AF860410

SAVE HARMLESS AGREEMENT

Solar Radiation Data Manual for Flat-Plate and Concentrating Collectors

National Renewable Energy Laboratory
1617 Cole Boulevard
Golden, Colorado 80401-3393
A national laboratory managed for the U.S. Department of Energy by the Midwest Research Institute under Contract No. DE-AC02-83CH-10093

Acknowledgements

Authors:	William Marion and Stephen Wilcox
Artists:	Joe Woodburn, Phyllis S. Kabins, and John Leedholm
Cover photographs:	Warren Gretz
Editor:	Mary Anne Dunlap

Preface

For designers and engineers of solar energy-related systems, the *Solar Radiation Data Manual for Flat-Plate and Concentrating Collectors* gives the solar resource available for various types of collectors for the United States and its territories. The data in the manual were modeled using hourly values of direct beam and diffuse horizontal solar radiation from the National Solar Radiation Data Base (NSRDB). The NSRDB contains modeled (93%) and measured (7%) global horizontal, diffuse horizontal, and direct beam solar radiation for 1961–1990.

This manual was produced by the National Renewable Energy Laboratory's (NREL's) Analytic Studies Division under the Solar Radiation Resource Assessment Project Task No. RA310102 and the Photovoltaic Solar Radiation Research Task No. PV360501. These tasks were funded and monitored by the Photovoltaics Branch of the Department of Energy's Office of Energy Efficiency and Renewable Energy.

The authors would like to acknowledge Charles Gay (UNISUN), Dave Menicucci (Sandia National Laboratories), Tom Ross (National Climatic Data Center), Frank Vignola (University of Oregon), and NREL staff members Carol Riordan, Dave Renné, Roland Hulstrom, Daryl Myers, Martin Rymes, and Tom Stoffel for their contributions to the manual and for their review.

In the early stages of developing the manual, we received suggestions and recommendations from more than 70 people on what they would like the manual to contain. This group consisted of designers, installers, manufacturers, consultants, university and national laboratory researchers, utility engineers, meteorologists, and state energy office staff. We appreciate their efforts and hope that this manual meets their expectations.

Approved for the
National Renewable Energy Laboratory

Thomas D. Bath, Director
Analytic Studies Division

Carol Riordan, Manager
Technology & Resource Assessment Branch

Daryl Myers, Project Manager
Photovoltaic Solar Radiation Research Task
Photovoltaics Division

Dave Renné, Project Manager
Solar Radiation Resource Assessment Project
Technology & Resource Assessment Branch

Table of Contents

Designers and engineers of solar energy conversion systems need solar resource information for different locations and types of collectors. Solar resource information provides data on how much solar energy is available to a collector and how it might vary from month to month and year to year.

This manual provides solar radiation values for common flat-plate and concentrating collectors for 239 stations in the United States and its territories. The solar radiation values are expressed as monthly and yearly averages for the period of 1961–1990. Minimum and maximum monthly and yearly averages are included to show the variability of a station's solar resource.

In addition to the solar radiation data, this manual contains tables listing climatic conditions such as average temperatures, average daily minimum and maximum temperatures, record minimum and maximum temperatures, average heating and cooling degree days, average relative humidity, and average wind speed.

The solar radiation and climatic data are presented in tables, with each station having its data presented on a single page. The pages are arranged alphabetically by the state or territory two-letter abbreviation, and within a state or territory the pages are arranged alphabetically by city or island.

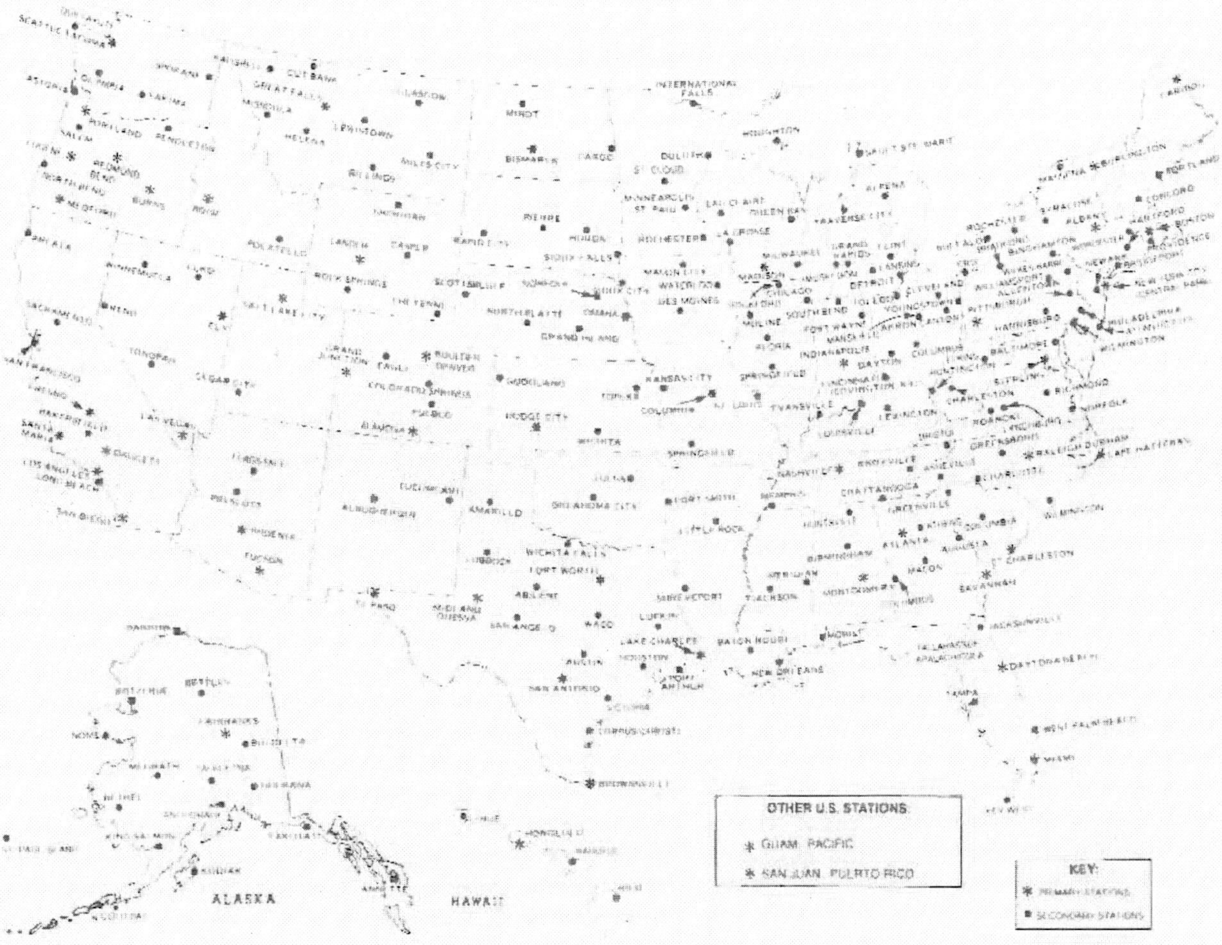

Map showing the location of the 239 stations in the National Solar Radiation Data Base, whose data were used to calculate values for this manual.

Technical Background

The solar radiation data in this manual are based on the National Solar Radiation Data Base (NSRDB), Version 1.1, which was completed in March 1994 by the National Renewable Energy Laboratory. Version 1.1 supersedes Version 1.0, which was completed in August 1992. Differences between the two versions are described briefly in the Appendix. The NSRDB contains hourly values of measured or modeled solar radiation and meteorological data for 239 stations for the 30-year period from 1961-1990. A complete description of the NSRDB and how it was produced is presented in its user's manual (NSRDB—Vol. 1, 1992).

There are two types of stations in the NSRDB: primary (denoted by asterisks on the station map) and secondary (denoted by dots on the station map). Primary stations, of which there are 56, measured solar radiation for a part (from 1 to 27 years) of the 30-year period. The remaining 183 stations made no solar radiation measurements and have modeled solar radiation data that are derived from meteorological data such as cloud cover. They are designated secondary stations. Both primary and secondary stations are National Weather Service stations that collected meteorological data for the period of 1961-1990.

Succeeding the older 1952-1975 SOLMET/ERSATZ data base, the NSRDB accounts for any recent climate changes and provides more accurate values of solar radiation for several reasons:

- More measured data
- Better model for estimating values (More than 90% of the solar radiation data in both data bases are modeled.)
- Improved instrument calibration methods
- Rigorous procedures for assessing quality of data.

A comparison of the old and new data bases provides the incentive for developing new solar radiation resource products such as this data manual. On an annual basis, 40% of the NSRDB and SOLMET/ERSATZ stations are in disagreement for global horizontal radiation by more than 5%, with some stations showing disagreement of up to 18% (Marion and Myers 1992). For direct beam radiation, 60% of the NSRDB and SOLMET/ERSATZ stations are in disagreement by more than 5%; some stations show disagreement of up to 33%. Disagreement between the two data bases is even greater when compared on a monthly basis. Most of the disagreement is attributed to differences in the instruments' calibration procedures and models.

This manual presents monthly and yearly average solar radiation values for various flat-plate and concentrating collectors to enable quick estimates of the incident solar energy for common collectors. The solar radiation values were computed using a model and NSRDB hourly values of direct beam and diffuse horizontal solar radiation. Climatic data were obtained from the NSRDB and from climatic data sets provided by the National Climatic Data Center, Asheville, North Carolina. The Appendix describes in more detail how this manual was produced.

References

NSRDB-Vol. 1 (1992). *User's Manual—National Solar Radiation Data Base (1961-1990). Version 1.0.* Asheville, NC: National Climatic Data Center.

Marion, W.; Myers, D. (1992). *A Comparison of Data from SOLMET/ERSATZ and the National Solar Radiation Data Base,* NREL/TP-463-5118, Golden, CO: National Renewable Energy Laboratory.

Interpreting the Data Tables

For each of the 239 stations, a data page describes the station location, presents average solar radiation values for flat-plate and concentrating collectors, and gives average climatic conditions. Except for mean atmospheric pressure, given in millibars, Standard International (SI) units are used. To convert to other units, use the conversion factor table on the inside back cover.

Station Description

Information at the top of each page describes the station.

- City and state in which the station is located
- Station Weather Bureau Army Navy (WBAN) identification number
- Latitude (degrees; north)
- Longitude (degrees; east or west)
- Elevation of station (meters)
- Mean atmospheric pressure of station (millibars)
- Type of station (primary or secondary).

Solar Radiation Data for Flat-Plate and Concentrating Collectors

For the period of 1961-1990, tables provide solar radiation data for flat-plate and concentrating collectors.

- Monthly and yearly averages of solar radiation ($kWh/m^2/day$)
- Minimum and maximum monthly and yearly averages of solar radiation ($kWh/m^2/day$)
- Uncertainty of solar radiation data (± %).

Minimum and maximum monthly and yearly averages are included to show the variability of a station's solar resource. The uncertainty of the data is presented in the table headings.

The manual includes data for the flat-plate and concentrating collectors described in the next few paragraphs.

Flat-plate collectors facing south at fixed tilt. Data are presented for five tilt angles from the horizontal: 0°, latitude minus 15°, latitude, latitude plus 15°, and 90°. Data for a tilt of 0°, referred to as global horizontal solar radiation, show how much solar radiation is received by a horizontal surface such as a solar pond.

Maximum yearly solar radiation can be achieved using a tilt angle approximately equal to a site's latitude. To optimize performance in the winter, the collector can be tilted 15° greater than the latitude; to optimize performance in the summer, the collector can be tilted 15° less than the latitude. Data for a tilt of 90° apply to collectors mounted vertically on south-facing walls and apply to south-facing windows for passive solar designs.

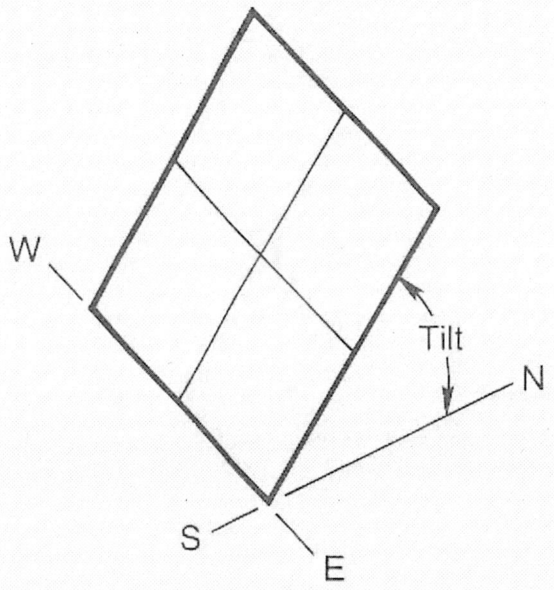

Flat-plate collector facing south at fixed tilt

Interpreting the Data Tables

One-axis tracking flat-plate collectors with axis oriented north-south. Data are presented for four different axis tilt angles from the horizontal: 0°, latitude minus 15°, latitude, and latitude plus 15°. These trackers pivot on their single axis to track the sun, facing east in the morning and west in the afternoon. Large collectors can use an axis tilt angle of 0° to minimize collector height and wind force. Small collectors can have their axis tilted up to increase the solar radiation on the collector. Just as for the flat-plate fixed tilt collector, the yearly and seasonal solar radiation can be optimized by the choice of tilt angle. The data presented assume continuous tracking of the sun throughout the day.

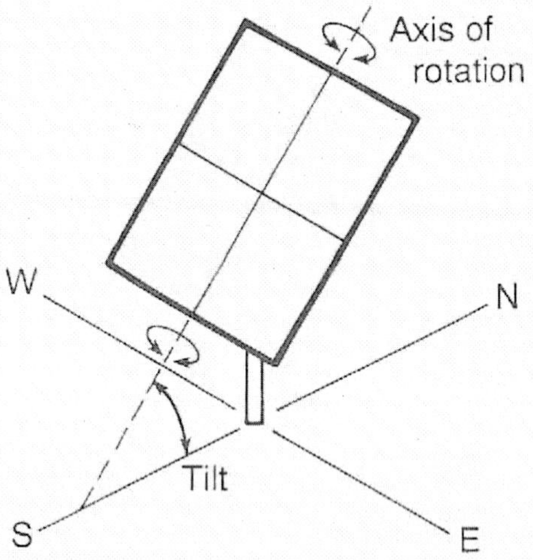

One-axis tracking flat-plate collector with axis oriented north-south

Two-axis tracking flat-plate collectors. Data for two-axis trackers represent the maximum solar radiation at a site available to a collector. Tracking the sun in both azimuth and elevation, these collectors keep the sun's rays normal to the collector surface.

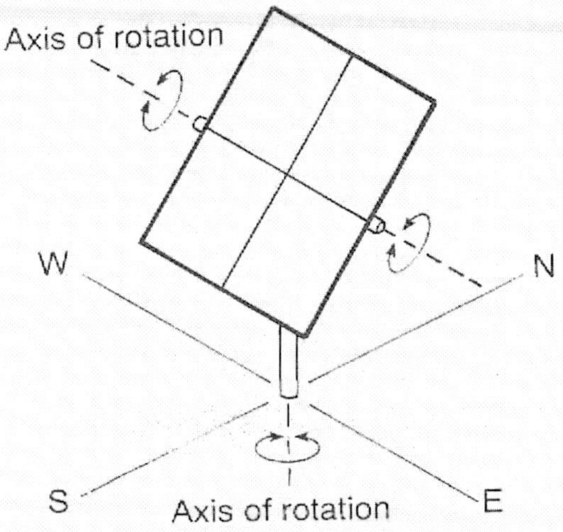

Two-axis tracking flat-plate collector

Concentrating collectors. Direct beam solar radiation data are presented for four concentrators: one-axis tracking parabolic troughs with a horizontal east-west axis, one-axis tracking parabolic troughs with a horizontal north-south axis, one-axis concentrators with the axis oriented north-south and tilted from the horizontal at an angle equal to the latitude, and two-axis tracking concentrator systems. Direct beam radiation comes in a direct line from the sun and is measured with instruments having a field-of-view of 5.7°. These instruments see only the sun's disk and a small portion of the sky surrounding the sun.

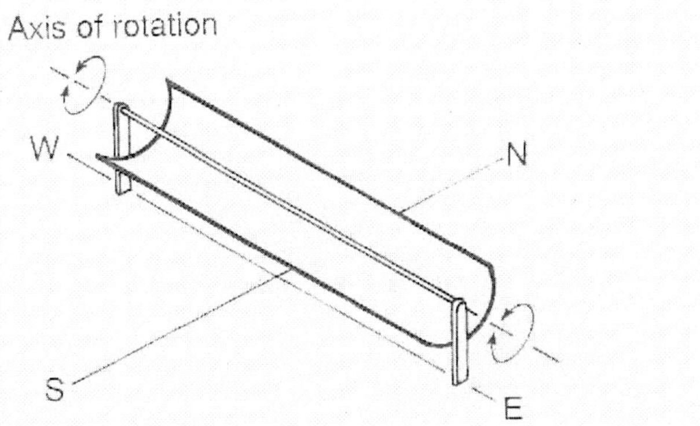

One-axis tracking parabolic trough with axis oriented east-west

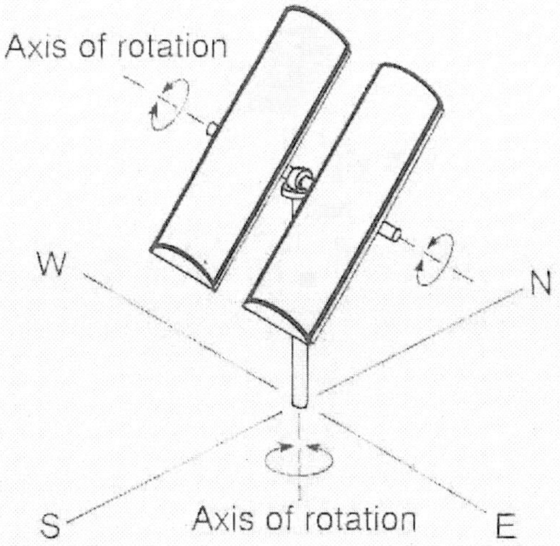

Two-axis tracking concentrator

Solar Radiation Graph

A graph at the top of each data page shows the variability of monthly and yearly solar radiation for a flat-plate collector facing south with a tilt equal to the station's latitude. For each month and year, 30 data values representing each year of the NSRDB are plotted along with the 1961–1990 averages for the months and year. The graph shows how the minimum and maximum values compare with the 1961–1990 average. It also shows the distribution of data points with respect to the average, minimum, and maximum values.

Climatic Conditions

A table shows average climatic conditions by listing monthly and yearly values for various parameters.

- Monthly and yearly average temperature (°C)
- Average daily minimum temperature (°C)
- Average daily maximum temperature (°C)
- Record minimum temperature (°C)
- Record maximum temperature (°C)
- Average heating degree days (HDD), base 18.3°C
- Average cooling degree days (CDD), base 18.3°C
- Average relative humidity (%)
- Average wind speed (m/s).

Degree days indicate heating and cooling requirements of buildings. They are defined as the difference between the average temperature for the day and a base temperature. If the average for the day (calculated by averaging the maximum and minimum temperature for the day) is less than the base value, then the difference is designated as heating degree days. If the average for the day is greater than the base value, the difference is designated as cooling degree days.

Other Data Formats

The data presented on the data pages in this manual, with the exception of the solar radiation graph, are also available on floppy disks in ASCII format. These data can be imported into popular spreadsheet programs. Also available on floppy disks are averages of solar radiation for each of the flat-plate and concentrating collectors for each month for the period of 1961-1990 (360 months). These data could be useful for identifying extended periods of low or high solar radiation or plotting graphs of monthly solar radiation for any of the flat-plate and concentrating collectors. The printed manual includes graphs only for flat-plate collectors tilted at an angle equal to the latitude.

To obtain either of these data sets on floppy disks, please contact the NREL Technical Inquiry Service at (303) 275-4099. "Readme" files, which describe the contents of the data sets, are included on the floppy disks.

Data Tables

Appendix - Methodology

Appendix - Methodology

This Appendix describes Version 1.1 revisions of the National Solar Radiation Data Base (NSRDB) and describes the method used to calculate the monthly and yearly averages of solar radiation for flat-plate and concentrating collectors. It also describes how the solar radiation data uncertainties were determined and how the climatic information was derived.

National Solar Radiation Data Base Version 1.1

This data manual is based on the NSRDB Version 1.1, completed in March 1994; the previous Version 1.0 was completed in August 1992. Version 1.1 corrects two types of errors discovered in Version 1.0: (1) for 23 stations, the wrong time zones were used, and data values were mismatched with their time stamp by 1 or 2 hours, and (2) for 8 stations that measured solar radiation, from 1 to 3 months per station had some hourly solar radiation values that were unrealistically low.

Version 1.1 corrects time zone errors for the following stations:

Anchorage, AK	McGrath, AK
Annette, AK	Nome, AK
Barrow, AK	St. Paul Island, AK
Bethel, AK	Talkeetna, AK
Bettles, AK	Honolulu, HI
Big Delta, AK	Lihue, HI
Cold Bay, AK	Evansville, IN
Fairbanks, AK	South Bend, IN
Gulkana, AK	Louisville, KY
King Salmon, AK	Lewistown, MT
Kodiak, AK	Ely, NV
Kotzebue, AK	

Version 1.1 replaces erroneous measured solar radiation data with modeled data for the following stations:

Montgomery, AL
Miami, FL
Great Falls, MT
Ely, NV
Albany, NY
Brownsville, TX
Seattle, WA
Lander, WY

Calculating Solar Radiation for Flat-Plate and Concentrating Collectors

The total solar radiation received by a flat-plate collector (I_c) is a combination of direct beam radiation (I_b), diffuse (sky) radiation (I_d), and radiation reflected from the surface in front of the collector (I_r):

$$I_c = I_b \cos\theta + I_d + I_r$$

where θ is the incident angle of the sun's rays to the collector. The incident angle is a function of the sun's postion in the sky and the orientation of the fixed or tracking collector. Algorithms presented by Menicucci and Fernandez (1988) were used to compute the incident angles for the various collectors. For tracking collectors, these algorithms also were used to compute collector tilt angles from the horizontal. Direct beam solar radiation hourly values from the National Solar Radiation Data Base (NSRDB) were used to determine the direct beam contribution ($I_b \cos\theta$) for each hour. Except for the first and last daylight hour, incident angles were calculated at the midpoint of the hour. For the first and last daylight hour, incident angles were calculated at the midpoint of the period during the hour when the sun was above the horizon.

The diffuse (sky) radiation, I_d, received by the collector was calculated by an anisotropic diffuse radiation model developed by Perez et al. (1990). The model determined the diffuse (sky) radiation for the collector using hourly values (from the NSRDB) of diffuse horizontal and direct beam solar radiation. Other inputs to the model included the sun's incident angle to the collector, the collector tilt angle from the horizontal, and the sun's zenith angle. The model is an improved and refined version of their original model that was recommended by the International Energy Agency for calculating diffuse radiation for tilted surfaces (Hay and McKay 1988).

The Perez model equation for diffuse sky radiation for a tilted surface is:

$$I_d = I_{dh} [\, 0.5(1 - F_1)(1 + \cos\beta) + F_1 a/b + F_2 \sin\beta \,]$$

where

I_{dh} = diffuse solar horizontal radiation

F_1 = circumsolar anisotropy coefficient, function of sky condition

F_2 = horizon/zenith anisotropy coefficient, function of sky condition

β = tilt of the collector from the horizontal

a = 0 or the cosine of the incident angle, whichever is greater

b = 0.087 or the cosine of the solar zenith angle, whichever is greater.

The model coefficients F_1 and F_2 are organized as an array of values that are selected for use depending on the solar zenith angle, the sky's clearness, and the sky's brightness. The manner in which this is done is described by Perez et al. (1990).

The ground-reflected radiation received by a collector is a function of the global horizontal radiation (I_h), the tilt of the collector from the horizontal (β), and the surface reflectivity or albedo (ρ):

$$I_r = 0.5\rho I_h (1 - \cos\beta).$$

Surface albedo was adjusted depending on the presence of snow cover, as indicated by the snow depth data in the NSRDB. If there was snow on the ground, the surface albedo was set to 0.6 (albedo for snow ranges from about 0.35 for old snow to 0.95 for dry new snow). If no snow was indicated, the surface albedo was set to 0.2, a nominal value for green vegetation and some soil types.

The concentrating collectors portrayed in the manual have small fields-of-view and do not receive diffuse (sky) radiation or ground-reflected radiation. Consequently, solar radiation for these concentrating collectors is solely a function of the direct beam radiation and the sun's incident angle to the collector. Solar radiation received by the concentrating collectors simplifies to

$$I_c = I_b \cos\theta.$$

For each station location, collector type, and collector orientation, hourly values of solar radiation received by the collectors were calculated. Monthly and yearly averages were then determined for the period of 1961–1990. For a few stations, monthly and yearly averages do not include data for 1989 or 1990 or both because NSRDB data did not include those station years. Stations with less than 30 years of NSRDB data, along with their period of record, are listed below:

Tucumcari, NM	1961–1988
Eagle, CO	1961–1988
Minot, ND	1961–1988
Miles City, MT	1961–1989
Cut Bank, MT	1961–1988
Burns, OR	1961–1988

Estimating the Uncertainty of Solar Radiation Data

The solar radiation values presented in the manual were calculated using improved models and data. The estimated data uncertainties assigned to the calculated values show how they might compare with true values. They were determined using the uncertainty method of Abernethy and Ringhiser (1985). This root-sum-square method defines an uncertainty, $\pm U_{RSS}$, in which 95% of the time, the true value will be within plus or minus the uncertainty of the calculated value.

$$U_{RSS} = [(tR)^2 + B^2]^{1/2}$$

where

t = student's T distribution factor (equals 2 for sample size greater than 30)

R = random error

B = bias error.

Random and Bias Errors. The two types of errors that contribute to uncertainties are random errors and bias errors. Random errors usually follow statistical distributions and result in values both above and below the true values. Random errors tend to cancel when individual values are used to determine an average. For example, a 30-year monthly average of solar radiation may use 10,800 hourly values (assuming 30 days per month and 12 hours of sunlight per day) to determine the average monthly solar radiation. The random error of the average is reduced by a factor of $10{,}800^{1/2}$, or approximately 100. Consequently, random error sources do not contribute significantly to the uncertainty of 30-year monthly averages.

Bias errors, however, are not reduced by averaging. Bias errors, which are often referred to as fixed or systematic errors, cause values to be in error by about the same amount and direction. The reason for bias errors, as well as their magnitude and direction, may be unknown; otherwise, corrections such as changes in the calibration factor can be made. When detailed information is not known about the bias errors, reasonable estimates of the bias error magnitude can be made using procedures similar to those described here.

For the monthly averages of solar radiation, we evaluated the three major bias errors: (1) errors in direct beam radiation incident on the collector caused by errors in NSRDB direct beam radiation data, (2) errors in diffuse radiation incident on the collector caused by errors in NSRDB diffuse horizontal radiation, and (3) errors in diffuse radiation incident on the collector caused by errors in modeling the diffuse solar radiation for the collector. Climate change could also bias monthly average solar radiation values but was not considered a major source of error for this work.

The root-sum-square of the individual bias errors yields the total bias error. Because the random error is negligible, the total bias error is the same as the total uncertainty of the monthly averages. Consequently, the uncertainty, $\pm U_{RSS}$, can be expressed as:

$$U_{RSS} = B = [B_b^2 + B_d^2 + B_m^2]^{1/2}$$

where

B_b = errors in collector direct beam radiation caused by errors in direct beam radiation data

B_d = errors in collector diffuse radiation caused by errors in diffuse horizontal radiation data

B_m = errors in total collector radiation caused by errors in modeling the diffuse solar radiation for the collector.

The bias errors for direct beam and diffuse horizontal radiation were extracted from the NSRDB daily statistic files for each station. The NSRDB daily statistic files include, among other information, 30-year averages and their uncertainties for direct beam and diffuse horizontal radiation. An integer number represents an uncertainty range. Examples of uncertainty ranges for the monthly averages are from 6% to 9%, from 9% to 13%, and from 13% to 18% of the monthly average.

For 30-year averages, most of the stations have direct beam radiation uncertainties in the 6% to 9% range and diffuse horizontal radiation uncertainties in the 9% to 13% range. The remaining stations have direct beam radiation uncertainties in the 9% to 13% range and diffuse horizontal radiation uncertainties in the 13% to 18% range. For the purpose of extracting the bias errors from the daily statistic files, a single integer value near the midpoint of the range was used (8% for the 6% to 9% range, 11% for the 9% to 13% range, and 16% for the 13% to 18% range).

The bias error for modeling the collector radiation is attributed to the diffuse solar radiation model because the direct beam component is considered an exact solution ($I_b\cos\theta$). An evaluation of the original Perez model by Hay and McKay (1988) provided information whereby we estimated the bias error to be about 5% of the total collector radiation for our applications.

The uncertainty, $\pm U_{RSS}$, can be expressed as a percentage of the total collector radiation by the following equation:

$$U_{RSS}\ (\pm\%) = 100\ [(w_b H_b)^2 + (w_{dh} H_d)^2 + (w_m H_c)^2]^{1/2}/H_c$$

where

- H_b = average monthly direct beam radiation incident on the collector
- H_d = average monthly diffuse radiation incident on the collector
- H_c = average monthly total radiation incident on the collector ($H_b + H_d$)
- w_b = percent bias uncertainty of average monthly beam radiation
- w_{dh} = percent bias uncertainty of average monthly diffuse horizontal radiation
- w_m = percent bias uncertainty of the solar radiation modeling for tilted surfaces

Uncertainty Values in Tables. Because of the large number of solar radiation values presented in the manual (546 per station), it was judged impractical with respect to space limitations to present uncertainty values for each solar radiation value. Rather, a simplifying assumption was made so that only one uncertainty value was presented for all flat-plate collectors. The assumption was that the direct beam radiation and diffuse radiation incident on the collector were of equal weight. The uncertainties of the diffuse horizontal and direct beam radiation have about the same value, so this assumption did not create large changes in calculated uncertainties for collector radiation.

Over a range of direct-beam-radiation-to-diffuse-radiation ratios (30/70 to 90/10), the assumption yielded uncertainties within 1% or 2% of that when calculated using the exact proportions of direct beam radiation and diffuse radiation (e.g., uncertainty of 8% or 10% instead of 9%). This was judged acceptable, considering that there are uncertainties associated with the uncertainty values used for the average monthly direct beam radiation, the average monthly diffuse horizontal radiation, and the solar radiation modeling for tilted surfaces. As a conservative measure, the calculated uncertainties were rounded to the next highest integer value.

For most of the stations in the data manual, uncertainties of 9% were assigned to the solar radiation data for flat-plate collectors. The few stations with higher uncertainties for direct beam and diffuse horizontal radiation were assigned uncertainties of 11%.

A separate value was assigned to the uncertainty for the direct beam radiation for concentrating collectors, which is only a function of the uncertainty (w_b) of the average monthly beam radiation. For most of the stations in the data manual, uncertainties of 8% were assigned to the solar radiation data for concentrating collectors. The few stations with higher uncertainties for direct beam radiation were assigned uncertainties of 11%.

Data values in the data manual are given to one significant figure by rounding the calculated value to the nearest tenth of a kWh/m^2. Consequently, the data values presented are within 0.05 kWh/m^2 of the calculated values. Because of the uncertainties of the data values, there is no benefit to expressing the data values to more than one significant figure.

Deriving Climatic Data

The climatic data presented in the manual were derived using both data from the National Solar Radiation Data Base (NSRDB) and from climatic data sets provided by the National Climatic Data Center (NCDC), Asheville, North Carolina (704) 271-4994.

Climatic data pertaining to average temperatures, average daily minimum temperature, average daily maximum temperature, average heating degree days base 18.3°C, and average cooling degree days base 18.3°C were extracted from NCDC's data tape, "1961-1990 Monthly Station Normals All Elements." This data tape includes temperature and degree day normals for about 4775 stations in the United States and its territories. The normals are averages computed by NCDC for the period of 1961-1990.

For this data set, NCDC used procedures, when possible, to estimate missing data and to correct for other inconsistencies by using data from neighboring stations. For one of the stations in this data manual, data were not available on NCDC's data tape. For this station, in Arcata, California, the averages were computed using NSRDB data, but no attempt was made to estimate missing data or to correct for other inconsistencies.

NSRDB data were used to calculate average relative humidity and average wind speed. Record minimum and maximum temperatures were obtained primarily form NCDC's data diskette, "Comparative Climatic Data Tables—1991." This data diskette contains, among other useful parameters, record minimum and maximum temperatures for about 90% of the stations in this manual and spans periods of record back to 1948 and earlier. For the remaining 10% of the stations, record minimum and maximum temperatures are based on NSRDB data.

References

Abernethy, R,; Ringhiser, B. (1985). "The History and Statistical Development of the New ASME-SAE-AIAA-ISO Measurement and Uncertainty Methodology." 20th AIAA/SAE/ASME Joint Propulsion Conference (July 1985) AIAA-85-1403. New York: American Institute of Astronautics and Aeronautics.

Hay, J.E.; McKay, D.C. (1988). *Final Report IEA Task IX—Calculation of Solar Irradiances for Inclined Surfaces: Verification of Models Which Use Hourly and Daily Data.* International Energy Agency Solar Heating and Cooling Programme.

Menicucci, D.; Fernandez, J.P. (1988). *User's Manual for PVFORM: A Photovoltaic System Simulation Program for Stand-Alone and Grid-Interactive Applications.* SAND85-0376, Albuquerque, NM: Sandia National Laboratories.

Perez, R.; Ineichen, P.; Seals, R.; Michalsky, J.; Stewart, R. (1990). "Modeling Daylight Availability and Irradiance Components from Direct and Global Irradiance." *Solar Energy*, 44(5), pp. 271-289.

Unit Conversion Factors

To Convert	Into	Multiply By
kilowatt-hours per square meter	megajoules per square meter	3.60
kilowatt-hours per square meter	Btus per square foot	317.2
kilowatt-hours per square meter	Langleys	86.04
kilowatt-hours per square meter	calories per square centimeter	86.04
meters	feet	3.281
meters per second	miles per hour	2.237
millibars	pascals	100.0
millibars	atmospheres	0.0009869
millibars	kilograms per square meter	10.20
millibars	pounds per square inch	0.0145
degrees Centigrade	degrees Fahrenheit	°C x 1.8 + 32
degrees (angle)	radians	0.017453
degree days (base 18.3°C)	degree days (base 65°F)	1.8

National Renewable Energy Laboratory
1617 Cole Boulevard
Golden, Colorado 80401-3393

NREL is a national laboratory of the U.S. Department of Energy

NREL/TP-463-5607
DE93018229
April 1994

Printed with a renewable source ink on
paper containing at least 50% wastepaper

M35-A0266

Anchorage, AK

WBAN NO. 26451

LATITUDE: 61.17° N
LONGITUDE: 150.02° W
ELEVATION: 35 meters
MEAN PRESSURE: 1004 millibars

STATION TYPE: Secondary

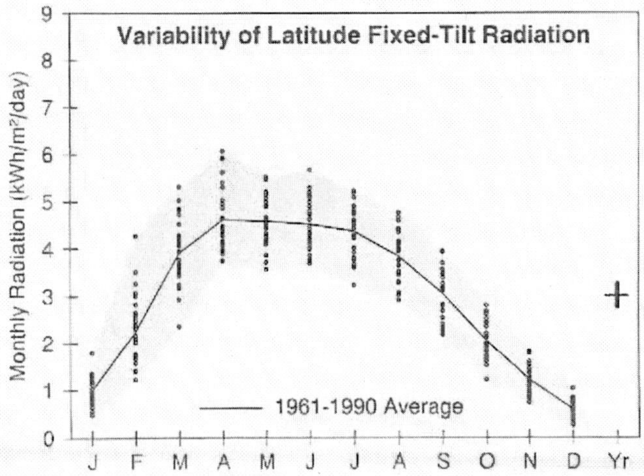

Solar Radiation for Flat-Plate Collectors Facing South at a Fixed Tilt (kWh/m²/day), Uncertainty ±9%

Tilt (°)		Jan	Feb	Mar	Apr	May	June	July	Aug	Sept	Oct	Nov	Dec	Year
0	Average	0.3	1.0	2.3	3.6	4.6	4.9	4.6	3.5	2.2	1.1	0.4	0.2	2.4
	Min/Max	0.2/0.4	0.7/1.3	1.7/2.6	3.2/4.5	3.8/5.3	4.1/6.0	3.7/5.4	2.9/4.2	1.9/2.6	0.9/1.4	0.3/0.5	0.1/0.2	2.2/2.5
Latitude -15	Average	0.9	2.1	3.8	4.7	4.9	5.0	4.8	4.1	3.1	2.0	1.1	0.5	3.1
	Min/Max	0.4/1.6	1.2/3.8	2.3/5.0	3.9/6.1	3.9/5.9	4.1/6.2	3.5/5.7	3.1/5.0	2.3/3.9	1.2/2.7	0.7/1.6	0.3/0.9	2.8/3.3
Latitude	Average	1.0	2.2	3.9	4.6	4.6	4.5	4.4	3.8	3.1	2.1	1.2	0.6	3.0
	Min/Max	0.5/1.8	1.2/4.3	2.4/5.3	3.7/6.1	3.6/5.5	3.7/5.7	3.2/5.2	2.9/4.8	2.2/4.0	1.2/2.8	0.7/1.8	0.3/1.0	2.8/3.2
Latitude +15	Average	1.0	2.3	3.9	4.3	4.0	3.9	3.8	3.4	2.9	2.0	1.3	0.6	2.8
	Min/Max	0.5/1.9	1.2/4.5	2.3/5.3	3.4/5.8	3.1/4.8	3.2/4.9	2.8/4.5	2.6/4.3	2.0/3.8	1.2/2.8	0.7/1.9	0.3/1.1	2.6/3.0
90	Average	1.0	2.3	3.7	3.8	3.3	3.2	3.1	2.9	2.6	1.9	1.3	0.6	2.5
	Min/Max	0.5/1.9	1.2/4.4	2.2/5.1	3.0/5.3	2.6/4.0	2.6/4.0	2.3/3.7	2.2/3.6	1.8/3.4	1.1/2.6	0.7/1.9	0.3/1.1	2.3/2.8

Solar Radiation for 1-Axis Tracking Flat-Plate Collectors with a North-South Axis (kWh/m²/day), Uncertainty ±9%

Axis Tilt (°)		Jan	Feb	Mar	Apr	May	June	July	Aug	Sept	Oct	Nov	Dec	Year
0	Average	0.5	1.5	3.5	5.2	6.1	6.3	6.0	4.7	3.1	1.6	0.7	0.3	3.3
	Min/Max	0.3/0.8	0.9/2.8	2.1/4.7	4.1/6.9	4.6/7.6	4.7/8.4	4.3/7.5	3.4/5.9	2.2/3.9	1.0/2.2	0.5/0.9	0.1/0.4	2.9/3.6
Latitude -15	Average	0.9	2.4	4.6	6.1	6.6	6.6	6.3	5.2	3.8	2.2	1.2	0.5	3.9
	Min/Max	0.5/1.7	1.3/4.6	2.6/6.5	4.8/8.3	4.8/8.3	4.8/8.9	4.3/8.0	3.6/6.7	2.5/4.9	1.3/3.2	0.7/1.8	0.3/0.9	3.4/4.2
Latitude	Average	1.0	2.5	4.8	6.1	6.4	6.3	6.1	5.1	3.8	2.3	1.3	0.6	3.9
	Min/Max	0.5/1.9	1.3/5.0	2.7/6.7	4.7/8.3	4.7/8.1	4.6/8.6	4.1/7.7	3.5/6.6	2.5/5.0	1.3/3.3	0.8/2.0	0.3/1.1	3.4/4.2
Latitude +15	Average	1.1	2.6	4.7	5.8	6.0	5.9	5.7	4.8	3.6	2.3	1.4	0.6	3.7
	Min/Max	0.5/2.0	1.3/5.1	2.6/6.7	4.5/8.1	4.3/7.6	4.2/8.1	3.8/7.3	3.3/6.2	2.4/4.8	1.3/3.3	0.8/2.1	0.3/1.2	3.3/4.1

Solar Radiation for 2-Axis Tracking Flat-Plate Collectors (kWh/m²/day), Uncertainty ±9%

Tracker		Jan	Feb	Mar	Apr	May	June	July	Aug	Sept	Oct	Nov	Dec	Year
2-Axis	Average	1.1	2.6	4.8	6.1	6.7	6.8	6.5	5.2	3.8	2.4	1.4	0.7	4.0
	Min/Max	0.5/2.1	1.3/5.1	2.7/6.7	4.8/8.3	5.0/8.4	5.0/9.2	4.5/8.2	3.7/6.8	2.6/5.0	1.3/3.3	0.8/2.1	0.3/1.2	3.6/4.4

Direct Beam Solar Radiation for Concentrating Collectors (kWh/m²/day), Uncertainty ±8%

Tracker		Jan	Feb	Mar	Apr	May	June	July	Aug	Sept	Oct	Nov	Dec	Year
1-Axis, E-W Horiz Axis	Average	0.6	1.3	2.2	2.6	2.8	2.8	2.6	2.1	1.7	1.2	0.8	0.4	1.8
	Min/Max	0.2/1.5	0.4/3.4	0.8/3.7	1.5/4.3	1.6/4.0	1.6/4.5	1.2/3.8	1.2/3.2	0.9/2.6	0.5/1.9	0.3/1.4	0.1/0.8	1.5/2.0
1-Axis, N-S Horiz Axis	Average	0.2	0.8	1.9	3.0	3.5	3.4	3.3	2.5	1.7	0.8	0.3	0.1	1.8
	Min/Max	0.1/0.5	0.2/2.0	0.6/3.3	1.7/4.9	2.0/5.0	1.8/5.5	1.5/4.8	1.3/3.7	0.8/2.5	0.3/1.4	0.1/0.6	0.0/0.2	1.4/2.1
1-Axis, N-S Tilt=Latitude	Average	0.6	1.5	2.8	3.6	3.7	3.5	3.4	2.9	2.3	1.4	0.8	0.4	2.3
	Min/Max	0.2/1.5	0.4/3.9	1.0/4.8	2.1/6.0	2.2/5.4	1.9/5.7	1.6/5.0	1.5/4.3	1.1/3.5	0.6/2.3	0.3/1.5	0.1/0.8	1.8/2.6
2-Axis	Average	0.7	1.5	2.8	3.7	4.0	3.8	3.7	3.0	2.3	1.4	0.9	0.4	2.3
	Min/Max	0.2/1.6	0.4/4.0	1.0/4.9	2.1/6.1	2.3/5.7	2.1/6.2	1.7/5.4	1.6/4.4	1.1/3.5	0.6/2.3	0.3/1.5	0.1/0.9	1.9/2.7

Average Climatic Conditions

Element	Jan	Feb	Mar	Apr	May	June	July	Aug	Sept	Oct	Nov	Dec	Year
Temperature (°C)	-9.5	-7.4	-3.5	2.1	8.1	12.4	14.7	13.5	9.1	1.4	-6.0	-8.7	2.2
Daily Minimum Temp	-13.1	-11.4	-7.7	-1.9	3.8	8.4	10.9	9.7	5.3	-1.8	-9.4	-12.2	-1.6
Daily Maximum Temp	-5.9	-3.4	0.6	6.0	12.4	16.4	18.4	17.2	12.9	4.7	-2.7	-5.3	5.9
Record Minimum Temp	-36.7	-32.2	-31.1	-20.0	-8.3	0.6	3.3	-0.6	-6.7	-20.6	-29.4	-34.4	-36.7
Record Maximum Temp	10.0	8.9	10.6	18.3	25.0	29.4	27.8	27.8	22.8	16.1	11.7	8.3	29.4
HDD, Base 18.3°C	863	720	677	487	317	177	114	150	277	523	730	839	5872
CDD, Base 18.3°C	0	0	0	0	0	0	0	0	0	0	0	0	0
Relative Humidity (%)	74	71	66	64	62	66	71	75	76	75	77	77	71
Wind Speed (m/s)	2.9	3.1	3.0	3.3	3.7	3.8	3.2	3.1	3.0	3.0	2.9	2.8	3.1

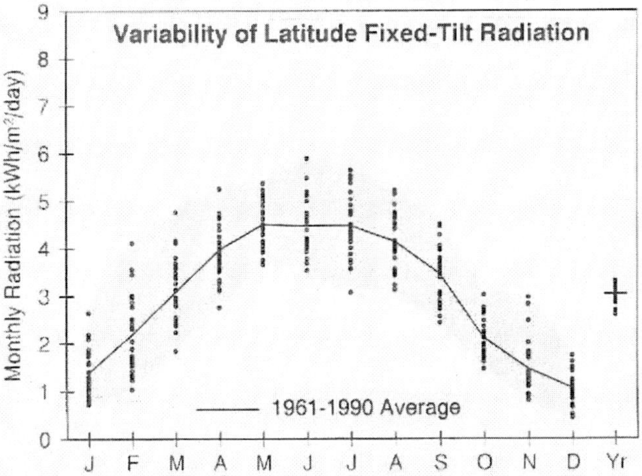

Annette, AK

WBAN NO. 25308

LATITUDE: 55.03° N
LONGITUDE: 131.57° W
ELEVATION: 34 meters
MEAN PRESSURE: 1009 millibars

STATION TYPE: Secondary

Solar Radiation for Flat-Plate Collectors Facing South at a Fixed Tilt (kWh/m²/day), Uncertainty ±9%

Tilt (°)		Jan	Feb	Mar	Apr	May	June	July	Aug	Sept	Oct	Nov	Dec	Year
0	Average	0.6	1.2	2.2	3.5	4.7	5.0	4.9	4.0	2.7	1.4	0.7	0.5	2.6
	Min/Max	0.4/0.9	0.9/1.6	1.6/3.0	2.7/4.4	4.0/5.4	4.1/6.4	3.5/5.9	3.2/4.8	2.2/3.3	1.3/1.8	0.6/1.1	0.3/0.6	2.4/2.8
Latitude -15	Average	1.2	2.1	3.1	4.1	4.9	4.9	4.9	4.4	3.5	2.1	1.3	0.9	3.1
	Min/Max	0.7/2.3	1.0/3.7	1.9/4.6	2.9/5.4	4.0/5.8	3.9/6.4	3.4/6.1	3.3/5.5	2.5/4.5	1.5/2.9	0.8/2.6	0.4/1.5	2.7/3.4
Latitude	Average	1.4	2.2	3.1	4.0	4.5	4.5	4.5	4.2	3.5	2.1	1.5	1.1	3.0
	Min/Max	0.7/2.7	1.0/4.1	1.8/4.8	2.8/5.3	3.7/5.4	3.5/5.9	3.1/5.7	3.1/5.2	2.4/4.5	1.5/3.0	0.8/3.0	0.4/1.7	2.6/3.3
Latitude +15	Average	1.4	2.2	3.0	3.6	4.0	3.9	3.9	3.7	3.2	2.1	1.5	1.1	2.8
	Min/Max	0.7/2.8	1.0/4.3	1.7/4.7	2.5/4.8	3.2/4.7	3.1/5.1	2.7/4.9	2.8/4.7	2.3/4.3	1.4/3.0	0.8/3.1	0.4/1.9	2.4/3.1
90	Average	1.4	2.1	2.6	2.9	3.0	2.8	2.9	2.9	2.7	1.8	1.4	1.1	2.3
	Min/Max	0.7/2.8	0.9/4.1	1.5/4.1	2.0/3.9	2.5/3.6	2.3/3.6	2.0/3.6	2.2/3.6	1.9/3.6	1.2/2.7	0.7/3.0	0.4/1.9	2.0/2.5

Solar Radiation for 1-Axis Tracking Flat-Plate Collectors with a North-South Axis (kWh/m²/day), Uncertainty ±9%

Axis Tilt (°)		Jan	Feb	Mar	Apr	May	June	July	Aug	Sept	Oct	Nov	Dec	Year
0	Average	0.9	1.8	3.1	4.7	6.1	6.4	6.3	5.3	3.7	1.9	1.0	0.6	3.5
	Min/Max	0.5/1.6	0.9/3.1	1.9/4.8	3.3/6.7	4.5/7.7	4.7/8.9	3.8/8.5	3.8/7.1	2.6/5.1	1.4/2.6	0.6/2.0	0.3/1.0	2.9/3.8
Latitude -15	Average	1.4	2.4	3.7	5.3	6.4	6.5	6.4	5.7	4.3	2.3	1.5	1.0	3.9
	Min/Max	0.7/2.7	1.1/4.6	2.1/6.0	3.5/7.5	4.6/8.2	4.7/9.3	3.8/8.9	4.0/7.8	2.9/6.1	1.6/3.4	0.8/3.1	0.4/1.7	3.3/4.3
Latitude	Average	1.5	2.5	3.7	5.2	6.2	6.2	6.2	5.6	4.3	2.4	1.6	1.1	3.9
	Min/Max	0.8/3.0	1.1/4.9	2.0/6.1	3.4/7.4	4.4/7.9	4.4/8.9	3.5/8.6	3.8/7.6	2.9/6.1	1.6/3.6	0.8/3.4	0.4/1.9	3.2/4.3
Latitude +15	Average	1.6	2.5	3.6	4.9	5.9	5.8	5.8	5.3	4.2	2.4	1.6	1.2	3.7
	Min/Max	0.8/3.2	1.0/5.1	2.0/6.0	3.2/7.2	4.1/7.5	4.1/8.4	3.3/8.1	3.6/7.2	2.7/5.9	1.5/3.6	0.8/3.5	0.4/2.0	3.1/4.1

Solar Radiation for 2-Axis Tracking Flat-Plate Collectors (kWh/m²/day), Uncertainty ±9%

Tracker		Jan	Feb	Mar	Apr	May	June	July	Aug	Sept	Oct	Nov	Dec	Year
2-Axis	Average	1.6	2.5	3.8	5.3	6.6	6.7	6.6	5.8	4.4	2.4	1.7	1.2	4.0
	Min/Max	0.8/3.2	1.1/5.1	2.1/6.1	3.5/7.6	4.7/8.3	4.9/9.5	3.9/9.1	4.0/7.8	2.9/6.1	1.6/3.6	0.8/3.5	0.4/2.0	3.4/4.4

Direct Beam Solar Radiation for Concentrating Collectors (kWh/m²/day), Uncertainty ±8%

Tracker		Jan	Feb	Mar	Apr	May	June	July	Aug	Sept	Oct	Nov	Dec	Year
1-Axis, E-W	Average	0.9	1.3	1.7	2.2	2.7	2.7	2.7	2.4	1.9	1.1	0.9	0.7	1.8
Horiz Axis	Min/Max	0.2/2.1	0.3/3.5	0.6/3.4	1.1/3.7	1.3/3.8	1.4/4.9	0.9/4.8	1.3/3.8	1.0/3.2	0.4/1.9	0.3/2.4	0.1/1.4	1.3/2.0
1-Axis, N-S	Average	0.4	0.9	1.6	2.7	3.5	3.4	3.5	3.0	2.1	0.9	0.5	0.3	1.9
Horiz Axis	Min/Max	0.1/1.1	0.2/2.4	0.6/3.4	1.3/4.7	1.6/5.0	1.8/6.3	1.1/6.2	1.6/4.9	1.1/3.6	0.3/1.5	0.2/1.4	0.1/0.6	1.4/2.3
1-Axis, N-S	Average	0.9	1.5	2.2	3.1	3.7	3.4	3.5	3.3	2.6	1.3	1.0	0.7	2.3
Tilt=Latitude	Min/Max	0.2/2.2	0.4/4.1	0.8/4.6	1.5/5.3	1.7/5.2	1.8/6.2	1.1/6.3	1.8/5.3	1.4/4.5	0.5/2.4	0.3/2.6	0.1/1.4	1.7/2.7
2-Axis	Average	1.0	1.6	2.2	3.1	3.9	3.7	3.8	3.4	2.6	1.3	1.0	0.7	2.4
	Min/Max	0.3/2.3	0.4/4.2	0.8/4.6	1.5/5.4	1.8/5.5	2.0/6.8	1.2/6.8	1.8/5.5	1.4/4.5	0.5/2.4	0.3/2.7	0.1/1.5	1.8/2.8

Average Climatic Conditions

Element	Jan	Feb	Mar	Apr	May	June	July	Aug	Sept	Oct	Nov	Dec	Year
Temperature (°C)	1.3	2.8	4.1	6.3	9.4	12.4	14.6	14.8	12.3	8.3	4.1	2.1	7.7
Daily Minimum Temp	-1.3	0.1	1.1	2.8	5.7	8.8	11.1	11.3	8.9	5.5	1.5	-0.3	4.6
Daily Maximum Temp	3.8	5.5	7.1	9.7	13.2	16.0	18.1	18.3	15.7	11.0	6.6	4.4	10.8
Record Minimum Temp	-17.2	-16.7	-17.2	-6.1	-0.6	2.8	4.4	4.4	0.6	-7.8	-19.4	-17.2	-19.4
Record Maximum Temp	16.1	17.2	16.7	27.8	31.1	31.7	31.7	32.2	27.8	21.7	19.4	16.7	32.2
HDD, Base 18.3°C	529	436	443	362	276	181	119	113	180	312	428	504	3882
CDD, Base 18.3°C	0	0	0	0	0	0	4	3	0	0	0	0	7
Relative Humidity (%)	78	78	76	75	75	77	79	81	82	83	80	80	79
Wind Speed (m/s)	4.9	5.0	4.4	4.4	3.8	3.7	3.4	3.4	3.7	4.8	5.0	4.8	4.2

Barrow, AK

WBAN NO. 27502

LATITUDE: 71.30° N
LONGITUDE: 156.78° W
ELEVATION: 4 meters
MEAN PRESSURE: 1016 millibars

STATION TYPE: Secondary

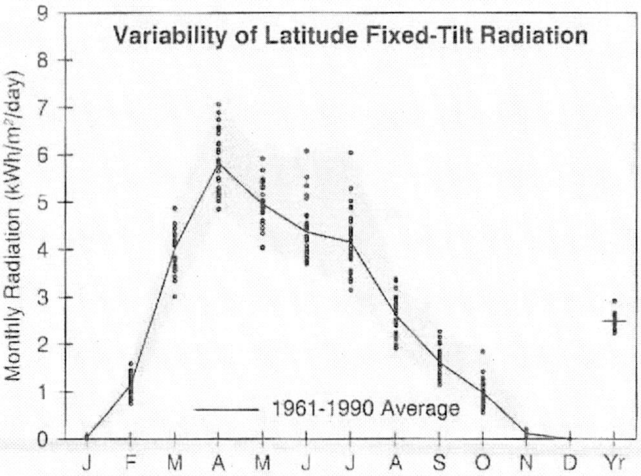

Solar Radiation for Flat-Plate Collectors Facing South at a Fixed Tilt (kWh/m²/day), Uncertainty ±9%

Tilt (°)		Jan	Feb	Mar	Apr	May	June	July	Aug	Sept	Oct	Nov	Dec	Year
0	Average	0.0	0.3	1.6	3.7	4.7	4.9	4.5	2.6	1.3	0.5	0.0	0.0	2.0
	Min/Max	0.0/0.0	0.2/0.3	1.4/1.7	3.3/4.1	4.0/5.3	4.4/6.2	3.7/5.9	2.2/3.0	1.2/1.5	0.4/0.6	0.0/0.0	0.0/0.0	1.9/2.1
Latitude -15	Average	0.0	1.1	3.8	5.8	5.2	4.8	4.6	2.8	1.7	0.9	0.1	0.0	2.6
	Min/Max	0.0/0.1	0.7/1.5	2.9/4.6	4.9/7.0	4.3/6.2	4.1/6.6	3.5/6.6	2.1/3.6	1.2/2.3	0.6/1.7	0.0/0.2	0.0/0.0	2.3/3.0
Latitude	Average	0.0	1.1	4.0	5.8	4.9	4.4	4.2	2.6	1.6	1.0	0.1	0.0	2.5
	Min/Max	0.0/0.1	0.8/1.6	3.0/4.9	4.8/7.1	4.0/5.9	3.7/6.1	3.1/6.0	1.9/3.4	1.1/2.3	0.6/1.8	0.0/0.2	0.0/0.0	2.2/2.9
Latitude +15	Average	0.0	1.2	4.0	5.6	4.6	3.8	3.6	2.3	1.5	1.0	0.1	0.0	2.3
	Min/Max	0.0/0.1	0.8/1.6	3.0/4.8	4.6/6.7	3.6/5.5	3.2/5.4	2.7/5.2	1.7/3.0	1.0/2.1	0.5/1.9	0.0/0.2	0.0/0.0	2.1/2.7
90	Average	0.0	1.2	3.9	5.4	4.5	3.6	3.4	2.2	1.4	1.0	0.1	0.0	2.2
	Min/Max	0.0/0.1	0.8/1.6	2.9/4.8	4.5/6.6	3.5/5.4	3.0/5.1	2.6/5.0	1.6/2.9	1.0/2.0	0.5/1.8	0.0/0.2	0.0/0.0	2.0/2.6

Solar Radiation for 1-Axis Tracking Flat-Plate Collectors with a North-South Axis (kWh/m²/day), Uncertainty ±9%

Axis Tilt (°)		Jan	Feb	Mar	Apr	May	June	July	Aug	Sept	Oct	Nov	Dec	Year
0	Average	0.0	0.5	3.2	6.5	6.3	6.4	6.0	3.2	1.6	0.7	0.0	0.0	2.9
	Min/Max	0.0/0.0	0.4/0.7	2.5/3.9	5.4/8.3	4.7/7.6	5.0/9.1	4.4/9.3	2.3/4.1	1.2/2.3	0.4/1.1	0.0/0.1	0.0/0.0	2.5/3.3
Latitude -15	Average	0.0	1.2	4.9	8.2	7.0	6.9	6.6	3.5	1.9	1.0	0.1	0.0	3.5
	Min/Max	0.0/0.1	0.8/1.6	3.6/6.0	6.8/10.6	5.1/8.7	5.2/10.3	4.6/10.6	2.3/4.8	1.3/2.9	0.6/2.0	0.0/0.2	0.0/0.0	3.0/4.1
Latitude	Average	0.0	1.2	5.0	8.2	6.9	6.7	6.4	3.4	1.9	1.1	0.1	0.0	3.4
	Min/Max	0.0/0.1	0.8/1.7	3.7/6.1	6.8/10.6	5.0/8.6	5.0/10.1	4.4/10.4	2.2/4.7	1.2/2.9	0.6/2.1	0.0/0.2	0.0/0.0	3.0/4.1
Latitude +15	Average	0.0	1.3	5.0	8.0	6.7	6.4	6.1	3.2	1.8	1.1	0.1	0.0	3.3
	Min/Max	0.0/0.1	0.8/1.8	3.7/6.1	6.7/10.4	4.8/8.3	4.8/9.7	4.1/10.0	2.0/4.4	1.2/2.8	0.6/2.1	0.0/0.2	0.0/0.0	2.9/4.0

Solar Radiation for 2-Axis Tracking Flat-Plate Collectors (kWh/m²/day), Uncertainty ±9%

Tracker		Jan	Feb	Mar	Apr	May	June	July	Aug	Sept	Oct	Nov	Dec	Year
2-Axis	Average	0.0	1.3	5.0	8.3	7.1	7.3	6.9	3.6	1.9	1.1	0.1	0.0	3.6
	Min/Max	0.0/0.1	0.8/1.8	3.7/6.1	6.9/10.6	5.2/8.8	5.5/10.7	4.8/11.1	2.4/4.9	1.3/3.0	0.6/2.1	0.0/0.2	0.0/0.0	3.1/4.2

Direct Beam Solar Radiation for Concentrating Collectors (kWh/m²/day), Uncertainty ±8%

Tracker		Jan	Feb	Mar	Apr	May	June	July	Aug	Sept	Oct	Nov	Dec	Year
1-Axis, E-W	Average	0.0	0.8	2.7	3.6	2.0	2.6	2.8	1.1	0.5	0.4	0.1	0.0	1.4
Horiz Axis	Min/Max	0.0/0.1	0.4/1.2	1.6/3.5	2.6/5.1	0.8/3.2	1.3/4.8	1.4/5.6	0.4/1.9	0.2/1.2	0.1/1.1	0.0/0.2	0.0/0.0	1.1/1.8
1-Axis, N-S	Average	0.0	0.3	2.0	3.9	2.4	3.0	3.2	1.2	0.5	0.2	0.0	0.0	1.4
Horiz Axis	Min/Max	0.0/0.0	0.2/0.5	1.3/2.7	2.7/5.8	0.9/3.8	1.5/5.4	1.5/6.4	0.5/2.2	0.1/1.2	0.1/0.6	0.0/0.0	0.0/0.0	1.0/1.8
1-Axis, N-S	Average	0.0	0.9	3.4	5.1	2.8	3.4	3.6	1.5	0.7	0.4	0.1	0.0	1.8
Tilt=Latitude	Min/Max	0.0/0.1	0.5/1.3	2.2/4.6	3.7/7.5	1.0/4.4	1.7/6.1	1.8/7.3	0.6/2.7	0.2/1.7	0.1/1.3	0.0/0.2	0.0/0.0	1.4/2.4
2-Axis	Average	0.0	0.9	3.4	5.2	2.9	3.7	3.9	1.6	0.7	0.5	0.1	0.0	1.9
	Min/Max	0.0/0.1	0.5/1.3	2.2/4.6	3.7/7.6	1.1/4.7	1.8/6.6	1.9/7.9	0.6/2.8	0.2/1.8	0.1/1.3	0.0/0.2	0.0/0.0	1.5/2.5

Average Climatic Conditions

Element	Jan	Feb	Mar	Apr	May	June	July	Aug	Sept	Oct	Nov	Dec	Year
Temperature (°C)	-25.2	-27.7	-26.2	-19.0	-7.1	1.1	4.1	3.3	-0.8	-10.3	-18.7	-24.0	-12.6
Daily Minimum Temp	-28.5	-30.9	-29.5	-22.8	-9.8	-1.3	0.9	0.7	-2.8	-12.9	-21.6	-27.3	-15.5
Daily Maximum Temp	-21.9	-24.3	-22.8	-15.2	-4.3	3.5	7.2	5.7	1.0	-7.7	-15.8	-20.7	-9.6
Record Minimum Temp	-47.2	-48.9	-46.7	-41.1	-28.3	-15.6	-5.6	-6.7	-17.2	-35.6	-40.0	-48.3	-48.9
Record Maximum Temp	2.2	2.2	0.6	5.6	7.2	21.7	25.6	24.4	16.7	6.1	3.9	1.1	25.6
HDD, Base 18.3°C	1350	1288	1379	1120	787	517	443	467	575	887	1112	1312	11237
CDD, Base 18.3°C	0	0	0	0	0	0	0	0	0	0	0	0	0
Relative Humidity (%)	73	70	71	77	87	89	88	91	91	86	79	74	81
Wind Speed (m/s)	5.5	5.0	5.1	5.2	5.3	5.3	5.2	5.6	6.0	5.8	5.6	5.2	5.4

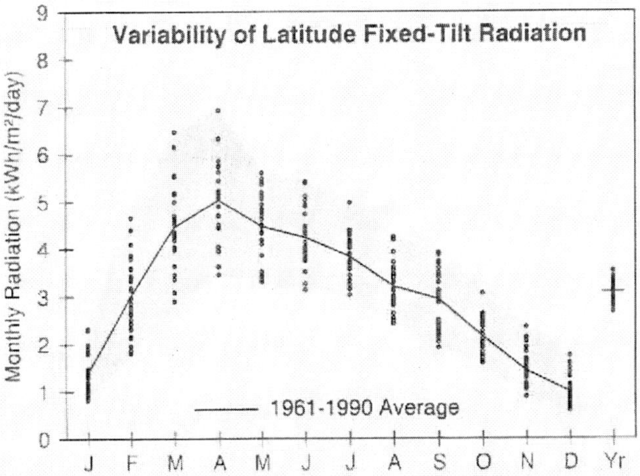

Bethel, AK

WBAN NO. 26615

LATITUDE: 60.78° N
LONGITUDE: 161.80° W
ELEVATION: 46 meters
MEAN PRESSURE: 1003 millibars

STATION TYPE: Secondary

Solar Radiation for Flat-Plate Collectors Facing South at a Fixed Tilt (kWh/m²/day), Uncertainty ±9%

Tilt (°)		Jan	Feb	Mar	Apr	May	June	July	Aug	Sept	Oct	Nov	Dec	Year
0	Average	0.4	1.1	2.5	3.9	4.5	4.8	4.3	3.2	2.2	1.2	0.5	0.2	2.4
	Min/Max	0.3/0.5	0.9/1.4	2.0/3.1	3.1/4.8	3.7/5.3	3.9/5.8	3.7/5.1	2.6/3.8	1.8/2.7	1.0/1.4	0.4/0.6	0.2/0.3	2.2/2.6
Latitude -15	Average	1.2	2.7	4.3	5.1	4.8	4.7	4.2	3.4	3.0	2.1	1.3	0.8	3.1
	Min/Max	0.7/2.0	1.7/4.2	2.8/6.1	3.6/6.8	3.6/5.9	3.5/5.9	3.4/5.4	2.6/4.5	2.0/3.9	1.6/2.9	0.8/2.1	0.5/1.5	2.7/3.5
Latitude	Average	1.4	3.0	4.5	5.0	4.5	4.2	3.8	3.2	2.9	2.2	1.4	1.0	3.1
	Min/Max	0.8/2.3	1.8/4.7	2.9/6.5	3.5/6.9	3.3/5.6	3.1/5.4	3.0/5.0	2.4/4.3	1.9/3.9	1.6/3.1	0.9/2.4	0.6/1.7	2.7/3.5
Latitude +15	Average	1.5	3.1	4.4	4.7	4.0	3.6	3.3	2.9	2.8	2.1	1.5	1.0	2.9
	Min/Max	0.9/2.5	1.8/4.9	2.8/6.5	3.2/6.7	2.9/5.2	2.7/4.7	2.6/4.3	2.1/3.8	1.8/3.7	1.5/3.1	0.9/2.5	0.6/1.9	2.5/3.4
90	Average	1.5	3.1	4.2	4.3	3.3	3.0	2.7	2.4	2.5	2.0	1.5	1.1	2.6
	Min/Max	0.8/2.5	1.8/4.8	2.7/6.2	2.7/6.1	2.4/4.6	2.2/3.8	2.2/3.5	1.8/3.2	1.5/3.3	1.4/2.9	0.9/2.5	0.6/1.9	2.3/3.1

Solar Radiation for 1-Axis Tracking Flat-Plate Collectors with a North-South Axis (kWh/m²/day), Uncertainty ±9%

Axis Tilt (°)		Jan	Feb	Mar	Apr	May	June	July	Aug	Sept	Oct	Nov	Dec	Year
0	Average	0.7	2.1	4.1	5.9	6.1	6.1	5.3	4.0	3.1	1.7	0.8	0.4	3.4
	Min/Max	0.4/1.1	1.4/3.2	2.7/6.0	4.0/8.3	4.5/8.0	4.2/8.3	3.8/7.1	3.0/5.4	2.0/4.2	1.3/2.4	0.6/1.3	0.3/0.7	2.9/3.8
Latitude -15	Average	1.3	3.3	5.5	6.9	6.6	6.3	5.5	4.3	3.7	2.4	1.4	0.9	4.0
	Min/Max	0.8/2.3	1.9/5.2	3.3/8.2	4.4/10.0	4.6/8.7	4.1/8.7	3.7/7.6	3.0/6.0	2.3/5.2	1.7/3.5	0.9/2.4	0.6/1.6	3.3/4.6
Latitude	Average	1.5	3.5	5.6	6.9	6.4	6.0	5.3	4.1	3.7	2.5	1.6	1.0	4.0
	Min/Max	0.9/2.5	2.0/5.6	3.4/8.5	4.4/10.0	4.4/8.5	3.8/8.4	3.5/7.4	2.9/5.9	2.2/5.2	1.8/3.7	0.9/2.6	0.6/1.9	3.3/4.6
Latitude +15	Average	1.6	3.6	5.6	6.7	6.0	5.6	4.9	3.9	3.5	2.5	1.6	1.1	3.9
	Min/Max	0.9/2.7	2.0/5.7	3.3/8.5	4.2/9.8	4.1/8.1	3.5/7.9	3.2/6.9	2.7/5.6	2.1/5.0	1.7/3.7	1.0/2.8	0.6/2.0	3.2/4.5

Solar Radiation for 2-Axis Tracking Flat-Plate Collectors (kWh/m²/day), Uncertainty ±9%

Tracker		Jan	Feb	Mar	Apr	May	June	July	Aug	Sept	Oct	Nov	Dec	Year
2-Axis	Average	1.6	3.6	5.6	6.9	6.7	6.5	5.7	4.3	3.7	2.5	1.6	1.1	4.2
	Min/Max	0.9/2.7	2.1/5.7	3.4/8.5	4.5/10.0	4.7/8.9	4.3/9.0	3.9/7.8	3.1/6.1	2.3/5.2	1.8/3.7	1.0/2.8	0.7/2.0	3.5/4.7

Direct Beam Solar Radiation for Concentrating Collectors (kWh/m²/day), Uncertainty ±8%

Tracker		Jan	Feb	Mar	Apr	May	June	July	Aug	Sept	Oct	Nov	Dec	Year
1-Axis, E-W Horiz Axis	Average	1.0	2.1	2.7	2.9	2.6	2.4	1.9	1.4	1.5	1.3	1.0	0.7	1.8
	Min/Max	0.4/2.0	0.8/4.0	1.0/5.0	0.9/4.8	1.3/4.1	0.7/4.0	0.6/3.6	0.6/2.8	0.6/2.5	0.7/2.2	0.4/1.9	0.3/1.5	1.3/2.2
1-Axis, N-S Horiz Axis	Average	0.4	1.3	2.5	3.4	3.4	3.1	2.5	1.7	1.6	0.9	0.4	0.2	1.8
	Min/Max	0.2/0.8	0.5/2.5	0.8/4.6	1.1/5.9	1.7/5.3	0.9/5.2	0.7/4.5	0.8/3.4	0.6/2.7	0.5/1.6	0.2/0.9	0.1/0.5	1.2/2.2
1-Axis, N-S Tilt=Latitude	Average	1.0	2.4	3.6	4.1	3.6	3.2	2.6	2.0	2.1	1.5	1.0	0.7	2.3
	Min/Max	0.4/2.0	0.9/4.6	1.3/6.7	1.3/7.1	1.8/5.7	1.0/5.3	0.8/4.7	0.9/4.0	0.8/3.5	0.9/2.7	0.5/2.0	0.3/1.5	1.6/2.9
2-Axis	Average	1.1	2.5	3.6	4.2	3.8	3.4	2.8	2.0	2.1	1.5	1.1	0.8	2.4
	Min/Max	0.4/2.2	0.9/4.7	1.3/6.7	1.3/7.2	1.9/6.0	1.1/5.7	0.8/5.1	0.9/4.1	0.8/3.5	0.9/2.7	0.5/2.1	0.3/1.6	1.7/3.0

Average Climatic Conditions

Element	Jan	Feb	Mar	Apr	May	June	July	Aug	Sept	Oct	Nov	Dec	Year
Temperature (°C)	-14.1	-14.4	-10.4	-4.7	4.4	10.3	12.8	11.6	7.3	-1.4	-8.4	-13.1	-1.7
Daily Minimum Temp	-17.6	-18.2	-14.8	-9.1	-0.2	5.7	8.7	7.9	3.5	-4.6	-11.7	-16.6	-5.6
Daily Maximum Temp	-10.6	-10.7	-6.1	-0.4	9.0	14.9	16.8	15.3	11.1	1.7	-5.2	-9.6	2.2
Record Minimum Temp	-44.4	-38.9	-39.4	-30.0	-15.6	-2.2	-0.6	-2.2	-7.8	-20.6	-31.1	-38.3	-44.4
Record Maximum Temp	8.9	7.8	7.8	12.8	26.1	30.0	28.3	28.9	22.2	13.3	8.9	7.2	30.0
HDD, Base 18.3°C	1004	918	891	690	432	242	172	208	330	613	803	973	7277
CDD, Base 18.3°C	0	0	0	0	0	0	0	0	0	0	0	0	0
Relative Humidity (%)	76	73	78	80	76	76	80	84	83	84	82	77	79
Wind Speed (m/s)	6.5	6.8	6.3	5.8	5.2	5.0	4.8	4.9	5.2	5.6	5.9	6.4	5.7

Bettles, AK

WBAN NO. 26533

LATITUDE: 66.92° N
LONGITUDE: 151.52° W
ELEVATION: 205 meters
MEAN PRESSURE: 988 millibars

STATION TYPE: Secondary

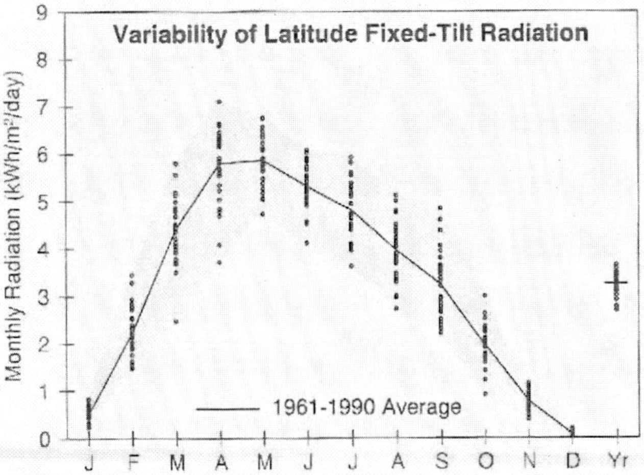

Solar Radiation for Flat-Plate Collectors Facing South at a Fixed Tilt (kWh/m²/day), Uncertainty ±9%

Tilt (°)		Jan	Feb	Mar	Apr	May	June	July	Aug	Sept	Oct	Nov	Dec	Year
0	Average	0.1	0.6	2.0	3.9	5.3	5.7	5.0	3.5	2.1	0.8	0.2	0.0	2.4
	Min/Max	0.1/0.1	0.4/0.7	1.5/2.2	2.8/4.4	4.6/6.0	4.7/6.3	4.1/5.8	2.7/4.2	1.7/2.6	0.6/1.0	0.1/0.2	0.0/0.0	2.1/2.6
Latitude -15	Average	0.4	2.0	4.2	5.8	6.2	5.8	5.2	4.2	3.3	1.9	0.7	0.1	3.3
	Min/Max	0.2/0.7	1.4/3.1	2.4/5.4	3.7/7.0	5.1/7.2	4.6/6.6	4.0/6.4	2.9/5.4	2.3/4.8	0.9/2.8	0.3/1.0	0.0/0.2	2.8/3.7
Latitude	Average	0.5	2.2	4.4	5.8	5.9	5.3	4.8	4.0	3.2	2.0	0.8	0.1	3.2
	Min/Max	0.2/0.8	1.5/3.4	2.5/5.8	3.7/7.1	4.7/6.8	4.1/6.1	3.6/5.9	2.7/5.1	2.2/4.8	0.9/3.0	0.4/1.1	0.0/0.2	2.7/3.6
Latitude +15	Average	0.5	2.3	4.4	5.5	5.2	4.5	4.1	3.5	3.0	2.0	0.8	0.1	3.0
	Min/Max	0.3/0.9	1.5/3.6	2.4/5.8	3.5/6.8	4.2/6.2	3.5/5.2	3.1/5.1	2.4/4.6	2.1/4.6	0.9/3.0	0.4/1.2	0.0/0.2	2.5/3.4
90	Average	0.5	2.3	4.3	5.3	4.8	4.1	3.7	3.2	2.9	1.9	0.8	0.1	2.8
	Min/Max	0.3/0.9	1.5/3.5	2.4/5.7	3.4/6.5	3.8/5.8	3.1/4.6	2.8/4.6	2.2/4.2	1.9/4.3	0.9/3.0	0.4/1.2	0.0/0.2	2.3/3.2

Solar Radiation for 1-Axis Tracking Flat-Plate Collectors with a North-South Axis (kWh/m²/day), Uncertainty ±9%

Axis Tilt (°)		Jan	Feb	Mar	Apr	May	June	July	Aug	Sept	Oct	Nov	Dec	Year
0	Average	0.1	1.3	3.8	6.6	8.3	8.3	7.1	5.0	3.2	1.4	0.3	0.0	3.8
	Min/Max	0.1/0.2	0.9/1.9	2.1/4.8	4.1/8.3	6.6/10.2	6.4/9.6	5.1/9.0	3.4/6.8	2.2/4.9	0.7/2.0	0.2/0.4	0.0/0.0	3.1/4.3
Latitude -15	Average	0.5	2.3	5.4	8.1	9.4	9.0	7.7	5.8	4.2	2.1	0.8	0.1	4.6
	Min/Max	0.2/0.8	1.5/3.6	2.8/7.2	4.9/10.4	7.3/11.6	6.7/10.5	5.3/9.9	3.7/7.9	2.7/6.6	0.9/3.3	0.4/1.1	0.0/0.2	3.7/5.3
Latitude	Average	0.5	2.5	5.5	8.1	9.2	8.8	7.5	5.6	4.2	2.2	0.8	0.1	4.6
	Min/Max	0.2/0.9	1.6/3.9	2.9/7.4	4.9/10.4	7.1/11.4	6.5/10.3	5.1/9.7	3.6/7.8	2.7/6.6	0.9/3.5	0.4/1.2	0.0/0.2	3.7/5.3
Latitude +15	Average	0.6	2.6	5.5	7.9	8.8	8.3	7.1	5.3	4.0	2.2	0.9	0.1	4.5
	Min/Max	0.3/0.9	1.6/4.0	2.9/7.4	4.8/10.3	6.8/11.0	6.2/9.8	4.7/9.2	3.4/7.4	2.6/6.4	0.9/3.5	0.4/1.3	0.0/0.2	3.5/5.1

Solar Radiation for 2-Axis Tracking Flat-Plate Collectors (kWh/m²/day), Uncertainty ±9%

Tracker		Jan	Feb	Mar	Apr	May	June	July	Aug	Sept	Oct	Nov	Dec	Year
2-Axis	Average	0.6	2.6	5.5	8.1	9.5	9.4	8.0	5.8	4.2	2.3	0.9	0.1	4.8
	Min/Max	0.3/0.9	1.6/4.0	2.9/7.4	4.9/10.4	7.5/11.9	7.1/11.0	5.5/10.3	3.8/8.0	2.8/6.6	1.0/3.5	0.4/1.3	0.0/0.2	3.8/5.5

Direct Beam Solar Radiation for Concentrating Collectors (kWh/m²/day), Uncertainty ±8%

Tracker		Jan	Feb	Mar	Apr	May	June	July	Aug	Sept	Oct	Nov	Dec	Year
1-Axis, E-W Horiz Axis	Average	0.5	1.6	2.9	3.5	4.2	4.2	3.4	2.4	2.0	1.2	0.6	0.1	2.2
	Min/Max	0.2/0.8	0.8/2.9	1.1/4.4	1.8/5.1	2.6/6.1	2.7/5.3	1.6/4.8	1.2/3.7	1.0/3.7	0.3/2.1	0.2/0.9	0.0/0.2	1.5/2.7
1-Axis, N-S Horiz Axis	Average	0.1	0.8	2.4	3.9	5.3	5.1	4.2	2.8	1.9	0.7	0.2	0.0	2.3
	Min/Max	0.0/0.2	0.4/1.5	0.8/3.6	2.0/5.9	3.2/7.7	3.2/6.5	2.0/5.9	1.4/4.6	0.9/3.7	0.2/1.2	0.1/0.3	0.0/0.0	1.5/2.8
1-Axis, N-S Tilt=Latitude	Average	0.5	1.8	3.8	5.0	5.9	5.5	4.5	3.3	2.7	1.3	0.6	0.1	2.9
	Min/Max	0.2/0.8	0.9/3.2	1.4/5.7	2.5/7.5	3.6/8.5	3.5/6.9	2.2/6.5	1.7/5.4	1.4/5.1	0.3/2.5	0.2/0.9	0.0/0.2	2.0/3.6
2-Axis	Average	0.5	1.8	3.8	5.1	6.2	6.0	4.8	3.5	2.7	1.4	0.6	0.1	3.0
	Min/Max	0.2/0.9	0.9/3.3	1.4/5.7	2.6/7.6	3.8/9.0	3.8/7.6	2.3/6.9	1.7/5.5	1.4/5.2	0.3/2.5	0.2/1.0	0.0/0.2	2.1/3.8

Average Climatic Conditions

Element	Jan	Feb	Mar	Apr	May	June	July	Aug	Sept	Oct	Nov	Dec	Year
Temperature (°C)	-24.8	-22.8	-16.2	-6.3	6.2	14.0	15.2	11.8	4.9	-7.5	-18.8	-22.9	-5.6
Daily Minimum Temp	-29.1	-27.9	-22.6	-12.4	0.8	8.3	9.6	6.6	0.2	-11.1	-22.7	-27.1	-10.6
Daily Maximum Temp	-20.5	-17.6	-9.8	-0.3	11.6	19.7	20.9	17.1	9.4	-3.9	-14.9	-18.8	-0.6
Record Minimum Temp	-56.7	-51.1	-48.9	-38.3	-23.3	-2.8	-1.7	-5.6	-16.7	-35.6	-49.4	-50.6	-56.7
Record Maximum Temp	8.9	4.4	6.7	13.3	30.0	33.3	33.9	30.6	26.1	11.7	7.2	3.3	33.9
HDD, Base 18.3°C	1338	1151	1069	738	376	132	107	207	403	801	1113	1279	8716
CDD, Base 18.3°C	0	0	0	0	0	0	10	6	0	0	0	0	16
Relative Humidity (%)	69	66	65	67	61	60	68	76	76	78	73	72	69
Wind Speed (m/s)	2.7	2.8	3.2	3.4	3.4	3.2	3.0	2.7	3.0	3.0	2.7	2.5	3.0

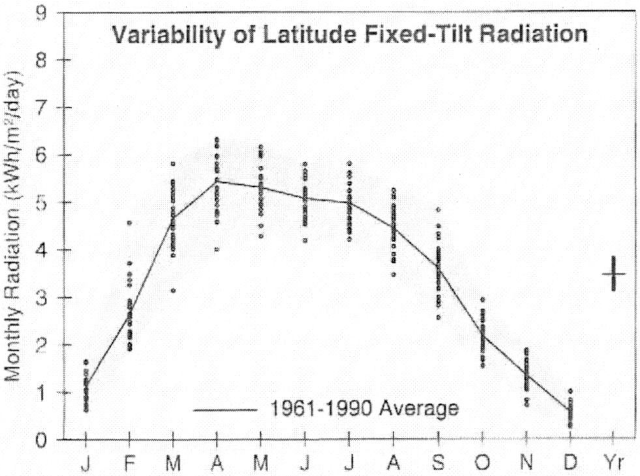

Big Delta, AK

WBAN NO. 26415

LATITUDE: 64.00° N
LONGITUDE: 145.73° W
ELEVATION: 388 meters
MEAN PRESSURE: 963 millibars

STATION TYPE: Secondary

Solar Radiation for Flat-Plate Collectors Facing South at a Fixed Tilt (kWh/m²/day), Uncertainty ±9%

Tilt (°)		Jan	Feb	Mar	Apr	May	June	July	Aug	Sept	Oct	Nov	Dec	Year
0	Average	0.2	0.8	2.3	3.9	5.1	5.5	5.2	3.9	2.4	1.1	0.3	0.1	2.6
	Min/Max	0.2/0.2	0.7/1.1	1.8/2.5	3.2/4.3	4.4/5.7	4.7/6.2	4.5/5.8	3.3/4.3	2.0/2.8	0.9/1.2	0.3/0.4	0.1/0.1	2.4/2.7
Latitude -15	Average	1.0	2.4	4.4	5.5	5.7	5.6	5.4	4.7	3.6	2.1	1.2	0.5	3.5
	Min/Max	0.6/1.4	1.8/4.1	3.0/5.5	4.1/6.3	4.6/6.6	4.6/6.4	4.6/6.3	3.7/5.5	2.6/4.8	1.5/2.7	0.6/1.6	0.2/0.9	3.2/3.8
Latitude	Average	1.1	2.6	4.6	5.4	5.3	5.1	5.0	4.4	3.6	2.2	1.3	0.6	3.4
	Min/Max	0.6/1.6	1.9/4.6	3.1/5.8	4.0/6.3	4.3/6.2	4.2/5.8	4.2/5.8	3.5/5.2	2.6/4.8	1.5/2.9	0.7/1.9	0.3/1.0	3.1/3.8
Latitude +15	Average	1.2	2.7	4.6	5.1	4.7	4.4	4.3	4.0	3.4	2.2	1.4	0.6	3.2
	Min/Max	0.6/1.8	2.0/4.8	3.1/5.8	3.7/6.1	3.7/5.4	3.6/5.0	3.6/5.0	3.1/4.7	2.4/4.6	1.5/2.9	0.7/2.0	0.3/1.1	2.9/3.5
90	Average	1.2	2.7	4.5	4.8	4.1	3.7	3.7	3.5	3.1	2.1	1.4	0.6	2.9
	Min/Max	0.6/1.8	1.9/4.7	3.0/5.6	3.3/5.7	3.2/4.8	3.1/4.2	3.1/4.3	2.7/4.2	2.2/4.3	1.4/2.8	0.7/2.0	0.3/1.1	2.7/3.3

Solar Radiation for 1-Axis Tracking Flat-Plate Collectors with a North-South Axis (kWh/m²/day), Uncertainty ±9%

Axis Tilt (°)		Jan	Feb	Mar	Apr	May	June	July	Aug	Sept	Oct	Nov	Dec	Year
0	Average	0.4	1.7	4.1	6.3	7.6	7.6	7.2	5.6	3.6	1.6	0.7	0.2	3.9
	Min/Max	0.3/0.6	1.2/2.8	2.7/5.2	4.7/7.5	6.0/9.3	5.9/9.3	5.8/8.9	4.4/6.7	2.6/4.8	1.2/2.1	0.4/0.9	0.1/0.3	3.5/4.3
Latitude -15	Average	1.1	2.8	5.7	7.6	8.3	8.1	7.8	6.4	4.6	2.4	1.3	0.5	4.7
	Min/Max	0.6/1.6	2.0/5.0	3.7/7.3	5.5/9.1	6.5/10.3	6.2/10.0	6.1/9.7	4.8/7.8	3.2/6.3	1.7/3.3	0.7/1.9	0.2/0.9	4.2/5.3
Latitude	Average	1.2	3.0	5.9	7.6	8.1	7.8	7.5	6.3	4.6	2.5	1.5	0.6	4.7
	Min/Max	0.6/1.8	2.1/5.4	3.8/7.6	5.4/9.1	6.3/10.1	5.9/9.7	5.9/9.4	4.7/7.7	3.2/6.3	1.7/3.4	0.8/2.0	0.3/1.0	4.2/5.3
Latitude +15	Average	1.3	3.1	5.9	7.3	7.7	7.4	7.1	5.9	4.5	2.5	1.5	0.6	4.6
	Min/Max	0.7/1.9	2.2/5.5	3.7/7.6	5.2/9.0	5.9/9.6	5.6/9.2	5.5/9.0	4.4/7.3	3.0/6.2	1.7/3.4	0.8/2.1	0.3/1.1	4.0/5.1

Solar Radiation for 2-Axis Tracking Flat-Plate Collectors (kWh/m²/day), Uncertainty ±9%

Tracker		Jan	Feb	Mar	Apr	May	June	July	Aug	Sept	Oct	Nov	Dec	Year
2-Axis	Average	1.3	3.1	5.9	7.6	8.5	8.4	8.0	6.5	4.7	2.5	1.5	0.6	4.9
	Min/Max	0.7/1.9	2.2/5.5	3.8/7.6	5.5/9.1	6.6/10.5	6.4/10.4	6.3/10.0	4.9/7.9	3.2/6.4	1.7/3.5	0.8/2.1	0.3/1.1	4.4/5.5

Direct Beam Solar Radiation for Concentrating Collectors (kWh/m²/day), Uncertainty ±8%

Tracker		Jan	Feb	Mar	Apr	May	June	July	Aug	Sept	Oct	Nov	Dec	Year
1-Axis, E-W Horiz Axis	Average	0.9	1.9	3.1	3.4	3.8	3.6	3.4	2.8	2.2	1.2	1.0	0.5	2.3
	Min/Max	0.4/1.5	0.9/4.0	1.7/4.4	2.1/4.4	2.4/5.1	2.3/5.0	2.2/5.0	1.7/3.8	1.2/3.5	0.6/2.1	0.4/1.6	0.2/1.0	1.9/2.8
1-Axis, N-S Horiz Axis	Average	0.3	1.0	2.7	3.9	4.8	4.5	4.3	3.3	2.2	0.8	0.4	0.1	2.4
	Min/Max	0.1/0.5	0.5/2.3	1.3/4.0	2.3/5.2	3.0/6.6	2.8/6.3	2.6/6.2	2.1/4.5	1.2/3.4	0.4/1.4	0.1/0.6	0.1/0.2	1.9/2.8
1-Axis, N-S Tilt=Latitude	Average	0.9	2.1	4.1	4.8	5.2	4.7	4.6	3.9	3.0	1.5	1.0	0.5	3.0
	Min/Max	0.4/1.5	1.1/4.6	2.1/5.9	2.9/6.4	3.3/7.2	2.9/6.6	2.8/6.6	2.4/5.3	1.7/4.7	0.8/2.5	0.4/1.6	0.2/0.9	2.5/3.6
2-Axis	Average	1.0	2.2	4.1	4.9	5.5	5.1	4.9	4.0	3.0	1.5	1.1	0.5	3.1
	Min/Max	0.4/1.6	1.1/4.7	2.1/5.9	2.9/6.5	3.5/7.6	3.2/7.2	3.0/7.1	2.5/5.4	1.7/4.7	0.8/2.5	0.4/1.7	0.2/1.0	2.6/3.8

Average Climatic Conditions

Element	Jan	Feb	Mar	Apr	May	June	July	Aug	Sept	Oct	Nov	Dec	Year
Temperature (°C)	-20.0	-16.6	-10.3	-0.9	8.2	13.8	15.7	13.0	6.9	-4.2	-14.7	-18.7	-2.3
Daily Minimum Temp	-23.9	-21.2	-16.4	-6.4	2.7	8.4	10.4	7.8	2.1	-7.9	-18.7	-22.5	-7.1
Daily Maximum Temp	-16.1	-11.9	-4.2	4.6	13.7	19.1	20.9	18.2	11.8	-0.4	-10.7	-14.8	2.5
Record Minimum Temp	-52.8	-51.1	-45.0	-38.3	-18.3	-1.1	0.0	-5.6	-18.9	-39.4	-43.9	-52.2	-52.8
Record Maximum Temp	8.9	10.6	14.4	22.2	32.2	33.3	32.8	31.1	26.1	18.9	11.1	12.8	33.3
HDD, Base 18.3°C	1188	977	887	577	313	139	88	171	343	698	992	1147	7519
CDD, Base 18.3°C	0	0	0	0	0	0	5	5	0	0	0	0	10
Relative Humidity (%)	67	68	64	59	52	59	66	68	70	77	73	71	66
Wind Speed (m/s)	5.4	4.7	4.1	3.8	3.8	3.5	2.9	3.2	3.8	3.9	4.6	5.2	4.1

Cold Bay, AK

WBAN NO. 25624

LATITUDE: 55.20° N
LONGITUDE: 162.72° W
ELEVATION: 29 meters
MEAN PRESSURE: 1003 millibars

STATION TYPE: Secondary

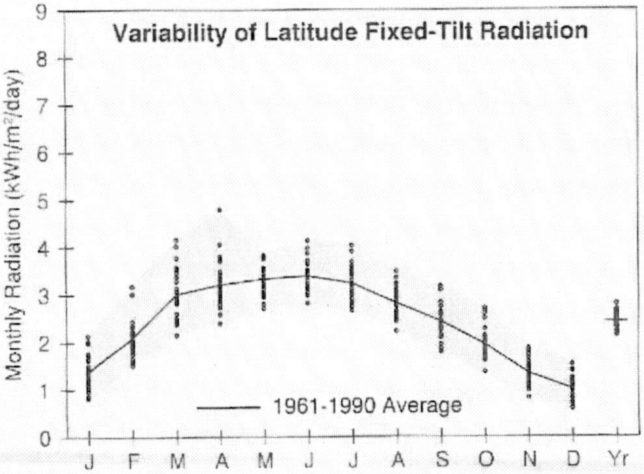

Solar Radiation for Flat-Plate Collectors Facing South at a Fixed Tilt (kWh/m²/day), Uncertainty ±9%

Tilt (°)		Jan	Feb	Mar	Apr	May	June	July	Aug	Sept	Oct	Nov	Dec	Year
0	Average	0.6	1.2	2.2	3.1	3.7	3.9	3.7	3.0	2.2	1.4	0.7	0.4	2.2
	Min/Max	0.5/0.8	1.0/1.5	1.8/2.8	2.6/4.1	3.1/4.1	3.3/4.6	3.2/4.3	2.5/3.5	1.8/2.6	1.2/1.6	0.5/0.9	0.3/0.5	2.0/2.5
Latitude -15	Average	1.2	2.0	3.0	3.3	3.6	3.7	3.5	3.0	2.5	1.9	1.2	0.9	2.5
	Min/Max	0.8/1.9	1.4/2.9	2.2/4.0	2.6/4.8	3.0/4.1	3.1/4.5	3.0/4.4	2.4/3.7	1.9/3.2	1.4/2.6	0.8/1.7	0.5/1.3	2.2/2.9
Latitude	Average	1.4	2.1	3.0	3.2	3.3	3.4	3.2	2.8	2.4	2.0	1.3	1.0	2.4
	Min/Max	0.8/2.1	1.5/3.2	2.2/4.2	2.4/4.8	2.7/3.8	2.8/4.1	2.7/4.0	2.2/3.5	1.8/3.2	1.4/2.7	0.8/1.9	0.6/1.5	2.1/2.8
Latitude +15	Average	1.4	2.1	2.9	2.9	2.9	2.9	2.8	2.5	2.2	1.9	1.4	1.1	2.3
	Min/Max	0.8/2.3	1.5/3.3	2.0/4.1	2.1/4.6	2.4/3.4	2.4/3.6	2.3/3.5	2.0/3.1	1.6/3.0	1.3/2.7	0.8/1.9	0.6/1.6	2.0/2.6
90	Average	1.4	2.0	2.6	2.4	2.3	2.2	2.2	2.0	1.9	1.7	1.3	1.0	1.9
	Min/Max	0.8/2.3	1.4/3.1	1.8/3.7	1.7/4.1	1.8/2.6	1.9/2.7	1.8/2.7	1.6/2.5	1.3/2.5	1.1/2.5	0.8/1.9	0.6/1.6	1.6/2.2

Solar Radiation for 1-Axis Tracking Flat-Plate Collectors with a North-South Axis (kWh/m²/day), Uncertainty ±9%

Axis Tilt (°)		Jan	Feb	Mar	Apr	May	June	July	Aug	Sept	Oct	Nov	Dec	Year
0	Average	0.9	1.7	2.9	3.6	4.2	4.3	4.0	3.3	2.6	1.7	0.9	0.6	2.6
	Min/Max	0.6/1.3	1.3/2.5	2.1/4.0	2.6/5.3	3.3/5.0	3.5/5.5	3.2/5.2	2.5/4.2	1.9/3.4	1.3/2.3	0.6/1.3	0.4/0.9	2.3/3.0
Latitude -15	Average	1.4	2.2	3.5	3.9	4.2	4.3	4.0	3.4	2.8	2.1	1.3	1.0	2.9
	Min/Max	0.8/2.2	1.6/3.5	2.4/5.0	2.7/5.9	3.2/5.1	3.4/5.6	3.2/5.3	2.5/4.4	2.0/3.9	1.5/3.0	0.8/1.9	0.6/1.5	2.5/3.3
Latitude	Average	1.5	2.3	3.6	3.8	4.1	4.1	3.8	3.2	2.8	2.2	1.4	1.1	2.8
	Min/Max	0.9/2.4	1.6/3.8	2.4/5.1	2.5/5.9	3.1/4.9	3.2/5.3	3.0/5.1	2.3/4.3	1.9/3.9	1.5/3.1	0.8/2.1	0.6/1.6	2.5/3.3
Latitude +15	Average	1.6	2.4	3.5	3.7	3.8	3.8	3.5	3.0	2.7	2.2	1.5	1.1	2.7
	Min/Max	0.9/2.5	1.6/3.8	2.3/5.1	2.4/5.8	2.8/4.6	3.0/4.9	2.8/4.7	2.1/4.0	1.8/3.7	1.4/3.1	0.9/2.1	0.6/1.8	2.4/3.2

Solar Radiation for 2-Axis Tracking Flat-Plate Collectors (kWh/m²/day), Uncertainty ±9%

Tracker		Jan	Feb	Mar	Apr	May	June	July	Aug	Sept	Oct	Nov	Dec	Year
2-Axis	Average	1.6	2.4	3.6	4.0	4.4	4.5	4.2	3.4	2.9	2.2	1.5	1.1	3.0
	Min/Max	0.9/2.5	1.6/3.8	2.4/5.2	2.7/5.9	3.3/5.2	3.6/5.7	3.3/5.5	2.5/4.5	2.0/3.9	1.5/3.2	0.9/2.2	0.6/1.8	2.6/3.4

Direct Beam Solar Radiation for Concentrating Collectors (kWh/m²/day), Uncertainty ±8%

Tracker		Jan	Feb	Mar	Apr	May	June	July	Aug	Sept	Oct	Nov	Dec	Year
1-Axis, E-W Horiz Axis	Average	0.8	1.0	1.4	1.1	1.2	1.1	1.0	0.8	0.9	0.9	0.7	0.6	1.0
	Min/Max	0.3/1.5	0.3/2.2	0.7/2.6	0.4/2.2	0.6/1.8	0.5/2.1	0.4/1.8	0.3/1.5	0.3/1.5	0.4/1.7	0.3/1.3	0.2/1.1	0.7/1.2
1-Axis, N-S Horiz Axis	Average	0.4	0.7	1.3	1.3	1.5	1.4	1.2	1.0	0.9	0.7	0.4	0.2	0.9
	Min/Max	0.1/0.8	0.2/1.6	0.6/2.4	0.4/2.5	0.7/2.3	0.6/2.6	0.4/2.3	0.3/1.8	0.3/1.6	0.3/1.3	0.2/0.7	0.1/0.5	0.7/1.1
1-Axis, N-S Tilt=Latitude	Average	0.8	1.2	1.8	1.6	1.6	1.4	1.2	1.1	1.2	1.1	0.7	0.6	1.2
	Min/Max	0.3/1.6	0.4/2.6	0.8/3.3	0.5/2.9	0.8/2.4	0.6/2.6	0.4/2.4	0.4/2.0	0.4/2.1	0.5/2.0	0.4/1.4	0.2/1.1	0.9/1.5
2-Axis	Average	0.9	1.2	1.8	1.6	1.6	1.5	1.3	1.1	1.2	1.1	0.8	0.6	1.2
	Min/Max	0.3/1.7	0.4/2.7	0.8/3.3	0.5/3.0	0.8/2.5	0.7/2.8	0.5/2.6	0.4/2.0	0.4/2.1	0.5/2.0	0.4/1.5	0.3/1.2	0.9/1.5

Average Climatic Conditions

Element	Jan	Feb	Mar	Apr	May	June	July	Aug	Sept	Oct	Nov	Dec	Year
Temperature (°C)	-1.9	-2.6	-1.2	0.7	4.2	7.6	10.3	10.8	8.7	4.2	1.3	-0.6	3.5
Daily Minimum Temp	-4.4	-5.1	-3.9	-1.9	1.6	4.9	7.8	8.4	6.2	1.6	-1.2	-3.0	0.9
Daily Maximum Temp	0.6	0.0	1.6	3.3	6.9	10.2	12.8	13.3	11.2	6.8	3.8	1.8	6.1
Record Minimum Temp	-22.2	-22.8	-25.0	-15.6	-7.8	-1.7	0.6	0.6	-2.8	-12.2	-17.2	-18.3	-25.0
Record Maximum Temp	10.0	10.0	13.3	15.6	19.4	20.6	25.0	25.6	24.4	20.6	15.0	12.2	25.6
HDD, Base 18.3°C	627	585	604	528	437	322	250	233	288	437	510	586	5407
CDD, Base 18.3°C	0	0	0	0	0	0	0	0	0	0	0	0	0
Relative Humidity (%)	84	84	83	82	82	84	87	88	85	82	83	84	84
Wind Speed (m/s)	8.0	8.0	7.7	8.0	7.3	7.1	7.0	7.2	7.4	7.6	7.8	7.8	7.6

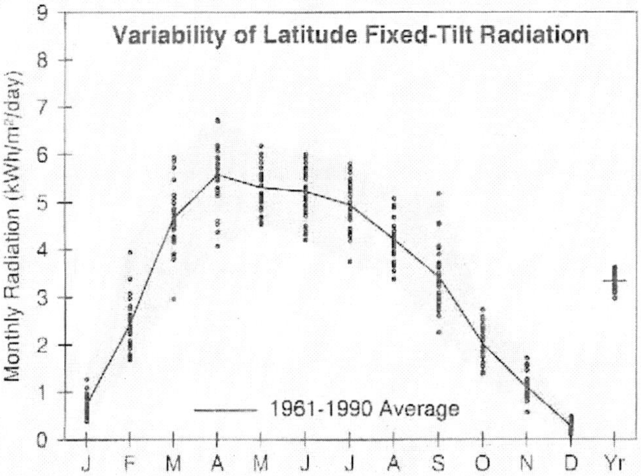

Fairbanks, AK

WBAN NO. 26411

LATITUDE: 64.82° N
LONGITUDE: 147.87° W
ELEVATION: 138 meters
MEAN PRESSURE: 994 millibars

STATION TYPE: Primary

Solar Radiation for Flat-Plate Collectors Facing South at a Fixed Tilt (kWh/m²/day), Uncertainty ±9%

Tilt (°)		Jan	Feb	Mar	Apr	May	June	July	Aug	Sept	Oct	Nov	Dec	Year
0	Average	0.1	0.8	2.3	4.0	5.1	5.6	5.1	3.7	2.3	1.0	0.3	0.0	2.5
	Min/Max	0.1/0.2	0.7/0.9	1.8/2.6	3.3/4.6	4.5/5.8	4.7/6.4	4.1/5.8	3.2/4.3	1.8/2.9	0.8/1.1	0.2/0.3	0.0/0.1	2.4/2.7
Latitude -15	Average	0.7	2.2	4.5	5.6	5.7	5.7	5.4	4.5	3.4	1.9	1.0	0.2	3.4
	Min/Max	0.3/1.1	1.6/3.5	2.9/5.6	4.2/6.7	4.9/6.6	4.6/6.6	4.1/6.3	3.6/5.3	2.3/5.1	1.4/2.6	0.5/1.5	0.1/0.4	3.1/3.7
Latitude	Average	0.7	2.4	4.7	5.6	5.3	5.2	4.9	4.2	3.4	2.0	1.1	0.3	3.3
	Min/Max	0.4/1.3	1.7/3.9	3.0/5.9	4.1/6.7	4.5/6.2	4.2/6.0	3.7/5.8	3.4/5.1	2.3/5.2	1.4/2.7	0.6/1.7	0.1/0.5	3.0/3.6
Latitude +15	Average	0.8	2.5	4.7	5.3	4.6	4.5	4.3	3.8	3.2	2.0	1.1	0.3	3.1
	Min/Max	0.4/1.3	1.7/4.1	2.9/5.9	3.7/6.4	4.0/5.4	3.6/5.1	3.2/5.0	3.0/4.6	2.1/4.9	1.4/2.8	0.6/1.8	0.1/0.5	2.7/3.4
90	Average	0.8	2.5	4.5	4.9	4.1	3.9	3.7	3.3	3.0	2.0	1.1	0.3	2.8
	Min/Max	0.4/1.4	1.7/4.1	2.8/5.8	3.4/6.1	3.5/4.8	3.1/4.4	2.8/4.4	2.7/4.1	1.9/4.6	1.3/2.7	0.6/1.8	0.1/0.5	2.5/3.1

Solar Radiation for 1-Axis Tracking Flat-Plate Collectors with a North-South Axis (kWh/m²/day), Uncertainty ±9%

Axis Tilt (°)		Jan	Feb	Mar	Apr	May	June	July	Aug	Sept	Oct	Nov	Dec	Year
0	Average	0.3	1.5	4.1	6.4	7.5	7.8	7.1	5.3	3.4	1.5	0.5	0.1	3.8
	Min/Max	0.2/0.5	1.1/2.3	2.5/5.2	4.7/7.8	6.1/9.0	6.0/9.4	5.0/8.6	4.0/6.6	2.4/5.2	1.1/1.9	0.3/0.8	0.0/0.1	3.3/4.2
Latitude -15	Average	0.7	2.6	5.7	7.7	8.2	8.3	7.6	6.0	4.4	2.2	1.1	0.2	4.6
	Min/Max	0.4/1.2	1.8/4.2	3.4/7.4	5.5/9.5	6.6/10.1	6.3/10.1	5.3/9.4	4.4/7.6	2.8/6.9	1.5/3.0	0.5/1.7	0.1/0.4	3.9/5.1
Latitude	Average	0.8	2.7	5.8	7.7	8.0	8.0	7.4	5.8	4.4	2.3	1.2	0.3	4.5
	Min/Max	0.4/1.3	1.9/4.6	3.5/7.7	5.4/9.6	6.4/9.8	6.0/9.8	5.0/9.1	4.3/7.4	2.8/6.9	1.5/3.2	0.6/1.9	0.1/0.5	3.9/5.0
Latitude +15	Average	0.8	2.8	5.8	7.4	7.6	7.6	6.9	5.5	4.2	2.3	1.2	0.3	4.4
	Min/Max	0.4/1.4	1.9/4.7	3.4/7.7	5.2/9.3	6.1/9.3	5.7/9.3	4.7/8.6	4.0/7.1	2.7/6.8	1.5/3.2	0.6/2.0	0.1/0.5	3.7/4.9

Solar Radiation for 2-Axis Tracking Flat-Plate Collectors (kWh/m²/day), Uncertainty ±9%

Tracker		Jan	Feb	Mar	Apr	May	June	July	Aug	Sept	Oct	Nov	Dec	Year
2-Axis	Average	0.8	2.8	5.8	7.7	8.4	8.7	7.9	6.0	4.4	2.3	1.2	0.3	4.7
	Min/Max	0.4/1.4	1.9/4.7	3.5/7.7	5.5/9.6	6.8/10.3	6.6/10.5	5.5/9.7	4.5/7.7	2.8/7.0	1.5/3.2	0.6/2.0	0.1/0.5	4.1/5.2

Direct Beam Solar Radiation for Concentrating Collectors (kWh/m²/day), Uncertainty ±8%

Tracker		Jan	Feb	Mar	Apr	May	June	July	Aug	Sept	Oct	Nov	Dec	Year
1-Axis, E-W	Average	0.6	1.6	2.9	3.2	3.6	3.7	3.3	2.5	2.0	1.1	0.7	0.3	2.1
Horiz Axis	Min/Max	0.2/1.1	0.8/3.3	1.2/4.4	1.6/4.6	2.3/4.9	2.3/4.8	1.8/4.6	1.5/3.5	1.0/3.8	0.5/1.8	0.2/1.4	0.1/0.5	1.7/2.6
1-Axis, N-S	Average	0.2	0.8	2.4	3.7	4.6	4.5	4.1	2.9	2.0	0.7	0.3	0.1	2.2
Horiz Axis	Min/Max	0.1/0.4	0.4/1.8	1.0/3.8	1.8/5.3	3.0/6.2	2.8/6.1	2.2/5.7	1.8/4.3	1.1/3.7	0.3/1.1	0.1/0.5	0.0/0.1	1.7/2.6
1-Axis, N-S	Average	0.6	1.8	3.8	4.6	5.0	4.8	4.3	3.4	2.8	1.3	0.8	0.3	2.8
Tilt=Latitude	Min/Max	0.2/1.1	0.9/3.7	1.5/5.8	2.3/6.6	3.2/6.8	3.0/6.4	2.3/6.1	2.1/5.0	1.5/5.2	0.6/2.1	0.2/1.4	0.1/0.5	2.1/3.3
2-Axis	Average	0.6	1.8	3.8	4.7	5.3	5.2	4.7	3.5	2.8	1.3	0.8	0.3	2.9
	Min/Max	0.2/1.2	0.9/3.8	1.5/5.8	2.3/6.7	3.4/7.2	3.2/6.9	2.5/6.6	2.1/5.1	1.5/5.2	0.6/2.2	0.3/1.5	0.1/0.5	2.2/3.5

Average Climatic Conditions

Element	Jan	Feb	Mar	Apr	May	June	July	Aug	Sept	Oct	Nov	Dec	Year
Temperature (°C)	-23.4	-19.8	-11.7	-0.7	9.2	15.4	16.9	13.8	7.5	-3.8	-16.3	-21.4	-2.8
Daily Minimum Temp	-28.1	-25.8	-18.7	-6.4	3.3	9.7	11.4	8.4	2.3	-7.7	-20.9	-26.0	-8.2
Daily Maximum Temp	-18.7	-13.8	-4.6	5.0	15.2	21.2	22.4	19.1	12.7	0.0	-11.7	-16.8	2.5
Record Minimum Temp	-51.7	-48.9	-45.0	-31.1	-18.3	-0.6	1.7	-2.8	-12.2	-32.8	-43.3	-52.2	-52.2
Record Maximum Temp	10.0	8.3	10.6	23.3	31.7	35.6	34.4	32.2	28.9	18.3	7.8	6.7	35.6
HDD, Base 18.3°C	1293	1067	930	572	282	101	68	148	325	687	1038	1232	7744
CDD, Base 18.3°C	0	0	0	0	0	14	25	7	0	0	0	0	47
Relative Humidity (%)	70	66	60	56	50	57	64	71	69	74	73	71	65
Wind Speed (m/s)	1.5	1.9	2.5	3.1	3.5	3.4	3.1	2.9	2.8	2.5	1.7	1.5	2.5

Gulkana, AK

WBAN NO. 26425

LATITUDE: 62.15° N
LONGITUDE: 145.45° W
ELEVATION: 481 meters
MEAN PRESSURE: 952 millibars

STATION TYPE: Secondary

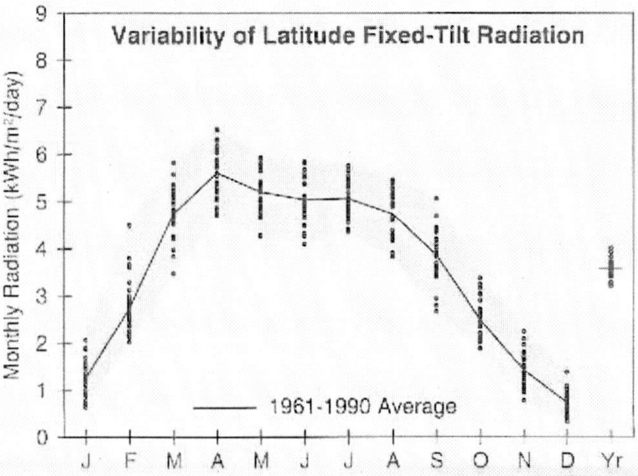

Solar Radiation for Flat-Plate Collectors Facing South at a Fixed Tilt (kWh/m²/day), Uncertainty ±9%

Tilt (°)		Jan	Feb	Mar	Apr	May	June	July	Aug	Sept	Oct	Nov	Dec	Year
0	Average	0.3	1.0	2.5	4.1	5.1	5.5	5.3	4.1	2.6	1.2	0.4	0.2	2.7
	Min/Max	0.2/0.3	0.9/1.2	2.1/2.8	3.7/4.6	4.3/5.6	4.7/6.3	4.6/5.9	3.5/4.6	2.1/3.0	1.1/1.4	0.4/0.6	0.1/0.2	2.5/2.9
Latitude -15	Average	1.1	2.5	4.5	5.6	5.6	5.5	5.5	5.0	3.9	2.4	1.3	0.6	3.6
	Min/Max	0.6/1.8	1.9/4.0	3.3/5.5	4.8/6.5	4.6/6.3	4.5/6.4	4.7/6.3	4.1/5.7	2.7/5.0	1.8/3.1	0.7/2.0	0.3/1.2	3.3/4.0
Latitude	Average	1.2	2.7	4.7	5.6	5.2	5.0	5.1	4.7	3.9	2.5	1.4	0.7	3.6
	Min/Max	0.6/2.1	2.0/4.5	3.5/5.8	4.7/6.5	4.3/5.9	4.1/5.8	4.3/5.8	3.8/5.4	2.7/5.1	1.9/3.4	0.8/2.2	0.3/1.4	3.2/4.0
Latitude +15	Average	1.3	2.8	4.7	5.3	4.6	4.3	4.4	4.3	3.6	2.5	1.5	0.8	3.3
	Min/Max	0.6/2.2	2.1/4.7	3.4/5.8	4.3/6.3	3.7/5.3	3.5/5.0	3.8/5.0	3.4/4.9	2.5/4.8	1.9/3.4	0.8/2.4	0.3/1.5	3.0/3.7
90	Average	1.3	2.8	4.5	4.8	3.9	3.6	3.7	3.7	3.3	2.4	1.5	0.8	3.0
	Min/Max	0.6/2.2	2.0/4.6	3.3/5.6	3.8/5.8	3.2/4.6	2.9/4.1	3.2/4.2	3.0/4.2	2.2/4.4	1.8/3.3	0.8/2.4	0.3/1.5	2.7/3.4

Solar Radiation for 1-Axis Tracking Flat-Plate Collectors with a North-South Axis (kWh/m²/day), Uncertainty ±9%

Axis Tilt (°)		Jan	Feb	Mar	Apr	May	June	July	Aug	Sept	Oct	Nov	Dec	Year
0	Average	0.6	1.8	4.3	6.6	7.5	7.7	7.4	6.0	4.0	1.9	0.8	0.3	4.1
	Min/Max	0.3/0.9	1.4/2.9	3.0/5.3	5.5/7.9	5.9/8.6	6.1/9.3	5.8/8.9	4.7/7.1	2.8/5.2	1.5/2.5	0.5/1.2	0.1/0.5	3.6/4.6
Latitude -15	Average	1.2	2.9	5.8	7.8	8.1	8.1	7.9	6.8	5.0	2.8	1.4	0.7	4.9
	Min/Max	0.6/2.0	2.1/4.9	4.0/7.3	6.5/9.4	6.4/9.4	6.2/9.8	6.1/9.6	5.2/8.1	3.3/6.7	2.0/3.8	0.8/2.3	0.3/1.3	4.3/5.5
Latitude	Average	1.3	3.1	6.0	7.8	7.9	7.8	7.6	6.7	5.0	2.9	1.5	0.8	4.9
	Min/Max	0.6/2.2	2.2/5.3	4.1/7.5	6.4/9.4	6.2/9.2	5.9/9.5	5.9/9.3	5.1/7.9	3.3/6.7	2.1/4.0	0.8/2.5	0.3/1.4	4.2/5.5
Latitude +15	Average	1.4	3.2	6.0	7.6	7.5	7.4	7.2	6.4	4.8	2.9	1.6	0.8	4.7
	Min/Max	0.7/2.4	2.3/5.4	4.1/7.5	6.2/9.1	5.8/8.9	5.6/9.0	5.5/8.8	4.8/7.6	3.1/6.6	2.1/4.0	0.8/2.6	0.3/1.5	4.1/5.4

Solar Radiation for 2-Axis Tracking Flat-Plate Collectors (kWh/m²/day), Uncertainty ±9%

Tracker		Jan	Feb	Mar	Apr	May	June	July	Aug	Sept	Oct	Nov	Dec	Year
2-Axis	Average	1.4	3.2	6.0	7.9	8.3	8.4	8.1	6.9	5.0	2.9	1.6	0.8	5.1
	Min/Max	0.7/2.4	2.3/5.4	4.1/7.5	6.5/9.5	6.5/9.6	6.5/10.2	6.3/9.9	5.3/8.1	3.3/6.8	2.1/4.0	0.8/2.6	0.3/1.6	4.4/5.7

Direct Beam Solar Radiation for Concentrating Collectors (kWh/m²/day), Uncertainty ±8%

Tracker		Jan	Feb	Mar	Apr	May	June	July	Aug	Sept	Oct	Nov	Dec	Year
1-Axis, E-W Horiz Axis	Average	0.9	1.8	2.9	3.5	3.6	3.6	3.5	3.0	2.4	1.5	0.9	0.6	2.4
	Min/Max	0.3/1.8	0.9/3.8	1.6/4.2	2.5/4.8	2.5/4.5	2.3/5.0	2.3/4.7	1.9/3.9	1.2/3.6	0.7/2.4	0.3/1.7	0.2/1.2	1.9/2.9
1-Axis, N-S Horiz Axis	Average	0.3	1.0	2.6	4.1	4.6	4.5	4.4	3.7	2.4	1.1	0.4	0.2	2.5
	Min/Max	0.1/0.7	0.5/2.3	1.3/3.8	2.9/5.9	3.1/5.8	3.0/6.3	2.7/6.1	2.3/4.7	1.2/3.6	0.5/1.6	0.1/0.8	0.0/0.4	2.0/3.0
1-Axis, N-S Tilt=Latitude	Average	0.9	2.0	3.9	5.0	5.0	4.7	4.7	4.3	3.3	1.8	1.0	0.5	3.1
	Min/Max	0.3/1.8	1.0/4.3	2.1/5.5	3.5/7.1	3.4/6.3	3.0/6.5	2.9/6.4	2.7/5.5	1.7/4.9	0.9/2.8	0.3/1.8	0.2/1.2	2.5/3.7
2-Axis	Average	1.0	2.1	3.9	5.1	5.3	5.1	5.0	4.4	3.3	1.8	1.0	0.6	3.2
	Min/Max	0.3/2.0	1.0/4.4	2.1/5.6	3.6/7.2	3.6/6.6	3.3/7.1	3.1/6.8	2.7/5.7	1.7/4.9	0.9/2.9	0.3/1.9	0.2/1.3	2.6/3.9

Average Climatic Conditions

Element	Jan	Feb	Mar	Apr	May	June	July	Aug	Sept	Oct	Nov	Dec	Year
Temperature (°C)	-20.9	-15.9	-9.2	-0.7	6.5	11.8	14.1	11.9	6.6	-2.6	-14.9	-19.6	-2.7
Daily Minimum Temp	-25.5	-21.7	-16.3	-6.7	0.3	5.7	8.0	5.7	0.8	-6.9	-19.2	-23.8	-8.3
Daily Maximum Temp	-16.4	-10.2	-2.2	5.3	12.7	17.9	20.2	18.2	12.3	1.8	-10.7	-15.2	2.8
Record Minimum Temp	-51.1	-53.9	-44.4	-41.1	-15.0	-2.2	-1.7	-6.7	-15.0	-30.6	-42.2	-50.0	-53.9
Record Maximum Temp	7.8	7.8	10.0	19.4	29.4	32.2	32.8	30.0	23.3	18.3	8.9	6.7	32.8
HDD, Base 18.3°C	1218	960	854	572	367	197	131	198	353	648	998	1174	7671
CDD, Base 18.3°C	0	0	0	0	0	0	0	0	0	0	0	0	0
Relative Humidity (%)	73	73	67	63	58	59	65	68	71	77	78	75	69
Wind Speed (m/s)	1.6	2.2	2.7	3.4	3.7	3.8	3.4	3.5	3.2	2.8	1.8	1.8	2.8

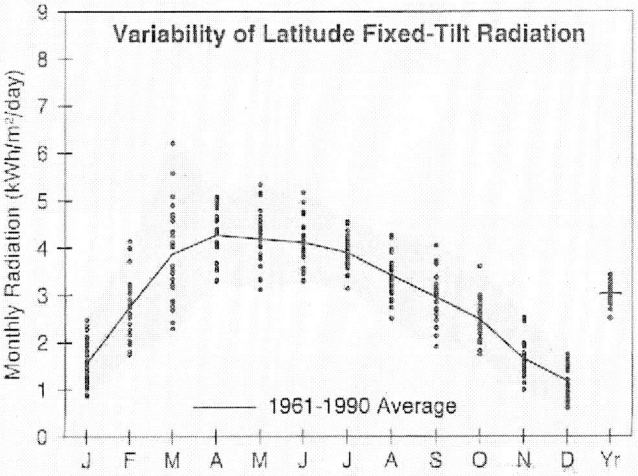

King Salmon, AK

WBAN NO. 25503

LATITUDE: 58.68° N
LONGITUDE: 156.65° W
ELEVATION: 15 meters
MEAN PRESSURE: 1005 millibars

STATION TYPE: Secondary

Solar Radiation for Flat-Plate Collectors Facing South at a Fixed Tilt (kWh/m²/day), Uncertainty ±9%

Tilt (°)		Jan	Feb	Mar	Apr	May	June	July	Aug	Sept	Oct	Nov	Dec	Year
0	Average	0.5	1.2	2.4	3.6	4.4	4.6	4.3	3.4	2.3	1.4	0.6	0.3	2.4
	Min/Max	0.4/0.6	0.9/1.5	1.8/3.1	3.0/4.0	3.5/5.3	3.9/5.5	3.7/4.9	2.7/4.0	1.9/2.8	1.2/1.7	0.5/0.7	0.2/0.4	2.2/2.6
Latitude -15	Average	1.4	2.5	3.7	4.4	4.5	4.5	4.3	3.6	3.0	2.4	1.5	1.0	3.1
	Min/Max	0.8/2.1	1.6/3.7	2.3/5.8	3.4/5.2	3.4/5.7	3.6/5.7	3.5/5.0	2.7/4.5	2.0/4.0	1.7/3.4	0.9/2.2	0.5/1.5	2.6/3.4
Latitude	Average	1.5	2.7	3.9	4.3	4.2	4.1	3.9	3.4	2.9	2.5	1.6	1.2	3.0
	Min/Max	0.9/2.5	1.7/4.1	2.3/6.2	3.3/5.1	3.1/5.3	3.3/5.2	3.1/4.5	2.5/4.3	1.9/4.0	1.7/3.6	1.0/2.5	0.6/1.7	2.5/3.4
Latitude +15	Average	1.6	2.8	3.8	3.9	3.7	3.6	3.4	3.0	2.8	2.5	1.7	1.3	2.8
	Min/Max	0.9/2.6	1.8/4.3	2.2/6.3	3.0/4.7	2.7/4.7	2.8/4.5	2.7/3.9	2.2/3.8	1.7/3.8	1.7/3.6	1.0/2.7	0.6/1.9	2.3/3.3
90	Average	1.6	2.7	3.5	3.4	3.0	2.8	2.7	2.5	2.4	2.3	1.7	1.3	2.5
	Min/Max	0.9/2.6	1.7/4.1	2.0/5.9	2.5/4.3	2.2/3.7	2.3/3.5	2.2/3.1	1.8/3.2	1.5/3.3	1.5/3.4	1.0/2.7	0.6/1.9	2.0/2.9

Solar Radiation for 1-Axis Tracking Flat-Plate Collectors with a North-South Axis (kWh/m²/day), Uncertainty ±9%

Axis Tilt (°)		Jan	Feb	Mar	Apr	May	June	July	Aug	Sept	Oct	Nov	Dec	Year
0	Average	0.8	2.0	3.7	5.0	5.7	5.7	5.2	4.2	3.1	2.0	1.0	0.6	3.3
	Min/Max	0.5/1.3	1.3/3.0	2.2/5.9	3.7/6.3	4.1/7.6	4.2/7.3	4.1/6.4	3.1/5.4	2.1/4.3	1.5/2.9	0.6/1.5	0.3/0.8	2.8/3.7
Latitude -15	Average	1.5	3.0	4.7	5.8	6.0	5.8	5.4	4.5	3.6	2.8	1.7	1.1	3.8
	Min/Max	0.8/2.5	1.8/4.6	2.6/7.8	4.1/7.3	4.1/8.2	4.2/7.6	4.1/6.6	3.2/5.9	2.3/5.2	1.9/4.1	1.0/2.6	0.6/1.7	3.2/4.4
Latitude	Average	1.7	3.2	4.8	5.7	5.8	5.6	5.2	4.3	3.6	2.9	1.8	1.3	3.8
	Min/Max	0.9/2.7	1.9/4.9	2.6/8.1	4.0/7.2	4.0/7.9	3.9/7.3	3.9/6.4	3.0/5.7	2.2/5.2	2.0/4.3	1.0/2.8	0.6/1.9	3.1/4.4
Latitude +15	Average	1.8	3.2	4.8	5.5	5.4	5.2	4.8	4.1	3.5	2.9	1.9	1.3	3.7
	Min/Max	0.9/2.9	2.0/5.1	2.5/8.1	3.8/7.0	3.7/7.5	3.6/6.8	3.6/6.0	2.8/5.4	2.1/5.1	1.9/4.3	1.1/3.0	0.6/2.0	3.0/4.3

Solar Radiation for 2-Axis Tracking Flat-Plate Collectors (kWh/m²/day), Uncertainty ±9%

Tracker		Jan	Feb	Mar	Apr	May	June	July	Aug	Sept	Oct	Nov	Dec	Year
2-Axis	Average	1.8	3.3	4.8	5.8	6.1	6.0	5.6	4.5	3.7	2.9	1.9	1.3	4.0
	Min/Max	0.9/2.9	2.0/5.1	2.7/8.1	4.1/7.3	4.3/8.3	4.3/7.8	4.2/6.8	3.2/5.9	2.3/5.3	2.0/4.3	1.1/3.0	0.6/2.0	3.3/4.6

Direct Beam Solar Radiation for Concentrating Collectors (kWh/m²/day), Uncertainty ±8%

Tracker		Jan	Feb	Mar	Apr	May	June	July	Aug	Sept	Oct	Nov	Dec	Year
1-Axis, E-W Horiz Axis	Average	1.1	1.9	2.3	2.4	2.3	2.1	1.9	1.5	1.4	1.5	1.1	0.9	1.7
	Min/Max	0.4/2.2	0.7/3.6	0.8/4.7	1.4/3.5	1.0/3.8	1.0/3.5	1.0/2.8	0.7/2.5	0.5/2.6	0.8/2.6	0.5/2.0	0.3/1.5	1.2/2.1
1-Axis, N-S Horiz Axis	Average	0.5	1.2	2.1	2.8	3.0	2.6	2.4	1.9	1.5	1.1	0.5	0.3	1.7
	Min/Max	0.2/1.0	0.4/2.4	0.8/4.4	1.6/4.3	1.4/4.9	1.2/4.3	1.2/3.5	0.9/3.0	0.6/2.7	0.6/1.9	0.2/1.0	0.1/0.6	1.2/2.0
1-Axis, N-S Tilt=Latitude	Average	1.2	2.1	3.0	3.4	3.1	2.7	2.4	2.1	2.0	1.8	1.2	0.9	2.2
	Min/Max	0.4/2.3	0.8/4.2	1.1/6.3	1.9/5.1	1.5/5.2	1.2/4.3	1.3/3.7	1.0/3.4	0.8/3.5	1.0/3.2	0.5/2.2	0.3/1.5	1.6/2.7
2-Axis	Average	1.2	2.2	3.0	3.4	3.3	2.9	2.6	2.2	2.0	1.8	1.2	0.9	2.2
	Min/Max	0.5/2.4	0.9/4.3	1.1/6.3	2.0/5.1	1.6/5.5	1.3/4.7	1.4/3.9	1.1/3.5	0.8/3.5	1.0/3.2	0.6/2.3	0.4/1.7	1.7/2.8

Average Climatic Conditions

Element	Jan	Feb	Mar	Apr	May	June	July	Aug	Sept	Oct	Nov	Dec	Year
Temperature (°C)	-9.5	-9.6	-5.3	-0.4	5.8	10.0	12.6	12.2	8.4	0.2	-5.4	-8.9	0.8
Daily Minimum Temp	-13.6	-14.1	-9.9	-4.8	1.1	5.2	8.0	8.0	4.3	-3.8	-9.4	-13.2	-3.5
Daily Maximum Temp	-5.5	-5.1	-0.7	3.9	10.5	14.8	17.2	16.3	12.6	4.3	-1.3	-4.8	5.2
Record Minimum Temp	-44.4	-40.6	-41.1	-26.1	-15.6	-1.7	0.6	-3.9	-9.4	-24.4	-33.3	-38.9	-44.4
Record Maximum Temp	11.7	13.9	13.3	18.3	25.0	31.1	30.0	28.9	23.3	19.4	13.3	8.9	31.1
HDD, Base 18.3°C	863	781	734	563	389	250	177	191	297	562	712	846	6364
CDD, Base 18.3°C	0	0	0	0	0	0	0	0	0	0	0	0	0
Relative Humidity (%)	76	73	73	72	70	74	78	80	79	79	81	77	76
Wind Speed (m/s)	4.9	5.1	5.1	5.0	5.0	4.9	4.5	4.7	4.7	4.6	4.7	4.9	4.8

Kodiak, AK

WBAN NO. 25501

LATITUDE: 57.75° N
LONGITUDE: 152.33° W
ELEVATION: 34 meters
MEAN PRESSURE: 1003 millibars

STATION TYPE: Secondary

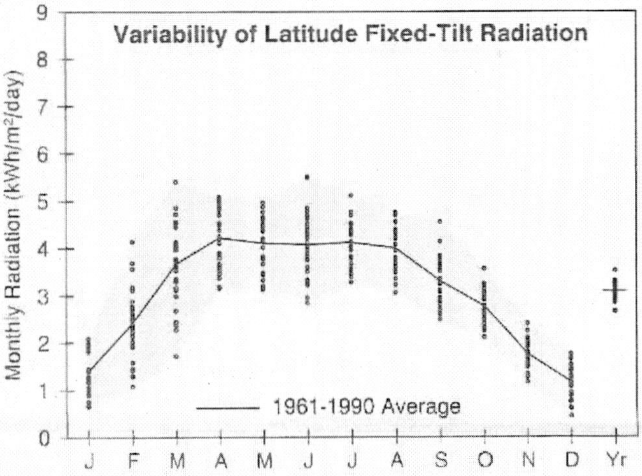

Solar Radiation for Flat-Plate Collectors Facing South at a Fixed Tilt (kWh/m²/day), Uncertainty ±9%

Tilt (°)		Jan	Feb	Mar	Apr	May	June	July	Aug	Sept	Oct	Nov	Dec	Year
0	Average	0.5	1.1	2.3	3.5	4.3	4.6	4.5	3.8	2.5	1.5	0.7	0.3	2.5
	Min/Max	0.3/0.6	0.8/1.5	1.4/3.0	2.9/4.0	3.4/4.9	3.4/6.1	3.7/5.4	3.1/4.4	2.1/3.1	1.2/1.6	0.5/0.8	0.2/0.4	2.2/2.7
Latitude -15	Average	1.2	2.2	3.5	4.3	4.4	4.5	4.5	4.2	3.3	2.6	1.6	1.0	3.1
	Min/Max	0.6/1.8	1.0/3.7	1.8/5.1	3.3/5.2	3.4/5.3	3.1/6.0	3.6/5.6	3.3/5.0	2.5/4.5	2.0/3.3	1.1/2.1	0.4/1.5	2.7/3.5
Latitude	Average	1.4	2.4	3.7	4.2	4.1	4.1	4.1	4.0	3.3	2.8	1.8	1.2	3.1
	Min/Max	0.6/2.1	1.1/4.1	1.7/5.4	3.2/5.1	3.1/5.0	2.8/5.5	3.3/5.1	3.1/4.8	2.5/4.6	2.1/3.6	1.2/2.4	0.4/1.8	2.7/3.5
Latitude +15	Average	1.5	2.5	3.6	3.9	3.6	3.5	3.6	3.6	3.1	2.8	1.8	1.2	2.9
	Min/Max	0.6/2.2	1.0/4.3	1.6/5.4	2.9/4.7	2.7/4.3	2.5/4.7	2.8/4.4	2.7/4.3	2.3/4.3	2.1/3.6	1.2/2.5	0.5/1.9	2.4/3.3
90	Average	1.4	2.4	3.3	3.3	2.9	2.7	2.8	2.9	2.7	2.6	1.8	1.2	2.5
	Min/Max	0.6/2.2	1.0/4.1	1.4/5.1	2.4/4.1	2.2/3.4	1.9/3.6	2.3/3.5	2.2/3.5	2.0/3.8	1.9/3.4	1.1/2.5	0.4/1.9	2.1/2.9

Solar Radiation for 1-Axis Tracking Flat-Plate Collectors with a North-South Axis (kWh/m²/day), Uncertainty ±9%

Axis Tilt (°)		Jan	Feb	Mar	Apr	May	June	July	Aug	Sept	Oct	Nov	Dec	Year
0	Average	0.8	1.8	3.5	5.0	5.6	5.8	5.8	5.1	3.5	2.3	1.1	0.6	3.4
	Min/Max	0.4/1.1	0.9/3.0	1.6/5.2	3.6/6.0	3.9/7.0	3.7/8.7	4.3/7.5	3.6/6.4	2.6/4.9	1.8/2.8	0.8/1.4	0.3/0.8	2.9/3.8
Latitude -15	Average	1.4	2.6	4.5	5.7	5.9	5.9	5.9	5.5	4.2	3.2	1.8	1.1	4.0
	Min/Max	0.6/2.1	1.1/4.6	1.9/6.7	4.0/7.0	4.0/7.4	3.6/9.0	4.4/7.9	3.9/7.0	3.0/6.0	2.4/4.1	1.2/2.5	0.4/1.7	3.4/4.5
Latitude	Average	1.5	2.8	4.6	5.7	5.7	5.7	5.7	5.4	4.2	3.3	2.0	1.2	4.0
	Min/Max	0.7/2.3	1.1/4.9	1.9/6.9	3.9/6.9	3.8/7.1	3.4/8.7	4.2/7.6	3.7/6.8	3.0/6.0	2.5/4.3	1.3/2.7	0.5/1.9	3.4/4.5
Latitude +15	Average	1.6	2.8	4.5	5.4	5.3	5.3	5.3	5.1	4.0	3.3	2.0	1.3	3.8
	Min/Max	0.7/2.4	1.1/5.1	1.8/6.9	3.7/6.7	3.5/6.7	3.2/8.3	3.9/7.2	3.5/6.5	2.9/5.9	2.5/4.3	1.3/2.8	0.5/2.0	3.2/4.4

Solar Radiation for 2-Axis Tracking Flat-Plate Collectors (kWh/m²/day), Uncertainty ±9%

Tracker		Jan	Feb	Mar	Apr	May	June	July	Aug	Sept	Oct	Nov	Dec	Year
2-Axis	Average	1.6	2.8	4.6	5.8	6.0	6.2	6.1	5.6	4.2	3.3	2.0	1.3	4.1
	Min/Max	0.7/2.4	1.1/5.1	1.9/6.9	4.0/7.0	4.1/7.5	3.8/9.4	4.5/8.1	3.9/7.1	3.0/6.1	2.5/4.3	1.3/2.8	0.5/2.0	3.5/4.7

Direct Beam Solar Radiation for Concentrating Collectors (kWh/m²/day), Uncertainty ±8%

Tracker		Jan	Feb	Mar	Apr	May	June	July	Aug	Sept	Oct	Nov	Dec	Year
1-Axis, E-W Horiz Axis	Average	1.0	1.5	2.2	2.5	2.3	2.3	2.3	2.3	1.8	1.8	1.3	0.9	1.9
	Min/Max	0.3/1.6	0.3/3.5	0.6/3.6	1.5/3.5	0.9/3.4	0.9/4.5	1.3/3.5	1.2/3.3	1.0/3.0	1.3/2.7	0.7/2.0	0.3/1.5	1.5/2.2
1-Axis, N-S Horiz Axis	Average	0.4	1.0	2.1	3.0	3.0	3.0	3.0	2.9	2.0	1.4	0.6	0.3	1.9
	Min/Max	0.1/0.8	0.2/2.4	0.5/3.4	1.8/4.1	1.1/4.3	1.2/5.9	1.6/4.6	1.4/4.3	1.1/3.3	0.9/2.0	0.4/1.0	0.1/0.6	1.5/2.2
1-Axis, N-S Tilt=Latitude	Average	1.0	1.8	3.0	3.5	3.2	3.0	3.1	3.2	2.5	2.2	1.4	0.9	2.4
	Min/Max	0.3/1.7	0.4/4.1	0.8/4.8	2.1/4.9	1.2/4.6	1.2/5.9	1.7/4.7	1.6/4.8	1.5/4.3	1.5/3.3	0.7/2.1	0.3/1.5	1.9/2.8
2-Axis	Average	1.1	1.8	3.0	3.6	3.4	3.3	3.3	3.3	2.6	2.3	1.4	0.9	2.5
	Min/Max	0.3/1.8	0.4/4.2	0.8/4.8	2.1/4.9	1.2/4.8	1.3/6.4	1.8/5.1	1.6/4.9	1.5/4.3	1.5/3.3	0.8/2.2	0.3/1.6	2.0/2.9

Average Climatic Conditions

Element	Jan	Feb	Mar	Apr	May	June	July	Aug	Sept	Oct	Nov	Dec	Year
Temperature (°C)	-1.2	-0.8	0.5	3.1	6.4	9.8	12.4	12.9	10.0	4.8	1.3	-0.7	4.9
Daily Minimum Temp	-4.1	-3.9	-2.8	-0.3	3.1	6.4	9.0	9.1	6.3	1.2	-1.8	-3.8	1.6
Daily Maximum Temp	1.7	2.3	3.8	6.5	9.7	13.1	15.8	16.6	13.7	8.4	4.4	2.4	8.2
Record Minimum Temp	-26.7	-24.4	-21.1	-13.9	-6.7	-1.1	2.8	2.2	-3.3	-12.2	-17.8	-18.3	-26.7
Record Maximum Temp	12.2	13.3	13.9	17.8	26.7	30.0	27.8	28.3	22.8	16.7	12.2	13.3	30.0
HDD, Base 18.3°C	604	537	553	458	371	257	183	169	250	418	510	589	4898
CDD, Base 18.3°C	0	0	0	0	0	0	0	0	0	0	0	0	0
Relative Humidity (%)	78	77	73	72	76	80	82	82	81	75	75	75	77
Wind Speed (m/s)	5.5	5.5	5.3	5.2	4.6	4.1	3.4	3.6	4.1	4.9	5.4	5.5	4.7

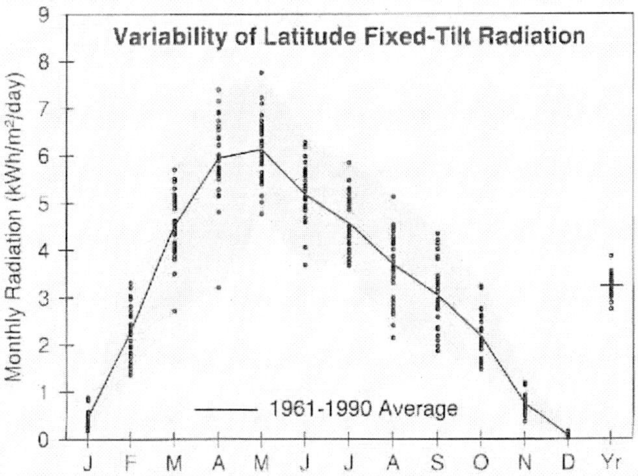

Kotzebue, AK

WBAN NO. 26616

LATITUDE: 66.87° N
LONGITUDE: 162.63° W
ELEVATION: 5 meters
MEAN PRESSURE: 1011 millibars

STATION TYPE: Secondary

Solar Radiation for Flat-Plate Collectors Facing South at a Fixed Tilt (kWh/m²/day), Uncertainty ±9%

Tilt (°)		Jan	Feb	Mar	Apr	May	June	July	Aug	Sept	Oct	Nov	Dec	Year
0	Average	0.1	0.6	2.1	4.1	5.5	5.5	4.8	3.3	2.0	0.9	0.2	0.0	2.4
	Min/Max	0.1/0.1	0.5/0.7	1.7/2.3	2.6/4.7	4.6/6.5	4.4/6.4	4.0/5.8	2.4/4.1	1.5/2.4	0.7/1.0	0.1/0.2	0.0/0.0	2.2/2.7
Latitude -15	Average	0.4	2.1	4.3	6.0	6.4	5.7	5.0	3.9	3.1	2.1	0.7	0.1	3.3
	Min/Max	0.2/0.8	1.3/3.0	2.7/5.4	3.3/7.3	5.1/8.0	4.1/6.8	4.1/6.4	2.3/5.4	1.9/4.3	1.4/3.0	0.3/1.0	0.0/0.2	2.8/3.9
Latitude	Average	0.5	2.3	4.5	6.0	6.1	5.2	4.6	3.7	3.1	2.2	0.7	0.1	3.2
	Min/Max	0.2/0.9	1.4/3.3	2.7/5.7	3.2/7.4	4.8/7.8	3.7/6.3	3.7/5.8	2.2/5.1	1.9/4.3	1.5/3.2	0.4/1.2	0.0/0.2	2.8/3.9
Latitude +15	Average	0.5	2.3	4.5	5.7	5.6	4.4	3.9	3.3	2.9	2.2	0.8	0.1	3.0
	Min/Max	0.2/0.9	1.4/3.4	2.7/5.7	3.0/7.1	4.3/7.2	3.2/5.6	3.2/5.0	1.9/4.6	1.7/4.1	1.5/3.3	0.4/1.2	0.0/0.2	2.6/3.6
90	Average	0.5	2.3	4.4	5.4	5.2	4.0	3.5	3.0	2.7	2.1	0.8	0.1	2.8
	Min/Max	0.2/0.9	1.4/3.4	2.6/5.5	2.9/6.7	4.0/6.7	2.8/5.1	2.8/4.5	1.7/4.2	1.6/3.9	1.4/3.2	0.4/1.2	0.0/0.2	2.4/3.4

Solar Radiation for 1-Axis Tracking Flat-Plate Collectors with a North-South Axis (kWh/m²/day), Uncertainty ±9%

Axis Tilt (°)		Jan	Feb	Mar	Apr	May	June	July	Aug	Sept	Oct	Nov	Dec	Year
0	Average	0.1	1.3	3.9	6.8	8.4	8.0	6.7	4.7	3.1	1.5	0.3	0.0	3.8
	Min/Max	0.1/0.3	0.8/1.8	2.3/4.8	3.5/8.6	6.3/11.4	5.3/9.8	5.0/9.3	2.7/6.7	1.9/4.5	1.0/2.2	0.2/0.5	0.0/0.0	3.1/4.6
Latitude -15	Average	0.4	2.4	5.5	8.4	9.5	8.7	7.3	5.3	3.9	2.4	0.7	0.1	4.6
	Min/Max	0.2/0.8	1.4/3.5	3.1/7.1	4.1/10.7	7.0/13.1	5.4/10.9	5.3/10.3	2.8/7.9	2.2/5.9	1.6/3.7	0.3/1.1	0.0/0.2	3.7/5.6
Latitude	Average	0.5	2.5	5.7	8.4	9.4	8.4	7.1	5.2	3.9	2.5	0.8	0.1	4.5
	Min/Max	0.2/0.9	1.5/3.8	3.2/7.3	4.1/10.7	6.8/13.1	5.2/10.6	5.1/10.0	2.6/7.7	2.2/5.9	1.6/3.8	0.4/1.3	0.0/0.2	3.7/5.6
Latitude +15	Average	0.5	2.6	5.7	8.2	9.0	8.0	6.7	4.9	3.8	2.5	0.8	0.1	4.4
	Min/Max	0.2/1.0	1.5/3.9	3.1/7.3	4.0/10.5	6.5/12.7	4.8/10.1	4.8/9.6	2.5/7.4	2.1/5.8	1.6/3.9	0.4/1.3	0.0/0.2	3.5/5.5

Solar Radiation for 2-Axis Tracking Flat-Plate Collectors (kWh/m²/day), Uncertainty ±9%

Tracker		Jan	Feb	Mar	Apr	May	June	July	Aug	Sept	Oct	Nov	Dec	Year
2-Axis	Average	0.5	2.6	5.7	8.4	9.6	9.0	7.6	5.4	4.0	2.5	0.8	0.1	4.7
	Min/Max	0.2/1.0	1.5/3.9	3.2/7.3	4.1/10.7	7.1/13.3	5.7/11.3	5.5/10.6	2.8/8.0	2.2/5.9	1.6/3.9	0.4/1.3	0.0/0.2	3.8/5.8

Direct Beam Solar Radiation for Concentrating Collectors (kWh/m²/day), Uncertainty ±8%

Tracker		Jan	Feb	Mar	Apr	May	June	July	Aug	Sept	Oct	Nov	Dec	Year
1-Axis, E-W Horiz Axis	Average	0.4	1.6	2.9	3.5	4.0	4.1	3.3	2.2	1.9	1.4	0.5	0.1	2.2
	Min/Max	0.1/0.9	0.7/2.8	1.0/4.2	1.2/5.1	2.5/6.5	1.8/5.8	2.0/5.4	0.6/3.8	0.7/3.2	0.7/2.5	0.2/1.0	0.0/0.2	1.5/3.0
1-Axis, N-S Horiz Axis	Average	0.1	0.8	2.4	4.1	5.0	5.0	4.0	2.6	1.8	0.9	0.2	0.0	2.2
	Min/Max	0.0/0.2	0.3/1.4	0.7/3.5	1.4/5.9	3.0/8.3	2.1/6.9	2.5/6.8	0.8/4.6	0.6/3.2	0.4/1.6	0.0/0.3	0.0/0.0	1.5/3.1
1-Axis, N-S Tilt=Latitude	Average	0.4	1.8	3.8	5.1	5.5	5.3	4.3	3.1	2.6	1.7	0.6	0.1	2.9
	Min/Max	0.1/0.9	0.8/3.1	1.2/5.4	1.7/7.4	3.3/9.2	2.3/7.4	2.7/7.3	0.9/5.5	0.9/4.5	0.9/3.0	0.2/1.0	0.0/0.2	1.9/3.9
2-Axis	Average	0.4	1.8	3.8	5.2	5.9	5.8	4.7	3.2	2.6	1.7	0.6	0.1	3.0
	Min/Max	0.1/0.9	0.8/3.2	1.2/5.4	1.8/7.6	3.5/9.7	2.5/8.1	2.9/7.9	0.9/5.7	0.9/4.5	0.9/3.1	0.2/1.1	0.0/0.2	2.0/4.1

Average Climatic Conditions

Element	Jan	Feb	Mar	Apr	May	June	July	Aug	Sept	Oct	Nov	Dec	Year
Temperature (°C)	-18.3	-20.4	-17.6	-11.5	-0.4	6.6	12.1	11.2	5.6	-5.1	-13.4	-18.3	-5.8
Daily Minimum Temp	-21.9	-24.4	-22.2	-16.5	-4.2	3.2	9.2	8.5	2.8	-7.7	-16.4	-21.9	-9.3
Daily Maximum Temp	-14.7	-16.5	-12.9	-6.5	3.3	9.8	15.1	13.9	8.3	-2.4	-10.5	-14.7	-2.3
Record Minimum Temp	-45.0	-46.7	-44.4	-42.2	-27.8	-6.7	-1.1	-1.1	-9.4	-28.3	-37.8	-43.9	-46.7
Record Maximum Temp	3.9	4.4	3.9	8.9	23.3	29.4	29.4	26.7	20.6	10.6	3.3	2.8	29.4
HDD, Base 18.3°C	1137	1086	1113	895	582	353	196	226	383	725	953	1135	8784
CDD, Base 18.3°C	0	0	0	0	0	0	3	4	0	0	0	0	7
Relative Humidity (%)	72	70	71	76	81	82	81	81	79	79	76	74	77
Wind Speed (m/s)	6.2	5.5	5.1	5.3	5.2	5.4	5.6	5.6	6.0	5.9	6.4	6.3	5.7

McGrath, AK

WBAN NO. 26510

LATITUDE: 62.97° N
LONGITUDE: 155.62° W
ELEVATION: 103 meters
MEAN PRESSURE: 997 millibars

STATION TYPE: Secondary

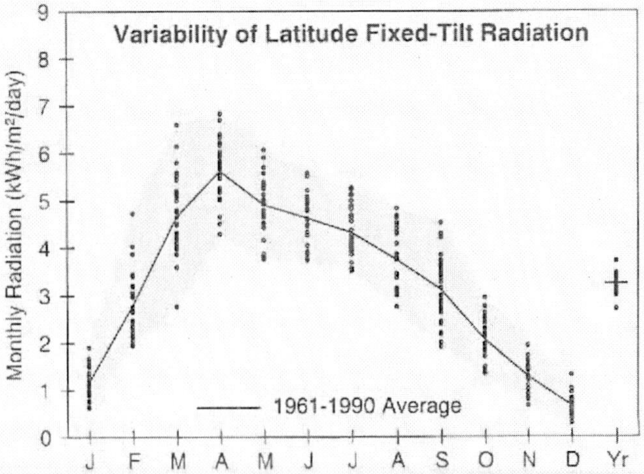

Solar Radiation for Flat-Plate Collectors Facing South at a Fixed Tilt (kWh/m²/day), Uncertainty ±9%

Tilt (°)		Jan	Feb	Mar	Apr	May	June	July	Aug	Sept	Oct	Nov	Dec	Year
0	Average	0.3	1.0	2.4	4.2	4.8	5.1	4.6	3.5	2.2	1.1	0.4	0.1	2.5
	Min/Max	0.2/0.3	0.8/1.2	1.9/2.9	3.5/4.7	3.9/5.7	4.3/5.9	4.0/5.4	2.8/4.1	1.7/2.8	0.9/1.3	0.3/0.4	0.1/0.1	2.2/2.8
Latitude -15	Average	1.0	2.5	4.4	5.6	5.3	5.1	4.7	4.0	3.1	2.0	1.1	0.6	3.3
	Min/Max	0.6/1.6	1.8/4.2	2.7/6.2	4.4/6.8	4.1/6.5	4.1/6.1	3.9/5.8	3.0/5.1	2.0/4.5	1.3/2.8	0.6/1.7	0.3/1.1	2.8/3.8
Latitude	Average	1.2	2.8	4.7	5.6	4.9	4.6	4.3	3.7	3.1	2.1	1.3	0.6	3.2
	Min/Max	0.6/1.9	1.9/4.7	2.8/6.6	4.3/6.8	3.8/6.1	3.7/5.6	3.5/5.3	2.8/4.8	1.9/4.5	1.3/2.9	0.7/1.9	0.3/1.3	2.7/3.7
Latitude +15	Average	1.2	2.9	4.6	5.4	4.4	4.0	3.7	3.3	2.9	2.1	1.3	0.7	3.0
	Min/Max	0.6/2.0	2.0/4.9	2.7/6.6	4.0/6.6	3.3/5.4	3.2/4.8	3.0/4.6	2.4/4.3	1.7/4.3	1.3/2.9	0.7/2.0	0.3/1.4	2.5/3.5
90	Average	1.2	2.8	4.5	5.0	3.8	3.3	3.2	2.9	2.7	2.0	1.3	0.7	2.8
	Min/Max	0.6/2.0	2.0/4.9	2.6/6.4	3.6/6.1	2.8/4.8	2.7/4.0	2.6/3.9	2.1/3.8	1.6/3.9	1.2/2.8	0.7/2.0	0.3/1.4	2.3/3.2

Solar Radiation for 1-Axis Tracking Flat-Plate Collectors with a North-South Axis (kWh/m²/day), Uncertainty ±9%

Axis Tilt (°)		Jan	Feb	Mar	Apr	May	June	July	Aug	Sept	Oct	Nov	Dec	Year
0	Average	0.5	1.8	4.2	6.5	6.9	6.9	6.2	4.7	3.2	1.6	0.7	0.2	3.6
	Min/Max	0.3/0.8	1.3/3.0	2.5/5.9	4.9/7.9	5.1/9.1	5.1/8.4	4.8/7.9	3.6/6.2	2.0/4.8	1.1/2.2	0.4/1.0	0.1/0.4	3.0/4.3
Latitude -15	Average	1.1	3.0	5.7	7.7	7.5	7.2	6.6	5.3	4.0	2.3	1.3	0.6	4.3
	Min/Max	0.6/1.8	2.0/5.2	3.1/8.3	5.6/9.6	5.5/10.1	5.2/9.0	4.9/8.5	3.8/7.1	2.2/6.1	1.5/3.3	0.6/1.9	0.3/1.2	3.5/5.2
Latitude	Average	1.2	3.2	5.9	7.7	7.3	7.0	6.3	5.1	4.0	2.4	1.4	0.7	4.3
	Min/Max	0.6/2.0	2.2/5.6	3.2/8.6	5.6/9.6	5.3/9.9	4.9/8.7	4.7/8.2	3.7/6.9	2.2/6.2	1.5/3.5	0.7/2.1	0.3/1.4	3.5/5.2
Latitude +15	Average	1.3	3.3	5.9	7.5	7.0	6.6	5.9	4.9	3.8	2.4	1.4	0.7	4.2
	Min/Max	0.7/2.2	2.2/5.7	3.2/8.6	5.5/9.4	5.0/9.4	4.6/8.3	4.4/7.8	3.4/6.6	2.0/6.0	1.5/3.5	0.7/2.2	0.3/1.5	3.3/5.1

Solar Radiation for 2-Axis Tracking Flat-Plate Collectors (kWh/m²/day), Uncertainty ±9%

Tracker		Jan	Feb	Mar	Apr	May	June	July	Aug	Sept	Oct	Nov	Dec	Year
2-Axis	Average	1.3	3.3	5.9	7.7	7.7	7.5	6.8	5.3	4.0	2.4	1.4	0.7	4.5
	Min/Max	0.7/2.2	2.2/5.7	3.2/8.6	5.7/9.6	5.6/10.3	5.4/9.4	5.1/8.7	3.8/7.2	2.2/6.2	1.5/3.5	0.7/2.2	0.3/1.5	3.6/5.4

Direct Beam Solar Radiation for Concentrating Collectors (kWh/m²/day), Uncertainty ±8%

Tracker		Jan	Feb	Mar	Apr	May	June	July	Aug	Sept	Oct	Nov	Dec	Year
1-Axis, E-W	Average	0.9	1.9	2.8	3.1	3.2	3.1	2.7	2.1	1.8	1.1	0.8	0.5	2.0
Horiz Axis	Min/Max	0.3/1.7	0.9/4.1	1.0/5.1	1.5/4.4	1.8/5.2	1.5/4.2	1.4/4.0	1.1/3.3	0.5/3.2	0.5/2.1	0.3/1.5	0.2/1.3	1.4/2.6
1-Axis, N-S	Average	0.3	1.1	2.5	3.6	4.1	3.9	3.4	2.5	1.8	0.8	0.3	0.1	2.0
Horiz Axis	Min/Max	0.1/0.6	0.5/2.4	0.8/4.5	1.7/5.1	2.4/6.7	1.9/5.3	1.8/5.0	1.5/3.9	0.5/3.4	0.3/1.5	0.1/0.6	0.0/0.3	1.4/2.8
1-Axis, N-S	Average	0.9	2.1	3.8	4.5	4.4	4.0	3.5	2.9	2.5	1.4	0.9	0.5	2.6
Tilt=Latitude	Min/Max	0.3/1.7	1.0/4.7	1.2/6.7	2.1/6.3	2.6/7.3	2.0/5.5	1.9/5.3	1.7/4.6	0.7/4.6	0.6/2.6	0.3/1.6	0.1/1.2	1.8/3.5
2-Axis	Average	0.9	2.2	3.8	4.5	4.7	4.4	3.8	3.0	2.5	1.4	0.9	0.5	2.7
	Min/Max	0.3/1.8	1.0/4.8	1.2/6.7	2.1/6.4	2.7/7.7	2.2/6.0	2.1/5.7	1.8/4.7	0.7/4.6	0.6/2.6	0.3/1.7	0.2/1.3	1.9/3.6

Average Climatic Conditions

Element	Jan	Feb	Mar	Apr	May	June	July	Aug	Sept	Oct	Nov	Dec	Year
Temperature (°C)	-22.6	-19.2	-12.1	-3.1	6.9	12.9	14.8	12.4	6.8	-4.1	-15.3	-21.1	-3.6
Daily Minimum Temp	-27.7	-25.7	-19.4	-8.9	1.4	7.3	9.4	7.2	1.9	-7.8	-19.8	-25.9	-9.0
Daily Maximum Temp	-17.6	-12.8	-4.8	2.8	12.5	18.6	20.2	17.5	11.6	-0.4	-10.9	-16.3	1.7
Record Minimum Temp	-59.4	-53.3	-46.1	-40.0	-18.9	-1.1	0.6	-3.9	-14.4	-33.3	-47.2	-55.0	-59.4
Record Maximum Temp	12.2	12.8	10.6	19.4	27.2	32.2	31.7	31.7	24.4	15.0	9.4	9.4	32.2
HDD, Base 18.3°C	1269	1052	944	642	353	162	115	188	347	694	1010	1223	7998
CDD, Base 18.3°C	0	0	0	0	0	0	7	3	0	0	0	0	10
Relative Humidity (%)	72	69	64	62	59	63	70	75	74	78	77	74	70
Wind Speed (m/s)	1.4	2.0	2.5	3.1	3.1	3.0	2.8	2.8	2.8	2.6	1.8	1.5	2.5

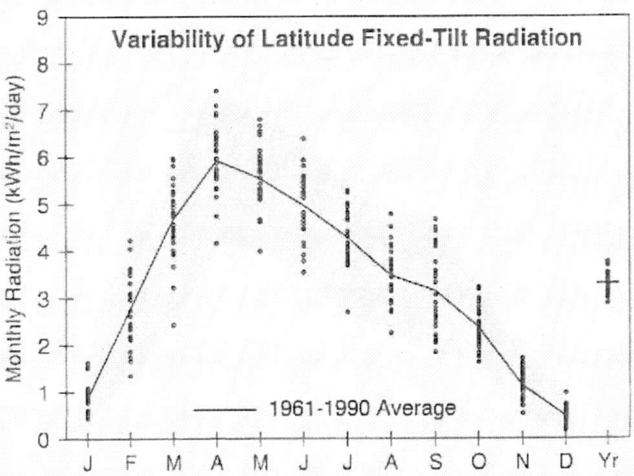

Nome, AK

WBAN NO. 26617

LATITUDE: 64.50° N
LONGITUDE: 165.43° W
ELEVATION: 7 meters
MEAN PRESSURE: 1009 millibars

STATION TYPE: Secondary

Solar Radiation for Flat-Plate Collectors Facing South at a Fixed Tilt (kWh/m²/day), Uncertainty ±9%

Tilt (°)		Jan	Feb	Mar	Apr	May	June	July	Aug	Sept	Oct	Nov	Dec	Year
0	Average	0.2	0.8	2.3	4.3	5.3	5.5	4.6	3.3	2.1	1.0	0.3	0.1	2.5
	Min/Max	0.1/0.2	0.7/1.0	1.7/2.7	3.6/4.8	4.0/6.2	4.2/6.8	3.3/5.5	2.5/4.1	1.6/2.8	0.8/1.2	0.2/0.3	0.0/0.1	2.3/2.7
Latitude -15	Average	0.8	2.5	4.5	5.9	5.9	5.5	4.7	3.7	3.2	2.2	1.0	0.4	3.4
	Min/Max	0.4/1.4	1.3/3.8	2.4/5.6	4.2/7.3	4.3/7.1	3.9/7.0	3.0/5.8	2.5/5.0	1.9/4.7	1.6/3.0	0.5/1.5	0.2/0.8	2.9/3.8
Latitude	Average	0.9	2.8	4.7	5.9	5.6	5.0	4.3	3.5	3.2	2.4	1.1	0.5	3.3
	Min/Max	0.4/1.6	1.4/4.2	2.4/6.0	4.2/7.4	4.0/6.8	3.5/6.4	2.7/5.3	2.3/4.8	1.9/4.7	1.6/3.2	0.5/1.7	0.2/1.0	2.9/3.8
Latitude +15	Average	0.9	2.9	4.7	5.6	5.0	4.3	3.7	3.1	3.0	2.4	1.2	0.5	3.1
	Min/Max	0.4/1.8	1.4/4.4	2.4/6.0	4.0/7.1	3.6/6.3	3.0/5.4	2.3/4.6	2.0/4.3	1.7/4.4	1.6/3.2	0.6/1.8	0.2/1.0	2.7/3.5
90	Average	0.9	2.8	4.5	5.3	4.5	3.7	3.2	2.7	2.7	2.3	1.2	0.5	2.9
	Min/Max	0.4/1.8	1.3/4.4	2.3/5.8	3.7/6.7	3.2/5.8	2.6/4.7	2.0/4.0	1.7/3.8	1.5/4.1	1.5/3.1	0.5/1.8	0.2/1.1	2.5/3.2

Solar Radiation for 1-Axis Tracking Flat-Plate Collectors with a North-South Axis (kWh/m²/day), Uncertainty ±9%

Axis Tilt (°)		Jan	Feb	Mar	Apr	May	June	July	Aug	Sept	Oct	Nov	Dec	Year
0	Average	0.3	1.7	4.2	6.9	7.9	7.8	6.3	4.5	3.3	1.8	0.5	0.1	3.8
	Min/Max	0.2/0.6	1.0/2.6	2.3/5.3	4.6/8.8	5.3/10.2	4.8/10.6	3.8/8.4	2.8/6.2	1.9/4.9	1.2/2.3	0.3/0.8	0.1/0.3	3.2/4.4
Latitude -15	Average	0.8	3.0	5.8	8.3	8.7	8.3	6.7	5.0	4.1	2.7	1.1	0.4	4.6
	Min/Max	0.4/1.6	1.4/4.6	2.8/7.5	5.2/10.8	5.6/11.4	4.8/11.4	3.8/9.0	2.9/7.1	2.2/6.3	1.8/3.7	0.5/1.7	0.2/0.9	3.8/5.4
Latitude	Average	0.9	3.2	5.9	8.3	8.5	8.0	6.4	4.8	4.1	2.8	1.2	0.5	4.6
	Min/Max	0.4/1.8	1.5/4.9	2.8/7.7	5.1/10.8	5.5/11.2	4.6/11.1	3.6/8.8	2.8/6.9	2.2/6.4	1.9/3.9	0.6/1.9	0.2/1.0	3.8/5.4
Latitude +15	Average	1.0	3.2	5.9	8.1	8.1	7.6	6.1	4.6	4.0	2.8	1.3	0.5	4.4
	Min/Max	0.5/1.9	1.5/5.1	2.8/7.7	5.0/10.6	5.2/10.7	4.3/10.6	3.3/8.4	2.6/6.6	2.1/6.2	1.8/3.9	0.6/1.9	0.2/1.1	3.7/5.3

Solar Radiation for 2-Axis Tracking Flat-Plate Collectors (kWh/m²/day), Uncertainty ±9%

Tracker		Jan	Feb	Mar	Apr	May	June	July	Aug	Sept	Oct	Nov	Dec	Year
2-Axis	Average	1.0	3.2	5.9	8.4	8.8	8.6	6.9	5.0	4.1	2.8	1.3	0.5	4.7
	Min/Max	0.5/1.9	1.5/5.1	2.8/7.7	5.2/10.8	5.8/11.7	5.1/11.9	3.9/9.4	3.0/7.1	2.2/6.4	1.9/3.9	0.6/2.0	0.2/1.1	4.0/5.6

Direct Beam Solar Radiation for Concentrating Collectors (kWh/m²/day), Uncertainty ±8%

Tracker		Jan	Feb	Mar	Apr	May	June	July	Aug	Sept	Oct	Nov	Dec	Year
1-Axis, E-W	Average	0.7	2.0	3.0	3.5	3.8	4.0	2.9	1.9	1.9	1.6	0.8	0.5	2.2
Horiz Axis	Min/Max	0.2/1.5	0.4/3.6	0.7/4.4	1.1/5.3	1.3/6.4	1.4/6.4	0.9/4.5	0.7/3.4	0.8/3.5	0.9/2.5	0.2/1.4	0.2/1.0	1.6/2.9
1-Axis, N-S	Average	0.2	1.1	2.6	4.2	4.8	5.0	3.6	2.4	2.0	1.1	0.3	0.1	2.3
Horiz Axis	Min/Max	0.1/0.5	0.2/2.0	0.6/4.0	1.2/6.3	1.6/8.1	1.7/8.0	1.3/5.8	0.9/4.3	0.8/3.6	0.6/1.7	0.1/0.6	0.0/0.2	1.6/3.0
1-Axis, N-S	Average	0.7	2.2	3.9	5.1	5.3	5.2	3.8	2.8	2.7	1.9	0.8	0.5	2.9
Tilt=Latitude	Min/Max	0.2/1.5	0.5/4.1	0.9/6.0	1.5/7.7	1.8/8.9	1.8/8.4	1.3/6.2	1.1/4.8	1.0/4.8	1.1/3.1	0.2/1.5	0.2/1.0	2.1/3.8
2-Axis	Average	0.7	2.3	3.9	5.2	5.6	5.7	4.1	2.9	2.7	2.0	0.9	0.5	3.0
	Min/Max	0.2/1.6	0.5/4.2	0.9/6.0	1.5/7.8	1.9/9.4	1.9/9.1	1.5/6.6	1.1/5.0	1.0/4.9	1.1/3.1	0.2/1.5	0.2/1.1	2.2/4.0

Average Climatic Conditions

Element	Jan	Feb	Mar	Apr	May	June	July	Aug	Sept	Oct	Nov	Dec	Year
Temperature (°C)	-13.9	-15.6	-13.0	-8.0	2.0	7.7	10.8	10.1	5.8	-2.2	-8.9	-13.7	-3.2
Daily Minimum Temp	-18.2	-20.3	-17.9	-12.4	-1.6	3.8	7.3	6.7	2.4	-5.4	-12.6	-18.1	-7.2
Daily Maximum Temp	-9.6	-11.0	-8.1	-3.6	5.6	11.6	14.3	13.4	9.3	0.9	-5.3	-9.4	0.7
Record Minimum Temp	-47.8	-41.1	-43.3	-34.4	-23.9	-5.0	-0.6	-2.8	-9.4	-23.3	-39.4	-40.6	-47.8
Record Maximum Temp	6.1	8.9	6.1	10.6	25.6	27.2	30.0	27.2	21.7	15.0	7.2	6.1	30.0
HDD, Base 18.3°C	999	951	971	790	506	318	235	255	375	637	818	994	7849
CDD, Base 18.3°C	0	0	0	0	0	0	0	0	0	0	0	0	0
Relative Humidity (%)	72	69	71	74	74	74	79	80	75	74	74	72	74
Wind Speed (m/s)	5.2	4.6	4.5	4.7	4.6	4.5	4.5	4.9	5.2	4.9	5.1	5.0	4.8

St Paul Is., AK

WBAN NO. 25713

LATITUDE: 57.15° N
LONGITUDE: 170.22° W
ELEVATION: 7 meters
MEAN PRESSURE: 1006 millibars

STATION TYPE: Secondary

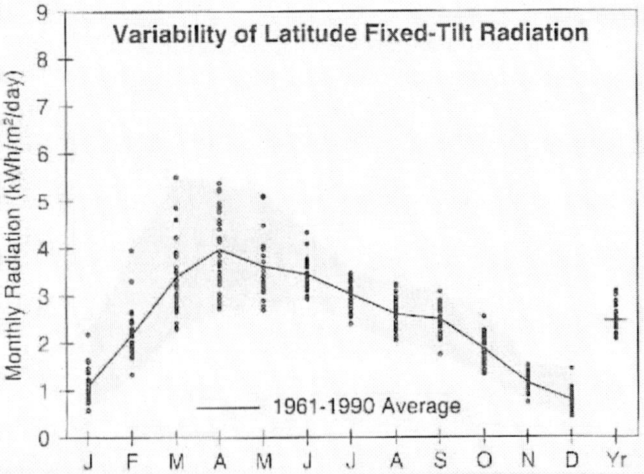

Solar Radiation for Flat-Plate Collectors Facing South at a Fixed Tilt (kWh/m²/day), Uncertainty ±9%

Tilt (°)		Jan	Feb	Mar	Apr	May	June	July	Aug	Sept	Oct	Nov	Dec	Year
0	Average	0.5	1.2	2.4	3.5	3.9	4.0	3.6	2.9	2.2	1.3	0.6	0.4	2.2
	Min/Max	0.4/0.6	0.9/1.6	1.9/3.1	2.7/4.4	3.2/5.1	3.5/4.7	3.0/3.9	2.4/3.4	1.8/2.4	1.1/1.5	0.5/0.7	0.3/0.5	2.0/2.5
Latitude -15	Average	1.0	2.0	3.3	4.1	3.9	3.8	3.3	2.8	2.6	1.8	1.1	0.7	2.5
	Min/Max	0.6/1.9	1.3/3.6	2.3/5.2	2.9/5.4	3.0/5.3	3.2/4.7	2.7/3.8	2.3/3.5	1.9/3.1	1.3/2.4	0.7/1.4	0.4/1.3	2.2/3.1
Latitude	Average	1.1	2.2	3.4	4.0	3.6	3.4	3.0	2.6	2.5	1.9	1.1	0.8	2.5
	Min/Max	0.6/2.2	1.3/4.0	2.3/5.5	2.7/5.4	2.7/5.1	2.9/4.3	2.4/3.5	2.0/3.2	1.8/3.1	1.3/2.5	0.7/1.5	0.4/1.4	2.1/3.1
Latitude +15	Average	1.1	2.2	3.3	3.7	3.2	3.0	2.6	2.3	2.3	1.8	1.2	0.8	2.3
	Min/Max	0.6/2.3	1.3/4.1	2.2/5.5	2.5/5.2	2.3/4.8	2.5/3.8	2.1/3.0	1.8/2.9	1.6/2.9	1.3/2.5	0.7/1.6	0.4/1.5	1.9/2.9
90	Average	1.1	2.1	3.1	3.3	2.6	2.3	2.1	1.9	2.0	1.7	1.1	0.8	2.0
	Min/Max	0.5/2.3	1.2/4.0	1.9/5.2	2.0/4.6	1.9/4.2	2.0/2.9	1.6/2.4	1.4/2.3	1.3/2.5	1.1/2.3	0.7/1.5	0.4/1.6	1.6/2.6

Solar Radiation for 1-Axis Tracking Flat-Plate Collectors with a North-South Axis (kWh/m²/day), Uncertainty ±9%

Axis Tilt (°)		Jan	Feb	Mar	Apr	May	June	July	Aug	Sept	Oct	Nov	Dec	Year
0	Average	0.7	1.7	3.3	4.5	4.4	4.4	3.7	3.0	2.6	1.6	0.8	0.5	2.6
	Min/Max	0.4/1.3	1.1/3.0	2.3/5.2	3.0/6.2	3.2/6.5	3.5/5.4	2.8/4.3	2.3/3.8	1.8/3.2	1.2/2.1	0.5/1.0	0.3/0.7	2.2/3.2
Latitude -15	Average	1.1	2.3	4.0	4.9	4.6	4.3	3.6	3.1	3.0	2.0	1.1	0.8	2.9
	Min/Max	0.6/2.2	1.4/4.4	2.7/6.8	3.2/6.9	3.2/6.9	3.4/5.6	2.7/4.4	2.3/4.0	1.9/3.8	1.4/2.8	0.7/1.5	0.4/1.4	2.4/3.7
Latitude	Average	1.2	2.5	4.1	4.9	4.4	4.1	3.4	2.9	2.9	2.1	1.2	0.8	2.9
	Min/Max	0.6/2.4	1.4/4.7	2.7/7.0	3.1/6.9	3.0/6.8	3.1/5.3	2.5/4.2	2.2/3.8	1.9/3.8	1.4/3.0	0.7/1.6	0.4/1.5	2.4/3.7
Latitude +15	Average	1.2	2.5	4.0	4.7	4.1	3.8	3.1	2.7	2.8	2.1	1.2	0.9	2.8
	Min/Max	0.6/2.6	1.4/4.8	2.6/7.0	2.9/6.8	2.7/6.5	2.9/4.9	2.3/3.9	2.0/3.6	1.8/3.6	1.4/2.9	0.7/1.7	0.4/1.6	2.3/3.6

Solar Radiation for 2-Axis Tracking Flat-Plate Collectors (kWh/m²/day), Uncertainty ±9%

Tracker		Jan	Feb	Mar	Apr	May	June	July	Aug	Sept	Oct	Nov	Dec	Year
2-Axis	Average	1.2	2.5	4.1	5.0	4.7	4.5	3.8	3.1	3.0	2.1	1.3	0.9	3.0
	Min/Max	0.6/2.6	1.5/4.8	2.7/7.0	3.2/7.0	3.3/7.0	3.5/5.8	2.8/4.6	2.4/4.0	1.9/3.8	1.5/3.0	0.8/1.7	0.4/1.7	2.5/3.9

Direct Beam Solar Radiation for Concentrating Collectors (kWh/m²/day), Uncertainty ±8%

Tracker		Jan	Feb	Mar	Apr	May	June	July	Aug	Sept	Oct	Nov	Dec	Year
1-Axis, E-W Horiz Axis	Average	0.5	1.1	1.5	1.4	1.2	1.1	0.7	0.6	1.0	0.9	0.6	0.4	0.9
	Min/Max	0.1/1.7	0.5/2.8	0.8/3.6	0.6/2.5	0.4/2.4	0.2/2.2	0.2/1.4	0.2/1.1	0.4/1.5	0.4/1.5	0.3/0.9	0.1/0.9	0.7/1.4
1-Axis, N-S Horiz Axis	Average	0.2	0.7	1.3	1.7	1.5	1.3	0.8	0.7	1.0	0.6	0.3	0.2	0.9
	Min/Max	0.1/0.8	0.4/2.0	0.7/3.3	0.7/3.0	0.6/3.0	0.2/2.5	0.2/1.7	0.3/1.4	0.4/1.6	0.3/1.1	0.1/0.5	0.1/0.4	0.7/1.4
1-Axis, N-S Tilt=Latitude	Average	0.6	1.2	1.9	2.0	1.5	1.3	0.8	0.8	1.3	1.0	0.6	0.4	1.1
	Min/Max	0.1/1.8	0.6/3.3	1.0/4.7	0.9/3.5	0.6/3.2	0.2/2.6	0.2/1.8	0.3/1.5	0.5/2.1	0.6/1.8	0.3/1.0	0.1/0.9	0.9/1.8
2-Axis	Average	0.6	1.3	1.9	2.0	1.6	1.4	0.9	0.9	1.3	1.1	0.6	0.4	1.2
	Min/Max	0.1/1.9	0.6/3.4	1.0/4.7	0.9/3.5	0.6/3.4	0.3/2.8	0.2/1.9	0.3/1.6	0.5/2.1	0.6/1.8	0.3/1.0	0.1/1.0	0.9/1.9

Average Climatic Conditions

Element	Jan	Feb	Mar	Apr	May	June	July	Aug	Sept	Oct	Nov	Dec	Year
Temperature (°C)	-3.0	-5.3	-4.3	-2.1	1.8	5.3	7.8	8.7	7.0	3.2	0.6	-1.7	1.5
Daily Minimum Temp	-5.3	-7.8	-7.0	-4.5	-0.4	2.8	5.8	6.9	4.6	0.9	-1.7	-4.0	-0.8
Daily Maximum Temp	-0.7	-2.8	-1.7	0.4	4.1	7.7	9.8	10.6	9.3	5.6	2.8	0.7	3.8
Record Minimum Temp	-32.2	-26.7	-28.3	-22.2	-13.3	-8.9	-2.2	-1.7	-3.9	-11.1	-15.6	-20.6	-32.2
Record Maximum Temp	8.9	6.7	10.0	8.9	14.4	16.7	17.2	18.9	16.1	12.2	10.0	11.1	18.9
HDD, Base 18.3°C	661	663	703	612	513	392	326	298	340	468	533	620	6128
CDD, Base 18.3°C	0	0	0	0	0	0	0	0	0	0	0	0	0
Relative Humidity (%)	84	84	85	85	87	90	93	93	88	82	82	83	86
Wind Speed (m/s)	8.6	8.9	8.3	8.0	6.8	6.1	5.5	6.1	7.0	8.0	8.4	8.8	7.5

Talkeetna, AK

WBAN NO. 26528

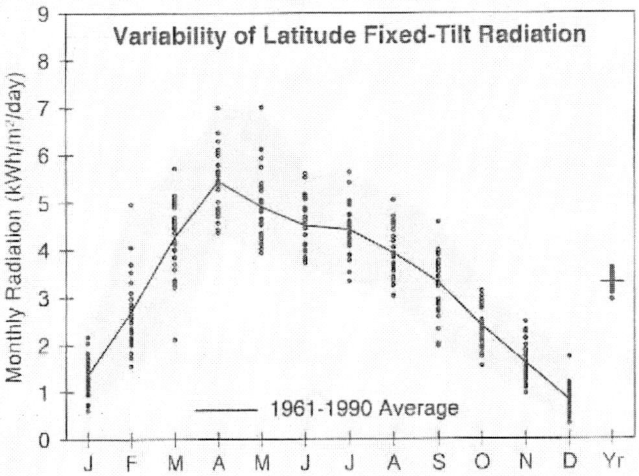

LATITUDE: 62.30° N
LONGITUDE: 150.10° W
ELEVATION: 105 meters
MEAN PRESSURE: 996 millibars

STATION TYPE: Secondary

Solar Radiation for Flat-Plate Collectors Facing South at a Fixed Tilt (kWh/m²/day), Uncertainty ±9%

Tilt (°)		Jan	Feb	Mar	Apr	May	June	July	Aug	Sept	Oct	Nov	Dec	Year
0	Average	0.3	1.0	2.3	4.1	4.8	5.0	4.7	3.6	2.3	1.2	0.4	0.2	2.5
	Min/Max	0.3/0.3	0.8/1.3	1.6/2.7	3.4/4.7	4.0/6.2	4.3/6.0	3.9/5.7	3.0/4.3	1.7/2.8	1.0/1.3	0.4/0.6	0.1/0.2	2.3/2.7
Latitude -15	Average	1.2	2.5	4.1	5.5	5.2	5.0	4.8	4.2	3.3	2.3	1.4	0.7	3.3
	Min/Max	0.6/1.9	1.5/4.4	2.1/5.4	4.4/6.9	4.3/7.3	4.1/6.1	3.7/6.1	3.3/5.3	2.0/4.5	1.5/2.9	0.9/2.2	0.3/1.5	3.0/3.6
Latitude	Average	1.3	2.7	4.3	5.4	4.9	4.5	4.4	3.9	3.3	2.4	1.6	0.8	3.3
	Min/Max	0.6/2.2	1.5/5.0	2.1/5.7	4.3/7.0	3.9/7.0	3.7/5.6	3.3/5.6	3.0/5.1	2.0/4.6	1.5/3.1	1.0/2.5	0.3/1.7	2.9/3.6
Latitude +15	Average	1.4	2.8	4.2	5.2	4.4	3.9	3.8	3.5	3.1	2.4	1.7	0.9	3.1
	Min/Max	0.6/2.3	1.6/5.2	2.0/5.7	4.0/6.7	3.4/6.4	3.2/4.8	2.9/4.9	2.7/4.5	1.8/4.3	1.5/3.2	1.0/2.6	0.3/1.9	2.7/3.4
90	Average	1.4	2.7	4.0	4.8	3.8	3.2	3.2	3.0	2.8	2.3	1.7	0.9	2.8
	Min/Max	0.6/2.3	1.5/5.1	1.8/5.5	3.5/6.2	2.9/5.8	2.7/4.0	2.4/4.1	2.3/3.9	1.6/4.0	1.4/3.1	1.0/2.6	0.3/1.9	2.5/3.1

Solar Radiation for 1-Axis Tracking Flat-Plate Collectors with a North-South Axis (kWh/m²/day), Uncertainty ±9%

Axis Tilt (°)		Jan	Feb	Mar	Apr	May	June	July	Aug	Sept	Oct	Nov	Dec	Year
0	Average	0.6	1.8	3.9	6.3	6.8	6.5	6.2	4.9	3.4	1.9	0.9	0.3	3.6
	Min/Max	0.3/0.9	1.2/3.2	2.0/5.2	5.0/8.5	5.3/9.4	5.2/8.6	4.7/8.4	3.6/6.6	2.0/4.6	1.3/2.3	0.6/1.3	0.1/0.6	3.2/4.1
Latitude -15	Average	1.3	2.9	5.2	7.5	7.3	6.8	6.6	5.5	4.2	2.7	1.6	0.8	4.4
	Min/Max	0.6/2.1	1.6/5.5	2.4/7.2	5.8/10.3	5.7/10.5	5.3/9.1	4.7/9.0	3.9/7.5	2.3/5.9	1.7/3.5	1.0/2.5	0.3/1.6	3.8/4.9
Latitude	Average	1.4	3.1	5.3	7.5	7.1	6.6	6.3	5.3	4.2	2.8	1.8	0.9	4.4
	Min/Max	0.6/2.4	1.7/5.9	2.4/7.4	5.8/10.3	5.5/10.4	5.1/8.8	4.4/8.8	3.7/7.3	2.3/6.0	1.7/3.7	1.0/2.8	0.3/1.8	3.8/4.9
Latitude +15	Average	1.5	3.2	5.3	7.3	6.8	6.2	5.9	5.0	4.1	2.8	1.8	0.9	4.2
	Min/Max	0.7/2.5	1.7/6.1	2.3/7.4	5.6/10.1	5.2/10.0	4.7/8.3	4.1/8.3	3.5/7.0	2.2/5.8	1.7/3.7	1.1/2.9	0.4/2.0	3.6/4.7

Solar Radiation for 2-Axis Tracking Flat-Plate Collectors (kWh/m²/day), Uncertainty ±9%

Tracker		Jan	Feb	Mar	Apr	May	June	July	Aug	Sept	Oct	Nov	Dec	Year
2-Axis	Average	1.5	3.2	5.3	7.5	7.5	7.1	6.8	5.5	4.2	2.8	1.8	0.9	4.5
	Min/Max	0.7/2.5	1.7/6.1	2.5/7.4	5.9/10.3	5.8/10.6	5.5/9.4	4.8/9.3	3.9/7.6	2.4/6.0	1.8/3.7	1.1/2.9	0.4/2.0	3.9/5.1

Direct Beam Solar Radiation for Concentrating Collectors (kWh/m²/day), Uncertainty ±8%

Tracker		Jan	Feb	Mar	Apr	May	June	July	Aug	Sept	Oct	Nov	Dec	Year
1-Axis, E-W Horiz Axis	Average	1.0	1.8	2.6	3.1	3.1	2.9	2.8	2.3	2.0	1.5	1.2	0.7	2.1
	Min/Max	0.2/1.9	0.4/4.4	0.9/4.2	1.7/5.1	1.5/4.8	1.6/4.5	1.2/4.5	1.2/3.7	0.8/3.2	0.6/2.3	0.4/2.0	0.2/1.7	1.7/2.4
1-Axis, N-S Horiz Axis	Average	0.4	1.1	2.3	3.6	3.9	3.6	3.4	2.7	2.0	1.0	0.5	0.2	2.1
	Min/Max	0.1/0.7	0.3/2.7	0.8/3.8	2.0/6.2	1.8/6.0	2.1/5.7	1.6/5.6	1.4/4.5	0.8/3.2	0.4/1.5	0.2/0.9	0.1/0.5	1.6/2.5
1-Axis, N-S Tilt=Latitude	Average	1.0	2.1	3.4	4.4	4.3	3.7	3.6	3.2	2.7	1.8	1.2	0.6	2.7
	Min/Max	0.2/2.0	0.5/5.1	1.2/5.6	2.5/7.5	2.0/6.5	2.1/5.9	1.6/5.9	1.6/5.2	1.0/4.4	0.8/2.7	0.5/2.2	0.2/1.6	2.1/3.1
2-Axis	Average	1.1	2.1	3.4	4.5	4.5	4.0	3.9	3.3	2.7	1.8	1.3	0.7	2.8
	Min/Max	0.3/2.1	0.5/5.2	1.2/5.6	2.5/7.6	2.1/6.9	2.3/6.4	1.8/6.4	1.6/5.3	1.0/4.4	0.8/2.8	0.5/2.3	0.2/1.8	2.2/3.3

Average Climatic Conditions

Element	Jan	Feb	Mar	Apr	May	June	July	Aug	Sept	Oct	Nov	Dec	Year
Temperature (°C)	-12.3	-9.6	-5.8	0.5	7.2	12.5	14.7	12.8	7.8	-0.4	-8.6	-11.5	0.6
Daily Minimum Temp	-17.4	-15.6	-12.3	-5.3	1.0	6.9	9.4	7.6	2.6	-4.9	-13.4	-16.2	-4.8
Daily Maximum Temp	-7.1	-3.7	0.8	6.3	13.3	18.0	19.8	18.1	13.1	4.1	-3.8	-6.8	6.0
Record Minimum Temp	-44.4	-43.3	-41.7	-38.3	-20.6	-2.2	0.6	-3.9	-9.4	-29.4	-40.6	-42.8	-44.4
Record Maximum Temp	7.2	10.6	11.7	20.6	27.8	32.8	32.2	29.4	25.6	20.0	10.0	8.3	32.8
HDD, Base 18.3°C	949	782	747	535	344	175	114	171	315	582	808	925	6448
CDD, Base 18.3°C	0	0	0	0	0	0	0	0	0	0	0	0	0
Relative Humidity (%)	72	72	69	68	66	70	75	81	82	81	76	75	74
Wind Speed (m/s)	2.9	2.7	2.7	2.5	2.5	2.6	2.4	2.0	1.9	2.1	2.7	2.8	2.5

Yakutat, AK

WBAN NO. 25339

LATITUDE: 59.52° N
LONGITUDE: 139.67° W
ELEVATION: 9 meters
MEAN PRESSURE: 1009 millibars

STATION TYPE: Secondary

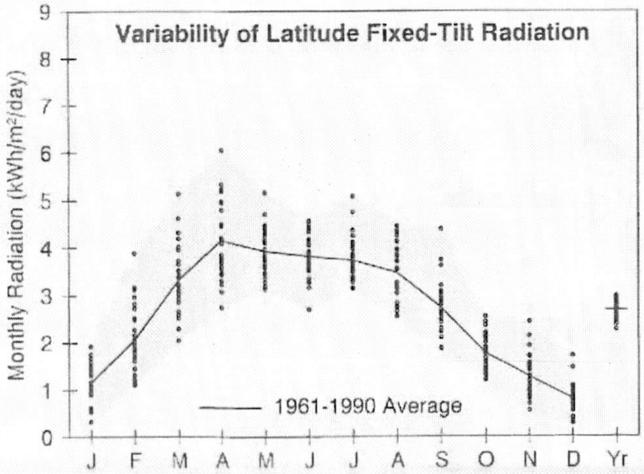

Solar Radiation for Flat-Plate Collectors Facing South at a Fixed Tilt (kWh/m²/day), Uncertainty ±9%

Tilt (°)		Jan	Feb	Mar	Apr	May	June	July	Aug	Sept	Oct	Nov	Dec	Year
0	Average	0.4	1.0	2.2	3.5	4.1	4.4	4.2	3.4	2.2	1.1	0.5	0.3	2.3
	Min/Max	0.2/0.5	0.7/1.4	1.6/2.8	2.6/4.7	3.5/4.9	3.3/5.1	3.6/5.3	2.7/4.1	1.7/3.0	0.9/1.3	0.4/0.7	0.2/0.3	2.0/2.5
Latitude -15	Average	1.0	1.9	3.2	4.2	4.2	4.2	4.1	3.7	2.8	1.7	1.2	0.7	2.8
	Min/Max	0.3/1.7	1.1/3.5	2.0/4.9	2.9/6.1	3.4/5.4	3.0/5.0	3.5/5.5	2.8/4.7	2.0/4.4	1.2/2.4	0.6/2.2	0.3/1.5	2.4/3.0
Latitude	Average	1.2	2.1	3.3	4.2	3.9	3.8	3.7	3.5	2.7	1.8	1.3	0.8	2.7
	Min/Max	0.3/1.9	1.1/3.9	2.0/5.1	2.7/6.1	3.1/5.2	2.7/4.6	3.1/5.1	2.5/4.4	1.9/4.4	1.2/2.5	0.6/2.4	0.3/1.7	2.3/3.0
Latitude +15	Average	1.2	2.1	3.3	3.9	3.5	3.3	3.2	3.1	2.5	1.7	1.3	0.8	2.5
	Min/Max	0.3/2.0	1.1/4.1	1.9/5.1	2.5/5.8	2.7/4.8	2.3/3.9	2.7/4.4	2.3/4.0	1.7/4.1	1.1/2.5	0.6/2.6	0.3/1.9	2.1/2.8
90	Average	1.2	2.0	3.1	3.5	2.8	2.6	2.6	2.6	2.2	1.6	1.3	0.8	2.2
	Min/Max	0.3/2.1	1.0/4.0	1.8/4.8	2.1/5.2	2.3/4.2	1.9/3.1	2.2/3.5	1.9/3.3	1.5/3.7	1.0/2.4	0.5/2.5	0.3/1.9	1.8/2.5

Solar Radiation for 1-Axis Tracking Flat-Plate Collectors with a North-South Axis (kWh/m²/day), Uncertainty ±9%

Axis Tilt (°)		Jan	Feb	Mar	Apr	May	June	July	Aug	Sept	Oct	Nov	Dec	Year
0	Average	0.6	1.5	3.2	4.9	5.3	5.3	5.0	4.3	2.9	1.5	0.8	0.4	3.0
	Min/Max	0.2/1.0	0.9/2.8	2.1/4.8	3.3/7.5	4.1/6.7	3.6/6.7	3.9/7.4	3.1/5.8	1.9/4.8	1.0/2.1	0.4/1.5	0.2/0.8	2.6/3.3
Latitude -15	Average	1.1	2.2	4.0	5.5	5.5	5.4	5.2	4.7	3.4	2.0	1.3	0.8	3.4
	Min/Max	0.3/1.9	1.1/4.3	2.4/6.3	3.6/8.6	4.2/7.2	3.5/6.9	3.9/7.9	3.3/6.4	2.1/5.9	1.3/2.9	0.5/2.5	0.3/1.6	2.9/3.8
Latitude	Average	1.2	2.3	4.0	5.5	5.3	5.1	4.9	4.5	3.4	2.0	1.4	0.8	3.4
	Min/Max	0.3/2.1	1.2/4.6	2.5/6.5	3.5/8.6	4.0/6.9	3.3/6.6	3.6/7.6	3.2/6.2	2.0/5.9	1.3/3.0	0.6/2.8	0.3/1.9	2.9/3.8
Latitude +15	Average	1.3	2.4	4.0	5.3	5.0	4.8	4.6	4.3	3.2	2.0	1.4	0.9	3.3
	Min/Max	0.3/2.2	1.2/4.7	2.4/6.5	3.3/8.4	3.7/6.7	3.1/6.2	3.3/7.2	2.9/5.9	1.9/5.7	1.3/3.0	0.6/2.9	0.3/2.0	2.8/3.7

Solar Radiation for 2-Axis Tracking Flat-Plate Collectors (kWh/m²/day), Uncertainty ±9%

Tracker		Jan	Feb	Mar	Apr	May	June	July	Aug	Sept	Oct	Nov	Dec	Year
2-Axis	Average	1.3	2.4	4.1	5.6	5.7	5.6	5.3	4.7	3.4	2.0	1.4	0.9	3.5
	Min/Max	0.3/2.2	1.2/4.7	2.5/6.5	3.6/8.6	4.3/7.3	3.7/7.2	4.0/8.1	3.3/6.4	2.1/5.9	1.3/3.0	0.6/2.9	0.3/2.0	3.0/3.9

Direct Beam Solar Radiation for Concentrating Collectors (kWh/m²/day), Uncertainty ±8%

Tracker		Jan	Feb	Mar	Apr	May	June	July	Aug	Sept	Oct	Nov	Dec	Year
1-Axis, E-W Horiz Axis	Average	0.7	1.1	1.6	2.0	2.0	1.9	1.8	1.7	1.3	0.9	0.8	0.5	1.4
	Min/Max	0.1/1.5	0.2/3.0	0.8/3.2	1.0/3.7	1.0/3.2	0.7/2.9	0.7/3.8	0.7/2.8	0.4/3.2	0.3/1.7	0.1/1.9	0.1/1.4	1.1/1.7
1-Axis, N-S Horiz Axis	Average	0.3	0.7	1.5	2.5	2.6	2.4	2.3	2.1	1.4	0.7	0.4	0.2	1.4
	Min/Max	0.1/0.6	0.2/1.9	0.7/3.0	1.3/4.7	1.3/4.0	0.9/3.7	0.9/4.8	0.8/3.5	0.4/3.5	0.3/1.3	0.1/1.0	0.0/0.5	1.2/1.8
1-Axis, N-S Tilt=Latitude	Average	0.7	1.3	2.2	2.9	2.8	2.4	2.3	2.4	1.9	1.1	0.8	0.5	1.8
	Min/Max	0.1/1.5	0.3/3.5	1.1/4.2	1.5/5.4	1.4/4.4	0.9/3.8	0.9/5.0	0.9/4.0	0.5/4.5	0.4/2.1	0.1/2.1	0.1/1.4	1.5/2.2
2-Axis	Average	0.8	1.3	2.2	3.0	2.9	2.6	2.5	2.5	1.9	1.1	0.9	0.5	1.9
	Min/Max	0.1/1.6	0.3/3.6	1.1/4.2	1.5/5.5	1.5/4.6	1.0/4.1	1.0/5.3	0.9/4.1	0.5/4.5	0.4/2.2	0.2/2.2	0.1/1.5	1.5/2.3

Average Climatic Conditions

Element	Jan	Feb	Mar	Apr	May	June	July	Aug	Sept	Oct	Nov	Dec	Year
Temperature (°C)	-3.8	-2.2	-0.6	2.4	6.3	9.7	12.0	11.8	9.0	4.9	-0.4	-2.7	3.9
Daily Minimum Temp	-7.4	-6.1	-4.6	-1.6	2.5	6.3	8.8	8.1	5.0	1.5	-3.8	-6.1	0.2
Daily Maximum Temp	-0.3	1.7	3.4	6.4	10.0	13.1	15.2	15.4	12.9	8.4	2.9	0.6	7.5
Record Minimum Temp	-30.0	-28.9	-28.9	-16.1	-6.1	-1.7	1.7	-1.7	-6.1	-14.4	-21.1	-31.1	-31.1
Record Maximum Temp	12.8	12.2	15.0	20.0	26.1	27.8	28.9	30.0	25.0	17.2	12.8	11.1	30.0
HDD, Base 18.3°C	687	576	586	478	374	260	196	203	280	415	562	653	5269
CDD, Base 18.3°C	0	0	0	0	0	0	0	0	0	0	0	0	0
Relative Humidity (%)	83	83	81	81	83	84	88	88	89	88	85	84	85
Wind Speed (m/s)	3.2	3.1	3.0	3.0	3.2	3.1	2.9	2.8	3.0	3.4	3.1	3.2	3.1

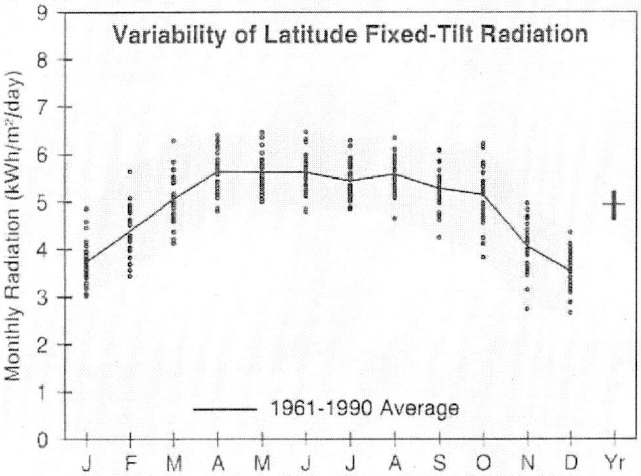

Birmingham, AL

WBAN NO. 13876

LATITUDE: 33.57° N
LONGITUDE: 86.75° W
ELEVATION: 192 meters
MEAN PRESSURE: 996 millibars

STATION TYPE: Secondary

Solar Radiation for Flat-Plate Collectors Facing South at a Fixed Tilt (kWh/m²/day), Uncertainty ±9%

Tilt (°)		Jan	Feb	Mar	Apr	May	June	July	Aug	Sept	Oct	Nov	Dec	Year
0	Average	2.5	3.3	4.4	5.5	6.0	6.2	5.9	5.6	4.8	4.0	2.8	2.3	4.4
	Min/Max	2.3/3.0	2.8/4.0	3.7/5.2	4.7/6.1	5.3/6.8	5.3/7.2	5.3/6.8	4.8/6.3	4.0/5.4	3.2/4.6	2.2/3.3	2.0/2.7	4.2/4.6
Latitude -15	Average	3.3	4.0	4.9	5.7	6.0	6.1	5.8	5.8	5.2	4.8	3.6	3.1	4.9
	Min/Max	2.8/4.2	3.2/5.1	4.1/6.0	4.9/6.5	5.3/6.9	5.2/7.0	5.2/6.8	4.8/6.5	4.3/6.0	3.7/5.7	2.6/4.3	2.4/3.7	4.6/5.1
Latitude	Average	3.7	4.4	5.1	5.6	5.6	5.6	5.4	5.6	5.3	5.2	4.1	3.5	4.9
	Min/Max	3.0/4.9	3.4/5.6	4.1/6.3	4.8/6.4	5.0/6.5	4.8/6.5	4.8/6.3	4.7/6.3	4.2/6.1	3.8/6.2	2.7/5.0	2.7/4.3	4.6/5.2
Latitude +15	Average	3.9	4.5	5.0	5.2	5.0	4.9	4.8	5.1	5.1	5.2	4.3	3.8	4.7
	Min/Max	3.1/5.3	3.5/5.9	4.0/6.2	4.5/5.9	4.4/5.7	4.2/5.6	4.3/5.5	4.2/5.8	4.0/5.9	3.8/6.4	2.8/5.3	2.8/4.7	4.5/5.0
90	Average	3.4	3.6	3.4	2.9	2.3	2.1	2.2	2.6	3.2	3.9	3.6	3.4	3.0
	Min/Max	2.6/4.7	2.7/4.8	2.7/4.2	2.6/3.2	2.2/2.5	1.9/2.2	2.1/2.3	2.3/2.9	2.5/3.7	2.8/4.9	2.2/4.6	2.4/4.3	2.9/3.3

Solar Radiation for 1-Axis Tracking Flat-Plate Collectors with a North-South Axis (kWh/m²/day), Uncertainty ±9%

Axis Tilt (°)		Jan	Feb	Mar	Apr	May	June	July	Aug	Sept	Oct	Nov	Dec	Year
0	Average	3.5	4.5	5.8	7.1	7.6	7.8	7.2	7.1	6.2	5.5	3.9	3.2	5.8
	Min/Max	2.8/4.6	3.4/5.8	4.5/7.4	5.8/8.3	6.4/9.2	6.3/9.3	6.2/8.7	5.6/8.3	4.8/7.2	4.0/6.7	2.6/4.8	2.4/3.9	5.4/6.1
Latitude -15	Average	4.1	5.0	6.2	7.3	7.6	7.7	7.2	7.2	6.5	6.1	4.5	3.8	6.1
	Min/Max	3.2/5.4	3.8/6.6	4.8/8.0	5.9/8.6	6.4/9.2	6.2/9.2	6.1/8.7	5.7/8.5	5.1/7.7	4.3/7.5	2.9/5.6	2.8/4.7	5.7/6.5
Latitude	Average	4.4	5.3	6.3	7.3	7.4	7.4	6.9	7.1	6.6	6.3	4.8	4.1	6.2
	Min/Max	3.4/5.9	4.0/7.1	4.9/8.2	5.9/8.6	6.2/9.0	6.0/8.8	5.9/8.4	5.6/8.3	5.1/7.8	4.5/7.9	3.0/6.1	3.0/5.2	5.7/6.6
Latitude +15	Average	4.6	5.4	6.2	7.0	6.9	6.9	6.5	6.7	6.4	6.4	5.0	4.3	6.0
	Min/Max	3.5/6.3	4.0/7.2	4.8/8.2	5.6/8.2	5.8/8.4	5.6/8.2	5.5/7.8	5.3/7.9	4.9/7.6	4.5/8.0	3.0/6.3	3.1/5.5	5.6/6.4

Solar Radiation for 2-Axis Tracking Flat-Plate Collectors (kWh/m²/day), Uncertainty ±9%

Tracker		Jan	Feb	Mar	Apr	May	June	July	Aug	Sept	Oct	Nov	Dec	Year
2-Axis	Average	4.6	5.4	6.3	7.4	7.7	7.8	7.3	7.2	6.6	6.4	5.0	4.4	6.4
	Min/Max	3.5/6.3	4.0/7.3	4.9/8.3	6.0/8.6	6.5/9.3	6.4/9.4	6.2/8.9	5.7/8.5	5.1/7.8	4.5/8.0	3.1/6.4	3.1/5.6	5.9/6.8

Direct Beam Solar Radiation for Concentrating Collectors (kWh/m²/day), Uncertainty ±8%

Tracker		Jan	Feb	Mar	Apr	May	June	July	Aug	Sept	Oct	Nov	Dec	Year
1-Axis, E-W Horiz Axis	Average	2.6	2.9	3.0	3.4	3.4	3.4	3.0	3.1	3.0	3.4	2.8	2.5	3.1
	Min/Max	1.6/4.2	1.8/4.4	1.9/4.6	2.5/4.4	2.5/4.7	2.4/4.7	2.3/4.2	2.0/4.1	1.9/3.9	1.9/4.7	1.1/4.0	1.4/3.6	2.7/3.5
1-Axis, N-S Horiz Axis	Average	2.2	2.9	3.6	4.5	4.5	4.5	3.9	4.0	3.6	3.6	2.4	2.0	3.5
	Min/Max	1.4/3.4	1.7/4.4	2.3/5.5	3.2/5.8	3.2/6.4	3.0/6.1	2.9/5.5	2.6/5.4	2.3/4.8	2.1/5.0	1.0/3.5	1.1/2.8	3.0/4.0
1-Axis, N-S Tilt=Latitude	Average	2.9	3.6	4.1	4.6	4.3	4.2	3.7	4.0	3.9	4.3	3.2	2.8	3.8
	Min/Max	1.8/4.7	2.1/5.4	2.6/6.2	3.3/5.9	3.1/6.1	2.8/5.7	2.7/5.2	2.6/5.4	2.5/5.2	2.5/5.9	1.3/4.6	1.6/3.9	3.3/4.3
2-Axis	Average	3.1	3.6	4.1	4.7	4.6	4.5	4.0	4.1	4.0	4.4	3.4	3.0	4.0
	Min/Max	2.0/5.0	2.2/5.6	2.6/6.2	3.3/6.0	3.3/6.5	3.1/6.2	2.9/5.6	2.7/5.5	2.5/5.2	2.5/6.0	1.4/4.8	1.7/4.2	3.5/4.5

Average Climatic Conditions

Element	Jan	Feb	Mar	Apr	May	June	July	Aug	Sept	Oct	Nov	Dec	Year
Temperature (°C)	5.3	7.6	12.3	16.7	20.8	24.6	26.6	26.1	23.0	16.9	11.7	7.3	16.6
Daily Minimum Temp	-0.4	1.4	5.7	9.6	14.3	18.4	20.8	20.4	17.2	10.1	5.3	1.6	10.4
Daily Maximum Temp	10.9	13.8	18.9	23.7	27.2	30.8	32.2	31.7	28.8	23.7	18.1	13.2	22.8
Record Minimum Temp	-21.1	-16.1	-11.7	-3.3	1.7	5.6	10.6	10.6	2.8	-2.8	-15.0	-17.2	-21.1
Record Maximum Temp	27.2	28.3	31.7	33.3	37.2	38.9	41.1	39.4	37.8	34.4	28.9	26.7	41.1
HDD, Base 18.3°C	405	300	194	75	17	0	0	0	4	82	202	341	1621
CDD, Base 18.3°C	0	0	8	25	93	188	255	241	144	39	4	0	998
Relative Humidity (%)	70	67	64	65	70	72	74	75	74	71	71	71	70
Wind Speed (m/s)	3.4	3.8	3.8	3.6	3.0	2.6	2.5	2.4	2.8	2.7	3.0	3.3	3.1

Huntsville, AL

WBAN NO. 03856

LATITUDE: 34.65° N
LONGITUDE: 86.77° W
ELEVATION: 190 meters
MEAN PRESSURE: 995 millibars

STATION TYPE: Secondary

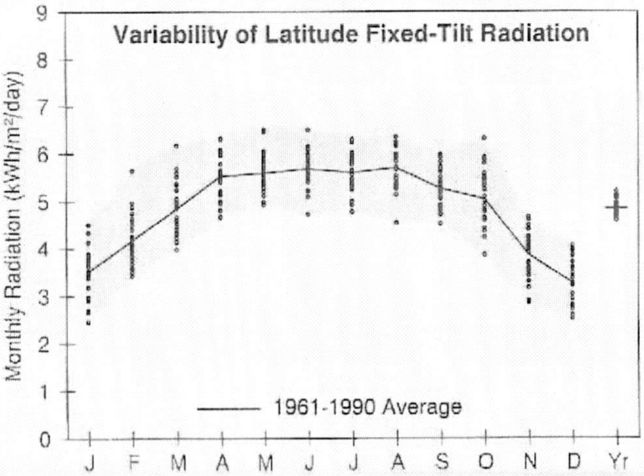

Solar Radiation for Flat-Plate Collectors Facing South at a Fixed Tilt (kWh/m²/day), Uncertainty ±9%

Tilt (°)		Jan	Feb	Mar	Apr	May	June	July	Aug	Sept	Oct	Nov	Dec	Year
0	Average	2.4	3.1	4.1	5.3	5.9	6.3	6.1	5.7	4.7	3.9	2.7	2.1	4.4
	Min/Max	1.9/2.8	2.7/3.9	3.5/5.1	4.5/6.0	5.3/6.9	5.2/7.2	5.2/6.9	4.7/6.3	4.2/5.3	3.2/4.6	2.2/3.0	1.9/2.5	4.2/4.6
Latitude -15	Average	3.1	3.8	4.7	5.6	5.9	6.2	6.0	5.9	5.2	4.7	3.5	2.9	4.8
	Min/Max	2.3/3.9	3.2/5.1	3.9/5.9	4.8/6.4	5.2/6.9	5.1/7.1	5.1/6.8	4.7/6.6	4.5/5.9	3.7/5.8	2.7/4.1	2.3/3.5	4.6/5.1
Latitude	Average	3.5	4.2	4.8	5.5	5.6	5.7	5.6	5.7	5.3	5.0	3.9	3.3	4.8
	Min/Max	2.5/4.5	3.4/5.6	4.0/6.2	4.7/6.3	4.9/6.5	4.7/6.5	4.8/6.3	4.5/6.3	4.5/6.0	3.9/6.3	2.9/4.7	2.5/4.1	4.6/5.2
Latitude +15	Average	3.7	4.3	4.7	5.1	5.0	4.9	4.9	5.2	5.0	5.1	4.1	3.5	4.6
	Min/Max	2.5/4.8	3.5/5.9	3.9/6.1	4.3/5.9	4.4/5.7	4.1/5.6	4.2/5.5	4.1/5.8	4.3/5.8	3.8/6.5	2.9/4.9	2.6/4.4	4.4/5.0
90	Average	3.2	3.4	3.3	2.9	2.4	2.1	2.2	2.7	3.2	3.9	3.4	3.1	3.0
	Min/Max	2.1/4.3	2.8/4.8	2.7/4.2	2.5/3.3	2.2/2.6	1.9/2.3	2.1/2.4	2.3/3.0	2.8/3.7	2.8/5.0	2.4/4.3	2.2/4.0	2.8/3.2

Solar Radiation for 1-Axis Tracking Flat-Plate Collectors with a North-South Axis (kWh/m²/day), Uncertainty ±9%

Axis Tilt (°)		Jan	Feb	Mar	Apr	May	June	July	Aug	Sept	Oct	Nov	Dec	Year
0	Average	3.2	4.2	5.5	7.0	7.6	8.0	7.6	7.3	6.1	5.3	3.6	2.9	5.7
	Min/Max	2.3/4.1	3.4/5.7	4.5/7.3	5.6/8.2	6.5/9.4	6.3/9.3	6.2/9.1	5.5/8.3	5.0/7.1	4.0/6.7	2.7/4.4	2.2/3.6	5.3/6.2
Latitude -15	Average	3.8	4.7	5.9	7.3	7.6	7.9	7.6	7.5	6.5	5.9	4.2	3.4	6.0
	Min/Max	2.6/4.9	3.8/6.6	4.8/7.9	5.8/8.6	6.5/9.4	6.2/9.3	6.2/9.0	5.6/8.5	5.3/7.6	4.4/7.6	3.0/5.2	2.6/4.4	5.6/6.6
Latitude	Average	4.1	5.0	6.0	7.2	7.4	7.6	7.3	7.3	6.6	6.2	4.5	3.8	6.1
	Min/Max	2.7/5.4	4.0/7.1	4.9/8.1	5.7/8.5	6.3/9.1	5.9/8.9	5.9/8.7	5.4/8.4	5.3/7.7	4.5/8.0	3.2/5.7	2.8/4.8	5.7/6.7
Latitude +15	Average	4.3	5.1	5.9	6.9	6.9	7.1	6.8	7.0	6.4	6.2	4.7	3.9	5.9
	Min/Max	2.8/5.7	4.0/7.2	4.8/8.0	5.5/8.2	5.9/8.6	5.5/8.3	5.5/8.1	5.1/8.0	5.2/7.6	4.5/8.1	3.3/5.9	2.9/5.1	5.5/6.5

Solar Radiation for 2-Axis Tracking Flat-Plate Collectors (kWh/m²/day), Uncertainty ±9%

Tracker		Jan	Feb	Mar	Apr	May	June	July	Aug	Sept	Oct	Nov	Dec	Year
2-Axis	Average	4.3	5.1	6.0	7.3	7.7	8.1	7.7	7.5	6.6	6.3	4.7	4.0	6.3
	Min/Max	2.8/5.8	4.1/7.3	4.9/8.1	5.8/8.6	6.6/9.5	6.3/9.4	6.3/9.2	5.6/8.6	5.3/7.7	4.6/8.1	3.3/5.9	2.9/5.2	5.8/6.9

Direct Beam Solar Radiation for Concentrating Collectors (kWh/m²/day), Uncertainty ±8%

Tracker		Jan	Feb	Mar	Apr	May	June	July	Aug	Sept	Oct	Nov	Dec	Year
1-Axis, E-W Horiz Axis	Average	2.4	2.7	2.9	3.4	3.5	3.7	3.5	3.4	3.1	3.4	2.7	2.3	3.1
	Min/Max	1.2/3.6	1.7/4.6	2.0/4.5	2.4/4.4	2.6/5.0	2.4/5.0	2.4/4.7	1.9/4.2	2.1/4.0	2.1/4.8	1.4/3.6	1.3/3.3	2.7/3.6
1-Axis, N-S Horiz Axis	Average	1.9	2.6	3.4	4.5	4.7	4.9	4.5	4.4	3.7	3.5	2.3	1.7	3.5
	Min/Max	1.0/2.9	1.7/4.4	2.4/5.4	3.0/5.8	3.5/6.7	3.1/6.4	3.1/6.1	2.5/5.5	2.5/4.9	2.2/5.1	1.2/3.1	1.0/2.5	3.1/4.1
1-Axis, N-S Tilt=Latitude	Average	2.7	3.3	3.8	4.7	4.5	4.5	4.2	4.5	4.1	4.2	3.0	2.5	3.8
	Min/Max	1.4/4.0	2.2/5.6	2.7/6.1	3.1/6.0	3.4/6.4	2.9/6.0	2.9/5.8	2.5/5.5	2.7/5.3	2.6/6.1	1.6/4.2	1.4/3.5	3.4/4.5
2-Axis	Average	2.9	3.4	3.8	4.7	4.7	4.9	4.5	4.6	4.1	4.3	3.2	2.7	4.0
	Min/Max	1.4/4.3	2.2/5.7	2.7/6.1	3.2/6.1	3.6/6.8	3.2/6.5	3.1/6.2	2.5/5.7	2.7/5.3	2.6/6.2	1.7/4.4	1.5/3.8	3.5/4.7

Average Climatic Conditions

Element	Jan	Feb	Mar	Apr	May	June	July	Aug	Sept	Oct	Nov	Dec	Year
Temperature (°C)	3.8	6.2	11.1	16.0	20.2	24.3	26.1	25.7	22.3	16.2	10.8	6.1	15.7
Daily Minimum Temp	-1.6	0.3	4.9	9.4	14.1	18.3	20.5	19.9	16.4	9.6	4.7	0.6	9.8
Daily Maximum Temp	9.0	11.9	17.1	22.5	26.3	30.3	31.7	31.6	28.2	22.8	16.9	11.4	21.7
Record Minimum Temp	-23.9	-14.4	-14.4	-3.3	2.2	7.2	12.2	12.2	4.4	-1.7	-9.4	-19.4	-23.9
Record Maximum Temp	24.4	26.1	28.9	32.2	33.9	38.3	38.9	39.4	38.3	32.2	28.3	25.0	39.4
HDD, Base 18.3°C	451	341	232	88	23	0	0	0	7	96	228	381	1846
CDD, Base 18.3°C	0	0	6	18	82	180	241	229	127	31	3	0	917
Relative Humidity (%)	72	68	66	63	69	71	74	74	74	70	71	72	70
Wind Speed (m/s)	4.1	4.4	4.5	4.2	3.5	3.0	2.8	2.7	3.1	3.3	3.8	4.2	3.6

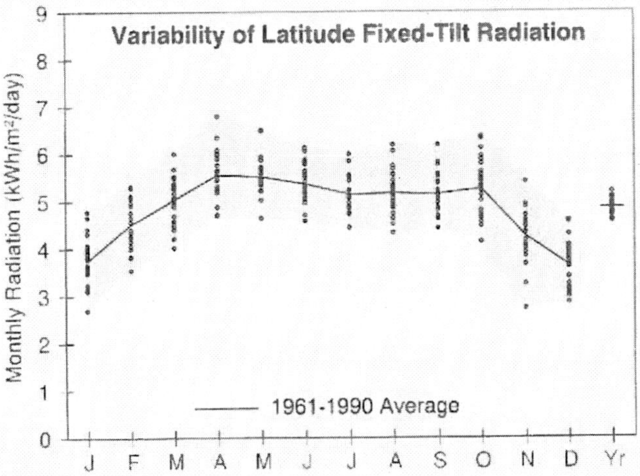

Mobile, AL

WBAN NO. 13894

LATITUDE: 30.68° N
LONGITUDE: 88.25° W
ELEVATION: 67 meters
MEAN PRESSURE: 1010 millibars

STATION TYPE: Secondary

Solar Radiation for Flat-Plate Collectors Facing South at a Fixed Tilt (kWh/m²/day), Uncertainty ±9%

Tilt (°)		Jan	Feb	Mar	Apr	May	June	July	Aug	Sept	Oct	Nov	Dec	Year
0	Average	2.7	3.5	4.4	5.4	5.9	5.9	5.6	5.2	4.7	4.2	3.1	2.5	4.4
	Min/Max	2.2/3.2	2.9/4.0	3.7/5.2	4.7/6.6	4.9/6.9	5.0/6.8	4.8/6.5	4.5/6.2	4.1/5.5	3.5/5.0	2.3/3.7	2.2/3.0	4.2/4.7
Latitude -15	Average	3.3	4.1	4.9	5.6	5.8	5.8	5.5	5.3	5.1	4.9	3.8	3.2	4.8
	Min/Max	2.5/4.1	3.3/4.8	4.0/5.8	4.8/6.9	4.9/6.9	4.9/6.6	4.7/6.4	4.5/6.4	4.4/6.0	3.9/5.9	2.5/4.7	2.6/3.9	4.5/5.1
Latitude	Average	3.7	4.5	5.0	5.6	5.5	5.4	5.1	5.2	5.1	5.2	4.3	3.6	4.9
	Min/Max	2.7/4.8	3.5/5.3	4.0/6.0	4.7/6.8	4.6/6.5	4.6/6.1	4.4/6.0	4.3/6.2	4.4/6.2	4.1/6.4	2.7/5.4	2.8/4.6	4.6/5.2
Latitude +15	Average	4.0	4.6	4.9	5.2	4.9	4.7	4.5	4.7	4.9	5.3	4.5	3.9	4.7
	Min/Max	2.7/5.1	3.6/5.5	3.9/5.9	4.4/6.3	4.2/5.7	4.0/5.3	3.9/5.2	4.0/5.6	4.2/6.0	4.1/6.5	2.8/5.8	3.0/5.0	4.4/5.0
90	Average	3.3	3.6	3.2	2.7	2.1	1.9	1.9	2.3	3.0	3.9	3.7	3.4	2.9
	Min/Max	2.2/4.5	2.8/4.3	2.5/3.8	2.3/3.1	2.0/2.3	1.8/2.0	1.8/2.1	2.1/2.6	2.5/3.6	2.9/4.8	2.1/4.9	2.5/4.5	2.7/3.1

Solar Radiation for 1-Axis Tracking Flat-Plate Collectors with a North-South Axis (kWh/m²/day), Uncertainty ±9%

Axis Tilt (°)		Jan	Feb	Mar	Apr	May	June	July	Aug	Sept	Oct	Nov	Dec	Year
0	Average	3.6	4.7	5.8	7.0	7.4	7.4	6.8	6.5	6.1	5.7	4.2	3.4	5.7
	Min/Max	2.6/4.6	3.6/5.8	4.5/7.1	5.7/9.1	5.8/9.2	6.0/8.8	5.6/8.3	5.2/8.1	5.0/7.4	4.4/7.1	2.7/5.4	2.7/4.4	5.3/6.2
Latitude -15	Average	4.1	5.2	6.1	7.2	7.4	7.3	6.7	6.6	6.3	6.2	4.7	3.9	6.0
	Min/Max	2.8/5.3	3.9/6.3	4.7/7.5	5.8/9.4	5.8/9.2	5.9/8.7	5.6/8.2	5.2/8.2	5.3/7.8	4.8/7.8	2.9/6.2	3.0/5.1	5.5/6.5
Latitude	Average	4.4	5.5	6.2	7.1	7.2	7.0	6.5	6.5	6.4	6.5	5.1	4.2	6.1
	Min/Max	2.9/5.8	4.1/6.7	4.7/7.7	5.7/9.4	5.6/8.9	5.7/8.3	5.3/7.9	5.1/8.1	5.3/8.0	4.9/8.1	3.0/6.7	3.2/5.6	5.6/6.6
Latitude +15	Average	4.6	5.5	6.2	6.9	6.7	6.5	6.1	6.2	6.2	6.6	5.3	4.4	5.9
	Min/Max	3.0/6.1	4.2/6.8	4.6/7.6	5.5/9.0	5.2/8.4	5.3/7.7	5.0/7.4	4.8/7.7	5.1/7.8	4.9/8.2	3.0/7.0	3.3/5.9	5.5/6.4

Solar Radiation for 2-Axis Tracking Flat-Plate Collectors (kWh/m²/day), Uncertainty ±9%

Tracker		Jan	Feb	Mar	Apr	May	June	July	Aug	Sept	Oct	Nov	Dec	Year
2-Axis	Average	4.6	5.6	6.3	7.2	7.5	7.4	6.8	6.6	6.4	6.6	5.3	4.5	6.2
	Min/Max	3.0/6.1	4.2/6.9	4.7/7.7	5.8/9.5	5.8/9.3	6.0/8.9	5.6/8.4	5.2/8.3	5.3/8.0	4.9/8.2	3.1/7.0	3.4/6.0	5.8/6.8

Direct Beam Solar Radiation for Concentrating Collectors (kWh/m²/day), Uncertainty ±8%

Tracker		Jan	Feb	Mar	Apr	May	June	July	Aug	Sept	Oct	Nov	Dec	Year
1-Axis, E-W Horiz Axis	Average	2.5	3.0	3.0	3.4	3.4	3.2	2.8	2.7	2.8	3.5	3.0	2.6	3.0
	Min/Max	1.1/3.8	2.0/4.0	1.8/4.2	2.3/5.2	2.0/4.9	2.2/4.3	2.0/4.1	1.6/4.1	1.8/4.2	2.3/4.9	1.2/4.4	1.5/3.8	2.6/3.5
1-Axis, N-S Horiz Axis	Average	2.2	3.0	3.6	4.5	4.4	4.2	3.6	3.6	3.6	3.8	2.7	2.1	3.4
	Min/Max	1.0/3.3	2.0/4.2	2.1/5.1	3.0/7.0	2.6/6.5	2.8/5.7	2.6/5.3	2.1/5.4	2.4/5.2	2.5/5.4	1.1/4.1	1.2/3.2	3.0/4.0
1-Axis, N-S Tilt=Latitude	Average	2.9	3.6	4.0	4.5	4.2	3.9	3.4	3.6	3.8	4.5	3.4	2.8	3.7
	Min/Max	1.3/4.3	2.4/5.0	2.3/5.6	3.0/7.1	2.5/6.3	2.6/5.3	2.4/4.9	2.1/5.3	2.5/5.6	2.9/6.3	1.4/5.1	1.6/4.2	3.2/4.3
2-Axis	Average	3.1	3.7	4.0	4.6	4.5	4.2	3.6	3.7	3.8	4.5	3.6	3.0	3.9
	Min/Max	1.3/4.6	2.5/5.2	2.3/5.6	3.1/7.2	2.6/6.6	2.8/5.7	2.6/5.3	2.2/5.5	2.5/5.6	2.9/6.3	1.5/5.4	1.8/4.6	3.3/4.5

Average Climatic Conditions

Element	Jan	Feb	Mar	Apr	May	June	July	Aug	Sept	Oct	Nov	Dec	Year
Temperature (°C)	9.9	11.8	15.8	19.9	23.6	26.9	27.9	27.7	25.5	20.2	15.4	11.7	19.7
Daily Minimum Temp	4.4	5.9	10.1	13.9	18.0	21.5	22.9	22.7	20.4	14.1	9.5	6.2	14.1
Daily Maximum Temp	15.4	17.6	21.6	25.8	29.2	32.2	32.9	32.5	30.5	26.4	21.3	17.2	25.2
Record Minimum Temp	-16.1	-11.7	-6.1	0.0	6.1	9.4	15.6	15.0	5.6	0.0	-5.6	-13.3	-16.1
Record Maximum Temp	28.9	27.8	32.2	34.4	37.8	38.9	40.0	38.9	37.2	33.9	30.6	27.2	40.0
HDD, Base 18.3°C	273	191	98	27	0	0	0	0	0	29	109	218	946
CDD, Base 18.3°C	13	8	21	73	164	257	298	289	215	87	22	12	1459
Relative Humidity (%)	72	70	71	70	71	73	76	78	75	71	73	74	73
Wind Speed (m/s)	4.5	4.7	4.7	4.6	3.9	3.4	3.0	2.9	3.4	3.6	4.1	4.4	3.9

Montgomery, AL

WBAN NO. 13895

LATITUDE: 32.30° N
LONGITUDE: 86.40° W
ELEVATION: 62 meters
MEAN PRESSURE: 1011 millibars

STATION TYPE: Primary

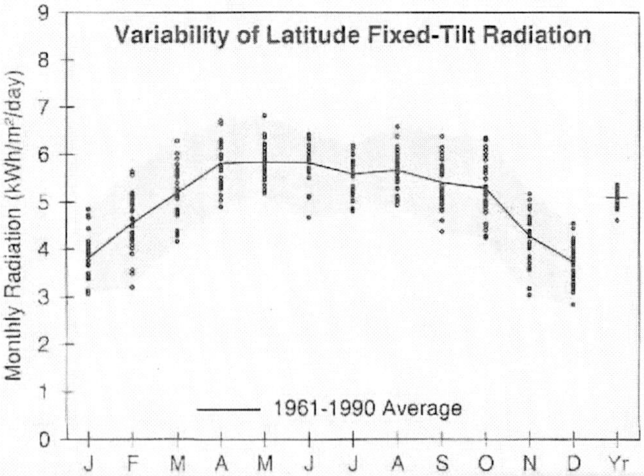

Solar Radiation for Flat-Plate Collectors Facing South at a Fixed Tilt (kWh/m²/day), Uncertainty ±9%

Tilt (°)		Jan	Feb	Mar	Apr	May	June	July	Aug	Sept	Oct	Nov	Dec	Year
0	Average	2.7	3.5	4.5	5.7	6.2	6.4	6.1	5.7	4.9	4.2	3.0	2.5	4.6
	Min/Max	2.2/3.2	2.6/4.1	3.7/5.3	4.8/6.4	5.5/7.2	5.2/7.1	5.2/6.7	5.0/6.6	4.1/5.7	3.5/4.9	2.4/3.5	2.1/2.9	4.2/4.9
Latitude -15	Average	3.4	4.2	5.0	5.9	6.2	6.3	6.0	5.9	5.3	4.9	3.8	3.3	5.0
	Min/Max	2.7/4.2	3.0/5.1	4.1/6.0	5.0/6.8	5.5/7.2	5.0/6.9	5.1/6.6	5.1/6.8	4.4/6.3	4.0/5.9	2.8/4.5	2.6/3.9	4.6/5.3
Latitude	Average	3.8	4.6	5.2	5.8	5.8	5.8	5.6	5.7	5.4	5.3	4.3	3.7	5.1
	Min/Max	3.1/4.8	3.2/5.6	4.2/6.3	4.9/6.7	5.2/6.8	4.7/6.4	4.8/6.2	4.9/6.6	4.4/6.4	4.3/6.3	3.0/5.2	2.8/4.6	4.6/5.4
Latitude +15	Average	4.0	4.7	5.1	5.4	5.2	5.1	4.9	5.2	5.2	5.4	4.5	4.0	4.9
	Min/Max	3.2/5.2	3.2/5.9	4.1/6.2	4.6/6.2	4.6/6.1	4.1/5.6	4.3/5.4	4.5/6.0	4.1/6.1	4.2/6.5	3.1/5.5	3.0/4.9	4.4/5.1
90	Average	3.5	3.7	3.4	2.9	2.3	2.1	2.2	2.6	3.2	4.0	3.8	3.5	3.1
	Min/Max	2.7/4.6	2.5/4.6	2.7/4.2	2.5/3.3	2.2/2.6	2.0/2.2	1.9/2.3	2.3/2.9	2.5/3.7	3.1/4.8	2.5/4.7	2.5/4.5	2.8/3.2

Solar Radiation for 1-Axis Tracking Flat-Plate Collectors with a North-South Axis (kWh/m²/day), Uncertainty ±9%

Axis Tilt (°)		Jan	Feb	Mar	Apr	May	June	July	Aug	Sept	Oct	Nov	Dec	Year
0	Average	3.6	4.7	5.9	7.3	7.8	7.9	7.3	7.1	6.3	5.7	4.1	3.4	5.9
	Min/Max	2.8/4.7	3.2/6.0	4.7/7.5	5.9/8.7	6.8/9.5	5.8/9.1	5.9/8.4	5.9/8.5	5.0/7.7	4.5/7.0	2.9/5.0	2.6/4.2	5.3/6.4
Latitude -15	Average	4.2	5.3	6.3	7.6	7.8	7.8	7.3	7.2	6.6	6.2	4.7	4.0	6.3
	Min/Max	3.2/5.5	3.5/6.7	5.0/8.0	6.1/9.0	6.8/9.6	5.7/9.0	5.9/8.3	6.0/8.7	5.2/8.1	4.8/7.7	3.2/5.8	3.0/5.0	5.6/6.7
Latitude	Average	4.5	5.5	6.5	7.5	7.6	7.5	7.0	7.1	6.7	6.5	5.1	4.4	6.3
	Min/Max	3.5/6.0	3.7/7.1	5.1/8.2	6.1/8.9	6.6/9.3	5.5/8.6	5.7/8.0	5.9/8.6	5.2/8.2	5.0/8.1	3.4/6.3	3.2/5.5	5.6/6.8
Latitude +15	Average	4.7	5.6	6.4	7.2	7.1	7.0	6.5	6.7	6.5	6.6	5.2	4.6	6.2
	Min/Max	3.6/6.3	3.7/7.3	5.0/8.2	5.8/8.6	6.2/8.8	5.1/8.0	5.3/7.5	5.6/8.1	5.0/8.0	5.0/8.2	3.4/6.5	3.3/5.8	5.5/6.6

Solar Radiation for 2-Axis Tracking Flat-Plate Collectors (kWh/m²/day), Uncertainty ±9%

Tracker		Jan	Feb	Mar	Apr	May	June	July	Aug	Sept	Oct	Nov	Dec	Year
2-Axis	Average	4.7	5.6	6.5	7.6	7.9	8.0	7.4	7.2	6.7	6.6	5.3	4.7	6.5
	Min/Max	3.7/6.3	3.8/7.3	5.1/8.3	6.2/9.1	6.9/9.7	5.8/9.2	6.0/8.5	6.1/8.7	5.2/8.2	5.1/8.2	3.4/6.6	3.3/5.9	5.8/7.0

Direct Beam Solar Radiation for Concentrating Collectors (kWh/m²/day), Uncertainty ±8%

Tracker		Jan	Feb	Mar	Apr	May	June	July	Aug	Sept	Oct	Nov	Dec	Year
1-Axis, E-W Horiz Axis	Average	2.7	3.0	3.1	3.6	3.5	3.5	3.0	3.0	3.0	3.5	3.0	2.7	3.1
	Min/Max	1.7/4.0	1.8/4.4	1.9/4.5	2.5/4.7	2.6/4.9	1.7/4.5	1.9/3.9	2.2/4.4	2.0/4.1	2.3/4.7	1.4/4.0	1.5/3.8	2.7/3.5
1-Axis, N-S Horiz Axis	Average	2.3	3.1	3.7	4.7	4.7	4.5	3.9	3.9	3.7	3.8	2.6	2.2	3.6
	Min/Max	1.5/3.5	1.7/4.5	2.3/5.5	3.1/6.2	3.5/6.5	2.2/5.8	2.4/5.0	2.8/5.6	2.4/5.1	2.4/5.2	1.2/3.6	1.2/3.0	3.0/4.0
1-Axis, N-S Tilt=Latitude	Average	3.0	3.7	4.2	4.8	4.5	4.2	3.7	3.9	3.9	4.4	3.4	3.0	3.9
	Min/Max	2.0/4.6	2.1/5.5	2.5/6.1	3.3/6.3	3.3/6.3	2.0/5.4	2.3/4.7	2.8/5.6	2.6/5.5	2.8/6.1	1.5/4.6	1.6/4.1	3.3/4.4
2-Axis	Average	3.2	3.8	4.2	4.9	4.7	4.5	3.9	4.0	3.9	4.5	3.6	3.2	4.0
	Min/Max	2.1/4.9	2.2/5.6	2.5/6.1	3.3/6.4	3.5/6.6	2.2/5.9	2.4/5.1	2.9/5.8	2.6/5.6	2.9/6.1	1.6/4.9	1.8/4.5	3.4/4.6

Average Climatic Conditions

Element	Jan	Feb	Mar	Apr	May	June	July	Aug	Sept	Oct	Nov	Dec	Year
Temperature (°C)	7.8	9.9	14.0	18.2	22.2	25.9	27.4	27.0	24.7	18.8	13.7	9.7	18.3
Daily Minimum Temp	2.1	3.8	7.6	11.6	16.0	19.9	21.9	21.6	18.9	11.8	7.0	3.7	12.2
Daily Maximum Temp	13.5	16.0	20.3	24.7	28.3	31.9	32.8	32.4	30.6	25.7	20.4	15.7	24.3
Record Minimum Temp	-17.8	-12.2	-7.2	-2.2	4.4	9.4	15.0	13.9	3.9	-3.3	-10.6	-15.0	-17.8
Record Maximum Temp	28.3	29.4	31.7	32.8	36.7	40.6	40.6	40.0	38.3	37.8	30.6	29.4	40.6
HDD, Base 18.3°C	330	237	146	43	3	0	0	0	0	51	152	273	1236
CDD, Base 18.3°C	4	0	12	38	122	228	281	269	192	64	12	6	1229
Relative Humidity (%)	70	67	66	67	71	72	76	76	74	71	72	71	71
Wind Speed (m/s)	3.4	3.6	3.7	3.3	2.7	2.5	2.5	2.3	2.6	2.5	2.9	3.3	3.0

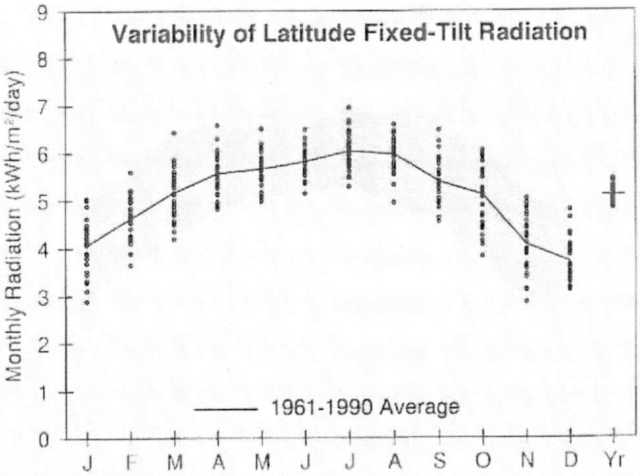

Fort Smith, AR

WBAN NO. 13964

LATITUDE: 35.33° N
LONGITUDE: 94.37° W
ELEVATION: 141 meters
MEAN PRESSURE: 1001 millibars

STATION TYPE: Secondary

Solar Radiation for Flat-Plate Collectors Facing South at a Fixed Tilt (kWh/m²/day), Uncertainty ±9%

Tilt (°)		Jan	Feb	Mar	Apr	May	June	July	Aug	Sept	Oct	Nov	Dec	Year
0	Average	2.6	3.4	4.4	5.4	6.0	6.5	6.6	6.0	4.8	3.9	2.8	2.3	4.6
	Min/Max	2.2/3.0	2.8/3.8	3.8/5.2	4.7/6.3	5.3/6.9	5.8/7.3	5.8/7.7	5.1/6.5	4.1/5.6	3.1/4.5	2.2/3.2	2.1/2.7	4.4/4.8
Latitude -15	Average	3.6	4.2	5.0	5.7	6.0	6.3	6.5	6.2	5.3	4.8	3.6	3.2	5.1
	Min/Max	2.7/4.3	3.4/5.0	4.1/6.1	4.9/6.7	5.3/6.9	5.6/7.1	5.7/7.5	5.2/6.8	4.5/6.3	3.6/5.6	2.7/4.4	2.8/4.1	4.8/5.4
Latitude	Average	4.1	4.6	5.2	5.6	5.7	5.8	6.0	6.0	5.4	5.1	4.1	3.7	5.1
	Min/Max	2.9/5.0	3.7/5.6	4.2/6.4	4.8/6.6	5.0/6.5	5.2/6.5	5.3/6.9	4.9/6.6	4.5/6.5	3.8/6.1	2.9/5.0	3.1/4.8	4.8/5.5
Latitude +15	Average	4.3	4.8	5.1	5.2	5.0	5.0	5.3	5.5	5.2	5.2	4.3	4.0	4.9
	Min/Max	3.0/5.4	3.7/5.8	4.0/6.4	4.5/6.1	4.4/5.8	4.5/5.6	4.6/6.0	4.5/6.0	4.3/6.3	3.8/6.2	2.9/5.4	3.3/5.2	4.6/5.2
90	Average	3.9	3.9	3.5	3.0	2.4	2.2	2.4	2.9	3.4	4.0	3.7	3.6	3.2
	Min/Max	2.5/4.9	2.9/4.8	2.8/4.5	2.6/3.4	2.2/2.7	2.1/2.3	2.2/2.5	2.5/3.1	2.9/4.0	2.8/4.8	2.4/4.7	2.9/4.8	3.0/3.5

Solar Radiation for 1-Axis Tracking Flat-Plate Collectors with a North-South Axis (kWh/m²/day), Uncertainty ±9%

Axis Tilt (°)		Jan	Feb	Mar	Apr	May	June	July	Aug	Sept	Oct	Nov	Dec	Year
0	Average	3.7	4.7	5.8	7.0	7.8	8.4	8.7	7.9	6.4	5.4	3.8	3.3	6.1
	Min/Max	2.6/4.6	3.6/5.8	4.7/7.5	6.0/8.8	6.5/9.3	7.2/9.8	7.5/10.6	6.3/8.9	5.3/8.0	4.0/6.5	2.7/4.8	2.7/4.2	5.7/6.6
Latitude -15	Average	4.4	5.3	6.3	7.3	7.8	8.3	8.7	8.2	6.8	6.1	4.5	3.9	6.5
	Min/Max	3.0/5.6	4.0/6.7	5.0/8.3	6.2/9.1	6.5/9.4	7.1/9.7	7.5/10.6	6.4/9.2	5.6/8.6	4.4/7.4	3.0/5.7	3.3/5.2	6.0/7.0
Latitude	Average	4.8	5.6	6.4	7.2	7.6	8.0	8.4	8.0	6.9	6.4	4.8	4.3	6.5
	Min/Max	3.2/6.2	4.2/7.1	5.0/8.5	6.1/9.1	6.3/9.1	6.8/9.3	7.2/10.2	6.3/9.1	5.7/8.7	4.5/7.7	3.2/6.2	3.6/5.8	6.1/7.1
Latitude +15	Average	5.0	5.7	6.4	7.0	7.1	7.5	7.8	7.6	6.7	6.4	5.0	4.6	6.4
	Min/Max	3.3/6.5	4.3/7.3	4.9/8.4	5.8/8.7	5.9/8.6	6.4/8.7	6.8/9.5	6.0/8.6	5.5/8.5	4.5/7.8	3.3/6.4	3.7/6.1	5.9/7.0

Solar Radiation for 2-Axis Tracking Flat-Plate Collectors (kWh/m²/day), Uncertainty ±9%

Tracker		Jan	Feb	Mar	Apr	May	June	July	Aug	Sept	Oct	Nov	Dec	Year
2-Axis	Average	5.1	5.7	6.5	7.3	7.9	8.5	8.8	8.2	6.9	6.4	5.0	4.6	6.8
	Min/Max	3.3/6.6	4.3/7.3	5.0/8.5	6.2/9.2	6.6/9.5	7.3/9.9	7.7/10.7	6.5/9.3	5.7/8.7	4.6/7.8	3.3/6.5	3.8/6.3	6.3/7.4

Direct Beam Solar Radiation for Concentrating Collectors (kWh/m²/day), Uncertainty ±8%

Tracker		Jan	Feb	Mar	Apr	May	June	July	Aug	Sept	Oct	Nov	Dec	Year
1-Axis, E-W Horiz Axis	Average	3.0	3.1	3.1	3.4	3.7	4.1	4.4	4.0	3.3	3.5	2.9	2.8	3.4
	Min/Max	1.5/4.3	2.0/4.4	1.9/4.8	2.4/4.9	2.5/4.9	3.0/5.3	3.4/6.1	2.5/4.8	2.4/4.7	2.1/4.6	1.4/4.1	2.1/4.2	3.0/4.0
1-Axis, N-S Horiz Axis	Average	2.4	3.0	3.7	4.4	4.9	5.4	5.8	5.3	4.1	3.7	2.5	2.1	3.9
	Min/Max	1.1/3.5	1.9/4.3	2.3/5.7	3.3/6.5	3.3/6.6	4.0/6.9	4.6/8.1	3.4/6.4	3.0/6.0	2.2/4.9	1.2/3.5	1.5/3.2	3.4/4.6
1-Axis, N-S Tilt=Latitude	Average	3.3	3.8	4.2	4.6	4.8	5.0	5.5	5.3	4.5	4.5	3.3	3.0	4.3
	Min/Max	1.6/4.8	2.4/5.5	2.7/6.4	3.4/6.7	3.2/6.4	3.7/6.5	4.3/7.7	3.3/6.4	3.2/6.5	2.6/6.0	1.6/4.7	2.2/4.6	3.7/5.0
2-Axis	Average	3.6	3.9	4.2	4.7	5.0	5.5	5.9	5.5	4.5	4.5	3.5	3.3	4.5
	Min/Max	1.7/5.2	2.5/5.6	2.7/6.5	3.4/6.8	3.4/6.7	4.1/7.1	4.6/8.3	3.5/6.6	3.2/6.5	2.7/6.0	1.7/5.0	2.4/4.9	3.9/5.2

Average Climatic Conditions

Element	Jan	Feb	Mar	Apr	May	June	July	Aug	Sept	Oct	Nov	Dec	Year
Temperature (°C)	2.7	5.5	10.9	16.3	20.6	24.9	27.5	27.0	23.1	16.8	10.3	4.6	15.9
Daily Minimum Temp	-3.6	-1.0	4.0	9.2	14.3	18.8	21.1	20.4	16.6	9.3	3.6	-1.5	9.3
Daily Maximum Temp	9.1	11.9	17.8	23.4	26.9	31.1	33.9	33.6	29.6	24.3	17.1	10.7	22.4
Record Minimum Temp	-23.3	-22.8	-13.9	-5.6	1.7	8.3	10.0	10.6	0.6	-5.6	-13.3	-20.6	-23.3
Record Maximum Temp	27.2	30.0	34.4	35.0	36.7	40.6	43.9	43.3	41.1	35.6	30.0	27.8	43.9
HDD, Base 18.3°C	484	359	237	84	20	0	0	0	7	76	240	426	1932
CDD, Base 18.3°C	0	0	6	24	91	198	284	269	151	29	0	0	1052
Relative Humidity (%)	69	68	64	64	71	71	69	69	72	69	70	71	69
Wind Speed (m/s)	3.6	3.8	4.2	4.0	3.4	2.9	2.8	2.8	3.0	3.0	3.5	3.7	3.4

Little Rock, AR

WBAN NO. 13963

LATITUDE: 34.73° N
LONGITUDE: 92.23° W
ELEVATION: 81 meters
MEAN PRESSURE: 1008 millibars

STATION TYPE: Secondary

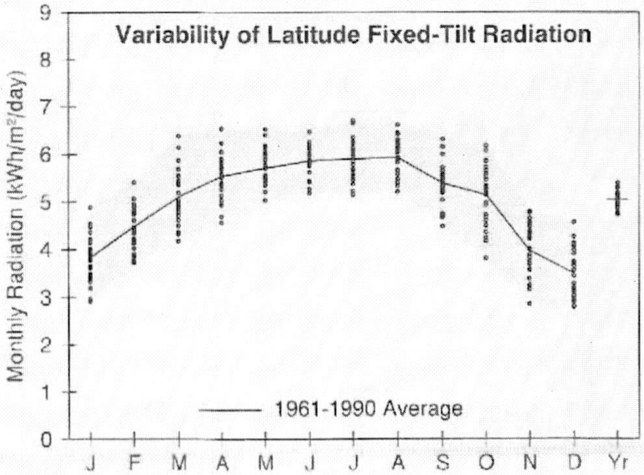

Solar Radiation for Flat-Plate Collectors Facing South at a Fixed Tilt ($kWh/m^2/day$), Uncertainty ±9%

Tilt (°)		Jan	Feb	Mar	Apr	May	June	July	Aug	Sept	Oct	Nov	Dec	Year
0	Average	2.5	3.3	4.3	5.3	6.1	6.5	6.4	5.9	4.8	3.9	2.7	2.2	4.5
	Min/Max	2.1/3.0	2.9/3.8	3.7/5.2	4.5/6.2	5.3/6.9	5.7/7.3	5.7/7.4	5.3/6.6	4.1/5.5	3.1/4.6	2.2/3.1	1.9/2.6	4.3/4.8
Latitude -15	Average	3.4	4.1	4.9	5.6	6.1	6.4	6.3	6.1	5.3	4.8	3.5	3.0	5.0
	Min/Max	2.7/4.2	3.5/4.9	4.1/6.1	4.7/6.6	5.3/6.9	5.6/7.1	5.5/7.3	5.4/6.8	4.5/6.2	3.6/5.7	2.7/4.2	2.5/3.9	4.7/5.3
Latitude	Average	3.8	4.5	5.1	5.5	5.7	5.9	5.9	5.9	5.4	5.1	4.0	3.5	5.0
	Min/Max	2.9/4.9	3.7/5.4	4.2/6.4	4.6/6.5	5.0/6.5	5.2/6.5	5.2/6.7	5.2/6.6	4.5/6.3	3.8/6.2	2.8/4.8	2.8/4.6	4.7/5.4
Latitude +15	Average	4.1	4.6	5.0	5.1	5.1	5.1	5.2	5.4	5.2	5.2	4.2	3.7	4.8
	Min/Max	3.0/5.3	3.8/5.6	4.0/6.3	4.2/6.1	4.5/5.8	4.5/5.6	4.5/5.8	4.7/6.0	4.3/6.1	3.8/6.3	2.9/5.1	2.9/5.0	4.5/5.2
90	Average	3.6	3.7	3.5	2.9	2.4	2.2	2.3	2.8	3.3	4.0	3.5	3.3	3.1
	Min/Max	2.6/4.8	3.0/4.6	2.8/4.4	2.5/3.3	2.2/2.6	2.1/2.3	2.1/2.4	2.5/3.1	2.8/3.9	2.8/4.8	2.4/4.4	2.6/4.6	2.9/3.4

Solar Radiation for 1-Axis Tracking Flat-Plate Collectors with a North-South Axis ($kWh/m^2/day$), Uncertainty ±9%

Axis Tilt (°)		Jan	Feb	Mar	Apr	May	June	July	Aug	Sept	Oct	Nov	Dec	Year
0	Average	3.5	4.5	5.7	7.0	7.8	8.4	8.3	7.8	6.4	5.4	3.7	3.1	6.0
	Min/Max	2.7/4.5	3.7/5.5	4.6/7.3	5.5/8.7	6.6/9.4	7.0/9.8	7.0/10.0	6.6/8.9	5.1/7.4	3.9/6.6	2.7/4.5	2.5/4.1	5.6/6.5
Latitude -15	Average	4.1	5.1	6.2	7.2	7.8	8.3	8.3	8.0	6.8	6.1	4.4	3.7	6.3
	Min/Max	3.1/5.4	4.1/6.4	4.9/8.0	5.7/9.1	6.7/9.4	7.0/9.7	7.0/9.9	6.8/9.1	5.4/8.0	4.3/7.4	3.0/5.3	2.9/5.0	5.9/6.9
Latitude	Average	4.5	5.4	6.4	7.2	7.6	8.0	8.0	7.8	6.8	6.4	4.7	4.1	6.4
	Min/Max	3.3/5.9	4.3/6.8	5.0/8.3	5.6/9.0	6.5/9.1	6.7/9.3	6.8/9.6	6.6/9.0	5.4/8.1	4.4/7.8	3.2/5.8	3.2/5.5	6.0/7.0
Latitude +15	Average	4.7	5.5	6.3	6.9	7.1	7.4	7.5	7.5	6.7	6.4	4.8	4.3	6.3
	Min/Max	3.4/6.3	4.4/6.9	4.9/8.2	5.3/8.6	6.1/8.6	6.2/8.7	6.3/9.0	6.3/8.5	5.2/7.9	4.4/7.9	3.2/6.0	3.3/5.8	5.8/6.8

Solar Radiation for 2-Axis Tracking Flat-Plate Collectors ($kWh/m^2/day$), Uncertainty ±9%

Tracker		Jan	Feb	Mar	Apr	May	June	July	Aug	Sept	Oct	Nov	Dec	Year
2-Axis	Average	4.7	5.5	6.4	7.3	7.9	8.5	8.4	8.0	6.9	6.4	4.9	4.3	6.6
	Min/Max	3.4/6.3	4.4/7.0	5.0/8.3	5.7/9.1	6.8/9.5	7.1/9.9	7.1/10.1	6.8/9.1	5.4/8.1	4.5/7.9	3.3/6.1	3.3/5.9	6.1/7.2

Direct Beam Solar Radiation for Concentrating Collectors ($kWh/m^2/day$), Uncertainty ±8%

Tracker		Jan	Feb	Mar	Apr	May	June	July	Aug	Sept	Oct	Nov	Dec	Year
1-Axis, E-W Horiz Axis	Average	2.7	3.0	3.1	3.4	3.6	4.0	4.0	3.8	3.2	3.5	2.8	2.5	3.3
	Min/Max	1.6/4.0	1.9/4.2	2.0/4.6	2.3/4.8	2.7/5.0	2.9/5.2	3.0/5.5	2.9/4.7	2.2/4.2	2.0/4.7	1.5/3.7	1.8/4.0	2.8/3.8
1-Axis, N-S Horiz Axis	Average	2.2	2.9	3.7	4.4	4.8	5.2	5.2	5.0	4.0	3.7	2.4	1.9	3.8
	Min/Max	1.3/3.3	1.8/4.1	2.3/5.4	2.9/6.4	3.7/6.6	3.8/6.9	3.9/7.3	3.8/6.1	2.6/5.2	2.1/5.0	1.2/3.2	1.3/3.1	3.3/4.4
1-Axis, N-S Tilt=Latitude	Average	3.0	3.6	4.2	4.6	4.6	4.9	4.9	5.0	4.4	4.4	3.2	2.7	4.1
	Min/Max	1.8/4.5	2.3/5.2	2.6/6.1	3.0/6.6	3.5/6.4	3.6/6.5	3.7/6.9	3.8/6.2	2.9/5.7	2.6/6.0	1.6/4.3	1.9/4.3	3.5/4.8
2-Axis	Average	3.3	3.7	4.2	4.7	4.9	5.3	5.3	5.1	4.4	4.5	3.3	3.0	4.3
	Min/Max	1.9/4.8	2.4/5.3	2.6/6.2	3.1/6.7	3.7/6.8	3.9/7.0	4.0/7.4	3.9/6.3	2.9/5.7	2.6/6.1	1.7/4.5	2.0/4.7	3.7/5.0

Average Climatic Conditions

Element	Jan	Feb	Mar	Apr	May	June	July	Aug	Sept	Oct	Nov	Dec	Year
Temperature (°C)	3.9	6.4	11.7	16.7	21.2	25.8	27.7	27.0	23.4	17.2	11.2	6.0	16.6
Daily Minimum Temp	-1.6	0.7	5.7	10.4	15.0	19.7	21.9	21.0	17.5	10.5	5.3	0.6	10.6
Daily Maximum Temp	9.4	12.2	17.8	23.0	27.4	31.8	33.6	33.0	29.2	23.9	17.1	11.4	22.5
Record Minimum Temp	-20.0	-20.6	-11.7	-2.2	4.4	7.8	12.2	11.1	2.8	-1.7	-8.3	-18.3	-20.6
Record Maximum Temp	28.3	29.4	32.8	35.0	36.7	40.6	44.4	42.2	41.1	36.1	30.0	26.7	44.4
HDD, Base 18.3°C	446	333	213	74	14	0	0	0	4	71	215	382	1753
CDD, Base 18.3°C	0	0	8	26	104	223	291	269	156	37	0	0	1114
Relative Humidity (%)	70	68	65	67	71	70	72	72	74	70	71	71	70
Wind Speed (m/s)	3.7	3.9	4.1	3.8	3.2	3.0	2.8	2.7	2.8	2.9	3.5	3.5	3.3

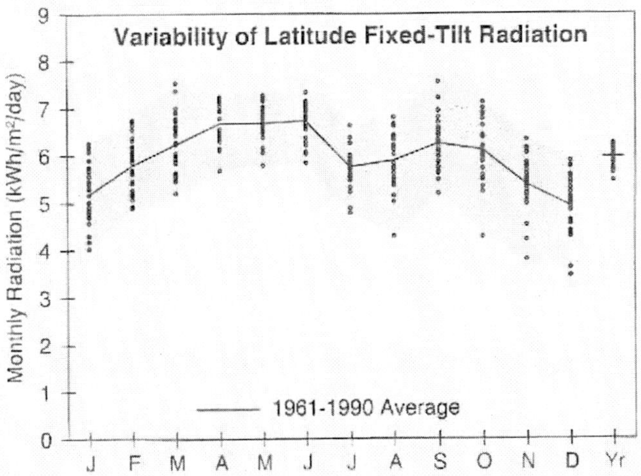

Flagstaff, AZ

WBAN NO. 03103

LATITUDE: 35.13° N
LONGITUDE: 111.67° W
ELEVATION: 2135 meters
MEAN PRESSURE: 788 millibars

STATION TYPE: Secondary

Solar Radiation for Flat-Plate Collectors Facing South at a Fixed Tilt (kWh/m²/day), Uncertainty ±9%

Tilt (°)		Jan	Feb	Mar	Apr	May	June	July	Aug	Sept	Oct	Nov	Dec	Year
0	Average	3.1	4.0	5.1	6.3	7.2	7.7	6.4	5.9	5.4	4.4	3.3	2.8	5.1
	Min/Max	2.7/3.5	3.5/4.5	4.3/6.0	5.4/6.8	6.2/7.8	6.6/8.4	5.2/7.4	4.4/6.8	4.6/6.4	3.4/5.0	2.6/3.7	2.2/3.2	4.8/5.4
Latitude -15	Average	4.4	5.2	5.9	6.8	7.2	7.4	6.2	6.1	6.1	5.6	4.7	4.2	5.8
	Min/Max	3.5/5.3	4.4/6.0	5.0/7.1	5.8/7.3	6.2/7.8	6.4/8.1	5.2/7.2	4.5/7.1	5.1/7.4	4.0/6.5	3.4/5.4	3.0/4.9	5.4/6.1
Latitude	Average	5.2	5.8	6.2	6.7	6.7	6.7	5.8	5.9	6.3	6.1	5.4	4.9	6.0
	Min/Max	4.0/6.3	4.9/6.8	5.2/7.5	5.7/7.2	5.8/7.3	5.8/7.3	4.8/6.6	4.3/6.8	5.2/7.6	4.3/7.1	3.8/6.3	3.5/5.9	5.5/6.2
Latitude +15	Average	5.6	6.1	6.2	6.2	5.9	5.7	5.0	5.4	6.0	6.3	5.8	5.4	5.8
	Min/Max	4.3/6.9	5.1/7.1	5.2/7.5	5.3/6.8	5.1/6.4	5.0/6.2	4.2/5.8	3.9/6.2	5.0/7.3	4.3/7.3	4.0/6.8	3.7/6.5	5.3/6.1
90	Average	5.3	5.3	4.6	3.6	2.6	2.2	2.2	2.8	3.9	4.9	5.1	5.1	4.0
	Min/Max	4.1/6.5	4.3/6.2	3.8/5.4	3.0/4.6	2.4/2.8	2.1/2.3	2.0/2.4	2.1/3.1	3.2/4.6	3.3/5.8	3.4/6.1	3.5/6.4	3.6/4.3

Solar Radiation for 1-Axis Tracking Flat-Plate Collectors with a North-South Axis (kWh/m²/day), Uncertainty ±9%

Axis Tilt (°)		Jan	Feb	Mar	Apr	May	June	July	Aug	Sept	Oct	Nov	Dec	Year
0	Average	4.8	6.0	7.4	9.1	10.1	10.8	8.6	8.2	7.9	6.8	5.2	4.4	7.5
	Min/Max	3.8/5.8	4.9/7.1	5.8/9.2	7.4/10.2	8.4/11.3	8.9/12.1	6.7/10.3	5.7/9.8	6.4/9.8	4.6/8.0	3.5/6.2	3.1/5.3	6.7/7.9
Latitude -15	Average	5.8	6.9	8.0	9.5	10.2	10.7	8.6	8.4	8.5	7.6	6.2	5.4	8.0
	Min/Max	4.4/7.1	5.6/8.2	6.3/10.1	7.7/10.6	8.5/11.4	8.9/12.0	6.7/10.3	5.8/10.0	6.9/10.6	5.1/9.1	4.2/7.4	3.7/6.4	7.1/8.4
Latitude	Average	6.4	7.4	8.3	9.4	9.9	10.3	8.3	8.3	8.6	8.0	6.7	6.0	8.1
	Min/Max	4.8/7.9	5.9/8.8	6.5/10.4	7.7/10.6	8.3/11.1	8.5/11.5	6.4/9.9	5.7/9.9	6.9/10.7	5.3/9.6	4.5/8.1	4.1/7.2	7.2/8.6
Latitude +15	Average	6.8	7.6	8.2	9.1	9.4	9.6	7.8	7.9	8.4	8.1	7.0	6.4	8.0
	Min/Max	5.1/8.4	6.1/9.0	6.4/10.3	7.4/10.2	7.8/10.5	7.9/10.8	6.0/9.3	5.5/9.4	6.8/10.5	5.3/9.7	4.6/8.5	4.3/7.7	7.1/8.5

Solar Radiation for 2-Axis Tracking Flat-Plate Collectors (kWh/m²/day), Uncertainty ±9%

Tracker		Jan	Feb	Mar	Apr	May	June	July	Aug	Sept	Oct	Nov	Dec	Year
2-Axis	Average	6.8	7.6	8.3	9.5	10.3	11.0	8.7	8.5	8.6	8.1	7.1	6.5	8.4
	Min/Max	5.1/8.5	6.1/9.0	6.5/10.4	7.8/10.7	8.6/11.6	9.1/12.3	6.8/10.5	5.9/10.1	6.9/10.7	5.3/9.7	4.6/8.5	4.3/7.9	7.5/8.9

Direct Beam Solar Radiation for Concentrating Collectors (kWh/m²/day), Uncertainty ±8%

Tracker		Jan	Feb	Mar	Apr	May	June	July	Aug	Sept	Oct	Nov	Dec	Year
1-Axis, E-W Horiz Axis	Average	4.4	4.4	4.4	5.1	5.7	6.3	4.4	4.3	4.6	4.9	4.6	4.3	4.8
	Min/Max	2.5/6.0	2.9/6.0	2.7/6.4	3.8/6.1	4.2/6.8	4.7/7.7	2.8/6.0	2.3/5.6	3.4/6.4	2.5/6.3	2.6/6.1	2.3/5.7	3.9/5.3
1-Axis, N-S Horiz Axis	Average	3.6	4.4	5.4	6.9	7.9	8.6	6.0	5.9	6.0	5.4	4.1	3.4	5.6
	Min/Max	2.1/5.0	2.8/5.9	3.3/7.8	5.1/8.4	5.8/9.4	6.3/10.3	3.7/8.1	3.3/7.7	4.4/8.4	2.7/6.9	2.2/5.4	1.8/4.5	4.6/6.2
1-Axis, N-S Tilt=Latitude	Average	4.9	5.5	6.1	7.2	7.6	8.0	5.7	5.9	6.5	6.4	5.4	4.7	6.2
	Min/Max	2.9/6.8	3.5/7.4	3.7/8.9	5.2/8.7	5.7/9.1	5.9/9.7	3.5/7.7	3.3/7.8	4.8/9.1	3.3/8.3	2.9/7.2	2.5/6.3	5.0/6.8
2-Axis	Average	5.3	5.7	6.1	7.3	8.1	8.7	6.1	6.1	6.6	6.5	5.7	5.1	6.4
	Min/Max	3.0/7.3	3.6/7.6	3.7/8.9	5.3/8.8	6.0/9.6	6.4/10.5	3.7/8.2	3.4/8.0	4.8/9.2	3.3/8.4	3.1/7.6	2.8/6.8	5.2/7.1

Average Climatic Conditions

Element	Jan	Feb	Mar	Apr	May	June	July	Aug	Sept	Oct	Nov	Dec	Year
Temperature (°C)	-1.8	-0.3	1.8	5.7	10.2	15.4	19.1	17.8	14.1	8.4	2.7	-1.3	7.7
Daily Minimum Temp	-9.3	-7.9	-5.9	-2.9	0.7	5.2	10.3	9.4	5.1	-0.6	-5.3	-9.0	-0.8
Daily Maximum Temp	5.7	7.4	9.6	14.3	19.7	25.7	27.7	26.3	22.9	17.4	10.6	6.3	16.1
Record Minimum Temp	-30.0	-30.6	-26.7	-18.9	-10.0	-5.6	0.0	-4.4	-5.0	-18.9	-25.0	-30.6	-30.6
Record Maximum Temp	18.9	21.7	22.8	26.7	30.6	35.6	36.1	33.3	32.2	29.4	23.3	20.0	36.1
HDD, Base 18.3°C	625	521	512	378	252	104	16	36	132	307	470	609	3962
CDD, Base 18.3°C	0	0	0	0	0	17	39	21	3	0	0	0	81
Relative Humidity (%)	62	60	56	47	40	33	51	58	55	53	57	61	53
Wind Speed (m/s)	2.9	2.8	3.1	3.3	3.2	3.0	2.3	2.0	2.3	2.4	2.8	2.6	2.7

Phoenix, AZ

WBAN NO. 23183

LATITUDE: 33.43° N
LONGITUDE: 112.02° W
ELEVATION: 339 meters
MEAN PRESSURE: 974 millibars

STATION TYPE: Primary

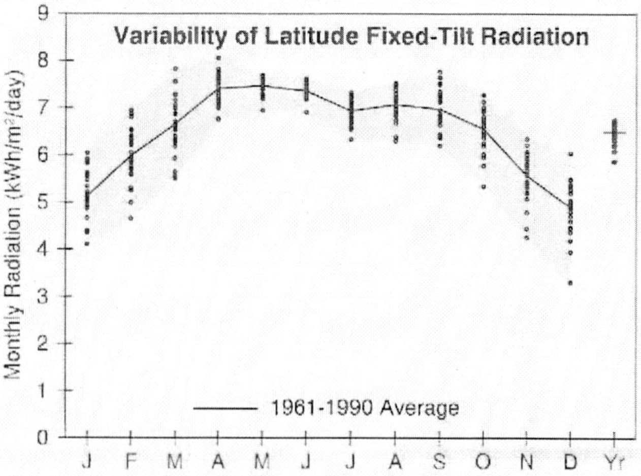

Solar Radiation for Flat-Plate Collectors Facing South at a Fixed Tilt (kWh/m²/day), Uncertainty ±9%

Tilt (°)		Jan	Feb	Mar	Apr	May	June	July	Aug	Sept	Oct	Nov	Dec	Year
0	Average	3.2	4.3	5.5	7.1	8.0	8.4	7.6	7.1	6.1	4.9	3.6	3.0	5.7
	Min/Max	2.8/3.7	3.5/4.8	4.7/6.4	6.4/7.8	7.5/8.3	7.8/8.7	6.9/8.1	6.3/7.5	5.4/6.7	4.2/5.3	3.0/4.0	2.2/3.5	5.3/5.9
Latitude -15	Average	4.4	5.4	6.4	7.5	8.0	8.1	7.5	7.3	6.8	6.0	4.9	4.2	6.4
	Min/Max	3.6/5.1	4.3/6.2	5.3/7.4	6.8/8.2	7.4/8.2	7.6/8.4	6.8/8.0	6.5/7.8	6.1/7.6	5.0/6.6	3.8/5.5	2.9/5.1	5.8/6.6
Latitude	Average	5.1	6.0	6.7	7.4	7.5	7.3	6.9	7.1	7.0	6.5	5.6	4.9	6.5
	Min/Max	4.1/6.0	4.7/6.9	5.5/7.8	6.8/8.1	6.9/7.7	6.9/7.6	6.3/7.3	6.3/7.5	6.2/7.8	5.3/7.3	4.3/6.3	3.3/6.0	5.9/6.7
Latitude +15	Average	5.5	6.2	6.6	6.9	6.6	6.3	6.0	6.4	6.7	6.7	5.9	5.3	6.3
	Min/Max	4.4/6.6	4.8/7.3	5.4/7.8	6.2/7.5	6.1/6.8	5.9/6.5	5.5/6.3	5.7/6.8	5.9/7.5	5.4/7.4	4.4/6.8	3.5/6.7	5.6/6.5
90	Average	4.9	5.0	4.5	3.7	2.7	2.3	2.4	3.1	4.2	5.1	5.1	4.8	4.0
	Min/Max	3.8/5.9	3.8/5.9	3.7/5.3	3.3/3.9	2.6/2.8	2.1/2.4	2.2/2.5	2.8/3.3	3.7/4.6	4.1/5.7	3.7/6.0	3.1/6.2	3.5/4.2

Solar Radiation for 1-Axis Tracking Flat-Plate Collectors with a North-South Axis (kWh/m²/day), Uncertainty ±9%

Axis Tilt (°)		Jan	Feb	Mar	Apr	May	June	July	Aug	Sept	Oct	Nov	Dec	Year
0	Average	4.7	6.2	7.8	9.9	11.0	11.4	10.0	9.6	8.6	7.1	5.3	4.4	8.0
	Min/Max	3.8/5.6	4.6/7.3	6.2/9.5	8.9/11.4	9.9/11.8	10.3/12.2	8.8/11.1	8.0/10.5	7.4/9.9	5.6/8.1	4.0/6.2	2.9/5.5	7.1/8.4
Latitude -15	Average	5.6	7.1	8.5	10.3	11.1	11.3	10.0	9.8	9.2	8.0	6.3	5.3	8.5
	Min/Max	4.4/6.7	5.2/8.3	6.7/10.3	9.2/11.9	9.9/11.8	10.2/12.1	8.8/11.1	8.2/10.8	7.9/10.5	6.2/9.1	4.6/7.3	3.4/6.7	7.6/8.9
Latitude	Average	6.2	7.5	8.7	10.3	10.7	10.8	9.6	9.6	9.3	8.4	6.8	5.8	8.6
	Min/Max	4.8/7.4	5.5/8.9	6.8/10.6	9.1/11.8	9.6/11.5	9.8/11.6	8.4/10.7	8.1/10.6	8.0/10.7	6.5/9.5	4.9/8.0	3.8/7.4	7.6/9.0
Latitude +15	Average	6.5	7.7	8.6	9.9	10.1	10.1	9.0	9.2	9.1	8.5	7.1	6.2	8.5
	Min/Max	5.0/7.8	5.6/9.1	6.8/10.5	8.8/11.4	9.1/10.8	9.1/10.8	7.9/10.0	7.7/10.1	7.8/10.5	6.6/9.7	5.1/8.3	4.0/7.9	7.5/8.9

Solar Radiation for 2-Axis Tracking Flat-Plate Collectors (kWh/m²/day), Uncertainty ±9%

Tracker		Jan	Feb	Mar	Apr	May	June	July	Aug	Sept	Oct	Nov	Dec	Year
2-Axis	Average	6.6	7.7	8.7	10.4	11.2	11.6	10.1	9.8	9.3	8.5	7.1	6.3	8.9
	Min/Max	5.1/7.9	5.6/9.1	6.9/10.6	9.2/11.9	10.1/12.0	10.5/12.4	8.9/11.3	8.3/10.9	8.0/10.7	6.6/9.7	5.1/8.4	4.0/8.1	7.9/9.4

Direct Beam Solar Radiation for Concentrating Collectors (kWh/m²/day), Uncertainty ±8%

Tracker		Jan	Feb	Mar	Apr	May	June	July	Aug	Sept	Oct	Nov	Dec	Year
1-Axis, E-W Horiz Axis	Average	4.2	4.6	4.8	5.7	6.4	6.8	5.6	5.3	5.2	5.0	4.5	4.1	5.2
	Min/Max	3.0/5.4	3.0/5.9	3.4/6.3	4.8/7.1	5.2/7.1	6.0/7.5	4.5/6.7	4.4/6.3	4.1/6.3	3.5/6.0	2.6/5.8	2.5/5.8	4.5/5.6
1-Axis, N-S Horiz Axis	Average	3.4	4.6	5.8	7.6	8.5	8.9	7.2	7.0	6.5	5.4	4.0	3.2	6.0
	Min/Max	2.4/4.5	2.8/5.9	4.0/7.7	6.5/9.8	7.0/9.7	7.9/10.0	5.7/8.8	5.6/8.4	5.1/8.1	3.6/6.6	2.2/5.1	1.8/4.5	5.2/6.5
1-Axis, N-S Tilt=Latitude	Average	4.7	5.7	6.5	7.8	8.2	8.3	6.8	7.0	7.1	6.5	5.2	4.5	6.5
	Min/Max	3.3/6.1	3.6/7.3	4.5/8.6	6.6/10.1	6.7/9.4	7.3/9.3	5.4/8.3	5.6/8.4	5.5/8.7	4.4/7.8	3.0/6.7	2.6/6.3	5.6/7.1
2-Axis	Average	5.0	5.8	6.5	8.0	8.7	9.1	7.3	7.2	7.1	6.5	5.5	4.8	6.8
	Min/Max	3.5/6.5	3.7/7.5	4.5/8.6	6.7/10.3	7.1/9.9	8.0/10.1	5.8/9.0	5.7/8.6	5.5/8.7	4.4/7.9	3.1/7.1	2.8/6.8	5.8/7.4

Average Climatic Conditions

Element	Jan	Feb	Mar	Apr	May	June	July	Aug	Sept	Oct	Nov	Dec	Year
Temperature (°C)	12.0	14.3	16.8	21.1	26.0	31.2	34.2	33.1	29.8	23.6	16.6	12.3	22.6
Daily Minimum Temp	5.1	7.1	9.3	12.9	17.7	22.7	27.2	26.2	22.7	16.0	9.4	5.4	15.2
Daily Maximum Temp	18.8	21.5	24.2	29.2	34.2	39.7	41.1	39.8	36.8	31.2	23.8	19.0	29.9
Record Minimum Temp	-8.3	-5.6	-3.9	0.0	4.4	10.0	16.1	15.6	8.3	1.1	-3.9	-5.6	-8.3
Record Maximum Temp	31.1	33.3	37.8	40.6	45.0	50.0	47.8	46.7	47.8	41.7	33.9	31.1	50.0
HDD, Base 18.3°C	201	126	101	42	4	0	0	0	0	9	74	192	750
CDD, Base 18.3°C	4	12	53	123	242	387	491	457	343	173	23	4	2312
Relative Humidity (%)	51	44	39	28	22	19	32	36	36	37	44	52	37
Wind Speed (m/s)	2.5	2.8	3.2	3.4	3.4	3.2	3.4	3.2	3.0	2.8	2.6	2.5	3.0

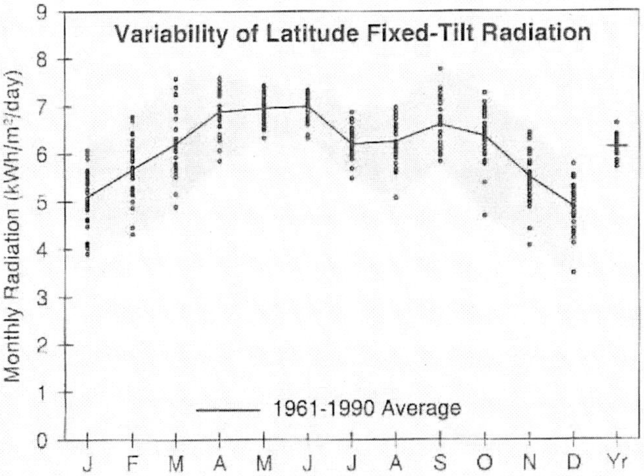

Prescott, AZ

WBAN NO. 23184

LATITUDE: 34.65° N
LONGITUDE: 112.43° W
ELEVATION: 1531 meters
MEAN PRESSURE: 847 millibars

STATION TYPE: Secondary

Solar Radiation for Flat-Plate Collectors Facing South at a Fixed Tilt (kWh/m²/day), Uncertainty ±9%

Tilt (°)		Jan	Feb	Mar	Apr	May	June	July	Aug	Sept	Oct	Nov	Dec	Year
0	Average	3.1	3.9	5.1	6.6	7.5	8.0	6.9	6.3	5.7	4.6	3.4	2.8	5.3
	Min/Max	2.6/3.5	3.2/4.5	4.2/6.1	5.7/7.1	6.8/8.0	7.2/8.5	6.1/7.7	5.1/7.0	5.1/6.7	3.6/5.2	2.8/3.9	2.2/3.2	5.1/5.7
Latitude -15	Average	4.4	5.1	5.9	7.0	7.5	7.7	6.7	6.5	6.5	5.8	4.8	4.1	6.0
	Min/Max	3.5/5.1	4.0/6.0	4.7/7.2	6.0/7.7	6.8/8.0	6.9/8.1	5.9/7.5	5.3/7.2	5.7/7.6	4.4/6.6	3.7/5.5	3.0/4.8	5.6/6.5
Latitude	Average	5.1	5.7	6.2	6.9	7.0	7.0	6.2	6.3	6.6	6.4	5.5	4.9	6.1
	Min/Max	3.9/6.1	4.3/6.8	4.9/7.6	5.8/7.6	6.3/7.4	6.3/7.3	5.5/6.9	5.1/7.0	5.8/7.8	4.7/7.3	4.1/6.4	3.5/5.8	5.7/6.6
Latitude +15	Average	5.5	5.9	6.1	6.4	6.1	6.0	5.4	5.7	6.4	6.5	5.9	5.4	5.9
	Min/Max	4.2/6.6	4.4/7.1	4.8/7.5	5.4/7.1	5.6/6.5	5.4/6.3	4.8/5.9	4.6/6.3	5.6/7.5	4.7/7.5	4.3/6.9	3.7/6.4	5.5/6.4
90	Average	5.0	4.9	4.3	3.5	2.6	2.3	2.3	2.9	4.0	5.0	5.2	5.0	3.9
	Min/Max	3.8/6.1	3.6/5.9	3.4/5.2	3.0/3.8	2.5/2.8	2.2/2.3	2.1/2.4	2.4/3.1	3.6/4.7	3.6/5.8	3.6/6.2	3.4/6.0	3.5/4.2

Solar Radiation for 1-Axis Tracking Flat-Plate Collectors with a North-South Axis (kWh/m²/day), Uncertainty ±9%

Axis Tilt (°)		Jan	Feb	Mar	Apr	May	June	July	Aug	Sept	Oct	Nov	Dec	Year
0	Average	4.8	5.9	7.4	9.4	10.6	11.3	9.3	8.8	8.4	7.1	5.4	4.4	7.7
	Min/Max	3.7/5.7	4.4/7.2	5.7/9.5	7.9/10.6	9.4/11.5	9.9/12.0	7.9/10.8	7.0/10.1	7.1/10.2	5.0/8.2	4.0/6.3	3.2/5.2	7.1/8.4
Latitude -15	Average	5.7	6.8	8.1	9.8	10.6	11.2	9.3	9.0	9.0	8.0	6.3	5.4	8.3
	Min/Max	4.3/6.9	5.0/8.3	6.2/10.3	8.2/11.0	9.4/11.6	9.8/11.9	7.8/10.7	7.1/10.4	7.6/10.9	5.6/9.3	4.6/7.5	3.8/6.4	7.5/9.0
Latitude	Average	6.3	7.2	8.3	9.8	10.3	10.7	8.9	8.8	9.1	8.4	6.9	6.0	8.4
	Min/Max	4.7/7.6	5.3/8.8	6.3/10.6	8.1/11.0	9.2/11.3	9.4/11.4	7.6/10.4	7.0/10.2	7.7/11.1	5.8/9.8	4.9/8.2	4.1/7.2	7.6/9.2
Latitude +15	Average	6.6	7.4	8.2	9.4	9.7	10.1	8.4	8.4	8.9	8.5	7.2	6.3	8.3
	Min/Max	4.9/8.1	5.3/9.1	6.2/10.6	7.7/10.6	8.6/10.6	8.8/10.7	7.1/9.7	6.7/9.8	7.5/10.9	5.8/9.9	5.1/8.6	4.3/7.6	7.5/9.0

Solar Radiation for 2-Axis Tracking Flat-Plate Collectors (kWh/m²/day), Uncertainty ±9%

Tracker		Jan	Feb	Mar	Apr	May	June	July	Aug	Sept	Oct	Nov	Dec	Year
2-Axis	Average	6.7	7.4	8.3	9.9	10.8	11.5	9.4	9.0	9.1	8.5	7.2	6.5	8.7
	Min/Max	4.9/8.2	5.3/9.1	6.3/10.6	8.2/11.1	9.6/11.8	10.0/12.2	8.0/11.0	7.2/10.4	7.7/11.1	5.8/9.9	5.1/8.7	4.4/7.8	7.9/9.5

Direct Beam Solar Radiation for Concentrating Collectors (kWh/m²/day), Uncertainty ±8%

Tracker		Jan	Feb	Mar	Apr	May	June	July	Aug	Sept	Oct	Nov	Dec	Year
1-Axis, E-W Horiz Axis	Average	4.4	4.5	4.5	5.4	6.1	6.8	5.0	4.7	5.0	5.2	4.8	4.4	5.1
	Min/Max	2.8/5.8	2.8/6.0	2.8/6.4	4.0/6.4	5.0/6.9	5.3/7.4	3.8/6.3	3.1/5.9	3.9/6.7	3.0/6.4	2.8/6.2	2.5/5.7	4.3/5.8
1-Axis, N-S Horiz Axis	Average	3.7	4.5	5.6	7.3	8.3	9.1	6.7	6.4	6.5	5.7	4.2	3.4	6.0
	Min/Max	2.3/4.8	2.8/6.1	3.5/8.1	5.5/8.9	6.9/9.5	7.2/10.1	5.1/8.6	4.4/8.1	4.9/8.9	3.2/7.1	2.5/5.5	2.0/4.4	5.1/6.8
1-Axis, N-S Tilt=Latitude	Average	5.0	5.6	6.3	7.6	8.0	8.5	6.3	6.4	7.1	6.8	5.6	4.8	6.5
	Min/Max	3.1/6.6	3.5/7.6	3.9/9.1	5.6/9.1	6.7/9.1	6.7/9.4	4.8/8.1	4.4/8.1	5.4/9.6	3.8/8.4	3.3/7.3	2.8/6.2	5.5/7.4
2-Axis	Average	5.4	5.8	6.3	7.7	8.5	9.3	6.8	6.6	7.1	6.9	5.9	5.2	6.8
	Min/Max	3.4/7.1	3.6/7.8	3.9/9.1	5.7/9.3	7.0/9.7	7.3/10.2	5.2/8.7	4.5/8.4	5.4/9.6	3.9/8.5	3.5/7.7	3.0/6.7	5.8/7.8

Average Climatic Conditions

Element	Jan	Feb	Mar	Apr	May	June	July	Aug	Sept	Oct	Nov	Dec	Year
Temperature (°C)	2.3	3.9	5.9	9.6	14.2	19.6	22.8	21.3	18.1	12.6	6.7	2.6	11.7
Daily Minimum Temp	-5.6	-4.4	-2.2	0.8	5.1	9.9	14.4	13.2	9.3	3.4	-1.9	-5.4	3.1
Daily Maximum Temp	10.2	12.2	14.1	18.3	23.3	29.2	31.2	29.4	26.7	21.8	15.3	10.6	20.2
Record Minimum Temp	-18.3	-18.3	-13.9	-6.1	-3.3	2.2	8.3	4.4	1.1	-7.2	-13.3	-22.8	-22.8
Record Maximum Temp	23.3	25.0	27.2	30.6	35.0	39.4	38.9	37.2	35.0	32.2	26.1	21.1	39.4
HDD, Base 18.3°C	496	403	384	262	138	32	0	0	41	182	350	489	2775
CDD, Base 18.3°C	0	0	0	0	11	68	139	96	32	4	0	0	351
Relative Humidity (%)	59	53	49	38	33	27	44	51	47	46	51	59	47
Wind Speed (m/s)	3.1	3.5	4.1	4.2	4.3	4.1	3.5	3.1	3.4	3.3	3.3	3.0	3.6

Tucson, AZ

WBAN NO. 23160

LATITUDE: 32.12° N
LONGITUDE: 110.93° W
ELEVATION: 779 meters
MEAN PRESSURE: 925 millibars

STATION TYPE: Primary

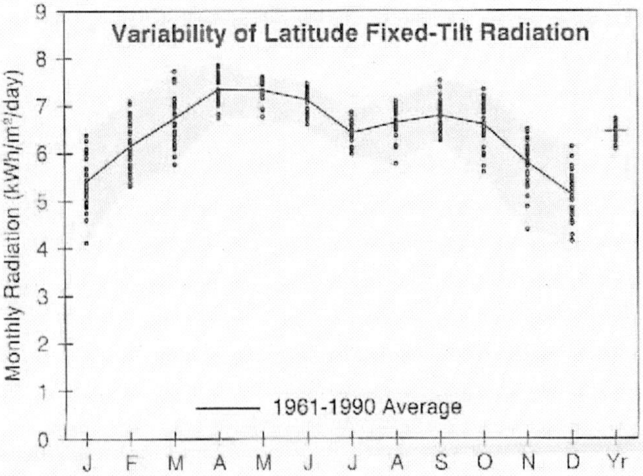

Solar Radiation for Flat-Plate Collectors Facing South at a Fixed Tilt (kWh/m²/day), Uncertainty ±9%

Tilt (°)		Jan	Feb	Mar	Apr	May	June	July	Aug	Sept	Oct	Nov	Dec	Year
0	Average	3.4	4.4	5.6	7.1	7.9	8.1	7.1	6.7	6.0	5.0	3.8	3.2	5.7
	Min/Max	2.8/3.9	3.9/4.9	5.0/6.3	6.5/7.6	7.3/8.2	7.4/8.6	6.5/7.6	5.8/7.2	5.5/6.6	4.4/5.4	3.1/4.2	2.7/3.6	5.5/5.9
Latitude -15	Average	4.6	5.5	6.4	7.5	7.8	7.8	6.9	6.9	6.6	6.1	5.0	4.3	6.3
	Min/Max	3.6/5.4	4.8/6.3	5.6/7.3	6.8/8.0	7.2/8.1	7.2/8.3	6.4/7.4	6.0/7.4	6.1/7.3	5.2/6.7	3.9/5.6	3.6/5.1	6.0/6.5
Latitude	Average	5.4	6.2	6.7	7.3	7.3	7.1	6.4	6.6	6.8	6.6	5.8	5.1	6.5
	Min/Max	4.1/6.4	5.3/7.1	5.8/7.7	6.7/7.8	6.8/7.6	6.6/7.5	6.0/6.9	5.8/7.1	6.3/7.5	5.6/7.3	4.4/6.5	4.1/6.1	6.1/6.7
Latitude +15	Average	5.9	6.4	6.6	6.8	6.4	6.1	5.6	6.0	6.6	6.8	6.2	5.6	6.3
	Min/Max	4.4/7.0	5.5/7.5	5.7/7.7	6.3/7.3	6.0/6.7	5.7/6.3	5.3/6.0	5.3/6.4	6.0/7.3	5.7/7.5	4.6/7.0	4.5/6.8	5.8/6.5
90	Average	5.2	5.1	4.4	3.5	2.5	2.1	2.2	2.8	4.0	5.1	5.3	5.1	3.9
	Min/Max	3.8/6.3	4.3/6.0	3.7/5.1	3.3/3.6	2.4/2.6	2.0/2.1	2.1/2.2	2.6/3.0	3.6/4.3	4.2/5.7	3.8/6.1	4.0/6.2	3.6/4.1

Solar Radiation for 1-Axis Tracking Flat-Plate Collectors with a North-South Axis (kWh/m²/day), Uncertainty ±9%

Axis Tilt (°)		Jan	Feb	Mar	Apr	May	June	July	Aug	Sept	Oct	Nov	Dec	Year
0	Average	5.2	6.6	8.1	10.1	11.0	11.2	9.2	9.0	8.5	7.4	5.8	4.8	8.1
	Min/Max	3.9/6.2	5.5/7.7	6.7/9.6	9.1/11.2	9.9/11.6	9.7/12.2	8.4/10.1	7.6/10.0	7.6/9.8	6.2/8.4	4.2/6.6	3.8/5.8	7.5/8.5
Latitude -15	Average	6.1	7.4	8.7	10.4	11.1	11.1	9.1	9.2	9.0	8.2	6.7	5.6	8.6
	Min/Max	4.5/7.3	6.2/8.7	7.2/10.3	9.4/11.6	9.9/11.6	9.6/12.0	8.4/10.1	7.7/10.2	8.1/10.4	6.8/9.3	4.8/7.6	4.5/6.9	7.9/9.0
Latitude	Average	6.7	7.8	9.0	10.4	10.7	10.6	8.8	9.1	9.1	8.6	7.3	6.2	8.7
	Min/Max	4.9/8.1	6.6/9.3	7.3/10.6	9.3/11.5	9.6/11.3	9.1/11.5	8.1/9.7	7.6/10.0	8.1/10.5	7.1/9.8	5.2/8.3	4.9/7.7	8.0/9.2
Latitude +15	Average	7.0	8.0	8.9	10.0	10.1	9.9	8.2	8.6	9.0	8.7	7.6	6.6	8.6
	Min/Max	5.1/8.5	6.7/9.6	7.3/10.6	9.0/11.1	9.1/10.6	8.5/10.8	7.5/9.1	7.2/9.5	7.9/10.3	7.2/10.0	5.3/8.7	5.2/8.1	7.9/9.0

Solar Radiation for 2-Axis Tracking Flat-Plate Collectors (kWh/m²/day), Uncertainty ±9%

Tracker		Jan	Feb	Mar	Apr	May	June	July	Aug	Sept	Oct	Nov	Dec	Year
2-Axis	Average	7.1	8.1	9.0	10.5	11.2	11.3	9.3	9.2	9.2	8.7	7.6	6.7	9.0
	Min/Max	5.2/8.6	6.7/9.6	7.3/10.7	9.4/11.6	10.1/11.8	9.8/12.3	8.5/10.2	7.8/10.2	8.1/10.5	7.2/10.0	5.4/8.7	5.3/8.3	8.3/9.5

Direct Beam Solar Radiation for Concentrating Collectors (kWh/m²/day), Uncertainty ±8%

Tracker		Jan	Feb	Mar	Apr	May	June	July	Aug	Sept	Oct	Nov	Dec	Year
1-Axis, E-W Horiz Axis	Average	4.7	5.0	5.1	5.9	6.5	6.7	5.0	5.0	5.1	5.3	5.0	4.5	5.3
	Min/Max	3.1/6.0	3.8/6.3	3.6/6.5	4.8/6.9	5.4/7.1	5.3/7.7	4.2/6.0	3.8/5.9	4.3/6.3	4.0/6.4	3.0/6.0	3.1/5.9	4.7/5.7
1-Axis, N-S Horiz Axis	Average	4.0	5.1	6.2	8.0	8.8	8.9	6.5	6.6	6.6	5.9	4.5	3.6	6.2
	Min/Max	2.6/5.2	3.8/6.6	4.4/8.1	6.5/9.5	7.4/9.6	6.9/10.4	5.4/7.8	5.0/7.9	5.4/8.2	4.4/7.2	2.6/5.5	2.5/4.8	5.4/6.8
1-Axis, N-S Tilt=Latitude	Average	5.3	6.2	6.9	8.2	8.5	8.3	6.2	6.6	7.0	6.9	5.8	4.9	6.7
	Min/Max	3.5/6.8	4.6/8.0	4.9/9.0	6.7/9.7	7.1/9.3	6.4/9.6	5.2/7.3	5.0/7.9	5.8/8.8	5.2/8.4	3.4/7.0	3.4/6.5	5.9/7.3
2-Axis	Average	5.7	6.3	6.9	8.3	9.0	9.0	6.6	6.8	7.1	7.0	6.1	5.4	7.0
	Min/Max	3.7/7.3	4.7/8.2	4.9/9.0	6.8/9.9	7.5/9.8	7.0/10.5	5.5/7.9	5.1/8.1	5.8/8.9	5.2/8.5	3.5/7.4	3.7/7.1	6.1/7.6

Average Climatic Conditions

Element	Jan	Feb	Mar	Apr	May	June	July	Aug	Sept	Oct	Nov	Dec	Year
Temperature (°C)	10.7	12.4	14.8	18.8	23.3	28.8	30.3	29.2	26.9	21.3	15.1	11.1	20.2
Daily Minimum Temp	3.7	5.0	7.0	10.2	14.4	19.9	23.1	22.3	19.7	13.7	7.6	4.3	12.6
Daily Maximum Temp	17.7	19.9	22.7	27.3	32.2	37.6	37.4	36.0	34.1	29.1	22.6	17.9	27.9
Record Minimum Temp	-8.9	-6.7	-6.7	-2.8	3.3	8.3	16.7	16.1	6.7	-3.3	-4.4	-8.9	-8.9
Record Maximum Temp	30.6	33.3	37.2	40.0	41.7	47.2	45.6	42.8	41.7	38.3	32.2	28.9	47.2
HDD, Base 18.3°C	236	168	127	54	4	0	0	0	0	15	104	224	932
CDD, Base 18.3°C	0	3	18	67	159	313	372	336	257	108	8	0	1641
Relative Humidity (%)	48	43	37	27	22	21	42	47	42	38	43	50	38
Wind Speed (m/s)	3.8	3.8	4.0	4.3	4.1	4.1	4.0	3.7	3.9	3.8	3.7	3.6	3.9

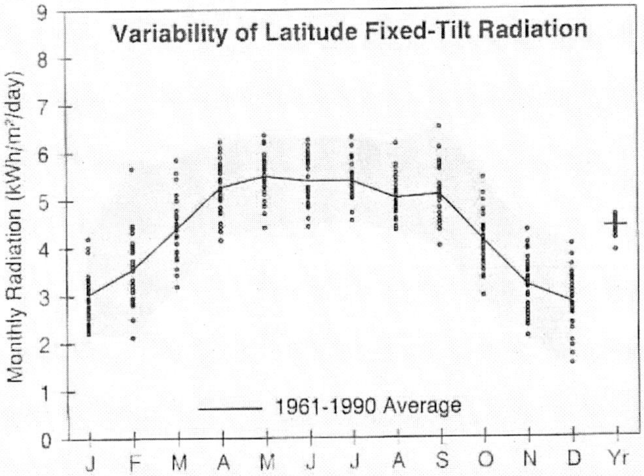

Arcata, CA

WBAN NO. 24283

LATITUDE: 40.98° N
LONGITUDE: 124.10° W
ELEVATION: 69 meters
MEAN PRESSURE: 1010 millibars

STATION TYPE: Secondary

Solar Radiation for Flat-Plate Collectors Facing South at a Fixed Tilt (kWh/m²/day), Uncertainty ±9%

Tilt (°)		Jan	Feb	Mar	Apr	May	June	July	Aug	Sept	Oct	Nov	Dec	Year
0	Average	1.8	2.5	3.6	5.0	5.8	6.0	5.9	5.0	4.4	3.1	2.0	1.6	3.9
	Min/Max	1.5/2.2	1.8/3.4	2.8/4.5	4.1/5.8	4.8/6.6	5.0/7.0	5.1/6.8	4.5/6.0	3.7/5.3	2.5/3.7	1.6/2.4	1.1/2.0	3.6/4.1
Latitude -15	Average	2.7	3.3	4.3	5.4	5.9	5.9	5.8	5.3	5.1	3.9	2.9	2.5	4.4
	Min/Max	2.0/3.6	2.1/5.1	3.2/5.6	4.3/6.3	4.7/6.8	4.8/6.8	4.9/6.8	4.6/6.4	4.1/6.4	2.9/5.1	2.0/3.8	1.4/3.5	4.0/4.6
Latitude	Average	3.0	3.5	4.4	5.3	5.5	5.4	5.4	5.0	5.1	4.1	3.2	2.8	4.4
	Min/Max	2.2/4.2	2.1/5.7	3.2/5.8	4.2/6.2	4.4/6.4	4.4/6.3	4.6/6.3	4.4/6.2	4.0/6.5	3.0/5.4	2.1/4.3	1.5/4.0	3.9/4.6
Latitude +15	Average	3.2	3.6	4.3	4.9	4.9	4.7	4.7	4.6	4.9	4.1	3.3	3.0	4.2
	Min/Max	2.3/4.5	2.1/5.9	3.0/5.8	3.8/5.8	3.9/5.6	3.9/5.4	4.0/5.5	4.0/5.6	3.8/6.3	2.9/5.5	2.1/4.6	1.5/4.4	3.6/4.4
90	Average	2.9	3.0	3.2	3.2	2.8	2.5	2.6	2.8	3.4	3.3	2.9	2.8	3.0
	Min/Max	2.0/4.2	1.7/5.1	2.3/4.4	2.5/3.7	2.3/3.1	2.3/2.8	2.3/2.9	2.5/3.4	2.7/4.4	2.3/4.5	1.8/4.2	1.3/4.2	2.5/3.1

Solar Radiation for 1-Axis Tracking Flat-Plate Collectors with a North-South Axis (kWh/m²/day), Uncertainty ±9%

Axis Tilt (°)		Jan	Feb	Mar	Apr	May	June	July	Aug	Sept	Oct	Nov	Dec	Year
0	Average	2.5	3.3	4.7	6.4	7.3	7.5	7.3	6.3	5.8	4.1	2.8	2.3	5.0
	Min/Max	1.9/3.5	1.9/5.4	3.2/6.5	4.8/7.9	5.5/8.8	5.7/9.2	5.9/9.0	5.2/8.0	4.4/7.7	3.0/5.6	1.9/3.8	1.3/3.2	4.5/5.4
Latitude -15	Average	3.2	3.9	5.3	6.8	7.5	7.5	7.4	6.5	6.4	4.8	3.4	2.9	5.5
	Min/Max	2.3/4.6	2.2/6.7	3.5/7.4	5.1/8.4	5.5/8.9	5.6/9.2	5.9/9.1	5.3/8.4	4.8/8.6	3.3/6.6	2.2/4.8	1.5/4.3	4.8/5.9
Latitude	Average	3.5	4.1	5.3	6.7	7.2	7.2	7.1	6.4	6.4	5.0	3.7	3.2	5.5
	Min/Max	2.4/5.0	2.2/7.1	3.5/7.6	4.9/8.4	5.3/8.7	5.3/8.8	5.6/8.8	5.2/8.2	4.7/8.7	3.3/6.9	2.3/5.2	1.6/4.7	4.7/5.9
Latitude +15	Average	3.6	4.2	5.2	6.4	6.8	6.7	6.6	6.0	6.3	5.0	3.8	3.4	5.3
	Min/Max	2.5/5.3	2.2/7.3	3.4/7.5	4.7/8.0	5.0/8.2	4.9/8.3	5.2/8.2	4.9/7.8	4.6/8.5	3.3/6.9	2.3/5.4	1.6/5.0	4.6/5.7

Solar Radiation for 2-Axis Tracking Flat-Plate Collectors (kWh/m²/day), Uncertainty ±9%

Tracker		Jan	Feb	Mar	Apr	May	June	July	Aug	Sept	Oct	Nov	Dec	Year
2-Axis	Average	3.6	4.2	5.4	6.8	7.5	7.7	7.5	6.5	6.5	5.0	3.8	3.4	5.7
	Min/Max	2.5/5.3	2.2/7.3	3.5/7.6	5.1/8.4	5.6/9.1	5.7/9.4	6.0/9.2	5.3/8.4	4.8/8.7	3.3/6.9	2.3/5.4	1.6/5.1	4.9/6.1

Direct Beam Solar Radiation for Concentrating Collectors (kWh/m²/day), Uncertainty ±8%

Tracker		Jan	Feb	Mar	Apr	May	June	July	Aug	Sept	Oct	Nov	Dec	Year
1-Axis, E-W Horiz Axis	Average	2.0	2.1	2.4	3.1	3.5	3.5	3.4	2.8	3.1	2.5	2.1	2.0	2.7
	Min/Max	1.2/3.6	0.7/4.8	1.2/4.3	1.7/4.4	1.9/4.7	1.8/4.9	2.0/4.9	1.7/4.5	1.8/4.9	1.2/4.1	0.8/3.4	0.6/3.4	2.2/3.1
1-Axis, N-S Horiz Axis	Average	1.5	1.8	2.7	3.9	4.5	4.5	4.4	3.6	3.7	2.5	1.6	1.3	3.0
	Min/Max	0.8/2.6	0.6/4.3	1.3/4.8	2.1/5.6	2.5/6.2	2.3/6.4	2.7/6.4	2.2/5.7	2.1/5.9	1.2/4.1	0.7/2.6	0.4/2.3	2.5/3.5
1-Axis, N-S Tilt=Latitude	Average	2.2	2.5	3.2	4.2	4.4	4.2	4.3	3.7	4.2	3.2	2.4	2.1	3.4
	Min/Max	1.3/4.0	0.8/5.8	1.5/5.7	2.3/5.9	2.4/6.1	2.2/6.0	2.5/6.2	2.3/5.9	2.4/6.7	1.5/5.2	0.9/3.9	0.6/3.6	2.7/3.9
2-Axis	Average	2.4	2.6	3.2	4.2	4.7	4.6	4.6	3.9	4.3	3.3	2.5	2.3	3.5
	Min/Max	1.3/4.3	0.8/5.9	1.5/5.7	2.3/6.1	2.5/6.5	2.3/6.5	2.7/6.6	2.3/6.1	2.4/6.7	1.5/5.3	1.0/4.1	0.7/3.9	2.8/4.0

Average Climatic Conditions

Element	Jan	Feb	Mar	Apr	May	June	July	Aug	Sept	Oct	Nov	Dec	Year
Temperature (°C)	7.8	8.7	8.9	9.5	11.2	12.8	13.6	13.8	13.2	11.7	9.9	8.0	10.8
Daily Minimum Temp	3.8	4.7	5.0	5.3	7.2	9.0	10.1	10.4	9.1	7.5	5.9	4.1	6.8
Daily Maximum Temp	12.2	12.7	12.7	13.4	14.7	16.3	17.2	17.4	17.7	16.1	14.2	12.2	14.7
Record Minimum Temp	-6.7	-5.0	-3.9	-1.7	0.0	1.7	2.8	4.4	0.6	-3.3	-2.8	-8.3	-8.3
Record Maximum Temp	23.9	25.6	23.9	25.6	31.1	27.2	27.8	31.7	32.2	29.4	24.4	23.3	32.2
HDD, Base 18.3°C	329	276	295	269	226	171	151	144	157	211	257	324	2809
CDD, Base 18.3°C	0	0	0	0	0	0	0	0	0	0	0	0	0
Relative Humidity (%)	82	81	82	81	83	85	85	87	87	87	84	83	84
Wind Speed (m/s)	2.9	3.2	3.5	3.4	3.4	3.0	2.8	2.4	2.3	2.3	2.8	2.8	2.9

Bakersfield, CA

WBAN NO. 23155

LATITUDE: 35.42° N
LONGITUDE: 119.05° W
ELEVATION: 150 meters
MEAN PRESSURE: 998 millibars

STATION TYPE: Secondary

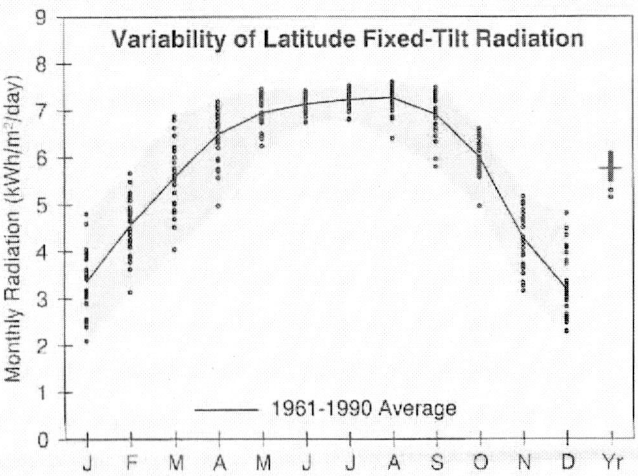

Solar Radiation for Flat-Plate Collectors Facing South at a Fixed Tilt (kWh/m²/day), Uncertainty ±9%

Tilt (°)		Jan	Feb	Mar	Apr	May	June	July	Aug	Sept	Oct	Nov	Dec	Year
0	Average	2.3	3.3	4.7	6.2	7.4	8.1	8.0	7.2	5.9	4.4	2.9	2.1	5.2
	Min/Max	1.8/2.9	2.6/3.9	3.6/5.5	4.8/6.8	6.7/7.9	7.6/8.4	7.4/8.4	6.4/7.6	5.1/6.4	3.8/4.8	2.4/3.3	1.8/2.7	4.8/5.4
Latitude -15	Average	3.0	4.2	5.4	6.6	7.4	7.8	7.8	7.5	6.8	5.5	3.8	2.8	5.7
	Min/Max	2.0/4.2	3.0/5.1	4.0/6.5	5.1/7.3	6.7/7.9	7.3/8.1	7.3/8.2	6.6/7.9	5.7/7.3	4.7/6.1	2.9/4.5	2.1/4.1	5.2/6.0
Latitude	Average	3.3	4.5	5.6	6.5	6.9	7.1	7.2	7.3	6.9	6.0	4.3	3.2	5.7
	Min/Max	2.1/4.8	3.1/5.7	4.0/6.9	5.0/7.2	6.2/7.4	6.7/7.4	6.8/7.5	6.4/7.6	5.8/7.5	5.0/6.6	3.2/5.2	2.3/4.8	5.2/6.1
Latitude +15	Average	3.5	4.7	5.5	6.0	6.1	6.1	6.2	6.6	6.7	6.1	4.5	3.4	5.4
	Min/Max	2.1/5.2	3.1/5.9	3.9/6.8	4.6/6.7	5.5/6.5	5.8/6.3	5.9/6.4	5.8/6.9	5.6/7.2	5.0/6.8	3.2/5.5	2.3/5.2	4.8/5.8
90	Average	3.0	3.8	3.8	3.4	2.8	2.4	2.6	3.3	4.3	4.7	3.9	3.0	3.4
	Min/Max	1.7/4.6	2.4/4.8	2.7/4.8	2.7/3.8	2.6/2.9	2.3/2.4	2.5/2.6	3.0/3.4	3.6/4.6	3.8/5.3	2.7/4.8	2.0/4.8	3.0/3.7

Solar Radiation for 1-Axis Tracking Flat-Plate Collectors with a North-South Axis (kWh/m²/day), Uncertainty ±9%

Axis Tilt (°)		Jan	Feb	Mar	Apr	May	June	July	Aug	Sept	Oct	Nov	Dec	Year
0	Average	3.0	4.5	6.3	8.4	10.1	11.0	11.1	10.1	8.5	6.3	3.9	2.7	7.2
	Min/Max	1.9/4.3	3.1/5.6	4.4/7.9	6.0/9.5	8.8/11.1	10.2/11.8	10.0/11.9	8.5/10.9	6.9/9.4	5.2/7.1	2.8/4.8	2.0/4.1	6.3/7.6
Latitude -15	Average	3.5	5.1	6.8	8.7	10.2	11.0	11.1	10.5	9.1	7.2	4.7	3.3	7.6
	Min/Max	2.1/5.2	3.4/6.5	4.7/8.7	6.2/9.9	8.8/11.2	10.1/11.7	10.0/11.9	8.7/11.2	7.4/10.1	5.8/8.0	3.3/5.7	2.3/5.1	6.7/8.1
Latitude	Average	3.8	5.4	7.0	8.7	9.9	10.5	10.7	10.3	9.2	7.5	5.0	3.6	7.6
	Min/Max	2.2/5.7	3.5/6.9	4.8/8.9	6.1/9.9	8.6/10.9	9.7/11.2	9.6/11.5	8.6/11.0	7.4/10.3	6.1/8.4	3.5/6.2	2.5/5.7	6.7/8.1
Latitude +15	Average	3.9	5.5	6.9	8.3	9.3	9.8	10.0	9.8	9.0	7.6	5.2	3.8	7.4
	Min/Max	2.2/6.0	3.5/7.1	4.7/8.9	5.9/9.5	8.1/10.3	9.1/10.4	9.0/10.8	8.1/10.5	7.3/10.1	6.1/8.5	3.6/6.5	2.5/6.0	6.4/7.9

Solar Radiation for 2-Axis Tracking Flat-Plate Collectors (kWh/m²/day), Uncertainty ±9%

Tracker		Jan	Feb	Mar	Apr	May	June	July	Aug	Sept	Oct	Nov	Dec	Year
2-Axis	Average	4.0	5.5	7.0	8.8	10.3	11.2	11.3	10.5	9.2	7.6	5.2	3.8	7.9
	Min/Max	2.2/6.1	3.5/7.1	4.8/9.0	6.2/10.0	8.9/11.4	10.4/11.9	10.1/12.1	8.8/11.3	7.5/10.3	6.2/8.5	3.6/6.5	2.5/6.1	6.9/8.4

Direct Beam Solar Radiation for Concentrating Collectors (kWh/m²/day), Uncertainty ±8%

Tracker		Jan	Feb	Mar	Apr	May	June	July	Aug	Sept	Oct	Nov	Dec	Year
1-Axis, E-W Horiz Axis	Average	2.1	2.9	3.5	4.5	5.6	6.5	6.5	5.9	5.2	4.4	3.0	2.0	4.3
	Min/Max	0.6/3.8	1.4/4.2	1.8/5.1	2.5/5.5	4.6/6.6	5.7/7.1	5.6/7.3	4.5/6.6	3.7/6.1	3.3/5.2	1.6/4.0	0.9/3.9	3.5/4.8
1-Axis, N-S Horiz Axis	Average	1.6	2.7	4.0	5.8	7.5	8.6	8.7	7.9	6.5	4.6	2.5	1.5	5.2
	Min/Max	0.5/3.0	1.3/3.9	2.1/5.9	3.2/7.2	6.0/8.8	7.5/9.5	7.3/9.9	5.8/8.9	4.6/7.7	3.4/5.5	1.3/3.3	0.7/2.9	4.1/5.6
1-Axis, N-S Tilt=Latitude	Average	2.2	3.5	4.6	6.0	7.2	8.0	8.2	7.9	7.1	5.5	3.4	2.2	5.5
	Min/Max	0.6/4.2	1.7/5.0	2.4/6.7	3.3/7.4	5.8/8.5	7.1/8.9	6.9/9.4	5.9/8.9	5.0/8.4	4.2/6.6	1.8/4.6	1.0/4.2	4.3/6.0
2-Axis	Average	2.4	3.5	4.6	6.1	7.7	8.7	8.9	8.2	7.1	5.6	3.5	2.3	5.7
	Min/Max	0.7/4.4	1.7/5.1	2.4/6.8	3.4/7.6	6.1/9.0	7.7/9.6	7.4/10.1	6.1/9.2	5.0/8.4	4.2/6.7	1.9/4.8	1.1/4.6	4.5/6.3

Average Climatic Conditions

Element	Jan	Feb	Mar	Apr	May	June	July	Aug	Sept	Oct	Nov	Dec	Year
Temperature (°C)	8.8	11.8	14.1	17.2	21.7	25.7	28.9	28.1	24.9	19.9	13.2	8.6	18.6
Daily Minimum Temp	3.7	5.9	7.7	10.1	14.1	17.8	20.9	20.3	17.5	12.7	7.1	3.5	11.8
Daily Maximum Temp	13.8	17.7	20.5	24.4	29.2	33.6	36.9	35.9	32.3	27.1	19.3	13.6	25.4
Record Minimum Temp	-6.7	-3.9	-0.6	1.1	2.8	7.2	11.1	11.1	7.2	-1.7	-2.2	-7.2	-7.2
Record Maximum Temp	27.8	30.6	33.3	38.3	41.7	45.6	46.1	44.4	44.4	39.4	32.8	28.3	46.1
HDD, Base 18.3°C	296	184	137	80	16	3	0	0	4	33	157	302	1212
CDD, Base 18.3°C	0	0	6	47	119	223	329	303	201	82	4	0	1314
Relative Humidity (%)	76	69	61	51	41	36	34	39	44	50	65	76	53
Wind Speed (m/s)	2.5	2.7	3.0	3.3	3.6	3.5	3.3	3.1	2.9	2.7	2.5	2.5	3.0

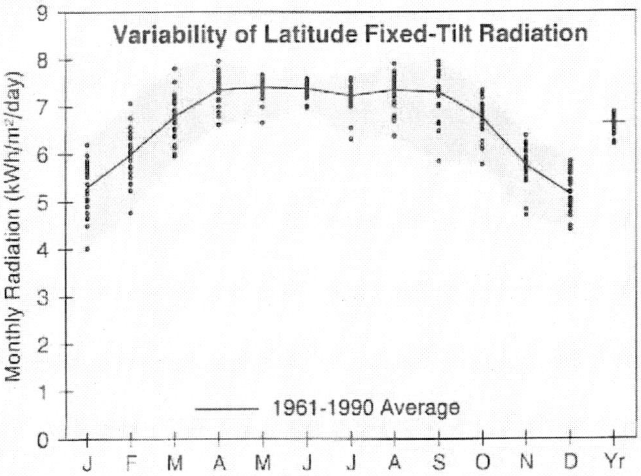

Daggett, CA

WBAN NO. 23161

LATITUDE: 34.87° N
LONGITUDE: 116.78° W
ELEVATION: 588 meters
MEAN PRESSURE: 947 millibars

STATION TYPE: Primary

Solar Radiation for Flat-Plate Collectors Facing South at a Fixed Tilt (kWh/m²/day), Uncertainty ±9%

Tilt (°)		Jan	Feb	Mar	Apr	May	June	July	Aug	Sept	Oct	Nov	Dec	Year
0	Average	3.2	4.2	5.5	7.0	7.9	8.4	8.0	7.3	6.3	4.9	3.6	2.9	5.8
	Min/Max	2.6/3.5	3.5/4.7	4.9/6.2	6.3/7.5	7.2/8.2	7.9/8.7	6.9/8.5	6.4/7.9	5.1/6.8	4.3/5.2	3.1/3.9	2.6/3.2	5.5/6.0
Latitude -15	Average	4.6	5.4	6.5	7.5	7.9	8.1	7.8	7.6	7.1	6.2	5.0	4.4	6.5
	Min/Max	3.5/5.3	4.4/6.3	5.7/7.4	6.7/8.1	7.1/8.2	7.7/8.4	6.8/8.3	6.6/8.2	5.7/7.8	5.3/6.7	4.2/5.5	3.8/4.9	6.1/6.7
Latitude	Average	5.3	6.0	6.8	7.4	7.4	7.4	7.2	7.3	7.3	6.8	5.8	5.2	6.6
	Min/Max	4.0/6.2	4.8/7.1	6.0/7.8	6.6/8.0	6.7/7.7	7.0/7.6	6.3/7.6	6.4/7.9	5.8/7.9	5.8/7.3	4.7/6.4	4.4/5.8	6.2/6.9
Latitude +15	Average	5.7	6.2	6.7	6.8	6.5	6.3	6.2	6.6	7.0	6.9	6.2	5.6	6.4
	Min/Max	4.3/6.8	4.9/7.4	5.8/7.8	6.1/7.4	5.8/6.7	6.0/6.4	5.5/6.5	5.8/7.1	5.6/7.7	5.9/7.5	5.0/6.9	4.8/6.4	5.9/6.6
90	Average	5.2	5.1	4.7	3.8	2.8	2.4	2.5	3.3	4.4	5.4	5.4	5.2	4.2
	Min/Max	3.7/6.2	3.9/6.2	4.0/5.4	3.4/4.0	2.6/2.9	2.2/2.5	2.3/2.6	3.0/3.4	3.6/4.8	4.5/5.9	4.3/6.1	4.3/6.0	3.8/4.3

Solar Radiation for 1-Axis Tracking Flat-Plate Collectors with a North-South Axis (kWh/m²/day), Uncertainty ±9%

Axis Tilt (°)		Jan	Feb	Mar	Apr	May	June	July	Aug	Sept	Oct	Nov	Dec	Year
0	Average	4.9	6.2	8.1	10.0	11.1	11.8	11.2	10.5	9.3	7.5	5.5	4.6	8.4
	Min/Max	3.7/5.7	4.6/7.4	7.0/9.6	8.6/11.0	9.9/11.6	10.9/12.4	9.0/12.2	8.5/11.6	7.0/10.4	6.2/8.2	4.4/6.3	3.9/5.2	7.7/8.8
Latitude -15	Average	5.9	7.1	8.8	10.4	11.2	11.7	11.1	10.8	10.0	8.4	6.6	5.6	9.0
	Min/Max	4.4/7.0	5.3/8.6	7.6/10.5	8.9/11.5	10.0/11.7	10.8/12.3	9.0/12.2	8.8/11.9	7.5/11.1	6.9/9.3	5.2/7.4	4.7/6.5	8.2/9.4
Latitude	Average	6.5	7.5	9.0	10.3	10.9	11.2	10.7	10.6	10.1	8.8	7.2	6.3	9.1
	Min/Max	4.8/7.7	5.6/9.1	7.8/10.8	8.8/11.4	9.7/11.4	10.4/11.8	8.7/11.8	8.6/11.7	7.6/11.3	7.3/9.7	5.6/8.1	5.2/7.2	8.3/9.5
Latitude +15	Average	6.8	7.7	8.9	10.0	10.3	10.5	10.1	10.1	9.9	8.9	7.5	6.6	8.9
	Min/Max	4.9/8.1	5.7/9.4	7.7/10.8	8.5/11.0	9.1/10.7	9.7/11.0	8.1/11.1	8.2/11.2	7.4/11.1	7.3/9.9	5.8/8.5	5.5/7.7	8.1/9.4

Solar Radiation for 2-Axis Tracking Flat-Plate Collectors (kWh/m²/day), Uncertainty ±9%

Tracker		Jan	Feb	Mar	Apr	May	June	July	Aug	Sept	Oct	Nov	Dec	Year
2-Axis	Average	6.9	7.7	9.0	10.4	11.3	12.0	11.4	10.8	10.1	9.0	7.5	6.8	9.4
	Min/Max	5.0/8.2	5.7/9.4	7.8/10.9	8.9/11.5	10.1/11.9	11.1/12.6	9.2/12.5	8.8/12.0	7.6/11.3	7.3/9.9	5.9/8.5	5.6/7.8	8.6/9.9

Direct Beam Solar Radiation for Concentrating Collectors (kWh/m²/day), Uncertainty ±8%

Tracker		Jan	Feb	Mar	Apr	May	June	July	Aug	Sept	Oct	Nov	Dec	Year
1-Axis, E-W Horiz Axis	Average	4.5	4.7	5.0	5.7	6.5	7.1	6.6	6.2	5.9	5.5	4.9	4.6	5.6
	Min/Max	2.9/5.8	3.0/6.2	3.9/6.5	4.4/6.8	5.6/7.0	6.1/7.8	4.6/7.6	4.6/7.2	4.2/6.8	4.1/6.3	3.4/5.9	3.5/5.6	4.8/6.0
1-Axis, N-S Horiz Axis	Average	3.7	4.6	6.1	7.7	8.7	9.6	8.8	8.4	7.5	6.0	4.3	3.5	6.6
	Min/Max	2.4/4.7	2.8/6.2	4.7/8.1	5.8/9.1	7.6/9.5	8.2/10.4	6.0/10.4	6.0/9.9	5.1/9.0	4.3/6.9	2.9/5.2	2.6/4.3	5.6/7.1
1-Axis, N-S Tilt=Latitude	Average	5.1	5.8	6.9	8.0	8.4	8.9	8.4	8.4	8.2	7.2	5.7	5.0	7.2
	Min/Max	3.3/6.5	3.6/7.8	5.3/9.1	6.0/9.4	7.3/9.2	7.7/9.8	5.7/9.8	6.1/9.9	5.6/9.7	5.2/8.3	3.9/6.9	3.7/6.1	6.1/7.7
2-Axis	Average	5.4	5.9	6.9	8.1	8.9	9.7	9.0	8.7	8.2	7.3	6.0	5.4	7.5
	Min/Max	3.5/7.0	3.7/8.0	5.3/9.1	6.1/9.6	7.7/9.7	8.4/10.6	6.1/10.5	6.2/10.2	5.6/9.8	5.3/8.4	4.1/7.2	4.1/6.6	6.3/8.0

Average Climatic Conditions

Element	Jan	Feb	Mar	Apr	May	June	July	Aug	Sept	Oct	Nov	Dec	Year
Temperature (°C)	9.3	12.1	14.6	18.2	23.1	28.2	31.6	30.6	26.5	20.7	13.8	9.2	19.8
Daily Minimum Temp	2.6	5.2	7.8	10.8	15.3	19.7	23.3	22.7	18.7	13.1	6.9	2.6	12.4
Daily Maximum Temp	15.9	18.9	21.4	25.6	30.8	36.6	39.9	38.6	34.3	28.2	20.8	15.8	27.2
Record Minimum Temp	-12.2	-6.7	-2.8	0.6	6.1	9.4	15.0	14.4	9.4	-1.1	-3.9	-12.2	-12.2
Record Maximum Temp	27.8	30.0	32.8	37.2	43.3	46.1	46.7	44.4	42.8	40.6	32.2	26.7	46.7
HDD, Base 18.3°C	281	178	129	67	8	0	0	0	0	22	139	282	1106
CDD, Base 18.3°C	0	0	13	62	154	295	412	381	245	94	4	0	1659
Relative Humidity (%)	47	42	38	32	29	24	24	28	31	33	40	46	35
Wind Speed (m/s)	3.5	4.5	5.9	6.3	6.8	6.4	5.5	4.9	4.7	4.2	4.3	3.8	5.1

Fresno, CA

WBAN NO. 93193

LATITUDE: 36.77° N
LONGITUDE: 119.72° W
ELEVATION: 100 meters
MEAN PRESSURE: 1004 millibars

STATION TYPE: Primary

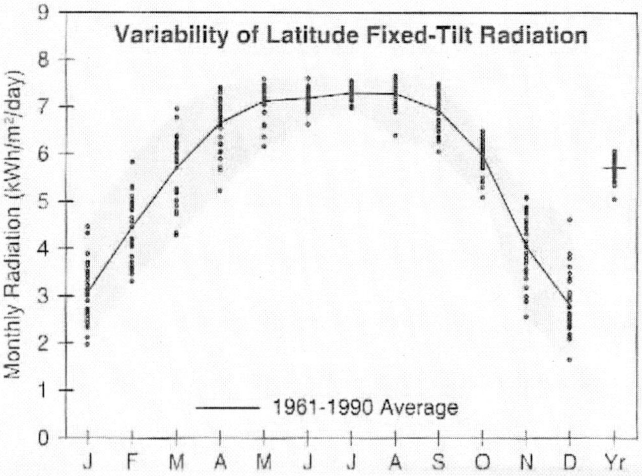

Solar Radiation for Flat-Plate Collectors Facing South at a Fixed Tilt (kWh/m²/day), Uncertainty ±9%

Tilt (°)		Jan	Feb	Mar	Apr	May	June	July	Aug	Sept	Oct	Nov	Dec	Year
0	Average	2.1	3.2	4.7	6.3	7.5	8.1	8.0	7.2	5.9	4.3	2.7	1.9	5.2
	Min/Max	1.7/2.7	2.5/3.9	3.7/5.5	5.1/7.0	6.6/8.0	7.4/8.7	7.6/8.4	6.3/7.6	5.2/6.3	3.9/4.6	2.0/3.2	1.4/2.6	4.7/5.4
Latitude -15	Average	2.8	4.1	5.5	6.8	7.6	7.8	7.9	7.5	6.8	5.5	3.6	2.5	5.7
	Min/Max	1.9/3.9	3.1/5.3	4.2/6.6	5.4/7.5	6.6/8.1	7.2/8.3	7.5/8.2	6.6/7.9	5.9/7.3	4.8/6.0	2.4/4.5	1.6/3.9	5.1/6.0
Latitude	Average	3.1	4.4	5.7	6.7	7.1	7.2	7.3	7.3	6.9	6.0	4.1	2.8	5.7
	Min/Max	2.0/4.5	3.3/5.8	4.3/6.9	5.2/7.4	6.2/7.6	6.6/7.6	7.0/7.6	6.4/7.7	6.1/7.5	5.1/6.5	2.6/5.1	1.7/4.6	5.1/6.1
Latitude +15	Average	3.2	4.5	5.6	6.2	6.3	6.1	6.3	6.6	6.7	6.1	4.2	3.0	5.4
	Min/Max	2.0/4.8	3.3/6.1	4.1/6.9	4.8/6.9	5.4/6.7	5.7/6.4	6.0/6.5	5.8/6.9	5.8/7.2	5.1/6.6	2.6/5.4	1.7/5.0	4.7/5.8
90	Average	2.8	3.7	4.0	3.6	3.0	2.6	2.7	3.5	4.4	4.8	3.7	2.6	3.4
	Min/Max	1.6/4.3	2.6/5.1	2.9/4.9	2.9/4.0	2.7/3.2	2.5/2.8	2.6/2.9	3.2/3.6	3.8/4.8	3.9/5.3	2.2/4.8	1.4/4.7	2.9/3.7

Solar Radiation for 1-Axis Tracking Flat-Plate Collectors with a North-South Axis (kWh/m²/day), Uncertainty ±9%

Axis Tilt (°)		Jan	Feb	Mar	Apr	May	June	July	Aug	Sept	Oct	Nov	Dec	Year
0	Average	2.7	4.3	6.4	8.5	10.2	11.0	11.2	10.1	8.3	6.2	3.7	2.4	7.1
	Min/Max	1.8/3.9	3.3/5.7	4.5/8.0	6.5/10.0	8.7/11.3	9.8/12.2	10.1/11.8	8.0/10.9	7.1/9.3	5.0/6.8	2.3/4.7	1.5/3.9	6.0/7.6
Latitude -15	Average	3.2	5.0	7.0	9.0	10.4	10.9	11.2	10.4	9.1	7.1	4.4	2.9	7.5
	Min/Max	2.0/4.8	3.7/6.7	4.9/8.9	6.8/10.5	8.8/11.5	9.8/12.1	10.2/11.8	8.3/11.3	7.7/10.0	5.7/7.8	2.6/5.6	1.6/4.9	6.3/8.1
Latitude	Average	3.4	5.2	7.2	8.9	10.1	10.5	10.8	10.3	9.2	7.4	4.7	3.1	7.6
	Min/Max	2.1/5.2	3.9/7.1	4.9/9.1	6.7/10.5	8.5/11.1	9.3/11.6	9.8/11.4	8.1/11.1	7.8/10.2	6.0/8.2	2.8/6.1	1.7/5.4	6.3/8.2
Latitude +15	Average	3.5	5.3	7.1	8.6	9.5	9.8	10.1	9.8	9.0	7.5	4.9	3.2	7.4
	Min/Max	2.0/5.5	4.0/7.3	4.8/9.0	6.4/10.1	8.0/10.5	8.7/10.9	9.1/10.7	7.7/10.6	7.6/10.0	6.0/8.3	2.8/6.4	1.7/5.8	6.0/8.0

Solar Radiation for 2-Axis Tracking Flat-Plate Collectors (kWh/m²/day), Uncertainty ±9%

Tracker		Jan	Feb	Mar	Apr	May	June	July	Aug	Sept	Oct	Nov	Dec	Year
2-Axis	Average	3.6	5.3	7.2	9.0	10.5	11.2	11.4	10.5	9.2	7.5	4.9	3.3	7.8
	Min/Max	2.1/5.5	4.0/7.3	4.9/9.1	6.8/10.6	8.9/11.6	9.9/12.4	10.3/12.1	8.3/11.4	7.8/10.2	6.0/8.3	2.8/6.4	1.7/5.8	6.5/8.4

Direct Beam Solar Radiation for Concentrating Collectors (kWh/m²/day), Uncertainty ±8%

Tracker		Jan	Feb	Mar	Apr	May	June	July	Aug	Sept	Oct	Nov	Dec	Year
1-Axis, E-W Horiz Axis	Average	1.8	2.7	3.6	4.6	5.7	6.4	6.6	5.9	5.1	4.3	2.7	1.7	4.3
	Min/Max	0.5/3.4	1.7/4.4	2.0/5.2	2.7/5.9	4.4/6.8	4.6/7.4	5.7/7.3	4.2/6.7	4.0/5.9	2.9/5.0	1.2/4.0	0.5/3.8	3.2/4.8
1-Axis, N-S Horiz Axis	Average	1.3	2.5	4.1	5.9	7.6	8.4	8.8	7.8	6.3	4.3	2.2	1.2	5.1
	Min/Max	0.4/2.6	1.5/4.1	2.1/6.0	3.6/7.9	5.8/9.0	5.8/10.0	7.2/9.9	5.2/9.0	4.8/7.4	2.8/5.2	0.9/3.3	0.4/2.7	3.7/5.8
1-Axis, N-S Tilt=Latitude	Average	1.9	3.3	4.7	6.2	7.3	7.9	8.4	7.9	6.9	5.4	3.1	1.8	5.4
	Min/Max	0.6/3.7	2.0/5.3	2.5/6.9	3.8/8.2	5.7/8.8	5.4/9.4	6.8/9.4	5.3/9.1	5.3/8.1	3.5/6.3	1.3/4.6	0.5/4.0	3.9/6.2
2-Axis	Average	2.1	3.4	4.8	6.3	7.8	8.6	9.0	8.1	7.0	5.4	3.2	1.9	5.6
	Min/Max	0.6/4.0	2.0/5.4	2.5/6.9	3.8/8.4	6.0/9.3	5.9/10.2	7.3/10.1	5.4/9.4	5.3/8.2	3.5/6.4	1.3/4.8	0.6/4.3	4.1/6.4

Average Climatic Conditions

Element	Jan	Feb	Mar	Apr	May	June	July	Aug	Sept	Oct	Nov	Dec	Year
Temperature (°C)	7.6	10.7	12.8	16.2	20.6	24.8	27.7	26.8	23.6	18.4	12.0	7.4	17.4
Daily Minimum Temp	3.0	4.7	6.3	8.5	12.1	15.8	18.4	17.7	14.9	10.4	5.8	2.8	10.1
Daily Maximum Temp	12.3	16.5	19.2	23.9	29.0	33.7	37.0	35.9	32.3	26.5	18.2	12.1	24.7
Record Minimum Temp	-7.2	-4.4	-3.3	0.0	2.2	6.7	10.0	9.4	2.8	-2.8	-3.3	-7.8	-7.8
Record Maximum Temp	25.6	26.7	32.2	37.8	41.7	43.3	44.4	43.9	43.9	38.9	31.7	24.4	44.4
HDD, Base 18.3°C	332	214	174	101	19	0	0	0	4	47	190	338	1420
CDD, Base 18.3°C	0	0	4	38	88	196	291	263	163	51	0	0	1093
Relative Humidity (%)	83	77	69	57	47	42	39	45	50	58	74	84	60
Wind Speed (m/s)	2.2	2.5	3.0	3.4	3.8	3.8	3.4	3.1	2.8	2.3	2.1	2.1	2.9

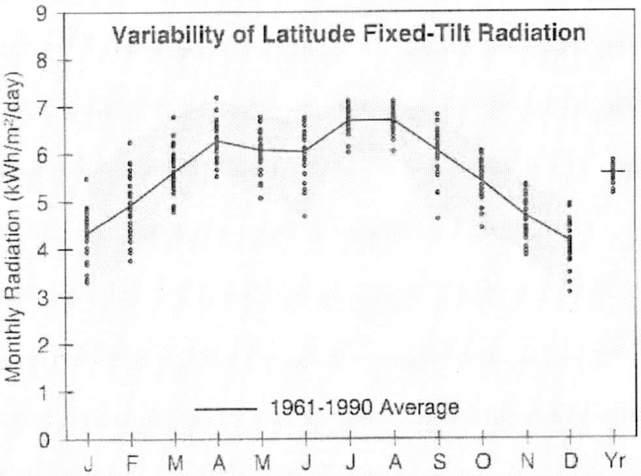

Long Beach, CA

WBAN NO. 23129

LATITUDE: 33.82° N
LONGITUDE: 118.15° W
ELEVATION: 17 meters
MEAN PRESSURE: 1014 millibars

STATION TYPE: Secondary

Solar Radiation for Flat-Plate Collectors Facing South at a Fixed Tilt (kWh/m²/day), Uncertainty ±9%

Tilt (°)		Jan	Feb	Mar	Apr	May	June	July	Aug	Sept	Oct	Nov	Dec	Year
0	Average	2.8	3.6	4.7	6.0	6.4	6.7	7.3	6.7	5.4	4.2	3.1	2.6	5.0
	Min/Max	2.3/3.3	2.9/4.3	4.1/5.6	5.3/6.8	5.4/7.2	5.2/7.5	6.5/7.7	6.1/7.1	4.3/5.9	3.6/4.6	2.8/3.5	2.1/2.9	4.6/5.2
Latitude -15	Average	3.8	4.5	5.4	6.4	6.4	6.5	7.2	6.9	6.0	5.0	4.1	3.6	5.5
	Min/Max	3.0/4.7	3.5/5.6	4.6/6.5	5.6/7.3	5.4/7.2	5.1/7.3	6.5/7.7	6.3/7.3	4.6/6.7	4.2/5.6	3.5/4.7	2.7/4.2	5.1/5.7
Latitude	Average	4.3	4.9	5.6	6.3	6.1	6.0	6.7	6.7	6.1	5.4	4.7	4.2	5.6
	Min/Max	3.3/5.5	3.8/6.3	4.8/6.8	5.5/7.2	5.1/6.8	4.7/6.8	6.0/7.1	6.1/7.1	4.6/6.8	4.4/6.1	3.9/5.3	3.1/4.9	5.2/5.9
Latitude +15	Average	4.6	5.1	5.5	5.8	5.4	5.2	5.8	6.1	5.8	5.5	5.0	4.5	5.4
	Min/Max	3.5/5.9	3.8/6.5	4.7/6.7	5.1/6.7	4.5/6.0	4.1/5.8	5.3/6.2	5.5/6.5	4.4/6.6	4.4/6.2	4.0/5.7	3.2/5.4	4.9/5.6
90	Average	4.1	4.1	3.8	3.2	2.5	2.2	2.4	3.0	3.7	4.2	4.3	4.1	3.4
	Min/Max	3.0/5.3	3.0/5.3	3.2/4.6	2.9/3.6	2.2/2.7	2.0/2.3	2.3/2.5	2.8/3.2	2.8/4.2	3.3/4.7	3.4/4.9	2.8/4.9	3.2/3.7

Solar Radiation for 1-Axis Tracking Flat-Plate Collectors with a North-South Axis (kWh/m²/day), Uncertainty ±9%

Axis Tilt (°)		Jan	Feb	Mar	Apr	May	June	July	Aug	Sept	Oct	Nov	Dec	Year
0	Average	3.9	4.9	6.3	7.9	8.1	8.4	9.3	8.7	7.1	5.7	4.4	3.7	6.5
	Min/Max	3.0/5.0	3.8/6.3	5.3/7.9	6.7/9.4	6.5/9.4	6.1/9.8	8.0/10.2	7.8/9.3	5.2/8.0	4.6/6.4	3.6/5.0	2.7/4.4	6.0/6.9
Latitude -15	Average	4.7	5.6	6.8	8.2	8.1	8.3	9.3	8.9	7.6	6.3	5.1	4.4	6.9
	Min/Max	3.5/6.0	4.2/7.3	5.7/8.6	7.0/9.8	6.5/9.4	6.0/9.7	8.0/10.2	7.9/9.6	5.5/8.6	5.1/7.2	4.1/5.9	3.2/5.3	6.3/7.3
Latitude	Average	5.1	5.9	7.0	8.1	7.9	7.9	8.9	8.8	7.6	6.6	5.6	4.9	7.0
	Min/Max	3.7/6.6	4.4/7.8	5.8/8.8	6.9/9.8	6.3/9.1	5.7/9.3	7.7/9.8	7.8/9.4	5.5/8.7	5.3/7.6	4.4/6.5	3.4/5.9	6.4/7.4
Latitude +15	Average	5.3	6.0	6.9	7.8	7.4	7.4	8.3	8.3	7.5	6.7	5.8	5.1	6.9
	Min/Max	3.9/7.0	4.5/8.0	5.7/8.7	6.6/9.4	5.9/8.6	5.3/8.7	7.2/9.1	7.4/8.9	5.3/8.5	5.3/7.6	4.6/6.8	3.6/6.3	6.2/7.3

Solar Radiation for 2-Axis Tracking Flat-Plate Collectors (kWh/m²/day), Uncertainty ±9%

Tracker		Jan	Feb	Mar	Apr	May	June	July	Aug	Sept	Oct	Nov	Dec	Year
2-Axis	Average	5.4	6.0	7.0	8.2	8.2	8.4	9.4	8.9	7.7	6.7	5.8	5.2	7.3
	Min/Max	3.9/7.1	4.5/8.0	5.8/8.8	7.0/9.8	6.6/9.5	6.1/9.9	8.1/10.3	8.0/9.6	5.5/8.7	5.3/7.6	4.6/6.8	3.6/6.4	6.6/7.7

Direct Beam Solar Radiation for Concentrating Collectors (kWh/m²/day), Uncertainty ±8%

Tracker		Jan	Feb	Mar	Apr	May	June	July	Aug	Sept	Oct	Nov	Dec	Year
1-Axis, E-W	Average	3.2	3.3	3.5	4.1	4.0	4.2	5.2	4.8	4.0	3.7	3.5	3.2	3.9
Horiz Axis	Min/Max	2.0/4.6	2.1/4.9	2.6/5.1	3.2/5.5	2.7/5.1	2.0/5.4	4.1/5.9	3.9/5.5	2.2/4.8	2.6/4.5	2.5/4.3	1.8/4.2	3.3/4.3
1-Axis, N-S	Average	2.6	3.2	4.1	5.3	5.2	5.4	6.6	6.2	4.8	3.8	2.9	2.4	4.4
Horiz Axis	Min/Max	1.6/3.7	1.9/4.7	3.0/6.0	4.0/7.1	3.5/6.6	2.7/7.0	5.1/7.6	5.1/7.0	2.7/5.8	2.8/4.7	2.1/3.7	1.4/3.2	3.7/4.7
1-Axis, N-S	Average	3.5	4.0	4.6	5.4	5.0	5.0	6.2	6.2	5.3	4.6	3.9	3.4	4.8
Tilt=Latitude	Min/Max	2.2/5.1	2.4/5.9	3.4/6.7	4.1/7.4	3.3/6.3	2.5/6.6	4.8/7.2	5.1/7.0	2.9/6.3	3.3/5.6	2.8/4.9	2.0/4.5	4.1/5.2
2-Axis	Average	3.8	4.1	4.6	5.5	5.3	5.5	6.7	6.4	5.3	4.6	4.1	3.7	5.0
	Min/Max	2.3/5.4	2.5/6.1	3.4/6.7	4.2/7.5	3.5/6.7	2.7/7.1	5.2/7.7	5.2/7.2	2.9/6.3	3.3/5.7	2.9/5.2	2.1/4.8	4.2/5.4

Average Climatic Conditions

Element	Jan	Feb	Mar	Apr	May	June	July	Aug	Sept	Oct	Nov	Dec	Year
Temperature (°C)	13.3	14.1	14.7	16.5	18.2	20.3	22.8	23.6	22.4	20.1	16.3	13.3	17.9
Daily Minimum Temp	7.2	8.3	9.4	11.0	13.5	15.4	17.4	18.2	17.1	14.3	10.2	7.2	12.4
Daily Maximum Temp	19.3	19.8	20.0	21.9	22.9	25.0	28.2	28.9	27.8	25.8	22.3	19.4	23.4
Record Minimum Temp	-3.9	0.6	0.6	3.3	4.4	8.3	10.6	11.1	10.0	3.9	1.1	-2.2	-3.9
Record Maximum Temp	32.8	32.8	36.7	40.6	39.4	42.8	41.7	40.6	43.3	43.9	38.3	33.3	43.9
HDD, Base 18.3°C	158	123	119	74	38	22	0	0	8	13	81	158	794
CDD, Base 18.3°C	0	3	7	19	35	80	142	164	132	67	19	0	667
Relative Humidity (%)	64	67	67	66	68	70	69	69	70	68	66	65	67
Wind Speed (m/s)	2.5	2.8	3.1	3.3	3.3	3.1	3.0	3.0	2.8	2.5	2.5	2.3	2.9

Los Angeles, CA

WBAN NO. 23174

LATITUDE: 33.93° N
LONGITUDE: 118.40° W
ELEVATION: 32 meters
MEAN PRESSURE: 1012 millibars

STATION TYPE: Primary

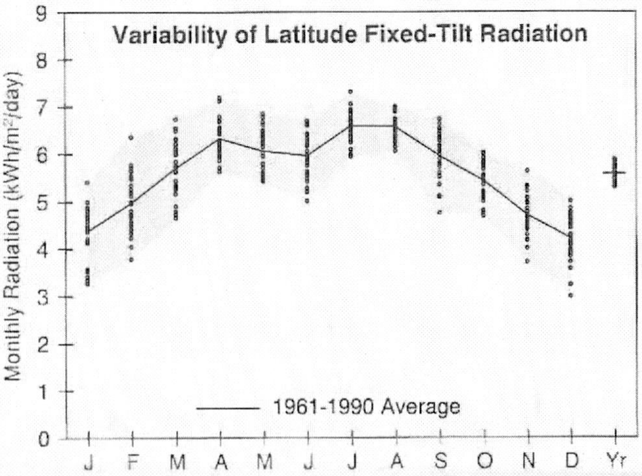

Solar Radiation for Flat-Plate Collectors Facing South at a Fixed Tilt (kWh/m²/day), Uncertainty ±9%

Tilt (°)		Jan	Feb	Mar	Apr	May	June	July	Aug	Sept	Oct	Nov	Dec	Year
0	Average	2.8	3.6	4.8	6.1	6.4	6.6	7.1	6.5	5.3	4.2	3.2	2.6	4.9
	Min/Max	2.3/3.3	3.0/4.4	4.0/5.6	5.5/6.8	5.7/7.2	5.6/7.7	6.4/8.0	6.1/7.0	4.4/5.8	3.8/4.5	2.7/3.6	2.1/3.0	4.7/5.1
Latitude -15	Average	3.8	4.5	5.5	6.4	6.4	6.4	7.1	6.8	5.9	5.0	4.2	3.6	5.5
	Min/Max	2.9/4.6	3.6/5.7	4.5/6.4	5.8/7.3	5.7/7.3	5.4/7.3	6.3/7.9	6.3/7.2	4.7/6.6	4.4/5.6	3.4/4.9	2.7/4.3	5.2/5.7
Latitude	Average	4.4	5.0	5.7	6.3	6.1	6.0	6.6	6.6	6.0	5.4	4.7	4.2	5.6
	Min/Max	3.3/5.4	3.8/6.4	4.7/6.7	5.6/7.2	5.4/6.8	5.0/6.7	5.9/7.3	6.1/7.0	4.8/6.7	4.7/6.0	3.7/5.6	3.0/5.0	5.3/5.9
Latitude +15	Average	4.7	5.1	5.6	5.9	5.4	5.2	5.8	6.0	5.7	5.5	5.0	4.5	5.4
	Min/Max	3.4/5.9	3.8/6.6	4.5/6.6	5.2/6.7	4.8/6.1	4.4/5.8	5.2/6.3	5.5/6.4	4.5/6.5	4.7/6.1	3.9/6.0	3.1/5.4	5.1/5.7
90	Average	4.1	4.1	3.8	3.3	2.5	2.2	2.4	3.0	3.6	4.2	4.3	4.1	3.5
	Min/Max	2.9/5.2	3.0/5.4	3.1/4.5	2.9/3.6	2.3/2.7	2.1/2.3	2.3/2.5	2.8/3.2	2.9/4.1	3.5/4.7	3.2/5.2	2.7/5.0	3.3/3.7

Solar Radiation for 1-Axis Tracking Flat-Plate Collectors with a North-South Axis (kWh/m²/day), Uncertainty ±9%

Axis Tilt (°)		Jan	Feb	Mar	Apr	May	June	July	Aug	Sept	Oct	Nov	Dec	Year
0	Average	3.9	5.0	6.4	7.9	8.0	8.1	9.0	8.3	6.9	5.6	4.4	3.7	6.4
	Min/Max	3.0/4.9	3.7/6.4	5.1/7.8	6.9/9.3	6.7/9.4	6.6/9.7	7.6/10.8	7.6/9.3	5.3/8.0	4.8/6.3	3.4/5.3	2.6/4.3	6.1/6.8
Latitude -15	Average	4.7	5.6	6.9	8.2	8.0	8.0	9.0	8.6	7.3	6.3	5.2	4.4	6.9
	Min/Max	3.5/5.9	4.1/7.4	5.4/8.5	7.1/9.7	6.8/9.4	6.5/9.6	7.6/10.8	7.8/9.5	5.6/8.5	5.4/7.0	3.9/6.3	3.0/5.3	6.5/7.3
Latitude	Average	5.1	6.0	7.1	8.2	7.8	7.7	8.7	8.4	7.4	6.6	5.6	4.9	7.0
	Min/Max	3.7/6.5	4.3/7.9	5.5/8.7	7.0/9.6	6.5/9.2	6.3/9.3	7.3/10.4	7.7/9.3	5.6/8.6	5.6/7.4	4.2/6.9	3.3/5.9	6.5/7.4
Latitude +15	Average	5.4	6.1	7.0	7.8	7.3	7.1	8.1	8.0	7.2	6.6	5.8	5.2	6.8
	Min/Max	3.8/6.8	4.3/8.1	5.4/8.6	6.7/9.3	6.1/8.6	5.8/8.6	6.8/9.8	7.3/8.9	5.4/8.4	5.6/7.5	4.4/7.2	3.4/6.3	6.4/7.3

Solar Radiation for 2-Axis Tracking Flat-Plate Collectors (kWh/m²/day), Uncertainty ±9%

Tracker		Jan	Feb	Mar	Apr	May	June	July	Aug	Sept	Oct	Nov	Dec	Year
2-Axis	Average	5.4	6.1	7.1	8.3	8.1	8.2	9.1	8.6	7.4	6.7	5.8	5.3	7.2
	Min/Max	3.9/6.9	4.4/8.1	5.6/8.7	7.1/9.7	6.8/9.6	6.7/9.9	7.7/11.0	7.9/9.5	5.7/8.6	5.6/7.5	4.4/7.2	3.5/6.4	6.7/7.7

Direct Beam Solar Radiation for Concentrating Collectors (kWh/m²/day), Uncertainty ±8%

Tracker		Jan	Feb	Mar	Apr	May	June	July	Aug	Sept	Oct	Nov	Dec	Year
1-Axis, E-W Horiz Axis	Average	3.2	3.4	3.6	4.1	3.9	3.9	4.8	4.5	3.8	3.6	3.5	3.2	3.8
	Min/Max	2.0/4.5	1.9/5.0	2.4/4.9	3.3/5.4	3.0/5.1	2.4/5.3	3.7/6.1	4.0/5.3	2.2/4.8	2.8/4.4	2.3/4.7	1.8/4.1	3.4/4.2
1-Axis, N-S Horiz Axis	Average	2.6	3.2	4.1	5.3	5.0	5.0	6.1	5.7	4.6	3.7	2.9	2.4	4.2
	Min/Max	1.6/3.6	1.8/4.8	2.8/5.8	4.2/7.0	3.7/6.6	3.2/6.9	4.6/8.0	4.9/6.8	2.7/5.8	2.9/4.5	1.9/4.0	1.3/3.1	3.8/4.7
1-Axis, N-S Tilt=Latitude	Average	3.6	4.1	4.7	5.5	4.8	4.6	5.8	5.7	5.0	4.5	3.9	3.5	4.6
	Min/Max	2.2/4.9	2.3/6.1	3.2/6.6	4.3/7.2	3.5/6.3	3.0/6.5	4.3/7.6	4.9/6.8	2.9/6.3	3.5/5.4	2.6/5.3	1.8/4.4	4.1/5.1
2-Axis	Average	3.8	4.2	4.7	5.5	5.1	5.0	6.2	5.9	5.0	4.6	4.1	3.8	4.8
	Min/Max	2.3/5.3	2.3/6.2	3.2/6.6	4.4/7.3	3.7/6.7	3.2/7.0	4.7/8.1	5.1/7.0	2.9/6.3	3.5/5.5	2.8/5.6	2.0/4.7	4.3/5.3

Average Climatic Conditions

Element	Jan	Feb	Mar	Apr	May	June	July	Aug	Sept	Oct	Nov	Dec	Year
Temperature (°C)	13.8	14.2	14.4	15.6	17.1	18.7	20.6	21.4	21.1	19.3	16.4	13.8	17.2
Daily Minimum Temp	8.8	9.6	10.3	11.6	13.5	15.3	17.1	17.9	17.3	15.1	11.6	8.8	13.1
Daily Maximum Temp	18.7	18.8	18.6	19.7	20.6	22.2	24.1	24.8	24.8	23.6	21.3	18.8	21.3
Record Minimum Temp	-5.0	0.0	1.1	3.9	6.1	8.9	9.4	10.6	8.3	5.0	1.1	0.0	-5.0
Record Maximum Temp	31.1	33.3	35.0	38.9	36.1	40.0	36.1	36.7	43.3	41.1	38.3	34.4	43.3
HDD, Base 18.3°C	143	119	124	88	53	30	5	3	12	18	71	143	810
CDD, Base 18.3°C	0	4	4	6	14	42	76	98	94	49	14	4	404
Relative Humidity (%)	63	68	71	71	74	76	77	77	74	71	65	63	71
Wind Speed (m/s)	3.1	3.4	3.8	4.0	3.9	3.8	3.7	3.7	3.5	3.3	3.2	3.0	3.5

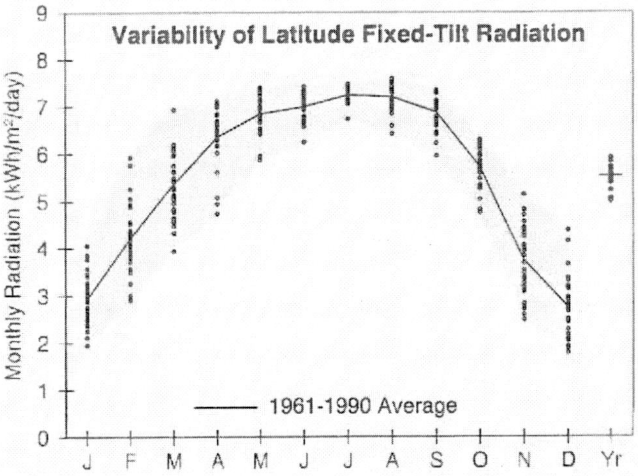

Sacramento, CA

WBAN NO. 23232

LATITUDE: 38.52° N
LONGITUDE: 121.50° W
ELEVATION: 8 meters
MEAN PRESSURE: 1015 millibars

STATION TYPE: Secondary

Solar Radiation for Flat-Plate Collectors Facing South at a Fixed Tilt (kWh/m²/day), Uncertainty ±9%

Tilt (°)		Jan	Feb	Mar	Apr	May	June	July	Aug	Sept	Oct	Nov	Dec	Year
0	Average	1.9	3.0	4.3	5.9	7.2	7.9	7.9	7.0	5.7	4.0	2.4	1.7	4.9
	Min/Max	1.6/2.4	2.3/3.8	3.4/5.4	4.6/6.6	6.2/7.8	7.0/8.4	7.4/8.2	6.3/7.4	5.1/6.1	3.5/4.3	1.9/3.0	1.4/2.3	4.6/5.1
Latitude -15	Average	2.6	3.9	5.2	6.5	7.3	7.6	7.8	7.5	6.7	5.3	3.3	2.4	5.5
	Min/Max	1.9/3.6	2.8/5.3	3.9/6.6	4.9/7.2	6.2/7.9	6.8/8.1	7.3/8.1	6.7/7.9	5.9/7.2	4.4/5.8	2.3/4.5	1.7/3.7	5.1/5.8
Latitude	Average	2.9	4.2	5.4	6.3	6.8	7.0	7.2	7.2	6.9	5.7	3.7	2.7	5.5
	Min/Max	2.0/4.1	2.9/5.9	3.9/6.9	4.7/7.1	5.9/7.4	6.2/7.4	6.7/7.5	6.4/7.6	5.9/7.3	4.8/6.3	2.5/5.1	1.8/4.4	5.0/5.9
Latitude +15	Average	3.1	4.3	5.2	5.9	6.0	6.0	6.3	6.5	6.6	5.8	3.9	2.9	5.2
	Min/Max	2.0/4.3	2.9/6.2	3.8/6.8	4.4/6.6	5.2/6.5	5.4/6.3	5.8/6.4	5.8/6.9	5.7/7.1	4.8/6.4	2.5/5.4	1.8/4.7	4.7/5.6
90	Average	2.7	3.6	3.8	3.6	3.0	2.7	2.9	3.6	4.5	4.6	3.4	2.6	3.4
	Min/Max	1.6/3.9	2.3/5.2	2.8/5.0	2.7/4.0	2.7/3.2	2.5/2.8	2.8/2.9	3.3/3.7	3.9/4.8	3.8/5.2	2.1/4.9	1.5/4.4	3.0/3.8

Solar Radiation for 1-Axis Tracking Flat-Plate Collectors with a North-South Axis (kWh/m²/day), Uncertainty ±9%

Axis Tilt (°)		Jan	Feb	Mar	Apr	May	June	July	Aug	Sept	Oct	Nov	Dec	Year
0	Average	2.5	4.0	5.9	8.1	9.9	10.8	11.1	10.0	8.3	5.8	3.3	2.2	6.8
	Min/Max	1.7/3.4	2.7/5.7	4.1/7.9	5.7/9.3	8.2/11.1	9.2/11.9	10.1/11.7	8.8/10.8	6.8/9.0	4.8/6.5	2.2/4.5	1.5/3.5	6.2/7.2
Latitude -15	Average	3.0	4.7	6.6	8.6	10.1	10.8	11.2	10.4	9.1	6.8	4.0	2.8	7.3
	Min/Max	2.0/4.3	3.0/6.8	4.5/8.9	5.9/9.8	8.3/11.3	9.2/11.9	10.2/11.8	9.1/11.2	7.5/9.9	5.6/7.6	2.6/5.6	1.8/4.6	6.5/7.8
Latitude	Average	3.3	4.9	6.7	8.5	9.8	10.3	10.8	10.2	9.2	7.1	4.3	3.0	7.4
	Min/Max	2.0/4.7	3.2/7.3	4.6/9.1	5.9/9.8	8.0/10.9	8.8/11.4	9.8/11.3	8.9/11.1	7.5/10.0	5.8/7.9	2.7/6.1	1.9/5.1	6.5/7.9
Latitude +15	Average	3.4	5.0	6.6	8.2	9.2	9.7	10.1	9.8	9.0	7.2	4.5	3.2	7.2
	Min/Max	2.1/4.9	3.2/7.4	4.5/9.0	5.6/9.4	7.6/10.3	8.2/10.6	9.2/10.6	8.5/10.6	7.4/9.8	5.8/8.0	2.7/6.4	1.9/5.4	6.3/7.7

Solar Radiation for 2-Axis Tracking Flat-Plate Collectors (kWh/m²/day), Uncertainty ±9%

Tracker		Jan	Feb	Mar	Apr	May	June	July	Aug	Sept	Oct	Nov	Dec	Year
2-Axis	Average	3.4	5.0	6.7	8.6	10.2	11.0	11.4	10.4	9.2	7.2	4.5	3.2	7.6
	Min/Max	2.1/5.0	3.2/7.4	4.6/9.1	6.0/9.9	8.4/11.4	9.4/12.2	10.4/12.0	9.1/11.3	7.6/10.0	5.8/8.1	2.8/6.4	1.9/5.5	6.7/8.1

Direct Beam Solar Radiation for Concentrating Collectors (kWh/m²/day), Uncertainty ±8%

Tracker		Jan	Feb	Mar	Apr	May	June	July	Aug	Sept	Oct	Nov	Dec	Year
1-Axis, E-W Horiz Axis	Average	1.8	2.6	3.4	4.3	5.5	6.2	6.6	5.8	5.2	4.1	2.5	1.7	4.1
	Min/Max	0.7/3.0	1.2/4.6	1.8/5.3	2.1/5.4	4.0/6.5	4.9/7.3	5.6/7.1	4.7/6.6	3.8/5.8	3.2/4.8	1.1/4.1	0.6/3.6	3.4/4.6
1-Axis, N-S Horiz Axis	Average	1.3	2.4	3.8	5.6	7.3	8.3	8.8	7.7	6.3	4.1	2.0	1.2	4.9
	Min/Max	0.5/2.2	1.0/4.2	2.0/6.1	2.8/7.0	5.3/8.7	6.4/9.8	7.4/9.6	6.3/8.8	4.4/7.2	3.1/4.9	0.9/3.2	0.4/2.5	4.1/5.3
1-Axis, N-S Tilt=Latitude	Average	1.9	3.1	4.4	5.9	7.1	7.8	8.4	7.9	7.1	5.2	2.8	1.8	5.3
	Min/Max	0.8/3.2	1.4/5.5	2.3/7.1	2.9/7.4	5.2/8.5	6.0/9.2	7.1/9.1	6.4/8.9	5.0/8.0	3.9/6.1	1.3/4.6	0.7/3.8	4.4/5.8
2-Axis	Average	2.1	3.2	4.4	6.0	7.5	8.4	9.0	8.1	7.1	5.3	2.9	1.9	5.5
	Min/Max	0.8/3.5	1.4/5.7	2.3/7.1	3.0/7.5	5.5/9.0	6.5/10.0	7.6/9.8	6.6/9.2	5.0/8.0	4.0/6.2	1.3/4.8	0.7/4.1	4.5/6.1

Average Climatic Conditions

Element	Jan	Feb	Mar	Apr	May	June	July	Aug	Sept	Oct	Nov	Dec	Year
Temperature (°C)	7.3	10.4	12.0	14.6	18.5	22.0	24.3	23.9	21.9	17.9	11.8	7.4	16.0
Daily Minimum Temp	3.2	5.2	6.2	7.5	10.2	12.9	14.5	14.4	13.2	10.2	6.3	3.2	8.9
Daily Maximum Temp	11.5	15.6	17.8	21.7	26.8	31.0	34.0	33.4	30.7	25.5	17.3	11.5	23.1
Record Minimum Temp	-5.0	-5.0	-3.3	0.0	2.2	5.0	8.9	9.4	6.1	2.2	-3.3	-7.8	-7.8
Record Maximum Temp	21.1	24.4	31.1	33.9	40.6	46.1	45.6	42.8	42.2	38.3	30.6	22.2	46.1
HDD, Base 18.3°C	341	222	198	128	44	7	0	0	9	43	195	339	1527
CDD, Base 18.3°C	0	0	0	16	49	117	184	174	117	29	0	0	687
Relative Humidity (%)	83	77	72	64	59	55	53	56	57	63	76	83	66
Wind Speed (m/s)	2.5	3.1	3.5	3.6	3.9	4.0	3.8	3.6	3.1	2.5	2.5	2.5	3.2

San Diego, CA

WBAN NO. 23188

LATITUDE: 32.73° N
LONGITUDE: 117.17° W
ELEVATION: 9 meters
MEAN PRESSURE: 1014 millibars

STATION TYPE: Primary

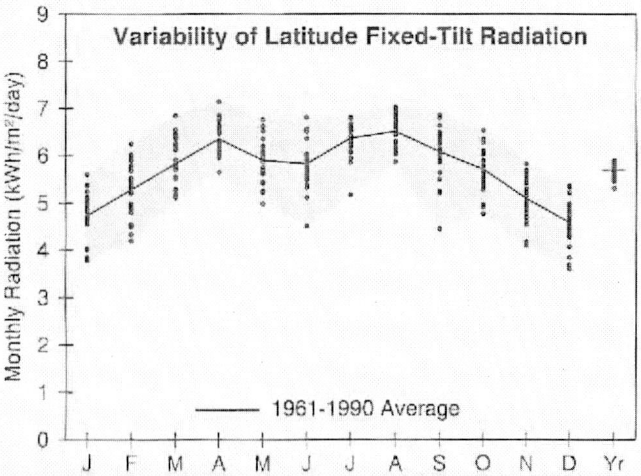

Solar Radiation for Flat-Plate Collectors Facing South at a Fixed Tilt (kWh/m²/day), Uncertainty ±9%

Tilt (°)		Jan	Feb	Mar	Apr	May	June	July	Aug	Sept	Oct	Nov	Dec	Year
0	Average	3.1	3.9	4.9	6.1	6.3	6.5	6.9	6.5	5.4	4.4	3.4	2.9	5.0
	Min/Max	2.6/3.5	3.3/4.4	4.4/5.7	5.5/6.8	5.3/7.2	5.0/7.6	5.6/7.4	5.9/7.0	4.2/6.0	3.8/4.9	3.0/3.7	2.4/3.2	4.7/5.2
Latitude -15	Average	4.1	4.8	5.6	6.4	6.3	6.3	6.8	6.7	6.0	5.3	4.5	3.9	5.6
	Min/Max	3.3/4.8	3.9/5.6	4.9/6.5	5.8/7.2	5.3/7.2	4.9/7.4	5.5/7.3	6.1/7.3	4.5/6.7	4.5/6.0	3.7/5.0	3.2/4.5	5.2/5.7
Latitude	Average	4.7	5.3	5.8	6.3	5.9	5.8	6.4	6.5	6.1	5.7	5.1	4.6	5.7
	Min/Max	3.8/5.6	4.2/6.3	5.1/6.8	5.7/7.1	5.0/6.8	4.5/6.8	5.2/6.8	5.9/7.0	4.4/6.9	4.8/6.5	4.1/5.8	3.6/5.4	5.3/5.9
Latitude +15	Average	5.1	5.5	5.7	5.9	5.2	5.1	5.6	5.9	5.8	5.8	5.4	5.0	5.5
	Min/Max	4.0/6.1	4.3/6.5	5.0/6.8	5.3/6.7	4.5/6.0	4.0/5.8	4.6/5.9	5.4/6.4	4.2/6.6	4.8/6.7	4.3/6.2	3.8/5.9	5.1/5.7
90	Average	4.5	4.3	3.9	3.2	2.4	2.1	2.2	2.9	3.6	4.4	4.6	4.5	3.5
	Min/Max	3.4/5.4	3.3/5.3	3.3/4.5	2.9/3.5	2.2/2.5	1.9/2.1	2.0/2.3	2.7/3.0	2.7/4.0	3.5/5.0	3.6/5.4	3.4/5.4	3.3/3.7

Solar Radiation for 1-Axis Tracking Flat-Plate Collectors with a North-South Axis (kWh/m²/day), Uncertainty ±9%

Axis Tilt (°)		Jan	Feb	Mar	Apr	May	June	July	Aug	Sept	Oct	Nov	Dec	Year
0	Average	4.4	5.4	6.7	8.0	7.7	7.8	8.8	8.4	7.2	6.1	4.9	4.2	6.6
	Min/Max	3.5/5.3	4.1/6.6	5.7/8.1	6.8/9.2	6.1/9.4	5.5/9.9	6.6/9.7	7.3/9.4	4.9/8.3	5.0/7.0	3.9/5.8	3.3/4.9	6.1/6.9
Latitude -15	Average	5.2	6.1	7.2	8.3	7.7	7.8	8.7	8.6	7.6	6.7	5.7	5.0	7.1
	Min/Max	4.1/6.3	4.6/7.4	6.1/8.7	7.1/9.6	6.1/9.4	5.4/9.8	6.6/9.7	7.4/9.6	5.2/8.8	5.5/7.8	4.4/6.7	3.8/5.9	6.4/7.3
Latitude	Average	5.7	6.5	7.4	8.2	7.5	7.4	8.4	8.4	7.7	7.1	6.2	5.5	7.2
	Min/Max	4.4/6.9	4.8/7.9	6.2/9.0	7.0/9.6	5.9/9.1	5.2/9.3	6.3/9.3	7.3/9.4	5.2/8.9	5.7/8.3	4.8/7.3	4.2/6.5	6.5/7.4
Latitude +15	Average	6.0	6.6	7.3	7.9	7.0	6.9	7.8	8.0	7.5	7.1	6.5	5.8	7.0
	Min/Max	4.6/7.3	4.9/8.1	6.1/8.9	6.7/9.2	5.5/8.6	4.8/8.6	5.9/8.7	6.9/8.9	5.0/8.7	5.7/8.4	4.9/7.6	4.4/6.9	6.4/7.3

Solar Radiation for 2-Axis Tracking Flat-Plate Collectors (kWh/m²/day), Uncertainty ±9%

Tracker		Jan	Feb	Mar	Apr	May	June	July	Aug	Sept	Oct	Nov	Dec	Year
2-Axis	Average	6.0	6.6	7.4	8.3	7.8	7.9	8.9	8.6	7.7	7.2	6.5	5.9	7.4
	Min/Max	4.6/7.3	4.9/8.1	6.2/9.0	7.1/9.6	6.2/9.5	5.5/10.0	6.7/9.8	7.4/9.6	5.2/8.9	5.8/8.4	5.0/7.7	4.4/7.1	6.8/7.7

Direct Beam Solar Radiation for Concentrating Collectors (kWh/m²/day), Uncertainty ±8%

Tracker		Jan	Feb	Mar	Apr	May	June	July	Aug	Sept	Oct	Nov	Dec	Year
1-Axis, E-W Horiz Axis	Average	3.8	3.8	3.9	4.3	3.9	4.0	4.9	4.8	4.2	4.1	4.0	3.8	4.1
	Min/Max	2.4/4.9	2.5/5.1	2.9/5.3	3.4/5.5	2.5/5.2	1.8/5.7	2.9/5.8	3.8/5.6	1.9/5.2	3.0/5.2	2.7/5.2	2.5/4.8	3.6/4.5
1-Axis, N-S Horiz Axis	Average	3.1	3.8	4.6	5.6	4.9	5.0	6.2	6.1	5.1	4.4	3.6	3.0	4.6
	Min/Max	2.0/4.1	2.3/5.1	3.4/6.3	4.3/7.1	3.1/6.8	2.3/7.3	3.6/7.4	4.7/7.2	2.3/6.3	3.2/5.5	2.4/4.6	2.0/3.8	3.9/5.0
1-Axis, N-S Tilt=Latitude	Average	4.2	4.7	5.1	5.7	4.7	4.7	5.8	6.1	5.5	5.2	4.6	4.1	5.0
	Min/Max	2.7/5.4	2.9/6.2	3.8/7.0	4.4/7.3	3.0/6.5	2.1/6.8	3.4/7.0	4.7/7.2	2.5/6.8	3.8/6.5	3.1/6.0	2.7/5.2	4.3/5.4
2-Axis	Average	4.5	4.8	5.1	5.8	5.0	5.1	6.2	6.3	5.5	5.3	4.9	4.5	5.3
	Min/Max	2.9/5.8	3.0/6.4	3.8/7.1	4.5/7.4	3.2/6.9	2.3/7.3	3.7/7.5	4.9/7.4	2.5/6.9	3.8/6.6	3.3/6.3	3.0/5.7	4.5/5.7

Average Climatic Conditions

Element	Jan	Feb	Mar	Apr	May	June	July	Aug	Sept	Oct	Nov	Dec	Year
Temperature (°C)	14.1	14.8	15.3	16.7	17.8	19.3	21.7	22.6	21.9	19.8	16.7	14.1	17.9
Daily Minimum Temp	9.4	10.4	11.6	13.1	15.1	16.6	18.7	19.6	18.7	16.1	12.2	9.3	14.2
Daily Maximum Temp	18.8	19.2	19.1	20.2	20.6	22.0	24.6	25.4	25.1	23.7	21.1	18.9	21.6
Record Minimum Temp	-1.7	2.2	3.9	5.0	8.9	10.6	12.8	13.9	10.6	6.1	3.3	1.1	-1.7
Record Maximum Temp	31.1	31.1	33.9	36.7	35.6	38.3	35.0	36.7	43.9	41.7	36.1	31.1	43.9
HDD, Base 18.3°C	136	105	98	63	41	28	7	0	11	13	61	135	698
CDD, Base 18.3°C	5	6	5	13	25	58	111	133	117	59	11	4	547
Relative Humidity (%)	63	66	67	67	71	74	75	74	73	69	66	64	69
Wind Speed (m/s)	2.7	3.0	3.5	3.7	3.7	3.6	3.5	3.5	3.4	3.1	2.8	2.6	3.3

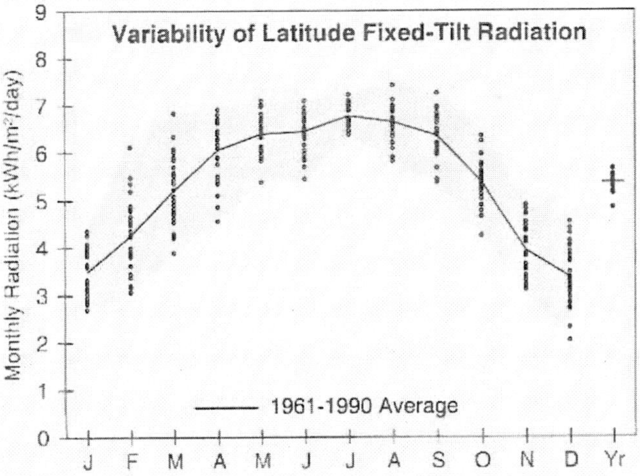

San Francisco, CA

WBAN NO. 23234

LATITUDE: 37.62° N
LONGITUDE: 122.38° W
ELEVATION: 5 meters
MEAN PRESSURE: 1017 millibars

STATION TYPE: Secondary

Solar Radiation for Flat-Plate Collectors Facing South at a Fixed Tilt (kWh/m²/day), Uncertainty ±9%

Tilt (°)		Jan	Feb	Mar	Apr	May	June	July	Aug	Sept	Oct	Nov	Dec	Year
0	Average	2.2	3.0	4.2	5.7	6.7	7.2	7.3	6.5	5.4	3.9	2.5	2.0	4.7
	Min/Max	1.8/2.5	2.3/3.9	3.3/5.3	4.4/6.4	5.7/7.4	6.1/7.9	6.9/7.9	5.7/7.3	4.7/6.0	3.2/4.4	2.1/2.9	1.4/2.4	4.4/4.9
Latitude -15	Average	3.1	3.9	5.0	6.2	6.8	7.0	7.3	6.9	6.2	5.0	3.5	2.9	5.3
	Min/Max	2.4/3.8	2.9/5.5	3.8/6.5	4.7/7.0	5.7/7.5	5.9/7.7	6.9/7.8	6.0/7.7	5.3/7.1	4.0/5.8	2.8/4.3	1.9/3.9	4.9/5.6
Latitude	Average	3.5	4.2	5.2	6.1	6.4	6.5	6.8	6.7	6.4	5.4	3.9	3.4	5.4
	Min/Max	2.7/4.3	3.1/6.1	3.9/6.8	4.6/6.9	5.4/7.1	5.4/7.1	6.4/7.2	5.8/7.4	5.4/7.3	4.3/6.4	3.1/4.9	2.0/4.6	4.9/5.7
Latitude +15	Average	3.7	4.4	5.1	5.6	5.7	5.6	5.9	6.1	6.1	5.5	4.1	3.6	5.1
	Min/Max	2.8/4.7	3.1/6.4	3.8/6.8	4.2/6.4	4.8/6.3	4.7/6.1	5.6/6.3	5.3/6.7	5.2/7.0	4.3/6.5	3.2/5.2	2.1/5.0	4.6/5.5
90	Average	3.3	3.6	3.7	3.4	2.8	2.5	2.7	3.3	4.1	4.3	3.6	3.3	3.4
	Min/Max	2.5/4.2	2.4/5.4	2.7/4.9	2.6/3.8	2.5/3.0	2.3/2.6	2.6/2.8	3.0/3.6	3.5/4.7	3.4/5.2	2.8/4.6	1.9/4.7	2.9/3.7

Solar Radiation for 1-Axis Tracking Flat-Plate Collectors with a North-South Axis (kWh/m²/day), Uncertainty ±9%

Axis Tilt (°)		Jan	Feb	Mar	Apr	May	June	July	Aug	Sept	Oct	Nov	Dec	Year
0	Average	3.0	4.1	5.7	7.6	8.8	9.3	9.7	8.7	7.4	5.5	3.5	2.8	6.3
	Min/Max	2.3/3.7	2.8/6.0	4.1/7.7	5.6/8.9	7.0/10.1	7.5/10.8	9.0/10.7	7.4/10.2	6.2/8.8	4.3/6.5	2.8/4.4	1.7/3.8	5.7/6.7
Latitude -15	Average	3.7	4.7	6.3	8.0	8.9	9.2	9.7	9.0	8.1	6.3	4.3	3.5	6.8
	Min/Max	2.8/4.6	3.3/7.1	4.5/8.6	5.8/9.4	7.1/10.3	7.4/10.7	9.0/10.8	7.7/10.6	6.7/9.6	4.8/7.6	3.3/5.4	2.0/4.8	6.1/7.2
Latitude	Average	4.0	5.0	6.5	8.0	8.7	8.9	9.4	8.8	8.2	6.6	4.6	3.9	6.9
	Min/Max	3.0/5.1	3.4/7.6	4.6/8.9	5.7/9.4	6.8/10.0	7.1/10.3	8.7/10.4	7.5/10.4	6.8/9.7	5.1/8.0	3.5/5.9	2.2/5.4	6.1/7.3
Latitude +15	Average	4.2	5.1	6.4	7.7	8.1	8.2	8.7	8.4	8.0	6.7	4.8	4.1	6.7
	Min/Max	3.1/5.4	3.4/7.8	4.5/8.8	5.5/9.0	6.4/9.4	6.6/9.6	8.1/9.7	7.1/9.9	6.6/9.5	5.1/8.1	3.6/6.1	2.3/5.7	5.9/7.1

Solar Radiation for 2-Axis Tracking Flat-Plate Collectors (kWh/m²/day), Uncertainty ±9%

Tracker		Jan	Feb	Mar	Apr	May	June	July	Aug	Sept	Oct	Nov	Dec	Year
2-Axis	Average	4.2	5.1	6.5	8.1	9.0	9.4	9.9	9.0	8.2	6.7	4.8	4.1	7.1
	Min/Max	3.1/5.4	3.4/7.8	4.6/8.9	5.8/9.5	7.2/10.4	7.6/11.0	9.2/11.0	7.7/10.6	6.8/9.7	5.1/8.1	3.7/6.2	2.3/5.8	6.3/7.5

Direct Beam Solar Radiation for Concentrating Collectors (kWh/m²/day), Uncertainty ±8%

Tracker		Jan	Feb	Mar	Apr	May	June	July	Aug	Sept	Oct	Nov	Dec	Year
1-Axis, E-W Horiz Axis	Average	2.5	2.8	3.3	4.0	4.7	5.1	5.5	4.9	4.6	3.9	2.8	2.5	3.9
	Min/Max	1.6/3.5	1.5/4.9	1.9/5.2	2.4/5.2	3.3/5.7	3.6/6.3	4.9/6.4	3.7/6.1	3.3/5.6	2.6/5.0	1.8/3.9	1.1/3.9	3.2/4.2
1-Axis, N-S Horiz Axis	Average	1.9	2.5	3.8	5.2	6.1	6.5	7.1	6.3	5.4	3.9	2.3	1.8	4.4
	Min/Max	1.2/2.6	1.3/4.6	2.1/6.0	3.1/6.7	4.2/7.5	4.6/8.3	6.2/8.3	4.7/8.0	4.0/6.8	2.5/5.0	1.5/3.2	0.8/2.8	3.7/4.7
1-Axis, N-S Tilt=Latitude	Average	2.7	3.3	4.4	5.4	5.9	6.1	6.8	6.4	6.0	4.8	3.2	2.7	4.8
	Min/Max	1.8/3.8	1.7/6.0	2.5/6.9	3.2/7.0	4.1/7.3	4.3/7.7	5.9/7.9	4.8/8.1	4.4/7.6	3.2/6.3	2.0/4.4	1.2/4.2	4.0/5.3
2-Axis	Average	2.9	3.4	4.4	5.5	6.3	6.7	7.3	6.6	6.1	4.9	3.3	2.9	5.0
	Min/Max	1.9/4.0	1.7/6.1	2.5/6.9	3.3/7.1	4.3/7.7	4.7/8.4	6.4/8.5	5.0/8.3	4.4/7.6	3.2/6.3	2.2/4.7	1.3/4.5	4.2/5.5

Average Climatic Conditions

Element	Jan	Feb	Mar	Apr	May	June	July	Aug	Sept	Oct	Nov	Dec	Year
Temperature (°C)	9.3	11.2	11.8	13.1	14.5	16.4	17.1	17.6	18.1	16.1	12.7	9.7	13.9
Daily Minimum Temp	5.4	7.2	7.7	8.4	9.8	11.4	12.2	12.8	12.9	11.0	8.4	5.9	9.4
Daily Maximum Temp	13.1	15.2	16.0	17.7	19.2	21.3	22.0	22.4	23.1	21.2	16.9	13.4	18.4
Record Minimum Temp	-4.4	-3.9	-1.1	-0.6	2.2	5.0	6.1	5.6	3.3	1.1	-3.9	-6.7	-6.7
Record Maximum Temp	22.2	25.6	29.4	33.3	36.1	41.1	40.6	36.7	39.4	37.2	29.4	23.9	41.1
HDD, Base 18.3°C	281	199	202	159	121	67	51	38	44	75	170	269	1676
CDD, Base 18.3°C	0	0	0	3	0	9	12	16	36	6	0	0	81
Relative Humidity (%)	78	76	73	71	71	72	73	74	72	72	75	77	74
Wind Speed (m/s)	3.3	4.0	4.8	5.5	6.2	6.2	6.2	5.7	5.0	4.3	3.7	3.5	4.9

Santa Maria, CA

WBAN NO. 23273

LATITUDE: 34.90° N
LONGITUDE: 120.45° W
ELEVATION: 72 meters
MEAN PRESSURE: 1008 millibars

STATION TYPE: Primary

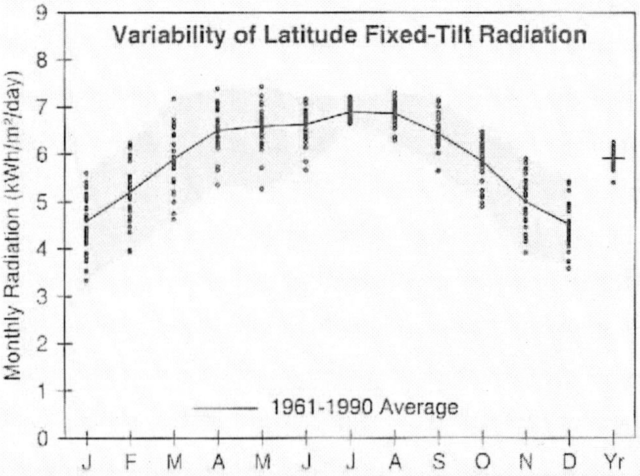

Solar Radiation for Flat-Plate Collectors Facing South at a Fixed Tilt (kWh/m²/day), Uncertainty ±9%

Tilt (°)		Jan	Feb	Mar	Apr	May	June	July	Aug	Sept	Oct	Nov	Dec	Year
0	Average	2.8	3.7	4.9	6.2	7.0	7.4	7.5	6.8	5.6	4.4	3.2	2.7	5.2
	Min/Max	2.3/3.3	3.0/4.2	4.0/5.8	5.3/6.9	5.6/7.9	6.3/8.0	7.1/7.9	6.3/7.2	5.0/6.1	3.8/4.7	2.7/3.6	2.3/3.0	4.8/5.4
Latitude -15	Average	4.0	4.7	5.7	6.6	7.0	7.2	7.4	7.1	6.3	5.4	4.4	3.9	5.8
	Min/Max	3.0/4.8	3.7/5.6	4.5/6.8	5.5/7.5	5.6/7.9	6.2/7.8	7.1/7.8	6.5/7.5	5.6/7.0	4.6/5.9	3.5/5.1	3.1/4.6	5.3/6.1
Latitude	Average	4.6	5.2	5.9	6.5	6.6	6.6	6.9	6.9	6.4	5.8	5.0	4.5	5.9
	Min/Max	3.3/5.6	3.9/6.2	4.6/7.2	5.3/7.4	5.3/7.4	5.7/7.2	6.6/7.2	6.3/7.3	5.6/7.1	4.9/6.5	3.9/5.9	3.6/5.4	5.4/6.2
Latitude +15	Average	4.9	5.4	5.8	6.0	5.8	5.7	6.0	6.2	6.2	6.0	5.3	4.9	5.7
	Min/Max	3.5/6.1	4.0/6.5	4.5/7.1	4.9/6.9	4.7/6.5	4.9/6.1	5.8/6.2	5.7/6.6	5.4/6.9	4.9/6.6	4.1/6.3	3.8/5.9	5.2/6.0
90	Average	4.4	4.4	4.0	3.4	2.7	2.3	2.5	3.2	3.9	4.6	4.6	4.5	3.7
	Min/Max	3.0/5.5	3.2/5.4	3.1/4.9	2.8/3.8	2.4/2.9	2.2/2.4	2.4/2.6	3.0/3.3	3.5/4.4	3.7/5.1	3.5/5.6	3.5/5.5	3.3/4.0

Solar Radiation for 1-Axis Tracking Flat-Plate Collectors with a North-South Axis (kWh/m²/day), Uncertainty ±9%

Axis Tilt (°)		Jan	Feb	Mar	Apr	May	June	July	Aug	Sept	Oct	Nov	Dec	Year
0	Average	4.2	5.2	6.7	8.3	9.0	9.4	9.6	8.8	7.5	6.1	4.7	4.0	7.0
	Min/Max	3.0/5.1	3.8/6.3	5.1/8.4	6.7/9.7	6.8/10.6	7.7/10.8	8.8/10.6	7.9/9.6	6.3/8.7	5.0/7.0	3.7/5.7	3.2/4.8	6.3/7.4
Latitude -15	Average	5.0	6.0	7.3	8.7	9.1	9.4	9.6	9.1	8.1	6.9	5.6	4.9	7.5
	Min/Max	3.5/6.2	4.3/7.3	5.5/9.2	6.9/10.2	6.9/10.8	7.6/10.7	8.8/10.6	8.1/9.9	6.7/9.4	5.6/7.9	4.3/6.7	3.8/5.9	6.7/7.9
Latitude	Average	5.5	6.4	7.5	8.6	8.8	9.0	9.2	8.9	8.2	7.3	6.0	5.4	7.6
	Min/Max	3.8/6.8	4.5/7.8	5.6/9.5	6.8/10.1	6.7/10.4	7.3/10.2	8.5/10.2	8.0/9.7	6.8/9.5	5.9/8.3	4.6/7.3	4.2/6.6	6.8/8.1
Latitude +15	Average	5.7	6.5	7.4	8.3	8.3	8.3	8.6	8.5	8.0	7.3	6.3	5.7	7.4
	Min/Max	4.0/7.2	4.5/8.0	5.5/9.4	6.4/9.8	6.3/9.8	6.8/9.5	7.9/9.5	7.6/9.2	6.6/9.3	5.9/8.4	4.7/7.7	4.4/7.0	6.6/7.9

Solar Radiation for 2-Axis Tracking Flat-Plate Collectors (kWh/m²/day), Uncertainty ±9%

Tracker		Jan	Feb	Mar	Apr	May	June	July	Aug	Sept	Oct	Nov	Dec	Year
2-Axis	Average	5.8	6.5	7.5	8.7	9.2	9.6	9.7	9.1	8.2	7.4	6.3	5.8	7.8
	Min/Max	4.0/7.3	4.6/8.0	5.6/9.5	6.9/10.2	6.9/10.9	7.8/10.9	8.9/10.8	8.1/9.9	6.8/9.5	5.9/8.4	4.8/7.7	4.4/7.2	7.0/8.4

Direct Beam Solar Radiation for Concentrating Collectors (kWh/m²/day), Uncertainty ±8%

Tracker		Jan	Feb	Mar	Apr	May	June	July	Aug	Sept	Oct	Nov	Dec	Year
1-Axis, E-W	Average	3.6	3.8	4.0	4.5	4.9	5.3	5.6	5.2	4.6	4.3	3.9	3.7	4.5
Horiz Axis	Min/Max	2.1/4.9	2.1/5.0	2.3/5.7	3.0/5.7	2.9/6.2	3.6/6.4	4.6/6.4	4.1/5.9	3.2/5.6	3.2/5.2	2.6/5.2	2.6/5.0	3.6/5.0
1-Axis, N-S	Average	2.9	3.6	4.6	5.9	6.4	6.8	7.1	6.5	5.5	4.5	3.4	2.8	5.0
Horiz Axis	Min/Max	1.6/3.9	2.0/4.8	2.7/6.7	4.0/7.5	3.8/8.2	4.5/8.3	5.7/8.3	5.0/7.5	3.7/7.0	3.3/5.5	2.2/4.5	2.0/3.8	4.1/5.5
1-Axis, N-S	Average	4.0	4.6	5.3	6.1	6.2	6.3	6.7	6.6	6.0	5.4	4.5	4.0	5.5
Tilt=Latitude	Min/Max	2.3/5.4	2.5/6.1	3.0/7.6	4.1/7.8	3.7/7.9	4.2/7.8	5.4/7.8	5.1/7.5	4.0/7.6	4.0/6.6	3.0/6.0	2.8/5.4	4.4/6.1
2-Axis	Average	4.3	4.7	5.3	6.2	6.5	6.9	7.2	6.8	6.1	5.5	4.8	4.4	5.7
	Min/Max	2.5/5.8	2.6/6.3	3.0/7.7	4.1/7.9	3.9/8.4	4.6/8.4	5.8/8.4	5.2/7.7	4.0/7.6	4.1/6.7	3.1/6.3	3.1/5.8	4.6/6.4

Average Climatic Conditions

Element	Jan	Feb	Mar	Apr	May	June	July	Aug	Sept	Oct	Nov	Dec	Year
Temperature (°C)	10.6	11.4	11.6	12.7	13.9	15.9	17.3	17.8	17.6	16.1	13.1	10.6	14.1
Daily Minimum Temp	3.5	4.7	5.2	5.8	8.0	10.2	11.6	12.2	11.3	8.9	5.7	3.2	7.5
Daily Maximum Temp	17.7	18.2	17.9	19.4	19.9	21.7	22.9	23.4	23.8	23.3	20.4	17.9	20.6
Record Minimum Temp	-6.7	-5.6	-4.4	-0.6	-0.6	2.2	6.1	6.1	2.2	-3.3	-3.9	-6.7	-6.7
Record Maximum Temp	30.0	30.6	35.0	39.4	37.8	38.9	40.0	39.4	39.4	42.2	33.9	32.2	42.2
HDD, Base 18.3°C	239	193	210	173	138	77	53	40	56	79	161	239	1658
CDD, Base 18.3°C	0	0	0	3	0	5	21	23	32	11	0	0	94
Relative Humidity (%)	70	73	76	75	76	76	77	78	78	73	71	70	75
Wind Speed (m/s)	2.8	3.2	3.7	3.8	4.0	3.7	3.2	3.0	2.8	2.8	3.0	2.8	3.2

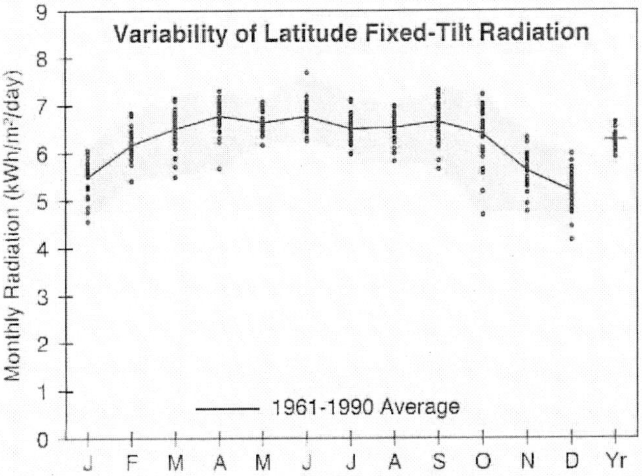

Alamosa, CO

WBAN NO. 23061

LATITUDE: 37.45° N
LONGITUDE: 105.87° W
ELEVATION: 2297 meters
MEAN PRESSURE: 772 millibars

STATION TYPE: Primary

Solar Radiation for Flat-Plate Collectors Facing South at a Fixed Tilt (kWh/m²/day), Uncertainty ±9%

Tilt (°)		Jan	Feb	Mar	Apr	May	June	July	Aug	Sept	Oct	Nov	Dec	Year
0	Average	3.0	4.0	5.2	6.4	7.1	7.7	7.2	6.5	5.6	4.5	3.3	2.7	5.3
	Min/Max	2.6/3.2	3.6/4.4	4.5/5.6	5.4/6.8	6.6/7.6	7.1/8.8	6.6/7.9	5.8/6.9	4.9/6.2	3.6/4.9	3.0/3.6	2.4/3.0	5.0/5.6
Latitude -15	Average	4.7	5.5	6.2	6.9	7.1	7.4	7.0	6.8	6.5	5.9	4.9	4.4	6.1
	Min/Max	3.9/5.1	4.9/6.1	5.3/6.8	5.8/7.4	6.6/7.6	6.8/8.4	6.5/7.7	6.1/7.3	5.6/7.2	4.4/6.6	4.2/5.5	3.6/4.9	5.8/6.5
Latitude	Average	5.5	6.2	6.5	6.8	6.6	6.8	6.5	6.5	6.7	6.4	5.6	5.2	6.3
	Min/Max	4.6/6.1	5.4/6.8	5.5/7.2	5.7/7.3	6.2/7.1	6.3/7.7	6.0/7.1	5.8/7.0	5.7/7.3	4.7/7.2	4.8/6.3	4.2/6.0	5.9/6.6
Latitude +15	Average	6.0	6.5	6.4	6.3	5.8	5.7	5.6	5.9	6.4	6.5	6.0	5.7	6.1
	Min/Max	4.9/6.6	5.6/7.3	5.4/7.1	5.3/6.8	5.4/6.2	5.3/6.5	5.2/6.2	5.3/6.3	5.4/7.0	4.7/7.4	5.0/6.8	4.5/6.7	5.7/6.5
90	Average	5.7	5.6	4.7	3.7	2.9	2.5	2.6	3.2	4.3	5.2	5.5	5.5	4.3
	Min/Max	4.6/6.2	4.8/6.7	4.0/5.2	3.2/4.1	2.7/3.0	2.4/2.7	2.4/2.8	2.9/3.4	3.7/4.7	3.7/5.9	4.4/6.1	4.2/6.7	3.9/4.5

Solar Radiation for 1-Axis Tracking Flat-Plate Collectors with a North-South Axis (kWh/m²/day), Uncertainty ±9%

Axis Tilt (°)		Jan	Feb	Mar	Apr	May	June	July	Aug	Sept	Oct	Nov	Dec	Year
0	Average	4.9	6.2	7.6	9.2	10.0	10.8	9.8	9.0	8.3	7.0	5.2	4.5	7.7
	Min/Max	3.9/5.4	5.3/7.1	6.1/8.6	7.4/10.2	8.8/10.9	9.7/12.3	8.9/11.3	7.8/10.0	6.5/9.2	5.0/8.0	4.3/5.9	3.5/5.2	7.1/8.3
Latitude -15	Average	6.1	7.4	8.4	9.6	10.1	10.7	9.8	9.4	9.0	8.0	6.4	5.7	8.4
	Min/Max	4.8/6.8	6.3/8.4	6.7/9.5	7.7/10.7	8.9/11.1	9.6/12.2	8.9/11.4	8.1/10.3	7.1/10.0	5.6/9.3	5.2/7.2	4.4/6.6	7.7/9.0
Latitude	Average	6.8	7.9	8.7	9.6	9.8	10.3	9.5	9.2	9.1	8.4	7.0	6.3	8.5
	Min/Max	5.3/7.6	6.7/9.0	6.9/9.8	7.6/10.7	8.7/10.8	9.2/11.7	8.6/11.0	7.9/10.2	7.1/10.2	5.8/9.7	5.6/7.8	4.8/7.4	7.8/9.2
Latitude +15	Average	7.2	8.1	8.6	9.2	9.3	9.6	8.9	8.8	9.0	8.5	7.3	6.7	8.4
	Min/Max	5.6/8.0	6.8/9.2	6.8/9.7	7.4/10.3	8.2/10.2	8.6/10.9	8.0/10.3	7.6/9.7	7.0/9.9	5.8/9.8	5.9/8.2	5.1/7.9	7.6/9.1

Solar Radiation for 2-Axis Tracking Flat-Plate Collectors (kWh/m²/day), Uncertainty ±9%

Tracker		Jan	Feb	Mar	Apr	May	June	July	Aug	Sept	Oct	Nov	Dec	Year
2-Axis	Average	7.2	8.1	8.7	9.7	10.2	11.0	10.0	9.4	9.2	8.5	7.3	6.8	8.8
	Min/Max	5.7/8.1	6.8/9.2	6.9/9.8	7.7/10.8	9.1/11.2	9.9/12.5	9.1/11.6	8.1/10.4	7.2/10.2	5.9/9.8	5.9/8.3	5.2/8.1	8.0/9.5

Direct Beam Solar Radiation for Concentrating Collectors (kWh/m²/day), Uncertainty ±8%

Tracker		Jan	Feb	Mar	Apr	May	June	July	Aug	Sept	Oct	Nov	Dec	Year
1-Axis, E-W Horiz Axis	Average	4.8	5.0	4.7	5.0	5.4	6.2	5.5	5.0	5.1	5.2	4.8	4.6	5.1
	Min/Max	3.5/5.7	3.8/6.0	3.3/5.7	3.3/5.9	4.3/6.3	5.1/7.4	4.7/6.9	4.0/5.9	3.5/6.3	2.8/6.4	3.5/5.8	3.2/5.6	4.4/5.7
1-Axis, N-S Horiz Axis	Average	3.7	4.8	5.7	6.9	7.4	8.4	7.4	6.7	6.5	5.5	4.0	3.4	5.9
	Min/Max	2.6/4.6	3.6/5.9	3.8/6.9	4.6/8.3	5.8/8.8	6.9/9.9	6.2/9.4	5.4/8.1	4.2/8.0	3.0/7.0	2.8/4.9	2.3/4.2	5.0/6.6
1-Axis, N-S Tilt=Latitude	Average	5.3	6.2	6.5	7.2	7.2	7.9	7.0	6.8	7.2	6.8	5.6	5.0	6.5
	Min/Max	3.8/6.4	4.7/7.6	4.4/7.9	4.8/8.6	5.7/8.5	6.4/9.3	5.9/8.9	5.5/8.1	4.7/8.8	3.7/8.5	4.0/6.8	3.4/6.1	5.5/7.4
2-Axis	Average	5.7	6.3	6.5	7.3	7.6	8.6	7.5	7.0	7.2	6.8	5.9	5.4	6.8
	Min/Max	4.1/6.9	4.8/7.8	4.4/8.0	4.9/8.7	6.0/9.0	7.0/10.1	6.3/9.6	5.6/8.4	4.7/8.8	3.7/8.6	4.2/7.1	3.7/6.6	5.7/7.7

Average Climatic Conditions

Element	Jan	Feb	Mar	Apr	May	June	July	Aug	Sept	Oct	Nov	Dec	Year
Temperature (°C)	-9.6	-5.3	0.2	5.1	10.2	15.2	18.3	16.9	12.7	6.4	-1.2	-8.1	5.1
Daily Minimum Temp	-19.9	-15.1	-9.0	-4.7	0.4	5.0	8.8	7.4	2.6	-4.1	-10.9	-18.1	-4.8
Daily Maximum Temp	0.7	4.4	9.3	14.8	20.0	25.4	27.8	26.2	22.6	16.9	8.6	1.9	14.9
Record Minimum Temp	-45.6	-37.2	-28.9	-21.1	-11.7	-4.4	1.1	-1.7	-9.4	-23.3	-34.4	-41.1	-45.6
Record Maximum Temp	16.7	18.9	22.8	26.7	29.4	33.9	35.6	32.2	30.6	27.2	21.7	16.1	35.6
HDD, Base 18.3°C	866	663	563	398	252	97	23	54	170	371	585	818	4861
CDD, Base 18.3°C	0	0	0	0	0	3	22	9	0	0	0	0	34
Relative Humidity (%)	71	67	57	48	47	46	56	60	56	55	64	70	58
Wind Speed (m/s)	3.4	3.4	4.7	5.1	4.8	4.5	3.6	3.4	2.7	3.4	4.0	2.6	3.8

Boulder, CO

WBAN NO. 94018

LATITUDE: 40.02° N
LONGITUDE: 105.25° W
ELEVATION: 1634 meters
MEAN PRESSURE: 836 millibars

STATION TYPE: Primary

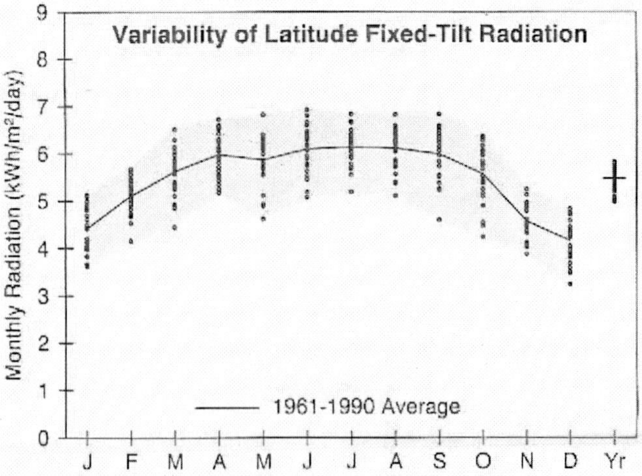

Solar Radiation for Flat-Plate Collectors Facing South at a Fixed Tilt (kWh/m²/day), Uncertainty ±9%

Tilt (°)		Jan	Feb	Mar	Apr	May	June	July	Aug	Sept	Oct	Nov	Dec	Year
0	Average	2.4	3.3	4.4	5.6	6.2	6.9	6.7	6.0	5.0	3.8	2.6	2.1	4.6
	Min/Max	2.1/2.7	2.8/3.5	3.7/5.0	4.8/6.1	5.1/7.2	5.7/7.8	5.6/7.4	5.2/6.6	4.0/5.5	3.1/4.2	2.3/2.8	1.9/2.3	4.3/4.8
Latitude -15	Average	3.8	4.6	5.4	6.1	6.2	6.6	6.6	6.3	5.9	5.1	4.0	3.5	5.4
	Min/Max	3.2/4.4	3.8/5.1	4.3/6.2	5.3/6.8	4.9/7.3	5.5/7.6	5.6/7.4	5.3/7.1	4.6/6.7	4.0/5.8	3.4/4.6	2.8/4.1	4.9/5.7
Latitude	Average	4.4	5.1	5.6	6.0	5.9	6.1	6.1	6.1	6.0	5.6	4.6	4.2	5.5
	Min/Max	3.6/5.1	4.2/5.7	4.4/6.5	5.2/6.7	4.6/6.8	5.1/6.9	5.2/6.8	5.1/6.8	4.6/6.8	4.2/6.4	3.9/5.2	3.2/4.8	5.0/5.8
Latitude +15	Average	4.8	5.3	5.6	5.6	5.2	5.2	5.3	5.5	5.8	5.7	4.8	4.5	5.3
	Min/Max	3.9/5.6	4.3/5.9	4.4/6.5	4.8/6.2	4.1/6.0	4.4/5.9	4.5/5.9	4.6/6.2	4.4/6.6	4.2/6.5	4.1/5.6	3.5/5.3	4.8/5.6
90	Average	4.5	4.6	4.3	3.6	2.8	2.6	2.7	3.2	4.0	4.6	4.4	4.3	3.8
	Min/Max	3.6/5.4	3.7/5.2	3.5/5.0	3.0/4.0	2.3/3.1	2.2/2.8	2.3/2.9	2.7/3.6	3.1/4.6	3.4/5.3	3.7/5.1	3.4/5.2	3.4/4.1

Solar Radiation for 1-Axis Tracking Flat-Plate Collectors with a North-South Axis (kWh/m²/day), Uncertainty ±9%

Axis Tilt (°)		Jan	Feb	Mar	Apr	May	June	July	Aug	Sept	Oct	Nov	Dec	Year
0	Average	3.7	4.9	6.2	7.6	8.2	9.1	9.0	8.2	7.1	5.7	4.0	3.3	6.4
	Min/Max	3.0/4.4	4.1/5.5	4.6/7.4	6.2/8.7	6.2/10.0	7.4/10.9	7.1/10.2	6.7/9.3	5.3/8.3	4.2/6.6	3.4/4.4	2.6/3.9	5.7/6.9
Latitude -15	Average	4.8	5.9	7.0	8.1	8.4	9.1	9.1	8.6	7.9	6.7	5.0	4.4	7.1
	Min/Max	3.8/5.6	4.8/6.7	5.1/8.4	6.6/9.2	6.3/10.2	7.4/10.9	7.1/10.3	7.0/9.8	5.8/9.2	4.8/7.8	4.2/5.7	3.3/5.2	6.2/7.6
Latitude	Average	5.2	6.2	7.2	8.0	8.1	8.8	8.7	8.4	7.9	7.1	5.5	4.9	7.2
	Min/Max	4.2/6.2	5.1/7.1	5.2/8.6	6.6/9.2	6.1/9.9	7.1/10.4	6.8/10.0	6.8/9.6	5.8/9.3	5.0/8.2	4.6/6.3	3.6/5.8	6.3/7.8
Latitude +15	Average	5.5	6.4	7.1	7.7	7.7	8.2	8.2	8.0	7.8	7.1	5.7	5.2	7.1
	Min/Max	4.4/6.6	5.2/7.3	5.2/8.6	6.3/8.9	5.8/9.4	6.6/9.8	6.4/9.3	6.5/9.2	5.6/9.1	5.0/8.3	4.8/6.6	3.8/6.2	6.1/7.6

Solar Radiation for 2-Axis Tracking Flat-Plate Collectors (kWh/m²/day), Uncertainty ±9%

Tracker		Jan	Feb	Mar	Apr	May	June	July	Aug	Sept	Oct	Nov	Dec	Year
2-Axis	Average	5.6	6.4	7.2	8.1	8.5	9.4	9.2	8.6	8.0	7.1	5.7	5.3	7.4
	Min/Max	4.5/6.7	5.2/7.3	5.2/8.6	6.7/9.3	6.4/10.4	7.6/11.1	7.2/10.5	7.0/9.8	5.8/9.3	5.1/8.3	4.8/6.6	3.9/6.3	6.5/8.0

Direct Beam Solar Radiation for Concentrating Collectors (kWh/m²/day), Uncertainty ±8%

Tracker		Jan	Feb	Mar	Apr	May	June	July	Aug	Sept	Oct	Nov	Dec	Year
1-Axis, E-W	Average	3.5	3.7	3.7	4.0	4.2	5.0	4.9	4.5	4.4	4.3	3.6	3.4	4.1
Horiz Axis	Min/Max	2.3/4.6	2.8/4.5	2.1/4.8	2.9/5.0	2.9/5.7	3.5/6.4	3.8/6.1	3.4/5.4	2.8/5.5	2.5/5.2	2.7/4.7	2.0/4.3	3.4/4.5
1-Axis, N-S	Average	2.6	3.4	4.2	5.3	5.6	6.6	6.5	6.0	5.4	4.3	2.8	2.3	4.6
Horiz Axis	Min/Max	1.6/3.4	2.5/4.2	2.2/5.7	3.6/6.4	3.8/7.6	4.8/8.5	4.8/8.1	4.5/7.1	3.4/6.7	2.4/5.3	2.2/3.6	1.3/3.0	3.7/5.1
1-Axis, N-S	Average	3.9	4.5	5.0	5.6	5.5	6.2	6.2	6.1	6.0	5.5	4.1	3.6	5.2
Tilt=Latitude	Min/Max	2.5/5.1	3.4/5.5	2.7/6.6	3.8/6.8	3.7/7.5	4.5/8.0	4.6/7.7	4.6/7.3	3.8/7.6	3.1/6.7	3.1/5.3	2.0/4.6	4.2/5.7
2-Axis	Average	4.1	4.6	5.0	5.7	5.8	6.8	6.7	6.3	6.1	5.6	4.3	4.0	5.4
	Min/Max	2.7/5.4	3.5/5.7	2.7/6.6	3.9/6.9	4.0/7.9	4.9/8.7	4.9/8.3	4.8/7.5	3.8/7.6	3.2/6.8	3.3/5.6	2.2/5.0	4.3/6.0

Average Climatic Conditions

Element	Jan	Feb	Mar	Apr	May	June	July	Aug	Sept	Oct	Nov	Dec	Year
Temperature (°C)	-1.3	0.8	3.9	9.0	14.0	19.4	23.1	21.9	16.8	10.8	3.9	-0.6	10.2
Daily Minimum Temp	-8.8	-6.6	-3.4	1.4	6.4	11.3	14.8	13.8	8.7	2.4	-3.7	-8.1	2.3
Daily Maximum Temp	6.2	8.1	11.2	16.6	21.6	27.4	31.2	29.9	24.9	19.1	11.4	6.9	17.9
Record Minimum Temp	-31.7	-34.4	-23.9	-18.9	-5.6	-1.1	6.1	5.0	-8.3	-16.1	-22.2	-31.7	-34.4
Record Maximum Temp	22.8	24.4	28.9	31.7	35.6	40.0	40.0	38.3	36.1	31.7	26.1	23.9	40.0
HDD, Base 18.3°C	608	492	448	280	141	39	0	0	80	238	433	586	3344
CDD, Base 18.3°C	0	0	0	0	6	71	148	113	35	4	0	0	377
Relative Humidity (%)	55	56	54	50	52	49	48	49	50	49	56	56	52
Wind Speed (m/s)	3.7	3.8	4.1	4.4	4.1	3.8	3.6	3.5	3.4	3.4	3.5	3.6	3.8

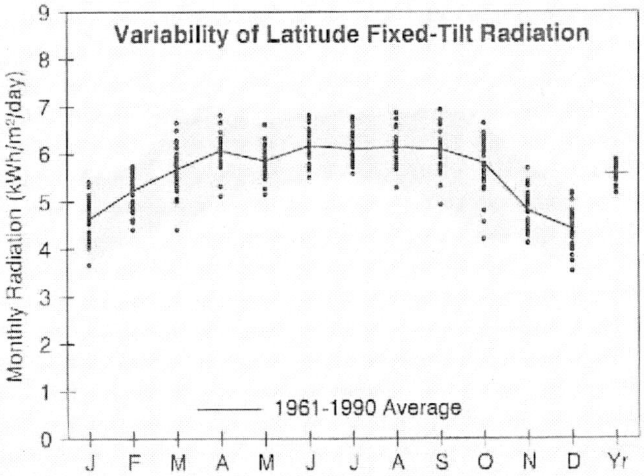

Colorado Springs, CO

WBAN NO. 93037

LATITUDE: 38.82° N
LONGITUDE: 104.72° W
ELEVATION: 1881 meters
MEAN PRESSURE: 811 millibars

STATION TYPE: Secondary

Solar Radiation for Flat-Plate Collectors Facing South at a Fixed Tilt (kWh/m²/day), Uncertainty ±9%

Tilt (°)		Jan	Feb	Mar	Apr	May	June	July	Aug	Sept	Oct	Nov	Dec	Year
0	Average	2.5	3.4	4.5	5.7	6.2	6.9	6.7	6.0	5.1	4.0	2.8	2.3	4.7
	Min/Max	2.2/2.8	3.0/3.7	3.7/5.1	4.8/6.3	5.5/7.0	6.2/7.7	6.1/7.4	5.2/6.7	4.3/5.8	3.2/4.5	2.5/3.1	2.0/2.5	4.5/4.9
Latitude -15	Average	4.0	4.7	5.5	6.2	6.2	6.7	6.6	6.4	6.0	5.4	4.2	3.7	5.5
	Min/Max	3.2/4.6	4.0/5.2	4.3/6.3	5.2/6.9	5.5/7.1	6.0/7.4	6.0/7.3	5.5/7.1	4.9/6.8	4.0/6.1	3.7/4.9	3.0/4.3	5.1/5.7
Latitude	Average	4.6	5.2	5.7	6.1	5.9	6.2	6.1	6.1	6.1	5.8	4.8	4.4	5.6
	Min/Max	3.7/5.4	4.4/5.7	4.4/6.7	5.1/6.8	5.2/6.6	5.5/6.8	5.6/6.8	5.3/6.9	4.9/6.9	4.2/6.6	4.1/5.7	3.5/5.2	5.2/5.9
Latitude +15	Average	5.0	5.4	5.6	5.6	5.2	5.3	5.3	5.6	5.9	5.9	5.1	4.8	5.4
	Min/Max	3.9/5.9	4.5/6.0	4.3/6.6	4.8/6.3	4.6/5.8	4.7/5.8	4.9/5.8	4.8/6.2	4.7/6.7	4.2/6.8	4.3/6.1	3.8/5.7	4.9/5.7
90	Average	4.7	4.6	4.2	3.5	2.7	2.5	2.6	3.2	4.0	4.8	4.6	4.6	3.8
	Min/Max	3.5/5.5	3.8/5.1	3.4/4.9	3.0/4.0	2.5/3.0	2.3/2.6	2.4/2.8	2.8/3.5	3.2/4.5	3.3/5.5	3.8/5.6	3.7/5.4	3.4/4.0

Solar Radiation for 1-Axis Tracking Flat-Plate Collectors with a North-South Axis (kWh/m²/day), Uncertainty ±9%

Axis Tilt (°)		Jan	Feb	Mar	Apr	May	June	July	Aug	Sept	Oct	Nov	Dec	Year
0	Average	4.0	5.1	6.4	7.9	8.3	9.4	9.0	8.3	7.4	6.1	4.3	3.7	6.7
	Min/Max	3.1/4.7	4.4/5.7	4.7/7.7	6.3/9.1	7.0/9.8	8.0/10.8	8.0/10.3	6.7/9.5	5.7/8.7	4.3/7.1	3.5/5.2	2.9/4.3	6.1/7.1
Latitude -15	Average	5.1	6.1	7.2	8.3	8.5	9.3	9.0	8.6	8.1	7.1	5.4	4.7	7.3
	Min/Max	3.8/6.0	5.1/6.8	5.2/8.6	6.6/9.7	7.1/10.0	7.9/10.7	8.0/10.3	7.0/9.9	6.2/9.5	4.9/8.3	4.4/6.4	3.7/5.6	6.6/7.7
Latitude	Average	5.6	6.5	7.4	8.3	8.2	9.0	8.7	8.5	8.1	7.4	5.8	5.3	7.4
	Min/Max	4.2/6.6	5.4/7.3	5.3/8.9	6.6/9.6	6.8/9.7	7.6/10.3	7.7/10.0	6.9/9.7	6.2/9.6	5.1/8.7	4.8/7.1	4.1/6.2	6.7/7.8
Latitude +15	Average	5.9	6.6	7.3	7.9	7.7	8.4	8.1	8.1	8.0	7.5	6.1	5.6	7.3
	Min/Max	4.4/7.0	5.5/7.5	5.2/8.8	6.3/9.3	6.4/9.1	7.1/9.6	7.2/9.4	6.5/9.3	6.1/9.4	5.1/8.8	5.0/7.4	4.3/6.6	6.5/7.7

Solar Radiation for 2-Axis Tracking Flat-Plate Collectors (kWh/m²/day), Uncertainty ±9%

Tracker		Jan	Feb	Mar	Apr	May	June	July	Aug	Sept	Oct	Nov	Dec	Year
2-Axis	Average	5.9	6.7	7.4	8.4	8.6	9.6	9.2	8.6	8.2	7.5	6.1	5.7	7.7
	Min/Max	4.4/7.1	5.5/7.5	5.3/8.9	6.7/9.7	7.2/10.1	8.1/11.0	8.2/10.5	7.0/9.9	6.3/9.6	5.1/8.8	5.0/7.4	4.3/6.8	6.9/8.1

Direct Beam Solar Radiation for Concentrating Collectors (kWh/m²/day), Uncertainty ±8%

Tracker		Jan	Feb	Mar	Apr	May	June	July	Aug	Sept	Oct	Nov	Dec	Year
1-Axis, E-W	Average	3.9	4.0	3.9	4.2	4.2	5.1	4.8	4.5	4.4	4.5	3.9	3.8	4.3
Horiz Axis	Min/Max	2.5/5.0	3.1/4.7	2.2/5.2	2.8/5.3	3.1/5.4	3.8/6.4	4.0/6.1	3.4/5.5	3.0/5.6	2.6/5.7	2.9/5.1	2.4/4.8	3.6/4.7
1-Axis, N-S	Average	2.9	3.7	4.5	5.6	5.7	6.8	6.4	6.0	5.5	4.7	3.2	2.7	4.8
Horiz Axis	Min/Max	1.8/3.8	2.8/4.5	2.4/6.0	3.7/7.1	4.2/7.4	5.1/8.5	5.2/8.1	4.4/7.4	3.6/7.1	2.6/5.9	2.3/4.2	1.6/3.5	4.1/5.3
1-Axis, N-S	Average	4.3	4.9	5.3	5.8	5.6	6.4	6.1	6.1	6.2	5.8	4.5	4.1	5.4
Tilt=Latitude	Min/Max	2.7/5.5	3.8/5.8	2.8/7.0	3.9/7.4	4.1/7.2	4.8/8.1	4.9/7.8	4.5/7.5	4.0/7.9	3.3/7.3	3.3/5.9	2.5/5.2	4.5/5.9
2-Axis	Average	4.6	5.0	5.3	5.9	5.9	7.0	6.6	6.2	6.2	5.9	4.8	4.4	5.6
	Min/Max	2.9/5.9	3.8/6.0	2.8/7.0	3.9/7.5	4.3/7.6	5.2/8.7	5.3/8.3	4.6/7.7	4.0/7.9	3.3/7.4	3.5/6.2	2.8/5.7	4.7/6.2

Average Climatic Conditions

Element	Jan	Feb	Mar	Apr	May	June	July	Aug	Sept	Oct	Nov	Dec	Year
Temperature (°C)	-1.8	0.0	2.9	8.0	13.0	18.3	21.6	20.2	15.8	9.9	3.2	-1.2	9.2
Daily Minimum Temp	-8.8	-7.1	-4.1	0.6	5.6	10.6	13.9	12.9	8.4	2.4	-3.9	-8.1	1.9
Daily Maximum Temp	5.2	7.0	10.0	15.4	20.4	26.1	29.1	27.4	23.1	17.5	10.4	5.7	16.4
Record Minimum Temp	-32.2	-32.8	-23.9	-19.4	-6.1	0.0	5.6	6.1	-5.6	-15.0	-22.2	-31.1	-32.8
Record Maximum Temp	22.2	24.4	27.2	28.3	33.9	37.8	37.8	37.2	34.4	30.0	25.6	25.0	37.8
HDD, Base 18.3°C	623	513	477	310	168	48	3	10	91	260	453	606	3564
CDD, Base 18.3°C	0	0	0	0	0	48	103	67	14	0	0	0	233
Relative Humidity (%)	51	50	50	46	49	48	50	54	51	46	50	51	50
Wind Speed (m/s)	4.1	4.3	4.7	5.1	4.8	4.3	4.0	3.8	4.1	4.2	4.2	4.0	4.3

Eagle, CO

WBAN NO. 23063

LATITUDE: 39.65° N
LONGITUDE: 106.92° W
ELEVATION: 1985 meters
MEAN PRESSURE: 802 millibars

STATION TYPE: Secondary

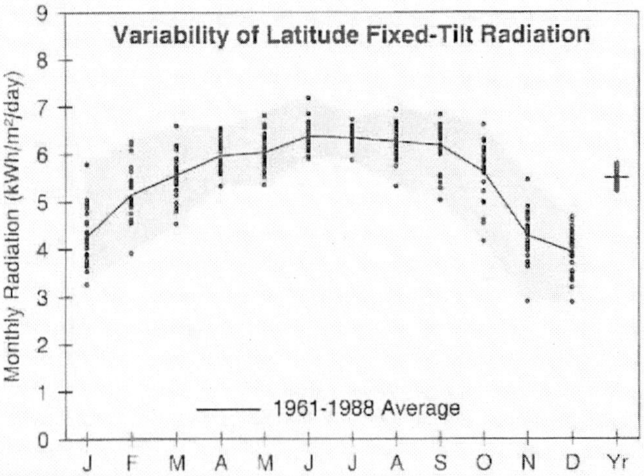

Solar Radiation for Flat-Plate Collectors Facing South at a Fixed Tilt (kWh/m²/day), Uncertainty ±9%

Tilt (°)		Jan	Feb	Mar	Apr	May	June	July	Aug	Sept	Oct	Nov	Dec	Year
0	Average	2.4	3.3	4.4	5.6	6.4	7.2	6.9	6.1	5.1	3.9	2.5	2.1	4.7
	Min/Max	2.0/2.9	2.7/3.9	3.8/5.0	5.0/6.0	5.6/7.2	6.6/8.2	6.4/7.4	5.3/6.8	4.4/5.6	3.1/4.4	2.0/3.0	1.8/2.3	4.5/4.9
Latitude -15	Average	3.7	4.6	5.3	6.1	6.4	7.0	6.9	6.5	6.1	5.2	3.8	3.4	5.4
	Min/Max	2.9/4.9	3.6/5.6	4.4/6.2	5.4/6.7	5.7/7.3	6.4/7.9	6.3/7.3	5.5/7.2	5.0/6.7	3.9/6.1	2.7/4.7	2.6/4.0	5.2/5.7
Latitude	Average	4.3	5.2	5.6	6.0	6.0	6.4	6.3	6.3	6.2	5.6	4.3	3.9	5.5
	Min/Max	3.3/5.8	3.9/6.3	4.5/6.6	5.3/6.6	5.4/6.8	5.9/7.2	5.9/6.7	5.3/6.9	5.0/6.8	4.2/6.6	2.9/5.5	2.9/4.7	5.2/5.8
Latitude +15	Average	4.6	5.4	5.5	5.6	5.3	5.5	5.5	5.7	5.9	5.7	4.5	4.3	5.3
	Min/Max	3.5/6.4	4.0/6.7	4.4/6.6	4.9/6.1	4.8/6.0	5.1/6.1	5.1/5.8	4.8/6.3	4.8/6.6	4.2/6.8	3.0/5.9	3.1/5.1	5.0/5.6
90	Average	4.4	4.8	4.3	3.5	2.8	2.6	2.7	3.3	4.1	4.7	4.1	4.1	3.8
	Min/Max	3.4/6.3	3.3/6.2	3.3/5.4	3.1/3.8	2.6/3.1	2.5/2.8	2.6/2.8	2.9/3.6	3.3/4.5	3.3/5.6	2.6/5.5	3.0/4.9	3.5/4.0

Solar Radiation for 1-Axis Tracking Flat-Plate Collectors with a North-South Axis (kWh/m²/day), Uncertainty ±9%

Axis Tilt (°)		Jan	Feb	Mar	Apr	May	June	July	Aug	Sept	Oct	Nov	Dec	Year
0	Average	3.6	5.0	6.2	7.7	8.7	9.9	9.6	8.6	7.5	5.9	3.8	3.2	6.7
	Min/Max	2.8/4.9	3.7/6.2	4.8/7.3	6.7/8.8	7.3/10.2	9.0/11.7	8.6/10.7	6.9/9.9	5.7/8.7	4.2/7.0	2.6/4.9	2.4/3.8	6.2/7.0
Latitude -15	Average	4.6	5.9	6.9	8.2	8.9	9.9	9.6	9.0	8.3	6.8	4.7	4.2	7.3
	Min/Max	3.5/6.4	4.4/7.4	5.3/8.3	7.1/9.3	7.4/10.4	9.0/11.7	8.6/10.8	7.2/10.3	6.2/9.5	4.8/8.2	3.1/6.2	3.0/5.1	6.7/7.7
Latitude	Average	5.1	6.3	7.1	8.1	8.6	9.5	9.3	8.8	8.4	7.2	5.1	4.6	7.3
	Min/Max	3.8/7.1	4.7/7.9	5.4/8.5	7.0/9.3	7.2/10.1	8.6/11.3	8.3/10.4	7.1/10.1	6.2/9.6	5.0/8.6	3.3/6.8	3.2/5.6	6.8/7.8
Latitude +15	Average	5.3	6.5	7.1	7.8	8.1	8.9	8.7	8.4	8.2	7.2	5.3	4.9	7.2
	Min/Max	4.0/7.6	4.7/8.2	5.3/8.6	6.7/8.9	6.8/9.6	8.1/10.6	7.8/9.8	6.7/9.7	6.1/9.4	5.0/8.7	3.4/7.1	3.4/6.0	6.6/7.7

Solar Radiation for 2-Axis Tracking Flat-Plate Collectors (kWh/m²/day), Uncertainty ±9%

Tracker		Jan	Feb	Mar	Apr	May	June	July	Aug	Sept	Oct	Nov	Dec	Year
2-Axis	Average	5.4	6.5	7.1	8.2	9.0	10.2	9.8	9.0	8.4	7.3	5.4	5.0	7.6
	Min/Max	4.0/7.7	4.7/8.2	5.4/8.6	7.1/9.4	7.5/10.5	9.2/12.0	8.8/11.0	7.3/10.3	6.3/9.6	5.1/8.7	3.4/7.1	3.4/6.1	7.0/8.1

Direct Beam Solar Radiation for Concentrating Collectors (kWh/m²/day), Uncertainty ±8%

Tracker		Jan	Feb	Mar	Apr	May	June	July	Aug	Sept	Oct	Nov	Dec	Year
1-Axis, E-W Horiz Axis	Average	3.1	3.6	3.5	4.0	4.4	5.4	5.1	4.6	4.5	4.3	3.3	3.1	4.1
	Min/Max	2.0/5.2	2.6/4.6	2.3/4.5	3.1/4.9	3.2/5.5	4.5/7.1	4.3/6.1	3.3/5.7	2.8/5.4	2.5/5.6	1.7/4.8	1.4/4.3	3.6/4.5
1-Axis, N-S Horiz Axis	Average	2.3	3.2	4.1	5.3	6.0	7.3	6.9	6.2	5.6	4.4	2.6	2.1	4.7
	Min/Max	1.4/3.9	2.3/4.3	2.5/5.1	4.1/6.6	4.3/7.6	6.0/9.6	5.8/8.4	4.3/7.8	3.3/7.0	2.5/5.7	1.4/3.8	0.9/3.0	4.1/5.1
1-Axis, N-S Tilt=Latitude	Average	3.4	4.3	4.8	5.6	5.8	6.9	6.6	6.3	6.3	5.5	3.7	3.3	5.2
	Min/Max	2.1/5.7	3.1/5.7	3.0/6.0	4.3/6.9	4.2/7.4	5.6/9.1	5.5/8.0	4.4/7.9	3.7/7.8	3.1/7.2	2.0/5.4	1.5/4.6	4.5/5.7
2-Axis	Average	3.7	4.4	4.8	5.7	6.2	7.5	7.1	6.5	6.3	5.6	3.9	3.5	5.4
	Min/Max	2.3/6.1	3.2/5.8	3.0/6.0	4.4/7.1	4.5/7.8	6.1/9.8	5.9/8.6	4.6/8.1	3.7/7.8	3.1/7.2	2.1/5.7	1.6/5.0	4.7/6.0

Average Climatic Conditions

Element	Jan	Feb	Mar	Apr	May	June	July	Aug	Sept	Oct	Nov	Dec	Year
Temperature (°C)	-7.4	-3.6	1.2	5.8	10.8	15.7	19.2	17.9	13.3	7.3	0.2	-6.3	6.2
Daily Minimum Temp	-15.7	-12.0	-6.8	-3.1	1.0	4.7	8.4	7.4	2.8	-2.9	-8.1	-14.4	-3.2
Daily Maximum Temp	1.0	4.8	9.1	14.8	20.6	26.7	30.0	28.5	23.9	17.4	8.4	1.6	15.6
Record Minimum Temp	-46.1	-34.4	-26.7	-19.4	-9.4	-3.3	-1.1	-2.2	-8.9	-16.1	-28.9	-35.0	-46.1
Record Maximum Temp	13.3	16.7	21.7	26.1	30.6	34.4	35.6	35.6	32.8	27.8	20.6	15.6	35.6
HDD, Base 18.3°C	797	614	532	375	234	91	14	40	153	343	545	764	4503
CDD, Base 18.3°C	0	0	0	0	0	11	42	28	3	0	0	0	84
Relative Humidity (%)	73	71	65	56	52	48	51	54	56	59	70	74	61
Wind Speed (m/s)	1.9	2.1	2.6	3.1	3.0	2.7	2.5	2.4	2.5	2.1	2.1	1.8	2.4

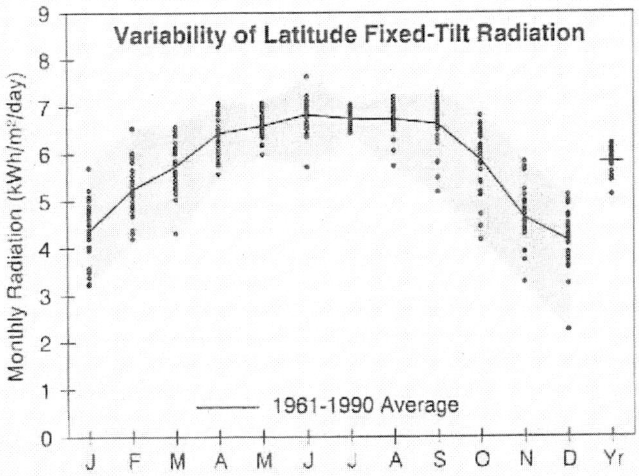

Grand Junction, CO

WBAN NO. 23066

LATITUDE: 39.12° N
LONGITUDE: 108.53° W
ELEVATION: 1475 meters
MEAN PRESSURE: 853 millibars

STATION TYPE: Primary

Solar Radiation for Flat-Plate Collectors Facing South at a Fixed Tilt (kWh/m²/day), Uncertainty ±9%

Tilt (°)		Jan	Feb	Mar	Apr	May	June	July	Aug	Sept	Oct	Nov	Dec	Year
0	Average	2.5	3.5	4.6	6.0	7.0	7.7	7.4	6.6	5.5	4.1	2.7	2.2	5.0
	Min/Max	2.1/2.9	3.0/4.1	3.8/5.1	5.3/6.5	6.3/7.5	6.4/8.7	7.1/7.7	5.7/7.1	4.5/6.0	3.1/4.5	2.2/3.2	1.5/2.5	4.6/5.2
Latitude -15	Average	3.8	4.7	5.5	6.5	7.0	7.5	7.3	7.0	6.5	5.4	4.1	3.6	5.7
	Min/Max	2.9/4.8	3.9/5.8	4.3/6.3	5.7/7.2	6.3/7.5	6.2/8.4	7.0/7.6	6.0/7.5	5.2/7.2	3.9/6.2	3.0/5.0	2.0/4.3	5.1/6.1
Latitude	Average	4.4	5.2	5.7	6.4	6.6	6.8	6.7	6.7	6.6	5.9	4.6	4.1	5.8
	Min/Max	3.2/5.7	4.2/6.6	4.3/6.6	5.6/7.1	6.0/7.1	5.7/7.6	6.4/7.0	5.7/7.2	5.2/7.3	4.2/6.8	3.3/5.8	2.3/5.1	5.1/6.2
Latitude +15	Average	4.7	5.4	5.6	6.0	5.8	5.8	5.8	6.1	6.4	6.0	4.9	4.5	5.6
	Min/Max	3.4/6.2	4.3/6.9	4.2/6.5	5.1/6.6	5.3/6.2	4.9/6.5	5.6/6.1	5.2/6.5	4.9/7.0	4.2/6.9	3.4/6.2	2.4/5.6	4.8/6.0
90	Average	4.4	4.7	4.2	3.7	3.0	2.7	2.8	3.5	4.4	4.8	4.4	4.3	3.9
	Min/Max	3.0/5.8	3.5/6.4	3.0/4.8	3.2/4.0	2.9/3.2	2.4/2.9	2.7/2.9	3.0/3.7	3.4/4.8	3.3/5.6	3.0/5.7	2.2/5.3	3.3/4.3

Solar Radiation for 1-Axis Tracking Flat-Plate Collectors with a North-South Axis (kWh/m²/day), Uncertainty ±9%

Axis Tilt (°)		Jan	Feb	Mar	Apr	May	June	July	Aug	Sept	Oct	Nov	Dec	Year
0	Average	3.7	5.1	6.4	8.2	9.5	10.6	10.1	9.2	8.1	6.1	4.2	3.4	7.0
	Min/Max	2.7/4.8	3.9/6.5	4.4/7.5	7.0/9.4	7.7/10.5	8.3/12.4	9.3/10.7	7.6/10.2	6.0/9.0	4.1/7.1	2.8/5.2	1.9/4.2	6.0/7.5
Latitude -15	Average	4.7	6.0	7.1	8.7	9.6	10.6	10.1	9.5	8.9	7.1	5.1	4.4	7.7
	Min/Max	3.4/6.2	4.6/7.8	4.9/8.4	7.3/10.0	7.9/10.7	8.3/12.3	9.4/10.7	7.9/10.6	6.5/9.9	4.8/8.4	3.4/6.6	2.3/5.5	6.4/8.2
Latitude	Average	5.2	6.4	7.3	8.7	9.3	10.1	9.8	9.4	9.0	7.5	5.6	4.9	7.8
	Min/Max	3.7/6.9	4.8/8.4	4.9/8.7	7.2/9.9	7.6/10.4	7.9/11.9	9.0/10.4	7.7/10.4	6.6/10.0	5.0/8.8	3.7/7.2	2.5/6.1	6.4/8.3
Latitude +15	Average	5.4	6.6	7.2	8.3	8.8	9.5	9.2	8.9	8.8	7.5	5.8	5.2	7.6
	Min/Max	3.8/7.3	4.9/8.6	4.8/8.6	6.9/9.6	7.2/9.8	7.4/11.1	8.4/9.7	7.4/10.0	6.4/9.8	5.0/8.9	3.8/7.5	2.6/6.5	6.3/8.2

Solar Radiation for 2-Axis Tracking Flat-Plate Collectors (kWh/m²/day), Uncertainty ±9%

Tracker		Jan	Feb	Mar	Apr	May	June	July	Aug	Sept	Oct	Nov	Dec	Year
2-Axis	Average	5.5	6.6	7.3	8.8	9.7	10.8	10.3	9.6	9.0	7.6	5.8	5.2	8.0
	Min/Max	3.8/7.4	4.9/8.6	5.0/8.7	7.3/10.0	8.0/10.9	8.4/12.7	9.5/10.9	7.9/10.7	6.6/10.0	5.0/8.9	3.8/7.5	2.6/6.6	6.7/8.6

Direct Beam Solar Radiation for Concentrating Collectors (kWh/m²/day), Uncertainty ±8%

Tracker		Jan	Feb	Mar	Apr	May	June	July	Aug	Sept	Oct	Nov	Dec	Year
1-Axis, E-W Horiz Axis	Average	3.3	3.7	3.7	4.3	5.0	5.9	5.5	5.1	4.9	4.5	3.6	3.4	4.4
	Min/Max	1.9/5.3	2.3/5.2	1.8/4.9	3.2/5.4	3.5/6.0	4.3/7.6	4.8/6.2	3.7/5.8	3.0/5.9	2.6/5.6	2.1/5.2	1.2/4.7	3.3/4.9
1-Axis, N-S Horiz Axis	Average	2.5	3.4	4.3	5.7	6.7	7.9	7.4	6.7	6.2	4.6	2.9	2.3	5.1
	Min/Max	1.5/4.0	2.0/4.9	1.9/5.7	4.0/7.2	4.5/8.2	5.6/10.2	6.3/8.2	4.7/7.9	3.7/7.3	2.5/5.8	1.6/4.2	0.8/3.3	3.7/5.6
1-Axis, N-S Tilt=Latitude	Average	3.6	4.5	5.0	6.0	6.5	7.5	7.0	6.9	6.9	5.8	4.1	3.6	5.6
	Min/Max	2.2/5.8	2.7/6.4	2.3/6.6	4.3/7.5	4.4/8.0	5.3/9.6	6.0/7.8	4.8/8.0	4.1/8.1	3.2/7.2	2.3/5.9	1.2/5.0	4.1/6.2
2-Axis	Average	3.9	4.7	5.0	6.1	6.9	8.1	7.6	7.1	6.9	5.8	4.4	3.9	5.9
	Min/Max	2.3/6.3	2.8/6.5	2.3/6.6	4.3/7.7	4.6/8.4	5.8/10.4	6.4/8.4	4.9/8.3	4.1/8.2	3.2/7.3	2.4/6.3	1.3/5.5	4.3/6.5

Average Climatic Conditions

Element	Jan	Feb	Mar	Apr	May	June	July	Aug	Sept	Oct	Nov	Dec	Year
Temperature (°C)	-3.9	1.4	6.3	11.2	16.7	22.4	26.0	24.7	19.4	12.6	4.7	-1.8	11.7
Daily Minimum Temp	-9.7	-4.7	-0.4	3.6	8.8	13.9	17.7	16.8	11.6	5.3	-1.4	-7.4	4.5
Daily Maximum Temp	1.9	7.4	13.1	18.8	24.4	30.9	34.2	32.5	27.3	19.8	10.8	3.7	18.8
Record Minimum Temp	-30.6	-27.8	-15.0	-11.7	-3.3	1.1	7.8	6.1	-1.7	-7.8	-18.9	-27.2	-30.6
Record Maximum Temp	15.6	20.0	27.2	29.4	35.0	40.6	40.6	39.4	36.7	31.1	23.9	17.8	40.6
HDD, Base 18.3°C	689	474	372	216	73	7	0	0	31	184	410	625	3082
CDD, Base 18.3°C	0	0	0	0	22	131	238	196	64	7	0	0	657
Relative Humidity (%)	70	60	50	40	36	29	34	37	39	46	58	68	47
Wind Speed (m/s)	2.5	2.9	3.6	4.1	4.2	4.2	4.1	3.9	3.8	3.4	3.0	2.5	3.5

Pueblo, CO

WBAN NO. 93058

LATITUDE: 38.28° N
LONGITUDE: 104.52° W
ELEVATION: 1439 meters
MEAN PRESSURE: 855 millibars

STATION TYPE: Secondary

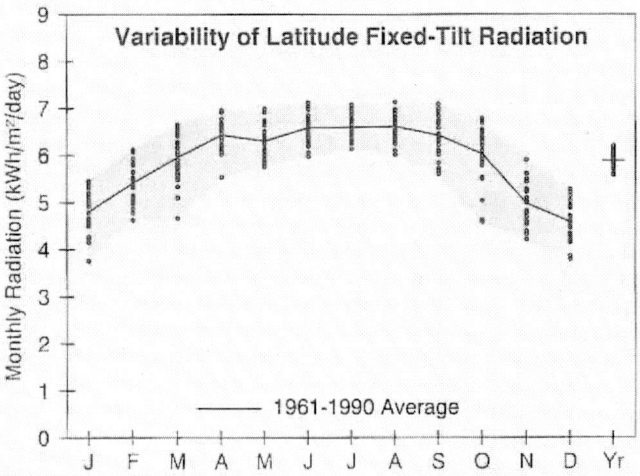

Solar Radiation for Flat-Plate Collectors Facing South at a Fixed Tilt (kWh/m²/day), Uncertainty ±9%

Tilt (°)		Jan	Feb	Mar	Apr	May	June	July	Aug	Sept	Oct	Nov	Dec	Year
0	Average	2.7	3.6	4.7	6.0	6.7	7.4	7.2	6.5	5.4	4.2	2.9	2.4	5.0
	Min/Max	2.3/2.9	3.2/3.9	3.9/5.2	5.2/6.5	6.1/7.3	6.7/8.1	6.7/7.8	6.0/7.0	4.9/5.9	3.4/4.6	2.7/3.3	2.2/2.6	4.8/5.2
Latitude -15	Average	4.1	4.9	5.7	6.5	6.7	7.2	7.1	6.9	6.3	5.6	4.4	3.9	5.8
	Min/Max	3.3/4.7	4.2/5.5	4.6/6.3	5.6/7.1	6.1/7.4	6.5/7.8	6.6/7.6	6.3/7.4	5.6/6.9	4.3/6.2	3.8/5.1	3.3/4.4	5.5/6.0
Latitude	Average	4.8	5.4	6.0	6.4	6.3	6.6	6.6	6.6	6.4	6.0	5.0	4.6	5.9
	Min/Max	3.8/5.5	4.6/6.1	4.7/6.7	5.5/6.9	5.8/7.0	6.0/7.1	6.1/7.1	6.0/7.1	5.6/7.1	4.6/6.8	4.2/5.9	3.8/5.3	5.6/6.2
Latitude +15	Average	5.2	5.6	5.9	6.0	5.6	5.6	5.7	6.0	6.2	6.2	5.3	5.0	5.7
	Min/Max	4.0/6.0	4.8/6.4	4.6/6.6	5.2/6.5	5.1/6.1	5.1/6.1	5.3/6.1	5.5/6.5	5.4/6.8	4.6/6.9	4.5/6.3	4.1/5.8	5.3/6.0
90	Average	4.9	4.8	4.3	3.6	2.8	2.6	2.7	3.3	4.2	4.9	4.8	4.8	4.0
	Min/Max	3.6/5.8	4.0/5.6	3.3/4.9	3.2/3.9	2.6/3.0	2.5/2.7	2.6/2.8	3.1/3.5	3.6/4.6	3.6/5.6	4.0/5.7	3.8/5.5	3.6/4.2

Solar Radiation for 1-Axis Tracking Flat-Plate Collectors with a North-South Axis (kWh/m²/day), Uncertainty ±9%

Axis Tilt (°)		Jan	Feb	Mar	Apr	May	June	July	Aug	Sept	Oct	Nov	Dec	Year
0	Average	4.2	5.3	6.8	8.4	9.0	10.0	9.7	8.9	7.8	6.4	4.5	3.8	7.1
	Min/Max	3.2/4.8	4.5/6.1	5.0/7.8	6.8/9.4	7.8/10.3	8.8/11.3	8.6/10.8	7.9/9.9	6.6/9.0	4.7/7.3	3.8/5.4	3.1/4.4	6.6/7.5
Latitude -15	Average	5.3	6.3	7.5	8.9	9.1	9.9	9.7	9.3	8.5	7.4	5.6	4.9	7.7
	Min/Max	3.9/6.1	5.3/7.2	5.5/8.6	7.2/9.9	7.9/10.5	8.8/11.2	8.7/10.9	8.2/10.3	7.1/9.8	5.4/8.5	4.7/6.7	4.0/5.7	7.2/8.2
Latitude	Average	5.8	6.7	7.7	8.8	8.8	9.5	9.4	9.1	8.6	7.7	6.1	5.5	7.8
	Min/Max	4.3/6.7	5.6/7.7	5.6/8.9	7.2/9.9	7.6/10.2	8.4/10.8	8.4/10.5	8.1/10.1	7.2/9.9	5.6/8.9	5.1/7.3	4.4/6.4	7.3/8.3
Latitude +15	Average	6.1	6.9	7.7	8.5	8.3	8.9	8.8	8.7	8.4	7.8	6.3	5.8	7.7
	Min/Max	4.5/7.1	5.7/7.9	5.5/8.8	6.9/9.5	7.2/9.6	7.8/10.1	7.8/9.8	7.7/9.7	7.0/9.7	5.6/9.0	5.3/7.7	4.6/6.8	7.1/8.2

Solar Radiation for 2-Axis Tracking Flat-Plate Collectors (kWh/m²/day), Uncertainty ±9%

Tracker		Jan	Feb	Mar	Apr	May	June	July	Aug	Sept	Oct	Nov	Dec	Year
2-Axis	Average	6.2	6.9	7.8	8.9	9.2	10.2	9.9	9.3	8.6	7.8	6.4	5.9	8.1
	Min/Max	4.6/7.2	5.7/7.9	5.7/8.9	7.3/10.0	8.0/10.6	9.0/11.5	8.8/11.1	8.3/10.3	7.2/9.9	5.6/9.0	5.3/7.7	4.7/6.9	7.5/8.6

Direct Beam Solar Radiation for Concentrating Collectors (kWh/m²/day), Uncertainty ±8%

Tracker		Jan	Feb	Mar	Apr	May	June	July	Aug	Sept	Oct	Nov	Dec	Year
1-Axis, E-W	Average	4.0	4.2	4.2	4.6	4.8	5.7	5.5	5.1	4.8	4.8	4.1	3.9	4.6
Horiz Axis	Min/Max	2.6/4.9	3.1/5.1	2.5/5.1	3.3/5.4	3.8/5.9	4.6/7.0	4.4/6.5	4.2/6.0	3.6/5.8	2.8/5.8	3.0/5.3	2.7/5.0	4.1/5.1
1-Axis, N-S	Average	3.1	3.9	4.9	6.1	6.4	7.5	7.2	6.6	5.9	4.9	3.3	2.8	5.2
Horiz Axis	Min/Max	1.9/3.8	2.9/4.8	2.8/6.1	4.3/7.3	4.9/8.0	6.0/9.3	5.7/8.7	5.4/7.9	4.4/7.4	2.9/6.1	2.5/4.4	1.9/3.6	4.6/5.7
1-Axis, N-S	Average	4.4	5.1	5.6	6.4	6.2	7.0	6.9	6.7	6.6	6.1	4.7	4.2	5.8
Tilt=Latitude	Min/Max	2.8/5.5	3.8/6.2	3.3/7.0	4.6/7.7	4.8/7.8	5.7/8.7	5.5/8.3	5.5/8.0	4.9/8.2	3.6/7.5	3.5/6.1	2.9/5.3	5.1/6.4
2-Axis	Average	4.8	5.2	5.7	6.5	6.6	7.6	7.4	6.9	6.6	6.2	4.9	4.6	6.1
	Min/Max	3.0/5.9	3.8/6.4	3.3/7.0	4.6/7.8	5.1/8.2	6.2/9.5	5.9/8.9	5.7/8.3	4.9/8.2	3.7/7.6	3.7/6.5	3.1/5.8	5.3/6.7

Average Climatic Conditions

Element	Jan	Feb	Mar	Apr	May	June	July	Aug	Sept	Oct	Nov	Dec	Year
Temperature (°C)	-1.2	1.7	5.4	11.0	16.2	21.7	25.1	23.5	18.8	12.0	4.7	-0.5	11.5
Daily Minimum Temp	-9.9	-6.9	-3.3	2.1	7.6	12.3	16.2	15.0	10.1	2.6	-4.3	-9.2	2.7
Daily Maximum Temp	7.4	10.4	14.0	19.9	24.7	30.9	33.9	32.1	27.4	21.4	13.8	8.2	20.3
Record Minimum Temp	-33.9	-35.0	-28.9	-16.7	-3.9	2.8	6.7	4.4	-2.8	-10.6	-27.2	-31.7	-35.0
Record Maximum Temp	25.6	27.2	30.0	33.9	36.7	42.2	41.1	40.0	37.8	34.4	28.9	27.8	42.2
HDD, Base 18.3°C	606	465	401	220	84	6	0	0	34	198	408	584	3007
CDD, Base 18.3°C	0	0	0	0	17	106	208	162	48	0	0	0	541
Relative Humidity (%)	60	56	51	45	47	46	49	52	50	48	57	60	52
Wind Speed (m/s)	3.7	3.9	4.6	5.1	4.9	4.5	4.2	3.9	3.9	3.7	3.7	3.6	4.2

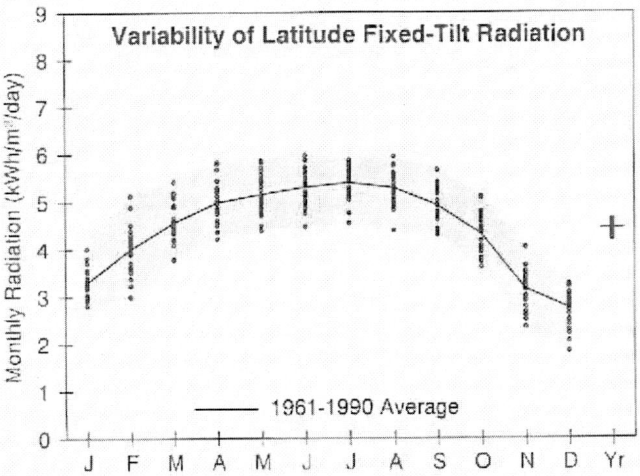

Bridgeport, CT

WBAN NO. 94702

LATITUDE: 41.17° N
LONGITUDE: 73.13° W
ELEVATION: 2 meters
MEAN PRESSURE: 1016 millibars

STATION TYPE: Secondary

Solar Radiation for Flat-Plate Collectors Facing South at a Fixed Tilt (kWh/m²/day), Uncertainty ±11%

Tilt (°)		Jan	Feb	Mar	Apr	May	June	July	Aug	Sept	Oct	Nov	Dec	Year
0	Average	1.9	2.7	3.7	4.7	5.4	5.9	5.8	5.2	4.2	3.1	2.0	1.6	3.8
	Min/Max	1.7/2.2	2.2/3.2	3.2/4.1	4.1/5.4	4.7/6.1	4.9/6.6	5.0/6.3	4.4/5.7	3.8/4.7	2.7/3.5	1.7/2.3	1.3/1.7	3.7/4.0
Latitude -15	Average	2.9	3.7	4.4	5.1	5.5	5.8	5.8	5.5	4.9	4.0	2.8	2.4	4.4
	Min/Max	2.5/3.5	2.8/4.6	3.7/5.2	4.3/5.9	4.7/6.2	4.8/6.5	4.9/6.3	4.6/6.2	4.3/5.6	3.5/4.7	2.2/3.6	1.7/2.8	4.2/4.6
Latitude	Average	3.3	4.0	4.6	5.0	5.2	5.3	5.4	5.3	4.9	4.3	3.1	2.8	4.4
	Min/Max	2.8/4.0	3.0/5.1	3.8/5.4	4.2/5.8	4.4/5.9	4.5/6.0	4.6/5.9	4.4/5.9	4.3/5.7	3.6/5.1	2.3/4.0	1.8/3.3	4.2/4.6
Latitude +15	Average	3.5	4.1	4.5	4.6	4.6	4.6	4.7	4.8	4.7	4.3	3.3	2.9	4.2
	Min/Max	3.0/4.3	3.0/5.3	3.6/5.4	3.9/5.4	3.9/5.2	3.9/5.2	4.0/5.1	4.0/5.4	4.1/5.4	3.6/5.2	2.4/4.3	1.9/3.5	4.0/4.4
90	Average	3.3	3.6	3.4	3.0	2.6	2.5	2.6	3.0	3.3	3.5	2.9	2.7	3.0
	Min/Max	2.7/4.2	2.5/4.7	2.7/4.2	2.6/3.6	2.3/2.9	2.2/2.7	2.3/2.8	2.5/3.3	2.9/3.9	2.9/4.3	2.1/3.9	1.7/3.3	2.9/3.2

Solar Radiation for 1-Axis Tracking Flat-Plate Collectors with a North-South Axis (kWh/m²/day), Uncertainty ±11%

Axis Tilt (°)		Jan	Feb	Mar	Apr	May	June	July	Aug	Sept	Oct	Nov	Dec	Year
0	Average	2.7	3.7	4.9	6.0	6.8	7.3	7.3	6.7	5.6	4.2	2.7	2.2	5.0
	Min/Max	2.3/3.3	2.8/4.8	4.0/5.9	4.8/7.3	5.8/8.0	5.7/8.4	6.0/8.1	5.3/7.6	4.8/6.4	3.6/5.0	2.0/3.5	1.5/2.6	4.8/5.2
Latitude -15	Average	3.5	4.5	5.5	6.4	6.9	7.3	7.4	7.0	6.1	5.0	3.4	2.8	5.5
	Min/Max	2.9/4.3	3.3/5.9	4.4/6.7	5.1/7.8	5.8/8.1	5.7/8.4	6.0/8.1	5.5/8.0	5.2/7.1	4.2/5.9	2.4/4.4	1.9/3.4	5.2/5.7
Latitude	Average	3.8	4.8	5.6	6.3	6.7	7.0	7.1	6.8	6.1	5.2	3.6	3.1	5.5
	Min/Max	3.2/4.7	3.4/6.3	4.5/6.9	5.0/7.8	5.6/7.9	5.4/8.1	5.7/7.8	5.4/7.8	5.2/7.2	4.3/6.2	2.5/4.8	2.0/3.7	5.2/5.7
Latitude +15	Average	4.0	4.9	5.5	6.1	6.3	6.5	6.6	6.5	6.0	5.2	3.7	3.3	5.4
	Min/Max	3.3/5.0	3.4/6.5	4.4/6.8	4.8/7.5	5.3/7.4	5.1/7.5	5.3/7.3	5.1/7.4	5.0/7.0	4.3/6.3	2.6/5.0	2.0/4.0	5.1/5.6

Solar Radiation for 2-Axis Tracking Flat-Plate Collectors (kWh/m²/day), Uncertainty ±11%

Tracker		Jan	Feb	Mar	Apr	May	June	July	Aug	Sept	Oct	Nov	Dec	Year
2-Axis	Average	4.0	4.9	5.6	6.4	7.0	7.5	7.5	7.0	6.2	5.3	3.8	3.3	5.7
	Min/Max	3.3/5.1	3.5/6.5	4.5/6.9	5.1/7.8	5.9/8.2	5.8/8.6	6.1/8.3	5.5/8.0	5.2/7.2	4.3/6.3	2.6/5.0	2.1/4.0	5.4/5.9

Direct Beam Solar Radiation for Concentrating Collectors (kWh/m²/day), Uncertainty ±11%

Tracker		Jan	Feb	Mar	Apr	May	June	July	Aug	Sept	Oct	Nov	Dec	Year
1-Axis, E-W Horiz Axis	Average	2.3	2.6	2.7	2.8	3.0	3.3	3.3	3.1	2.9	2.8	2.1	1.9	2.7
	Min/Max	1.8/2.9	1.5/3.9	1.9/3.7	1.8/3.9	2.1/3.9	2.2/4.3	2.2/4.0	2.1/4.0	2.1/3.8	2.0/3.6	1.2/3.1	1.0/2.5	2.5/2.9
1-Axis, N-S Horiz Axis	Average	1.6	2.2	3.0	3.6	3.9	4.3	4.3	4.0	3.4	2.7	1.6	1.3	3.0
	Min/Max	1.3/2.1	1.4/3.4	2.1/4.1	2.2/4.9	2.8/5.1	2.7/5.5	2.7/5.1	2.6/5.1	2.5/4.3	2.0/3.4	0.9/2.4	0.6/1.6	2.7/3.2
1-Axis, N-S Tilt=Latitude	Average	2.5	3.0	3.6	3.8	3.9	4.0	4.1	4.1	3.9	3.5	2.4	2.0	3.4
	Min/Max	2.0/3.2	1.9/4.7	2.5/4.8	2.4/5.3	2.8/5.0	2.6/5.2	2.6/4.9	2.7/5.2	2.8/5.0	2.6/4.5	1.3/3.5	1.0/2.7	3.1/3.6
2-Axis	Average	2.7	3.1	3.6	3.9	4.1	4.4	4.4	4.2	3.9	3.5	2.5	2.2	3.5
	Min/Max	2.1/3.4	1.9/4.8	2.5/4.8	2.4/5.4	3.0/5.3	2.8/5.6	2.8/5.3	2.8/5.4	2.9/5.0	2.6/4.5	1.4/3.7	1.1/2.9	3.2/3.8

Average Climatic Conditions

Element	Jan	Feb	Mar	Apr	May	June	July	Aug	Sept	Oct	Nov	Dec	Year
Temperature (°C)	-1.7	-0.9	3.7	9.0	14.7	19.8	23.2	22.8	18.8	13.1	7.6	1.3	10.9
Daily Minimum Temp	-5.6	-4.9	-0.6	4.3	10.0	15.1	18.7	18.4	14.2	8.4	3.4	-2.4	6.6
Daily Maximum Temp	2.2	3.1	7.9	13.7	19.3	24.4	27.6	27.2	23.4	17.8	11.7	5.0	15.3
Record Minimum Temp	-21.7	-20.6	-15.6	-7.8	-0.6	5.0	9.4	6.7	2.2	-3.3	-8.9	-20.0	-21.7
Record Maximum Temp	18.3	19.4	28.9	32.8	33.3	35.6	39.4	36.7	37.2	29.4	25.6	18.3	39.4
HDD, Base 18.3°C	622	538	454	280	122	10	0	0	30	168	323	529	3076
CDD, Base 18.3°C	0	0	0	0	8	53	150	139	45	6	0	0	402
Relative Humidity (%)	66	66	65	64	70	73	73	74	74	70	70	69	70
Wind Speed (m/s)	5.9	6.1	6.1	5.9	5.2	4.8	4.6	4.7	5.2	5.3	5.8	5.8	5.5

Hartford, CT

WBAN NO. 14740

LATITUDE: 41.93° N
LONGITUDE: 72.68° W
ELEVATION: 55 meters
MEAN PRESSURE: 1010 millibars

STATION TYPE: Secondary

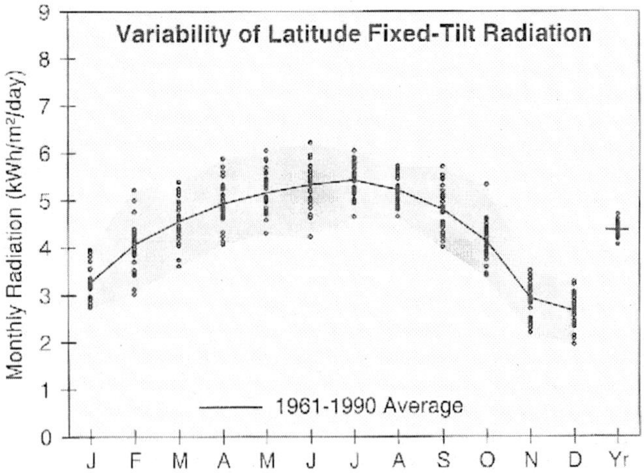

Solar Radiation for Flat-Plate Collectors Facing South at a Fixed Tilt (kWh/m²/day), Uncertainty ±11%

Tilt (°)		Jan	Feb	Mar	Apr	May	June	July	Aug	Sept	Oct	Nov	Dec	Year
0	Average	1.9	2.7	3.7	4.6	5.4	5.9	5.9	5.1	4.1	3.0	1.9	1.5	3.8
	Min/Max	1.7/2.2	2.2/3.2	3.1/4.1	4.0/5.4	4.6/6.3	4.7/6.8	5.1/6.5	4.7/5.6	3.6/4.7	2.6/3.6	1.5/2.1	1.3/1.7	3.6/4.0
Latitude -15	Average	2.9	3.7	4.4	5.1	5.5	5.8	5.8	5.4	4.8	3.9	2.7	2.3	4.4
	Min/Max	2.5/3.5	2.8/4.7	3.6/5.2	4.2/6.0	4.6/6.4	4.6/6.8	5.0/6.5	4.9/5.9	4.0/5.6	3.3/4.9	2.0/3.2	1.8/2.8	4.1/4.7
Latitude	Average	3.3	4.1	4.6	4.9	5.2	5.3	5.4	5.2	4.8	4.1	2.9	2.7	4.4
	Min/Max	2.8/4.0	3.0/5.2	3.6/5.4	4.1/5.9	4.3/6.1	4.2/6.2	4.7/6.1	4.7/5.7	4.0/5.7	3.4/5.3	2.2/3.5	2.0/3.3	4.1/4.7
Latitude +15	Average	3.5	4.2	4.5	4.6	4.6	4.6	4.8	4.7	4.6	4.2	3.0	2.8	4.2
	Min/Max	2.9/4.3	3.0/5.5	3.5/5.4	3.7/5.5	3.8/5.4	3.7/5.4	4.1/5.3	4.2/5.2	3.8/5.5	3.4/5.4	2.2/3.7	2.0/3.5	3.9/4.5
90	Average	3.3	3.8	3.5	3.0	2.7	2.6	2.7	3.0	3.3	3.4	2.7	2.7	3.0
	Min/Max	2.7/4.2	2.6/5.1	2.6/4.4	2.5/3.6	2.3/3.1	2.2/2.9	2.4/2.9	2.7/3.3	2.7/3.9	2.7/4.5	2.0/3.3	1.9/3.4	2.8/3.4

Solar Radiation for 1-Axis Tracking Flat-Plate Collectors with a North-South Axis (kWh/m²/day), Uncertainty ±11%

Axis Tilt (°)		Jan	Feb	Mar	Apr	May	June	July	Aug	Sept	Oct	Nov	Dec	Year
0	Average	2.7	3.8	4.9	6.0	6.8	7.3	7.3	6.6	5.4	4.0	2.5	2.1	4.9
	Min/Max	2.2/3.2	2.8/4.9	3.8/5.8	4.7/7.2	5.5/8.0	5.4/8.8	5.9/8.3	5.8/7.3	4.4/6.4	3.3/5.2	1.9/3.0	1.5/2.5	4.6/5.3
Latitude -15	Average	3.4	4.5	5.5	6.3	6.9	7.3	7.4	6.9	5.9	4.7	3.1	2.7	5.4
	Min/Max	2.8/4.2	3.3/6.0	4.2/6.6	5.0/7.7	5.5/8.3	5.4/8.9	5.9/8.4	6.0/7.7	4.7/7.1	3.8/6.3	2.3/3.8	2.0/3.4	5.0/5.8
Latitude	Average	3.7	4.8	5.6	6.2	6.7	7.0	7.1	6.7	6.0	4.9	3.3	3.0	5.4
	Min/Max	3.1/4.6	3.5/6.4	4.2/6.8	4.9/7.7	5.3/8.0	5.1/8.5	5.7/8.1	5.8/7.6	4.7/7.2	4.0/6.6	2.5/4.1	2.1/3.8	5.0/5.9
Latitude +15	Average	3.9	4.9	5.5	6.0	6.2	6.5	6.6	6.4	5.8	5.0	3.4	3.1	5.3
	Min/Max	3.2/4.9	3.5/6.6	4.1/6.7	4.7/7.4	5.0/7.5	4.7/7.9	5.3/7.6	5.5/7.2	4.6/7.0	3.9/6.6	2.5/4.2	2.2/4.0	4.9/5.8

Solar Radiation for 2-Axis Tracking Flat-Plate Collectors (kWh/m²/day), Uncertainty ±11%

Tracker		Jan	Feb	Mar	Apr	May	June	July	Aug	Sept	Oct	Nov	Dec	Year
2-Axis	Average	4.0	4.9	5.6	6.3	7.0	7.5	7.5	6.9	6.0	5.0	3.4	3.2	5.6
	Min/Max	3.2/4.9	3.5/6.6	4.2/6.8	5.0/7.8	5.6/8.3	5.5/9.1	6.0/8.6	6.0/7.7	4.8/7.2	4.0/6.7	2.5/4.2	2.2/4.0	5.2/6.1

Direct Beam Solar Radiation for Concentrating Collectors (kWh/m²/day), Uncertainty ±11%

Tracker		Jan	Feb	Mar	Apr	May	June	July	Aug	Sept	Oct	Nov	Dec	Year
1-Axis, E-W Horiz Axis	Average	2.1	2.5	2.5	2.7	2.9	3.2	3.2	2.9	2.7	2.6	1.8	1.7	2.6
	Min/Max	1.6/2.7	1.6/3.7	1.7/3.5	1.8/3.7	1.8/3.9	1.6/4.4	2.0/3.9	2.2/3.5	1.8/3.6	1.8/3.9	1.1/2.5	1.0/2.5	2.3/2.9
1-Axis, N-S Horiz Axis	Average	1.5	2.1	2.8	3.4	3.8	4.1	4.1	3.8	3.2	2.4	1.3	1.1	2.8
	Min/Max	1.1/1.8	1.4/3.2	1.9/3.7	2.2/4.7	2.4/5.0	2.1/5.7	2.6/5.1	2.8/4.7	2.1/4.2	1.7/3.8	0.8/1.9	0.6/1.6	2.5/3.1
1-Axis, N-S Tilt=Latitude	Average	2.3	2.9	3.3	3.7	3.7	3.9	4.0	3.9	3.6	3.2	2.0	1.8	3.2
	Min/Max	1.7/2.9	2.0/4.4	2.2/4.5	2.4/5.0	2.4/4.9	2.0/5.4	2.5/4.9	2.9/4.8	2.4/4.8	2.2/4.9	1.2/2.8	1.0/2.6	2.8/3.5
2-Axis	Average	2.5	3.0	3.3	3.7	3.9	4.2	4.3	4.0	3.6	3.2	2.1	2.0	3.3
	Min/Max	1.9/3.1	2.0/4.5	2.2/4.5	2.4/5.1	2.5/5.2	2.2/5.9	2.6/5.2	3.0/4.9	2.4/4.8	2.2/4.9	1.3/2.9	1.1/2.8	3.0/3.7

Average Climatic Conditions

Element	Jan	Feb	Mar	Apr	May	June	July	Aug	Sept	Oct	Nov	Dec	Year
Temperature (°C)	-4.1	-2.5	3.1	9.3	15.3	20.3	23.2	22.0	17.4	11.2	5.5	-1.4	9.9
Daily Minimum Temp	-9.0	-7.4	-2.2	3.1	8.7	13.8	16.8	15.8	11.0	4.8	0.4	-5.9	4.2
Daily Maximum Temp	0.7	2.4	8.2	15.5	22.0	26.7	29.4	28.2	23.8	17.6	10.6	3.1	15.7
Record Minimum Temp	-32.2	-29.4	-21.1	-12.8	-2.2	2.8	6.7	2.2	-1.1	-8.3	-17.2	-25.6	-32.2
Record Maximum Temp	18.3	22.8	30.6	35.6	36.1	37.8	38.9	38.3	37.2	32.8	27.2	23.3	38.9
HDD, Base 18.3°C	696	583	474	272	108	11	0	3	53	221	385	612	3417
CDD, Base 18.3°C	0	0	0	0	15	69	150	117	25	0	0	0	376
Relative Humidity (%)	64	63	60	58	63	67	68	71	73	69	68	68	66
Wind Speed (m/s)	3.8	4.0	4.3	4.4	3.8	3.5	3.3	3.1	3.1	3.3	3.7	3.7	3.7

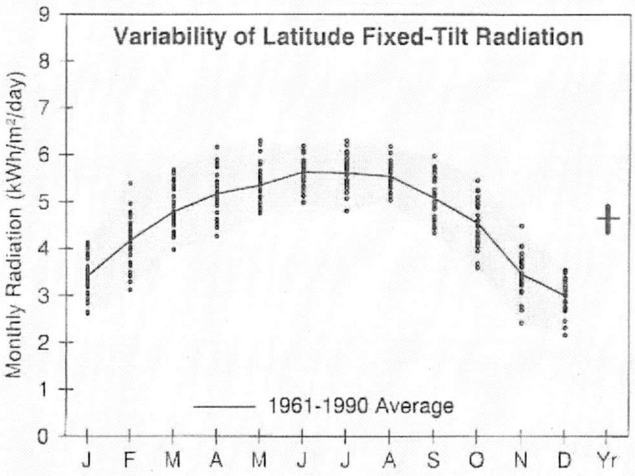

Wilmington, DE

WBAN NO. 13781

LATITUDE: 39.67° N
LONGITUDE: 75.60° W
ELEVATION: 24 meters
MEAN PRESSURE: 1015 millibars

STATION TYPE: Secondary

Solar Radiation for Flat-Plate Collectors Facing South at a Fixed Tilt (kWh/m²/day), Uncertainty ±11%

Tilt (°)		Jan	Feb	Mar	Apr	May	June	July	Aug	Sept	Oct	Nov	Dec	Year
0	Average	2.0	2.9	3.9	4.9	5.6	6.2	6.1	5.4	4.4	3.3	2.2	1.7	4.1
	Min/Max	1.7/2.3	2.4/3.5	3.4/4.4	4.2/5.7	5.1/6.6	5.5/6.9	5.3/6.8	5.0/6.0	3.9/4.9	2.8/3.8	1.8/2.6	1.5/2.0	3.8/4.2
Latitude -15	Average	3.0	3.8	4.6	5.3	5.7	6.1	6.0	5.8	5.0	4.3	3.1	2.6	4.6
	Min/Max	2.4/3.6	2.9/4.8	3.9/5.4	4.4/6.3	5.1/6.7	5.4/6.7	5.2/6.8	5.2/6.4	4.3/5.8	3.5/5.1	2.3/3.9	2.0/3.1	4.3/4.9
Latitude	Average	3.4	4.2	4.8	5.2	5.4	5.6	5.6	5.5	5.1	4.5	3.5	3.0	4.6
	Min/Max	2.6/4.1	3.1/5.4	4.0/5.7	4.3/6.2	4.8/6.3	5.0/6.2	4.8/6.3	5.0/6.2	4.3/6.0	3.6/5.4	2.4/4.5	2.2/3.5	4.3/4.9
Latitude +15	Average	3.6	4.3	4.7	4.8	4.7	4.9	4.9	5.0	4.9	4.6	3.6	3.2	4.4
	Min/Max	2.8/4.4	3.2/5.7	3.8/5.6	3.9/5.7	4.2/5.6	4.3/5.3	4.2/5.5	4.6/5.6	4.1/5.8	3.6/5.5	2.5/4.8	2.2/3.8	4.1/4.7
90	Average	3.4	3.7	3.5	3.0	2.6	2.5	2.6	3.0	3.4	3.7	3.2	3.0	3.1
	Min/Max	2.4/4.2	2.7/5.2	2.8/4.2	2.5/3.6	2.4/3.0	2.3/2.6	2.4/2.8	2.8/3.3	2.8/4.1	2.8/4.4	2.1/4.3	2.0/3.7	2.9/3.3

Solar Radiation for 1-Axis Tracking Flat-Plate Collectors with a North-South Axis (kWh/m²/day), Uncertainty ±11%

Axis Tilt (°)		Jan	Feb	Mar	Apr	May	June	July	Aug	Sept	Oct	Nov	Dec	Year
0	Average	2.9	4.0	5.3	6.4	7.1	7.8	7.7	7.0	5.8	4.6	3.0	2.5	5.3
	Min/Max	2.3/3.5	3.0/5.3	4.2/6.3	5.1/8.0	6.2/8.7	6.5/9.1	6.2/8.7	6.2/8.0	4.9/6.8	3.6/5.6	2.2/3.9	1.9/2.9	5.0/5.7
Latitude -15	Average	3.6	4.7	5.9	6.7	7.2	7.8	7.7	7.3	6.3	5.3	3.7	3.1	5.8
	Min/Max	2.7/4.5	3.5/6.3	4.7/7.1	5.3/8.5	6.2/8.8	6.5/9.1	6.2/8.8	6.5/8.4	5.3/7.6	4.1/6.6	2.5/4.9	2.2/3.7	5.4/6.2
Latitude	Average	4.0	5.0	6.0	6.7	7.0	7.5	7.4	7.2	6.3	5.5	4.0	3.4	5.8
	Min/Max	3.0/4.9	3.6/6.7	4.7/7.3	5.2/8.4	6.0/8.6	6.2/8.7	6.0/8.5	6.3/8.2	5.3/7.7	4.2/6.8	2.7/5.3	2.4/4.1	5.4/6.3
Latitude +15	Average	4.1	5.1	5.9	6.4	6.6	7.0	6.9	6.8	6.2	5.5	4.1	3.6	5.7
	Min/Max	3.1/5.2	3.6/6.9	4.6/7.2	4.9/8.1	5.6/8.1	5.8/8.1	5.6/7.9	6.0/7.8	5.1/7.5	4.2/6.9	2.7/5.6	2.4/4.3	5.2/6.1

Solar Radiation for 2-Axis Tracking Flat-Plate Collectors (kWh/m²/day), Uncertainty ±11%

Tracker		Jan	Feb	Mar	Apr	May	June	July	Aug	Sept	Oct	Nov	Dec	Year
2-Axis	Average	4.2	5.1	6.0	6.8	7.3	8.0	7.8	7.3	6.4	5.6	4.2	3.6	6.0
	Min/Max	3.1/5.2	3.7/6.9	4.7/7.3	5.3/8.5	6.3/8.9	6.7/9.3	6.3/8.9	6.5/8.4	5.3/7.7	4.2/6.9	2.8/5.6	2.5/4.4	5.6/6.5

Direct Beam Solar Radiation for Concentrating Collectors (kWh/m²/day), Uncertainty ±11%

Tracker		Jan	Feb	Mar	Apr	May	June	July	Aug	Sept	Oct	Nov	Dec	Year
1-Axis, E-W Horiz Axis	Average	2.4	2.7	2.9	3.0	3.2	3.6	3.5	3.3	3.0	2.9	2.3	2.1	2.9
	Min/Max	1.5/3.3	1.5/3.8	1.9/4.0	1.7/4.2	2.4/4.4	2.6/4.7	2.3/4.4	2.7/4.2	2.1/4.0	1.9/4.1	1.2/3.6	1.1/2.8	2.6/3.3
1-Axis, N-S Horiz Axis	Average	1.8	2.5	3.3	3.9	4.2	4.7	4.6	4.2	3.5	2.9	1.8	1.4	3.2
	Min/Max	1.1/2.4	1.4/3.5	2.1/4.5	2.3/5.6	3.2/5.8	3.3/6.2	3.0/5.7	3.4/5.3	2.5/4.7	1.9/4.1	1.0/2.8	0.8/1.9	2.8/3.7
1-Axis, N-S Tilt=Latitude	Average	2.6	3.3	3.9	4.1	4.1	4.4	4.4	4.3	4.0	3.7	2.6	2.2	3.6
	Min/Max	1.7/3.6	1.8/4.6	2.5/5.3	2.4/5.8	3.1/5.7	3.1/5.8	2.8/5.4	3.5/5.5	2.8/5.3	2.3/5.1	1.4/4.1	1.2/3.0	3.2/4.1
2-Axis	Average	2.8	3.4	3.9	4.2	4.3	4.8	4.7	4.5	4.0	3.7	2.8	2.4	3.8
	Min/Max	1.8/3.8	1.9/4.8	2.5/5.3	2.4/5.9	3.3/6.0	3.4/6.3	3.0/5.8	3.6/5.6	2.8/5.3	2.4/5.2	1.5/4.3	1.3/3.2	3.3/4.3

Average Climatic Conditions

Element	Jan	Feb	Mar	Apr	May	June	July	Aug	Sept	Oct	Nov	Dec	Year
Temperature (°C)	-0.8	0.8	5.9	11.2	16.9	21.9	24.7	23.9	20.0	13.4	7.9	2.1	12.3
Daily Minimum Temp	-5.3	-4.0	0.6	5.4	11.2	16.4	19.5	18.8	14.6	7.6	2.8	-2.4	7.1
Daily Maximum Temp	3.7	5.5	11.2	17.0	22.7	27.4	29.8	28.9	25.4	19.2	13.1	6.6	17.6
Record Minimum Temp	-25.6	-21.1	-16.7	-7.8	-1.1	5.0	8.9	6.1	2.2	-4.4	-10.0	-21.7	-25.6
Record Maximum Temp	23.9	25.6	30.0	34.4	35.0	37.2	38.9	38.3	37.8	32.8	29.4	23.3	38.9
HDD, Base 18.3°C	592	492	384	213	68	0	0	0	19	160	312	503	2743
CDD, Base 18.3°C	0	0	0	0	24	111	196	172	69	8	0	0	581
Relative Humidity (%)	68	65	62	62	67	69	70	72	72	71	69	69	68
Wind Speed (m/s)	4.5	4.6	5.1	4.9	4.2	3.8	3.5	3.4	3.6	3.7	4.2	4.3	4.2

Daytona Beach, FL

WBAN NO. 12834

LATITUDE: 29.18° N
LONGITUDE: 81.05° W
ELEVATION: 12 meters
MEAN PRESSURE: 1017 millibars

STATION TYPE: Primary

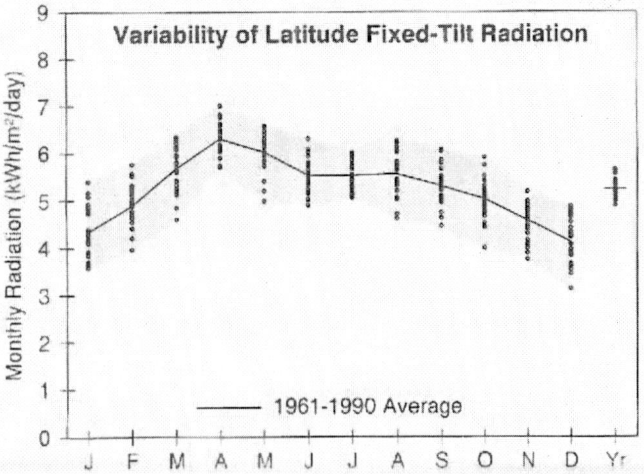

Solar Radiation for Flat-Plate Collectors Facing South at a Fixed Tilt (kWh/m²/day), Uncertainty ±9%

Tilt (°)		Jan	Feb	Mar	Apr	May	June	July	Aug	Sept	Oct	Nov	Dec	Year
0	Average	3.1	3.9	5.0	6.2	6.4	6.1	6.0	5.7	4.9	4.2	3.4	2.9	4.8
	Min/Max	2.7/3.7	3.2/4.4	4.2/5.5	5.6/6.8	5.3/7.0	5.4/7.0	5.5/6.6	4.8/6.3	4.3/5.5	3.5/4.8	2.9/3.7	2.4/3.3	4.6/5.1
Latitude -15	Average	3.8	4.5	5.5	6.4	6.4	6.0	5.9	5.8	5.2	4.7	4.1	3.6	5.2
	Min/Max	3.2/4.6	3.7/5.2	4.5/6.1	5.8/7.1	5.3/7.0	5.3/6.8	5.4/6.4	4.8/6.5	4.5/6.0	3.8/5.5	3.4/4.6	2.8/4.1	4.8/5.5
Latitude	Average	4.3	4.9	5.7	6.3	6.0	5.5	5.5	5.6	5.3	5.0	4.6	4.1	5.2
	Min/Max	3.6/5.4	4.0/5.8	4.6/6.3	5.7/7.0	5.0/6.6	4.9/6.3	5.1/6.0	4.6/6.3	4.5/6.1	4.0/5.9	3.8/5.2	3.1/4.9	4.9/5.7
Latitude +15	Average	4.6	5.1	5.6	5.9	5.4	4.8	4.9	5.1	5.1	5.1	4.8	4.4	5.1
	Min/Max	3.8/5.8	4.1/6.0	4.5/6.3	5.3/6.5	4.5/5.8	4.3/5.5	4.5/5.3	4.2/5.7	4.3/5.9	4.0/6.0	3.9/5.6	3.3/5.3	4.7/5.5
90	Average	3.9	3.8	3.6	2.9	2.1	1.8	1.9	2.4	3.0	3.6	4.0	3.9	3.1
	Min/Max	3.1/5.1	3.1/4.7	2.9/4.0	2.7/3.1	2.0/2.2	1.6/1.9	1.8/2.0	2.0/2.6	2.5/3.4	2.7/4.3	3.1/4.7	2.8/4.7	2.8/3.3

Solar Radiation for 1-Axis Tracking Flat-Plate Collectors with a North-South Axis (kWh/m²/day), Uncertainty ±9%

Axis Tilt (°)		Jan	Feb	Mar	Apr	May	June	July	Aug	Sept	Oct	Nov	Dec	Year
0	Average	4.3	5.2	6.6	8.1	8.2	7.5	7.4	7.1	6.2	5.5	4.6	3.9	6.2
	Min/Max	3.5/5.4	4.2/6.2	5.2/7.6	7.1/9.5	6.3/9.2	6.5/8.9	6.6/8.3	5.8/8.3	5.0/7.3	4.2/6.6	3.8/5.3	3.0/4.7	5.7/6.8
Latitude -15	Average	4.8	5.7	7.0	8.3	8.2	7.4	7.4	7.2	6.5	5.9	5.1	4.4	6.5
	Min/Max	3.8/6.1	4.6/6.8	5.5/8.0	7.3/9.7	6.3/9.2	6.4/8.8	6.5/8.2	5.9/8.4	5.2/7.6	4.5/7.1	4.1/5.9	3.3/5.4	6.0/7.1
Latitude	Average	5.2	6.0	7.2	8.3	7.9	7.1	7.1	7.0	6.5	6.1	5.5	4.9	6.6
	Min/Max	4.1/6.7	4.8/7.3	5.6/8.3	7.2/9.7	6.1/8.9	6.1/8.4	6.2/7.9	5.8/8.3	5.2/7.7	4.6/7.4	4.4/6.4	3.6/5.9	6.0/7.2
Latitude +15	Average	5.4	6.1	7.1	8.0	7.4	6.6	6.6	6.7	6.4	6.2	5.7	5.1	6.4
	Min/Max	4.3/7.0	4.9/7.4	5.5/8.2	6.9/9.3	5.7/8.4	5.7/7.8	5.8/7.4	5.5/7.9	5.0/7.5	4.6/7.5	4.5/6.7	3.7/6.3	5.9/7.1

Solar Radiation for 2-Axis Tracking Flat-Plate Collectors (kWh/m²/day), Uncertainty ±9%

Tracker		Jan	Feb	Mar	Apr	May	June	July	Aug	Sept	Oct	Nov	Dec	Year
2-Axis	Average	5.5	6.1	7.2	8.4	8.2	7.5	7.5	7.2	6.5	6.2	5.7	5.2	6.8
	Min/Max	4.3/7.1	4.9/7.5	5.6/8.3	7.3/9.8	6.3/9.3	6.5/9.0	6.6/8.4	5.9/8.4	5.2/7.7	4.6/7.5	4.5/6.8	3.7/6.4	6.2/7.5

Direct Beam Solar Radiation for Concentrating Collectors (kWh/m²/day), Uncertainty ±8%

Tracker		Jan	Feb	Mar	Apr	May	June	July	Aug	Sept	Oct	Nov	Dec	Year
1-Axis, E-W Horiz Axis	Average	3.1	3.3	3.6	4.2	3.9	3.3	3.3	3.1	2.8	3.1	3.2	3.0	3.3
	Min/Max	2.1/4.5	2.4/4.4	2.4/4.6	3.3/5.5	2.2/4.8	2.4/4.6	2.4/4.2	2.1/4.0	1.7/3.6	1.8/4.1	2.2/4.1	1.8/4.1	2.9/3.9
1-Axis, N-S Horiz Axis	Average	2.7	3.4	4.4	5.5	5.1	4.3	4.2	4.0	3.5	3.3	3.0	2.5	3.8
	Min/Max	1.9/4.0	2.5/4.5	2.9/5.6	4.3/7.3	2.9/6.3	3.1/5.8	3.2/5.3	2.8/5.3	2.1/4.6	2.0/4.5	2.0/3.8	1.5/3.5	3.3/4.5
1-Axis, N-S Tilt=Latitude	Average	3.5	4.1	4.8	5.6	4.9	4.0	4.0	4.0	3.7	3.9	3.7	3.3	4.1
	Min/Max	2.4/5.1	3.0/5.4	3.2/6.1	4.4/7.4	2.8/6.0	2.9/5.4	3.0/5.0	2.8/5.3	2.2/4.8	2.3/5.2	2.5/4.7	1.9/4.6	3.5/4.8
2-Axis	Average	3.7	4.2	4.8	5.7	5.2	4.3	4.3	4.1	3.8	3.9	3.9	3.6	4.3
	Min/Max	2.6/5.4	3.0/5.6	3.2/6.1	4.4/7.6	2.9/6.4	3.1/5.9	3.2/5.3	2.9/5.4	2.2/4.8	2.3/5.3	2.7/5.0	2.1/4.9	3.7/5.0

Average Climatic Conditions

Element	Jan	Feb	Mar	Apr	May	June	July	Aug	Sept	Oct	Nov	Dec	Year
Temperature (°C)	14.2	15.0	17.9	20.7	23.7	26.3	27.3	27.2	26.3	23.0	18.8	15.6	21.3
Daily Minimum Temp	8.3	9.1	12.2	14.8	18.3	21.6	22.5	22.7	22.2	18.4	13.5	9.8	16.1
Daily Maximum Temp	20.0	20.8	23.8	26.7	29.2	31.1	32.1	31.7	30.4	27.5	24.2	21.3	26.6
Record Minimum Temp	-9.4	-4.4	-3.3	1.7	6.7	11.1	15.6	18.3	11.1	5.0	-2.8	-7.2	-9.4
Record Maximum Temp	30.6	31.7	32.8	35.6	37.8	38.9	38.9	37.8	37.2	35.0	31.7	31.1	38.9
HDD, Base 18.3°C	157	114	62	12	0	0	0	0	0	0	46	115	505
CDD, Base 18.3°C	28	21	50	83	167	240	279	276	240	147	61	31	1622
Relative Humidity (%)	75	72	71	69	72	77	78	80	79	75	76	76	75
Wind Speed (m/s)	3.8	4.1	4.2	4.1	3.8	3.4	3.2	3.0	3.5	4.0	3.7	3.6	3.7

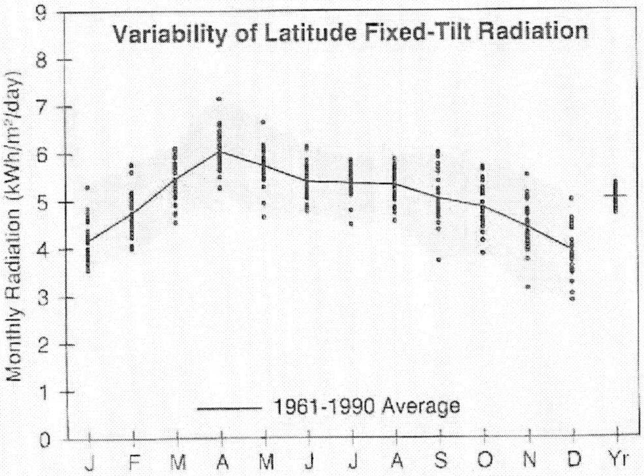

Jacksonville, FL

WBAN NO. 13889

LATITUDE: 30.50° N
LONGITUDE: 81.70° W
ELEVATION: 9 meters
MEAN PRESSURE: 1017 millibars

STATION TYPE: Secondary

Solar Radiation for Flat-Plate Collectors Facing South at a Fixed Tilt (kWh/m²/day), Uncertainty ±9%

Tilt (°)		Jan	Feb	Mar	Apr	May	June	July	Aug	Sept	Oct	Nov	Dec	Year
0	Average	2.9	3.7	4.7	5.9	6.1	6.0	5.8	5.4	4.6	4.0	3.2	2.7	4.6
	Min/Max	2.6/3.5	3.2/4.3	4.0/5.2	5.1/6.9	5.0/7.1	5.3/6.8	4.9/6.4	4.7/5.9	3.6/5.4	3.3/4.5	2.5/3.8	2.2/3.2	4.4/4.8
Latitude -15	Average	3.6	4.3	5.2	6.1	6.1	5.8	5.7	5.5	5.0	4.5	3.9	3.4	4.9
	Min/Max	3.2/4.5	3.7/5.2	4.4/5.8	5.3/7.2	4.9/7.1	5.1/6.7	4.8/6.3	4.7/6.1	3.7/5.9	3.7/5.3	2.9/4.8	2.6/4.2	4.7/5.2
Latitude	Average	4.2	4.7	5.5	6.0	5.7	5.4	5.4	5.3	5.0	4.9	4.4	3.9	5.0
	Min/Max	3.5/5.3	4.0/5.8	4.5/6.1	5.3/7.1	4.7/6.7	4.8/6.1	4.5/5.8	4.6/5.9	3.7/6.0	3.9/5.7	3.1/5.5	2.9/5.0	4.7/5.3
Latitude +15	Average	4.4	4.9	5.4	5.6	5.1	4.7	4.7	4.9	4.8	4.9	4.7	4.2	4.9
	Min/Max	3.7/5.8	4.1/6.0	4.4/6.1	4.9/6.7	4.2/5.9	4.2/5.3	4.0/5.1	4.2/5.4	3.5/5.8	3.8/5.8	3.2/5.9	3.0/5.5	4.5/5.2
90	Average	3.8	3.8	3.5	2.9	2.1	1.8	1.9	2.3	2.9	3.6	3.9	3.7	3.0
	Min/Max	3.1/5.0	3.1/4.7	2.9/4.0	2.6/3.3	2.0/2.3	1.8/1.9	1.8/2.0	2.1/2.5	2.1/3.5	2.7/4.2	2.6/5.0	2.5/4.9	2.7/3.2

Solar Radiation for 1-Axis Tracking Flat-Plate Collectors with a North-South Axis (kWh/m²/day), Uncertainty ±9%

Axis Tilt (°)		Jan	Feb	Mar	Apr	May	June	July	Aug	Sept	Oct	Nov	Dec	Year
0	Average	4.0	5.0	6.3	7.8	7.8	7.4	7.2	6.8	5.9	5.3	4.4	3.7	6.0
	Min/Max	3.3/5.2	4.1/6.2	5.2/7.3	6.5/9.6	5.8/9.5	6.2/8.9	5.9/8.2	5.5/7.7	4.2/7.4	4.1/6.3	3.0/5.5	2.7/4.7	5.5/6.4
Latitude -15	Average	4.6	5.5	6.7	8.0	7.8	7.3	7.2	6.9	6.2	5.7	4.9	4.2	6.3
	Min/Max	3.8/6.0	4.5/6.9	5.4/7.8	6.7/9.9	5.8/9.5	6.1/8.8	5.8/8.1	5.6/7.8	4.3/7.8	4.3/6.8	3.3/6.3	3.0/5.5	5.8/6.7
Latitude	Average	5.0	5.8	6.9	8.0	7.6	7.0	6.9	6.8	6.2	6.0	5.3	4.6	6.3
	Min/Max	4.1/6.6	4.7/7.3	5.5/8.0	6.6/9.8	5.6/9.2	5.9/8.5	5.6/7.8	5.5/7.7	4.3/7.9	4.5/7.2	3.5/6.8	3.2/6.1	5.8/6.8
Latitude +15	Average	5.2	5.9	6.8	7.6	7.1	6.5	6.5	6.4	6.1	6.0	5.5	4.9	6.2
	Min/Max	4.2/6.9	4.8/7.5	5.5/8.0	6.4/9.4	5.2/8.7	5.5/7.9	5.3/7.3	5.2/7.3	4.2/7.7	4.4/7.3	3.6/7.1	3.3/6.4	5.7/6.6

Solar Radiation for 2-Axis Tracking Flat-Plate Collectors (kWh/m²/day), Uncertainty ±9%

Tracker		Jan	Feb	Mar	Apr	May	June	July	Aug	Sept	Oct	Nov	Dec	Year
2-Axis	Average	5.2	5.9	6.9	8.0	7.9	7.5	7.3	6.9	6.3	6.0	5.5	4.9	6.5
	Min/Max	4.3/7.0	4.8/7.5	5.6/8.1	6.7/9.9	5.8/9.6	6.3/9.0	5.9/8.3	5.6/7.9	4.3/7.9	4.5/7.3	3.6/7.2	3.4/6.6	6.0/7.0

Direct Beam Solar Radiation for Concentrating Collectors (kWh/m²/day), Uncertainty ±8%

Tracker		Jan	Feb	Mar	Apr	May	June	July	Aug	Sept	Oct	Nov	Dec	Year
1-Axis, E-W Horiz Axis	Average	3.1	3.2	3.5	4.0	3.7	3.3	3.2	3.0	2.7	3.0	3.1	2.9	3.2
	Min/Max	2.2/4.5	2.3/4.5	2.5/4.4	3.0/5.5	2.0/5.0	2.3/4.5	2.1/4.1	1.9/3.7	1.4/3.8	1.8/4.0	1.5/4.5	1.5/4.4	2.8/3.6
1-Axis, N-S Horiz Axis	Average	2.6	3.3	4.2	5.3	4.9	4.3	4.2	3.9	3.4	3.3	2.8	2.4	3.7
	Min/Max	1.8/3.9	2.3/4.7	3.0/5.4	3.9/7.3	2.6/6.7	3.0/5.9	2.8/5.3	2.5/4.9	1.8/5.0	2.0/4.4	1.4/4.1	1.3/3.6	3.2/4.1
1-Axis, N-S Tilt=Latitude	Average	3.4	4.0	4.7	5.4	4.7	4.0	3.9	3.8	3.6	3.8	3.6	3.2	4.0
	Min/Max	2.4/5.1	2.8/5.6	3.3/6.0	4.0/7.5	2.5/6.4	2.7/5.4	2.7/5.0	2.5/4.9	1.9/5.3	2.3/5.1	1.8/5.2	1.7/4.8	3.4/4.4
2-Axis	Average	3.7	4.1	4.7	5.4	5.0	4.4	4.2	4.0	3.6	3.9	3.8	3.4	4.2
	Min/Max	2.6/5.5	2.9/5.8	3.3/6.0	4.0/7.6	2.7/6.8	3.0/5.9	2.9/5.3	2.5/5.0	1.9/5.3	2.3/5.2	1.9/5.4	1.8/5.2	3.6/4.6

Average Climatic Conditions

Element	Jan	Feb	Mar	Apr	May	June	July	Aug	Sept	Oct	Nov	Dec	Year
Temperature (°C)	11.3	12.9	16.2	19.4	23.0	26.2	27.6	27.3	25.6	21.0	16.6	12.8	20.0
Daily Minimum Temp	4.7	6.3	9.6	12.7	16.7	20.6	22.2	22.1	20.6	15.2	10.1	6.3	13.9
Daily Maximum Temp	17.9	19.4	22.8	26.2	29.3	31.8	33.0	32.6	30.7	26.8	23.1	19.3	26.1
Record Minimum Temp	-13.9	-7.2	-5.0	1.1	7.2	8.3	16.1	17.2	8.9	2.2	-6.1	-11.7	-13.9
Record Maximum Temp	29.4	31.1	32.8	35.0	37.8	39.4	40.6	38.9	37.8	35.6	31.1	28.9	40.6
HDD, Base 18.3°C	234	164	94	21	0	0	0	0	0	17	83	184	797
CDD, Base 18.3°C	17	12	27	54	144	235	286	279	218	99	31	14	1417
Relative Humidity (%)	75	72	71	69	73	77	78	80	81	79	78	77	76
Wind Speed (m/s)	3.6	3.9	3.9	3.7	3.4	3.2	3.0	2.8	3.1	3.4	3.3	3.3	3.4

Key West, FL

WBAN NO. 12836

LATITUDE: 24.55° N
LONGITUDE: 81.75° W
ELEVATION: 1 meters
MEAN PRESSURE: 1016 millibars

STATION TYPE: Secondary

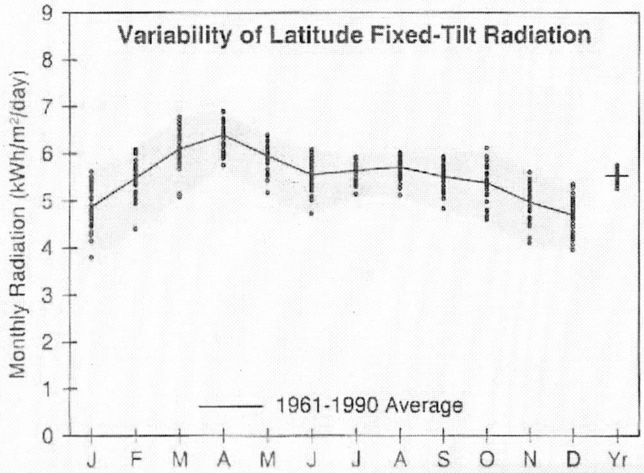

Solar Radiation for Flat-Plate Collectors Facing South at a Fixed Tilt (kWh/m²/day), Uncertainty ±9%

Tilt (°)		Jan	Feb	Mar	Apr	May	June	July	Aug	Sept	Oct	Nov	Dec	Year
0	Average	3.7	4.4	5.5	6.3	6.3	6.1	6.1	5.8	5.2	4.6	3.8	3.4	5.1
	Min/Max	3.0/4.1	3.7/4.9	4.6/6.0	5.7/6.9	5.5/6.8	5.2/6.7	5.6/6.4	5.2/6.2	4.6/5.5	4.0/5.1	3.3/4.3	3.0/3.8	4.9/5.3
Latitude -15	Average	4.2	4.9	5.8	6.5	6.3	6.0	6.0	5.9	5.4	5.0	4.4	4.0	5.4
	Min/Max	3.4/4.8	4.0/5.4	4.9/6.4	5.8/7.0	5.4/6.8	5.1/6.6	5.5/6.3	5.3/6.2	4.8/5.8	4.3/5.6	3.7/4.9	3.4/4.5	5.1/5.6
Latitude	Average	4.9	5.5	6.1	6.4	6.0	5.5	5.6	5.7	5.5	5.4	5.0	4.7	5.5
	Min/Max	3.8/5.6	4.4/6.1	5.1/6.8	5.8/6.9	5.2/6.4	4.7/6.1	5.1/5.9	5.1/6.0	4.8/5.9	4.6/6.1	4.1/5.6	4.0/5.3	5.2/5.8
Latitude +15	Average	5.3	5.7	6.0	6.0	5.3	4.8	5.0	5.2	5.3	5.5	5.3	5.1	5.4
	Min/Max	4.0/6.1	4.5/6.4	5.0/6.7	5.4/6.4	4.6/5.7	4.2/5.3	4.6/5.2	4.7/5.5	4.6/5.7	4.6/6.3	4.3/6.0	4.2/5.9	5.1/5.6
90	Average	4.3	4.2	3.5	2.5	1.7	1.5	1.6	2.1	2.8	3.7	4.2	4.3	3.0
	Min/Max	3.2/5.2	3.3/4.7	2.9/3.9	2.3/2.6	1.7/1.8	1.5/1.6	1.5/1.7	1.9/2.1	2.4/3.0	3.0/4.3	3.3/4.8	3.5/5.1	2.8/3.2

Solar Radiation for 1-Axis Tracking Flat-Plate Collectors with a North-South Axis (kWh/m²/day), Uncertainty ±9%

Axis Tilt (°)		Jan	Feb	Mar	Apr	May	June	July	Aug	Sept	Oct	Nov	Dec	Year
0	Average	5.1	6.1	7.4	8.4	8.1	7.5	7.5	7.3	6.6	6.0	5.2	4.8	6.7
	Min/Max	3.8/6.0	4.7/7.0	5.8/8.6	7.4/9.6	6.6/8.9	6.1/8.6	6.8/8.1	6.4/7.8	5.6/7.2	5.0/7.1	4.2/6.0	4.0/5.6	6.2/7.0
Latitude -15	Average	5.5	6.5	7.7	8.5	8.1	7.5	7.5	7.3	6.7	6.3	5.6	5.2	6.9
	Min/Max	4.1/6.5	4.9/7.4	6.0/8.9	7.5/9.7	6.6/8.8	6.0/8.5	6.7/8.1	6.4/7.9	5.7/7.4	5.2/7.4	4.5/6.4	4.3/6.1	6.4/7.2
Latitude	Average	6.0	6.9	7.9	8.5	7.8	7.1	7.2	7.2	6.8	6.6	6.1	5.7	7.0
	Min/Max	4.4/7.2	5.2/7.9	6.2/9.1	7.4/9.6	6.4/8.6	5.8/8.1	6.4/7.8	6.3/7.7	5.8/7.5	5.4/7.8	4.8/7.0	4.7/6.7	6.5/7.4
Latitude +15	Average	6.3	7.1	7.8	8.2	7.3	6.6	6.7	6.8	6.6	6.7	6.3	6.0	6.9
	Min/Max	4.6/7.6	5.3/8.1	6.1/9.1	7.1/9.3	6.0/8.0	5.4/7.5	6.0/7.2	6.0/7.3	5.6/7.3	5.4/7.9	5.0/7.3	4.9/7.1	6.4/7.3

Solar Radiation for 2-Axis Tracking Flat-Plate Collectors (kWh/m²/day), Uncertainty ±9%

Tracker		Jan	Feb	Mar	Apr	May	June	July	Aug	Sept	Oct	Nov	Dec	Year
2-Axis	Average	6.4	7.1	7.9	8.6	8.1	7.6	7.6	7.3	6.8	6.7	6.4	6.2	7.2
	Min/Max	4.6/7.7	5.3/8.1	6.2/9.1	7.5/9.7	6.6/8.9	6.1/8.6	6.8/8.2	6.5/7.9	5.8/7.5	5.5/7.9	5.0/7.3	5.0/7.3	6.7/7.6

Direct Beam Solar Radiation for Concentrating Collectors (kWh/m²/day), Uncertainty ±8%

Tracker		Jan	Feb	Mar	Apr	May	June	July	Aug	Sept	Oct	Nov	Dec	Year
1-Axis, E-W	Average	3.8	4.1	4.3	4.5	4.1	3.6	3.5	3.4	3.1	3.5	3.6	3.7	3.8
Horiz Axis	Min/Max	2.3/4.9	2.7/5.1	2.9/5.3	3.6/5.5	2.8/4.7	2.5/4.5	2.9/4.0	2.8/3.9	2.4/3.7	2.4/4.5	2.5/4.5	2.6/4.7	3.3/4.1
1-Axis, N-S	Average	3.6	4.4	5.3	6.0	5.3	4.5	4.5	4.3	3.9	3.9	3.5	3.3	4.4
Horiz Axis	Min/Max	2.1/4.6	2.8/5.6	3.5/6.7	4.7/7.5	3.6/6.2	3.1/5.7	3.6/5.1	3.6/5.0	3.0/4.7	2.8/5.1	2.4/4.4	2.3/4.3	3.8/4.8
1-Axis, N-S	Average	4.3	5.0	5.7	6.0	5.0	4.2	4.2	4.3	4.1	4.4	4.2	4.1	4.6
Tilt=Latitude	Min/Max	2.6/5.6	3.2/6.3	3.8/7.2	4.7/7.5	3.5/5.9	2.9/5.3	3.4/4.8	3.5/4.9	3.1/4.9	3.1/5.7	2.9/5.2	2.9/5.2	4.0/5.1
2-Axis	Average	4.6	5.2	5.7	6.1	5.3	4.6	4.5	4.4	4.1	4.4	4.4	4.4	4.8
	Min/Max	2.8/6.0	3.3/6.5	3.8/7.2	4.8/7.6	3.7/6.3	3.1/5.8	3.7/5.2	3.6/5.0	3.2/4.9	3.1/5.8	3.0/5.5	3.1/5.7	4.2/5.3

Average Climatic Conditions

Element	Jan	Feb	Mar	Apr	May	June	July	Aug	Sept	Oct	Nov	Dec	Year
Temperature (°C)	21.1	21.4	23.2	25.0	27.0	28.4	29.1	29.1	28.5	26.7	24.2	21.9	25.4
Daily Minimum Temp	18.3	18.7	20.6	22.3	24.5	25.8	26.4	26.3	25.8	24.2	21.8	19.3	22.8
Daily Maximum Temp	23.8	24.1	25.9	27.6	29.5	30.9	31.7	31.8	31.1	29.1	26.7	24.5	28.1
Record Minimum Temp	5.0	7.8	8.3	8.9	18.3	20.0	20.6	20.0	20.6	15.6	9.4	6.7	5.0
Record Maximum Temp	30.0	29.4	30.6	32.2	32.8	34.4	35.0	35.0	34.4	33.9	31.7	30.0	35.0
HDD, Base 18.3°C	24	17	4	0	0	0	0	0	0	0	0	11	56
CDD, Base 18.3°C	109	102	155	200	269	302	334	332	305	258	177	123	2666
Relative Humidity (%)	76	74	73	70	72	74	72	73	75	75	76	76	74
Wind Speed (m/s)	5.2	5.4	5.5	5.3	4.8	4.5	4.3	4.1	4.3	5.2	5.5	5.3	5.0

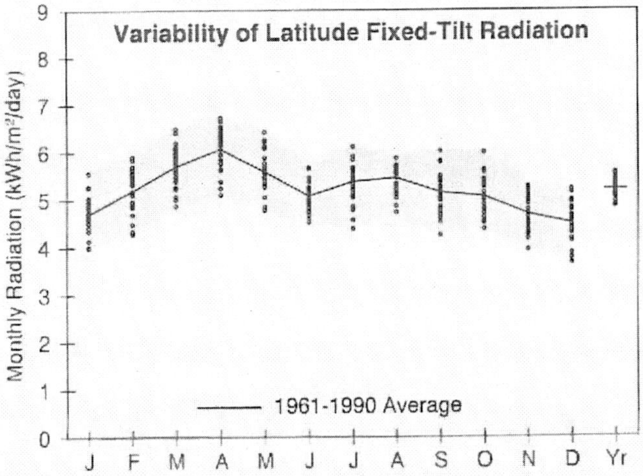

Miami, FL

WBAN NO. 12839

LATITUDE: 25.80° N
LONGITUDE: 80.27° W
ELEVATION: 2 meters
MEAN PRESSURE: 1017 millibars

STATION TYPE: Primary

Solar Radiation for Flat-Plate Collectors Facing South at a Fixed Tilt (kWh/m²/day), Uncertainty ±9%

Tilt (°)		Jan	Feb	Mar	Apr	May	June	July	Aug	Sept	Oct	Nov	Dec	Year
0	Average	3.5	4.2	5.2	6.0	6.0	5.6	5.8	5.6	4.9	4.4	3.7	3.3	4.8
	Min/Max	3.1/4.0	3.6/4.7	4.5/5.8	5.1/6.7	5.1/6.9	4.9/6.3	4.7/6.7	4.9/6.0	4.1/5.7	3.9/4.9	3.2/4.0	2.8/3.7	4.5/5.1
Latitude -15	Average	4.1	4.7	5.5	6.2	5.9	5.5	5.7	5.6	5.1	4.7	4.2	3.9	5.1
	Min/Max	3.5/4.8	4.0/5.3	4.7/6.2	5.2/6.8	5.0/6.8	4.8/6.1	4.7/6.5	4.9/6.1	4.2/5.9	4.2/5.5	3.6/4.6	3.2/4.4	4.8/5.4
Latitude	Average	4.7	5.2	5.7	6.1	5.6	5.1	5.4	5.5	5.1	5.1	4.7	4.5	5.2
	Min/Max	4.0/5.6	4.3/5.9	4.9/6.5	5.1/6.7	4.8/6.4	4.5/5.7	4.4/6.1	4.7/5.9	4.2/6.0	4.4/6.0	3.9/5.3	3.7/5.2	4.8/5.5
Latitude +15	Average	5.0	5.4	5.6	5.7	5.0	4.5	4.8	5.0	4.9	5.1	4.9	4.9	5.1
	Min/Max	4.2/6.0	4.4/6.2	4.8/6.4	4.8/6.3	4.3/5.7	4.0/5.0	3.9/5.4	4.3/5.4	4.1/5.8	4.4/6.1	4.1/5.6	3.9/5.7	4.7/5.4
90	Average	4.1	3.9	3.4	2.6	1.9	1.6	1.7	2.1	2.7	3.5	3.9	4.1	3.0
	Min/Max	3.4/5.1	3.1/4.6	2.9/3.9	2.3/2.8	1.6/2.0	1.5/1.8	1.6/2.0	2.0/2.2	2.3/3.1	3.0/4.2	3.2/4.5	3.2/4.9	2.7/3.1

Solar Radiation for 1-Axis Tracking Flat-Plate Collectors with a North-South Axis (kWh/m²/day), Uncertainty ±9%

Axis Tilt (°)		Jan	Feb	Mar	Apr	May	June	July	Aug	Sept	Oct	Nov	Dec	Year
0	Average	4.8	5.7	6.8	7.8	7.4	6.7	7.0	6.8	6.0	5.6	4.8	4.5	6.2
	Min/Max	4.0/5.8	4.5/6.7	5.6/7.8	6.5/9.1	6.1/8.9	5.6/7.7	5.3/8.2	5.8/7.5	4.7/7.1	4.7/6.8	4.0/5.5	3.6/5.3	5.6/6.7
Latitude -15	Average	5.2	6.1	7.0	7.9	7.4	6.6	7.0	6.9	6.2	5.9	5.2	4.9	6.4
	Min/Max	4.3/6.3	4.8/7.1	5.8/8.2	6.6/9.2	6.1/8.9	5.5/7.6	5.3/8.2	5.8/7.6	4.8/7.3	4.9/7.2	4.3/6.0	3.9/5.8	5.8/6.9
Latitude	Average	5.7	6.4	7.2	7.8	7.2	6.3	6.7	6.7	6.2	6.1	5.6	5.4	6.5
	Min/Max	4.6/6.9	5.0/7.6	5.9/8.4	6.5/9.2	5.9/8.6	5.3/7.3	5.1/7.9	5.7/7.4	4.8/7.4	5.1/7.6	4.6/6.5	4.2/6.4	5.8/7.0
Latitude +15	Average	5.9	6.5	7.1	7.5	6.7	5.9	6.3	6.4	6.1	6.2	5.8	5.7	6.3
	Min/Max	4.8/7.3	5.0/7.8	5.9/8.3	6.3/8.8	5.5/8.1	4.9/6.8	4.8/7.3	5.4/7.1	4.7/7.3	5.1/7.7	4.7/6.7	4.4/6.8	5.7/6.9

Solar Radiation for 2-Axis Tracking Flat-Plate Collectors (kWh/m²/day), Uncertainty ±9%

Tracker		Jan	Feb	Mar	Apr	May	June	July	Aug	Sept	Oct	Nov	Dec	Year
2-Axis	Average	6.0	6.6	7.2	7.9	7.4	6.7	7.1	6.9	6.2	6.2	5.9	5.8	6.7
	Min/Max	4.8/7.4	5.1/7.8	6.0/8.4	6.6/9.3	6.1/8.9	5.6/7.8	5.4/8.3	5.8/7.6	4.8/7.4	5.1/7.7	4.7/6.8	4.5/6.9	6.0/7.2

Direct Beam Solar Radiation for Concentrating Collectors (kWh/m²/day), Uncertainty ±8%

Tracker		Jan	Feb	Mar	Apr	May	June	July	Aug	Sept	Oct	Nov	Dec	Year
1-Axis, E-W Horiz Axis	Average	3.3	3.4	3.4	3.7	3.2	2.7	2.9	2.8	2.5	2.8	3.1	3.2	3.1
	Min/Max	2.3/4.6	2.1/4.3	2.5/4.4	2.6/4.7	2.2/4.5	1.8/3.4	1.2/3.8	1.8/3.3	1.5/3.5	1.9/4.0	2.1/3.8	2.0/4.4	2.4/3.6
1-Axis, N-S Horiz Axis	Average	3.1	3.7	4.3	4.9	4.2	3.4	3.7	3.6	3.2	3.2	2.9	2.8	3.6
	Min/Max	2.0/4.3	2.2/4.8	3.1/5.6	3.5/6.5	2.8/5.9	2.2/4.3	1.6/4.9	2.4/4.3	1.9/4.4	2.1/4.6	2.0/3.7	1.7/3.8	2.8/4.2
1-Axis, N-S Tilt=Latitude	Average	3.8	4.3	4.6	4.9	4.0	3.2	3.5	3.5	3.3	3.6	3.5	3.6	3.8
	Min/Max	2.5/5.3	2.6/5.5	3.3/6.0	3.5/6.5	2.7/5.7	2.1/4.0	1.5/4.6	2.3/4.3	1.9/4.6	2.3/5.2	2.4/4.4	2.1/4.8	3.0/4.4
2-Axis	Average	4.1	4.4	4.6	5.0	4.3	3.4	3.7	3.6	3.3	3.7	3.7	3.9	4.0
	Min/Max	2.7/5.6	2.6/5.7	3.3/6.0	3.5/6.6	2.8/6.0	2.2/4.3	1.6/4.9	2.4/4.4	2.0/4.6	2.4/5.2	2.5/4.7	2.3/5.2	3.1/4.6

Average Climatic Conditions

Element	Jan	Feb	Mar	Apr	May	June	July	Aug	Sept	Oct	Nov	Dec	Year
Temperature (°C)	19.6	20.3	22.1	24.0	25.9	27.4	28.1	28.2	27.7	25.7	23.1	20.6	24.4
Daily Minimum Temp	15.1	15.8	17.9	19.9	22.3	23.9	24.6	24.8	24.4	22.3	19.3	16.4	20.6
Daily Maximum Temp	24.0	24.7	26.2	28.0	29.6	30.9	31.7	31.7	31.0	29.2	26.9	24.8	28.2
Record Minimum Temp	-1.1	0.0	0.0	7.8	11.7	15.6	20.6	20.0	20.0	10.6	3.9	-1.1	-1.1
Record Maximum Temp	31.1	31.7	33.3	35.6	35.0	36.7	36.7	36.7	36.1	35.0	31.7	30.6	36.7
HDD, Base 18.3°C	49	28	8	0	0	0	0	0	0	0	3	23	111
CDD, Base 18.3°C	87	83	123	170	236	273	303	307	282	229	147	93	2332
Relative Humidity (%)	73	71	69	67	72	76	75	76	78	75	74	72	73
Wind Speed (m/s)	4.3	4.7	4.9	4.8	4.4	3.8	3.7	3.7	3.8	4.4	4.5	4.3	4.2

Tallahassee, FL

WBAN NO. 93805

LATITUDE: 30.38° N
LONGITUDE: 84.37° W
ELEVATION: 21 meters
MEAN PRESSURE: 1016 millibars

STATION TYPE: Primary

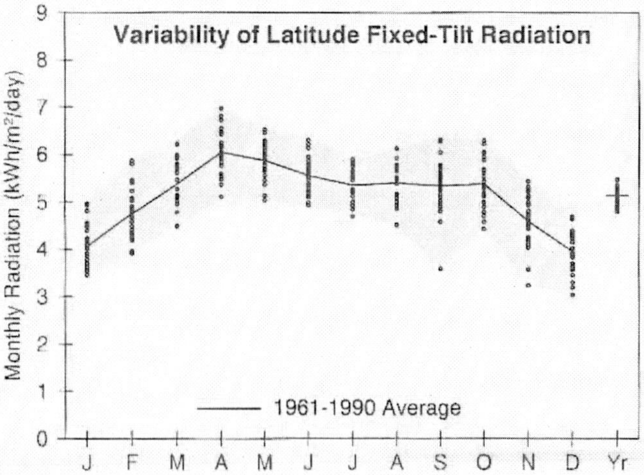

Solar Radiation for Flat-Plate Collectors Facing South at a Fixed Tilt (kWh/m²/day), Uncertainty ±9%

Tilt (°)		Jan	Feb	Mar	Apr	May	June	July	Aug	Sept	Oct	Nov	Dec	Year
0	Average	2.9	3.7	4.7	5.9	6.3	6.1	5.8	5.5	4.9	4.3	3.3	2.7	4.7
	Min/Max	2.5/3.4	3.1/4.4	4.0/5.3	5.1/6.7	5.5/6.9	5.4/7.0	5.2/6.4	4.6/6.2	3.5/5.7	3.7/5.0	2.6/3.8	2.3/3.1	4.4/5.0
Latitude -15	Average	3.6	4.3	5.2	6.1	6.2	6.0	5.7	5.6	5.3	5.0	4.1	3.4	5.0
	Min/Max	3.1/4.3	3.6/5.3	4.3/5.9	5.2/7.1	5.3/6.9	5.3/6.8	5.0/6.3	4.7/6.3	3.6/6.2	4.2/5.8	3.0/4.7	2.7/4.0	4.7/5.4
Latitude	Average	4.0	4.7	5.4	6.0	5.9	5.6	5.4	5.4	5.3	5.4	4.6	4.0	5.1
	Min/Max	3.5/5.0	3.9/5.9	4.5/6.2	5.1/7.0	5.0/6.5	4.9/6.3	4.7/5.9	4.5/6.1	3.6/6.3	4.4/6.3	3.2/5.4	3.0/4.7	4.8/5.5
Latitude +15	Average	4.3	4.9	5.3	5.6	5.2	4.9	4.7	4.9	5.1	5.5	4.8	4.2	5.0
	Min/Max	3.6/5.3	4.0/6.1	4.3/6.1	4.7/6.5	4.5/5.8	4.3/5.5	4.2/5.2	4.1/5.6	3.4/6.1	4.4/6.4	3.3/5.8	3.2/5.1	4.6/5.3
90	Average	3.6	3.7	3.4	2.9	2.2	1.9	2.0	2.4	3.1	4.0	4.0	3.7	3.1
	Min/Max	3.0/4.7	3.1/4.8	2.8/4.0	2.5/3.2	1.9/2.4	1.7/2.0	1.8/2.1	2.1/2.5	2.0/3.6	3.2/4.7	2.6/4.9	2.7/4.5	2.9/3.3

Solar Radiation for 1-Axis Tracking Flat-Plate Collectors with a North-South Axis (kWh/m²/day), Uncertainty ±9%

Axis Tilt (°)		Jan	Feb	Mar	Apr	May	June	July	Aug	Sept	Oct	Nov	Dec	Year
0	Average	3.9	5.0	6.2	7.7	7.9	7.5	7.1	6.8	6.2	5.9	4.5	3.7	6.0
	Min/Max	3.3/4.9	4.0/6.3	5.0/7.2	6.3/9.2	6.4/9.0	6.5/8.9	6.0/8.1	5.5/8.1	4.1/7.7	4.7/7.0	3.1/5.4	2.9/4.4	5.6/6.6
Latitude -15	Average	4.4	5.5	6.6	7.9	7.9	7.4	7.1	6.9	6.5	6.4	5.1	4.2	6.3
	Min/Max	3.7/5.6	4.4/6.9	5.3/7.7	6.4/9.5	6.4/9.0	6.5/8.8	6.0/8.1	5.6/8.2	4.2/8.0	5.1/7.7	3.4/6.2	3.2/5.1	5.8/6.9
Latitude	Average	4.8	5.8	6.7	7.8	7.7	7.1	6.8	6.8	6.6	6.7	5.5	4.6	6.4
	Min/Max	4.0/6.1	4.7/7.3	5.3/7.9	6.3/9.4	6.2/8.7	6.2/8.4	5.7/7.8	5.4/8.0	4.2/8.1	5.3/8.0	3.6/6.7	3.4/5.6	5.9/7.0
Latitude +15	Average	5.0	5.9	6.6	7.5	7.2	6.6	6.3	6.4	6.4	6.7	5.7	4.9	6.3
	Min/Max	4.1/6.4	4.7/7.5	5.2/7.8	6.1/9.1	5.8/8.2	5.8/7.8	5.3/7.3	5.2/7.7	4.1/7.9	5.3/8.1	3.7/7.0	3.5/5.9	5.8/6.8

Solar Radiation for 2-Axis Tracking Flat-Plate Collectors (kWh/m²/day), Uncertainty ±9%

Tracker		Jan	Feb	Mar	Apr	May	June	July	Aug	Sept	Oct	Nov	Dec	Year
2-Axis	Average	5.1	5.9	6.7	7.9	8.0	7.6	7.2	6.9	6.6	6.7	5.7	5.0	6.6
	Min/Max	4.1/6.5	4.8/7.5	5.3/7.9	6.4/9.5	6.5/9.1	6.6/9.0	6.1/8.2	5.6/8.2	4.2/8.1	5.3/8.1	3.7/7.0	3.6/6.1	6.1/7.2

Direct Beam Solar Radiation for Concentrating Collectors (kWh/m²/day), Uncertainty ±8%

Tracker		Jan	Feb	Mar	Apr	May	June	July	Aug	Sept	Oct	Nov	Dec	Year
1-Axis, E-W Horiz Axis	Average	2.9	3.2	3.3	3.8	3.7	3.3	2.9	2.8	2.9	3.5	3.3	2.9	3.2
	Min/Max	2.0/4.1	2.3/4.6	2.0/4.2	2.8/5.2	2.4/4.5	2.4/4.4	2.0/3.8	1.9/4.0	1.5/4.1	2.3/4.8	1.6/4.4	1.7/3.8	2.8/3.7
1-Axis, N-S Horiz Axis	Average	2.5	3.2	4.0	5.0	4.8	4.2	3.8	3.7	3.6	3.9	3.0	2.4	3.7
	Min/Max	1.7/3.5	2.3/4.8	2.5/5.0	3.7/6.9	3.1/5.8	3.2/5.7	2.6/5.0	2.5/5.2	1.9/5.2	2.5/5.3	1.5/4.0	1.4/3.1	3.3/4.3
1-Axis, N-S Tilt=Latitude	Average	3.2	3.9	4.4	5.1	4.6	3.9	3.6	3.7	3.8	4.5	3.8	3.1	4.0
	Min/Max	2.2/4.6	2.8/5.8	2.8/5.6	3.7/7.0	2.9/5.6	2.9/5.3	2.4/4.7	2.5/5.1	2.0/5.5	3.0/6.1	1.9/5.1	1.9/4.2	3.5/4.7
2-Axis	Average	3.5	4.0	4.4	5.2	4.9	4.3	3.8	3.8	3.8	4.6	4.0	3.4	4.1
	Min/Max	2.3/4.9	2.9/5.9	2.8/5.6	3.8/7.1	3.1/5.9	3.2/5.7	2.6/5.0	2.5/5.3	2.0/5.5	3.0/6.2	1.9/5.4	2.0/4.5	3.7/4.9

Average Climatic Conditions

Element	Jan	Feb	Mar	Apr	May	June	July	Aug	Sept	Oct	Nov	Dec	Year
Temperature (°C)	10.3	11.8	15.7	19.1	23.1	26.4	27.4	27.4	25.7	20.4	15.4	11.8	19.6
Daily Minimum Temp	3.4	4.5	8.2	11.2	16.0	20.3	21.7	21.9	20.0	13.3	7.9	4.6	12.8
Daily Maximum Temp	17.2	19.1	23.1	26.9	30.2	32.7	32.9	32.8	31.4	27.5	22.8	18.8	26.3
Record Minimum Temp	-14.4	-10.0	-6.7	-1.7	1.1	7.8	13.9	15.6	4.4	-1.1	-10.6	-12.2	-14.4
Record Maximum Temp	27.8	29.4	32.2	35.0	36.7	39.4	39.4	38.9	37.2	34.4	31.1	28.9	39.4
HDD, Base 18.3°C	262	191	106	33	0	0	0	0	0	27	113	216	947
CDD, Base 18.3°C	12	7	23	54	148	243	281	281	222	91	24	13	1399
Relative Humidity (%)	75	72	72	70	72	76	80	81	78	74	76	76	75
Wind Speed (m/s)	3.0	3.3	3.4	3.2	2.9	2.5	2.3	2.3	2.7	2.9	2.8	2.9	2.8

Tampa, FL

WBAN NO. 12842

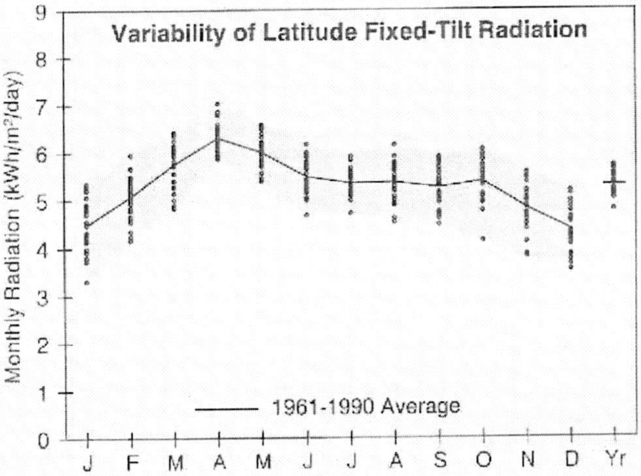

LATITUDE: 27.97° N
LONGITUDE: 82.53° W
ELEVATION: 3 meters
MEAN PRESSURE: 1018 millibars

STATION TYPE: Secondary

Solar Radiation for Flat-Plate Collectors Facing South at a Fixed Tilt (kWh/m²/day), Uncertainty ±9%

Tilt (°)		Jan	Feb	Mar	Apr	May	June	July	Aug	Sept	Oct	Nov	Dec	Year
0	Average	3.2	4.0	5.1	6.2	6.4	6.1	5.8	5.5	4.9	4.4	3.6	3.1	4.9
	Min/Max	2.6/3.7	3.4/4.5	4.3/5.6	5.8/6.8	5.7/7.1	5.1/6.9	5.1/6.4	4.6/6.3	4.3/5.4	3.6/4.9	3.0/4.0	2.7/3.5	4.5/5.2
Latitude -15	Average	3.9	4.6	5.5	6.4	6.4	5.9	5.7	5.5	5.2	5.0	4.2	3.8	5.2
	Min/Max	3.0/4.6	3.8/5.3	4.7/6.1	5.9/7.1	5.7/7.0	5.0/6.7	5.0/6.3	4.7/6.3	4.5/5.8	4.0/5.6	3.4/4.8	3.1/4.4	4.7/5.5
Latitude	Average	4.5	5.1	5.8	6.3	6.0	5.5	5.3	5.4	5.2	5.4	4.8	4.4	5.3
	Min/Max	3.3/5.3	4.2/5.9	4.8/6.4	5.9/7.0	5.4/6.6	4.7/6.2	4.7/5.9	4.5/6.2	4.5/5.9	4.1/6.1	3.8/5.6	3.5/5.2	4.8/5.7
Latitude +15	Average	4.8	5.3	5.7	5.9	5.3	4.8	4.7	4.9	5.0	5.5	5.1	4.7	5.1
	Min/Max	3.5/5.8	4.3/6.2	4.7/6.4	5.4/6.5	4.8/5.8	4.1/5.3	4.2/5.2	4.2/5.6	4.3/5.7	4.1/6.2	4.0/6.0	3.7/5.7	4.6/5.5
90	Average	4.0	4.0	3.5	2.8	2.0	1.7	1.8	2.2	2.9	3.9	4.2	4.1	3.1
	Min/Max	2.8/5.0	3.2/4.8	3.0/4.0	2.6/3.0	1.9/2.1	1.6/1.8	1.7/1.9	2.0/2.4	2.4/3.2	2.8/4.4	3.1/5.0	3.1/5.0	2.7/3.3

Solar Radiation for 1-Axis Tracking Flat-Plate Collectors with a North-South Axis (kWh/m²/day), Uncertainty ±9%

Axis Tilt (°)		Jan	Feb	Mar	Apr	May	June	July	Aug	Sept	Oct	Nov	Dec	Year
0	Average	4.5	5.5	6.8	8.3	8.3	7.5	7.1	6.8	6.2	6.0	4.9	4.2	6.3
	Min/Max	3.3/5.4	4.3/6.6	5.5/7.9	7.5/9.4	7.1/9.4	6.0/8.8	6.0/8.1	5.6/8.1	5.2/7.2	4.4/6.8	3.9/5.8	3.3/5.2	5.6/6.9
Latitude -15	Average	5.0	5.9	7.2	8.4	8.2	7.4	7.1	6.9	6.5	6.4	5.4	4.8	6.6
	Min/Max	3.5/6.0	4.6/7.1	5.8/8.3	7.7/9.6	7.1/9.3	5.9/8.7	6.0/8.0	5.7/8.2	5.3/7.5	4.7/7.3	4.2/6.4	3.7/5.8	5.8/7.2
Latitude	Average	5.4	6.3	7.3	8.4	8.0	7.1	6.8	6.8	6.5	6.7	5.9	5.2	6.7
	Min/Max	3.8/6.6	4.9/7.6	5.9/8.5	7.6/9.6	6.9/9.1	5.7/8.4	5.7/7.7	5.6/8.0	5.3/7.6	4.8/7.7	4.5/6.9	4.0/6.4	5.9/7.3
Latitude +15	Average	5.6	6.4	7.3	8.1	7.5	6.6	6.3	6.4	6.4	6.8	6.1	5.5	6.6
	Min/Max	3.9/7.0	4.9/7.8	5.8/8.5	7.3/9.2	6.4/8.5	5.3/7.8	5.3/7.2	5.3/7.6	5.2/7.4	4.8/7.8	4.6/7.3	4.2/6.8	5.8/7.2

Solar Radiation for 2-Axis Tracking Flat-Plate Collectors (kWh/m²/day), Uncertainty ±9%

Tracker		Jan	Feb	Mar	Apr	May	June	July	Aug	Sept	Oct	Nov	Dec	Year
2-Axis	Average	5.7	6.4	7.4	8.5	8.3	7.6	7.2	6.9	6.5	6.8	6.1	5.6	6.9
	Min/Max	3.9/7.1	5.0/7.8	5.9/8.5	7.7/9.7	7.2/9.4	6.0/8.9	6.0/8.1	5.7/8.2	5.4/7.6	4.9/7.8	4.6/7.3	4.3/7.0	6.1/7.6

Direct Beam Solar Radiation for Concentrating Collectors (kWh/m²/day), Uncertainty ±8%

Tracker		Jan	Feb	Mar	Apr	May	June	July	Aug	Sept	Oct	Nov	Dec	Year
1-Axis, E-W Horiz Axis	Average	3.3	3.6	3.8	4.3	4.1	3.4	3.1	3.0	2.9	3.5	3.5	3.4	3.5
	Min/Max	1.9/4.6	2.4/4.8	2.6/4.8	3.5/5.3	3.1/5.0	2.2/4.6	2.3/3.9	2.0/4.0	2.0/3.7	2.0/4.4	2.3/4.6	2.2/4.5	2.8/4.1
1-Axis, N-S Horiz Axis	Average	3.0	3.7	4.7	5.7	5.4	4.4	4.0	3.9	3.6	3.9	3.3	2.9	4.0
	Min/Max	1.6/4.1	2.4/5.1	3.2/5.9	4.7/7.1	4.1/6.7	2.8/5.9	3.0/5.0	2.7/5.3	2.5/4.8	2.3/4.9	2.2/4.3	1.8/3.9	3.2/4.7
1-Axis, N-S Tilt=Latitude	Average	3.8	4.4	5.1	5.8	5.1	4.1	3.8	3.8	3.8	4.5	4.1	3.7	4.3
	Min/Max	2.1/5.2	2.8/6.0	3.5/6.5	4.8/7.2	3.9/6.3	2.6/5.5	2.8/4.7	2.7/5.2	2.7/5.0	2.6/5.6	2.8/5.3	2.3/5.0	3.4/5.1
2-Axis	Average	4.0	4.5	5.1	5.9	5.4	4.5	4.0	3.9	3.8	4.5	4.3	4.0	4.5
	Min/Max	2.2/5.5	2.9/6.2	3.5/6.5	4.8/7.3	4.1/6.7	2.8/6.0	3.0/5.1	2.7/5.4	2.7/5.0	2.6/5.7	2.9/5.6	2.5/5.4	3.6/5.3

Average Climatic Conditions

Element	Jan	Feb	Mar	Apr	May	June	July	Aug	Sept	Oct	Nov	Dec	Year
Temperature (°C)	15.5	16.4	19.1	21.8	25.1	27.2	27.8	27.8	27.2	23.8	19.8	16.8	22.4
Daily Minimum Temp	10.0	10.9	13.6	16.0	19.7	22.7	23.6	23.6	22.7	18.4	14.0	11.3	17.2
Daily Maximum Temp	21.0	21.9	24.8	27.6	30.7	31.9	32.3	32.3	31.7	29.1	25.4	22.3	27.6
Record Minimum Temp	-6.1	-4.4	-1.7	4.4	9.4	11.7	17.2	19.4	13.9	4.4	-5.0	-7.8	-7.8
Record Maximum Temp	30.0	31.1	32.8	33.9	36.7	37.2	36.1	36.7	35.6	34.4	32.2	30.0	37.2
HDD, Base 18.3°C	130	89	47	4	0	0	0	0	0	0	39	95	403
CDD, Base 18.3°C	42	34	71	107	210	267	294	294	267	171	82	47	1887
Relative Humidity (%)	75	73	72	69	70	74	77	78	77	74	75	75	74
Wind Speed (m/s)	3.9	4.1	4.2	4.1	3.9	3.6	3.3	3.1	3.4	3.8	3.8	3.8	3.8

West Palm Beach, FL

WBAN NO. 12844

LATITUDE: 26.68° N
LONGITUDE: 80.10° W
ELEVATION: 6 meters
MEAN PRESSURE: 1017 millibars

STATION TYPE: Secondary

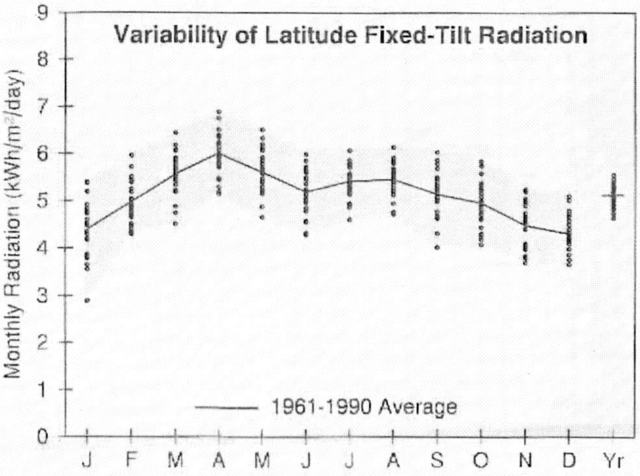

Solar Radiation for Flat-Plate Collectors Facing South at a Fixed Tilt (kWh/m²/day), Uncertainty ±9%

Tilt (°)		Jan	Feb	Mar	Apr	May	June	July	Aug	Sept	Oct	Nov	Dec	Year
0	Average	3.3	4.0	5.0	5.9	6.0	5.7	5.9	5.6	4.8	4.2	3.4	3.1	4.7
	Min/Max	2.4/3.8	3.5/4.6	4.1/5.7	5.1/6.8	4.9/7.0	4.7/6.6	5.0/6.6	4.8/6.3	3.9/5.6	3.6/4.8	3.0/3.9	2.8/3.5	4.4/5.1
Latitude -15	Average	3.8	4.5	5.3	6.1	5.9	5.6	5.8	5.6	5.1	4.6	4.0	3.7	5.0
	Min/Max	2.7/4.6	4.0/5.3	4.4/6.1	5.2/7.0	4.9/6.9	4.6/6.5	4.9/6.5	4.9/6.3	4.0/5.9	3.9/5.4	3.3/4.6	3.2/4.3	4.6/5.4
Latitude	Average	4.4	5.0	5.6	6.0	5.6	5.2	5.4	5.4	5.1	4.9	4.5	4.3	5.1
	Min/Max	2.9/5.4	4.3/6.0	4.5/6.4	5.1/6.9	4.7/6.5	4.3/6.0	4.6/6.1	4.7/6.1	4.0/6.0	4.1/5.8	3.7/5.2	3.7/5.1	4.6/5.6
Latitude +15	Average	4.7	5.1	5.5	5.6	5.0	4.5	4.8	5.0	4.9	5.0	4.7	4.7	5.0
	Min/Max	3.0/5.9	4.4/6.2	4.4/6.4	4.8/6.4	4.2/5.8	3.8/5.2	4.1/5.3	4.3/5.6	3.8/5.8	4.1/6.0	3.8/5.6	3.9/5.6	4.5/5.4
90	Average	3.9	3.8	3.4	2.6	1.9	1.6	1.7	2.2	2.7	3.5	3.8	4.0	2.9
	Min/Max	2.3/5.0	3.2/4.7	2.7/3.9	2.3/2.8	1.7/2.0	1.6/1.7	1.6/1.8	2.0/2.3	2.2/3.2	2.8/4.2	3.0/4.5	3.3/4.9	2.6/3.2

Solar Radiation for 1-Axis Tracking Flat-Plate Collectors with a North-South Axis (kWh/m²/day), Uncertainty ±9%

Axis Tilt (°)		Jan	Feb	Mar	Apr	May	June	July	Aug	Sept	Oct	Nov	Dec	Year
0	Average	4.5	5.5	6.6	7.8	7.5	6.9	7.2	6.9	6.0	5.4	4.6	4.3	6.1
	Min/Max	2.9/5.6	4.5/6.7	5.1/7.8	6.5/9.3	5.8/9.1	5.2/8.4	5.7/8.4	5.8/8.1	4.6/7.3	4.3/6.6	3.7/5.4	3.5/5.1	5.4/6.7
Latitude -15	Average	4.9	5.8	6.9	7.9	7.5	6.8	7.2	7.0	6.2	5.8	5.0	4.7	6.3
	Min/Max	3.0/6.2	4.8/7.2	5.3/8.2	6.6/9.5	5.8/9.1	5.2/8.3	5.7/8.4	5.8/8.2	4.7/7.5	4.6/7.0	4.0/6.0	3.9/5.7	5.6/6.9
Latitude	Average	5.3	6.2	7.1	7.9	7.3	6.6	6.9	6.8	6.3	6.0	5.4	5.2	6.4
	Min/Max	3.2/6.8	5.1/7.7	5.3/8.4	6.5/9.4	5.6/8.8	5.0/7.9	5.5/8.1	5.7/8.1	4.7/7.6	4.7/7.4	4.3/6.5	4.2/6.3	5.6/7.1
Latitude +15	Average	5.6	6.3	7.0	7.6	6.9	6.1	6.4	6.5	6.1	6.1	5.6	5.4	6.3
	Min/Max	3.3/7.1	5.2/7.8	5.3/8.4	6.3/9.1	5.3/8.3	4.6/7.4	5.1/7.5	5.4/7.7	4.6/7.4	4.7/7.5	4.4/6.7	4.4/6.7	5.5/6.9

Solar Radiation for 2-Axis Tracking Flat-Plate Collectors (kWh/m²/day), Uncertainty ±9%

Tracker		Jan	Feb	Mar	Apr	May	June	July	Aug	Sept	Oct	Nov	Dec	Year
2-Axis	Average	5.6	6.3	7.1	8.0	7.6	7.0	7.3	7.0	6.3	6.1	5.6	5.5	6.6
	Min/Max	3.3/7.2	5.2/7.9	5.4/8.4	6.6/9.5	5.9/9.2	5.3/8.5	5.8/8.5	5.8/8.2	4.7/7.6	4.8/7.5	4.4/6.8	4.5/6.8	5.8/7.3

Direct Beam Solar Radiation for Concentrating Collectors (kWh/m²/day), Uncertainty ±8%

Tracker		Jan	Feb	Mar	Apr	May	June	July	Aug	Sept	Oct	Nov	Dec	Year
1-Axis, E-W	Average	3.1	3.3	3.5	3.8	3.5	3.0	3.2	3.0	2.7	2.9	3.0	3.2	3.2
Horiz Axis	Min/Max	1.3/4.6	2.4/4.6	2.1/4.7	2.7/5.2	2.1/4.8	1.5/4.1	1.9/4.2	2.0/4.0	1.6/3.8	2.0/4.0	2.0/4.0	2.3/4.3	2.6/3.7
1-Axis, N-S	Average	2.9	3.6	4.3	5.1	4.6	3.8	4.0	3.9	3.4	3.2	2.9	2.7	3.7
Horiz Axis	Min/Max	1.2/4.2	2.5/4.9	2.6/5.8	3.6/7.0	2.8/6.3	1.9/5.2	2.5/5.4	2.6/5.2	2.1/4.7	2.2/4.5	1.9/3.8	2.0/3.7	3.0/4.4
1-Axis, N-S	Average	3.6	4.2	4.7	5.2	4.4	3.5	3.8	3.8	3.5	3.7	3.5	3.5	3.9
Tilt=Latitude	Min/Max	1.4/5.2	2.9/5.7	2.8/6.3	3.6/7.0	2.6/6.0	1.8/4.8	2.3/5.0	2.6/5.2	2.2/4.9	2.5/5.1	2.3/4.7	2.5/4.7	3.2/4.6
2-Axis	Average	3.8	4.3	4.7	5.2	4.6	3.8	4.1	3.9	3.5	3.7	3.7	3.8	4.1
	Min/Max	1.5/5.6	3.0/5.9	2.8/6.3	3.7/7.1	2.8/6.3	2.0/5.3	2.5/5.4	2.7/5.3	2.2/5.0	2.5/5.1	2.5/4.9	2.7/5.1	3.3/4.8

Average Climatic Conditions

Element	Jan	Feb	Mar	Apr	May	June	July	Aug	Sept	Oct	Nov	Dec	Year
Temperature (°C)	18.4	19.0	21.1	23.0	25.3	27.0	27.9	28.1	27.6	25.4	22.4	19.7	23.7
Daily Minimum Temp	13.2	13.6	16.2	18.2	20.9	22.8	23.6	23.9	23.7	21.5	18.1	14.8	19.2
Daily Maximum Temp	23.6	24.4	26.0	27.8	29.8	31.2	32.2	32.2	31.4	29.4	26.7	24.4	28.3
Record Minimum Temp	-2.8	0.0	-1.1	6.1	11.7	16.1	18.9	18.3	18.9	7.8	2.2	-2.2	-2.8
Record Maximum Temp	31.7	32.2	34.4	37.2	35.6	36.7	38.3	36.7	36.1	35.0	32.8	32.2	38.3
HDD, Base 18.3°C	68	47	15	0	0	0	0	0	0	0	8	41	179
CDD, Base 18.3°C	69	66	101	140	217	260	296	302	277	221	130	83	2162
Relative Humidity (%)	73	72	70	67	71	77	77	76	77	74	73	73	73
Wind Speed (m/s)	4.6	4.8	5.1	4.9	4.6	3.9	3.7	3.7	3.9	4.8	4.8	4.6	4.4

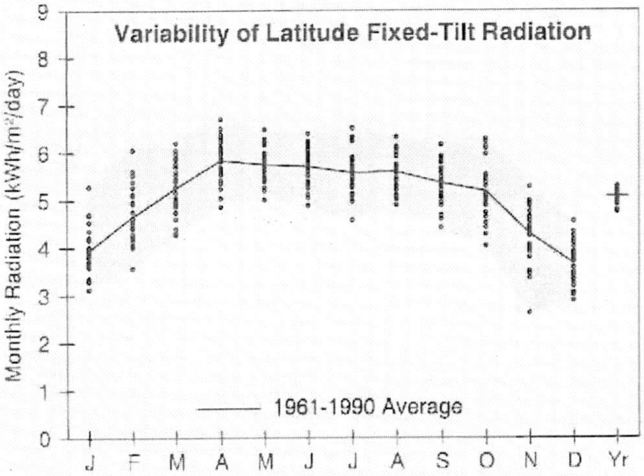

Athens, GA

WBAN NO. 13873

LATITUDE: 33.95° N
LONGITUDE: 83.32° W
ELEVATION: 244 meters
MEAN PRESSURE: 989 millibars

STATION TYPE: Secondary

Solar Radiation for Flat-Plate Collectors Facing South at a Fixed Tilt (kWh/m²/day), Uncertainty ±9%

Tilt (°)		Jan	Feb	Mar	Apr	May	June	July	Aug	Sept	Oct	Nov	Dec	Year
0	Average	2.6	3.4	4.5	5.6	6.1	6.4	6.1	5.6	4.8	4.0	2.9	2.4	4.5
	Min/Max	2.3/3.2	2.8/4.2	3.8/5.1	4.8/6.4	5.3/6.9	5.4/7.1	5.0/7.1	5.0/6.3	4.2/5.4	3.3/4.7	2.1/3.4	2.0/2.7	4.3/4.7
Latitude -15	Average	3.5	4.2	5.1	5.9	6.1	6.2	6.0	5.8	5.3	4.8	3.8	3.2	5.0
	Min/Max	2.8/4.5	3.3/5.4	4.2/5.9	5.0/6.8	5.3/6.9	5.3/7.0	4.9/7.0	5.1/6.6	4.4/6.1	3.8/5.8	2.5/4.6	2.6/3.9	4.7/5.2
Latitude	Average	3.9	4.6	5.2	5.8	5.7	5.7	5.6	5.6	5.4	5.2	4.3	3.7	5.1
	Min/Max	3.1/5.3	3.6/6.1	4.3/6.2	4.9/6.7	5.0/6.5	4.9/6.4	4.6/6.5	4.9/6.3	4.4/6.2	4.0/6.3	2.6/5.3	2.9/4.6	4.8/5.3
Latitude +15	Average	4.2	4.8	5.2	5.4	5.1	5.0	4.9	5.1	5.1	5.3	4.5	3.9	4.9
	Min/Max	3.3/5.7	3.6/6.3	4.1/6.1	4.5/6.3	4.5/5.8	4.3/5.5	4.1/5.7	4.5/5.8	4.2/6.0	4.0/6.4	2.7/5.6	3.0/5.0	4.5/5.1
90	Average	3.7	3.8	3.5	3.0	2.4	2.1	2.2	2.7	3.3	4.0	3.9	3.5	3.2
	Min/Max	2.7/5.2	2.9/5.2	2.8/4.2	2.6/3.5	2.2/2.6	2.0/2.2	2.0/2.4	2.4/2.9	2.6/3.8	3.0/4.9	2.1/4.9	2.7/4.5	2.9/3.4

Solar Radiation for 1-Axis Tracking Flat-Plate Collectors with a North-South Axis (kWh/m²/day), Uncertainty ±9%

Axis Tilt (°)		Jan	Feb	Mar	Apr	May	June	July	Aug	Sept	Oct	Nov	Dec	Year
0	Average	3.7	4.8	6.0	7.5	7.8	8.0	7.6	7.1	6.3	5.5	4.1	3.3	6.0
	Min/Max	2.9/4.9	3.6/6.4	4.7/7.2	5.9/8.7	6.4/9.1	6.4/9.3	5.8/9.2	6.0/8.4	5.0/7.4	4.2/6.8	2.5/5.0	2.5/4.1	5.5/6.3
Latitude -15	Average	4.3	5.4	6.5	7.7	7.8	7.9	7.5	7.3	6.6	6.1	4.7	3.9	6.3
	Min/Max	3.3/5.9	4.0/7.3	5.1/7.8	6.1/9.1	6.4/9.2	6.3/9.2	5.8/9.1	6.1/8.6	5.3/7.9	4.6/7.7	2.7/5.9	3.0/4.9	5.9/6.7
Latitude	Average	4.7	5.7	6.6	7.7	7.6	7.6	7.3	7.1	6.7	6.4	5.1	4.3	6.4
	Min/Max	3.5/6.5	4.2/7.8	5.1/8.1	6.0/9.0	6.2/8.9	6.1/8.8	5.5/8.8	6.0/8.5	5.3/8.0	4.8/8.1	2.9/6.4	3.2/5.5	5.9/6.8
Latitude +15	Average	4.9	5.8	6.5	7.4	7.1	7.0	6.8	6.8	6.5	6.5	5.3	4.5	6.3
	Min/Max	3.7/6.8	4.2/8.0	5.0/8.0	5.7/8.7	5.8/8.4	5.6/8.2	5.1/8.2	5.7/8.1	5.1/7.8	4.8/8.1	2.9/6.7	3.4/5.8	5.8/6.6

Solar Radiation for 2-Axis Tracking Flat-Plate Collectors (kWh/m²/day), Uncertainty ±9%

Tracker		Jan	Feb	Mar	Apr	May	June	July	Aug	Sept	Oct	Nov	Dec	Year
2-Axis	Average	4.9	5.8	6.6	7.8	7.9	8.1	7.6	7.3	6.7	6.5	5.3	4.6	6.6
	Min/Max	3.7/6.9	4.3/8.0	5.1/8.1	6.1/9.1	6.5/9.3	6.5/9.4	5.8/9.3	6.2/8.7	5.3/8.0	4.8/8.1	2.9/6.8	3.4/5.9	6.1/7.0

Direct Beam Solar Radiation for Concentrating Collectors (kWh/m²/day), Uncertainty ±8%

Tracker		Jan	Feb	Mar	Apr	May	June	July	Aug	Sept	Oct	Nov	Dec	Year
1-Axis, E-W Horiz Axis	Average	2.9	3.2	3.3	3.7	3.6	3.7	3.4	3.2	3.1	3.5	3.1	2.8	3.3
	Min/Max	1.9/4.7	2.1/5.1	2.1/4.4	2.5/4.7	2.5/4.8	2.3/4.8	2.1/4.7	2.4/4.2	2.0/4.1	2.2/4.9	1.2/4.2	1.8/4.0	2.9/3.7
1-Axis, N-S Horiz Axis	Average	2.4	3.2	3.9	4.9	4.8	4.8	4.4	4.2	3.8	3.7	2.7	2.1	3.7
	Min/Max	1.5/3.9	2.0/5.1	2.4/5.2	3.3/6.4	3.2/6.4	3.0/6.3	2.7/6.1	3.1/5.6	2.4/5.1	2.4/5.2	1.0/3.7	1.3/3.1	3.3/4.2
1-Axis, N-S Tilt=Latitude	Average	3.3	4.0	4.4	5.1	4.6	4.5	4.1	4.2	4.1	4.4	3.6	3.0	4.1
	Min/Max	2.1/5.2	2.5/6.3	2.8/5.9	3.4/6.6	3.1/6.1	2.8/5.8	2.5/5.8	3.1/5.5	2.6/5.6	2.8/6.2	1.4/4.9	1.9/4.3	3.6/4.6
2-Axis	Average	3.5	4.1	4.5	5.2	4.9	4.8	4.4	4.3	4.1	4.5	3.7	3.3	4.3
	Min/Max	2.2/5.6	2.6/6.5	2.8/5.9	3.4/6.7	3.3/6.5	3.1/6.4	2.7/6.2	3.2/5.7	2.6/5.6	2.8/6.3	1.5/5.1	2.1/4.7	3.8/4.8

Average Climatic Conditions

Element	Jan	Feb	Mar	Apr	May	June	July	Aug	Sept	Oct	Nov	Dec	Year
Temperature (°C)	5.4	7.5	12.0	16.4	20.7	24.7	26.4	25.8	22.7	16.8	11.9	7.3	16.5
Daily Minimum Temp	0.0	1.5	5.6	9.9	14.5	18.8	20.8	20.4	17.2	10.4	5.7	1.7	10.6
Daily Maximum Temp	10.9	13.5	18.4	23.0	26.9	30.6	32.0	31.2	28.2	23.1	18.0	12.8	22.4
Record Minimum Temp	-20.0	-15.0	-11.7	-2.8	2.8	7.2	12.8	12.2	2.2	-4.4	-13.9	-16.7	-20.0
Record Maximum Temp	26.7	27.2	31.1	33.9	36.1	40.0	40.0	41.7	37.2	36.7	30.0	26.1	41.7
HDD, Base 18.3°C	399	303	199	73	15	0	0	0	0	78	197	343	1607
CDD, Base 18.3°C	0	0	3	16	89	190	252	233	133	29	3	0	949
Relative Humidity (%)	68	64	64	63	70	71	75	77	76	71	69	70	70
Wind Speed (m/s)	3.8	4.0	3.9	3.7	3.1	2.9	2.8	2.6	2.9	3.0	3.3	3.6	3.3

Atlanta, GA

WBAN NO. 13874

LATITUDE: 33.65° N
LONGITUDE: 84.43° W
ELEVATION: 315 meters
MEAN PRESSURE: 981 millibars

STATION TYPE: Primary

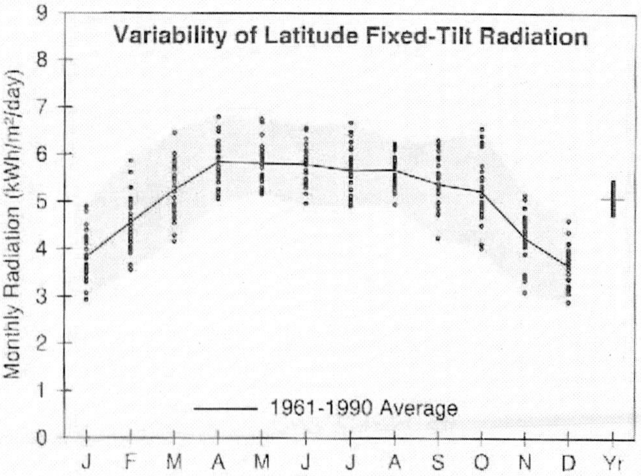

Solar Radiation for Flat-Plate Collectors Facing South at a Fixed Tilt (kWh/m²/day), Uncertainty ±9%

Tilt (°)		Jan	Feb	Mar	Apr	May	June	July	Aug	Sept	Oct	Nov	Dec	Year
0	Average	2.6	3.4	4.5	5.7	6.2	6.4	6.2	5.7	4.8	4.1	2.9	2.4	4.6
	Min/Max	2.2/3.0	2.9/4.1	3.7/5.3	5.0/6.5	5.5/7.2	5.5/7.3	5.3/7.3	5.0/6.3	4.0/5.6	3.4/4.9	2.4/3.4	2.0/2.8	4.3/4.9
Latitude -15	Average	3.4	4.2	5.1	6.0	6.2	6.3	6.1	5.9	5.3	4.9	3.8	3.2	5.0
	Min/Max	2.7/4.2	3.3/5.3	4.1/6.2	5.2/6.9	5.5/7.2	5.4/7.1	5.3/7.2	5.1/6.4	4.3/6.2	3.8/6.0	2.9/4.5	2.6/3.9	4.7/5.4
Latitude	Average	3.8	4.6	5.3	5.8	5.8	5.8	5.7	5.7	5.4	5.2	4.2	3.7	5.1
	Min/Max	2.9/4.9	3.6/5.9	4.2/6.5	5.1/6.8	5.2/6.8	5.0/6.6	4.9/6.7	4.9/6.2	4.2/6.3	4.0/6.6	3.1/5.2	2.9/4.6	4.8/5.4
Latitude +15	Average	4.1	4.7	5.1	5.4	5.2	5.1	5.0	5.2	5.1	5.3	4.5	3.9	4.9
	Min/Max	3.0/5.3	3.6/6.1	4.0/6.4	4.7/6.3	4.6/6.0	4.3/5.7	4.4/5.9	4.5/5.7	4.0/6.1	4.0/6.7	3.2/5.5	3.0/5.0	4.6/5.2
90	Average	3.5	3.7	3.5	3.0	2.4	2.2	2.2	2.7	3.2	4.0	3.8	3.5	3.1
	Min/Max	2.5/4.8	2.8/5.0	2.7/4.3	2.6/3.4	2.2/2.6	2.0/2.3	2.1/2.6	2.4/2.9	2.5/3.8	2.9/5.1	2.6/4.8	2.6/4.6	2.9/3.4

Solar Radiation for 1-Axis Tracking Flat-Plate Collectors with a North-South Axis (kWh/m²/day), Uncertainty ±9%

Axis Tilt (°)		Jan	Feb	Mar	Apr	May	June	July	Aug	Sept	Oct	Nov	Dec	Year
0	Average	3.6	4.7	6.0	7.4	7.9	8.0	7.6	7.2	6.3	5.6	4.0	3.3	6.0
	Min/Max	2.8/4.6	3.5/6.2	4.7/7.6	6.3/8.9	6.7/9.6	6.6/9.5	6.1/9.2	6.1/8.1	4.8/7.5	4.2/7.1	2.9/5.0	2.6/4.1	5.5/6.5
Latitude -15	Average	4.2	5.3	6.5	7.7	7.9	8.0	7.6	7.4	6.6	6.2	4.7	3.9	6.3
	Min/Max	3.1/5.4	3.9/7.0	4.9/8.3	6.4/9.2	6.7/9.7	6.5/9.4	6.1/9.2	6.2/8.3	5.1/8.0	4.6/8.0	3.2/5.8	3.0/4.9	5.8/6.9
Latitude	Average	4.5	5.5	6.6	7.6	7.7	7.6	7.3	7.2	6.7	6.4	5.0	4.3	6.4
	Min/Max	3.3/6.0	4.0/7.4	5.0/8.5	6.3/9.2	6.5/9.4	6.2/9.0	5.8/8.8	6.0/8.1	5.0/8.1	4.8/8.4	3.4/6.3	3.3/5.5	5.9/7.0
Latitude +15	Average	4.7	5.6	6.5	7.3	7.2	7.1	6.8	6.9	6.5	6.5	5.2	4.5	6.2
	Min/Max	3.4/6.3	4.1/7.6	4.9/8.4	6.1/8.8	6.1/8.9	5.8/8.4	5.4/8.3	5.7/7.8	4.9/7.9	4.7/8.5	3.5/6.6	3.4/5.8	5.7/6.8

Solar Radiation for 2-Axis Tracking Flat-Plate Collectors (kWh/m²/day), Uncertainty ±9%

Tracker		Jan	Feb	Mar	Apr	May	June	July	Aug	Sept	Oct	Nov	Dec	Year
2-Axis	Average	4.8	5.7	6.6	7.7	8.0	8.1	7.7	7.4	6.7	6.5	5.3	4.5	6.6
	Min/Max	3.4/6.4	4.1/7.7	5.0/8.5	6.4/9.3	6.8/9.8	6.6/9.6	6.1/9.3	6.2/8.3	5.1/8.1	4.8/8.5	3.5/6.6	3.4/5.9	6.1/7.2

Direct Beam Solar Radiation for Concentrating Collectors (kWh/m²/day), Uncertainty ±8%

Tracker		Jan	Feb	Mar	Apr	May	June	July	Aug	Sept	Oct	Nov	Dec	Year
1-Axis, E-W Horiz Axis	Average	2.8	3.1	3.3	3.7	3.7	3.7	3.3	3.1	3.0	3.5	3.0	2.7	3.2
	Min/Max	1.7/4.3	1.8/4.7	2.0/4.8	2.7/4.9	2.8/5.1	2.5/4.8	2.1/4.6	2.3/3.9	1.9/4.1	2.1/5.1	1.6/4.1	1.7/3.9	2.8/3.7
1-Axis, N-S Horiz Axis	Average	2.3	3.1	3.9	4.9	4.9	4.7	4.3	4.1	3.7	3.7	2.6	2.1	3.7
	Min/Max	1.4/3.6	1.8/4.7	2.4/5.8	3.5/6.4	3.7/6.9	3.3/6.3	2.6/6.0	2.9/5.2	2.4/5.1	2.3/5.4	1.3/3.6	1.3/3.0	3.2/4.3
1-Axis, N-S Tilt=Latitude	Average	3.1	3.8	4.4	5.0	4.7	4.4	4.0	4.1	4.0	4.4	3.5	2.9	4.0
	Min/Max	1.9/4.8	2.2/5.9	2.7/6.5	3.6/6.6	3.5/6.6	3.0/5.8	2.5/5.7	2.9/5.2	2.5/5.6	2.7/6.5	1.8/4.8	1.9/4.2	3.5/4.6
2-Axis	Average	3.3	3.9	4.4	5.1	4.9	4.8	4.3	4.2	4.0	4.5	3.6	3.2	4.2
	Min/Max	2.0/5.2	2.3/6.0	2.7/6.5	3.7/6.7	3.7/7.0	3.3/6.3	2.7/6.1	3.0/5.3	2.5/5.6	2.7/6.5	1.9/5.0	2.0/4.6	3.6/4.8

Average Climatic Conditions

Element	Jan	Feb	Mar	Apr	May	June	July	Aug	Sept	Oct	Nov	Dec	Year
Temperature (°C)	5.0	7.1	11.9	16.4	20.7	24.4	26.0	25.6	22.6	16.8	11.7	6.9	16.3
Daily Minimum Temp	-0.3	1.4	5.8	10.1	14.8	19.0	20.8	20.6	17.5	11.1	6.0	1.7	10.7
Daily Maximum Temp	10.2	12.8	17.9	22.6	26.4	29.9	31.1	30.6	27.7	22.6	17.4	12.2	21.8
Record Minimum Temp	-22.2	-15.0	-12.2	-3.3	2.8	7.8	11.7	12.8	2.2	-2.2	-16.1	-17.8	-22.2
Record Maximum Temp	26.1	26.7	29.4	33.9	35.0	38.3	40.6	38.9	36.7	35.0	28.9	26.1	40.6
HDD, Base 18.3°C	413	314	203	77	15	0	0	0	6	77	204	353	1662
CDD, Base 18.3°C	0	0	4	18	87	183	238	226	134	30	6	0	926
Relative Humidity (%)	68	63	62	61	67	70	74	75	74	68	68	68	68
Wind Speed (m/s)	4.4	4.6	4.5	4.3	3.8	3.5	3.4	3.2	3.5	3.7	3.9	4.2	3.9

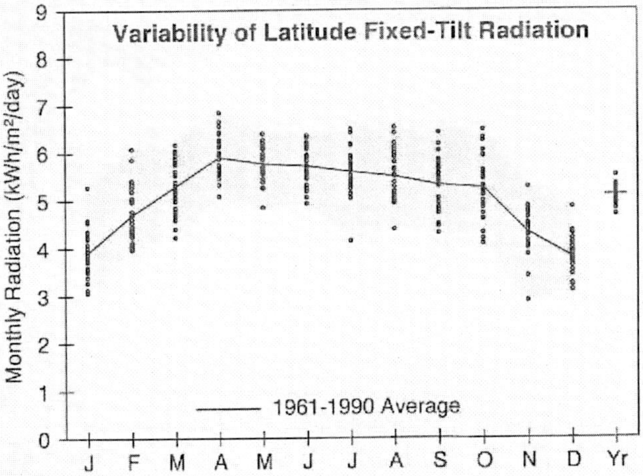

Augusta, GA

WBAN NO. 03820

LATITUDE: 33.37° N
LONGITUDE: 81.97° W
ELEVATION: 45 meters
MEAN PRESSURE: 1013 millibars

STATION TYPE: Secondary

Solar Radiation for Flat-Plate Collectors Facing South at a Fixed Tilt (kWh/m²/day), Uncertainty ±9%

Tilt (°)		Jan	Feb	Mar	Apr	May	June	July	Aug	Sept	Oct	Nov	Dec	Year
0	Average	2.6	3.5	4.5	5.7	6.1	6.3	6.1	5.5	4.8	4.1	3.0	2.4	4.6
	Min/Max	2.3/3.3	3.1/4.2	3.8/5.1	5.0/6.5	5.2/6.8	5.4/7.1	4.5/7.1	4.5/6.5	4.1/5.6	3.4/4.8	2.3/3.4	2.2/2.9	4.2/4.9
Latitude -15	Average	3.4	4.3	5.1	6.0	6.1	6.2	6.0	5.7	5.2	4.9	3.8	3.3	5.0
	Min/Max	2.8/4.5	3.7/5.4	4.2/5.9	5.2/6.9	5.1/6.8	5.3/6.9	4.4/7.0	4.6/6.8	4.3/6.3	3.9/5.9	2.7/4.6	2.8/4.1	4.6/5.4
Latitude	Average	3.9	4.7	5.3	5.9	5.8	5.7	5.6	5.5	5.3	5.3	4.3	3.8	5.1
	Min/Max	3.1/5.3	4.0/6.1	4.2/6.2	5.1/6.8	4.8/6.4	4.9/6.4	4.1/6.5	4.4/6.5	4.3/6.4	4.1/6.5	2.9/5.3	3.1/4.8	4.7/5.5
Latitude +15	Average	4.1	4.8	5.2	5.5	5.1	5.0	4.9	5.0	5.1	5.3	4.6	4.1	4.9
	Min/Max	3.2/5.7	4.1/6.4	4.1/6.1	4.7/6.4	4.3/5.7	4.3/5.5	3.7/5.7	4.0/5.9	4.1/6.2	4.1/6.6	2.9/5.6	3.3/5.3	4.5/5.3
90	Average	3.6	3.8	3.5	3.0	2.4	2.1	2.2	2.6	3.2	4.0	3.9	3.6	3.2
	Min/Max	2.7/5.1	3.2/5.2	2.8/4.2	2.7/3.4	2.2/2.5	2.0/2.2	1.9/2.3	2.2/3.0	2.5/3.9	3.0/5.0	2.4/4.8	2.8/4.8	2.9/3.5

Solar Radiation for 1-Axis Tracking Flat-Plate Collectors with a North-South Axis (kWh/m²/day), Uncertainty ±9%

Axis Tilt (°)		Jan	Feb	Mar	Apr	May	June	July	Aug	Sept	Oct	Nov	Dec	Year
0	Average	3.6	4.8	6.1	7.6	7.8	7.9	7.5	6.9	6.2	5.6	4.1	3.4	6.0
	Min/Max	2.8/5.0	4.0/6.4	4.6/7.1	6.2/9.2	6.0/9.0	6.4/9.2	5.1/9.3	5.3/8.7	4.9/7.7	4.2/7.1	2.7/5.1	2.7/4.4	5.4/6.5
Latitude -15	Average	4.2	5.4	6.5	7.8	7.8	7.8	7.5	7.1	6.5	6.2	4.8	4.0	6.3
	Min/Max	3.3/5.9	4.4/7.3	4.9/7.7	6.4/9.6	6.0/9.1	6.3/9.1	5.0/9.3	5.4/8.9	5.1/8.2	4.6/7.9	3.0/5.9	3.2/5.3	5.7/6.9
Latitude	Average	4.6	5.7	6.7	7.8	7.6	7.5	7.2	7.0	6.6	6.5	5.2	4.4	6.4
	Min/Max	3.5/6.5	4.7/7.8	4.9/7.9	6.3/9.5	5.8/8.8	6.1/8.8	4.8/8.9	5.2/8.7	5.1/8.3	4.8/8.3	3.2/6.4	3.5/5.8	5.8/7.0
Latitude +15	Average	4.8	5.8	6.6	7.5	7.1	7.0	6.7	6.6	6.4	6.5	5.3	4.7	6.3
	Min/Max	3.6/6.8	4.8/7.9	4.8/7.9	6.0/9.2	5.4/8.3	5.6/8.1	4.5/8.3	5.0/8.3	4.9/8.1	4.8/8.4	3.2/6.7	3.7/6.2	5.6/6.9

Solar Radiation for 2-Axis Tracking Flat-Plate Collectors (kWh/m²/day), Uncertainty ±9%

Tracker		Jan	Feb	Mar	Apr	May	June	July	Aug	Sept	Oct	Nov	Dec	Year
2-Axis	Average	4.9	5.9	6.7	7.9	7.9	8.0	7.6	7.1	6.6	6.6	5.4	4.8	6.6
	Min/Max	3.6/6.9	4.8/8.0	4.9/8.0	6.4/9.6	6.1/9.2	6.4/9.3	5.1/9.4	5.4/8.9	5.1/8.3	4.8/8.4	3.3/6.8	3.7/6.3	5.9/7.3

Direct Beam Solar Radiation for Concentrating Collectors (kWh/m²/day), Uncertainty ±8%

Tracker		Jan	Feb	Mar	Apr	May	June	July	Aug	Sept	Oct	Nov	Dec	Year
1-Axis, E-W Horiz Axis	Average	2.8	3.2	3.3	3.8	3.7	3.7	3.4	3.1	3.0	3.5	3.1	2.9	3.3
	Min/Max	1.8/4.7	2.4/5.0	1.9/4.4	2.7/5.2	2.2/4.8	2.3/4.9	1.5/5.0	1.8/4.6	2.0/4.3	2.2/5.1	1.4/4.3	1.9/4.3	2.8/3.8
1-Axis, N-S Horiz Axis	Average	2.4	3.2	3.9	5.0	4.8	4.8	4.3	4.0	3.7	3.8	2.7	2.3	3.7
	Min/Max	1.5/3.9	2.3/5.0	2.2/5.2	3.5/7.0	2.8/6.3	3.0/6.3	1.9/6.6	2.3/6.0	2.4/5.3	2.3/5.5	1.2/3.8	1.5/3.3	3.1/4.3
1-Axis, N-S Tilt=Latitude	Average	3.2	4.0	4.4	5.2	4.7	4.4	4.1	4.0	4.0	4.5	3.6	3.1	4.1
	Min/Max	2.0/5.2	2.9/6.2	2.5/5.9	3.6/7.2	2.7/6.1	2.8/5.9	1.8/6.2	2.3/6.0	2.6/5.8	2.7/6.5	1.6/4.9	2.1/4.6	3.4/4.8
2-Axis	Average	3.4	4.1	4.5	5.2	4.9	4.8	4.4	4.1	4.0	4.5	3.8	3.4	4.3
	Min/Max	2.1/5.6	3.0/6.4	2.5/5.9	3.7/7.3	2.8/6.4	3.0/6.4	1.9/6.7	2.4/6.2	2.6/5.8	2.8/6.6	1.7/5.2	2.3/5.0	3.6/5.0

Average Climatic Conditions

Element	Jan	Feb	Mar	Apr	May	June	July	Aug	Sept	Oct	Nov	Dec	Year
Temperature (°C)	6.6	8.6	13.1	17.1	21.5	25.3	27.1	26.5	23.6	17.7	12.8	8.4	17.3
Daily Minimum Temp	0.0	1.5	5.7	9.2	14.2	18.7	21.1	20.6	17.3	10.2	5.3	1.6	10.4
Daily Maximum Temp	13.2	15.6	20.4	24.8	28.7	31.8	33.2	32.4	29.8	25.1	20.2	15.3	24.2
Record Minimum Temp	-18.3	-12.8	-11.1	-3.3	1.7	8.3	12.8	12.2	2.2	-5.6	-9.4	-15.0	-18.3
Record Maximum Temp	26.7	30.0	31.1	35.6	37.2	40.6	41.7	42.2	38.3	36.1	32.2	27.8	42.2
HDD, Base 18.3°C	363	274	169	59	7	0	0	0	0	72	174	307	1425
CDD, Base 18.3°C	0	0	6	21	104	208	272	253	158	51	8	0	1082
Relative Humidity (%)	70	66	65	65	70	71	74	76	76	73	72	72	71
Wind Speed (m/s)	3.2	3.4	3.5	3.4	2.9	2.8	2.7	2.4	2.5	2.5	2.8	3.0	2.9

Columbus, GA

WBAN NO. 93842

LATITUDE: 32.52° N
LONGITUDE: 84.95° W
ELEVATION: 136 meters
MEAN PRESSURE: 1004 millibars

STATION TYPE: Secondary

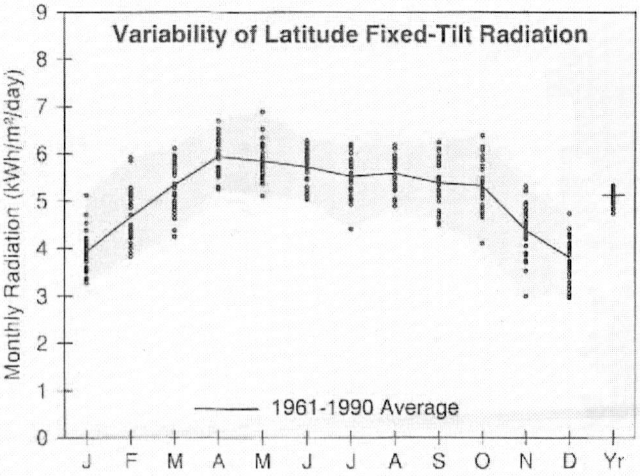

Solar Radiation for Flat-Plate Collectors Facing South at a Fixed Tilt (kWh/m²/day), Uncertainty ±9%

Tilt (°)		Jan	Feb	Mar	Apr	May	June	July	Aug	Sept	Oct	Nov	Dec	Year
0	Average	2.7	3.5	4.6	5.7	6.2	6.4	6.0	5.6	4.9	4.2	3.1	2.5	4.6
	Min/Max	2.4/3.2	3.0/4.2	3.9/5.1	5.1/6.4	5.5/7.3	5.6/7.0	4.8/6.8	5.0/6.2	4.3/5.5	3.5/4.9	2.4/3.5	2.1/2.9	4.3/4.8
Latitude -15	Average	3.5	4.3	5.2	6.0	6.2	6.2	5.9	5.8	5.3	4.9	3.9	3.3	5.0
	Min/Max	3.0/4.4	3.6/5.3	4.2/5.8	5.3/6.8	5.4/7.3	5.4/6.8	4.7/6.7	5.1/6.4	4.5/6.1	3.9/5.9	2.8/4.6	2.7/4.0	4.7/5.2
Latitude	Average	3.9	4.7	5.3	5.9	5.8	5.7	5.5	5.6	5.4	5.3	4.4	3.8	5.1
	Min/Max	3.3/5.1	3.8/5.9	4.3/6.1	5.2/6.7	5.1/6.9	5.0/6.3	4.4/6.2	4.9/6.2	4.5/6.2	4.1/6.4	3.0/5.3	3.0/4.7	4.7/5.3
Latitude +15	Average	4.2	4.8	5.2	5.5	5.2	5.0	4.9	5.1	5.2	5.4	4.6	4.1	4.9
	Min/Max	3.4/5.5	3.9/6.2	4.1/6.0	4.9/6.2	4.6/6.1	4.4/5.4	3.9/5.4	4.5/5.6	4.3/6.0	4.1/6.5	3.1/5.7	3.1/5.2	4.5/5.1
90	Average	3.6	3.8	3.5	3.0	2.3	2.1	2.1	2.6	3.2	4.0	3.9	3.6	3.1
	Min/Max	2.9/4.9	3.0/5.0	2.8/4.0	2.7/3.3	2.2/2.5	1.9/2.1	1.9/2.3	2.3/2.8	2.6/3.8	3.0/4.9	2.5/4.9	2.7/4.7	2.9/3.3

Solar Radiation for 1-Axis Tracking Flat-Plate Collectors with a North-South Axis (kWh/m²/day), Uncertainty ±9%

Axis Tilt (°)		Jan	Feb	Mar	Apr	May	June	July	Aug	Sept	Oct	Nov	Dec	Year
0	Average	3.7	4.9	6.1	7.6	7.9	7.9	7.4	7.1	6.3	5.7	4.2	3.5	6.0
	Min/Max	3.1/4.9	3.9/6.2	4.7/7.2	6.4/8.8	6.7/10.0	6.7/8.9	5.3/8.7	6.0/7.9	5.2/7.4	4.3/7.0	2.8/5.2	2.7/4.4	5.5/6.3
Latitude -15	Average	4.3	5.4	6.6	7.8	8.0	7.8	7.3	7.2	6.6	6.3	4.8	4.1	6.4
	Min/Max	3.5/5.8	4.3/7.0	5.0/7.8	6.6/9.1	6.7/10.1	6.6/8.8	5.3/8.7	6.1/8.1	5.4/7.8	4.7/7.8	3.1/6.0	3.1/5.2	5.8/6.7
Latitude	Average	4.7	5.7	6.7	7.8	7.7	7.5	7.1	7.1	6.7	6.6	5.2	4.5	6.4
	Min/Max	3.7/6.3	4.5/7.5	5.1/8.0	6.5/9.1	6.4/9.8	6.3/8.4	5.0/8.3	6.0/8.0	5.4/7.9	4.9/8.2	3.3/6.6	3.3/5.8	5.8/6.8
Latitude +15	Average	4.8	5.8	6.6	7.5	7.3	7.0	6.6	6.7	6.5	6.6	5.4	4.7	6.3
	Min/Max	3.8/6.6	4.5/7.7	4.9/7.9	6.2/8.7	6.1/9.2	5.9/7.8	4.7/7.8	5.6/7.6	5.2/7.8	4.8/8.3	3.4/6.8	3.5/6.1	5.7/6.6

Solar Radiation for 2-Axis Tracking Flat-Plate Collectors (kWh/m²/day), Uncertainty ±9%

Tracker		Jan	Feb	Mar	Apr	May	June	July	Aug	Sept	Oct	Nov	Dec	Year
2-Axis	Average	4.9	5.8	6.7	7.9	8.1	8.0	7.4	7.3	6.7	6.7	5.4	4.8	6.6
	Min/Max	3.9/6.7	4.6/7.7	5.1/8.0	6.6/9.2	6.7/10.2	6.8/8.9	5.3/8.8	6.1/8.1	5.4/8.0	4.9/8.3	3.4/6.9	3.5/6.2	6.0/7.0

Direct Beam Solar Radiation for Concentrating Collectors (kWh/m²/day), Uncertainty ±8%

Tracker		Jan	Feb	Mar	Apr	May	June	July	Aug	Sept	Oct	Nov	Dec	Year
1-Axis, E-W	Average	2.8	3.2	3.4	3.8	3.7	3.6	3.2	3.1	3.0	3.6	3.1	2.9	3.3
Horiz Axis	Min/Max	1.9/4.4	2.2/4.8	2.0/4.4	2.8/4.8	2.6/5.4	2.6/4.4	1.7/4.3	2.3/3.9	2.1/4.1	2.2/4.9	1.4/4.4	1.8/4.2	2.8/3.5
1-Axis, N-S	Average	2.4	3.2	4.0	5.0	4.9	4.7	4.1	4.1	3.8	3.8	2.8	2.3	3.8
Horiz Axis	Min/Max	1.6/3.7	2.2/4.8	2.3/5.3	3.6/6.4	3.5/7.2	3.3/5.7	2.1/5.6	3.0/5.0	2.7/5.0	2.4/5.3	1.2/3.9	1.4/3.4	3.2/4.1
1-Axis, N-S	Average	3.2	3.9	4.5	5.2	4.7	4.4	3.9	4.1	4.1	4.5	3.6	3.1	4.1
Tilt=Latitude	Min/Max	2.1/4.9	2.7/5.9	2.6/5.9	3.8/6.6	3.3/7.0	3.1/5.3	2.0/5.3	3.0/5.0	2.8/5.4	2.8/6.3	1.6/5.1	1.9/4.6	3.5/4.4
2-Axis	Average	3.4	4.0	4.5	5.2	5.0	4.8	4.2	4.2	4.1	4.6	3.8	3.4	4.3
	Min/Max	2.3/5.3	2.7/6.1	2.6/5.9	3.8/6.7	3.5/7.3	3.4/5.7	2.1/5.7	3.1/5.1	2.9/5.4	2.8/6.4	1.7/5.4	2.0/5.0	3.6/4.6

Average Climatic Conditions

Element	Jan	Feb	Mar	Apr	May	June	July	Aug	Sept	Oct	Nov	Dec	Year
Temperature (°C)	7.6	9.6	14.0	18.2	22.4	26.4	27.7	27.4	24.7	18.9	13.7	9.2	18.3
Daily Minimum Temp	1.8	3.1	7.3	11.2	16.1	20.2	22.1	21.9	19.1	12.3	7.2	3.4	12.2
Daily Maximum Temp	13.4	16.1	20.6	25.2	28.7	32.4	33.2	32.9	30.1	25.3	20.1	15.0	24.4
Record Minimum Temp	-18.9	-11.7	-8.9	-2.2	3.9	6.7	15.0	13.9	3.3	-4.4	-12.2	-15.6	-18.9
Record Maximum Temp	28.3	28.3	31.7	33.9	36.1	40.0	40.0	39.4	37.8	35.6	30.0	27.8	40.0
HDD, Base 18.3°C	337	244	144	42	4	0	0	0	0	48	149	287	1256
CDD, Base 18.3°C	5	0	10	39	130	242	291	282	190	65	11	4	1269
Relative Humidity (%)	71	68	67	65	69	70	75	75	75	71	72	73	71
Wind Speed (m/s)	3.3	3.6	3.5	3.4	3.0	2.8	2.7	2.5	2.9	3.0	3.0	3.1	3.0

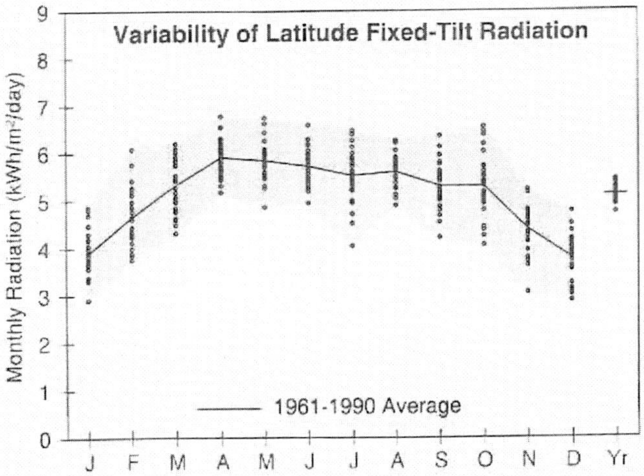

Macon, GA

WBAN NO. 03813

LATITUDE: 32.70° N
LONGITUDE: 83.65° W
ELEVATION: 110 meters
MEAN PRESSURE: 1005 millibars

STATION TYPE: Secondary

Solar Radiation for Flat-Plate Collectors Facing South at a Fixed Tilt (kWh/m²/day), Uncertainty ±9%

Tilt (°)		Jan	Feb	Mar	Apr	May	June	July	Aug	Sept	Oct	Nov	Dec	Year
0	Average	2.7	3.5	4.6	5.7	6.2	6.3	6.0	5.6	4.8	4.1	3.0	2.5	4.6
	Min/Max	2.2/3.1	3.0/4.3	3.9/5.2	5.1/6.5	5.2/7.2	5.4/7.3	4.4/7.0	5.0/6.3	4.0/5.6	3.4/4.9	2.4/3.5	2.1/2.9	4.3/4.9
Latitude -15	Average	3.4	4.3	5.1	6.0	6.2	6.2	5.9	5.8	5.2	4.9	3.9	3.3	5.0
	Min/Max	2.7/4.2	3.5/5.4	4.2/5.9	5.3/6.9	5.2/7.2	5.3/7.2	4.3/6.9	5.1/6.5	4.2/6.2	3.9/6.0	2.8/4.6	2.6/4.0	4.7/5.3
Latitude	Average	3.9	4.7	5.3	5.9	5.8	5.7	5.5	5.6	5.3	5.3	4.4	3.8	5.1
	Min/Max	2.9/4.9	3.8/6.1	4.3/6.2	5.2/6.8	4.9/6.7	4.9/6.6	4.0/6.5	4.9/6.3	4.2/6.3	4.0/6.5	3.0/5.2	2.9/4.8	4.7/5.4
Latitude +15	Average	4.1	4.8	5.2	5.5	5.2	5.0	4.9	5.1	5.1	5.4	4.6	4.0	4.9
	Min/Max	3.0/5.2	3.8/6.4	4.2/6.1	4.8/6.3	4.4/6.0	4.4/5.7	3.6/5.7	4.5/5.7	4.0/6.1	4.0/6.7	3.1/5.6	3.0/5.2	4.6/5.2
90	Average	3.6	3.8	3.5	3.0	2.3	2.1	2.1	2.6	3.2	4.0	3.9	3.6	3.1
	Min/Max	2.5/4.7	3.0/5.1	2.8/4.1	2.6/3.3	2.2/2.5	2.0/2.2	1.8/2.3	2.3/2.8	2.5/3.8	2.9/5.0	2.5/4.8	2.6/4.7	2.9/3.3

Solar Radiation for 1-Axis Tracking Flat-Plate Collectors with a North-South Axis (kWh/m²/day), Uncertainty ±9%

Axis Tilt (°)		Jan	Feb	Mar	Apr	May	June	July	Aug	Sept	Oct	Nov	Dec	Year
0	Average	3.7	4.8	6.1	7.6	7.9	7.9	7.3	7.1	6.2	5.7	4.2	3.4	6.0
	Min/Max	2.7/4.6	3.8/6.4	4.8/7.3	6.4/9.0	6.1/9.6	6.3/9.3	4.8/9.1	6.0/8.3	4.7/7.6	4.3/7.1	2.9/5.0	2.6/4.3	5.5/6.5
Latitude -15	Average	4.2	5.4	6.5	7.8	7.9	7.8	7.3	7.2	6.5	6.3	4.8	4.0	6.3
	Min/Max	3.0/5.4	4.2/7.3	5.1/7.9	6.6/9.3	6.1/9.6	6.2/9.2	4.8/9.1	6.1/8.4	4.9/8.1	4.7/7.9	3.2/5.8	3.0/5.2	5.8/6.8
Latitude	Average	4.6	5.7	6.7	7.7	7.7	7.5	7.0	7.1	6.6	6.5	5.2	4.4	6.4
	Min/Max	3.3/5.9	4.4/7.8	5.2/8.1	6.5/9.3	5.9/9.3	5.9/8.8	4.6/8.7	6.0/8.3	4.8/8.2	4.8/8.3	3.4/6.3	3.2/5.7	5.8/6.9
Latitude +15	Average	4.8	5.8	6.6	7.4	7.2	6.9	6.6	6.7	6.4	6.6	5.4	4.6	6.3
	Min/Max	3.3/6.2	4.4/8.0	5.1/8.1	6.2/8.9	5.6/8.8	5.5/8.2	4.3/8.1	5.7/7.9	4.7/8.0	4.8/8.4	3.5/6.6	3.3/6.1	5.7/6.8

Solar Radiation for 2-Axis Tracking Flat-Plate Collectors (kWh/m²/day), Uncertainty ±9%

Tracker		Jan	Feb	Mar	Apr	May	June	July	Aug	Sept	Oct	Nov	Dec	Year
2-Axis	Average	4.8	5.8	6.7	7.8	8.0	8.0	7.4	7.2	6.6	6.6	5.5	4.7	6.6
	Min/Max	3.4/6.3	4.5/8.0	5.2/8.1	6.6/9.4	6.2/9.7	6.3/9.4	4.9/9.2	6.1/8.5	4.9/8.2	4.8/8.4	3.5/6.7	3.4/6.2	6.0/7.1

Direct Beam Solar Radiation for Concentrating Collectors (kWh/m²/day), Uncertainty ±8%

Tracker		Jan	Feb	Mar	Apr	May	June	July	Aug	Sept	Oct	Nov	Dec	Year
1-Axis, E-W Horiz Axis	Average	2.8	3.2	3.3	3.8	3.7	3.6	3.2	3.1	2.9	3.6	3.2	2.8	3.3
	Min/Max	1.6/4.0	2.0/5.0	2.0/4.6	2.9/5.0	2.2/5.1	2.2/4.8	1.4/4.8	2.1/4.2	1.6/4.2	2.1/5.1	1.5/4.2	1.6/4.2	2.9/3.7
1-Axis, N-S Horiz Axis	Average	2.4	3.2	4.0	5.0	4.9	4.7	4.1	4.1	3.6	3.8	2.8	2.2	3.7
	Min/Max	1.3/3.4	2.0/5.1	2.4/5.5	3.8/6.7	2.9/6.8	2.8/6.1	1.8/6.2	2.8/5.5	2.0/5.2	2.3/5.5	1.3/3.7	1.3/3.3	3.2/4.3
1-Axis, N-S Tilt=Latitude	Average	3.1	3.9	4.5	5.1	4.7	4.3	3.9	4.1	3.9	4.5	3.6	3.1	4.1
	Min/Max	1.8/4.5	2.5/6.2	2.7/6.2	3.8/6.8	2.8/6.5	2.6/5.7	1.7/5.8	2.8/5.5	2.1/5.7	2.8/6.5	1.7/4.8	1.7/4.5	3.5/4.6
2-Axis	Average	3.4	4.0	4.5	5.2	5.0	4.7	4.2	4.2	4.0	4.6	3.8	3.3	4.2
	Min/Max	1.9/4.8	2.6/6.4	2.7/6.2	3.9/6.9	2.9/6.9	2.8/6.2	1.8/6.3	2.9/5.7	2.1/5.7	2.8/6.6	1.8/5.0	1.9/4.9	3.7/4.8

Average Climatic Conditions

Element	Jan	Feb	Mar	Apr	May	June	July	Aug	Sept	Oct	Nov	Dec	Year
Temperature (°C)	7.4	9.3	13.8	17.9	22.2	25.8	27.3	26.9	24.1	18.2	13.3	9.3	18.0
Daily Minimum Temp	1.2	2.7	6.8	10.5	15.2	19.3	21.4	21.0	17.9	10.9	6.2	2.8	11.3
Daily Maximum Temp	13.7	16.0	20.8	25.4	29.2	32.3	33.3	32.8	30.2	25.5	20.4	15.7	24.6
Record Minimum Temp	-21.1	-12.8	-10.0	-1.7	4.4	7.8	12.2	12.8	1.7	-3.3	-12.2	-15.0	-21.1
Record Maximum Temp	28.9	29.4	35.0	35.6	37.2	41.1	42.2	40.6	38.9	37.8	31.1	27.8	42.2
HDD, Base 18.3°C	342	252	148	46	4	0	0	0	3	59	159	284	1297
CDD, Base 18.3°C	4	0	8	34	123	225	279	265	174	56	9	3	1181
Relative Humidity (%)	70	67	67	65	69	71	74	76	76	71	71	71	71
Wind Speed (m/s)	3.4	3.8	3.8	3.7	3.2	3.1	3.0	2.8	2.9	2.9	3.0	3.3	3.2

Savannah, GA

WBAN NO. 03822

LATITUDE: 32.13° N
LONGITUDE: 81.20° W
ELEVATION: 16 meters
MEAN PRESSURE: 1017 millibars

STATION TYPE: Primary

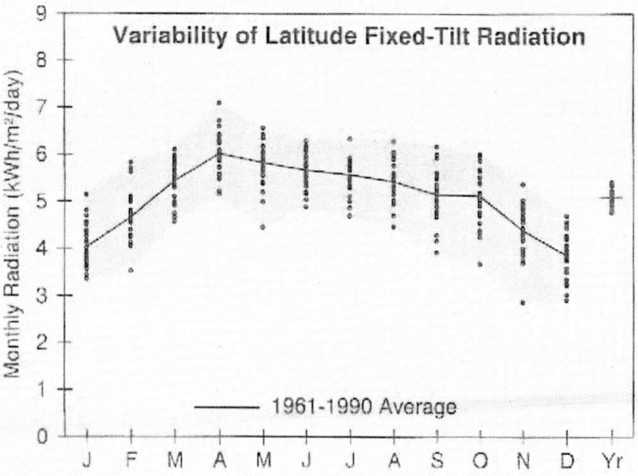

Solar Radiation for Flat-Plate Collectors Facing South at a Fixed Tilt (kWh/m²/day), Uncertainty ±9%

Tilt (°)		Jan	Feb	Mar	Apr	May	June	July	Aug	Sept	Oct	Nov	Dec	Year
0	Average	2.8	3.5	4.7	5.8	6.2	6.3	6.1	5.5	4.7	4.1	3.1	2.6	4.6
	Min/Max	2.4/3.3	2.7/4.2	4.1/5.2	5.0/6.8	4.8/7.0	5.4/7.0	5.1/6.9	4.6/6.3	3.7/5.5	3.1/4.6	2.2/3.5	2.0/2.9	4.4/4.8
Latitude -15	Average	3.5	4.3	5.2	6.1	6.2	6.1	6.0	5.6	5.1	4.8	3.9	3.4	5.0
	Min/Max	3.0/4.4	3.3/5.3	4.5/5.8	5.3/7.2	4.7/7.0	5.3/6.8	5.0/6.8	4.6/6.5	3.9/6.0	3.5/5.6	2.6/4.7	2.6/4.0	4.7/5.3
Latitude	Average	4.0	4.7	5.4	6.0	5.8	5.7	5.6	5.4	5.1	5.1	4.4	3.9	5.1
	Min/Max	3.4/5.1	3.5/5.8	4.6/6.1	5.2/7.1	4.4/6.6	4.9/6.3	4.7/6.3	4.5/6.3	3.9/6.2	3.7/6.0	2.9/5.4	2.9/4.7	4.8/5.4
Latitude +15	Average	4.3	4.8	5.3	5.6	5.2	4.9	4.9	5.0	4.9	5.2	4.6	4.2	4.9
	Min/Max	3.5/5.6	3.6/6.1	4.4/6.0	4.8/6.6	4.0/5.8	4.3/5.4	4.2/5.6	4.1/5.7	3.7/5.9	3.7/6.1	2.9/5.8	3.1/5.1	4.6/5.2
90	Average	3.7	3.8	3.6	3.0	2.3	2.0	2.1	2.5	3.0	3.8	3.9	3.7	3.1
	Min/Max	3.0/4.9	2.8/4.8	3.0/4.0	2.6/3.4	2.0/2.5	1.9/2.1	1.9/2.2	2.1/2.8	2.3/3.7	2.6/4.6	2.4/4.9	2.7/4.6	2.9/3.3

Solar Radiation for 1-Axis Tracking Flat-Plate Collectors with a North-South Axis (kWh/m²/day), Uncertainty ±9%

Axis Tilt (°)		Jan	Feb	Mar	Apr	May	June	July	Aug	Sept	Oct	Nov	Dec	Year
0	Average	3.8	4.8	6.2	7.7	7.9	7.8	7.4	6.8	6.0	5.5	4.3	3.5	6.0
	Min/Max	3.2/4.9	3.5/6.1	5.1/7.2	6.3/9.3	5.5/9.2	6.3/8.8	5.9/8.6	5.5/8.3	4.5/7.5	3.9/6.6	2.8/5.2	2.7/4.3	5.5/6.4
Latitude -15	Average	4.4	5.4	6.7	8.0	7.9	7.7	7.4	6.9	6.3	6.0	4.9	4.1	6.3
	Min/Max	3.6/5.7	3.9/6.9	5.4/7.7	6.5/9.7	5.5/9.2	6.3/8.7	5.9/8.6	5.6/8.5	4.7/7.9	4.2/7.3	3.1/6.1	3.1/5.1	5.8/6.7
Latitude	Average	4.8	5.7	6.8	7.9	7.7	7.4	7.1	6.8	6.3	6.3	5.3	4.5	6.4
	Min/Max	3.9/6.3	4.2/7.3	5.5/7.9	6.4/9.6	5.3/8.9	6.0/8.4	5.6/8.3	5.5/8.3	4.7/8.0	4.4/7.7	3.2/6.6	3.3/5.6	5.8/6.8
Latitude +15	Average	5.0	5.8	6.7	7.6	7.2	6.8	6.6	6.5	6.2	6.3	5.4	4.8	6.2
	Min/Max	4.0/6.6	4.2/7.5	5.4/7.8	6.2/9.2	5.0/8.4	5.6/7.8	5.2/7.7	5.2/7.9	4.5/7.8	4.4/7.8	3.3/6.9	3.5/6.0	5.7/6.7

Solar Radiation for 2-Axis Tracking Flat-Plate Collectors (kWh/m²/day), Uncertainty ±9%

Tracker		Jan	Feb	Mar	Apr	May	June	July	Aug	Sept	Oct	Nov	Dec	Year
2-Axis	Average	5.0	5.8	6.8	8.0	8.0	7.8	7.5	7.0	6.4	6.4	5.5	4.9	6.6
	Min/Max	4.0/6.7	4.3/7.5	5.6/7.9	6.5/9.7	5.6/9.3	6.4/8.9	5.9/8.7	5.6/8.5	4.7/8.0	4.4/7.8	3.3/7.0	3.5/6.1	6.0/7.0

Direct Beam Solar Radiation for Concentrating Collectors (kWh/m²/day), Uncertainty ±8%

Tracker		Jan	Feb	Mar	Apr	May	June	July	Aug	Sept	Oct	Nov	Dec	Year
1-Axis, E-W	Average	2.9	3.1	3.4	3.9	3.6	3.5	3.2	2.9	2.7	3.3	3.2	2.9	3.2
Horiz Axis	Min/Max	2.0/4.3	2.2/4.4	2.4/4.3	2.7/5.4	1.6/4.8	2.2/4.4	2.1/4.2	2.0/4.0	1.6/4.0	2.0/4.5	1.5/4.7	1.7/4.1	2.7/3.7
1-Axis, N-S	Average	2.5	3.1	4.1	5.1	4.8	4.5	4.1	3.7	3.4	3.6	2.8	2.3	3.7
Horiz Axis	Min/Max	1.7/3.6	2.1/4.5	2.9/5.2	3.5/7.1	2.1/6.3	2.8/5.7	2.7/5.4	2.6/5.3	2.0/5.0	2.2/5.0	1.3/4.2	1.3/3.2	3.1/4.1
1-Axis, N-S	Average	3.3	3.9	4.6	5.3	4.6	4.2	3.9	3.7	3.6	4.2	3.6	3.1	4.0
Tilt=Latitude	Min/Max	2.3/4.8	2.7/5.5	3.3/5.7	3.6/7.2	2.0/6.1	2.6/5.3	2.6/5.1	2.6/5.3	2.2/5.4	2.6/5.8	1.7/5.4	1.8/4.4	3.3/4.5
2-Axis	Average	3.5	4.0	4.6	5.3	4.9	4.5	4.2	3.8	3.6	4.2	3.8	3.4	4.2
	Min/Max	2.5/5.2	2.7/5.7	3.3/5.8	3.7/7.3	2.2/6.4	2.9/5.8	2.8/5.5	2.7/5.5	2.2/5.4	2.6/5.9	1.8/5.7	2.0/4.8	3.5/4.7

Average Climatic Conditions

Element	Jan	Feb	Mar	Apr	May	June	July	Aug	Sept	Oct	Nov	Dec	Year
Temperature (°C)	9.4	11.0	15.1	18.9	23.1	26.2	27.7	27.2	24.8	19.6	15.1	10.9	19.1
Daily Minimum Temp	3.4	5.1	9.1	12.5	17.2	20.7	22.4	22.3	19.9	13.8	8.9	5.0	13.3
Daily Maximum Temp	15.4	16.9	21.2	25.3	28.9	31.6	32.8	32.1	29.6	25.3	21.1	16.8	24.7
Record Minimum Temp	-16.1	-10.0	-6.7	0.0	3.9	10.6	16.1	13.9	6.1	-2.2	-9.4	-12.8	-16.1
Record Maximum Temp	28.9	30.0	32.8	35.0	37.8	40.0	40.6	40.0	36.7	36.1	31.7	28.3	40.6
HDD, Base 18.3°C	287	210	113	26	0	0	0	0	0	35	118	237	1026
CDD, Base 18.3°C	9	5	13	43	148	235	289	276	193	75	20	8	1314
Relative Humidity (%)	70	67	67	65	70	74	76	79	78	73	72	71	72
Wind Speed (m/s)	3.8	4.1	4.1	3.9	3.4	3.3	3.2	2.9	3.2	3.3	3.4	3.4	3.5

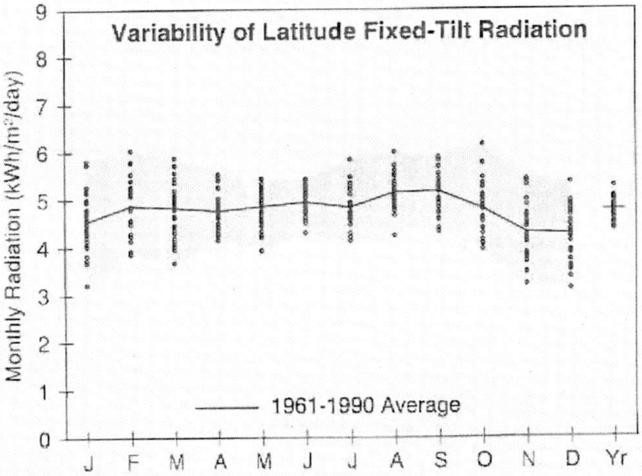

Hilo, HI

WBAN NO. 21504

LATITUDE: 19.72° N
LONGITUDE: 155.07° W
ELEVATION: 11 meters
MEAN PRESSURE: 1015 millibars

STATION TYPE: Secondary

Solar Radiation for Flat-Plate Collectors Facing South at a Fixed Tilt (kWh/m²/day), Uncertainty ±9%

Tilt (°)		Jan	Feb	Mar	Apr	May	June	July	Aug	Sept	Oct	Nov	Dec	Year
0	Average	3.8	4.3	4.6	4.8	5.2	5.4	5.2	5.3	5.0	4.3	3.7	3.5	4.6
	Min/Max	2.8/4.6	3.5/5.1	3.5/5.5	4.2/5.6	4.2/5.8	4.6/5.9	4.4/6.4	4.4/6.2	4.2/5.7	3.7/5.4	2.9/4.5	2.8/4.2	4.2/5.0
Latitude -15	Average	4.0	4.4	4.7	4.8	5.1	5.3	5.1	5.3	5.1	4.5	3.9	3.7	4.7
	Min/Max	2.9/4.9	3.6/5.4	3.6/5.6	4.2/5.6	4.1/5.8	4.6/5.8	4.4/6.3	4.4/6.2	4.3/5.8	3.8/5.6	3.0/4.7	2.9/4.5	4.3/5.1
Latitude	Average	4.5	4.9	4.8	4.7	4.8	4.9	4.8	5.1	5.2	4.8	4.3	4.3	4.8
	Min/Max	3.2/5.8	3.8/6.0	3.7/5.9	4.1/5.5	3.9/5.4	4.3/5.4	4.1/5.8	4.2/6.0	4.3/5.9	4.0/6.2	3.2/5.4	3.1/5.3	4.4/5.3
Latitude +15	Average	4.9	5.0	4.7	4.4	4.3	4.3	4.3	4.7	5.0	4.8	4.5	4.6	4.6
	Min/Max	3.3/6.3	3.9/6.3	3.6/5.8	3.9/5.2	3.5/4.8	3.8/4.7	3.7/5.1	3.9/5.4	4.1/5.7	4.0/6.3	3.3/5.8	3.2/5.9	4.2/5.1
90	Average	3.8	3.4	2.6	1.8	1.4	1.4	1.4	1.6	2.3	3.0	3.3	3.7	2.5
	Min/Max	2.5/5.1	2.6/4.3	2.0/3.1	1.7/2.0	1.4/1.5	1.3/1.5	1.3/1.5	1.5/1.7	2.0/2.5	2.5/3.9	2.3/4.4	2.4/4.8	2.1/2.7

Solar Radiation for 1-Axis Tracking Flat-Plate Collectors with a North-South Axis (kWh/m²/day), Uncertainty ±9%

Axis Tilt (°)		Jan	Feb	Mar	Apr	May	June	July	Aug	Sept	Oct	Nov	Dec	Year
0	Average	4.9	5.5	5.6	5.8	6.2	6.5	6.3	6.6	6.3	5.5	4.6	4.5	5.7
	Min/Max	3.3/6.5	4.1/7.1	3.9/7.0	4.8/7.2	4.7/7.3	5.3/7.4	5.1/8.1	5.1/8.0	5.0/7.6	4.4/7.4	3.3/6.0	3.1/5.9	5.1/6.5
Latitude -15	Average	5.1	5.6	5.7	5.8	6.2	6.5	6.2	6.6	6.4	5.6	4.8	4.7	5.8
	Min/Max	3.3/6.8	4.2/7.3	4.0/7.1	4.8/7.2	4.7/7.2	5.3/7.3	5.1/8.1	5.1/8.0	5.1/7.7	4.5/7.5	3.3/6.2	3.2/6.1	5.2/6.6
Latitude	Average	5.5	5.9	5.8	5.7	6.0	6.2	6.0	6.5	6.4	5.8	5.1	5.1	5.8
	Min/Max	3.6/7.4	4.3/7.7	4.0/7.3	4.8/7.2	4.5/7.0	5.1/7.0	4.9/7.8	5.0/7.9	5.1/7.7	4.6/7.9	3.5/6.7	3.4/6.7	5.2/6.7
Latitude +15	Average	5.7	6.0	5.8	5.5	5.6	5.8	5.6	6.1	6.3	5.9	5.3	5.4	5.7
	Min/Max	3.7/7.8	4.4/7.9	4.0/7.2	4.6/6.9	4.3/6.6	4.7/6.5	4.6/7.2	4.7/7.5	5.0/7.6	4.6/8.0	3.6/7.0	3.5/7.1	5.1/6.6

Solar Radiation for 2-Axis Tracking Flat-Plate Collectors (kWh/m²/day), Uncertainty ±9%

Tracker		Jan	Feb	Mar	Apr	May	June	July	Aug	Sept	Oct	Nov	Dec	Year
2-Axis	Average	5.8	6.1	5.9	5.8	6.2	6.6	6.3	6.6	6.4	5.9	5.3	5.4	6.0
	Min/Max	3.7/7.9	4.4/8.0	4.0/7.3	4.8/7.3	4.7/7.3	5.4/7.5	5.2/8.2	5.1/8.1	5.1/7.8	4.7/8.0	3.6/7.1	3.5/7.3	5.4/6.9

Direct Beam Solar Radiation for Concentrating Collectors (kWh/m²/day), Uncertainty ±8%

Tracker		Jan	Feb	Mar	Apr	May	June	July	Aug	Sept	Oct	Nov	Dec	Year
1-Axis, E-W Horiz Axis	Average	3.2	3.1	2.5	2.3	2.7	3.0	2.8	3.0	3.0	2.8	2.6	2.9	2.8
	Min/Max	1.5/5.0	1.5/4.7	1.4/3.8	1.5/3.4	1.5/3.6	2.0/3.9	1.7/4.5	1.9/4.4	1.9/4.0	1.8/4.6	1.4/4.2	1.2/4.5	2.4/3.6
1-Axis, N-S Horiz Axis	Average	3.1	3.5	3.2	3.0	3.4	3.8	3.5	4.0	3.9	3.3	2.7	2.8	3.3
	Min/Max	1.4/5.1	1.7/5.3	1.7/4.7	1.9/4.6	1.9/4.7	2.5/4.9	2.1/5.8	2.4/5.7	2.5/5.3	2.1/5.4	1.4/4.3	1.1/4.3	2.8/4.3
1-Axis, N-S Tilt=Latitude	Average	3.6	3.8	3.4	3.0	3.3	3.5	3.3	3.9	4.0	3.5	3.1	3.3	3.5
	Min/Max	1.6/5.8	1.9/5.9	1.8/5.0	1.9/4.6	1.8/4.4	2.4/4.6	2.0/5.4	2.4/5.5	2.5/5.4	2.3/5.9	1.6/4.9	1.3/5.1	2.9/4.5
2-Axis	Average	3.9	3.9	3.4	3.1	3.5	3.8	3.6	4.0	4.0	3.6	3.2	3.5	3.6
	Min/Max	1.8/6.2	2.0/6.1	1.8/5.0	2.0/4.6	1.9/4.7	2.6/5.0	2.2/5.8	2.5/5.7	2.5/5.4	2.3/5.9	1.6/5.2	1.4/5.5	3.1/4.7

Average Climatic Conditions

Element	Jan	Feb	Mar	Apr	May	June	July	Aug	Sept	Oct	Nov	Dec	Year
Temperature (°C)	22.1	22.1	22.2	22.6	23.3	24.0	24.3	24.6	24.6	24.3	23.4	22.4	23.3
Daily Minimum Temp	17.6	17.6	18.0	18.6	19.2	19.8	20.3	20.5	20.3	20.1	19.3	18.2	19.1
Daily Maximum Temp	26.6	26.6	26.4	26.6	27.3	28.2	28.3	28.7	28.8	28.4	27.4	26.7	27.5
Record Minimum Temp	12.2	11.7	12.2	13.3	14.4	15.6	16.7	17.2	16.1	16.7	14.4	12.8	11.7
Record Maximum Temp	32.8	33.3	33.9	31.7	34.4	32.2	31.7	33.9	33.3	32.8	32.2	33.9	34.4
HDD, Base 18.3°C	0	0	0	0	0	0	0	0	0	0	0	0	0
CDD, Base 18.3°C	116	104	121	128	153	170	186	194	187	184	153	127	1824
Relative Humidity (%)	77	76	78	80	79	77	80	80	79	80	80	79	79
Wind Speed (m/s)	3.5	3.6	3.6	3.6	3.5	3.4	3.3	3.2	3.2	3.2	3.3	3.4	3.4

Honolulu, HI

WBAN NO. 22521

LATITUDE: 21.33° N
LONGITUDE: 157.92° W
ELEVATION: 5 meters
MEAN PRESSURE: 1016 millibars

STATION TYPE: Primary

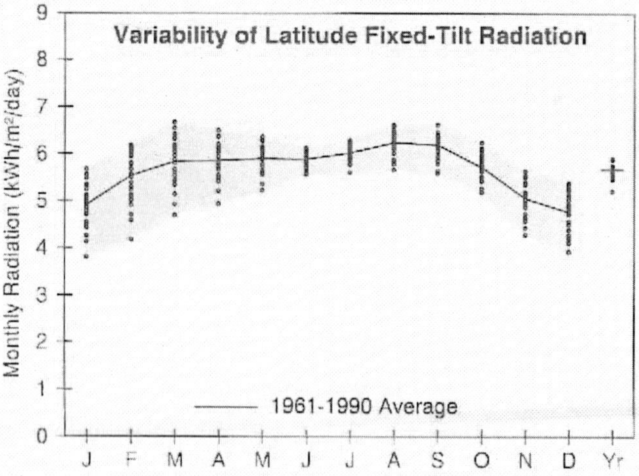

Solar Radiation for Flat-Plate Collectors Facing South at a Fixed Tilt (kWh/m²/day), Uncertainty ±9%

Tilt (°)		Jan	Feb	Mar	Apr	May	June	July	Aug	Sept	Oct	Nov	Dec	Year
0	Average	3.9	4.7	5.4	5.9	6.4	6.5	6.6	6.5	5.9	5.0	4.1	3.7	5.4
	Min/Max	3.2/4.4	3.7/5.2	4.4/6.1	5.0/6.5	5.6/6.9	6.2/6.9	6.1/7.0	5.8/6.9	5.4/6.3	4.6/5.4	3.7/4.5	3.2/4.1	5.0/5.6
Latitude -15	Average	4.3	5.0	5.6	5.9	6.3	6.4	6.5	6.5	6.1	5.3	4.5	4.1	5.5
	Min/Max	3.4/4.8	3.9/5.5	4.6/6.3	5.1/6.6	5.6/6.8	6.1/6.7	6.0/6.8	5.9/6.9	5.5/6.5	4.9/5.7	3.9/4.9	3.5/4.6	5.1/5.8
Latitude	Average	4.9	5.5	5.8	5.9	5.9	5.9	6.0	6.2	6.2	5.7	5.1	4.8	5.7
	Min/Max	3.8/5.7	4.2/6.2	4.7/6.7	4.9/6.5	5.2/6.4	5.6/6.1	5.6/6.3	5.7/6.6	5.6/6.6	5.2/6.2	4.3/5.6	3.9/5.4	5.2/5.9
Latitude +15	Average	5.3	5.8	5.8	5.5	5.3	5.1	5.3	5.7	6.0	5.8	5.4	5.2	5.5
	Min/Max	4.0/6.2	4.3/6.5	4.6/6.6	4.6/6.1	4.7/5.6	4.8/5.3	4.9/5.4	5.2/6.0	5.4/6.4	5.3/6.4	4.5/6.0	4.2/5.9	5.0/5.7
90	Average	4.2	4.0	3.2	2.2	1.5	1.4	1.4	1.8	2.8	3.7	4.1	4.3	2.9
	Min/Max	3.1/5.0	2.9/4.6	2.6/3.6	1.9/2.3	1.5/1.6	1.4/1.5	1.4/1.5	1.8/1.9	2.6/3.0	3.3/4.1	3.3/4.7	3.3/4.9	2.6/3.1

Solar Radiation for 1-Axis Tracking Flat-Plate Collectors with a North-South Axis (kWh/m²/day), Uncertainty ±9%

Axis Tilt (°)		Jan	Feb	Mar	Apr	May	June	July	Aug	Sept	Oct	Nov	Dec	Year
0	Average	5.3	6.4	7.2	7.7	8.3	8.5	8.7	8.7	8.0	6.8	5.6	5.1	7.2
	Min/Max	4.1/6.3	4.7/7.3	5.4/8.4	6.2/8.9	6.9/9.4	7.9/9.3	7.9/9.3	7.4/9.4	7.0/8.7	6.0/7.6	4.5/6.2	4.0/5.9	6.4/7.6
Latitude -15	Average	5.6	6.6	7.3	7.7	8.3	8.5	8.6	8.7	8.1	7.0	5.8	5.3	7.3
	Min/Max	4.2/6.6	4.8/7.6	5.5/8.6	6.2/9.0	6.9/9.3	7.8/9.2	7.9/9.3	7.4/9.4	7.1/8.9	6.2/7.8	4.7/6.5	4.2/6.2	6.5/7.7
Latitude	Average	6.1	7.0	7.5	7.7	8.0	8.1	8.3	8.5	8.2	7.3	6.2	5.9	7.4
	Min/Max	4.5/7.3	5.0/8.1	5.7/8.9	6.1/8.9	6.7/9.0	7.5/8.8	7.6/8.9	7.3/9.2	7.2/9.0	6.4/8.2	5.0/7.0	4.6/6.8	6.6/7.8
Latitude +15	Average	6.3	7.2	7.5	7.4	7.5	7.5	7.7	8.1	8.0	7.4	6.5	6.2	7.3
	Min/Max	4.7/7.6	5.1/8.3	5.6/8.8	5.9/8.6	6.3/8.5	7.0/8.2	7.1/8.3	6.9/8.8	7.0/8.8	6.5/8.3	5.2/7.3	4.8/7.2	6.4/7.7

Solar Radiation for 2-Axis Tracking Flat-Plate Collectors (kWh/m²/day), Uncertainty ±9%

Tracker		Jan	Feb	Mar	Apr	May	June	July	Aug	Sept	Oct	Nov	Dec	Year
2-Axis	Average	6.4	7.2	7.5	7.8	8.3	8.6	8.8	8.7	8.2	7.4	6.5	6.3	7.7
	Min/Max	4.7/7.8	5.1/8.3	5.7/8.9	6.2/9.0	7.0/9.4	8.0/9.4	8.0/9.4	7.4/9.4	7.2/9.0	6.5/8.3	5.2/7.4	4.8/7.3	6.8/8.1

Direct Beam Solar Radiation for Concentrating Collectors (kWh/m²/day), Uncertainty ±8%

Tracker		Jan	Feb	Mar	Apr	May	June	July	Aug	Sept	Oct	Nov	Dec	Year
1-Axis, E-W Horiz Axis	Average	3.5	3.8	3.7	3.6	4.1	4.4	4.6	4.4	4.1	3.8	3.5	3.5	3.9
	Min/Max	2.2/4.8	2.1/4.8	2.3/4.8	2.5/4.7	3.0/5.1	3.9/5.2	3.8/5.3	3.3/5.3	3.2/4.7	2.9/4.5	2.3/4.4	2.2/4.4	3.2/4.3
1-Axis, N-S Horiz Axis	Average	3.5	4.3	4.7	4.9	5.5	5.7	6.0	6.0	5.5	4.5	3.6	3.3	4.8
	Min/Max	2.1/4.7	2.4/5.5	2.9/6.2	3.3/6.4	3.9/6.9	5.0/6.8	5.0/7.0	4.4/7.1	4.3/6.3	3.5/5.5	2.3/4.4	2.0/4.2	3.9/5.3
1-Axis, N-S Tilt=Latitude	Average	4.1	4.8	5.0	4.9	5.2	5.4	5.7	5.9	5.6	4.9	4.1	4.0	4.9
	Min/Max	2.5/5.5	2.7/6.1	3.1/6.5	3.3/6.4	3.7/6.6	4.7/6.4	4.7/6.6	4.3/6.9	4.4/6.5	3.8/5.9	2.7/5.1	2.4/5.0	4.0/5.4
2-Axis	Average	4.3	4.9	5.0	4.9	5.5	5.8	6.1	6.0	5.6	5.0	4.3	4.3	5.2
	Min/Max	2.7/5.9	2.8/6.3	3.1/6.6	3.4/6.5	3.9/7.0	5.1/6.9	5.1/7.1	4.4/7.1	4.4/6.5	3.9/6.0	2.8/5.4	2.6/5.4	4.2/5.6

Average Climatic Conditions

Element	Jan	Feb	Mar	Apr	May	June	July	Aug	Sept	Oct	Nov	Dec	Year
Temperature (°C)	22.7	22.8	23.6	24.3	25.3	26.3	26.9	27.4	27.2	26.4	25.1	23.4	25.1
Daily Minimum Temp	18.7	18.6	19.6	20.4	21.3	22.3	23.1	23.4	23.1	22.4	21.3	19.4	21.1
Daily Maximum Temp	26.7	26.9	27.6	28.2	29.3	30.3	30.8	31.5	31.4	30.5	28.9	27.3	29.1
Record Minimum Temp	11.7	11.7	12.8	13.9	15.6	18.3	18.9	19.4	18.9	17.8	13.9	12.2	11.7
Record Maximum Temp	30.6	31.1	31.1	31.7	33.9	33.3	33.3	33.9	34.4	34.4	33.9	31.7	34.4
HDD, Base 18.3°C	0	0	0	0	0	0	0	0	0	0	0	0	0
CDD, Base 18.3°C	136	124	162	180	216	240	267	282	267	252	203	157	2486
Relative Humidity (%)	73	71	69	67	66	64	65	64	66	68	70	72	68
Wind Speed (m/s)	4.3	4.3	5.0	5.1	5.1	5.5	5.7	5.4	4.9	4.6	4.6	4.3	4.9

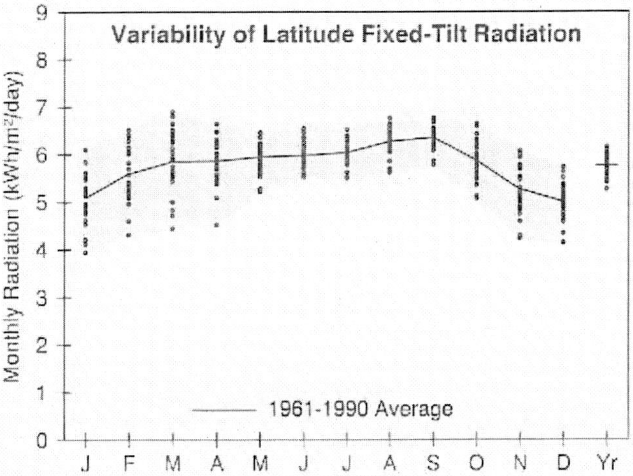

Kahului, HI

WBAN NO. 22516

LATITUDE: 20.90° N
LONGITUDE: 156.43° W
ELEVATION: 15 meters
MEAN PRESSURE: 1013 millibars

STATION TYPE: Secondary

Solar Radiation for Flat-Plate Collectors Facing South at a Fixed Tilt (kWh/m²/day), Uncertainty ±9%

Tilt (°)		Jan	Feb	Mar	Apr	May	June	July	Aug	Sept	Oct	Nov	Dec	Year
0	Average	4.0	4.7	5.4	5.9	6.4	6.7	6.7	6.5	6.1	5.1	4.3	3.9	5.5
	Min/Max	3.3/4.7	3.8/5.4	4.2/6.3	4.6/6.7	5.6/7.0	6.2/7.4	6.0/7.3	5.8/7.1	5.6/6.5	4.5/5.7	3.6/4.8	3.3/4.3	5.0/5.8
Latitude -15	Average	4.4	5.0	5.6	5.9	6.4	6.6	6.6	6.5	6.2	5.4	4.6	4.2	5.6
	Min/Max	3.5/5.1	4.0/5.8	4.3/6.5	4.6/6.7	5.6/6.9	6.0/7.3	5.9/7.1	5.8/7.1	5.7/6.6	4.7/6.1	3.8/5.2	3.6/4.8	5.2/6.0
Latitude	Average	5.1	5.6	5.9	5.9	5.9	6.0	6.0	6.3	6.3	5.9	5.3	5.0	5.8
	Min/Max	3.9/6.1	4.3/6.5	4.4/6.9	4.5/6.6	5.2/6.5	5.5/6.6	5.5/6.5	5.6/6.8	5.8/6.8	5.1/6.6	4.2/6.1	4.1/5.7	5.3/6.2
Latitude +15	Average	5.5	5.8	5.8	5.5	5.3	5.1	5.2	5.7	6.1	6.0	5.6	5.5	5.6
	Min/Max	4.2/6.7	4.4/6.8	4.3/6.9	4.2/6.2	4.6/5.7	4.8/5.5	4.8/5.6	5.1/6.1	5.6/6.5	5.2/6.8	4.4/6.6	4.5/6.3	5.1/6.0
90	Average	4.4	4.0	3.1	2.1	1.4	1.3	1.3	1.7	2.8	3.8	4.3	4.5	2.9
	Min/Max	3.3/5.4	3.0/4.8	2.5/3.7	1.8/2.2	1.4/1.5	1.2/1.4	1.2/1.4	1.6/1.8	2.6/2.9	3.3/4.3	3.3/5.1	3.6/5.3	2.6/3.1

Solar Radiation for 1-Axis Tracking Flat-Plate Collectors with a North-South Axis (kWh/m²/day), Uncertainty ±9%

Axis Tilt (°)		Jan	Feb	Mar	Apr	May	June	July	Aug	Sept	Oct	Nov	Dec	Year
0	Average	5.6	6.5	7.3	7.8	8.6	9.0	9.0	8.9	8.4	7.1	5.9	5.4	7.5
	Min/Max	4.3/6.9	4.7/7.8	5.2/9.0	5.6/9.3	7.2/9.7	8.1/10.5	7.8/10.0	7.5/9.9	7.4/9.2	6.0/8.3	4.5/7.0	4.4/6.4	6.6/8.1
Latitude -15	Average	5.9	6.8	7.5	7.9	8.5	8.9	8.9	8.9	8.5	7.3	6.2	5.7	7.6
	Min/Max	4.4/7.3	4.8/8.1	5.3/9.2	5.6/9.3	7.1/9.7	8.0/10.4	7.8/9.9	7.5/9.9	7.5/9.3	6.2/8.5	4.7/7.3	4.6/6.7	6.7/8.2
Latitude	Average	6.4	7.2	7.7	7.8	8.3	8.5	8.6	8.7	8.6	7.6	6.7	6.2	7.7
	Min/Max	4.8/8.0	5.1/8.6	5.4/9.4	5.5/9.3	6.9/9.4	7.7/9.9	7.5/9.5	7.3/9.8	7.5/9.4	6.4/8.9	5.0/7.9	5.0/7.4	6.8/8.3
Latitude +15	Average	6.7	7.3	7.6	7.5	7.8	7.9	8.0	8.3	8.4	7.7	6.9	6.6	7.6
	Min/Max	4.9/8.4	5.2/8.9	5.3/9.4	5.3/8.9	6.5/8.8	7.1/9.2	7.0/8.9	7.0/9.3	7.4/9.3	6.5/9.1	5.2/8.3	5.3/7.8	6.7/8.2

Solar Radiation for 2-Axis Tracking Flat-Plate Collectors (kWh/m²/day), Uncertainty ±9%

Tracker		Jan	Feb	Mar	Apr	May	June	July	Aug	Sept	Oct	Nov	Dec	Year
2-Axis	Average	6.8	7.4	7.7	7.9	8.7	9.1	9.1	9.0	8.6	7.7	7.0	6.7	8.0
	Min/Max	5.0/8.5	5.2/8.9	5.4/9.5	5.6/9.4	7.2/9.8	8.2/10.7	7.9/10.1	7.5/10.0	7.6/9.4	6.5/9.1	5.2/8.4	5.4/8.0	7.0/8.6

Direct Beam Solar Radiation for Concentrating Collectors (kWh/m²/day), Uncertainty ±8%

Tracker		Jan	Feb	Mar	Apr	May	June	July	Aug	Sept	Oct	Nov	Dec	Year
1-Axis, E-W Horiz Axis	Average	4.0	4.1	4.0	4.0	4.6	5.1	5.1	4.9	4.7	4.3	4.0	4.1	4.4
	Min/Max	2.5/5.7	2.3/5.6	2.1/5.5	2.4/5.3	3.3/5.7	4.2/6.6	3.8/6.1	3.7/5.9	3.8/5.4	3.2/5.5	2.4/5.3	2.8/5.3	3.6/5.0
1-Axis, N-S Horiz Axis	Average	4.0	4.7	5.2	5.4	6.2	6.7	6.7	6.7	6.3	5.2	4.2	3.9	5.4
	Min/Max	2.5/5.7	2.5/6.4	2.7/7.3	3.2/7.2	4.3/7.8	5.4/8.8	5.1/8.1	4.9/8.2	5.2/7.4	3.8/6.7	2.5/5.6	2.7/5.1	4.4/6.2
1-Axis, N-S Tilt=Latitude	Average	4.6	5.2	5.4	5.4	5.9	6.2	6.4	6.6	6.5	5.6	4.8	4.6	5.6
	Min/Max	3.0/6.6	2.8/7.1	2.9/7.6	3.1/7.2	4.1/7.4	5.1/8.2	4.8/7.6	4.8/8.0	5.3/7.6	4.2/7.2	2.8/6.4	3.2/5.9	4.5/6.4
2-Axis	Average	5.0	5.4	5.5	5.5	6.3	6.8	6.8	6.8	6.5	5.7	5.1	5.0	5.8
	Min/Max	3.2/7.1	2.9/7.3	2.9/7.7	3.2/7.3	4.3/7.8	5.5/8.9	5.1/8.2	4.9/8.3	5.3/7.6	4.2/7.3	3.0/6.7	3.5/6.4	4.7/6.7

Average Climatic Conditions

Element	Jan	Feb	Mar	Apr	May	June	July	Aug	Sept	Oct	Nov	Dec	Year
Temperature (°C)	22.1	22.1	22.8	23.4	24.3	25.3	25.9	26.3	26.0	25.5	24.3	22.8	24.2
Daily Minimum Temp	17.6	17.4	18.1	18.9	19.4	20.6	21.4	21.6	21.0	20.7	19.8	18.4	19.6
Daily Maximum Temp	26.6	26.7	27.3	27.9	29.0	29.9	30.4	30.9	30.9	30.2	28.7	27.3	28.8
Record Minimum Temp	8.9	10.0	12.8	12.2	13.9	14.4	14.4	16.1	15.6	14.4	12.8	11.1	8.9
Record Maximum Temp	31.7	31.1	32.2	32.8	33.3	33.9	34.4	35.6	35.0	35.6	33.9	32.2	35.6
HDD, Base 18.3°C	0	0	0	0	0	0	0	0	0	0	0	0	0
CDD, Base 18.3°C	116	106	138	153	184	208	236	246	230	222	178	139	2157
Relative Humidity (%)	75	74	72	71	70	68	69	69	69	71	73	74	71
Wind Speed (m/s)	5.2	5.4	5.7	6.3	6.7	7.1	7.4	7.1	6.2	5.5	5.7	5.0	6.1

Lihue, HI

WBAN NO. 22536

LATITUDE: 21.98° N
LONGITUDE: 159.35° W
ELEVATION: 45 meters
MEAN PRESSURE: 1012 millibars

STATION TYPE: Secondary

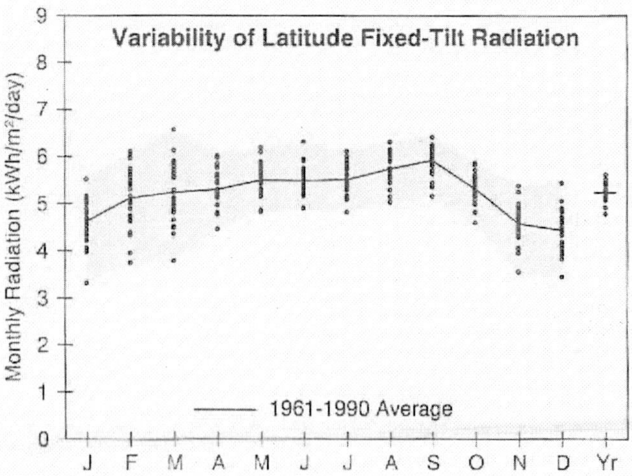

Solar Radiation for Flat-Plate Collectors Facing South at a Fixed Tilt (kWh/m²/day), Uncertainty ±9%

Tilt (°)		Jan	Feb	Mar	Apr	May	June	July	Aug	Sept	Oct	Nov	Dec	Year
0	Average	3.7	4.3	4.9	5.3	5.9	6.1	6.0	5.9	5.6	4.7	3.8	3.5	5.0
	Min/Max	2.8/4.3	3.4/5.0	3.6/6.0	4.5/6.0	5.1/6.7	5.4/7.1	5.2/6.7	5.2/6.5	5.0/6.1	4.1/5.1	3.1/4.3	2.9/4.1	4.6/5.3
Latitude -15	Average	4.0	4.6	5.0	5.4	5.8	5.9	5.9	5.9	5.8	4.9	4.1	3.8	5.1
	Min/Max	3.0/4.7	3.5/5.4	3.7/6.2	4.6/6.1	5.1/6.6	5.3/6.9	5.1/6.6	5.2/6.5	5.1/6.3	4.3/5.4	3.3/4.7	3.1/4.6	4.7/5.5
Latitude	Average	4.6	5.1	5.2	5.3	5.5	5.5	5.5	5.7	5.9	5.3	4.6	4.4	5.2
	Min/Max	3.3/5.5	3.7/6.1	3.8/6.6	4.5/6.0	4.8/6.2	4.9/6.3	4.8/6.1	5.0/6.3	5.2/6.4	4.6/5.9	3.5/5.4	3.4/5.4	4.8/5.6
Latitude +15	Average	5.0	5.3	5.2	4.9	4.9	4.8	4.8	5.2	5.7	5.4	4.8	4.8	5.1
	Min/Max	3.5/6.0	3.8/6.4	3.7/6.5	4.2/5.6	4.3/5.5	4.3/5.4	4.3/5.3	4.6/5.8	4.9/6.2	4.6/6.0	3.7/5.8	3.6/6.0	4.6/5.4
90	Average	3.9	3.7	2.9	2.1	1.5	1.4	1.4	1.8	2.7	3.5	3.7	3.9	2.7
	Min/Max	2.6/4.9	2.5/4.6	2.2/3.6	1.9/2.2	1.5/1.5	1.3/1.5	1.4/1.5	1.7/1.9	2.5/2.9	2.9/3.9	2.7/4.5	2.8/5.1	2.5/2.9

Solar Radiation for 1-Axis Tracking Flat-Plate Collectors with a North-South Axis (kWh/m²/day), Uncertainty ±9%

Axis Tilt (°)		Jan	Feb	Mar	Apr	May	June	July	Aug	Sept	Oct	Nov	Dec	Year
0	Average	4.9	5.8	6.2	6.7	7.4	7.7	7.6	7.6	7.5	6.1	4.9	4.6	6.4
	Min/Max	3.4/6.1	4.0/7.2	4.1/8.3	5.4/7.8	6.1/8.7	6.4/9.5	6.2/8.7	6.3/8.8	6.2/8.2	5.1/6.9	3.6/5.9	3.5/5.9	5.7/7.0
Latitude -15	Average	5.2	6.0	6.4	6.7	7.4	7.6	7.5	7.7	7.6	6.3	5.1	4.9	6.5
	Min/Max	3.5/6.4	4.1/7.5	4.2/8.5	5.4/7.9	6.1/8.7	6.3/9.4	6.2/8.6	6.3/8.8	6.3/8.3	5.3/7.2	3.8/6.2	3.6/6.3	5.8/7.1
Latitude	Average	5.6	6.4	6.5	6.7	7.2	7.3	7.2	7.5	7.7	6.6	5.5	5.4	6.6
	Min/Max	3.8/7.0	4.3/8.0	4.3/8.7	5.3/7.8	5.9/8.4	6.1/9.0	6.0/8.3	6.2/8.6	6.3/8.5	5.5/7.5	4.0/6.7	3.9/6.9	5.8/7.2
Latitude +15	Average	5.9	6.5	6.5	6.4	6.7	6.7	6.7	7.2	7.5	6.7	5.7	5.6	6.5
	Min/Max	3.9/7.4	4.3/8.2	4.2/8.7	5.1/7.6	5.5/7.9	5.6/8.3	5.6/7.7	5.9/8.2	6.2/8.3	5.5/7.6	4.1/7.0	4.1/7.3	5.7/7.1

Solar Radiation for 2-Axis Tracking Flat-Plate Collectors (kWh/m²/day), Uncertainty ±9%

Tracker		Jan	Feb	Mar	Apr	May	June	July	Aug	Sept	Oct	Nov	Dec	Year
2-Axis	Average	6.0	6.5	6.5	6.7	7.5	7.8	7.6	7.7	7.7	6.7	5.7	5.7	6.8
	Min/Max	3.9/7.5	4.3/8.2	4.3/8.7	5.4/7.9	6.2/8.8	6.4/9.6	6.3/8.8	6.3/8.8	6.3/8.5	5.5/7.6	4.1/7.0	4.1/7.5	6.0/7.5

Direct Beam Solar Radiation for Concentrating Collectors (kWh/m²/day), Uncertainty ±8%

Tracker		Jan	Feb	Mar	Apr	May	June	July	Aug	Sept	Oct	Nov	Dec	Year
1-Axis, E-W	Average	3.3	3.4	3.1	3.0	3.7	3.9	3.8	3.9	3.9	3.5	3.0	3.2	3.5
Horiz Axis	Min/Max	1.7/4.7	1.4/4.9	1.4/5.0	2.0/4.1	2.6/4.8	2.7/5.7	2.6/4.9	2.7/4.8	2.7/4.7	2.6/4.3	1.6/4.2	1.7/4.8	2.8/4.1
1-Axis, N-S	Average	3.2	3.8	3.9	4.0	4.8	5.0	4.9	5.1	5.1	4.1	3.0	3.0	4.2
Horiz Axis	Min/Max	1.6/4.6	1.6/5.6	1.6/6.3	2.6/5.4	3.3/6.3	3.4/7.4	3.3/6.3	3.5/6.5	3.5/6.0	3.0/5.0	1.6/4.2	1.6/4.5	3.3/4.9
1-Axis, N-S	Average	3.8	4.3	4.1	4.0	4.5	4.7	4.6	5.0	5.3	4.4	3.5	3.6	4.3
Tilt=Latitude	Min/Max	1.9/5.4	1.8/6.2	1.7/6.7	2.6/5.4	3.1/6.0	3.2/6.9	3.1/5.9	3.5/6.3	3.6/6.2	3.3/5.5	1.8/4.9	1.9/5.4	3.4/5.0
2-Axis	Average	4.0	4.4	4.1	4.0	4.8	5.1	4.9	5.2	5.3	4.5	3.7	3.9	4.5
	Min/Max	2.0/5.8	1.8/6.4	1.8/6.7	2.7/5.5	3.3/6.3	3.5/7.5	3.3/6.4	3.6/6.5	3.6/6.3	3.4/5.6	1.9/5.2	2.1/5.8	3.6/5.2

Average Climatic Conditions

Element	Jan	Feb	Mar	Apr	May	June	July	Aug	Sept	Oct	Nov	Dec	Year
Temperature (°C)	22.0	22.0	22.6	23.3	24.3	25.4	26.1	26.4	26.2	25.3	24.2	22.7	24.2
Daily Minimum Temp	18.5	18.4	19.4	20.3	21.3	22.5	23.2	23.4	23.1	22.2	21.3	19.5	21.1
Daily Maximum Temp	25.5	25.6	25.8	26.4	27.3	28.4	28.9	29.3	29.3	28.4	27.1	25.9	27.3
Record Minimum Temp	10.0	11.1	10.6	13.3	15.0	16.1	16.7	18.9	18.3	16.7	13.9	11.1	10.0
Record Maximum Temp	29.4	30.0	31.1	31.1	31.1	31.7	31.7	32.2	31.7	32.2	30.6	30.0	32.2
HDD, Base 18.3°C	0	0	0	0	0	0	0	0	0	0	0	0	0
CDD, Base 18.3°C	116	104	133	150	186	213	239	250	237	217	175	136	2157
Relative Humidity (%)	76	75	75	75	75	73	74	74	74	76	77	76	75
Wind Speed (m/s)	5.2	5.3	5.9	6.4	6.1	6.1	6.4	5.9	5.4	5.3	5.7	5.2	5.8

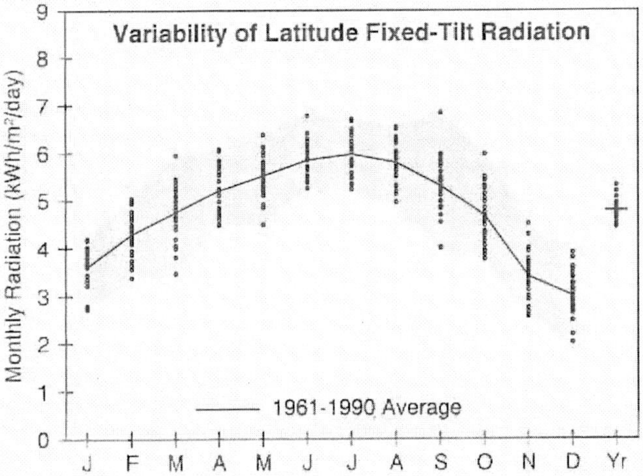

Des Moines, IA

WBAN NO. 14933

LATITUDE: 41.53° N
LONGITUDE: 93.65° W
ELEVATION: 294 meters
MEAN PRESSURE: 982 millibars

STATION TYPE: Secondary

Solar Radiation for Flat-Plate Collectors Facing South at a Fixed Tilt ($kWh/m^2/day$), Uncertainty ±9%

Tilt (°)		Jan	Feb	Mar	Apr	May	June	July	Aug	Sept	Oct	Nov	Dec	Year
0	Average	2.0	2.8	3.8	4.9	5.8	6.5	6.5	5.7	4.4	3.2	2.1	1.7	4.1
	Min/Max	1.7/2.2	2.5/3.2	3.0/4.4	4.3/5.6	4.8/6.7	5.8/7.5	5.8/7.3	4.9/6.3	3.6/5.5	2.8/3.9	1.7/2.4	1.4/2.0	3.9/4.5
Latitude -15	Average	3.2	3.9	4.6	5.3	5.9	6.4	6.5	6.0	5.2	4.4	3.1	2.6	4.8
	Min/Max	2.4/3.6	3.2/4.6	3.4/5.7	4.6/6.2	4.8/6.8	5.7/7.4	5.7/7.3	5.2/6.8	4.0/6.7	3.6/5.5	2.4/4.0	1.9/3.3	4.4/5.3
Latitude	Average	3.6	4.3	4.7	5.2	5.5	5.8	6.0	5.8	5.3	4.7	3.4	3.0	4.8
	Min/Max	2.7/4.2	3.4/5.0	3.5/5.9	4.5/6.1	4.5/6.4	5.3/6.8	5.2/6.7	5.0/6.5	4.0/6.8	3.8/6.0	2.6/4.5	2.0/3.9	4.4/5.3
Latitude +15	Average	3.9	4.4	4.7	4.8	4.9	5.1	5.2	5.3	5.1	4.7	3.6	3.2	4.6
	Min/Max	2.8/4.6	3.5/5.2	3.3/5.9	4.1/5.6	4.0/5.6	4.6/5.8	4.6/5.8	4.5/5.9	3.8/6.6	3.8/6.1	2.6/4.8	2.1/4.3	4.2/5.1
90	Average	3.7	4.0	3.6	3.2	2.8	2.7	2.8	3.2	3.6	3.9	3.2	3.1	3.3
	Min/Max	2.6/4.4	3.0/4.8	2.5/4.5	2.7/3.6	2.4/3.1	2.5/2.9	2.5/3.0	2.8/3.6	2.7/4.7	3.0/5.1	2.4/4.4	1.9/4.2	3.0/3.6

Solar Radiation for 1-Axis Tracking Flat-Plate Collectors with a North-South Axis ($kWh/m^2/day$), Uncertainty ±9%

Axis Tilt (°)		Jan	Feb	Mar	Apr	May	June	July	Aug	Sept	Oct	Nov	Dec	Year
0	Average	3.0	4.0	5.1	6.5	7.6	8.6	8.6	7.7	6.2	4.7	2.9	2.4	5.6
	Min/Max	2.2/3.4	3.1/4.8	3.7/6.5	5.4/7.8	6.0/9.1	7.2/10.3	7.3/10.1	6.4/8.8	4.5/8.3	3.7/6.1	2.3/3.9	1.7/3.1	5.2/6.4
Latitude -15	Average	3.8	4.9	5.8	6.9	7.7	8.6	8.7	8.0	6.8	5.5	3.7	3.1	6.1
	Min/Max	2.8/4.5	3.7/5.9	4.1/7.4	5.7/8.3	6.0/9.4	7.2/10.3	7.4/10.2	6.7/9.2	4.9/9.2	4.3/7.2	2.7/5.0	2.0/4.1	5.6/7.0
Latitude	Average	4.2	5.2	5.9	6.8	7.5	8.2	8.4	7.9	6.8	5.8	4.0	3.4	6.2
	Min/Max	3.0/5.0	3.9/6.3	4.1/7.7	5.6/8.3	5.8/9.1	6.9/9.9	7.1/9.9	6.5/9.1	4.8/9.3	4.4/7.6	2.9/5.4	2.2/4.6	5.7/7.1
Latitude +15	Average	4.4	5.3	5.8	6.5	7.1	7.7	7.9	7.5	6.7	5.8	4.1	3.6	6.0
	Min/Max	3.1/5.3	3.9/6.4	4.0/7.6	5.3/7.9	5.5/8.6	6.4/9.3	6.7/9.3	6.2/8.6	4.7/9.1	4.4/7.7	2.9/5.6	2.3/4.9	5.5/6.9

Solar Radiation for 2-Axis Tracking Flat-Plate Collectors ($kWh/m^2/day$), Uncertainty ±9%

Tracker		Jan	Feb	Mar	Apr	May	June	July	Aug	Sept	Oct	Nov	Dec	Year
2-Axis	Average	4.5	5.3	5.9	6.9	7.8	8.8	8.9	8.1	6.9	5.8	4.1	3.7	6.4
	Min/Max	3.1/5.3	3.9/6.4	4.1/7.7	5.7/8.4	6.1/9.5	7.3/10.6	7.5/10.4	6.7/9.3	4.9/9.3	4.5/7.7	3.0/5.7	2.3/4.9	5.9/7.3

Direct Beam Solar Radiation for Concentrating Collectors ($kWh/m^2/day$), Uncertainty ±8%

Tracker		Jan	Feb	Mar	Apr	May	June	July	Aug	Sept	Oct	Nov	Dec	Year
1-Axis, E-W Horiz Axis	Average	2.6	2.8	2.8	3.1	3.6	4.2	4.3	3.9	3.4	3.2	2.4	2.1	3.2
	Min/Max	1.4/3.3	1.7/3.8	1.6/4.4	2.2/4.4	2.2/5.0	3.0/5.9	3.2/5.5	2.8/4.8	1.9/5.4	2.3/4.7	1.2/3.6	1.0/3.0	2.7/4.0
1-Axis, N-S Horiz Axis	Average	1.8	2.4	3.1	4.0	4.7	5.6	5.7	5.1	4.1	3.2	1.8	1.4	3.6
	Min/Max	1.0/2.3	1.5/3.4	1.7/4.8	2.9/5.7	3.0/6.6	3.8/7.7	4.2/7.4	3.7/6.4	2.3/6.5	2.2/4.7	0.9/2.8	0.6/2.0	3.1/4.5
1-Axis, N-S Tilt=Latitude	Average	2.8	3.3	3.7	4.3	4.7	5.3	5.5	5.3	4.7	4.1	2.7	2.2	4.0
	Min/Max	1.5/3.5	2.0/4.6	2.1/5.8	3.1/6.0	3.0/6.5	3.6/7.3	4.1/7.1	3.8/6.6	2.6/7.4	2.8/6.0	1.3/4.1	1.0/3.2	3.5/5.1
2-Axis	Average	3.0	3.4	3.7	4.4	4.9	5.7	5.9	5.4	4.7	4.1	2.8	2.4	4.2
	Min/Max	1.6/3.8	2.1/4.7	2.1/5.8	3.2/6.1	3.2/6.8	3.9/7.9	4.3/7.6	3.9/6.8	2.6/7.4	2.9/6.0	1.4/4.3	1.1/3.5	3.6/5.3

Average Climatic Conditions

Element	Jan	Feb	Mar	Apr	May	June	July	Aug	Sept	Oct	Nov	Dec	Year
Temperature (°C)	-7.0	-4.1	2.9	10.5	16.8	22.1	24.8	23.3	18.4	11.9	3.9	-4.2	9.9
Daily Minimum Temp	-11.8	-9.1	-2.4	4.4	10.8	16.2	19.2	17.6	12.5	5.9	-1.2	-8.8	4.4
Daily Maximum Temp	-2.2	0.9	8.3	16.6	22.8	27.9	30.4	29.0	24.2	17.9	8.9	0.3	15.4
Record Minimum Temp	-31.1	-28.9	-30.0	-12.8	-1.1	3.3	8.3	4.4	-3.3	-10.0	-20.0	-30.0	-31.1
Record Maximum Temp	18.3	22.8	32.8	33.9	36.7	39.4	40.6	42.2	38.3	35.0	24.4	20.6	42.2
HDD, Base 18.3°C	786	627	477	238	92	6	0	6	39	207	433	699	3609
CDD, Base 18.3°C	0	0	0	3	45	119	200	159	41	8	0	0	576
Relative Humidity (%)	71	71	68	63	63	65	68	70	71	67	71	75	68
Wind Speed (m/s)	5.1	5.0	5.5	5.5	4.8	4.4	3.9	3.8	4.1	4.4	4.8	4.9	4.7

Mason City, IA

WBAN NO. 14940

LATITUDE: 43.15° N
LONGITUDE: 93.33° W
ELEVATION: 373 meters
MEAN PRESSURE: 972 millibars

STATION TYPE: Secondary

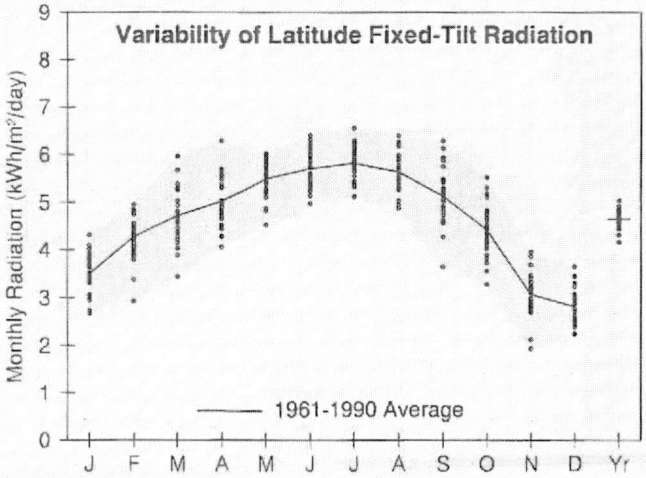

Solar Radiation for Flat-Plate Collectors Facing South at a Fixed Tilt (kWh/m²/day), Uncertainty ±9%

Tilt (°)		Jan	Feb	Mar	Apr	May	June	July	Aug	Sept	Oct	Nov	Dec	Year
0	Average	1.9	2.7	3.7	4.7	5.8	6.3	6.3	5.5	4.3	3.0	1.8	1.5	4.0
	Min/Max	1.6/2.2	2.2/3.1	3.0/4.3	3.9/5.6	4.8/6.3	5.6/7.1	5.6/7.1	4.8/6.2	3.4/5.0	2.4/3.6	1.5/2.1	1.3/1.8	3.7/4.2
Latitude -15	Average	3.1	3.9	4.6	5.1	5.9	6.2	6.3	5.9	5.1	4.1	2.8	2.5	4.6
	Min/Max	2.4/3.7	2.8/4.4	3.4/5.7	4.2/6.4	4.8/6.4	5.4/7.0	5.5/7.1	5.1/6.7	3.7/6.2	3.1/5.1	1.8/3.5	2.0/3.1	4.2/5.0
Latitude	Average	3.5	4.3	4.7	5.0	5.5	5.7	5.8	5.6	5.1	4.4	3.1	2.8	4.6
	Min/Max	2.7/4.3	2.9/4.9	3.4/6.0	4.1/6.3	4.5/6.0	5.0/6.4	5.1/6.6	4.9/6.4	3.7/6.3	3.3/5.5	1.9/4.0	2.2/3.7	4.2/5.0
Latitude +15	Average	3.8	4.4	4.7	4.6	4.9	4.9	5.1	5.1	4.9	4.4	3.2	3.0	4.4
	Min/Max	2.8/4.7	3.0/5.2	3.3/6.0	3.8/5.8	4.0/5.3	4.3/5.5	4.5/5.7	4.4/5.8	3.4/6.1	3.3/5.6	2.0/4.2	2.3/4.0	3.9/4.8
90	Average	3.6	4.1	3.8	3.2	2.9	2.7	2.9	3.2	3.6	3.7	2.9	2.9	3.3
	Min/Max	2.7/4.4	2.7/4.9	2.6/5.3	2.7/3.9	2.5/3.1	2.5/3.0	2.6/3.1	2.8/3.6	2.5/4.5	2.7/4.7	1.7/3.8	2.2/4.0	2.9/3.5

Solar Radiation for 1-Axis Tracking Flat-Plate Collectors with a North-South Axis (kWh/m²/day), Uncertainty ±9%

Axis Tilt (°)		Jan	Feb	Mar	Apr	May	June	July	Aug	Sept	Oct	Nov	Dec	Year
0	Average	2.8	3.9	5.0	6.2	7.5	8.3	8.4	7.4	5.9	4.3	2.6	2.2	5.4
	Min/Max	2.2/3.4	2.7/4.6	3.5/6.5	4.7/7.9	5.9/8.5	7.1/9.6	7.3/10.0	6.2/8.6	4.1/7.4	3.2/5.4	1.7/3.3	1.7/2.8	4.8/5.9
Latitude -15	Average	3.7	4.8	5.7	6.6	7.7	8.3	8.5	7.7	6.5	5.1	3.3	2.9	5.9
	Min/Max	2.8/4.7	3.1/5.7	3.9/7.5	5.0/8.5	6.0/8.7	7.0/9.6	7.3/10.1	6.5/9.1	4.4/8.3	3.7/6.6	2.0/4.3	2.2/3.8	5.2/6.5
Latitude	Average	4.1	5.1	5.8	6.5	7.5	8.0	8.2	7.6	6.6	5.4	3.5	3.2	6.0
	Min/Max	3.0/5.1	3.2/6.1	3.9/7.7	4.9/8.4	5.8/8.5	6.7/9.2	7.0/9.8	6.4/9.0	4.3/8.4	3.8/6.9	2.1/4.7	2.5/4.2	5.2/6.6
Latitude +15	Average	4.3	5.2	5.8	6.2	7.0	7.5	7.7	7.2	6.4	5.4	3.6	3.3	5.8
	Min/Max	3.1/5.4	3.3/6.2	3.8/7.7	4.7/8.1	5.5/8.0	6.3/8.6	6.6/9.2	6.0/8.6	4.2/8.3	3.8/7.0	2.1/4.8	2.6/4.5	5.1/6.4

Solar Radiation for 2-Axis Tracking Flat-Plate Collectors (kWh/m²/day), Uncertainty ±9%

Tracker		Jan	Feb	Mar	Apr	May	June	July	Aug	Sept	Oct	Nov	Dec	Year
2-Axis	Average	4.3	5.2	5.8	6.6	7.8	8.5	8.6	7.8	6.6	5.4	3.6	3.4	6.2
	Min/Max	3.2/5.5	3.3/6.2	3.9/7.7	5.0/8.5	6.1/8.9	7.2/9.9	7.4/10.3	6.6/9.2	4.4/8.4	3.9/7.0	2.1/4.9	2.6/4.6	5.4/6.8

Direct Beam Solar Radiation for Concentrating Collectors (kWh/m²/day), Uncertainty ±8%

Tracker		Jan	Feb	Mar	Apr	May	June	July	Aug	Sept	Oct	Nov	Dec	Year
1-Axis, E-W Horiz Axis	Average	2.4	2.6	2.6	2.9	3.5	4.0	4.1	3.6	3.2	3.0	2.0	1.9	3.0
	Min/Max	1.5/3.7	1.0/3.7	1.3/4.3	1.6/4.5	2.3/4.3	3.0/5.1	3.1/5.3	2.7/4.8	1.5/4.6	1.9/4.1	0.8/3.0	1.3/2.7	2.4/3.6
1-Axis, N-S Horiz Axis	Average	1.6	2.3	2.9	3.7	4.7	5.2	5.4	4.8	3.8	2.8	1.5	1.2	3.3
	Min/Max	1.0/2.6	0.9/3.2	1.4/4.8	2.1/5.7	3.0/5.8	3.9/6.7	4.2/7.2	3.5/6.3	1.9/5.5	1.7/3.9	0.5/2.3	0.8/1.7	2.7/4.0
1-Axis, N-S Tilt=Latitude	Average	2.6	3.2	3.5	4.0	4.6	5.0	5.2	4.9	4.4	3.7	2.3	2.0	3.8
	Min/Max	1.6/4.1	1.2/4.4	1.7/5.8	2.3/6.1	3.0/5.7	3.7/6.4	4.0/6.9	3.7/6.5	2.1/6.4	2.3/5.2	0.9/3.4	1.3/2.8	3.0/4.5
2-Axis	Average	2.8	3.2	3.5	4.0	4.9	5.4	5.6	5.1	4.4	3.7	2.4	2.1	3.9
	Min/Max	1.7/4.4	1.3/4.5	1.7/5.8	2.3/6.2	3.2/6.0	4.1/6.9	4.3/7.4	3.8/6.7	2.1/6.4	2.3/5.2	0.9/3.6	1.4/3.0	3.1/4.7

Average Climatic Conditions

Element	Jan	Feb	Mar	Apr	May	June	July	Aug	Sept	Oct	Nov	Dec	Year
Temperature (°C)	-10.4	-7.4	-0.3	7.9	14.8	20.2	22.5	21.0	15.9	9.6	1.1	-7.6	7.3
Daily Minimum Temp	-15.4	-12.3	-5.1	2.0	8.2	13.8	16.4	14.8	9.6	3.3	-3.9	-12.1	1.6
Daily Maximum Temp	-5.5	-2.5	4.4	13.9	21.4	26.6	28.7	27.2	22.3	15.8	6.1	-3.1	12.9
Record Minimum Temp	-36.1	-31.1	-32.8	-14.4	-5.0	3.3	6.7	3.9	-3.3	-11.1	-26.7	-32.2	-36.1
Record Maximum Temp	15.0	18.9	28.3	33.9	34.4	39.4	39.4	37.8	36.1	35.0	22.8	18.3	39.4
HDD, Base 18.3°C	892	720	579	312	129	22	6	15	83	276	517	804	4354
CDD, Base 18.3°C	0	0	0	0	21	78	134	97	11	4	0	0	346
Relative Humidity (%)	74	76	76	68	65	67	73	76	76	72	78	79	73
Wind Speed (m/s)	5.9	5.5	5.9	5.9	5.4	4.9	4.0	3.8	4.3	4.9	5.4	5.6	5.1

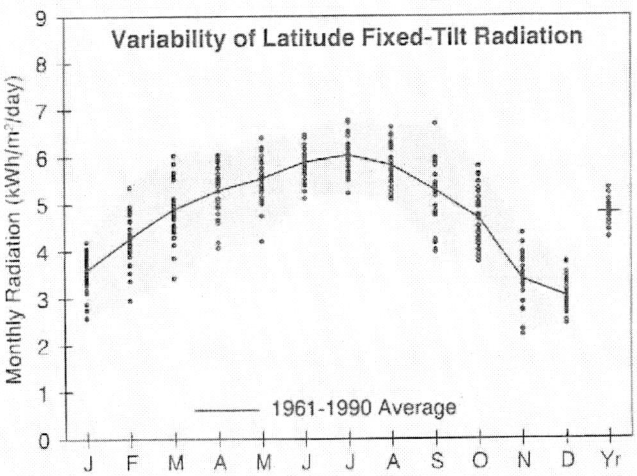

Sioux City, IA

WBAN NO. 14943

LATITUDE: 42.40° N
LONGITUDE: 96.38° W
ELEVATION: 336 meters
MEAN PRESSURE: 976 millibars

STATION TYPE: Secondary

Solar Radiation for Flat-Plate Collectors Facing South at a Fixed Tilt (kWh/m²/day), Uncertainty ±9%

Tilt (°)		Jan	Feb	Mar	Apr	May	June	July	Aug	Sept	Oct	Nov	Dec	Year
0	Average	1.9	2.8	3.8	4.9	5.8	6.6	6.5	5.7	4.4	3.2	2.0	1.6	4.1
	Min/Max	1.6/2.1	2.3/3.2	3.0/4.5	3.9/5.5	4.5/6.7	5.7/7.2	5.8/7.3	5.0/6.3	3.6/5.3	2.8/3.8	1.6/2.3	1.4/1.8	3.8/4.4
Latitude -15	Average	3.1	3.9	4.7	5.4	5.9	6.4	6.5	6.1	5.2	4.4	3.0	2.6	4.8
	Min/Max	2.3/3.6	2.8/4.8	3.4/5.8	4.2/6.2	4.5/6.8	5.6/7.1	5.7/7.3	5.3/6.9	4.0/6.5	3.6/5.3	2.1/3.8	2.2/3.2	4.3/5.3
Latitude	Average	3.6	4.3	4.9	5.3	5.5	5.9	6.0	5.8	5.3	4.7	3.4	3.0	4.8
	Min/Max	2.6/4.2	3.0/5.4	3.4/6.0	4.1/6.0	4.2/6.4	5.1/6.5	5.2/6.8	5.1/6.6	4.0/6.7	3.8/5.8	2.2/4.3	2.4/3.8	4.3/5.3
Latitude +15	Average	3.9	4.4	4.8	4.9	4.9	5.1	5.3	5.3	5.1	4.7	3.5	3.2	4.6
	Min/Max	2.7/4.5	3.0/5.6	3.3/5.9	3.7/5.6	3.7/5.7	4.4/5.6	4.6/5.9	4.6/6.0	3.7/6.4	3.7/5.9	2.3/4.6	2.6/4.1	4.0/5.1
90	Average	3.7	4.0	3.8	3.3	2.9	2.7	2.9	3.3	3.7	3.9	3.2	3.1	3.4
	Min/Max	2.6/4.3	2.7/5.0	2.5/4.8	2.5/3.7	2.3/3.2	2.5/2.9	2.5/3.2	2.9/3.7	2.7/4.7	3.0/4.9	2.0/4.2	2.4/3.9	2.9/3.7

Solar Radiation for 1-Axis Tracking Flat-Plate Collectors with a North-South Axis (kWh/m²/day), Uncertainty ±9%

Axis Tilt (°)		Jan	Feb	Mar	Apr	May	June	July	Aug	Sept	Oct	Nov	Dec	Year
0	Average	2.9	4.0	5.3	6.6	7.6	8.6	8.7	7.7	6.2	4.6	2.9	2.4	5.6
	Min/Max	2.1/3.4	2.7/5.0	3.6/6.6	4.8/7.8	5.4/9.1	7.2/9.9	7.6/10.1	6.4/8.9	4.5/8.0	3.8/5.8	1.9/3.7	1.9/2.9	5.0/6.3
Latitude -15	Average	3.8	4.8	6.0	7.0	7.8	8.7	8.9	8.1	6.8	5.5	3.7	3.1	6.2
	Min/Max	2.7/4.4	3.1/6.2	3.9/7.6	5.0/8.4	5.5/9.4	7.2/9.9	7.6/10.3	6.7/9.4	4.9/9.0	4.4/7.0	2.3/4.8	2.5/3.9	5.4/7.0
Latitude	Average	4.2	5.1	6.1	6.9	7.6	8.3	8.5	8.0	6.9	5.8	3.9	3.5	6.2
	Min/Max	2.9/4.9	3.2/6.7	4.0/7.8	4.9/8.3	5.3/9.1	6.8/9.6	7.3/9.9	6.5/9.2	4.9/9.1	4.5/7.3	2.4/5.2	2.7/4.4	5.4/7.0
Latitude +15	Average	4.4	5.2	6.0	6.6	7.1	7.8	8.0	7.6	6.7	5.8	4.1	3.6	6.1
	Min/Max	3.0/5.2	3.3/6.8	3.9/7.7	4.7/8.0	5.0/8.6	6.4/8.9	6.9/9.3	6.2/8.8	4.7/8.9	4.5/7.4	2.4/5.4	2.8/4.7	5.2/6.9

Solar Radiation for 2-Axis Tracking Flat-Plate Collectors (kWh/m²/day), Uncertainty ±9%

Tracker		Jan	Feb	Mar	Apr	May	June	July	Aug	Sept	Oct	Nov	Dec	Year
2-Axis	Average	4.4	5.2	6.1	7.0	7.9	8.9	9.0	8.1	6.9	5.8	4.1	3.7	6.4
	Min/Max	3.1/5.2	3.3/6.8	4.0/7.8	5.1/8.4	5.6/9.5	7.3/10.2	7.8/10.5	6.7/9.4	4.9/9.1	4.5/7.4	2.5/5.4	2.9/4.7	5.6/7.3

Direct Beam Solar Radiation for Concentrating Collectors (kWh/m²/day), Uncertainty ±8%

Tracker		Jan	Feb	Mar	Apr	May	June	July	Aug	Sept	Oct	Nov	Dec	Year
1-Axis, E-W Horiz Axis	Average	2.6	2.8	2.9	3.3	3.6	4.3	4.5	4.0	3.5	3.3	2.4	2.1	3.3
	Min/Max	1.2/3.5	1.1/4.1	1.5/4.3	1.7/4.3	1.9/4.9	3.1/5.3	3.5/5.6	2.8/4.9	1.9/5.2	2.2/4.5	1.0/3.5	1.4/3.1	2.5/4.0
1-Axis, N-S Horiz Axis	Average	1.8	2.4	3.2	4.2	4.8	5.7	5.9	5.2	4.1	3.2	1.8	1.4	3.6
	Min/Max	0.9/2.4	0.9/3.7	1.7/4.8	2.3/5.6	2.5/6.5	4.1/7.1	4.7/7.4	3.6/6.4	2.3/6.3	2.2/4.4	0.7/2.7	0.9/2.0	2.9/4.4
1-Axis, N-S Tilt=Latitude	Average	2.8	3.3	3.9	4.5	4.7	5.4	5.7	5.4	4.7	4.1	2.7	2.3	4.1
	Min/Max	1.3/3.7	1.3/5.0	2.0/5.8	2.4/6.0	2.5/6.4	3.9/6.7	4.5/7.2	3.8/6.7	2.7/7.2	2.8/5.7	1.1/4.0	1.5/3.3	3.2/5.0
2-Axis	Average	3.0	3.4	3.9	4.5	5.0	5.9	6.1	5.6	4.7	4.1	2.8	2.5	4.3
	Min/Max	1.4/4.0	1.3/5.1	2.0/5.8	2.5/6.1	2.7/6.8	4.2/7.3	4.8/7.7	3.9/6.8	2.7/7.2	2.8/5.8	1.2/4.2	1.6/3.6	3.3/5.2

Average Climatic Conditions

Element	Jan	Feb	Mar	Apr	May	June	July	Aug	Sept	Oct	Nov	Dec	Year
Temperature (°C)	-7.9	-4.7	2.1	10.1	16.4	21.6	24.3	22.7	17.4	11.0	2.4	-5.7	9.2
Daily Minimum Temp	-13.5	-10.1	-3.5	3.5	10.0	15.3	18.2	16.7	11.1	4.2	-3.1	-10.7	3.2
Daily Maximum Temp	-2.4	0.7	7.7	16.8	22.9	27.8	30.3	28.6	23.8	17.8	8.0	-0.6	15.1
Record Minimum Temp	-32.2	-32.2	-30.0	-18.9	-3.9	3.3	5.6	2.8	-4.4	-10.6	-22.8	-31.1	-32.2
Record Maximum Temp	21.1	21.7	32.8	36.1	38.9	42.2	41.7	40.0	38.9	33.9	27.2	20.0	42.2
HDD, Base 18.3°C	814	644	503	247	95	11	0	7	57	232	477	744	3829
CDD, Base 18.3°C	0	0	0	0	36	107	184	141	31	4	0	0	504
Relative Humidity (%)	72	72	70	62	62	65	69	72	71	66	72	76	69
Wind Speed (m/s)	5.2	5.1	5.7	5.9	5.4	4.8	4.2	4.1	4.4	4.7	5.0	5.1	5.0

Waterloo, IA

WBAN NO. 94910

LATITUDE: 42.55° N
LONGITUDE: 92.40° W
ELEVATION: 265 meters
MEAN PRESSURE: 985 millibars

STATION TYPE: Secondary

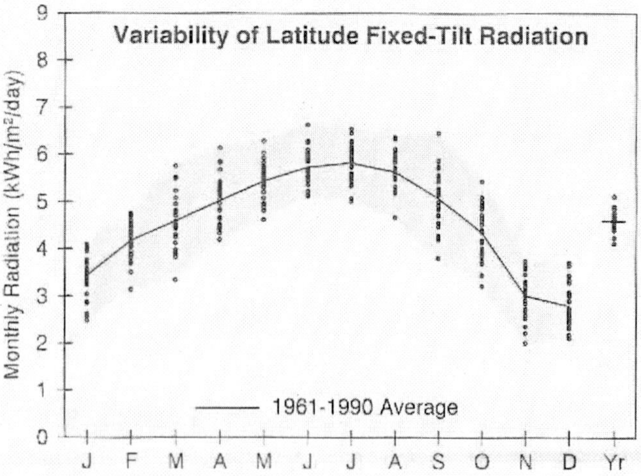

Solar Radiation for Flat-Plate Collectors Facing South at a Fixed Tilt (kWh/m²/day), Uncertainty ±9%

Tilt (°)		Jan	Feb	Mar	Apr	May	June	July	Aug	Sept	Oct	Nov	Dec	Year
0	Average	1.9	2.7	3.6	4.7	5.7	6.4	6.3	5.5	4.3	3.0	1.9	1.5	4.0
	Min/Max	1.5/2.2	2.3/3.0	3.0/4.3	4.1/5.6	4.9/6.5	5.7/7.4	5.5/7.0	4.7/6.1	3.5/5.1	2.4/3.6	1.5/2.1	1.3/1.9	3.7/4.4
Latitude -15	Average	3.0	3.8	4.4	5.1	5.8	6.2	6.3	5.9	5.0	4.1	2.7	2.5	4.6
	Min/Max	2.2/3.5	3.0/4.3	3.3/5.5	4.3/6.2	4.9/6.7	5.6/7.2	5.4/7.1	4.9/6.6	3.8/6.3	3.1/5.0	1.9/3.3	1.9/3.2	4.2/5.1
Latitude	Average	3.4	4.2	4.6	5.0	5.4	5.7	5.8	5.6	5.1	4.3	3.0	2.8	4.6
	Min/Max	2.5/4.1	3.1/4.8	3.3/5.8	4.2/6.1	4.6/6.3	5.1/6.6	5.0/6.5	4.7/6.4	3.8/6.4	3.2/5.4	2.0/3.7	2.1/3.7	4.1/5.1
Latitude +15	Average	3.7	4.3	4.5	4.6	4.8	5.0	5.1	5.1	4.8	4.4	3.1	3.0	4.4
	Min/Max	2.6/4.4	3.2/5.0	3.2/5.7	3.8/5.7	4.1/5.5	4.4/5.7	4.4/5.7	4.2/5.8	3.6/6.2	3.2/5.5	2.0/3.9	2.2/4.0	3.9/4.9
90	Average	3.5	3.9	3.6	3.1	2.8	2.7	2.8	3.2	3.5	3.6	2.8	2.9	3.2
	Min/Max	2.4/4.3	2.8/4.6	2.5/4.4	2.6/3.8	2.5/3.2	2.5/3.0	2.5/3.1	2.7/3.5	2.6/4.5	2.6/4.6	1.7/3.6	2.1/3.9	2.8/3.5

Solar Radiation for 1-Axis Tracking Flat-Plate Collectors with a North-South Axis (kWh/m²/day), Uncertainty ±9%

Axis Tilt (°)		Jan	Feb	Mar	Apr	May	June	July	Aug	Sept	Oct	Nov	Dec	Year
0	Average	2.8	3.9	4.9	6.2	7.4	8.2	8.3	7.4	5.9	4.3	2.6	2.2	5.3
	Min/Max	2.0/3.3	2.8/4.5	3.4/6.2	5.0/7.9	5.9/8.7	7.2/10.1	6.9/9.8	6.0/8.6	4.3/7.6	3.1/5.4	1.7/3.1	1.6/2.9	4.8/6.1
Latitude -15	Average	3.6	4.7	5.5	6.6	7.5	8.3	8.4	7.7	6.5	5.1	3.2	2.9	5.8
	Min/Max	2.5/4.4	3.3/5.5	3.8/7.2	5.3/8.4	6.0/9.0	7.1/10.1	7.0/9.9	6.2/9.0	4.6/8.5	3.6/6.5	2.0/4.0	2.1/3.8	5.1/6.7
Latitude	Average	4.0	5.0	5.6	6.5	7.3	7.9	8.1	7.6	6.5	5.3	3.4	3.2	5.9
	Min/Max	2.8/4.8	3.4/5.8	3.8/7.4	5.2/8.4	5.8/8.7	6.8/9.8	6.7/9.6	6.0/8.9	4.6/8.6	3.7/6.8	2.1/4.4	2.3/4.3	5.1/6.8
Latitude +15	Average	4.2	5.1	5.6	6.2	6.9	7.4	7.6	7.2	6.3	5.3	3.5	3.3	5.7
	Min/Max	2.8/5.1	3.5/5.9	3.7/7.3	4.9/8.1	5.5/8.2	6.4/9.1	6.3/9.0	5.7/8.4	4.4/8.4	3.7/6.9	2.2/4.5	2.4/4.6	5.0/6.6

Solar Radiation for 2-Axis Tracking Flat-Plate Collectors (kWh/m²/day), Uncertainty ±9%

Tracker		Jan	Feb	Mar	Apr	May	June	July	Aug	Sept	Oct	Nov	Dec	Year
2-Axis	Average	4.2	5.1	5.6	6.6	7.6	8.5	8.6	7.8	6.5	5.3	3.6	3.4	6.1
	Min/Max	2.9/5.1	3.5/5.9	3.8/7.4	5.3/8.5	6.1/9.1	7.3/10.4	7.1/10.1	6.2/9.1	4.6/8.6	3.7/6.9	2.2/4.6	2.4/4.6	5.3/7.0

Direct Beam Solar Radiation for Concentrating Collectors (kWh/m²/day), Uncertainty ±8%

Tracker		Jan	Feb	Mar	Apr	May	June	July	Aug	Sept	Oct	Nov	Dec	Year
1-Axis, E-W	Average	2.3	2.6	2.5	2.9	3.4	3.9	4.0	3.6	3.1	2.9	1.9	1.9	2.9
Horiz Axis	Min/Max	1.4/3.3	1.3/3.5	1.1/4.0	1.9/4.5	2.2/4.5	2.9/5.6	2.7/5.2	2.4/4.7	1.7/4.8	1.7/4.1	0.8/2.7	1.2/3.0	2.3/3.7
1-Axis, N-S	Average	1.6	2.2	2.8	3.7	4.5	5.1	5.3	4.7	3.7	2.8	1.4	1.2	3.3
Horiz Axis	Min/Max	1.0/2.3	1.0/3.0	1.2/4.4	2.5/5.8	2.9/5.9	3.8/7.4	3.6/7.0	3.2/6.2	2.0/5.7	1.6/4.0	0.6/2.0	0.7/1.9	2.6/4.2
1-Axis, N-S	Average	2.5	3.1	3.4	3.9	4.4	4.9	5.1	4.9	4.3	3.6	2.2	2.0	3.7
Tilt=Latitude	Min/Max	1.5/3.7	1.5/4.2	1.5/5.3	2.6/6.2	2.8/5.8	3.6/7.0	3.5/6.8	3.3/6.4	2.3/6.6	2.1/5.2	0.9/3.1	1.2/3.1	2.9/4.7
2-Axis	Average	2.7	3.1	3.4	4.0	4.7	5.3	5.5	5.0	4.3	3.6	2.3	2.1	3.8
	Min/Max	1.7/3.9	1.5/4.3	1.5/5.3	2.7/6.3	3.0/6.1	3.9/7.6	3.7/7.3	3.4/6.6	2.3/6.6	2.1/5.2	1.0/3.2	1.3/3.4	3.0/4.9

Average Climatic Conditions

Element	Jan	Feb	Mar	Apr	May	June	July	Aug	Sept	Oct	Nov	Dec	Year
Temperature (°C)	-9.7	-6.6	1.2	9.1	15.7	20.6	22.8	21.3	16.6	10.1	2.1	-6.5	8.1
Daily Minimum Temp	-14.8	-11.9	-3.9	2.8	9.3	14.2	16.8	15.0	10.0	3.6	-3.2	-11.3	2.2
Daily Maximum Temp	-4.6	-1.4	6.3	15.2	22.0	26.9	28.8	27.5	23.2	16.6	7.3	-1.7	13.8
Record Minimum Temp	-35.0	-33.9	-36.7	-20.0	-3.9	3.3	5.6	3.3	-5.6	-11.7	-27.2	-32.8	-36.7
Record Maximum Temp	14.4	18.9	30.6	37.8	34.4	39.4	40.6	40.6	36.7	35.0	26.1	18.3	40.6
HDD, Base 18.3°C	868	698	531	278	113	15	6	16	71	261	488	770	4114
CDD, Base 18.3°C	0	0	0	0	31	83	145	108	19	4	0	0	390
Relative Humidity (%)	73	74	73	66	66	68	72	74	74	70	75	77	72
Wind Speed (m/s)	5.2	5.0	5.6	5.6	4.9	4.4	3.8	3.8	4.1	4.5	5.0	5.1	4.7

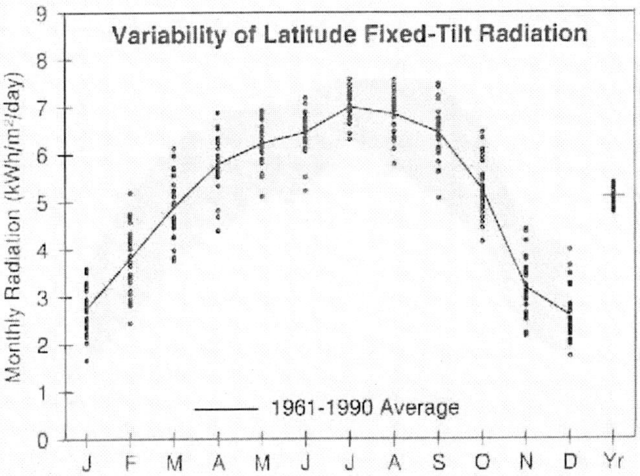

Boise, ID

WBAN NO. 24131

LATITUDE: 43.57° N
LONGITUDE: 116.22° W
ELEVATION: 874 meters
MEAN PRESSURE: 916 millibars

STATION TYPE: Primary

Solar Radiation for Flat-Plate Collectors Facing South at a Fixed Tilt (kWh/m²/day), Uncertainty ±9%

Tilt (°)		Jan	Feb	Mar	Apr	May	June	July	Aug	Sept	Oct	Nov	Dec	Year
0	Average	1.6	2.5	3.8	5.3	6.5	7.2	7.6	6.6	5.1	3.4	1.9	1.4	4.4
	Min/Max	1.3/2.0	1.9/3.0	3.1/4.4	4.2/6.1	5.5/7.2	5.9/8.0	6.9/8.3	5.6/7.2	4.3/5.8	2.9/3.9	1.4/2.3	1.2/1.8	4.2/4.6
Latitude -15	Average	2.5	3.5	4.7	5.9	6.7	7.1	7.6	7.1	6.3	4.9	2.9	2.3	5.1
	Min/Max	1.6/3.2	2.3/4.7	3.7/5.8	4.5/7.0	5.5/7.4	5.7/7.9	6.8/8.2	6.0/7.9	5.0/7.3	3.9/5.9	2.0/3.9	1.6/3.4	4.8/5.4
Latitude	Average	2.8	3.8	4.9	5.8	6.2	6.5	7.0	6.8	6.5	5.2	3.2	2.6	5.1
	Min/Max	1.7/3.6	2.5/5.2	3.8/6.1	4.4/6.9	5.1/6.9	5.2/7.2	6.3/7.6	5.8/7.6	5.1/7.5	4.2/6.5	2.2/4.4	1.7/4.0	4.8/5.4
Latitude +15	Average	2.9	3.9	4.8	5.4	5.5	5.5	6.0	6.2	6.2	5.3	3.3	2.8	4.8
	Min/Max	1.7/3.9	2.5/5.4	3.6/6.1	4.0/6.4	4.5/6.1	4.5/6.2	5.5/6.5	5.3/6.8	4.8/7.2	4.2/6.6	2.2/4.7	1.8/4.3	4.5/5.1
90	Average	2.7	3.4	3.7	3.6	3.2	3.0	3.3	3.8	4.6	4.5	3.0	2.6	3.5
	Min/Max	1.5/3.8	2.1/4.9	2.8/4.8	2.8/4.3	2.7/3.5	2.6/3.2	3.0/3.5	3.3/4.2	3.5/5.3	3.5/5.6	1.9/4.3	1.7/4.2	3.2/3.8

Solar Radiation for 1-Axis Tracking Flat-Plate Collectors with a North-South Axis (kWh/m²/day), Uncertainty ±9%

Axis Tilt (°)		Jan	Feb	Mar	Apr	May	June	July	Aug	Sept	Oct	Nov	Dec	Year
0	Average	2.2	3.5	5.4	7.4	9.0	10.0	10.9	9.5	7.8	5.2	2.7	2.0	6.3
	Min/Max	1.5/2.9	2.3/4.8	3.9/6.7	5.3/8.8	7.1/10.3	7.6/11.4	9.5/12.2	7.6/10.9	6.0/9.3	4.1/6.3	1.8/3.6	1.4/2.9	5.8/6.7
Latitude -15	Average	2.9	4.3	6.1	7.9	9.2	10.0	11.0	10.1	8.7	6.3	3.4	2.7	6.9
	Min/Max	1.7/3.8	2.7/6.0	4.4/7.7	5.5/9.6	7.3/10.6	7.6/11.4	9.6/12.4	8.0/11.6	6.6/10.5	4.8/7.8	2.3/4.8	1.7/4.1	6.3/7.3
Latitude	Average	3.1	4.5	6.2	7.8	9.0	9.6	10.7	9.9	8.8	6.6	3.7	2.9	6.9
	Min/Max	1.8/4.2	2.8/6.4	4.4/8.0	5.4/9.5	7.0/10.3	7.3/11.0	9.3/12.0	7.9/11.4	6.7/10.6	5.0/8.3	2.4/5.2	1.8/4.6	6.3/7.4
Latitude +15	Average	3.2	4.6	6.1	7.5	8.5	9.0	10.0	9.5	8.6	6.7	3.8	3.1	6.7
	Min/Max	1.8/4.4	2.8/6.6	4.3/7.9	5.2/9.2	6.6/9.8	6.8/10.3	8.7/11.3	7.5/10.9	6.5/10.4	5.0/8.4	2.4/5.5	1.9/4.9	6.1/7.2

Solar Radiation for 2-Axis Tracking Flat-Plate Collectors (kWh/m²/day), Uncertainty ±9%

Tracker		Jan	Feb	Mar	Apr	May	June	July	Aug	Sept	Oct	Nov	Dec	Year
2-Axis	Average	3.3	4.6	6.2	7.9	9.4	10.3	11.3	10.1	8.8	6.7	3.8	3.1	7.1
	Min/Max	1.8/4.4	2.8/6.6	4.4/8.0	5.6/9.6	7.4/10.8	7.8/11.8	9.8/12.7	8.1/11.6	6.7/10.6	5.1/8.4	2.4/5.5	1.9/5.0	6.5/7.6

Direct Beam Solar Radiation for Concentrating Collectors (kWh/m²/day), Uncertainty ±8%

Tracker		Jan	Feb	Mar	Apr	May	June	July	Aug	Sept	Oct	Nov	Dec	Year
1-Axis, E-W Horiz Axis	Average	1.7	2.4	3.0	3.8	4.7	5.5	6.4	5.6	5.0	4.0	2.2	1.7	3.8
	Min/Max	0.4/2.8	1.0/3.9	1.8/4.5	1.8/5.2	3.1/5.9	3.4/6.9	5.1/7.6	3.9/7.0	3.3/6.4	2.6/5.7	0.9/3.6	0.6/3.4	3.3/4.3
1-Axis, N-S Horiz Axis	Average	1.1	2.1	3.4	4.9	6.3	7.3	8.7	7.5	6.1	3.9	1.6	1.1	4.5
	Min/Max	0.3/1.9	0.9/3.4	1.9/5.1	2.4/6.7	4.2/8.0	4.5/9.2	6.8/10.4	5.1/9.4	4.0/7.9	2.5/5.4	0.7/2.6	0.4/2.1	3.8/5.0
1-Axis, N-S Tilt=Latitude	Average	1.8	2.9	4.1	5.3	6.3	7.0	8.4	7.7	6.9	5.0	2.5	1.8	5.0
	Min/Max	0.5/3.0	1.2/4.7	2.4/6.1	2.6/7.2	4.1/7.9	4.3/8.8	6.6/10.0	5.3/9.7	4.6/9.1	3.3/7.2	1.1/4.0	0.6/3.6	4.2/5.6
2-Axis	Average	1.9	3.0	4.1	5.4	6.6	7.6	9.0	8.0	7.0	5.1	2.6	2.0	5.2
	Min/Max	0.5/3.2	1.3/4.8	2.4/6.1	2.6/7.3	4.4/8.3	4.7/9.5	7.1/10.7	5.5/10.0	4.6/9.1	3.3/7.3	1.1/4.3	0.7/3.9	4.4/5.8

Average Climatic Conditions

Element	Jan	Feb	Mar	Apr	May	June	July	Aug	Sept	Oct	Nov	Dec	Year
Temperature (°C)	-1.7	2.2	5.8	9.5	14.2	19.2	23.3	22.5	17.0	11.0	4.4	-1.1	10.5
Daily Minimum Temp	-5.8	-2.5	-0.1	2.6	6.6	11.2	14.3	13.8	9.0	3.9	-0.5	-5.3	3.9
Daily Maximum Temp	2.4	6.8	11.6	16.3	21.7	27.2	32.3	31.2	25.0	18.1	9.3	3.2	17.1
Record Minimum Temp	-27.2	-26.1	-14.4	-7.2	-5.6	-0.6	1.7	2.8	-5.0	-11.7	-19.4	-31.7	-31.7
Record Maximum Temp	17.2	21.1	27.2	33.3	36.7	42.8	43.9	43.3	38.9	32.8	23.3	18.3	43.9
HDD, Base 18.3°C	620	453	389	265	134	42	3	11	89	230	418	601	3256
CDD, Base 18.3°C	0	0	0	0	5	67	158	140	49	0	0	0	419
Relative Humidity (%)	75	70	59	52	49	45	36	37	45	54	69	75	55
Wind Speed (m/s)	3.2	3.8	4.2	4.3	4.2	3.9	3.6	3.5	3.5	3.4	3.6	3.3	3.7

Pocatello, ID

WBAN NO. 24156

LATITUDE: 42.92° N
LONGITUDE: 112.60° W
ELEVATION: 1365 meters
MEAN PRESSURE: 864 millibars

STATION TYPE: Secondary

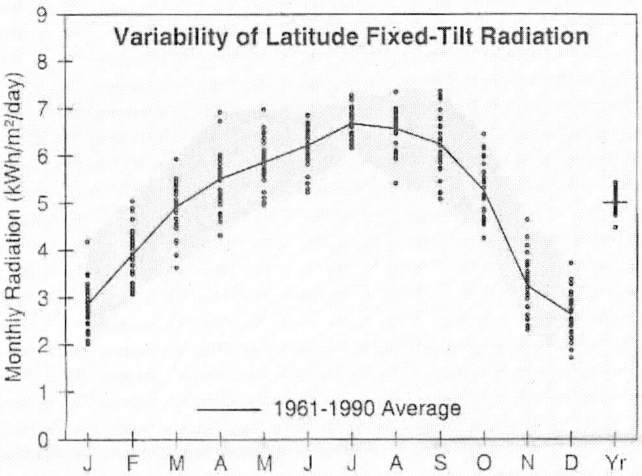

Solar Radiation for Flat-Plate Collectors Facing South at a Fixed Tilt (kWh/m²/day), Uncertainty ±9%

Tilt (°)		Jan	Feb	Mar	Apr	May	June	July	Aug	Sept	Oct	Nov	Dec	Year
0	Average	1.7	2.6	3.8	5.1	6.2	7.0	7.3	6.3	5.0	3.5	2.0	1.5	4.3
	Min/Max	1.4/2.1	2.3/3.0	3.0/4.3	4.2/6.2	5.2/7.3	5.8/7.8	6.7/8.0	5.3/7.0	4.2/5.7	3.0/4.0	1.6/2.4	1.2/1.8	4.0/4.6
Latitude -15	Average	2.6	3.6	4.7	5.6	6.3	6.8	7.3	6.8	6.1	4.9	2.9	2.3	5.0
	Min/Max	1.9/3.6	2.9/4.6	3.6/5.7	4.5/7.0	5.3/7.5	5.7/7.6	6.7/7.9	5.7/7.6	5.0/7.2	4.0/5.9	2.2/4.1	1.6/3.2	4.5/5.4
Latitude	Average	2.9	3.9	4.9	5.5	5.9	6.2	6.7	6.6	6.2	5.3	3.2	2.6	5.0
	Min/Max	2.0/4.2	3.1/5.0	3.6/5.9	4.3/6.9	4.9/7.0	5.2/6.9	6.2/7.3	5.4/7.3	5.1/7.4	4.3/6.4	2.3/4.6	1.7/3.7	4.5/5.4
Latitude +15	Average	3.0	4.0	4.8	5.1	5.2	5.4	5.8	6.0	6.0	5.3	3.4	2.8	4.7
	Min/Max	2.1/4.5	3.1/5.2	3.5/5.9	4.0/6.4	4.4/6.2	4.5/5.9	5.4/6.3	4.9/6.7	4.8/7.1	4.3/6.6	2.3/4.9	1.8/4.0	4.2/5.2
90	Average	2.9	3.6	3.8	3.4	3.0	2.8	3.1	3.7	4.4	4.5	3.1	2.7	3.4
	Min/Max	1.9/4.3	2.6/4.6	2.7/4.6	2.7/4.3	2.7/3.5	2.5/3.0	2.9/3.3	3.1/4.1	3.5/5.2	3.5/5.6	2.0/4.5	1.7/3.8	3.0/3.8

Solar Radiation for 1-Axis Tracking Flat-Plate Collectors with a North-South Axis (kWh/m²/day), Uncertainty ±9%

Axis Tilt (°)		Jan	Feb	Mar	Apr	May	June	July	Aug	Sept	Oct	Nov	Dec	Year
0	Average	2.4	3.6	5.3	7.0	8.4	9.6	10.3	9.1	7.5	5.3	2.8	2.1	6.1
	Min/Max	1.7/3.3	2.8/4.7	3.7/6.4	5.3/9.1	6.7/10.3	7.5/11.2	9.1/11.6	7.1/10.4	6.0/9.0	4.2/6.5	2.0/3.8	1.4/2.8	5.4/6.6
Latitude -15	Average	3.0	4.4	6.1	7.5	8.6	9.6	10.4	9.6	8.4	6.3	3.5	2.7	6.7
	Min/Max	2.1/4.5	3.3/5.8	4.1/7.4	5.6/9.8	6.8/10.6	7.5/11.2	9.2/11.8	7.4/11.1	6.7/10.1	4.9/7.9	2.4/5.1	1.7/3.9	5.8/7.3
Latitude	Average	3.3	4.6	6.2	7.4	8.3	9.3	10.1	9.5	8.5	6.6	3.8	3.0	6.7
	Min/Max	2.2/4.9	3.5/6.2	4.1/7.6	5.5/9.8	6.6/10.3	7.2/10.8	8.9/11.4	7.3/10.9	6.7/10.3	5.1/8.3	2.5/5.5	1.8/4.3	5.8/7.3
Latitude +15	Average	3.4	4.7	6.1	7.1	7.8	8.7	9.5	9.0	8.3	6.7	3.9	3.1	6.5
	Min/Max	2.3/5.2	3.5/6.3	4.1/7.6	5.3/9.4	6.2/9.7	6.8/10.1	8.3/10.7	6.9/10.4	6.5/10.1	5.2/8.4	2.6/5.7	1.9/4.6	5.7/7.1

Solar Radiation for 2-Axis Tracking Flat-Plate Collectors (kWh/m²/day), Uncertainty ±9%

Tracker		Jan	Feb	Mar	Apr	May	June	July	Aug	Sept	Oct	Nov	Dec	Year
2-Axis	Average	3.4	4.7	6.2	7.5	8.7	9.9	10.7	9.7	8.5	6.7	3.9	3.2	6.9
	Min/Max	2.3/5.2	3.5/6.3	4.2/7.7	5.6/9.9	6.9/10.7	7.7/11.5	9.4/12.0	7.5/11.1	6.7/10.3	5.2/8.4	2.6/5.8	1.9/4.6	6.0/7.5

Direct Beam Solar Radiation for Concentrating Collectors (kWh/m²/day), Uncertainty ±8%

Tracker		Jan	Feb	Mar	Apr	May	June	July	Aug	Sept	Oct	Nov	Dec	Year
1-Axis, E-W Horiz Axis	Average	1.7	2.4	2.9	3.5	4.3	5.3	6.0	5.2	4.7	3.9	2.1	1.7	3.6
	Min/Max	0.8/3.4	1.4/3.8	1.5/4.2	2.1/5.3	2.8/5.9	3.5/6.7	4.9/7.2	3.5/6.5	3.4/6.2	2.7/5.4	1.0/3.8	0.5/3.0	3.0/4.2
1-Axis, N-S Horiz Axis	Average	1.2	2.1	3.3	4.5	5.7	7.0	8.0	7.0	5.7	3.9	1.6	1.1	4.3
	Min/Max	0.6/2.4	1.2/3.4	1.6/4.6	2.7/6.9	3.9/7.9	4.6/9.0	6.4/9.7	4.6/8.7	4.0/7.7	2.6/5.3	0.7/2.8	0.3/1.9	3.5/4.9
1-Axis, N-S Tilt=Latitude	Average	1.8	2.9	4.0	4.8	5.6	6.6	7.7	7.2	6.5	5.0	2.4	1.8	4.7
	Min/Max	0.9/3.7	1.6/4.7	1.9/5.6	2.9/7.5	3.8/7.8	4.4/8.6	6.2/9.3	4.7/8.9	4.6/8.7	3.4/6.9	1.1/4.2	0.6/3.2	3.8/5.4
2-Axis	Average	2.0	3.0	4.0	4.9	6.0	7.2	8.3	7.4	6.5	5.1	2.5	1.9	4.9
	Min/Max	0.9/4.0	1.7/4.8	1.9/5.6	3.0/7.6	4.0/8.3	4.8/9.3	6.7/10.0	4.9/9.2	4.6/8.8	3.5/7.0	1.2/4.4	0.6/3.4	3.9/5.7

Average Climatic Conditions

Element	Jan	Feb	Mar	Apr	May	June	July	Aug	Sept	Oct	Nov	Dec	Year
Temperature (°C)	-4.8	-1.6	2.4	7.2	12.0	17.1	21.4	20.4	15.0	8.9	2.0	-4.0	8.0
Daily Minimum Temp	-9.8	-6.8	-3.4	0.2	4.2	8.5	11.7	10.5	6.0	0.8	-3.3	-9.0	0.8
Daily Maximum Temp	0.1	3.6	8.2	14.2	19.7	25.6	31.2	30.2	23.9	16.9	7.3	0.9	15.2
Record Minimum Temp	-34.4	-36.1	-24.4	-9.4	-6.7	-1.1	1.1	0.0	-7.2	-12.2	-25.0	-33.9	-36.1
Record Maximum Temp	13.9	16.7	23.9	29.4	33.9	39.4	38.9	40.0	36.7	31.1	21.7	15.0	40.0
HDD, Base 18.3°C	718	558	494	335	196	69	5	16	121	293	490	692	3989
CDD, Base 18.3°C	0	0	0	0	0	31	102	80	21	0	0	0	234
Relative Humidity (%)	75	72	65	54	52	49	41	41	47	54	69	75	58
Wind Speed (m/s)	4.5	4.6	4.9	5.2	4.8	4.3	4.0	3.9	4.0	4.1	4.7	4.3	4.5

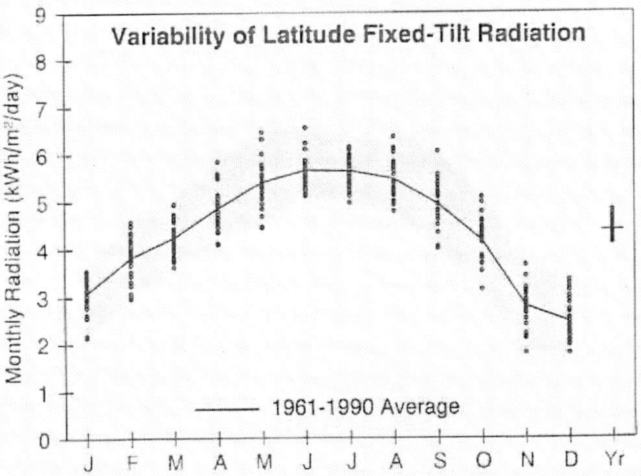

Chicago, IL

WBAN NO. 94846

LATITUDE: 41.78° N
LONGITUDE: 87.75° W
ELEVATION: 190 meters
MEAN PRESSURE: 992 millibars

STATION TYPE: Secondary

Solar Radiation for Flat-Plate Collectors Facing South at a Fixed Tilt (kWh/m²/day), Uncertainty ±9%

Tilt (°)		Jan	Feb	Mar	Apr	May	June	July	Aug	Sept	Oct	Nov	Dec	Year
0	Average	1.8	2.6	3.5	4.6	5.7	6.3	6.1	5.4	4.2	3.0	1.8	1.5	3.9
	Min/Max	1.5/2.1	2.2/3.0	3.1/3.9	4.0/5.3	4.8/6.8	5.7/7.2	5.4/6.6	4.8/6.1	3.6/4.9	2.5/3.5	1.4/2.1	1.2/1.7	3.7/4.2
Latitude -15	Average	2.7	3.5	4.1	5.0	5.8	6.1	6.1	5.7	4.9	3.9	2.5	2.2	4.4
	Min/Max	2.0/3.1	2.8/4.2	3.6/4.8	4.2/5.9	4.8/6.9	5.5/7.1	5.4/6.6	5.0/6.6	4.0/5.9	3.0/4.7	1.7/3.3	1.6/2.9	4.2/4.8
Latitude	Average	3.1	3.8	4.2	4.9	5.4	5.7	5.6	5.5	4.9	4.2	2.8	2.4	4.4
	Min/Max	2.1/3.5	3.0/4.6	3.6/4.9	4.1/5.8	4.4/6.5	5.1/6.6	5.0/6.2	4.8/6.4	4.0/6.1	3.1/5.1	1.8/3.7	1.8/3.3	4.1/4.8
Latitude +15	Average	3.3	3.9	4.1	4.5	4.8	4.9	4.9	5.0	4.7	4.2	2.9	2.6	4.1
	Min/Max	2.2/3.8	3.0/4.7	3.5/4.9	3.8/5.4	4.0/5.7	4.4/5.7	4.3/5.4	4.3/5.8	3.8/5.8	3.1/5.2	1.8/3.8	1.8/3.6	3.8/4.5
90	Average	3.1	3.5	3.2	3.0	2.8	2.6	2.7	3.1	3.4	3.4	2.5	2.4	3.0
	Min/Max	2.0/3.7	2.5/4.3	2.7/3.8	2.5/3.6	2.4/3.2	2.5/3.0	2.5/2.9	2.8/3.5	2.7/4.2	2.5/4.3	1.5/3.5	1.7/3.5	2.7/3.3

Solar Radiation for 1-Axis Tracking Flat-Plate Collectors with a North-South Axis (kWh/m²/day), Uncertainty ±9%

Axis Tilt (°)		Jan	Feb	Mar	Apr	May	June	July	Aug	Sept	Oct	Nov	Dec	Year
0	Average	2.5	3.5	4.5	5.9	7.3	8.0	7.9	7.0	5.6	4.1	2.4	1.9	5.1
	Min/Max	1.7/2.9	2.7/4.2	3.9/5.4	4.7/7.3	5.8/9.1	7.1/9.6	6.9/8.9	5.9/8.5	4.4/6.9	3.0/5.0	1.6/3.0	1.5/2.6	4.7/5.7
Latitude -15	Average	3.2	4.2	5.0	6.2	7.4	8.0	7.9	7.3	6.2	4.8	2.9	2.5	5.5
	Min/Max	2.2/3.7	3.2/5.2	4.3/6.1	5.0/7.8	5.9/9.4	7.1/9.6	6.9/9.0	6.1/9.0	4.8/7.7	3.5/6.0	1.8/3.9	1.8/3.4	5.1/6.2
Latitude	Average	3.5	4.5	5.1	6.1	7.2	7.7	7.7	7.2	6.2	5.0	3.1	2.7	5.5
	Min/Max	2.3/4.1	3.3/5.5	4.3/6.2	4.9/7.8	5.7/9.1	6.8/9.2	6.7/8.7	6.0/8.8	4.8/7.8	3.6/6.3	1.9/4.2	1.9/3.8	5.1/6.2
Latitude +15	Average	3.7	4.5	5.0	5.9	6.8	7.2	7.2	6.8	6.0	5.0	3.2	2.9	5.3
	Min/Max	2.4/4.3	3.3/5.6	4.2/6.1	4.6/7.4	5.4/8.6	6.3/8.6	6.2/8.1	5.7/8.4	4.6/7.6	3.5/6.3	1.9/4.4	2.0/4.1	4.9/6.0

Solar Radiation for 2-Axis Tracking Flat-Plate Collectors (kWh/m²/day), Uncertainty ±9%

Tracker		Jan	Feb	Mar	Apr	May	June	July	Aug	Sept	Oct	Nov	Dec	Year
2-Axis	Average	3.7	4.6	5.1	6.2	7.5	8.2	8.1	7.4	6.2	5.0	3.2	2.9	5.7
	Min/Max	2.4/4.3	3.4/5.6	4.3/6.2	5.0/7.8	6.0/9.5	7.2/9.8	7.1/9.2	6.2/9.0	4.8/7.8	3.6/6.3	1.9/4.4	2.0/4.1	5.2/6.4

Direct Beam Solar Radiation for Concentrating Collectors (kWh/m²/day), Uncertainty ±8%

Tracker		Jan	Feb	Mar	Apr	May	June	July	Aug	Sept	Oct	Nov	Dec	Year
1-Axis, E-W Horiz Axis	Average	2.0	2.2	2.2	2.6	3.3	3.7	3.6	3.3	2.9	2.6	1.7	1.5	2.6
	Min/Max	1.1/2.5	1.3/3.3	1.6/3.1	1.8/3.8	2.1/4.9	3.0/4.9	2.8/4.4	2.3/4.4	1.9/4.1	1.5/3.6	0.7/2.6	0.9/2.5	2.3/3.2
1-Axis, N-S Horiz Axis	Average	1.3	1.9	2.4	3.3	4.3	4.8	4.8	4.2	3.4	2.5	1.3	1.0	2.9
	Min/Max	0.7/1.7	1.1/2.8	1.6/3.4	2.2/4.8	2.8/6.4	4.0/6.4	3.8/5.9	3.0/5.9	2.2/4.7	1.4/3.5	0.5/1.9	0.6/1.6	2.6/3.6
1-Axis, N-S Tilt=Latitude	Average	2.1	2.6	2.9	3.6	4.3	4.6	4.6	4.3	3.9	3.2	1.9	1.6	3.3
	Min/Max	1.2/2.7	1.5/3.9	2.0/4.1	2.4/5.2	2.8/6.3	3.7/6.1	3.7/5.6	3.1/6.1	2.5/5.4	1.9/4.5	0.8/2.9	0.9/2.6	2.9/4.0
2-Axis	Average	2.3	2.7	2.9	3.6	4.5	5.0	4.9	4.5	3.9	3.3	2.0	1.7	3.4
	Min/Max	1.3/2.9	1.6/4.0	2.0/4.1	2.4/5.2	3.0/6.7	4.1/6.6	3.9/6.1	3.2/6.2	2.5/5.4	1.9/4.5	0.8/3.0	1.0/2.8	3.0/4.2

Average Climatic Conditions

Element	Jan	Feb	Mar	Apr	May	June	July	Aug	Sept	Oct	Nov	Dec	Year
Temperature (°C)	-6.1	-3.7	2.9	9.2	14.9	20.3	22.9	22.1	18.0	11.6	4.4	-3.0	9.4
Daily Minimum Temp	-10.6	-8.2	-1.9	3.7	8.7	14.2	17.0	16.4	12.2	5.7	-0.2	-7.2	4.2
Daily Maximum Temp	-1.7	0.8	7.7	14.8	21.2	26.4	28.7	27.7	23.8	17.4	9.1	1.1	14.8
Record Minimum Temp	-32.8	-27.2	-22.2	-13.9	-4.4	2.2	4.4	5.0	-2.2	-8.3	-17.2	-31.7	-32.8
Record Maximum Temp	18.3	21.7	31.1	32.8	33.9	40.0	38.9	38.3	37.2	32.8	25.6	21.7	40.0
HDD, Base 18.3°C	758	616	479	273	131	19	3	11	47	217	417	661	3631
CDD, Base 18.3°C	0	0	0	0	26	79	144	126	37	7	0	0	418
Relative Humidity (%)	72	72	70	65	64	66	68	71	71	69	73	76	70
Wind Speed (m/s)	5.2	5.1	5.4	5.3	4.6	4.2	3.7	3.6	4.0	4.5	4.8	4.9	4.6

Moline, IL

WBAN NO. 14923

LATITUDE: 41.45° N
LONGITUDE: 90.52° W
ELEVATION: 181 meters
MEAN PRESSURE: 995 millibars

STATION TYPE: Secondary

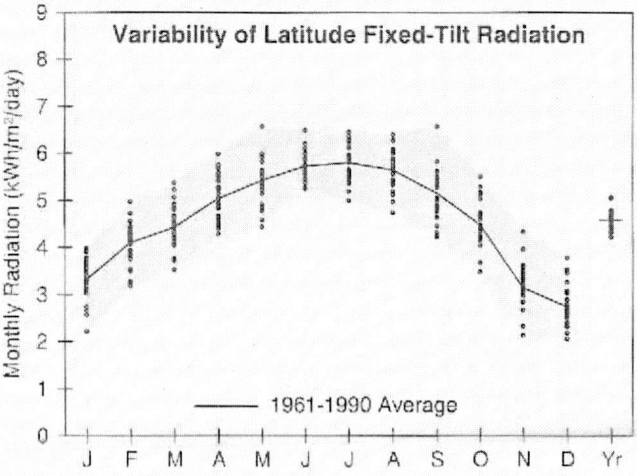

Solar Radiation for Flat-Plate Collectors Facing South at a Fixed Tilt (kWh/m²/day), Uncertainty ±9%

Tilt (°)		Jan	Feb	Mar	Apr	May	June	July	Aug	Sept	Oct	Nov	Dec	Year
0	Average	1.9	2.7	3.6	4.7	5.7	6.4	6.3	5.5	4.3	3.2	2.0	1.6	4.0
	Min/Max	1.5/2.2	2.4/3.1	3.1/4.1	4.2/5.5	4.7/6.9	5.8/7.2	5.4/6.9	4.7/6.1	3.8/5.3	2.6/3.7	1.6/2.4	1.3/1.9	3.8/4.4
Latitude -15	Average	2.9	3.8	4.3	5.1	5.8	6.2	6.2	5.9	5.1	4.2	2.8	2.4	4.6
	Min/Max	2.0/3.5	3.0/4.5	3.5/5.2	4.4/6.1	4.7/7.0	5.7/7.1	5.4/6.9	4.9/6.6	4.2/6.4	3.3/5.1	2.0/3.8	1.9/3.2	4.3/5.0
Latitude	Average	3.3	4.1	4.4	5.0	5.4	5.7	5.8	5.6	5.1	4.5	3.1	2.7	4.6
	Min/Max	2.2/4.0	3.2/5.0	3.5/5.4	4.3/6.0	4.4/6.6	5.3/6.5	5.0/6.4	4.7/6.4	4.2/6.6	3.5/5.5	2.1/4.3	2.1/3.8	4.2/5.1
Latitude +15	Average	3.5	4.2	4.3	4.6	4.8	4.9	5.1	5.1	4.9	4.5	3.3	2.9	4.4
	Min/Max	2.3/4.3	3.2/5.1	3.4/5.3	3.9/5.6	3.9/5.8	4.5/5.6	4.4/5.6	4.3/5.8	4.0/6.3	3.5/5.6	2.1/4.6	2.2/4.1	4.0/4.8
90	Average	3.4	3.8	3.3	3.1	2.7	2.6	2.7	3.1	3.5	3.7	2.9	2.8	3.1
	Min/Max	2.1/4.2	2.9/4.4	2.6/4.1	2.6/3.6	2.4/3.2	2.5/2.9	2.5/3.0	2.7/3.5	2.8/4.5	2.8/4.6	1.8/4.2	1.9/4.0	2.8/3.5

Solar Radiation for 1-Axis Tracking Flat-Plate Collectors with a North-South Axis (kWh/m²/day), Uncertainty ±9%

Axis Tilt (°)		Jan	Feb	Mar	Apr	May	June	July	Aug	Sept	Oct	Nov	Dec	Year
0	Average	2.8	3.8	4.8	6.2	7.4	8.3	8.3	7.4	6.0	4.5	2.7	2.2	5.4
	Min/Max	1.8/3.3	3.0/4.7	3.7/6.0	5.1/7.6	5.7/9.4	7.2/9.7	6.9/9.4	6.1/8.5	4.8/7.8	3.5/5.4	1.9/3.7	1.6/3.0	5.0/6.1
Latitude -15	Average	3.5	4.6	5.3	6.5	7.6	8.3	8.3	7.7	6.6	5.2	3.4	2.8	5.8
	Min/Max	2.3/4.3	3.5/5.7	4.0/6.8	5.4/8.1	5.8/9.7	7.2/9.7	6.9/9.5	6.3/8.9	5.2/8.7	4.0/6.5	2.2/4.8	2.1/4.0	5.3/6.6
Latitude	Average	3.8	4.9	5.4	6.5	7.3	7.9	8.0	7.6	6.6	5.5	3.6	3.1	5.9
	Min/Max	2.4/4.7	3.7/6.1	4.1/7.0	5.3/8.0	5.6/9.4	6.9/9.4	6.7/9.1	6.1/8.8	5.2/8.8	4.1/6.8	2.3/5.2	2.2/4.4	5.3/6.6
Latitude +15	Average	4.0	5.0	5.3	6.2	6.9	7.4	7.5	7.2	6.4	5.5	3.7	3.3	5.7
	Min/Max	2.5/4.9	3.7/6.2	4.0/6.9	5.0/7.7	5.2/8.9	6.4/8.7	6.3/8.6	5.8/8.4	5.0/8.6	4.1/6.9	2.3/5.4	2.3/4.7	5.1/6.5

Solar Radiation for 2-Axis Tracking Flat-Plate Collectors (kWh/m²/day), Uncertainty ±9%

Tracker		Jan	Feb	Mar	Apr	May	June	July	Aug	Sept	Oct	Nov	Dec	Year
2-Axis	Average	4.1	5.0	5.4	6.6	7.7	8.5	8.5	7.7	6.6	5.5	3.7	3.3	6.1
	Min/Max	2.5/5.0	3.7/6.2	4.1/7.0	5.4/8.1	5.9/9.8	7.4/10.0	7.1/9.6	6.3/8.9	5.2/8.8	4.1/6.9	2.3/5.4	2.3/4.8	5.5/6.9

Direct Beam Solar Radiation for Concentrating Collectors (kWh/m²/day), Uncertainty ±8%

Tracker		Jan	Feb	Mar	Apr	May	June	July	Aug	Sept	Oct	Nov	Dec	Year
1-Axis, E-W	Average	2.3	2.6	2.4	2.9	3.4	4.0	4.0	3.6	3.2	3.0	2.1	1.8	2.9
Horiz Axis	Min/Max	1.1/2.9	1.5/3.7	1.4/3.6	2.0/4.1	2.0/5.1	3.1/5.2	2.8/4.9	2.6/4.5	2.1/4.9	1.9/4.1	0.8/3.4	1.2/3.0	2.5/3.6
1-Axis, N-S	Average	1.6	2.2	2.7	3.7	4.6	5.2	5.3	4.7	3.9	2.9	1.6	1.2	3.3
Horiz Axis	Min/Max	0.8/2.1	1.3/3.3	1.5/4.2	2.7/5.3	2.7/6.8	4.0/6.8	3.8/6.5	3.5/5.9	2.5/6.0	1.9/3.9	0.6/2.6	0.8/2.0	2.8/4.1
1-Axis, N-S	Average	2.4	3.1	3.3	3.9	4.5	4.9	5.1	4.9	4.4	3.7	2.3	1.9	3.7
Tilt=Latitude	Min/Max	1.2/3.2	1.8/4.5	1.8/4.9	2.9/5.6	2.6/6.7	3.8/6.5	3.7/6.3	3.5/6.0	2.9/6.8	2.4/5.1	0.9/3.8	1.3/3.2	3.1/4.6
2-Axis	Average	2.6	3.2	3.3	4.0	4.7	5.3	5.4	5.0	4.4	3.8	2.4	2.1	3.9
	Min/Max	1.3/3.5	1.8/4.6	1.8/5.0	2.9/5.7	2.8/7.0	4.1/7.0	3.9/6.7	3.7/6.2	2.9/6.8	2.5/5.1	0.9/4.0	1.4/3.5	3.3/4.8

Average Climatic Conditions

Element	Jan	Feb	Mar	Apr	May	June	July	Aug	Sept	Oct	Nov	Dec	Year
Temperature (°C)	-6.7	-4.0	3.0	10.2	16.3	21.7	24.0	22.6	18.2	11.7	4.2	-3.7	9.8
Daily Minimum Temp	-11.5	-8.9	-2.2	4.1	9.9	15.3	18.1	16.6	11.8	5.4	-0.7	-8.1	4.2
Daily Maximum Temp	-2.0	0.9	8.2	16.3	22.7	28.1	29.9	28.6	24.4	17.9	9.1	0.8	15.4
Record Minimum Temp	-32.8	-31.7	-28.3	-13.9	-3.3	3.9	7.8	4.4	-4.4	-8.9	-22.8	-31.1	-32.8
Record Maximum Temp	20.6	21.7	31.1	33.9	40.0	40.0	40.6	41.1	37.8	33.3	26.7	20.6	41.1
HDD, Base 18.3°C	777	626	476	243	104	6	0	6	41	213	423	682	3597
CDD, Base 18.3°C	0	0	0	0	42	107	176	139	36	7	0	0	506
Relative Humidity (%)	70	70	68	64	65	66	70	73	73	68	71	74	69
Wind Speed (m/s)	5.1	4.8	5.5	5.4	4.6	4.2	3.5	3.4	3.7	4.3	4.8	4.9	4.5

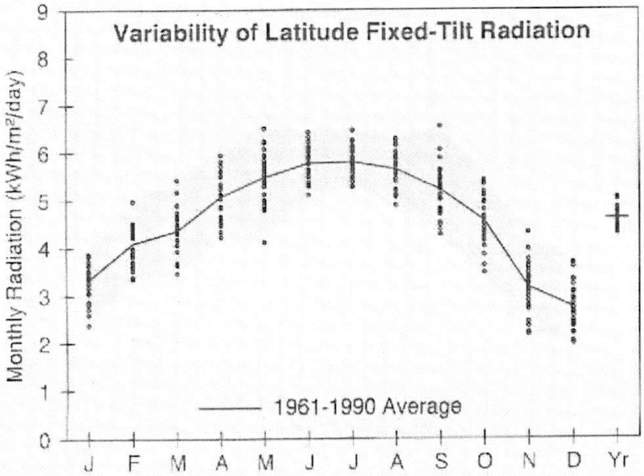

Peoria, IL

WBAN NO. 14842

LATITUDE: 40.67° N
LONGITUDE: 89.68° W
ELEVATION: 199 meters
MEAN PRESSURE: 993 millibars

STATION TYPE: Secondary

Solar Radiation for Flat-Plate Collectors Facing South at a Fixed Tilt (kWh/m²/day), Uncertainty ±9%

Tilt (°)		Jan	Feb	Mar	Apr	May	June	July	Aug	Sept	Oct	Nov	Dec	Year
0	Average	2.0	2.8	3.6	4.8	5.8	6.4	6.3	5.5	4.4	3.2	2.0	1.6	4.0
	Min/Max	1.6/2.2	2.5/3.2	3.0/4.2	4.1/5.6	4.4/6.8	5.8/7.2	5.8/6.9	4.8/6.1	3.8/5.3	2.7/3.7	1.6/2.5	1.3/1.9	3.8/4.4
Latitude -15	Average	2.9	3.7	4.2	5.2	5.8	6.3	6.2	5.9	5.2	4.3	2.9	2.4	4.6
	Min/Max	2.2/3.4	3.1/4.5	3.4/5.2	4.3/6.1	4.4/6.9	5.6/7.0	5.7/6.9	5.1/6.5	4.3/6.4	3.3/5.0	2.0/3.8	1.8/3.2	4.3/5.0
Latitude	Average	3.3	4.1	4.4	5.1	5.5	5.8	5.8	5.7	5.2	4.6	3.2	2.7	4.6
	Min/Max	2.4/3.9	3.3/5.0	3.5/5.4	4.2/5.9	4.1/6.5	5.1/6.4	5.3/6.5	4.9/6.3	4.3/6.5	3.5/5.4	2.2/4.3	2.0/3.7	4.3/5.0
Latitude +15	Average	3.5	4.2	4.2	4.7	4.8	5.0	5.1	5.1	5.0	4.6	3.3	2.9	4.4
	Min/Max	2.5/4.1	3.4/5.1	3.3/5.3	3.9/5.5	3.7/5.7	4.4/5.5	4.6/5.6	4.4/5.7	4.0/6.3	3.4/5.5	2.2/4.6	2.0/4.0	4.0/4.8
90	Average	3.3	3.7	3.2	3.0	2.7	2.6	2.7	3.1	3.5	3.7	2.9	2.7	3.1
	Min/Max	2.2/4.0	3.0/4.4	2.6/4.0	2.5/3.5	2.2/3.1	2.3/2.8	2.5/3.0	2.7/3.4	2.8/4.4	2.7/4.5	1.9/4.1	1.8/3.9	2.8/3.4

Solar Radiation for 1-Axis Tracking Flat-Plate Collectors with a North-South Axis (kWh/m²/day), Uncertainty ±9%

Axis Tilt (°)		Jan	Feb	Mar	Apr	May	June	July	Aug	Sept	Oct	Nov	Dec	Year
0	Average	2.8	3.8	4.7	6.2	7.5	8.3	8.2	7.4	6.0	4.6	2.7	2.2	5.4
	Min/Max	2.0/3.3	3.2/4.7	3.7/5.9	5.1/7.6	5.3/9.4	7.4/9.7	7.3/9.1	6.2/8.2	4.9/7.8	3.4/5.5	1.9/3.7	1.6/2.9	5.0/6.1
Latitude -15	Average	3.5	4.6	5.2	6.6	7.6	8.3	8.3	7.7	6.6	5.3	3.4	2.8	5.8
	Min/Max	2.4/4.1	3.7/5.7	4.0/6.7	5.3/8.1	5.3/9.6	7.4/9.7	7.3/9.3	6.4/8.6	5.3/8.6	3.9/6.4	2.2/4.7	2.0/3.8	5.4/6.6
Latitude	Average	3.8	4.8	5.3	6.5	7.4	8.0	8.0	7.5	6.7	5.6	3.6	3.1	5.9
	Min/Max	2.6/4.5	3.9/6.1	4.0/6.8	5.2/8.0	5.2/9.3	7.1/9.3	7.1/8.9	6.2/8.5	5.3/8.7	4.0/6.8	2.3/5.1	2.1/4.3	5.4/6.6
Latitude +15	Average	4.0	4.9	5.2	6.2	6.9	7.4	7.5	7.2	6.5	5.6	3.7	3.2	5.7
	Min/Max	2.7/4.8	3.9/6.2	4.0/6.8	5.0/7.7	4.8/8.8	6.6/8.7	6.6/8.4	5.9/8.1	5.1/8.5	4.0/6.8	2.4/5.3	2.2/4.5	5.2/6.5

Solar Radiation for 2-Axis Tracking Flat-Plate Collectors (kWh/m²/day), Uncertainty ±9%

Tracker		Jan	Feb	Mar	Apr	May	June	July	Aug	Sept	Oct	Nov	Dec	Year
2-Axis	Average	4.0	5.0	5.3	6.6	7.7	8.5	8.4	7.7	6.7	5.6	3.8	3.3	6.1
	Min/Max	2.7/4.8	3.9/6.2	4.1/6.9	5.3/8.1	5.4/9.7	7.6/9.9	7.5/9.4	6.4/8.6	5.3/8.7	4.1/6.8	2.4/5.3	2.2/4.6	5.6/6.8

Direct Beam Solar Radiation for Concentrating Collectors (kWh/m²/day), Uncertainty ±8%

Tracker		Jan	Feb	Mar	Apr	May	June	July	Aug	Sept	Oct	Nov	Dec	Year
1-Axis, E-W	Average	2.2	2.5	2.3	2.9	3.4	4.0	3.9	3.6	3.2	3.0	2.1	1.8	2.9
Horiz Axis	Min/Max	1.3/2.9	1.6/3.7	1.3/3.5	2.0/4.0	1.8/5.0	3.1/5.1	3.2/4.8	2.6/4.2	2.2/4.8	1.9/4.0	0.9/3.3	1.0/3.0	2.5/3.5
1-Axis, N-S	Average	1.6	2.2	2.6	3.7	4.6	5.2	5.2	4.7	3.9	3.0	1.6	1.2	3.3
Horiz Axis	Min/Max	0.9/2.1	1.4/3.3	1.5/3.9	2.6/5.2	2.4/6.7	4.2/6.7	4.2/6.4	3.4/5.6	2.7/5.8	1.8/4.0	0.7/2.5	0.6/2.0	2.8/4.1
1-Axis, N-S	Average	2.4	3.0	3.1	3.9	4.5	4.9	5.0	4.8	4.4	3.8	2.3	1.9	3.7
Tilt=Latitude	Min/Max	1.3/3.2	1.9/4.4	1.8/4.7	2.7/5.6	2.4/6.5	3.9/6.4	4.0/6.1	3.5/5.7	3.0/6.6	2.3/5.1	1.0/3.7	1.0/3.1	3.2/4.5
2-Axis	Average	2.6	3.1	3.1	4.0	4.7	5.4	5.3	4.9	4.4	3.9	2.4	2.1	3.8
	Min/Max	1.4/3.4	2.0/4.6	1.8/4.7	2.8/5.6	2.5/6.9	4.3/6.9	4.3/6.5	3.6/5.9	3.0/6.6	2.4/5.1	1.0/3.8	1.1/3.4	3.3/4.7

Average Climatic Conditions

Element	Jan	Feb	Mar	Apr	May	June	July	Aug	Sept	Oct	Nov	Dec	Year
Temperature (°C)	-5.8	-3.2	3.9	10.8	16.6	21.9	24.2	22.8	18.9	12.2	5.1	-2.8	10.4
Daily Minimum Temp	-10.4	-7.9	-1.2	4.9	10.5	15.9	18.6	17.3	12.9	6.2	0.3	-7.1	5.0
Daily Maximum Temp	-1.2	1.6	8.9	16.7	22.7	27.9	29.8	28.4	24.9	18.2	9.9	1.4	15.8
Record Minimum Temp	-31.7	-27.8	-23.3	-10.0	-3.9	3.9	8.3	5.0	-3.3	-7.2	-18.9	-30.6	-31.7
Record Maximum Temp	21.1	22.2	30.0	33.3	33.9	40.6	39.4	39.4	37.8	32.2	27.2	21.7	40.6
HDD, Base 18.3°C	747	602	448	227	102	6	0	5	30	198	397	654	3416
CDD, Base 18.3°C	0	0	0	0	48	114	181	145	48	8	0	0	546
Relative Humidity (%)	74	74	71	65	66	67	72	74	73	70	74	78	71
Wind Speed (m/s)	4.9	4.8	5.3	5.1	4.3	3.8	3.4	3.3	3.6	4.1	4.7	4.7	4.3

Rockford, IL

WBAN NO. 94822

LATITUDE: 42.20° N
LONGITUDE: 89.10° W
ELEVATION: 221 meters
MEAN PRESSURE: 990 millibars

STATION TYPE: Secondary

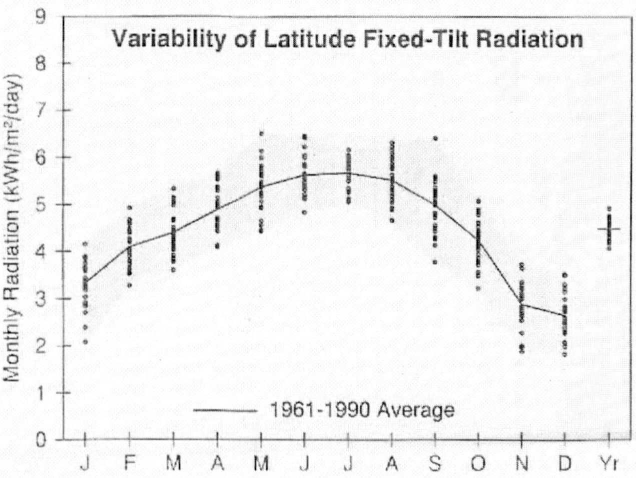

Solar Radiation for Flat-Plate Collectors Facing South at a Fixed Tilt (kWh/m²/day), Uncertainty ±9%

Tilt (°)		Jan	Feb	Mar	Apr	May	June	July	Aug	Sept	Oct	Nov	Dec	Year
0	Average	1.9	2.7	3.5	4.6	5.7	6.3	6.1	5.4	4.2	3.0	1.8	1.5	3.9
	Min/Max	1.4/2.2	2.4/3.1	3.1/4.1	4.0/5.2	4.7/6.8	5.4/7.2	5.6/6.7	4.6/6.0	3.5/5.1	2.5/3.4	1.5/2.1	1.2/1.8	3.7/4.2
Latitude -15	Average	2.9	3.8	4.3	5.0	5.7	6.1	6.1	5.7	5.0	4.0	2.6	2.3	4.5
	Min/Max	1.9/3.6	3.1/4.5	3.5/5.1	4.2/5.8	4.7/6.9	5.3/7.0	5.4/6.7	4.9/6.5	3.8/6.3	3.1/4.7	1.8/3.3	1.7/3.0	4.1/4.9
Latitude	Average	3.3	4.1	4.4	4.9	5.4	5.6	5.7	5.5	5.0	4.2	2.9	2.6	4.5
	Min/Max	2.1/4.2	3.3/4.9	3.6/5.3	4.1/5.7	4.4/6.5	4.8/6.5	5.1/6.2	4.7/6.3	3.8/6.4	3.2/5.1	1.9/3.7	1.8/3.5	4.1/4.9
Latitude +15	Average	3.6	4.2	4.3	4.5	4.8	4.9	4.9	5.0	4.8	4.3	3.0	2.8	4.3
	Min/Max	2.1/4.5	3.3/5.1	3.5/5.2	3.8/5.3	3.9/5.7	4.2/5.6	4.4/5.4	4.2/5.7	3.5/6.2	3.2/5.1	1.9/3.9	1.9/3.8	3.8/4.7
90	Average	3.4	3.8	3.4	3.0	2.8	2.7	2.8	3.1	3.4	3.5	2.7	2.7	3.1
	Min/Max	1.9/4.3	3.0/4.5	2.8/4.2	2.5/3.5	2.4/3.2	2.4/2.9	2.5/2.9	2.6/3.5	2.5/4.5	2.5/4.3	1.6/3.6	1.7/3.8	2.7/3.4

Solar Radiation for 1-Axis Tracking Flat-Plate Collectors with a North-South Axis (kWh/m²/day), Uncertainty ±9%

Axis Tilt (°)		Jan	Feb	Mar	Apr	May	June	July	Aug	Sept	Oct	Nov	Dec	Year
0	Average	2.7	3.8	4.7	6.0	7.3	8.1	8.0	7.1	5.7	4.2	2.5	2.1	5.2
	Min/Max	1.7/3.3	3.0/4.6	3.8/5.8	4.8/7.2	5.8/9.3	6.8/9.5	6.9/9.1	5.9/8.4	4.3/7.4	3.2/5.0	1.7/3.2	1.5/2.7	4.7/5.8
Latitude -15	Average	3.5	4.6	5.3	6.3	7.4	8.1	8.1	7.4	6.3	4.9	3.1	2.7	5.7
	Min/Max	2.1/4.4	3.5/5.7	4.2/6.6	5.0/7.6	5.8/9.6	6.8/9.6	7.0/9.2	6.2/8.8	4.7/8.3	3.7/6.0	1.9/4.1	1.8/3.6	5.1/6.4
Latitude	Average	3.8	4.9	5.4	6.2	7.2	7.7	7.8	7.3	6.3	5.1	3.3	3.0	5.7
	Min/Max	2.3/4.8	3.7/6.0	4.2/6.8	4.9/7.5	5.6/9.3	6.5/9.2	6.7/8.8	6.0/8.7	4.6/8.5	3.8/6.3	2.0/4.4	2.0/4.1	5.1/6.4
Latitude +15	Average	4.0	5.0	5.3	6.0	6.8	7.2	7.3	6.9	6.2	5.1	3.4	3.1	5.5
	Min/Max	2.3/5.1	3.7/6.2	4.1/6.7	4.7/7.2	5.3/8.8	6.1/8.6	6.3/8.3	5.7/8.3	4.4/8.3	3.7/6.3	2.0/4.6	2.0/4.3	4.9/6.3

Solar Radiation for 2-Axis Tracking Flat-Plate Collectors (kWh/m²/day), Uncertainty ±9%

Tracker		Jan	Feb	Mar	Apr	May	June	July	Aug	Sept	Oct	Nov	Dec	Year
2-Axis	Average	4.1	5.0	5.4	6.3	7.5	8.3	8.2	7.5	6.4	5.2	3.4	3.2	5.9
	Min/Max	2.3/5.2	3.7/6.2	4.2/6.8	5.1/7.6	5.9/9.7	7.0/9.8	7.1/9.3	6.2/8.9	4.7/8.5	3.8/6.4	2.0/4.6	2.0/4.4	5.2/6.6

Direct Beam Solar Radiation for Concentrating Collectors (kWh/m²/day), Uncertainty ±8%

Tracker		Jan	Feb	Mar	Apr	May	June	July	Aug	Sept	Oct	Nov	Dec	Year
1-Axis, E-W Horiz Axis	Average	2.2	2.5	2.4	2.7	3.3	3.8	3.8	3.4	3.0	2.7	1.8	1.7	2.8
	Min/Max	1.0/3.1	1.5/3.7	1.5/3.5	1.8/3.7	2.1/5.1	2.8/5.1	2.8/4.6	2.5/4.5	1.8/4.7	1.7/3.6	0.7/2.8	0.9/2.8	2.3/3.4
1-Axis, N-S Horiz Axis	Average	1.5	2.2	2.6	3.4	4.4	4.9	5.0	4.4	3.6	2.6	1.4	1.1	3.1
	Min/Max	0.7/2.1	1.2/3.3	1.6/3.8	2.3/4.9	2.7/6.8	3.7/6.6	3.7/6.2	3.2/5.8	2.2/5.5	1.7/3.5	0.6/2.1	0.6/1.8	2.6/3.8
1-Axis, N-S Tilt=Latitude	Average	2.4	3.0	3.1	3.7	4.3	4.7	4.8	4.5	4.1	3.4	2.0	1.8	3.5
	Min/Max	1.1/3.3	1.8/4.5	1.9/4.6	2.5/5.2	2.7/6.7	3.5/6.3	3.6/5.9	3.3/6.0	2.5/6.3	2.2/4.5	0.8/3.1	1.0/2.9	2.9/4.3
2-Axis	Average	2.6	3.1	3.1	3.8	4.6	5.1	5.1	4.7	4.1	3.5	2.1	2.0	3.6
	Min/Max	1.1/3.6	1.8/4.6	1.9/4.6	2.5/5.3	2.8/7.1	3.8/6.8	3.8/6.3	3.3/6.2	2.5/6.3	2.2/4.6	0.9/3.3	1.0/3.2	3.0/4.5

Average Climatic Conditions

Element	Jan	Feb	Mar	Apr	May	June	July	Aug	Sept	Oct	Nov	Dec	Year
Temperature (°C)	-7.7	-5.1	1.7	8.7	15.1	20.4	22.9	21.6	17.3	10.8	3.3	-4.4	8.7
Daily Minimum Temp	-12.3	-9.9	-3.3	2.7	8.6	14.1	16.9	15.7	11.1	4.8	-1.4	-8.8	3.2
Daily Maximum Temp	-3.0	-0.2	6.6	14.7	21.6	26.8	28.8	27.4	23.4	16.8	7.9	-0.1	14.2
Record Minimum Temp	-32.8	-30.0	-23.9	-15.0	-4.4	3.3	6.1	5.0	-2.8	-9.4	-23.3	-31.1	-32.8
Record Maximum Temp	17.2	20.0	29.4	32.8	35.0	38.3	39.4	40.0	38.9	32.2	24.4	18.9	40.0
HDD, Base 18.3°C	806	655	517	288	133	13	3	8	55	237	450	706	3872
CDD, Base 18.3°C	0	0	0	0	31	77	144	110	23	5	0	0	390
Relative Humidity (%)	75	74	72	66	65	66	71	74	74	71	75	78	72
Wind Speed (m/s)	5.0	4.8	5.4	5.3	4.7	4.2	3.7	3.5	3.8	4.3	4.8	4.8	4.5

Springfield, IL

WBAN NO. 93822

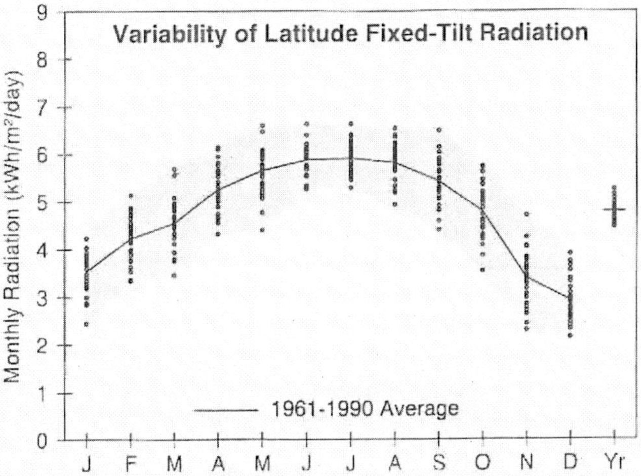

LATITUDE: 39.83° N
LONGITUDE: 89.67° W
ELEVATION: 187 meters
MEAN PRESSURE: 995 millibars

STATION TYPE: Secondary

Solar Radiation for Flat-Plate Collectors Facing South at a Fixed Tilt (kWh/m²/day), Uncertainty ±9%

Tilt (°)		Jan	Feb	Mar	Apr	May	June	July	Aug	Sept	Oct	Nov	Dec	Year
0	Average	2.1	2.9	3.7	5.0	6.0	6.5	6.4	5.7	4.6	3.4	2.2	1.7	4.2
	Min/Max	1.6/2.3	2.5/3.3	3.1/4.4	4.3/5.7	4.7/6.9	5.9/7.4	5.8/7.1	4.9/6.4	3.9/5.3	2.8/3.9	1.8/2.7	1.4/2.1	4.0/4.6
Latitude -15	Average	3.1	3.9	4.4	5.4	6.0	6.4	6.4	6.0	5.4	4.5	3.1	2.6	4.8
	Min/Max	2.2/3.7	3.1/4.7	3.4/5.4	4.5/6.2	4.7/7.0	5.7/7.2	5.7/7.1	5.1/6.8	4.4/6.3	3.4/5.3	2.2/4.1	2.0/3.3	4.5/5.2
Latitude	Average	3.5	4.2	4.5	5.3	5.6	5.9	5.9	5.8	5.4	4.8	3.4	2.9	4.8
	Min/Max	2.5/4.2	3.3/5.1	3.5/5.7	4.3/6.1	4.4/6.6	5.3/6.6	5.3/6.6	4.9/6.5	4.4/6.5	3.5/5.7	2.3/4.7	2.2/3.9	4.5/5.3
Latitude +15	Average	3.8	4.4	4.4	4.9	5.0	5.1	5.1	5.3	5.2	4.8	3.5	3.1	4.5
	Min/Max	2.5/4.6	3.4/5.3	3.3/5.6	4.0/5.7	3.9/5.8	4.6/5.7	4.6/5.8	4.5/5.9	4.2/6.2	3.5/5.8	2.3/5.0	2.2/4.2	4.2/5.0
90	Average	3.5	3.8	3.3	3.1	2.7	2.5	2.7	3.1	3.6	3.9	3.1	2.9	3.2
	Min/Max	2.3/4.4	2.9/4.6	2.6/4.4	2.5/3.6	2.3/3.0	2.4/2.7	2.5/2.9	2.7/3.4	2.9/4.4	2.8/4.7	2.0/4.5	2.0/4.1	2.9/3.5

Solar Radiation for 1-Axis Tracking Flat-Plate Collectors with a North-South Axis (kWh/m²/day), Uncertainty ±9%

Axis Tilt (°)		Jan	Feb	Mar	Apr	May	June	July	Aug	Sept	Oct	Nov	Dec	Year
0	Average	3.0	4.0	4.9	6.5	7.8	8.5	8.5	7.7	6.3	4.8	3.0	2.4	5.6
	Min/Max	2.0/3.5	3.2/4.9	3.7/6.3	5.2/7.9	5.7/9.5	7.3/10.0	7.5/9.5	6.2/8.7	5.0/7.7	3.6/5.8	2.1/4.1	1.7/3.1	5.2/6.3
Latitude -15	Average	3.8	4.8	5.5	6.9	7.9	8.5	8.5	8.0	6.9	5.6	3.7	3.0	6.1
	Min/Max	2.5/4.5	3.7/5.9	3.9/7.1	5.4/8.4	5.8/9.7	7.3/10.0	7.5/9.6	6.4/9.1	5.4/8.5	4.0/6.8	2.4/5.2	2.1/4.1	5.6/6.9
Latitude	Average	4.1	5.1	5.6	6.8	7.6	8.2	8.2	7.8	7.0	5.9	3.9	3.3	6.1
	Min/Max	2.7/5.0	3.9/6.3	4.0/7.3	5.3/8.4	5.6/9.4	7.0/9.6	7.2/9.2	6.3/9.0	5.4/8.6	4.2/7.2	2.5/5.6	2.3/4.5	5.6/6.9
Latitude +15	Average	4.3	5.2	5.5	6.5	7.2	7.6	7.7	7.5	6.8	5.9	4.1	3.5	6.0
	Min/Max	2.8/5.3	3.9/6.4	3.9/7.2	5.0/8.0	5.2/8.9	6.6/8.9	6.8/8.6	5.9/8.6	5.3/8.4	4.1/7.2	2.5/5.8	2.4/4.8	5.4/6.8

Solar Radiation for 2-Axis Tracking Flat-Plate Collectors (kWh/m²/day), Uncertainty ±9%

Tracker		Jan	Feb	Mar	Apr	May	June	July	Aug	Sept	Oct	Nov	Dec	Year
2-Axis	Average	4.4	5.2	5.6	6.9	8.0	8.7	8.7	8.0	7.0	5.9	4.1	3.5	6.3
	Min/Max	2.8/5.3	3.9/6.4	4.0/7.3	5.4/8.4	5.8/9.8	7.5/10.2	7.6/9.7	6.4/9.2	5.5/8.6	4.2/7.3	2.5/5.9	2.4/4.9	5.8/7.2

Direct Beam Solar Radiation for Concentrating Collectors (kWh/m²/day), Uncertainty ±8%

Tracker		Jan	Feb	Mar	Apr	May	June	July	Aug	Sept	Oct	Nov	Dec	Year
1-Axis, E-W	Average	2.5	2.7	2.5	3.1	3.7	4.2	4.1	3.8	3.5	3.3	2.3	2.0	3.1
Horiz Axis	Min/Max	1.4/3.2	1.6/3.9	1.1/3.9	2.1/4.2	2.2/5.2	3.1/5.5	3.3/5.0	2.6/4.7	2.3/4.7	2.0/4.3	0.9/3.7	1.2/3.0	2.6/3.8
1-Axis, N-S	Average	1.8	2.4	2.8	4.0	4.9	5.5	5.5	5.0	4.2	3.2	1.8	1.4	3.6
Horiz Axis	Min/Max	0.9/2.3	1.5/3.5	1.3/4.4	2.7/5.6	2.9/6.9	4.1/7.2	4.4/6.6	3.4/6.3	2.8/5.7	2.0/4.3	0.7/2.9	0.8/2.0	3.0/4.4
1-Axis, N-S	Average	2.7	3.3	3.4	4.2	4.8	5.2	5.3	5.1	4.7	4.1	2.6	2.1	4.0
Tilt=Latitude	Min/Max	1.5/3.5	2.0/4.7	1.5/5.2	2.8/5.9	2.9/6.8	3.9/6.8	4.2/6.3	3.5/6.4	3.2/6.4	2.5/5.4	1.0/4.2	1.3/3.2	3.3/4.9
2-Axis	Average	2.9	3.3	3.4	4.3	5.1	5.6	5.6	5.3	4.7	4.2	2.7	2.3	4.1
	Min/Max	1.6/3.7	2.0/4.8	1.5/5.2	2.8/6.0	3.0/7.2	4.2/7.4	4.5/6.8	3.6/6.6	3.2/6.4	2.5/5.5	1.1/4.4	1.4/3.4	3.5/5.1

Average Climatic Conditions

Element	Jan	Feb	Mar	Apr	May	June	July	Aug	Sept	Oct	Nov	Dec	Year
Temperature (°C)	-4.3	-1.8	4.9	11.8	17.5	22.7	24.7	23.2	19.6	13.1	6.1	-1.3	11.3
Daily Minimum Temp	-8.9	-6.6	-0.2	5.8	11.3	16.6	18.9	17.4	13.3	6.9	1.1	-5.6	5.8
Daily Maximum Temp	0.3	2.9	10.0	17.7	23.7	28.8	30.5	29.0	25.9	19.3	10.9	2.9	16.8
Record Minimum Temp	-29.4	-30.0	-24.4	-7.2	-2.2	4.4	8.9	6.1	0.0	-8.3	-19.4	-29.4	-30.0
Record Maximum Temp	21.7	23.3	30.6	32.2	35.0	39.4	44.4	39.4	38.3	33.9	28.3	23.3	44.4
HDD, Base 18.3°C	703	564	417	201	92	4	0	4	24	174	368	608	3160
CDD, Base 18.3°C	0	0	0	4	67	136	198	154	63	12	0	0	634
Relative Humidity (%)	73	74	71	65	66	67	70	74	72	68	74	78	71
Wind Speed (m/s)	5.5	5.3	5.9	5.6	4.8	4.2	3.7	3.5	4.0	4.6	5.2	5.4	4.8

Evansville, IN

WBAN NO. 93817

LATITUDE: 38.05° N
LONGITUDE: 87.53° W
ELEVATION: 118 meters
MEAN PRESSURE: 1004 millibars

STATION TYPE: Secondary

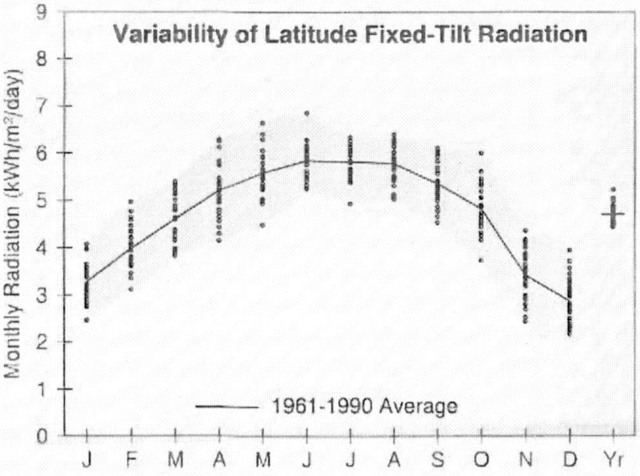

Solar Radiation for Flat-Plate Collectors Facing South at a Fixed Tilt (kWh/m²/day), Uncertainty ±9%

Tilt (°)		Jan	Feb	Mar	Apr	May	June	July	Aug	Sept	Oct	Nov	Dec	Year
0	Average	2.1	2.9	3.8	5.0	5.9	6.5	6.3	5.7	4.6	3.5	2.3	1.8	4.2
	Min/Max	1.8/2.5	2.5/3.3	3.4/4.3	4.1/5.9	4.8/7.0	5.8/7.7	5.3/6.9	5.0/6.3	4.1/5.2	2.9/4.2	1.9/2.7	1.5/2.2	4.0/4.6
Latitude -15	Average	2.9	3.7	4.5	5.3	5.9	6.3	6.2	6.0	5.3	4.5	3.1	2.5	4.7
	Min/Max	2.3/3.5	3.0/4.5	3.8/5.2	4.3/6.4	4.8/7.1	5.7/7.5	5.3/6.8	5.2/6.6	4.5/6.0	3.5/5.5	2.3/3.8	2.0/3.4	4.4/5.2
Latitude	Average	3.3	4.0	4.6	5.2	5.6	5.8	5.8	5.8	5.3	4.8	3.4	2.9	4.7
	Min/Max	2.5/4.1	3.1/5.0	3.8/5.4	4.1/6.3	4.5/6.6	5.2/6.8	4.9/6.3	5.0/6.4	4.5/6.1	3.7/6.0	2.4/4.3	2.2/4.0	4.4/5.2
Latitude +15	Average	3.5	4.1	4.5	4.8	4.9	5.0	5.1	5.3	5.1	4.9	3.6	3.0	4.5
	Min/Max	2.5/4.4	3.1/5.2	3.7/5.3	3.8/5.8	4.0/5.8	4.6/5.8	4.3/5.5	4.6/5.8	4.3/5.9	3.7/6.1	2.5/4.6	2.3/4.3	4.2/5.0
90	Average	3.1	3.4	3.3	3.0	2.6	2.4	2.5	3.0	3.5	3.8	3.1	2.8	3.0
	Min/Max	2.2/4.1	2.6/4.5	2.7/3.9	2.4/3.6	2.2/2.9	2.3/2.6	2.3/2.7	2.6/3.3	2.9/4.0	2.9/4.9	2.0/4.1	2.0/4.0	2.8/3.3

Solar Radiation for 1-Axis Tracking Flat-Plate Collectors with a North-South Axis (kWh/m²/day), Uncertainty ±9%

Axis Tilt (°)		Jan	Feb	Mar	Apr	May	June	July	Aug	Sept	Oct	Nov	Dec	Year
0	Average	2.9	3.9	5.1	6.5	7.7	8.4	8.2	7.6	6.2	4.9	3.1	2.4	5.6
	Min/Max	2.2/3.6	3.1/4.8	4.1/6.1	5.0/8.2	5.8/9.6	7.2/10.4	6.5/9.1	6.4/8.5	5.2/7.3	3.7/6.3	2.3/3.9	1.8/3.3	5.2/6.4
Latitude -15	Average	3.5	4.5	5.6	6.8	7.7	8.4	8.2	7.8	6.7	5.7	3.7	3.0	6.0
	Min/Max	2.6/4.4	3.4/5.7	4.4/6.8	5.2/8.6	5.9/9.8	7.1/10.4	6.5/9.1	6.6/8.8	5.6/7.9	4.2/7.2	2.6/4.8	2.2/4.2	5.6/6.8
Latitude	Average	3.8	4.7	5.7	6.7	7.5	8.1	7.9	7.7	6.8	5.9	4.0	3.3	6.0
	Min/Max	2.8/4.8	3.6/6.1	4.5/6.9	5.1/8.5	5.7/9.5	6.8/10.0	6.3/8.8	6.4/8.6	5.6/8.0	4.3/7.6	2.7/5.2	2.3/4.6	5.6/6.9
Latitude +15	Average	3.9	4.8	5.6	6.4	7.1	7.5	7.4	7.3	6.6	5.9	4.1	3.4	5.9
	Min/Max	2.8/5.0	3.6/6.2	4.4/6.8	4.9/8.2	5.3/9.0	6.3/9.3	5.9/8.3	6.1/8.2	5.4/7.9	4.3/7.7	2.7/5.4	2.4/4.9	5.5/6.7

Solar Radiation for 2-Axis Tracking Flat-Plate Collectors (kWh/m²/day), Uncertainty ±9%

Tracker		Jan	Feb	Mar	Apr	May	June	July	Aug	Sept	Oct	Nov	Dec	Year
2-Axis	Average	4.0	4.8	5.7	6.8	7.8	8.6	8.4	7.9	6.8	6.0	4.1	3.5	6.2
	Min/Max	2.8/5.1	3.6/6.2	4.5/6.9	5.2/8.6	5.9/9.9	7.3/10.6	6.7/9.3	6.6/8.8	5.6/8.1	4.4/7.7	2.8/5.4	2.4/5.0	5.8/7.1

Direct Beam Solar Radiation for Concentrating Collectors (kWh/m²/day), Uncertainty ±8%

Tracker		Jan	Feb	Mar	Apr	May	June	July	Aug	Sept	Oct	Nov	Dec	Year
1-Axis, E-W Horiz Axis	Average	2.2	2.4	2.6	3.0	3.6	4.1	3.9	3.6	3.3	3.2	2.3	1.9	3.0
	Min/Max	1.3/3.2	1.4/3.5	1.7/3.6	1.8/4.4	2.1/5.3	2.8/5.9	2.5/4.8	2.6/4.5	2.4/4.2	2.0/4.6	1.2/3.3	1.1/3.2	2.7/3.8
1-Axis, N-S Horiz Axis	Average	1.6	2.3	3.0	4.0	4.8	5.4	5.2	4.8	4.0	3.3	1.8	1.4	3.5
	Min/Max	1.0/2.4	1.4/3.2	1.9/4.2	2.3/5.8	2.9/7.1	3.7/7.8	3.4/6.3	3.5/5.9	2.9/5.3	2.0/4.7	1.0/2.7	0.8/2.3	3.1/4.4
1-Axis, N-S Tilt=Latitude	Average	2.4	3.0	3.5	4.2	4.6	5.0	4.9	4.9	4.4	4.1	2.6	2.1	3.8
	Min/Max	1.4/3.5	1.8/4.2	2.2/4.9	2.4/6.1	2.8/6.9	3.5/7.3	3.2/6.0	3.5/6.0	3.3/5.8	2.5/5.8	1.4/3.8	1.2/3.5	3.4/4.8
2-Axis	Average	2.6	3.1	3.5	4.2	4.9	5.5	5.3	5.0	4.4	4.1	2.7	2.2	4.0
	Min/Max	1.5/3.7	1.8/4.3	2.2/4.9	2.4/6.2	3.0/7.3	3.8/7.9	3.4/6.4	3.6/6.2	3.3/5.9	2.5/5.9	1.4/3.9	1.3/3.8	3.6/5.0

Average Climatic Conditions

Element	Jan	Feb	Mar	Apr	May	June	July	Aug	Sept	Oct	Nov	Dec	Year
Temperature (°C)	-1.1	1.3	7.7	13.4	18.6	23.8	25.8	24.5	20.7	14.0	7.9	1.8	13.2
Daily Minimum Temp	-6.0	-3.9	2.1	7.2	12.3	17.4	19.7	18.3	14.2	7.1	2.5	-2.9	7.3
Daily Maximum Temp	3.8	6.5	13.3	19.7	24.9	30.1	31.7	30.7	27.1	20.9	13.3	6.4	19.1
Record Minimum Temp	-29.4	-30.6	-22.8	-5.0	-2.2	5.0	8.3	6.1	-0.6	-6.1	-19.4	-26.1	-30.6
Record Maximum Temp	24.4	26.1	28.9	32.8	35.0	40.0	40.6	38.9	39.4	34.4	28.3	25.0	40.6
HDD, Base 18.3°C	601	476	331	152	63	0	0	0	18	148	313	513	2616
CDD, Base 18.3°C	0	0	0	5	72	163	231	191	88	14	0	0	764
Relative Humidity (%)	72	71	68	65	68	68	71	73	73	69	71	74	70
Wind Speed (m/s)	4.0	4.0	4.3	4.1	3.4	3.1	2.7	2.5	2.7	3.0	3.6	3.8	3.4

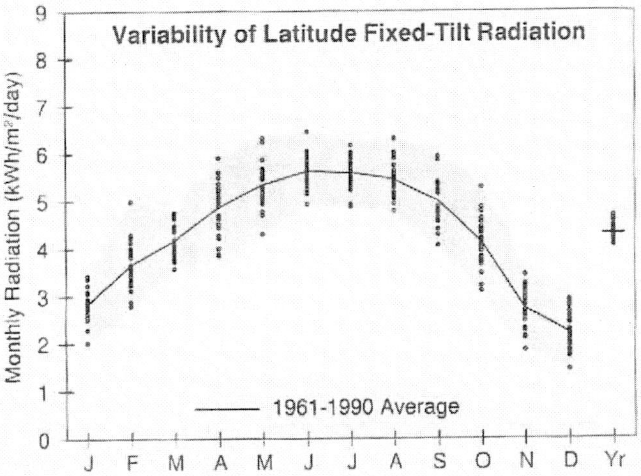

Fort Wayne, IN

WBAN NO. 14827

LATITUDE: 41.00° N
LONGITUDE: 85.20° W
ELEVATION: 252 meters
MEAN PRESSURE: 987 millibars

STATION TYPE: Secondary

Solar Radiation for Flat-Plate Collectors Facing South at a Fixed Tilt (kWh/m²/day), Uncertainty ±9%

Tilt (°)		Jan	Feb	Mar	Apr	May	June	July	Aug	Sept	Oct	Nov	Dec	Year
0	Average	1.8	2.6	3.5	4.6	5.6	6.2	6.1	5.3	4.3	3.0	1.8	1.4	3.9
	Min/Max	1.4/2.0	2.2/3.3	3.2/3.9	3.8/5.4	4.6/6.7	5.5/7.2	5.3/6.7	4.8/6.1	3.7/4.9	2.5/3.6	1.5/2.1	1.1/1.7	3.7/4.2
Latitude -15	Average	2.5	3.4	4.1	5.0	5.7	6.1	6.0	5.7	5.0	3.9	2.5	2.0	4.3
	Min/Max	1.9/3.0	2.7/4.5	3.6/4.6	4.0/6.0	4.6/6.8	5.3/7.0	5.2/6.7	5.0/6.5	4.1/5.8	3.0/4.9	1.8/3.1	1.4/2.5	4.1/4.7
Latitude	Average	2.8	3.7	4.2	4.9	5.4	5.6	5.6	5.4	5.0	4.1	2.7	2.2	4.3
	Min/Max	2.0/3.4	2.8/5.0	3.6/4.8	3.8/5.9	4.3/6.3	4.9/6.5	4.9/6.2	4.8/6.3	4.1/5.9	3.1/5.3	1.9/3.4	1.4/2.9	4.1/4.7
Latitude +15	Average	3.0	3.8	4.1	4.5	4.8	4.9	4.9	4.9	4.8	4.1	2.8	2.3	4.1
	Min/Max	2.1/3.7	2.8/5.2	3.5/4.6	3.5/5.5	3.8/5.6	4.3/5.6	4.3/5.4	4.3/5.7	3.8/5.7	3.0/5.4	1.8/3.6	1.5/3.1	3.8/4.4
90	Average	2.8	3.3	3.1	2.9	2.7	2.6	2.7	3.0	3.4	3.4	2.5	2.2	2.9
	Min/Max	1.8/3.5	2.4/4.8	2.7/3.8	2.3/3.6	2.3/3.1	2.4/2.8	2.4/2.8	2.7/3.5	2.7/4.1	2.4/4.4	1.5/3.3	1.3/3.0	2.7/3.1

Solar Radiation for 1-Axis Tracking Flat-Plate Collectors with a North-South Axis (kWh/m²/day), Uncertainty ±9%

Axis Tilt (°)		Jan	Feb	Mar	Apr	May	June	July	Aug	Sept	Oct	Nov	Dec	Year
0	Average	2.4	3.4	4.4	5.9	7.2	7.9	7.8	7.0	5.7	4.1	2.4	1.8	5.0
	Min/Max	1.6/2.8	2.6/4.6	3.8/5.2	4.6/7.2	5.5/9.0	6.5/9.6	6.7/8.8	5.9/8.5	4.6/6.7	3.0/5.3	1.6/2.9	1.2/2.3	4.7/5.6
Latitude -15	Average	2.9	4.0	4.9	6.2	7.3	7.9	7.8	7.3	6.2	4.8	2.9	2.3	5.4
	Min/Max	2.0/3.6	3.0/5.5	4.1/5.8	4.7/7.7	5.6/9.2	6.5/9.6	6.7/8.8	6.2/8.9	4.9/7.4	3.5/6.3	1.9/3.7	1.4/3.0	5.1/6.0
Latitude	Average	3.2	4.3	5.0	6.2	7.1	7.6	7.6	7.1	6.3	4.9	3.1	2.4	5.4
	Min/Max	2.1/3.9	3.1/5.9	4.1/5.9	4.6/7.7	5.4/8.9	6.2/9.2	6.4/8.5	6.0/8.7	4.9/7.5	3.5/6.6	2.0/4.0	1.5/3.3	5.0/6.0
Latitude +15	Average	3.3	4.3	4.9	5.9	6.7	7.1	7.1	6.8	6.1	5.0	3.1	2.5	5.2
	Min/Max	2.2/4.1	3.1/6.0	4.1/5.8	4.4/7.3	5.1/8.4	5.8/8.6	6.0/8.0	5.7/8.3	4.7/7.4	3.5/6.6	2.0/4.1	1.5/3.5	4.9/5.9

Solar Radiation for 2-Axis Tracking Flat-Plate Collectors (kWh/m²/day), Uncertainty ±9%

Tracker		Jan	Feb	Mar	Apr	May	June	July	Aug	Sept	Oct	Nov	Dec	Year
2-Axis	Average	3.4	4.4	5.0	6.3	7.4	8.1	8.0	7.3	6.3	5.0	3.2	2.6	5.6
	Min/Max	2.2/4.2	3.2/6.0	4.2/5.9	4.7/7.7	5.7/9.3	6.6/9.8	6.8/9.0	6.2/8.9	4.9/7.6	3.5/6.6	2.0/4.2	1.5/3.5	5.2/6.2

Direct Beam Solar Radiation for Concentrating Collectors (kWh/m²/day), Uncertainty ±8%

Tracker		Jan	Feb	Mar	Apr	May	June	July	Aug	Sept	Oct	Nov	Dec	Year
1-Axis, E-W	Average	1.7	2.0	2.1	2.6	3.2	3.6	3.5	3.2	2.9	2.5	1.6	1.2	2.5
Horiz Axis	Min/Max	0.9/2.5	1.2/2.8	1.2/2.8	1.5/3.7	1.9/4.7	2.5/5.0	2.5/4.4	2.4/4.3	1.9/3.8	1.4/3.8	0.8/2.4	0.5/2.0	2.2/3.1
1-Axis, N-S	Average	1.2	1.8	2.3	3.4	4.2	4.7	4.7	4.2	3.4	2.4	1.2	0.8	2.9
Horiz Axis	Min/Max	0.6/1.8	1.0/2.5	1.4/3.2	1.9/4.8	2.6/6.2	3.2/6.5	3.4/5.8	3.1/5.7	2.3/4.5	1.4/3.7	0.6/1.8	0.3/1.3	2.5/3.5
1-Axis, N-S	Average	1.8	2.4	2.8	3.6	4.1	4.4	4.5	4.3	3.9	3.1	1.8	1.3	3.2
Tilt=Latitude	Min/Max	1.0/2.7	1.4/3.3	1.6/3.8	2.0/5.1	2.5/6.1	3.0/6.2	3.3/5.5	3.1/5.9	2.6/5.1	1.8/4.8	0.8/2.7	0.5/2.0	2.8/3.9
2-Axis	Average	1.9	2.5	2.8	3.6	4.4	4.8	4.8	4.4	3.9	3.2	1.9	1.4	3.3
	Min/Max	1.0/2.9	1.4/3.4	1.6/3.8	2.1/5.2	2.7/6.4	3.3/6.7	3.5/5.9	3.2/6.1	2.6/5.1	1.8/4.8	0.9/2.8	0.6/2.2	2.9/4.1

Average Climatic Conditions

Element	Jan	Feb	Mar	Apr	May	June	July	Aug	Sept	Oct	Nov	Dec	Year
Temperature (°C)	-5.1	-3.4	3.1	9.6	15.7	21.2	23.3	22.0	18.3	11.6	5.2	-1.9	9.9
Daily Minimum Temp	-9.3	-7.9	-1.8	3.6	9.5	15.2	17.3	16.1	12.3	5.8	0.8	-5.8	4.7
Daily Maximum Temp	-0.9	1.1	7.9	15.4	21.8	27.2	29.2	27.9	24.2	17.3	9.5	1.9	15.2
Record Minimum Temp	-30.0	-27.8	-23.3	-13.9	-2.8	3.3	6.7	3.3	-1.7	-7.2	-18.3	-27.8	-30.0
Record Maximum Temp	20.6	20.6	27.8	31.1	34.4	41.1	39.4	38.3	37.8	32.2	26.1	21.7	41.1
HDD, Base 18.3°C	725	608	472	263	121	8	0	6	45	216	395	627	3485
CDD, Base 18.3°C	0	0	0	0	38	95	155	119	43	7	0	0	458
Relative Humidity (%)	76	74	72	66	65	66	69	73	73	72	76	79	72
Wind Speed (m/s)	5.3	4.9	5.4	5.1	4.4	4.1	3.6	3.4	3.6	4.1	4.7	5.0	4.5

Indianapolis, IN

WBAN NO. 93819

LATITUDE: 39.73° N
LONGITUDE: 86.28° W
ELEVATION: 246 meters
MEAN PRESSURE: 988 millibars

STATION TYPE: Primary

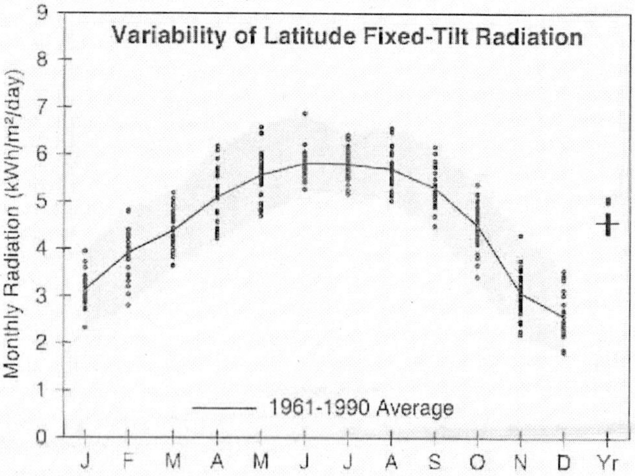

Solar Radiation for Flat-Plate Collectors Facing South at a Fixed Tilt (kWh/m²/day), Uncertainty ±9%

Tilt (°)		Jan	Feb	Mar	Apr	May	June	July	Aug	Sept	Oct	Nov	Dec	Year
0	Average	2.0	2.8	3.7	4.9	5.9	6.5	6.3	5.6	4.6	3.3	2.1	1.6	4.1
	Min/Max	1.6/2.3	2.3/3.3	3.2/4.1	4.1/5.7	5.1/6.9	5.9/7.7	5.6/7.0	5.0/6.4	4.0/5.2	2.8/3.8	1.7/2.5	1.3/2.0	3.9/4.5
Latitude -15	Average	2.8	3.6	4.3	5.2	5.9	6.3	6.2	5.9	5.2	4.2	2.8	2.3	4.6
	Min/Max	2.2/3.5	2.7/4.4	3.6/5.0	4.3/6.3	5.0/7.0	5.7/7.5	5.6/6.9	5.2/6.8	4.5/6.1	3.3/5.0	2.1/3.8	1.7/3.1	4.4/5.1
Latitude	Average	3.1	3.9	4.4	5.1	5.6	5.8	5.8	5.7	5.3	4.5	3.1	2.6	4.6
	Min/Max	2.3/4.0	2.8/4.8	3.6/5.2	4.2/6.2	4.7/6.6	5.3/6.9	5.2/6.4	5.0/6.6	4.5/6.2	3.4/5.4	2.2/4.3	1.8/3.5	4.3/5.1
Latitude +15	Average	3.3	4.0	4.3	4.7	4.9	5.0	5.1	5.2	5.1	4.5	3.2	2.7	4.3
	Min/Max	2.4/4.3	2.8/5.0	3.5/5.1	3.9/5.7	4.2/5.8	4.6/5.9	4.5/5.6	4.6/5.9	4.3/5.9	3.4/5.4	2.2/4.5	1.9/3.8	4.1/4.8
90	Average	3.0	3.5	3.2	3.0	2.7	2.6	2.7	3.1	3.5	3.6	2.8	2.5	3.0
	Min/Max	2.1/4.1	2.4/4.6	2.5/3.8	2.5/3.6	2.4/3.1	2.4/2.8	2.5/2.9	2.7/3.5	2.9/4.1	2.6/4.4	1.8/4.1	1.6/3.7	2.8/3.3

Solar Radiation for 1-Axis Tracking Flat-Plate Collectors with a North-South Axis (kWh/m²/day), Uncertainty ±9%

Axis Tilt (°)		Jan	Feb	Mar	Apr	May	June	July	Aug	Sept	Oct	Nov	Dec	Year
0	Average	2.7	3.7	4.7	6.2	7.5	8.2	8.1	7.4	6.1	4.5	2.7	2.1	5.3
	Min/Max	2.0/3.4	2.8/4.6	3.8/5.7	4.9/7.9	6.1/9.1	7.2/10.4	6.9/9.2	6.4/8.9	5.1/7.2	3.4/5.4	1.9/3.8	1.5/2.8	5.1/6.1
Latitude -15	Average	3.3	4.3	5.2	6.5	7.6	8.2	8.1	7.7	6.6	5.2	3.3	2.7	5.7
	Min/Max	2.4/4.3	3.1/5.5	4.2/6.3	5.1/8.4	6.1/9.3	7.2/10.4	6.9/9.3	6.6/9.3	5.5/7.9	3.8/6.4	2.2/4.7	1.8/3.7	5.4/6.6
Latitude	Average	3.6	4.6	5.3	6.5	7.4	7.9	7.8	7.5	6.7	5.5	3.5	2.9	5.8
	Min/Max	2.5/4.7	3.2/5.8	4.2/6.4	4.9/8.4	5.9/9.0	6.9/10.0	6.7/9.0	6.4/9.1	5.5/8.0	3.9/6.7	2.3/5.1	1.9/4.1	5.4/6.6
Latitude +15	Average	3.7	4.7	5.2	6.2	6.9	7.3	7.3	7.1	6.5	5.5	3.6	3.0	5.6
	Min/Max	2.6/4.9	3.2/5.9	4.1/6.4	4.7/8.1	5.6/8.5	6.4/9.3	6.2/8.4	6.1/8.7	5.3/7.8	3.9/6.7	2.4/5.3	2.0/4.3	5.3/6.4

Solar Radiation for 2-Axis Tracking Flat-Plate Collectors (kWh/m²/day), Uncertainty ±9%

Tracker		Jan	Feb	Mar	Apr	May	June	July	Aug	Sept	Oct	Nov	Dec	Year
2-Axis	Average	3.7	4.7	5.3	6.6	7.7	8.4	8.3	7.7	6.7	5.5	3.6	3.1	5.9
	Min/Max	2.6/4.9	3.2/5.9	4.2/6.5	5.1/8.5	6.2/9.4	7.3/10.6	7.1/9.5	6.6/9.4	5.6/8.1	4.0/6.7	2.4/5.3	2.0/4.4	5.6/6.8

Direct Beam Solar Radiation for Concentrating Collectors (kWh/m²/day), Uncertainty ±8%

Tracker		Jan	Feb	Mar	Apr	May	June	July	Aug	Sept	Oct	Nov	Dec	Year
1-Axis, E-W	Average	2.0	2.3	2.3	2.8	3.4	3.7	3.7	3.4	3.1	2.9	1.9	1.6	2.7
Horiz Axis	Min/Max	1.1/3.0	1.1/3.4	1.4/3.2	1.7/4.2	2.2/4.8	3.0/5.8	2.7/4.8	2.6/4.7	2.3/4.3	1.6/3.8	0.9/3.3	0.8/2.7	2.5/3.5
1-Axis, N-S	Average	1.4	2.0	2.5	3.6	4.4	4.9	4.8	4.5	3.8	2.8	1.5	1.1	3.1
Horiz Axis	Min/Max	0.8/2.2	1.0/3.1	1.5/3.6	2.1/5.5	2.9/6.4	3.9/7.6	3.6/6.3	3.5/6.3	2.8/5.1	1.6/3.8	0.7/2.6	0.5/1.8	2.8/4.0
1-Axis, N-S	Average	2.1	2.7	3.0	3.8	4.3	4.6	4.6	4.6	4.2	3.6	2.1	1.7	3.4
Tilt=Latitude	Min/Max	1.2/3.3	1.3/4.1	1.8/4.2	2.2/5.8	2.8/6.2	3.6/7.1	3.5/6.1	3.5/6.4	3.1/5.8	2.1/4.8	1.0/3.7	0.8/2.8	3.1/4.4
2-Axis	Average	2.3	2.8	3.0	3.9	4.6	5.0	5.0	4.7	4.2	3.6	2.2	1.8	3.6
	Min/Max	1.3/3.5	1.4/4.2	1.9/4.2	2.3/5.9	3.0/6.6	4.0/7.8	3.7/6.5	3.6/6.7	3.2/5.8	2.1/4.8	1.1/3.9	0.9/3.0	3.2/4.6

Average Climatic Conditions

Element	Jan	Feb	Mar	Apr	May	June	July	Aug	Sept	Oct	Nov	Dec	Year
Temperature (°C)	-3.6	-1.3	5.2	11.3	17.1	22.2	24.1	22.9	19.2	12.6	6.1	-0.6	11.3
Daily Minimum Temp	-8.2	-6.2	-0.1	5.3	10.9	16.1	18.4	17.1	13.1	6.4	1.2	-4.9	5.8
Daily Maximum Temp	0.9	3.5	10.5	17.4	23.2	28.2	29.7	28.7	25.3	18.8	11.1	3.6	16.7
Record Minimum Temp	-30.0	-29.4	-21.7	-8.9	-2.2	3.9	6.7	5.0	-2.2	-8.3	-18.9	-30.6	-30.6
Record Maximum Temp	21.7	23.3	29.4	31.7	33.9	38.9	40.0	38.9	37.8	32.2	27.2	23.3	40.0
HDD, Base 18.3°C	681	551	407	210	92	3	0	3	32	188	367	587	3119
CDD, Base 18.3°C	0	0	0	0	53	118	179	144	59	10	0	0	563
Relative Humidity (%)	75	74	70	66	67	68	73	75	74	72	76	78	72
Wind Speed (m/s)	4.7	4.6	5.1	4.7	4.1	3.7	3.3	3.1	3.4	3.8	4.4	4.6	4.1

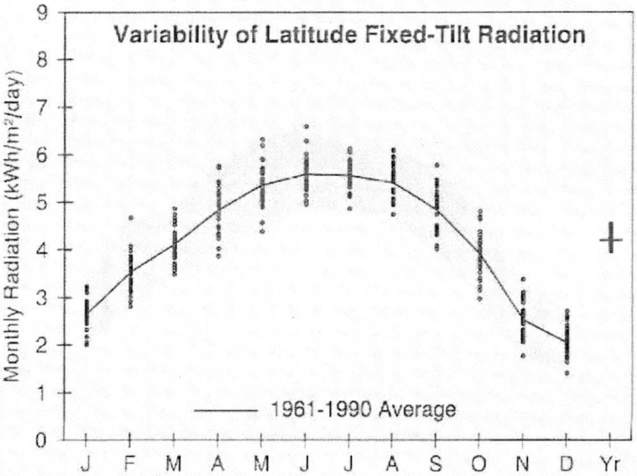

South Bend, IN

WBAN NO. 14848

LATITUDE: 41.70° N
LONGITUDE: 86.32° W
ELEVATION: 236 meters
MEAN PRESSURE: 988 millibars

STATION TYPE: Secondary

Solar Radiation for Flat-Plate Collectors Facing South at a Fixed Tilt (kWh/m²/day), Uncertainty ±9%

Tilt (°)		Jan	Feb	Mar	Apr	May	June	July	Aug	Sept	Oct	Nov	Dec	Year
0	Average	1.7	2.5	3.4	4.6	5.6	6.2	6.0	5.3	4.1	2.9	1.7	1.4	3.8
	Min/Max	1.5/2.0	2.2/3.2	3.0/3.9	3.8/5.3	4.7/6.6	5.5/7.3	5.3/6.6	4.8/5.9	3.6/4.8	2.4/3.3	1.4/2.0	1.1/1.6	3.6/4.1
Latitude -15	Average	2.4	3.3	4.0	5.0	5.7	6.1	6.0	5.6	4.8	3.7	2.3	1.9	4.2
	Min/Max	1.9/2.9	2.7/4.3	3.5/4.7	4.0/5.8	4.7/6.7	5.4/7.2	5.2/6.6	4.9/6.3	4.0/5.7	2.9/4.5	1.7/3.0	1.3/2.4	4.0/4.6
Latitude	Average	2.7	3.5	4.1	4.8	5.4	5.6	5.6	5.4	4.8	3.9	2.5	2.0	4.2
	Min/Max	2.0/3.2	2.8/4.7	3.5/4.9	3.9/5.8	4.4/6.3	4.9/6.6	4.9/6.1	4.7/6.1	4.0/5.8	3.0/4.8	1.8/3.4	1.4/2.7	4.0/4.5
Latitude +15	Average	2.8	3.6	4.0	4.5	4.7	4.8	4.9	4.9	4.6	3.9	2.6	2.1	4.0
	Min/Max	2.0/3.4	2.8/4.9	3.3/4.8	3.5/5.4	3.9/5.6	4.3/5.7	4.3/5.3	4.3/5.6	3.8/5.6	2.9/4.8	1.8/3.5	1.4/2.9	3.7/4.3
90	Average	2.6	3.2	3.1	3.0	2.8	2.6	2.7	3.0	3.3	3.2	2.3	2.0	2.8
	Min/Max	1.8/3.3	2.5/4.5	2.5/3.9	2.4/3.5	2.3/3.1	2.4/2.9	2.4/2.9	2.7/3.4	2.7/4.0	2.3/4.0	1.5/3.2	1.2/2.8	2.6/3.1

Solar Radiation for 1-Axis Tracking Flat-Plate Collectors with a North-South Axis (kWh/m²/day), Uncertainty ±9%

Axis Tilt (°)		Jan	Feb	Mar	Apr	May	June	July	Aug	Sept	Oct	Nov	Dec	Year
0	Average	2.2	3.3	4.3	5.9	7.2	7.9	7.7	6.9	5.5	3.8	2.2	1.7	4.9
	Min/Max	1.6/2.7	2.6/4.3	3.6/5.2	4.5/7.2	5.7/8.9	6.7/9.7	6.7/8.7	5.8/8.0	4.5/6.7	2.9/4.7	1.6/2.9	1.2/2.1	4.6/5.4
Latitude -15	Average	2.7	3.9	4.8	6.2	7.3	7.9	7.8	7.2	6.0	4.5	2.6	2.1	5.3
	Min/Max	2.0/3.3	3.0/5.1	4.0/5.8	4.7/7.7	5.8/9.1	6.6/9.8	6.7/8.8	6.0/8.4	4.9/7.5	3.3/5.6	1.8/3.6	1.4/2.8	4.9/5.9
Latitude	Average	3.0	4.1	4.9	6.1	7.1	7.6	7.5	7.1	6.0	4.7	2.8	2.2	5.3
	Min/Max	2.1/3.6	3.1/5.4	4.0/6.0	4.6/7.6	5.5/8.9	6.3/9.4	6.4/8.5	5.9/8.3	4.9/7.6	3.4/5.9	1.8/3.9	1.5/3.0	4.9/5.9
Latitude +15	Average	3.1	4.1	4.8	5.8	6.6	7.0	7.0	6.7	5.9	4.7	2.9	2.3	5.1
	Min/Max	2.2/3.8	3.2/5.6	3.9/5.9	4.3/7.3	5.2/8.3	5.9/8.8	6.0/8.0	5.6/7.9	4.7/7.4	3.3/5.9	1.8/4.1	1.5/3.2	4.7/5.7

Solar Radiation for 2-Axis Tracking Flat-Plate Collectors (kWh/m²/day), Uncertainty ±9%

Tracker		Jan	Feb	Mar	Apr	May	June	July	Aug	Sept	Oct	Nov	Dec	Year
2-Axis	Average	3.1	4.2	4.9	6.2	7.4	8.1	7.9	7.3	6.1	4.7	2.9	2.3	5.4
	Min/Max	2.2/3.8	3.2/5.6	4.1/6.0	4.7/7.7	5.8/9.2	6.8/10.0	6.8/9.0	6.1/8.4	4.9/7.6	3.4/5.9	1.9/4.1	1.5/3.3	5.1/6.1

Direct Beam Solar Radiation for Concentrating Collectors (kWh/m²/day), Uncertainty ±8%

Tracker		Jan	Feb	Mar	Apr	May	June	July	Aug	Sept	Oct	Nov	Dec	Year
1-Axis, E-W Horiz Axis	Average	1.4	1.8	2.0	2.6	3.2	3.6	3.5	3.2	2.7	2.3	1.4	1.0	2.4
	Min/Max	0.8/2.1	1.1/2.4	1.2/2.7	1.5/3.7	2.1/4.6	2.6/5.1	2.7/4.5	2.3/4.0	1.8/3.8	1.3/3.3	0.6/2.3	0.4/1.8	2.1/3.0
1-Axis, N-S Horiz Axis	Average	0.9	1.5	2.2	3.3	4.2	4.7	4.6	4.1	3.2	2.2	1.0	0.6	2.7
	Min/Max	0.5/1.5	0.9/2.0	1.4/3.0	1.9/4.7	2.8/6.2	3.4/6.7	3.6/5.9	3.0/5.3	2.2/4.6	1.2/3.2	0.4/1.7	0.3/1.2	2.4/3.4
1-Axis, N-S Tilt=Latitude	Average	1.5	2.1	2.6	3.5	4.2	4.5	4.4	4.2	3.7	2.9	1.5	1.0	3.0
	Min/Max	0.9/2.3	1.3/2.9	1.6/3.6	2.0/5.1	2.8/6.1	3.2/6.3	3.5/5.6	3.0/5.4	2.4/5.3	1.6/4.1	0.7/2.6	0.5/1.9	2.6/3.7
2-Axis	Average	1.6	2.2	2.6	3.6	4.4	4.8	4.8	4.4	3.7	2.9	1.6	1.1	3.1
	Min/Max	0.9/2.5	1.3/2.9	1.6/3.6	2.1/5.1	2.9/6.4	3.5/6.9	3.7/6.1	3.1/5.6	2.5/5.3	1.6/4.2	0.7/2.7	0.5/2.1	2.7/3.9

Average Climatic Conditions

Element	Jan	Feb	Mar	Apr	May	June	July	Aug	Sept	Oct	Nov	Dec	Year
Temperature (°C)	-4.8	-3.1	3.0	9.3	15.2	20.6	22.7	21.6	17.7	11.4	4.9	-1.7	9.7
Daily Minimum Temp	-8.8	-7.4	-1.6	3.7	9.3	14.8	17.2	16.2	12.1	6.0	0.8	-5.4	4.7
Daily Maximum Temp	-0.9	1.2	7.6	14.8	21.1	26.4	28.3	27.1	23.4	16.8	9.2	1.9	14.7
Record Minimum Temp	-30.0	-27.2	-25.0	-11.7	-4.4	1.7	6.7	4.4	-1.7	-6.7	-21.7	-26.7	-30.0
Record Maximum Temp	20.0	20.6	29.4	32.8	35.0	40.0	38.3	39.4	37.2	33.3	27.8	21.1	40.0
HDD, Base 18.3°C	718	601	476	272	129	16	4	12	47	219	402	622	3517
CDD, Base 18.3°C	0	0	0	0	33	84	139	114	29	6	0	0	404
Relative Humidity (%)	77	75	71	66	66	67	70	73	74	72	76	79	72
Wind Speed (m/s)	5.3	5.0	5.3	5.1	4.4	4.0	3.6	3.4	3.7	4.2	4.8	5.1	4.5

Dodge City, KS

WBAN NO. 13985

LATITUDE: 37.77° N
LONGITUDE: 99.97° W
ELEVATION: 787 meters
MEAN PRESSURE: 925 millibars

STATION TYPE: Primary

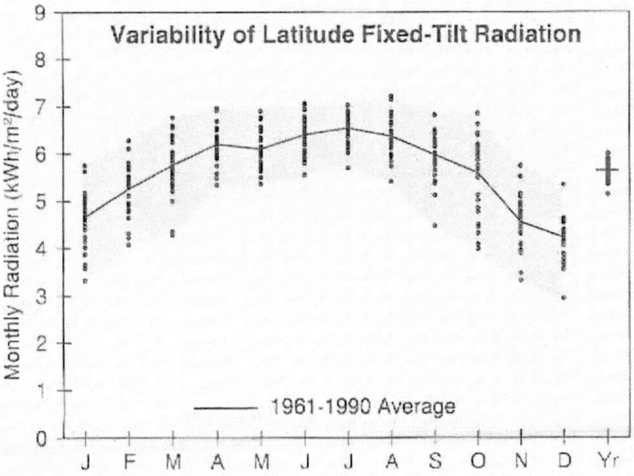

Solar Radiation for Flat-Plate Collectors Facing South at a Fixed Tilt (kWh/m²/day), Uncertainty ±9%

Tilt (°)		Jan	Feb	Mar	Apr	May	June	July	Aug	Sept	Oct	Nov	Dec	Year
0	Average	2.7	3.6	4.7	5.9	6.5	7.2	7.2	6.3	5.1	4.0	2.8	2.4	4.9
	Min/Max	2.2/3.1	3.0/4.1	3.7/5.3	5.2/6.5	5.7/7.4	6.2/8.0	6.3/7.7	5.4/7.1	4.1/5.8	3.2/4.7	2.2/3.3	1.9/2.7	4.6/5.1
Latitude -15	Average	4.0	4.8	5.5	6.3	6.5	7.0	7.1	6.6	5.9	5.2	4.0	3.6	5.5
	Min/Max	3.0/4.9	3.8/5.6	4.2/6.4	5.5/7.1	5.7/7.4	6.0/7.7	6.1/7.6	5.6/7.5	4.5/6.7	3.8/6.3	3.0/5.0	2.6/4.5	5.1/5.9
Latitude	Average	4.6	5.3	5.8	6.2	6.1	6.4	6.5	6.4	6.0	5.6	4.6	4.2	5.6
	Min/Max	3.3/5.8	4.1/6.3	4.3/6.8	5.3/7.0	5.4/6.9	5.6/7.1	5.7/7.0	5.4/7.2	4.5/6.8	4.0/6.8	3.3/5.7	2.9/5.3	5.1/6.0
Latitude +15	Average	5.0	5.5	5.7	5.8	5.4	5.5	5.7	5.8	5.7	5.7	4.8	4.6	5.4
	Min/Max	3.5/6.3	4.2/6.6	4.2/6.7	4.9/6.5	4.8/6.1	4.8/6.1	5.0/6.1	4.9/6.5	4.2/6.6	4.0/7.0	3.5/6.2	3.1/5.8	4.9/5.8
90	Average	4.6	4.6	4.2	3.5	2.7	2.5	2.7	3.2	3.9	4.5	4.3	4.3	3.7
	Min/Max	3.2/5.8	3.5/5.6	3.1/4.9	3.1/3.9	2.5/3.0	2.3/2.6	2.5/2.8	2.8/3.5	2.8/4.4	3.0/5.6	3.0/5.5	2.8/5.5	3.3/4.0

Solar Radiation for 1-Axis Tracking Flat-Plate Collectors with a North-South Axis (kWh/m²/day), Uncertainty ±9%

Axis Tilt (°)		Jan	Feb	Mar	Apr	May	June	July	Aug	Sept	Oct	Nov	Dec	Year
0	Average	4.1	5.2	6.5	8.0	8.5	9.5	9.5	8.5	7.1	5.8	4.1	3.6	6.7
	Min/Max	2.9/5.1	3.9/6.2	4.6/7.6	6.6/8.9	7.0/10.1	7.8/11.0	8.0/10.4	6.9/9.8	5.2/8.2	4.1/7.2	3.0/5.2	2.5/4.5	6.1/7.2
Latitude -15	Average	5.1	6.1	7.2	8.4	8.7	9.4	9.6	8.8	7.7	6.7	5.0	4.5	7.3
	Min/Max	3.5/6.4	4.5/7.4	5.0/8.5	6.9/9.4	7.1/10.2	7.7/10.9	8.0/10.5	7.1/10.2	5.5/8.9	4.6/8.4	3.6/6.5	3.1/5.8	6.6/7.8
Latitude	Average	5.6	6.5	7.4	8.3	8.4	9.0	9.2	8.7	7.8	7.0	5.5	5.0	7.4
	Min/Max	3.8/7.1	4.7/7.9	5.1/8.8	6.8/9.4	6.9/9.9	7.4/10.5	7.7/10.1	7.0/10.0	5.5/9.0	4.7/8.8	3.9/7.1	3.3/6.5	6.6/8.0
Latitude +15	Average	5.8	6.6	7.3	8.0	7.9	8.4	8.6	8.2	7.6	7.1	5.7	5.3	7.2
	Min/Max	4.0/7.5	4.8/8.1	5.0/8.7	6.5/9.0	6.4/9.3	6.9/9.8	7.2/9.5	6.6/9.6	5.3/8.9	4.7/8.9	4.0/7.4	3.5/6.9	6.4/7.8

Solar Radiation for 2-Axis Tracking Flat-Plate Collectors (kWh/m²/day), Uncertainty ±9%

Tracker		Jan	Feb	Mar	Apr	May	June	July	Aug	Sept	Oct	Nov	Dec	Year
2-Axis	Average	5.9	6.6	7.4	8.4	8.8	9.7	9.7	8.9	7.8	7.1	5.7	5.4	7.6
	Min/Max	4.0/7.6	4.8/8.1	5.1/8.8	6.9/9.5	7.2/10.3	7.9/11.2	8.2/10.7	7.2/10.2	5.6/9.1	4.8/8.9	4.0/7.4	3.5/7.0	6.8/8.3

Direct Beam Solar Radiation for Concentrating Collectors (kWh/m²/day), Uncertainty ±8%

Tracker		Jan	Feb	Mar	Apr	May	June	July	Aug	Sept	Oct	Nov	Dec	Year
1-Axis, E-W Horiz Axis	Average	3.7	3.8	3.8	4.2	4.3	5.1	5.2	4.5	4.1	4.1	3.5	3.4	4.1
	Min/Max	2.1/5.4	2.2/5.3	2.0/5.0	2.9/5.2	3.1/5.6	3.5/6.5	3.9/6.0	3.3/5.8	2.3/5.0	2.1/5.6	2.1/5.1	1.9/5.0	3.4/4.7
1-Axis, N-S Horiz Axis	Average	2.8	3.6	4.4	5.5	5.8	6.6	6.8	6.0	5.0	4.2	2.9	2.5	4.7
	Min/Max	1.6/4.2	2.0/4.9	2.2/5.7	3.8/6.7	4.0/7.6	4.6/8.6	5.2/7.9	4.3/7.5	2.9/6.1	2.2/5.7	1.8/4.2	1.4/3.7	3.9/5.2
1-Axis, N-S Tilt=Latitude	Average	4.1	4.7	5.1	5.8	5.6	6.2	6.5	6.1	5.5	5.2	4.0	3.7	5.2
	Min/Max	2.3/6.0	2.6/6.4	2.6/6.6	4.0/7.1	3.9/7.4	4.3/8.1	4.9/7.6	4.3/7.6	3.2/6.8	2.7/7.1	2.5/5.9	2.0/5.4	4.3/5.9
2-Axis	Average	4.4	4.8	5.1	5.9	5.9	6.8	7.0	6.2	5.5	5.3	4.2	4.0	5.4
	Min/Max	2.5/6.4	2.7/6.6	2.6/6.7	4.0/7.2	4.1/7.8	4.7/8.8	5.3/8.1	4.5/7.8	3.2/6.8	2.7/7.2	2.6/6.2	2.2/5.9	4.5/6.1

Average Climatic Conditions

Element	Jan	Feb	Mar	Apr	May	June	July	Aug	Sept	Oct	Nov	Dec	Year
Temperature (°C)	-1.2	1.7	6.4	12.6	17.9	23.6	26.8	25.7	20.6	14.2	6.2	0.3	12.9
Daily Minimum Temp	-7.7	-5.2	-0.9	5.2	11.1	16.6	19.7	18.7	13.7	6.8	-0.5	-6.0	5.9
Daily Maximum Temp	5.3	8.5	13.7	20.0	24.6	30.6	33.9	32.6	27.4	21.5	12.9	6.6	19.8
Record Minimum Temp	-25.0	-26.1	-26.1	-9.4	-3.3	5.0	7.8	8.3	-1.7	-8.3	-17.8	-29.4	-29.4
Record Maximum Temp	26.7	29.4	33.9	37.8	38.9	42.2	42.8	41.7	41.1	35.6	32.8	30.0	42.8
HDD, Base 18.3°C	606	467	371	179	66	4	0	0	22	141	363	560	2778
CDD, Base 18.3°C	0	0	0	8	52	161	263	230	88	12	0	0	814
Relative Humidity (%)	66	65	61	58	63	60	55	58	62	58	64	67	61
Wind Speed (m/s)	5.9	6.0	6.7	6.7	6.3	6.0	5.8	5.5	5.8	5.8	5.9	5.9	6.0

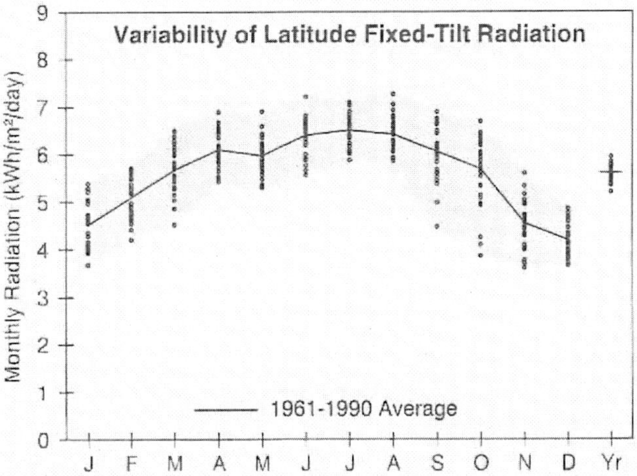

Goodland, KS

WBAN NO. 23065

LATITUDE: 39.37° N
LONGITUDE: 101.70° W
ELEVATION: 1124 meters
MEAN PRESSURE: 888 millibars

STATION TYPE: Secondary

Solar Radiation for Flat-Plate Collectors Facing South at a Fixed Tilt (kWh/m²/day), Uncertainty ±9%

Tilt (°)		Jan	Feb	Mar	Apr	May	June	July	Aug	Sept	Oct	Nov	Dec	Year
0	Average	2.5	3.3	4.5	5.7	6.3	7.2	7.1	6.3	5.1	3.9	2.7	2.2	4.7
	Min/Max	2.2/2.8	2.9/3.6	3.8/4.9	5.2/6.3	5.7/7.3	6.2/8.1	6.4/7.8	5.8/7.1	4.0/5.7	3.1/4.4	2.3/3.1	2.0/2.4	4.5/5.0
Latitude -15	Average	3.9	4.6	5.4	6.2	6.4	7.0	7.0	6.7	6.0	5.3	4.0	3.6	5.5
	Min/Max	3.2/4.6	3.9/5.1	4.4/6.2	5.6/7.0	5.7/7.3	6.1/7.9	6.3/7.7	6.1/7.5	4.5/6.7	3.7/6.1	3.3/4.9	3.2/4.1	5.2/5.8
Latitude	Average	4.5	5.1	5.7	6.1	6.0	6.4	6.5	6.4	6.1	5.7	4.6	4.2	5.6
	Min/Max	3.7/5.4	4.2/5.7	4.5/6.5	5.4/6.9	5.3/6.9	5.6/7.2	5.9/7.1	5.9/7.3	4.5/6.9	3.9/6.7	3.6/5.6	3.7/4.8	5.2/5.9
Latitude +15	Average	4.9	5.3	5.6	5.7	5.3	5.5	5.6	5.8	5.8	5.8	4.8	4.6	5.4
	Min/Max	3.9/5.9	4.3/6.0	4.4/6.5	5.0/6.4	4.7/6.1	4.8/6.1	5.1/6.1	5.3/6.6	4.2/6.6	3.8/6.9	3.8/6.0	3.9/5.3	5.0/5.7
90	Average	4.6	4.6	4.3	3.6	2.8	2.6	2.8	3.3	4.0	4.7	4.4	4.4	3.8
	Min/Max	3.6/5.5	3.7/5.3	3.3/5.0	3.2/3.9	2.5/3.2	2.4/2.8	2.6/3.0	3.0/3.7	2.9/4.6	3.0/5.6	3.3/5.4	3.7/5.1	3.5/4.1

Solar Radiation for 1-Axis Tracking Flat-Plate Collectors with a North-South Axis (kWh/m²/day), Uncertainty ±9%

Axis Tilt (°)		Jan	Feb	Mar	Apr	May	June	July	Aug	Sept	Oct	Nov	Dec	Year
0	Average	3.9	4.9	6.4	7.9	8.4	9.6	9.6	8.7	7.3	5.9	4.1	3.4	6.7
	Min/Max	3.2/4.6	4.1/5.6	4.9/7.4	6.8/9.1	7.3/10.0	8.0/11.3	8.4/11.0	7.7/10.1	5.3/8.5	4.0/7.1	3.2/4.9	3.0/4.0	6.2/7.1
Latitude -15	Average	4.9	5.9	7.1	8.3	8.6	9.6	9.7	9.1	8.0	6.9	5.0	4.5	7.3
	Min/Max	3.9/6.0	4.8/6.7	5.4/8.4	7.1/9.7	7.4/10.2	7.9/11.3	8.4/11.1	8.0/10.5	5.7/9.3	4.5/8.3	3.9/6.2	3.9/5.2	6.8/7.8
Latitude	Average	5.4	6.3	7.3	8.3	8.3	9.2	9.4	8.9	8.0	7.2	5.5	4.9	7.4
	Min/Max	4.3/6.6	5.1/7.2	5.5/8.6	7.1/9.6	7.2/9.9	7.6/10.9	8.1/10.7	7.8/10.4	5.7/9.4	4.6/8.8	4.2/6.8	4.3/5.8	6.8/7.9
Latitude +15	Average	5.7	6.4	7.2	8.0	7.8	8.6	8.8	8.5	7.9	7.3	5.7	5.3	7.3
	Min/Max	4.5/7.0	5.1/7.4	5.4/8.6	6.8/9.3	6.8/9.3	7.1/10.1	7.6/10.1	7.4/9.9	5.5/9.2	4.6/8.9	4.3/7.1	4.5/6.2	6.7/7.8

Solar Radiation for 2-Axis Tracking Flat-Plate Collectors (kWh/m²/day), Uncertainty ±9%

Tracker		Jan	Feb	Mar	Apr	May	June	July	Aug	Sept	Oct	Nov	Dec	Year
2-Axis	Average	5.7	6.4	7.3	8.4	8.7	9.9	9.9	9.1	8.1	7.3	5.7	5.3	7.7
	Min/Max	4.5/7.1	5.2/7.4	5.5/8.6	7.2/9.8	7.5/10.3	8.1/11.6	8.6/11.3	8.0/10.6	5.7/9.4	4.7/8.9	4.3/7.2	4.6/6.3	7.1/8.2

Direct Beam Solar Radiation for Concentrating Collectors (kWh/m²/day), Uncertainty ±8%

Tracker		Jan	Feb	Mar	Apr	May	June	July	Aug	Sept	Oct	Nov	Dec	Year
1-Axis, E-W	Average	3.6	3.7	3.7	4.2	4.3	5.4	5.4	4.9	4.4	4.4	3.6	3.5	4.2
Horiz Axis	Min/Max	2.5/4.8	2.7/4.6	2.3/4.8	3.1/5.3	3.3/5.7	3.8/7.0	4.3/6.7	3.9/6.2	2.5/5.5	2.0/5.7	2.4/4.8	2.8/4.4	3.7/4.7
1-Axis, N-S	Average	2.7	3.4	4.3	5.5	5.8	7.1	7.1	6.4	5.4	4.4	2.9	2.4	4.8
Horiz Axis	Min/Max	1.9/3.6	2.5/4.3	2.5/5.6	4.1/7.1	4.5/7.6	4.9/9.2	5.7/9.0	5.1/8.2	3.1/6.8	2.1/5.9	1.9/3.8	1.9/3.2	4.2/5.3
1-Axis, N-S	Average	4.0	4.5	5.0	5.8	5.7	6.6	6.8	6.6	6.0	5.6	4.1	3.7	5.4
Tilt=Latitude	Min/Max	2.8/5.3	3.3/5.7	3.0/6.5	4.3/7.4	4.4/7.4	4.7/8.7	5.4/8.6	5.2/8.3	3.5/7.6	2.6/7.3	2.7/5.5	3.0/4.8	4.7/5.9
2-Axis	Average	4.2	4.6	5.1	5.9	6.0	7.2	7.3	6.8	6.0	5.6	4.3	4.0	5.6
	Min/Max	3.0/5.7	3.4/5.8	3.0/6.6	4.4/7.5	4.6/7.8	5.1/9.4	5.8/9.3	5.3/8.6	3.5/7.7	2.6/7.4	2.8/5.8	3.2/5.2	4.9/6.2

Average Climatic Conditions

Element	Jan	Feb	Mar	Apr	May	June	July	Aug	Sept	Oct	Nov	Dec	Year
Temperature (°C)	-2.0	0.1	3.9	9.7	15.0	20.7	24.2	22.9	17.8	11.2	3.7	-1.4	10.5
Daily Minimum Temp	-9.3	-7.2	-3.7	1.8	7.6	12.9	16.3	15.1	10.0	3.1	-3.5	-8.3	2.9
Daily Maximum Temp	5.3	7.4	11.5	17.6	22.4	28.6	32.1	30.7	25.6	19.2	10.9	5.6	18.1
Record Minimum Temp	-32.2	-30.0	-28.9	-17.8	-6.1	-0.6	5.6	3.3	-7.2	-12.2	-24.4	-32.8	-32.8
Record Maximum Temp	26.1	27.2	31.7	35.6	37.2	42.8	43.9	43.3	40.6	35.6	30.6	28.3	43.9
HDD, Base 18.3°C	631	510	446	258	117	18	0	6	60	222	438	612	3319
CDD, Base 18.3°C	0	0	0	0	14	90	183	147	43	0	0	0	477
Relative Humidity (%)	66	66	63	59	64	60	58	60	60	57	65	66	62
Wind Speed (m/s)	5.7	5.6	6.4	6.6	6.1	5.7	5.5	5.1	5.4	5.3	5.4	5.5	5.7

Topeka, KS

WBAN NO. 13996

LATITUDE: 39.07° N
LONGITUDE: 95.63° W
ELEVATION: 270 meters
MEAN PRESSURE: 985 millibars

STATION TYPE: Secondary

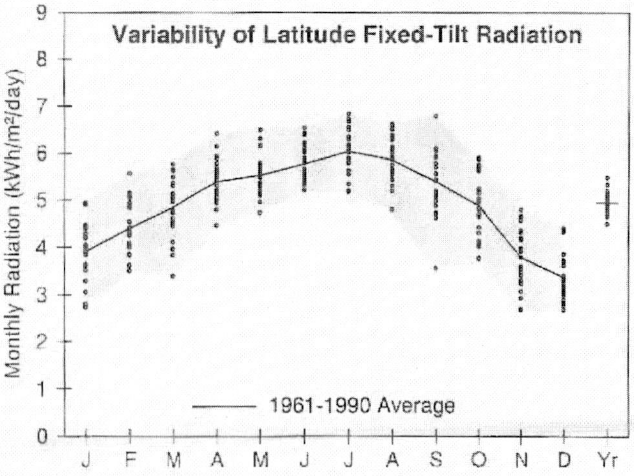

Solar Radiation for Flat-Plate Collectors Facing South at a Fixed Tilt (kWh/m²/day), Uncertainty ±9%

Tilt (°)		Jan	Feb	Mar	Apr	May	June	July	Aug	Sept	Oct	Nov	Dec	Year
0	Average	2.3	3.0	4.0	5.1	5.9	6.5	6.6	5.8	4.6	3.5	2.4	1.9	4.3
	Min/Max	1.8/2.6	2.6/3.6	3.0/4.6	4.3/6.0	5.1/6.8	5.8/7.3	5.7/7.5	4.8/6.4	3.3/5.6	2.9/4.1	1.9/2.8	1.7/2.3	4.1/4.7
Latitude -15	Average	3.4	4.0	4.7	5.5	5.9	6.3	6.5	6.1	5.3	4.6	3.4	2.9	4.9
	Min/Max	2.5/4.2	3.3/5.0	3.4/5.6	4.6/6.5	5.1/6.9	5.7/7.2	5.6/7.4	5.0/6.9	3.6/6.6	3.6/5.4	2.5/4.2	2.4/3.8	4.5/5.4
Latitude	Average	3.9	4.4	4.9	5.4	5.5	5.8	6.0	5.9	5.4	4.9	3.8	3.4	4.9
	Min/Max	2.7/4.9	3.5/5.6	3.4/5.8	4.5/6.4	4.7/6.5	5.2/6.5	5.2/6.8	4.8/6.6	3.6/6.8	3.8/5.9	2.7/4.8	2.7/4.4	4.5/5.5
Latitude +15	Average	4.2	4.5	4.7	5.0	4.9	5.0	5.3	5.3	5.2	4.9	4.0	3.6	4.7
	Min/Max	2.9/5.3	3.5/5.8	3.3/5.7	4.1/5.9	4.2/5.8	4.6/5.6	4.5/6.0	4.3/6.0	3.4/6.6	3.7/6.0	2.8/5.1	2.8/4.8	4.3/5.3
90	Average	3.9	3.9	3.5	3.1	2.6	2.5	2.6	3.1	3.6	4.0	3.5	3.4	3.3
	Min/Max	2.6/4.9	2.9/5.0	2.4/4.2	2.6/3.6	2.4/3.0	2.3/2.6	2.3/2.9	2.6/3.4	2.3/4.5	2.9/4.9	2.4/4.6	2.5/4.6	2.9/3.7

Solar Radiation for 1-Axis Tracking Flat-Plate Collectors with a North-South Axis (kWh/m²/day), Uncertainty ±9%

Axis Tilt (°)		Jan	Feb	Mar	Apr	May	June	July	Aug	Sept	Oct	Nov	Dec	Year
0	Average	3.4	4.3	5.4	6.7	7.6	8.4	8.7	7.8	6.3	5.0	3.4	2.8	5.8
	Min/Max	2.3/4.2	3.3/5.5	3.7/6.4	5.3/8.3	6.3/9.1	7.2/10.0	7.2/10.3	6.1/9.0	3.9/8.3	3.8/6.2	2.4/4.3	2.3/3.6	5.4/6.6
Latitude -15	Average	4.2	5.0	5.9	7.1	7.7	8.4	8.8	8.1	6.9	5.8	4.1	3.5	6.3
	Min/Max	2.8/5.3	3.9/6.6	4.0/7.2	5.5/8.8	6.4/9.2	7.2/10.0	7.2/10.3	6.3/9.3	4.2/9.1	4.3/7.2	2.8/5.3	2.8/4.7	5.7/7.2
Latitude	Average	4.6	5.3	6.1	7.0	7.5	8.0	8.5	8.0	7.0	6.1	4.5	3.9	6.4
	Min/Max	3.0/5.9	4.1/7.0	4.0/7.3	5.4/8.8	6.2/9.0	6.9/9.6	7.0/10.0	6.1/9.2	4.1/9.2	4.4/7.6	3.0/5.8	3.0/5.2	5.7/7.2
Latitude +15	Average	4.8	5.4	6.0	6.7	7.0	7.5	7.9	7.6	6.8	6.1	4.6	4.1	6.2
	Min/Max	3.2/6.2	4.1/7.2	3.9/7.3	5.2/8.4	5.8/8.4	6.4/9.0	6.5/9.4	5.8/8.8	4.0/9.0	4.4/7.6	3.1/6.0	3.1/5.5	5.5/7.1

Solar Radiation for 2-Axis Tracking Flat-Plate Collectors (kWh/m²/day), Uncertainty ±9%

Tracker		Jan	Feb	Mar	Apr	May	June	July	Aug	Sept	Oct	Nov	Dec	Year
2-Axis	Average	4.9	5.5	6.1	7.1	7.8	8.5	8.9	8.2	7.0	6.1	4.7	4.2	6.6
	Min/Max	3.2/6.3	4.1/7.2	4.1/7.3	5.5/8.9	6.4/9.3	7.4/10.3	7.4/10.5	6.3/9.4	4.2/9.2	4.4/7.7	3.1/6.1	3.2/5.6	5.9/7.5

Direct Beam Solar Radiation for Concentrating Collectors (kWh/m²/day), Uncertainty ±8%

Tracker		Jan	Feb	Mar	Apr	May	June	July	Aug	Sept	Oct	Nov	Dec	Year
1-Axis, E-W	Average	2.9	3.0	2.9	3.3	3.6	4.1	4.4	4.0	3.4	3.4	2.7	2.5	3.4
Horiz Axis	Min/Max	1.5/4.3	1.8/4.4	1.5/4.0	2.2/4.8	2.4/4.9	3.0/5.5	3.1/5.8	2.5/4.9	1.5/5.2	2.0/4.6	1.3/3.9	1.7/3.8	2.8/4.1
1-Axis, N-S	Average	2.2	2.7	3.3	4.3	4.7	5.4	5.8	5.2	4.2	3.4	2.2	1.8	3.8
Horiz Axis	Min/Max	1.1/3.2	1.6/4.2	1.8/4.5	2.8/6.2	3.3/6.4	4.0/7.4	4.1/7.6	3.3/6.6	1.7/6.4	2.0/4.7	1.1/3.1	1.2/2.7	3.2/4.7
1-Axis, N-S	Average	3.2	3.6	3.9	4.5	4.6	5.1	5.6	5.3	4.7	4.3	3.1	2.7	4.2
Tilt=Latitude	Min/Max	1.7/4.7	2.1/5.4	2.0/5.3	2.9/6.5	3.2/6.3	3.7/6.9	4.0/7.3	3.3/6.7	1.9/7.1	2.6/5.9	1.5/4.4	1.8/4.1	3.5/5.2
2-Axis	Average	3.5	3.7	3.9	4.6	4.9	5.5	6.0	5.5	4.7	4.4	3.3	2.9	4.4
	Min/Max	1.8/5.1	2.2/5.6	2.0/5.3	3.0/6.6	3.4/6.6	4.0/7.5	4.2/7.8	3.5/6.9	1.9/7.1	2.6/6.0	1.6/4.7	2.0/4.4	3.7/5.4

Average Climatic Conditions

Element	Jan	Feb	Mar	Apr	May	June	July	Aug	Sept	Oct	Nov	Dec	Year
Temperature (°C)	-2.9	0.1	6.4	12.7	18.1	23.1	25.8	24.6	19.9	13.5	6.1	-0.7	12.2
Daily Minimum Temp	-8.7	-5.7	0.1	6.0	11.8	17.2	19.8	18.3	13.2	6.4	0.0	-6.1	6.0
Daily Maximum Temp	2.8	5.9	12.8	19.4	24.3	29.0	31.8	30.8	26.5	20.6	12.2	4.7	18.4
Record Minimum Temp	-28.9	-30.6	-21.7	-12.2	-3.3	6.1	6.1	5.0	-1.7	-7.2	-16.7	-32.2	-32.2
Record Maximum Temp	22.8	28.9	31.7	35.0	36.1	41.7	43.3	43.3	42.8	35.6	29.4	22.8	43.3
HDD, Base 18.3°C	659	510	368	173	67	3	0	0	31	157	367	589	2925
CDD, Base 18.3°C	0	0	0	5	61	146	233	195	78	7	0	0	724
Relative Humidity (%)	70	69	66	64	69	71	70	71	73	68	71	72	69
Wind Speed (m/s)	4.4	4.5	5.3	5.2	4.6	4.2	3.8	3.6	3.7	3.9	4.3	4.2	4.3

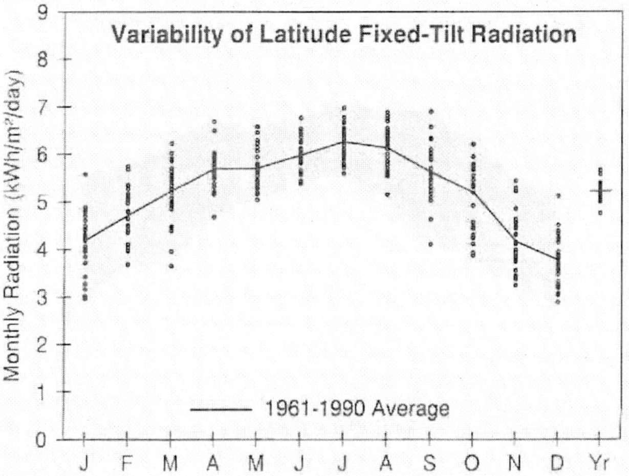

Wichita, KS

WBAN NO. 03928

LATITUDE: 37.65° N
LONGITUDE: 97.42° W
ELEVATION: 408 meters
MEAN PRESSURE: 968 millibars

STATION TYPE: Secondary

Solar Radiation for Flat-Plate Collectors Facing South at a Fixed Tilt (kWh/m²/day), Uncertainty ±9%

Tilt (°)		Jan	Feb	Mar	Apr	May	June	July	Aug	Sept	Oct	Nov	Dec	Year
0	Average	2.5	3.3	4.3	5.4	6.1	6.7	6.8	6.1	4.9	3.8	2.6	2.2	4.6
	Min/Max	2.0/3.0	2.8/3.8	3.2/4.9	4.6/6.2	5.4/7.0	6.0/7.6	6.1/7.7	5.2/6.8	3.8/5.8	3.1/4.4	2.2/3.1	1.9/2.6	4.3/4.8
Latitude -15	Average	3.6	4.3	5.0	5.8	6.1	6.5	6.8	6.4	5.5	4.8	3.7	3.3	5.2
	Min/Max	2.7/4.8	3.4/5.2	3.6/5.9	4.8/6.8	5.4/7.0	5.8/7.4	6.0/7.6	5.4/7.1	4.1/6.7	3.7/5.7	2.9/4.7	2.6/4.3	4.8/5.5
Latitude	Average	4.2	4.7	5.2	5.7	5.7	6.0	6.3	6.1	5.6	5.2	4.2	3.8	5.2
	Min/Max	3.0/5.6	3.7/5.7	3.7/6.2	4.7/6.7	5.0/6.6	5.4/6.8	5.6/7.0	5.1/6.9	4.1/6.9	3.9/6.2	3.2/5.4	2.9/5.1	4.8/5.7
Latitude +15	Average	4.5	4.9	5.1	5.3	5.1	5.2	5.5	5.6	5.4	5.3	4.4	4.1	5.0
	Min/Max	3.1/6.1	3.7/6.0	3.5/6.1	4.3/6.2	4.5/5.8	4.7/5.8	4.9/6.0	4.7/6.3	3.9/6.7	3.8/6.3	3.4/5.8	3.1/5.6	4.5/5.5
90	Average	4.1	4.1	3.7	3.2	2.6	2.4	2.6	3.1	3.6	4.2	3.9	3.8	3.4
	Min/Max	2.7/5.6	3.2/5.0	2.5/4.4	2.6/3.7	2.4/2.9	2.3/2.6	2.4/2.7	2.7/3.4	2.6/4.5	2.9/5.1	3.0/5.2	2.8/5.3	3.0/3.8

Solar Radiation for 1-Axis Tracking Flat-Plate Collectors with a North-South Axis (kWh/m²/day), Uncertainty ±9%

Axis Tilt (°)		Jan	Feb	Mar	Apr	May	June	July	Aug	Sept	Oct	Nov	Dec	Year
0	Average	3.7	4.6	5.8	7.2	7.9	8.8	9.1	8.2	6.7	5.4	3.8	3.2	6.2
	Min/Max	2.6/4.9	3.5/5.7	4.1/7.1	5.7/8.8	6.9/9.4	7.4/10.3	8.0/10.6	6.6/9.6	4.7/8.4	4.0/6.7	3.0/4.9	2.5/4.3	5.7/6.8
Latitude -15	Average	4.5	5.4	6.4	7.6	8.0	8.7	9.1	8.5	7.2	6.2	4.6	4.0	6.7
	Min/Max	3.1/6.2	4.0/6.7	4.4/7.9	5.9/9.3	6.9/9.6	7.4/10.3	8.0/10.6	6.8/9.9	5.0/9.2	4.4/7.7	3.5/6.1	3.0/5.5	6.1/7.4
Latitude	Average	5.0	5.7	6.6	7.5	7.8	8.4	8.8	8.4	7.3	6.5	5.0	4.4	6.8
	Min/Max	3.3/6.8	4.2/7.2	4.4/8.1	5.8/9.2	6.7/9.3	7.1/9.8	7.7/10.2	6.6/9.8	5.0/9.3	4.6/8.0	3.7/6.6	3.3/6.2	6.1/7.5
Latitude +15	Average	5.2	5.8	6.5	7.2	7.3	7.8	8.3	8.0	7.1	6.5	5.1	4.7	6.6
	Min/Max	3.4/7.2	4.3/7.3	4.3/8.0	5.6/8.9	6.3/8.7	6.6/9.2	7.2/9.6	6.3/9.3	4.8/9.1	4.5/8.1	3.8/6.9	3.4/6.6	5.9/7.3

Solar Radiation for 2-Axis Tracking Flat-Plate Collectors (kWh/m²/day), Uncertainty ±9%

Tracker		Jan	Feb	Mar	Apr	May	June	July	Aug	Sept	Oct	Nov	Dec	Year
2-Axis	Average	5.3	5.9	6.6	7.6	8.1	8.9	9.3	8.6	7.3	6.6	5.2	4.7	7.0
	Min/Max	3.5/7.3	4.3/7.3	4.4/8.1	5.9/9.3	7.0/9.7	7.5/10.5	8.2/10.8	6.8/10.0	5.0/9.3	4.6/8.1	3.9/7.0	3.5/6.7	6.3/7.7

Direct Beam Solar Radiation for Concentrating Collectors (kWh/m²/day), Uncertainty ±8%

Tracker		Jan	Feb	Mar	Apr	May	June	July	Aug	Sept	Oct	Nov	Dec	Year
1-Axis, E-W Horiz Axis	Average	3.2	3.2	3.3	3.6	3.8	4.5	4.7	4.3	3.7	3.7	3.1	3.0	3.7
	Min/Max	1.6/5.1	1.8/4.5	1.7/4.5	2.5/5.1	2.9/5.1	3.3/5.7	3.8/6.2	3.0/5.5	1.9/5.3	2.0/5.0	2.0/4.7	1.9/4.8	3.1/4.3
1-Axis, N-S Horiz Axis	Average	2.5	3.0	3.8	4.7	5.1	5.9	6.3	5.7	4.5	3.8	2.5	2.1	4.2
	Min/Max	1.2/4.0	1.7/4.2	2.0/5.2	3.2/6.7	4.0/6.9	4.3/7.6	5.0/8.2	3.9/7.4	2.4/6.5	2.0/5.2	1.6/3.8	1.4/3.5	3.6/4.9
1-Axis, N-S Tilt=Latitude	Average	3.5	4.0	4.4	5.0	4.9	5.5	6.0	5.8	5.0	4.7	3.5	3.2	4.6
	Min/Max	1.8/5.7	2.2/5.5	2.3/6.0	3.3/7.0	3.8/6.7	4.1/7.1	4.8/7.8	3.9/7.5	2.7/7.2	2.5/6.4	2.3/5.3	2.0/5.2	3.9/5.5
2-Axis	Average	3.8	4.1	4.4	5.0	5.2	6.0	6.4	6.0	5.0	4.8	3.7	3.4	4.8
	Min/Max	1.9/6.1	2.2/5.6	2.3/6.0	3.4/7.1	4.1/7.1	4.4/7.7	5.1/8.4	4.0/7.7	2.7/7.2	2.5/6.5	2.4/5.6	2.2/5.6	4.0/5.7

Average Climatic Conditions

Element	Jan	Feb	Mar	Apr	May	June	July	Aug	Sept	Oct	Nov	Dec	Year
Temperature (°C)	-1.4	1.6	7.4	13.6	18.7	24.3	27.4	26.3	21.3	14.8	7.1	0.6	13.4
Daily Minimum Temp	-7.1	-4.6	0.9	6.9	12.4	18.1	21.1	19.9	15.1	8.1	1.1	-5.0	7.2
Daily Maximum Temp	4.3	7.7	14.0	20.2	24.9	30.4	33.8	32.6	27.4	21.4	12.9	6.1	19.7
Record Minimum Temp	-24.4	-29.4	-18.9	-9.4	-0.6	6.1	10.6	8.9	-0.6	-6.1	-17.2	-26.7	-29.4
Record Maximum Temp	23.9	28.9	31.7	35.6	37.8	43.3	45.0	43.3	41.7	35.0	29.4	28.3	45.0
HDD, Base 18.3°C	612	470	338	154	57	3	0	0	16	123	338	551	2662
CDD, Base 18.3°C	0	0	0	11	67	181	282	246	104	12	0	0	904
Relative Humidity (%)	70	68	64	63	67	64	59	61	67	65	70	72	66
Wind Speed (m/s)	5.4	5.5	6.2	6.1	5.4	5.2	5.1	4.9	5.1	5.2	5.4	5.4	5.4

Covington, KY

WBAN NO. 93814

LATITUDE: 39.07° N
LONGITUDE: 84.67° W
ELEVATION: 271 meters
MEAN PRESSURE: 986 millibars

STATION TYPE: Secondary

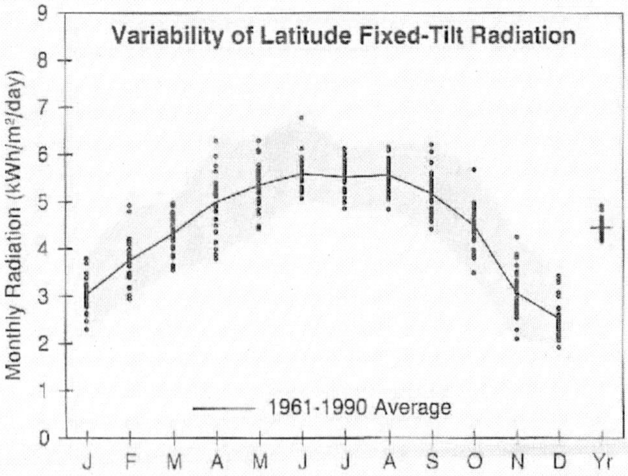

Solar Radiation for Flat-Plate Collectors Facing South at a Fixed Tilt (kWh/m²/day), Uncertainty ±9%

Tilt (°)		Jan	Feb	Mar	Apr	May	June	July	Aug	Sept	Oct	Nov	Dec	Year
0	Average	1.9	2.7	3.6	4.8	5.7	6.2	6.0	5.5	4.5	3.3	2.1	1.6	4.0
	Min/Max	1.6/2.3	2.3/3.3	3.1/4.1	3.8/5.8	4.8/6.6	5.6/7.6	5.3/6.6	4.9/6.0	4.0/5.2	2.8/4.0	1.7/2.5	1.4/1.9	3.8/4.4
Latitude -15	Average	2.7	3.5	4.2	5.1	5.7	6.1	5.9	5.8	5.1	4.2	2.8	2.3	4.5
	Min/Max	2.1/3.3	2.8/4.4	3.5/4.8	3.9/6.4	4.7/6.7	5.5/7.4	5.2/6.6	5.1/6.4	4.4/6.1	3.3/5.3	2.0/3.8	1.8/3.0	4.2/4.9
Latitude	Average	3.0	3.7	4.3	5.0	5.4	5.6	5.5	5.6	5.2	4.5	3.1	2.5	4.5
	Min/Max	2.3/3.8	2.9/4.9	3.5/5.0	3.8/6.3	4.4/6.3	5.1/6.8	4.8/6.1	4.8/6.2	4.4/6.2	3.5/5.7	2.1/4.2	1.9/3.4	4.2/4.9
Latitude +15	Average	3.2	3.8	4.2	4.6	4.8	4.9	4.8	5.1	4.9	4.5	3.2	2.7	4.2
	Min/Max	2.4/4.1	2.9/5.2	3.4/4.9	3.5/5.8	3.9/5.6	4.4/5.8	4.3/5.3	4.4/5.6	4.2/6.0	3.5/5.8	2.1/4.5	2.0/3.7	3.9/4.7
90	Average	2.9	3.3	3.1	2.9	2.6	2.4	2.5	3.0	3.4	3.6	2.7	2.4	2.9
	Min/Max	2.0/3.9	2.5/4.8	2.5/3.6	2.2/3.6	2.2/2.9	2.3/2.7	2.3/2.7	2.6/3.3	2.8/4.1	2.7/4.6	1.7/4.0	1.7/3.4	2.7/3.2

Solar Radiation for 1-Axis Tracking Flat-Plate Collectors with a North-South Axis (kWh/m²/day), Uncertainty ±9%

Axis Tilt (°)		Jan	Feb	Mar	Apr	May	June	July	Aug	Sept	Oct	Nov	Dec	Year
0	Average	2.6	3.6	4.7	6.1	7.2	7.9	7.7	7.2	6.0	4.5	2.7	2.1	5.2
	Min/Max	2.0/3.3	2.8/4.7	3.8/5.6	4.5/8.0	5.7/8.7	7.1/10.2	6.5/8.8	6.1/8.1	5.1/7.3	3.5/5.7	1.9/3.8	1.6/2.8	4.8/5.9
Latitude -15	Average	3.2	4.2	5.1	6.4	7.3	7.9	7.7	7.5	6.5	5.2	3.3	2.6	5.6
	Min/Max	2.4/4.1	3.1/5.6	4.1/6.2	4.6/8.5	5.8/8.9	7.0/10.2	6.5/8.8	6.3/8.4	5.4/8.0	3.9/6.7	2.1/4.7	1.9/3.6	5.2/6.4
Latitude	Average	3.5	4.4	5.2	6.4	7.1	7.6	7.4	7.3	6.5	5.4	3.5	2.8	5.6
	Min/Max	2.5/4.5	3.2/5.9	4.2/6.3	4.5/8.4	5.5/8.6	6.7/9.8	6.2/8.5	6.2/8.3	5.4/8.1	4.0/7.0	2.2/5.1	2.0/3.9	5.2/6.4
Latitude +15	Average	3.6	4.5	5.1	6.1	6.7	7.1	7.0	7.0	6.4	5.5	3.6	2.9	5.4
	Min/Max	2.6/4.7	3.3/6.1	4.1/6.2	4.3/8.1	5.2/8.1	6.3/9.1	5.8/8.0	5.8/7.9	5.2/8.0	4.0/7.1	2.3/5.3	2.1/4.2	5.0/6.2

Solar Radiation for 2-Axis Tracking Flat-Plate Collectors (kWh/m²/day), Uncertainty ±9%

Tracker		Jan	Feb	Mar	Apr	May	June	July	Aug	Sept	Oct	Nov	Dec	Year
2-Axis	Average	3.6	4.5	5.2	6.5	7.4	8.1	7.9	7.5	6.6	5.5	3.6	3.0	5.8
	Min/Max	2.6/4.7	3.3/6.1	4.2/6.3	4.6/8.5	5.8/9.0	7.2/10.4	6.6/8.9	6.3/8.5	5.4/8.1	4.1/7.1	2.3/5.3	2.1/4.2	5.3/6.6

Direct Beam Solar Radiation for Concentrating Collectors (kWh/m²/day), Uncertainty ±8%

Tracker		Jan	Feb	Mar	Apr	May	June	July	Aug	Sept	Oct	Nov	Dec	Year
1-Axis, E-W	Average	1.9	2.2	2.3	2.8	3.2	3.6	3.5	3.3	3.1	2.9	1.9	1.6	2.7
Horiz Axis	Min/Max	1.2/2.8	1.2/3.4	1.5/3.1	1.3/4.3	2.0/4.6	2.9/5.7	2.5/4.4	2.5/4.3	2.1/4.3	1.8/4.1	0.8/3.2	0.9/2.6	2.3/3.4
1-Axis, N-S	Average	1.4	2.0	2.6	3.6	4.3	4.8	4.6	4.4	3.7	2.9	1.5	1.1	3.1
Horiz Axis	Min/Max	0.9/2.1	1.0/3.2	1.6/3.5	1.7/5.6	2.8/6.0	3.9/7.4	3.3/5.8	3.3/5.7	2.6/5.1	1.8/4.1	0.6/2.6	0.6/1.8	2.7/3.9
1-Axis, N-S	Average	2.1	2.6	3.0	3.8	4.2	4.5	4.4	4.5	4.2	3.6	2.1	1.7	3.4
Tilt=Latitude	Min/Max	1.3/3.1	1.4/4.2	2.0/4.1	1.8/5.9	2.7/5.9	3.6/7.0	3.1/5.6	3.3/5.7	2.9/5.8	2.3/5.1	0.9/3.7	0.9/2.8	2.9/4.3
2-Axis	Average	2.2	2.7	3.1	3.8	4.4	4.9	4.7	4.6	4.2	3.7	2.2	1.8	3.5
	Min/Max	1.4/3.3	1.4/4.3	2.0/4.1	1.8/6.0	2.9/6.2	4.0/7.6	3.3/6.0	3.4/5.9	2.9/5.8	2.3/5.2	1.0/3.9	1.0/3.0	3.0/4.5

Average Climatic Conditions

Element	Jan	Feb	Mar	Apr	May	June	July	Aug	Sept	Oct	Nov	Dec	Year
Temperature (°C)	-2.2	-0.1	6.1	11.8	17.2	21.7	23.9	23.1	19.6	12.8	6.8	0.8	11.8
Daily Minimum Temp	-6.9	-5.2	0.6	5.7	11.0	15.6	18.2	17.2	13.7	6.8	1.8	-3.7	6.2
Daily Maximum Temp	2.6	4.9	11.7	17.9	23.3	27.8	29.7	28.9	25.5	18.9	11.8	5.3	17.3
Record Minimum Temp	-31.7	-23.3	-23.3	-7.2	-2.8	5.0	8.9	6.7	1.7	-8.3	-16.7	-28.9	-31.7
Record Maximum Temp	20.0	22.8	28.3	31.1	32.8	38.3	38.9	38.3	36.1	30.6	26.7	23.9	38.9
HDD, Base 18.3°C	636	517	379	197	84	6	0	0	28	182	345	543	2916
CDD, Base 18.3°C	0	0	0	0	48	106	174	148	67	11	0	0	553
Relative Humidity (%)	72	70	67	63	67	69	72	72	73	69	71	74	70
Wind Speed (m/s)	4.7	4.5	4.9	4.6	3.8	3.5	3.1	3.1	3.2	3.6	4.2	4.5	4.0

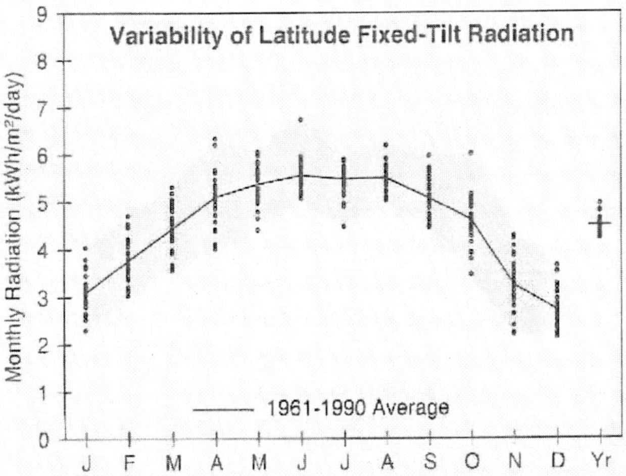

Lexington, KY

WBAN NO. 93820

LATITUDE: 38.03° N
LONGITUDE: 84.60° W
ELEVATION: 301 meters
MEAN PRESSURE: 982 millibars

STATION TYPE: Secondary

Solar Radiation for Flat-Plate Collectors Facing South at a Fixed Tilt (kWh/m²/day), Uncertainty ±9%

Tilt (°)		Jan	Feb	Mar	Apr	May	June	July	Aug	Sept	Oct	Nov	Dec	Year
0	Average	2.0	2.8	3.7	4.9	5.7	6.2	6.0	5.5	4.4	3.4	2.2	1.7	4.1
	Min/Max	1.7/2.3	2.5/3.2	3.2/4.4	4.0/5.9	4.7/6.4	5.6/7.5	4.9/6.4	5.0/6.1	4.0/5.1	2.8/4.2	1.8/2.6	1.5/2.1	3.8/4.4
Latitude -15	Average	2.8	3.5	4.3	5.2	5.7	6.0	5.9	5.7	5.0	4.3	2.9	2.4	4.5
	Min/Max	2.1/3.5	2.9/4.3	3.5/5.2	4.1/6.4	4.7/6.5	5.5/7.4	4.8/6.4	5.2/6.4	4.4/5.8	3.3/5.6	2.1/3.8	2.0/3.1	4.2/4.9
Latitude	Average	3.1	3.8	4.4	5.1	5.4	5.6	5.5	5.5	5.1	4.6	3.2	2.7	4.5
	Min/Max	2.3/4.0	3.0/4.7	3.6/5.3	4.0/6.3	4.4/6.1	5.1/6.7	4.5/5.9	5.0/6.2	4.4/6.0	3.5/6.0	2.2/4.3	2.1/3.6	4.2/4.9
Latitude +15	Average	3.3	3.8	4.3	4.7	4.8	4.8	4.8	5.0	4.8	4.6	3.4	2.9	4.3
	Min/Max	2.3/4.3	3.0/4.9	3.4/5.2	3.7/5.9	3.9/5.4	4.4/5.8	4.0/5.2	4.6/5.6	4.2/5.8	3.4/6.1	2.2/4.5	2.2/3.9	4.0/4.7
90	Average	2.9	3.2	3.1	2.9	2.5	2.4	2.4	2.9	3.3	3.7	2.9	2.6	2.9
	Min/Max	2.0/4.1	2.5/4.3	2.5/3.8	2.3/3.6	2.2/2.8	2.3/2.6	2.1/2.6	2.6/3.2	2.8/3.9	2.6/4.9	1.8/4.0	1.9/3.7	2.7/3.2

Solar Radiation for 1-Axis Tracking Flat-Plate Collectors with a North-South Axis (kWh/m²/day), Uncertainty ±9%

Axis Tilt (°)		Jan	Feb	Mar	Apr	May	June	July	Aug	Sept	Oct	Nov	Dec	Year
0	Average	2.7	3.7	4.8	6.3	7.3	7.8	7.6	7.1	5.9	4.7	2.9	2.3	5.3
	Min/Max	2.1/3.5	2.9/4.7	3.8/6.1	4.6/8.2	5.6/8.7	6.8/10.2	5.9/8.4	6.4/8.3	5.1/7.2	3.4/6.4	2.0/3.9	1.9/3.0	4.9/6.0
Latitude -15	Average	3.3	4.2	5.3	6.6	7.3	7.8	7.6	7.4	6.4	5.4	3.5	2.8	5.6
	Min/Max	2.4/4.4	3.2/5.4	4.1/6.7	4.8/8.6	5.6/8.8	6.8/10.1	5.8/8.5	6.6/8.6	5.4/7.8	3.8/7.3	2.3/4.7	2.2/3.8	5.2/6.4
Latitude	Average	3.5	4.4	5.4	6.5	7.1	7.5	7.3	7.2	6.4	5.6	3.7	3.1	5.7
	Min/Max	2.5/4.8	3.3/5.8	4.1/6.8	4.7/8.6	5.4/8.5	6.5/9.7	5.6/8.1	6.4/8.5	5.4/7.9	4.0/7.7	2.4/5.1	2.4/4.2	5.2/6.4
Latitude +15	Average	3.7	4.5	5.3	6.2	6.7	7.0	6.9	6.9	6.2	5.6	3.8	3.2	5.5
	Min/Max	2.6/5.1	3.3/5.9	4.0/6.7	4.5/8.3	5.0/8.0	6.0/9.1	5.2/7.6	6.1/8.1	5.2/7.8	3.9/7.8	2.4/5.3	2.4/4.4	5.1/6.3

Solar Radiation for 2-Axis Tracking Flat-Plate Collectors (kWh/m²/day), Uncertainty ±9%

Tracker		Jan	Feb	Mar	Apr	May	June	July	Aug	Sept	Oct	Nov	Dec	Year
2-Axis	Average	3.7	4.5	5.4	6.6	7.4	8.0	7.7	7.4	6.4	5.7	3.9	3.2	5.8
	Min/Max	2.6/5.1	3.4/5.9	4.1/6.8	4.8/8.7	5.7/8.9	6.9/10.4	5.9/8.6	6.6/8.7	5.4/7.9	4.0/7.8	2.4/5.3	2.5/4.5	5.4/6.6

Direct Beam Solar Radiation for Concentrating Collectors (kWh/m²/day), Uncertainty ±8%

Tracker		Jan	Feb	Mar	Apr	May	June	July	Aug	Sept	Oct	Nov	Dec	Year
1-Axis, E-W Horiz Axis	Average	1.9	2.2	2.4	2.9	3.2	3.5	3.4	3.3	2.9	3.0	2.0	1.7	2.7
	Min/Max	1.1/3.2	1.3/3.3	1.4/3.5	1.4/4.4	2.0/4.3	2.7/5.6	2.0/4.1	2.7/4.3	2.0/3.9	1.7/4.6	0.9/3.2	1.0/2.8	2.4/3.4
1-Axis, N-S Horiz Axis	Average	1.5	2.0	2.8	3.8	4.3	4.7	4.5	4.3	3.6	3.0	1.7	1.2	3.1
	Min/Max	0.9/2.4	1.2/3.1	1.6/4.2	1.8/5.8	2.7/5.9	3.5/7.4	2.7/5.4	3.5/5.7	2.5/4.9	1.6/4.8	0.7/2.7	0.8/2.0	2.7/3.9
1-Axis, N-S Tilt=Latitude	Average	2.1	2.6	3.2	3.9	4.2	4.4	4.2	4.4	4.0	3.8	2.3	1.8	3.4
	Min/Max	1.2/3.5	1.5/4.0	1.9/4.8	1.9/6.1	2.6/5.7	3.3/6.9	2.6/5.2	3.5/5.8	2.8/5.5	2.0/5.9	1.0/3.7	1.1/3.0	3.0/4.3
2-Axis	Average	2.3	2.7	3.2	4.0	4.4	4.7	4.5	4.5	4.0	3.8	2.4	2.0	3.6
	Min/Max	1.3/3.8	1.6/4.2	1.9/4.8	2.0/6.2	2.7/6.0	3.6/7.5	2.8/5.6	3.6/6.0	2.8/5.5	2.1/6.0	1.1/3.9	1.2/3.3	3.1/4.4

Average Climatic Conditions

Element	Jan	Feb	Mar	Apr	May	June	July	Aug	Sept	Oct	Nov	Dec	Year
Temperature (°C)	-0.7	1.4	7.4	12.7	17.8	22.3	24.3	23.7	20.1	13.7	7.8	2.2	12.7
Daily Minimum Temp	-5.3	-3.7	1.8	6.8	11.9	16.4	18.7	18.0	14.4	7.8	2.8	-2.4	7.3
Daily Maximum Temp	3.9	6.4	12.9	18.6	23.5	28.2	29.9	29.4	25.7	19.6	12.7	6.8	18.2
Record Minimum Temp	-29.4	-26.1	-18.9	-7.8	-3.3	3.9	8.3	5.6	1.7	-6.7	-19.4	-28.3	-29.4
Record Maximum Temp	24.4	24.4	28.3	31.1	33.3	38.3	39.4	39.4	39.4	32.8	28.3	23.9	39.4
HDD, Base 18.3°C	589	474	339	173	75	3	0	0	26	159	317	501	2657
CDD, Base 18.3°C	0	0	0	3	58	123	186	167	79	17	0	0	633
Relative Humidity (%)	73	71	66	63	68	69	72	73	73	69	72	74	70
Wind Speed (m/s)	4.4	4.4	4.7	4.4	3.7	3.4	3.0	3.0	3.1	3.4	4.1	4.4	3.8

Louisville, KY

WBAN NO. 93821

LATITUDE: 38.18° N
LONGITUDE: 85.73° W
ELEVATION: 149 meters
MEAN PRESSURE: 1000 millibars

STATION TYPE: Secondary

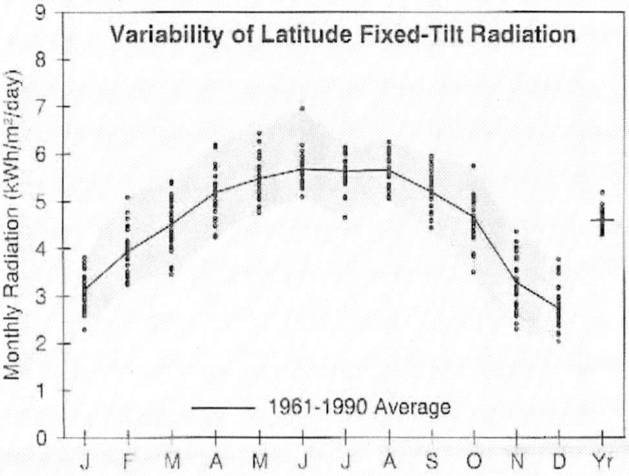

Solar Radiation for Flat-Plate Collectors Facing South at a Fixed Tilt (kWh/m²/day), Uncertainty ±9%

Tilt (°)		Jan	Feb	Mar	Apr	May	June	July	Aug	Sept	Oct	Nov	Dec	Year
0	Average	2.0	2.8	3.8	5.0	5.8	6.3	6.1	5.6	4.5	3.5	2.2	1.7	4.1
	Min/Max	1.7/2.3	2.5/3.4	3.1/4.4	4.2/5.8	5.1/6.8	5.7/7.8	5.1/6.7	5.1/6.1	4.0/5.1	2.8/4.1	1.8/2.7	1.5/2.1	3.9/4.6
Latitude -15	Average	2.8	3.6	4.4	5.3	5.8	6.2	6.0	5.9	5.1	4.4	3.0	2.4	4.6
	Min/Max	2.1/3.3	3.1/4.6	3.4/5.2	4.4/6.3	5.1/6.8	5.5/7.6	5.0/6.6	5.3/6.5	4.4/5.8	3.4/5.3	2.2/3.8	1.9/3.3	4.3/5.1
Latitude	Average	3.1	3.9	4.5	5.2	5.5	5.7	5.6	5.7	5.2	4.7	3.3	2.7	4.6
	Min/Max	2.3/3.8	3.2/5.1	3.5/5.4	4.2/6.2	4.8/6.4	5.1/6.9	4.6/6.1	5.1/6.3	4.4/5.9	3.5/5.7	2.3/4.3	2.0/3.8	4.3/5.2
Latitude +15	Average	3.3	4.0	4.4	4.8	4.9	4.9	4.9	5.1	5.0	4.7	3.4	2.9	4.4
	Min/Max	2.3/4.1	3.3/5.3	3.3/5.3	3.9/5.8	4.2/5.7	4.4/6.0	4.1/5.4	4.6/5.7	4.2/5.7	3.5/5.8	2.3/4.6	2.1/4.1	4.1/4.9
90	Average	3.0	3.4	3.2	3.0	2.6	2.4	2.5	2.9	3.4	3.7	2.9	2.6	3.0
	Min/Max	2.0/3.8	2.7/4.9	2.4/3.9	2.4/3.5	2.3/2.9	2.3/2.7	2.2/2.7	2.7/3.2	2.8/3.9	2.7/4.6	1.9/4.1	1.8/3.8	2.8/3.3

Solar Radiation for 1-Axis Tracking Flat-Plate Collectors with a North-South Axis (kWh/m²/day), Uncertainty ±9%

Axis Tilt (°)		Jan	Feb	Mar	Apr	May	June	July	Aug	Sept	Oct	Nov	Dec	Year
0	Average	2.7	3.8	4.9	6.4	7.4	8.0	7.8	7.3	6.0	4.7	2.9	2.3	5.4
	Min/Max	2.0/3.3	3.1/4.9	3.7/6.1	5.0/8.0	6.2/9.2	7.0/10.5	6.2/8.9	6.5/8.2	5.0/7.0	3.5/5.9	2.1/3.8	1.8/3.1	5.0/6.2
Latitude -15	Average	3.3	4.4	5.4	6.7	7.5	8.0	7.8	7.5	6.5	5.4	3.5	2.8	5.7
	Min/Max	2.4/4.1	3.5/5.8	4.0/6.7	5.2/8.4	6.3/9.4	7.0/10.4	6.2/8.9	6.7/8.5	5.3/7.7	4.0/6.8	2.4/4.7	2.1/3.9	5.3/6.7
Latitude	Average	3.6	4.6	5.5	6.6	7.2	7.7	7.5	7.4	6.6	5.7	3.8	3.1	5.8
	Min/Max	2.5/4.5	3.7/6.2	4.0/6.9	5.1/8.4	6.1/9.1	6.7/10.0	6.0/8.6	6.5/8.4	5.3/7.8	4.1/7.2	2.5/5.2	2.2/4.3	5.3/6.7
Latitude +15	Average	3.7	4.7	5.4	6.3	6.8	7.2	7.1	7.0	6.4	5.7	3.9	3.2	5.6
	Min/Max	2.5/4.7	3.7/6.4	3.9/6.8	4.8/8.1	5.7/8.6	6.2/9.4	5.6/8.1	6.2/8.0	5.2/7.6	4.0/7.2	2.5/5.3	2.3/4.6	5.2/6.6

Solar Radiation for 2-Axis Tracking Flat-Plate Collectors (kWh/m²/day), Uncertainty ±9%

Tracker		Jan	Feb	Mar	Apr	May	June	July	Aug	Sept	Oct	Nov	Dec	Year
2-Axis	Average	3.8	4.7	5.5	6.7	7.6	8.2	8.0	7.6	6.6	5.7	3.9	3.2	6.0
	Min/Max	2.5/4.7	3.7/6.4	4.0/6.9	5.2/8.5	6.3/9.5	7.1/10.7	6.3/9.1	6.7/8.6	5.4/7.8	4.1/7.3	2.5/5.4	2.3/4.7	5.5/7.0

Direct Beam Solar Radiation for Concentrating Collectors (kWh/m²/day), Uncertainty ±8%

Tracker		Jan	Feb	Mar	Apr	May	June	July	Aug	Sept	Oct	Nov	Dec	Year
1-Axis, E-W Horiz Axis	Average	2.0	2.4	2.5	3.0	3.4	3.7	3.5	3.4	3.1	3.0	2.1	1.7	2.8
	Min/Max	1.0/2.8	1.5/3.3	1.4/3.6	1.6/4.3	2.3/4.9	2.7/5.9	2.3/4.5	2.7/4.3	2.2/4.0	1.7/4.2	1.0/3.3	1.0/3.0	2.4/3.7
1-Axis, N-S Horiz Axis	Average	1.5	2.2	2.8	3.8	4.4	4.9	4.7	4.5	3.7	3.0	1.7	1.2	3.2
	Min/Max	0.8/2.1	1.4/3.1	1.6/4.2	2.1/5.6	3.0/6.5	3.6/7.8	3.1/5.9	3.6/5.5	2.6/4.9	1.7/4.3	0.8/2.7	0.7/2.0	2.8/4.2
1-Axis, N-S Tilt=Latitude	Average	2.2	2.9	3.3	4.0	4.3	4.6	4.4	4.5	4.1	3.8	2.3	1.9	3.5
	Min/Max	1.1/3.1	1.9/4.1	1.9/4.8	2.3/5.9	2.9/6.3	3.4/7.3	2.9/5.7	3.6/5.6	2.9/5.4	2.1/5.3	1.1/3.8	1.1/3.1	3.0/4.6
2-Axis	Average	2.3	2.9	3.3	4.1	4.6	4.9	4.8	4.7	4.2	3.8	2.5	2.0	3.7
	Min/Max	1.2/3.3	1.9/4.2	1.9/4.8	2.3/6.0	3.1/6.7	3.7/7.9	3.2/6.1	3.7/5.8	2.9/5.4	2.2/5.4	1.2/4.0	1.2/3.4	3.2/4.8

Average Climatic Conditions

Element	Jan	Feb	Mar	Apr	May	June	July	Aug	Sept	Oct	Nov	Dec	Year
Temperature (°C)	-0.2	2.1	7.9	13.5	18.5	22.9	25.1	24.3	20.8	14.2	8.4	2.7	13.4
Daily Minimum Temp	-4.9	-3.1	2.3	7.4	12.6	17.2	19.6	18.8	14.8	7.7	2.9	-1.9	7.8
Daily Maximum Temp	4.6	7.1	13.5	19.6	24.4	28.6	30.6	29.8	26.8	20.7	13.8	7.3	18.9
Record Minimum Temp	-28.9	-28.3	-18.3	-5.6	-0.6	5.6	10.0	7.8	0.6	-5.0	-18.3	-26.1	-28.9
Record Maximum Temp	25.0	25.0	30.0	32.8	35.0	38.9	40.6	38.3	40.0	33.3	28.9	24.4	40.6
HDD, Base 18.3°C	573	456	322	152	58	3	0	0	20	141	298	484	2508
CDD, Base 18.3°C	0	0	0	7	64	140	210	186	95	14	0	0	716
Relative Humidity (%)	69	68	64	61	67	69	71	72	73	70	69	70	69
Wind Speed (m/s)	4.2	4.2	4.5	4.3	3.6	3.4	3.0	3.0	3.0	3.3	3.9	4.2	3.7

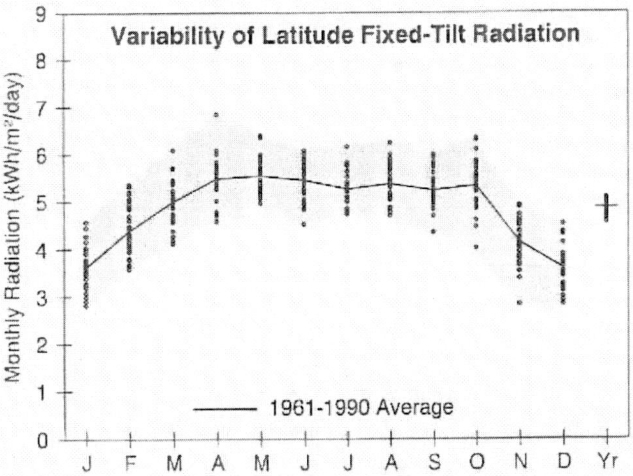

Baton Rouge, LA

WBAN NO. 13970

LATITUDE: 30.53° N
LONGITUDE: 91.15° W
ELEVATION: 23 meters
MEAN PRESSURE: 1015 millibars

STATION TYPE: Secondary

Solar Radiation for Flat-Plate Collectors Facing South at a Fixed Tilt (kWh/m²/day), Uncertainty ±9%

Tilt (°)		Jan	Feb	Mar	Apr	May	June	July	Aug	Sept	Oct	Nov	Dec	Year
0	Average	2.6	3.5	4.4	5.4	5.9	6.0	5.7	5.4	4.8	4.3	3.0	2.5	4.5
	Min/Max	2.2/3.1	3.0/4.1	3.8/5.2	4.6/6.6	5.3/6.9	5.0/6.8	5.1/6.7	4.8/6.3	4.0/5.4	3.4/4.9	2.4/3.4	2.2/3.0	4.2/4.6
Latitude -15	Average	3.2	4.0	4.8	5.6	5.9	5.9	5.6	5.6	5.2	5.0	3.7	3.1	4.8
	Min/Max	2.6/4.0	3.4/4.8	4.0/5.8	4.7/6.9	5.2/6.8	4.9/6.6	5.0/6.6	4.9/6.4	4.3/5.9	3.8/5.8	2.7/4.3	2.6/3.9	4.5/5.0
Latitude	Average	3.6	4.4	5.0	5.5	5.5	5.5	5.3	5.4	5.2	5.3	4.1	3.6	4.9
	Min/Max	2.8/4.6	3.6/5.4	4.1/6.1	4.6/6.8	5.0/6.4	4.5/6.1	4.7/6.2	4.7/6.2	4.3/6.0	4.0/6.3	2.8/4.9	2.8/4.5	4.6/5.1
Latitude +15	Average	3.8	4.5	4.9	5.1	4.9	4.7	4.6	4.9	5.0	5.4	4.4	3.8	4.7
	Min/Max	2.9/4.9	3.6/5.6	4.0/6.0	4.3/6.3	4.4/5.6	3.9/5.2	4.2/5.4	4.3/5.7	4.2/5.8	4.0/6.5	2.9/5.2	3.0/4.9	4.4/4.9
90	Average	3.2	3.4	3.2	2.7	2.1	1.9	1.9	2.4	3.0	3.9	3.6	3.3	2.9
	Min/Max	2.4/4.3	2.7/4.3	2.6/3.9	2.3/3.1	2.0/2.3	1.7/1.9	1.8/2.1	2.1/2.6	2.6/3.5	2.8/4.7	2.2/4.4	2.5/4.4	2.7/3.0

Solar Radiation for 1-Axis Tracking Flat-Plate Collectors with a North-South Axis (kWh/m²/day), Uncertainty ±9%

Axis Tilt (°)		Jan	Feb	Mar	Apr	May	June	July	Aug	Sept	Oct	Nov	Dec	Year
0	Average	3.5	4.6	5.7	6.9	7.4	7.6	7.0	6.8	6.2	5.8	4.1	3.4	5.8
	Min/Max	2.7/4.4	3.7/5.7	4.6/7.3	5.3/9.2	6.3/9.1	6.2/9.0	6.0/8.4	5.8/8.2	4.9/7.3	4.3/7.2	2.7/4.9	2.7/4.3	5.4/6.0
Latitude -15	Average	3.9	5.0	6.1	7.1	7.4	7.5	7.0	6.9	6.5	6.3	4.6	3.8	6.0
	Min/Max	3.0/5.1	4.0/6.3	4.8/7.8	5.4/9.5	6.3/9.1	6.1/8.9	6.0/8.4	5.9/8.4	5.1/7.6	4.6/7.8	2.9/5.6	3.0/4.9	5.6/6.3
Latitude	Average	4.2	5.3	6.2	7.0	7.2	7.2	6.7	6.8	6.5	6.6	4.9	4.2	6.1
	Min/Max	3.2/5.5	4.1/6.7	4.9/8.0	5.4/9.4	6.1/8.9	5.9/8.5	5.7/8.1	5.8/8.2	5.1/7.7	4.7/8.2	3.1/6.0	3.2/5.4	5.6/6.4
Latitude +15	Average	4.4	5.4	6.1	6.7	6.8	6.7	6.3	6.4	6.4	6.7	5.1	4.4	5.9
	Min/Max	3.2/5.8	4.2/6.8	4.8/7.9	5.1/9.1	5.7/8.3	5.5/7.9	5.3/7.6	5.5/7.8	5.0/7.6	4.7/8.3	3.1/6.3	3.3/5.8	5.5/6.2

Solar Radiation for 2-Axis Tracking Flat-Plate Collectors (kWh/m²/day), Uncertainty ±9%

Tracker		Jan	Feb	Mar	Apr	May	June	July	Aug	Sept	Oct	Nov	Dec	Year
2-Axis	Average	4.4	5.4	6.2	7.1	7.5	7.6	7.1	6.9	6.6	6.7	5.1	4.5	6.3
	Min/Max	3.3/5.9	4.2/6.8	4.9/8.0	5.5/9.5	6.4/9.3	6.2/9.1	6.0/8.5	5.9/8.4	5.2/7.7	4.7/8.3	3.1/6.3	3.3/5.9	5.8/6.6

Direct Beam Solar Radiation for Concentrating Collectors (kWh/m²/day), Uncertainty ±8%

Tracker		Jan	Feb	Mar	Apr	May	June	July	Aug	Sept	Oct	Nov	Dec	Year
1-Axis, E-W	Average	2.4	2.8	3.0	3.3	3.4	3.4	3.1	3.0	3.0	3.6	2.8	2.6	3.0
Horiz Axis	Min/Max	1.5/3.7	1.8/4.0	1.9/4.4	2.2/5.3	2.4/4.9	2.4/4.5	2.2/4.2	2.3/4.3	2.0/3.9	2.1/5.0	1.2/3.8	1.6/3.8	2.7/3.3
1-Axis, N-S	Average	2.1	2.9	3.6	4.3	4.5	4.4	3.9	4.0	3.7	3.9	2.6	2.1	3.5
Horiz Axis	Min/Max	1.3/3.2	1.9/4.1	2.3/5.4	2.7/7.0	3.2/6.5	3.2/5.9	2.8/5.4	2.9/5.6	2.4/5.0	2.3/5.5	1.1/3.5	1.3/3.1	3.1/3.8
1-Axis, N-S	Average	2.7	3.5	4.0	4.4	4.3	4.1	3.7	3.9	4.0	4.6	3.3	2.8	3.8
Tilt=Latitude	Min/Max	1.7/4.1	2.3/4.9	2.5/6.0	2.7/7.2	3.0/6.3	3.0/5.5	2.7/5.1	2.9/5.6	2.6/5.4	2.7/6.4	1.4/4.4	1.7/4.1	3.3/4.1
2-Axis	Average	2.9	3.6	4.0	4.5	4.6	4.5	4.0	4.0	4.0	4.6	3.4	3.0	3.9
	Min/Max	1.8/4.4	2.4/5.0	2.5/6.0	2.8/7.3	3.2/6.6	3.2/6.0	2.9/5.5	3.0/5.8	2.6/5.4	2.7/6.5	1.4/4.6	1.9/4.5	3.5/4.3

Average Climatic Conditions

Element	Jan	Feb	Mar	Apr	May	June	July	Aug	Sept	Oct	Nov	Dec	Year
Temperature (°C)	9.9	11.7	16.3	20.5	24.1	26.9	27.9	27.7	25.6	20.3	15.6	11.6	19.8
Daily Minimum Temp	4.2	5.8	10.1	14.4	18.2	21.3	22.9	22.6	20.4	13.9	9.4	5.8	14.1
Daily Maximum Temp	15.4	17.6	22.4	26.6	29.8	32.5	33.0	32.8	30.8	26.7	21.6	17.4	25.6
Record Minimum Temp	-12.8	-6.7	-6.7	0.0	6.7	11.7	14.4	15.0	6.1	0.0	-6.1	-13.3	-13.3
Record Maximum Temp	27.8	29.4	32.8	33.3	36.7	39.4	38.3	38.9	37.2	34.4	30.6	29.4	39.4
HDD, Base 18.3°C	272	192	94	15	0	0	0	0	0	27	109	218	927
CDD, Base 18.3°C	11	7	31	80	177	258	298	291	218	88	26	9	1494
Relative Humidity (%)	74	71	70	71	72	74	77	78	77	73	75	75	74
Wind Speed (m/s)	3.8	4.0	4.1	3.9	3.4	2.9	2.6	2.4	2.9	2.9	3.4	3.6	3.3

Lake Charles, LA

WBAN NO. 03937

LATITUDE: 30.12° N
LONGITUDE: 93.22° W
ELEVATION: 3 meters
MEAN PRESSURE: 1016 millibars

STATION TYPE: Primary

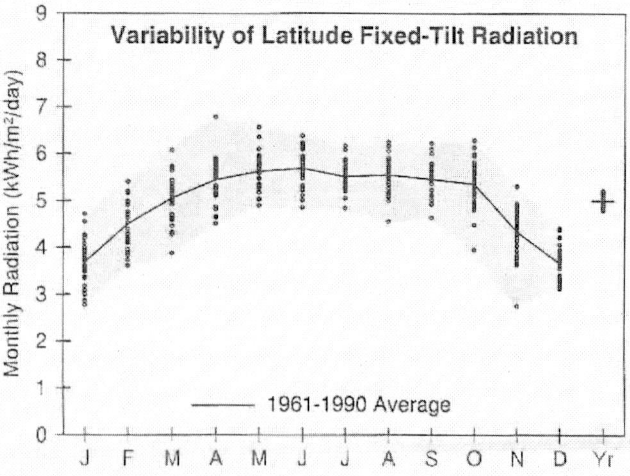

Solar Radiation for Flat-Plate Collectors Facing South at a Fixed Tilt (kWh/m²/day), Uncertainty ±9%

Tilt (°)		Jan	Feb	Mar	Apr	May	June	July	Aug	Sept	Oct	Nov	Dec	Year
0	Average	2.7	3.6	4.5	5.4	6.0	6.3	6.0	5.6	5.0	4.3	3.2	2.6	4.6
	Min/Max	2.1/3.3	3.1/4.1	3.6/5.2	4.5/6.6	5.2/7.1	5.4/7.2	5.3/6.8	4.7/6.3	4.3/5.6	3.4/5.0	2.4/3.7	2.3/3.0	4.5/4.8
Latitude -15	Average	3.3	4.1	4.9	5.5	6.0	6.2	5.9	5.7	5.4	5.0	3.9	3.2	4.9
	Min/Max	2.5/4.1	3.4/4.9	3.8/5.8	4.6/6.9	5.2/7.0	5.2/6.9	5.2/6.6	4.7/6.5	4.6/6.1	3.8/5.8	2.6/4.6	2.8/3.8	4.7/5.1
Latitude	Average	3.7	4.5	5.1	5.4	5.6	5.7	5.5	5.6	5.4	5.4	4.3	3.7	5.0
	Min/Max	2.8/4.7	3.6/5.4	3.9/6.1	4.5/6.8	4.9/6.6	4.8/6.4	4.8/6.2	4.6/6.2	4.6/6.2	3.9/6.3	2.8/5.3	3.1/4.4	4.8/5.2
Latitude +15	Average	3.9	4.6	4.9	5.1	5.0	5.0	4.9	5.1	5.2	5.4	4.6	3.9	4.8
	Min/Max	2.9/5.1	3.6/5.6	3.8/6.0	4.2/6.3	4.4/5.8	4.2/5.5	4.3/5.4	4.2/5.7	4.4/6.0	3.9/6.4	2.8/5.7	3.3/4.8	4.6/5.0
90	Average	3.2	3.5	3.2	2.6	2.1	1.9	2.0	2.4	3.1	3.9	3.7	3.4	2.9
	Min/Max	2.4/4.3	2.7/4.3	2.4/3.9	2.3/3.1	2.0/2.3	1.8/2.0	1.8/2.0	2.1/2.6	2.7/3.6	2.7/4.6	2.1/4.8	2.8/4.2	2.8/3.1

Solar Radiation for 1-Axis Tracking Flat-Plate Collectors with a North-South Axis (kWh/m²/day), Uncertainty ±9%

Axis Tilt (°)		Jan	Feb	Mar	Apr	May	June	July	Aug	Sept	Oct	Nov	Dec	Year
0	Average	3.6	4.7	5.8	6.8	7.5	7.9	7.4	7.1	6.4	5.8	4.3	3.4	5.9
	Min/Max	2.6/4.6	3.8/5.8	4.3/7.4	5.3/9.0	6.3/9.4	6.5/9.2	6.3/8.7	5.6/8.2	5.3/7.5	4.2/7.1	2.6/5.3	2.8/4.2	5.5/6.2
Latitude -15	Average	4.0	5.2	6.1	6.9	7.5	7.8	7.3	7.2	6.7	6.3	4.8	3.9	6.1
	Min/Max	2.9/5.2	4.1/6.4	4.5/7.8	5.4/9.3	6.3/9.4	6.4/9.1	6.3/8.6	5.6/8.4	5.5/7.9	4.4/7.7	2.8/6.0	3.2/4.8	5.8/6.5
Latitude	Average	4.3	5.4	6.2	6.9	7.2	7.5	7.0	7.0	6.8	6.6	5.2	4.3	6.2
	Min/Max	3.1/5.7	4.2/6.8	4.6/8.0	5.3/9.2	6.0/9.1	6.2/8.7	6.1/8.3	5.5/8.2	5.6/8.0	4.6/8.1	2.9/6.6	3.5/5.3	5.9/6.6
Latitude +15	Average	4.5	5.5	6.1	6.6	6.8	6.9	6.6	6.7	6.6	6.6	5.3	4.5	6.1
	Min/Max	3.2/6.0	4.2/6.9	4.5/8.0	5.1/8.8	5.7/8.5	5.7/8.1	5.7/7.8	5.2/7.8	5.4/7.8	4.5/8.1	3.0/6.8	3.6/5.6	5.7/6.4

Solar Radiation for 2-Axis Tracking Flat-Plate Collectors (kWh/m²/day), Uncertainty ±9%

Tracker		Jan	Feb	Mar	Apr	May	June	July	Aug	Sept	Oct	Nov	Dec	Year
2-Axis	Average	4.5	5.5	6.2	7.0	7.5	7.9	7.4	7.2	6.8	6.7	5.4	4.5	6.4
	Min/Max	3.3/6.0	4.3/6.9	4.6/8.1	5.4/9.3	6.3/9.5	6.6/9.3	6.4/8.8	5.7/8.4	5.6/8.0	4.6/8.2	3.0/6.9	3.7/5.7	6.0/6.8

Direct Beam Solar Radiation for Concentrating Collectors (kWh/m²/day), Uncertainty ±8%

Tracker		Jan	Feb	Mar	Apr	May	June	July	Aug	Sept	Oct	Nov	Dec	Year
1-Axis, E-W	Average	2.4	2.9	2.9	3.1	3.3	3.6	3.2	3.0	3.1	3.5	3.0	2.6	3.0
Horiz Axis	Min/Max	1.4/3.8	1.7/4.0	1.8/4.3	1.9/5.0	2.4/5.0	2.6/4.6	2.5/4.3	2.0/3.9	2.1/4.0	1.9/4.8	1.0/4.5	1.8/3.6	2.8/3.4
1-Axis, N-S	Average	2.1	3.0	3.5	4.1	4.3	4.6	4.1	4.0	3.8	3.8	2.8	2.1	3.5
Horiz Axis	Min/Max	1.3/3.3	1.9/4.2	2.1/5.4	2.5/6.7	3.1/6.6	3.4/6.0	3.2/5.6	2.6/5.2	2.7/5.2	2.1/5.3	0.9/4.1	1.4/3.0	3.1/4.0
1-Axis, N-S	Average	2.7	3.6	3.9	4.1	4.1	4.3	3.9	3.9	4.1	4.4	3.5	2.8	3.8
Tilt=Latitude	Min/Max	1.6/4.3	2.2/5.0	2.3/5.9	2.5/6.8	2.9/6.3	3.1/5.6	3.0/5.3	2.5/5.1	2.8/5.5	2.4/6.1	1.1/5.2	1.9/3.9	3.3/4.3
2-Axis	Average	2.9	3.7	3.9	4.2	4.4	4.7	4.1	4.1	4.1	4.5	3.7	3.0	3.9
	Min/Max	1.7/4.6	2.3/5.1	2.3/5.9	2.6/6.9	3.1/6.7	3.4/6.1	3.3/5.7	2.6/5.3	2.8/5.5	2.4/6.2	1.2/5.4	2.1/4.3	3.5/4.5

Average Climatic Conditions

Element	Jan	Feb	Mar	Apr	May	June	July	Aug	Sept	Oct	Nov	Dec	Year
Temperature (°C)	10.2	12.1	15.9	20.2	23.8	26.8	27.9	27.7	25.4	20.6	15.8	11.9	19.9
Daily Minimum Temp	5.1	6.7	10.4	14.8	18.7	22.0	23.1	22.8	20.3	14.4	10.0	6.4	14.6
Daily Maximum Temp	15.4	17.4	21.5	25.5	28.9	31.7	32.7	32.7	30.4	26.8	21.6	17.4	25.2
Record Minimum Temp	-9.4	-5.6	-3.9	1.1	9.4	13.3	16.1	16.1	8.3	0.0	-5.0	-11.7	-11.7
Record Maximum Temp	27.8	28.3	30.0	35.0	35.6	37.2	38.9	38.3	36.7	33.3	30.6	27.8	38.9
HDD, Base 18.3°C	259	182	99	22	0	0	0	0	0	19	107	208	898
CDD, Base 18.3°C	8	7	25	77	171	255	296	291	212	90	31	10	1472
Relative Humidity (%)	79	77	76	76	77	78	80	80	79	76	77	79	78
Wind Speed (m/s)	4.4	4.6	4.6	4.5	3.9	3.4	2.9	2.7	3.2	3.4	4.1	4.4	3.8

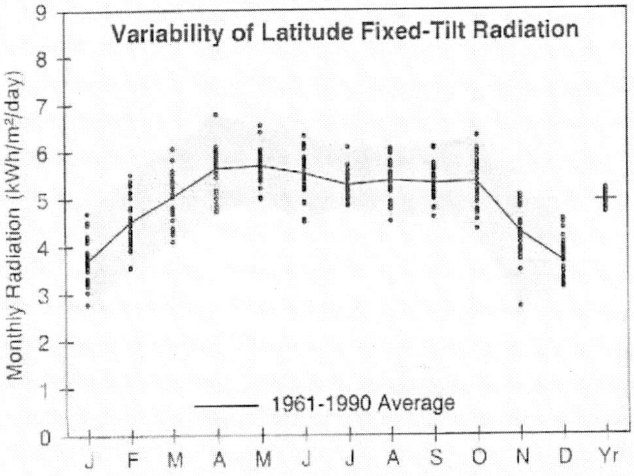

New Orleans, LA

WBAN NO. 12916

LATITUDE: 29.98° N
LONGITUDE: 90.25° W
ELEVATION: 3 meters
MEAN PRESSURE: 1017 millibars

STATION TYPE: Secondary

Solar Radiation for Flat-Plate Collectors Facing South at a Fixed Tilt (kWh/m²/day), Uncertainty ±9%

Tilt (°)		Jan	Feb	Mar	Apr	May	June	July	Aug	Sept	Oct	Nov	Dec	Year
0	Average	2.7	3.6	4.5	5.5	6.1	6.1	5.7	5.5	4.9	4.3	3.1	2.6	4.6
	Min/Max	2.3/3.2	3.0/4.2	3.8/5.2	4.7/6.6	5.3/7.0	5.0/7.1	5.3/6.6	4.7/6.1	4.3/5.5	3.7/5.0	2.3/3.6	2.3/3.0	4.3/4.8
Latitude -15	Average	3.3	4.2	4.9	5.7	6.0	6.0	5.7	5.6	5.3	5.0	3.8	3.2	4.9
	Min/Max	2.6/4.1	3.3/5.0	4.0/5.8	4.8/6.9	5.3/7.0	4.8/6.9	5.2/6.5	4.7/6.3	4.6/6.0	4.1/5.8	2.5/4.4	2.8/3.9	4.6/5.1
Latitude	Average	3.7	4.5	5.0	5.6	5.7	5.5	5.3	5.4	5.3	5.3	4.3	3.7	5.0
	Min/Max	2.8/4.7	3.5/5.5	4.1/6.1	4.7/6.8	5.0/6.6	4.5/6.3	4.8/6.1	4.5/6.1	4.6/6.1	4.3/6.3	2.7/5.1	3.1/4.6	4.7/5.2
Latitude +15	Average	3.9	4.6	4.9	5.3	5.1	4.8	4.7	4.9	5.1	5.4	4.5	3.9	4.8
	Min/Max	2.8/5.1	3.6/5.7	4.0/6.0	4.4/6.3	4.4/5.8	4.0/5.4	4.3/5.3	4.1/5.5	4.4/5.9	4.3/6.5	2.7/5.4	3.3/4.9	4.5/5.0
90	Average	3.3	3.5	3.2	2.7	2.1	1.8	1.9	2.3	3.0	3.9	3.7	3.4	2.9
	Min/Max	2.3/4.4	2.7/4.4	2.6/3.9	2.3/3.0	1.9/2.2	1.7/1.9	1.8/2.0	2.1/2.5	2.7/3.5	3.0/4.7	2.1/4.5	2.7/4.4	2.7/3.0

Solar Radiation for 1-Axis Tracking Flat-Plate Collectors with a North-South Axis (kWh/m²/day), Uncertainty ±9%

Axis Tilt (°)		Jan	Feb	Mar	Apr	May	June	July	Aug	Sept	Oct	Nov	Dec	Year
0	Average	3.6	4.7	5.8	7.1	7.7	7.6	7.0	6.8	6.3	5.8	4.2	3.5	5.8
	Min/Max	2.7/4.6	3.6/6.0	4.5/7.2	5.7/9.1	6.5/9.3	5.8/9.3	6.3/8.3	5.5/7.9	5.2/7.4	4.6/7.1	2.6/5.0	2.8/4.3	5.5/6.2
Latitude -15	Average	4.0	5.2	6.1	7.3	7.7	7.5	6.9	6.9	6.6	6.3	4.7	3.9	6.1
	Min/Max	3.0/5.2	3.9/6.6	4.8/7.6	5.8/9.3	6.5/9.3	5.8/9.1	6.2/8.2	5.5/8.0	5.4/7.7	4.9/7.8	2.8/5.7	3.2/5.0	5.7/6.5
Latitude	Average	4.3	5.5	6.3	7.3	7.4	7.2	6.7	6.7	6.6	6.6	5.1	4.3	6.2
	Min/Max	3.1/5.7	4.1/7.0	4.8/7.8	5.8/9.3	6.3/9.0	5.5/8.8	6.0/7.9	5.4/7.9	5.4/7.8	5.1/8.1	2.9/6.1	3.4/5.5	5.8/6.6
Latitude +15	Average	4.5	5.6	6.2	7.0	7.0	6.7	6.2	6.4	6.5	6.7	5.3	4.5	6.0
	Min/Max	3.2/6.0	4.1/7.1	4.7/7.8	5.5/8.9	5.9/8.5	5.1/8.1	5.6/7.4	5.1/7.5	5.3/7.6	5.1/8.2	2.9/6.4	3.6/5.8	5.6/6.4

Solar Radiation for 2-Axis Tracking Flat-Plate Collectors (kWh/m²/day), Uncertainty ±9%

Tracker		Jan	Feb	Mar	Apr	May	June	July	Aug	Sept	Oct	Nov	Dec	Year
2-Axis	Average	4.5	5.6	6.3	7.3	7.8	7.7	7.0	6.9	6.6	6.7	5.3	4.5	6.4
	Min/Max	3.2/6.1	4.2/7.2	4.8/7.9	5.9/9.4	6.6/9.4	5.8/9.4	6.3/8.4	5.5/8.1	5.4/7.8	5.2/8.2	2.9/6.4	3.6/5.9	5.9/6.8

Direct Beam Solar Radiation for Concentrating Collectors (kWh/m²/day), Uncertainty ±8%

Tracker		Jan	Feb	Mar	Apr	May	June	July	Aug	Sept	Oct	Nov	Dec	Year
1-Axis, E-W Horiz Axis	Average	2.4	2.9	3.0	3.5	3.6	3.5	3.0	3.0	3.0	3.6	2.9	2.6	3.1
	Min/Max	1.2/3.8	1.8/4.2	1.9/4.2	2.2/5.1	2.7/5.0	2.0/4.9	2.4/4.2	1.8/3.9	2.1/4.0	2.4/4.9	1.0/4.0	1.8/3.7	2.8/3.5
1-Axis, N-S Horiz Axis	Average	2.1	3.0	3.6	4.5	4.8	4.5	3.9	3.9	3.8	3.9	2.7	2.1	3.6
	Min/Max	1.1/3.3	1.9/4.4	2.2/5.1	2.9/6.8	3.5/6.7	2.6/6.3	3.1/5.3	2.3/5.1	2.5/5.0	2.6/5.4	0.9/3.7	1.4/3.1	3.1/4.0
1-Axis, N-S Tilt=Latitude	Average	2.7	3.6	4.0	4.6	4.6	4.2	3.6	3.8	4.0	4.5	3.4	2.8	3.8
	Min/Max	1.4/4.2	2.3/5.3	2.5/5.6	3.0/6.9	3.4/6.4	2.4/5.9	2.9/5.0	2.3/5.1	2.7/5.3	3.0/6.3	1.1/4.6	1.9/4.1	3.4/4.3
2-Axis	Average	2.9	3.7	4.0	4.7	4.8	4.5	3.9	3.9	4.1	4.6	3.6	3.0	4.0
	Min/Max	1.5/4.6	2.3/5.4	2.5/5.7	3.0/7.1	3.6/6.7	2.6/6.4	3.2/5.3	2.4/5.2	2.7/5.3	3.0/6.3	1.2/4.9	2.0/4.4	3.5/4.5

Average Climatic Conditions

Element	Jan	Feb	Mar	Apr	May	June	July	Aug	Sept	Oct	Nov	Dec	Year
Temperature (°C)	10.7	12.4	16.4	20.3	23.8	26.7	27.7	27.5	25.6	20.6	16.2	12.5	20.1
Daily Minimum Temp	5.4	6.9	10.9	14.7	18.4	21.6	22.8	22.7	20.8	14.8	10.6	7.1	14.7
Daily Maximum Temp	16.0	17.8	22.0	25.8	29.1	31.8	32.6	32.3	30.3	26.3	21.7	17.9	25.3
Record Minimum Temp	-10.0	-7.2	-3.9	0.0	5.0	10.0	15.6	15.6	5.6	1.7	-4.4	-11.7	-11.7
Record Maximum Temp	28.3	29.4	31.7	33.3	35.6	37.8	38.3	38.9	38.3	33.3	30.6	28.9	38.9
HDD, Base 18.3°C	250	176	90	16	0	0	0	0	0	17	99	194	841
CDD, Base 18.3°C	14	9	31	74	169	250	291	284	218	87	34	13	1475
Relative Humidity (%)	76	73	73	73	74	76	79	79	78	75	77	77	76
Wind Speed (m/s)	4.0	4.3	4.2	4.1	3.6	3.0	2.6	2.6	3.1	3.3	3.8	4.0	3.5

Shreveport, LA

WBAN NO. 13957

LATITUDE: 32.47° N
LONGITUDE: 93.82° W
ELEVATION: 79 meters
MEAN PRESSURE: 1008 millibars

STATION TYPE: Secondary

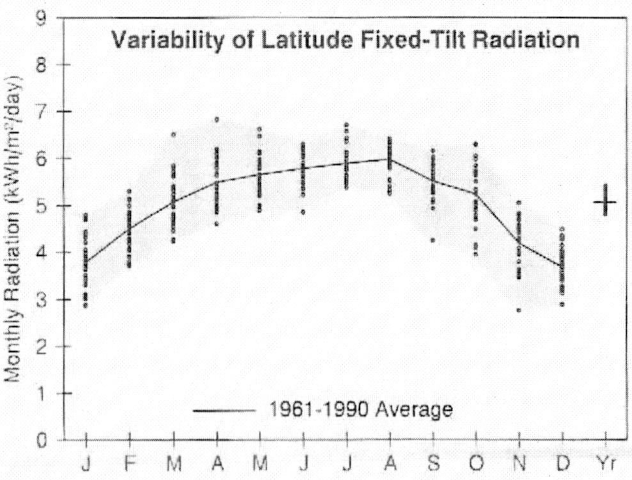

Solar Radiation for Flat-Plate Collectors Facing South at a Fixed Tilt (kWh/m²/day), Uncertainty ±9%

Tilt (°)		Jan	Feb	Mar	Apr	May	June	July	Aug	Sept	Oct	Nov	Dec	Year
0	Average	2.6	3.4	4.4	5.4	6.0	6.4	6.4	6.0	5.0	4.1	3.0	2.5	4.6
	Min/Max	2.2/3.1	3.0/3.9	3.8/5.4	4.6/6.6	5.2/7.0	5.4/7.1	5.8/7.4	5.3/6.5	4.0/5.5	3.3/4.8	2.3/3.4	2.1/2.8	4.4/4.9
Latitude -15	Average	3.4	4.1	4.9	5.6	6.0	6.3	6.3	6.2	5.4	4.9	3.7	3.2	5.0
	Min/Max	2.7/4.2	3.5/4.8	4.1/6.2	4.7/6.9	5.2/7.0	5.3/6.8	5.8/7.2	5.4/6.7	4.2/6.0	3.8/5.8	2.6/4.4	2.6/3.8	4.8/5.3
Latitude	Average	3.8	4.5	5.1	5.5	5.7	5.8	5.9	6.0	5.5	5.2	4.2	3.7	5.1
	Min/Max	2.9/4.8	3.7/5.3	4.2/6.5	4.6/6.8	4.9/6.6	4.8/6.3	5.4/6.7	5.3/6.4	4.2/6.2	3.9/6.3	2.8/5.1	2.9/4.5	4.8/5.4
Latitude +15	Average	4.0	4.6	5.0	5.1	5.0	5.0	5.2	5.5	5.3	5.3	4.4	3.9	4.9
	Min/Max	3.0/5.2	3.7/5.5	4.1/6.4	4.3/6.3	4.4/5.8	4.2/5.4	4.8/5.8	4.8/5.9	4.0/5.9	3.9/6.4	2.8/5.4	3.0/4.9	4.6/5.2
90	Average	3.4	3.6	3.3	2.8	2.3	2.0	2.2	2.7	3.3	3.9	3.7	3.5	3.1
	Min/Max	2.4/4.5	2.8/4.4	2.7/4.3	2.4/3.3	2.1/2.5	1.9/2.1	2.0/2.3	2.4/2.8	2.5/3.7	2.8/4.8	2.2/4.6	2.6/4.4	2.9/3.3

Solar Radiation for 1-Axis Tracking Flat-Plate Collectors with a North-South Axis (kWh/m²/day), Uncertainty ±9%

Axis Tilt (°)		Jan	Feb	Mar	Apr	May	June	July	Aug	Sept	Oct	Nov	Dec	Year
0	Average	3.6	4.6	5.8	6.9	7.6	8.1	8.2	7.8	6.5	5.6	4.1	3.4	6.0
	Min/Max	2.7/4.6	3.8/5.6	4.7/7.6	5.6/9.1	6.2/9.3	6.6/9.2	7.1/9.8	6.6/8.7	4.8/7.4	4.1/7.0	2.6/4.9	2.6/4.2	5.6/6.5
Latitude -15	Average	4.1	5.2	6.2	7.1	7.6	8.1	8.2	8.0	6.9	6.2	4.6	3.9	6.3
	Min/Max	3.0/5.3	4.1/6.3	4.9/8.2	5.7/9.4	6.2/9.4	6.5/9.1	7.1/9.8	6.7/8.8	5.0/7.8	4.4/7.8	2.9/5.7	3.0/4.9	5.9/6.9
Latitude	Average	4.4	5.4	6.3	7.0	7.4	7.7	7.9	7.8	7.0	6.5	5.0	4.3	6.4
	Min/Max	3.2/5.8	4.3/6.6	5.0/8.5	5.6/9.4	6.0/9.1	6.2/8.7	6.8/9.4	6.6/8.7	5.0/7.9	4.6/8.1	3.0/6.2	3.2/5.4	5.9/6.9
Latitude +15	Average	4.6	5.5	6.2	6.8	7.0	7.2	7.3	7.4	6.8	6.5	5.2	4.5	6.3
	Min/Max	3.3/6.1	4.3/6.8	4.9/8.4	5.4/9.0	5.6/8.5	5.8/8.1	6.4/8.8	6.3/8.3	4.9/7.8	4.6/8.2	3.0/6.4	3.3/5.7	5.8/6.8

Solar Radiation for 2-Axis Tracking Flat-Plate Collectors (kWh/m²/day), Uncertainty ±9%

Tracker		Jan	Feb	Mar	Apr	May	June	July	Aug	Sept	Oct	Nov	Dec	Year
2-Axis	Average	4.7	5.5	6.3	7.1	7.7	8.2	8.3	8.0	7.0	6.6	5.2	4.6	6.6
	Min/Max	3.3/6.2	4.3/6.8	5.0/8.5	5.7/9.5	6.3/9.5	6.7/9.3	7.2/10.0	6.8/8.9	5.1/8.0	4.6/8.2	3.1/6.5	3.4/5.8	6.1/7.2

Direct Beam Solar Radiation for Concentrating Collectors (kWh/m²/day), Uncertainty ±8%

Tracker		Jan	Feb	Mar	Apr	May	June	July	Aug	Sept	Oct	Nov	Dec	Year
1-Axis, E-W Horiz Axis	Average	2.6	2.9	3.0	3.3	3.5	3.8	3.9	3.7	3.2	3.5	2.9	2.7	3.3
	Min/Max	1.4/3.9	1.9/4.0	1.9/4.8	2.2/5.2	2.2/5.0	2.7/4.8	3.1/5.3	2.8/4.5	1.8/4.1	2.0/4.9	1.2/4.0	1.6/3.7	2.8/3.8
1-Axis, N-S Horiz Axis	Average	2.2	3.0	3.6	4.3	4.6	5.0	5.1	4.9	4.1	3.8	2.6	2.1	3.8
	Min/Max	1.2/3.3	2.0/4.0	2.3/5.7	2.9/6.9	2.9/6.6	3.5/6.2	4.0/7.0	3.6/6.0	2.2/5.1	2.1/5.3	1.0/3.6	1.3/3.0	3.2/4.4
1-Axis, N-S Tilt=Latitude	Average	2.9	3.6	4.1	4.4	4.4	4.6	4.8	4.9	4.4	4.5	3.4	2.9	4.1
	Min/Max	1.6/4.3	2.4/4.9	2.6/6.4	3.0/7.0	2.8/6.4	3.3/5.8	3.8/6.6	3.6/6.0	2.4/5.5	2.5/6.3	1.4/4.6	1.7/4.1	3.5/4.7
2-Axis	Average	3.1	3.7	4.1	4.5	4.7	5.0	5.2	5.0	4.4	4.5	3.6	3.1	4.2
	Min/Max	1.8/4.6	2.4/5.0	2.6/6.4	3.0/7.2	3.0/6.7	3.5/6.3	4.1/7.1	3.7/6.2	2.4/5.5	2.5/6.4	1.4/4.9	1.9/4.4	3.6/4.9

Average Climatic Conditions

Element	Jan	Feb	Mar	Apr	May	June	July	Aug	Sept	Oct	Nov	Dec	Year
Temperature (°C)	7.3	9.6	14.2	18.7	22.6	26.3	28.2	27.9	24.8	19.2	13.7	8.9	18.4
Daily Minimum Temp	1.6	3.3	7.7	12.3	16.7	20.6	22.4	21.8	18.9	12.4	7.4	2.9	12.3
Daily Maximum Temp	13.0	15.9	20.7	25.1	28.4	32.1	33.9	33.9	30.7	25.9	20.0	14.7	24.6
Record Minimum Temp	-16.1	-11.1	-6.7	-0.6	5.6	11.1	14.4	12.2	5.6	-1.7	-8.9	-15.0	-16.1
Record Maximum Temp	28.9	31.7	33.3	34.4	35.0	38.3	41.1	41.7	39.4	36.1	31.1	28.9	41.7
HDD, Base 18.3°C	346	249	146	38	0	0	0	0	0	35	147	297	1258
CDD, Base 18.3°C	3	4	17	48	133	240	305	296	195	61	8	4	1316
Relative Humidity (%)	73	70	68	70	73	73	72	72	74	72	74	74	72
Wind Speed (m/s)	3.9	4.2	4.4	4.2	3.7	3.3	3.1	2.9	3.2	3.2	3.8	3.9	3.6

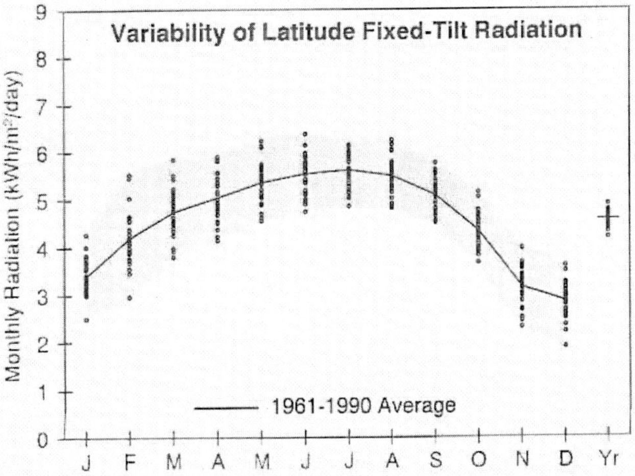

Boston, MA

WBAN NO. 14739

LATITUDE: 42.37° N
LONGITUDE: 71.03° W
ELEVATION: 5 meters
MEAN PRESSURE: 1015 millibars

STATION TYPE: Primary

Solar Radiation for Flat-Plate Collectors Facing South at a Fixed Tilt (kWh/m²/day), Uncertainty ±9%

Tilt (°)		Jan	Feb	Mar	Apr	May	June	July	Aug	Sept	Oct	Nov	Dec	Year
0	Average	1.9	2.7	3.7	4.7	5.6	6.1	6.1	5.4	4.3	3.0	1.9	1.5	3.9
	Min/Max	1.5/2.2	2.2/3.3	3.2/4.4	4.0/5.4	4.8/6.4	5.3/7.0	5.2/6.6	4.8/6.0	3.9/4.7	2.7/3.5	1.5/2.2	1.2/1.8	3.7/4.1
Latitude -15	Average	3.0	3.8	4.6	5.2	5.7	6.0	6.0	5.7	5.0	4.1	2.8	2.5	4.5
	Min/Max	2.2/3.7	2.8/5.0	3.7/5.6	4.3/6.0	4.8/6.6	5.2/6.9	5.2/6.6	5.0/6.5	4.5/5.7	3.5/4.8	2.1/3.5	1.7/3.1	4.2/4.8
Latitude	Average	3.4	4.2	4.7	5.0	5.3	5.5	5.6	5.5	5.1	4.3	3.1	2.9	4.6
	Min/Max	2.5/4.3	3.0/5.5	3.8/5.8	4.1/5.9	4.6/6.2	4.7/6.4	4.8/6.1	4.8/6.2	4.5/5.8	3.7/5.1	2.3/4.0	1.9/3.6	4.2/4.9
Latitude +15	Average	3.6	4.3	4.6	4.7	4.7	4.8	4.9	5.0	4.9	4.4	3.3	3.1	4.4
	Min/Max	2.6/4.6	3.0/5.8	3.7/5.8	3.8/5.5	4.1/5.5	4.1/5.5	4.3/5.3	4.4/5.7	4.3/5.5	3.7/5.2	2.4/4.2	2.0/3.9	4.0/4.7
90	Average	3.4	3.9	3.7	3.1	2.8	2.6	2.8	3.1	3.5	3.6	3.0	2.9	3.2
	Min/Max	2.4/4.5	2.6/5.3	2.8/4.9	2.6/3.7	2.5/3.2	2.4/2.9	2.5/3.0	2.8/3.5	3.1/4.0	3.0/4.3	2.1/3.8	1.8/3.8	2.9/3.5

Solar Radiation for 1-Axis Tracking Flat-Plate Collectors with a North-South Axis (kWh/m²/day), Uncertainty ±9%

Axis Tilt (°)		Jan	Feb	Mar	Apr	May	June	July	Aug	Sept	Oct	Nov	Dec	Year
0	Average	2.7	3.8	5.1	6.2	7.1	7.7	7.7	7.0	5.7	4.2	2.6	2.2	5.2
	Min/Max	2.0/3.5	2.8/5.1	3.9/6.2	5.0/7.4	5.7/8.6	6.4/9.2	6.1/8.8	5.8/8.0	5.0/6.7	3.5/5.0	2.0/3.3	1.5/2.7	4.8/5.5
Latitude -15	Average	3.6	4.7	5.7	6.5	7.3	7.7	7.8	7.3	6.3	5.0	3.4	2.9	5.7
	Min/Max	2.6/4.6	3.2/6.3	4.3/7.1	5.2/7.9	5.8/8.8	6.3/9.3	6.2/8.9	6.1/8.4	5.5/7.4	4.1/6.0	2.4/4.3	1.9/3.7	5.2/6.1
Latitude	Average	3.9	5.0	5.9	6.5	7.1	7.4	7.5	7.1	6.4	5.2	3.6	3.2	5.7
	Min/Max	2.8/5.0	3.4/6.8	4.4/7.3	5.1/7.8	5.6/8.5	6.1/8.9	5.9/8.6	5.9/8.3	5.5/7.5	4.3/6.3	2.6/4.7	2.0/4.1	5.2/6.2
Latitude +15	Average	4.1	5.1	5.8	6.2	6.6	6.9	7.0	6.8	6.2	5.2	3.7	3.4	5.6
	Min/Max	2.9/5.3	3.4/7.0	4.3/7.3	4.9/7.5	5.2/8.0	5.6/8.3	5.5/8.0	5.6/7.9	5.4/7.3	4.3/6.3	2.6/4.8	2.1/4.4	5.0/6.1

Solar Radiation for 2-Axis Tracking Flat-Plate Collectors (kWh/m²/day), Uncertainty ±9%

Tracker		Jan	Feb	Mar	Apr	May	June	July	Aug	Sept	Oct	Nov	Dec	Year
2-Axis	Average	4.1	5.1	5.9	6.6	7.4	7.9	7.9	7.3	6.4	5.3	3.8	3.4	5.9
	Min/Max	2.9/5.3	3.4/7.0	4.4/7.3	5.2/7.9	5.8/8.9	6.5/9.5	6.3/9.1	6.1/8.5	5.6/7.5	4.3/6.3	2.6/4.9	2.1/4.4	5.4/6.4

Direct Beam Solar Radiation for Concentrating Collectors (kWh/m²/day), Uncertainty ±8%

Tracker		Jan	Feb	Mar	Apr	May	June	July	Aug	Sept	Oct	Nov	Dec	Year
1-Axis, E-W Horiz Axis	Average	2.4	2.7	2.8	2.9	3.2	3.5	3.5	3.3	3.1	2.8	2.2	2.1	2.9
	Min/Max	1.6/3.1	1.5/4.2	1.9/3.6	1.9/4.0	2.0/4.4	2.5/4.7	2.3/4.5	2.2/4.2	2.5/3.9	2.1/3.6	1.3/3.1	1.0/2.8	2.4/3.2
1-Axis, N-S Horiz Axis	Average	1.7	2.3	3.1	3.7	4.2	4.5	4.6	4.2	3.6	2.7	1.6	1.3	3.1
	Min/Max	1.0/2.1	1.3/3.6	2.0/3.9	2.5/5.0	2.6/5.8	3.2/6.1	2.8/5.9	2.8/5.3	2.8/4.6	2.0/3.4	0.9/2.3	0.6/1.8	2.7/3.4
1-Axis, N-S Tilt=Latitude	Average	2.6	3.2	3.7	4.0	4.2	4.3	4.4	4.3	4.1	3.5	2.4	2.2	3.6
	Min/Max	1.7/3.4	1.7/5.0	2.5/4.8	2.6/5.4	2.6/5.7	3.0/5.8	2.7/5.7	2.9/5.5	3.3/5.3	2.6/4.4	1.4/3.4	1.0/2.9	3.0/3.9
2-Axis	Average	2.8	3.3	3.7	4.0	4.4	4.7	4.7	4.5	4.1	3.5	2.5	2.3	3.7
	Min/Max	1.8/3.6	1.8/5.1	2.5/4.8	2.7/5.5	2.7/6.0	3.3/6.3	2.9/6.1	3.0/5.7	3.3/5.3	2.7/4.5	1.5/3.6	1.1/3.2	3.2/4.1

Average Climatic Conditions

Element	Jan	Feb	Mar	Apr	May	June	July	Aug	Sept	Oct	Nov	Dec	Year
Temperature (°C)	-1.9	-0.9	3.7	8.9	14.6	19.8	23.1	22.2	18.2	12.7	7.4	0.9	10.7
Daily Minimum Temp	-5.8	-5.0	-0.4	4.6	9.9	15.1	18.4	17.8	13.8	8.3	3.5	-2.9	6.4
Daily Maximum Temp	2.1	3.1	7.7	13.3	19.2	24.6	27.7	26.6	22.7	17.1	11.2	4.7	15.0
Record Minimum Temp	-24.4	-20.0	-14.4	-8.9	1.1	7.2	10.0	8.3	3.3	-2.2	-9.4	-21.7	-24.4
Record Maximum Temp	17.2	21.1	27.2	34.4	35.0	37.8	38.9	38.9	37.8	32.2	25.6	22.8	38.9
HDD, Base 18.3°C	627	540	454	282	123	18	0	3	40	178	328	541	3134
CDD, Base 18.3°C	0	0	0	0	6	63	147	122	37	3	0	0	377
Relative Humidity (%)	62	62	63	63	67	68	68	71	72	68	67	65	67
Wind Speed (m/s)	6.2	6.1	6.1	5.9	5.5	5.1	4.9	4.8	5.1	5.4	5.8	6.1	5.6

Worchester, MA

WBAN NO. 94746

LATITUDE: 42.27° N
LONGITUDE: 71.87° W
ELEVATION: 301 meters
MEAN PRESSURE: 979 millibars

STATION TYPE: Secondary

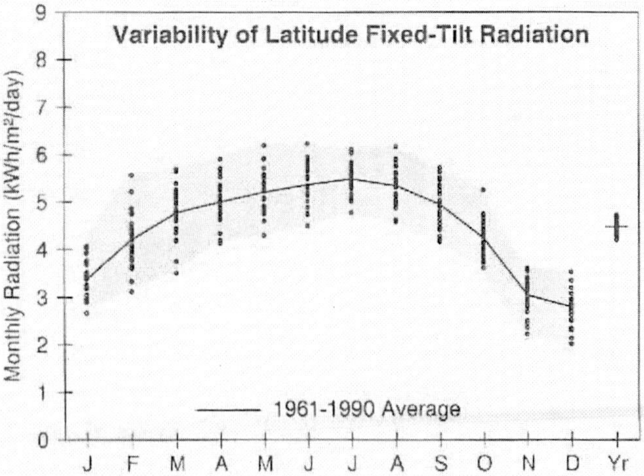

Solar Radiation for Flat-Plate Collectors Facing South at a Fixed Tilt (kWh/m²/day), Uncertainty ±11%

Tilt (°)		Jan	Feb	Mar	Apr	May	June	July	Aug	Sept	Oct	Nov	Dec	Year
0	Average	1.9	2.8	3.8	4.7	5.5	6.0	5.9	5.2	4.2	3.0	1.9	1.5	3.9
	Min/Max	1.6/2.2	2.2/3.4	3.0/4.3	4.0/5.4	4.6/6.4	5.0/6.8	5.3/6.6	4.6/5.9	3.7/4.7	2.7/3.5	1.5/2.1	1.3/1.8	3.7/4.0
Latitude -15	Average	3.0	3.8	4.6	5.1	5.5	5.8	5.9	5.6	4.9	4.0	2.8	2.4	4.5
	Min/Max	2.4/3.5	2.9/5.0	3.5/5.4	4.2/6.0	4.6/6.6	4.9/6.8	5.2/6.6	4.8/6.4	4.2/5.7	3.5/4.9	2.0/3.2	1.8/3.0	4.2/4.7
Latitude	Average	3.4	4.2	4.8	5.0	5.2	5.4	5.5	5.3	5.0	4.3	3.0	2.8	4.5
	Min/Max	2.7/4.1	3.1/5.6	3.5/5.7	4.1/5.9	4.3/6.2	4.5/6.2	4.8/6.1	4.6/6.2	4.2/5.7	3.6/5.3	2.2/3.6	2.0/3.5	4.2/4.7
Latitude +15	Average	3.6	4.4	4.7	4.6	4.6	4.6	4.8	4.8	4.7	4.3	3.2	3.0	4.3
	Min/Max	2.8/4.4	3.2/5.9	3.4/5.7	3.8/5.5	3.8/5.5	3.9/5.4	4.2/5.3	4.2/5.6	4.0/5.5	3.6/5.3	2.3/3.8	2.1/3.8	4.0/4.5
90	Average	3.5	4.0	3.8	3.1	2.7	2.6	2.7	3.0	3.4	3.5	2.9	2.8	3.2
	Min/Max	2.7/4.3	2.8/5.5	2.6/4.9	2.5/3.6	2.3/3.2	2.3/2.9	2.5/2.9	2.7/3.5	2.8/4.0	2.9/4.4	2.0/3.5	2.0/3.7	3.0/3.4

Solar Radiation for 1-Axis Tracking Flat-Plate Collectors with a North-South Axis (kWh/m²/day), Uncertainty ±11%

Axis Tilt (°)		Jan	Feb	Mar	Apr	May	June	July	Aug	Sept	Oct	Nov	Dec	Year
0	Average	2.7	3.9	5.1	6.1	6.9	7.5	7.5	6.8	5.6	4.1	2.6	2.2	5.1
	Min/Max	2.2/3.3	3.0/5.2	3.7/6.1	4.7/7.4	5.4/8.4	5.9/8.9	6.3/8.7	5.7/7.9	4.7/6.5	3.5/5.1	1.9/3.0	1.6/2.7	4.8/5.4
Latitude -15	Average	3.5	4.7	5.7	6.4	7.0	7.5	7.6	7.1	6.2	4.9	3.2	2.9	5.6
	Min/Max	2.8/4.3	3.5/6.4	4.1/6.9	4.9/8.0	5.5/8.6	5.8/8.9	6.3/8.8	5.8/8.4	5.1/7.3	4.1/6.1	2.3/3.8	2.0/3.7	5.2/5.9
Latitude	Average	3.9	5.0	5.9	6.4	6.8	7.2	7.3	6.9	6.2	5.1	3.5	3.1	5.6
	Min/Max	3.0/4.8	3.6/6.8	4.1/7.1	4.8/7.9	5.3/8.4	5.6/8.6	6.1/8.5	5.7/8.2	5.1/7.3	4.2/6.4	2.5/4.2	2.2/4.1	5.2/6.0
Latitude +15	Average	4.1	5.1	5.8	6.1	6.4	6.7	6.8	6.6	6.0	5.1	3.6	3.3	5.5
	Min/Max	3.2/5.0	3.7/7.0	4.0/7.1	4.6/7.6	4.9/7.9	5.2/8.0	5.7/7.9	5.4/7.8	4.9/7.2	4.2/6.5	2.5/4.3	2.3/4.3	5.1/5.8

Solar Radiation for 2-Axis Tracking Flat-Plate Collectors (kWh/m²/day), Uncertainty ±11%

Tracker		Jan	Feb	Mar	Apr	May	June	July	Aug	Sept	Oct	Nov	Dec	Year
2-Axis	Average	4.1	5.1	5.9	6.5	7.1	7.7	7.7	7.1	6.2	5.2	3.6	3.4	5.8
	Min/Max	3.2/5.1	3.7/7.0	4.1/7.1	4.9/8.0	5.6/8.7	6.0/9.1	6.5/8.9	5.9/8.4	5.1/7.3	4.2/6.5	2.5/4.3	2.3/4.4	5.4/6.2

Direct Beam Solar Radiation for Concentrating Collectors (kWh/m²/day), Uncertainty ±11%

Tracker		Jan	Feb	Mar	Apr	May	June	July	Aug	Sept	Oct	Nov	Dec	Year
1-Axis, E-W Horiz Axis	Average	2.2	2.5	2.6	2.8	3.0	3.3	3.3	3.1	2.9	2.7	2.0	1.9	2.7
	Min/Max	1.5/2.9	1.6/3.9	1.7/3.4	1.7/3.9	1.9/4.3	1.9/4.4	2.3/4.4	2.1/4.1	2.0/3.8	1.9/3.8	1.1/2.7	0.9/2.7	2.3/3.0
1-Axis, N-S Horiz Axis	Average	1.5	2.2	2.9	3.5	3.9	4.3	4.3	4.0	3.4	2.5	1.4	1.2	2.9
	Min/Max	1.0/2.0	1.4/3.4	1.8/3.7	2.0/5.0	2.5/5.5	2.5/5.7	3.0/5.7	2.8/5.3	2.3/4.4	1.8/3.6	0.8/2.0	0.6/1.7	2.6/3.2
1-Axis, N-S Tilt=Latitude	Average	2.4	3.0	3.5	3.8	3.8	4.1	4.1	4.1	3.9	3.3	2.2	2.0	3.3
	Min/Max	1.6/3.2	2.0/4.6	2.2/4.5	2.2/5.4	2.5/5.4	2.4/5.4	2.9/5.5	2.8/5.4	2.7/5.1	2.4/4.6	1.3/3.0	1.0/2.8	2.9/3.7
2-Axis	Average	2.6	3.1	3.5	3.8	4.1	4.4	4.5	4.2	3.9	3.4	2.3	2.1	3.5
	Min/Max	1.7/3.4	2.0/4.8	2.2/4.5	2.3/5.4	2.6/5.7	2.6/5.8	3.1/5.9	2.9/5.6	2.7/5.1	2.4/4.7	1.3/3.1	1.0/3.1	3.0/3.9

Average Climatic Conditions

Element	Jan	Feb	Mar	Apr	May	June	July	Aug	Sept	Oct	Nov	Dec	Year
Temperature (°C)	-5.1	-4.0	0.9	6.9	13.1	17.9	20.9	20.0	15.7	10.0	4.1	-2.6	8.2
Daily Minimum Temp	-9.4	-8.6	-3.9	1.5	7.3	12.2	15.6	14.8	10.3	4.7	-0.4	-6.6	3.1
Daily Maximum Temp	-0.7	0.6	5.8	12.2	18.8	23.6	26.3	25.2	20.9	15.3	8.6	1.5	13.2
Record Minimum Temp	-28.3	-24.4	-20.0	-11.7	-2.2	2.2	6.1	3.3	-1.1	-6.7	-14.4	-25.0	-28.3
Record Maximum Temp	15.6	19.4	27.2	32.8	33.3	34.4	35.6	35.6	32.8	29.4	25.6	21.1	35.6
HDD, Base 18.3°C	727	626	539	343	167	38	3	14	87	258	427	648	3877
CDD, Base 18.3°C	0	0	0	0	3	25	84	66	7	0	0	0	185
Relative Humidity (%)	65	64	63	60	64	69	70	73	74	69	70	70	68
Wind Speed (m/s)	4.9	4.8	4.8	4.6	4.1	3.8	3.5	3.4	3.6	4.0	4.5	4.6	4.2

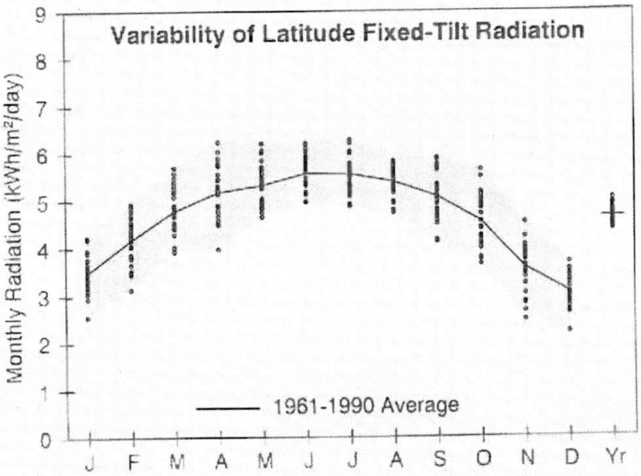

Baltimore, MD

WBAN NO. 93721

LATITUDE: 39.18° N
LONGITUDE: 76.67° W
ELEVATION: 47 meters
MEAN PRESSURE: 1012 millibars

STATION TYPE: Secondary

Solar Radiation for Flat-Plate Collectors Facing South at a Fixed Tilt (kWh/m²/day), Uncertainty ±9%

Tilt (°)		Jan	Feb	Mar	Apr	May	June	July	Aug	Sept	Oct	Nov	Dec	Year
0	Average	2.1	2.9	3.9	4.9	5.6	6.2	6.0	5.3	4.4	3.3	2.2	1.8	4.0
	Min/Max	1.7/2.4	2.4/3.2	3.4/4.5	3.9/5.8	5.0/6.5	5.5/6.9	5.3/6.8	4.7/5.7	3.7/5.0	2.9/3.9	1.8/2.6	1.5/2.0	3.8/4.3
Latitude -15	Average	3.1	3.8	4.6	5.3	5.7	6.0	6.0	5.6	5.0	4.3	3.2	2.7	4.6
	Min/Max	2.3/3.7	3.0/4.4	3.8/5.4	4.1/6.3	5.0/6.6	5.4/6.7	5.2/6.7	4.9/6.0	4.1/5.8	3.5/5.2	2.3/4.0	2.0/3.2	4.3/4.9
Latitude	Average	3.5	4.2	4.8	5.2	5.3	5.6	5.5	5.4	5.1	4.6	3.6	3.1	4.6
	Min/Max	2.5/4.2	3.1/4.9	3.9/5.7	4.0/6.2	4.7/6.2	5.0/6.2	4.9/6.3	4.7/5.8	4.1/5.9	3.6/5.6	2.5/4.5	2.2/3.7	4.4/5.0
Latitude +15	Average	3.7	4.3	4.7	4.8	4.7	4.8	4.9	4.9	4.8	4.6	3.7	3.3	4.4
	Min/Max	2.6/4.6	3.2/5.1	3.8/5.6	3.7/5.8	4.1/5.5	4.3/5.3	4.3/5.5	4.3/5.3	3.9/5.7	3.6/5.7	2.5/4.8	2.3/4.0	4.1/4.8
90	Average	3.4	3.7	3.5	3.0	2.6	2.4	2.5	2.9	3.3	3.7	3.3	3.0	3.1
	Min/Max	2.3/4.4	2.7/4.5	2.8/4.2	2.4/3.6	2.3/2.9	2.2/2.6	2.3/2.8	2.6/3.1	2.7/3.9	2.8/4.6	2.2/4.3	2.0/3.7	2.9/3.4

Solar Radiation for 1-Axis Tracking Flat-Plate Collectors with a North-South Axis (kWh/m²/day), Uncertainty ±9%

Axis Tilt (°)		Jan	Feb	Mar	Apr	May	June	July	Aug	Sept	Oct	Nov	Dec	Year
0	Average	3.0	4.0	5.3	6.4	7.1	7.8	7.7	6.9	5.8	4.6	3.2	2.5	5.4
	Min/Max	2.3/3.7	3.0/4.8	4.2/6.4	4.8/8.1	6.1/8.5	6.5/9.0	6.3/8.9	5.7/7.5	4.7/6.9	3.6/5.7	2.2/4.0	1.9/3.0	5.0/5.9
Latitude -15	Average	3.7	4.7	5.9	6.8	7.2	7.8	7.7	7.1	6.3	5.3	3.9	3.2	5.8
	Min/Max	2.7/4.6	3.4/5.7	4.6/7.2	4.9/8.6	6.1/8.6	6.5/9.0	6.3/9.0	5.9/7.8	5.1/7.5	4.2/6.7	2.6/5.0	2.3/3.9	5.4/6.4
Latitude	Average	4.1	5.0	6.0	6.7	7.0	7.5	7.4	7.0	6.4	5.5	4.2	3.5	5.9
	Min/Max	2.9/5.1	3.6/6.1	4.7/7.3	4.9/8.5	5.9/8.4	6.2/8.6	6.1/8.6	5.8/7.7	5.1/7.6	4.3/7.0	2.8/5.4	2.4/4.3	5.4/6.4
Latitude +15	Average	4.3	5.1	5.9	6.4	6.6	7.0	6.9	6.6	6.2	5.6	4.3	3.7	5.7
	Min/Max	3.0/5.3	3.6/6.3	4.6/7.3	4.7/8.2	5.5/7.9	5.8/8.1	5.6/8.1	5.4/7.3	4.9/7.4	4.3/7.1	2.8/5.6	2.5/4.6	5.3/6.3

Solar Radiation for 2-Axis Tracking Flat-Plate Collectors (kWh/m²/day), Uncertainty ±9%

Tracker		Jan	Feb	Mar	Apr	May	June	July	Aug	Sept	Oct	Nov	Dec	Year
2-Axis	Average	4.3	5.1	6.0	6.8	7.3	8.0	7.8	7.2	6.4	5.6	4.3	3.7	6.0
	Min/Max	3.0/5.4	3.6/6.3	4.7/7.3	5.0/8.6	6.2/8.7	6.7/9.2	6.4/9.1	5.9/7.8	5.1/7.6	4.3/7.1	2.8/5.7	2.5/4.6	5.6/6.7

Direct Beam Solar Radiation for Concentrating Collectors (kWh/m²/day), Uncertainty ±8%

Tracker		Jan	Feb	Mar	Apr	May	June	July	Aug	Sept	Oct	Nov	Dec	Year
1-Axis, E-W Horiz Axis	Average	2.5	2.8	3.0	3.1	3.3	3.7	3.6	3.2	3.0	3.0	2.5	2.2	3.0
	Min/Max	1.4/3.5	1.5/3.7	1.8/4.1	1.6/4.4	2.5/4.4	2.7/4.6	2.5/4.6	2.3/3.8	2.0/3.9	2.0/4.2	1.3/3.6	1.2/3.0	2.7/3.5
1-Axis, N-S Horiz Axis	Average	1.9	2.5	3.4	4.0	4.3	4.8	4.7	4.2	3.6	3.0	2.0	1.5	3.3
	Min/Max	1.1/2.5	1.4/3.4	2.1/4.7	2.2/5.8	3.3/5.8	3.4/6.1	3.1/6.0	2.9/5.0	2.5/4.7	1.9/4.2	1.0/2.9	0.8/2.1	2.9/3.9
1-Axis, N-S Tilt=Latitude	Average	2.8	3.4	3.9	4.2	4.2	4.5	4.5	4.2	4.1	3.8	2.8	2.4	3.7
	Min/Max	1.6/3.8	1.8/4.5	2.4/5.4	2.3/6.1	3.2/5.6	3.2/5.7	3.0/5.7	3.0/5.0	2.8/5.2	2.5/5.3	1.4/4.1	1.3/3.2	3.3/4.4
2-Axis	Average	3.0	3.4	4.0	4.3	4.4	4.9	4.8	4.4	4.1	3.8	3.0	2.6	3.9
	Min/Max	1.7/4.1	1.9/4.6	2.4/5.5	2.4/6.2	3.4/5.9	3.5/6.2	3.2/6.1	3.1/5.2	2.8/5.3	2.5/5.3	1.5/4.3	1.4/3.5	3.5/4.6

Average Climatic Conditions

Element	Jan	Feb	Mar	Apr	May	June	July	Aug	Sept	Oct	Nov	Dec	Year
Temperature (°C)	-0.1	1.6	6.7	11.9	17.4	22.5	25.0	24.2	20.3	13.7	8.2	2.6	12.8
Daily Minimum Temp	-4.8	-3.4	1.2	5.8	11.4	16.6	19.3	18.7	14.7	7.7	2.8	-2.1	7.3
Daily Maximum Temp	4.6	6.5	12.2	17.9	23.4	28.4	30.7	29.7	25.8	19.6	13.6	7.3	18.3
Record Minimum Temp	-21.7	-19.4	-14.4	-6.7	0.0	4.4	10.0	7.2	1.7	-3.9	-10.6	-17.8	-21.7
Record Maximum Temp	23.9	26.1	30.6	34.4	36.7	37.8	40.0	40.6	37.8	33.3	28.3	25.0	40.6
HDD, Base 18.3°C	572	470	360	193	60	0	0	0	16	153	303	487	2615
CDD, Base 18.3°C	0	0	0	0	33	126	207	183	74	9	0	0	632
Relative Humidity (%)	63	61	59	59	66	68	69	71	71	69	67	66	66
Wind Speed (m/s)	4.2	4.4	4.6	4.6	3.9	3.6	3.4	3.4	3.5	3.7	4.0	4.1	3.9

Caribou, ME

WBAN NO. 14607

LATITUDE: 46.87° N
LONGITUDE: 68.02° W
ELEVATION: 190 meters
MEAN PRESSURE: 991 millibars

STATION TYPE: Primary

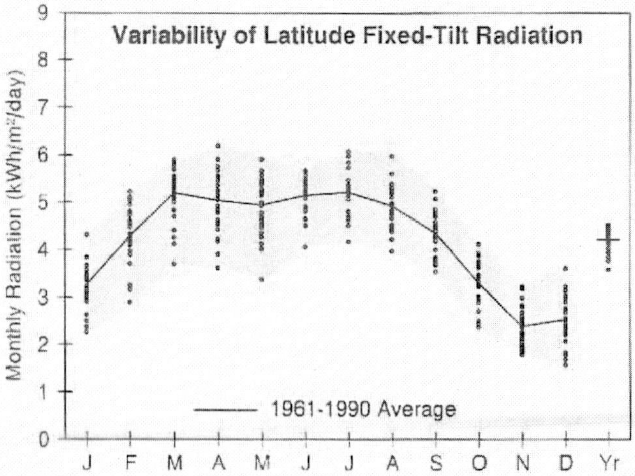

Solar Radiation for Flat-Plate Collectors Facing South at a Fixed Tilt (kWh/m²/day), Uncertainty ±9%

Tilt (°)		Jan	Feb	Mar	Apr	May	June	July	Aug	Sept	Oct	Nov	Dec	Year
0	Average	1.6	2.6	3.8	4.6	5.2	5.7	5.6	4.8	3.6	2.3	1.4	1.2	3.6
	Min/Max	1.3/1.8	1.9/2.9	3.0/4.3	3.5/5.4	3.6/6.2	4.7/6.4	4.6/6.5	4.0/5.7	3.1/4.2	1.8/2.7	1.1/1.7	0.9/1.5	3.1/3.7
Latitude -15	Average	2.9	3.9	5.0	5.1	5.3	5.6	5.6	5.2	4.3	3.1	2.2	2.2	4.2
	Min/Max	2.0/3.7	2.7/4.7	3.6/5.6	3.8/6.2	3.6/6.3	4.4/6.2	4.5/6.6	4.2/6.2	3.6/5.2	2.3/3.9	1.6/2.9	1.4/3.1	3.6/4.5
Latitude	Average	3.3	4.3	5.2	5.0	4.9	5.1	5.2	4.9	4.3	3.3	2.4	2.5	4.2
	Min/Max	2.3/4.3	2.9/5.2	3.7/5.9	3.6/6.2	3.4/5.9	4.1/5.7	4.2/6.1	4.0/6.0	3.5/5.2	2.4/4.1	1.8/3.2	1.6/3.6	3.6/4.5
Latitude +15	Average	3.5	4.5	5.2	4.7	4.4	4.5	4.6	4.5	4.1	3.3	2.5	2.7	4.0
	Min/Max	2.4/4.7	3.0/5.5	3.6/5.9	3.3/5.9	3.0/5.2	3.5/4.9	3.6/5.3	3.6/5.4	3.3/5.0	2.3/4.2	1.8/3.4	1.6/3.9	3.4/4.3
90	Average	3.4	4.2	4.6	3.7	2.9	2.8	2.9	3.0	3.2	2.8	2.3	2.7	3.2
	Min/Max	2.3/4.7	2.8/5.2	3.1/5.3	2.4/4.8	2.0/3.4	2.3/3.0	2.3/3.2	2.5/3.7	2.5/3.8	2.0/3.6	1.6/3.2	1.5/3.9	2.6/3.5

Solar Radiation for 1-Axis Tracking Flat-Plate Collectors with a North-South Axis (kWh/m²/day), Uncertainty ±9%

Axis Tilt (°)		Jan	Feb	Mar	Apr	May	June	July	Aug	Sept	Oct	Nov	Dec	Year
0	Average	2.5	3.8	5.4	6.1	6.8	7.3	7.3	6.4	5.0	3.1	1.9	1.9	4.8
	Min/Max	1.8/3.2	2.7/4.6	3.7/6.2	4.1/7.6	4.2/8.1	5.4/8.6	5.8/9.2	5.1/7.8	3.9/5.8	2.3/4.0	1.5/2.6	1.1/2.6	4.0/5.1
Latitude -15	Average	3.4	4.8	6.3	6.6	7.0	7.4	7.4	6.7	5.5	3.8	2.5	2.6	5.3
	Min/Max	2.4/4.6	3.3/5.9	4.3/7.3	4.4/8.3	4.2/8.3	5.4/8.6	5.8/9.4	5.4/8.3	4.3/6.7	2.7/4.8	1.9/3.5	1.5/3.8	4.4/5.7
Latitude	Average	3.7	5.1	6.5	6.5	6.7	7.1	7.2	6.6	5.6	3.9	2.7	2.8	5.4
	Min/Max	2.6/5.1	3.4/6.4	4.3/7.5	4.3/8.3	4.1/8.1	5.1/8.3	5.6/9.1	5.2/8.2	4.3/6.7	2.8/5.1	2.0/3.8	1.7/4.2	4.4/5.8
Latitude +15	Average	3.9	5.2	6.5	6.3	6.3	6.6	6.7	6.3	5.4	3.9	2.8	3.0	5.2
	Min/Max	2.7/5.4	3.5/6.5	4.3/7.5	4.1/8.1	3.8/7.6	4.8/7.8	5.3/8.6	5.0/7.8	4.2/6.5	2.8/5.1	2.0/3.9	1.7/4.4	4.2/5.7

Solar Radiation for 2-Axis Tracking Flat-Plate Collectors (kWh/m²/day), Uncertainty ±9%

Tracker		Jan	Feb	Mar	Apr	May	June	July	Aug	Sept	Oct	Nov	Dec	Year
2-Axis	Average	4.0	5.2	6.5	6.6	7.1	7.6	7.6	6.8	5.6	4.0	2.8	3.0	5.6
	Min/Max	2.7/5.4	3.5/6.5	4.3/7.5	4.4/8.3	4.3/8.4	5.6/8.9	6.0/9.6	5.4/8.4	4.3/6.7	2.8/5.1	2.0/3.9	1.7/4.5	4.5/6.0

Direct Beam Solar Radiation for Concentrating Collectors (kWh/m²/day), Uncertainty ±8%

Tracker		Jan	Feb	Mar	Apr	May	June	July	Aug	Sept	Oct	Nov	Dec	Year
1-Axis, E-W	Average	2.1	2.5	2.7	2.6	2.9	3.2	3.2	2.9	2.5	1.9	1.4	1.6	2.5
Horiz Axis	Min/Max	1.1/3.5	1.4/3.6	1.5/3.7	1.4/3.6	1.5/4.0	1.7/4.1	2.4/4.7	2.2/4.1	1.7/3.4	1.0/2.8	0.8/2.5	0.8/2.7	1.9/2.8
1-Axis, N-S	Average	1.3	2.1	2.9	3.3	3.9	4.2	4.3	3.8	3.0	1.8	0.9	0.9	2.7
Horiz Axis	Min/Max	0.7/2.2	1.1/2.9	1.5/4.0	1.7/4.5	1.8/5.4	2.3/5.5	3.2/6.4	2.9/5.3	2.0/4.0	0.9/2.7	0.5/1.8	0.4/1.6	2.0/3.0
1-Axis, N-S	Average	2.3	3.0	3.7	3.6	3.9	4.0	4.2	4.0	3.5	2.4	1.5	1.7	3.1
Tilt=Latitude	Min/Max	1.3/3.7	1.6/4.3	1.9/4.9	1.9/5.0	1.8/5.4	2.2/5.3	3.1/6.2	3.1/5.6	2.3/4.7	1.3/3.6	0.9/2.8	0.8/2.9	2.3/3.5
2-Axis	Average	2.4	3.1	3.7	3.7	4.1	4.4	4.5	4.1	3.5	2.4	1.6	1.8	3.3
	Min/Max	1.3/4.0	1.7/4.4	2.0/4.9	1.9/5.1	1.9/5.7	2.3/5.8	3.3/6.7	3.2/5.8	2.3/4.7	1.3/3.7	0.9/3.0	0.9/3.1	2.4/3.7

Average Climatic Conditions

Element	Jan	Feb	Mar	Apr	May	June	July	Aug	Sept	Oct	Nov	Dec	Year
Temperature (°C)	-12.8	-11.2	-4.1	3.3	10.5	15.8	18.6	17.1	12.1	6.2	-0.7	-9.6	3.8
Daily Minimum Temp	-18.7	-17.4	-9.5	-1.7	4.5	9.5	12.5	11.2	6.2	1.3	-4.6	-14.7	-1.8
Daily Maximum Temp	-7.0	-5.0	1.3	8.2	16.5	22.2	24.7	23.1	17.8	11.1	3.1	-4.4	9.3
Record Minimum Temp	-35.6	-40.6	-28.9	-18.9	-7.8	-1.1	2.2	1.1	-5.0	-10.0	-20.6	-35.0	-40.6
Record Maximum Temp	11.1	11.1	22.8	30.0	35.6	35.6	35.0	35.0	32.8	26.1	20.0	14.4	35.6
HDD, Base 18.3°C	966	826	696	452	243	79	34	64	191	376	572	864	5362
CDD, Base 18.3°C	0	0	0	0	0	4	42	26	0	0	0	0	73
Relative Humidity (%)	72	70	69	69	66	70	75	78	78	77	80	77	73
Wind Speed (m/s)	5.1	5.0	5.3	5.1	4.8	4.4	4.1	3.8	4.2	4.5	4.7	4.9	4.6

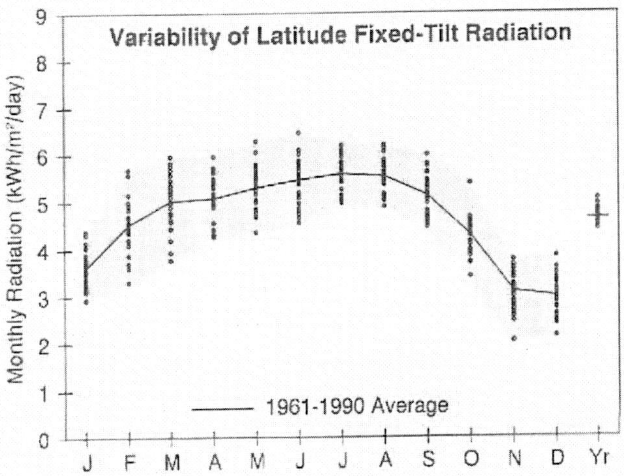

Portland, ME

WBAN NO. 14764

LATITUDE: 43.65° N
LONGITUDE: 70.32° W
ELEVATION: 19 meters
MEAN PRESSURE: 1013 millibars

STATION TYPE: Secondary

Solar Radiation for Flat-Plate Collectors Facing South at a Fixed Tilt (kWh/m²/day), Uncertainty ±11%

Tilt (°)		Jan	Feb	Mar	Apr	May	June	July	Aug	Sept	Oct	Nov	Dec	Year
0	Average	1.9	2.8	3.8	4.7	5.6	6.1	6.0	5.4	4.2	2.9	1.8	1.5	3.9
	Min/Max	1.7/2.2	2.2/3.3	3.1/4.5	4.1/5.4	4.6/6.6	5.2/7.2	5.4/6.7	4.8/5.9	3.9/4.8	2.5/3.5	1.4/2.0	1.2/1.8	3.7/4.2
Latitude -15	Average	3.1	4.1	4.8	5.2	5.7	6.0	6.0	5.8	5.1	4.0	2.8	2.6	4.6
	Min/Max	2.6/3.8	3.1/5.1	3.7/5.7	4.4/6.1	4.7/6.7	5.0/7.1	5.3/6.7	5.1/6.4	4.5/5.9	3.2/5.0	1.9/3.3	1.9/3.3	4.4/5.0
Latitude	Average	3.6	4.5	5.0	5.1	5.3	5.5	5.6	5.5	5.1	4.3	3.1	3.0	4.6
	Min/Max	2.9/4.4	3.3/5.7	3.8/6.0	4.3/6.0	4.4/6.3	4.6/6.5	5.0/6.2	4.9/6.2	4.4/6.0	3.4/5.4	2.0/3.8	2.1/3.8	4.4/5.0
Latitude +15	Average	3.9	4.7	5.0	4.7	4.7	4.7	4.9	5.0	4.9	4.3	3.2	3.2	4.4
	Min/Max	3.1/4.8	3.4/6.0	3.7/6.0	3.9/5.6	3.9/5.5	4.0/5.6	4.4/5.4	4.4/5.6	4.2/5.8	3.4/5.4	2.1/4.0	2.3/4.2	4.2/4.8
90	Average	3.8	4.4	4.1	3.2	2.8	2.7	2.8	3.2	3.6	3.6	2.9	3.1	3.4
	Min/Max	3.0/4.7	3.0/5.6	2.9/5.2	2.7/4.0	2.4/3.2	2.3/3.0	2.6/3.0	2.8/3.6	3.1/4.3	2.8/4.6	1.8/3.6	2.1/4.2	3.1/3.7

Solar Radiation for 1-Axis Tracking Flat-Plate Collectors with a North-South Axis (kWh/m²/day), Uncertainty ±11%

Axis Tilt (°)		Jan	Feb	Mar	Apr	May	June	July	Aug	Sept	Oct	Nov	Dec	Year
0	Average	2.9	4.1	5.4	6.2	7.2	7.8	7.8	7.2	5.8	4.2	2.6	2.3	5.3
	Min/Max	2.4/3.5	3.0/5.1	4.0/6.4	5.1/7.5	5.7/9.0	6.1/9.7	6.4/9.0	6.3/8.2	5.0/6.8	3.3/5.3	1.7/3.1	1.6/2.9	5.0/5.8
Latitude -15	Average	3.8	5.1	6.1	6.7	7.4	7.9	7.9	7.5	6.5	5.0	3.3	3.1	5.9
	Min/Max	3.1/4.7	3.6/6.5	4.5/7.3	5.4/8.0	5.8/9.3	6.1/9.7	6.4/9.1	6.6/8.7	5.5/7.6	3.8/6.4	2.1/4.0	2.1/4.0	5.5/6.4
Latitude	Average	4.2	5.4	6.3	6.6	7.1	7.5	7.6	7.4	6.6	5.2	3.6	3.4	5.9
	Min/Max	3.4/5.2	3.8/6.9	4.6/7.6	5.3/8.0	5.5/9.0	5.8/9.4	6.2/8.8	6.4/8.6	5.5/7.7	4.0/6.7	2.2/4.4	2.3/4.4	5.5/6.5
Latitude +15	Average	4.4	5.6	6.2	6.3	6.7	7.0	7.1	7.0	6.4	5.2	3.7	3.6	5.8
	Min/Max	3.5/5.5	3.9/7.2	4.5/7.6	5.0/7.7	5.2/8.5	5.4/8.8	5.8/8.3	6.1/8.1	5.3/7.5	4.0/6.8	2.3/4.6	2.4/4.7	5.4/6.4

Solar Radiation for 2-Axis Tracking Flat-Plate Collectors (kWh/m²/day), Uncertainty ±11%

Tracker		Jan	Feb	Mar	Apr	May	June	July	Aug	Sept	Oct	Nov	Dec	Year
2-Axis	Average	4.5	5.6	6.3	6.7	7.5	8.1	8.1	7.6	6.6	5.3	3.7	3.6	6.1
	Min/Max	3.6/5.5	3.9/7.2	4.6/7.6	5.4/8.1	5.8/9.4	6.3/10.0	6.6/9.3	6.6/8.8	5.6/7.7	4.0/6.8	2.3/4.6	2.5/4.8	5.7/6.7

Direct Beam Solar Radiation for Concentrating Collectors (kWh/m²/day), Uncertainty ±11%

Tracker		Jan	Feb	Mar	Apr	May	June	July	Aug	Sept	Oct	Nov	Dec	Year
1-Axis, E-W Horiz Axis	Average	2.5	2.9	2.9	3.0	3.3	3.7	3.8	3.6	3.2	2.9	2.1	2.1	3.0
	Min/Max	1.8/3.2	2.0/4.1	1.8/3.8	2.1/4.0	2.2/4.9	2.3/5.2	2.6/4.6	2.9/4.4	2.4/4.1	1.9/4.1	1.1/2.8	1.2/2.9	2.8/3.4
1-Axis, N-S Horiz Axis	Average	1.7	2.5	3.2	3.8	4.4	4.9	4.9	4.6	3.8	2.7	1.5	1.3	3.3
	Min/Max	1.2/2.2	1.6/3.4	2.0/4.1	2.6/5.1	2.9/6.4	2.9/6.8	3.3/6.1	3.8/5.8	2.8/4.9	1.7/4.0	0.8/2.1	0.7/1.8	3.0/3.7
1-Axis, N-S Tilt=Latitude	Average	2.8	3.5	3.9	4.1	4.3	4.6	4.7	4.8	4.4	3.6	2.4	2.2	3.8
	Min/Max	1.9/3.5	2.3/4.8	2.4/5.0	2.8/5.5	2.9/6.4	2.8/6.5	3.2/5.9	3.9/6.0	3.2/5.7	2.3/5.2	1.2/3.2	1.3/3.1	3.4/4.3
2-Axis	Average	2.9	3.6	3.9	4.2	4.6	5.0	5.1	4.9	4.4	3.6	2.5	2.4	3.9
	Min/Max	2.1/3.8	2.4/5.0	2.4/5.0	2.9/5.6	3.1/6.7	3.0/7.0	3.4/6.3	4.0/6.2	3.3/5.7	2.4/5.3	1.3/3.3	1.4/3.3	3.6/4.5

Average Climatic Conditions

Element	Jan	Feb	Mar	Apr	May	June	July	Aug	Sept	Oct	Nov	Dec	Year
Temperature (°C)	-6.2	-4.8	0.6	6.3	11.8	16.9	20.3	19.6	15.1	9.2	3.7	-3.1	7.4
Daily Minimum Temp	-11.4	-10.3	-4.2	1.2	6.3	11.2	14.6	13.9	9.4	3.5	-0.9	-7.9	2.1
Daily Maximum Temp	-0.9	0.6	5.2	11.3	17.3	22.6	26.0	25.2	20.7	14.8	8.3	1.7	12.7
Record Minimum Temp	-32.2	-39.4	-29.4	-13.3	-5.0	0.6	4.4	0.6	-5.0	-9.4	-16.1	-29.4	-39.4
Record Maximum Temp	17.8	17.8	30.0	29.4	34.4	36.7	37.2	39.4	35.0	31.1	23.3	20.6	39.4
HDD, Base 18.3°C	761	649	551	362	202	56	6	22	105	284	438	663	4099
CDD, Base 18.3°C	0	0	0	0	0	12	68	62	7	0	0	0	149
Relative Humidity (%)	67	65	66	67	71	75	75	76	77	74	73	70	71
Wind Speed (m/s)	4.1	4.1	4.4	4.4	4.1	3.8	3.4	3.4	3.5	3.7	4.0	3.9	3.9

Alpena, MI

WBAN NO. 94849

LATITUDE: 45.07° N
LONGITUDE: 83.57° W
ELEVATION: 210 meters
MEAN PRESSURE: 990 millibars

STATION TYPE: Secondary

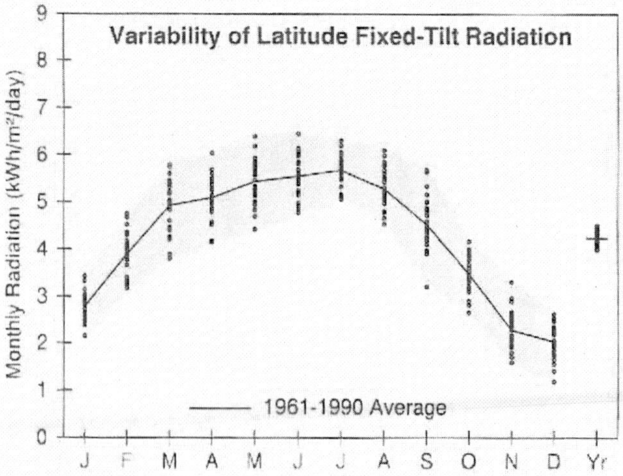

Solar Radiation for Flat-Plate Collectors Facing South at a Fixed Tilt (kWh/m²/day), Uncertainty ±9%

Tilt (°)		Jan	Feb	Mar	Apr	May	June	July	Aug	Sept	Oct	Nov	Dec	Year
0	Average	1.6	2.5	3.7	4.7	5.7	6.2	6.1	5.1	3.8	2.5	1.5	1.2	3.7
	Min/Max	1.4/1.8	2.2/2.9	3.1/4.3	4.0/5.3	4.7/6.6	5.4/7.2	5.5/6.8	4.5/5.8	3.0/4.4	2.2/2.8	1.2/1.8	0.9/1.4	3.6/3.9
Latitude -15	Average	2.5	3.6	4.7	5.2	5.8	6.1	6.1	5.5	4.5	3.3	2.1	1.8	4.3
	Min/Max	2.0/3.0	3.0/4.3	3.7/5.5	4.3/6.1	4.8/6.8	5.2/7.1	5.5/6.8	4.7/6.3	3.3/5.6	2.6/3.9	1.5/3.0	1.1/2.3	4.1/4.5
Latitude	Average	2.8	3.9	4.9	5.1	5.4	5.5	5.7	5.3	4.5	3.5	2.3	2.0	4.2
	Min/Max	2.2/3.4	3.2/4.8	3.8/5.8	4.2/6.0	4.4/6.4	4.8/6.4	5.1/6.3	4.5/6.1	3.2/5.7	2.7/4.2	1.6/3.3	1.2/2.6	4.0/4.5
Latitude +15	Average	2.9	4.0	4.9	4.7	4.8	4.8	4.9	4.8	4.3	3.5	2.3	2.1	4.0
	Min/Max	2.3/3.7	3.3/5.0	3.7/5.8	3.8/5.7	3.9/5.7	4.1/5.6	4.4/5.5	4.1/5.5	3.0/5.5	2.6/4.2	1.6/3.5	1.2/2.8	3.8/4.3
90	Average	2.9	3.8	4.2	3.4	3.0	2.8	2.9	3.1	3.2	2.9	2.1	2.1	3.0
	Min/Max	2.2/3.6	3.0/4.7	3.0/5.1	2.7/4.2	2.5/3.4	2.5/3.1	2.7/3.2	2.7/3.6	2.2/4.2	2.1/3.5	1.4/3.2	1.1/2.8	2.8/3.3

Solar Radiation for 1-Axis Tracking Flat-Plate Collectors with a North-South Axis (kWh/m²/day), Uncertainty ±9%

Axis Tilt (°)		Jan	Feb	Mar	Apr	May	June	July	Aug	Sept	Oct	Nov	Dec	Year
0	Average	2.2	3.6	5.2	6.3	7.6	8.2	8.2	7.0	5.2	3.4	1.9	1.6	5.0
	Min/Max	1.8/2.8	2.8/4.3	4.0/6.1	5.0/7.4	5.8/9.2	6.8/10.0	7.1/9.3	5.9/8.4	3.5/6.5	2.6/4.1	1.4/2.7	1.0/2.0	4.7/5.4
Latitude -15	Average	2.9	4.3	6.0	6.8	7.8	8.2	8.4	7.4	5.7	4.0	2.4	2.1	5.5
	Min/Max	2.2/3.7	3.4/5.4	4.5/7.1	5.3/8.1	5.9/9.6	6.8/10.1	7.2/9.5	6.2/8.9	3.8/7.4	3.0/4.9	1.6/3.6	1.2/2.7	5.2/5.9
Latitude	Average	3.1	4.6	6.1	6.7	7.5	7.9	8.1	7.2	5.8	4.2	2.5	2.2	5.5
	Min/Max	2.4/4.0	3.6/5.7	4.6/7.3	5.2/8.1	5.7/9.3	6.6/9.7	7.0/9.2	6.0/8.8	3.8/7.5	3.0/5.2	1.7/3.8	1.2/2.9	5.1/5.9
Latitude +15	Average	3.3	4.7	6.1	6.4	7.1	7.4	7.6	6.9	5.6	4.2	2.6	2.3	5.4
	Min/Max	2.5/4.2	3.6/5.8	4.5/7.3	4.9/7.8	5.4/8.8	6.1/9.1	6.5/8.6	5.7/8.4	3.6/7.3	3.0/5.2	1.7/4.0	1.3/3.1	5.0/5.7

Solar Radiation for 2-Axis Tracking Flat-Plate Collectors (kWh/m²/day), Uncertainty ±9%

Tracker		Jan	Feb	Mar	Apr	May	June	July	Aug	Sept	Oct	Nov	Dec	Year
2-Axis	Average	3.3	4.7	6.1	6.8	7.9	8.5	8.6	7.4	5.8	4.2	2.6	2.4	5.7
	Min/Max	2.5/4.2	3.6/5.8	4.6/7.3	5.3/8.1	6.0/9.7	7.0/10.4	7.4/9.7	6.2/9.0	3.8/7.5	3.0/5.2	1.7/4.0	1.3/3.1	5.3/6.1

Direct Beam Solar Radiation for Concentrating Collectors (kWh/m²/day), Uncertainty ±8%

Tracker		Jan	Feb	Mar	Apr	May	June	July	Aug	Sept	Oct	Nov	Dec	Year
1-Axis, E-W Horiz Axis	Average	1.5	2.1	2.7	3.1	3.7	4.0	4.1	3.4	2.7	2.1	1.2	1.0	2.6
	Min/Max	0.8/2.3	1.4/2.8	1.9/3.5	2.0/4.0	2.2/5.2	2.7/5.6	3.2/5.0	2.5/4.6	1.3/4.1	1.0/2.8	0.6/2.4	0.4/1.5	2.3/3.0
1-Axis, N-S Horiz Axis	Average	1.0	1.7	2.9	3.9	4.9	5.3	5.5	4.5	3.2	1.9	0.9	0.6	3.0
	Min/Max	0.5/1.5	1.1/2.4	2.0/4.0	2.5/5.1	2.9/6.9	3.7/7.4	4.2/6.6	3.3/6.1	1.6/4.8	1.0/2.7	0.4/1.8	0.3/1.0	2.7/3.5
1-Axis, N-S Tilt=Latitude	Average	1.6	2.5	3.6	4.3	4.9	5.1	5.3	4.7	3.7	2.6	1.4	1.0	3.4
	Min/Max	0.9/2.5	1.6/3.4	2.5/4.8	2.7/5.6	2.9/6.8	3.5/7.1	4.1/6.4	3.4/6.4	1.8/5.6	1.3/3.6	0.7/2.7	0.4/1.6	3.0/3.9
2-Axis	Average	1.7	2.5	3.6	4.3	5.2	5.5	5.7	4.9	3.7	2.6	1.5	1.1	3.5
	Min/Max	0.9/2.7	1.6/3.5	2.5/4.8	2.8/5.6	3.0/7.2	3.8/7.7	4.4/6.8	3.5/6.6	1.8/5.6	1.3/3.6	0.8/2.8	0.5/1.8	3.1/4.1

Average Climatic Conditions

Element	Jan	Feb	Mar	Apr	May	June	July	Aug	Sept	Oct	Nov	Dec	Year
Temperature (°C)	-8.0	-7.6	-2.2	4.9	11.1	16.3	19.5	18.2	14.1	8.3	2.1	-4.6	6.0
Daily Minimum Temp	-12.9	-13.2	-7.8	-1.1	4.0	8.9	12.2	11.4	7.9	2.8	-2.0	-8.6	0.2
Daily Maximum Temp	-3.1	-2.1	3.3	10.9	18.2	23.6	26.8	24.9	20.2	13.8	6.1	-0.6	11.8
Record Minimum Temp	-33.3	-38.3	-32.8	-17.8	-6.7	-2.2	1.1	-1.1	-3.9	-8.9	-18.3	-27.8	-38.3
Record Maximum Temp	11.1	18.3	24.4	32.2	34.4	36.7	38.9	38.9	34.4	31.1	24.4	18.3	38.9
HDD, Base 18.3°C	816	727	637	402	228	81	24	44	132	312	488	711	4602
CDD, Base 18.3°C	0	0	0	0	4	19	60	41	3	0	0	0	128
Relative Humidity (%)	76	73	72	67	66	70	71	76	78	76	78	79	73
Wind Speed (m/s)	4.1	3.9	4.2	4.2	3.8	3.5	3.3	3.1	3.3	3.6	4.0	3.9	3.8

Detroit, MI

WBAN NO. 94847

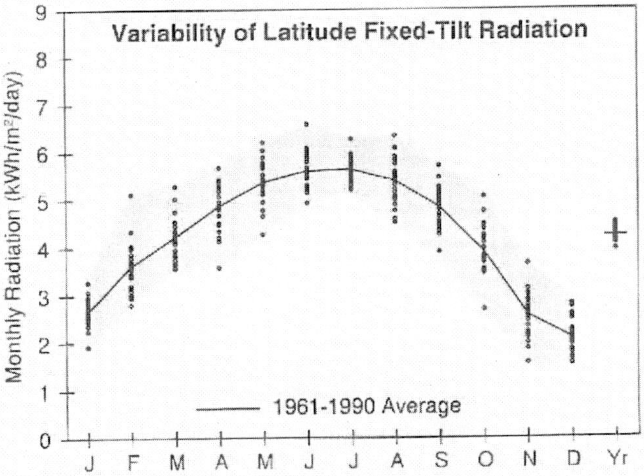

LATITUDE: 42.42° N
LONGITUDE: 83.02° W
ELEVATION: 191 meters
MEAN PRESSURE: 992 millibars

STATION TYPE: Secondary

Solar Radiation for Flat-Plate Collectors Facing South at a Fixed Tilt (kWh/m²/day), Uncertainty ±9%

Tilt (°)		Jan	Feb	Mar	Apr	May	June	July	Aug	Sept	Oct	Nov	Dec	Year
0	Average	1.6	2.5	3.4	4.6	5.6	6.2	6.1	5.3	4.1	2.8	1.7	1.3	3.8
	Min/Max	1.3/1.9	2.1/3.2	3.1/4.0	3.6/5.2	4.6/6.5	5.5/7.3	5.7/6.8	4.6/6.1	3.5/4.7	2.2/3.4	1.3/2.1	1.1/1.5	3.6/4.0
Latitude -15	Average	2.4	3.3	4.1	5.0	5.7	6.1	6.1	5.6	4.8	3.7	2.4	1.9	4.3
	Min/Max	1.8/2.9	2.7/4.6	3.5/5.1	3.7/5.8	4.6/6.7	5.4/7.2	5.6/6.8	4.8/6.6	3.9/5.6	2.6/4.7	1.5/3.3	1.4/2.5	4.0/4.5
Latitude	Average	2.7	3.6	4.2	4.9	5.4	5.6	5.6	5.4	4.8	3.9	2.6	2.1	4.2
	Min/Max	1.9/3.3	2.8/5.1	3.6/5.3	3.6/5.7	4.3/6.2	4.9/6.6	5.2/6.3	4.5/6.3	3.9/5.7	2.7/5.1	1.6/3.7	1.5/2.8	4.0/4.5
Latitude +15	Average	2.8	3.7	4.1	4.5	4.8	4.9	4.9	4.9	4.6	3.9	2.6	2.2	4.0
	Min/Max	2.0/3.5	2.8/5.4	3.5/5.2	3.3/5.3	3.8/5.5	4.3/5.7	4.6/5.5	4.1/5.8	3.7/5.5	2.6/5.1	1.5/3.8	1.6/3.0	3.7/4.2
90	Average	2.6	3.3	3.2	3.0	2.8	2.7	2.8	3.1	3.3	3.2	2.3	2.0	2.9
	Min/Max	1.8/3.4	2.5/5.0	2.7/4.1	2.2/3.6	2.3/3.1	2.5/3.0	2.6/3.0	2.6/3.6	2.6/4.0	2.1/4.2	1.3/3.5	1.4/2.9	2.7/3.1

Solar Radiation for 1-Axis Tracking Flat-Plate Collectors with a North-South Axis (kWh/m²/day), Uncertainty ±9%

Axis Tilt (°)		Jan	Feb	Mar	Apr	May	June	July	Aug	Sept	Oct	Nov	Dec	Year
0	Average	2.2	3.3	4.5	5.9	7.2	8.0	7.9	6.9	5.5	3.8	2.2	1.7	4.9
	Min/Max	1.6/2.7	2.6/4.6	3.8/5.7	4.3/7.1	5.6/9.0	6.7/9.8	7.1/9.1	5.8/8.3	4.3/6.4	2.6/5.0	1.4/3.1	1.2/2.2	4.6/5.3
Latitude -15	Average	2.8	4.0	5.0	6.3	7.4	8.0	8.0	7.3	6.0	4.5	2.7	2.1	5.3
	Min/Max	1.9/3.5	3.0/5.7	4.1/6.5	4.4/7.5	5.7/9.2	6.7/9.8	7.2/9.3	6.0/8.8	4.7/7.2	3.0/5.9	1.6/4.0	1.5/2.9	4.9/5.8
Latitude	Average	3.0	4.2	5.1	6.2	7.1	7.7	7.7	7.1	6.1	4.7	2.9	2.3	5.3
	Min/Max	2.0/3.8	3.2/6.1	4.2/6.6	4.3/7.5	5.5/8.9	6.4/9.4	6.9/9.0	5.8/8.6	4.7/7.3	3.0/6.2	1.6/4.3	1.6/3.2	4.9/5.8
Latitude +15	Average	3.1	4.3	5.0	5.9	6.7	7.1	7.2	6.8	5.9	4.7	2.9	2.4	5.2
	Min/Max	2.1/4.0	3.2/6.3	4.1/6.6	4.1/7.2	5.1/8.4	5.9/8.8	6.4/8.4	5.5/8.2	4.5/7.1	3.0/6.3	1.6/4.4	1.7/3.3	4.8/5.6

Solar Radiation for 2-Axis Tracking Flat-Plate Collectors (kWh/m²/day), Uncertainty ±9%

Tracker		Jan	Feb	Mar	Apr	May	June	July	Aug	Sept	Oct	Nov	Dec	Year
2-Axis	Average	3.1	4.3	5.1	6.3	7.5	8.2	8.2	7.3	6.1	4.7	3.0	2.4	5.5
	Min/Max	2.1/4.0	3.2/6.3	4.2/6.7	4.5/7.6	5.8/9.3	6.8/10.1	7.3/9.4	6.0/8.8	4.7/7.3	3.0/6.3	1.6/4.4	1.7/3.4	5.1/6.0

Direct Beam Solar Radiation for Concentrating Collectors (kWh/m²/day), Uncertainty ±8%

Tracker		Jan	Feb	Mar	Apr	May	June	July	Aug	Sept	Oct	Nov	Dec	Year
1-Axis, E-W Horiz Axis	Average	1.5	2.0	2.2	2.7	3.2	3.7	3.6	3.2	2.8	2.3	1.5	1.1	2.5
	Min/Max	0.9/2.1	1.2/3.3	1.2/3.5	1.2/3.6	2.0/4.6	2.6/5.1	2.9/4.6	2.3/4.4	1.7/3.7	1.1/3.5	0.4/2.6	0.6/1.9	2.2/2.9
1-Axis, N-S Horiz Axis	Average	1.0	1.7	2.4	3.4	4.3	4.8	4.8	4.2	3.3	2.2	1.1	0.7	2.8
	Min/Max	0.6/1.4	1.0/2.7	1.4/3.8	1.7/4.6	2.7/6.2	3.4/6.7	3.9/6.2	3.1/5.7	2.0/4.3	1.0/3.4	0.3/2.0	0.4/1.2	2.5/3.3
1-Axis, N-S Tilt=Latitude	Average	1.6	2.4	2.9	3.7	4.2	4.6	4.6	4.3	3.7	2.9	1.7	1.2	3.2
	Min/Max	1.0/2.3	1.4/3.9	1.6/4.6	1.8/4.9	2.6/6.1	3.2/6.4	3.7/5.9	3.2/5.9	2.3/4.9	1.3/4.4	0.5/3.0	0.6/2.0	2.7/3.6
2-Axis	Average	1.8	2.5	2.9	3.7	4.4	5.0	5.0	4.5	3.8	3.0	1.7	1.3	3.3
	Min/Max	1.0/2.5	1.4/4.0	1.6/4.6	1.8/5.0	2.8/6.4	3.5/6.9	4.0/6.4	3.3/6.1	2.3/4.9	1.4/4.5	0.5/3.1	0.7/2.2	2.9/3.8

Average Climatic Conditions

Element	Jan	Feb	Mar	Apr	May	June	July	Aug	Sept	Oct	Nov	Dec	Year
Temperature (°C)	-5.1	-3.7	2.1	8.5	14.7	19.8	22.4	21.4	17.3	10.7	4.6	-2.1	9.2
Daily Minimum Temp	-9.1	-8.0	-2.8	2.7	8.4	13.5	16.3	15.3	11.4	4.9	0.1	-5.9	3.9
Daily Maximum Temp	-0.9	0.7	6.9	14.3	20.9	26.1	28.5	27.4	23.3	16.4	8.9	1.8	14.5
Record Minimum Temp	-29.4	-26.1	-20.0	-12.2	-3.9	2.2	5.0	3.3	-1.7	-8.3	-12.8	-23.3	-29.4
Record Maximum Temp	16.7	18.3	27.2	31.7	33.9	40.0	38.9	37.8	36.7	32.8	25.0	20.0	40.0
HDD, Base 18.3°C	725	616	504	295	135	21	0	9	57	242	413	632	3649
CDD, Base 18.3°C	0	0	0	0	21	64	128	103	27	4	0	0	348
Relative Humidity (%)	75	73	70	66	65	67	69	72	73	72	75	77	71
Wind Speed (m/s)	5.4	5.1	5.3	5.2	4.6	4.2	3.8	3.7	3.9	4.4	5.0	5.1	4.6

Flint, MI

WBAN NO. 14826

LATITUDE: 42.97° N
LONGITUDE: 83.73° W
ELEVATION: 233 meters
MEAN PRESSURE: 988 millibars

STATION TYPE: Secondary

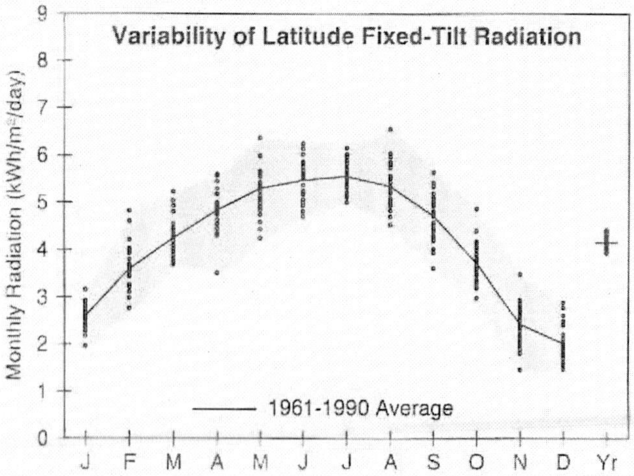

Solar Radiation for Flat-Plate Collectors Facing South at a Fixed Tilt (kWh/m²/day), Uncertainty ±9%

Tilt (°)		Jan	Feb	Mar	Apr	May	June	July	Aug	Sept	Oct	Nov	Dec	Year
0	Average	1.6	2.5	3.4	4.6	5.6	6.1	6.0	5.2	4.0	2.7	1.6	1.3	3.7
	Min/Max	1.4/1.8	2.1/3.0	3.1/4.0	3.5/5.1	4.6/6.6	5.2/6.9	5.4/6.6	4.5/6.3	3.3/4.6	2.4/3.3	1.3/2.0	1.1/1.6	3.6/3.9
Latitude -15	Average	2.3	3.3	4.1	5.0	5.7	6.0	6.0	5.6	4.7	3.5	2.2	1.8	4.2
	Min/Max	1.8/2.8	2.6/4.4	3.6/5.0	3.7/5.7	4.6/6.8	5.1/6.8	5.4/6.6	4.7/6.8	3.7/5.5	2.9/4.5	1.4/3.1	1.4/2.5	4.0/4.4
Latitude	Average	2.6	3.6	4.2	4.8	5.3	5.5	5.5	5.3	4.7	3.7	2.4	2.0	4.1
	Min/Max	2.0/3.2	2.8/4.8	3.7/5.2	3.5/5.6	4.2/6.4	4.7/6.2	5.0/6.2	4.5/6.6	3.6/5.6	3.0/4.9	1.5/3.5	1.5/2.9	3.9/4.4
Latitude +15	Average	2.7	3.7	4.1	4.5	4.7	4.7	4.8	4.8	4.5	3.7	2.5	2.1	3.9
	Min/Max	2.0/3.4	2.8/5.1	3.6/5.1	3.2/5.2	3.8/5.6	4.1/5.4	4.4/5.4	4.1/5.9	3.4/5.4	2.9/4.9	1.4/3.6	1.5/3.1	3.7/4.2
90	Average	2.6	3.3	3.3	3.0	2.8	2.7	2.8	3.1	3.3	3.0	2.2	2.0	2.8
	Min/Max	1.9/3.3	2.4/4.7	2.8/4.1	2.2/3.6	2.3/3.3	2.4/2.9	2.5/3.0	2.6/3.7	2.4/4.0	2.4/4.1	1.1/3.3	1.3/3.0	2.6/3.1

Solar Radiation for 1-Axis Tracking Flat-Plate Collectors with a North-South Axis (kWh/m²/day), Uncertainty ±9%

Axis Tilt (°)		Jan	Feb	Mar	Apr	May	June	July	Aug	Sept	Oct	Nov	Dec	Year
0	Average	2.1	3.3	4.5	5.9	7.1	7.8	7.8	6.9	5.3	3.6	2.1	1.6	4.9
	Min/Max	1.6/2.5	2.5/4.3	3.8/5.6	4.2/6.9	5.5/9.0	6.2/9.2	6.9/8.9	5.6/8.8	4.0/6.4	2.8/4.7	1.3/3.0	1.2/2.2	4.6/5.2
Latitude -15	Average	2.7	4.0	5.0	6.3	7.3	7.8	7.9	7.2	5.9	4.3	2.5	2.0	5.2
	Min/Max	2.0/3.3	2.9/5.3	4.2/6.4	4.4/7.4	5.5/9.3	6.2/9.3	7.0/9.0	5.8/9.3	4.3/7.2	3.2/5.7	1.5/3.8	1.5/2.9	4.9/5.6
Latitude	Average	2.9	4.2	5.1	6.2	7.1	7.5	7.6	7.0	5.9	4.4	2.7	2.2	5.2
	Min/Max	2.1/3.6	3.0/5.7	4.3/6.6	4.3/7.3	5.3/9.0	5.9/8.9	6.7/8.7	5.7/9.1	4.3/7.3	3.3/5.9	1.5/4.1	1.5/3.2	4.9/5.7
Latitude +15	Average	3.0	4.3	5.1	5.9	6.6	7.0	7.2	6.7	5.7	4.4	2.8	2.3	5.1
	Min/Max	2.2/3.8	3.0/5.9	4.2/6.5	4.1/7.0	5.0/8.5	5.5/8.3	6.3/8.2	5.4/8.7	4.1/7.1	3.3/6.0	1.5/4.2	1.6/3.4	4.7/5.5

Solar Radiation for 2-Axis Tracking Flat-Plate Collectors (kWh/m²/day), Uncertainty ±9%

Tracker		Jan	Feb	Mar	Apr	May	June	July	Aug	Sept	Oct	Nov	Dec	Year
2-Axis	Average	3.1	4.3	5.1	6.3	7.4	8.0	8.1	7.2	6.0	4.5	2.8	2.3	5.4
	Min/Max	2.2/3.8	3.1/5.9	4.3/6.6	4.4/7.4	5.6/9.4	6.4/9.5	7.1/9.2	5.8/9.3	4.3/7.3	3.3/6.0	1.5/4.2	1.6/3.5	5.0/5.8

Direct Beam Solar Radiation for Concentrating Collectors (kWh/m²/day), Uncertainty ±8%

Tracker		Jan	Feb	Mar	Apr	May	June	July	Aug	Sept	Oct	Nov	Dec	Year
1-Axis, E-W	Average	1.4	1.9	2.2	2.6	3.2	3.5	3.6	3.2	2.7	2.2	1.3	1.0	2.4
Horiz Axis	Min/Max	0.8/1.9	1.0/2.9	1.4/3.4	1.2/3.5	1.8/4.7	2.2/4.6	2.9/4.4	2.3/4.8	1.5/3.6	1.3/3.4	0.3/2.5	0.4/1.9	2.1/2.7
1-Axis, N-S	Average	1.0	1.6	2.4	3.4	4.2	4.6	4.7	4.1	3.2	2.1	1.0	0.6	2.7
Horiz Axis	Min/Max	0.5/1.2	0.8/2.4	1.6/3.7	1.7/4.4	2.4/6.2	2.9/6.0	3.9/5.8	3.0/6.2	1.8/4.3	1.2/3.3	0.2/1.9	0.3/1.2	2.4/3.1
1-Axis, N-S	Average	1.5	2.3	2.9	3.6	4.1	4.4	4.6	4.3	3.6	2.7	1.5	1.1	3.0
Tilt=Latitude	Min/Max	0.9/2.0	1.2/3.4	1.9/4.5	1.8/4.8	2.4/6.1	2.8/5.7	3.7/5.6	3.0/6.4	2.1/4.9	1.6/4.3	0.3/2.8	0.4/2.0	2.7/3.5
2-Axis	Average	1.6	2.4	2.9	3.7	4.4	4.7	4.9	4.4	3.7	2.7	1.5	1.2	3.2
	Min/Max	0.9/2.1	1.2/3.5	1.9/4.5	1.8/4.9	2.5/6.5	3.0/6.2	4.0/6.0	3.1/6.6	2.1/4.9	1.6/4.3	0.3/2.9	0.5/2.1	2.8/3.6

Average Climatic Conditions

Element	Jan	Feb	Mar	Apr	May	June	July	Aug	Sept	Oct	Nov	Dec	Year
Temperature (°C)	-5.8	-4.7	1.1	7.8	13.8	18.8	21.4	20.3	16.3	10.1	4.1	-2.7	8.4
Daily Minimum Temp	-9.9	-9.1	-3.6	2.2	7.6	12.6	15.4	14.4	10.6	4.7	0.1	-6.3	3.2
Daily Maximum Temp	-1.8	-0.4	5.7	13.3	20.0	24.9	27.5	26.2	22.0	15.4	8.1	1.0	13.5
Record Minimum Temp	-31.7	-30.0	-24.4	-14.4	-5.6	0.6	4.4	2.8	-3.3	-7.2	-14.4	-24.4	-31.7
Record Maximum Temp	15.6	17.2	25.6	30.6	33.9	38.3	38.3	36.7	33.9	31.7	24.4	19.4	38.3
HDD, Base 18.3°C	749	646	534	317	162	36	6	14	76	260	427	651	3877
CDD, Base 18.3°C	0	0	0	0	21	51	102	76	16	3	0	0	268
Relative Humidity (%)	75	73	70	66	66	68	70	73	76	73	76	77	72
Wind Speed (m/s)	5.1	4.8	5.1	4.9	4.4	4.0	3.6	3.4	3.8	4.2	4.8	4.9	4.4

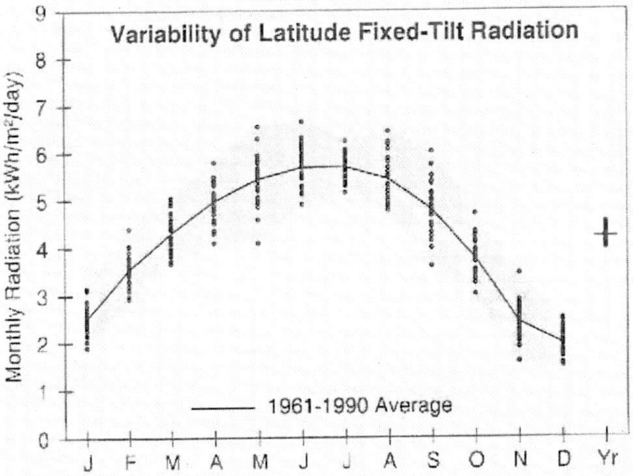

Grand Rapids, MI

WBAN NO. 94860

LATITUDE: 42.88° N
LONGITUDE: 85.52° W
ELEVATION: 245 meters
MEAN PRESSURE: 987 millibars

STATION TYPE: Secondary

Solar Radiation for Flat-Plate Collectors Facing South at a Fixed Tilt (kWh/m²/day), Uncertainty ±9%

Tilt (°)		Jan	Feb	Mar	Apr	May	June	July	Aug	Sept	Oct	Nov	Dec	Year
0	Average	1.6	2.5	3.5	4.7	5.7	6.3	6.2	5.3	4.1	2.7	1.6	1.3	3.8
	Min/Max	1.4/1.9	2.2/2.9	3.1/3.9	4.0/5.3	4.4/6.8	5.5/7.4	5.6/6.7	4.7/6.1	3.3/4.8	2.4/3.2	1.4/2.0	1.1/1.5	3.6/4.0
Latitude -15	Average	2.3	3.3	4.2	5.1	5.8	6.2	6.1	5.7	4.8	3.6	2.2	1.8	4.3
	Min/Max	1.8/2.8	2.8/4.0	3.6/4.9	4.2/5.9	4.4/7.0	5.3/7.3	5.5/6.7	5.0/6.7	3.7/5.9	2.9/4.4	1.5/3.1	1.4/2.2	4.0/4.5
Latitude	Average	2.5	3.5	4.3	5.0	5.4	5.7	5.7	5.4	4.8	3.8	2.4	2.0	4.2
	Min/Max	1.9/3.2	2.9/4.4	3.7/5.0	4.1/5.8	4.1/6.5	4.9/6.7	5.1/6.2	4.8/6.4	3.6/6.0	3.0/4.7	1.6/3.4	1.5/2.5	4.0/4.5
Latitude +15	Average	2.6	3.6	4.2	4.6	4.8	4.9	5.0	4.9	4.6	3.8	2.5	2.0	4.0
	Min/Max	2.0/3.3	3.0/4.6	3.6/4.9	3.8/5.4	3.6/5.8	4.3/5.7	4.5/5.4	4.3/5.8	3.4/5.8	3.0/4.7	1.5/3.6	1.5/2.7	3.8/4.3
90	Average	2.5	3.3	3.4	3.1	2.9	2.7	2.8	3.1	3.3	3.1	2.2	1.9	2.9
	Min/Max	1.8/3.2	2.7/4.3	2.8/3.9	2.6/3.6	2.3/3.3	2.5/3.0	2.6/3.0	2.8/3.7	2.4/4.3	2.4/3.9	1.3/3.3	1.4/2.5	2.7/3.1

Solar Radiation for 1-Axis Tracking Flat-Plate Collectors with a North-South Axis (kWh/m²/day), Uncertainty ±9%

Axis Tilt (°)		Jan	Feb	Mar	Apr	May	June	July	Aug	Sept	Oct	Nov	Dec	Year
0	Average	2.1	3.3	4.6	6.1	7.4	8.2	8.1	7.1	5.5	3.7	2.1	1.6	5.0
	Min/Max	1.6/2.5	2.7/4.0	3.9/5.3	4.9/7.1	5.3/9.1	6.8/10.0	7.2/9.0	6.0/8.6	4.1/6.9	3.0/4.6	1.4/2.9	1.3/2.0	4.7/5.4
Latitude -15	Average	2.6	3.9	5.1	6.5	7.6	8.2	8.2	7.4	6.0	4.3	2.5	2.0	5.4
	Min/Max	1.9/3.3	3.2/4.9	4.2/6.1	5.1/7.6	5.3/9.4	6.7/10.1	7.2/9.2	6.3/9.1	4.4/7.8	3.4/5.5	1.6/3.7	1.5/2.5	5.0/5.8
Latitude	Average	2.8	4.1	5.2	6.4	7.3	7.9	7.9	7.3	6.1	4.5	2.7	2.1	5.4
	Min/Max	2.0/3.5	3.3/5.2	4.3/6.3	5.0/7.6	5.2/9.1	6.4/9.7	6.9/8.8	6.1/8.9	4.4/7.8	3.5/5.8	1.6/4.0	1.5/2.8	5.0/5.8
Latitude +15	Average	2.9	4.2	5.1	6.1	6.9	7.4	7.4	6.9	5.9	4.5	2.7	2.2	5.2
	Min/Max	2.1/3.7	3.3/5.3	4.2/6.2	4.8/7.3	4.8/8.6	6.0/9.1	6.5/8.3	5.8/8.5	4.2/7.7	3.4/5.8	1.6/4.1	1.5/2.9	4.9/5.7

Solar Radiation for 2-Axis Tracking Flat-Plate Collectors (kWh/m²/day), Uncertainty ±9%

Tracker		Jan	Feb	Mar	Apr	May	June	July	Aug	Sept	Oct	Nov	Dec	Year
2-Axis	Average	2.9	4.2	5.2	6.5	7.7	8.4	8.3	7.5	6.1	4.5	2.8	2.2	5.5
	Min/Max	2.1/3.8	3.3/5.3	4.3/6.3	5.1/7.7	5.4/9.4	6.9/10.3	7.3/9.3	6.3/9.1	4.4/7.9	3.5/5.8	1.6/4.2	1.5/3.0	5.2/6.0

Direct Beam Solar Radiation for Concentrating Collectors (kWh/m²/day), Uncertainty ±8%

Tracker		Jan	Feb	Mar	Apr	May	June	July	Aug	Sept	Oct	Nov	Dec	Year
1-Axis, E-W Horiz Axis	Average	1.2	1.8	2.2	2.8	3.4	3.9	3.8	3.3	2.8	2.2	1.3	0.9	2.5
	Min/Max	0.7/2.0	1.1/2.4	1.4/3.2	1.8/3.7	1.8/4.8	2.6/5.4	3.0/4.8	2.4/4.6	1.5/4.2	1.4/3.3	0.4/2.4	0.2/1.6	2.2/2.8
1-Axis, N-S Horiz Axis	Average	0.8	1.5	2.4	3.5	4.5	5.1	5.0	4.4	3.3	2.1	1.0	0.6	2.9
	Min/Max	0.5/1.3	0.9/2.0	1.6/3.4	2.3/4.6	2.4/6.3	3.5/7.2	4.0/6.2	3.2/6.0	1.9/4.9	1.3/3.2	0.3/1.8	0.1/1.0	2.6/3.3
1-Axis, N-S Tilt=Latitude	Average	1.3	2.1	2.9	3.8	4.5	4.9	4.8	4.5	3.8	2.8	1.4	0.9	3.2
	Min/Max	0.7/2.1	1.3/2.8	1.9/4.2	2.5/5.0	2.4/6.2	3.3/6.8	3.8/6.0	3.3/6.2	2.2/5.7	1.7/4.1	0.5/2.7	0.2/1.6	2.8/3.6
2-Axis	Average	1.4	2.2	2.9	3.8	4.7	5.3	5.2	4.7	3.8	2.8	1.5	1.0	3.3
	Min/Max	0.8/2.2	1.3/2.9	1.9/4.2	2.5/5.0	2.5/6.5	3.6/7.4	4.1/6.4	3.4/6.4	2.2/5.7	1.8/4.1	0.5/2.8	0.2/1.8	3.0/3.8

Average Climatic Conditions

Element	Jan	Feb	Mar	Apr	May	June	July	Aug	Sept	Oct	Nov	Dec	Year
Temperature (°C)	-5.7	-4.6	1.2	7.8	14.2	19.4	22.0	20.8	16.1	9.7	3.3	-2.7	8.4
Daily Minimum Temp	-9.6	-9.0	-3.7	1.9	7.6	12.9	15.8	14.7	9.9	3.9	-1.0	-6.3	3.1
Daily Maximum Temp	-1.7	-0.2	6.0	13.7	20.7	25.9	28.2	26.9	22.2	15.4	7.7	0.8	13.8
Record Minimum Temp	-29.4	-28.3	-22.2	-16.1	-5.6	0.6	5.0	3.9	-2.8	-7.8	-15.0	-27.8	-29.4
Record Maximum Temp	16.7	19.4	25.6	31.1	33.3	36.7	37.8	37.8	33.9	30.6	25.0	19.4	37.8
HDD, Base 18.3°C	744	642	532	317	152	28	0	10	77	269	450	653	3874
CDD, Base 18.3°C	0	0	0	0	22	61	116	87	11	0	0	0	297
Relative Humidity (%)	77	75	71	66	65	67	69	73	76	74	77	79	72
Wind Speed (m/s)	5.1	4.7	5.1	4.9	4.3	4.0	3.7	3.5	3.7	4.0	4.6	4.8	4.4

Houghton, MI

WBAN NO. 94814

LATITUDE: 47.17° N
LONGITUDE: 88.50° W
ELEVATION: 329 meters
MEAN PRESSURE: 974 millibars

STATION TYPE: Secondary

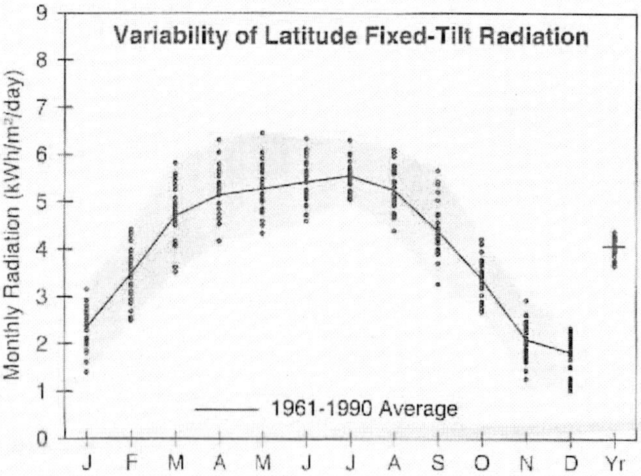

Solar Radiation for Flat-Plate Collectors Facing South at a Fixed Tilt (kWh/m²/day), Uncertainty ±9%

Tilt (°)		Jan	Feb	Mar	Apr	May	June	July	Aug	Sept	Oct	Nov	Dec	Year
0	Average	1.3	2.2	3.5	4.6	5.5	6.0	6.0	5.0	3.6	2.3	1.3	1.1	3.6
	Min/Max	1.0/1.6	1.7/2.6	2.8/4.1	4.0/5.3	4.6/6.6	5.2/7.0	5.5/6.7	4.3/5.7	3.0/4.3	2.0/2.7	1.1/1.6	0.7/1.3	3.3/3.8
Latitude -15	Average	2.1	3.2	4.5	5.2	5.6	5.9	6.0	5.5	4.4	3.2	1.9	1.6	4.1
	Min/Max	1.3/2.8	2.4/4.0	3.5/5.5	4.3/6.3	4.7/6.9	5.0/6.9	5.4/6.8	4.6/6.3	3.3/5.6	2.6/3.9	1.3/2.6	1.0/2.1	3.7/4.4
Latitude	Average	2.3	3.5	4.7	5.2	5.3	5.4	5.6	5.2	4.4	3.4	2.1	1.8	4.1
	Min/Max	1.4/3.2	2.5/4.4	3.5/5.8	4.2/6.3	4.3/6.5	4.6/6.3	5.1/6.3	4.4/6.1	3.3/5.7	2.7/4.2	1.3/2.9	1.0/2.3	3.7/4.4
Latitude +15	Average	2.5	3.6	4.7	4.8	4.7	4.7	4.8	4.7	4.2	3.4	2.1	1.9	3.8
	Min/Max	1.4/3.4	2.5/4.6	3.4/5.9	3.8/6.1	3.8/5.7	4.0/5.4	4.4/5.5	4.0/5.5	3.1/5.4	2.7/4.3	1.2/3.1	1.0/2.5	3.4/4.1
90	Average	2.4	3.3	4.1	3.7	3.0	2.9	3.0	3.2	3.2	2.9	2.0	1.8	3.0
	Min/Max	1.2/3.3	2.2/4.4	2.7/5.3	2.7/5.0	2.5/3.6	2.5/3.2	2.8/3.3	2.8/3.8	2.3/4.3	2.3/3.7	1.0/2.8	0.9/2.5	2.5/3.2

Solar Radiation for 1-Axis Tracking Flat-Plate Collectors with a North-South Axis (kWh/m²/day), Uncertainty ±9%

Axis Tilt (°)		Jan	Feb	Mar	Apr	May	June	July	Aug	Sept	Oct	Nov	Dec	Year
0	Average	1.8	3.1	4.9	6.3	7.4	8.1	8.1	6.9	5.0	3.2	1.7	1.4	4.8
	Min/Max	1.1/2.4	2.2/4.0	3.7/6.2	5.0/7.4	5.8/9.4	6.6/10.0	7.2/9.3	5.7/8.5	3.6/6.3	2.5/4.1	1.1/2.4	0.8/1.7	4.4/5.2
Latitude -15	Average	2.4	3.8	5.7	6.8	7.6	8.2	8.3	7.4	5.6	3.9	2.2	1.8	5.3
	Min/Max	1.4/3.3	2.7/5.0	4.2/7.2	5.4/8.3	6.0/9.8	6.6/10.1	7.3/9.6	6.0/9.1	3.9/7.3	3.0/5.0	1.3/3.2	1.0/2.3	4.8/5.8
Latitude	Average	2.6	4.1	5.8	6.8	7.4	7.8	8.0	7.2	5.6	4.0	2.3	2.0	5.3
	Min/Max	1.5/3.6	2.8/5.3	4.3/7.4	5.3/8.3	5.8/9.6	6.3/9.8	7.0/9.3	5.9/8.9	3.9/7.4	3.0/5.2	1.3/3.4	1.0/2.6	4.7/5.8
Latitude +15	Average	2.7	4.1	5.8	6.6	7.0	7.3	7.5	6.9	5.5	4.0	2.4	2.1	5.2
	Min/Max	1.5/3.8	2.8/5.5	4.2/7.4	5.0/8.2	5.4/9.1	5.9/9.2	6.6/8.7	5.6/8.5	3.7/7.2	3.0/5.3	1.3/3.5	1.0/2.7	4.5/5.6

Solar Radiation for 2-Axis Tracking Flat-Plate Collectors (kWh/m²/day), Uncertainty ±9%

Tracker		Jan	Feb	Mar	Apr	May	June	July	Aug	Sept	Oct	Nov	Dec	Year
2-Axis	Average	2.7	4.1	5.8	6.9	7.7	8.4	8.4	7.4	5.7	4.1	2.4	2.1	5.5
	Min/Max	1.5/3.8	2.8/5.5	4.3/7.4	5.4/8.3	6.1/10.0	6.8/10.4	7.4/9.8	6.1/9.1	3.9/7.4	3.0/5.3	1.3/3.5	1.1/2.8	4.9/6.0

Direct Beam Solar Radiation for Concentrating Collectors (kWh/m²/day), Uncertainty ±8%

Tracker		Jan	Feb	Mar	Apr	May	June	July	Aug	Sept	Oct	Nov	Dec	Year
1-Axis, E-W	Average	1.2	1.8	2.4	2.9	3.6	3.9	4.0	3.4	2.6	2.0	1.1	0.8	2.5
Horiz Axis	Min/Max	0.5/2.0	1.1/2.7	1.6/3.4	2.0/3.7	2.3/5.3	2.8/5.4	3.1/5.1	2.4/4.6	1.5/4.0	1.1/3.0	0.3/2.0	0.3/1.3	2.2/2.9
1-Axis, N-S	Average	0.8	1.5	2.6	3.7	4.7	5.3	5.3	4.5	3.1	1.9	0.7	0.5	2.9
Horiz Axis	Min/Max	0.3/1.3	0.8/2.3	1.7/3.7	2.6/4.6	3.1/7.0	3.6/7.2	4.2/6.7	3.2/6.2	1.7/4.5	1.0/2.8	0.2/1.4	0.2/0.8	2.5/3.3
1-Axis, N-S	Average	1.3	2.2	3.2	4.1	4.8	5.1	5.2	4.7	3.6	2.6	1.2	0.9	3.2
Tilt=Latitude	Min/Max	0.6/2.2	1.3/3.3	2.1/4.6	2.9/5.1	3.1/7.0	3.5/6.9	4.1/6.5	3.4/6.5	2.0/5.4	1.4/3.8	0.3/2.3	0.3/1.4	2.8/3.7
2-Axis	Average	1.4	2.2	3.2	4.1	5.0	5.5	5.6	4.9	3.6	2.6	1.3	0.9	3.4
	Min/Max	0.6/2.4	1.3/3.4	2.1/4.6	2.9/5.2	3.3/7.4	3.8/7.6	4.4/7.0	3.5/6.7	2.0/5.4	1.4/3.9	0.4/2.4	0.3/1.5	3.0/3.9

Average Climatic Conditions

Element	Jan	Feb	Mar	Apr	May	June	July	Aug	Sept	Oct	Nov	Dec	Year
Temperature (°C)	-8.4	-7.7	-2.1	5.7	12.3	17.1	19.7	18.4	14.1	8.2	1.8	-5.2	6.2
Daily Minimum Temp	-13.1	-13.2	-7.7	-0.2	5.6	10.3	13.0	12.1	8.3	3.1	-2.0	-9.1	0.6
Daily Maximum Temp	-3.7	-2.2	3.4	11.5	18.9	23.8	26.4	24.7	19.9	13.2	5.6	-1.4	11.7
Record Minimum Temp	-32.2	-36.7	-30.6	-16.1	-6.1	-1.7	0.6	-1.7	-6.1	-8.9	-18.9	-29.4	-36.7
Record Maximum Temp	11.7	15.0	24.4	30.0	32.2	35.0	36.7	34.4	33.3	29.4	21.1	17.2	36.7
HDD, Base 18.3°C	828	728	634	380	199	67	19	39	130	315	497	730	4566
CDD, Base 18.3°C	0	0	0	0	11	28	62	41	0	0	0	0	142
Relative Humidity (%)	79	76	74	68	65	69	71	75	78	77	81	82	75
Wind Speed (m/s)	4.5	4.1	4.4	4.5	4.1	3.8	3.4	3.3	3.5	3.9	4.3	4.3	4.0

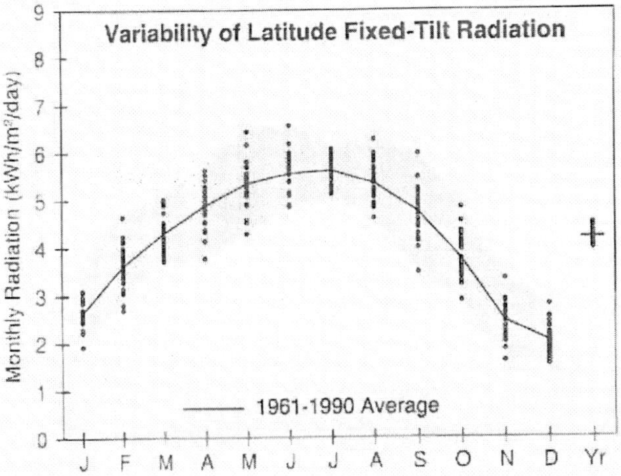

Lansing, MI

WBAN NO. 14836

LATITUDE: 42.78° N
LONGITUDE: 84.60° W
ELEVATION: 256 meters
MEAN PRESSURE: 985 millibars

STATION TYPE: Secondary

Solar Radiation for Flat-Plate Collectors Facing South at a Fixed Tilt (kWh/m²/day), Uncertainty ±9%

Tilt (°)		Jan	Feb	Mar	Apr	May	June	July	Aug	Sept	Oct	Nov	Dec	Year
0	Average	1.6	2.5	3.5	4.6	5.6	6.2	6.1	5.2	4.0	2.7	1.7	1.3	3.8
	Min/Max	1.4/1.9	2.1/3.0	3.2/3.9	3.8/5.1	4.6/6.7	5.4/7.3	5.6/6.5	4.7/6.0	3.2/4.8	2.3/3.3	1.4/2.0	1.1/1.6	3.6/4.0
Latitude -15	Average	2.4	3.3	4.2	5.0	5.7	6.0	6.1	5.6	4.7	3.6	2.3	1.8	4.2
	Min/Max	1.8/2.8	2.6/4.2	3.7/4.8	3.9/5.7	4.6/6.9	5.3/7.2	5.5/6.5	4.8/6.5	3.5/5.8	2.8/4.5	1.6/3.0	1.4/2.5	4.0/4.5
Latitude	Average	2.6	3.6	4.3	4.9	5.3	5.6	5.6	5.4	4.8	3.8	2.4	2.0	4.2
	Min/Max	1.9/3.1	2.7/4.6	3.7/5.0	3.8/5.6	4.3/6.4	4.8/6.6	5.1/6.1	4.6/6.3	3.5/6.0	2.9/4.8	1.6/3.3	1.5/2.8	4.0/4.5
Latitude +15	Average	2.8	3.7	4.2	4.5	4.7	4.8	4.9	4.9	4.5	3.7	2.5	2.1	4.0
	Min/Max	2.0/3.3	2.7/4.8	3.6/4.9	3.5/5.2	3.8/5.7	4.2/5.7	4.5/5.3	4.2/5.7	3.3/5.8	2.8/4.9	1.6/3.5	1.6/3.0	3.7/4.3
90	Average	2.6	3.4	3.4	3.1	2.8	2.7	2.8	3.1	3.3	3.1	2.2	2.0	2.9
	Min/Max	1.8/3.2	2.4/4.5	2.8/4.2	2.4/3.6	2.3/3.3	2.5/3.0	2.6/3.0	2.7/3.6	2.3/4.2	2.3/4.1	1.3/3.1	1.4/2.9	2.7/3.1

Solar Radiation for 1-Axis Tracking Flat-Plate Collectors with a North-South Axis (kWh/m²/day), Uncertainty ±9%

Axis Tilt (°)		Jan	Feb	Mar	Apr	May	June	July	Aug	Sept	Oct	Nov	Dec	Year
0	Average	2.2	3.3	4.6	5.9	7.2	7.9	7.9	6.9	5.4	3.7	2.1	1.6	4.9
	Min/Max	1.6/2.5	2.5/4.1	3.9/5.3	4.6/6.9	5.5/9.1	6.5/9.8	7.1/8.8	5.9/8.4	3.9/6.8	2.8/4.8	1.4/2.8	1.3/2.2	4.6/5.3
Latitude -15	Average	2.7	4.0	5.1	6.3	7.4	7.9	8.0	7.2	5.9	4.3	2.6	2.1	5.3
	Min/Max	2.0/3.2	2.9/5.1	4.3/6.1	4.8/7.4	5.6/9.4	6.5/9.9	7.1/8.9	6.2/8.8	4.2/7.7	3.2/5.7	1.6/3.6	1.5/2.8	5.0/5.7
Latitude	Average	2.9	4.2	5.2	6.2	7.1	7.6	7.7	7.1	6.0	4.5	2.7	2.2	5.3
	Min/Max	2.1/3.5	3.0/5.4	4.3/6.2	4.7/7.4	5.4/9.1	6.2/9.5	6.8/8.6	6.0/8.7	4.2/7.8	3.2/6.0	1.6/3.8	1.6/3.1	5.0/5.8
Latitude +15	Average	3.1	4.3	5.1	6.0	6.7	7.1	7.2	6.7	5.8	4.5	2.8	2.3	5.1
	Min/Max	2.1/3.7	3.0/5.6	4.2/6.1	4.4/7.1	5.0/8.6	5.7/8.9	6.4/8.0	5.7/8.3	4.0/7.6	3.2/6.0	1.6/4.0	1.6/3.3	4.8/5.6

Solar Radiation for 2-Axis Tracking Flat-Plate Collectors (kWh/m²/day), Uncertainty ±9%

Tracker		Jan	Feb	Mar	Apr	May	June	July	Aug	Sept	Oct	Nov	Dec	Year
2-Axis	Average	3.1	4.3	5.2	6.3	7.5	8.1	8.1	7.3	6.0	4.5	2.8	2.3	5.5
	Min/Max	2.1/3.7	3.0/5.6	4.4/6.2	4.8/7.5	5.7/9.5	6.6/10.1	7.2/9.0	6.2/8.9	4.2/7.8	3.3/6.1	1.6/4.0	1.7/3.3	5.1/6.0

Direct Beam Solar Radiation for Concentrating Collectors (kWh/m²/day), Uncertainty ±8%

Tracker		Jan	Feb	Mar	Apr	May	June	July	Aug	Sept	Oct	Nov	Dec	Year
1-Axis, E-W Horiz Axis	Average	1.4	1.9	2.2	2.7	3.3	3.6	3.6	3.2	2.7	2.2	1.3	1.0	2.4
	Min/Max	0.8/1.9	1.0/2.6	1.6/3.1	1.4/3.6	2.0/4.8	2.4/5.2	2.8/4.3	2.3/4.3	1.4/4.1	1.2/3.4	0.4/2.3	0.4/1.7	2.1/2.8
1-Axis, N-S Horiz Axis	Average	0.9	1.6	2.4	3.4	4.3	4.8	4.8	4.2	3.2	2.1	1.0	0.6	2.8
	Min/Max	0.6/1.3	0.8/2.2	1.7/3.4	1.9/4.5	2.6/6.3	3.1/6.8	3.7/5.7	3.1/5.7	1.7/4.7	1.1/3.3	0.3/1.7	0.3/1.0	2.4/3.3
1-Axis, N-S Tilt=Latitude	Average	1.5	2.3	2.9	3.7	4.2	4.5	4.6	4.3	3.7	2.7	1.5	1.0	3.1
	Min/Max	0.9/2.1	1.1/3.0	2.1/4.2	2.0/4.9	2.6/6.2	2.9/6.5	3.6/5.5	3.2/5.9	1.9/5.5	1.5/4.3	0.4/2.6	0.4/1.8	2.7/3.6
2-Axis	Average	1.6	2.4	2.9	3.7	4.5	4.9	5.0	4.4	3.7	2.8	1.6	1.1	3.2
	Min/Max	0.9/2.2	1.2/3.1	2.1/4.2	2.1/5.0	2.7/6.6	3.2/7.1	3.8/5.9	3.3/6.1	2.0/5.5	1.5/4.4	0.4/2.7	0.5/1.9	2.8/3.7

Average Climatic Conditions

Element	Jan	Feb	Mar	Apr	May	June	July	Aug	Sept	Oct	Nov	Dec	Year
Temperature (°C)	-6.2	-5.1	0.9	7.7	13.9	19.1	21.6	20.4	16.1	9.6	3.5	-3.1	8.2
Daily Minimum Temp	-10.4	-9.9	-4.2	1.7	7.2	12.5	15.0	13.8	9.9	4.0	-0.7	-7.1	2.7
Daily Maximum Temp	-1.9	-0.2	5.9	13.7	20.6	25.7	28.1	26.9	22.2	15.2	7.7	0.8	13.7
Record Minimum Temp	-33.9	-31.1	-26.1	-18.9	-7.2	-1.1	2.8	1.7	-5.6	-9.4	-15.6	-27.2	-33.9
Record Maximum Temp	18.9	17.8	25.6	30.0	34.4	37.2	37.8	37.8	36.1	31.7	25.0	18.9	37.8
HDD, Base 18.3°C	759	655	541	318	157	33	4	14	82	271	445	665	3945
CDD, Base 18.3°C	0	0	0	0	19	57	104	78	13	0	0	0	272
Relative Humidity (%)	79	76	73	68	67	69	71	75	77	76	79	81	74
Wind Speed (m/s)	5.4	4.9	5.2	5.1	4.5	4.1	3.6	3.4	3.7	4.1	4.8	5.0	4.5

Muskegon, MI

WBAN NO. 14840

LATITUDE: 43.17° N
LONGITUDE: 86.25° W
ELEVATION: 191 meters
MEAN PRESSURE: 993 millibars

STATION TYPE: Secondary

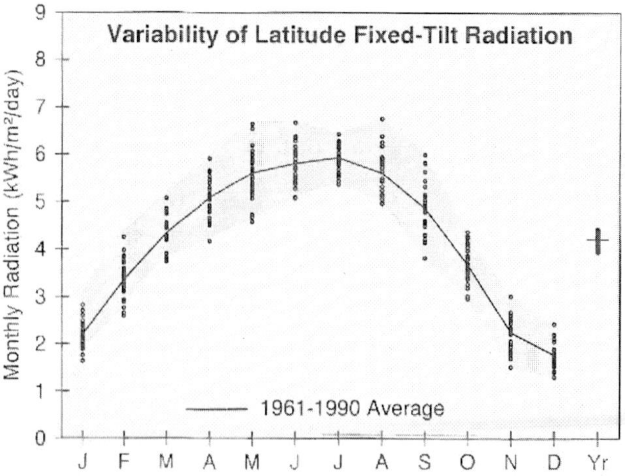

Solar Radiation for Flat-Plate Collectors Facing South at a Fixed Tilt (kWh/m²/day), Uncertainty ±9%

Tilt (°)		Jan	Feb	Mar	Apr	May	June	July	Aug	Sept	Oct	Nov	Dec	Year
0	Average	1.6	2.4	3.5	4.7	5.9	6.4	6.4	5.4	4.1	2.7	1.6	1.2	3.8
	Min/Max	1.3/1.8	2.1/2.9	3.1/4.0	4.0/5.4	4.9/6.9	5.7/7.4	5.8/6.9	4.9/6.4	3.4/4.8	2.3/3.0	1.3/1.9	1.0/1.4	3.6/4.0
Latitude -15	Average	2.0	3.1	4.2	5.2	6.0	6.3	6.4	5.8	4.8	3.5	2.1	1.6	4.3
	Min/Max	1.6/2.5	2.5/3.9	3.7/4.9	4.3/6.0	4.9/7.1	5.5/7.3	5.8/6.9	5.2/7.0	3.8/5.9	2.9/4.1	1.5/2.7	1.3/2.1	4.0/4.5
Latitude	Average	2.2	3.3	4.3	5.1	5.6	5.8	5.9	5.6	4.9	3.7	2.2	1.8	4.2
	Min/Max	1.6/2.8	2.6/4.3	3.7/5.1	4.2/5.9	4.6/6.6	5.1/6.7	5.4/6.4	4.9/6.8	3.8/6.0	3.0/4.3	1.5/3.0	1.3/2.4	3.9/4.4
Latitude +15	Average	2.3	3.4	4.2	4.7	5.0	5.0	5.2	5.1	4.6	3.7	2.3	1.8	3.9
	Min/Max	1.6/3.0	2.6/4.4	3.6/5.1	3.8/5.5	4.1/5.9	4.4/5.7	4.7/5.6	4.5/6.1	3.6/5.8	2.9/4.4	1.5/3.1	1.3/2.6	3.7/4.2
90	Average	2.1	3.1	3.4	3.2	2.9	2.8	2.9	3.2	3.4	3.0	2.0	1.7	2.8
	Min/Max	1.5/2.8	2.3/4.1	2.8/4.4	2.6/3.7	2.5/3.4	2.5/3.0	2.7/3.1	2.8/3.8	2.6/4.3	2.3/3.6	1.2/2.8	1.1/2.4	2.6/3.0

Solar Radiation for 1-Axis Tracking Flat-Plate Collectors with a North-South Axis (kWh/m²/day), Uncertainty ±9%

Axis Tilt (°)		Jan	Feb	Mar	Apr	May	June	July	Aug	Sept	Oct	Nov	Dec	Year
0	Average	1.9	3.1	4.6	6.2	7.6	8.3	8.4	7.3	5.5	3.6	1.9	1.5	5.0
	Min/Max	1.4/2.3	2.4/3.9	3.9/5.4	4.9/7.3	5.9/9.3	6.9/10.1	7.5/9.2	6.2/9.1	4.2/6.8	2.9/4.3	1.3/2.6	1.1/1.9	4.6/5.4
Latitude -15	Average	2.3	3.7	5.1	6.6	7.8	8.3	8.5	7.6	6.1	4.2	2.3	1.8	5.4
	Min/Max	1.6/3.0	2.7/4.7	4.3/6.2	5.2/7.9	6.0/9.5	6.9/10.2	7.6/9.4	6.5/9.6	4.6/7.6	3.3/5.1	1.5/3.2	1.3/2.5	4.9/5.7
Latitude	Average	2.4	3.8	5.2	6.6	7.6	8.0	8.2	7.5	6.1	4.4	2.5	1.9	5.3
	Min/Max	1.7/3.2	2.8/5.0	4.3/6.3	5.1/7.8	5.8/9.3	6.6/9.8	7.3/9.0	6.3/9.4	4.5/7.7	3.3/5.3	1.5/3.5	1.3/2.7	4.9/5.7
Latitude +15	Average	2.5	3.9	5.2	6.3	7.1	7.5	7.7	7.1	6.0	4.3	2.5	2.0	5.2
	Min/Max	1.7/3.4	2.9/5.1	4.3/6.3	4.8/7.5	5.4/8.7	6.1/9.2	6.8/8.5	6.0/9.0	4.4/7.6	3.3/5.3	1.5/3.6	1.3/2.8	4.7/5.5

Solar Radiation for 2-Axis Tracking Flat-Plate Collectors (kWh/m²/day), Uncertainty ±9%

Tracker		Jan	Feb	Mar	Apr	May	June	July	Aug	Sept	Oct	Nov	Dec	Year
2-Axis	Average	2.5	3.9	5.3	6.7	7.9	8.5	8.7	7.7	6.2	4.4	2.5	2.0	5.5
	Min/Max	1.7/3.4	2.9/5.1	4.4/6.3	5.2/7.9	6.1/9.6	7.1/10.4	7.7/9.5	6.5/9.6	4.6/7.8	3.4/5.3	1.5/3.6	1.3/2.9	5.0/5.9

Direct Beam Solar Radiation for Concentrating Collectors (kWh/m²/day), Uncertainty ±8%

Tracker		Jan	Feb	Mar	Apr	May	June	July	Aug	Sept	Oct	Nov	Dec	Year
1-Axis, E-W Horiz Axis	Average	0.8	1.5	2.2	3.0	3.7	4.1	4.2	3.6	2.9	2.1	1.1	0.7	2.5
	Min/Max	0.3/1.8	0.7/2.4	1.2/3.3	1.9/3.9	2.3/5.0	2.9/5.6	3.5/5.0	2.7/5.1	1.8/4.2	1.3/2.8	0.4/1.9	0.1/1.5	2.1/2.8
1-Axis, N-S Horiz Axis	Average	0.5	1.3	2.4	3.8	4.8	5.3	5.5	4.7	3.4	2.0	0.8	0.4	2.9
	Min/Max	0.2/1.2	0.6/2.0	1.4/3.5	2.4/4.9	2.9/6.6	3.8/7.3	4.5/6.6	3.5/6.7	2.0/4.8	1.2/2.8	0.3/1.4	0.1/1.0	2.5/3.3
1-Axis, N-S Tilt=Latitude	Average	0.8	1.8	2.9	4.0	4.7	5.0	5.3	4.8	3.9	2.7	1.2	0.7	3.2
	Min/Max	0.3/1.9	0.9/2.8	1.7/4.3	2.6/5.3	2.9/6.5	3.6/6.9	4.4/6.4	3.6/6.9	2.3/5.6	1.6/3.6	0.4/2.1	0.1/1.6	2.7/3.6
2-Axis	Average	0.9	1.8	2.9	4.1	5.0	5.5	5.7	5.0	3.9	2.7	1.3	0.8	3.3
	Min/Max	0.4/2.1	0.9/2.8	1.7/4.3	2.7/5.4	3.0/6.8	3.9/7.5	4.7/6.8	3.7/7.1	2.4/5.7	1.6/3.6	0.4/2.3	0.2/1.7	2.8/3.7

Average Climatic Conditions

Element	Jan	Feb	Mar	Apr	May	June	July	Aug	Sept	Oct	Nov	Dec	Year
Temperature (°C)	-4.8	-4.2	0.7	7.2	13.3	18.3	21.3	20.3	16.2	10.2	4.1	-1.9	8.4
Daily Minimum Temp	-7.9	-7.8	-3.7	1.9	7.3	12.4	15.7	14.9	10.8	5.3	0.4	-4.8	3.7
Daily Maximum Temp	-1.8	-0.7	5.1	12.5	19.2	24.2	26.8	25.6	21.6	15.0	7.8	0.9	13.1
Record Minimum Temp	-25.0	-25.6	-23.3	-17.2	-5.6	-0.6	4.4	2.2	-2.8	-6.1	-25.6	-26.1	-26.1
Record Maximum Temp	17.2	16.7	26.7	30.0	33.9	36.1	35.6	37.2	35.0	28.3	24.4	17.8	37.2
HDD, Base 18.3°C	718	632	546	333	172	43	6	14	74	253	427	629	3847
CDD, Base 18.3°C	0	0	0	0	15	43	97	74	11	0	0	0	239
Relative Humidity (%)	78	75	71	65	64	68	71	75	76	74	75	79	73
Wind Speed (m/s)	5.6	5.1	5.4	5.3	4.5	4.2	4.0	3.9	4.2	4.7	5.3	5.4	4.8

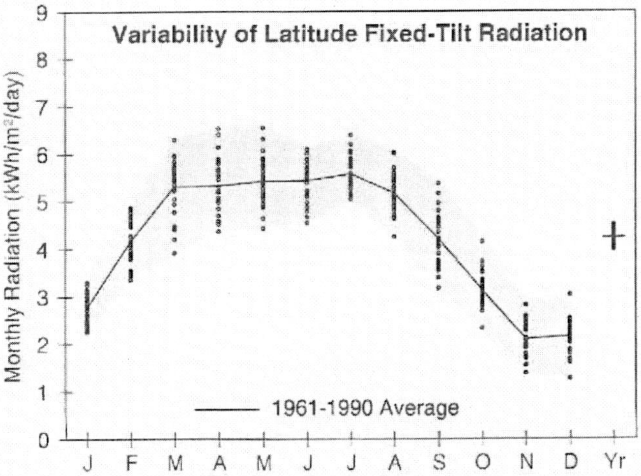

Sault Ste. Marie, MI

WBAN NO. 14847

LATITUDE: 46.47° N
LONGITUDE: 84.37° W
ELEVATION: 221 meters
MEAN PRESSURE: 989 millibars

STATION TYPE: Secondary

Solar Radiation for Flat-Plate Collectors Facing South at a Fixed Tilt (kWh/m²/day), Uncertainty ±9%

Tilt (°)		Jan	Feb	Mar	Apr	May	June	July	Aug	Sept	Oct	Nov	Dec	Year
0	Average	1.6	2.6	3.9	4.8	5.7	6.1	6.0	5.0	3.5	2.2	1.4	1.2	3.7
	Min/Max	1.4/1.7	2.3/2.9	3.1/4.5	4.1/5.7	4.7/6.7	5.1/6.7	5.5/6.8	4.2/5.7	2.9/4.3	2.0/2.7	1.1/1.6	0.9/1.5	3.5/3.9
Latitude -15	Average	2.5	3.8	5.1	5.4	5.8	5.9	6.0	5.4	4.2	3.0	1.9	1.9	4.3
	Min/Max	2.1/2.9	3.1/4.4	3.8/6.0	4.5/6.5	4.8/7.0	5.0/6.7	5.5/6.9	4.5/6.3	3.2/5.3	2.3/3.9	1.3/2.5	1.2/2.6	4.0/4.5
Latitude	Average	2.8	4.2	5.3	5.3	5.4	5.4	5.6	5.2	4.2	3.1	2.1	2.2	4.2
	Min/Max	2.3/3.3	3.4/4.9	3.9/6.3	4.4/6.5	4.4/6.6	4.6/6.1	5.0/6.4	4.3/6.0	3.2/5.4	2.3/4.2	1.4/2.8	1.3/3.0	4.0/4.5
Latitude +15	Average	2.9	4.3	5.3	5.0	4.8	4.7	4.9	4.7	4.0	3.1	2.1	2.3	4.0
	Min/Max	2.4/3.5	3.5/5.1	3.8/6.3	4.0/6.2	3.9/5.8	4.0/5.3	4.4/5.6	3.8/5.5	3.0/5.1	2.3/4.2	1.4/2.9	1.3/3.3	3.8/4.3
90	Average	2.9	4.1	4.7	3.7	3.0	2.8	3.0	3.2	3.0	2.6	2.0	2.2	3.1
	Min/Max	2.3/3.5	3.2/4.8	3.2/5.6	3.0/4.9	2.6/3.6	2.5/3.1	2.7/3.3	2.6/3.7	2.2/3.9	1.9/3.6	1.2/2.7	1.2/3.3	2.8/3.3

Solar Radiation for 1-Axis Tracking Flat-Plate Collectors with a North-South Axis (kWh/m²/day), Uncertainty ±9%

Axis Tilt (°)		Jan	Feb	Mar	Apr	May	June	July	Aug	Sept	Oct	Nov	Dec	Year
0	Average	2.1	3.7	5.5	6.6	7.6	8.0	8.1	6.8	4.8	3.0	1.7	1.6	5.0
	Min/Max	1.8/2.5	2.9/4.3	4.1/6.6	5.4/8.0	5.8/9.6	6.3/9.4	7.1/9.6	5.3/8.1	3.5/6.2	2.3/4.0	1.2/2.3	1.0/2.2	4.6/5.3
Latitude -15	Average	2.9	4.6	6.4	7.1	7.9	8.1	8.3	7.2	5.3	3.6	2.2	2.2	5.5
	Min/Max	2.3/3.5	3.6/5.5	4.7/7.8	5.8/8.8	5.9/10.0	6.3/9.4	7.2/9.8	5.6/8.6	3.8/7.0	2.6/4.9	1.4/3.0	1.3/3.1	5.1/5.9
Latitude	Average	3.1	4.9	6.6	7.1	7.6	7.8	8.0	7.0	5.3	3.7	2.3	2.4	5.5
	Min/Max	2.5/3.8	3.8/5.8	4.7/8.0	5.7/8.8	5.7/9.7	6.1/9.1	6.9/9.5	5.4/8.5	3.8/7.1	2.6/5.1	1.5/3.3	1.3/3.4	5.1/5.9
Latitude +15	Average	3.2	5.0	6.6	6.8	7.2	7.3	7.5	6.7	5.2	3.7	2.4	2.5	5.3
	Min/Max	2.6/4.0	3.9/6.0	4.7/8.0	5.5/8.6	5.3/9.2	5.7/8.5	6.5/8.9	5.1/8.1	3.6/6.9	2.6/5.1	1.5/3.3	1.4/3.6	4.9/5.7

Solar Radiation for 2-Axis Tracking Flat-Plate Collectors (kWh/m²/day), Uncertainty ±9%

Tracker		Jan	Feb	Mar	Apr	May	June	July	Aug	Sept	Oct	Nov	Dec	Year
2-Axis	Average	3.3	5.0	6.6	7.1	8.0	8.3	8.5	7.2	5.4	3.7	2.4	2.5	5.7
	Min/Max	2.6/4.0	3.9/6.0	4.8/8.0	5.8/8.8	5.9/10.1	6.5/9.7	7.4/10.0	5.6/8.7	3.8/7.1	2.6/5.1	1.5/3.4	1.4/3.7	5.2/6.1

Direct Beam Solar Radiation for Concentrating Collectors (kWh/m²/day), Uncertainty ±8%

Tracker		Jan	Feb	Mar	Apr	May	June	July	Aug	Sept	Oct	Nov	Dec	Year
1-Axis, E-W Horiz Axis	Average	1.4	2.3	2.8	3.2	3.8	4.0	4.1	3.4	2.4	1.7	1.0	1.1	2.6
	Min/Max	0.8/2.0	1.3/3.1	1.6/3.7	2.3/4.2	2.2/5.5	2.5/5.1	3.3/5.4	2.2/4.6	1.3/3.8	0.8/2.8	0.3/1.9	0.5/2.0	2.3/2.9
1-Axis, N-S Horiz Axis	Average	0.9	1.8	3.0	4.1	5.0	5.3	5.5	4.4	2.8	1.6	0.7	0.6	3.0
	Min/Max	0.5/1.3	1.1/2.5	1.7/4.0	3.0/5.3	2.8/7.3	3.4/6.8	4.4/7.1	2.8/5.9	1.6/4.4	0.8/2.6	0.2/1.4	0.3/1.1	2.6/3.3
1-Axis, N-S Tilt=Latitude	Average	1.5	2.7	3.7	4.4	5.0	5.0	5.3	4.6	3.3	2.2	1.2	1.1	3.3
	Min/Max	0.8/2.2	1.6/3.7	2.1/5.0	3.3/5.9	2.8/7.3	3.2/6.5	4.3/6.9	3.0/6.2	1.8/5.1	1.0/3.5	0.4/2.2	0.5/2.0	2.9/3.7
2-Axis	Average	1.6	2.7	3.7	4.5	5.3	5.5	5.7	4.7	3.3	2.2	1.2	1.2	3.5
	Min/Max	0.9/2.3	1.7/3.8	2.1/5.0	3.3/5.9	3.0/7.7	3.5/7.1	4.6/7.4	3.1/6.4	1.8/5.1	1.0/3.6	0.4/2.3	0.5/2.2	3.0/3.9

Average Climatic Conditions

Element	Jan	Feb	Mar	Apr	May	June	July	Aug	Sept	Oct	Nov	Dec	Year
Temperature (°C)	-10.6	-10.0	-4.4	3.4	10.3	14.4	17.7	17.0	12.8	7.4	0.6	-7.2	4.3
Daily Minimum Temp	-15.2	-15.1	-9.3	-2.0	3.6	7.5	10.7	10.7	6.8	2.3	-3.4	-11.2	-1.2
Daily Maximum Temp	-6.1	-4.9	0.4	8.9	17.0	21.4	24.6	23.2	18.8	12.4	4.4	-3.2	9.8
Record Minimum Temp	-37.8	-37.2	-31.1	-18.9	-7.8	-3.3	2.2	-1.7	-3.9	-8.9	-23.3	-31.7	-37.8
Record Maximum Temp	7.2	8.3	23.9	29.4	31.7	33.9	36.1	36.7	35.0	26.7	19.4	15.6	36.7
HDD, Base 18.3°C	897	793	706	447	254	124	51	71	167	339	533	792	5176
CDD, Base 18.3°C	0	0	0	0	4	8	31	30	0	0	0	0	73
Relative Humidity (%)	77	75	75	70	68	75	76	80	82	80	82	81	77
Wind Speed (m/s)	4.2	3.9	4.3	4.4	4.1	3.7	3.4	3.3	3.6	3.8	4.2	4.1	3.9

Traverse City, MI

WBAN NO. 14850

LATITUDE: 44.73° N
LONGITUDE: 85.58° W
ELEVATION: 192 meters
MEAN PRESSURE: 993 millibars

STATION TYPE: Secondary

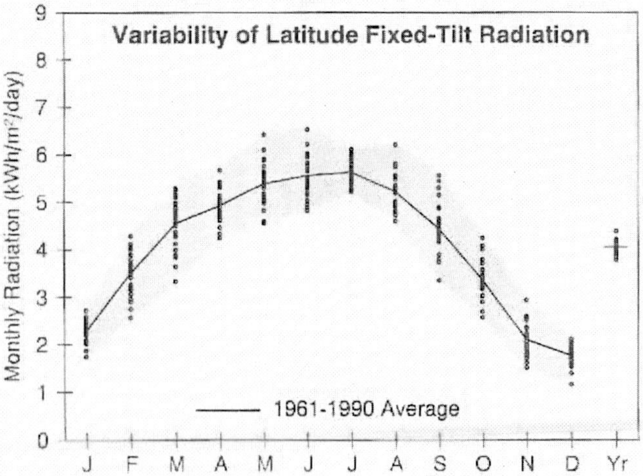

Solar Radiation for Flat-Plate Collectors Facing South at a Fixed Tilt (kWh/m²/day), Uncertainty ±9%

Tilt (°)		Jan	Feb	Mar	Apr	May	June	July	Aug	Sept	Oct	Nov	Dec	Year
0	Average	1.5	2.4	3.5	4.6	5.6	6.2	6.1	5.1	3.7	2.4	1.4	1.2	3.6
	Min/Max	1.3/1.7	1.9/2.8	2.8/4.0	4.1/5.1	4.8/6.6	5.4/7.2	5.7/6.6	4.6/5.9	3.1/4.4	2.0/2.9	1.2/1.7	0.9/1.4	3.5/3.8
Latitude -15	Average	2.1	3.3	4.4	5.0	5.8	6.1	6.1	5.4	4.4	3.2	2.0	1.6	4.1
	Min/Max	1.6/2.4	2.4/3.9	3.3/5.0	4.4/5.8	4.9/6.8	5.3/7.1	5.6/6.6	4.8/6.4	3.4/5.5	2.5/4.0	1.5/2.7	1.1/1.9	3.9/4.4
Latitude	Average	2.3	3.5	4.6	4.9	5.4	5.6	5.6	5.2	4.4	3.4	2.1	1.8	4.1
	Min/Max	1.7/2.7	2.6/4.3	3.3/5.3	4.2/5.7	4.6/6.4	4.8/6.5	5.2/6.1	4.6/6.2	3.3/5.6	2.6/4.2	1.5/2.9	1.1/2.1	3.8/4.4
Latitude +15	Average	2.4	3.6	4.5	4.6	4.8	4.8	4.9	4.7	4.2	3.3	2.1	1.8	3.8
	Min/Max	1.8/2.9	2.6/4.4	3.2/5.3	3.9/5.3	4.0/5.7	4.2/5.6	4.5/5.3	4.2/5.6	3.1/5.3	2.5/4.3	1.5/3.1	1.1/2.2	3.5/4.1
90	Average	2.2	3.3	3.8	3.2	2.9	2.8	2.9	3.1	3.1	2.8	1.9	1.7	2.8
	Min/Max	1.7/2.8	2.3/4.1	2.5/4.7	2.7/3.7	2.6/3.4	2.5/3.1	2.7/3.1	2.8/3.7	2.3/4.0	2.0/3.6	1.3/2.8	1.0/2.2	2.5/3.1

Solar Radiation for 1-Axis Tracking Flat-Plate Collectors with a North-South Axis (kWh/m²/day), Uncertainty ±9%

Axis Tilt (°)		Jan	Feb	Mar	Apr	May	June	July	Aug	Sept	Oct	Nov	Dec	Year
0	Average	1.9	3.2	4.8	6.1	7.4	8.1	8.1	6.8	5.0	3.2	1.8	1.4	4.8
	Min/Max	1.4/2.2	2.3/3.9	3.4/5.6	5.2/6.9	5.9/9.0	6.8/9.9	7.4/8.9	5.9/8.4	3.7/6.3	2.5/4.2	1.3/2.5	0.9/1.7	4.4/5.2
Latitude -15	Average	2.3	3.8	5.4	6.5	7.6	8.1	8.2	7.1	5.6	3.8	2.2	1.8	5.2
	Min/Max	1.7/2.8	2.7/4.8	3.8/6.5	5.5/7.5	6.0/9.4	6.8/10.0	7.5/9.1	6.2/8.9	4.0/7.1	2.9/5.0	1.5/3.2	1.1/2.1	4.8/5.7
Latitude	Average	2.5	4.0	5.6	6.4	7.4	7.8	7.9	7.0	5.6	3.9	2.3	1.9	5.2
	Min/Max	1.8/3.1	2.8/5.0	3.8/6.7	5.4/7.5	5.8/9.1	6.5/9.6	7.2/8.8	6.0/8.7	3.9/7.2	3.0/5.2	1.6/3.4	1.2/2.3	4.7/5.7
Latitude +15	Average	2.6	4.1	5.5	6.2	7.0	7.3	7.4	6.6	5.4	3.9	2.4	2.0	5.0
	Min/Max	1.9/3.2	2.9/5.2	3.8/6.7	5.2/7.2	5.4/8.6	6.1/9.0	6.8/8.2	5.7/8.3	3.7/7.1	2.9/5.2	1.6/3.5	1.2/2.4	4.6/5.5

Solar Radiation for 2-Axis Tracking Flat-Plate Collectors (kWh/m²/day), Uncertainty ±9%

Tracker		Jan	Feb	Mar	Apr	May	June	July	Aug	Sept	Oct	Nov	Dec	Year
2-Axis	Average	2.6	4.1	5.6	6.5	7.7	8.3	8.3	7.2	5.6	4.0	2.4	2.0	5.4
	Min/Max	1.9/3.2	2.9/5.2	3.8/6.7	5.6/7.6	6.1/9.5	7.0/10.3	7.7/9.3	6.2/8.9	4.0/7.2	3.0/5.3	1.6/3.5	1.2/2.5	4.9/5.9

Direct Beam Solar Radiation for Concentrating Collectors (kWh/m²/day), Uncertainty ±8%

Tracker		Jan	Feb	Mar	Apr	May	June	July	Aug	Sept	Oct	Nov	Dec	Year
1-Axis, E-W	Average	0.9	1.6	2.3	2.9	3.5	3.9	3.9	3.2	2.5	1.9	1.0	0.7	2.4
Horiz Axis	Min/Max	0.5/1.4	0.9/2.5	1.3/3.4	2.1/3.7	2.3/4.9	2.8/5.6	3.3/4.6	2.4/4.4	1.2/3.8	1.1/2.9	0.4/2.0	0.2/1.2	2.1/2.7
1-Axis, N-S	Average	0.5	1.3	2.5	3.7	4.7	5.1	5.2	4.2	2.9	1.7	0.7	0.4	2.7
Horiz Axis	Min/Max	0.3/0.9	0.8/2.1	1.4/3.7	2.7/4.6	2.9/6.3	3.6/7.3	4.3/6.1	3.2/5.8	1.5/4.3	1.0/2.8	0.3/1.5	0.1/0.7	2.4/3.2
1-Axis, N-S	Average	0.9	1.9	3.0	4.0	4.6	4.9	5.0	4.3	3.4	2.3	1.1	0.7	3.0
Tilt=Latitude	Min/Max	0.6/1.5	1.1/3.0	1.7/4.6	2.9/5.0	2.9/6.3	3.5/7.0	4.2/5.9	3.3/6.1	1.7/5.1	1.4/3.7	0.5/2.3	0.2/1.2	2.6/3.5
2-Axis	Average	1.0	1.9	3.0	4.0	4.9	5.3	5.4	4.5	3.4	2.4	1.2	0.7	3.2
	Min/Max	0.6/1.6	1.1/3.1	1.7/4.6	3.0/5.1	3.1/6.7	3.8/7.6	4.5/6.4	3.4/6.3	1.7/5.1	1.4/3.7	0.5/2.4	0.3/1.3	2.8/3.7

Average Climatic Conditions

Element	Jan	Feb	Mar	Apr	May	June	July	Aug	Sept	Oct	Nov	Dec	Year
Temperature (°C)	-6.8	-6.8	-1.4	5.8	12.1	17.5	20.7	19.5	15.3	9.3	2.9	-3.7	7.1
Daily Minimum Temp	-10.4	-11.6	-6.6	-0.1	5.0	10.6	14.1	13.3	9.6	4.2	-0.8	-6.9	1.7
Daily Maximum Temp	-3.2	-2.1	3.6	11.8	19.1	24.4	27.4	25.7	20.9	14.4	6.6	-0.4	12.3
Record Minimum Temp	-28.9	-37.8	-31.7	-15.0	-6.1	-1.7	0.0	0.0	-2.2	-5.6	-15.0	-26.1	-37.8
Record Maximum Temp	11.7	14.4	25.6	32.2	33.3	37.2	37.2	36.7	33.3	31.1	23.3	17.2	37.2
HDD, Base 18.3°C	778	703	613	375	206	64	14	32	96	279	463	682	4305
CDD, Base 18.3°C	0	0	0	0	11	39	88	68	4	0	0	0	209
Relative Humidity (%)	77	74	71	66	66	69	68	73	76	74	76	79	72
Wind Speed (m/s)	4.6	4.2	4.2	4.3	3.8	3.7	3.4	3.4	3.6	3.8	4.2	4.3	3.9

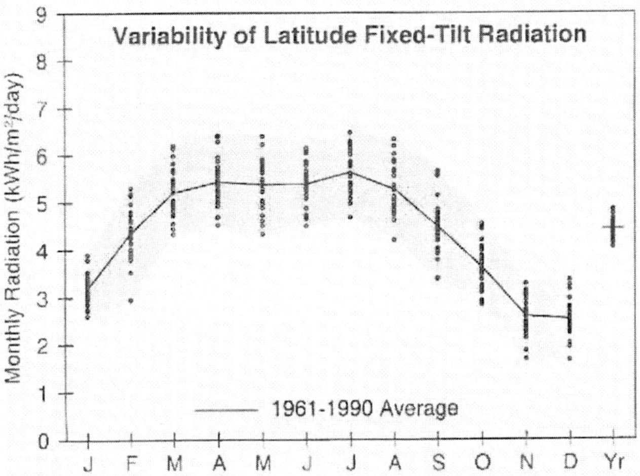

Duluth, MN

WBAN NO. 14913

LATITUDE: 46.83° N
LONGITUDE: 92.18° W
ELEVATION: 432 meters
MEAN PRESSURE: 964 millibars

STATION TYPE: Secondary

Solar Radiation for Flat-Plate Collectors Facing South at a Fixed Tilt (kWh/m²/day), Uncertainty ±9%

Tilt (°)		Jan	Feb	Mar	Apr	May	June	July	Aug	Sept	Oct	Nov	Dec	Year
0	Average	1.6	2.6	3.8	4.8	5.6	6.0	6.1	5.1	3.7	2.5	1.5	1.2	3.7
	Min/Max	1.5/1.8	2.1/2.9	3.3/4.4	4.2/5.5	4.6/6.6	5.1/6.8	5.1/6.8	4.2/5.9	3.0/4.4	2.1/2.9	1.3/1.7	1.0/1.5	3.5/4.0
Latitude -15	Average	2.8	4.0	5.0	5.5	5.7	5.9	6.1	5.5	4.5	3.4	2.4	2.2	4.4
	Min/Max	2.3/3.4	2.8/4.7	4.2/5.9	4.7/6.5	4.6/6.8	4.9/6.7	5.1/7.0	4.4/6.6	3.4/5.6	2.7/4.2	1.6/2.9	1.5/2.9	4.1/4.8
Latitude	Average	3.2	4.4	5.2	5.4	5.4	5.4	5.6	5.3	4.5	3.6	2.6	2.5	4.4
	Min/Max	2.6/3.9	3.0/5.3	4.3/6.2	4.5/6.4	4.3/6.4	4.5/6.1	4.7/6.5	4.2/6.3	3.4/5.7	2.8/4.5	1.7/3.3	1.7/3.4	4.0/4.8
Latitude +15	Average	3.4	4.5	5.2	5.1	4.7	4.7	4.9	4.8	4.3	3.6	2.7	2.7	4.2
	Min/Max	2.8/4.2	3.0/5.6	4.3/6.2	4.2/6.1	3.8/5.6	3.9/5.3	4.1/5.6	3.8/5.7	3.2/5.4	2.8/4.6	1.7/3.4	1.8/3.7	3.9/4.6
90	Average	3.4	4.3	4.6	3.9	3.0	2.9	3.0	3.2	3.3	3.1	2.5	2.7	3.3
	Min/Max	2.7/4.2	2.8/5.3	3.8/5.6	3.0/4.8	2.5/3.5	2.5/3.1	2.6/3.4	2.6/3.9	2.4/4.2	2.3/3.9	1.5/3.2	1.7/3.7	3.0/3.6

Solar Radiation for 1-Axis Tracking Flat-Plate Collectors with a North-South Axis (kWh/m²/day), Uncertainty ±9%

Axis Tilt (°)		Jan	Feb	Mar	Apr	May	June	July	Aug	Sept	Oct	Nov	Dec	Year
0	Average	2.4	3.8	5.4	6.6	7.4	7.8	8.1	6.9	5.1	3.4	2.1	1.9	5.1
	Min/Max	2.0/2.9	2.6/4.6	4.4/6.4	5.5/8.1	5.4/9.1	6.2/9.4	6.6/9.4	5.4/8.5	3.8/6.4	2.7/4.3	1.5/2.7	1.2/2.4	4.6/5.7
Latitude -15	Average	3.3	4.9	6.3	7.2	7.6	7.9	8.3	7.3	5.7	4.2	2.7	2.6	5.7
	Min/Max	2.7/4.1	3.2/6.0	5.1/7.6	6.0/8.9	5.6/9.5	6.2/9.5	6.7/9.7	5.6/9.1	4.1/7.3	3.2/5.3	1.7/3.6	1.7/3.5	5.1/6.4
Latitude	Average	3.7	5.2	6.5	7.1	7.4	7.6	8.0	7.2	5.7	4.3	2.9	2.9	5.7
	Min/Max	2.9/4.5	3.3/6.4	5.2/7.8	5.9/8.9	5.4/9.2	5.9/9.1	6.4/9.4	5.5/8.9	4.1/7.4	3.2/5.6	1.8/3.8	1.8/3.9	5.1/6.4
Latitude +15	Average	3.8	5.3	6.5	6.9	7.0	7.1	7.5	6.8	5.5	4.3	3.0	3.0	5.6
	Min/Max	3.0/4.8	3.3/6.6	5.1/7.8	5.7/8.6	5.0/8.7	5.5/8.5	6.0/8.8	5.2/8.5	3.9/7.2	3.2/5.6	1.8/4.0	1.9/4.1	5.0/6.3

Solar Radiation for 2-Axis Tracking Flat-Plate Collectors (kWh/m²/day), Uncertainty ±9%

Tracker		Jan	Feb	Mar	Apr	May	June	July	Aug	Sept	Oct	Nov	Dec	Year
2-Axis	Average	3.9	5.3	6.5	7.2	7.8	8.1	8.5	7.4	5.7	4.4	3.1	3.1	5.9
	Min/Max	3.1/4.8	3.4/6.6	5.2/7.8	6.0/9.0	5.7/9.6	6.4/9.7	6.8/9.9	5.7/9.1	4.1/7.4	3.3/5.6	1.8/4.0	1.9/4.2	5.3/6.6

Direct Beam Solar Radiation for Concentrating Collectors (kWh/m²/day), Uncertainty ±8%

Tracker		Jan	Feb	Mar	Apr	May	June	July	Aug	Sept	Oct	Nov	Dec	Year
1-Axis, E-W Horiz Axis	Average	2.1	2.6	2.7	3.1	3.5	3.6	3.9	3.4	2.7	2.2	1.6	1.6	2.8
	Min/Max	1.4/3.0	1.0/3.7	1.9/3.6	2.1/4.7	2.1/5.1	2.1/4.9	2.7/5.2	2.2/4.6	1.4/4.0	1.3/3.2	0.5/2.4	0.8/2.6	2.3/3.4
1-Axis, N-S Horiz Axis	Average	1.3	2.1	2.9	4.0	4.7	4.8	5.2	4.4	3.1	2.0	1.1	0.9	3.0
	Min/Max	0.9/1.9	0.8/3.0	1.9/3.9	2.7/6.0	2.6/6.6	2.8/6.5	3.6/6.8	2.8/6.1	1.7/4.6	1.2/2.9	0.3/1.7	0.5/1.5	2.6/3.8
1-Axis, N-S Tilt=Latitude	Average	2.2	3.1	3.6	4.4	4.7	4.6	5.1	4.6	3.6	2.8	1.8	1.7	3.5
	Min/Max	1.5/3.2	1.2/4.4	2.5/4.9	2.9/6.6	2.7/6.6	2.7/6.3	3.5/6.7	3.0/6.4	1.9/5.4	1.6/4.0	0.6/2.7	0.9/2.7	2.9/4.3
2-Axis	Average	2.4	3.2	3.6	4.4	4.9	5.0	5.5	4.8	3.7	2.8	1.8	1.8	3.7
	Min/Max	1.6/3.4	1.3/4.5	2.5/4.9	3.0/6.7	2.8/7.0	3.0/6.8	3.8/7.2	3.1/6.6	2.0/5.4	1.6/4.1	0.6/2.8	0.9/2.9	3.1/4.5

Average Climatic Conditions

Element	Jan	Feb	Mar	Apr	May	June	July	Aug	Sept	Oct	Nov	Dec	Year
Temperature (°C)	-13.9	-10.9	-4.2	3.7	10.4	15.4	18.9	17.6	12.3	6.5	-2.0	-10.7	3.6
Daily Minimum Temp	-19.0	-16.2	-9.1	-1.7	4.2	9.2	12.8	11.8	6.9	1.7	-5.8	-15.1	-1.7
Daily Maximum Temp	-8.8	-5.7	0.5	9.0	16.6	21.7	25.1	23.3	17.7	11.3	1.8	-6.3	8.8
Record Minimum Temp	-39.4	-36.1	-33.9	-20.6	-8.3	-2.8	1.7	0.0	-5.6	-13.3	-30.6	-36.7	-39.4
Record Maximum Temp	11.1	12.8	25.6	31.1	32.2	33.9	36.1	36.1	35.0	30.0	21.1	12.8	36.1
HDD, Base 18.3°C	999	820	699	440	247	94	33	63	183	367	610	899	5454
CDD, Base 18.3°C	0	0	0	0	0	8	52	40	0	0	0	0	100
Relative Humidity (%)	72	70	69	64	63	69	71	75	76	71	75	76	71
Wind Speed (m/s)	5.1	4.8	5.3	5.2	4.9	4.5	4.0	4.0	4.4	4.6	4.9	4.8	4.7

International Falls, MN

WBAN NO. 14918

LATITUDE: 48.57° N
LONGITUDE: 93.38° W
ELEVATION: 361 meters
MEAN PRESSURE: 972 millibars

STATION TYPE: Secondary

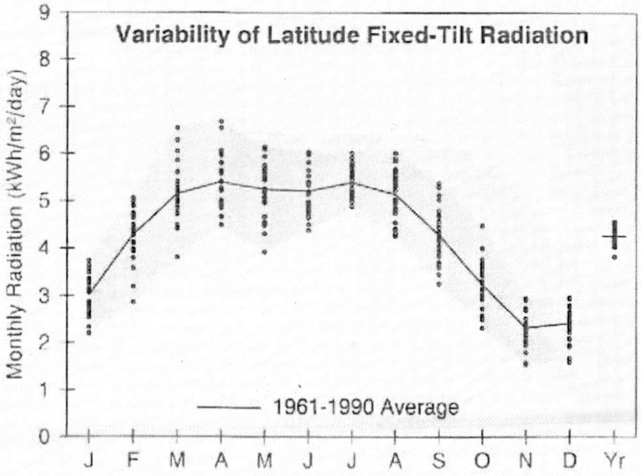

Solar Radiation for Flat-Plate Collectors Facing South at a Fixed Tilt (kWh/m²/day), Uncertainty ±9%

Tilt (°)		Jan	Feb	Mar	Apr	May	June	July	Aug	Sept	Oct	Nov	Dec	Year
0	Average	1.4	2.4	3.7	4.8	5.5	5.8	5.8	4.9	3.5	2.2	1.4	1.1	3.6
	Min/Max	1.2/1.6	1.9/2.7	3.0/4.3	4.1/5.6	4.3/6.3	5.0/6.7	5.3/6.4	4.3/5.6	2.8/4.1	1.9/2.7	1.2/1.5	0.8/1.3	3.3/3.8
Latitude -15	Average	2.6	3.9	4.9	5.5	5.6	5.7	5.9	5.4	4.3	3.1	2.1	2.1	4.3
	Min/Max	2.0/3.2	2.7/4.6	3.7/6.2	4.6/6.8	4.2/6.6	4.8/6.6	5.3/6.5	4.5/6.3	3.3/5.3	2.3/4.2	1.5/2.6	1.4/2.5	3.9/4.6
Latitude	Average	3.0	4.3	5.1	5.4	5.2	5.2	5.4	5.1	4.3	3.2	2.3	2.4	4.3
	Min/Max	2.2/3.7	2.9/5.1	3.8/6.5	4.5/6.7	3.9/6.2	4.4/6.0	4.9/6.0	4.3/6.0	3.3/5.4	2.3/4.5	1.5/3.0	1.6/3.0	3.8/4.6
Latitude +15	Average	3.2	4.5	5.1	5.1	4.6	4.5	4.7	4.6	4.1	3.2	2.4	2.6	4.0
	Min/Max	2.3/4.0	2.9/5.3	3.7/6.6	4.1/6.2	3.4/5.4	3.8/5.2	4.3/5.2	3.8/5.4	3.0/5.1	2.2/4.5	1.5/3.1	1.7/3.2	3.6/4.3
90	Average	3.2	4.3	4.6	3.9	3.1	2.9	3.0	3.3	3.2	2.8	2.2	2.6	3.2
	Min/Max	2.3/4.0	2.8/5.1	3.1/5.9	3.0/5.0	2.3/3.5	2.5/3.3	2.8/3.3	2.7/3.8	2.4/4.1	1.9/4.0	1.4/3.0	1.6/3.2	2.8/3.5

Solar Radiation for 1-Axis Tracking Flat-Plate Collectors with a North-South Axis (kWh/m²/day), Uncertainty ±9%

Axis Tilt (°)		Jan	Feb	Mar	Apr	May	June	July	Aug	Sept	Oct	Nov	Dec	Year
0	Average	2.2	3.7	5.3	6.7	7.4	7.7	7.9	6.8	4.8	3.1	1.8	1.7	4.9
	Min/Max	1.6/2.7	2.5/4.3	3.9/6.7	5.6/8.5	5.3/9.0	6.2/9.3	6.7/8.9	5.5/8.0	3.6/6.1	2.2/4.2	1.3/2.3	1.1/2.1	4.5/5.4
Latitude -15	Average	3.1	4.8	6.3	7.3	7.6	7.7	8.1	7.3	5.5	3.7	2.4	2.4	5.5
	Min/Max	2.2/3.8	3.1/5.7	4.5/8.0	6.0/9.5	5.4/9.4	6.2/9.4	6.8/9.2	5.7/8.7	4.0/7.0	2.5/5.3	1.5/3.1	1.6/3.0	4.9/6.1
Latitude	Average	3.4	5.1	6.4	7.2	7.4	7.4	7.8	7.1	5.5	3.8	2.6	2.7	5.5
	Min/Max	2.4/4.3	3.2/6.1	4.6/8.3	5.9/9.4	5.2/9.1	5.9/9.1	6.6/8.9	5.6/8.5	4.0/7.1	2.6/5.6	1.6/3.4	1.7/3.3	4.9/6.1
Latitude +15	Average	3.6	5.2	6.4	7.0	7.0	7.0	7.4	6.8	5.3	3.8	2.7	2.9	5.4
	Min/Max	2.5/4.5	3.3/6.3	4.5/8.3	5.7/9.1	4.9/8.6	5.6/8.5	6.1/8.4	5.3/8.1	3.8/6.9	2.5/5.6	1.6/3.5	1.8/3.6	4.7/5.9

Solar Radiation for 2-Axis Tracking Flat-Plate Collectors (kWh/m²/day), Uncertainty ±9%

Tracker		Jan	Feb	Mar	Apr	May	June	July	Aug	Sept	Oct	Nov	Dec	Year
2-Axis	Average	3.6	5.2	6.4	7.3	7.7	8.0	8.3	7.3	5.5	3.9	2.7	2.9	5.7
	Min/Max	2.6/4.6	3.3/6.3	4.6/8.3	6.0/9.5	5.5/9.5	6.4/9.7	7.0/9.4	5.8/8.7	4.0/7.1	2.6/5.6	1.6/3.5	1.8/3.6	5.1/6.3

Direct Beam Solar Radiation for Concentrating Collectors (kWh/m²/day), Uncertainty ±8%

Tracker		Jan	Feb	Mar	Apr	May	June	July	Aug	Sept	Oct	Nov	Dec	Year
1-Axis, E-W	Average	1.9	2.6	2.8	3.2	3.5	3.5	3.7	3.3	2.5	1.9	1.3	1.6	2.6
Horiz Axis	Min/Max	1.1/2.7	1.2/3.5	1.8/4.1	2.2/5.1	1.8/4.9	2.2/4.9	2.7/4.6	2.1/4.4	1.4/3.6	0.8/3.3	0.4/2.0	0.8/2.2	2.2/3.1
1-Axis, N-S	Average	1.2	2.1	2.9	4.1	4.6	4.6	5.0	4.3	2.9	1.7	0.9	0.8	2.9
Horiz Axis	Min/Max	0.7/1.6	0.9/2.8	1.9/4.3	2.7/6.4	2.5/6.5	3.0/6.6	3.6/6.2	2.8/5.8	1.7/4.2	0.8/3.0	0.3/1.3	0.4/1.2	2.5/3.5
1-Axis, N-S	Average	2.1	3.1	3.7	4.5	4.7	4.5	4.9	4.5	3.4	2.4	1.4	1.6	3.4
Tilt=Latitude	Min/Max	1.2/2.8	1.4/4.2	2.4/5.4	3.0/7.2	2.5/6.5	2.9/6.3	3.5/6.1	3.0/6.1	2.0/5.0	1.0/4.2	0.5/2.2	0.9/2.2	2.9/4.0
2-Axis	Average	2.2	3.2	3.7	4.6	4.9	4.9	5.3	4.7	3.5	2.4	1.5	1.7	3.6
	Min/Max	1.3/3.0	1.5/4.3	2.4/5.5	3.1/7.3	2.6/6.9	3.2/6.9	3.8/6.5	3.1/6.3	2.0/5.0	1.1/4.3	0.5/2.3	0.9/2.4	3.0/4.1

Average Climatic Conditions

Element	Jan	Feb	Mar	Apr	May	June	July	Aug	Sept	Oct	Nov	Dec	Year
Temperature (°C)	-17.2	-13.5	-5.5	3.9	11.2	16.3	19.3	17.6	11.9	5.8	-3.9	-13.8	2.7
Daily Minimum Temp	-23.3	-20.0	-11.4	-2.3	4.2	9.7	12.6	10.9	5.8	0.5	-8.3	-19.0	-3.4
Daily Maximum Temp	-11.2	-7.1	0.4	10.1	18.1	23.0	26.0	24.2	17.9	11.0	0.4	-8.6	8.7
Record Minimum Temp	-43.3	-42.2	-38.9	-25.6	-11.7	-5.0	1.7	-1.1	-6.7	-16.7	-35.6	-40.6	-43.3
Record Maximum Temp	8.9	12.8	24.4	33.9	35.0	36.7	36.7	35.0	35.0	31.1	22.8	13.9	36.7
HDD, Base 18.3°C	1102	891	739	433	232	87	31	64	193	389	668	996	5826
CDD, Base 18.3°C	0	0	0	0	10	27	59	42	0	0	0	0	138
Relative Humidity (%)	71	67	66	61	61	68	71	75	77	74	79	77	71
Wind Speed (m/s)	3.9	3.8	4.2	4.3	4.1	3.7	3.3	3.4	3.7	4.0	4.1	3.9	3.9

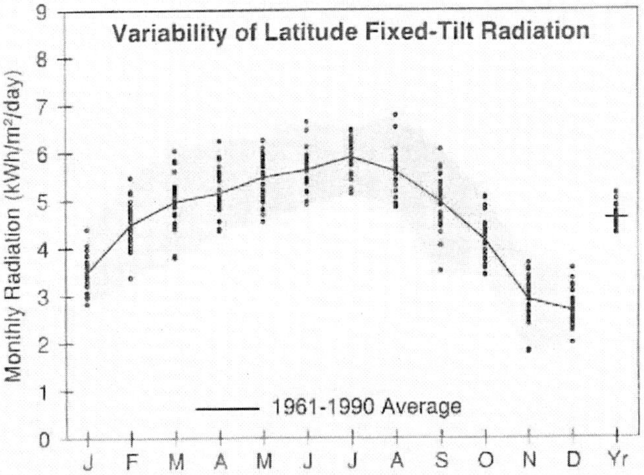

Minneapolis, MN

WBAN NO. 14922

LATITUDE: 44.88° N
LONGITUDE: 93.22° W
ELEVATION: 255 meters
MEAN PRESSURE: 986 millibars

STATION TYPE: Secondary

Solar Radiation for Flat-Plate Collectors Facing South at a Fixed Tilt (kWh/m²/day), Uncertainty ±9%

Tilt (°)		Jan	Feb	Mar	Apr	May	June	July	Aug	Sept	Oct	Nov	Dec	Year
0	Average	1.8	2.7	3.8	4.7	5.7	6.3	6.3	5.4	4.1	2.8	1.7	1.4	3.9
	Min/Max	1.5/2.0	2.4/3.1	3.1/4.3	4.1/5.5	4.9/6.5	5.5/7.4	5.6/6.9	4.8/6.4	3.3/4.8	2.4/3.2	1.3/2.0	1.1/1.7	3.7/4.2
Latitude -15	Average	3.1	4.1	4.8	5.3	5.8	6.1	6.4	5.8	4.9	3.9	2.6	2.3	4.6
	Min/Max	2.5/3.8	3.2/4.9	3.7/5.7	4.4/6.3	4.9/6.7	5.3/7.3	5.6/7.0	5.1/7.0	3.6/6.0	3.3/4.7	1.7/3.3	1.8/3.1	4.3/5.1
Latitude	Average	3.5	4.5	5.0	5.1	5.5	5.6	5.9	5.6	5.0	4.1	2.9	2.7	4.6
	Min/Max	2.8/4.4	3.4/5.5	3.8/6.0	4.4/6.2	4.5/6.3	4.9/6.6	5.1/6.5	4.8/6.8	3.5/6.1	3.4/5.1	1.8/3.7	2.0/3.6	4.3/5.1
Latitude +15	Average	3.8	4.7	4.9	4.8	4.8	4.9	5.1	5.1	4.7	4.2	3.0	2.9	4.4
	Min/Max	3.0/4.8	3.5/5.8	3.7/6.1	4.0/5.8	4.0/5.5	4.3/5.7	4.5/5.6	4.4/6.2	3.3/5.8	3.4/5.1	1.8/3.9	2.1/3.9	4.0/4.9
90	Average	3.7	4.4	4.1	3.3	3.0	2.8	3.0	3.3	3.5	3.5	2.8	2.8	3.3
	Min/Max	2.8/4.8	3.2/5.4	2.9/5.3	2.8/4.0	2.5/3.3	2.5/3.2	2.7/3.3	2.9/4.0	2.4/4.4	2.8/4.3	1.6/3.6	2.0/3.9	3.0/3.7

Solar Radiation for 1-Axis Tracking Flat-Plate Collectors with a North-South Axis (kWh/m²/day), Uncertainty ±9%

Axis Tilt (°)		Jan	Feb	Mar	Apr	May	June	July	Aug	Sept	Oct	Nov	Dec	Year
0	Average	2.7	4.0	5.2	6.3	7.5	8.2	8.5	7.4	5.7	4.0	2.4	2.0	5.3
	Min/Max	2.2/3.5	3.0/4.9	3.9/6.4	5.0/7.8	6.2/8.9	6.9/10.4	7.3/9.6	6.3/9.1	4.0/7.0	3.2/4.9	1.5/3.1	1.5/2.6	4.9/6.1
Latitude -15	Average	3.7	5.0	6.0	6.8	7.8	8.3	8.7	7.8	6.3	4.8	3.1	2.7	5.9
	Min/Max	3.0/4.7	3.7/6.2	4.4/7.5	5.3/8.5	6.3/9.2	6.9/10.5	7.4/9.9	6.6/9.8	4.3/7.9	3.8/6.0	1.8/4.0	2.0/3.7	5.4/6.8
Latitude	Average	4.0	5.4	6.2	6.7	7.5	8.0	8.4	7.7	6.4	5.0	3.3	3.0	6.0
	Min/Max	3.2/5.2	3.8/6.7	4.4/7.7	5.3/8.5	6.1/8.9	6.6/10.1	7.1/9.5	6.4/9.6	4.3/8.0	4.0/6.3	1.9/4.3	2.2/4.1	5.4/6.8
Latitude +15	Average	4.3	5.5	6.1	6.5	7.1	7.4	7.9	7.3	6.2	5.0	3.4	3.2	5.8
	Min/Max	3.3/5.5	3.9/6.9	4.3/7.7	5.1/8.1	5.7/8.4	6.2/9.5	6.7/8.9	6.1/9.1	4.1/7.8	4.0/6.4	1.9/4.5	2.3/4.3	5.2/6.7

Solar Radiation for 2-Axis Tracking Flat-Plate Collectors (kWh/m²/day), Uncertainty ±9%

Tracker		Jan	Feb	Mar	Apr	May	June	July	Aug	Sept	Oct	Nov	Dec	Year
2-Axis	Average	4.3	5.5	6.2	6.8	7.9	8.5	8.9	7.9	6.4	5.1	3.4	3.2	6.2
	Min/Max	3.4/5.6	3.9/6.9	4.4/7.7	5.3/8.6	6.4/9.3	7.1/10.8	7.5/10.0	6.6/9.8	4.3/8.0	4.0/6.4	2.0/4.5	2.3/4.4	5.6/7.1

Direct Beam Solar Radiation for Concentrating Collectors (kWh/m²/day), Uncertainty ±8%

Tracker		Jan	Feb	Mar	Apr	May	June	July	Aug	Sept	Oct	Nov	Dec	Year
1-Axis, E-W	Average	2.4	2.8	2.8	3.1	3.6	4.0	4.3	3.7	3.1	2.7	1.9	1.8	3.0
Horiz Axis	Min/Max	1.4/3.5	1.4/4.0	1.7/4.0	1.7/4.5	2.4/4.7	2.7/5.8	3.3/5.3	2.8/5.3	1.5/4.2	1.9/3.8	0.7/2.7	1.1/2.6	2.5/3.7
1-Axis, N-S	Average	1.6	2.4	3.1	3.9	4.8	5.2	5.6	4.9	3.7	2.5	1.3	1.1	3.3
Horiz Axis	Min/Max	0.9/2.3	1.2/3.3	1.8/4.4	2.1/5.6	3.2/6.2	3.6/7.7	4.4/6.9	3.7/6.8	1.8/5.0	1.7/3.6	0.5/2.0	0.7/1.6	2.8/4.2
1-Axis, N-S	Average	2.6	3.4	3.8	4.3	4.7	5.0	5.5	5.1	4.3	3.4	2.1	1.9	3.8
Tilt=Latitude	Min/Max	1.5/3.8	1.7/4.7	2.2/5.4	2.3/6.1	3.2/6.1	3.5/7.4	4.2/6.7	3.8/7.1	2.1/5.8	2.3/4.8	0.8/3.0	1.2/2.8	3.1/4.8
2-Axis	Average	2.8	3.5	3.8	4.3	5.0	5.4	5.9	5.2	4.3	3.5	2.2	2.0	4.0
	Min/Max	1.6/4.1	1.7/4.9	2.2/5.4	2.3/6.2	3.3/6.5	3.8/8.0	4.5/7.2	4.0/7.3	2.1/5.8	2.4/4.8	0.8/3.2	1.3/3.0	3.3/5.0

Average Climatic Conditions

Element	Jan	Feb	Mar	Apr	May	June	July	Aug	Sept	Oct	Nov	Dec	Year
Temperature (°C)	-11.2	-7.8	-0.6	8.0	14.7	20.1	23.1	21.4	15.8	9.3	0.7	-7.8	7.2
Daily Minimum Temp	-16.2	-12.7	-5.2	2.3	8.7	14.2	17.3	15.7	10.2	3.8	-3.8	-12.1	1.8
Daily Maximum Temp	-6.3	-3.0	4.0	13.6	20.8	26.0	28.9	27.1	21.5	14.9	5.0	-3.6	12.4
Record Minimum Temp	-36.7	-33.3	-35.6	-16.7	-7.8	1.1	6.1	3.9	-3.3	-9.4	-27.2	-33.9	-36.7
Record Maximum Temp	14.4	15.6	28.3	35.0	35.6	38.9	40.6	38.9	36.7	31.7	23.9	17.2	40.6
HDD, Base 18.3°C	916	733	586	310	136	23	6	12	93	279	530	811	4434
CDD, Base 18.3°C	0	0	0	0	24	76	154	107	18	0	0	0	379
Relative Humidity (%)	70	70	67	60	60	64	65	68	71	68	73	74	67
Wind Speed (m/s)	4.7	4.6	5.1	5.4	4.8	4.6	4.2	4.1	4.3	4.6	4.7	4.6	4.6

Rochester, MN

WBAN NO. 14925

LATITUDE: 43.92° N
LONGITUDE: 92.50° W
ELEVATION: 402 meters
MEAN PRESSURE: 968 millibars

STATION TYPE: Secondary

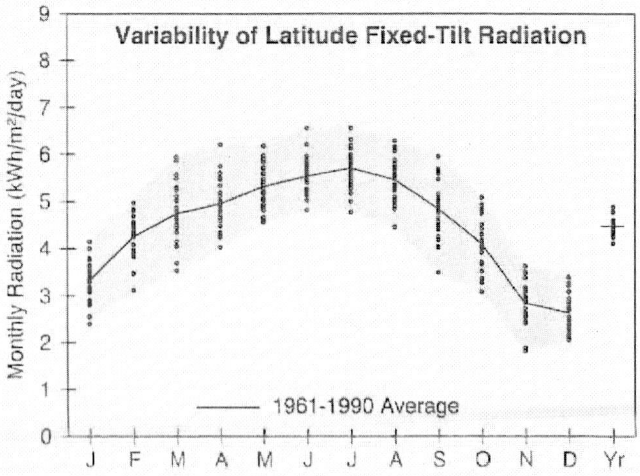

Solar Radiation for Flat-Plate Collectors Facing South at a Fixed Tilt (kWh/m²/day), Uncertainty ±9%

Tilt (°)		Jan	Feb	Mar	Apr	May	June	July	Aug	Sept	Oct	Nov	Dec	Year
0	Average	1.8	2.7	3.7	4.6	5.6	6.2	6.2	5.3	4.0	2.8	1.7	1.4	3.8
	Min/Max	1.5/2.1	2.3/3.0	3.0/4.3	3.9/5.5	4.8/6.5	5.4/7.3	5.3/7.0	4.5/6.0	3.3/4.7	2.3/3.3	1.4/2.0	1.2/1.7	3.7/4.1
Latitude -15	Average	2.9	3.9	4.6	5.1	5.7	6.0	6.2	5.7	4.8	3.8	2.6	2.3	4.5
	Min/Max	2.2/3.6	2.9/4.5	3.5/5.6	4.1/6.3	4.9/6.6	5.3/7.1	5.2/7.1	4.7/6.5	3.6/5.8	3.0/4.7	1.7/3.2	1.8/2.9	4.1/4.9
Latitude	Average	3.3	4.2	4.7	4.9	5.3	5.5	5.7	5.5	4.8	4.1	2.8	2.6	4.5
	Min/Max	2.4/4.2	3.1/5.0	3.5/5.9	4.0/6.2	4.6/6.2	4.8/6.6	4.8/6.6	4.4/6.3	3.5/5.9	3.1/5.1	1.8/3.6	2.0/3.4	4.1/4.9
Latitude +15	Average	3.6	4.4	4.7	4.6	4.7	4.8	5.0	4.9	4.6	4.1	2.9	2.8	4.3
	Min/Max	2.5/4.5	3.2/5.2	3.4/6.0	3.8/5.8	4.1/5.4	4.2/5.6	4.2/5.7	4.0/5.7	3.3/5.7	3.0/5.1	1.8/3.8	2.2/3.7	3.9/4.6
90	Average	3.5	4.1	3.9	3.2	2.8	2.7	2.9	3.2	3.4	3.4	2.7	2.7	3.2
	Min/Max	2.4/4.4	2.9/4.9	2.7/5.3	2.7/3.9	2.6/3.2	2.5/3.1	2.5/3.2	2.6/3.6	2.4/4.2	2.5/4.3	1.5/3.5	2.1/3.7	2.9/3.5

Solar Radiation for 1-Axis Tracking Flat-Plate Collectors with a North-South Axis (kWh/m²/day), Uncertainty ±9%

Axis Tilt (°)		Jan	Feb	Mar	Apr	May	June	July	Aug	Sept	Oct	Nov	Dec	Year
0	Average	2.7	3.9	5.0	6.1	7.3	8.0	8.2	7.1	5.5	4.0	2.4	2.0	5.2
	Min/Max	1.9/3.3	2.8/4.5	3.6/6.3	4.7/7.7	5.8/8.9	6.8/9.9	6.7/9.6	5.7/8.3	3.9/6.9	2.9/4.9	1.5/3.0	1.6/2.5	4.8/5.8
Latitude -15	Average	3.5	4.8	5.7	6.5	7.5	8.0	8.3	7.5	6.1	4.7	3.0	2.7	5.7
	Min/Max	2.5/4.4	3.3/5.7	4.0/7.3	5.0/8.4	5.9/9.1	6.8/10.0	6.7/9.7	5.9/8.8	4.2/7.7	3.4/6.0	1.8/3.9	2.1/3.5	5.2/6.4
Latitude	Average	3.8	5.1	5.8	6.4	7.2	7.7	8.0	7.4	6.2	4.9	3.2	2.9	5.7
	Min/Max	2.7/4.9	3.5/6.1	4.1/7.6	4.9/8.3	5.7/8.9	6.5/9.6	6.5/9.4	5.7/8.7	4.2/7.8	3.5/6.3	1.9/4.3	2.3/3.9	5.2/6.4
Latitude +15	Average	4.0	5.2	5.8	6.1	6.8	7.2	7.5	7.0	6.0	5.0	3.3	3.1	5.6
	Min/Max	2.8/5.2	3.5/6.2	4.0/7.6	4.7/8.0	5.3/8.4	6.1/9.0	6.1/8.8	5.4/8.3	4.0/7.6	3.5/6.3	1.9/4.4	2.4/4.1	5.0/6.3

Solar Radiation for 2-Axis Tracking Flat-Plate Collectors (kWh/m²/day), Uncertainty ±9%

Tracker		Jan	Feb	Mar	Apr	May	June	July	Aug	Sept	Oct	Nov	Dec	Year
2-Axis	Average	4.1	5.2	5.8	6.5	7.6	8.2	8.5	7.6	6.2	5.0	3.3	3.1	5.9
	Min/Max	2.8/5.2	3.6/6.2	4.1/7.6	5.0/8.4	6.0/9.3	7.0/10.3	6.9/9.9	5.9/8.9	4.2/7.8	3.6/6.3	1.9/4.4	2.4/4.2	5.3/6.6

Direct Beam Solar Radiation for Concentrating Collectors (kWh/m²/day), Uncertainty ±8%

Tracker		Jan	Feb	Mar	Apr	May	June	July	Aug	Sept	Oct	Nov	Dec	Year
1-Axis, E-W	Average	2.2	2.6	2.6	2.8	3.4	3.7	3.9	3.5	2.9	2.6	1.8	1.7	2.8
Horiz Axis	Min/Max	1.2/3.1	1.2/3.5	1.4/3.6	1.6/4.4	2.2/4.7	2.6/5.4	2.6/5.0	2.2/4.5	1.4/4.2	1.7/3.8	0.6/2.7	1.1/2.4	2.3/3.4
1-Axis, N-S	Average	1.5	2.2	2.8	3.6	4.4	4.9	5.2	4.6	3.5	2.5	1.3	1.0	3.1
Horiz Axis	Min/Max	0.8/2.1	0.9/3.0	1.5/3.9	2.0/5.5	2.8/6.3	3.6/7.1	3.6/6.6	3.0/5.9	1.7/5.0	1.5/3.6	0.4/2.0	0.7/1.5	2.6/3.8
1-Axis, N-S	Average	2.4	3.1	3.4	3.9	4.4	4.7	5.0	4.7	4.0	3.3	2.0	1.8	3.6
Tilt=Latitude	Min/Max	1.2/3.3	1.4/4.3	1.8/4.8	2.2/6.0	2.8/6.2	3.4/6.8	3.5/6.4	3.1/6.1	2.0/5.7	2.1/4.7	0.6/3.0	1.1/2.5	2.9/4.3
2-Axis	Average	2.6	3.2	3.4	3.9	4.7	5.1	5.4	4.9	4.0	3.4	2.1	1.9	3.7
	Min/Max	1.3/3.5	1.4/4.4	1.8/4.8	2.2/6.1	3.0/6.5	3.7/7.4	3.7/6.8	3.2/6.3	2.0/5.8	2.1/4.8	0.7/3.1	1.2/2.8	3.1/4.5

Average Climatic Conditions

Element	Jan	Feb	Mar	Apr	May	June	July	Aug	Sept	Oct	Nov	Dec	Year
Temperature (°C)	-11.4	-8.3	-1.2	7.2	13.8	19.2	21.6	20.1	15.1	8.8	0.3	-8.2	6.4
Daily Minimum Temp	-16.3	-13.3	-5.9	1.4	7.5	12.9	15.6	14.2	9.2	3.1	-4.2	-12.6	0.9
Daily Maximum Temp	-6.6	-3.3	3.4	12.9	20.1	25.4	27.7	26.0	21.0	14.6	4.9	-3.9	11.8
Record Minimum Temp	-35.6	-33.9	-35.0	-15.0	-6.1	1.7	5.6	2.8	-5.0	-11.7	-28.9	-36.1	-36.1
Record Maximum Temp	12.8	17.2	26.1	32.8	33.3	38.3	38.9	37.2	35.0	31.1	22.8	16.7	38.9
HDD, Base 18.3°C	922	745	606	335	157	31	9	18	103	294	540	823	4583
CDD, Base 18.3°C	0	0	0	0	16	56	111	74	6	0	0	0	262
Relative Humidity (%)	77	76	75	68	67	69	72	75	75	72	77	80	74
Wind Speed (m/s)	6.6	6.3	6.6	6.6	6.0	5.7	4.9	4.9	5.4	5.9	6.2	6.3	5.9

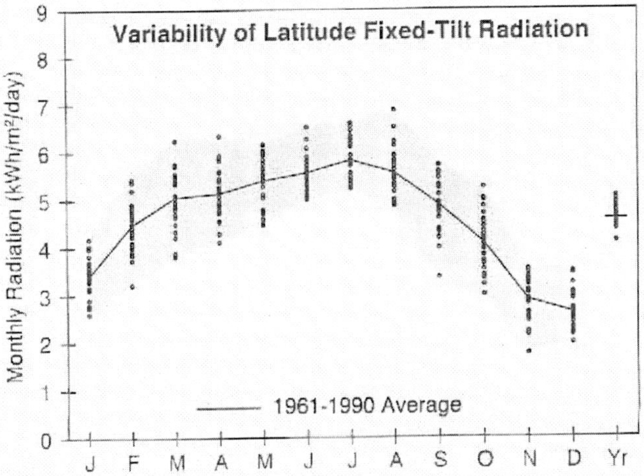

Saint Cloud, MN

WBAN NO. 14926

LATITUDE: 45.55° N
LONGITUDE: 94.07° W
ELEVATION: 313 meters
MEAN PRESSURE: 978 millibars

STATION TYPE: Secondary

Solar Radiation for Flat-Plate Collectors Facing South at a Fixed Tilt (kWh/m²/day), Uncertainty ±9%

Tilt (°)		Jan	Feb	Mar	Apr	May	June	July	Aug	Sept	Oct	Nov	Dec	Year
0	Average	1.7	2.7	3.8	4.7	5.6	6.2	6.3	5.4	4.0	2.7	1.7	1.3	3.8
	Min/Max	1.4/1.9	2.2/3.0	3.1/4.4	4.0/5.6	4.7/6.4	5.6/7.2	5.6/7.1	4.8/6.5	3.2/4.5	2.2/3.2	1.3/1.9	1.1/1.6	3.6/4.2
Latitude -15	Average	3.0	4.1	4.8	5.3	5.8	6.1	6.3	5.8	4.8	3.8	2.6	2.3	4.6
	Min/Max	2.3/3.6	3.0/4.8	3.7/5.9	4.3/6.4	4.8/6.6	5.5/7.1	5.6/7.1	5.1/7.1	3.4/5.6	2.9/4.9	1.7/3.1	1.8/3.0	4.1/5.0
Latitude	Average	3.4	4.5	5.0	5.1	5.4	5.6	5.8	5.6	4.9	4.1	2.9	2.6	4.6
	Min/Max	2.6/4.2	3.2/5.4	3.8/6.2	4.1/6.3	4.5/6.2	5.0/6.5	5.2/6.6	4.9/6.9	3.4/5.7	3.0/5.3	1.8/3.5	2.0/3.5	4.1/5.1
Latitude +15	Average	3.7	4.7	5.0	4.7	4.8	4.8	5.1	5.0	4.6	4.1	3.0	2.8	4.4
	Min/Max	2.7/4.5	3.3/5.7	3.7/6.3	3.8/5.9	4.0/5.4	4.3/5.6	4.6/5.7	4.4/6.2	3.1/5.5	3.0/5.3	1.8/3.7	2.1/3.8	3.9/4.8
90	Average	3.6	4.4	4.3	3.4	3.0	2.8	3.0	3.3	3.5	3.5	2.8	2.7	3.3
	Min/Max	2.6/4.4	3.1/5.4	3.0/5.6	2.7/4.1	2.6/3.3	2.6/3.2	2.8/3.4	2.9/4.0	2.3/4.2	2.5/4.6	1.6/3.4	2.0/3.8	2.9/3.7

Solar Radiation for 1-Axis Tracking Flat-Plate Collectors with a North-South Axis (kWh/m²/day), Uncertainty ±9%

Axis Tilt (°)		Jan	Feb	Mar	Apr	May	June	July	Aug	Sept	Oct	Nov	Dec	Year
0	Average	2.6	4.0	5.3	6.4	7.5	8.2	8.5	7.4	5.6	3.9	2.4	1.9	5.3
	Min/Max	2.0/3.1	2.8/4.8	3.9/6.5	4.8/8.1	5.6/8.8	7.1/10.1	7.5/10.0	6.2/9.7	3.8/6.7	2.8/5.1	1.5/2.9	1.5/2.5	4.7/5.9
Latitude -15	Average	3.6	5.0	6.2	6.8	7.7	8.2	8.7	7.8	6.3	4.8	3.1	2.7	5.9
	Min/Max	2.7/4.4	3.4/6.2	4.4/7.7	5.1/8.8	5.8/9.1	7.1/10.2	7.6/10.2	6.5/10.3	4.1/7.6	3.3/6.3	1.8/3.8	2.0/3.6	5.2/6.7
Latitude	Average	3.9	5.4	6.3	6.8	7.4	7.9	8.4	7.7	6.3	5.0	3.3	2.9	5.9
	Min/Max	3.0/4.9	3.6/6.7	4.5/7.9	5.0/8.8	5.6/8.9	6.8/9.8	7.3/9.8	6.3/10.1	4.0/7.7	3.5/6.6	1.9/4.1	2.2/4.0	5.2/6.7
Latitude +15	Average	4.1	5.5	6.3	6.5	7.0	7.4	7.9	7.3	6.1	5.0	3.4	3.1	5.8
	Min/Max	3.1/5.2	3.7/6.9	4.4/7.9	4.8/8.4	5.2/8.4	6.3/9.2	6.9/9.3	6.0/9.7	3.9/7.5	3.4/6.6	1.9/4.3	2.2/4.3	5.0/6.6

Solar Radiation for 2-Axis Tracking Flat-Plate Collectors (kWh/m²/day), Uncertainty ±9%

Tracker		Jan	Feb	Mar	Apr	May	June	July	Aug	Sept	Oct	Nov	Dec	Year
2-Axis	Average	4.2	5.5	6.3	6.9	7.8	8.5	8.9	7.9	6.3	5.0	3.4	3.1	6.1
	Min/Max	3.1/5.2	3.7/6.9	4.5/7.9	5.1/8.9	5.9/9.2	7.3/10.4	7.8/10.4	6.5/10.4	4.1/7.7	3.5/6.6	1.9/4.3	2.3/4.3	5.4/7.0

Direct Beam Solar Radiation for Concentrating Collectors (kWh/m²/day), Uncertainty ±8%

Tracker		Jan	Feb	Mar	Apr	May	June	July	Aug	Sept	Oct	Nov	Dec	Year
1-Axis, E-W Horiz Axis	Average	2.4	2.9	2.9	3.1	3.5	3.9	4.2	3.7	3.0	2.7	1.9	1.8	3.0
	Min/Max	1.4/3.6	1.4/4.1	1.9/4.2	1.3/4.7	2.2/4.6	2.9/5.4	3.4/5.3	2.7/5.7	1.4/4.1	1.6/4.1	0.7/2.7	1.0/2.8	2.4/3.7
1-Axis, N-S Horiz Axis	Average	1.5	2.4	3.1	3.9	4.7	5.2	5.6	4.9	3.6	2.5	1.3	1.0	3.3
	Min/Max	0.9/2.3	1.1/3.4	1.9/4.5	1.6/6.0	2.8/6.1	3.8/7.3	4.4/7.1	3.5/7.6	1.7/4.9	1.5/3.8	0.4/1.9	0.6/1.7	2.7/4.1
1-Axis, N-S Tilt=Latitude	Average	2.5	3.4	3.9	4.3	4.7	4.9	5.5	5.1	4.2	3.4	2.1	1.8	3.8
	Min/Max	1.5/3.9	1.6/4.9	2.4/5.6	1.8/6.6	2.9/6.1	3.7/6.9	4.3/6.9	3.6/7.9	1.9/5.6	2.0/5.1	0.7/3.0	1.1/3.0	3.1/4.7
2-Axis	Average	2.7	3.5	3.9	4.3	4.9	5.4	5.9	5.3	4.2	3.4	2.2	2.0	4.0
	Min/Max	1.6/4.2	1.7/5.0	2.4/5.6	1.8/6.7	3.0/6.5	4.0/7.5	4.6/7.4	3.8/8.2	1.9/5.7	2.0/5.2	0.8/3.2	1.2/3.2	3.2/4.9

Average Climatic Conditions

Element	Jan	Feb	Mar	Apr	May	June	July	Aug	Sept	Oct	Nov	Dec	Year
Temperature (°C)	-13.3	-9.8	-2.4	6.4	13.3	18.2	21.2	19.5	14.1	7.7	-1.3	-9.9	5.3
Daily Minimum Temp	-19.1	-15.7	-8.0	0.0	6.3	11.2	14.2	12.7	7.4	1.3	-6.5	-15.0	-0.9
Daily Maximum Temp	-7.5	-4.0	3.1	12.8	20.2	25.2	28.1	26.3	20.6	14.1	3.9	-4.9	11.5
Record Minimum Temp	-41.7	-37.2	-35.6	-19.4	-7.2	0.6	4.4	0.6	-7.8	-15.0	-28.9	-40.6	-41.7
Record Maximum Temp	12.8	12.8	26.1	35.6	36.1	38.9	39.4	39.4	36.7	32.2	23.3	15.6	39.4
HDD, Base 18.3°C	980	789	644	358	171	48	13	30	133	329	588	877	4960
CDD, Base 18.3°C	0	0	0	0	14	45	101	66	5	0	0	0	231
Relative Humidity (%)	72	70	70	64	63	68	69	72	74	70	75	76	70
Wind Speed (m/s)	4.7	3.7	5.0	4.7	4.4	3.7	3.2	3.3	3.4	3.9	4.2	4.4	4.1

Columbia, MO

WBAN NO. 03945

LATITUDE: 38.82° N
LONGITUDE: 92.22° W
ELEVATION: 270 meters
MEAN PRESSURE: 986 millibars

STATION TYPE: Primary

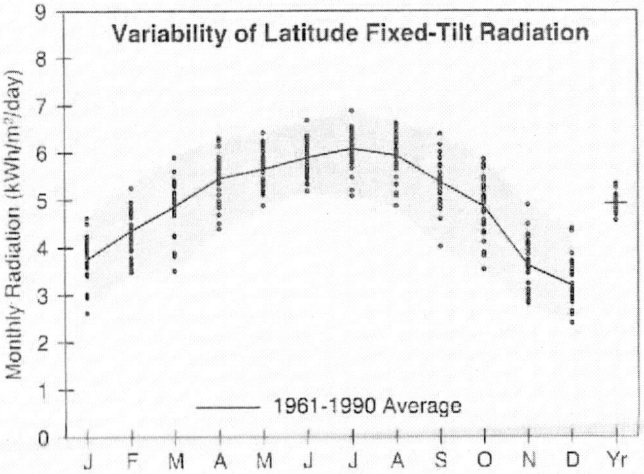

Solar Radiation for Flat-Plate Collectors Facing South at a Fixed Tilt (kWh/m²/day), Uncertainty ±9%

Tilt (°)		Jan	Feb	Mar	Apr	May	June	July	Aug	Sept	Oct	Nov	Dec	Year
0	Average	2.2	3.0	4.0	5.2	6.0	6.6	6.6	5.9	4.6	3.5	2.3	1.9	4.3
	Min/Max	1.8/2.6	2.6/3.4	3.2/4.7	4.3/5.9	5.2/6.8	5.8/7.5	5.5/7.5	5.0/6.5	3.7/5.3	2.9/4.1	2.0/2.8	1.5/2.3	4.1/4.6
Latitude -15	Average	3.3	4.0	4.7	5.6	6.0	6.4	6.6	6.2	5.3	4.5	3.2	2.8	4.9
	Min/Max	2.4/4.0	3.3/4.7	3.5/5.7	4.5/6.4	5.2/6.8	5.6/7.3	5.5/7.4	5.1/6.8	4.1/6.3	3.4/5.4	2.6/4.3	2.2/3.7	4.6/5.3
Latitude	Average	3.8	4.4	4.9	5.5	5.6	5.9	6.1	5.9	5.4	4.9	3.6	3.2	4.9
	Min/Max	2.6/4.6	3.5/5.3	3.5/5.9	4.4/6.3	4.9/6.4	5.2/6.7	5.1/6.9	4.9/6.6	4.0/6.4	3.5/5.9	2.8/4.9	2.4/4.4	4.6/5.3
Latitude +15	Average	4.0	4.5	4.8	5.1	5.0	5.1	5.3	5.4	5.1	4.9	3.8	3.4	4.7
	Min/Max	2.7/5.0	3.5/5.5	3.4/5.8	4.0/5.8	4.3/5.7	4.5/5.7	4.5/6.0	4.4/6.0	3.8/6.2	3.5/6.0	2.9/5.2	2.5/4.8	4.3/5.1
90	Average	3.7	3.9	3.5	3.2	2.7	2.5	2.7	3.1	3.5	3.9	3.3	3.2	3.3
	Min/Max	2.4/4.7	3.0/4.9	2.6/4.3	2.5/3.6	2.4/2.9	2.3/2.7	2.4/2.9	2.6/3.5	2.6/4.2	2.7/4.8	2.5/4.7	2.2/4.6	2.9/3.6

Solar Radiation for 1-Axis Tracking Flat-Plate Collectors with a North-South Axis (kWh/m²/day), Uncertainty ±9%

Axis Tilt (°)		Jan	Feb	Mar	Apr	May	June	July	Aug	Sept	Oct	Nov	Dec	Year
0	Average	3.2	4.2	5.3	6.8	7.7	8.5	8.7	7.8	6.3	5.0	3.2	2.7	5.8
	Min/Max	2.2/4.0	3.4/5.1	3.8/6.6	5.2/8.3	6.6/9.2	7.2/10.0	6.7/10.0	6.3/8.9	4.4/7.6	3.6/6.0	2.5/4.3	2.0/3.7	5.4/6.3
Latitude -15	Average	4.0	4.9	5.9	7.2	7.8	8.4	8.7	8.0	6.8	5.7	3.9	3.3	6.2
	Min/Max	2.7/5.0	3.8/6.1	4.1/7.4	5.4/8.7	6.6/9.3	7.2/10.0	6.7/10.1	6.4/9.2	4.7/8.3	4.0/7.0	3.0/5.4	2.4/4.7	5.8/6.9
Latitude	Average	4.4	5.2	6.0	7.1	7.6	8.1	8.4	7.9	6.9	6.0	4.2	3.7	6.3
	Min/Max	2.9/5.5	4.0/6.5	4.1/7.6	5.3/8.6	6.4/9.1	6.9/9.6	6.4/9.7	6.3/9.1	4.7/8.4	4.1/7.3	3.2/5.9	2.6/5.2	5.8/6.9
Latitude +15	Average	4.6	5.3	5.9	6.8	7.1	7.6	7.8	7.5	6.7	6.0	4.4	3.9	6.1
	Min/Max	3.0/5.8	4.1/6.7	4.0/7.5	5.0/8.3	6.0/8.5	6.4/8.9	6.0/9.1	5.9/8.6	4.6/8.2	4.1/7.4	3.2/6.1	2.7/5.6	5.6/6.8

Solar Radiation for 2-Axis Tracking Flat-Plate Collectors (kWh/m²/day), Uncertainty ±9%

Tracker		Jan	Feb	Mar	Apr	May	June	July	Aug	Sept	Oct	Nov	Dec	Year
2-Axis	Average	4.6	5.3	6.0	7.2	7.9	8.6	8.8	8.1	6.9	6.0	4.4	3.9	6.5
	Min/Max	3.0/5.9	4.1/6.7	4.1/7.6	5.4/8.7	6.7/9.4	7.3/10.2	6.8/10.3	6.4/9.3	4.8/8.4	4.1/7.4	3.2/6.2	2.8/5.7	5.9/7.2

Direct Beam Solar Radiation for Concentrating Collectors (kWh/m²/day), Uncertainty ±8%

Tracker		Jan	Feb	Mar	Apr	May	June	July	Aug	Sept	Oct	Nov	Dec	Year
1-Axis, E-W Horiz Axis	Average	2.7	2.8	2.9	3.3	3.6	4.0	4.2	3.8	3.3	3.3	2.5	2.3	3.2
	Min/Max	1.5/3.9	1.8/3.8	1.2/4.2	2.0/4.3	2.6/4.8	2.9/5.4	2.4/5.4	2.6/4.6	1.7/4.4	1.7/4.4	1.4/3.9	1.3/3.8	2.8/3.8
1-Axis, N-S Horiz Axis	Average	2.0	2.6	3.3	4.3	4.8	5.3	5.6	5.0	4.0	3.3	2.0	1.6	3.6
	Min/Max	1.0/2.9	1.7/3.5	1.4/4.7	2.6/5.8	3.5/6.4	3.9/7.1	3.2/7.1	3.4/6.1	2.0/5.3	1.8/4.4	1.1/3.1	0.9/2.6	3.2/4.3
1-Axis, N-S Tilt=Latitude	Average	3.0	3.4	3.8	4.5	4.6	5.0	5.3	5.1	4.5	4.2	2.8	2.5	4.1
	Min/Max	1.5/4.3	2.2/4.6	1.6/5.5	2.7/6.1	3.5/6.3	3.7/6.7	3.0/6.8	3.4/6.2	2.2/6.0	2.2/5.5	1.6/4.5	1.4/4.0	3.5/4.8
2-Axis	Average	3.2	3.5	3.8	4.6	4.9	5.4	5.7	5.2	4.5	4.2	3.0	2.7	4.2
	Min/Max	1.7/4.6	2.2/4.8	1.6/5.6	2.8/6.2	3.6/6.6	4.0/7.3	3.3/7.3	3.5/6.4	2.2/6.0	2.2/5.6	1.6/4.7	1.5/4.4	3.6/5.0

Average Climatic Conditions

Element	Jan	Feb	Mar	Apr	May	June	July	Aug	Sept	Oct	Nov	Dec	Year
Temperature (°C)	-2.4	0.1	6.2	12.6	17.6	22.3	25.2	24.0	19.9	13.6	6.7	-0.1	12.2
Daily Minimum Temp	-7.5	-5.1	0.6	6.5	11.7	16.2	19.0	17.7	13.9	7.5	1.4	-4.9	6.4
Daily Maximum Temp	2.6	5.2	11.8	18.7	23.4	28.2	31.4	30.4	26.0	19.8	12.0	4.6	17.8
Record Minimum Temp	-28.3	-26.1	-20.6	-7.2	-1.7	6.1	8.9	5.6	0.0	-5.6	-17.8	-28.9	-28.9
Record Maximum Temp	23.3	27.8	29.4	32.2	32.2	39.4	43.9	43.3	38.3	33.9	28.3	24.4	43.9
HDD, Base 18.3°C	644	512	376	176	76	6	0	0	29	158	348	572	2896
CDD, Base 18.3°C	0	0	0	4	52	124	213	178	77	12	0	0	661
Relative Humidity (%)	71	71	67	63	69	71	69	70	71	68	71	74	69
Wind Speed (m/s)	5.0	5.0	5.5	5.3	4.3	3.9	3.7	3.6	3.9	4.4	4.8	4.9	4.5

Kansas City, MO

WBAN NO. 03947

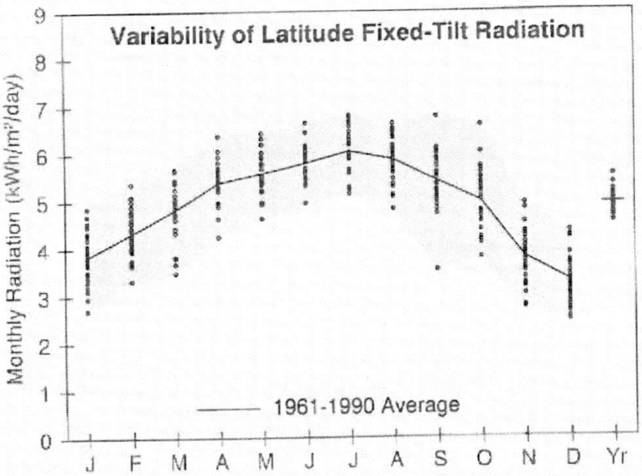

LATITUDE: 39.30° N
LONGITUDE: 94.72° W
ELEVATION: 315 meters
MEAN PRESSURE: 984 millibars

STATION TYPE: Secondary

Solar Radiation for Flat-Plate Collectors Facing South at a Fixed Tilt (kWh/m²/day), Uncertainty ±9%

Tilt (°)		Jan	Feb	Mar	Apr	May	June	July	Aug	Sept	Oct	Nov	Dec	Year
0	Average	2.2	3.0	3.9	5.1	5.9	6.5	6.6	5.8	4.6	3.6	2.3	1.9	4.3
	Min/Max	1.8/2.6	2.5/3.5	3.1/4.5	4.2/5.9	4.9/6.8	5.5/7.4	5.7/7.4	4.9/6.4	3.3/5.6	2.9/4.4	1.9/2.8	1.6/2.2	4.1/4.7
Latitude -15	Average	3.4	4.0	4.7	5.5	5.9	6.3	6.5	6.1	5.3	4.6	3.4	2.9	4.9
	Min/Max	2.4/4.2	3.1/4.9	3.5/5.4	4.4/6.5	4.9/6.8	5.4/7.3	5.6/7.4	5.1/6.9	3.6/6.6	3.6/6.0	2.5/4.3	2.2/3.7	4.6/5.4
Latitude	Average	3.8	4.3	4.8	5.4	5.6	5.8	6.0	5.9	5.4	5.0	3.8	3.3	4.9
	Min/Max	2.7/4.8	3.3/5.4	3.5/5.7	4.2/6.4	4.6/6.4	4.9/6.6	5.2/6.8	4.8/6.6	3.5/6.8	3.8/6.6	2.7/4.9	2.5/4.3	4.5/5.5
Latitude +15	Average	4.1	4.4	4.7	5.0	4.9	5.0	5.3	5.3	5.2	5.0	4.0	3.5	4.7
	Min/Max	2.8/5.2	3.3/5.6	3.3/5.6	3.9/5.9	4.1/5.7	4.3/5.7	4.5/5.9	4.4/6.0	3.3/6.5	3.8/6.7	2.8/5.3	2.6/4.7	4.3/5.3
90	Average	3.8	3.8	3.5	3.1	2.7	2.5	2.7	3.1	3.6	4.0	3.5	3.3	3.3
	Min/Max	2.5/4.9	2.7/4.8	2.4/4.2	2.5/3.6	2.3/2.9	2.3/2.7	2.4/2.9	2.6/3.5	2.3/4.5	3.0/5.5	2.4/4.8	2.3/4.5	2.9/3.7

Solar Radiation for 1-Axis Tracking Flat-Plate Collectors with a North-South Axis (kWh/m²/day), Uncertainty ±9%

Axis Tilt (°)		Jan	Feb	Mar	Apr	May	June	July	Aug	Sept	Oct	Nov	Dec	Year
0	Average	3.3	4.2	5.3	6.7	7.7	8.4	8.7	7.8	6.4	5.1	3.4	2.7	5.8
	Min/Max	2.3/4.1	3.2/5.2	3.8/6.5	5.0/8.4	6.1/9.2	6.8/10.2	7.2/10.3	6.1/8.9	4.0/8.2	3.9/6.8	2.5/4.4	2.1/3.5	5.3/6.6
Latitude -15	Average	4.1	4.9	5.9	7.1	7.8	8.4	8.8	8.1	6.9	5.9	4.1	3.5	6.3
	Min/Max	2.8/5.3	3.7/6.2	4.1/7.3	5.2/8.8	6.1/9.4	6.8/10.1	7.3/10.4	6.3/9.3	4.2/9.0	4.4/8.0	3.0/5.5	2.5/4.6	5.8/7.2
Latitude	Average	4.5	5.2	6.0	7.0	7.5	8.1	8.5	8.0	7.0	6.2	4.5	3.8	6.4
	Min/Max	3.0/5.8	3.9/6.6	4.1/7.4	5.1/8.8	5.9/9.1	6.5/9.7	7.0/10.0	6.1/9.2	4.2/9.1	4.6/8.4	3.2/5.9	2.7/5.1	5.8/7.3
Latitude +15	Average	4.7	5.3	5.9	6.7	7.1	7.5	7.9	7.6	6.8	6.2	4.6	4.0	6.2
	Min/Max	3.1/6.1	3.9/6.8	4.0/7.3	4.8/8.4	5.6/8.6	6.0/9.1	6.6/9.4	5.8/8.7	4.0/9.0	4.6/8.5	3.2/6.2	2.8/5.4	5.6/7.1

Solar Radiation for 2-Axis Tracking Flat-Plate Collectors (kWh/m²/day), Uncertainty ±9%

Tracker		Jan	Feb	Mar	Apr	May	June	July	Aug	Sept	Oct	Nov	Dec	Year
2-Axis	Average	4.8	5.3	6.0	7.1	7.9	8.6	8.9	8.2	7.0	6.2	4.7	4.1	6.6
	Min/Max	3.1/6.2	3.9/6.8	4.2/7.4	5.2/8.9	6.2/9.5	6.9/10.4	7.4/10.6	6.3/9.4	4.2/9.1	4.6/8.5	3.3/6.2	2.9/5.5	5.9/7.5

Direct Beam Solar Radiation for Concentrating Collectors (kWh/m²/day), Uncertainty ±8%

Tracker		Jan	Feb	Mar	Apr	May	June	July	Aug	Sept	Oct	Nov	Dec	Year
1-Axis, E-W Horiz Axis	Average	2.9	2.9	2.9	3.3	3.6	4.1	4.4	4.0	3.5	3.5	2.8	2.5	3.4
	Min/Max	1.5/4.2	1.8/4.2	1.5/4.0	1.9/4.7	2.3/5.0	2.7/5.7	3.1/5.7	2.5/4.9	1.4/5.2	2.2/5.3	1.6/4.1	1.5/3.8	2.8/4.2
1-Axis, N-S Horiz Axis	Average	2.1	2.6	3.3	4.3	4.8	5.5	5.8	5.3	4.3	3.5	2.2	1.7	3.8
	Min/Max	1.0/3.1	1.7/3.9	1.6/4.7	2.4/6.2	3.0/6.6	3.5/7.5	4.1/7.7	3.3/6.4	1.7/6.3	2.2/5.3	1.3/3.2	1.0/2.6	3.2/4.7
1-Axis, N-S Tilt=Latitude	Average	3.2	3.5	3.9	4.5	4.7	5.1	5.6	5.3	4.8	4.5	3.1	2.6	4.2
	Min/Max	1.6/4.6	2.2/5.1	1.9/5.5	2.5/6.5	3.0/6.4	3.3/7.1	3.9/7.3	3.3/6.5	1.9/7.1	2.8/6.7	1.9/4.6	1.6/4.0	3.6/5.3
2-Axis	Average	3.4	3.6	3.9	4.6	5.0	5.6	6.0	5.5	4.8	4.5	3.3	2.8	4.4
	Min/Max	1.7/4.9	2.3/5.3	2.0/5.5	2.5/6.6	3.1/6.8	3.6/7.7	4.2/7.9	3.5/6.7	1.9/7.1	2.8/6.8	2.0/4.8	1.7/4.3	3.7/5.5

Average Climatic Conditions

Element	Jan	Feb	Mar	Apr	May	June	July	Aug	Sept	Oct	Nov	Dec	Year
Temperature (°C)	-3.5	-0.4	5.9	12.5	17.8	22.9	25.8	24.5	19.7	13.7	6.2	-0.9	12.0
Daily Minimum Temp	-8.5	-5.7	0.3	6.6	12.2	17.3	20.1	18.7	13.8	7.6	0.9	-5.6	6.5
Daily Maximum Temp	1.5	4.8	11.6	18.4	23.5	28.5	31.5	30.2	25.6	19.7	11.4	3.8	17.6
Record Minimum Temp	-27.2	-28.3	-23.3	-11.1	-1.1	5.6	11.1	6.1	0.6	-6.1	-17.2	-30.6	-30.6
Record Maximum Temp	20.6	24.4	30.0	33.9	33.3	40.6	41.7	42.8	38.9	33.3	27.8	21.1	42.8
HDD, Base 18.3°C	677	526	384	181	75	4	0	3	31	155	365	596	2996
CDD, Base 18.3°C	0	0	0	6	59	141	233	194	73	10	0	0	716
Relative Humidity (%)	68	67	64	61	66	68	66	68	70	64	68	71	67
Wind Speed (m/s)	4.9	4.9	5.5	5.4	4.6	4.3	4.1	4.0	4.1	4.5	4.8	4.8	4.6

Springfield, MO

WBAN NO. 13995

LATITUDE: 37.23° N
LONGITUDE: 93.38° W
ELEVATION: 387 meters
MEAN PRESSURE: 972 millibars

STATION TYPE: Secondary

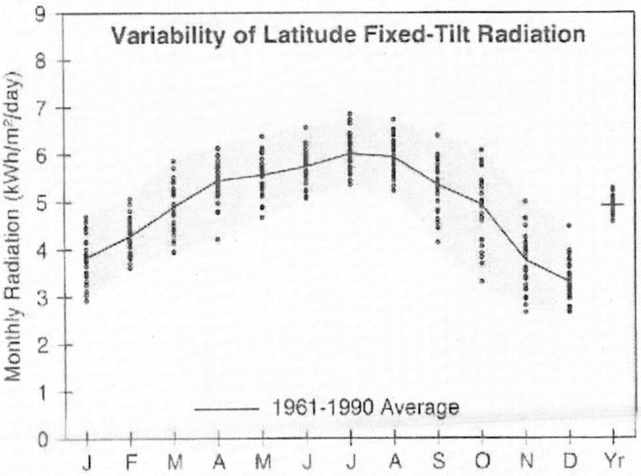

Solar Radiation for Flat-Plate Collectors Facing South at a Fixed Tilt (kWh/m²/day), Uncertainty ±9%

Tilt (°)		Jan	Feb	Mar	Apr	May	June	July	Aug	Sept	Oct	Nov	Dec	Year
0	Average	2.4	3.1	4.1	5.2	5.9	6.4	6.6	5.9	4.7	3.7	2.5	2.0	4.4
	Min/Max	2.0/2.7	2.8/3.4	3.4/4.7	4.2/5.8	5.0/6.7	5.7/7.4	5.9/7.5	5.3/6.6	3.8/5.4	2.7/4.3	2.0/3.0	1.8/2.4	4.2/4.6
Latitude -15	Average	3.4	3.9	4.7	5.6	5.9	6.2	6.5	6.2	5.3	4.6	3.4	2.9	4.9
	Min/Max	2.6/4.1	3.4/4.6	3.9/5.6	4.3/6.2	5.0/6.8	5.5/7.2	5.8/7.4	5.4/7.0	4.1/6.3	3.2/5.6	2.5/4.4	2.4/3.8	4.6/5.2
Latitude	Average	3.8	4.3	4.9	5.4	5.6	5.7	6.0	5.9	5.4	4.9	3.8	3.3	4.9
	Min/Max	2.9/4.7	3.6/5.1	3.9/5.9	4.2/6.1	4.7/6.4	5.1/6.6	5.3/6.8	5.2/6.7	4.1/6.4	3.3/6.1	2.7/5.0	2.7/4.5	4.6/5.3
Latitude +15	Average	4.1	4.4	4.8	5.0	4.9	5.0	5.2	5.4	5.1	5.0	3.9	3.5	4.7
	Min/Max	3.1/5.1	3.7/5.3	3.8/5.8	3.9/5.7	4.2/5.7	4.4/5.6	4.7/5.9	4.7/6.1	3.9/6.2	3.3/6.2	2.7/5.3	2.8/4.9	4.3/5.1
90	Average	3.7	3.7	3.5	3.0	2.5	2.3	2.5	3.0	3.4	3.9	3.4	3.3	3.2
	Min/Max	2.7/4.6	3.0/4.5	2.7/4.2	2.4/3.4	2.2/2.8	2.1/2.5	2.3/2.7	2.6/3.3	2.6/4.1	2.5/4.9	2.2/4.8	2.5/4.6	2.8/3.5

Solar Radiation for 1-Axis Tracking Flat-Plate Collectors with a North-South Axis (kWh/m²/day), Uncertainty ±9%

Axis Tilt (°)		Jan	Feb	Mar	Apr	May	June	July	Aug	Sept	Oct	Nov	Dec	Year
0	Average	3.4	4.2	5.4	6.8	7.6	8.3	8.7	7.9	6.3	5.1	3.5	2.8	5.9
	Min/Max	2.5/4.1	3.5/4.9	4.2/6.6	4.9/8.0	6.1/9.0	6.9/9.9	7.6/10.3	6.7/9.4	4.8/7.8	3.4/6.5	2.5/4.6	2.3/3.8	5.5/6.3
Latitude -15	Average	4.1	4.9	6.0	7.1	7.7	8.2	8.7	8.2	6.8	5.9	4.1	3.5	6.3
	Min/Max	3.0/5.1	4.0/5.8	4.5/7.3	5.1/8.4	6.1/9.1	6.9/9.8	7.6/10.3	6.9/9.7	5.1/8.5	3.8/7.4	2.8/5.6	2.7/4.8	5.8/6.8
Latitude	Average	4.5	5.1	6.1	7.1	7.5	7.9	8.4	8.0	6.9	6.1	4.4	3.8	6.3
	Min/Max	3.2/5.6	4.2/6.2	4.5/7.5	5.0/8.4	5.9/8.8	6.6/9.4	7.3/10.0	6.8/9.5	5.1/8.6	3.9/7.8	3.0/6.1	3.0/5.3	5.8/6.9
Latitude +15	Average	4.7	5.2	6.0	6.8	7.0	7.4	7.9	7.6	6.7	6.2	4.6	4.0	6.2
	Min/Max	3.4/5.9	4.2/6.4	4.4/7.4	4.8/8.0	5.5/8.3	6.2/8.8	6.8/9.3	6.4/9.1	4.9/8.4	3.8/7.9	3.0/6.4	3.1/5.7	5.6/6.7

Solar Radiation for 2-Axis Tracking Flat-Plate Collectors (kWh/m²/day), Uncertainty ±9%

Tracker		Jan	Feb	Mar	Apr	May	June	July	Aug	Sept	Oct	Nov	Dec	Year
2-Axis	Average	4.7	5.2	6.1	7.2	7.8	8.4	8.9	8.2	6.9	6.2	4.6	4.1	6.5
	Min/Max	3.4/5.9	4.2/6.4	4.6/7.5	5.1/8.4	6.2/9.2	7.0/10.1	7.7/10.5	6.9/9.7	5.1/8.6	3.9/7.9	3.0/6.4	3.1/5.7	6.0/7.1

Direct Beam Solar Radiation for Concentrating Collectors (kWh/m²/day), Uncertainty ±8%

Tracker		Jan	Feb	Mar	Apr	May	June	July	Aug	Sept	Oct	Nov	Dec	Year
1-Axis, E-W Horiz Axis	Average	2.8	2.7	2.9	3.3	3.6	4.0	4.3	4.0	3.3	3.4	2.7	2.4	3.3
	Min/Max	1.7/3.9	1.9/3.8	1.7/4.1	1.9/4.4	2.5/4.7	2.8/5.4	3.4/5.8	3.1/5.2	1.9/4.6	1.7/4.8	1.4/4.1	1.4/3.8	2.8/3.8
1-Axis, N-S Horiz Axis	Average	2.1	2.6	3.4	4.4	4.8	5.3	5.8	5.3	4.1	3.5	2.2	1.7	3.8
	Min/Max	1.2/3.0	1.8/3.5	1.9/4.6	2.4/5.8	3.3/6.2	3.7/7.2	4.5/7.8	4.1/6.9	2.5/5.8	1.8/5.0	1.2/3.5	0.9/2.8	3.3/4.3
1-Axis, N-S Tilt=Latitude	Average	3.1	3.3	3.9	4.6	4.6	5.0	5.5	5.3	4.6	4.3	3.0	2.6	4.2
	Min/Max	1.8/4.3	2.3/4.6	2.2/5.4	2.5/6.0	3.2/6.0	3.5/6.7	4.3/7.4	4.1/7.0	2.7/6.3	2.2/6.1	1.6/4.8	1.4/4.1	3.6/4.8
2-Axis	Average	3.3	3.4	3.9	4.6	4.9	5.4	5.9	5.5	4.6	4.4	3.2	2.8	4.3
	Min/Max	1.9/4.6	2.4/4.8	2.2/5.4	2.5/6.1	3.4/6.4	3.8/7.3	4.6/7.9	4.2/7.2	2.7/6.4	2.2/6.2	1.7/5.0	1.5/4.5	3.7/5.0

Average Climatic Conditions

Element	Jan	Feb	Mar	Apr	May	June	July	Aug	Sept	Oct	Nov	Dec	Year
Temperature (°C)	-0.5	2.1	7.8	13.3	18.1	22.9	25.6	24.9	20.6	14.3	7.8	1.8	13.2
Daily Minimum Temp	-6.4	-3.9	1.3	6.7	11.8	16.6	19.2	18.3	14.3	7.7	1.9	-3.7	7.0
Daily Maximum Temp	5.4	7.9	14.1	19.9	24.4	29.1	32.0	31.4	26.8	21.0	13.7	7.4	19.4
Record Minimum Temp	-25.0	-27.2	-19.4	-7.8	-1.1	5.6	6.7	6.7	-0.6	-6.1	-15.6	-26.7	-27.2
Record Maximum Temp	24.4	27.2	30.6	33.9	33.9	38.3	45.0	41.1	40.0	33.9	26.7	25.0	45.0
HDD, Base 18.3°C	584	456	327	156	61	3	0	0	24	138	317	512	2577
CDD, Base 18.3°C	0	0	0	6	54	139	226	203	91	14	0	0	733
Relative Humidity (%)	68	69	65	65	71	72	70	70	73	68	70	71	69
Wind Speed (m/s)	4.9	4.9	5.4	5.0	4.2	3.8	3.4	3.5	3.8	4.2	4.6	4.8	4.4

St. Louis, MO

WBAN NO. 13994

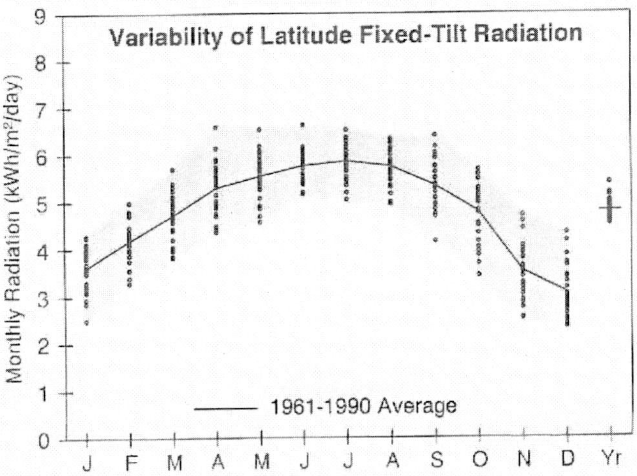

LATITUDE: 38.75° N
LONGITUDE: 90.38° W
ELEVATION: 172 meters
MEAN PRESSURE: 997 millibars

STATION TYPE: Secondary

Solar Radiation for Flat-Plate Collectors Facing South at a Fixed Tilt (kWh/m²/day), Uncertainty ±9%

Tilt (°)		Jan	Feb	Mar	Apr	May	June	July	Aug	Sept	Oct	Nov	Dec	Year
0	Average	2.2	2.9	3.9	5.0	5.9	6.4	6.4	5.7	4.6	3.5	2.3	1.8	4.2
	Min/Max	1.7/2.5	2.5/3.4	3.3/4.5	4.3/6.1	4.9/6.9	5.8/7.4	5.5/7.1	4.9/6.2	3.8/5.3	2.8/3.9	1.9/2.7	1.6/2.3	4.0/4.7
Latitude -15	Average	3.2	3.8	4.6	5.4	5.9	6.3	6.3	6.0	5.3	4.5	3.2	2.7	4.8
	Min/Max	2.3/3.7	3.1/4.5	3.8/5.5	4.5/6.7	4.9/6.9	5.6/7.2	5.4/7.0	5.1/6.6	4.2/6.3	3.3/5.2	2.4/4.1	2.1/3.7	4.5/5.3
Latitude	Average	3.6	4.2	4.7	5.3	5.6	5.8	5.9	5.7	5.3	4.8	3.5	3.1	4.8
	Min/Max	2.5/4.3	3.3/5.0	3.8/5.7	4.4/6.6	4.6/6.5	5.2/6.6	5.0/6.5	4.9/6.3	4.2/6.4	3.4/5.7	2.5/4.7	2.3/4.3	4.5/5.4
Latitude +15	Average	3.8	4.3	4.6	4.9	4.9	5.0	5.1	5.2	5.1	4.8	3.7	3.3	4.6
	Min/Max	2.6/4.6	3.3/5.2	3.7/5.6	4.0/6.1	4.1/5.8	4.5/5.7	4.4/5.7	4.5/5.8	4.0/6.2	3.4/5.8	2.6/5.0	2.4/4.7	4.3/5.1
90	Average	3.5	3.7	3.4	3.1	2.6	2.4	2.6	3.0	3.5	3.8	3.2	3.0	3.2
	Min/Max	2.3/4.3	2.7/4.6	2.7/4.1	2.5/3.7	2.3/2.9	2.3/2.6	2.3/2.7	2.6/3.3	2.7/4.2	2.6/4.7	2.2/4.5	2.2/4.5	2.8/3.5

Solar Radiation for 1-Axis Tracking Flat-Plate Collectors with a North-South Axis (kWh/m²/day), Uncertainty ±9%

Axis Tilt (°)		Jan	Feb	Mar	Apr	May	June	July	Aug	Sept	Oct	Nov	Dec	Year
0	Average	3.1	4.0	5.1	6.6	7.6	8.3	8.3	7.6	6.2	4.9	3.1	2.5	5.6
	Min/Max	2.2/3.7	3.1/4.8	4.0/6.4	5.2/8.5	6.1/9.4	7.3/10.0	7.2/9.5	6.3/8.4	4.7/7.8	3.4/5.9	2.3/4.2	2.0/3.5	5.3/6.5
Latitude -15	Average	3.8	4.7	5.7	6.9	7.7	8.3	8.4	7.8	6.8	5.6	3.8	3.2	6.1
	Min/Max	2.6/4.6	3.6/5.6	4.3/7.2	5.4/9.0	6.1/9.6	7.2/10.0	7.2/9.5	6.4/8.8	5.0/8.6	3.9/6.8	2.7/5.2	2.4/4.6	5.7/7.0
Latitude	Average	4.2	5.0	5.8	6.8	7.5	7.9	8.1	7.7	6.8	5.9	4.1	3.5	6.1
	Min/Max	2.8/5.1	3.7/6.0	4.4/7.3	5.3/8.9	5.9/9.3	6.9/9.6	6.9/9.2	6.3/8.6	5.0/8.7	4.0/7.2	2.8/5.7	2.6/5.1	5.7/7.1
Latitude +15	Average	4.4	5.1	5.7	6.5	7.0	7.4	7.6	7.3	6.7	5.9	4.2	3.7	6.0
	Min/Max	2.8/5.3	3.8/6.2	4.3/7.3	5.1/8.6	5.5/8.8	6.5/8.9	6.5/8.6	6.0/8.2	4.8/8.5	4.0/7.2	2.8/5.9	2.7/5.4	5.5/6.9

Solar Radiation for 2-Axis Tracking Flat-Plate Collectors (kWh/m²/day), Uncertainty ±9%

Tracker		Jan	Feb	Mar	Apr	May	June	July	Aug	Sept	Oct	Nov	Dec	Year
2-Axis	Average	4.4	5.1	5.8	6.9	7.8	8.5	8.5	7.9	6.9	5.9	4.3	3.7	6.3
	Min/Max	2.9/5.4	3.8/6.2	4.4/7.4	5.4/9.0	6.2/9.7	7.4/10.2	7.3/9.7	6.5/8.8	5.1/8.7	4.0/7.3	2.9/5.9	2.7/5.5	5.9/7.3

Direct Beam Solar Radiation for Concentrating Collectors (kWh/m²/day), Uncertainty ±8%

Tracker		Jan	Feb	Mar	Apr	May	June	July	Aug	Sept	Oct	Nov	Dec	Year
1-Axis, E-W Horiz Axis	Average	2.5	2.6	2.7	3.1	3.5	3.9	4.0	3.7	3.3	3.2	2.4	2.1	3.1
	Min/Max	1.3/3.4	1.7/3.4	1.7/4.0	1.9/4.8	2.3/5.1	3.2/5.5	2.8/5.1	2.6/4.5	2.0/4.7	1.8/4.2	1.2/3.8	1.3/3.6	2.8/3.9
1-Axis, N-S Horiz Axis	Average	1.9	2.4	3.0	4.0	4.7	5.2	5.3	4.9	4.0	3.3	1.9	1.5	3.5
	Min/Max	0.9/2.6	1.5/3.2	1.9/4.6	2.4/6.2	3.1/6.8	4.3/7.2	3.9/6.7	3.5/6.0	2.3/5.9	1.8/4.3	1.0/3.1	0.9/2.5	3.1/4.5
1-Axis, N-S Tilt=Latitude	Average	2.7	3.1	3.6	4.3	4.6	4.9	5.1	4.9	4.5	4.1	2.7	2.2	3.9
	Min/Max	1.4/3.8	2.0/4.2	2.2/5.3	2.5/6.5	3.1/6.7	4.0/6.8	3.7/6.4	3.5/6.1	2.6/6.6	2.3/5.4	1.4/4.3	1.4/3.8	3.5/5.0
2-Axis	Average	2.9	3.2	3.6	4.3	4.8	5.3	5.5	5.1	4.5	4.1	2.9	2.4	4.1
	Min/Max	1.5/4.0	2.0/4.2	2.2/5.3	2.6/6.6	3.2/7.1	4.3/7.4	4.0/6.9	3.6/6.3	2.6/6.6	2.3/5.4	1.4/4.6	1.5/4.2	3.6/5.2

Average Climatic Conditions

Element	Jan	Feb	Mar	Apr	May	June	July	Aug	Sept	Oct	Nov	Dec	Year
Temperature (°C)	-1.5	1.1	7.3	13.7	18.9	24.1	26.6	25.3	21.2	14.7	7.9	1.1	13.4
Daily Minimum Temp	-6.2	-3.8	1.9	8.0	13.3	18.7	21.3	19.9	15.8	9.1	3.2	-3.3	8.2
Daily Maximum Temp	3.2	5.9	12.6	19.4	24.5	29.6	31.8	30.7	26.6	20.3	12.6	5.4	18.6
Record Minimum Temp	-27.8	-23.3	-20.6	-5.6	-0.6	6.1	10.6	8.3	2.2	-5.0	-17.2	-26.7	-27.8
Record Maximum Temp	24.4	29.4	31.7	33.9	33.9	38.9	41.7	41.7	40.0	34.4	29.4	24.4	41.7
HDD, Base 18.3°C	615	484	343	148	62	0	0	0	12	132	313	536	2643
CDD, Base 18.3°C	0	0	0	9	81	173	255	217	98	18	0	0	852
Relative Humidity (%)	73	72	68	63	66	67	68	70	72	69	72	76	70
Wind Speed (m/s)	4.9	4.8	5.3	5.2	4.3	4.1	3.8	3.6	3.8	4.1	4.6	4.8	4.4

Jackson, MS

WBAN NO. 03940

LATITUDE: 32.32° N
LONGITUDE: 90.08° W
ELEVATION: 101 meters
MEAN PRESSURE: 1006 millibars

STATION TYPE: Secondary

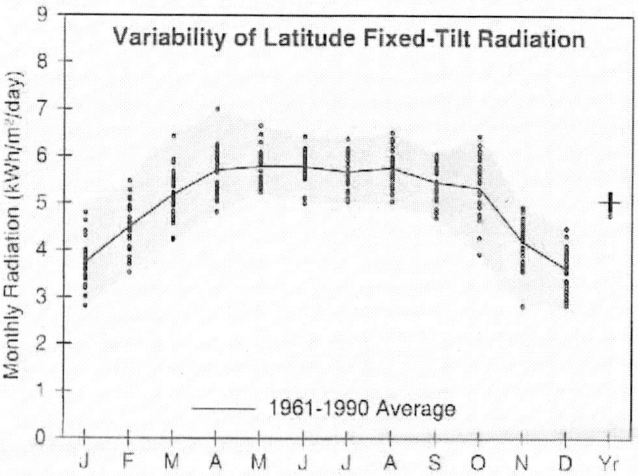

Solar Radiation for Flat-Plate Collectors Facing South at a Fixed Tilt (kWh/m²/day), Uncertainty ±9%

Tilt (°)		Jan	Feb	Mar	Apr	May	June	July	Aug	Sept	Oct	Nov	Dec	Year
0	Average	2.6	3.5	4.5	5.5	6.1	6.4	6.2	5.8	4.9	4.2	3.0	2.4	4.6
	Min/Max	2.2/3.1	2.9/4.0	3.8/5.4	4.8/6.7	5.6/7.1	5.5/7.2	5.4/6.9	5.1/6.5	4.3/5.4	3.3/4.9	2.3/3.3	2.1/2.8	4.3/4.8
Latitude -15	Average	3.3	4.1	5.0	5.8	6.1	6.3	6.1	6.0	5.4	5.0	3.7	3.2	5.0
	Min/Max	2.6/4.1	3.3/4.9	4.1/6.1	4.9/7.1	5.5/7.1	5.4/7.0	5.4/6.8	5.2/6.7	4.7/5.9	3.8/5.9	2.7/4.3	2.6/3.8	4.7/5.2
Latitude	Average	3.7	4.5	5.2	5.7	5.8	5.8	5.7	5.8	5.5	5.3	4.2	3.6	5.1
	Min/Max	2.8/4.8	3.5/5.5	4.2/6.4	4.8/7.0	5.2/6.7	5.0/6.4	5.0/6.4	5.0/6.5	4.7/6.0	3.9/6.4	2.8/4.9	2.8/4.5	4.8/5.2
Latitude +15	Average	4.0	4.6	5.1	5.3	5.1	5.0	5.0	5.3	5.2	5.4	4.4	3.9	4.9
	Min/Max	2.9/5.2	3.5/5.7	4.1/6.3	4.5/6.5	4.7/5.9	4.4/5.5	4.4/5.6	4.6/5.9	4.5/5.8	3.9/6.6	2.9/5.2	3.0/4.8	4.6/5.0
90	Average	3.4	3.6	3.4	2.9	2.3	2.0	2.1	2.6	3.2	4.0	3.7	3.4	3.0
	Min/Max	2.4/4.6	2.7/4.6	2.7/4.2	2.5/3.4	2.1/2.5	1.9/2.1	2.0/2.2	2.4/2.9	2.8/3.6	2.8/4.9	2.3/4.4	2.5/4.4	2.9/3.2

Solar Radiation for 1-Axis Tracking Flat-Plate Collectors with a North-South Axis (kWh/m²/day), Uncertainty ±9%

Axis Tilt (°)		Jan	Feb	Mar	Apr	May	June	July	Aug	Sept	Oct	Nov	Dec	Year
0	Average	3.6	4.7	5.9	7.2	7.8	8.1	7.7	7.4	6.4	5.7	4.1	3.3	6.0
	Min/Max	2.6/4.6	3.6/5.8	4.8/7.6	5.9/9.4	6.8/9.5	6.5/9.4	6.5/9.1	6.1/8.6	5.4/7.2	4.2/7.0	2.7/4.8	2.6/4.1	5.5/6.3
Latitude -15	Average	4.1	5.2	6.4	7.4	7.8	8.0	7.7	7.5	6.8	6.3	4.6	3.9	6.3
	Min/Max	2.9/5.3	3.9/6.5	5.0/8.1	6.0/9.7	6.8/9.5	6.5/9.3	6.4/9.0	6.2/8.8	5.7/7.6	4.5/7.8	2.9/5.5	2.9/4.8	5.8/6.6
Latitude	Average	4.4	5.4	6.5	7.3	7.6	7.7	7.4	7.4	6.8	6.6	5.0	4.2	6.4
	Min/Max	3.1/5.8	4.1/6.9	5.1/8.4	6.0/9.6	6.6/9.3	6.2/8.9	6.2/8.7	6.1/8.7	5.7/7.7	4.7/8.2	3.1/6.0	3.2/5.3	5.9/6.7
Latitude +15	Average	4.6	5.5	6.4	7.1	7.1	7.1	6.9	7.0	6.7	6.6	5.2	4.4	6.2
	Min/Max	3.2/6.1	4.1/7.0	5.0/8.3	5.7/9.3	6.2/8.7	5.8/8.3	5.8/8.1	5.8/8.3	5.6/7.6	4.6/8.3	3.1/6.2	3.3/5.6	5.8/6.5

Solar Radiation for 2-Axis Tracking Flat-Plate Collectors (kWh/m²/day), Uncertainty ±9%

Tracker		Jan	Feb	Mar	Apr	May	June	July	Aug	Sept	Oct	Nov	Dec	Year
2-Axis	Average	4.6	5.6	6.5	7.4	7.9	8.2	7.8	7.6	6.8	6.7	5.2	4.5	6.6
	Min/Max	3.2/6.2	4.1/7.1	5.1/8.4	6.1/9.8	6.9/9.6	6.6/9.5	6.5/9.2	6.3/8.8	5.7/7.8	4.7/8.3	3.1/6.3	3.3/5.7	6.1/6.9

Direct Beam Solar Radiation for Concentrating Collectors (kWh/m²/day), Uncertainty ±8%

Tracker		Jan	Feb	Mar	Apr	May	June	July	Aug	Sept	Oct	Nov	Dec	Year
1-Axis, E-W Horiz Axis	Average	2.6	3.0	3.2	3.5	3.7	3.8	3.5	3.4	3.2	3.6	2.9	2.6	3.2
	Min/Max	1.4/4.0	1.8/4.2	2.1/4.8	2.5/5.4	2.9/5.0	2.5/4.9	2.5/4.5	2.3/4.5	2.3/3.9	2.1/4.9	1.3/3.9	1.5/3.7	2.9/3.5
1-Axis, N-S Horiz Axis	Average	2.2	3.0	3.8	4.6	4.8	4.9	4.5	4.5	3.9	3.9	2.6	2.1	3.7
	Min/Max	1.1/3.4	1.8/4.2	2.6/5.8	3.3/7.2	3.8/6.7	3.2/6.4	3.2/5.9	3.0/5.9	2.9/4.9	2.3/5.2	1.1/3.4	1.2/2.9	3.3/4.1
1-Axis, N-S Tilt=Latitude	Average	2.9	3.7	4.3	4.7	4.6	4.6	4.2	4.5	4.3	4.6	3.4	2.8	4.0
	Min/Max	1.5/4.5	2.2/5.2	2.9/6.4	3.4/7.4	3.6/6.5	3.0/6.0	3.0/5.6	3.0/5.8	3.1/5.3	2.7/6.2	1.4/4.4	1.6/4.0	3.6/4.4
2-Axis	Average	3.1	3.8	4.3	4.8	4.9	4.9	4.6	4.6	4.3	4.6	3.6	3.1	4.2
	Min/Max	1.6/4.8	2.2/5.3	2.9/6.5	3.4/7.5	3.8/6.8	3.2/6.5	3.3/6.0	3.1/6.0	3.1/5.3	2.7/6.3	1.5/4.7	1.8/4.4	3.7/4.6

Average Climatic Conditions

Element	Jan	Feb	Mar	Apr	May	June	July	Aug	Sept	Oct	Nov	Dec	Year
Temperature (°C)	6.7	8.8	13.7	18.1	22.2	26.0	27.5	27.2	24.4	18.2	13.2	8.8	17.9
Daily Minimum Temp	0.4	2.1	6.7	11.1	15.6	19.5	21.4	20.9	17.6	10.2	5.7	2.3	11.1
Daily Maximum Temp	13.1	15.6	20.7	25.2	28.9	32.6	33.6	33.3	31.1	26.2	20.7	15.3	24.7
Record Minimum Temp	-16.7	-11.7	-9.4	-2.8	3.3	8.3	10.6	12.8	1.7	-1.7	-8.3	-15.6	-16.7
Record Maximum Temp	27.8	29.4	31.7	34.4	37.2	40.6	41.1	38.9	40.0	35.0	31.1	28.9	41.1
HDD, Base 18.3°C	364	269	158	48	4	0	0	0	3	58	164	301	1371
CDD, Base 18.3°C	4	3	16	42	124	230	284	274	185	52	11	5	1231
Relative Humidity (%)	76	73	71	71	73	74	77	77	77	74	76	76	75
Wind Speed (m/s)	3.7	3.8	4.0	3.7	3.2	2.8	2.6	2.5	2.9	2.9	3.4	3.8	3.3

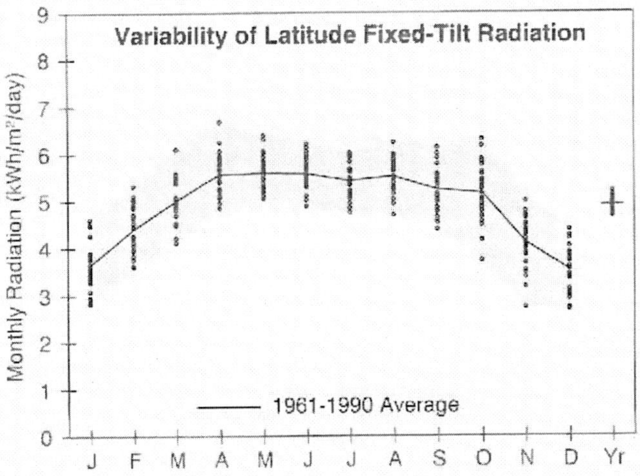

Meridian, MS

WBAN NO. 13865

LATITUDE: 32.33° N
LONGITUDE: 88.75° W
ELEVATION: 94 meters
MEAN PRESSURE: 1007 millibars

STATION TYPE: Secondary

Solar Radiation for Flat-Plate Collectors Facing South at a Fixed Tilt (kWh/m²/day), Uncertainty ±9%

Tilt (°)		Jan	Feb	Mar	Apr	May	June	July	Aug	Sept	Oct	Nov	Dec	Year
0	Average	2.6	3.4	4.4	5.4	5.9	6.2	5.9	5.6	4.8	4.1	2.9	2.4	4.5
	Min/Max	2.2/3.0	3.0/3.9	3.7/5.1	4.8/6.4	5.4/6.8	5.3/6.9	5.2/6.5	4.8/6.3	4.1/5.5	3.2/4.8	2.3/3.4	2.0/2.8	4.2/4.7
Latitude -15	Average	3.2	4.1	4.8	5.7	5.9	6.0	5.8	5.7	5.2	4.8	3.7	3.1	4.8
	Min/Max	2.6/4.0	3.4/4.8	4.1/5.8	4.9/6.8	5.4/6.8	5.2/6.7	5.1/6.4	4.9/6.4	4.4/6.0	3.6/5.8	2.6/4.4	2.5/3.8	4.6/5.1
Latitude	Average	3.6	4.4	5.0	5.6	5.6	5.6	5.4	5.5	5.2	5.2	4.1	3.5	4.9
	Min/Max	2.8/4.6	3.6/5.3	4.1/6.1	4.8/6.7	5.1/6.4	4.9/6.2	4.8/6.0	4.7/6.2	4.4/6.1	3.7/6.3	2.7/5.0	2.7/4.4	4.7/5.2
Latitude +15	Average	3.8	4.5	4.9	5.2	5.0	4.9	4.8	5.0	5.0	5.2	4.3	3.8	4.7
	Min/Max	2.9/5.0	3.6/5.5	4.0/6.0	4.5/6.2	4.5/5.7	4.3/5.4	4.2/5.3	4.3/5.7	4.2/5.9	3.7/6.4	2.8/5.3	2.8/4.8	4.5/5.0
90	Average	3.3	3.5	3.3	2.8	2.2	2.0	2.1	2.6	3.1	3.9	3.6	3.3	3.0
	Min/Max	2.4/4.4	2.8/4.4	2.7/4.0	2.5/3.3	2.1/2.4	1.9/2.1	2.0/2.2	2.3/2.8	2.6/3.7	2.6/4.8	2.2/4.5	2.3/4.3	2.8/3.2

Solar Radiation for 1-Axis Tracking Flat-Plate Collectors with a North-South Axis (kWh/m²/day), Uncertainty ±9%

Axis Tilt (°)		Jan	Feb	Mar	Apr	May	June	July	Aug	Sept	Oct	Nov	Dec	Year
0	Average	3.5	4.6	5.7	7.0	7.5	7.7	7.2	7.0	6.1	5.5	4.0	3.2	5.7
	Min/Max	2.7/4.4	3.6/5.6	4.6/7.1	5.8/8.9	6.5/8.9	6.3/9.0	6.2/8.4	5.6/8.1	4.9/7.3	3.8/6.9	2.6/4.8	2.5/4.0	5.4/6.2
Latitude -15	Average	4.0	5.1	6.1	7.2	7.5	7.6	7.2	7.1	6.5	6.1	4.5	3.8	6.0
	Min/Max	3.0/5.1	4.0/6.3	4.8/7.7	5.9/9.1	6.5/8.9	6.2/8.9	6.1/8.4	5.7/8.2	5.1/7.8	4.1/7.6	2.8/5.6	2.8/4.7	5.6/6.5
Latitude	Average	4.3	5.3	6.2	7.1	7.2	7.3	6.9	7.0	6.5	6.4	4.9	4.1	6.1
	Min/Max	3.1/5.6	4.1/6.7	4.9/7.8	5.9/9.1	6.3/8.6	5.9/8.5	5.9/8.1	5.6/8.1	5.1/7.8	4.3/8.0	3.0/6.1	3.0/5.2	5.7/6.6
Latitude +15	Average	4.4	5.4	6.1	6.8	6.8	6.8	6.5	6.6	6.3	6.4	5.0	4.3	6.0
	Min/Max	3.2/5.9	4.2/6.8	4.8/7.8	5.6/8.7	5.9/8.1	5.5/7.9	5.5/7.6	5.3/7.7	4.9/7.7	4.2/8.1	3.0/6.3	3.0/5.5	5.6/6.5

Solar Radiation for 2-Axis Tracking Flat-Plate Collectors (kWh/m²/day), Uncertainty ±9%

Tracker		Jan	Feb	Mar	Apr	May	June	July	Aug	Sept	Oct	Nov	Dec	Year
2-Axis	Average	4.5	5.4	6.2	7.2	7.6	7.8	7.3	7.1	6.5	6.4	5.1	4.4	6.3
	Min/Max	3.2/6.0	4.2/6.9	4.9/7.9	6.0/9.2	6.6/9.0	6.3/9.1	6.2/8.5	5.7/8.3	5.1/7.9	4.3/8.1	3.0/6.3	3.1/5.6	5.9/6.8

Direct Beam Solar Radiation for Concentrating Collectors (kWh/m²/day), Uncertainty ±8%

Tracker		Jan	Feb	Mar	Apr	May	June	July	Aug	Sept	Oct	Nov	Dec	Year
1-Axis, E-W Horiz Axis	Average	2.5	2.8	2.9	3.3	3.3	3.4	3.1	3.0	2.9	3.4	2.8	2.5	3.0
	Min/Max	1.3/3.7	1.8/4.1	1.9/4.3	2.4/4.9	2.5/4.5	2.3/4.4	2.1/4.0	1.9/3.9	1.9/3.9	1.7/4.7	1.2/3.9	1.4/3.5	2.7/3.4
1-Axis, N-S Horiz Axis	Average	2.1	2.9	3.5	4.4	4.4	4.4	4.0	3.9	3.6	3.6	2.5	2.0	3.4
	Min/Max	1.1/3.2	1.7/4.1	2.3/5.1	3.1/6.5	3.3/6.0	2.9/5.8	2.8/5.3	2.5/5.1	2.3/4.8	1.8/5.1	1.0/3.4	1.0/2.8	3.0/4.0
1-Axis, N-S Tilt=Latitude	Average	2.8	3.5	3.9	4.5	4.2	4.1	3.8	3.9	3.9	4.3	3.2	2.7	3.7
	Min/Max	1.5/4.2	2.2/5.1	2.6/5.7	3.2/6.7	3.1/5.8	2.7/5.4	2.6/5.0	2.5/5.1	2.5/5.2	2.1/6.0	1.3/4.5	1.4/3.8	3.3/4.3
2-Axis	Average	2.9	3.6	3.9	4.5	4.5	4.4	4.0	4.0	3.9	4.3	3.4	2.9	3.9
	Min/Max	1.6/4.5	2.2/5.2	2.6/5.7	3.3/6.8	3.3/6.1	2.9/5.9	2.8/5.4	2.6/5.2	2.5/5.3	2.2/6.1	1.4/4.7	1.6/4.2	3.5/4.5

Average Climatic Conditions

Element	Jan	Feb	Mar	Apr	May	June	July	Aug	Sept	Oct	Nov	Dec	Year
Temperature (°C)	7.2	9.4	13.7	17.8	21.8	25.6	27.2	27.0	24.1	17.8	13.1	9.1	17.8
Daily Minimum Temp	0.8	2.6	6.4	10.5	14.9	18.9	21.1	20.7	17.7	10.3	5.8	2.6	11.0
Daily Maximum Temp	13.6	16.2	20.9	25.2	28.7	32.2	33.4	33.2	30.5	25.4	20.3	15.6	24.6
Record Minimum Temp	-17.8	-13.3	-9.4	-2.2	3.3	5.6	12.8	11.7	1.1	-4.4	-8.9	-16.7	-17.8
Record Maximum Temp	28.3	29.4	32.2	35.0	37.2	40.0	41.7	40.0	40.6	36.1	30.6	27.8	41.7
HDD, Base 18.3°C	349	253	162	55	8	0	0	0	3	68	167	292	1358
CDD, Base 18.3°C	4	3	17	40	117	218	276	269	177	53	9	6	1188
Relative Humidity (%)	74	71	70	71	73	74	77	77	76	75	75	75	74
Wind Speed (m/s)	3.2	3.4	3.5	3.2	2.6	2.3	2.1	2.1	2.4	2.3	2.8	3.2	2.7

Billings, MT

WBAN NO. 24033

LATITUDE: 45.80° N
LONGITUDE: 108.53° W
ELEVATION: 1088 meters
MEAN PRESSURE: 891 millibars

STATION TYPE: Secondary

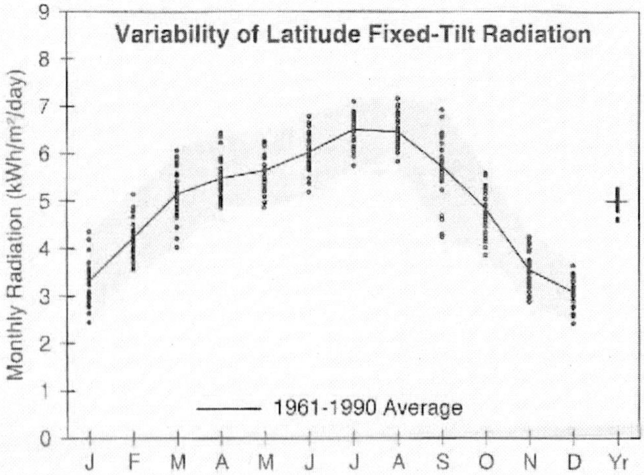

Solar Radiation for Flat-Plate Collectors Facing South at a Fixed Tilt (kWh/m²/day), Uncertainty ±9%

Tilt (°)		Jan	Feb	Mar	Apr	May	June	July	Aug	Sept	Oct	Nov	Dec	Year
0	Average	1.7	2.6	3.8	5.0	5.9	6.7	7.0	6.1	4.5	3.1	1.9	1.4	4.1
	Min/Max	1.5/2.0	2.3/2.8	3.2/4.2	4.5/5.7	5.2/6.5	5.8/7.5	6.1/7.6	5.5/6.7	3.6/5.3	2.7/3.4	1.7/2.1	1.3/1.6	3.9/4.3
Latitude -15	Average	2.9	3.9	4.9	5.6	6.0	6.6	7.1	6.7	5.6	4.5	3.2	2.7	5.0
	Min/Max	2.2/3.8	3.3/4.6	4.0/5.8	5.0/6.6	5.2/6.7	5.7/7.4	6.2/7.7	6.1/7.4	4.2/6.8	3.7/5.2	2.6/3.8	2.1/3.1	4.6/5.2
Latitude	Average	3.3	4.2	5.1	5.5	5.6	6.0	6.5	6.5	5.7	4.8	3.5	3.1	5.0
	Min/Max	2.5/4.4	3.6/5.2	4.0/6.1	4.9/6.4	4.9/6.3	5.2/6.8	5.7/7.1	5.8/7.2	4.2/6.9	3.9/5.6	2.9/4.3	2.4/3.6	4.6/5.3
Latitude +15	Average	3.5	4.4	5.1	5.1	5.0	5.2	5.7	5.9	5.5	4.9	3.7	3.3	4.8
	Min/Max	2.6/4.7	3.7/5.4	3.9/6.0	4.5/6.0	4.3/5.5	4.5/5.8	5.0/6.2	5.3/6.5	4.0/6.7	3.9/5.7	3.0/4.5	2.6/4.0	4.4/5.0
90	Average	3.4	4.0	4.2	3.6	3.1	3.0	3.3	3.8	4.2	4.2	3.5	3.3	3.6
	Min/Max	2.4/4.7	3.2/4.9	3.1/5.0	3.2/4.2	2.8/3.4	2.7/3.3	3.0/3.6	3.5/4.2	3.1/5.1	3.3/4.9	2.8/4.3	2.5/3.9	3.3/3.9

Solar Radiation for 1-Axis Tracking Flat-Plate Collectors with a North-South Axis (kWh/m²/day), Uncertainty ±9%

Axis Tilt (°)		Jan	Feb	Mar	Apr	May	June	July	Aug	Sept	Oct	Nov	Dec	Year
0	Average	2.5	3.7	5.4	6.8	7.9	9.1	9.8	8.8	6.7	4.6	2.9	2.2	5.9
	Min/Max	2.0/3.3	3.1/4.5	4.1/6.5	5.8/8.4	6.7/9.1	7.5/10.7	8.0/10.9	7.6/10.1	4.7/8.3	3.7/5.4	2.3/3.5	1.8/2.6	5.3/6.3
Latitude -15	Average	3.5	4.7	6.3	7.4	8.2	9.2	10.1	9.4	7.6	5.7	3.8	3.2	6.6
	Min/Max	2.5/4.6	3.9/5.9	4.7/7.7	6.3/9.1	6.8/9.4	7.5/10.9	8.2/11.1	8.1/10.8	5.2/9.5	4.4/6.7	3.0/4.7	2.4/3.8	5.9/7.0
Latitude	Average	3.8	5.0	6.4	7.3	7.9	8.8	9.7	9.2	7.6	6.0	4.1	3.5	6.6
	Min/Max	2.7/5.1	4.1/6.3	4.8/7.9	6.2/9.1	6.6/9.1	7.2/10.5	7.9/10.8	8.0/10.6	5.3/9.6	4.6/7.0	3.3/5.1	2.7/4.2	5.9/7.0
Latitude +15	Average	4.0	5.1	6.4	7.0	7.5	8.3	9.1	8.8	7.5	6.0	4.3	3.7	6.5
	Min/Max	2.8/5.4	4.2/6.4	4.7/7.8	5.9/8.7	6.2/8.6	6.7/9.8	7.4/10.1	7.6/10.2	5.1/9.4	4.6/7.1	3.4/5.3	2.8/4.4	5.8/6.9

Solar Radiation for 2-Axis Tracking Flat-Plate Collectors (kWh/m²/day), Uncertainty ±9%

Tracker		Jan	Feb	Mar	Apr	May	June	July	Aug	Sept	Oct	Nov	Dec	Year
2-Axis	Average	4.0	5.1	6.5	7.4	8.3	9.4	10.3	9.4	7.7	6.1	4.3	3.8	6.9
	Min/Max	2.8/5.4	4.2/6.4	4.8/7.9	6.3/9.2	6.9/9.5	7.7/11.1	8.4/11.4	8.2/10.9	5.3/9.6	4.7/7.1	3.4/5.3	2.9/4.5	6.1/7.3

Direct Beam Solar Radiation for Concentrating Collectors (kWh/m²/day), Uncertainty ±8%

Tracker		Jan	Feb	Mar	Apr	May	June	July	Aug	Sept	Oct	Nov	Dec	Year
1-Axis, E-W	Average	2.3	2.7	3.1	3.4	3.9	4.8	5.6	5.1	4.1	3.5	2.5	2.2	3.6
Horiz Axis	Min/Max	1.2/3.6	1.8/3.9	2.0/4.5	2.4/4.8	2.7/5.0	3.5/6.3	4.2/6.5	4.1/6.3	2.3/5.7	2.3/4.4	1.7/3.4	1.5/3.1	3.0/4.0
1-Axis, N-S	Average	1.4	2.2	3.3	4.3	5.2	6.4	7.4	6.7	4.9	3.3	1.8	1.3	4.0
Horiz Axis	Min/Max	0.8/2.3	1.5/3.2	2.1/4.8	3.1/6.3	3.7/6.6	4.6/8.4	5.4/8.6	5.3/8.4	2.7/6.8	2.1/4.2	1.2/2.5	0.8/1.8	3.3/4.5
1-Axis, N-S	Average	2.4	3.2	4.1	4.7	5.2	6.1	7.2	6.9	5.7	4.4	2.8	2.3	4.6
Tilt=Latitude	Min/Max	1.3/3.9	2.2/4.7	2.6/6.0	3.4/6.8	3.7/6.6	4.4/8.0	5.3/8.4	5.5/8.7	3.1/7.9	2.8/5.6	1.9/3.9	1.5/3.2	3.8/5.1
2-Axis	Average	2.6	3.3	4.1	4.8	5.5	6.6	7.8	7.2	5.7	4.4	2.9	2.5	4.8
	Min/Max	1.4/4.2	2.3/4.8	2.6/6.1	3.4/6.9	3.9/6.9	4.8/8.7	5.6/9.0	5.7/9.0	3.1/8.0	2.9/5.7	2.0/4.1	1.6/3.5	3.9/5.3

Average Climatic Conditions

Element	Jan	Feb	Mar	Apr	May	June	July	Aug	Sept	Oct	Nov	Dec	Year
Temperature (°C)	-5.1	-1.7	1.9	7.6	12.8	18.2	22.5	21.5	15.1	9.5	1.7	-3.6	8.4
Daily Minimum Temp	-10.2	-7.0	-3.8	1.1	6.3	11.1	14.6	13.7	8.1	3.1	-3.6	-8.6	2.1
Daily Maximum Temp	-0.1	3.7	7.7	13.9	19.3	25.3	30.4	29.3	22.0	15.9	6.9	1.3	14.6
Record Minimum Temp	-34.4	-38.9	-28.3	-20.6	-10.0	0.0	5.0	4.4	-5.6	-21.7	-30.0	-35.6	-38.9
Record Maximum Temp	20.0	22.2	26.1	33.3	35.6	40.6	41.1	40.6	39.4	32.2	25.0	20.6	41.1
HDD, Base 18.3°C	727	560	508	323	176	66	7	23	134	277	498	681	3980
CDD, Base 18.3°C	0	0	0	0	3	63	136	122	36	3	0	0	362
Relative Humidity (%)	60	59	58	54	55	53	46	44	52	53	59	61	54
Wind Speed (m/s)	6.3	5.5	5.0	5.1	4.7	4.3	4.2	4.2	4.4	4.9	5.2	5.9	5.0

Cut Bank, MT

WBAN NO. 24137

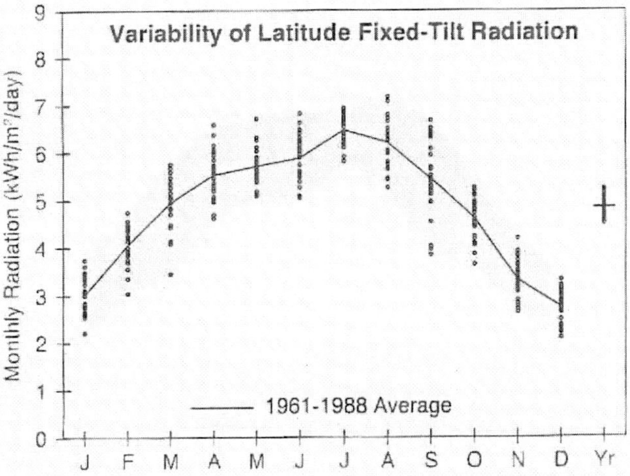

LATITUDE: 48.60° N
LONGITUDE: 112.37° W
ELEVATION: 1170 meters
MEAN PRESSURE: 881 millibars

STATION TYPE: Secondary

Solar Radiation for Flat-Plate Collectors Facing South at a Fixed Tilt (kWh/m²/day), Uncertainty ±9%

Tilt (°)		Jan	Feb	Mar	Apr	May	June	July	Aug	Sept	Oct	Nov	Dec	Year
0	Average	1.4	2.2	3.5	4.9	5.9	6.6	6.9	5.8	4.2	2.8	1.6	1.1	3.9
	Min/Max	1.2/1.6	1.9/2.5	2.8/3.9	4.2/5.6	5.3/6.8	5.7/7.6	6.3/7.4	5.1/6.6	3.3/4.8	2.4/3.0	1.4/1.8	1.0/1.3	3.7/4.1
Latitude -15	Average	2.6	3.7	4.8	5.6	6.1	6.5	7.0	6.5	5.4	4.3	2.9	2.4	4.8
	Min/Max	2.0/3.2	2.8/4.3	3.4/5.5	4.8/6.7	5.5/7.2	5.5/7.5	6.3/7.5	5.5/7.5	3.9/6.5	3.5/4.9	2.4/3.7	1.8/2.8	4.5/5.2
Latitude	Average	3.0	4.0	4.9	5.5	5.7	5.9	6.5	6.2	5.4	4.6	3.3	2.7	4.8
	Min/Max	2.2/3.7	3.0/4.7	3.4/5.8	4.6/6.6	5.1/6.7	5.0/6.8	5.8/6.9	5.3/7.2	3.8/6.7	3.6/5.3	2.6/4.2	2.1/3.3	4.5/5.2
Latitude +15	Average	3.2	4.2	4.9	5.1	5.1	5.1	5.6	5.6	5.2	4.6	3.5	2.9	4.6
	Min/Max	2.3/4.1	3.1/4.9	3.3/5.7	4.3/6.1	4.5/5.9	4.4/5.9	5.1/6.0	4.8/6.5	3.6/6.4	3.6/5.3	2.7/4.4	2.2/3.6	4.2/5.0
90	Average	3.1	3.9	4.1	3.8	3.4	3.2	3.6	3.9	4.1	4.1	3.3	2.9	3.6
	Min/Max	2.2/4.0	2.8/4.5	2.7/4.8	3.2/4.6	3.0/3.8	2.8/3.6	3.2/3.8	3.3/4.5	2.8/5.1	3.2/4.7	2.5/4.2	2.2/3.5	3.3/3.9

Solar Radiation for 1-Axis Tracking Flat-Plate Collectors with a North-South Axis (kWh/m²/day), Uncertainty ±9%

Axis Tilt (°)		Jan	Feb	Mar	Apr	May	June	July	Aug	Sept	Oct	Nov	Dec	Year
0	Average	2.2	3.4	5.1	6.9	8.1	9.0	9.8	8.4	6.2	4.3	2.5	1.8	5.7
	Min/Max	1.7/2.7	2.6/4.0	3.5/6.1	5.5/8.5	7.1/9.8	7.3/11.1	8.6/10.7	6.9/10.1	4.2/7.7	3.4/5.0	2.0/3.2	1.4/2.2	5.2/6.1
Latitude -15	Average	3.1	4.5	6.1	7.6	8.4	9.1	10.1	9.0	7.1	5.4	3.5	2.8	6.4
	Min/Max	2.3/4.0	3.3/5.4	4.0/7.3	5.9/9.4	7.3/10.3	7.4/11.3	8.8/11.1	7.3/10.9	4.7/9.0	4.2/6.3	2.8/4.5	2.1/3.4	5.8/7.0
Latitude	Average	3.4	4.8	6.2	7.5	8.2	8.7	9.8	8.9	7.2	5.7	3.8	3.1	6.5
	Min/Max	2.5/4.4	3.5/5.7	4.0/7.5	5.9/9.4	7.1/10.0	7.1/10.9	8.5/10.7	7.2/10.7	4.7/9.1	4.3/6.6	3.0/4.9	2.3/3.8	5.9/7.0
Latitude +15	Average	3.6	4.9	6.2	7.2	7.7	8.2	9.2	8.4	7.0	5.7	4.0	3.2	6.3
	Min/Max	2.5/4.6	3.5/5.9	3.9/7.4	5.6/9.1	6.7/9.4	6.6/10.3	8.0/10.1	6.8/10.2	4.6/8.9	4.3/6.7	3.1/5.2	2.4/4.0	5.7/6.9

Solar Radiation for 2-Axis Tracking Flat-Plate Collectors (kWh/m²/day), Uncertainty ±9%

Tracker		Jan	Feb	Mar	Apr	May	June	July	Aug	Sept	Oct	Nov	Dec	Year
2-Axis	Average	3.6	4.9	6.2	7.6	8.6	9.3	10.3	9.1	7.2	5.7	4.0	3.3	6.7
	Min/Max	2.6/4.7	3.5/5.9	4.1/7.5	6.0/9.5	7.4/10.4	7.6/11.7	9.0/11.3	7.4/10.9	4.8/9.1	4.4/6.7	3.1/5.2	2.4/4.1	6.1/7.3

Direct Beam Solar Radiation for Concentrating Collectors (kWh/m²/day), Uncertainty ±8%

Tracker		Jan	Feb	Mar	Apr	May	June	July	Aug	Sept	Oct	Nov	Dec	Year
1-Axis, E-W Horiz Axis	Average	2.2	2.7	3.1	3.4	3.9	4.4	5.4	4.6	3.7	3.2	2.4	2.1	3.4
	Min/Max	1.1/3.0	1.8/3.7	1.6/4.1	2.3/5.0	3.0/5.5	3.0/6.4	4.2/6.2	3.3/6.1	2.0/5.2	2.1/4.0	1.6/3.5	1.2/2.8	3.0/4.0
1-Axis, N-S Horiz Axis	Average	1.3	2.1	3.2	4.4	5.2	5.9	7.2	6.0	4.3	2.9	1.6	1.1	3.8
	Min/Max	0.7/1.8	1.4/2.8	1.6/4.4	2.8/6.4	4.0/7.3	4.1/8.7	5.5/8.2	4.3/8.1	2.2/6.0	1.9/3.7	1.1/2.3	0.7/1.6	3.2/4.4
1-Axis, N-S Tilt=Latitude	Average	2.3	3.2	4.1	4.8	5.3	5.7	7.0	6.4	5.1	4.1	2.7	2.1	4.4
	Min/Max	1.2/3.2	2.1/4.3	2.0/5.5	3.1/7.1	4.0/7.3	4.0/8.4	5.4/8.1	4.6/8.5	2.7/7.2	2.6/5.1	1.8/3.9	1.3/2.9	3.8/5.1
2-Axis	Average	2.5	3.3	4.1	4.9	5.6	6.2	7.6	6.6	5.1	4.1	2.8	2.3	4.6
	Min/Max	1.3/3.5	2.2/4.4	2.0/5.5	3.2/7.2	4.3/7.8	4.3/9.1	5.8/8.7	4.7/8.8	2.7/7.2	2.7/5.2	1.9/4.1	1.4/3.2	3.9/5.3

Average Climatic Conditions

Element	Jan	Feb	Mar	Apr	May	June	July	Aug	Sept	Oct	Nov	Dec	Year
Temperature (°C)	-8.1	-4.9	-1.4	4.6	9.9	14.6	17.9	17.4	11.8	6.7	-1.4	-6.7	5.1
Daily Minimum Temp	-13.7	-10.8	-7.3	-2.1	3.1	7.5	9.7	9.2	4.1	-0.4	-7.2	-12.2	-1.7
Daily Maximum Temp	-2.5	1.0	4.6	11.2	16.8	21.7	26.1	25.6	19.4	13.7	4.3	-1.2	11.7
Record Minimum Temp	-37.2	-34.4	-31.1	-18.9	-10.6	-1.1	1.7	-0.6	-6.1	-21.1	-30.6	-39.4	-39.4
Record Maximum Temp	19.4	18.9	22.2	29.4	31.7	34.4	36.7	41.1	35.0	30.0	25.0	16.1	41.1
HDD, Base 18.3°C	818	650	612	412	260	125	43	86	209	362	593	777	4947
CDD, Base 18.3°C	0	0	0	0	0	13	31	57	13	0	0	0	114
Relative Humidity (%)	66	66	64	57	55	56	51	51	56	56	66	68	59
Wind Speed (m/s)	6.5	6.0	5.6	6.0	5.5	5.1	4.8	4.6	4.9	5.5	5.5	5.9	5.5

Glasgow, MT

WBAN NO. 94008

LATITUDE: 48.22° N
LONGITUDE: 106.62° W
ELEVATION: 700 meters
MEAN PRESSURE: 933 millibars

STATION TYPE: Secondary

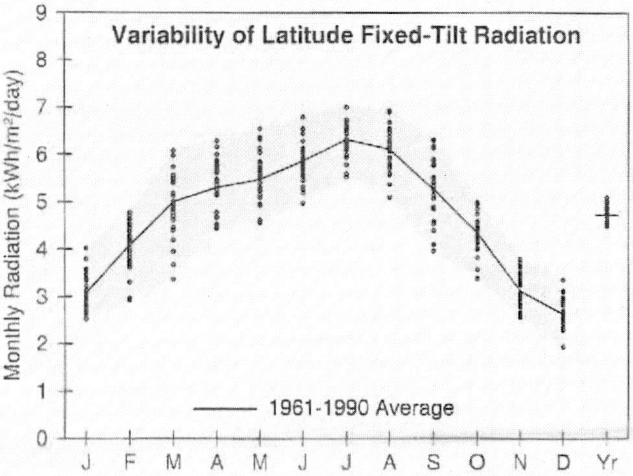

Solar Radiation for Flat-Plate Collectors Facing South at a Fixed Tilt (kWh/m²/day), Uncertainty ±9%

Tilt (°)		Jan	Feb	Mar	Apr	May	June	July	Aug	Sept	Oct	Nov	Dec	Year
0	Average	1.5	2.3	3.6	4.7	5.7	6.5	6.7	5.7	4.1	2.7	1.6	1.2	3.9
	Min/Max	1.3/1.7	1.9/2.7	2.8/4.2	4.1/5.4	4.8/6.7	5.6/7.4	5.9/7.4	5.0/6.4	3.3/4.7	2.3/2.9	1.4/1.7	1.0/1.3	3.7/4.1
Latitude -15	Average	2.7	3.7	4.8	5.4	5.8	6.4	6.8	6.4	5.2	4.1	2.8	2.3	4.7
	Min/Max	2.3/3.5	2.7/4.3	3.3/5.8	4.6/6.4	4.9/7.0	5.4/7.4	6.0/7.6	5.3/7.2	4.0/6.2	3.2/4.6	2.3/3.3	1.7/2.9	4.5/5.1
Latitude	Average	3.1	4.1	5.0	5.3	5.5	5.9	6.3	6.1	5.3	4.3	3.1	2.6	4.7
	Min/Max	2.5/4.0	2.9/4.8	3.4/6.1	4.4/6.3	4.6/6.5	5.0/6.8	5.5/7.0	5.1/6.9	4.0/6.3	3.4/5.0	2.6/3.8	1.9/3.3	4.5/5.1
Latitude +15	Average	3.3	4.2	4.9	4.9	4.8	5.0	5.5	5.5	5.0	4.4	3.3	2.8	4.5
	Min/Max	2.7/4.3	3.0/5.0	3.2/6.1	4.1/5.8	4.0/5.8	4.3/5.8	4.8/6.0	4.6/6.3	3.8/6.1	3.4/5.0	2.7/4.0	2.0/3.7	4.2/4.8
90	Average	3.2	3.9	4.2	3.6	3.2	3.1	3.4	3.8	3.9	3.8	3.1	2.8	3.5
	Min/Max	2.6/4.3	2.7/4.7	2.6/5.5	3.0/4.2	2.7/3.7	2.8/3.5	3.1/3.7	3.2/4.3	2.9/4.8	2.9/4.4	2.5/3.8	2.0/3.7	3.3/3.8

Solar Radiation for 1-Axis Tracking Flat-Plate Collectors with a North-South Axis (kWh/m²/day), Uncertainty ±9%

Axis Tilt (°)		Jan	Feb	Mar	Apr	May	June	July	Aug	Sept	Oct	Nov	Dec	Year
0	Average	2.3	3.5	5.2	6.6	7.7	8.9	9.5	8.2	6.0	4.1	2.4	1.8	5.5
	Min/Max	1.8/2.9	2.5/4.1	3.5/6.3	5.1/8.2	6.1/9.6	7.1/10.3	8.0/10.7	6.9/9.6	4.4/7.4	3.1/4.7	2.0/2.9	1.4/2.3	5.2/6.0
Latitude -15	Average	3.2	4.5	6.1	7.2	8.0	9.0	9.8	8.9	6.9	5.1	3.3	2.7	6.2
	Min/Max	2.6/4.2	3.2/5.4	4.0/7.5	5.5/9.0	6.3/10.1	7.1/10.5	8.2/11.1	7.3/10.4	5.0/8.6	3.8/5.9	2.7/4.1	1.9/3.5	5.8/6.8
Latitude	Average	3.5	4.8	6.3	7.2	7.7	8.6	9.4	8.7	6.9	5.3	3.6	3.0	6.3
	Min/Max	2.8/4.7	3.3/5.8	4.0/7.8	5.4/9.0	6.1/9.8	6.8/10.1	7.9/10.7	7.1/10.2	5.0/8.7	4.0/6.2	2.9/4.4	2.1/3.8	5.8/6.9
Latitude +15	Average	3.7	4.9	6.2	6.9	7.3	8.1	8.9	8.3	6.8	5.3	3.7	3.1	6.1
	Min/Max	2.9/4.9	3.4/5.9	3.9/7.8	5.2/8.6	5.7/9.3	6.4/9.5	7.4/10.1	6.8/9.7	4.8/8.5	4.0/6.2	3.0/4.6	2.2/4.1	5.7/6.7

Solar Radiation for 2-Axis Tracking Flat-Plate Collectors (kWh/m²/day), Uncertainty ±9%

Tracker		Jan	Feb	Mar	Apr	May	June	July	Aug	Sept	Oct	Nov	Dec	Year
2-Axis	Average	3.7	5.0	6.3	7.3	8.1	9.2	10.0	8.9	7.0	5.4	3.7	3.2	6.5
	Min/Max	3.0/5.0	3.4/5.9	4.0/7.8	5.5/9.1	6.4/10.2	7.3/10.8	8.4/11.3	7.3/10.4	5.0/8.7	4.0/6.3	3.0/4.7	2.2/4.2	6.0/7.1

Direct Beam Solar Radiation for Concentrating Collectors (kWh/m²/day), Uncertainty ±8%

Tracker		Jan	Feb	Mar	Apr	May	June	July	Aug	Sept	Oct	Nov	Dec	Year
1-Axis, E-W Horiz Axis	Average	2.0	2.6	2.9	3.3	3.7	4.5	5.2	4.5	3.5	3.0	2.2	1.8	3.3
	Min/Max	1.3/3.0	1.4/3.5	1.5/4.1	1.8/4.8	2.3/5.3	2.9/6.0	3.9/6.3	3.2/5.8	2.1/4.9	2.0/3.8	1.5/3.0	0.9/2.6	2.9/3.9
1-Axis, N-S Horiz Axis	Average	1.2	2.0	3.1	4.2	4.9	5.9	6.8	5.9	4.1	2.7	1.5	1.0	3.6
	Min/Max	0.8/1.8	1.1/2.8	1.6/4.3	2.2/6.2	3.1/7.1	3.8/7.8	5.2/8.3	4.3/7.6	2.4/5.8	1.7/3.4	1.0/2.0	0.5/1.5	3.2/4.3
1-Axis, N-S Tilt=Latitude	Average	2.2	3.0	3.9	4.6	4.9	5.7	6.7	6.2	4.9	3.8	2.4	1.9	4.2
	Min/Max	1.4/3.2	1.7/4.2	2.0/5.4	2.4/6.8	3.2/7.2	3.7/7.5	5.1/8.1	4.5/8.0	2.9/6.8	2.4/4.7	1.6/3.4	1.0/2.7	3.7/4.9
2-Axis	Average	2.3	3.1	3.9	4.7	5.2	6.2	7.2	6.4	4.9	3.8	2.6	2.0	4.4
	Min/Max	1.5/3.4	1.7/4.3	2.0/5.4	2.5/6.9	3.3/7.6	4.0/8.1	5.4/8.7	4.7/8.3	2.9/6.8	2.5/4.8	1.7/3.5	1.0/3.0	3.9/5.2

Average Climatic Conditions

Element	Jan	Feb	Mar	Apr	May	June	July	Aug	Sept	Oct	Nov	Dec	Year
Temperature (°C)	-11.9	-8.1	-1.4	6.8	12.8	18.0	21.6	20.8	14.1	7.8	-1.6	-9.3	5.8
Daily Minimum Temp	-17.1	-13.4	-7.2	0.0	5.8	10.7	13.7	12.9	6.7	0.7	-7.3	-14.7	-0.8
Daily Maximum Temp	-6.7	-2.7	4.3	13.6	19.6	25.3	29.3	28.6	21.3	14.8	4.2	-3.9	12.3
Record Minimum Temp	-43.9	-38.3	-32.8	-19.4	-6.7	0.6	5.0	2.8	-6.7	-21.1	-32.2	-38.9	-43.9
Record Maximum Temp	12.8	18.9	25.0	32.8	38.9	42.2	40.0	42.2	39.4	32.2	23.9	15.0	42.2
HDD, Base 18.3°C	937	739	613	345	179	67	11	39	148	326	597	858	4858
CDD, Base 18.3°C	0	0	0	0	7	57	111	116	20	0	0	0	310
Relative Humidity (%)	73	75	70	58	56	55	51	48	55	59	70	75	62
Wind Speed (m/s)	4.4	4.6	5.1	5.4	5.5	4.9	4.9	4.9	5.0	4.7	4.3	4.5	4.8

Great Falls, MT

WBAN NO. 24143

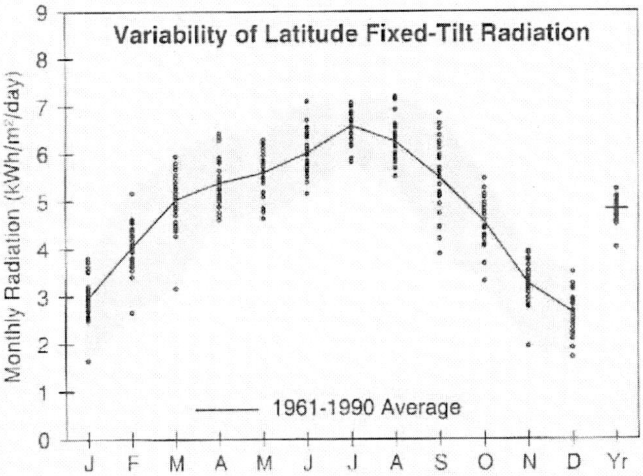

LATITUDE: 47.48° N
LONGITUDE: 111.37° W
ELEVATION: 1116 meters
MEAN PRESSURE: 888 millibars

STATION TYPE: Primary

Solar Radiation for Flat-Plate Collectors Facing South at a Fixed Tilt (kWh/m²/day), Uncertainty ±9%

Tilt (°)		Jan	Feb	Mar	Apr	May	June	July	Aug	Sept	Oct	Nov	Dec	Year
0	Average	1.4	2.4	3.7	4.9	5.8	6.7	7.1	5.9	4.3	2.8	1.7	1.2	4.0
	Min/Max	1.1/1.7	2.2/2.8	2.9/4.1	4.3/5.7	4.9/6.5	5.7/7.9	6.3/7.6	5.3/6.7	3.4/5.1	2.4/3.2	1.3/1.9	0.9/1.4	3.7/4.2
Latitude -15	Average	2.6	3.7	4.8	5.5	6.0	6.6	7.2	6.5	5.4	4.3	2.9	2.3	4.8
	Min/Max	1.5/3.3	2.6/4.7	3.2/5.6	4.8/6.6	5.0/6.7	5.6/7.8	6.3/7.7	5.8/7.5	3.9/6.7	3.2/5.1	1.8/3.5	1.5/3.0	4.1/5.2
Latitude	Average	3.0	4.0	5.0	5.4	5.6	6.0	6.6	6.3	5.5	4.6	3.3	2.7	4.8
	Min/Max	1.6/3.8	2.7/5.2	3.2/5.9	4.6/6.4	4.6/6.3	5.2/7.1	5.8/7.1	5.5/7.2	3.9/6.9	3.3/5.5	2.0/4.0	1.7/3.5	4.0/5.3
Latitude +15	Average	3.2	4.2	5.0	5.0	4.9	5.2	5.7	5.7	5.3	4.6	3.4	2.9	4.6
	Min/Max	1.7/4.1	2.6/5.4	3.0/5.9	4.3/6.0	4.1/5.6	4.5/6.1	5.1/6.2	5.0/6.5	3.7/6.6	3.3/5.6	2.0/4.2	1.8/3.8	3.7/5.0
90	Average	3.1	3.8	4.2	3.7	3.2	3.2	3.5	3.9	4.1	4.0	3.2	2.8	3.5
	Min/Max	1.5/4.1	2.3/5.0	2.4/5.2	3.1/4.4	2.7/3.6	2.8/3.6	3.2/3.8	3.4/4.4	2.8/5.1	2.8/4.9	1.8/3.9	1.7/3.7	2.8/3.9

Solar Radiation for 1-Axis Tracking Flat-Plate Collectors with a North-South Axis (kWh/m²/day), Uncertainty ±9%

Axis Tilt (°)		Jan	Feb	Mar	Apr	May	June	July	Aug	Sept	Oct	Nov	Dec	Year
0	Average	2.2	3.5	5.2	6.6	7.8	9.0	9.9	8.4	6.3	4.3	2.5	1.8	5.6
	Min/Max	1.3/2.8	2.4/4.7	3.2/6.4	5.6/8.3	6.0/9.1	7.1/11.4	8.4/11.0	7.2/10.2	4.3/7.7	2.9/5.2	1.5/3.1	1.3/2.4	4.7/6.2
Latitude -15	Average	3.1	4.5	6.1	7.2	8.1	9.1	10.2	9.0	7.2	5.3	3.5	2.7	6.3
	Min/Max	1.7/4.0	2.8/6.1	3.5/7.4	6.0/9.1	6.2/9.4	7.2/11.6	8.6/11.3	7.7/10.9	4.7/8.9	3.6/6.6	2.0/4.3	1.7/3.6	5.1/7.0
Latitude	Average	3.4	4.8	6.3	7.2	7.8	8.7	9.8	8.9	7.3	5.6	3.7	3.0	6.4
	Min/Max	1.8/4.4	2.9/6.5	3.5/7.6	6.0/9.1	6.0/9.2	6.9/11.1	8.3/10.9	7.5/10.8	4.7/9.1	3.7/6.9	2.1/4.6	1.9/4.0	5.0/7.1
Latitude +15	Average	3.5	4.9	6.2	6.9	7.4	8.2	9.2	8.4	7.1	5.6	3.9	3.2	6.2
	Min/Max	1.8/4.6	2.8/6.6	3.4/7.6	5.7/8.7	5.6/8.7	6.4/10.5	7.8/10.3	7.2/10.3	4.6/8.9	3.7/7.0	2.1/4.8	2.0/4.3	4.8/6.9

Solar Radiation for 2-Axis Tracking Flat-Plate Collectors (kWh/m²/day), Uncertainty ±9%

Tracker		Jan	Feb	Mar	Apr	May	June	July	Aug	Sept	Oct	Nov	Dec	Year
2-Axis	Average	3.6	4.9	6.3	7.3	8.2	9.3	10.4	9.1	7.3	5.6	3.9	3.2	6.6
	Min/Max	1.8/4.7	2.9/6.6	3.5/7.6	6.1/9.1	6.3/9.6	7.4/11.9	8.8/11.6	7.7/11.0	4.8/9.1	3.7/7.0	2.2/4.9	2.0/4.3	5.2/7.3

Direct Beam Solar Radiation for Concentrating Collectors (kWh/m²/day), Uncertainty ±8%

Tracker		Jan	Feb	Mar	Apr	May	June	July	Aug	Sept	Oct	Nov	Dec	Year
1-Axis, E-W Horiz Axis	Average	2.0	2.6	3.0	3.2	3.7	4.5	5.4	4.6	3.7	3.2	2.3	1.9	3.3
	Min/Max	0.5/2.9	0.4/3.9	0.8/4.0	2.1/4.7	2.4/4.7	3.0/6.8	4.1/6.3	3.7/6.2	1.9/5.3	1.7/4.4	0.9/3.2	1.0/2.9	2.2/3.9
1-Axis, N-S Horiz Axis	Average	1.2	2.0	3.1	4.0	4.9	5.9	7.2	6.0	4.4	2.9	1.5	1.1	3.7
	Min/Max	0.3/1.8	0.3/3.4	0.8/4.2	2.7/6.2	3.1/6.3	3.8/9.0	5.5/8.5	4.8/8.3	2.1/6.0	1.4/4.1	0.6/2.1	0.5/1.7	2.5/4.3
1-Axis, N-S Tilt=Latitude	Average	2.1	3.0	3.9	4.4	4.9	5.7	7.1	6.4	5.2	4.0	2.5	2.0	4.3
	Min/Max	0.6/3.1	0.5/4.8	1.0/5.3	3.0/6.7	3.1/6.3	3.7/8.7	5.4/8.3	5.0/8.7	2.5/7.2	2.0/5.6	1.0/3.5	1.0/3.0	2.8/5.0
2-Axis	Average	2.3	3.1	3.9	4.5	5.2	6.2	7.6	6.6	5.2	4.0	2.7	2.1	4.5
	Min/Max	0.6/3.3	0.5/4.9	1.0/5.3	3.0/6.8	3.3/6.7	4.0/9.4	5.8/8.9	5.2/9.0	2.6/7.2	2.0/5.6	1.0/3.7	1.1/3.3	2.9/5.3

Average Climatic Conditions

Element	Jan	Feb	Mar	Apr	May	June	July	Aug	Sept	Oct	Nov	Dec	Year
Temperature (°C)	-6.0	-2.6	0.7	6.4	11.7	16.4	20.1	19.4	13.7	8.6	1.1	-4.5	7.1
Daily Minimum Temp	-11.3	-8.2	-5.1	-0.1	4.9	9.2	11.8	11.2	6.4	2.1	-4.3	-9.7	0.6
Daily Maximum Temp	-0.8	3.1	6.5	12.9	18.4	23.7	28.5	27.6	20.9	15.2	6.4	0.6	13.6
Record Minimum Temp	-38.3	-37.2	-33.9	-21.1	-9.4	-0.6	4.4	1.7	-6.1	-23.9	-31.7	-41.7	-41.7
Record Maximum Temp	16.7	20.0	25.6	31.7	33.9	38.3	40.6	41.1	36.7	32.8	24.4	20.6	41.1
HDD, Base 18.3°C	754	585	546	357	207	87	21	51	166	302	518	708	4301
CDD, Base 18.3°C	0	0	0	0	0	30	76	83	26	0	0	0	216
Relative Humidity (%)	63	62	60	54	54	52	45	46	53	53	59	63	55
Wind Speed (m/s)	6.4	5.9	5.4	5.5	4.9	4.6	4.3	4.3	4.7	5.5	6.2	6.4	5.4

Helena, MT

WBAN NO. 24144

LATITUDE: 46.60° N
LONGITUDE: 112.00° W
ELEVATION: 1188 meters
MEAN PRESSURE: 881 millibars

STATION TYPE: Secondary

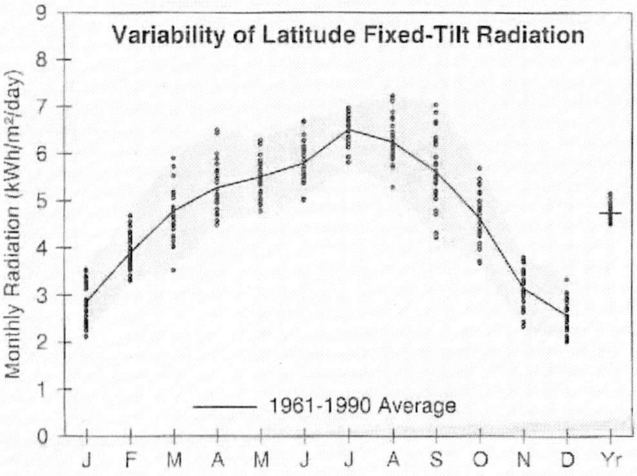

Solar Radiation for Flat-Plate Collectors Facing South at a Fixed Tilt (kWh/m²/day), Uncertainty ±9%

Tilt (°)		Jan	Feb	Mar	Apr	May	June	July	Aug	Sept	Oct	Nov	Dec	Year
0	Average	1.5	2.3	3.5	4.8	5.8	6.5	7.0	5.9	4.4	2.9	1.7	1.2	4.0
	Min/Max	1.2/1.7	2.1/2.7	2.9/4.2	4.2/5.6	5.1/6.5	5.6/7.5	6.3/7.6	5.1/6.8	3.6/5.2	2.5/3.3	1.5/1.9	1.1/1.4	3.8/4.2
Latitude -15	Average	2.5	3.6	4.6	5.4	5.9	6.4	7.1	6.5	5.5	4.3	2.8	2.2	4.7
	Min/Max	1.9/3.1	3.1/4.3	3.5/5.6	4.6/6.6	5.1/6.7	5.5/7.4	6.3/7.6	5.5/7.5	4.2/6.9	3.5/5.3	2.2/3.3	1.8/2.8	4.5/5.1
Latitude	Average	2.8	3.9	4.8	5.3	5.5	5.8	6.5	6.2	5.6	4.6	3.1	2.6	4.7
	Min/Max	2.1/3.5	3.3/4.7	3.5/5.9	4.5/6.5	4.8/6.3	5.0/6.7	5.8/7.0	5.3/7.2	4.2/7.0	3.7/5.7	2.3/3.8	2.0/3.3	4.5/5.2
Latitude +15	Average	3.0	4.0	4.7	4.9	4.9	5.0	5.7	5.7	5.4	4.7	3.3	2.8	4.5
	Min/Max	2.2/3.8	3.4/4.9	3.4/5.9	4.2/6.1	4.2/5.6	4.3/5.7	5.1/6.0	4.8/6.6	4.0/6.8	3.7/5.8	2.4/4.0	2.1/3.6	4.3/4.9
90	Average	2.9	3.7	3.9	3.5	3.1	3.0	3.4	3.8	4.1	4.0	3.1	2.7	3.4
	Min/Max	2.1/3.8	3.0/4.6	2.7/5.2	3.0/4.3	2.8/3.5	2.6/3.3	3.1/3.6	3.2/4.3	3.1/5.2	3.1/5.1	2.3/3.8	2.0/3.6	3.2/3.8

Solar Radiation for 1-Axis Tracking Flat-Plate Collectors with a North-South Axis (kWh/m²/day), Uncertainty ±9%

Axis Tilt (°)		Jan	Feb	Mar	Apr	May	June	July	Aug	Sept	Oct	Nov	Dec	Year
0	Average	2.2	3.4	5.0	6.5	7.8	8.8	10.0	8.6	6.5	4.4	2.5	1.8	5.6
	Min/Max	1.6/2.7	3.0/4.1	3.7/6.2	5.4/8.4	6.4/9.2	7.2/11.0	8.5/11.3	7.2/10.2	4.8/8.3	3.5/5.6	1.9/3.0	1.4/2.3	5.2/6.1
Latitude -15	Average	3.0	4.3	5.8	7.1	8.1	8.9	10.2	9.2	7.4	5.5	3.4	2.6	6.3
	Min/Max	2.1/3.7	3.7/5.2	4.1/7.3	5.8/9.2	6.6/9.6	7.3/11.1	8.7/11.5	7.6/11.0	5.3/9.6	4.2/6.9	2.4/4.1	2.0/3.4	5.8/6.9
Latitude	Average	3.2	4.6	6.0	7.0	7.8	8.6	9.9	9.0	7.5	5.7	3.6	2.9	6.3
	Min/Max	2.3/4.1	3.9/5.6	4.2/7.6	5.7/9.1	6.3/9.3	7.0/10.7	8.4/11.1	7.4/10.8	5.3/9.7	4.4/7.3	2.6/4.5	2.2/3.8	5.9/7.0
Latitude +15	Average	3.4	4.7	5.9	6.7	7.4	8.0	9.3	8.6	7.3	5.7	3.8	3.0	6.2
	Min/Max	2.4/4.3	3.9/5.7	4.1/7.6	5.4/8.8	5.9/8.8	6.5/10.1	7.9/10.5	7.0/10.3	5.1/9.5	4.4/7.4	2.7/4.7	2.3/4.1	5.7/6.8

Solar Radiation for 2-Axis Tracking Flat-Plate Collectors (kWh/m²/day), Uncertainty ±9%

Tracker		Jan	Feb	Mar	Apr	May	June	July	Aug	Sept	Oct	Nov	Dec	Year
2-Axis	Average	3.4	4.7	6.0	7.1	8.2	9.2	10.4	9.2	7.5	5.8	3.8	3.1	6.5
	Min/Max	2.4/4.4	3.9/5.7	4.2/7.6	5.8/9.2	6.7/9.7	7.5/11.5	8.9/11.8	7.6/11.1	5.3/9.7	4.4/7.4	2.7/4.7	2.3/4.1	6.1/7.2

Direct Beam Solar Radiation for Concentrating Collectors (kWh/m²/day), Uncertainty ±8%

Tracker		Jan	Feb	Mar	Apr	May	June	July	Aug	Sept	Oct	Nov	Dec	Year
1-Axis, E-W	Average	1.9	2.5	2.8	3.2	3.8	4.5	5.7	4.8	4.0	3.3	2.2	1.8	3.4
Horiz Axis	Min/Max	1.1/2.7	1.6/3.4	1.6/3.8	2.1/4.9	2.6/5.0	3.1/6.5	4.3/6.7	3.6/6.4	2.3/5.8	2.2/4.7	1.0/3.0	1.1/2.6	3.0/4.0
1-Axis, N-S	Average	1.2	2.0	3.0	4.1	5.0	6.0	7.6	6.4	4.7	3.0	1.5	1.0	3.8
Horiz Axis	Min/Max	0.7/1.7	1.3/2.8	1.7/4.2	2.7/6.3	3.4/6.7	4.1/8.8	5.6/9.1	4.8/8.5	2.7/6.8	2.0/4.4	0.6/2.1	0.6/1.5	3.3/4.5
1-Axis, N-S	Average	2.0	3.0	3.8	4.5	5.0	5.8	7.4	6.7	5.5	4.1	2.4	1.8	4.3
Tilt=Latitude	Min/Max	1.2/3.0	1.9/4.1	2.1/5.2	2.9/6.9	3.4/6.7	4.0/8.4	5.5/8.9	5.0/8.9	3.1/8.1	2.7/6.0	1.1/3.3	1.2/2.7	3.8/5.1
2-Axis	Average	2.1	3.0	3.8	4.5	5.3	6.3	7.9	6.9	5.6	4.2	2.5	2.0	4.5
	Min/Max	1.3/3.2	1.9/4.2	2.1/5.2	3.0/7.0	3.6/7.1	4.3/9.2	5.9/9.6	5.2/9.2	3.1/8.1	2.8/6.0	1.1/3.5	1.3/2.9	4.0/5.3

Average Climatic Conditions

Element	Jan	Feb	Mar	Apr	May	June	July	Aug	Sept	Oct	Nov	Dec	Year
Temperature (°C)	-6.9	-3.1	0.9	6.3	11.4	16.7	20.7	19.7	13.0	7.3	-0.2	-6.0	6.7
Daily Minimum Temp	-12.4	-8.9	-5.4	-0.8	4.2	9.1	11.9	10.9	5.0	-0.2	-6.3	-11.6	-0.4
Daily Maximum Temp	-1.3	2.7	7.1	13.4	18.6	24.3	29.4	28.4	21.0	14.7	5.8	-0.4	13.7
Record Minimum Temp	-41.1	-36.1	-34.4	-17.2	-8.3	-1.1	2.2	0.0	-7.8	-22.2	-39.4	-38.9	-41.1
Record Maximum Temp	16.7	20.0	25.0	29.4	33.3	37.8	38.9	40.6	37.2	29.4	22.2	17.8	40.6
HDD, Base 18.3°C	782	601	541	360	216	76	19	36	178	343	557	754	4462
CDD, Base 18.3°C	0	0	0	0	0	28	91	77	18	0	0	0	214
Relative Humidity (%)	67	65	60	54	54	52	46	47	55	59	65	69	58
Wind Speed (m/s)	3.1	3.4	3.7	4.1	3.9	3.7	3.5	3.2	3.2	3.1	3.1	2.8	3.4

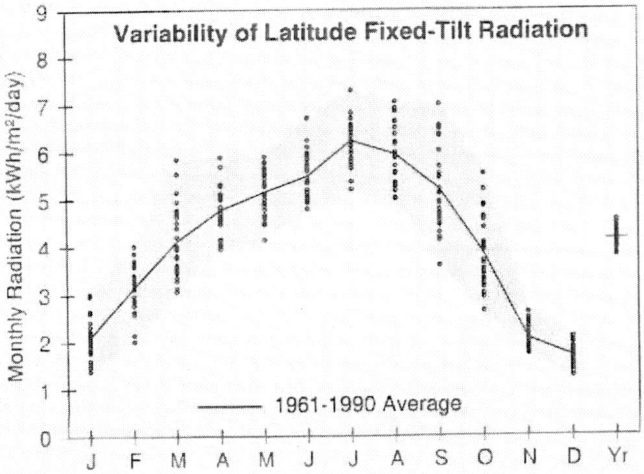

Kalispell, MT

WBAN NO. 24146

LATITUDE: 48.30° N
LONGITUDE: 114.27° W
ELEVATION: 904 meters
MEAN PRESSURE: 912 millibars

STATION TYPE: Secondary

Solar Radiation for Flat-Plate Collectors Facing South at a Fixed Tilt (kWh/m²/day), Uncertainty ±9%

Tilt (°)		Jan	Feb	Mar	Apr	May	June	July	Aug	Sept	Oct	Nov	Dec	Year
0	Average	1.2	2.0	3.1	4.3	5.4	6.1	6.7	5.6	4.0	2.5	1.3	1.0	3.6
	Min/Max	0.9/1.5	1.5/2.3	2.5/4.0	3.7/5.1	4.5/6.1	5.4/7.5	5.6/7.8	4.9/6.5	3.2/5.1	2.0/3.2	1.2/1.4	0.8/1.1	3.4/3.8
Latitude -15	Average	1.9	2.9	4.0	4.9	5.5	6.0	6.7	6.2	5.1	3.6	1.9	1.5	4.2
	Min/Max	1.3/2.7	1.9/3.7	3.0/5.5	4.1/6.0	4.5/6.3	5.3/7.4	5.6/7.9	5.3/7.3	3.7/6.8	2.6/5.1	1.6/2.4	1.2/1.9	3.9/4.6
Latitude	Average	2.1	3.1	4.1	4.8	5.1	5.5	6.2	5.9	5.2	3.8	2.0	1.7	4.1
	Min/Max	1.4/3.0	2.0/4.0	3.1/5.8	4.0/5.9	4.1/5.9	4.8/6.7	5.2/7.3	5.0/7.1	3.6/7.0	2.6/5.5	1.7/2.6	1.2/2.1	3.8/4.5
Latitude +15	Average	2.2	3.2	4.0	4.4	4.6	4.7	5.4	5.4	5.0	3.8	2.1	1.7	3.9
	Min/Max	1.4/3.2	2.0/4.2	2.9/5.9	3.6/5.5	3.7/5.2	4.2/5.8	4.6/6.3	4.5/6.4	3.4/6.7	2.6/5.6	1.7/2.7	1.3/2.2	3.5/4.3
90	Average	2.1	3.0	3.4	3.2	3.0	3.0	3.4	3.7	3.9	3.3	1.9	1.7	3.0
	Min/Max	1.3/3.2	1.7/4.0	2.4/5.3	2.7/4.0	2.5/3.4	2.7/3.5	3.0/3.9	3.1/4.4	2.6/5.3	2.2/4.9	1.5/2.5	1.1/2.2	2.7/3.3

Solar Radiation for 1-Axis Tracking Flat-Plate Collectors with a North-South Axis (kWh/m²/day), Uncertainty ±9%

Axis Tilt (°)		Jan	Feb	Mar	Apr	May	June	July	Aug	Sept	Oct	Nov	Dec	Year
0	Average	1.6	2.8	4.3	5.9	7.2	8.2	9.5	8.1	6.0	3.6	1.7	1.3	5.0
	Min/Max	1.1/2.3	1.9/3.5	3.2/6.0	4.7/7.4	5.6/8.6	6.9/10.8	7.2/11.5	6.5/9.9	4.0/8.3	2.6/5.2	1.4/2.0	1.0/1.5	4.5/5.5
Latitude -15	Average	2.1	3.5	4.9	6.4	7.5	8.3	9.7	8.7	6.8	4.5	2.1	1.7	5.5
	Min/Max	1.4/3.1	2.2/4.5	3.6/7.2	5.0/8.1	5.6/8.9	6.9/11.0	7.4/11.9	6.9/10.6	4.4/9.7	3.0/6.7	1.8/2.7	1.2/2.1	5.0/6.1
Latitude	Average	2.3	3.6	5.0	6.3	7.2	8.0	9.4	8.5	6.9	4.7	2.3	1.8	5.5
	Min/Max	1.5/3.4	2.3/4.8	3.6/7.4	4.9/8.1	5.4/8.7	6.7/10.6	7.1/11.5	6.8/10.5	4.4/9.8	3.0/7.0	1.9/2.9	1.3/2.3	4.9/6.2
Latitude +15	Average	2.4	3.7	5.0	6.0	6.8	7.5	8.9	8.1	6.7	4.7	2.3	1.9	5.3
	Min/Max	1.5/3.6	2.2/4.9	3.5/7.4	4.7/7.8	5.1/8.2	6.2/10.0	6.7/10.9	6.4/10.0	4.2/9.6	3.0/7.0	1.9/3.0	1.3/2.4	4.8/6.0

Solar Radiation for 2-Axis Tracking Flat-Plate Collectors (kWh/m²/day), Uncertainty ±9%

Tracker		Jan	Feb	Mar	Apr	May	June	July	Aug	Sept	Oct	Nov	Dec	Year
2-Axis	Average	2.4	3.7	5.1	6.4	7.6	8.6	10.0	8.7	6.9	4.7	2.3	1.9	5.7
	Min/Max	1.5/3.6	2.3/4.9	3.7/7.4	5.0/8.2	5.8/9.1	7.2/11.3	7.5/12.1	7.0/10.7	4.4/9.8	3.0/7.1	1.9/3.0	1.3/2.4	5.1/6.3

Direct Beam Solar Radiation for Concentrating Collectors (kWh/m²/day), Uncertainty ±8%

Tracker		Jan	Feb	Mar	Apr	May	June	July	Aug	Sept	Oct	Nov	Dec	Year
1-Axis, E-W Horiz Axis	Average	0.9	1.6	2.2	2.7	3.3	4.0	5.2	4.4	3.6	2.5	1.0	0.7	2.7
	Min/Max	0.1/2.0	0.5/2.6	1.2/3.7	1.8/4.1	1.9/4.4	2.8/6.3	3.4/7.0	3.0/6.0	1.8/5.8	1.2/4.4	0.6/1.6	0.3/1.2	2.3/3.2
1-Axis, N-S Horiz Axis	Average	0.6	1.2	2.3	3.5	4.4	5.3	6.9	5.8	4.2	2.3	0.7	0.4	3.2
	Min/Max	0.1/1.2	0.4/2.1	1.2/3.8	2.2/5.2	2.6/6.0	3.7/8.4	4.4/9.3	3.9/8.0	2.0/6.9	1.1/4.0	0.4/1.2	0.2/0.7	2.7/3.7
1-Axis, N-S Tilt=Latitude	Average	1.0	1.9	2.9	3.8	4.4	5.1	6.8	6.2	5.0	3.2	1.1	0.8	3.5
	Min/Max	0.1/2.1	0.6/3.1	1.6/4.9	2.5/5.7	2.6/6.0	3.6/8.1	4.3/9.1	4.2/8.5	2.4/8.1	1.5/5.6	0.6/1.9	0.4/1.2	3.0/4.2
2-Axis	Average	1.1	1.9	2.9	3.9	4.7	5.6	7.3	6.4	5.0	3.2	1.2	0.8	3.7
	Min/Max	0.1/2.3	0.7/3.2	1.6/4.9	2.5/5.8	2.7/6.3	3.9/8.8	4.6/9.8	4.3/8.7	2.4/8.2	1.5/5.6	0.6/1.9	0.4/1.3	3.1/4.4

Average Climatic Conditions

Element	Jan	Feb	Mar	Apr	May	June	July	Aug	Sept	Oct	Nov	Dec	Year
Temperature (°C)	-6.4	-3.0	0.9	6.2	10.7	14.3	17.6	17.2	11.8	5.5	-0.6	-5.2	5.8
Daily Minimum Temp	-10.7	-7.7	-4.5	-0.5	3.6	6.7	8.4	7.9	3.7	-1.4	-4.6	-9.2	-0.7
Daily Maximum Temp	-2.1	1.7	6.3	12.9	17.9	21.9	26.7	26.3	19.8	12.4	3.5	-1.2	12.2
Record Minimum Temp	-38.9	-37.8	-33.9	-12.2	-7.2	-2.2	-0.6	-0.6	-8.9	-19.4	-33.3	-37.2	-38.9
Record Maximum Temp	11.7	13.9	22.2	28.9	34.4	35.6	40.0	40.6	37.2	27.8	18.3	13.9	40.6
HDD, Base 18.3°C	767	597	539	363	236	131	54	71	204	398	567	728	4654
CDD, Base 18.3°C	0	0	0	0	0	11	30	35	7	0	0	0	83
Relative Humidity (%)	78	76	70	61	61	62	57	57	65	73	79	81	68
Wind Speed (m/s)	2.6	2.6	3.1	3.6	3.4	3.1	3.0	2.9	2.9	2.3	2.6	2.3	2.9

Lewistown, MT

WBAN NO. 24036

LATITUDE: 47.05° N
LONGITUDE: 109.45° W
ELEVATION: 1264 meters
MEAN PRESSURE: 872 millibars

STATION TYPE: Secondary

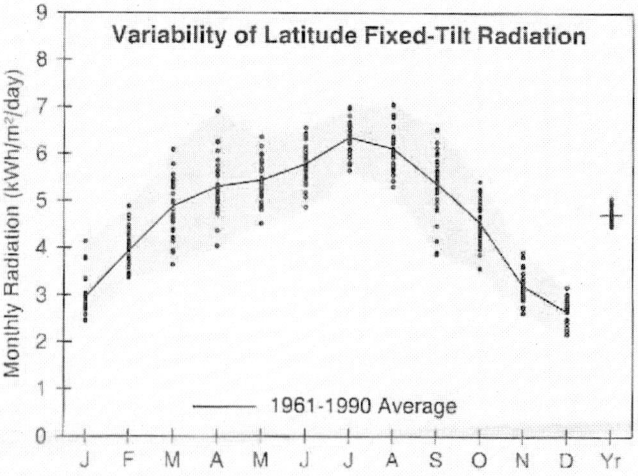

Solar Radiation for Flat-Plate Collectors Facing South at a Fixed Tilt (kWh/m²/day), Uncertainty ±9%

Tilt (°)		Jan	Feb	Mar	Apr	May	June	July	Aug	Sept	Oct	Nov	Dec	Year
0	Average	1.5	2.3	3.6	4.8	5.7	6.4	6.8	5.8	4.2	2.8	1.7	1.2	3.9
	Min/Max	1.3/1.8	2.1/2.7	3.0/4.2	3.9/5.9	4.8/6.5	5.5/7.3	6.1/7.4	5.1/6.6	3.3/4.9	2.5/3.2	1.5/1.9	1.1/1.4	3.7/4.1
Latitude -15	Average	2.6	3.6	4.7	5.4	5.8	6.3	6.9	6.4	5.3	4.2	2.9	2.3	4.7
	Min/Max	2.2/3.6	3.1/4.4	3.6/5.8	4.2/7.0	4.8/6.8	5.3/7.2	6.1/7.6	5.5/7.3	3.9/6.4	3.4/5.0	2.4/3.4	1.9/2.7	4.5/5.0
Latitude	Average	3.0	3.9	4.9	5.3	5.4	5.8	6.4	6.1	5.4	4.5	3.2	2.7	4.7
	Min/Max	2.5/4.1	3.4/4.9	3.6/6.1	4.0/6.9	4.5/6.4	4.9/6.6	5.7/7.0	5.3/7.1	3.9/6.5	3.6/5.4	2.6/3.9	2.2/3.2	4.5/5.0
Latitude +15	Average	3.2	4.1	4.8	4.9	4.8	5.0	5.5	5.5	5.1	4.6	3.4	2.8	4.5
	Min/Max	2.6/4.5	3.4/5.1	3.5/6.2	3.7/6.4	4.0/5.6	4.2/5.6	4.9/6.1	4.8/6.4	3.7/6.3	3.6/5.5	2.7/4.2	2.3/3.4	4.2/4.8
90	Average	3.1	3.8	4.1	3.7	3.1	3.0	3.4	3.7	4.0	4.0	3.2	2.8	3.5
	Min/Max	2.5/4.3	3.0/4.8	2.9/5.5	2.8/4.6	2.6/3.6	2.6/3.3	3.1/3.6	3.3/4.3	2.8/4.9	3.1/4.8	2.5/3.9	2.2/3.3	3.3/3.8

Solar Radiation for 1-Axis Tracking Flat-Plate Collectors with a North-South Axis (kWh/m²/day), Uncertainty ±9%

Axis Tilt (°)		Jan	Feb	Mar	Apr	May	June	July	Aug	Sept	Oct	Nov	Dec	Year
0	Average	2.2	3.5	5.1	6.6	7.6	8.7	9.6	8.3	6.2	4.3	2.5	1.9	5.6
	Min/Max	1.9/3.0	3.0/4.3	3.8/6.4	4.8/9.1	6.2/9.4	6.9/10.5	8.2/10.8	7.1/9.9	4.3/7.6	3.4/5.1	2.0/3.1	1.6/2.2	5.2/6.0
Latitude -15	Average	3.1	4.4	5.9	7.2	7.9	8.8	9.8	8.9	7.0	5.3	3.4	2.7	6.2
	Min/Max	2.5/4.3	3.7/5.6	4.3/7.6	5.1/10.0	6.4/9.8	6.9/10.6	8.4/11.1	7.5/10.6	4.8/8.8	4.1/6.5	2.7/4.3	2.2/3.2	5.8/6.7
Latitude	Average	3.4	4.7	6.1	7.1	7.7	8.5	9.5	8.7	7.1	5.6	3.7	3.0	6.3
	Min/Max	2.7/4.8	3.9/6.0	4.3/7.9	5.0/9.9	6.2/9.5	6.6/10.3	8.1/10.8	7.4/10.5	4.8/8.9	4.3/6.8	2.9/4.7	2.4/3.6	5.8/6.7
Latitude +15	Average	3.5	4.8	6.0	6.8	7.2	7.9	8.9	8.3	6.9	5.6	3.8	3.2	6.1
	Min/Max	2.8/5.1	4.0/6.2	4.3/7.9	4.8/9.6	5.8/9.0	6.2/9.6	7.6/10.1	7.0/10.0	4.7/8.7	4.3/6.8	3.0/4.8	2.5/3.8	5.7/6.6

Solar Radiation for 2-Axis Tracking Flat-Plate Collectors (kWh/m²/day), Uncertainty ±9%

Tracker		Jan	Feb	Mar	Apr	May	June	July	Aug	Sept	Oct	Nov	Dec	Year
2-Axis	Average	3.6	4.8	6.1	7.2	8.0	9.0	10.0	8.9	7.1	5.6	3.9	3.2	6.5
	Min/Max	2.9/5.2	4.0/6.2	4.4/7.9	5.1/10.0	6.5/9.9	7.1/10.9	8.6/11.3	7.6/10.7	4.8/8.9	4.3/6.8	3.0/4.9	2.5/3.9	6.0/7.0

Direct Beam Solar Radiation for Concentrating Collectors (kWh/m²/day), Uncertainty ±8%

Tracker		Jan	Feb	Mar	Apr	May	June	July	Aug	Sept	Oct	Nov	Dec	Year
1-Axis, E-W	Average	2.0	2.5	2.8	3.2	3.6	4.4	5.3	4.6	3.6	3.2	2.2	1.9	3.3
Horiz Axis	Min/Max	1.2/3.7	1.6/3.7	1.6/4.0	1.6/5.4	2.3/5.1	2.8/6.0	4.0/6.5	3.5/6.0	2.0/5.0	1.9/4.2	1.7/3.1	1.3/2.6	2.9/3.8
1-Axis, N-S	Average	1.2	2.0	3.0	4.0	4.8	5.8	7.0	6.0	4.3	2.9	1.5	1.1	3.7
Horiz Axis	Min/Max	0.7/2.3	1.4/3.1	1.7/4.3	2.1/7.0	3.2/6.8	3.8/8.0	5.3/8.6	4.6/8.0	2.3/5.9	1.8/3.9	1.1/2.2	0.7/1.5	3.2/4.3
1-Axis, N-S	Average	2.1	3.0	3.8	4.4	4.8	5.6	6.9	6.3	5.1	4.0	2.5	1.9	4.2
Tilt=Latitude	Min/Max	1.3/3.9	2.0/4.5	2.2/5.3	2.3/7.7	3.2/6.8	3.6/7.7	5.2/8.4	4.8/8.4	2.7/7.0	2.5/5.3	1.9/3.5	1.4/2.7	3.7/4.9
2-Axis	Average	2.2	3.0	3.8	4.5	5.1	6.1	7.4	6.5	5.1	4.0	2.6	2.1	4.4
	Min/Max	1.3/4.2	2.0/4.6	2.2/5.4	2.3/7.8	3.3/7.2	3.9/8.4	5.6/9.0	5.0/8.7	2.7/7.0	2.5/5.4	2.0/3.7	1.5/2.9	3.9/5.1

Average Climatic Conditions

Element	Jan	Feb	Mar	Apr	May	June	July	Aug	Sept	Oct	Nov	Dec	Year
Temperature (°C)	-6.6	-3.9	-0.8	4.9	9.9	14.7	18.5	18.2	12.3	7.2	0.1	-5.2	5.8
Daily Minimum Temp	-12.7	-9.9	-6.9	-1.8	2.9	7.3	9.9	9.5	4.4	-0.2	-6.4	-11.3	-1.3
Daily Maximum Temp	-0.6	2.2	5.2	11.6	16.9	22.2	27.1	26.8	20.1	14.6	6.4	0.9	12.8
Record Minimum Temp	-37.2	-36.1	-31.1	-20.6	-8.9	-4.4	1.7	0.6	-8.3	-20.6	-33.3	-41.1	-41.1
Record Maximum Temp	22.8	18.9	22.2	28.0	31.0	36.7	36.7	39.4	36.1	29.4	23.9	23.9	39.4
HDD, Base 18.3°C	773	622	594	403	260	123	42	73	197	344	548	730	4711
CDD, Base 18.3°C	0	0	0	0	0	14	47	67	16	0	0	0	144
Relative Humidity (%)	68	69	68	62	63	64	55	53	59	59	65	67	63
Wind Speed (m/s)	5.0	4.6	4.3	4.6	4.3	3.8	3.8	3.7	3.8	4.0	4.3	4.6	4.2

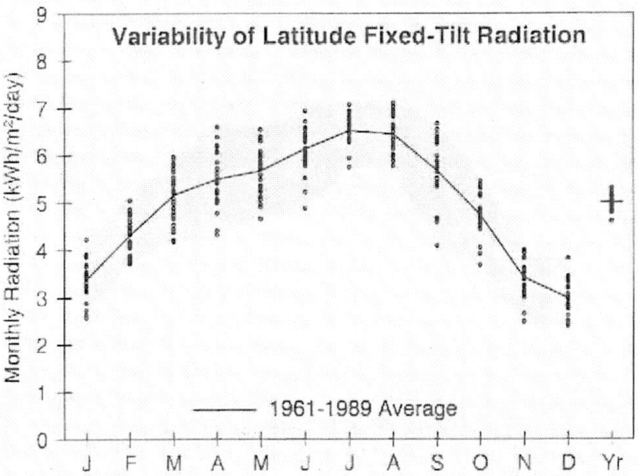

Miles City, MT

WBAN NO. 24037

LATITUDE: 46.43° N
LONGITUDE: 105.87° W
ELEVATION: 803 meters
MEAN PRESSURE: 922 millibars

STATION TYPE: Secondary

Solar Radiation for Flat-Plate Collectors Facing South at a Fixed Tilt (kWh/m²/day), Uncertainty ±9%

Tilt (°)		Jan	Feb	Mar	Apr	May	June	July	Aug	Sept	Oct	Nov	Dec	Year
0	Average	1.7	2.6	3.8	4.9	5.9	6.8	7.0	6.0	4.4	3.0	1.8	1.4	4.1
	Min/Max	1.4/1.9	2.3/2.8	3.3/4.2	4.1/5.8	4.9/6.8	5.5/7.7	6.2/7.5	5.5/6.6	3.5/5.0	2.5/3.2	1.5/2.0	1.2/1.5	3.8/4.4
Latitude -15	Average	3.0	3.9	5.0	5.6	6.0	6.7	7.1	6.7	5.6	4.4	3.0	2.6	5.0
	Min/Max	2.3/3.7	3.4/4.6	4.1/5.7	4.5/6.7	5.0/7.0	5.3/7.6	6.2/7.6	6.0/7.4	4.1/6.5	3.5/5.0	2.3/3.5	2.1/3.3	4.6/5.3
Latitude	Average	3.4	4.3	5.2	5.5	5.7	6.1	6.5	6.5	5.7	4.7	3.4	3.0	5.0
	Min/Max	2.6/4.2	3.7/5.1	4.2/6.0	4.3/6.6	4.7/6.5	4.9/6.9	5.8/7.1	5.8/7.1	4.1/6.7	3.7/5.4	2.5/4.0	2.4/3.8	4.6/5.3
Latitude +15	Average	3.6	4.5	5.1	5.1	5.0	5.3	5.6	5.8	5.4	4.8	3.6	3.2	4.8
	Min/Max	2.7/4.6	3.9/5.3	4.1/5.9	4.0/6.1	4.1/5.7	4.2/5.9	5.0/6.1	5.2/6.4	3.9/6.4	3.7/5.5	2.6/4.3	2.5/4.2	4.4/5.0
90	Average	3.5	4.1	4.3	3.7	3.2	3.1	3.4	3.9	4.2	4.1	3.4	3.2	3.7
	Min/Max	2.6/4.5	3.4/4.8	3.3/5.2	2.9/4.3	2.7/3.6	2.6/3.4	3.0/3.6	3.5/4.2	2.9/4.9	3.1/4.8	2.3/4.1	2.4/4.1	3.4/3.9

Solar Radiation for 1-Axis Tracking Flat-Plate Collectors with a North-South Axis (kWh/m²/day), Uncertainty ±9%

Axis Tilt (°)		Jan	Feb	Mar	Apr	May	June	July	Aug	Sept	Oct	Nov	Dec	Year
0	Average	2.6	3.8	5.4	6.9	8.0	9.3	9.9	8.8	6.6	4.5	2.7	2.1	5.9
	Min/Max	2.0/3.2	3.2/4.4	4.3/6.3	5.2/8.6	6.6/9.6	6.7/10.9	8.6/10.9	7.5/10.0	4.6/7.9	3.5/5.2	2.0/3.2	1.7/2.7	5.3/6.4
Latitude -15	Average	3.5	4.8	6.3	7.5	8.3	9.4	10.1	9.4	7.5	5.6	3.6	3.0	6.6
	Min/Max	2.7/4.4	4.1/5.7	4.9/7.5	5.5/9.4	6.7/9.9	6.8/11.0	8.8/11.2	8.0/10.7	5.1/9.1	4.2/6.5	2.6/4.4	2.4/4.0	6.0/7.2
Latitude	Average	3.9	5.1	6.5	7.4	8.0	9.1	9.8	9.2	7.6	5.9	4.0	3.4	6.7
	Min/Max	2.9/4.9	4.3/6.1	5.0/7.7	5.4/9.3	6.5/9.7	6.5/10.6	8.5/10.8	7.9/10.6	5.1/9.2	4.4/6.8	2.8/4.8	2.7/4.4	6.0/7.2
Latitude +15	Average	4.1	5.3	6.4	7.1	7.6	8.5	9.2	8.8	7.4	5.9	4.1	3.6	6.5
	Min/Max	3.0/5.2	4.4/6.3	4.9/7.7	5.2/9.0	6.2/9.1	6.0/10.0	7.9/10.2	7.5/10.1	4.9/9.0	4.4/6.9	2.8/4.9	2.8/4.7	5.8/7.1

Solar Radiation for 2-Axis Tracking Flat-Plate Collectors (kWh/m²/day), Uncertainty ±9%

Tracker		Jan	Feb	Mar	Apr	May	June	July	Aug	Sept	Oct	Nov	Dec	Year
2-Axis	Average	4.1	5.3	6.5	7.5	8.4	9.7	10.3	9.4	7.6	5.9	4.1	3.6	6.9
	Min/Max	3.0/5.2	4.4/6.3	5.0/7.7	5.5/9.4	6.8/10.1	6.9/11.3	8.9/11.4	8.1/10.8	5.1/9.2	4.4/6.9	2.9/5.0	2.8/4.8	6.2/7.5

Direct Beam Solar Radiation for Concentrating Collectors (kWh/m²/day), Uncertainty ±8%

Tracker		Jan	Feb	Mar	Apr	May	June	July	Aug	Sept	Oct	Nov	Dec	Year
1-Axis, E-W	Average	2.3	2.7	3.1	3.4	3.9	4.9	5.5	5.0	4.0	3.4	2.4	2.1	3.6
Horiz Axis	Min/Max	1.2/3.2	2.0/3.9	2.0/4.5	1.7/4.9	2.7/5.3	2.7/6.4	4.4/6.5	4.0/6.1	2.1/5.2	2.2/4.3	1.5/3.2	1.4/3.3	3.0/4.2
1-Axis, N-S	Average	1.4	2.2	3.3	4.4	5.2	6.5	7.4	6.5	4.8	3.1	1.7	1.2	4.0
Horiz Axis	Min/Max	0.7/2.1	1.6/3.2	2.1/4.7	2.2/6.4	3.7/7.0	3.5/8.4	5.9/8.8	5.1/8.1	2.5/6.3	2.0/4.0	1.0/2.3	0.8/1.9	3.3/4.7
1-Axis, N-S	Average	2.5	3.2	4.2	4.8	5.2	6.3	7.2	6.8	5.6	4.3	2.7	2.2	4.6
Tilt=Latitude	Min/Max	1.3/3.5	2.3/4.6	2.6/5.9	2.5/6.9	3.7/7.0	3.4/8.1	5.7/8.6	5.3/8.5	2.9/7.3	2.7/5.4	1.6/3.6	1.4/3.4	3.9/5.4
2-Axis	Average	2.6	3.3	4.2	4.9	5.5	6.8	7.7	7.1	5.6	4.3	2.8	2.4	4.8
	Min/Max	1.4/3.7	2.4/4.8	2.6/5.9	2.5/7.1	3.9/7.4	3.7/8.8	6.2/9.2	5.5/8.8	2.9/7.3	2.8/5.4	1.7/3.8	1.5/3.7	4.0/5.6

Average Climatic Conditions

Element	Jan	Feb	Mar	Apr	May	June	July	Aug	Sept	Oct	Nov	Dec	Year
Temperature (°C)	-8.9	-4.8	0.7	7.8	13.8	19.4	23.7	22.5	15.4	8.9	0.1	-7.1	7.6
Daily Minimum Temp	-14.3	-10.6	-5.4	1.0	7.0	12.3	15.9	14.7	8.1	1.9	-5.7	-12.6	1.1
Daily Maximum Temp	-3.5	0.8	6.8	14.5	20.6	26.6	31.6	30.3	22.8	15.8	5.8	-1.7	14.2
Record Minimum Temp	-37.2	-37.8	-30.6	-15.0	-5.6	1.7	6.7	2.2	-5.0	-17.2	-31.7	-40.0	-40.0
Record Maximum Temp	14.4	19.4	26.7	32.8	37.8	41.1	42.2	40.6	40.0	33.9	22.2	18.9	42.2
HDD, Base 18.3°C	844	649	546	317	148	47	3	22	127	293	548	789	4331
CDD, Base 18.3°C	0	0	0	0	8	80	170	151	40	0	0	0	449
Relative Humidity (%)	72	72	68	58	57	55	47	45	54	60	69	73	61
Wind Speed (m/s)	4.6	4.5	4.9	5.2	5.1	4.6	4.4	4.4	4.5	4.5	4.2	4.4	4.6

Missoula, MT

WBAN NO. 24153

LATITUDE: 46.92° N
LONGITUDE: 114.08° W
ELEVATION: 972 meters
MEAN PRESSURE: 905 millibars

STATION TYPE: Secondary

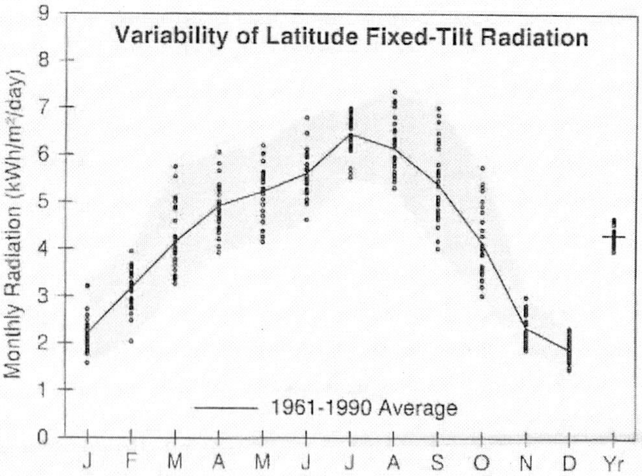

Solar Radiation for Flat-Plate Collectors Facing South at a Fixed Tilt (kWh/m²/day), Uncertainty ±9%

Tilt (°)		Jan	Feb	Mar	Apr	May	June	July	Aug	Sept	Oct	Nov	Dec	Year
0	Average	1.3	2.1	3.2	4.5	5.5	6.3	6.9	5.8	4.2	2.7	1.4	1.1	3.8
	Min/Max	1.1/1.6	1.6/2.5	2.7/4.0	3.8/5.4	4.4/6.3	5.2/7.6	5.9/7.5	5.1/6.8	3.4/5.2	2.3/3.4	1.3/1.6	1.0/1.3	3.5/4.0
Latitude -15	Average	2.0	3.0	4.0	5.0	5.6	6.1	7.0	6.4	5.3	3.9	2.1	1.7	4.4
	Min/Max	1.5/2.8	2.0/3.6	3.2/5.4	4.1/6.2	4.5/6.6	5.1/7.5	6.0/7.6	5.6/7.6	4.0/6.8	2.9/5.3	1.8/2.7	1.4/2.0	4.0/4.7
Latitude	Average	2.2	3.2	4.1	4.9	5.2	5.6	6.4	6.1	5.4	4.1	2.3	1.9	4.3
	Min/Max	1.6/3.2	2.0/3.9	3.3/5.7	3.9/6.1	4.2/6.2	4.6/6.8	5.5/7.0	5.3/7.3	4.0/7.0	3.0/5.7	1.9/3.0	1.5/2.3	4.0/4.7
Latitude +15	Average	2.3	3.2	4.0	4.5	4.6	4.8	5.6	5.5	5.1	4.1	2.4	2.0	4.0
	Min/Max	1.6/3.4	2.0/4.1	3.1/5.8	3.6/5.6	3.7/5.5	4.0/5.8	4.8/6.1	4.8/6.6	3.8/6.7	3.0/5.8	1.9/3.1	1.5/2.5	3.7/4.4
90	Average	2.2	2.9	3.3	3.3	3.0	2.9	3.4	3.7	4.0	3.6	2.2	1.9	3.0
	Min/Max	1.5/3.4	1.7/3.8	2.5/5.1	2.6/4.0	2.4/3.5	2.5/3.4	3.0/3.6	3.3/4.4	2.9/5.2	2.5/5.1	1.7/2.9	1.4/2.4	2.8/3.4

Solar Radiation for 1-Axis Tracking Flat-Plate Collectors with a North-South Axis (kWh/m²/day), Uncertainty ±9%

Axis Tilt (°)		Jan	Feb	Mar	Apr	May	June	July	Aug	Sept	Oct	Nov	Dec	Year
0	Average	1.7	2.8	4.3	6.1	7.3	8.4	9.9	8.5	6.2	4.0	1.9	1.4	5.2
	Min/Max	1.3/2.5	1.8/3.4	3.5/6.0	4.7/7.8	5.4/8.8	6.6/10.9	8.0/11.1	7.0/10.3	4.4/8.4	3.0/5.6	1.5/2.4	1.1/1.7	4.8/5.7
Latitude -15	Average	2.3	3.5	5.0	6.5	7.5	8.5	10.2	9.0	7.1	4.9	2.5	1.9	5.7
	Min/Max	1.6/3.4	2.1/4.3	3.9/7.1	5.0/8.6	5.6/9.2	6.6/11.0	8.1/11.4	7.4/11.1	4.9/9.7	3.5/7.0	1.9/3.2	1.5/2.4	5.2/6.3
Latitude	Average	2.4	3.7	5.1	6.5	7.3	8.2	9.9	8.9	7.1	5.1	2.6	2.0	5.7
	Min/Max	1.6/3.7	2.2/4.6	3.9/7.3	4.9/8.5	5.3/8.9	6.4/10.6	7.9/11.0	7.3/11.0	4.9/9.8	3.5/7.3	2.0/3.4	1.5/2.6	5.2/6.3
Latitude +15	Average	2.5	3.7	5.0	6.2	6.9	7.7	9.3	8.5	7.0	5.1	2.7	2.1	5.6
	Min/Max	1.7/3.9	2.2/4.7	3.8/7.3	4.7/8.1	5.0/8.4	5.9/10.0	7.4/10.4	6.9/10.5	4.7/9.6	3.5/7.4	2.1/3.5	1.6/2.7	5.0/6.1

Solar Radiation for 2-Axis Tracking Flat-Plate Collectors (kWh/m²/day), Uncertainty ±9%

Tracker		Jan	Feb	Mar	Apr	May	June	July	Aug	Sept	Oct	Nov	Dec	Year
2-Axis	Average	2.6	3.8	5.1	6.6	7.6	8.7	10.4	9.1	7.2	5.1	2.7	2.1	5.9
	Min/Max	1.7/3.9	2.2/4.7	3.9/7.3	5.0/8.6	5.6/9.3	6.8/11.4	8.3/11.6	7.5/11.2	4.9/9.8	3.6/7.4	2.1/3.6	1.6/2.8	5.4/6.5

Direct Beam Solar Radiation for Concentrating Collectors (kWh/m²/day), Uncertainty ±8%

Tracker		Jan	Feb	Mar	Apr	May	June	July	Aug	Sept	Oct	Nov	Dec	Year
1-Axis, E-W Horiz Axis	Average	1.1	1.7	2.3	2.9	3.4	4.2	5.6	4.7	3.8	2.8	1.3	1.0	2.9
	Min/Max	0.2/2.1	0.6/2.5	1.3/3.7	1.8/4.4	1.7/4.7	2.9/6.5	3.9/6.6	3.5/6.4	2.1/5.8	1.5/4.7	0.7/2.0	0.5/1.6	2.5/3.4
1-Axis, N-S Horiz Axis	Average	0.7	1.4	2.4	3.7	4.5	5.6	7.6	6.3	4.4	2.6	0.9	0.6	3.4
	Min/Max	0.1/1.4	0.5/2.1	1.4/3.9	2.3/5.7	2.3/6.3	3.8/8.7	5.2/8.9	4.6/8.6	2.4/7.0	1.5/4.4	0.5/1.4	0.3/0.9	2.9/3.9
1-Axis, N-S Tilt=Latitude	Average	1.2	2.0	3.0	4.0	4.5	5.4	7.4	6.6	5.2	3.6	1.5	1.0	3.8
	Min/Max	0.2/2.3	0.7/3.0	1.8/4.9	2.5/6.2	2.3/6.3	3.7/8.3	5.1/8.7	4.8/9.0	2.8/8.2	2.0/5.9	0.7/2.3	0.6/1.6	3.2/4.4
2-Axis	Average	1.2	2.1	3.0	4.1	4.8	5.9	7.9	6.8	5.2	3.6	1.6	1.1	4.0
	Min/Max	0.3/2.5	0.8/3.1	1.8/5.0	2.5/6.3	2.4/6.6	4.0/9.1	5.5/9.4	5.0/9.3	2.8/8.2	2.0/6.0	0.8/2.4	0.6/1.8	3.4/4.6

Average Climatic Conditions

Element	Jan	Feb	Mar	Apr	May	June	July	Aug	Sept	Oct	Nov	Dec	Year
Temperature (°C)	-5.2	-1.6	2.1	6.8	11.0	15.6	19.3	18.8	13.2	6.8	0.2	-4.8	6.8
Daily Minimum Temp	-9.2	-6.2	-3.9	-0.6	3.3	7.8	10.1	9.6	4.7	-0.4	-4.3	-8.7	0.2
Daily Maximum Temp	-1.1	3.0	8.1	14.2	18.7	23.3	28.6	27.9	21.6	13.9	4.8	-1.0	13.5
Record Minimum Temp	-36.1	-32.8	-25.0	-10.0	-6.1	-0.6	-0.6	0.0	-6.7	-17.8	-30.6	-34.4	-36.1
Record Maximum Temp	15.0	15.6	23.9	30.6	35.0	36.7	40.6	40.6	37.2	29.4	18.9	15.6	40.6
HDD, Base 18.3°C	728	557	503	347	227	99	36	45	168	358	543	717	4329
CDD, Base 18.3°C	0	0	0	0	0	16	67	59	13	0	0	0	156
Relative Humidity (%)	81	78	70	61	62	61	52	53	63	71	80	84	68
Wind Speed (m/s)	2.4	2.6	3.0	3.5	3.4	3.3	3.3	3.0	2.8	2.3	2.5	2.2	2.9

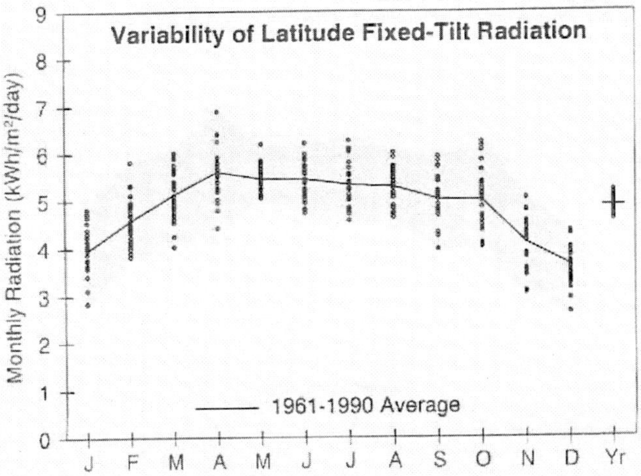

Asheville, NC

WBAN NO. 03812

LATITUDE: 35.43° N
LONGITUDE: 82.53° W
ELEVATION: 661 meters
MEAN PRESSURE: 942 millibars

STATION TYPE: Secondary

Solar Radiation for Flat-Plate Collectors Facing South at a Fixed Tilt (kWh/m²/day), Uncertainty ±9%

Tilt (°)		Jan	Feb	Mar	Apr	May	June	July	Aug	Sept	Oct	Nov	Dec	Year
0	Average	2.5	3.3	4.3	5.4	5.8	6.0	5.8	5.3	4.5	3.8	2.7	2.2	4.3
	Min/Max	2.0/2.9	2.9/3.9	3.5/4.9	4.3/6.5	5.4/6.6	5.3/6.9	5.0/6.8	4.7/5.9	3.7/5.1	3.2/4.5	2.2/3.1	1.8/2.5	4.1/4.6
Latitude -15	Average	3.5	4.2	4.9	5.7	5.8	5.9	5.7	5.5	5.0	4.6	3.6	3.1	4.8
	Min/Max	2.6/4.1	3.5/5.2	3.9/5.8	4.5/7.0	5.4/6.6	5.1/6.8	4.9/6.7	4.8/6.2	4.0/5.8	3.8/5.7	2.8/4.4	2.3/3.7	4.5/5.1
Latitude	Average	3.9	4.6	5.1	5.6	5.4	5.4	5.3	5.3	5.0	5.0	4.1	3.6	4.9
	Min/Max	2.8/4.8	3.8/5.8	4.0/6.0	4.4/6.9	5.1/6.2	4.7/6.2	4.6/6.3	4.6/6.0	4.0/5.9	4.0/6.2	3.0/5.0	2.6/4.3	4.6/5.2
Latitude +15	Average	4.2	4.7	5.0	5.2	4.8	4.7	4.7	4.8	4.8	5.0	4.3	3.9	4.7
	Min/Max	3.0/5.2	3.9/6.1	3.9/5.9	4.1/6.4	4.5/5.5	4.1/5.4	4.0/5.5	4.2/5.5	3.8/5.7	4.0/6.4	3.1/5.4	2.8/4.7	4.4/5.0
90	Average	3.8	3.9	3.6	3.0	2.4	2.1	2.2	2.6	3.1	3.9	3.7	3.5	3.2
	Min/Max	2.6/4.7	3.2/5.1	2.7/4.2	2.4/3.6	2.2/2.5	2.0/2.3	2.0/2.5	2.3/2.9	2.4/3.7	3.0/5.0	2.6/4.7	2.4/4.4	2.9/3.4

Solar Radiation for 1-Axis Tracking Flat-Plate Collectors with a North-South Axis (kWh/m²/day), Uncertainty ±9%

Axis Tilt (°)		Jan	Feb	Mar	Apr	May	June	July	Aug	Sept	Oct	Nov	Dec	Year
0	Average	3.6	4.6	5.8	7.1	7.3	7.5	7.1	6.5	5.7	5.2	3.8	3.2	5.6
	Min/Max	2.6/4.3	3.8/5.9	4.4/6.9	5.4/9.1	6.5/8.7	6.2/9.1	5.9/8.8	5.6/7.7	4.4/6.8	4.2/6.5	2.8/4.7	2.3/3.8	5.2/6.1
Latitude -15	Average	4.3	5.3	6.3	7.4	7.3	7.4	7.1	6.7	6.1	5.8	4.5	3.8	6.0
	Min/Max	3.0/5.3	4.3/6.9	4.7/7.6	5.6/9.5	6.6/8.8	6.2/9.1	5.9/8.8	5.7/7.9	4.7/7.4	4.6/7.4	3.2/5.6	2.7/4.7	5.6/6.6
Latitude	Average	4.7	5.6	6.4	7.3	7.1	7.1	6.8	6.6	6.1	6.1	4.9	4.2	6.1
	Min/Max	3.2/5.8	4.5/7.3	4.8/7.8	5.6/9.5	6.4/8.5	5.9/8.7	5.7/8.4	5.6/7.8	4.7/7.5	4.8/7.8	3.4/6.1	2.9/5.2	5.6/6.7
Latitude +15	Average	4.9	5.7	6.4	7.1	6.7	6.6	6.4	6.2	6.0	6.1	5.0	4.5	6.0
	Min/Max	3.3/6.1	4.6/7.5	4.7/7.7	5.3/9.1	6.0/8.0	5.5/8.1	5.3/7.9	5.3/7.4	4.5/7.3	4.8/7.9	3.5/6.4	3.1/5.5	5.5/6.5

Solar Radiation for 2-Axis Tracking Flat-Plate Collectors (kWh/m²/day), Uncertainty ±9%

Tracker		Jan	Feb	Mar	Apr	May	June	July	Aug	Sept	Oct	Nov	Dec	Year
2-Axis	Average	4.9	5.7	6.4	7.4	7.4	7.6	7.2	6.7	6.1	6.2	5.1	4.5	6.3
	Min/Max	3.4/6.2	4.6/7.6	4.8/7.8	5.7/9.6	6.6/8.9	6.3/9.3	6.0/8.9	5.8/8.0	4.7/7.5	4.8/7.9	3.5/6.4	3.1/5.6	5.8/6.9

Direct Beam Solar Radiation for Concentrating Collectors (kWh/m²/day), Uncertainty ±8%

Tracker		Jan	Feb	Mar	Apr	May	June	July	Aug	Sept	Oct	Nov	Dec	Year
1-Axis, E-W Horiz Axis	Average	3.0	3.2	3.3	3.6	3.5	3.5	3.2	3.0	2.9	3.4	3.0	2.8	3.2
	Min/Max	1.8/4.1	2.3/4.9	2.2/4.4	2.3/5.3	2.7/4.6	2.4/4.9	2.3/4.4	2.3/4.0	1.9/3.9	2.3/4.9	1.8/4.2	1.6/3.8	2.8/3.7
1-Axis, N-S Horiz Axis	Average	2.4	3.1	3.8	4.7	4.5	4.5	4.1	3.8	3.4	3.5	2.5	2.1	3.5
	Min/Max	1.4/3.3	2.2/4.7	2.5/5.2	3.1/6.9	3.5/6.1	3.1/6.4	2.9/5.7	2.8/5.1	2.1/4.7	2.4/5.0	1.5/3.5	1.2/2.9	3.1/4.2
1-Axis, N-S Tilt=Latitude	Average	3.3	3.9	4.4	4.9	4.4	4.2	3.9	3.8	3.8	4.3	3.5	3.0	3.9
	Min/Max	1.9/4.6	2.8/6.0	2.9/5.9	3.2/7.2	3.4/5.9	2.9/6.0	2.7/5.4	2.8/5.1	2.4/5.2	2.9/6.1	2.1/4.7	1.7/4.1	3.4/4.6
2-Axis	Average	3.5	4.0	4.4	5.0	4.6	4.5	4.2	3.9	3.8	4.3	3.6	3.3	4.1
	Min/Max	2.1/4.9	2.9/6.1	2.9/5.9	3.3/7.3	3.5/6.2	3.1/6.5	2.9/5.8	2.9/5.3	2.4/5.2	2.9/6.1	2.2/5.0	1.9/4.5	3.6/4.8

Average Climatic Conditions

Element	Jan	Feb	Mar	Apr	May	June	July	Aug	Sept	Oct	Nov	Dec	Year
Temperature (°C)	2.1	3.7	8.6	12.9	17.2	20.8	22.7	22.2	19.0	13.3	8.6	4.2	12.9
Daily Minimum Temp	-4.0	-2.6	1.9	5.9	10.5	14.6	17.1	16.6	13.1	6.4	2.1	-1.9	6.6
Daily Maximum Temp	8.1	10.0	15.1	19.9	23.9	26.9	28.3	27.8	24.9	20.2	15.2	10.2	19.2
Record Minimum Temp	-26.7	-18.9	-12.8	-5.6	-2.2	1.7	6.7	5.6	-1.1	-6.1	-13.3	-21.7	-26.7
Record Maximum Temp	25.6	25.0	28.3	31.7	32.8	35.6	35.6	37.8	33.3	30.0	27.2	25.6	37.8
HDD, Base 18.3°C	504	409	303	166	65	10	3	3	36	164	292	439	2393
CDD, Base 18.3°C	0	0	0	0	31	83	137	123	56	7	0	0	437
Relative Humidity (%)	72	70	68	65	75	79	81	83	84	78	74	74	75
Wind Speed (m/s)	4.3	4.3	4.2	4.1	3.2	2.6	2.5	2.4	2.5	3.0	3.7	4.0	3.4

Cape Hatteras, NC

WBAN NO. 93729

LATITUDE: 35.27° N
LONGITUDE: 75.55° W
ELEVATION: 2 meters
MEAN PRESSURE: 1018 millibars

STATION TYPE: Primary

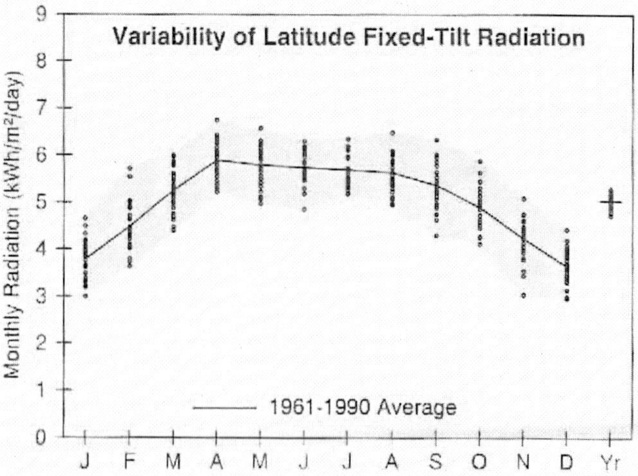

Solar Radiation for Flat-Plate Collectors Facing South at a Fixed Tilt (kWh/m²/day), Uncertainty ±9%

Tilt (°)		Jan	Feb	Mar	Apr	May	June	July	Aug	Sept	Oct	Nov	Dec	Year
0	Average	2.4	3.3	4.4	5.6	6.1	6.4	6.2	5.6	4.8	3.7	2.8	2.2	4.5
	Min/Max	2.1/2.8	2.8/3.9	3.8/5.0	5.0/6.4	5.3/6.9	5.4/7.0	5.6/6.9	5.0/6.4	4.0/5.4	3.3/4.3	2.2/3.2	2.0/2.5	4.2/4.7
Latitude -15	Average	3.3	4.1	5.1	6.0	6.1	6.2	6.1	5.8	5.3	4.6	3.7	3.2	5.0
	Min/Max	2.7/4.0	3.4/5.1	4.3/5.8	5.3/6.8	5.3/7.0	5.2/6.8	5.5/6.8	5.1/6.7	4.3/6.2	3.9/5.4	2.8/4.4	2.6/3.8	4.7/5.2
Latitude	Average	3.8	4.5	5.2	5.9	5.8	5.7	5.7	5.6	5.4	4.9	4.2	3.6	5.0
	Min/Max	3.0/4.7	3.6/5.7	4.4/6.0	5.2/6.7	5.0/6.6	4.8/6.3	5.2/6.3	4.9/6.5	4.3/6.3	4.1/5.9	3.0/5.1	3.0/4.4	4.7/5.3
Latitude +15	Average	4.0	4.6	5.2	5.5	5.1	5.0	5.0	5.1	5.1	4.9	4.5	3.9	4.8
	Min/Max	3.1/5.0	3.7/6.0	4.3/5.9	4.8/6.3	4.4/5.8	4.3/5.4	4.6/5.5	4.5/5.9	4.1/6.1	4.1/6.0	3.1/5.4	3.1/4.8	4.5/5.1
90	Average	3.6	3.7	3.6	3.1	2.5	2.2	2.3	2.7	3.3	3.8	3.9	3.6	3.2
	Min/Max	2.7/4.5	3.0/5.0	3.0/4.1	2.8/3.5	2.3/2.7	2.1/2.3	2.2/2.5	2.4/3.0	2.7/4.0	3.1/4.7	2.6/4.8	2.8/4.4	2.9/3.3

Solar Radiation for 1-Axis Tracking Flat-Plate Collectors with a North-South Axis (kWh/m²/day), Uncertainty ±9%

Axis Tilt (°)		Jan	Feb	Mar	Apr	May	June	July	Aug	Sept	Oct	Nov	Dec	Year
0	Average	3.4	4.5	5.9	7.5	7.8	8.0	7.7	7.1	6.2	5.1	3.9	3.2	5.9
	Min/Max	2.8/4.2	3.7/5.8	4.9/6.9	6.5/8.8	6.4/9.1	6.3/9.0	6.7/8.9	6.2/8.6	5.0/7.6	4.2/6.1	2.8/4.7	2.5/3.9	5.5/6.3
Latitude -15	Average	4.1	5.1	6.5	7.8	7.9	7.9	7.7	7.3	6.6	5.7	4.6	3.9	6.3
	Min/Max	3.2/5.1	4.1/6.8	5.3/7.6	6.8/9.2	6.4/9.2	6.2/8.9	6.7/8.9	6.3/8.8	5.2/8.1	4.7/7.0	3.2/5.6	3.0/4.8	5.9/6.7
Latitude	Average	4.5	5.4	6.6	7.7	7.6	7.6	7.4	7.2	6.7	6.0	5.0	4.3	6.3
	Min/Max	3.5/5.6	4.3/7.2	5.4/7.8	6.7/9.1	6.2/8.9	6.0/8.6	6.4/8.5	6.2/8.6	5.2/8.2	4.8/7.3	3.4/6.1	3.3/5.3	5.9/6.8
Latitude +15	Average	4.7	5.5	6.5	7.4	7.2	7.0	6.9	6.8	6.5	6.0	5.2	4.5	6.2
	Min/Max	3.6/5.9	4.3/7.4	5.3/7.7	6.4/8.8	5.8/8.4	5.5/8.0	6.0/8.0	5.9/8.2	5.1/8.1	4.8/7.4	3.5/6.4	3.5/5.6	5.8/6.6

Solar Radiation for 2-Axis Tracking Flat-Plate Collectors (kWh/m²/day), Uncertainty ±9%

Tracker		Jan	Feb	Mar	Apr	May	June	July	Aug	Sept	Oct	Nov	Dec	Year
2-Axis	Average	4.7	5.5	6.6	7.8	7.9	8.1	7.8	7.4	6.7	6.0	5.2	4.6	6.5
	Min/Max	3.6/6.0	4.4/7.4	5.4/7.8	6.8/9.2	6.5/9.3	6.3/9.1	6.8/9.0	6.3/8.8	5.3/8.3	4.9/7.4	3.5/6.4	3.5/5.7	6.1/7.0

Direct Beam Solar Radiation for Concentrating Collectors (kWh/m²/day), Uncertainty ±8%

Tracker		Jan	Feb	Mar	Apr	May	June	July	Aug	Sept	Oct	Nov	Dec	Year
1-Axis, E-W Horiz Axis	Average	2.8	3.1	3.4	3.9	3.8	3.8	3.6	3.4	3.2	3.2	3.1	2.8	3.4
	Min/Max	1.8/4.0	2.2/4.6	2.4/4.4	3.1/4.9	2.7/4.9	2.3/4.7	2.9/4.6	2.5/4.6	1.9/4.5	2.3/4.3	1.8/4.2	1.9/3.8	3.0/3.7
1-Axis, N-S Horiz Axis	Average	2.3	3.0	4.0	5.1	5.0	4.9	4.6	4.3	3.9	3.3	2.6	2.1	3.8
	Min/Max	1.5/3.2	2.0/4.5	2.8/5.2	4.0/6.5	3.5/6.4	3.0/6.1	3.6/6.0	3.2/6.0	2.4/5.4	2.3/4.4	1.4/3.5	1.3/2.9	3.4/4.2
1-Axis, N-S Tilt=Latitude	Average	3.1	3.8	4.6	5.3	4.8	4.6	4.4	4.4	4.2	4.0	3.5	3.0	4.2
	Min/Max	2.1/4.4	2.6/5.7	3.2/5.9	4.2/6.7	3.4/6.2	2.8/5.7	3.4/5.7	3.2/6.0	2.6/6.0	2.9/5.4	2.0/4.7	2.0/4.1	3.7/4.6
2-Axis	Average	3.4	3.9	4.6	5.4	5.1	5.0	4.7	4.5	4.3	4.1	3.7	3.3	4.3
	Min/Max	2.2/4.7	2.7/5.8	3.2/6.0	4.3/6.8	3.6/6.6	3.0/6.2	3.7/6.1	3.3/6.2	2.6/6.0	2.9/5.5	2.1/4.9	2.1/4.4	3.9/4.8

Average Climatic Conditions

Element	Jan	Feb	Mar	Apr	May	June	July	Aug	Sept	Oct	Nov	Dec	Year
Temperature (°C)	7.0	7.5	10.9	14.9	19.4	23.4	25.7	25.8	23.4	18.6	13.9	9.4	16.7
Daily Minimum Temp	2.6	3.1	6.4	10.4	15.3	19.7	22.1	22.2	19.8	14.6	9.6	5.1	12.6
Daily Maximum Temp	11.3	11.9	15.3	19.4	23.6	27.1	29.2	29.3	27.1	22.4	18.2	13.8	20.7
Record Minimum Temp	-14.4	-10.0	-7.2	-3.3	3.9	6.7	12.2	13.3	7.2	0.0	-5.6	-11.1	-14.4
Record Maximum Temp	23.9	24.4	27.2	31.7	32.8	35.0	35.0	34.4	33.3	31.7	27.2	25.6	35.0
HDD, Base 18.3°C	355	303	234	109	21	0	0	0	0	50	147	279	1499
CDD, Base 18.3°C	3	0	3	7	56	153	229	231	153	57	14	4	910
Relative Humidity (%)	75	74	74	72	77	79	81	81	78	76	75	75	77
Wind Speed (m/s)	5.2	5.3	5.3	5.2	4.7	4.7	4.4	4.2	4.7	4.9	4.9	5.0	4.9

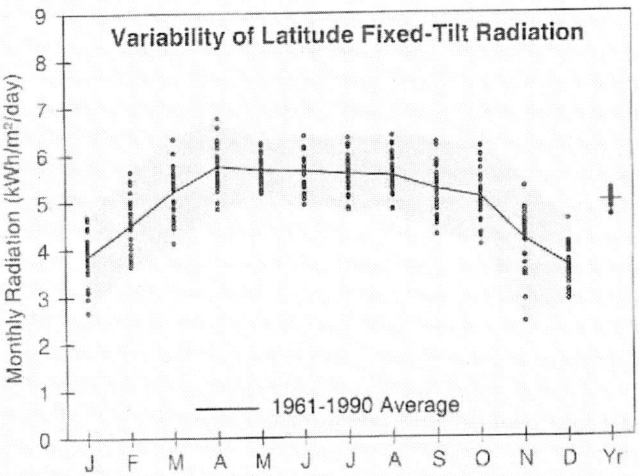

Charlotte, NC

WBAN NO. 13881

LATITUDE: 35.22° N
LONGITUDE: 80.93° W
ELEVATION: 234 meters
MEAN PRESSURE: 991 millibars

STATION TYPE: Secondary

Solar Radiation for Flat-Plate Collectors Facing South at a Fixed Tilt (kWh/m²/day), Uncertainty ±9%

Tilt (°)		Jan	Feb	Mar	Apr	May	June	July	Aug	Sept	Oct	Nov	Dec	Year
0	Average	2.5	3.3	4.4	5.5	6.0	6.3	6.1	5.6	4.7	3.9	2.8	2.3	4.4
	Min/Max	2.0/2.8	2.8/3.8	3.6/5.2	4.7/6.4	5.5/6.6	5.5/7.1	5.3/6.9	4.9/6.3	4.0/5.4	3.3/4.5	2.0/3.3	2.0/2.6	4.2/4.6
Latitude -15	Average	3.4	4.1	5.0	5.9	6.0	6.1	6.0	5.8	5.2	4.7	3.7	3.1	4.9
	Min/Max	2.5/4.0	3.4/5.1	4.1/6.1	4.9/6.9	5.5/6.6	5.3/6.9	5.2/6.8	5.0/6.6	4.3/6.1	3.9/5.7	2.3/4.6	2.6/3.9	4.6/5.1
Latitude	Average	3.8	4.5	5.2	5.7	5.7	5.7	5.6	5.6	5.3	5.1	4.2	3.6	5.0
	Min/Max	2.7/4.7	3.6/5.7	4.1/6.3	4.8/6.8	5.2/6.2	4.9/6.4	4.8/6.3	4.8/6.4	4.3/6.3	4.1/6.2	2.4/5.3	2.9/4.6	4.7/5.2
Latitude +15	Average	4.1	4.6	5.1	5.3	5.0	4.9	4.9	5.1	5.0	5.2	4.4	3.9	4.8
	Min/Max	2.8/5.1	3.7/5.9	4.0/6.3	4.5/6.3	4.6/5.6	4.3/5.5	4.3/5.6	4.4/5.8	4.1/6.0	4.1/6.3	2.5/5.7	3.1/5.0	4.5/5.1
90	Average	3.6	3.8	3.6	3.1	2.5	2.2	2.3	2.7	3.3	4.0	3.8	3.5	3.2
	Min/Max	2.3/4.6	3.0/4.9	2.8/4.4	2.7/3.6	2.3/2.7	2.1/2.3	2.1/2.5	2.4/3.0	2.6/4.0	3.0/4.9	2.0/5.0	2.7/4.6	2.9/3.4

Solar Radiation for 1-Axis Tracking Flat-Plate Collectors with a North-South Axis (kWh/m²/day), Uncertainty ±9%

Axis Tilt (°)		Jan	Feb	Mar	Apr	May	June	July	Aug	Sept	Oct	Nov	Dec	Year
0	Average	3.5	4.6	5.9	7.3	7.6	7.9	7.5	7.1	6.1	5.3	3.9	3.2	5.8
	Min/Max	2.5/4.2	3.6/5.8	4.6/7.5	6.0/8.9	6.8/8.6	6.6/9.2	6.2/8.9	5.9/8.5	4.9/7.4	4.3/6.6	2.3/5.0	2.5/4.0	5.4/6.2
Latitude -15	Average	4.2	5.2	6.4	7.6	7.7	7.8	7.5	7.3	6.5	6.0	4.6	3.8	6.2
	Min/Max	2.8/5.1	4.0/6.7	4.9/8.1	6.2/9.3	6.8/8.7	6.6/9.1	6.2/8.9	6.0/8.8	5.2/8.0	4.7/7.5	2.6/6.0	3.0/4.9	5.7/6.6
Latitude	Average	4.5	5.5	6.6	7.5	7.4	7.5	7.3	7.1	6.6	6.3	5.0	4.2	6.3
	Min/Max	3.0/5.6	4.2/7.1	4.9/8.4	6.1/9.3	6.6/8.4	6.3/8.8	5.9/8.6	5.9/8.6	5.2/8.1	4.8/7.8	2.7/6.5	3.2/5.5	5.8/6.6
Latitude +15	Average	4.7	5.6	6.5	7.2	7.0	6.9	6.8	6.8	6.4	6.3	5.2	4.5	6.2
	Min/Max	3.1/5.9	4.3/7.3	4.8/8.3	5.9/8.9	6.2/7.9	5.9/8.1	5.5/8.0	5.6/8.2	5.0/8.0	4.8/7.9	2.7/6.8	3.4/5.8	5.6/6.5

Solar Radiation for 2-Axis Tracking Flat-Plate Collectors (kWh/m²/day), Uncertainty ±9%

Tracker		Jan	Feb	Mar	Apr	May	June	July	Aug	Sept	Oct	Nov	Dec	Year
2-Axis	Average	4.8	5.6	6.6	7.6	7.7	8.0	7.7	7.3	6.6	6.3	5.2	4.5	6.5
	Min/Max	3.1/6.0	4.3/7.3	5.0/8.4	6.2/9.4	6.9/8.8	6.7/9.3	6.3/9.0	6.0/8.8	5.2/8.1	4.9/7.9	2.7/6.8	3.4/5.9	5.9/6.9

Direct Beam Solar Radiation for Concentrating Collectors (kWh/m²/day), Uncertainty ±8%

Tracker		Jan	Feb	Mar	Apr	May	June	July	Aug	Sept	Oct	Nov	Dec	Year
1-Axis, E-W Horiz Axis	Average	2.8	3.1	3.3	3.7	3.6	3.6	3.4	3.2	3.0	3.4	3.1	2.8	3.3
	Min/Max	1.4/4.0	2.0/4.5	2.1/4.7	2.5/5.1	2.7/4.3	2.6/5.0	2.5/4.5	2.3/4.4	2.1/4.3	2.2/4.7	1.0/4.4	1.8/4.0	2.8/3.5
1-Axis, N-S Horiz Axis	Average	2.3	3.0	3.9	4.8	4.7	4.7	4.4	4.2	3.7	3.6	2.6	2.1	3.7
	Min/Max	1.1/3.2	1.9/4.4	2.4/5.6	3.3/6.7	3.6/5.7	3.4/6.4	3.1/5.9	2.9/5.7	2.6/5.2	2.3/5.0	0.9/3.7	1.3/3.0	3.2/4.1
1-Axis, N-S Tilt=Latitude	Average	3.2	3.8	4.4	5.0	4.5	4.4	4.2	4.2	4.1	4.3	3.5	3.0	4.0
	Min/Max	1.6/4.4	2.4/5.6	2.8/6.3	3.5/6.9	3.5/5.5	3.2/6.0	3.0/5.6	2.9/5.7	2.8/5.8	2.8/6.0	1.2/5.0	1.8/4.3	3.5/4.4
2-Axis	Average	3.4	3.9	4.4	5.1	4.8	4.8	4.5	4.4	4.1	4.4	3.7	3.2	4.2
	Min/Max	1.7/4.7	2.5/5.7	2.8/6.3	3.5/7.0	3.7/5.8	3.5/6.5	3.2/6.0	3.0/5.9	2.8/5.8	2.8/6.1	1.3/5.3	2.0/4.6	3.6/4.6

Average Climatic Conditions

Element	Jan	Feb	Mar	Apr	May	June	July	Aug	Sept	Oct	Nov	Dec	Year
Temperature (°C)	4.1	5.8	10.5	15.2	19.7	24.3	26.3	25.7	22.4	16.3	11.2	5.9	15.6
Daily Minimum Temp	-1.3	-0.1	4.1	8.6	13.6	18.7	20.9	20.5	17.2	10.3	5.3	0.4	9.8
Daily Maximum Temp	9.4	11.7	16.8	21.8	25.7	29.9	31.6	30.9	27.7	22.2	17.0	11.3	21.3
Record Minimum Temp	-20.6	-15.0	-15.6	-4.4	0.0	7.2	11.7	11.7	3.9	-4.4	-11.7	-16.7	-20.6
Record Maximum Temp	25.6	27.2	32.2	33.9	37.8	39.4	39.4	39.4	40.0	36.7	29.4	25.0	40.0
HDD, Base 18.3°C	443	350	243	102	23	0	0	0	3	89	217	386	1856
CDD, Base 18.3°C	0	0	0	8	64	178	246	229	127	26	0	0	879
Relative Humidity (%)	66	62	61	59	67	70	72	73	73	70	68	67	67
Wind Speed (m/s)	3.4	3.6	3.9	3.8	3.4	3.1	2.9	2.8	2.9	2.9	3.1	3.3	3.3

Greensboro, NC

WBAN NO. 13723

LATITUDE: 36.08° N
LONGITUDE: 79.95° W
ELEVATION: 270 meters
MEAN PRESSURE: 986 millibars

STATION TYPE: Secondary

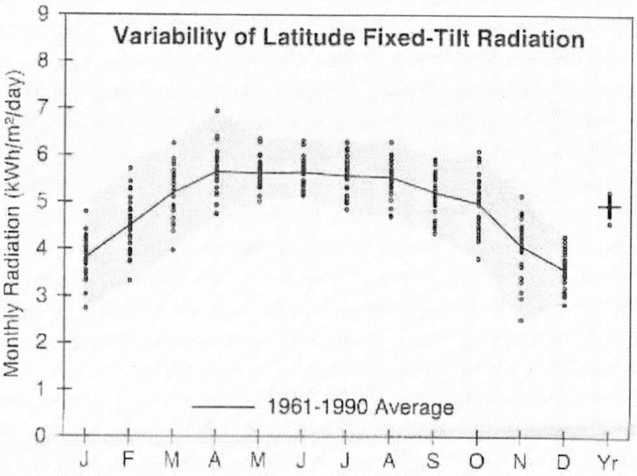

Solar Radiation for Flat-Plate Collectors Facing South at a Fixed Tilt (kWh/m²/day), Uncertainty ±9%

Tilt (°)		Jan	Feb	Mar	Apr	May	June	July	Aug	Sept	Oct	Nov	Dec	Year
0	Average	2.4	3.2	4.3	5.4	6.0	6.3	6.1	5.5	4.6	3.7	2.7	2.2	4.4
	Min/Max	2.0/2.8	2.6/3.8	3.5/5.0	4.6/6.5	5.3/6.7	5.7/7.0	5.3/6.8	4.7/6.2	4.0/5.2	3.1/4.4	2.0/3.1	1.9/2.4	4.1/4.5
Latitude -15	Average	3.3	4.1	5.0	5.8	6.0	6.1	6.0	5.7	5.2	4.6	3.6	3.1	4.9
	Min/Max	2.5/4.1	3.1/5.1	3.9/6.0	4.8/7.0	5.3/6.7	5.6/6.9	5.2/6.8	4.9/6.5	4.3/5.8	3.6/5.6	2.4/4.5	2.5/3.7	4.5/5.1
Latitude	Average	3.8	4.5	5.2	5.7	5.6	5.6	5.6	5.5	5.2	5.0	4.1	3.6	5.0
	Min/Max	2.7/4.8	3.3/5.7	4.0/6.3	4.7/6.9	5.0/6.3	5.1/6.3	4.8/6.3	4.7/6.3	4.3/5.9	3.8/6.1	2.5/5.1	2.8/4.3	4.6/5.2
Latitude +15	Average	4.1	4.6	5.1	5.2	5.0	4.9	4.9	5.1	5.0	5.0	4.3	3.8	4.8
	Min/Max	2.8/5.2	3.3/6.0	3.8/6.2	4.4/6.4	4.5/5.6	4.5/5.5	4.3/5.5	4.3/5.8	4.1/5.7	3.8/6.2	2.6/5.5	3.0/4.7	4.4/5.0
90	Average	3.7	3.8	3.6	3.1	2.5	2.3	2.3	2.8	3.3	3.9	3.7	3.5	3.2
	Min/Max	2.4/4.7	2.7/5.0	2.7/4.5	2.7/3.7	2.3/2.7	2.1/2.4	2.2/2.5	2.4/3.1	2.7/3.8	2.9/4.8	2.1/4.8	2.7/4.3	2.9/3.4

Solar Radiation for 1-Axis Tracking Flat-Plate Collectors with a North-South Axis (kWh/m²/day), Uncertainty ±9%

Axis Tilt (°)		Jan	Feb	Mar	Apr	May	June	July	Aug	Sept	Oct	Nov	Dec	Year
0	Average	3.4	4.5	5.9	7.2	7.6	7.9	7.6	7.1	6.1	5.2	3.8	3.1	5.8
	Min/Max	2.5/4.3	3.4/5.8	4.4/7.1	5.7/9.2	6.6/8.9	6.9/9.2	6.5/8.9	5.7/8.1	4.9/7.2	4.0/6.5	2.3/4.8	2.4/3.7	5.3/6.1
Latitude -15	Average	4.1	5.2	6.4	7.5	7.6	7.8	7.6	7.3	6.5	5.9	4.5	3.8	6.2
	Min/Max	2.9/5.3	3.8/6.8	4.8/7.8	5.9/9.7	6.6/8.9	6.9/9.1	6.4/8.9	5.9/8.4	5.2/7.7	4.4/7.4	2.6/5.8	2.9/4.6	5.6/6.6
Latitude	Average	4.5	5.5	6.6	7.4	7.4	7.5	7.3	7.1	6.5	6.1	4.8	4.2	6.3
	Min/Max	3.1/5.8	3.9/7.2	4.8/8.0	5.8/9.6	6.4/8.7	6.6/8.8	6.2/8.5	5.7/8.3	5.2/7.8	4.6/7.8	2.8/6.3	3.2/5.1	5.6/6.7
Latitude +15	Average	4.7	5.6	6.5	7.1	7.0	7.0	6.8	6.8	6.4	6.2	5.0	4.4	6.1
	Min/Max	3.2/6.1	3.9/7.4	4.7/8.0	5.6/9.3	6.0/8.2	6.1/8.1	5.8/8.0	5.4/7.9	5.1/7.6	4.6/7.9	2.8/6.6	3.3/5.4	5.5/6.6

Solar Radiation for 2-Axis Tracking Flat-Plate Collectors (kWh/m²/day), Uncertainty ±9%

Tracker		Jan	Feb	Mar	Apr	May	June	July	Aug	Sept	Oct	Nov	Dec	Year
2-Axis	Average	4.8	5.6	6.6	7.5	7.7	8.0	7.7	7.3	6.6	6.2	5.0	4.5	6.5
	Min/Max	3.2/6.2	4.0/7.4	4.8/8.1	5.9/9.7	6.7/9.1	7.0/9.3	6.6/9.0	5.9/8.4	5.3/7.8	4.6/7.9	2.8/6.6	3.3/5.4	5.8/6.9

Direct Beam Solar Radiation for Concentrating Collectors (kWh/m²/day), Uncertainty ±8%

Tracker		Jan	Feb	Mar	Apr	May	June	July	Aug	Sept	Oct	Nov	Dec	Year
1-Axis, E-W Horiz Axis	Average	2.8	3.1	3.3	3.6	3.5	3.6	3.4	3.2	3.0	3.4	3.0	2.7	3.2
	Min/Max	1.6/4.1	1.7/4.6	2.0/4.3	2.3/5.2	2.8/4.5	2.7/4.7	2.4/4.3	2.2/4.1	2.1/3.9	2.1/4.7	1.2/4.2	1.8/3.6	2.7/3.6
1-Axis, N-S Horiz Axis	Average	2.2	3.0	3.9	4.7	4.6	4.7	4.4	4.2	3.7	3.5	2.5	2.0	3.6
	Min/Max	1.3/3.3	1.7/4.4	2.4/5.0	3.0/7.0	3.6/6.0	3.6/6.1	3.2/5.7	2.8/5.3	2.5/4.9	2.3/4.9	1.0/3.5	1.3/2.7	3.0/4.0
1-Axis, N-S Tilt=Latitude	Average	3.1	3.8	4.5	4.9	4.5	4.4	4.2	4.3	4.1	4.2	3.4	2.9	4.0
	Min/Max	1.7/4.6	2.2/5.7	2.8/5.8	3.2/7.2	3.5/5.8	3.4/5.7	3.0/5.4	2.9/5.4	2.8/5.4	2.8/6.0	1.4/4.8	1.9/3.9	3.4/4.5
2-Axis	Average	3.3	3.9	4.5	5.0	4.8	4.8	4.5	4.4	4.1	4.3	3.5	3.2	4.2
	Min/Max	1.9/4.9	2.2/5.8	2.8/5.8	3.2/7.4	3.7/6.2	3.7/6.2	3.3/5.8	2.9/5.5	2.8/5.4	2.8/6.1	1.4/5.1	2.0/4.2	3.5/4.7

Average Climatic Conditions

Element	Jan	Feb	Mar	Apr	May	June	July	Aug	Sept	Oct	Nov	Dec	Year
Temperature (°C)	2.6	4.4	9.3	14.2	18.8	22.8	24.9	24.3	20.9	14.8	9.7	4.7	14.3
Daily Minimum Temp	-3.0	-1.5	3.0	7.6	12.6	17.0	19.4	18.8	15.3	8.4	3.6	-0.8	8.4
Daily Maximum Temp	8.2	10.4	15.7	20.9	25.0	28.7	30.5	29.7	26.6	21.1	15.8	10.3	20.2
Record Minimum Temp	-22.2	-20.0	-15.0	-6.1	0.0	5.6	8.9	7.2	1.7	-6.7	-12.2	-17.8	-22.2
Record Maximum Temp	25.6	27.2	32.2	34.4	36.7	38.9	38.9	39.4	37.8	35.0	29.4	25.6	39.4
HDD, Base 18.3°C	487	389	279	131	43	3	0	0	9	126	258	422	2147
CDD, Base 18.3°C	0	0	0	7	59	138	205	184	87	15	0	0	696
Relative Humidity (%)	67	64	63	61	70	73	75	76	76	72	69	69	70
Wind Speed (m/s)	3.4	3.6	3.8	3.7	3.2	3.0	2.8	2.7	2.8	2.9	3.2	3.4	3.2

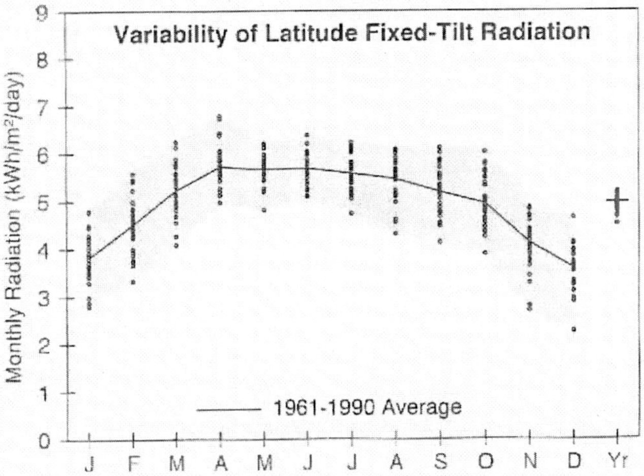

Raleigh, NC

WBAN NO. 13722

LATITUDE: 35.87° N
LONGITUDE: 78.78° W
ELEVATION: 134 meters
MEAN PRESSURE: 1002 millibars

STATION TYPE: Primary

Solar Radiation for Flat-Plate Collectors Facing South at a Fixed Tilt (kWh/m²/day), Uncertainty ±9%

Tilt (°)		Jan	Feb	Mar	Apr	May	June	July	Aug	Sept	Oct	Nov	Dec	Year
0	Average	2.4	3.2	4.4	5.5	6.0	6.3	6.1	5.5	4.6	3.8	2.7	2.2	4.4
	Min/Max	2.0/2.8	2.5/3.8	3.6/5.1	4.8/6.4	5.1/6.6	5.7/7.1	5.2/6.8	4.3/6.1	3.8/5.3	3.2/4.4	2.0/3.0	1.6/2.6	4.0/4.6
Latitude -15	Average	3.4	4.1	5.0	5.8	6.0	6.2	6.0	5.7	5.1	4.6	3.7	3.1	4.9
	Min/Max	2.6/4.1	3.1/5.0	4.0/6.0	5.1/6.9	5.1/6.6	5.5/6.9	5.1/6.7	4.5/6.3	4.1/6.0	3.7/5.6	2.5/4.3	2.1/4.0	4.5/5.1
Latitude	Average	3.8	4.5	5.2	5.7	5.7	5.7	5.6	5.5	5.2	4.9	4.1	3.6	5.0
	Min/Max	2.8/4.8	3.3/5.6	4.1/6.2	5.0/6.8	4.8/6.2	5.1/6.4	4.7/6.2	4.3/6.1	4.1/6.1	3.9/6.0	2.7/4.8	2.3/4.6	4.5/5.2
Latitude +15	Average	4.1	4.6	5.1	5.3	5.0	4.9	4.9	5.0	5.0	5.0	4.3	3.8	4.8
	Min/Max	2.9/5.2	3.4/5.8	3.9/6.1	4.6/6.3	4.3/5.5	4.5/5.5	4.2/5.5	3.9/5.6	3.9/5.9	3.8/6.1	2.8/5.2	2.3/5.1	4.3/5.0
90	Average	3.6	3.8	3.6	3.1	2.5	2.3	2.4	2.8	3.3	3.8	3.7	3.5	3.2
	Min/Max	2.5/4.7	2.8/4.8	2.8/4.3	2.7/3.6	2.3/2.7	2.2/2.4	2.1/2.6	2.2/3.0	2.5/3.8	2.9/4.8	2.3/4.5	2.0/4.7	2.9/3.4

Solar Radiation for 1-Axis Tracking Flat-Plate Collectors with a North-South Axis (kWh/m²/day), Uncertainty ±9%

Axis Tilt (°)		Jan	Feb	Mar	Apr	May	June	July	Aug	Sept	Oct	Nov	Dec	Year
0	Average	3.4	4.5	5.9	7.2	7.6	7.9	7.5	6.8	6.0	5.1	3.8	3.1	5.7
	Min/Max	2.5/4.3	3.3/5.8	4.5/7.2	6.1/8.9	6.2/8.6	6.8/9.2	6.1/8.7	5.3/7.8	4.6/7.3	4.0/6.4	2.5/4.5	2.0/4.0	5.2/6.0
Latitude -15	Average	4.1	5.2	6.4	7.5	7.6	7.8	7.5	7.0	6.4	5.8	4.5	3.8	6.1
	Min/Max	2.9/5.3	3.7/6.7	4.8/7.9	6.4/9.3	6.3/8.7	6.7/9.1	6.1/8.7	5.5/8.0	4.8/7.8	4.4/7.2	2.8/5.4	2.3/5.0	5.5/6.4
Latitude	Average	4.5	5.5	6.5	7.5	7.4	7.5	7.2	6.9	6.4	6.0	4.9	4.2	6.2
	Min/Max	3.1/5.8	3.9/7.1	4.8/8.1	6.3/9.2	6.1/8.5	6.4/8.7	5.9/8.4	5.4/7.9	4.8/7.9	4.5/7.6	3.0/5.9	2.5/5.6	5.5/6.5
Latitude +15	Average	4.7	5.6	6.5	7.2	7.0	7.0	6.7	6.5	6.3	6.1	5.0	4.4	6.1
	Min/Max	3.2/6.1	4.0/7.2	4.7/8.0	6.0/8.9	5.7/8.0	6.0/8.1	5.5/7.8	5.1/7.5	4.7/7.7	4.5/7.7	3.1/6.1	2.6/5.9	5.4/6.4

Solar Radiation for 2-Axis Tracking Flat-Plate Collectors (kWh/m²/day), Uncertainty ±9%

Tracker		Jan	Feb	Mar	Apr	May	June	July	Aug	Sept	Oct	Nov	Dec	Year
2-Axis	Average	4.7	5.6	6.6	7.5	7.7	8.0	7.6	7.1	6.5	6.1	5.1	4.4	6.4
	Min/Max	3.3/6.2	4.0/7.3	4.8/8.1	6.4/9.3	6.3/8.8	6.8/9.3	6.2/8.9	5.5/8.0	4.8/7.9	4.5/7.7	3.1/6.2	2.6/6.0	5.7/6.7

Direct Beam Solar Radiation for Concentrating Collectors (kWh/m²/day), Uncertainty ±8%

Tracker		Jan	Feb	Mar	Apr	May	June	July	Aug	Sept	Oct	Nov	Dec	Year
1-Axis, E-W Horiz Axis	Average	2.8	3.0	3.3	3.5	3.4	3.5	3.2	2.9	2.9	3.2	3.0	2.7	3.1
	Min/Max	1.5/4.1	1.9/4.4	1.9/4.5	2.6/4.9	2.3/4.3	2.4/4.6	2.1/4.3	1.8/3.7	1.8/4.0	1.9/4.6	1.4/3.8	1.1/4.0	2.7/3.4
1-Axis, N-S Horiz Axis	Average	2.2	2.9	3.8	4.6	4.5	4.6	4.1	3.8	3.5	3.3	2.5	2.0	3.5
	Min/Max	1.3/3.2	1.8/4.3	2.2/5.3	3.4/6.5	3.1/5.7	3.2/6.0	2.8/5.7	2.3/4.9	2.1/4.9	2.0/4.7	1.1/3.3	0.9/3.0	3.0/3.8
1-Axis, N-S Tilt=Latitude	Average	3.1	3.7	4.4	4.8	4.4	4.3	3.9	3.8	3.9	4.1	3.4	2.9	3.9
	Min/Max	1.7/4.5	2.3/5.5	2.6/6.0	3.6/6.8	3.0/5.5	3.0/5.6	2.7/5.4	2.3/4.9	2.3/5.4	2.4/5.8	1.5/4.4	1.3/4.3	3.3/4.2
2-Axis	Average	3.3	3.8	4.4	4.9	4.6	4.6	4.2	3.9	3.9	4.1	3.5	3.1	4.0
	Min/Max	1.9/4.8	2.4/5.6	2.6/6.1	3.6/6.9	3.2/5.8	3.3/6.1	2.9/5.8	2.4/5.1	2.3/5.4	2.4/5.9	1.6/4.7	1.4/4.7	3.4/4.4

Average Climatic Conditions

Element	Jan	Feb	Mar	Apr	May	June	July	Aug	Sept	Oct	Nov	Dec	Year
Temperature (°C)	3.8	5.6	10.2	15.0	19.4	23.5	25.6	25.1	21.7	15.6	10.7	5.9	15.2
Daily Minimum Temp	-1.8	-0.4	3.7	7.9	12.9	17.6	20.1	19.7	16.2	9.1	4.3	0.2	9.1
Daily Maximum Temp	9.4	11.4	16.7	22.1	25.9	29.4	31.1	30.4	27.3	22.0	17.0	11.5	21.2
Record Minimum Temp	-22.8	-15.0	-11.7	-5.0	-0.6	3.3	8.9	7.8	2.8	-7.2	-11.7	-15.6	-22.8
Record Maximum Temp	26.1	28.9	33.3	35.0	36.1	40.0	40.6	40.6	40.0	36.7	31.1	26.1	40.6
HDD, Base 18.3°C	449	358	254	107	26	0	0	0	5	105	230	386	1921
CDD, Base 18.3°C	0	0	3	7	61	155	226	208	107	21	0	0	787
Relative Humidity (%)	67	64	63	62	71	74	76	78	77	73	69	69	70
Wind Speed (m/s)	3.8	3.9	4.1	3.9	3.4	3.1	2.9	2.8	3.0	3.1	3.4	3.6	3.4

Wilmington, NC

WBAN NO. 13748

LATITUDE: 34.27° N
LONGITUDE: 77.90° W
ELEVATION: 9 meters
MEAN PRESSURE: 1017 millibars

STATION TYPE: Secondary

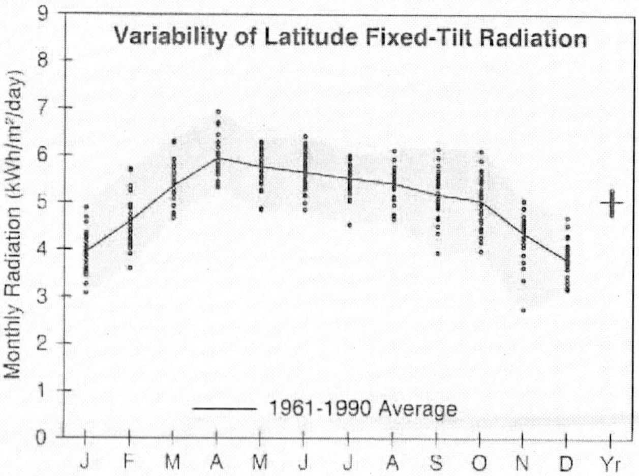

Solar Radiation for Flat-Plate Collectors Facing South at a Fixed Tilt (kWh/m²/day), Uncertainty ±9%

Tilt (°)		Jan	Feb	Mar	Apr	May	June	July	Aug	Sept	Oct	Nov	Dec	Year
0	Average	2.6	3.4	4.5	5.7	6.1	6.3	6.0	5.4	4.6	3.9	2.9	2.4	4.5
	Min/Max	2.2/3.0	2.8/4.0	4.1/5.2	5.2/6.6	5.2/6.6	5.4/7.1	4.9/6.5	4.7/6.1	3.7/5.4	3.3/4.5	2.1/3.3	2.1/2.8	4.2/4.7
Latitude -15	Average	3.5	4.2	5.2	6.0	6.1	6.1	5.9	5.6	5.1	4.7	3.9	3.3	5.0
	Min/Max	2.8/4.2	3.4/5.1	4.6/6.0	5.4/7.0	5.1/6.7	5.2/7.0	4.8/6.4	4.8/6.3	4.0/6.0	3.8/5.6	2.5/4.4	2.8/4.0	4.7/5.2
Latitude	Average	4.0	4.6	5.4	5.9	5.8	5.6	5.5	5.4	5.2	5.0	4.4	3.8	5.0
	Min/Max	3.1/4.9	3.6/5.7	4.7/6.3	5.3/6.9	4.8/6.3	4.8/6.4	4.5/6.0	4.7/6.1	3.9/6.1	4.0/6.1	2.8/5.1	3.2/4.7	4.8/5.3
Latitude +15	Average	4.2	4.7	5.3	5.5	5.1	4.9	4.8	4.9	5.0	5.1	4.6	4.1	4.9
	Min/Max	3.2/5.3	3.7/6.0	4.5/6.2	4.9/6.4	4.3/5.6	4.3/5.5	4.0/5.3	4.3/5.6	3.7/5.9	4.0/6.2	2.8/5.4	3.4/5.1	4.6/5.1
90	Average	3.7	3.8	3.6	3.1	2.4	2.1	2.2	2.6	3.2	3.9	3.9	3.7	3.2
	Min/Max	2.8/4.7	2.9/4.9	3.1/4.3	2.8/3.6	2.2/2.6	2.0/2.3	2.0/2.3	2.3/2.8	2.4/3.8	2.9/4.8	2.3/4.7	3.0/4.7	3.0/3.3

Solar Radiation for 1-Axis Tracking Flat-Plate Collectors with a North-South Axis (kWh/m²/day), Uncertainty ±9%

Axis Tilt (°)		Jan	Feb	Mar	Apr	May	June	July	Aug	Sept	Oct	Nov	Dec	Year
0	Average	3.7	4.7	6.1	7.5	7.8	7.8	7.5	6.8	6.0	5.3	4.1	3.4	5.9
	Min/Max	2.8/4.5	3.7/5.9	5.2/7.3	6.5/9.0	6.1/8.6	6.4/9.3	5.8/8.4	5.6/8.1	4.3/7.3	4.1/6.6	2.6/4.8	2.7/4.2	5.5/6.3
Latitude -15	Average	4.3	5.3	6.6	7.8	7.8	7.8	7.4	7.0	6.4	5.9	4.8	4.0	6.3
	Min/Max	3.3/5.4	4.1/6.8	5.6/7.9	6.7/9.4	6.1/8.7	6.3/9.2	5.8/8.3	5.7/8.3	4.5/7.8	4.5/7.4	2.9/5.7	3.3/5.1	5.8/6.7
Latitude	Average	4.7	5.6	6.7	7.8	7.6	7.4	7.2	6.9	6.4	6.2	5.2	4.5	6.3
	Min/Max	3.5/5.9	4.3/7.3	5.7/8.2	6.7/9.3	5.9/8.4	6.1/8.8	5.5/8.0	5.6/8.1	4.5/7.9	4.7/7.8	3.1/6.2	3.6/5.6	5.9/6.7
Latitude +15	Average	4.9	5.7	6.6	7.5	7.1	6.9	6.7	6.5	6.3	6.2	5.4	4.7	6.2
	Min/Max	3.7/6.2	4.3/7.4	5.6/8.1	6.4/9.0	5.5/7.9	5.6/8.2	5.2/7.5	5.3/7.7	4.4/7.7	4.7/7.9	3.1/6.4	3.7/6.0	5.7/6.6

Solar Radiation for 2-Axis Tracking Flat-Plate Collectors (kWh/m²/day), Uncertainty ±9%

Tracker		Jan	Feb	Mar	Apr	May	June	July	Aug	Sept	Oct	Nov	Dec	Year
2-Axis	Average	4.9	5.7	6.7	7.9	7.9	7.9	7.6	7.0	6.5	6.3	5.4	4.8	6.6
	Min/Max	3.7/6.3	4.4/7.4	5.7/8.2	6.8/9.4	6.2/8.8	6.5/9.4	5.8/8.5	5.7/8.3	4.6/7.9	4.7/7.9	3.1/6.5	3.8/6.1	6.1/7.0

Direct Beam Solar Radiation for Concentrating Collectors (kWh/m²/day), Uncertainty ±8%

Tracker		Jan	Feb	Mar	Apr	May	June	July	Aug	Sept	Oct	Nov	Dec	Year
1-Axis, E-W	Average	2.9	3.1	3.4	3.8	3.6	3.6	3.3	3.0	2.9	3.3	3.2	2.9	3.3
Horiz Axis	Min/Max	1.9/4.2	2.0/4.5	2.6/4.5	3.0/5.1	2.3/4.4	2.3/4.8	2.0/4.2	2.0/4.2	1.5/4.0	2.1/4.6	1.4/4.1	2.1/4.0	2.9/3.6
1-Axis, N-S	Average	2.4	3.1	4.0	5.0	4.8	4.6	4.3	3.9	3.5	3.5	2.7	2.2	3.7
Horiz Axis	Min/Max	1.5/3.4	2.0/4.5	3.0/5.3	3.8/6.6	3.0/5.8	3.0/6.3	2.7/5.4	2.6/5.5	1.9/4.9	2.2/4.9	1.2/3.5	1.5/3.1	3.2/4.1
1-Axis, N-S	Average	3.3	3.9	4.6	5.2	4.6	4.3	4.0	3.9	3.9	4.2	3.6	3.1	4.0
Tilt=Latitude	Min/Max	2.1/4.6	2.5/5.6	3.4/6.1	3.9/6.8	2.9/5.6	2.8/5.9	2.5/5.1	2.6/5.5	2.0/5.3	2.6/5.9	1.6/4.7	2.2/4.4	3.5/4.5
2-Axis	Average	3.5	4.0	4.6	5.2	4.9	4.7	4.3	4.0	3.9	4.2	3.8	3.4	4.2
	Min/Max	2.2/5.0	2.6/5.8	3.4/6.1	4.0/6.9	3.0/5.9	3.1/6.4	2.7/5.5	2.7/5.7	2.0/5.4	2.7/6.0	1.7/4.9	2.3/4.7	3.7/4.7

Average Climatic Conditions

Element	Jan	Feb	Mar	Apr	May	June	July	Aug	Sept	Oct	Nov	Dec	Year
Temperature (°C)	7.2	8.5	12.4	16.8	21.2	24.7	26.7	26.3	24.1	18.5	13.9	9.2	17.4
Daily Minimum Temp	1.3	2.4	6.2	10.3	15.2	19.7	22.1	21.7	18.5	12.1	7.1	3.1	11.6
Daily Maximum Temp	12.9	14.5	18.7	23.3	27.1	29.7	31.4	30.9	29.6	24.9	20.6	15.2	23.2
Record Minimum Temp	-15.0	-11.7	-12.8	-1.1	3.3	8.9	12.8	12.8	6.7	-2.8	-6.7	-17.8	-17.8
Record Maximum Temp	27.8	29.4	31.7	35.0	36.7	40.0	38.9	38.9	36.7	35.0	30.6	27.2	40.0
HDD, Base 18.3°C	351	276	188	64	6	0	0	0	0	57	144	288	1372
CDD, Base 18.3°C	4	0	6	19	93	192	260	248	172	62	11	4	1070
Relative Humidity (%)	71	68	69	67	74	76	78	81	80	76	73	72	74
Wind Speed (m/s)	3.9	4.2	4.3	4.4	3.9	3.7	3.4	3.1	3.4	3.4	3.5	3.7	3.8

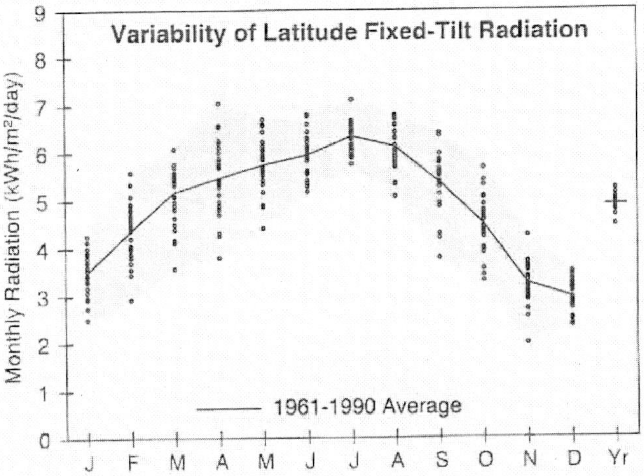

Bismarck, ND

WBAN NO. 24011

LATITUDE: 46.77° N
LONGITUDE: 100.75° W
ELEVATION: 502 meters
MEAN PRESSURE: 956 millibars

STATION TYPE: Primary

Solar Radiation for Flat-Plate Collectors Facing South at a Fixed Tilt (kWh/m²/day), Uncertainty ±9%

Tilt (°)		Jan	Feb	Mar	Apr	May	June	July	Aug	Sept	Oct	Nov	Dec	Year
0	Average	1.7	2.6	3.8	4.9	6.0	6.6	6.8	5.8	4.2	2.8	1.7	1.4	4.0
	Min/Max	1.4/1.9	2.0/3.0	3.0/4.4	3.6/6.1	4.7/6.8	5.8/7.6	6.2/7.6	4.7/6.4	3.3/4.8	2.3/3.3	1.4/2.0	1.2/1.5	3.9/4.2
Latitude -15	Average	3.1	4.0	5.0	5.6	6.1	6.5	6.8	6.4	5.3	4.2	2.9	2.6	4.9
	Min/Max	2.2/3.7	2.8/5.0	3.5/5.8	3.9/7.2	4.7/7.1	5.6/7.4	6.2/7.7	5.3/7.1	3.8/6.3	3.2/5.2	1.9/3.8	2.1/3.0	4.5/5.2
Latitude	Average	3.5	4.4	5.2	5.5	5.7	5.9	6.3	6.1	5.4	4.5	3.2	3.0	4.9
	Min/Max	2.5/4.3	2.9/5.6	3.6/6.1	3.8/7.0	4.4/6.7	5.2/6.8	5.8/7.1	5.1/6.8	3.8/6.4	3.3/5.7	2.0/4.3	2.3/3.5	4.5/5.2
Latitude +15	Average	3.7	4.5	5.1	5.1	5.1	5.1	5.5	5.5	5.1	4.5	3.4	3.2	4.7
	Min/Max	2.6/4.6	3.0/5.9	3.5/6.1	3.5/6.5	3.9/5.9	4.5/5.8	5.0/6.2	4.6/6.1	3.6/6.2	3.3/5.8	2.0/4.5	2.5/3.8	4.2/5.0
90	Average	3.7	4.2	4.3	3.7	3.3	3.1	3.4	3.7	3.9	3.9	3.2	3.1	3.6
	Min/Max	2.5/4.6	2.7/5.6	2.8/5.4	2.5/4.6	2.6/4.0	2.8/3.4	3.0/3.7	3.1/4.1	2.7/4.7	2.8/5.0	1.8/4.2	2.4/3.8	3.2/3.9

Solar Radiation for 1-Axis Tracking Flat-Plate Collectors with a North-South Axis (kWh/m²/day), Uncertainty ±9%

Axis Tilt (°)		Jan	Feb	Mar	Apr	May	June	July	Aug	Sept	Oct	Nov	Dec	Year
0	Average	2.6	3.8	5.4	6.7	8.0	8.8	9.3	8.1	6.1	4.2	2.6	2.1	5.7
	Min/Max	1.9/3.2	2.5/4.8	3.7/6.4	4.7/9.0	5.7/9.6	7.2/10.5	8.3/10.6	6.8/9.2	4.3/7.5	3.2/5.3	1.7/3.3	1.6/2.5	5.2/6.1
Latitude -15	Average	3.6	4.9	6.3	7.3	8.3	8.9	9.5	8.7	7.0	5.3	3.4	3.0	6.4
	Min/Max	2.5/4.5	3.1/6.3	4.2/7.4	5.1/9.9	5.8/9.9	7.2/10.7	8.5/10.9	7.2/9.9	4.7/8.6	3.8/6.7	2.0/4.6	2.4/3.6	5.7/6.9
Latitude	Average	4.0	5.2	6.4	7.3	8.0	8.6	9.2	8.5	7.0	5.5	3.7	3.3	6.4
	Min/Max	2.8/4.9	3.3/6.7	4.2/7.7	5.0/9.8	5.6/9.6	6.9/10.3	8.1/10.5	7.1/9.7	4.7/8.7	3.9/7.1	2.2/5.0	2.6/4.0	5.7/7.0
Latitude +15	Average	4.2	5.3	6.4	7.0	7.6	8.0	8.6	8.1	6.8	5.5	3.9	3.5	6.3
	Min/Max	2.9/5.2	3.3/6.9	4.1/7.7	4.8/9.4	5.3/9.1	6.5/9.6	7.6/9.9	6.8/9.3	4.5/8.5	3.9/7.1	2.2/5.2	2.7/4.3	5.6/6.8

Solar Radiation for 2-Axis Tracking Flat-Plate Collectors (kWh/m²/day), Uncertainty ±9%

Tracker		Jan	Feb	Mar	Apr	May	June	July	Aug	Sept	Oct	Nov	Dec	Year
2-Axis	Average	4.2	5.3	6.5	7.4	8.4	9.2	9.7	8.7	7.0	5.5	3.9	3.6	6.6
	Min/Max	2.9/5.3	3.3/6.9	4.2/7.7	5.1/9.9	5.9/10.1	7.4/10.9	8.6/11.1	7.3/9.9	4.7/8.7	3.9/7.1	2.2/5.2	2.8/4.3	5.9/7.2

Direct Beam Solar Radiation for Concentrating Collectors (kWh/m²/day), Uncertainty ±8%

Tracker		Jan	Feb	Mar	Apr	May	June	July	Aug	Sept	Oct	Nov	Dec	Year
1-Axis, E-W Horiz Axis	Average	2.4	2.7	3.0	3.4	3.9	4.5	5.0	4.4	3.7	3.1	2.2	2.1	3.4
	Min/Max	1.5/3.1	1.3/3.9	1.5/4.1	1.4/5.4	2.1/5.1	3.1/5.9	4.0/6.2	3.3/5.3	1.8/5.1	1.8/4.4	0.8/3.4	1.3/2.8	2.8/3.8
1-Axis, N-S Horiz Axis	Average	1.5	2.1	3.2	4.3	5.2	5.9	6.5	5.7	4.3	2.9	1.5	1.2	3.7
	Min/Max	0.9/2.0	1.0/3.1	1.6/4.3	1.8/6.7	2.8/6.8	4.0/7.8	5.2/8.0	4.3/6.9	2.1/6.0	1.7/4.0	0.5/2.4	0.8/1.6	3.1/4.2
1-Axis, N-S Tilt=Latitude	Average	2.6	3.2	4.0	4.7	5.2	5.7	6.4	6.0	5.0	3.9	2.5	2.1	4.3
	Min/Max	1.6/3.4	1.5/4.6	2.0/5.4	2.0/7.4	2.8/6.8	3.8/7.4	5.1/7.9	4.5/7.3	2.5/7.0	2.3/5.4	0.8/3.8	1.4/2.9	3.5/4.9
2-Axis	Average	2.7	3.3	4.0	4.8	5.5	6.2	6.8	6.2	5.0	4.0	2.6	2.3	4.5
	Min/Max	1.7/3.6	1.5/4.7	2.0/5.5	2.0/7.5	2.9/7.2	4.2/8.1	5.5/8.4	4.7/7.5	2.5/7.1	2.3/5.5	0.9/4.0	1.5/3.2	3.7/5.1

Average Climatic Conditions

Element	Jan	Feb	Mar	Apr	May	June	July	Aug	Sept	Oct	Nov	Dec	Year
Temperature (°C)	-12.7	-9.1	-2.1	6.1	12.8	18.0	21.3	20.2	13.9	7.6	-1.9	-10.0	5.3
Daily Minimum Temp	-18.7	-14.9	-7.9	-0.6	5.7	10.9	13.6	12.2	6.2	0.3	-7.9	-15.9	-1.4
Daily Maximum Temp	-6.6	-3.1	3.6	12.7	19.9	25.1	29.1	28.2	21.6	14.8	4.1	-4.2	12.1
Record Minimum Temp	-42.2	-39.4	-35.0	-24.4	-9.4	-1.1	1.7	0.6	-11.7	-23.3	-34.4	-41.7	-42.2
Record Maximum Temp	16.7	20.0	27.2	33.9	36.7	41.7	42.8	42.8	40.6	35.0	23.9	18.3	42.8
HDD, Base 18.3°C	961	767	634	367	180	64	8	38	146	332	607	878	4982
CDD, Base 18.3°C	0	0	0	0	8	54	102	95	12	0	0	0	271
Relative Humidity (%)	71	72	70	62	60	65	62	61	64	64	72	75	66
Wind Speed (m/s)	4.4	4.2	4.6	5.0	4.8	4.2	3.8	3.9	4.1	4.3	4.0	4.1	4.3

Fargo, ND

WBAN NO. 14914

LATITUDE: 46.90° N
LONGITUDE: 96.80° W
ELEVATION: 274 meters
MEAN PRESSURE: 983 millibars

STATION TYPE: Secondary

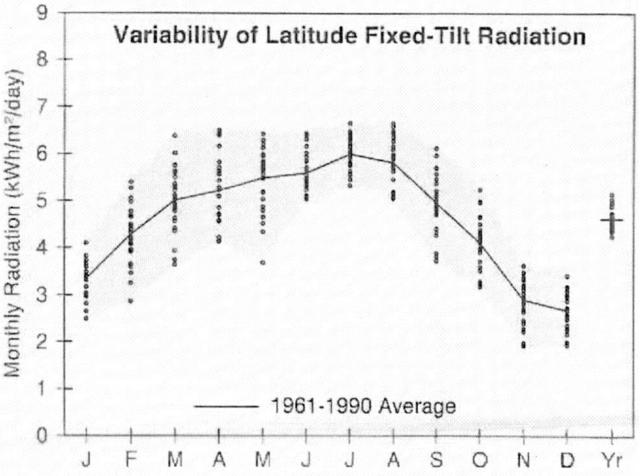

Solar Radiation for Flat-Plate Collectors Facing South at a Fixed Tilt (kWh/m²/day), Uncertainty ±9%

Tilt (°)		Jan	Feb	Mar	Apr	May	June	July	Aug	Sept	Oct	Nov	Dec	Year
0	Average	1.6	2.5	3.7	4.7	5.7	6.2	6.4	5.5	4.0	2.7	1.6	1.3	3.8
	Min/Max	1.3/1.8	1.9/2.9	3.0/4.5	3.8/5.7	4.0/6.6	5.7/7.1	5.8/7.1	4.8/6.2	3.2/4.6	2.2/3.1	1.3/1.8	1.0/1.5	3.7/4.2
Latitude -15	Average	2.9	3.9	4.8	5.3	5.9	6.1	6.5	6.1	4.9	3.9	2.6	2.3	4.6
	Min/Max	2.2/3.5	2.7/4.8	3.6/6.0	4.2/6.6	4.0/6.9	5.5/7.1	5.8/7.2	5.3/6.9	3.8/6.0	3.0/4.8	1.8/3.2	1.7/2.9	4.3/5.1
Latitude	Average	3.4	4.3	5.0	5.2	5.5	5.6	6.0	5.8	5.0	4.1	2.9	2.7	4.6
	Min/Max	2.5/4.1	2.9/5.4	3.6/6.4	4.1/6.5	3.7/6.4	5.0/6.4	5.3/6.7	5.1/6.7	3.7/6.1	3.2/5.2	1.9/3.6	1.9/3.4	4.3/5.1
Latitude +15	Average	3.6	4.5	5.0	4.8	4.9	4.8	5.2	5.3	4.7	4.1	3.0	2.9	4.4
	Min/Max	2.6/4.4	2.9/5.7	3.5/6.4	3.8/6.0	3.3/5.7	4.4/5.6	4.7/5.8	4.6/6.0	3.5/5.9	3.2/5.3	2.0/3.8	2.0/3.7	4.0/4.9
90	Average	3.5	4.2	4.3	3.5	3.1	2.9	3.2	3.6	3.6	3.6	2.8	2.8	3.4
	Min/Max	2.5/4.4	2.5/5.4	2.9/5.7	2.8/4.3	2.2/3.6	2.7/3.3	2.9/3.5	3.1/4.0	2.7/4.6	2.7/4.6	1.8/3.7	1.9/3.7	3.0/3.8

Solar Radiation for 1-Axis Tracking Flat-Plate Collectors with a North-South Axis (kWh/m²/day), Uncertainty ±9%

Axis Tilt (°)		Jan	Feb	Mar	Apr	May	June	July	Aug	Sept	Oct	Nov	Dec	Year
0	Average	2.5	3.8	5.2	6.4	7.7	8.3	8.9	7.8	5.7	3.9	2.3	1.9	5.4
	Min/Max	1.8/3.1	2.4/4.8	3.6/6.7	4.7/8.4	4.7/9.3	7.1/10.0	7.8/10.1	6.5/9.3	4.3/7.0	3.0/5.1	1.6/2.9	1.4/2.4	4.9/6.0
Latitude -15	Average	3.5	4.8	6.1	7.0	7.9	8.4	9.1	8.3	6.4	4.8	3.1	2.7	6.0
	Min/Max	2.5/4.4	3.0/6.2	4.1/7.9	5.0/9.2	4.7/9.7	7.1/10.2	8.0/10.3	6.9/9.9	4.7/8.1	3.6/6.3	2.0/3.9	1.9/3.5	5.5/6.8
Latitude	Average	3.8	5.1	6.2	6.9	7.7	8.1	8.8	8.1	6.5	5.0	3.3	3.0	6.1
	Min/Max	2.8/4.8	3.2/6.6	4.2/8.1	5.0/9.1	4.5/9.4	6.8/9.8	7.7/10.0	6.8/9.8	4.7/8.2	3.8/6.6	2.1/4.2	2.1/3.9	5.5/6.9
Latitude +15	Average	4.1	5.2	6.2	6.7	7.2	7.6	8.2	7.7	6.3	5.0	3.4	3.2	5.9
	Min/Max	2.9/5.1	3.2/6.8	4.1/8.1	4.8/8.8	4.3/8.9	6.4/9.2	7.2/9.4	6.4/9.4	4.5/8.0	3.8/6.7	2.1/4.4	2.2/4.2	5.3/6.7

Solar Radiation for 2-Axis Tracking Flat-Plate Collectors (kWh/m²/day), Uncertainty ±9%

Tracker		Jan	Feb	Mar	Apr	May	June	July	Aug	Sept	Oct	Nov	Dec	Year
2-Axis	Average	4.1	5.2	6.3	7.0	8.0	8.7	9.3	8.3	6.5	5.1	3.5	3.2	6.3
	Min/Max	2.9/5.1	3.2/6.8	4.2/8.1	5.1/9.2	4.8/9.8	7.3/10.4	8.1/10.6	6.9/10.0	4.7/8.2	3.8/6.7	2.1/4.4	2.2/4.2	5.6/7.1

Direct Beam Solar Radiation for Concentrating Collectors (kWh/m²/day), Uncertainty ±8%

Tracker		Jan	Feb	Mar	Apr	May	June	July	Aug	Sept	Oct	Nov	Dec	Year
1-Axis, E-W Horiz Axis	Average	2.3	2.6	2.8	3.1	3.7	4.0	4.5	4.1	3.2	2.8	1.9	1.8	3.1
	Min/Max	1.4/3.1	1.5/3.7	1.6/3.9	1.6/4.8	1.3/5.1	3.0/5.6	3.6/5.5	3.1/5.4	1.8/4.5	1.8/4.1	0.8/2.7	1.0/2.5	2.6/3.7
1-Axis, N-S Horiz Axis	Average	1.4	2.1	3.0	4.0	4.9	5.4	6.0	5.3	3.7	2.5	1.3	1.0	3.4
	Min/Max	0.9/2.0	1.1/3.0	1.6/4.1	1.9/6.2	1.8/6.7	3.9/7.4	4.8/7.4	4.0/7.2	2.1/5.2	1.6/3.8	0.5/1.9	0.6/1.5	2.8/4.1
1-Axis, N-S Tilt=Latitude	Average	2.4	3.1	3.7	4.4	4.9	5.1	5.9	5.6	4.4	3.5	2.1	1.9	3.9
	Min/Max	1.5/3.3	1.7/4.4	2.0/5.2	2.2/6.8	1.8/6.7	3.8/7.1	4.7/7.2	4.2/7.5	2.5/6.2	2.2/5.2	0.9/3.0	1.0/2.6	3.3/4.7
2-Axis	Average	2.6	3.2	3.7	4.5	5.2	5.6	6.3	5.8	4.4	3.5	2.2	2.0	4.1
	Min/Max	1.6/3.6	1.8/4.6	2.0/5.2	2.2/6.9	1.9/7.1	4.1/7.7	5.0/7.7	4.3/7.7	2.5/6.2	2.3/5.2	0.9/3.2	1.1/2.9	3.5/4.9

Average Climatic Conditions

Element	Jan	Feb	Mar	Apr	May	June	July	Aug	Sept	Oct	Nov	Dec	Year
Temperature (°C)	-14.5	-11.1	-3.4	6.1	13.4	18.6	21.7	20.4	14.3	7.6	-2.2	-11.3	5.0
Daily Minimum Temp	-19.8	-16.3	-8.2	0.1	6.6	12.0	14.9	13.6	7.7	1.4	-7.0	-16.1	-0.9
Daily Maximum Temp	-9.2	-6.1	1.4	12.1	20.3	25.2	28.6	27.4	20.8	13.7	2.7	-6.6	10.8
Record Minimum Temp	-37.2	-36.7	-30.6	-21.7	-6.7	-1.1	2.2	0.6	-7.2	-13.9	-31.1	-35.6	-37.2
Record Maximum Temp	11.1	18.9	25.6	37.8	36.7	37.2	41.1	41.1	38.9	33.9	23.3	13.9	41.1
HDD, Base 18.3°C	1018	824	673	367	171	52	11	27	133	332	615	919	5141
CDD, Base 18.3°C	0	0	0	0	19	60	116	92	11	0	0	0	298
Relative Humidity (%)	73	75	76	66	60	66	67	67	69	69	76	76	70
Wind Speed (m/s)	5.5	5.4	5.6	5.8	5.4	4.9	4.5	4.6	4.9	5.3	5.2	5.1	5.2

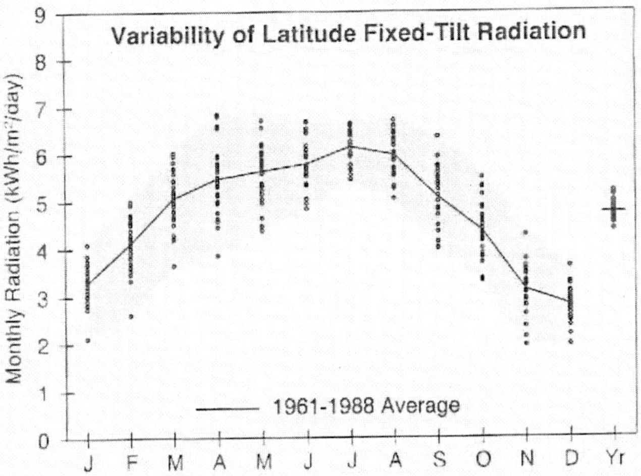

Minot, ND

WBAN NO. 24013

LATITUDE: 48.27° N
LONGITUDE: 101.28° W
ELEVATION: 522 meters
MEAN PRESSURE: 953 millibars

STATION TYPE: Secondary

Solar Radiation for Flat-Plate Collectors Facing South at a Fixed Tilt (kWh/m²/day), Uncertainty ±9%

Tilt (°)		Jan	Feb	Mar	Apr	May	June	July	Aug	Sept	Oct	Nov	Dec	Year
0	Average	1.5	2.4	3.6	4.9	5.8	6.4	6.6	5.6	4.0	2.7	1.6	1.2	3.9
	Min/Max	1.2/1.7	1.8/2.7	3.0/4.2	3.9/5.9	4.7/6.8	5.4/7.4	5.9/7.0	4.8/6.2	3.3/4.7	2.2/3.1	1.3/1.9	1.0/1.4	3.7/4.1
Latitude -15	Average	2.9	3.8	4.9	5.6	6.0	6.3	6.6	6.2	5.0	4.1	2.8	2.4	4.7
	Min/Max	1.9/3.5	2.5/4.5	3.6/5.7	4.0/6.9	4.7/7.2	5.3/7.3	5.9/7.2	5.3/7.0	4.0/6.2	3.2/5.1	1.8/3.8	1.7/3.1	4.4/5.1
Latitude	Average	3.3	4.1	5.1	5.5	5.6	5.8	6.1	6.0	5.1	4.4	3.1	2.8	4.7
	Min/Max	2.1/4.1	2.6/5.0	3.6/6.0	3.8/6.8	4.4/6.7	4.8/6.7	5.4/6.6	5.0/6.7	4.0/6.3	3.3/5.5	1.9/4.3	2.0/3.6	4.4/5.2
Latitude +15	Average	3.5	4.3	5.0	5.1	5.0	5.0	5.3	5.4	4.8	4.4	3.3	3.0	4.5
	Min/Max	2.2/4.4	2.6/5.2	3.6/6.1	3.5/6.3	3.9/5.9	4.2/5.7	4.7/5.8	4.6/6.1	3.8/6.1	3.3/5.6	2.0/4.5	2.1/3.9	4.1/4.9
90	Average	3.5	4.0	4.3	3.8	3.3	3.1	3.4	3.7	3.8	3.8	3.1	3.0	3.6
	Min/Max	2.1/4.4	2.4/4.9	3.1/5.4	2.5/4.7	2.6/3.8	2.7/3.5	3.0/3.6	3.2/4.2	2.9/4.8	2.8/4.9	1.8/4.3	2.0/4.0	3.2/3.9

Solar Radiation for 1-Axis Tracking Flat-Plate Collectors with a North-South Axis (kWh/m²/day), Uncertainty ±9%

Axis Tilt (°)		Jan	Feb	Mar	Apr	May	June	July	Aug	Sept	Oct	Nov	Dec	Year
0	Average	2.4	3.6	5.3	6.8	8.0	8.7	9.2	8.1	5.8	4.1	2.4	1.9	5.5
	Min/Max	1.6/3.0	2.2/4.3	3.7/6.2	4.8/8.9	5.8/9.9	7.0/10.6	8.0/10.1	6.6/9.4	4.4/7.3	3.0/5.1	1.6/3.3	1.4/2.5	5.1/6.1
Latitude -15	Average	3.4	4.6	6.2	7.4	8.3	8.9	9.5	8.6	6.6	5.1	3.3	2.8	6.2
	Min/Max	2.1/4.3	2.8/5.7	4.3/7.4	5.0/9.8	6.0/10.3	7.0/10.7	8.2/10.4	7.1/10.1	4.9/8.5	3.7/6.6	2.0/4.6	1.9/3.7	5.7/6.9
Latitude	Average	3.8	4.9	6.4	7.4	8.1	8.5	9.2	8.5	6.7	5.3	3.6	3.1	6.3
	Min/Max	2.3/4.8	2.9/6.1	4.3/7.6	4.9/9.8	5.8/10.1	6.7/10.4	7.9/10.1	6.9/10.0	4.9/8.6	3.9/6.9	2.1/5.1	2.1/4.2	5.7/7.0
Latitude +15	Average	4.0	5.0	6.3	7.1	7.6	8.0	8.6	8.1	6.5	5.4	3.7	3.3	6.1
	Min/Max	2.4/5.1	2.9/6.3	4.3/7.6	4.6/9.5	5.4/9.5	6.3/9.7	7.4/9.5	6.6/9.5	4.8/8.4	3.8/6.9	2.2/5.3	2.2/4.4	5.5/6.8

Solar Radiation for 2-Axis Tracking Flat-Plate Collectors (kWh/m²/day), Uncertainty ±9%

Tracker		Jan	Feb	Mar	Apr	May	June	July	Aug	Sept	Oct	Nov	Dec	Year
2-Axis	Average	4.0	5.0	6.4	7.5	8.4	9.1	9.7	8.7	6.7	5.4	3.7	3.4	6.5
	Min/Max	2.4/5.1	2.9/6.3	4.3/7.6	5.0/9.9	6.1/10.5	7.2/11.0	8.3/10.6	7.1/10.2	5.0/8.6	3.9/7.0	2.2/5.3	2.3/4.5	5.9/7.2

Direct Beam Solar Radiation for Concentrating Collectors (kWh/m²/day), Uncertainty ±8%

Tracker		Jan	Feb	Mar	Apr	May	June	July	Aug	Sept	Oct	Nov	Dec	Year
1-Axis, E-W Horiz Axis	Average	2.3	2.5	3.0	3.4	3.9	4.4	4.9	4.3	3.4	3.0	2.2	2.0	3.3
	Min/Max	1.2/3.2	1.1/3.8	1.4/4.1	1.5/5.2	2.2/5.4	2.7/6.0	3.9/5.7	3.1/5.4	2.1/4.8	1.8/4.3	0.9/3.6	1.2/2.8	2.8/3.9
1-Axis, N-S Horiz Axis	Average	1.4	2.0	3.1	4.3	5.2	5.9	6.5	5.7	3.9	2.7	1.5	1.1	3.6
	Min/Max	0.7/2.0	0.9/3.1	1.5/4.3	2.1/6.7	2.9/7.3	3.7/8.0	5.2/7.7	4.1/7.2	2.4/5.6	1.6/3.9	0.6/2.5	0.6/1.6	3.1/4.3
1-Axis, N-S Tilt=Latitude	Average	2.4	3.0	4.0	4.8	5.2	5.6	6.4	6.0	4.7	3.8	2.4	2.1	4.2
	Min/Max	1.2/3.5	1.3/4.6	1.9/5.5	2.3/7.3	3.0/7.3	3.5/7.7	5.1/7.5	4.4/7.6	2.8/6.7	2.3/5.4	1.0/4.1	1.2/3.0	3.6/5.0
2-Axis	Average	2.6	3.1	4.0	4.8	5.5	6.1	6.9	6.2	4.7	3.8	2.5	2.3	4.4
	Min/Max	1.3/3.7	1.4/4.7	1.9/5.5	2.3/7.4	3.1/7.8	3.8/8.4	5.4/8.1	4.5/7.8	2.8/6.7	2.3/5.5	1.0/4.3	1.3/3.2	3.7/5.2

Average Climatic Conditions

Element	Jan	Feb	Mar	Apr	May	June	July	Aug	Sept	Oct	Nov	Dec	Year
Temperature (°C)	-12.8	-9.6	-2.9	5.7	12.4	17.9	21.1	19.9	13.5	7.3	-2.3	-10.2	5.0
Daily Minimum Temp	-17.6	-14.3	-7.8	-0.3	6.0	11.6	14.3	12.9	7.2	1.4	-6.9	-14.8	-0.7
Daily Maximum Temp	-8.0	-4.8	1.9	11.6	18.8	24.2	27.8	26.9	19.8	13.2	2.3	-5.7	10.7
Record Minimum Temp	-36.1	-35.0	-33.9	-20.6	-7.2	0.6	5.6	1.1	-6.1	-14.4	-28.3	-37.8	-37.8
Record Maximum Temp	13.3	17.8	25.0	32.2	37.2	38.3	40.6	39.4	39.4	34.4	23.3	13.3	40.6
HDD, Base 18.3°C	964	781	658	380	194	69	11	47	159	341	618	885	5107
CDD, Base 18.3°C	0	0	0	0	12	56	95	97	14	0	0	0	273
Relative Humidity (%)	68	71	71	62	58	61	60	60	64	64	71	71	65
Wind Speed (m/s)	6.1	5.6	5.7	5.8	5.6	5.1	4.7	4.9	5.3	5.6	5.3	5.6	5.4

Grand Island, NE

WBAN NO. 14935

LATITUDE: 40.97° N
LONGITUDE: 98.32° W
ELEVATION: 566 meters
MEAN PRESSURE: 950 millibars

STATION TYPE: Secondary

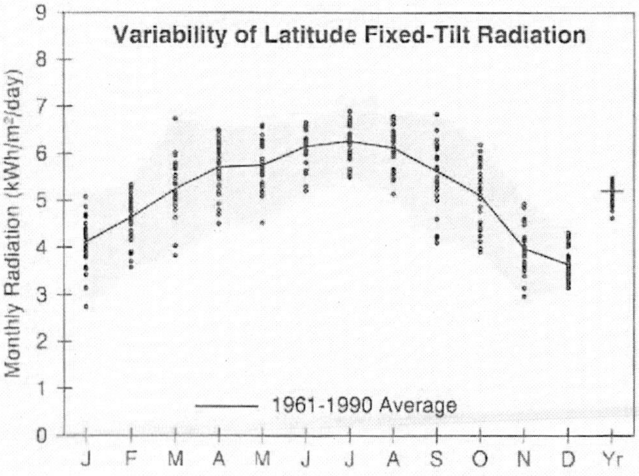

Solar Radiation for Flat-Plate Collectors Facing South at a Fixed Tilt (kWh/m²/day), Uncertainty ±9%

Tilt (°)		Jan	Feb	Mar	Apr	May	June	July	Aug	Sept	Oct	Nov	Dec	Year
0	Average	2.2	3.0	4.1	5.3	6.1	6.9	6.8	6.0	4.7	3.5	2.3	1.9	4.4
	Min/Max	1.7/2.5	2.6/3.3	3.3/5.0	4.4/6.0	4.8/6.9	5.9/7.5	6.0/7.5	5.1/6.6	3.7/5.5	2.9/4.1	2.0/2.7	1.8/2.1	4.1/4.6
Latitude -15	Average	3.6	4.2	5.0	5.8	6.1	6.7	6.8	6.4	5.5	4.7	3.5	3.1	5.1
	Min/Max	2.5/4.3	3.3/4.8	3.8/6.4	4.6/6.6	4.8/7.1	5.7/7.3	5.9/7.5	5.3/7.1	4.1/6.7	3.7/5.7	2.7/4.3	2.8/3.7	4.6/5.4
Latitude	Average	4.1	4.6	5.2	5.7	5.7	6.2	6.3	6.1	5.6	5.1	4.0	3.6	5.2
	Min/Max	2.7/5.1	3.6/5.3	3.8/6.7	4.5/6.5	4.5/6.6	5.2/6.7	5.5/6.9	5.1/6.8	4.1/6.8	3.9/6.2	3.0/4.9	3.2/4.3	4.6/5.5
Latitude +15	Average	4.4	4.8	5.2	5.3	5.1	5.3	5.5	5.6	5.4	5.2	4.2	3.9	5.0
	Min/Max	2.9/5.5	3.7/5.5	3.7/6.7	4.2/6.0	4.0/5.8	4.5/5.7	4.8/6.0	4.7/6.2	3.9/6.6	3.9/6.3	3.1/5.2	3.4/4.7	4.4/5.2
90	Average	4.2	4.3	4.0	3.4	2.9	2.7	2.8	3.3	3.8	4.2	3.8	3.8	3.6
	Min/Max	2.7/5.1	3.2/4.8	2.8/5.1	2.7/3.9	2.4/3.2	2.4/2.8	2.6/3.0	2.9/3.7	2.7/4.7	3.2/5.2	2.8/4.8	3.2/4.6	3.1/3.8

Solar Radiation for 1-Axis Tracking Flat-Plate Collectors with a North-South Axis (kWh/m²/day), Uncertainty ±9%

Axis Tilt (°)		Jan	Feb	Mar	Apr	May	June	July	Aug	Sept	Oct	Nov	Dec	Year
0	Average	3.4	4.4	5.8	7.3	8.0	9.2	9.2	8.2	6.7	5.2	3.5	2.9	6.1
	Min/Max	2.3/4.2	3.3/5.2	4.1/7.5	5.4/8.6	6.0/9.4	7.5/10.3	7.8/10.5	6.4/9.3	4.8/8.3	3.9/6.3	2.5/4.3	2.5/3.4	5.5/6.6
Latitude -15	Average	4.4	5.3	6.5	7.7	8.2	9.2	9.3	8.6	7.3	6.1	4.3	3.8	6.7
	Min/Max	2.8/5.5	3.8/6.3	4.6/8.6	5.6/9.1	6.1/9.6	7.5/10.3	7.8/10.6	6.7/9.8	5.1/9.2	4.5/7.5	3.1/5.4	3.3/4.5	5.9/7.2
Latitude	Average	4.9	5.6	6.6	7.6	7.9	8.8	9.0	8.5	7.4	6.4	4.7	4.2	6.8
	Min/Max	3.1/6.1	4.1/6.7	4.6/8.8	5.5/9.1	5.9/9.4	7.2/9.9	7.6/10.2	6.6/9.6	5.1/9.4	4.6/7.9	3.3/5.9	3.6/5.0	5.9/7.3
Latitude +15	Average	5.1	5.8	6.6	7.3	7.4	8.2	8.4	8.0	7.2	6.4	4.9	4.5	6.7
	Min/Max	3.2/6.4	4.1/6.8	4.5/8.7	5.3/8.7	5.5/8.8	6.7/9.3	7.1/9.6	6.2/9.2	5.0/9.2	4.7/8.0	3.4/6.2	3.8/5.4	5.8/7.1

Solar Radiation for 2-Axis Tracking Flat-Plate Collectors (kWh/m²/day), Uncertainty ±9%

Tracker		Jan	Feb	Mar	Apr	May	June	July	Aug	Sept	Oct	Nov	Dec	Year
2-Axis	Average	5.2	5.8	6.6	7.7	8.3	9.4	9.5	8.6	7.4	6.4	4.9	4.5	7.0
	Min/Max	3.2/6.5	4.2/6.8	4.6/8.8	5.6/9.2	6.1/9.8	7.6/10.5	8.0/10.8	6.8/9.8	5.2/9.4	4.7/8.0	3.5/6.2	3.9/5.5	6.1/7.5

Direct Beam Solar Radiation for Concentrating Collectors (kWh/m²/day), Uncertainty ±8%

Tracker		Jan	Feb	Mar	Apr	May	June	July	Aug	Sept	Oct	Nov	Dec	Year
1-Axis, E-W Horiz Axis	Average	3.2	3.2	3.3	3.7	3.9	4.8	4.9	4.4	3.8	3.7	2.9	2.8	3.7
	Min/Max	1.5/4.5	1.9/4.1	1.7/5.1	2.0/4.8	2.3/5.1	3.3/5.8	3.7/6.0	2.9/5.4	2.1/5.3	2.3/4.9	1.8/4.1	2.1/3.7	2.9/4.2
1-Axis, N-S Horiz Axis	Average	2.3	2.8	3.7	4.8	5.2	6.3	6.5	5.8	4.7	3.7	2.3	1.9	4.2
	Min/Max	1.0/3.3	1.5/3.8	1.8/5.7	2.6/6.4	3.1/6.8	4.4/7.7	4.9/8.0	3.8/7.2	2.6/6.5	2.3/4.9	1.3/3.1	1.4/2.5	3.4/4.7
1-Axis, N-S Tilt=Latitude	Average	3.5	3.8	4.4	5.1	5.1	6.0	6.3	6.0	5.3	4.7	3.3	3.0	4.7
	Min/Max	1.6/5.0	2.2/5.1	2.2/6.8	2.8/6.8	3.1/6.6	4.2/7.3	4.7/7.7	3.8/7.3	2.9/7.4	2.9/6.3	2.0/4.6	2.2/3.9	3.7/5.3
2-Axis	Average	3.7	3.9	4.4	5.2	5.4	6.5	6.7	6.1	5.3	4.7	3.5	3.3	4.9
	Min/Max	1.8/5.3	2.2/5.2	2.2/6.9	2.8/6.9	3.2/7.0	4.6/7.9	5.1/8.2	4.0/7.6	2.9/7.4	3.0/6.3	2.1/4.8	2.4/4.2	3.9/5.5

Average Climatic Conditions

Element	Jan	Feb	Mar	Apr	May	June	July	Aug	Sept	Oct	Nov	Dec	Year
Temperature (°C)	-5.6	-2.6	3.2	10.4	16.3	21.9	24.8	23.3	17.7	11.3	3.1	-3.7	10.0
Daily Minimum Temp	-11.6	-8.7	-3.2	3.4	9.7	15.2	18.2	16.7	10.9	4.0	-3.2	-9.5	3.5
Daily Maximum Temp	0.3	3.4	9.6	17.5	22.9	28.7	31.4	29.9	24.6	18.6	9.3	2.1	16.6
Record Minimum Temp	-33.3	-28.3	-29.4	-13.9	-5.0	3.3	5.6	4.4	-5.0	-8.9	-23.9	-32.2	-33.3
Record Maximum Temp	24.4	25.0	32.2	35.6	38.3	41.7	42.8	43.3	40.0	35.6	27.8	24.4	43.3
HDD, Base 18.3°C	742	587	470	239	95	10	0	7	59	219	457	682	3567
CDD, Base 18.3°C	0	0	0	0	33	118	202	160	41	0	0	0	554
Relative Humidity (%)	69	70	66	61	64	63	65	67	68	63	69	71	66
Wind Speed (m/s)	5.4	5.3	6.0	6.2	5.6	5.2	4.6	4.5	4.8	4.9	5.2	5.2	5.2

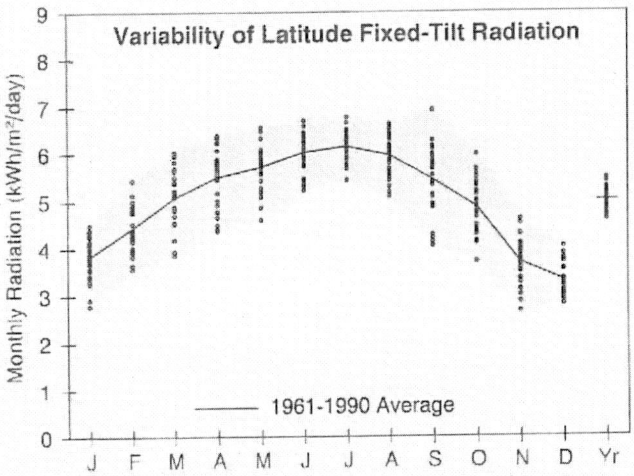

Norfolk, NE

WBAN NO. 14941

LATITUDE: 41.98° N
LONGITUDE: 97.43° W
ELEVATION: 471 meters
MEAN PRESSURE: 960 millibars

STATION TYPE: Secondary

Solar Radiation for Flat-Plate Collectors Facing South at a Fixed Tilt (kWh/m²/day), Uncertainty ±9%

Tilt (°)		Jan	Feb	Mar	Apr	May	June	July	Aug	Sept	Oct	Nov	Dec	Year
0	Average	2.1	2.9	4.0	5.1	6.0	6.7	6.7	5.8	4.5	3.3	2.2	1.7	4.3
	Min/Max	1.7/2.3	2.5/3.3	3.3/4.6	4.2/5.8	4.9/6.8	5.8/7.5	6.0/7.3	5.1/6.3	3.6/5.5	2.8/3.9	1.8/2.5	1.6/2.0	4.0/4.5
Latitude -15	Average	3.3	4.0	4.9	5.6	6.1	6.6	6.7	6.2	5.4	4.5	3.3	2.9	5.0
	Min/Max	2.5/3.9	3.3/4.9	3.8/5.8	4.5/6.5	4.9/7.0	5.7/7.3	5.9/7.3	5.3/6.9	4.1/6.7	3.5/5.5	2.4/4.0	2.5/3.4	4.6/5.4
Latitude	Average	3.8	4.4	5.1	5.5	5.7	6.0	6.2	6.0	5.4	4.9	3.7	3.3	5.0
	Min/Max	2.8/4.5	3.6/5.4	3.8/6.0	4.4/6.4	4.6/6.6	5.2/6.7	5.4/6.8	5.1/6.6	4.0/6.9	3.7/6.0	2.7/4.6	2.8/4.0	4.6/5.4
Latitude +15	Average	4.1	4.6	5.0	5.1	5.1	5.2	5.4	5.4	5.2	4.9	3.9	3.6	4.8
	Min/Max	3.0/4.9	3.7/5.7	3.7/5.9	4.1/5.9	4.1/5.8	4.5/5.8	4.7/5.9	4.6/6.0	3.8/6.7	3.7/6.1	2.7/4.9	3.0/4.3	4.3/5.2
90	Average	4.0	4.1	4.0	3.4	2.9	2.7	2.9	3.3	3.7	4.0	3.5	3.4	3.5
	Min/Max	2.8/4.8	3.2/5.1	2.8/4.8	2.8/3.9	2.5/3.2	2.5/2.9	2.6/3.1	2.9/3.7	2.7/4.8	3.0/5.1	2.4/4.5	2.9/4.2	3.1/3.8

Solar Radiation for 1-Axis Tracking Flat-Plate Collectors with a North-South Axis (kWh/m²/day), Uncertainty ±9%

Axis Tilt (°)		Jan	Feb	Mar	Apr	May	June	July	Aug	Sept	Oct	Nov	Dec	Year
0	Average	3.1	4.1	5.5	6.9	7.9	8.9	9.0	7.9	6.4	4.9	3.1	2.6	5.9
	Min/Max	2.3/3.7	3.2/5.2	4.2/6.6	5.3/8.3	6.1/9.3	7.3/10.2	7.9/10.2	6.6/8.9	4.6/8.3	3.8/6.0	2.3/4.0	2.3/3.1	5.4/6.5
Latitude -15	Average	4.1	5.0	6.2	7.3	8.1	8.9	9.1	8.3	7.0	5.8	4.0	3.4	6.5
	Min/Max	2.9/4.8	3.8/6.4	4.6/7.5	5.6/8.9	6.2/9.6	7.3/10.3	8.0/10.3	6.9/9.4	4.9/9.3	4.3/7.2	2.8/5.1	2.9/4.2	5.8/7.2
Latitude	Average	4.5	5.3	6.4	7.3	7.9	8.6	8.8	8.2	7.1	6.0	4.3	3.8	6.5
	Min/Max	3.1/5.3	4.1/6.8	4.6/7.7	5.5/8.8	6.0/9.3	7.0/9.9	7.7/9.9	6.7/9.3	4.9/9.4	4.4/7.6	3.0/5.5	3.2/4.7	5.8/7.2
Latitude +15	Average	4.7	5.4	6.3	7.0	7.4	8.0	8.2	7.8	6.9	6.0	4.5	4.0	6.4
	Min/Max	3.3/5.6	4.2/7.0	4.5/7.7	5.3/8.5	5.7/8.8	6.5/9.2	7.2/9.3	6.4/8.8	4.8/9.2	4.4/7.6	3.0/5.8	3.3/4.9	5.7/7.1

Solar Radiation for 2-Axis Tracking Flat-Plate Collectors (kWh/m²/day), Uncertainty ±9%

Tracker		Jan	Feb	Mar	Apr	May	June	July	Aug	Sept	Oct	Nov	Dec	Year
2-Axis	Average	4.8	5.4	6.4	7.4	8.2	9.2	9.3	8.4	7.1	6.1	4.5	4.1	6.7
	Min/Max	3.3/5.7	4.2/7.0	4.6/7.7	5.6/8.9	6.3/9.7	7.5/10.5	8.1/10.5	6.9/9.4	5.0/9.4	4.5/7.7	3.0/5.8	3.4/5.0	6.0/7.5

Direct Beam Solar Radiation for Concentrating Collectors (kWh/m²/day), Uncertainty ±8%

Tracker		Jan	Feb	Mar	Apr	May	June	July	Aug	Sept	Oct	Nov	Dec	Year
1-Axis, E-W Horiz Axis	Average	2.8	2.9	3.0	3.4	3.8	4.6	4.7	4.1	3.6	3.4	2.6	2.4	3.5
	Min/Max	1.4/3.7	1.8/4.2	1.4/4.3	1.8/4.6	2.5/5.0	3.3/5.9	3.8/5.7	3.0/5.1	2.0/5.4	2.1/4.7	1.5/3.7	1.7/3.4	2.9/4.1
1-Axis, N-S Horiz Axis	Average	2.0	2.5	3.4	4.4	5.1	6.0	6.2	5.5	4.3	3.3	2.0	1.6	3.9
	Min/Max	1.0/2.6	1.5/3.8	1.6/4.8	2.5/6.0	3.3/6.7	4.2/7.7	4.9/7.6	3.9/6.7	2.4/6.5	2.1/4.6	1.1/2.9	1.1/2.2	3.3/4.6
1-Axis, N-S Tilt=Latitude	Average	3.1	3.5	4.1	4.7	5.0	5.7	6.0	5.6	4.9	4.3	3.0	2.5	4.4
	Min/Max	1.5/4.1	2.1/5.1	1.9/5.7	2.6/6.4	3.3/6.6	4.0/7.3	4.7/7.3	4.0/6.9	2.7/7.4	2.7/5.9	1.6/4.2	1.8/3.6	3.6/5.2
2-Axis	Average	3.3	3.6	4.1	4.8	5.3	6.2	6.4	5.8	5.0	4.4	3.1	2.8	4.6
	Min/Max	1.6/4.3	2.2/5.2	1.9/5.8	2.7/6.5	3.4/6.9	4.3/7.9	5.1/7.8	4.1/7.1	2.8/7.5	2.7/6.0	1.7/4.4	1.9/3.9	3.8/5.4

Average Climatic Conditions

Element	Jan	Feb	Mar	Apr	May	June	July	Aug	Sept	Oct	Nov	Dec	Year
Temperature (°C)	-7.2	-4.2	2.1	9.8	16.1	21.3	24.1	22.6	17.3	10.8	2.3	-5.1	9.2
Daily Minimum Temp	-13.1	-9.9	-3.9	3.1	9.6	14.8	17.8	16.2	10.6	3.7	-3.6	-10.7	2.9
Daily Maximum Temp	-1.3	1.5	7.9	16.4	22.6	27.8	30.4	28.8	23.9	17.8	8.3	0.4	15.4
Record Minimum Temp	-32.8	-32.2	-28.9	-16.7	-4.4	3.3	5.6	4.4	-3.3	-10.6	-26.1	-34.4	-34.4
Record Maximum Temp	21.7	22.8	31.1	35.0	39.4	41.1	45.0	41.7	38.3	35.0	27.8	21.7	45.0
HDD, Base 18.3°C	792	632	504	255	102	12	0	9	69	237	480	727	3818
CDD, Base 18.3°C	0	0	0	0	31	101	179	139	37	0	0	0	487
Relative Humidity (%)	68	70	68	61	62	62	65	68	67	64	69	72	66
Wind Speed (m/s)	5.6	5.3	6.0	6.1	5.5	4.9	4.4	4.4	4.8	5.0	5.4	5.4	5.2

North Platte, NE

WBAN NO. 24023

LATITUDE: 41.13° N
LONGITUDE: 100.68° W
ELEVATION: 849 meters
MEAN PRESSURE: 917 millibars

STATION TYPE: Secondary

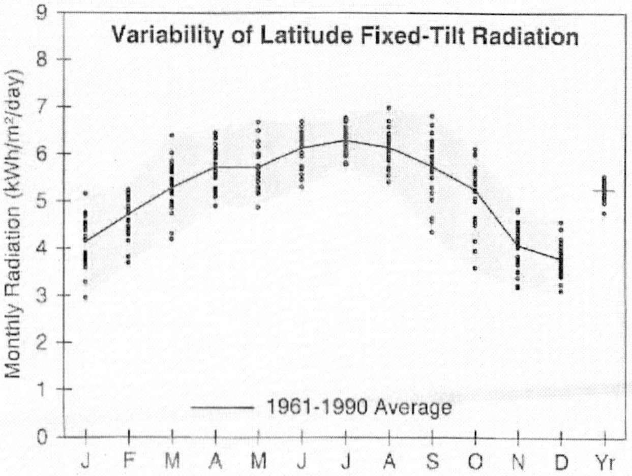

Solar Radiation for Flat-Plate Collectors Facing South at a Fixed Tilt (kWh/m²/day), Uncertainty ±9%

Tilt (°)		Jan	Feb	Mar	Apr	May	June	July	Aug	Sept	Oct	Nov	Dec	Year
0	Average	2.2	3.1	4.2	5.3	6.0	6.8	6.8	6.0	4.8	3.6	2.4	1.9	4.4
	Min/Max	1.8/2.5	2.6/3.3	3.5/4.8	4.7/5.9	5.3/7.0	5.9/7.5	6.2/7.3	5.3/6.7	3.9/5.5	2.7/4.0	2.0/2.6	1.8/2.2	4.2/4.6
Latitude -15	Average	3.6	4.3	5.1	5.8	6.1	6.7	6.8	6.4	5.7	4.9	3.6	3.3	5.2
	Min/Max	2.6/4.4	3.4/4.7	4.1/6.1	5.0/6.6	5.2/7.1	5.8/7.3	6.2/7.3	5.6/7.2	4.4/6.7	3.4/5.6	2.9/4.3	2.7/3.9	4.8/5.4
Latitude	Average	4.1	4.7	5.3	5.7	5.7	6.1	6.3	6.1	5.7	5.3	4.1	3.8	5.3
	Min/Max	3.0/5.2	3.7/5.2	4.2/6.4	4.9/6.4	4.9/6.7	5.3/6.7	5.8/6.8	5.4/7.0	4.3/6.8	3.6/6.1	3.2/4.8	3.1/4.6	4.8/5.5
Latitude +15	Average	4.4	4.9	5.2	5.3	5.1	5.3	5.5	5.6	5.5	5.3	4.3	4.1	5.0
	Min/Max	3.1/5.6	3.8/5.4	4.1/6.3	4.6/6.0	4.3/5.9	4.6/5.7	5.0/5.9	4.9/6.3	4.1/6.6	3.6/6.2	3.3/5.2	3.3/5.0	4.5/5.3
90	Average	4.2	4.3	4.0	3.5	2.8	2.7	2.9	3.3	3.9	4.4	3.9	3.9	3.7
	Min/Max	2.9/5.3	3.2/4.9	3.1/4.8	3.0/3.8	2.5/3.2	2.4/2.8	2.7/3.0	3.0/3.7	2.9/4.7	2.9/5.2	3.0/4.7	3.2/4.8	3.2/3.9

Solar Radiation for 1-Axis Tracking Flat-Plate Collectors with a North-South Axis (kWh/m²/day), Uncertainty ±9%

Axis Tilt (°)		Jan	Feb	Mar	Apr	May	June	July	Aug	Sept	Oct	Nov	Dec	Year
0	Average	3.4	4.5	5.8	7.3	7.9	9.1	9.2	8.2	6.8	5.3	3.5	3.0	6.2
	Min/Max	2.4/4.2	3.4/5.0	4.3/7.2	5.9/8.5	6.7/9.6	7.6/10.2	7.9/10.1	6.9/9.6	5.1/8.3	3.6/6.3	2.8/4.1	2.5/3.6	5.6/6.5
Latitude -15	Average	4.4	5.4	6.5	7.8	8.1	9.1	9.3	8.6	7.5	6.3	4.5	4.0	6.8
	Min/Max	3.0/5.6	4.1/6.1	4.8/8.2	6.2/9.1	6.7/9.8	7.6/10.3	8.0/10.2	7.2/10.1	5.5/9.2	4.1/7.5	3.4/5.3	3.2/4.8	6.0/7.2
Latitude	Average	4.9	5.7	6.7	7.7	7.8	8.8	9.0	8.5	7.5	6.6	4.8	4.4	6.9
	Min/Max	3.3/6.2	4.3/6.5	4.9/8.4	6.2/9.0	6.5/9.5	7.3/9.9	7.7/9.9	7.1/10.0	5.4/9.4	4.3/7.8	3.7/5.8	3.5/5.4	6.1/7.3
Latitude +15	Average	5.1	5.9	6.6	7.4	7.4	8.2	8.4	8.1	7.4	6.6	5.0	4.7	6.7
	Min/Max	3.4/6.5	4.3/6.7	4.8/8.3	5.9/8.6	6.1/9.0	6.8/9.2	7.2/9.3	6.7/9.5	5.3/9.2	4.3/7.9	3.8/6.1	3.7/5.7	5.9/7.1

Solar Radiation for 2-Axis Tracking Flat-Plate Collectors (kWh/m²/day), Uncertainty ±9%

Tracker		Jan	Feb	Mar	Apr	May	June	July	Aug	Sept	Oct	Nov	Dec	Year
2-Axis	Average	5.2	5.9	6.7	7.8	8.2	9.4	9.5	8.7	7.6	6.7	5.0	4.7	7.1
	Min/Max	3.5/6.6	4.4/6.7	4.9/8.4	6.3/9.1	6.8/9.9	7.7/10.5	8.2/10.4	7.3/10.2	5.5/9.4	4.3/7.9	3.8/6.1	3.8/5.8	6.3/7.5

Direct Beam Solar Radiation for Concentrating Collectors (kWh/m²/day), Uncertainty ±8%

Tracker		Jan	Feb	Mar	Apr	May	June	July	Aug	Sept	Oct	Nov	Dec	Year
1-Axis, E-W	Average	3.1	3.2	3.3	3.7	3.9	4.9	5.0	4.5	4.0	3.8	3.0	3.0	3.8
Horiz Axis	Min/Max	1.7/4.6	2.1/4.1	1.9/4.7	2.5/4.7	2.6/5.3	3.5/5.9	4.0/5.8	3.3/5.8	2.3/5.4	2.0/4.9	1.7/4.0	2.0/4.1	3.1/4.2
1-Axis, N-S	Average	2.2	2.9	3.7	4.9	5.2	6.4	6.6	5.9	4.8	3.8	2.3	2.0	4.2
Horiz Axis	Min/Max	1.1/3.3	1.8/3.7	2.0/5.4	3.2/6.3	3.5/7.1	4.5/7.8	5.1/7.7	4.3/7.6	2.8/6.7	1.9/4.9	1.3/3.0	1.3/2.8	3.4/4.6
1-Axis, N-S	Average	3.4	3.9	4.4	5.2	5.1	6.0	6.3	6.0	5.5	4.9	3.4	3.2	4.8
Tilt=Latitude	Min/Max	1.8/5.0	2.5/5.0	2.5/6.4	3.4/6.7	3.5/6.9	4.3/7.3	4.9/7.3	4.4/7.8	3.2/7.5	2.5/6.2	1.9/4.5	2.1/4.4	3.8/5.2
2-Axis	Average	3.7	4.0	4.4	5.3	5.4	6.6	6.8	6.2	5.5	5.0	3.6	3.4	5.0
	Min/Max	2.0/5.4	2.6/5.2	2.5/6.4	3.5/6.8	3.7/7.3	4.7/8.0	5.3/7.9	4.5/8.0	3.2/7.6	2.5/6.3	2.0/4.7	2.3/4.8	4.0/5.5

Average Climatic Conditions

Element	Jan	Feb	Mar	Apr	May	June	July	Aug	Sept	Oct	Nov	Dec	Year
Temperature (°C)	-5.8	-2.4	2.5	9.0	14.6	19.9	23.3	22.1	16.3	9.8	1.9	-4.3	8.9
Daily Minimum Temp	-13.0	-9.8	-5.0	1.1	7.0	12.3	15.6	14.2	7.9	0.8	-5.9	-11.6	1.1
Daily Maximum Temp	1.4	4.9	9.9	16.9	22.1	27.6	31.0	30.0	24.8	18.7	9.7	2.9	16.7
Record Minimum Temp	-30.6	-30.0	-30.0	-13.9	-7.2	-1.7	4.4	1.7	-8.3	-12.2	-25.0	-36.7	-36.7
Record Maximum Temp	22.8	26.1	30.0	34.4	36.1	41.7	44.4	40.6	38.9	34.4	27.8	23.9	44.4
HDD, Base 18.3°C	747	582	491	280	130	27	0	6	88	263	493	703	3811
CDD, Base 18.3°C	0	0	0	0	13	76	157	123	28	0	0	0	396
Relative Humidity (%)	69	68	65	60	63	64	63	64	64	62	67	69	65
Wind Speed (m/s)	4.2	4.3	5.2	5.6	5.1	4.5	4.3	4.1	4.3	4.2	4.2	4.1	4.5

Omaha, NE

WBAN NO. 94918

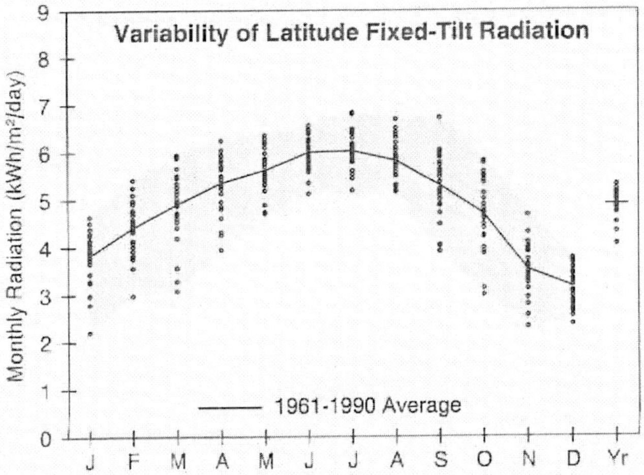

LATITUDE: 41.37° N
LONGITUDE: 96.52° W
ELEVATION: 404 meters
MEAN PRESSURE: 976 millibars

STATION TYPE: Primary

Solar Radiation for Flat-Plate Collectors Facing South at a Fixed Tilt (kWh/m²/day), Uncertainty ±9%

Tilt (°)		Jan	Feb	Mar	Apr	May	June	July	Aug	Sept	Oct	Nov	Dec	Year
0	Average	2.1	2.9	3.9	5.0	5.9	6.7	6.6	5.7	4.5	3.3	2.1	1.7	4.2
	Min/Max	1.6/2.4	2.3/3.3	2.7/4.6	3.9/5.7	5.0/6.7	5.7/7.4	5.7/7.4	5.1/6.5	3.6/5.4	2.3/3.9	1.6/2.5	1.5/1.9	3.7/4.4
Latitude -15	Average	3.3	4.0	4.7	5.5	6.0	6.5	6.5	6.1	5.2	4.4	3.2	2.7	4.9
	Min/Max	2.0/4.0	2.8/4.9	3.0/5.7	4.1/6.3	5.0/6.8	5.6/7.2	5.6/7.4	5.4/6.9	4.0/6.6	2.9/5.4	2.1/4.1	2.1/3.2	4.1/5.2
Latitude	Average	3.8	4.4	4.9	5.3	5.6	6.0	6.0	5.8	5.3	4.7	3.5	3.2	4.9
	Min/Max	2.2/4.6	3.0/5.4	3.1/5.9	3.9/6.3	4.7/6.4	5.1/6.6	5.2/6.8	5.2/6.7	3.9/6.7	3.0/5.8	2.3/4.7	2.4/3.8	4.0/5.3
Latitude +15	Average	4.1	4.6	4.8	5.0	5.0	5.2	5.3	5.3	5.1	4.7	3.7	3.4	4.7
	Min/Max	2.3/5.0	3.0/5.6	3.0/5.8	3.6/5.8	4.2/5.6	4.5/5.7	4.6/6.0	4.7/6.1	3.7/6.5	3.0/5.9	2.4/5.0	2.5/4.1	3.8/5.1
90	Average	3.9	4.1	3.8	3.2	2.8	2.7	2.8	3.2	3.6	3.9	3.3	3.2	3.4
	Min/Max	2.1/4.7	2.6/4.9	2.3/4.7	2.5/3.9	2.4/3.1	2.4/2.9	2.5/3.1	2.9/3.7	2.6/4.6	2.3/4.8	2.1/4.5	2.3/4.0	2.7/3.7

Solar Radiation for 1-Axis Tracking Flat-Plate Collectors with a North-South Axis (kWh/m²/day), Uncertainty ±9%

Axis Tilt (°)		Jan	Feb	Mar	Apr	May	June	July	Aug	Sept	Oct	Nov	Dec	Year
0	Average	3.1	4.1	5.3	6.6	7.6	8.7	8.5	7.6	6.1	4.7	3.0	2.5	5.7
	Min/Max	1.8/3.8	2.7/5.0	3.1/6.6	4.5/8.1	6.1/9.0	6.9/10.0	7.2/10.0	6.5/8.7	4.5/7.9	2.9/5.8	2.0/4.0	1.9/3.0	4.6/6.2
Latitude -15	Average	4.1	5.0	6.0	7.0	7.8	8.7	8.6	8.0	6.7	5.5	3.8	3.3	6.2
	Min/Max	2.2/5.0	3.2/6.2	3.4/7.5	4.7/8.7	6.2/9.2	6.9/10.1	7.2/10.1	6.8/9.2	4.8/8.8	3.3/6.9	2.4/5.2	2.4/4.0	4.9/6.8
Latitude	Average	4.5	5.3	6.1	7.0	7.5	8.3	8.3	7.8	6.7	5.7	4.1	3.6	6.3
	Min/Max	2.4/5.5	3.3/6.6	3.4/7.7	4.6/8.6	5.9/8.9	6.6/9.7	6.9/9.8	6.7/9.0	4.8/8.9	3.4/7.3	2.6/5.6	2.6/4.4	4.9/6.9
Latitude +15	Average	4.7	5.4	6.0	6.7	7.1	7.8	7.8	7.4	6.6	5.8	4.2	3.8	6.1
	Min/Max	2.5/5.8	3.3/6.8	3.3/7.6	4.4/8.3	5.6/8.4	6.1/9.0	6.5/9.2	6.3/8.6	4.6/8.7	3.4/7.3	2.7/5.8	2.7/4.7	4.7/6.7

Solar Radiation for 2-Axis Tracking Flat-Plate Collectors (kWh/m²/day), Uncertainty ±9%

Tracker		Jan	Feb	Mar	Apr	May	June	July	Aug	Sept	Oct	Nov	Dec	Year
2-Axis	Average	4.7	5.4	6.1	7.1	7.9	8.9	8.8	8.0	6.8	5.8	4.3	3.9	6.5
	Min/Max	2.5/5.9	3.3/6.8	3.4/7.7	4.7/8.7	6.2/9.3	7.0/10.3	7.3/10.3	6.8/9.2	4.8/8.9	3.5/7.3	2.7/5.9	2.8/4.7	5.1/7.1

Direct Beam Solar Radiation for Concentrating Collectors (kWh/m²/day), Uncertainty ±8%

Tracker		Jan	Feb	Mar	Apr	May	June	July	Aug	Sept	Oct	Nov	Dec	Year
1-Axis, E-W Horiz Axis	Average	2.8	2.9	3.0	3.3	3.6	4.2	4.2	3.8	3.4	3.2	2.5	2.3	3.3
	Min/Max	0.8/3.9	1.2/4.3	1.0/4.2	1.6/4.3	2.1/4.8	2.9/5.6	3.0/5.4	2.8/4.7	1.8/5.1	1.5/4.4	1.1/3.8	1.4/3.1	2.2/3.7
1-Axis, N-S Horiz Axis	Average	2.0	2.5	3.3	4.2	4.7	5.5	5.5	4.9	4.0	3.1	1.9	1.5	3.6
	Min/Max	0.6/2.8	1.0/3.8	1.0/4.7	1.9/5.7	2.8/6.3	3.7/7.4	4.0/7.1	3.7/6.2	2.3/6.0	1.5/4.3	0.8/2.9	0.9/2.0	2.4/4.1
1-Axis, N-S Tilt=Latitude	Average	3.0	3.5	3.9	4.4	4.6	5.2	5.3	5.1	4.5	4.0	2.8	2.4	4.1
	Min/Max	0.9/4.3	1.4/5.2	1.2/5.6	2.0/6.1	2.8/6.2	3.5/7.0	3.8/6.8	3.8/6.4	2.5/6.8	1.9/5.5	1.2/4.3	1.5/3.3	2.7/4.7
2-Axis	Average	3.3	3.6	3.9	4.5	4.9	5.7	5.7	5.2	4.6	4.1	2.9	2.6	4.2
	Min/Max	0.9/4.6	1.4/5.3	1.2/5.6	2.1/6.2	2.9/6.6	3.8/7.6	4.1/7.3	3.9/6.6	2.5/6.9	1.9/5.5	1.3/4.5	1.6/3.5	2.8/4.9

Average Climatic Conditions

Element	Jan	Feb	Mar	Apr	May	June	July	Aug	Sept	Oct	Nov	Dec	Year
Temperature (°C)	-6.4	-3.4	3.2	10.8	16.9	22.2	24.7	23.3	18.2	12.0	3.7	-4.2	10.1
Daily Minimum Temp	-11.6	-8.6	-2.3	4.6	11.0	16.3	19.2	17.7	12.6	6.1	-1.3	-8.9	4.6
Daily Maximum Temp	-1.3	1.7	8.7	16.9	22.7	28.0	30.3	28.9	23.8	17.8	8.7	0.5	15.6
Record Minimum Temp	-30.6	-29.4	-26.7	-15.0	-2.8	3.3	6.7	6.1	-3.9	-10.6	-22.8	-30.6	-30.6
Record Maximum Temp	20.6	25.6	31.7	36.1	37.2	40.6	45.6	43.3	40.0	35.6	26.7	22.2	45.6
HDD, Base 18.3°C	767	608	470	230	86	6	0	6	50	202	438	699	3563
CDD, Base 18.3°C	0	0	0	3	42	121	198	159	47	6	0	0	576
Relative Humidity (%)	71	71	67	61	64	65	68	70	71	67	71	73	68
Wind Speed (m/s)	4.6	4.6	5.1	5.2	4.4	4.0	3.7	3.6	3.9	4.1	4.5	4.5	4.3

Scottsbluff, NE

WBAN NO. 24028

LATITUDE: 41.87° N
LONGITUDE: 103.60° W
ELEVATION: 1206 meters
MEAN PRESSURE: 879 millibars

STATION TYPE: Secondary

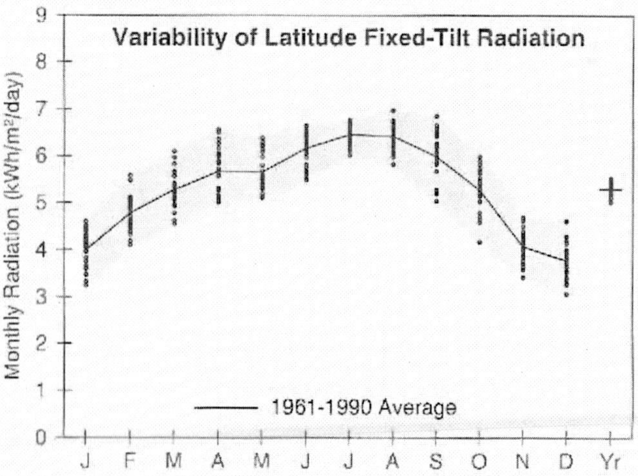

Solar Radiation for Flat-Plate Collectors Facing South at a Fixed Tilt (kWh/m²/day), Uncertainty ±9%

Tilt (°)		Jan	Feb	Mar	Apr	May	June	July	Aug	Sept	Oct	Nov	Dec	Year
0	Average	2.1	3.0	4.1	5.3	6.0	6.9	7.0	6.2	4.9	3.5	2.3	1.9	4.4
	Min/Max	1.9/2.3	2.7/3.3	3.7/4.5	4.7/5.9	5.4/6.7	6.1/7.4	6.5/7.3	5.6/6.7	4.3/5.5	3.0/3.9	2.1/2.5	1.6/2.1	4.3/4.6
Latitude -15	Average	3.5	4.3	5.1	5.8	6.0	6.7	7.0	6.7	5.9	4.9	3.6	3.2	5.2
	Min/Max	2.9/4.0	3.8/5.0	4.4/5.8	5.1/6.7	5.4/6.8	5.9/7.2	6.5/7.3	6.0/7.2	5.0/6.7	4.0/5.5	3.1/4.1	2.7/3.9	5.0/5.4
Latitude	Average	4.0	4.8	5.3	5.7	5.7	6.2	6.5	6.4	6.0	5.3	4.1	3.8	5.3
	Min/Max	3.2/4.6	4.1/5.6	4.5/6.1	5.0/6.6	5.1/6.4	5.5/6.6	6.0/6.8	5.8/7.0	5.0/6.8	4.2/6.0	3.4/4.7	3.1/4.6	5.0/5.5
Latitude +15	Average	4.3	4.9	5.2	5.3	5.0	5.3	5.6	5.8	5.7	5.3	4.3	4.1	5.1
	Min/Max	3.4/5.0	4.2/5.8	4.4/6.0	4.6/6.1	4.5/5.6	4.7/5.7	5.2/5.9	5.3/6.3	4.8/6.6	4.2/6.1	3.5/5.0	3.3/5.1	4.8/5.3
90	Average	4.1	4.4	4.1	3.5	2.9	2.7	3.0	3.5	4.1	4.4	3.9	3.9	3.7
	Min/Max	3.1/4.8	3.6/5.2	3.5/4.7	3.1/4.1	2.6/3.1	2.6/2.9	2.8/3.1	3.3/3.8	3.4/4.7	3.4/5.1	3.2/4.6	3.1/5.0	3.5/3.9

Solar Radiation for 1-Axis Tracking Flat-Plate Collectors with a North-South Axis (kWh/m²/day), Uncertainty ±9%

Axis Tilt (°)		Jan	Feb	Mar	Apr	May	June	July	Aug	Sept	Oct	Nov	Dec	Year
0	Average	3.3	4.5	5.8	7.2	7.9	9.2	9.5	8.6	7.1	5.3	3.5	2.9	6.2
	Min/Max	2.5/3.8	3.8/5.3	4.8/6.7	6.1/8.5	6.8/9.1	7.6/10.4	8.7/10.1	7.4/9.6	5.6/8.4	4.1/6.1	3.0/4.0	2.4/3.6	5.9/6.5
Latitude -15	Average	4.3	5.5	6.5	7.6	8.1	9.2	9.6	9.1	7.9	6.3	4.4	3.9	6.9
	Min/Max	3.3/5.0	4.7/6.5	5.4/7.7	6.5/9.1	6.9/9.3	7.6/10.4	8.8/10.2	7.8/10.1	6.2/9.3	4.8/7.3	3.7/5.2	3.1/4.9	6.5/7.2
Latitude	Average	4.7	5.8	6.7	7.6	7.8	8.9	9.2	8.9	7.9	6.6	4.8	4.4	6.9
	Min/Max	3.6/5.5	4.9/7.0	5.6/7.9	6.4/9.1	6.7/9.0	7.3/10.0	8.4/9.9	7.6/10.0	6.3/9.5	5.0/7.7	4.0/5.6	3.5/5.5	6.5/7.3
Latitude +15	Average	4.9	5.9	6.6	7.3	7.4	8.3	8.7	8.5	7.8	6.7	5.0	4.6	6.8
	Min/Max	3.8/5.8	5.0/7.2	5.5/7.8	6.2/8.8	6.3/8.5	6.8/9.3	7.9/9.3	7.2/9.5	6.1/9.3	5.0/7.8	4.1/5.8	3.7/5.8	6.4/7.2

Solar Radiation for 2-Axis Tracking Flat-Plate Collectors (kWh/m²/day), Uncertainty ±9%

Tracker		Jan	Feb	Mar	Apr	May	June	July	Aug	Sept	Oct	Nov	Dec	Year
2-Axis	Average	5.0	5.9	6.7	7.7	8.2	9.4	9.8	9.1	8.0	6.7	5.0	4.7	7.2
	Min/Max	3.8/5.9	5.0/7.2	5.6/7.9	6.5/9.2	7.0/9.4	7.8/10.6	8.9/10.4	7.8/10.2	6.3/9.5	5.0/7.8	4.2/5.9	3.7/5.9	6.7/7.6

Direct Beam Solar Radiation for Concentrating Collectors (kWh/m²/day), Uncertainty ±8%

Tracker		Jan	Feb	Mar	Apr	May	June	July	Aug	Sept	Oct	Nov	Dec	Year
1-Axis, E-W Horiz Axis	Average	3.0	3.3	3.3	3.6	3.9	5.0	5.4	4.9	4.3	3.9	3.0	3.0	3.9
	Min/Max	2.2/4.0	2.7/4.4	2.4/4.5	2.8/4.9	2.9/4.9	3.7/5.9	4.7/5.9	4.0/5.8	3.0/5.5	2.4/4.8	2.2/3.9	2.2/4.2	3.5/4.2
1-Axis, N-S Horiz Axis	Average	2.1	3.0	3.7	4.7	5.2	6.6	7.0	6.4	5.2	3.8	2.3	1.9	4.3
	Min/Max	1.4/2.9	2.3/4.0	2.6/5.0	3.5/6.4	3.8/6.5	4.7/7.8	6.1/7.8	5.0/7.6	3.5/6.8	2.4/4.9	1.7/3.0	1.4/2.8	3.9/4.7
1-Axis, N-S Tilt=Latitude	Average	3.3	4.0	4.4	5.0	5.1	6.2	6.7	6.6	6.0	5.0	3.4	3.1	4.9
	Min/Max	2.3/4.4	3.2/5.4	3.2/6.0	3.8/6.9	3.8/6.4	4.5/7.4	5.8/7.5	5.2/7.9	4.0/7.7	3.1/6.2	2.5/4.4	2.3/4.4	4.4/5.3
2-Axis	Average	3.5	4.2	4.4	5.1	5.4	6.8	7.2	6.8	6.0	5.0	3.6	3.4	5.1
	Min/Max	2.5/4.7	3.3/5.6	3.2/6.0	3.9/7.0	4.0/6.8	4.9/8.0	6.3/8.1	5.3/8.1	4.0/7.8	3.1/6.3	2.6/4.6	2.5/4.8	4.6/5.6

Average Climatic Conditions

Element	Jan	Feb	Mar	Apr	May	June	July	Aug	Sept	Oct	Nov	Dec	Year
Temperature (°C)	-3.9	-1.0	2.3	8.1	13.6	19.6	23.4	22.0	16.3	9.8	2.4	-3.2	9.1
Daily Minimum Temp	-11.2	-8.4	-5.5	-0.3	5.4	11.5	14.8	13.3	7.6	0.9	-5.3	-10.6	1.0
Daily Maximum Temp	3.3	6.4	10.2	16.3	21.6	27.8	32.1	30.7	25.1	18.6	10.1	4.2	17.2
Record Minimum Temp	-35.6	-33.3	-32.8	-22.2	-9.4	-1.1	4.4	3.9	-7.2	-21.1	-25.0	-41.1	-41.1
Record Maximum Temp	23.3	25.0	30.6	33.3	35.0	41.1	42.8	40.0	38.3	33.3	26.7	25.0	42.8
HDD, Base 18.3°C	691	541	496	308	154	34	0	5	97	265	478	668	3738
CDD, Base 18.3°C	0	0	0	0	7	72	161	119	37	0	0	0	396
Relative Humidity (%)	64	63	60	56	58	57	56	58	57	56	62	65	59
Wind Speed (m/s)	5.1	5.1	5.5	5.7	5.3	4.7	4.2	4.0	4.2	4.3	4.5	4.7	4.7

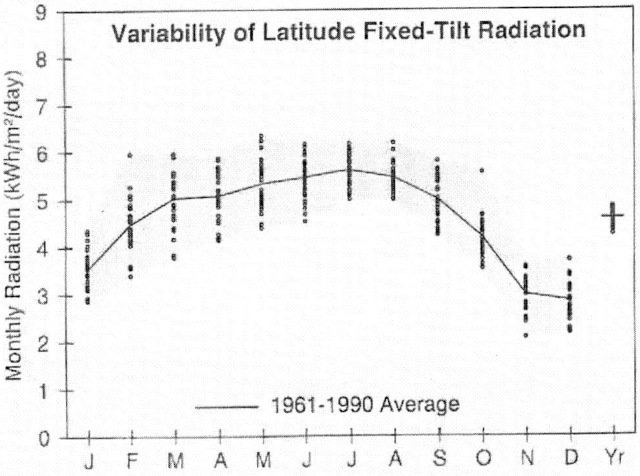

Concord, NH

WBAN NO. 14745

LATITUDE: 43.20° N
LONGITUDE: 71.50° W
ELEVATION: 105 meters
MEAN PRESSURE: 1003 millibars

STATION TYPE: Secondary

Solar Radiation for Flat-Plate Collectors Facing South at a Fixed Tilt (kWh/m²/day), Uncertainty ±11%

Tilt (°)		Jan	Feb	Mar	Apr	May	June	July	Aug	Sept	Oct	Nov	Dec	Year
0	Average	1.9	2.8	3.9	4.7	5.6	6.1	6.1	5.3	4.2	2.9	1.8	1.5	3.9
	Min/Max	1.6/2.2	2.3/3.4	3.2/4.5	4.0/5.3	4.7/6.6	5.1/6.9	5.5/6.7	4.9/5.9	3.7/4.7	2.5/3.6	1.4/2.0	1.3/1.7	3.7/4.1
Latitude -15	Average	3.1	4.1	4.8	5.2	5.7	5.9	6.0	5.7	5.0	3.9	2.7	2.5	4.6
	Min/Max	2.6/3.8	3.1/5.3	3.7/5.7	4.3/6.0	4.7/6.8	4.9/6.7	5.4/6.6	5.2/6.4	4.3/5.7	3.3/5.2	1.9/3.2	1.9/3.2	4.2/4.8
Latitude	Average	3.5	4.5	5.0	5.1	5.3	5.4	5.6	5.5	5.0	4.2	3.0	2.8	4.6
	Min/Max	2.9/4.3	3.4/6.0	3.8/6.0	4.1/5.9	4.4/6.3	4.5/6.2	5.0/6.2	5.0/6.2	4.2/5.8	3.5/5.6	2.1/3.6	2.2/3.7	4.2/4.8
Latitude +15	Average	3.8	4.7	5.0	4.7	4.7	4.7	4.9	4.9	4.8	4.2	3.1	3.1	4.4
	Min/Max	3.0/4.7	3.5/6.3	3.7/6.0	3.8/5.4	3.9/5.6	4.0/5.3	4.4/5.4	4.5/5.6	4.0/5.6	3.5/5.7	2.1/3.8	2.3/4.1	4.0/4.6
90	Average	3.7	4.3	4.1	3.2	2.8	2.6	2.8	3.1	3.5	3.5	2.8	3.0	3.3
	Min/Max	2.9/4.6	3.3/5.9	2.9/5.2	2.6/3.8	2.4/3.3	2.3/2.9	2.5/3.0	2.9/3.6	2.9/4.1	2.9/4.7	1.8/3.4	2.1/4.0	3.0/3.5

Solar Radiation for 1-Axis Tracking Flat-Plate Collectors with a North-South Axis (kWh/m²/day), Uncertainty ±11%

Axis Tilt (°)		Jan	Feb	Mar	Apr	May	June	July	Aug	Sept	Oct	Nov	Dec	Year
0	Average	2.8	4.1	5.4	6.3	7.2	7.9	7.9	7.1	5.6	4.1	2.5	2.2	5.3
	Min/Max	2.2/3.5	3.2/5.5	4.0/6.6	4.9/7.4	5.5/8.9	6.0/9.4	6.8/8.9	6.3/8.1	4.7/6.7	3.5/5.5	1.7/3.0	1.6/2.8	4.8/5.6
Latitude -15	Average	3.7	5.0	6.1	6.7	7.4	7.9	7.9	7.4	6.3	4.8	3.2	2.9	5.8
	Min/Max	3.0/4.6	3.8/6.9	4.4/7.4	5.1/8.0	5.6/9.2	6.0/9.4	6.9/9.0	6.6/8.5	5.2/7.5	4.1/6.7	2.1/3.8	2.2/3.9	5.3/6.2
Latitude	Average	4.1	5.4	6.3	6.6	7.1	7.5	7.7	7.3	6.3	5.0	3.4	3.2	5.8
	Min/Max	3.3/5.1	4.0/7.4	4.5/7.6	5.0/7.9	5.4/8.9	5.7/9.1	6.7/8.7	6.4/8.4	5.2/7.5	4.2/7.0	2.3/4.2	2.4/4.3	5.3/6.2
Latitude +15	Average	4.3	5.5	6.2	6.3	6.7	7.0	7.2	6.9	6.1	5.1	3.5	3.4	5.7
	Min/Max	3.4/5.4	4.1/7.6	4.4/7.6	4.8/7.6	5.0/8.4	5.3/8.5	6.2/8.1	6.1/8.0	5.0/7.4	4.2/7.0	2.3/4.3	2.5/4.6	5.2/6.1

Solar Radiation for 2-Axis Tracking Flat-Plate Collectors (kWh/m²/day), Uncertainty ±11%

Tracker		Jan	Feb	Mar	Apr	May	June	July	Aug	Sept	Oct	Nov	Dec	Year
2-Axis	Average	4.3	5.5	6.3	6.7	7.5	8.1	8.1	7.4	6.3	5.1	3.5	3.5	6.0
	Min/Max	3.4/5.4	4.1/7.6	4.5/7.6	5.1/8.0	5.7/9.3	6.2/9.7	7.1/9.2	6.6/8.6	5.2/7.6	4.2/7.0	2.3/4.3	2.5/4.7	5.5/6.4

Direct Beam Solar Radiation for Concentrating Collectors (kWh/m²/day), Uncertainty ±11%

Tracker		Jan	Feb	Mar	Apr	May	June	July	Aug	Sept	Oct	Nov	Dec	Year
1-Axis, E-W Horiz Axis	Average	2.4	2.7	2.8	3.0	3.3	3.6	3.7	3.4	3.0	2.7	1.9	1.9	2.9
	Min/Max	1.6/3.2	1.8/4.4	1.5/3.7	1.9/3.9	1.9/4.7	2.1/4.8	3.0/4.5	2.7/4.2	2.2/4.0	2.0/4.3	1.1/2.6	0.9/2.9	2.4/3.2
1-Axis, N-S Horiz Axis	Average	1.6	2.3	3.1	3.8	4.4	4.8	4.9	4.4	3.5	2.5	1.4	1.2	3.2
	Min/Max	1.1/2.1	1.6/3.9	1.7/4.0	2.4/5.1	2.5/6.2	2.8/6.5	3.9/5.9	3.5/5.5	2.6/4.7	1.9/4.1	0.8/1.9	0.6/1.8	2.6/3.6
1-Axis, N-S Tilt=Latitude	Average	2.6	3.3	3.8	4.1	4.3	4.6	4.7	4.6	4.1	3.4	2.2	2.0	3.6
	Min/Max	1.7/3.4	2.2/5.4	2.1/4.8	2.6/5.4	2.5/6.1	2.6/6.2	3.8/5.7	3.7/5.7	3.0/5.4	2.5/5.4	1.2/2.9	1.0/3.0	3.0/4.0
2-Axis	Average	2.8	3.4	3.8	4.2	4.6	5.0	5.1	4.7	4.1	3.4	2.3	2.2	3.8
	Min/Max	1.8/3.6	2.3/5.5	2.1/4.8	2.6/5.5	2.6/6.5	2.9/6.7	4.1/6.1	3.8/5.9	3.0/5.4	2.6/5.4	1.3/3.1	1.1/3.3	3.1/4.2

Average Climatic Conditions

Element	Jan	Feb	Mar	Apr	May	June	July	Aug	Sept	Oct	Nov	Dec	Year
Temperature (°C)	-7.4	-5.7	0.2	6.6	12.9	17.9	20.8	19.6	14.9	8.8	2.8	-4.3	7.3
Daily Minimum Temp	-13.7	-12.0	-5.5	-0.3	5.2	10.7	13.6	12.6	7.8	1.6	-2.8	-9.8	0.6
Daily Maximum Temp	-1.2	0.6	6.0	13.5	20.5	25.2	28.0	26.6	22.0	15.9	8.4	1.2	13.9
Record Minimum Temp	-36.1	-38.3	-26.7	-13.3	-6.1	-1.1	1.7	-1.7	-6.1	-12.2	-20.6	-30.0	-38.3
Record Maximum Temp	20.0	18.9	29.4	35.0	36.1	36.7	38.9	38.3	36.7	32.2	26.7	20.0	38.9
HDD, Base 18.3°C	799	672	562	352	173	39	7	22	109	296	465	701	4197
CDD, Base 18.3°C	0	0	0	0	4	26	85	62	6	0	0	0	182
Relative Humidity (%)	68	66	65	62	65	71	72	75	76	73	73	72	70
Wind Speed (m/s)	3.0	3.2	3.5	3.5	3.0	2.9	2.5	2.4	2.4	2.5	3.0	3.0	3.0

Atlantic City, NJ

WBAN NO. 93730

LATITUDE: 39.45° N
LONGITUDE: 74.57° W
ELEVATION: 20 meters
MEAN PRESSURE: 1015 millibars

STATION TYPE: Secondary

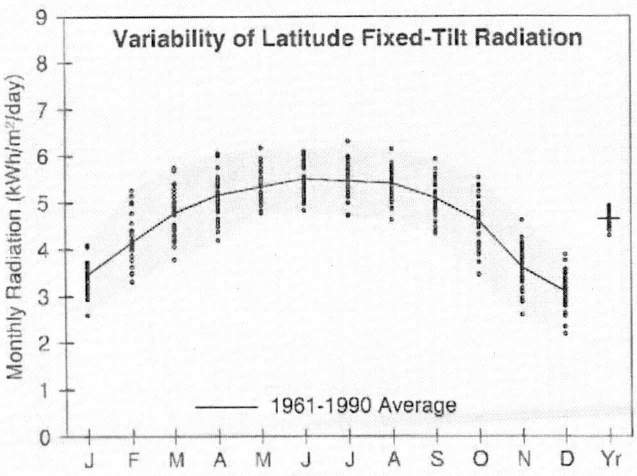

Solar Radiation for Flat-Plate Collectors Facing South at a Fixed Tilt (kWh/m²/day), Uncertainty ±9%

Tilt (°)		Jan	Feb	Mar	Apr	May	June	July	Aug	Sept	Oct	Nov	Dec	Year
0	Average	2.0	2.8	3.9	4.9	5.6	6.1	5.9	5.3	4.4	3.3	2.2	1.8	4.0
	Min/Max	1.7/2.3	2.4/3.4	3.3/4.5	4.1/5.6	5.1/6.4	5.4/6.8	5.1/6.8	4.7/6.0	3.9/4.9	2.8/3.8	1.9/2.6	1.5/2.0	3.8/4.3
Latitude -15	Average	3.0	3.8	4.6	5.3	5.7	6.0	5.9	5.6	5.0	4.3	3.2	2.7	4.6
	Min/Max	2.4/3.5	3.1/4.7	3.7/5.5	4.3/6.1	5.1/6.6	5.2/6.6	5.1/6.8	4.8/6.4	4.3/5.8	3.3/5.1	2.4/4.0	2.0/3.3	4.3/4.9
Latitude	Average	3.5	4.1	4.8	5.2	5.3	5.5	5.5	5.4	5.1	4.6	3.6	3.1	4.6
	Min/Max	2.6/4.1	3.3/5.3	3.8/5.8	4.2/6.1	4.8/6.2	4.8/6.1	4.7/6.3	4.6/6.2	4.3/5.9	3.5/5.5	2.6/4.6	2.2/3.9	4.3/4.9
Latitude +15	Average	3.7	4.3	4.7	4.8	4.7	4.8	4.8	4.9	4.9	4.6	3.8	3.3	4.4
	Min/Max	2.7/4.4	3.4/5.5	3.6/5.7	3.9/5.6	4.2/5.5	4.2/5.3	4.2/5.5	4.2/5.6	4.1/5.7	3.4/5.6	2.7/4.9	2.3/4.2	4.1/4.7
90	Average	3.4	3.7	3.5	3.0	2.6	2.4	2.5	2.9	3.4	3.7	3.3	3.1	3.1
	Min/Max	2.4/4.2	2.8/5.0	2.7/4.2	2.5/3.5	2.4/2.9	2.2/2.6	2.3/2.8	2.5/3.3	2.8/4.0	2.7/4.5	2.3/4.4	2.0/4.0	2.8/3.3

Solar Radiation for 1-Axis Tracking Flat-Plate Collectors with a North-South Axis (kWh/m²/day), Uncertainty ±9%

Axis Tilt (°)		Jan	Feb	Mar	Apr	May	June	July	Aug	Sept	Oct	Nov	Dec	Year
0	Average	3.0	4.0	5.2	6.4	7.1	7.7	7.5	6.9	5.9	4.7	3.2	2.5	5.3
	Min/Max	2.3/3.6	3.2/5.1	4.1/6.4	5.0/7.7	6.1/8.5	6.5/8.8	6.0/8.9	5.6/8.0	4.9/7.0	3.5/5.7	2.4/4.1	1.9/3.2	4.9/5.7
Latitude -15	Average	3.7	4.7	5.8	6.7	7.2	7.7	7.5	7.2	6.4	5.4	3.9	3.2	5.8
	Min/Max	2.8/4.5	3.7/6.1	4.4/7.2	5.2/8.2	6.2/8.7	6.5/8.8	6.0/8.9	5.8/8.3	5.2/7.7	3.9/6.6	2.8/5.1	2.2/4.1	5.3/6.2
Latitude	Average	4.0	5.0	5.9	6.6	7.0	7.4	7.2	7.0	6.4	5.6	4.2	3.6	5.8
	Min/Max	2.9/4.9	3.8/6.5	4.5/7.4	5.1/8.1	6.0/8.4	6.2/8.4	5.8/8.6	5.7/8.2	5.2/7.8	4.0/6.9	2.9/5.6	2.4/4.6	5.3/6.3
Latitude +15	Average	4.2	5.1	5.9	6.4	6.5	6.9	6.8	6.7	6.3	5.7	4.4	3.7	5.7
	Min/Max	3.0/5.1	3.9/6.7	4.3/7.3	4.9/7.8	5.6/7.9	5.8/7.8	5.4/8.1	5.4/7.8	5.1/7.6	4.0/7.0	3.0/5.8	2.5/4.8	5.2/6.1

Solar Radiation for 2-Axis Tracking Flat-Plate Collectors (kWh/m²/day), Uncertainty ±9%

Tracker		Jan	Feb	Mar	Apr	May	June	July	Aug	Sept	Oct	Nov	Dec	Year
2-Axis	Average	4.3	5.1	6.0	6.7	7.3	7.9	7.6	7.2	6.4	5.7	4.4	3.8	6.0
	Min/Max	3.0/5.2	3.9/6.7	4.5/7.4	5.2/8.2	6.3/8.8	6.6/9.0	6.1/9.1	5.8/8.4	5.3/7.8	4.0/7.0	3.0/5.8	2.5/4.9	5.5/6.5

Direct Beam Solar Radiation for Concentrating Collectors (kWh/m²/day), Uncertainty ±8%

Tracker		Jan	Feb	Mar	Apr	May	June	July	Aug	Sept	Oct	Nov	Dec	Year
1-Axis, E-W Horiz Axis	Average	2.5	2.7	2.9	3.0	3.2	3.6	3.4	3.2	3.0	3.1	2.5	2.3	3.0
	Min/Max	1.5/3.3	1.7/3.8	1.7/4.1	1.7/4.2	2.3/4.4	2.6/4.4	2.2/4.5	2.2/4.2	2.1/4.0	1.8/4.2	1.3/3.8	1.2/3.2	2.5/3.3
1-Axis, N-S Horiz Axis	Average	1.8	2.5	3.3	3.9	4.2	4.6	4.4	4.2	3.6	3.0	2.0	1.6	3.3
	Min/Max	1.1/2.5	1.6/3.5	1.9/4.6	2.2/5.4	3.0/5.7	3.3/5.8	2.8/5.8	2.8/5.4	2.5/4.8	1.8/4.2	1.0/3.0	0.8/2.2	2.8/3.7
1-Axis, N-S Tilt=Latitude	Average	2.7	3.3	3.9	4.1	4.1	4.4	4.2	4.2	4.1	3.9	2.9	2.4	3.7
	Min/Max	1.7/3.7	2.1/4.6	2.3/5.4	2.4/5.7	3.0/5.6	3.1/5.4	2.7/5.6	2.9/5.5	2.8/5.4	2.3/5.3	1.5/4.3	1.3/3.4	3.2/4.1
2-Axis	Average	2.9	3.4	3.9	4.2	4.3	4.7	4.5	4.4	4.1	3.9	3.0	2.6	3.8
	Min/Max	1.8/3.9	2.1/4.8	2.3/5.4	2.4/5.8	3.1/5.9	3.4/5.9	2.8/6.0	3.0/5.7	2.8/5.4	2.3/5.3	1.6/4.5	1.4/3.7	3.3/4.3

Average Climatic Conditions

Element	Jan	Feb	Mar	Apr	May	June	July	Aug	Sept	Oct	Nov	Dec	Year
Temperature (°C)	-0.6	0.6	5.3	10.0	15.8	20.8	23.7	23.0	18.9	12.7	7.7	2.1	11.7
Daily Minimum Temp	-5.9	-4.7	-0.4	4.1	9.8	14.8	18.2	17.5	13.1	6.5	2.1	-3.2	6.0
Daily Maximum Temp	4.7	5.8	10.9	15.9	21.8	26.7	29.2	28.5	24.8	18.9	13.2	7.4	17.3
Record Minimum Temp	-23.3	-23.9	-15.0	-11.1	-3.9	2.8	5.6	4.4	0.0	-6.7	-12.2	-21.7	-23.9
Record Maximum Temp	25.6	23.9	30.6	34.4	37.2	41.1	40.0	38.9	37.2	32.2	28.9	23.9	41.1
HDD, Base 18.3°C	587	498	405	250	93	7	0	0	29	180	320	503	2872
CDD, Base 18.3°C	0	0	0	0	14	80	167	144	48	6	0	0	459
Relative Humidity (%)	70	69	67	66	71	73	74	76	76	75	73	71	72
Wind Speed (m/s)	4.6	4.8	5.1	5.2	4.5	4.0	3.7	3.5	3.7	3.9	4.6	4.6	4.3

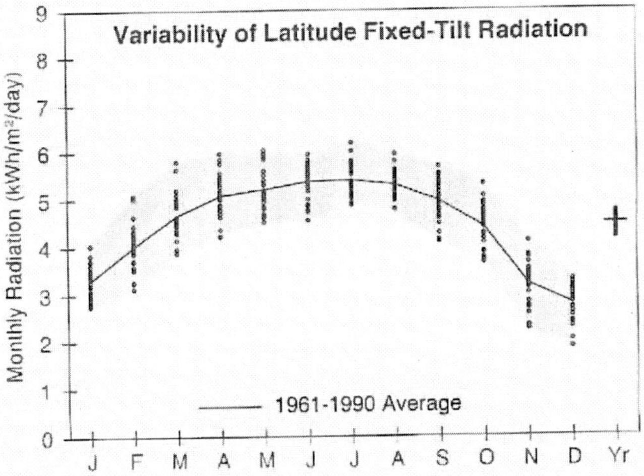

Newark, NJ

WBAN NO. 14734

LATITUDE: 40.70° N
LONGITUDE: 74.17° W
ELEVATION: 9 meters
MEAN PRESSURE: 1016 millibars

STATION TYPE: Secondary

Solar Radiation for Flat-Plate Collectors Facing South at a Fixed Tilt (kWh/m²/day), Uncertainty ±11%

Tilt (°)		Jan	Feb	Mar	Apr	May	June	July	Aug	Sept	Oct	Nov	Dec	Year
0	Average	1.9	2.7	3.8	4.8	5.5	6.0	5.9	5.2	4.3	3.2	2.0	1.6	3.9
	Min/Max	1.7/2.2	2.3/3.2	3.3/4.4	4.1/5.5	4.8/6.3	5.1/6.6	5.3/6.7	4.8/5.8	3.7/4.7	2.8/3.7	1.7/2.4	1.3/1.8	3.7/4.1
Latitude -15	Average	2.9	3.7	4.5	5.2	5.5	5.8	5.8	5.5	4.9	4.1	2.9	2.4	4.4
	Min/Max	2.5/3.5	2.9/4.6	3.8/5.6	4.3/6.1	4.8/6.4	4.9/6.4	5.3/6.7	5.0/6.2	4.1/5.6	3.5/4.9	2.1/3.6	1.7/2.8	4.2/4.7
Latitude	Average	3.3	4.0	4.6	5.1	5.2	5.4	5.4	5.3	5.0	4.4	3.2	2.8	4.5
	Min/Max	2.8/4.0	3.1/5.1	3.9/5.8	4.2/6.0	4.5/6.0	4.6/5.9	4.9/6.2	4.8/5.9	4.1/5.7	3.7/5.3	2.3/4.1	1.9/3.3	4.2/4.7
Latitude +15	Average	3.5	4.1	4.5	4.7	4.6	4.7	4.7	4.8	4.7	4.4	3.3	3.0	4.3
	Min/Max	2.9/4.3	3.1/5.3	3.7/5.7	3.9/5.5	4.0/5.3	4.0/5.2	4.3/5.4	4.3/5.4	3.9/5.5	3.6/5.4	2.3/4.3	1.9/3.5	4.0/4.5
90	Average	3.3	3.6	3.4	3.0	2.6	2.5	2.6	2.9	3.3	3.6	2.9	2.8	3.0
	Min/Max	2.6/4.1	2.6/4.8	2.8/4.3	2.5/3.6	2.4/3.0	2.2/2.7	2.4/2.9	2.7/3.3	2.7/3.9	2.9/4.4	1.9/3.9	1.7/3.3	2.8/3.2

Solar Radiation for 1-Axis Tracking Flat-Plate Collectors with a North-South Axis (kWh/m²/day), Uncertainty ±11%

Axis Tilt (°)		Jan	Feb	Mar	Apr	May	June	July	Aug	Sept	Oct	Nov	Dec	Year
0	Average	2.7	3.7	5.0	6.2	6.9	7.4	7.3	6.7	5.6	4.4	2.7	2.2	5.1
	Min/Max	2.3/3.4	2.9/4.7	4.1/6.4	5.0/7.4	5.7/8.2	5.9/8.3	6.4/8.5	5.8/7.5	4.5/6.3	3.6/5.3	2.0/3.5	1.5/2.6	4.7/5.4
Latitude -15	Average	3.5	4.5	5.6	6.5	7.0	7.4	7.4	6.9	6.1	5.1	3.4	2.9	5.5
	Min/Max	2.9/4.3	3.4/5.8	4.5/7.2	5.2/7.9	5.8/8.3	5.8/8.3	6.4/8.6	6.0/7.9	4.9/7.0	4.2/6.2	2.3/4.4	1.9/3.4	5.1/5.9
Latitude	Average	3.8	4.7	5.7	6.5	6.7	7.1	7.1	6.8	6.2	5.3	3.7	3.1	5.6
	Min/Max	3.1/4.7	3.5/6.2	4.6/7.4	5.1/7.8	5.6/8.1	5.6/8.0	6.2/8.3	5.9/7.7	4.9/7.1	4.3/6.5	2.4/4.8	2.0/3.8	5.1/5.9
Latitude +15	Average	4.0	4.8	5.6	6.2	6.3	6.6	6.6	6.5	6.0	5.3	3.8	3.3	5.4
	Min/Max	3.2/5.0	3.6/6.3	4.4/7.3	4.9/7.5	5.3/7.6	5.2/7.4	5.8/7.7	5.6/7.3	4.7/6.9	4.3/6.6	2.5/5.0	2.0/4.0	5.0/5.8

Solar Radiation for 2-Axis Tracking Flat-Plate Collectors (kWh/m²/day), Uncertainty ±11%

Tracker		Jan	Feb	Mar	Apr	May	June	July	Aug	Sept	Oct	Nov	Dec	Year
2-Axis	Average	4.0	4.8	5.7	6.6	7.0	7.5	7.5	7.0	6.2	5.4	3.8	3.3	5.7
	Min/Max	3.2/5.0	3.6/6.3	4.6/7.4	5.2/7.9	5.9/8.4	6.0/8.5	6.6/8.8	6.1/7.9	4.9/7.1	4.4/6.6	2.5/5.0	2.0/4.0	5.3/6.1

Direct Beam Solar Radiation for Concentrating Collectors (kWh/m²/day), Uncertainty ±11%

Tracker		Jan	Feb	Mar	Apr	May	June	July	Aug	Sept	Oct	Nov	Dec	Year
1-Axis, E-W Horiz Axis	Average	2.3	2.5	2.7	2.9	3.0	3.2	3.2	3.0	2.8	2.8	2.1	1.9	2.7
	Min/Max	1.7/3.0	1.6/3.7	2.0/4.1	1.8/3.9	2.0/4.0	2.2/3.9	2.4/4.2	2.3/3.8	2.0/3.5	2.0/3.8	1.0/3.1	0.9/2.5	2.4/3.0
1-Axis, N-S Horiz Axis	Average	1.6	2.2	3.0	3.7	3.9	4.2	4.2	3.8	3.3	2.7	1.6	1.3	3.0
	Min/Max	1.2/2.2	1.4/3.2	2.1/4.6	2.4/4.9	2.7/5.2	2.8/5.1	3.0/5.4	2.9/4.8	2.3/4.1	2.0/3.7	0.8/2.3	0.6/1.6	2.6/3.3
1-Axis, N-S Tilt=Latitude	Average	2.4	3.0	3.6	3.9	3.8	3.9	4.0	3.9	3.8	3.5	2.3	2.0	3.4
	Min/Max	1.8/3.3	1.9/4.4	2.5/5.5	2.5/5.2	2.6/5.1	2.7/4.8	2.9/5.2	3.0/4.9	2.6/4.7	2.5/4.7	1.1/3.4	1.0/2.6	2.9/3.7
2-Axis	Average	2.6	3.1	3.6	4.0	4.0	4.3	4.3	4.1	3.8	3.5	2.5	2.2	3.5
	Min/Max	2.0/3.5	2.0/4.5	2.5/5.5	2.6/5.3	2.8/5.4	2.9/5.3	3.1/5.6	3.1/5.1	2.6/4.7	2.6/4.7	1.2/3.6	1.1/2.8	3.1/3.9

Average Climatic Conditions

Element	Jan	Feb	Mar	Apr	May	June	July	Aug	Sept	Oct	Nov	Dec	Year
Temperature (°C)	-0.8	0.6	5.6	11.3	17.1	22.6	25.4	24.7	20.4	14.2	8.5	2.2	12.7
Daily Minimum Temp	-4.8	-3.7	0.8	5.9	11.8	17.1	20.3	19.7	15.5	9.0	4.0	-1.6	7.8
Daily Maximum Temp	3.2	4.7	10.4	16.6	22.4	27.9	30.6	29.7	25.3	19.3	13.0	6.1	17.4
Record Minimum Temp	-22.2	-21.7	-14.4	-8.9	0.6	6.1	11.1	7.2	1.7	-2.2	-9.4	-18.3	-22.2
Record Maximum Temp	23.3	24.4	31.7	34.4	36.7	38.9	40.6	39.4	40.6	33.3	29.4	22.2	40.6
HDD, Base 18.3°C	592	498	394	212	71	0	0	0	14	140	295	499	2716
CDD, Base 18.3°C	0	0	0	0	33	129	221	196	78	11	0	0	667
Relative Humidity (%)	66	63	60	58	62	63	64	66	68	66	66	67	64
Wind Speed (m/s)	4.9	5.1	5.4	5.2	4.6	4.3	4.0	4.0	4.2	4.2	4.7	4.8	4.6

Albuquerque, NM

WBAN NO. 23050

LATITUDE: 35.05° N
LONGITUDE: 106.62° W
ELEVATION: 1619 meters
MEAN PRESSURE: 838 millibars

STATION TYPE: Primary

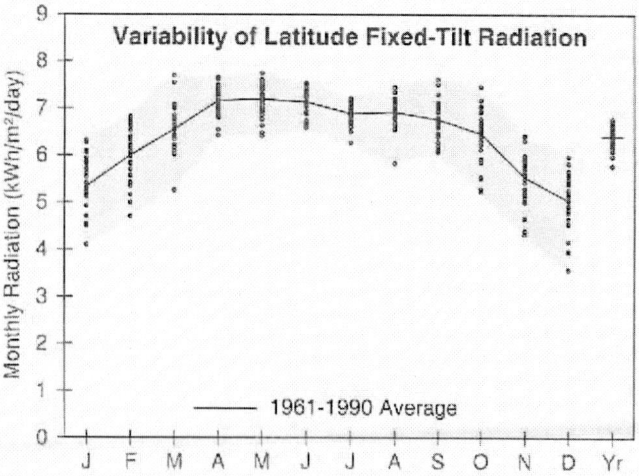

Solar Radiation for Flat-Plate Collectors Facing South at a Fixed Tilt (kWh/m²/day), Uncertainty ±9%

Tilt (°)		Jan	Feb	Mar	Apr	May	June	July	Aug	Sept	Oct	Nov	Dec	Year
0	Average	3.2	4.2	5.4	6.8	7.7	8.1	7.5	6.9	5.9	4.7	3.5	2.9	5.6
	Min/Max	2.6/3.7	3.4/4.7	4.4/6.2	6.2/7.3	6.9/8.3	7.4/8.7	6.8/7.9	5.9/7.4	5.4/6.6	4.0/5.3	2.8/3.9	2.2/3.3	5.1/5.9
Latitude -15	Average	4.6	5.4	6.3	7.3	7.7	7.8	7.4	7.2	6.6	5.9	4.8	4.3	6.3
	Min/Max	3.6/5.4	4.3/6.1	5.1/7.3	6.5/7.8	6.8/8.3	7.2/8.3	6.7/7.8	6.1/7.7	6.0/7.4	4.9/6.8	3.8/5.5	3.1/5.0	5.7/6.6
Latitude	Average	5.3	6.0	6.5	7.2	7.2	7.1	6.9	6.9	6.8	6.5	5.5	5.0	6.4
	Min/Max	4.1/6.3	4.7/6.8	5.3/7.7	6.4/7.7	6.4/7.7	6.6/7.5	6.3/7.2	5.8/7.4	6.1/7.6	5.2/7.4	4.3/6.4	3.6/6.0	5.8/6.8
Latitude +15	Average	5.8	6.2	6.5	6.6	6.3	6.1	6.0	6.3	6.5	6.6	5.9	5.5	6.2
	Min/Max	4.4/6.9	4.8/7.1	5.1/7.6	5.9/7.1	5.7/6.8	5.6/6.4	5.5/6.3	5.3/6.7	5.8/7.3	5.3/7.7	4.6/6.9	3.8/6.6	5.5/6.5
90	Average	5.2	5.1	4.5	3.7	2.8	2.4	2.5	3.2	4.2	5.1	5.2	5.1	4.1
	Min/Max	3.9/6.4	3.9/5.8	3.5/5.4	3.4/4.0	2.5/3.0	2.2/2.5	2.3/2.7	2.8/3.4	3.7/4.6	4.0/6.0	3.9/6.2	3.5/6.2	3.5/4.4

Solar Radiation for 1-Axis Tracking Flat-Plate Collectors with a North-South Axis (kWh/m²/day), Uncertainty ±9%

Axis Tilt (°)		Jan	Feb	Mar	Apr	May	June	July	Aug	Sept	Oct	Nov	Dec	Year
0	Average	4.9	6.2	7.6	9.6	10.6	10.9	9.9	9.2	8.3	7.0	5.3	4.5	7.8
	Min/Max	3.7/5.9	4.7/7.1	5.7/9.3	8.4/10.6	9.0/11.8	9.5/12.1	8.5/10.8	7.4/10.3	7.0/9.6	5.5/8.4	4.1/6.3	3.2/5.4	6.9/8.3
Latitude -15	Average	5.9	7.1	8.3	10.0	10.6	10.8	9.9	9.5	8.8	7.9	6.3	5.5	8.4
	Min/Max	4.4/7.2	5.4/8.1	6.2/10.2	8.7/11.0	9.1/11.9	9.5/12.0	8.5/10.8	7.6/10.6	7.5/10.4	6.1/9.5	4.8/7.5	3.8/6.7	7.4/8.9
Latitude	Average	6.5	7.5	8.6	9.9	10.3	10.4	9.5	9.3	9.0	8.3	6.8	6.1	8.5
	Min/Max	4.8/7.9	5.7/8.7	6.4/10.5	8.7/10.9	8.8/11.6	9.1/11.5	8.1/10.4	7.5/10.4	7.6/10.5	6.3/10.0	5.2/8.1	4.2/7.4	7.5/9.1
Latitude +15	Average	6.9	7.7	8.5	9.5	9.7	9.7	8.9	8.9	8.8	8.4	7.1	6.5	8.4
	Min/Max	5.1/8.4	5.8/8.9	6.3/10.4	8.3/10.6	8.3/10.9	8.4/10.7	7.6/9.8	7.1/9.9	7.4/10.3	6.4/10.1	5.3/8.5	4.4/7.9	7.3/8.9

Solar Radiation for 2-Axis Tracking Flat-Plate Collectors (kWh/m²/day), Uncertainty ±9%

Tracker		Jan	Feb	Mar	Apr	May	June	July	Aug	Sept	Oct	Nov	Dec	Year
2-Axis	Average	6.9	7.7	8.6	10.0	10.8	11.1	10.0	9.5	9.0	8.4	7.2	6.6	8.8
	Min/Max	5.1/8.5	5.8/8.9	6.4/10.5	8.8/11.1	9.2/12.1	9.7/12.3	8.6/11.0	7.7/10.6	7.6/10.5	6.4/10.1	5.3/8.6	4.4/8.1	7.7/9.4

Direct Beam Solar Radiation for Concentrating Collectors (kWh/m²/day), Uncertainty ±8%

Tracker		Jan	Feb	Mar	Apr	May	June	July	Aug	Sept	Oct	Nov	Dec	Year
1-Axis, E-W	Average	4.5	4.6	4.6	5.3	5.9	6.3	5.5	5.2	4.9	5.1	4.6	4.4	5.1
Horiz Axis	Min/Max	3.1/5.8	3.0/5.7	3.0/6.2	4.3/6.1	4.7/7.1	5.3/7.4	4.3/6.4	3.7/6.0	3.9/6.2	3.3/6.5	2.8/5.9	2.6/5.7	4.3/5.5
1-Axis, N-S	Average	3.7	4.6	5.6	7.2	8.0	8.4	7.2	6.7	6.2	5.4	4.0	3.4	5.9
Horiz Axis	Min/Max	2.5/4.8	3.0/5.7	3.4/7.6	5.7/8.4	6.1/9.7	6.8/10.0	5.4/8.4	4.7/7.9	4.7/7.9	3.4/7.2	2.4/5.2	2.0/4.5	4.8/6.4
1-Axis, N-S	Average	5.1	5.7	6.3	7.4	7.7	7.8	6.8	6.8	6.7	6.5	5.4	4.8	6.4
Tilt=Latitude	Min/Max	3.4/6.6	3.7/7.2	3.9/8.6	5.8/8.6	5.9/9.3	6.4/9.4	5.1/7.9	4.7/8.0	5.1/8.6	4.1/8.6	3.2/6.9	2.8/6.3	5.3/7.0
2-Axis	Average	5.4	5.9	6.3	7.5	8.1	8.5	7.3	7.0	6.8	6.6	5.7	5.2	6.7
	Min/Max	3.7/7.1	3.8/7.3	3.9/8.6	5.9/8.8	6.2/9.9	6.9/10.2	5.5/8.5	4.9/8.2	5.2/8.7	4.2/8.6	3.4/7.3	3.1/6.8	5.6/7.3

Average Climatic Conditions

Element	Jan	Feb	Mar	Apr	May	June	July	Aug	Sept	Oct	Nov	Dec	Year
Temperature (°C)	1.2	4.4	8.3	12.9	17.9	23.4	25.8	24.4	20.3	13.9	6.8	1.8	13.4
Daily Minimum Temp	-5.7	-3.1	0.1	4.2	9.2	14.6	18.0	17.0	12.9	6.1	-0.4	-4.9	5.7
Daily Maximum Temp	8.2	11.9	16.3	21.6	26.5	32.2	33.6	31.7	27.7	21.7	14.1	8.6	21.2
Record Minimum Temp	-27.2	-20.6	-13.3	-7.2	-2.2	4.4	11.1	11.1	2.8	-6.1	-21.7	-21.7	-27.2
Record Maximum Temp	20.6	24.4	29.4	31.7	36.7	40.6	40.6	38.3	37.8	32.8	25.0	22.2	40.6
HDD, Base 18.3°C	531	389	312	167	49	0	0	0	10	144	345	512	2458
CDD, Base 18.3°C	0	0	0	4	36	155	233	188	70	6	0	0	691
Relative Humidity (%)	56	50	40	33	31	30	42	47	48	45	50	57	44
Wind Speed (m/s)	3.7	3.9	4.5	4.9	4.8	4.5	4.0	3.8	3.8	3.6	3.6	3.5	4.1

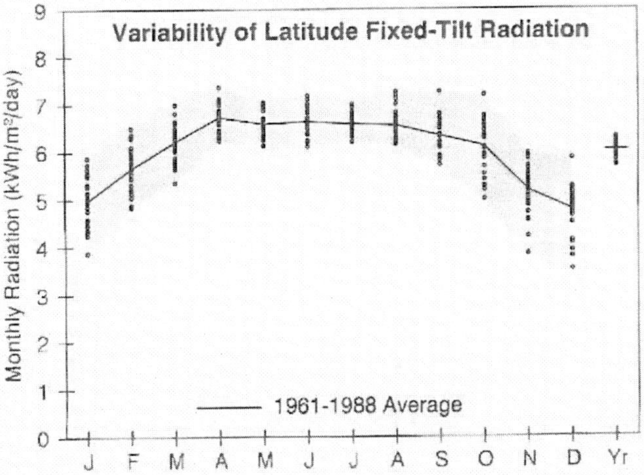

Tucumcari, NM

WBAN NO. 23048

LATITUDE: 35.18° N
LONGITUDE: 103.60° W
ELEVATION: 1231 meters
MEAN PRESSURE: 878 millibars

STATION TYPE: Secondary

Solar Radiation for Flat-Plate Collectors Facing South at a Fixed Tilt (kWh/m²/day), Uncertainty ±9%

Tilt (°)		Jan	Feb	Mar	Apr	May	June	July	Aug	Sept	Oct	Nov	Dec	Year
0	Average	3.0	3.9	5.1	6.4	7.0	7.5	7.2	6.5	5.5	4.5	3.3	2.7	5.2
	Min/Max	2.5/3.3	3.5/4.3	4.5/5.6	5.9/6.9	6.5/7.5	6.8/8.1	6.7/7.7	6.1/7.3	5.1/6.2	3.8/5.1	2.7/3.6	2.2/3.1	5.0/5.4
Latitude -15	Average	4.3	5.1	5.9	6.8	7.0	7.2	7.1	6.8	6.2	5.6	4.5	4.0	5.9
	Min/Max	3.4/5.0	4.4/5.8	5.2/6.6	6.3/7.4	6.5/7.5	6.6/7.8	6.7/7.6	6.4/7.5	5.7/7.1	4.7/6.5	3.5/5.2	3.1/4.9	5.6/6.1
Latitude	Average	5.0	5.6	6.2	6.7	6.6	6.6	6.6	6.5	6.3	6.1	5.2	4.8	6.0
	Min/Max	3.9/5.9	4.8/6.5	5.3/7.0	6.2/7.3	6.1/7.0	6.1/7.2	6.2/7.0	6.2/7.3	5.7/7.3	5.0/7.2	3.8/5.9	3.5/5.8	5.7/6.3
Latitude +15	Average	5.4	5.9	6.1	6.2	5.8	5.7	5.7	5.9	6.1	6.2	5.5	5.2	5.8
	Min/Max	4.1/6.4	5.0/6.8	5.3/6.9	5.8/6.8	5.4/6.2	5.3/6.1	5.4/6.0	5.6/6.6	5.5/7.0	5.0/7.4	4.0/6.4	3.7/6.4	5.5/6.1
90	Average	4.9	4.8	4.3	3.5	2.7	2.3	2.4	3.1	3.9	4.8	4.8	4.8	3.9
	Min/Max	3.7/5.9	4.0/5.6	3.7/4.9	3.3/3.8	2.5/2.8	2.2/2.4	2.4/2.5	2.9/3.3	3.5/4.5	3.8/5.8	3.4/5.7	3.3/6.1	3.5/4.1

Solar Radiation for 1-Axis Tracking Flat-Plate Collectors with a North-South Axis (kWh/m²/day), Uncertainty ±9%

Axis Tilt (°)		Jan	Feb	Mar	Apr	May	June	July	Aug	Sept	Oct	Nov	Dec	Year
0	Average	4.6	5.8	7.2	8.9	9.4	10.0	9.6	8.9	7.7	6.6	4.9	4.2	7.3
	Min/Max	3.5/5.4	4.8/6.7	5.9/8.3	7.9/9.9	8.4/10.5	8.7/11.2	8.7/10.6	8.1/10.3	6.9/9.1	5.3/8.0	3.6/5.7	3.0/5.2	6.8/7.8
Latitude -15	Average	5.5	6.6	7.9	9.3	9.5	9.9	9.6	9.1	8.3	7.5	5.9	5.2	7.9
	Min/Max	4.2/6.6	5.5/7.8	6.5/9.1	8.2/10.3	8.5/10.6	8.6/11.1	8.7/10.6	8.3/10.6	7.4/9.8	5.9/9.0	4.2/6.8	3.7/6.5	7.3/8.3
Latitude	Average	6.0	7.1	8.1	9.2	9.3	9.5	9.3	9.0	8.4	7.9	6.4	5.7	8.0
	Min/Max	4.6/7.3	5.8/8.3	6.6/9.4	8.1/10.3	8.3/10.3	8.2/10.7	8.4/10.2	8.2/10.4	7.4/9.9	6.2/9.5	4.5/7.4	4.0/7.2	7.4/8.5
Latitude +15	Average	6.3	7.2	8.0	8.9	8.7	8.9	8.7	8.5	8.2	7.9	6.6	6.1	7.8
	Min/Max	4.8/7.7	5.9/8.5	6.5/9.3	7.8/9.9	7.8/9.7	7.7/10.0	7.8/9.6	7.8/9.9	7.2/9.7	6.2/9.6	4.7/7.7	4.2/7.7	7.2/8.3

Solar Radiation for 2-Axis Tracking Flat-Plate Collectors (kWh/m²/day), Uncertainty ±9%

Tracker		Jan	Feb	Mar	Apr	May	June	July	Aug	Sept	Oct	Nov	Dec	Year
2-Axis	Average	6.4	7.2	8.1	9.3	9.6	10.1	9.8	9.2	8.4	8.0	6.7	6.2	8.3
	Min/Max	4.8/7.8	5.9/8.6	6.6/9.4	8.2/10.4	8.6/10.7	8.8/11.4	8.8/10.8	8.4/10.6	7.4/9.9	6.2/9.6	4.7/7.8	4.3/7.8	7.6/8.8

Direct Beam Solar Radiation for Concentrating Collectors (kWh/m²/day), Uncertainty ±8%

Tracker		Jan	Feb	Mar	Apr	May	June	July	Aug	Sept	Oct	Nov	Dec	Year
1-Axis, E-W Horiz Axis	Average	4.1	4.3	4.4	4.9	5.1	5.6	5.3	4.8	4.5	4.7	4.2	4.1	4.7
	Min/Max	2.6/5.4	3.1/5.4	3.1/5.4	4.0/5.8	4.2/6.1	4.4/6.7	4.4/6.1	4.1/6.1	3.8/5.8	3.3/6.2	2.5/5.2	2.4/5.7	4.1/5.1
1-Axis, N-S Horiz Axis	Average	3.3	4.2	5.2	6.6	6.8	7.4	6.9	6.3	5.7	5.1	3.7	3.1	5.4
	Min/Max	2.1/4.4	3.0/5.4	3.6/6.4	5.2/7.8	5.6/8.2	5.6/8.9	5.7/8.2	5.4/8.3	4.6/7.3	3.5/6.7	2.1/4.5	1.8/4.4	4.7/5.9
1-Axis, N-S Tilt=Latitude	Average	4.6	5.3	6.0	6.8	6.6	6.9	6.6	6.4	6.2	6.1	4.9	4.4	5.9
	Min/Max	2.9/6.0	3.8/6.8	4.1/7.3	5.4/8.0	5.4/7.9	5.3/8.4	5.4/7.7	5.5/8.3	5.1/8.0	4.2/8.1	2.9/6.1	2.6/6.2	5.1/6.5
2-Axis	Average	4.9	5.5	6.0	6.9	7.0	7.5	7.0	6.6	6.2	6.2	5.1	4.8	6.1
	Min/Max	3.1/6.4	3.9/6.9	4.1/7.3	5.5/8.2	5.7/8.4	5.7/9.1	5.8/8.3	5.6/8.5	5.1/8.0	4.3/8.2	3.0/6.4	2.8/6.7	5.3/6.7

Average Climatic Conditions

Element	Jan	Feb	Mar	Apr	May	June	July	Aug	Sept	Oct	Nov	Dec	Year
Temperature (°C)	2.9	5.0	9.1	14.1	18.8	23.6	25.8	24.6	20.6	15.1	8.8	3.4	14.3
Daily Minimum Temp	-5.7	-3.6	0.3	5.4	10.3	15.3	17.9	16.7	12.8	6.4	0.6	-4.8	5.9
Daily Maximum Temp	11.6	13.6	17.9	22.8	27.2	31.9	33.7	32.3	28.3	23.7	16.9	11.7	22.6
Record Minimum Temp	-27.8	-19.0	-12.8	-9.4	0.0	6.1	11.7	10.6	1.7	-3.9	-16.1	-23.9	-27.8
Record Maximum Temp	25.6	27.2	36.7	34.4	36.7	42.8	40.0	38.9	38.3	36.1	30.0	25.6	42.8
HDD, Base 18.3°C	477	373	286	136	31	4	0	0	8	110	287	462	2173
CDD, Base 18.3°C	0	0	0	9	45	162	231	193	76	10	0	0	726
Relative Humidity (%)	58	56	48	43	46	47	52	56	58	52	56	58	53
Wind Speed (m/s)	4.3	4.4	4.9	5.1	4.8	4.6	4.3	3.9	4.1	4.0	4.3	4.2	4.4

Elko, NV

WBAN NO. 24121

LATITUDE: 40.83° N
LONGITUDE: 115.78° W
ELEVATION: 1547 meters
MEAN PRESSURE: 845 millibars

STATION TYPE: Secondary

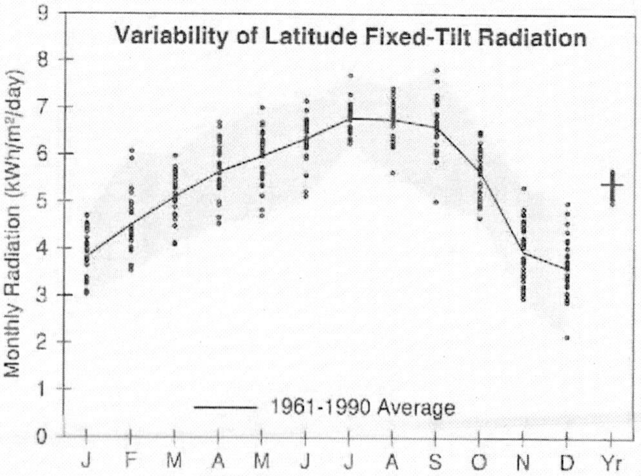

Solar Radiation for Flat-Plate Collectors Facing South at a Fixed Tilt (kWh/m²/day), Uncertainty ±9%

Tilt (°)		Jan	Feb	Mar	Apr	May	June	July	Aug	Sept	Oct	Nov	Dec	Year
0	Average	2.1	2.9	4.0	5.3	6.3	7.1	7.4	6.6	5.4	3.8	2.3	1.9	4.6
	Min/Max	1.8/2.4	2.5/3.7	3.4/4.6	4.3/6.1	5.0/7.3	5.8/8.1	6.8/8.5	5.5/7.2	4.3/6.2	3.3/4.2	1.9/2.8	1.5/2.3	4.4/4.8
Latitude -15	Average	3.3	4.1	4.9	5.7	6.4	6.9	7.3	7.0	6.5	5.2	3.5	3.1	5.3
	Min/Max	2.7/4.0	3.3/5.4	4.0/5.7	4.6/6.8	5.0/7.5	5.6/7.8	6.7/8.4	5.8/7.7	5.0/7.6	4.4/5.9	2.7/4.6	2.0/4.2	5.0/5.6
Latitude	Average	3.8	4.5	5.1	5.6	6.0	6.3	6.8	6.8	6.6	5.7	4.0	3.6	5.4
	Min/Max	3.0/4.7	3.5/6.1	4.1/6.0	4.5/6.7	4.7/7.0	5.1/7.1	6.2/7.7	5.6/7.4	5.0/7.8	4.7/6.5	3.0/5.3	2.2/5.0	5.0/5.7
Latitude +15	Average	4.1	4.7	5.0	5.2	5.3	5.4	5.9	6.1	6.4	5.8	4.2	3.9	5.2
	Min/Max	3.2/5.1	3.6/6.4	4.0/5.9	4.2/6.2	4.2/6.2	4.4/6.1	5.4/6.6	5.1/6.7	4.8/7.5	4.7/6.7	3.1/5.7	2.3/5.4	4.7/5.4
90	Average	3.9	4.1	3.8	3.4	2.9	2.7	3.0	3.6	4.5	4.7	3.8	3.7	3.7
	Min/Max	3.0/4.9	3.1/6.0	3.0/4.5	2.8/4.0	2.4/3.3	2.3/2.9	2.8/3.2	3.1/3.9	3.3/5.3	3.9/5.5	2.8/5.2	2.1/5.2	3.3/3.9

Solar Radiation for 1-Axis Tracking Flat-Plate Collectors with a North-South Axis (kWh/m²/day), Uncertainty ±9%

Axis Tilt (°)		Jan	Feb	Mar	Apr	May	June	July	Aug	Sept	Oct	Nov	Dec	Year
0	Average	3.2	4.3	5.7	7.2	8.6	9.8	10.4	9.4	8.1	5.9	3.5	2.9	6.6
	Min/Max	2.5/3.9	3.4/5.8	4.4/6.8	5.4/8.8	6.6/10.4	7.5/11.7	9.2/12.4	7.5/10.8	5.8/9.9	4.8/6.9	2.7/4.6	1.8/3.9	6.0/7.0
Latitude -15	Average	4.1	5.2	6.4	7.7	8.8	9.8	10.5	9.9	8.9	6.9	4.3	3.8	7.2
	Min/Max	3.2/5.1	4.0/7.1	4.8/7.7	5.7/9.4	6.6/10.7	7.5/11.7	9.3/12.6	7.8/11.3	6.4/10.9	5.6/8.1	3.2/6.0	2.2/5.3	6.5/7.7
Latitude	Average	4.5	5.5	6.5	7.6	8.5	9.5	10.2	9.7	9.0	7.2	4.7	4.2	7.3
	Min/Max	3.5/5.7	4.2/7.6	4.9/7.9	5.6/9.4	6.4/10.4	7.2/11.2	8.9/12.1	7.7/11.1	6.4/11.1	5.8/8.6	3.4/6.5	2.3/5.9	6.5/7.7
Latitude +15	Average	4.7	5.6	6.4	7.3	8.0	8.9	9.6	9.3	8.9	7.3	4.9	4.5	7.1
	Min/Max	3.6/6.0	4.2/7.8	4.8/7.8	5.4/9.0	6.0/9.8	6.7/10.6	8.4/11.4	7.3/10.6	6.2/10.9	5.8/8.6	3.5/6.8	2.4/6.3	6.4/7.6

Solar Radiation for 2-Axis Tracking Flat-Plate Collectors (kWh/m²/day), Uncertainty ±9%

Tracker		Jan	Feb	Mar	Apr	May	June	July	Aug	Sept	Oct	Nov	Dec	Year
2-Axis	Average	4.8	5.6	6.5	7.7	8.9	10.1	10.7	9.9	9.1	7.3	4.9	4.5	7.5
	Min/Max	3.7/6.1	4.3/7.8	4.9/7.9	5.7/9.4	6.7/10.8	7.7/12.0	9.4/12.8	7.9/11.3	6.4/11.1	5.8/8.6	3.6/6.8	2.5/6.4	6.8/8.0

Direct Beam Solar Radiation for Concentrating Collectors (kWh/m²/day), Uncertainty ±8%

Tracker		Jan	Feb	Mar	Apr	May	June	July	Aug	Sept	Oct	Nov	Dec	Year
1-Axis, E-W	Average	2.8	3.1	3.2	3.6	4.4	5.4	6.0	5.4	5.1	4.4	3.0	2.8	4.1
Horiz Axis	Min/Max	1.7/4.2	1.8/5.0	1.9/4.2	2.0/5.1	2.5/6.0	3.5/7.0	4.9/7.9	3.7/6.5	3.1/6.8	3.2/5.5	1.7/4.7	1.0/4.7	3.5/4.5
1-Axis, N-S	Average	2.1	2.8	3.7	4.8	5.9	7.2	8.1	7.3	6.3	4.5	2.3	1.9	4.8
Horiz Axis	Min/Max	1.2/3.1	1.6/4.6	2.2/4.9	2.7/6.7	3.4/8.1	4.8/9.5	6.5/10.7	5.0/8.9	3.8/8.6	3.2/5.7	1.3/3.7	0.6/3.2	4.0/5.3
1-Axis, N-S	Average	3.1	3.8	4.4	5.1	5.8	6.8	7.8	7.4	7.1	5.7	3.4	3.0	5.3
Tilt=Latitude	Min/Max	1.9/4.6	2.2/6.2	2.6/5.8	2.8/7.1	3.3/8.0	4.5/9.0	6.2/10.3	5.1/9.1	4.2/9.6	4.1/7.2	1.9/5.4	1.0/5.0	4.4/5.9
2-Axis	Average	3.3	3.9	4.4	5.2	6.2	7.4	8.3	7.7	7.2	5.8	3.6	3.3	5.5
	Min/Max	2.0/5.0	2.3/6.3	2.6/5.8	2.9/7.2	3.5/8.4	4.9/9.8	6.7/11.0	5.3/9.4	4.2/9.7	4.1/7.3	2.0/5.7	1.1/5.4	4.6/6.1

Average Climatic Conditions

Element	Jan	Feb	Mar	Apr	May	June	July	Aug	Sept	Oct	Nov	Dec	Year
Temperature (°C)	-3.8	-0.3	3.1	6.8	11.7	16.9	21.5	20.4	14.8	8.7	2.1	-3.5	8.2
Daily Minimum Temp	-10.3	-6.7	-3.9	-1.4	2.7	7.0	10.2	9.2	3.8	-1.3	-5.3	-10.0	-0.5
Daily Maximum Temp	2.6	6.1	10.1	15.1	20.8	26.8	32.8	31.4	25.7	18.8	9.5	3.0	16.9
Record Minimum Temp	-41.7	-38.3	-22.8	-18.9	-12.2	-5.0	-1.1	-4.4	-12.8	-16.1	-24.4	-38.9	-41.7
Record Maximum Temp	17.8	21.1	25.0	30.0	33.3	40.0	41.7	41.7	37.2	31.1	25.6	17.8	41.7
HDD, Base 18.3°C	687	521	472	345	205	77	10	23	131	298	487	677	3932
CDD, Base 18.3°C	0	0	0	0	0	33	108	87	26	0	0	0	254
Relative Humidity (%)	72	70	62	53	48	44	33	35	41	50	65	71	53
Wind Speed (m/s)	2.3	2.6	3.0	3.2	3.1	2.9	2.8	2.7	2.5	2.3	2.4	2.2	2.7

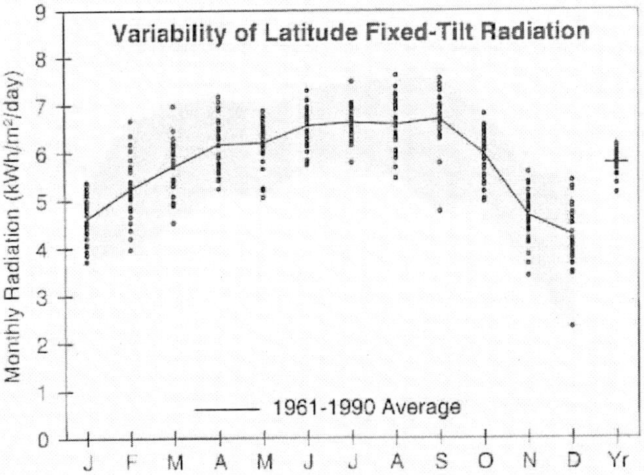

Ely, NV

WBAN NO. 23154

LATITUDE: 39.28° N
LONGITUDE: 114.85° W
ELEVATION: 1906 meters
MEAN PRESSURE: 809 millibars

STATION TYPE: Primary

Solar Radiation for Flat-Plate Collectors Facing South at a Fixed Tilt (kWh/m²/day), Uncertainty ±9%

Tilt (°)		Jan	Feb	Mar	Apr	May	June	July	Aug	Sept	Oct	Nov	Dec	Year
0	Average	2.6	3.4	4.5	5.8	6.6	7.5	7.3	6.5	5.6	4.1	2.8	2.2	4.9
	Min/Max	2.2/2.9	2.8/4.1	3.8/5.3	5.0/6.6	5.5/7.4	6.5/8.3	6.3/8.3	5.5/7.5	4.2/6.2	3.6/4.6	2.3/3.1	1.6/2.6	4.5/5.1
Latitude -15	Average	4.0	4.7	5.5	6.3	6.6	7.2	7.2	6.9	6.6	5.5	4.1	3.6	5.7
	Min/Max	3.3/4.6	3.6/5.9	4.4/6.6	5.3/7.3	5.4/7.4	6.3/8.0	6.3/8.2	5.7/7.9	4.8/7.4	4.7/6.2	3.1/4.9	2.1/4.6	5.1/6.0
Latitude	Average	4.6	5.2	5.7	6.2	6.2	6.6	6.6	6.6	6.7	6.0	4.7	4.3	5.8
	Min/Max	3.7/5.4	4.0/6.7	4.5/7.0	5.2/7.2	5.1/6.9	5.8/7.3	5.8/7.5	5.5/7.6	4.8/7.6	5.0/6.8	3.4/5.6	2.3/5.4	5.1/6.2
Latitude +15	Average	5.0	5.5	5.6	5.7	5.5	5.6	5.7	6.0	6.5	6.1	5.0	4.7	5.6
	Min/Max	4.0/5.9	4.1/7.1	4.4/7.0	4.9/6.7	4.5/6.1	5.0/6.3	5.1/6.5	4.9/6.9	4.5/7.3	5.0/7.0	3.6/6.0	2.5/5.9	4.9/6.0
90	Average	4.8	4.8	4.3	3.6	2.9	2.6	2.8	3.4	4.4	4.9	4.5	4.5	4.0
	Min/Max	3.7/5.7	3.6/6.3	3.3/5.7	3.2/4.1	2.5/3.1	2.4/2.8	2.6/3.0	2.9/3.8	3.1/5.0	4.0/5.7	3.2/5.5	2.3/5.7	3.4/4.3

Solar Radiation for 1-Axis Tracking Flat-Plate Collectors with a North-South Axis (kWh/m²/day), Uncertainty ±9%

Axis Tilt (°)		Jan	Feb	Mar	Apr	May	June	July	Aug	Sept	Oct	Nov	Dec	Year
0	Average	4.0	5.1	6.5	8.0	9.0	10.3	10.2	9.3	8.3	6.3	4.2	3.5	7.1
	Min/Max	3.0/4.7	4.0/6.7	4.8/8.1	6.2/9.7	7.3/10.4	8.5/11.8	8.3/12.0	7.4/11.2	5.6/9.8	5.1/7.1	3.0/5.1	2.0/4.4	6.2/7.5
Latitude -15	Average	5.0	6.1	7.2	8.5	9.2	10.3	10.3	9.6	9.1	7.3	5.2	4.6	7.7
	Min/Max	3.8/6.0	4.6/8.0	5.3/9.1	6.6/10.3	7.3/10.6	8.5/11.7	8.3/12.1	7.7/11.6	6.0/10.7	5.9/8.4	3.7/6.4	2.4/5.9	6.7/8.2
Latitude	Average	5.5	6.5	7.4	8.4	8.9	9.9	9.9	9.5	9.2	7.7	5.6	5.1	7.8
	Min/Max	4.2/6.6	4.9/8.6	5.4/9.4	6.5/10.3	7.1/10.3	8.1/11.3	8.0/11.6	7.5/11.4	6.1/10.9	6.1/8.8	3.9/6.9	2.6/6.6	6.7/8.3
Latitude +15	Average	5.8	6.6	7.3	8.1	8.4	9.2	9.3	9.0	9.0	7.7	5.9	5.4	7.7
	Min/Max	4.4/7.0	5.0/8.9	5.3/9.4	6.2/9.9	6.7/9.8	7.6/10.6	7.5/11.0	7.2/10.9	5.9/10.6	6.2/8.9	4.0/7.3	2.7/7.0	6.5/8.2

Solar Radiation for 2-Axis Tracking Flat-Plate Collectors (kWh/m²/day), Uncertainty ±9%

Tracker		Jan	Feb	Mar	Apr	May	June	July	Aug	Sept	Oct	Nov	Dec	Year
2-Axis	Average	5.9	6.6	7.4	8.5	9.3	10.5	10.5	9.7	9.2	7.8	5.9	5.5	8.1
	Min/Max	4.4/7.1	5.0/8.9	5.5/9.4	6.6/10.4	7.5/10.8	8.7/12.0	8.5/12.3	7.7/11.7	6.1/10.9	6.2/8.9	4.1/7.3	2.7/7.1	6.9/8.6

Direct Beam Solar Radiation for Concentrating Collectors (kWh/m²/day), Uncertainty ±8%

Tracker		Jan	Feb	Mar	Apr	May	June	July	Aug	Sept	Oct	Nov	Dec	Year
1-Axis, E-W Horiz Axis	Average	3.6	3.8	3.7	4.2	4.6	5.7	5.7	5.1	5.2	4.7	3.7	3.5	4.5
	Min/Max	2.4/4.8	2.3/5.7	2.3/5.0	2.7/5.7	3.0/5.8	4.1/7.0	4.2/7.3	3.5/6.9	2.8/6.4	3.5/5.8	2.1/5.0	1.2/5.3	3.5/5.0
1-Axis, N-S Horiz Axis	Average	2.7	3.5	4.4	5.5	6.3	7.7	7.8	7.0	6.5	4.8	3.0	2.5	5.1
	Min/Max	1.6/3.6	2.1/5.4	2.5/5.9	3.4/7.7	4.3/8.0	5.4/9.5	5.5/9.9	4.9/9.4	3.5/8.4	3.5/5.9	1.6/4.1	0.8/3.7	4.1/5.6
1-Axis, N-S Tilt=Latitude	Average	4.0	4.6	5.1	5.8	6.1	7.2	7.4	7.1	7.3	6.1	4.2	3.8	5.7
	Min/Max	2.6/5.3	2.8/7.1	3.0/6.9	3.6/8.1	4.2/7.8	5.1/8.9	5.3/9.5	5.0/9.5	3.9/9.3	4.4/7.4	2.3/5.8	1.2/5.7	4.5/6.3
2-Axis	Average	4.3	4.7	5.1	5.9	6.5	7.9	8.0	7.3	7.3	6.1	4.4	4.1	6.0
	Min/Max	2.7/5.7	2.9/7.3	3.0/6.9	3.6/8.2	4.4/8.3	5.6/9.7	5.7/10.2	5.1/9.9	3.9/9.3	4.5/7.5	2.4/6.1	1.4/6.2	4.6/6.6

Average Climatic Conditions

Element	Jan	Feb	Mar	Apr	May	June	July	Aug	Sept	Oct	Nov	Dec	Year
Temperature (°C)	-4.1	-1.4	1.4	5.3	10.3	15.3	19.7	18.6	13.5	7.7	1.2	-3.6	7.0
Daily Minimum Temp	-12.6	-9.2	-6.3	-3.3	0.9	4.8	8.9	8.1	2.9	-2.1	-7.2	-11.9	-2.2
Daily Maximum Temp	4.3	6.4	9.1	13.9	19.6	25.7	30.6	29.1	24.0	17.5	9.6	4.8	16.2
Record Minimum Temp	-32.8	-34.4	-25.0	-20.6	-13.9	-7.8	-1.1	-4.4	-9.4	-19.4	-26.1	-33.9	-34.4
Record Maximum Temp	20.0	19.4	22.8	27.2	31.7	37.2	37.8	36.1	33.9	28.9	23.9	19.4	37.8
HDD, Base 18.3°C	696	552	523	392	250	104	8	34	152	331	513	678	4234
CDD, Base 18.3°C	0	0	0	0	0	14	51	43	7	0	0	0	116
Relative Humidity (%)	65	64	59	52	47	39	35	39	42	51	60	64	51
Wind Speed (m/s)	4.1	4.2	4.5	4.6	4.6	4.5	4.5	4.5	4.3	4.1	4.1	4.1	4.3

Las Vegas, NV

WBAN NO. 23169

LATITUDE: 36.08° N
LONGITUDE: 115.17° W
ELEVATION: 664 meters
MEAN PRESSURE: 938 millibars

STATION TYPE: Primary

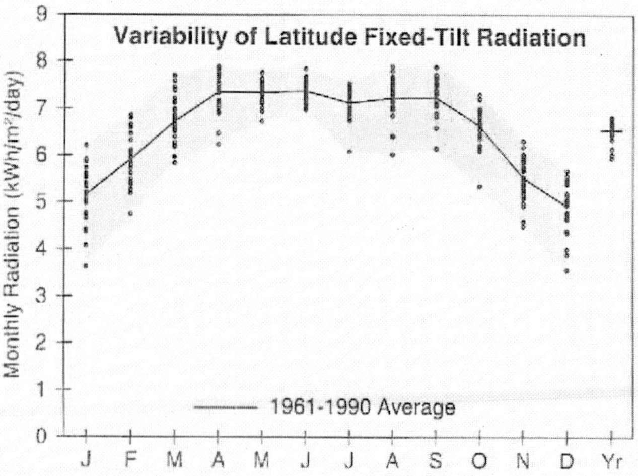

Solar Radiation for Flat-Plate Collectors Facing South at a Fixed Tilt (kWh/m²/day), Uncertainty ±9%

Tilt (°)		Jan	Feb	Mar	Apr	May	June	July	Aug	Sept	Oct	Nov	Dec	Year
0	Average	3.0	4.0	5.4	6.9	7.8	8.4	7.9	7.2	6.2	4.7	3.4	2.8	5.7
	Min/Max	2.3/3.4	3.4/4.5	4.8/6.1	6.1/7.4	7.2/8.3	7.8/8.9	6.6/8.4	6.0/7.8	5.4/6.7	4.0/5.1	3.0/3.7	2.2/3.1	5.3/5.8
Latitude -15	Average	4.4	5.3	6.4	7.5	7.8	8.1	7.7	7.5	7.1	6.1	4.8	4.2	6.4
	Min/Max	3.2/5.3	4.4/6.1	5.6/7.3	6.4/8.0	7.2/8.3	7.6/8.6	6.5/8.2	6.2/8.2	6.1/7.7	5.0/6.7	4.0/5.4	3.1/4.8	5.9/6.7
Latitude	Average	5.1	5.9	6.7	7.4	7.3	7.4	7.1	7.2	7.2	6.6	5.5	4.9	6.5
	Min/Max	3.6/6.2	4.8/6.8	5.8/7.7	6.2/7.9	6.7/7.8	7.0/7.8	6.1/7.5	6.0/7.9	6.1/7.9	5.3/7.3	4.5/6.3	3.6/5.7	5.9/6.8
Latitude +15	Average	5.6	6.1	6.6	6.8	6.5	6.3	6.2	6.5	7.0	6.8	5.9	5.4	6.3
	Min/Max	3.8/6.8	4.9/7.2	5.7/7.6	5.7/7.3	5.9/6.8	6.0/6.7	5.3/6.5	5.5/7.2	5.9/7.6	5.4/7.5	4.7/6.8	3.8/6.2	5.7/6.6
90	Average	5.0	5.1	4.7	3.9	3.0	2.6	2.6	3.4	4.5	5.3	5.2	5.0	4.2
	Min/Max	3.4/6.2	4.0/6.0	4.0/5.4	3.3/4.1	2.8/3.1	2.4/2.7	2.4/2.7	2.9/3.7	3.8/4.9	4.2/5.9	4.1/6.1	3.5/5.8	3.7/4.5

Solar Radiation for 1-Axis Tracking Flat-Plate Collectors with a North-South Axis (kWh/m²/day), Uncertainty ±9%

Axis Tilt (°)		Jan	Feb	Mar	Apr	May	June	July	Aug	Sept	Oct	Nov	Dec	Year
0	Average	4.6	5.9	7.8	9.8	10.9	11.6	10.8	10.1	9.0	7.1	5.1	4.2	8.1
	Min/Max	3.3/5.6	4.6/7.0	6.2/9.3	8.2/10.9	9.8/11.9	10.5/12.5	8.2/11.8	8.1/11.3	7.1/10.1	5.7/7.9	4.1/6.0	3.0/4.9	7.2/8.5
Latitude -15	Average	5.7	6.9	8.5	10.3	11.0	11.5	10.8	10.4	9.7	8.1	6.2	5.3	8.7
	Min/Max	3.9/6.9	5.3/8.2	6.9/10.2	8.5/11.4	9.9/12.0	10.4/12.4	8.2/11.8	8.4/11.7	7.7/10.9	6.4/9.1	4.9/7.3	3.7/6.2	7.7/9.1
Latitude	Average	6.2	7.3	8.8	10.2	10.6	11.1	10.4	10.3	9.8	8.6	6.7	5.9	8.8
	Min/Max	4.3/7.7	5.6/8.7	7.1/10.5	8.4/11.3	9.6/11.6	10.0/11.9	7.8/11.4	8.2/11.5	7.8/11.1	6.7/9.5	5.3/7.9	4.1/6.9	7.8/9.3
Latitude +15	Average	6.5	7.5	8.7	9.8	10.0	10.3	9.8	9.8	9.6	8.7	7.0	6.2	8.7
	Min/Max	4.4/8.1	5.7/9.0	7.0/10.4	8.1/10.9	9.0/11.0	9.3/11.1	7.3/10.7	7.8/11.0	7.6/10.9	6.7/9.7	5.4/8.3	4.3/7.3	7.6/9.1

Solar Radiation for 2-Axis Tracking Flat-Plate Collectors (kWh/m²/day), Uncertainty ±9%

Tracker		Jan	Feb	Mar	Apr	May	June	July	Aug	Sept	Oct	Nov	Dec	Year
2-Axis	Average	6.6	7.5	8.8	10.3	11.1	11.8	11.0	10.5	9.8	8.7	7.1	6.3	9.1
	Min/Max	4.5/8.2	5.7/9.0	7.1/10.5	8.5/11.4	10.0/12.1	10.6/12.7	8.3/12.1	8.4/11.8	7.8/11.1	6.7/9.7	5.5/8.3	4.3/7.4	8.1/9.6

Direct Beam Solar Radiation for Concentrating Collectors (kWh/m²/day), Uncertainty ±8%

Tracker		Jan	Feb	Mar	Apr	May	June	July	Aug	Sept	Oct	Nov	Dec	Year
1-Axis, E-W Horiz Axis	Average	4.3	4.5	4.8	5.5	6.1	6.8	6.1	5.8	5.6	5.3	4.6	4.3	5.3
	Min/Max	2.7/5.7	3.0/5.8	3.8/6.3	4.1/6.4	5.2/7.1	5.7/7.6	4.4/7.1	4.4/6.9	4.0/6.6	3.5/6.1	3.1/5.7	2.7/5.3	4.5/5.8
1-Axis, N-S Horiz Axis	Average	3.4	4.4	5.7	7.4	8.2	9.1	8.2	7.8	7.1	5.6	3.9	3.2	6.2
	Min/Max	2.1/4.6	2.8/5.7	4.2/7.7	5.5/8.7	7.0/9.6	7.5/10.3	5.5/9.6	5.8/9.4	4.9/8.5	3.7/6.5	2.5/4.9	1.9/4.0	5.2/6.7
1-Axis, N-S Tilt=Latitude	Average	4.8	5.6	6.5	7.7	8.0	8.5	7.8	7.8	7.8	6.8	5.2	4.6	6.8
	Min/Max	3.0/6.4	3.6/7.2	4.9/8.8	5.6/9.0	6.8/9.3	7.1/9.6	5.2/9.1	5.8/9.4	5.4/9.3	4.5/7.9	3.5/6.7	2.8/5.8	5.7/7.4
2-Axis	Average	5.2	5.7	6.6	7.8	8.4	9.2	8.4	8.1	7.8	6.9	5.5	5.0	7.1
	Min/Max	3.2/6.9	3.7/7.4	4.9/8.8	5.8/9.2	7.2/9.8	7.7/10.4	5.6/9.7	6.0/9.7	5.4/9.4	4.5/8.0	3.7/7.0	3.1/6.3	6.0/7.7

Average Climatic Conditions

Element	Jan	Feb	Mar	Apr	May	June	July	Aug	Sept	Oct	Nov	Dec	Year
Temperature (°C)	7.5	10.6	13.5	17.8	23.3	29.4	32.8	31.5	26.9	20.2	12.8	7.6	19.5
Daily Minimum Temp	0.9	3.8	6.6	10.4	15.7	20.8	24.6	23.4	19.0	12.4	5.9	1.1	12.1
Daily Maximum Temp	14.1	17.4	20.4	25.3	31.0	37.9	41.1	39.6	34.8	27.8	19.7	14.2	26.9
Record Minimum Temp	-13.3	-8.9	-5.0	-0.6	4.4	9.4	15.6	13.3	7.8	-3.3	-6.1	-11.7	-13.3
Record Maximum Temp	25.0	30.6	32.8	37.2	42.8	46.1	46.7	46.7	45.0	39.4	30.6	25.0	46.7
HDD, Base 18.3°C	336	216	162	79	8	0	0	0	0	34	169	332	1337
CDD, Base 18.3°C	0	0	12	64	163	332	449	408	258	91	0	0	1778
Relative Humidity (%)	45	40	33	25	21	16	21	26	25	29	37	45	30
Wind Speed (m/s)	3.6	4.1	4.9	5.1	5.2	5.1	4.9	4.5	4.3	3.8	3.8	3.4	4.4

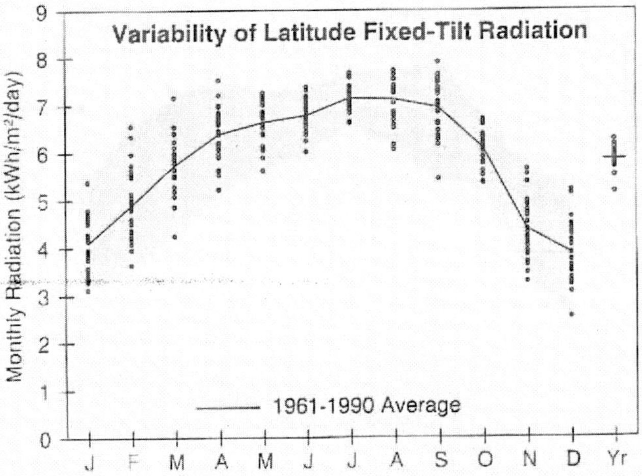

Reno, NV

WBAN NO. 23185

LATITUDE: 39.50° N
LONGITUDE: 119.78° W
ELEVATION: 1341 meters
MEAN PRESSURE: 867 millibars

STATION TYPE: Secondary

Solar Radiation for Flat-Plate Collectors Facing South at a Fixed Tilt (kWh/m²/day), Uncertainty ±9%

Tilt (°)		Jan	Feb	Mar	Apr	May	June	July	Aug	Sept	Oct	Nov	Dec	Year
0	Average	2.3	3.2	4.5	5.9	7.0	7.6	7.8	6.9	5.7	4.1	2.6	2.1	5.0
	Min/Max	2.0/2.7	2.6/4.0	3.6/5.4	4.9/6.9	6.0/7.6	6.8/8.4	7.3/8.5	6.0/7.5	4.7/6.4	3.7/4.5	2.1/3.1	1.6/2.5	4.6/5.3
Latitude -15	Average	3.6	4.4	5.5	6.5	7.1	7.4	7.7	7.4	6.8	5.6	3.9	3.3	5.8
	Min/Max	2.8/4.6	3.4/5.8	4.2/6.8	5.3/7.6	6.0/7.7	6.5/8.1	7.2/8.3	6.3/8.0	5.4/7.7	4.9/6.1	3.0/4.9	2.3/4.4	5.2/6.1
Latitude	Average	4.1	4.9	5.7	6.4	6.6	6.8	7.1	7.1	6.9	6.1	4.4	3.9	5.8
	Min/Max	3.1/5.4	3.6/6.6	4.2/7.2	5.2/7.5	5.6/7.3	6.0/7.4	6.6/7.7	6.1/7.7	5.4/7.9	5.3/6.7	3.3/5.6	2.5/5.2	5.2/6.3
Latitude +15	Average	4.4	5.1	5.6	5.9	5.8	5.8	6.1	6.4	6.7	6.2	4.6	4.2	5.6
	Min/Max	3.3/5.8	3.7/6.9	4.1/7.1	4.8/7.0	5.0/6.4	5.2/6.3	5.7/6.6	5.5/7.0	5.2/7.6	5.4/6.9	3.4/6.0	2.7/5.7	4.9/6.0
90	Average	4.1	4.3	4.2	3.7	3.0	2.7	2.9	3.6	4.6	5.0	4.2	4.0	3.9
	Min/Max	3.0/5.5	3.0/5.9	3.1/5.3	3.1/4.3	2.7/3.3	2.5/2.8	2.8/3.1	3.2/3.9	3.6/5.2	4.3/5.6	3.0/5.5	2.4/5.4	3.3/4.3

Solar Radiation for 1-Axis Tracking Flat-Plate Collectors with a North-South Axis (kWh/m²/day), Uncertainty ±9%

Axis Tilt (°)		Jan	Feb	Mar	Apr	May	June	July	Aug	Sept	Oct	Nov	Dec	Year
0	Average	3.5	4.7	6.4	8.3	9.7	10.5	11.1	10.0	8.6	6.3	3.9	3.2	7.2
	Min/Max	2.7/4.6	3.3/6.4	4.4/8.4	6.2/10.1	7.8/10.9	9.0/11.9	9.9/12.2	8.2/11.2	6.4/10.0	5.4/7.0	2.9/5.1	2.0/4.2	6.3/7.7
Latitude -15	Average	4.4	5.6	7.2	8.8	9.8	10.5	11.1	10.4	9.4	7.4	4.8	4.1	7.8
	Min/Max	3.3/6.0	3.9/7.8	4.9/9.4	6.6/10.8	7.9/11.1	8.9/11.8	9.9/12.3	8.5/11.6	6.9/11.1	6.3/8.3	3.5/6.4	2.6/5.6	6.7/8.4
Latitude	Average	4.8	6.0	7.4	8.7	9.5	10.1	10.8	10.3	9.5	7.8	5.2	4.5	7.9
	Min/Max	3.5/6.6	4.2/8.3	5.0/9.7	6.5/10.8	7.7/10.8	8.6/11.4	9.6/11.9	8.4/11.4	7.0/11.2	6.6/8.7	3.8/7.0	2.8/6.3	6.8/8.5
Latitude +15	Average	5.1	6.1	7.3	8.4	9.0	9.4	10.1	9.8	9.3	7.9	5.4	4.8	7.7
	Min/Max	3.7/7.0	4.2/8.5	4.9/9.6	6.3/10.4	7.2/10.2	8.0/10.7	9.0/11.1	7.9/10.9	6.8/11.0	6.6/8.8	3.9/7.3	2.9/6.7	6.6/8.4

Solar Radiation for 2-Axis Tracking Flat-Plate Collectors (kWh/m²/day), Uncertainty ±9%

Tracker		Jan	Feb	Mar	Apr	May	June	July	Aug	Sept	Oct	Nov	Dec	Year
2-Axis	Average	5.1	6.1	7.4	8.8	10.0	10.8	11.4	10.5	9.5	7.9	5.5	4.9	8.2
	Min/Max	3.7/7.1	4.2/8.5	5.0/9.7	6.6/10.9	8.0/11.2	9.2/12.2	10.1/12.5	8.6/11.7	7.0/11.2	6.7/8.8	3.9/7.3	2.9/6.8	7.0/8.9

Direct Beam Solar Radiation for Concentrating Collectors (kWh/m²/day), Uncertainty ±8%

Tracker		Jan	Feb	Mar	Apr	May	June	July	Aug	Sept	Oct	Nov	Dec	Year
1-Axis, E-W Horiz Axis	Average	3.2	3.5	3.9	4.4	5.2	6.0	6.6	5.9	5.4	4.8	3.4	3.1	4.6
	Min/Max	1.8/5.1	2.0/5.6	2.1/5.8	2.7/6.1	3.5/6.3	4.6/7.2	5.3/7.7	4.2/6.9	3.5/6.9	3.7/5.6	2.0/5.0	1.5/4.8	3.6/5.2
1-Axis, N-S Horiz Axis	Average	2.3	3.2	4.5	5.8	7.0	8.0	8.8	7.9	6.8	4.9	2.7	2.2	5.4
	Min/Max	1.3/3.8	1.8/5.2	2.3/6.8	3.5/8.1	4.8/8.5	6.1/9.7	7.1/10.3	5.6/9.4	4.2/8.8	3.7/5.7	1.6/4.1	1.0/3.4	4.2/6.0
1-Axis, N-S Tilt=Latitude	Average	3.5	4.3	5.2	6.2	6.9	7.5	8.4	8.0	7.6	6.1	3.8	3.3	5.9
	Min/Max	2.0/5.6	2.4/6.9	2.7/7.9	3.7/8.6	4.6/8.3	5.8/9.1	6.8/9.9	5.7/9.5	4.7/9.7	4.7/7.2	2.3/5.8	1.6/5.2	4.6/6.7
2-Axis	Average	3.7	4.4	5.3	6.3	7.3	8.2	9.1	8.3	7.6	6.2	4.0	3.6	6.2
	Min/Max	2.1/6.0	2.4/7.0	2.7/8.0	3.7/8.7	4.9/8.8	6.3/9.9	7.3/10.6	5.8/9.8	4.7/9.8	4.7/7.3	2.4/6.0	1.7/5.6	4.8/7.0

Average Climatic Conditions

Element	Jan	Feb	Mar	Apr	May	June	July	Aug	Sept	Oct	Nov	Dec	Year
Temperature (°C)	0.5	3.3	6.0	9.2	13.6	18.4	22.0	20.9	15.8	10.4	4.6	0.4	10.4
Daily Minimum Temp	-6.3	-4.3	-1.6	0.7	4.5	8.3	10.7	9.8	5.2	0.5	-2.9	-6.7	1.5
Daily Maximum Temp	7.3	10.9	13.5	17.6	22.7	28.4	33.3	32.0	26.4	20.3	12.1	7.5	19.3
Record Minimum Temp	-26.7	-26.7	-18.9	-10.6	-7.8	-3.9	0.6	-4.4	-6.7	-13.3	-17.2	-26.7	-26.7
Record Maximum Temp	21.1	23.9	28.3	31.7	35.6	39.4	40.0	40.6	38.3	32.8	25.0	21.1	40.6
HDD, Base 18.3°C	553	420	382	273	152	42	7	11	99	244	412	556	3152
CDD, Base 18.3°C	0	0	0	0	6	44	121	90	22	0	0	0	282
Relative Humidity (%)	68	60	53	46	43	40	36	39	44	51	61	68	51
Wind Speed (m/s)	2.3	2.8	3.7	3.8	3.8	3.5	3.4	3.1	2.8	2.5	2.7	2.5	3.1

Tonopah, NV

WBAN NO. 23153

LATITUDE: 38.07° N
LONGITUDE: 117.13° W
ELEVATION: 1653 meters
MEAN PRESSURE: 834 millibars

STATION TYPE: Secondary

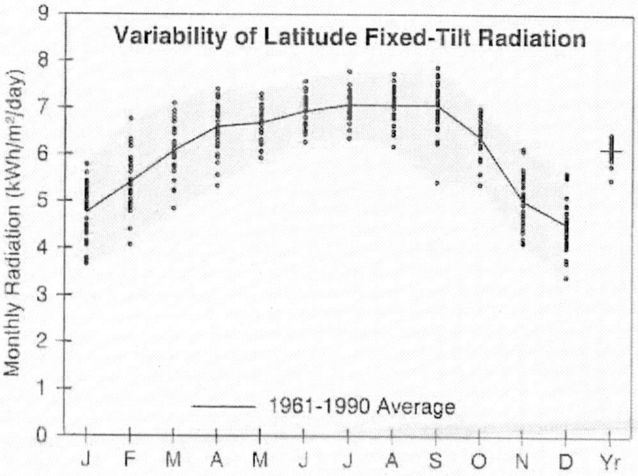

Solar Radiation for Flat-Plate Collectors Facing South at a Fixed Tilt (kWh/m²/day), Uncertainty ±9%

Tilt (°)		Jan	Feb	Mar	Apr	May	June	July	Aug	Sept	Oct	Nov	Dec	Year
0	Average	2.7	3.6	4.8	6.2	7.1	7.9	7.8	7.0	5.9	4.4	3.0	2.4	5.2
	Min/Max	2.3/3.0	2.9/4.2	4.0/5.5	5.2/6.8	6.3/7.8	7.1/8.6	6.9/8.6	6.1/7.6	4.7/6.5	3.9/4.7	2.6/3.4	2.0/2.8	4.8/5.4
Latitude -15	Average	4.1	4.9	5.8	6.7	7.1	7.6	7.7	7.3	6.9	5.9	4.4	3.8	6.0
	Min/Max	3.2/4.9	3.8/6.0	4.7/6.7	5.5/7.5	6.3/7.8	6.9/8.3	6.8/8.4	6.4/8.0	5.4/7.7	5.0/6.4	3.7/5.3	3.0/4.7	5.4/6.3
Latitude	Average	4.8	5.4	6.1	6.6	6.7	6.9	7.1	7.1	7.1	6.4	5.0	4.5	6.1
	Min/Max	3.7/5.8	4.1/6.8	4.8/7.1	5.3/7.4	5.9/7.3	6.3/7.6	6.3/7.8	6.2/7.7	5.4/7.9	5.4/7.0	4.1/6.1	3.4/5.6	5.5/6.4
Latitude +15	Average	5.1	5.6	6.0	6.1	5.9	5.9	6.1	6.4	6.8	6.5	5.3	4.9	5.9
	Min/Max	3.9/6.3	4.2/7.1	4.7/7.0	4.9/6.9	5.2/6.4	5.4/6.4	5.5/6.7	5.6/7.0	5.2/7.6	5.4/7.2	4.3/6.6	3.6/6.2	5.2/6.2
90	Average	4.7	4.8	4.4	3.7	2.9	2.6	2.8	3.5	4.6	5.2	4.8	4.6	4.0
	Min/Max	3.5/5.9	3.4/6.0	3.4/5.1	3.0/4.1	2.7/3.1	2.4/2.7	2.6/2.9	3.2/3.7	3.5/5.1	4.3/5.8	3.8/6.0	3.3/5.9	3.5/4.3

Solar Radiation for 1-Axis Tracking Flat-Plate Collectors with a North-South Axis (kWh/m²/day), Uncertainty ±9%

Axis Tilt (°)		Jan	Feb	Mar	Apr	May	June	July	Aug	Sept	Oct	Nov	Dec	Year
0	Average	4.2	5.4	7.0	8.7	9.8	10.9	10.9	10.0	8.8	6.8	4.6	3.8	7.6
	Min/Max	3.2/5.1	3.8/6.8	5.4/8.6	6.8/10.0	8.4/11.2	9.7/12.2	9.3/12.6	8.4/11.3	6.4/10.1	5.8/7.5	3.7/5.7	2.8/4.7	6.7/8.0
Latitude -15	Average	5.2	6.4	7.8	9.2	10.0	10.9	11.0	10.3	9.6	7.9	5.6	4.9	8.2
	Min/Max	3.9/6.5	4.5/8.1	5.9/9.5	7.1/10.6	8.5/11.3	9.7/12.2	9.3/12.6	8.7/11.7	6.9/11.0	6.6/8.7	4.5/7.0	3.5/6.1	7.2/8.7
Latitude	Average	5.7	6.8	8.0	9.2	9.7	10.4	10.6	10.2	9.8	8.3	6.1	5.4	8.4
	Min/Max	4.3/7.2	4.8/8.7	6.0/9.8	7.0/10.5	8.3/11.0	9.3/11.7	8.9/12.2	8.5/11.5	6.9/11.2	6.9/9.1	4.9/7.7	3.9/6.8	7.2/8.8
Latitude +15	Average	6.0	6.9	7.9	8.8	9.1	9.8	9.9	9.7	9.6	8.4	6.4	5.7	8.2
	Min/Max	4.5/7.6	4.8/8.9	5.9/9.7	6.7/10.1	7.8/10.4	8.7/10.9	8.4/11.4	8.1/11.0	6.7/11.0	6.9/9.2	5.1/8.0	4.1/7.3	7.1/8.7

Solar Radiation for 2-Axis Tracking Flat-Plate Collectors (kWh/m²/day), Uncertainty ±9%

Tracker		Jan	Feb	Mar	Apr	May	June	July	Aug	Sept	Oct	Nov	Dec	Year
2-Axis	Average	6.1	6.9	8.0	9.3	10.1	11.1	11.2	10.4	9.8	8.4	6.4	5.8	8.6
	Min/Max	4.5/7.7	4.8/8.9	6.0/9.8	7.1/10.6	8.6/11.5	9.9/12.5	9.4/12.9	8.7/11.8	6.9/11.2	6.9/9.2	5.1/8.1	4.2/7.4	7.5/9.2

Direct Beam Solar Radiation for Concentrating Collectors (kWh/m²/day), Uncertainty ±8%

Tracker		Jan	Feb	Mar	Apr	May	June	July	Aug	Sept	Oct	Nov	Dec	Year
1-Axis, E-W Horiz Axis	Average	4.0	4.1	4.3	4.8	5.3	6.3	6.4	5.8	5.6	5.2	4.1	3.9	5.0
	Min/Max	2.5/5.6	2.5/5.9	2.7/5.7	3.0/6.0	4.0/6.4	5.1/7.6	5.0/7.9	4.3/6.9	3.2/6.8	3.9/5.9	3.0/5.7	2.4/5.5	4.0/5.5
1-Axis, N-S Horiz Axis	Average	3.1	3.9	5.1	6.4	7.2	8.4	8.6	7.8	7.1	5.4	3.4	2.9	5.8
	Min/Max	1.9/4.3	2.2/5.7	3.2/6.9	4.0/8.0	5.4/8.8	6.8/10.2	6.5/10.7	5.8/9.5	4.0/8.8	4.2/6.3	2.4/4.7	1.7/4.0	4.7/6.3
1-Axis, N-S Tilt=Latitude	Average	4.4	5.1	5.9	6.7	7.0	7.9	8.2	7.9	7.8	6.7	4.8	4.3	6.4
	Min/Max	2.8/6.2	2.9/7.4	3.7/8.0	4.1/8.4	5.3/8.5	6.4/9.6	6.2/10.2	5.9/9.6	4.4/9.7	5.1/7.7	3.4/6.6	2.5/6.0	5.1/7.1
2-Axis	Average	4.7	5.2	5.9	6.8	7.4	8.6	8.8	8.1	7.9	6.8	5.0	4.6	6.7
	Min/Max	3.0/6.7	3.0/7.6	3.7/8.0	4.2/8.5	5.6/9.0	7.0/10.4	6.7/11.0	6.1/9.9	4.4/9.7	5.2/7.8	3.5/6.9	2.8/6.5	5.3/7.4

Average Climatic Conditions

Element	Jan	Feb	Mar	Apr	May	June	July	Aug	Sept	Oct	Nov	Dec	Year
Temperature (°C)	-0.5	2.4	5.0	8.9	14.1	19.5	23.2	21.9	17.3	11.4	4.3	-0.1	10.6
Daily Minimum Temp	-7.7	-4.7	-2.6	0.6	5.4	10.3	13.6	12.6	8.2	2.7	-3.2	-7.2	2.3
Daily Maximum Temp	6.6	9.6	12.5	17.2	22.8	28.6	32.8	31.3	26.4	20.1	11.8	6.9	18.9
Record Minimum Temp	-25.0	-22.2	-15.0	-7.8	-6.7	-1.1	4.4	5.0	-3.9	-9.4	-15.0	-23.3	-25.0
Record Maximum Temp	17.8	23.3	25.6	28.9	33.9	37.8	38.3	37.8	34.4	31.7	24.4	17.2	38.3
HDD, Base 18.3°C	584	445	413	288	142	33	0	6	65	218	420	572	3185
CDD, Base 18.3°C	0	0	0	4	11	68	152	118	35	4	0	0	392
Relative Humidity (%)	62	57	48	39	34	27	26	29	32	38	52	59	42
Wind Speed (m/s)	3.7	4.1	4.8	4.9	4.6	4.3	4.0	3.8	3.9	3.8	3.8	3.7	4.1

Winnemucca, NV

WBAN NO. 24128

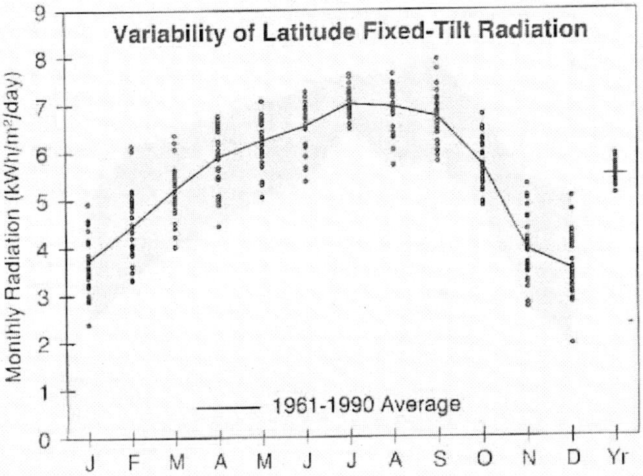

LATITUDE: 40.90° N
LONGITUDE: 117.80° W
ELEVATION: 1323 meters
MEAN PRESSURE: 869 millibars

STATION TYPE: Secondary

Solar Radiation for Flat-Plate Collectors Facing South at a Fixed Tilt (kWh/m²/day), Uncertainty ±9%

Tilt (°)		Jan	Feb	Mar	Apr	May	June	July	Aug	Sept	Oct	Nov	Dec	Year
0	Average	2.1	2.9	4.1	5.5	6.6	7.4	7.7	6.7	5.5	3.8	2.3	1.9	4.7
	Min/Max	1.6/2.4	2.5/3.6	3.4/4.8	4.3/6.1	5.4/7.4	6.1/8.2	7.1/8.4	5.6/7.4	4.9/6.3	3.4/4.3	1.9/2.8	1.4/2.3	4.5/5.0
Latitude -15	Average	3.3	4.1	5.0	6.0	6.7	7.1	7.6	7.2	6.6	5.3	3.5	3.0	5.5
	Min/Max	2.2/4.2	3.1/5.5	4.0/6.1	4.6/6.9	5.4/7.5	5.9/8.0	7.0/8.3	5.9/7.9	5.7/7.7	4.5/6.2	2.5/4.6	1.8/4.2	5.1/5.8
Latitude	Average	3.7	4.5	5.2	5.9	6.2	6.5	7.0	6.9	6.7	5.7	3.9	3.5	5.5
	Min/Max	2.4/4.9	3.3/6.1	4.0/6.4	4.4/6.8	5.1/7.1	5.4/7.3	6.5/7.6	5.7/7.6	5.8/7.9	4.8/6.8	2.7/5.3	1.9/5.0	5.1/5.9
Latitude +15	Average	4.0	4.6	5.1	5.5	5.5	5.6	6.0	6.3	6.5	5.8	4.1	3.8	5.2
	Min/Max	2.5/5.3	3.3/6.4	3.9/6.3	4.1/6.3	4.5/6.2	4.6/6.2	5.6/6.6	5.2/6.9	5.5/7.7	4.9/6.9	2.8/5.7	2.0/5.5	4.8/5.7
90	Average	3.7	4.0	3.9	3.5	3.0	2.8	3.0	3.7	4.6	4.8	3.7	3.6	3.7
	Min/Max	2.2/5.0	2.8/5.6	2.9/4.7	2.8/4.0	2.6/3.3	2.5/3.0	2.9/3.2	3.2/4.0	3.9/5.4	4.0/5.8	2.4/5.2	1.8/5.3	3.3/4.1

Solar Radiation for 1-Axis Tracking Flat-Plate Collectors with a North-South Axis (kWh/m²/day), Uncertainty ±9%

Axis Tilt (°)		Jan	Feb	Mar	Apr	May	June	July	Aug	Sept	Oct	Nov	Dec	Year
0	Average	3.1	4.2	5.8	7.6	9.0	10.1	10.8	9.7	8.2	5.9	3.4	2.8	6.8
	Min/Max	2.0/4.1	3.2/5.9	4.4/7.2	5.3/8.8	6.9/10.6	7.9/11.7	9.6/12.2	7.5/11.1	6.8/10.1	4.9/7.1	2.4/4.6	1.6/4.0	6.2/7.3
Latitude -15	Average	4.0	5.1	6.5	8.0	9.2	10.1	10.9	10.2	9.1	7.0	4.3	3.7	7.4
	Min/Max	2.5/5.4	3.7/7.3	4.8/8.2	5.5/9.4	7.0/10.8	7.9/11.7	9.8/12.3	7.9/11.7	7.4/11.2	5.8/8.4	2.9/5.9	1.9/5.4	6.7/8.0
Latitude	Average	4.4	5.4	6.7	8.0	8.9	9.7	10.6	10.0	9.2	7.3	4.7	4.1	7.4
	Min/Max	2.6/6.0	3.9/7.8	4.9/8.4	5.4/9.4	6.7/10.5	7.6/11.3	9.4/11.9	7.7/11.5	7.5/11.3	6.1/8.9	3.0/6.4	2.1/6.0	6.7/8.1
Latitude +15	Average	4.6	5.5	6.6	7.7	8.4	9.1	9.9	9.6	9.0	7.4	4.8	4.3	7.2
	Min/Max	2.7/6.3	3.9/8.0	4.8/8.3	5.2/9.0	6.3/9.9	7.1/10.6	8.8/11.2	7.3/11.0	7.3/11.1	6.1/9.0	3.1/6.7	2.1/6.4	6.5/7.9

Solar Radiation for 2-Axis Tracking Flat-Plate Collectors (kWh/m²/day), Uncertainty ±9%

Tracker		Jan	Feb	Mar	Apr	May	June	July	Aug	Sept	Oct	Nov	Dec	Year
2-Axis	Average	4.6	5.5	6.7	8.1	9.3	10.4	11.2	10.2	9.2	7.4	4.9	4.4	7.7
	Min/Max	2.7/6.4	3.9/8.0	4.9/8.4	5.6/9.5	7.1/10.9	8.1/12.1	9.9/12.6	7.9/11.7	7.5/11.3	6.1/9.0	3.1/6.8	2.2/6.5	7.0/8.4

Direct Beam Solar Radiation for Concentrating Collectors (kWh/m²/day), Uncertainty ±8%

Tracker		Jan	Feb	Mar	Apr	May	June	July	Aug	Sept	Oct	Nov	Dec	Year
1-Axis, E-W Horiz Axis	Average	2.8	3.0	3.3	3.8	4.6	5.6	6.3	5.6	5.2	4.4	2.9	2.7	4.2
	Min/Max	1.1/4.5	1.6/5.2	2.0/4.7	1.8/5.0	2.7/6.0	3.6/7.1	5.2/7.5	3.6/6.8	3.8/6.9	3.3/5.8	1.3/4.6	0.8/4.7	3.6/4.8
1-Axis, N-S Horiz Axis	Average	2.0	2.7	3.8	5.1	6.2	7.4	8.5	7.5	6.5	4.5	2.3	1.8	4.9
	Min/Max	0.8/3.3	1.4/4.8	2.3/5.4	2.3/6.6	3.7/8.2	4.9/9.5	6.9/10.2	4.9/9.4	4.6/8.8	3.2/5.9	1.0/3.5	0.5/3.2	4.2/5.5
1-Axis, N-S Tilt=Latitude	Average	3.0	3.7	4.5	5.4	6.1	7.0	8.1	7.7	7.3	5.7	3.3	2.9	5.4
	Min/Max	1.2/4.9	2.0/6.4	2.7/6.4	2.5/7.0	3.6/8.0	4.6/9.0	6.7/9.8	5.0/9.6	5.2/9.9	4.1/7.5	1.5/5.2	0.8/5.0	4.6/6.1
2-Axis	Average	3.2	3.8	4.5	5.5	6.4	7.6	8.7	8.0	7.3	5.7	3.5	3.2	5.6
	Min/Max	1.3/5.3	2.0/6.6	2.7/6.4	2.5/7.1	3.8/8.4	5.0/9.8	7.1/10.5	5.1/9.9	5.2/9.9	4.2/7.6	1.6/5.5	0.9/5.5	4.8/6.4

Average Climatic Conditions

Element	Jan	Feb	Mar	Apr	May	June	July	Aug	Sept	Oct	Nov	Dec	Year
Temperature (°C)	-1.4	2.2	4.5	7.9	13.0	18.1	22.3	20.9	15.3	9.3	3.2	-1.2	9.5
Daily Minimum Temp	-8.5	-5.1	-3.8	-1.4	3.2	7.7	10.7	9.3	3.8	-1.5	-4.8	-8.3	0.1
Daily Maximum Temp	5.6	9.4	12.8	17.2	22.8	28.4	33.9	32.6	26.7	20.1	11.3	5.9	18.9
Record Minimum Temp	-31.1	-33.3	-19.4	-14.4	-11.1	-5.0	-1.7	-2.2	-11.1	-13.9	-22.2	-38.3	-38.3
Record Maximum Temp	20.0	23.3	27.2	32.2	35.6	41.1	41.1	42.2	39.4	32.8	25.0	19.4	42.2
HDD, Base 18.3°C	613	453	429	313	171	58	3	16	113	281	453	606	3508
CDD, Base 18.3°C	0	0	0	0	5	49	127	96	21	0	0	0	299
Relative Humidity (%)	69	63	56	47	42	38	28	30	36	47	63	70	49
Wind Speed (m/s)	3.2	3.6	3.9	4.0	4.0	3.9	3.8	3.6	3.6	3.4	3.5	3.4	3.7

Albany, NY

WBAN NO. 14735

LATITUDE: 42.75° N
LONGITUDE: 73.80° W
ELEVATION: 89 meters
MEAN PRESSURE: 1006 millibars

STATION TYPE: Primary

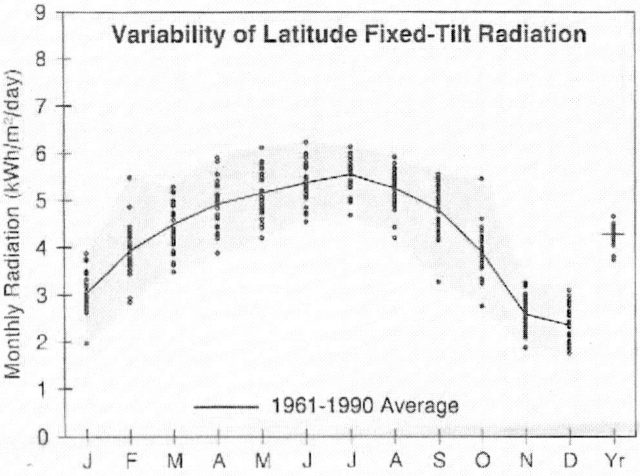

Solar Radiation for Flat-Plate Collectors Facing South at a Fixed Tilt (kWh/m²/day), Uncertainty ±9%

Tilt (°)		Jan	Feb	Mar	Apr	May	June	July	Aug	Sept	Oct	Nov	Dec	Year
0	Average	1.8	2.6	3.6	4.7	5.5	6.0	6.1	5.2	4.1	2.8	1.7	1.4	3.8
	Min/Max	1.2/2.1	2.1/3.3	3.1/4.1	3.9/5.4	4.5/6.4	5.1/6.9	5.4/6.6	4.6/5.8	3.2/4.6	2.2/3.6	1.4/2.0	1.2/1.6	3.6/4.1
Latitude -15	Average	2.7	3.6	4.4	5.0	5.5	5.8	6.0	5.5	4.8	3.7	2.4	2.1	4.3
	Min/Max	1.8/3.4	2.6/4.9	3.4/5.1	4.0/6.0	4.5/6.5	4.9/6.8	5.0/6.6	4.4/6.2	3.3/5.5	2.6/5.1	1.8/2.9	1.6/2.7	3.7/4.7
Latitude	Average	3.0	3.9	4.5	4.9	5.1	5.4	5.5	5.2	4.8	3.9	2.6	2.4	4.3
	Min/Max	2.0/3.9	2.8/5.5	3.5/5.3	3.9/5.9	4.2/6.1	4.5/6.2	4.7/6.1	4.2/5.9	3.3/5.6	2.8/5.5	1.9/3.3	1.8/3.1	3.7/4.7
Latitude +15	Average	3.2	4.1	4.4	4.5	4.6	4.6	4.8	4.8	4.6	3.9	2.7	2.5	4.1
	Min/Max	2.1/4.2	2.9/5.8	3.4/5.2	3.6/5.5	3.8/5.4	4.0/5.4	4.1/5.3	3.8/5.4	3.1/5.3	2.7/5.5	1.9/3.4	1.9/3.3	3.5/4.4
90	Average	3.1	3.7	3.5	3.1	2.7	2.6	2.8	3.0	3.3	3.2	2.4	2.4	3.0
	Min/Max	2.0/4.1	2.5/5.4	2.6/4.2	2.4/3.8	2.3/3.1	2.2/2.9	2.3/3.0	2.4/3.3	2.3/3.9	2.3/4.6	1.6/3.0	1.8/3.2	2.6/3.3

Solar Radiation for 1-Axis Tracking Flat-Plate Collectors with a North-South Axis (kWh/m²/day), Uncertainty ±9%

Axis Tilt (°)		Jan	Feb	Mar	Apr	May	June	July	Aug	Sept	Oct	Nov	Dec	Year
0	Average	2.5	3.6	4.8	6.0	6.9	7.5	7.7	6.7	5.4	3.8	2.2	1.9	4.9
	Min/Max	1.6/3.2	2.7/5.2	3.8/5.7	4.6/7.4	5.2/8.5	5.8/8.9	6.5/8.7	5.3/8.0	3.6/6.3	2.6/5.3	1.6/2.7	1.4/2.4	4.4/5.4
Latitude -15	Average	3.2	4.4	5.4	6.4	7.0	7.5	7.7	7.0	6.0	4.5	2.7	2.4	5.4
	Min/Max	2.1/4.1	3.2/6.4	4.3/6.4	4.7/7.9	5.3/8.7	5.8/8.9	6.6/8.8	5.6/8.3	3.9/7.0	3.1/6.4	1.9/3.5	1.8/3.2	4.8/5.9
Latitude	Average	3.5	4.6	5.5	6.3	6.8	7.2	7.5	6.9	6.0	4.6	2.9	2.6	5.4
	Min/Max	2.2/4.5	3.3/6.8	4.3/6.6	4.6/7.8	5.1/8.4	5.5/8.6	6.3/8.5	5.4/8.2	3.9/7.0	3.2/6.8	2.0/3.7	1.9/3.6	4.8/5.9
Latitude +15	Average	3.6	4.7	5.4	6.0	6.4	6.7	7.0	6.5	5.9	4.6	3.0	2.7	5.2
	Min/Max	2.3/4.8	3.4/7.0	4.2/6.6	4.4/7.6	4.7/7.9	5.1/8.0	5.9/7.9	5.2/7.8	3.8/6.9	3.2/6.8	2.0/3.8	2.0/3.8	4.7/5.7

Solar Radiation for 2-Axis Tracking Flat-Plate Collectors (kWh/m²/day), Uncertainty ±9%

Tracker		Jan	Feb	Mar	Apr	May	June	July	Aug	Sept	Oct	Nov	Dec	Year
2-Axis	Average	3.7	4.7	5.5	6.4	7.1	7.7	7.9	7.1	6.1	4.7	3.0	2.8	5.6
	Min/Max	2.4/4.8	3.4/7.0	4.3/6.6	4.7/7.9	5.3/8.8	5.9/9.1	6.7/9.0	5.6/8.4	4.0/7.1	3.2/6.8	2.1/3.8	2.0/3.8	5.0/6.1

Direct Beam Solar Radiation for Concentrating Collectors (kWh/m²/day), Uncertainty ±8%

Tracker		Jan	Feb	Mar	Apr	May	June	July	Aug	Sept	Oct	Nov	Dec	Year
1-Axis, E-W	Average	1.9	2.2	2.4	2.7	3.0	3.3	3.4	3.0	2.8	2.3	1.5	1.4	2.5
Horiz Axis	Min/Max	1.1/2.6	1.5/3.8	1.6/3.3	1.5/3.7	1.7/4.3	2.0/4.4	2.4/4.3	2.3/3.9	1.7/3.6	1.5/4.0	0.8/2.1	0.7/2.3	2.2/2.8
1-Axis, N-S	Average	1.3	1.9	2.7	3.5	3.9	4.4	4.5	3.9	3.3	2.2	1.1	0.9	2.8
Horiz Axis	Min/Max	0.8/1.8	1.3/3.4	1.8/3.6	2.0/4.8	2.1/5.6	2.5/5.8	3.1/5.6	3.0/5.2	2.0/4.2	1.4/3.8	0.6/1.6	0.4/1.5	2.5/3.2
1-Axis, N-S	Average	2.0	2.7	3.2	3.7	3.9	4.1	4.3	4.1	3.7	2.9	1.6	1.4	3.2
Tilt=Latitude	Min/Max	1.2/2.8	1.8/4.6	2.2/4.3	2.1/5.1	2.1/5.5	2.4/5.5	3.0/5.4	3.1/5.3	2.3/4.8	1.9/5.0	0.9/2.4	0.8/2.4	2.8/3.6
2-Axis	Average	2.2	2.7	3.2	3.8	4.1	4.5	4.7	4.2	3.8	3.0	1.7	1.6	3.3
	Min/Max	1.3/3.0	1.8/4.7	2.2/4.3	2.1/5.2	2.2/5.8	2.6/5.9	3.2/5.8	3.2/5.5	2.3/4.8	1.9/5.1	0.9/2.5	0.8/2.6	2.9/3.7

Average Climatic Conditions

Element	Jan	Feb	Mar	Apr	May	June	July	Aug	Sept	Oct	Nov	Dec	Year
Temperature (°C)	-6.3	-4.7	1.3	8.0	14.2	19.4	22.1	20.9	16.3	10.1	4.3	-3.1	8.6
Daily Minimum Temp	-11.7	-10.1	-4.2	1.7	7.4	12.6	15.3	14.3	9.7	3.7	-0.7	-7.7	2.6
Daily Maximum Temp	-1.0	0.7	6.7	14.2	20.9	26.1	28.9	27.4	22.9	16.6	9.3	1.6	14.5
Record Minimum Temp	-33.3	-29.4	-29.4	-12.2	-3.3	2.2	4.4	1.1	-4.4	-8.9	-15.0	-30.0	-33.3
Record Maximum Temp	16.7	19.4	30.0	33.3	34.4	37.2	37.8	37.2	37.8	31.7	27.8	21.7	37.8
HDD, Base 18.3°C	764	646	529	310	137	19	0	7	78	255	422	663	3830
CDD, Base 18.3°C	0	0	0	0	10	51	118	86	17	0	0	0	282
Relative Humidity (%)	71	68	65	61	66	70	71	74	76	72	73	74	70
Wind Speed (m/s)	4.2	4.4	4.7	4.6	3.9	3.6	3.3	3.1	3.3	3.4	4.1	4.2	3.9

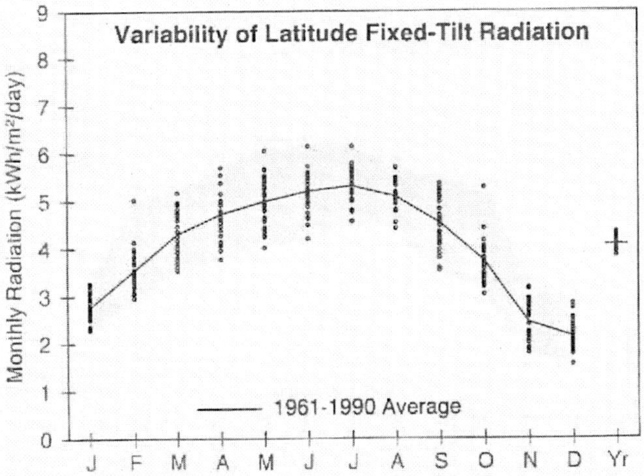

Binghamton, NY

WBAN NO. 04725

LATITUDE: 42.22° N
LONGITUDE: 75.98° W
ELEVATION: 499 meters
MEAN PRESSURE: 957 millibars

STATION TYPE: Secondary

Solar Radiation for Flat-Plate Collectors Facing South at a Fixed Tilt (kWh/m²/day), Uncertainty ±11%

Tilt (°)		Jan	Feb	Mar	Apr	May	June	July	Aug	Sept	Oct	Nov	Dec	Year
0	Average	1.7	2.5	3.5	4.5	5.3	5.8	5.8	5.0	3.9	2.7	1.7	1.4	3.7
	Min/Max	1.5/2.0	2.2/3.1	3.1/4.0	3.8/5.3	4.4/6.3	4.8/6.8	5.0/6.6	4.4/5.5	3.2/4.4	2.4/3.5	1.4/2.0	1.2/1.6	3.5/3.8
Latitude -15	Average	2.5	3.3	4.2	4.8	5.3	5.6	5.7	5.3	4.5	3.5	2.3	1.9	4.1
	Min/Max	2.1/2.9	2.8/4.5	3.5/5.0	3.9/5.8	4.3/6.5	4.6/6.7	4.9/6.6	4.6/5.9	3.6/5.3	2.9/4.9	1.7/2.9	1.5/2.5	3.9/4.3
Latitude	Average	2.8	3.5	4.3	4.7	5.0	5.2	5.3	5.1	4.5	3.7	2.4	2.1	4.1
	Min/Max	2.3/3.3	3.0/5.0	3.5/5.2	3.8/5.7	4.0/6.1	4.2/6.2	4.6/6.2	4.4/5.7	3.5/5.4	3.0/5.3	1.8/3.2	1.5/2.8	3.8/4.3
Latitude +15	Average	2.9	3.6	4.2	4.4	4.4	4.5	4.6	4.6	4.3	3.7	2.5	2.2	3.8
	Min/Max	2.4/3.5	3.0/5.3	3.4/5.1	3.5/5.3	3.6/5.3	3.7/5.3	4.0/5.4	4.0/5.2	3.3/5.1	3.0/5.3	1.8/3.3	1.5/3.0	3.6/4.1
90	Average	2.8	3.3	3.4	2.9	2.6	2.5	2.6	2.9	3.1	3.0	2.2	2.1	2.8
	Min/Max	2.2/3.4	2.7/4.9	2.6/4.2	2.5/3.6	2.2/3.0	2.2/2.8	2.3/3.0	2.6/3.3	2.4/3.7	2.4/4.4	1.5/3.0	1.4/2.9	2.6/3.0

Solar Radiation for 1-Axis Tracking Flat-Plate Collectors with a North-South Axis (kWh/m²/day), Uncertainty ±11%

Axis Tilt (°)		Jan	Feb	Mar	Apr	May	June	July	Aug	Sept	Oct	Nov	Dec	Year
0	Average	2.3	3.3	4.6	5.8	6.7	7.3	7.4	6.5	5.2	3.7	2.1	1.7	4.7
	Min/Max	2.0/2.7	2.8/4.7	3.8/5.7	4.5/7.2	5.0/8.6	5.4/8.9	6.1/8.6	5.6/7.5	3.9/6.2	3.0/5.3	1.6/2.8	1.3/2.2	4.4/5.1
Latitude -15	Average	2.9	3.9	5.1	6.1	6.8	7.3	7.5	6.8	5.7	4.3	2.6	2.2	5.1
	Min/Max	2.4/3.5	3.3/5.8	4.2/6.5	4.7/7.6	5.1/8.8	5.3/8.9	6.1/8.8	5.8/7.9	4.2/6.9	3.4/6.3	1.8/3.5	1.5/2.9	4.7/5.5
Latitude	Average	3.1	4.1	5.2	6.0	6.6	7.0	7.2	6.7	5.7	4.4	2.7	2.3	5.1
	Min/Max	2.5/3.8	3.4/6.2	4.2/6.6	4.6/7.6	4.8/8.6	5.1/8.5	5.9/8.4	5.6/7.8	4.2/6.9	3.5/6.6	1.9/3.7	1.6/3.2	4.7/5.5
Latitude +15	Average	3.2	4.2	5.2	5.7	6.2	6.5	6.7	6.3	5.5	4.4	2.8	2.4	4.9
	Min/Max	2.6/4.0	3.4/6.3	4.1/6.6	4.4/7.3	4.5/8.1	4.7/8.0	5.5/7.9	5.3/7.4	4.0/6.8	3.5/6.7	1.9/3.8	1.6/3.4	4.6/5.3

Solar Radiation for 2-Axis Tracking Flat-Plate Collectors (kWh/m²/day), Uncertainty ±11%

Tracker		Jan	Feb	Mar	Apr	May	June	July	Aug	Sept	Oct	Nov	Dec	Year
2-Axis	Average	3.3	4.2	5.2	6.1	6.9	7.5	7.6	6.8	5.7	4.5	2.8	2.4	5.3
	Min/Max	2.6/4.0	3.5/6.3	4.3/6.7	4.7/7.7	5.1/8.9	5.5/9.1	6.2/8.9	5.8/7.9	4.2/6.9	3.6/6.7	1.9/3.8	1.6/3.4	4.9/5.7

Direct Beam Solar Radiation for Concentrating Collectors (kWh/m²/day), Uncertainty ±11%

Tracker		Jan	Feb	Mar	Apr	May	June	July	Aug	Sept	Oct	Nov	Dec	Year
1-Axis, E-W	Average	1.4	1.7	2.1	2.5	2.8	3.1	3.2	2.9	2.5	2.2	1.2	1.0	2.2
Horiz Axis	Min/Max	0.9/2.0	1.1/3.4	1.4/3.3	1.3/3.6	1.4/4.4	1.5/4.6	2.1/4.3	2.0/3.8	1.4/3.4	1.4/3.9	0.5/2.0	0.3/1.8	1.8/2.6
1-Axis, N-S	Average	1.0	1.5	2.4	3.2	3.8	4.1	4.3	3.8	3.0	2.1	0.9	0.7	2.6
Horiz Axis	Min/Max	0.7/1.4	1.0/3.0	1.6/3.8	1.8/4.6	1.9/5.9	2.0/5.9	2.8/5.6	2.7/4.9	1.6/4.0	1.4/3.8	0.4/1.6	0.2/1.1	2.1/3.0
1-Axis, N-S	Average	1.6	2.1	2.8	3.4	3.7	3.9	4.1	3.9	3.4	2.7	1.4	1.1	2.8
Tilt=Latitude	Min/Max	1.0/2.2	1.4/4.1	1.9/4.6	1.9/4.9	1.9/5.8	1.9/5.6	2.7/5.4	2.8/5.1	1.9/4.6	1.8/4.9	0.6/2.4	0.3/1.9	2.4/3.3
2-Axis	Average	1.7	2.1	2.8	3.5	3.9	4.3	4.4	4.0	3.4	2.8	1.5	1.2	3.0
	Min/Max	1.1/2.3	1.5/4.2	1.9/4.6	2.0/5.0	2.0/6.1	2.0/6.1	2.9/5.8	2.9/5.2	1.9/4.6	1.8/5.0	0.6/2.5	0.3/2.0	2.4/3.5

Average Climatic Conditions

Element	Jan	Feb	Mar	Apr	May	June	July	Aug	Sept	Oct	Nov	Dec	Year
Temperature (°C)	-6.1	-5.2	0.3	6.9	13.2	17.9	20.7	19.6	15.3	9.3	3.5	-3.1	7.7
Daily Minimum Temp	-9.8	-9.3	-4.1	1.8	7.9	12.6	15.4	14.4	10.3	4.6	-0.2	-6.5	3.1
Daily Maximum Temp	-2.3	-1.1	4.7	11.9	18.6	23.2	25.9	24.7	20.3	14.1	7.2	0.4	12.3
Record Minimum Temp	-28.9	-26.1	-21.1	-12.8	-3.9	0.6	3.9	2.8	-3.9	-8.3	-16.1	-27.8	-28.9
Record Maximum Temp	17.2	18.9	27.8	31.1	31.7	34.4	36.7	34.4	35.6	27.8	25.0	18.3	36.7
HDD, Base 18.3°C	756	658	558	343	169	41	9	19	100	279	445	663	4041
CDD, Base 18.3°C	0	0	0	0	11	27	82	57	10	0	0	0	187
Relative Humidity (%)	74	72	69	65	67	72	72	75	78	74	76	78	73
Wind Speed (m/s)	5.0	4.9	5.2	5.1	4.5	4.1	3.7	3.7	3.9	4.3	4.9	5.0	4.5

Buffalo, NY

WBAN NO. 14733

LATITUDE: 42.93° N
LONGITUDE: 78.73° W
ELEVATION: 215 meters
MEAN PRESSURE: 991 millibars

STATION TYPE: Secondary

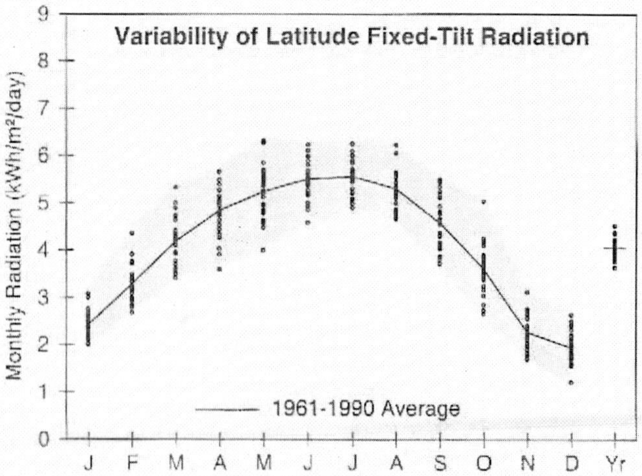

Solar Radiation for Flat-Plate Collectors Facing South at a Fixed Tilt (kWh/m²/day), Uncertainty ±9%

Tilt (°)		Jan	Feb	Mar	Apr	May	June	July	Aug	Sept	Oct	Nov	Dec	Year
0	Average	1.6	2.4	3.4	4.5	5.5	6.1	6.0	5.2	3.9	2.6	1.6	1.3	3.7
	Min/Max	1.4/1.8	2.0/2.8	3.0/4.0	3.6/5.2	4.3/6.5	5.1/6.9	5.4/6.7	4.6/5.9	3.4/4.5	2.2/3.3	1.3/1.9	1.1/1.5	3.4/4.0
Latitude -15	Average	2.2	3.1	4.1	5.0	5.6	6.0	6.0	5.5	4.6	3.4	2.1	1.8	4.1
	Min/Max	1.9/2.7	2.5/4.0	3.4/5.1	3.7/5.8	4.3/6.7	5.0/6.8	5.3/6.7	4.8/6.4	3.8/5.4	2.6/4.7	1.6/2.8	1.2/2.3	3.7/4.5
Latitude	Average	2.4	3.3	4.2	4.8	5.2	5.5	5.5	5.3	4.6	3.6	2.3	1.9	4.1
	Min/Max	2.0/3.1	2.7/4.4	3.4/5.3	3.6/5.7	4.0/6.3	4.6/6.2	4.9/6.2	4.7/6.2	3.7/5.5	2.7/5.0	1.7/3.1	1.2/2.6	3.6/4.5
Latitude +15	Average	2.5	3.4	4.1	4.5	4.6	4.8	4.8	4.8	4.3	3.6	2.3	2.0	3.8
	Min/Max	2.1/3.3	2.7/4.6	3.3/5.2	3.3/5.3	3.5/5.6	4.0/5.4	4.3/5.4	4.2/5.7	3.5/5.3	2.6/5.1	1.7/3.2	1.2/2.8	3.4/4.3
90	Average	2.4	3.0	3.3	3.0	2.7	2.6	2.8	3.0	3.2	3.0	2.0	1.9	2.7
	Min/Max	1.9/3.2	2.3/4.2	2.5/4.1	2.3/3.6	2.2/3.2	2.3/2.9	2.5/3.0	2.7/3.6	2.5/3.8	2.1/4.2	1.4/2.9	1.0/2.7	2.4/3.1

Solar Radiation for 1-Axis Tracking Flat-Plate Collectors with a North-South Axis (kWh/m²/day), Uncertainty ±9%

Axis Tilt (°)		Jan	Feb	Mar	Apr	May	June	July	Aug	Sept	Oct	Nov	Dec	Year
0	Average	2.0	3.1	4.4	5.9	7.1	7.9	7.9	6.8	5.2	3.5	1.9	1.6	4.8
	Min/Max	1.7/2.5	2.5/4.1	3.6/5.7	4.3/7.2	5.1/8.9	6.1/9.4	6.8/8.9	5.9/8.1	4.2/6.4	2.6/5.0	1.5/2.6	1.1/2.0	4.3/5.4
Latitude -15	Average	2.5	3.6	5.0	6.3	7.2	7.9	7.9	7.2	5.8	4.1	2.4	1.9	5.2
	Min/Max	2.1/3.2	2.9/5.0	4.0/6.6	4.4/7.7	5.1/9.1	6.1/9.4	6.8/9.0	6.1/8.5	4.5/7.1	2.9/6.0	1.7/3.3	1.2/2.7	4.6/5.8
Latitude	Average	2.7	3.8	5.1	6.2	7.0	7.6	7.7	7.0	5.8	4.3	2.5	2.1	5.1
	Min/Max	2.2/3.5	3.0/5.3	4.0/6.7	4.3/7.7	4.9/8.9	5.9/9.1	6.5/8.7	6.0/8.4	4.4/7.2	3.0/6.2	1.8/3.5	1.2/2.9	4.5/5.8
Latitude +15	Average	2.8	3.8	5.0	6.0	6.6	7.1	7.2	6.7	5.6	4.3	2.5	2.2	5.0
	Min/Max	2.3/3.7	3.1/5.4	3.9/6.6	4.2/7.4	4.6/8.4	5.4/8.5	6.1/8.2	5.7/8.0	4.3/7.0	2.9/6.3	1.8/3.7	1.2/3.1	4.4/5.7

Solar Radiation for 2-Axis Tracking Flat-Plate Collectors (kWh/m²/day), Uncertainty ±9%

Tracker		Jan	Feb	Mar	Apr	May	June	July	Aug	Sept	Oct	Nov	Dec	Year
2-Axis	Average	2.8	3.8	5.1	6.3	7.3	8.1	8.1	7.2	5.8	4.3	2.6	2.2	5.3
	Min/Max	2.3/3.7	3.1/5.4	4.0/6.7	4.5/7.8	5.2/9.2	6.3/9.6	6.9/9.2	6.2/8.6	4.5/7.2	3.0/6.3	1.8/3.7	1.2/3.1	4.7/6.0

Direct Beam Solar Radiation for Concentrating Collectors (kWh/m²/day), Uncertainty ±8%

Tracker		Jan	Feb	Mar	Apr	May	June	July	Aug	Sept	Oct	Nov	Dec	Year
1-Axis, E-W Horiz Axis	Average	1.1	1.5	2.1	2.7	3.3	3.7	3.7	3.2	2.6	2.1	1.1	0.9	2.3
	Min/Max	0.6/1.7	1.0/2.8	1.4/3.6	1.4/3.7	1.7/4.7	2.4/4.9	2.8/4.7	2.4/4.3	1.6/3.6	1.0/3.6	0.6/2.1	0.1/1.5	1.8/2.9
1-Axis, N-S Horiz Axis	Average	0.7	1.3	2.3	3.5	4.3	4.9	4.9	4.2	3.1	2.0	0.8	0.6	2.7
	Min/Max	0.3/1.1	0.9/2.5	1.6/4.0	1.8/4.8	2.2/6.2	3.1/6.5	3.7/6.1	3.1/5.4	2.0/4.4	1.0/3.5	0.4/1.5	0.1/0.9	2.2/3.4
1-Axis, N-S Tilt=Latitude	Average	1.2	1.8	2.8	3.7	4.2	4.6	4.7	4.3	3.5	2.6	1.3	0.9	3.0
	Min/Max	0.6/1.8	1.2/3.4	1.9/4.8	1.9/5.1	2.2/6.1	2.9/6.1	3.6/5.9	3.2/5.7	2.2/5.0	1.3/4.6	0.6/2.3	0.2/1.6	2.3/3.7
2-Axis	Average	1.3	1.8	2.8	3.8	4.5	5.0	5.0	4.4	3.5	2.6	1.3	1.0	3.1
	Min/Max	0.6/2.0	1.3/3.5	1.9/4.8	1.9/5.2	2.3/6.4	3.2/6.7	3.8/6.3	3.3/5.8	2.2/5.0	1.3/4.6	0.7/2.4	0.2/1.7	2.4/3.9

Average Climatic Conditions

Element	Jan	Feb	Mar	Apr	May	June	July	Aug	Sept	Oct	Nov	Dec	Year
Temperature (°C)	-4.7	-4.2	1.0	7.3	13.7	18.8	21.7	20.6	16.6	10.6	4.7	-1.6	8.7
Daily Minimum Temp	-8.3	-8.1	-3.4	2.3	8.3	13.6	16.6	15.6	11.7	5.9	1.1	-5.1	4.2
Daily Maximum Temp	-1.0	-0.2	5.4	12.3	18.9	24.1	26.8	25.5	21.6	15.2	8.4	1.8	13.2
Record Minimum Temp	-26.7	-28.9	-21.7	-11.1	-3.3	1.7	6.1	3.3	0.0	-6.7	-12.8	-23.3	-28.9
Record Maximum Temp	22.2	18.3	27.2	34.4	32.2	35.6	36.1	37.2	36.7	30.6	26.7	23.3	37.2
HDD, Base 18.3°C	713	630	537	330	155	33	3	9	72	239	408	618	3748
CDD, Base 18.3°C	0	0	0	0	11	48	108	78	21	0	0	0	265
Relative Humidity (%)	76	76	73	68	67	69	68	72	74	73	76	78	72
Wind Speed (m/s)	6.2	5.6	5.5	5.3	4.8	4.6	4.3	4.1	4.2	4.6	5.3	5.6	5.0

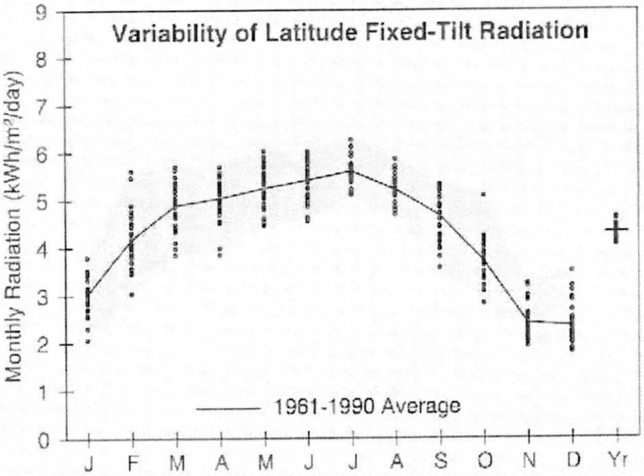

Massena, NY

WBAN NO. 94725

LATITUDE: 44.93° N
LONGITUDE: 74.85° W
ELEVATION: 63 meters
MEAN PRESSURE: 1008 millibars

STATION TYPE: Secondary

Solar Radiation for Flat-Plate Collectors Facing South at a Fixed Tilt (kWh/m²/day), Uncertainty ±9%

Tilt (°)		Jan	Feb	Mar	Apr	May	June	July	Aug	Sept	Oct	Nov	Dec	Year
0	Average	1.7	2.6	3.7	4.6	5.5	6.0	6.1	5.1	3.9	2.6	1.5	1.3	3.7
	Min/Max	1.3/1.9	2.1/3.2	3.2/4.0	3.8/5.1	4.7/6.3	5.2/6.7	5.5/6.7	4.7/5.6	3.2/4.3	2.2/3.2	1.3/1.8	1.1/1.6	3.6/3.9
Latitude -15	Average	2.7	3.8	4.7	5.2	5.6	5.9	6.0	5.4	4.6	3.5	2.2	2.1	4.3
	Min/Max	1.9/3.3	2.8/5.0	3.8/5.4	4.0/5.8	4.8/6.4	5.0/6.6	5.5/6.8	4.9/6.1	3.6/5.3	2.7/4.7	1.8/2.9	1.6/3.0	4.1/4.6
Latitude	Average	3.0	4.2	4.9	5.0	5.2	5.4	5.6	5.2	4.7	3.7	2.4	2.3	4.3
	Min/Max	2.1/3.8	3.0/5.6	3.8/5.7	3.8/5.7	4.5/6.0	4.6/6.0	5.1/6.3	4.7/5.9	3.6/5.3	2.8/5.1	1.9/3.2	1.8/3.5	4.1/4.6
Latitude +15	Average	3.2	4.3	4.8	4.7	4.6	4.7	4.9	4.7	4.4	3.7	2.5	2.5	4.1
	Min/Max	2.2/4.1	3.1/5.9	3.7/5.8	3.5/5.3	3.9/5.3	4.0/5.2	4.5/5.4	4.2/5.3	3.3/5.1	2.8/5.2	2.0/3.4	1.9/3.8	3.8/4.4
90	Average	3.1	4.0	4.1	3.3	2.9	2.7	2.9	3.1	3.3	3.1	2.2	2.4	3.1
	Min/Max	2.0/4.0	2.8/5.6	3.0/5.1	2.5/4.0	2.5/3.2	2.4/3.0	2.7/3.1	2.8/3.5	2.5/3.8	2.3/4.4	1.7/3.2	1.8/3.8	2.8/3.4

Solar Radiation for 1-Axis Tracking Flat-Plate Collectors with a North-South Axis (kWh/m²/day), Uncertainty ±9%

Axis Tilt (°)		Jan	Feb	Mar	Apr	May	June	July	Aug	Sept	Oct	Nov	Dec	Year
0	Average	2.4	3.8	5.2	6.3	7.3	7.9	8.1	6.9	5.4	3.6	2.0	1.8	5.1
	Min/Max	1.6/3.0	2.8/5.2	4.1/6.1	4.6/7.2	5.8/8.7	6.3/9.1	6.9/9.3	6.1/7.8	4.1/6.2	2.8/4.9	1.6/2.7	1.4/2.6	4.7/5.4
Latitude -15	Average	3.2	4.7	6.0	6.7	7.4	8.0	8.2	7.2	6.0	4.3	2.5	2.4	5.6
	Min/Max	2.1/4.0	3.3/6.5	4.6/7.1	4.9/7.8	5.9/9.0	6.3/9.1	7.1/9.5	6.4/8.3	4.4/7.0	3.2/6.0	2.0/3.5	1.8/3.6	5.1/6.0
Latitude	Average	3.4	5.0	6.1	6.6	7.2	7.6	7.9	7.1	6.0	4.5	2.7	2.6	5.6
	Min/Max	2.3/4.4	3.5/7.0	4.6/7.3	4.8/7.7	5.7/8.7	6.0/8.8	6.8/9.2	6.2/8.2	4.4/7.0	3.3/6.3	2.1/3.8	2.0/4.0	5.1/6.1
Latitude +15	Average	3.6	5.1	6.1	6.4	6.8	7.2	7.4	6.7	5.8	4.5	2.8	2.8	5.4
	Min/Max	2.3/4.7	3.5/7.2	4.5/7.3	4.5/7.4	5.4/8.2	5.6/8.2	6.3/8.6	5.9/7.8	4.2/6.8	3.2/6.3	2.1/3.9	2.0/4.3	5.0/5.9

Solar Radiation for 2-Axis Tracking Flat-Plate Collectors (kWh/m²/day), Uncertainty ±9%

Tracker		Jan	Feb	Mar	Apr	May	June	July	Aug	Sept	Oct	Nov	Dec	Year
2-Axis	Average	3.6	5.1	6.1	6.7	7.6	8.2	8.4	7.3	6.0	4.5	2.8	2.8	5.8
	Min/Max	2.3/4.7	3.6/7.2	4.6/7.3	4.9/7.8	6.0/9.1	6.4/9.4	7.2/9.7	6.4/8.4	4.4/7.0	3.3/6.4	2.1/3.9	2.0/4.4	5.3/6.3

Direct Beam Solar Radiation for Concentrating Collectors (kWh/m²/day), Uncertainty ±8%

Tracker		Jan	Feb	Mar	Apr	May	June	July	Aug	Sept	Oct	Nov	Dec	Year
1-Axis, E-W Horiz Axis	Average	1.8	2.5	2.8	3.0	3.3	3.7	3.9	3.2	2.7	2.2	1.3	1.4	2.7
	Min/Max	1.1/2.7	1.5/4.1	1.8/3.8	1.8/3.7	2.2/4.6	2.3/4.7	3.0/5.0	2.6/4.2	1.7/3.6	1.3/3.7	0.8/2.2	0.7/2.7	2.4/3.0
1-Axis, N-S Horiz Axis	Average	1.2	2.1	3.0	3.8	4.5	4.9	5.2	4.2	3.3	2.1	1.0	0.8	3.0
	Min/Max	0.7/1.8	1.2/3.6	2.0/4.1	2.2/4.8	3.0/6.1	3.1/6.2	3.9/6.7	3.4/5.4	2.0/4.3	1.3/3.5	0.5/1.6	0.4/1.6	2.7/3.5
1-Axis, N-S Tilt=Latitude	Average	2.0	3.0	3.7	4.1	4.4	4.7	5.0	4.4	3.8	2.8	1.5	1.4	3.4
	Min/Max	1.2/2.9	1.7/5.0	2.4/5.0	2.4/5.2	3.0/6.1	2.9/6.0	3.8/6.5	3.5/5.7	2.4/5.0	1.6/4.6	0.9/2.5	0.8/2.8	3.1/3.9
2-Axis	Average	2.1	3.1	3.7	4.2	4.7	5.1	5.4	4.5	3.8	2.9	1.6	1.6	3.6
	Min/Max	1.3/3.1	1.8/5.1	2.5/5.0	2.5/5.3	3.1/6.4	3.2/6.5	4.1/6.9	3.6/5.8	2.4/5.0	1.7/4.7	0.9/2.6	0.8/3.0	3.2/4.1

Average Climatic Conditions

Element	Jan	Feb	Mar	Apr	May	June	July	Aug	Sept	Oct	Nov	Dec	Year
Temperature (°C)	-9.8	-8.6	-2.1	5.9	12.9	17.7	20.7	19.2	14.4	8.3	1.8	-6.6	6.2
Daily Minimum Temp	-15.4	-14.2	-7.2	0.3	6.6	11.4	14.3	13.1	8.4	2.8	-2.4	-11.3	0.6
Daily Maximum Temp	-4.3	-3.1	3.1	11.5	19.2	24.1	27.0	25.3	20.4	13.8	6.0	-1.8	11.8
Record Minimum Temp	-38.9	-36.7	-31.1	-13.9	-5.6	-0.6	3.3	1.7	-2.8	-8.9	-20.6	-35.6	-38.9
Record Maximum Temp	15.0	16.1	27.8	31.7	34.4	33.9	35.6	37.8	32.8	27.8	23.9	18.3	37.8
HDD, Base 18.3°C	873	754	632	372	177	41	9	29	123	310	495	772	4586
CDD, Base 18.3°C	0	0	0	0	8	22	81	55	6	0	0	0	172
Relative Humidity (%)	76	74	73	68	68	72	72	76	79	77	81	82	75
Wind Speed (m/s)	4.2	4.1	4.2	4.3	3.8	3.4	3.1	2.9	3.0	3.6	4.0	4.0	3.7

New York City, NY

WBAN NO. 94728

LATITUDE: 40.78° N
LONGITUDE: 73.97° W
ELEVATION: 57 meters
MEAN PRESSURE: 1016 millibars

STATION TYPE: Primary

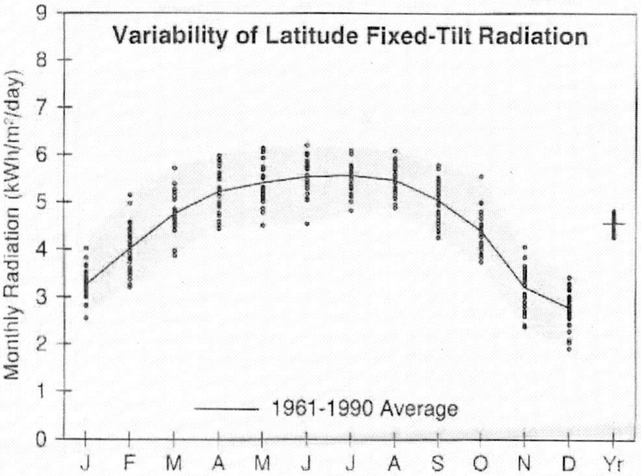

Solar Radiation for Flat-Plate Collectors Facing South at a Fixed Tilt (kWh/m²/day), Uncertainty ±9%

Tilt (°)		Jan	Feb	Mar	Apr	May	June	July	Aug	Sept	Oct	Nov	Dec	Year
0	Average	1.9	2.7	3.8	4.9	5.7	6.1	6.0	5.4	4.3	3.2	2.0	1.6	4.0
	Min/Max	1.7/2.2	2.3/3.3	3.3/4.4	4.3/5.5	4.8/6.5	5.0/6.8	5.3/6.6	4.8/5.9	3.9/4.9	2.9/3.8	1.7/2.4	1.4/1.8	3.7/4.2
Latitude -15	Average	2.9	3.7	4.6	5.3	5.8	6.0	6.0	5.7	5.0	4.1	2.9	2.4	4.5
	Min/Max	2.3/3.5	3.0/4.6	3.8/5.5	4.6/6.1	4.8/6.5	4.9/6.7	5.2/6.6	5.1/6.3	4.3/5.7	3.6/5.1	2.2/3.6	1.8/3.0	4.3/4.8
Latitude	Average	3.2	4.0	4.8	5.2	5.4	5.5	5.6	5.5	5.0	4.4	3.2	2.8	4.6
	Min/Max	2.5/4.0	3.2/5.2	3.9/5.7	4.4/6.0	4.5/6.1	4.5/6.2	4.8/6.1	4.9/6.1	4.3/5.8	3.7/5.6	2.4/4.1	1.9/3.4	4.3/4.8
Latitude +15	Average	3.4	4.1	4.6	4.8	4.8	4.8	4.9	5.0	4.8	4.4	3.3	3.0	4.3
	Min/Max	2.7/4.3	3.3/5.4	3.7/5.6	4.1/5.6	4.0/5.4	4.0/5.3	4.2/5.3	4.4/5.5	4.0/5.5	3.7/5.6	2.4/4.3	2.0/3.7	4.0/4.6
90	Average	3.2	3.6	3.5	3.1	2.7	2.6	2.7	3.0	3.4	3.6	3.0	2.7	3.1
	Min/Max	2.4/4.2	2.8/4.9	2.8/4.3	2.7/3.7	2.4/3.1	2.3/2.8	2.4/2.9	2.8/3.3	2.8/3.9	3.0/4.6	2.1/3.8	1.7/3.5	2.9/3.4

Solar Radiation for 1-Axis Tracking Flat-Plate Collectors with a North-South Axis (kWh/m²/day), Uncertainty ±9%

Axis Tilt (°)		Jan	Feb	Mar	Apr	May	June	July	Aug	Sept	Oct	Nov	Dec	Year
0	Average	2.7	3.7	5.1	6.3	7.1	7.5	7.5	6.8	5.7	4.3	2.7	2.2	5.1
	Min/Max	2.1/3.3	3.0/4.9	4.1/6.2	5.2/7.4	5.5/8.4	5.8/8.8	6.2/8.4	5.8/7.7	4.6/6.6	3.7/5.6	2.0/3.4	1.6/2.7	4.7/5.4
Latitude -15	Average	3.4	4.5	5.7	6.7	7.2	7.5	7.5	7.1	6.2	5.1	3.4	2.8	5.6
	Min/Max	2.6/4.3	3.5/5.9	4.5/7.1	5.4/7.9	5.5/8.6	5.8/8.8	6.2/8.5	6.0/8.0	5.0/7.2	4.2/6.6	2.4/4.4	1.9/3.6	5.1/5.9
Latitude	Average	3.7	4.7	5.8	6.6	6.9	7.2	7.2	6.9	6.2	5.3	3.7	3.1	5.6
	Min/Max	2.8/4.7	3.7/6.3	4.5/7.3	5.3/7.9	5.3/8.3	5.5/8.4	5.9/8.2	5.9/7.9	5.0/7.3	4.3/6.9	2.5/4.7	2.0/3.9	5.2/6.0
Latitude +15	Average	3.9	4.8	5.7	6.3	6.5	6.7	6.7	6.6	6.0	5.3	3.8	3.3	5.5
	Min/Max	2.9/5.0	3.8/6.4	4.4/7.2	5.1/7.6	5.0/7.8	5.1/7.8	5.5/7.6	5.6/7.5	4.8/7.1	4.3/7.0	2.6/4.9	2.1/4.2	5.0/5.9

Solar Radiation for 2-Axis Tracking Flat-Plate Collectors (kWh/m²/day), Uncertainty ±9%

Tracker		Jan	Feb	Mar	Apr	May	June	July	Aug	Sept	Oct	Nov	Dec	Year
2-Axis	Average	3.9	4.8	5.8	6.7	7.3	7.7	7.6	7.1	6.2	5.3	3.8	3.3	5.8
	Min/Max	2.9/5.0	3.8/6.4	4.5/7.3	5.4/8.0	5.6/8.7	5.9/9.0	6.3/8.6	6.1/8.1	5.0/7.3	4.4/7.0	2.6/4.9	2.1/4.2	5.4/6.2

Direct Beam Solar Radiation for Concentrating Collectors (kWh/m²/day), Uncertainty ±8%

Tracker		Jan	Feb	Mar	Apr	May	June	July	Aug	Sept	Oct	Nov	Dec	Year
1-Axis, E-W Horiz Axis	Average	2.2	2.5	2.7	2.9	3.0	3.2	3.2	3.0	2.8	2.7	2.1	1.9	2.7
	Min/Max	1.4/2.9	1.7/3.5	1.7/3.9	2.0/3.8	1.8/4.2	1.9/4.4	2.2/4.0	2.2/3.9	1.9/3.5	2.0/4.1	1.1/3.0	0.9/2.7	2.3/3.0
1-Axis, N-S Horiz Axis	Average	1.5	2.2	3.0	3.7	3.9	4.1	4.1	3.8	3.3	2.6	1.5	1.2	2.9
	Min/Max	0.9/2.1	1.4/3.1	1.8/4.3	2.4/4.9	2.2/5.4	2.4/5.6	2.7/5.2	2.8/4.9	2.2/4.2	1.9/4.0	0.8/2.3	0.6/1.8	2.4/3.3
1-Axis, N-S Tilt=Latitude	Average	2.3	2.9	3.6	3.9	3.9	3.9	3.9	3.9	3.7	3.4	2.3	2.0	3.3
	Min/Max	1.5/3.2	2.0/4.2	2.2/5.2	2.6/5.2	2.2/5.3	2.3/5.3	2.6/5.0	2.8/5.0	2.5/4.8	2.4/5.1	1.2/3.3	0.9/2.8	2.8/3.7
2-Axis	Average	2.5	3.0	3.6	4.0	4.1	4.2	4.2	4.1	3.8	3.4	2.4	2.1	3.5
	Min/Max	1.6/3.4	2.1/4.3	2.2/5.2	2.6/5.3	2.3/5.7	2.5/5.8	2.8/5.3	2.9/5.2	2.5/4.8	2.5/5.1	1.3/3.5	1.0/3.1	2.9/3.8

Average Climatic Conditions

Element	Jan	Feb	Mar	Apr	May	June	July	Aug	Sept	Oct	Nov	Dec	Year
Temperature (°C)	-0.3	0.9	5.8	11.4	17.1	22.0	24.9	24.2	20.1	14.2	8.7	2.6	12.6
Daily Minimum Temp	-3.7	-2.8	1.6	6.6	12.1	17.2	20.2	19.6	15.6	9.8	5.1	-0.7	8.4
Daily Maximum Temp	3.1	4.6	10.0	16.2	22.1	26.7	29.6	28.7	24.6	18.5	12.2	5.8	16.8
Record Minimum Temp	-21.1	-26.1	-16.1	-11.1	0.0	6.7	11.1	10.0	3.9	-2.2	-15.0	-25.0	-26.1
Record Maximum Temp	22.2	23.9	30.0	35.6	37.2	38.3	41.1	40.0	38.9	34.4	28.9	22.2	41.1
HDD, Base 18.3°C	577	488	389	208	69	0	0	0	19	139	290	489	2669
CDD, Base 18.3°C	0	0	0	0	30	113	203	181	72	9	0	0	609
Relative Humidity (%)	61	60	60	59	64	65	65	67	67	65	64	63	63
Wind Speed (m/s)	6.1	6.1	6.1	5.8	5.2	4.9	4.7	4.7	5.0	5.1	5.8	6.0	5.5

Rochester, NY

WBAN NO. 14768

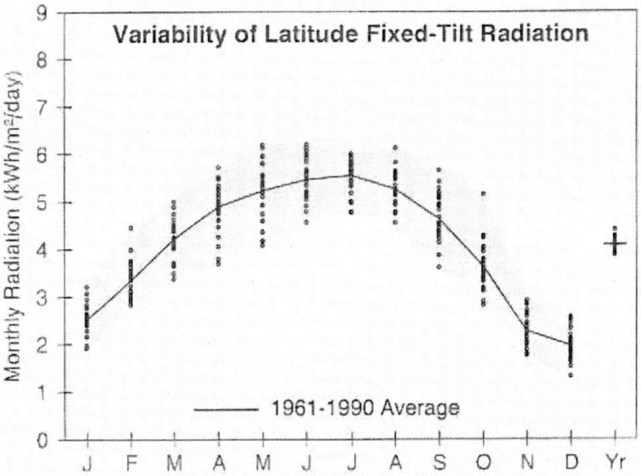

LATITUDE: 43.12° N
LONGITUDE: 77.67° W
ELEVATION: 169 meters
MEAN PRESSURE: 996 millibars

STATION TYPE: Secondary

Solar Radiation for Flat-Plate Collectors Facing South at a Fixed Tilt (kWh/m²/day), Uncertainty ±9%

Tilt (°)		Jan	Feb	Mar	Apr	May	June	July	Aug	Sept	Oct	Nov	Dec	Year
0	Average	1.6	2.4	3.4	4.6	5.5	6.1	6.0	5.2	4.0	2.7	1.6	1.3	3.7
	Min/Max	1.4/1.9	2.1/2.8	3.0/3.8	3.7/5.2	4.5/6.4	5.1/6.9	5.2/6.5	4.6/5.8	3.3/4.6	2.3/3.4	1.4/1.8	1.1/1.5	3.5/3.9
Latitude -15	Average	2.3	3.1	4.1	5.0	5.6	5.9	6.0	5.5	4.6	3.4	2.1	1.8	4.1
	Min/Max	1.8/2.9	2.7/4.0	3.4/4.8	3.9/5.8	4.4/6.6	5.0/6.8	5.2/6.5	4.8/6.3	3.7/5.6	2.8/4.8	1.7/2.6	1.3/2.3	3.9/4.4
Latitude	Average	2.5	3.3	4.2	4.9	5.2	5.4	5.5	5.3	4.6	3.6	2.3	2.0	4.1
	Min/Max	1.9/3.2	2.8/4.4	3.4/5.0	3.7/5.7	4.1/6.2	4.6/6.2	4.8/6.0	4.6/6.1	3.6/5.7	2.8/5.2	1.8/2.9	1.3/2.6	3.9/4.4
Latitude +15	Average	2.6	3.4	4.1	4.5	4.6	4.7	4.8	4.8	4.4	3.6	2.3	2.0	3.8
	Min/Max	2.0/3.4	2.9/4.7	3.3/4.9	3.4/5.3	3.6/5.4	4.0/5.3	4.2/5.2	4.1/5.6	3.4/5.4	2.8/5.2	1.7/3.0	1.3/2.7	3.6/4.2
90	Average	2.5	3.1	3.3	3.1	2.7	2.6	2.8	3.0	3.2	3.0	2.0	1.9	2.8
	Min/Max	1.8/3.3	2.5/4.3	2.5/4.0	2.3/3.5	2.3/3.1	2.3/2.9	2.4/2.9	2.6/3.5	2.4/4.0	2.2/4.4	1.5/2.8	1.1/2.7	2.6/3.0

Solar Radiation for 1-Axis Tracking Flat-Plate Collectors with a North-South Axis (kWh/m²/day), Uncertainty ±9%

Axis Tilt (°)		Jan	Feb	Mar	Apr	May	June	July	Aug	Sept	Oct	Nov	Dec	Year
0	Average	2.1	3.1	4.5	6.0	7.1	7.9	8.0	6.9	5.3	3.6	2.0	1.6	4.8
	Min/Max	1.6/2.6	2.6/4.2	3.7/5.5	4.4/7.1	5.3/8.8	6.3/9.3	6.6/8.7	5.9/8.0	4.1/6.7	2.8/5.2	1.6/2.5	1.2/2.0	4.6/5.3
Latitude -15	Average	2.6	3.7	5.0	6.4	7.3	7.9	8.0	7.2	5.8	4.2	2.4	2.0	5.2
	Min/Max	1.9/3.3	3.0/5.2	4.0/6.2	4.6/7.7	5.3/9.0	6.3/9.3	6.6/8.7	6.1/8.5	4.4/7.4	3.1/6.2	1.8/3.1	1.3/2.6	4.9/5.7
Latitude	Average	2.8	3.9	5.1	6.3	7.1	7.6	7.7	7.0	5.9	4.3	2.5	2.1	5.2
	Min/Max	2.1/3.6	3.1/5.5	4.0/6.4	4.5/7.6	5.1/8.7	6.0/8.9	6.3/8.4	5.9/8.4	4.4/7.5	3.2/6.5	1.8/3.3	1.3/2.9	4.9/5.7
Latitude +15	Average	2.9	3.9	5.0	6.0	6.6	7.1	7.3	6.7	5.7	4.3	2.6	2.2	5.0
	Min/Max	2.1/3.8	3.2/5.6	4.0/6.3	4.3/7.3	4.7/8.2	5.6/8.4	5.9/7.9	5.6/8.0	4.2/7.3	3.2/6.5	1.8/3.4	1.3/3.0	4.8/5.6

Solar Radiation for 2-Axis Tracking Flat-Plate Collectors (kWh/m²/day), Uncertainty ±9%

Tracker		Jan	Feb	Mar	Apr	May	June	July	Aug	Sept	Oct	Nov	Dec	Year
2-Axis	Average	2.9	3.9	5.1	6.4	7.4	8.1	8.2	7.2	5.9	4.3	2.6	2.2	5.4
	Min/Max	2.1/3.8	3.2/5.6	4.1/6.4	4.6/7.7	5.3/9.1	6.4/9.5	6.7/8.9	6.1/8.5	4.4/7.5	3.2/6.5	1.9/3.5	1.3/3.0	5.1/5.9

Direct Beam Solar Radiation for Concentrating Collectors (kWh/m²/day), Uncertainty ±8%

Tracker		Jan	Feb	Mar	Apr	May	June	July	Aug	Sept	Oct	Nov	Dec	Year
1-Axis, E-W	Average	1.2	1.5	2.1	2.8	3.3	3.7	3.7	3.2	2.6	2.1	1.1	0.9	2.4
Horiz Axis	Min/Max	0.8/1.8	1.0/2.9	1.4/3.4	1.4/3.7	1.5/4.8	2.3/4.7	2.5/4.4	2.4/4.2	1.6/3.8	1.2/3.8	0.5/1.8	0.1/1.5	2.1/2.8
1-Axis, N-S	Average	0.8	1.3	2.3	3.6	4.3	4.9	5.0	4.2	3.2	2.0	0.8	0.6	2.8
Horiz Axis	Min/Max	0.5/1.2	0.8/2.6	1.6/3.8	1.8/4.7	2.1/6.3	3.1/6.3	3.4/5.8	3.2/5.4	2.0/4.7	1.1/3.8	0.4/1.4	0.1/0.9	2.5/3.2
1-Axis, N-S	Average	1.3	1.8	2.8	3.8	4.3	4.6	4.8	4.3	3.6	2.6	1.3	0.9	3.0
Tilt=Latitude	Min/Max	0.8/1.9	1.2/3.5	1.9/4.5	1.9/5.1	2.1/6.2	3.0/6.0	3.3/5.5	3.3/5.6	2.2/5.3	1.5/4.9	0.6/2.0	0.1/1.5	2.8/3.6
2-Axis	Average	1.4	1.9	2.8	3.9	4.5	5.0	5.2	4.5	3.6	2.7	1.3	1.0	3.2
	Min/Max	0.9/2.1	1.2/3.6	1.9/4.5	1.9/5.2	2.2/6.5	3.2/6.5	3.5/5.9	3.4/5.8	2.2/5.3	1.5/4.9	0.6/2.2	0.1/1.6	2.9/3.7

Average Climatic Conditions

Element	Jan	Feb	Mar	Apr	May	June	July	Aug	Sept	Oct	Nov	Dec	Year
Temperature (°C)	-4.7	-4.1	1.3	7.7	13.9	18.4	21.2	20.0	16.5	10.6	4.7	-1.6	8.7
Daily Minimum Temp	-8.7	-8.6	-3.5	2.2	7.9	12.4	15.3	14.3	10.9	5.3	0.7	-5.3	3.6
Daily Maximum Temp	-0.6	0.3	5.9	13.3	19.9	24.3	27.1	25.6	22.1	15.8	8.8	2.1	13.7
Record Minimum Temp	-26.7	-28.3	-21.1	-10.6	-3.3	1.7	5.6	2.2	-2.2	-6.7	-15.0	-26.7	-28.3
Record Maximum Temp	23.3	19.4	28.9	33.9	34.4	37.8	36.7	37.2	37.2	32.8	27.2	22.2	37.8
HDD, Base 18.3°C	713	628	529	318	150	34	6	18	76	242	408	618	3741
CDD, Base 18.3°C	0	0	0	0	14	36	95	70	21	0	0	0	236
Relative Humidity (%)	74	74	71	67	67	69	70	74	77	74	76	78	73
Wind Speed (m/s)	5.4	5.0	5.0	5.1	4.3	4.0	3.7	3.6	3.7	4.0	4.6	4.9	4.4

Syracuse, NY

WBAN NO. 14771

LATITUDE: 43.12° N
LONGITUDE: 76.12° W
ELEVATION: 124 meters
MEAN PRESSURE: 1001 millibars

STATION TYPE: Secondary

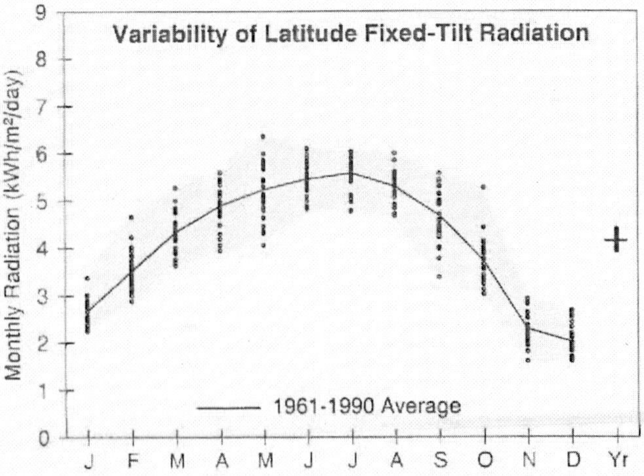

Solar Radiation for Flat-Plate Collectors Facing South at a Fixed Tilt (kWh/m²/day), Uncertainty ±9%

Tilt (°)		Jan	Feb	Mar	Apr	May	June	July	Aug	Sept	Oct	Nov	Dec	Year
0	Average	1.7	2.5	3.5	4.6	5.5	6.1	6.0	5.2	4.0	2.7	1.6	1.3	3.7
	Min/Max	1.5/1.9	2.1/2.9	3.1/4.0	3.9/5.1	4.4/6.6	5.4/6.8	5.3/6.5	4.7/5.7	3.1/4.5	2.3/3.4	1.3/1.8	1.1/1.5	3.6/3.9
Latitude -15	Average	2.4	3.3	4.2	5.0	5.6	5.9	6.0	5.5	4.7	3.5	2.1	1.8	4.2
	Min/Max	2.0/3.0	2.7/4.2	3.6/5.1	4.1/5.7	4.3/6.8	5.2/6.6	5.2/6.5	4.9/6.2	3.4/5.5	2.9/4.9	1.5/2.6	1.5/2.4	4.0/4.4
Latitude	Average	2.7	3.5	4.3	4.9	5.2	5.4	5.6	5.3	4.7	3.7	2.3	2.0	4.1
	Min/Max	2.2/3.4	2.9/4.7	3.6/5.3	3.9/5.6	4.1/6.3	4.8/6.1	4.8/6.0	4.7/6.0	3.4/5.6	3.0/5.3	1.6/2.9	1.6/2.7	3.9/4.4
Latitude +15	Average	2.8	3.6	4.2	4.5	4.6	4.7	4.9	4.8	4.5	3.7	2.3	2.1	3.9
	Min/Max	2.3/3.6	2.9/4.9	3.5/5.2	3.6/5.2	3.6/5.6	4.2/5.2	4.2/5.3	4.2/5.5	3.2/5.3	3.0/5.3	1.6/3.0	1.6/2.9	3.7/4.1
90	Average	2.7	3.3	3.5	3.1	2.7	2.6	2.8	3.1	3.2	3.1	2.0	2.0	2.8
	Min/Max	2.2/3.5	2.5/4.5	2.7/4.2	2.5/3.5	2.3/3.2	2.4/2.9	2.5/3.0	2.7/3.5	2.3/3.9	2.5/4.5	1.3/2.8	1.5/2.8	2.6/3.1

Solar Radiation for 1-Axis Tracking Flat-Plate Collectors with a North-South Axis (kWh/m²/day), Uncertainty ±9%

Axis Tilt (°)		Jan	Feb	Mar	Apr	May	June	July	Aug	Sept	Oct	Nov	Dec	Year
0	Average	2.2	3.2	4.6	6.0	7.1	7.9	8.0	6.9	5.4	3.6	2.0	1.6	4.9
	Min/Max	1.9/2.8	2.7/4.4	3.8/5.8	4.6/7.2	5.0/8.9	6.6/9.1	6.7/8.8	6.1/7.9	3.8/6.4	3.0/5.2	1.4/2.5	1.3/2.1	4.6/5.2
Latitude -15	Average	2.8	3.9	5.2	6.4	7.3	7.9	8.0	7.2	5.9	4.3	2.4	2.0	5.3
	Min/Max	2.3/3.5	3.1/5.3	4.2/6.6	4.8/7.7	5.0/9.1	6.6/9.1	6.7/8.9	6.3/8.4	4.1/7.1	3.5/6.2	1.6/3.1	1.6/2.7	5.0/5.6
Latitude	Average	3.0	4.1	5.3	6.3	7.0	7.6	7.8	7.1	5.9	4.4	2.5	2.2	5.3
	Min/Max	2.5/3.9	3.3/5.7	4.3/6.8	4.7/7.6	4.8/8.9	6.3/8.7	6.4/8.6	6.2/8.2	4.1/7.2	3.6/6.5	1.7/3.3	1.7/3.0	5.0/5.7
Latitude +15	Average	3.1	4.1	5.2	6.1	6.6	7.1	7.3	6.7	5.8	4.4	2.6	2.3	5.1
	Min/Max	2.5/4.1	3.3/5.8	4.1/6.7	4.5/7.3	4.5/8.4	5.8/8.2	6.0/8.1	5.9/7.8	3.9/7.0	3.5/6.6	1.7/3.4	1.7/3.2	4.8/5.5

Solar Radiation for 2-Axis Tracking Flat-Plate Collectors (kWh/m²/day), Uncertainty ±9%

Tracker		Jan	Feb	Mar	Apr	May	June	July	Aug	Sept	Oct	Nov	Dec	Year
2-Axis	Average	3.1	4.1	5.3	6.4	7.4	8.1	8.2	7.2	6.0	4.5	2.6	2.3	5.4
	Min/Max	2.5/4.1	3.3/5.8	4.3/6.8	4.8/7.7	5.1/9.3	6.7/9.3	6.8/9.1	6.4/8.4	4.1/7.2	3.6/6.6	1.7/3.5	1.8/3.2	5.2/5.8

Direct Beam Solar Radiation for Concentrating Collectors (kWh/m²/day), Uncertainty ±8%

Tracker		Jan	Feb	Mar	Apr	May	June	July	Aug	Sept	Oct	Nov	Dec	Year
1-Axis, E-W Horiz Axis	Average	1.4	1.7	2.2	2.8	3.3	3.7	3.8	3.2	2.7	2.2	1.1	1.0	2.4
	Min/Max	0.6/2.0	0.9/3.0	1.5/3.6	1.7/3.7	1.4/4.8	2.5/4.7	2.6/4.6	2.5/4.1	1.4/3.7	1.5/3.9	0.5/1.9	0.4/1.8	2.1/2.8
1-Axis, N-S Horiz Axis	Average	0.9	1.4	2.4	3.6	4.3	4.9	5.0	4.2	3.2	2.1	0.8	0.6	2.8
	Min/Max	0.4/1.3	0.7/2.7	1.7/4.0	2.0/4.8	1.9/6.3	3.4/6.3	3.5/6.2	3.4/5.3	1.7/4.4	1.4/3.7	0.3/1.4	0.3/1.1	2.5/3.3
1-Axis, N-S Tilt=Latitude	Average	1.5	2.0	2.9	3.9	4.3	4.6	4.9	4.4	3.7	2.8	1.3	1.0	3.1
	Min/Max	0.7/2.1	1.0/3.7	2.1/4.8	2.2/5.2	1.9/6.2	3.2/5.9	3.4/6.0	3.5/5.6	1.9/5.1	1.8/4.9	0.5/2.1	0.4/1.9	2.7/3.6
2-Axis	Average	1.6	2.1	2.9	3.9	4.5	5.0	5.2	4.5	3.7	2.8	1.3	1.1	3.2
	Min/Max	0.7/2.3	1.0/3.8	2.1/4.8	2.2/5.2	2.0/6.5	3.5/6.5	3.7/6.4	3.6/5.7	1.9/5.1	1.9/4.9	0.6/2.2	0.5/2.0	2.8/3.7

Average Climatic Conditions

Element	Jan	Feb	Mar	Apr	May	June	July	Aug	Sept	Oct	Nov	Dec	Year
Temperature (°C)	-5.3	-4.4	1.1	7.6	13.9	18.5	21.3	20.2	16.4	10.4	4.7	-2.1	8.6
Daily Minimum Temp	-9.9	-9.2	-3.8	1.9	7.8	12.1	15.0	14.3	10.8	5.1	0.6	-6.1	3.2
Daily Maximum Temp	-0.8	0.3	5.9	13.3	20.2	24.8	27.6	26.1	22.0	15.7	8.9	1.9	13.8
Record Minimum Temp	-32.2	-32.2	-26.7	-12.8	-3.9	1.7	7.2	4.4	-2.2	-7.2	-15.0	-30.0	-32.2
Record Maximum Temp	21.1	20.6	30.6	33.3	35.6	36.7	36.1	36.1	36.1	30.6	27.2	21.1	36.7
HDD, Base 18.3°C	734	638	536	322	149	34	6	16	77	246	408	632	3797
CDD, Base 18.3°C	0	0	0	0	13	39	99	74	19	0	0	0	243
Relative Humidity (%)	73	72	70	65	67	70	71	75	76	74	75	77	72
Wind Speed (m/s)	4.8	4.8	4.8	4.6	4.0	3.6	3.5	3.4	3.6	3.8	4.5	4.7	4.2

Akron, OH

WBAN NO. 14895

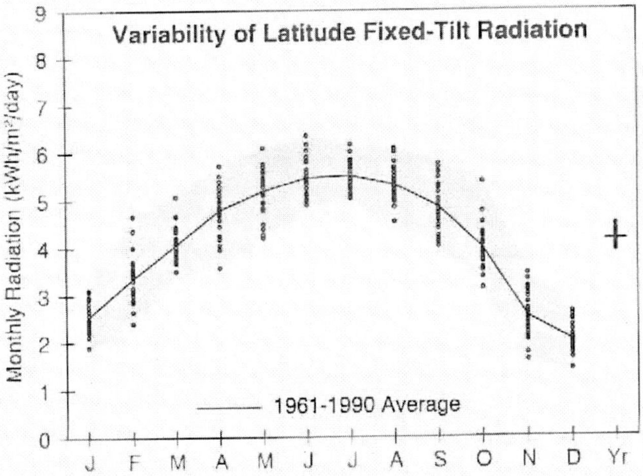

LATITUDE: 40.92° N
LONGITUDE: 81.43° W
ELEVATION: 377 meters
MEAN PRESSURE: 973 millibars

STATION TYPE: Secondary

Solar Radiation for Flat-Plate Collectors Facing South at a Fixed Tilt (kWh/m²/day), Uncertainty ±9%

Tilt (°)		Jan	Feb	Mar	Apr	May	June	July	Aug	Sept	Oct	Nov	Dec	Year
0	Average	1.7	2.4	3.4	4.6	5.5	6.1	6.0	5.2	4.2	2.9	1.8	1.4	3.8
	Min/Max	1.5/1.9	2.0/3.1	3.0/4.0	3.6/5.3	4.5/6.4	5.4/7.1	5.5/6.7	4.6/5.9	3.6/4.8	2.5/3.7	1.4/2.1	1.2/1.6	3.6/4.0
Latitude -15	Average	2.3	3.1	4.0	4.9	5.5	5.9	5.9	5.5	4.8	3.8	2.3	1.8	4.2
	Min/Max	1.8/2.8	2.3/4.2	3.5/4.9	3.7/5.8	4.5/6.5	5.3/6.9	5.4/6.7	4.7/6.3	4.0/5.7	3.0/5.0	1.6/3.1	1.4/2.3	4.0/4.5
Latitude	Average	2.5	3.4	4.1	4.8	5.2	5.5	5.5	5.3	4.8	4.0	2.5	2.0	4.1
	Min/Max	1.9/3.1	2.4/4.7	3.5/5.1	3.6/5.7	4.2/6.1	4.9/6.3	5.0/6.2	4.5/6.1	4.0/5.8	3.1/5.4	1.6/3.4	1.4/2.6	3.9/4.4
Latitude +15	Average	2.7	3.4	3.9	4.4	4.6	4.7	4.8	4.8	4.6	4.0	2.6	2.1	3.9
	Min/Max	1.9/3.3	2.4/4.9	3.4/5.0	3.3/5.3	3.7/5.4	4.3/5.5	4.4/5.4	4.1/5.5	3.8/5.5	3.1/5.4	1.6/3.6	1.4/2.8	3.7/4.2
90	Average	2.5	3.0	3.0	2.9	2.6	2.5	2.6	3.0	3.3	3.2	2.2	1.9	2.7
	Min/Max	1.8/3.2	2.0/4.5	2.5/3.8	2.2/3.4	2.2/3.0	2.3/2.8	2.4/2.8	2.6/3.4	2.7/3.9	2.4/4.5	1.3/3.2	1.2/2.7	2.6/3.0

Solar Radiation for 1-Axis Tracking Flat-Plate Collectors with a North-South Axis (kWh/m²/day), Uncertainty ±9%

Axis Tilt (°)		Jan	Feb	Mar	Apr	May	June	July	Aug	Sept	Oct	Nov	Dec	Year
0	Average	2.2	3.2	4.3	5.8	6.9	7.6	7.6	6.8	5.5	3.9	2.2	1.7	4.8
	Min/Max	1.7/2.6	2.3/4.3	3.8/5.4	4.2/7.2	5.3/8.5	6.6/9.6	6.9/8.9	5.6/8.0	4.5/6.6	3.1/5.3	1.4/2.9	1.2/2.1	4.5/5.2
Latitude -15	Average	2.6	3.7	4.8	6.1	7.0	7.6	7.7	7.0	6.0	4.5	2.7	2.0	5.2
	Min/Max	1.9/3.3	2.5/5.2	4.2/6.2	4.3/7.6	5.3/8.7	6.6/9.6	6.9/9.0	5.7/8.4	4.8/7.3	3.5/6.3	1.6/3.7	1.4/2.7	4.8/5.6
Latitude	Average	2.8	3.9	4.8	6.0	6.8	7.3	7.4	6.9	6.0	4.7	2.8	2.2	5.1
	Min/Max	2.0/3.6	2.6/5.5	4.2/6.3	4.2/7.6	5.1/8.4	6.3/9.2	6.7/8.7	5.6/8.2	4.8/7.4	3.6/6.6	1.7/3.9	1.4/2.9	4.8/5.6
Latitude +15	Average	2.9	3.9	4.7	5.8	6.3	6.8	6.9	6.6	5.9	4.7	2.9	2.3	5.0
	Min/Max	2.0/3.7	2.6/5.7	4.1/6.2	4.0/7.3	4.8/7.9	5.8/8.6	6.2/8.2	5.3/7.8	4.7/7.2	3.5/6.7	1.6/4.1	1.4/3.1	4.6/5.5

Solar Radiation for 2-Axis Tracking Flat-Plate Collectors (kWh/m²/day), Uncertainty ±9%

Tracker		Jan	Feb	Mar	Apr	May	June	July	Aug	Sept	Oct	Nov	Dec	Year
2-Axis	Average	3.0	3.9	4.8	6.1	7.1	7.8	7.8	7.1	6.1	4.8	2.9	2.3	5.3
	Min/Max	2.1/3.8	2.6/5.7	4.2/6.3	4.3/7.7	5.4/8.8	6.7/9.8	7.0/9.2	5.8/8.4	4.9/7.4	3.6/6.7	1.7/4.1	1.4/3.1	5.0/5.8

Direct Beam Solar Radiation for Concentrating Collectors (kWh/m²/day), Uncertainty ±8%

Tracker		Jan	Feb	Mar	Apr	May	June	July	Aug	Sept	Oct	Nov	Dec	Year
1-Axis, E-W Horiz Axis	Average	1.3	1.7	2.0	2.5	2.9	3.4	3.4	3.0	2.7	2.3	1.4	1.0	2.3
	Min/Max	0.5/2.0	0.7/2.8	1.3/3.2	1.2/3.6	1.8/4.2	2.6/4.9	2.7/4.5	2.0/4.1	1.9/3.7	1.5/3.7	0.5/2.4	0.3/1.6	2.0/2.7
1-Axis, N-S Horiz Axis	Average	0.9	1.5	2.2	3.2	3.9	4.4	4.4	3.9	3.2	2.3	1.0	0.6	2.6
	Min/Max	0.3/1.4	0.7/2.5	1.5/3.5	1.6/4.7	2.2/5.6	3.4/6.5	3.5/5.9	2.6/5.4	2.2/4.4	1.4/3.6	0.4/1.8	0.2/1.0	2.3/3.1
1-Axis, N-S Tilt=Latitude	Average	1.4	2.0	2.6	3.4	3.8	4.2	4.3	4.0	3.7	2.9	1.5	1.0	2.9
	Min/Max	0.6/2.2	0.9/3.4	1.8/4.2	1.6/5.0	2.2/5.5	3.2/6.2	3.4/5.7	2.7/5.5	2.5/5.0	1.8/4.6	0.5/2.6	0.3/1.7	2.5/3.4
2-Axis	Average	1.5	2.1	2.6	3.5	4.0	4.5	4.6	4.2	3.7	2.9	1.6	1.1	3.0
	Min/Max	0.6/2.3	0.9/3.5	1.8/4.2	1.7/5.1	2.3/5.8	3.5/6.7	3.7/6.1	2.8/5.7	2.5/5.0	1.9/4.7	0.6/2.8	0.3/1.8	2.7/3.5

Average Climatic Conditions

Element	Jan	Feb	Mar	Apr	May	June	July	Aug	Sept	Oct	Nov	Dec	Year
Temperature (°C)	-4.0	-2.6	3.3	9.2	15.0	19.9	22.2	21.3	17.6	11.3	5.6	-0.7	9.8
Daily Minimum Temp	-8.4	-7.3	-1.9	3.3	9.0	13.9	16.4	15.6	12.1	5.9	1.2	-4.7	4.6
Daily Maximum Temp	0.3	2.2	8.5	15.1	20.9	25.8	27.9	26.9	23.2	16.7	9.8	3.2	15.1
Record Minimum Temp	-31.1	-25.0	-19.4	-12.2	-4.4	0.0	6.1	5.0	0.0	-6.7	-18.3	-26.7	-31.1
Record Maximum Temp	21.1	20.0	27.2	31.1	33.9	37.8	38.3	36.7	37.2	30.0	26.7	24.4	38.3
HDD, Base 18.3°C	692	585	465	275	124	21	0	7	57	222	383	591	3422
CDD, Base 18.3°C	0	0	0	0	21	67	121	98	36	4	0	0	347
Relative Humidity (%)	73	72	68	64	66	68	70	73	74	70	72	75	70
Wind Speed (m/s)	5.2	4.9	5.1	4.8	4.1	3.7	3.4	3.3	3.6	4.1	4.8	5.1	4.3

Cleveland, OH

WBAN NO. 14820

LATITUDE: 41.40° N
LONGITUDE: 81.85° W
ELEVATION: 245 meters
MEAN PRESSURE: 988 millibars

STATION TYPE: Secondary

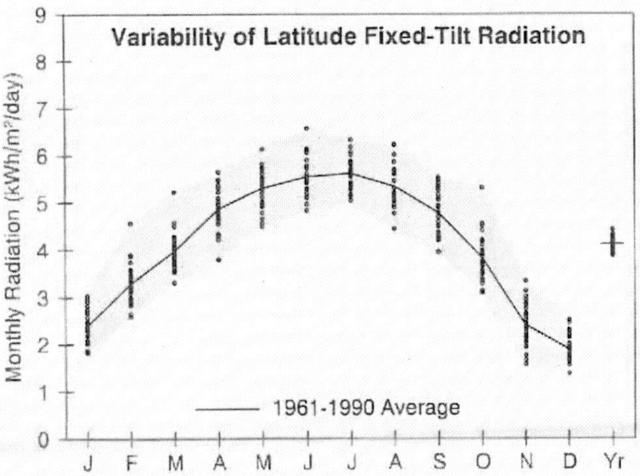

Solar Radiation for Flat-Plate Collectors Facing South at a Fixed Tilt (kWh/m²/day), Uncertainty ±9%

Tilt (°)		Jan	Feb	Mar	Apr	May	June	July	Aug	Sept	Oct	Nov	Dec	Year
0	Average	1.6	2.4	3.3	4.6	5.6	6.2	6.1	5.3	4.1	2.8	1.7	1.3	3.8
	Min/Max	1.4/1.9	2.1/3.0	2.9/4.1	3.8/5.2	4.8/6.4	5.4/7.3	5.5/6.9	4.5/6.0	3.5/4.6	2.4/3.6	1.3/2.1	1.2/1.5	3.6/3.9
Latitude -15	Average	2.2	3.1	3.9	5.0	5.6	6.0	6.1	5.6	4.7	3.6	2.2	1.7	4.2
	Min/Max	1.8/2.7	2.5/4.2	3.3/5.0	3.9/5.8	4.8/6.6	5.3/7.2	5.4/6.8	4.7/6.5	4.0/5.4	3.0/4.9	1.5/3.0	1.3/2.2	3.9/4.4
Latitude	Average	2.4	3.3	4.0	4.9	5.3	5.5	5.6	5.3	4.8	3.8	2.4	1.9	4.1
	Min/Max	1.8/3.0	2.6/4.6	3.3/5.2	3.8/5.7	4.5/6.1	4.8/6.6	5.0/6.3	4.4/6.2	4.0/5.5	3.1/5.3	1.6/3.3	1.4/2.5	3.9/4.4
Latitude +15	Average	2.5	3.3	3.9	4.5	4.7	4.8	4.9	4.9	4.6	3.8	2.4	2.0	3.9
	Min/Max	1.9/3.2	2.6/4.8	3.2/5.1	3.5/5.3	4.0/5.4	4.2/5.7	4.4/5.5	4.0/5.7	3.8/5.3	3.1/5.4	1.5/3.5	1.4/2.7	3.6/4.2
90	Average	2.3	2.9	3.0	3.0	2.7	2.6	2.7	3.0	3.2	3.1	2.1	1.8	2.7
	Min/Max	1.7/3.0	2.3/4.4	2.4/4.0	2.3/3.5	2.4/3.0	2.3/2.9	2.5/2.9	2.5/3.4	2.7/3.8	2.4/4.4	1.2/3.1	1.2/2.6	2.5/3.0

Solar Radiation for 1-Axis Tracking Flat-Plate Collectors with a North-South Axis (kWh/m²/day), Uncertainty ±9%

Axis Tilt (°)		Jan	Feb	Mar	Apr	May	June	July	Aug	Sept	Oct	Nov	Dec	Year
0	Average	2.0	3.1	4.2	5.9	7.0	7.8	7.8	6.8	5.4	3.7	2.1	1.6	4.8
	Min/Max	1.6/2.5	2.4/4.1	3.5/5.6	4.4/6.8	5.6/8.7	6.6/9.6	6.8/9.2	5.4/8.3	4.3/6.3	3.0/5.3	1.4/2.8	1.2/2.0	4.5/5.1
Latitude -15	Average	2.5	3.6	4.7	6.2	7.2	7.8	7.9	7.1	5.9	4.3	2.5	1.9	5.1
	Min/Max	1.9/3.2	2.8/5.0	3.8/6.3	4.6/7.3	5.7/8.9	6.5/9.7	6.8/9.3	5.6/8.7	4.7/7.0	3.4/6.3	1.6/3.5	1.4/2.5	4.8/5.6
Latitude	Average	2.6	3.8	4.7	6.1	6.9	7.5	7.6	7.0	5.9	4.5	2.6	2.0	5.1
	Min/Max	1.9/3.5	2.8/5.3	3.8/6.5	4.5/7.3	5.5/8.6	6.3/9.3	6.6/9.0	5.4/8.6	4.7/7.1	3.5/6.5	1.6/3.8	1.4/2.8	4.8/5.6
Latitude +15	Average	2.7	3.8	4.6	5.9	6.5	7.0	7.1	6.6	5.8	4.5	2.7	2.1	4.9
	Min/Max	2.0/3.6	2.9/5.5	3.7/6.4	4.3/6.9	5.1/8.1	5.8/8.7	6.2/8.4	5.1/8.2	4.5/6.9	3.5/6.6	1.6/4.0	1.4/2.9	4.6/5.4

Solar Radiation for 2-Axis Tracking Flat-Plate Collectors (kWh/m²/day), Uncertainty ±9%

Tracker		Jan	Feb	Mar	Apr	May	June	July	Aug	Sept	Oct	Nov	Dec	Year
2-Axis	Average	2.8	3.8	4.7	6.2	7.3	8.0	8.0	7.2	6.0	4.5	2.7	2.1	5.3
	Min/Max	2.0/3.6	2.9/5.5	3.8/6.5	4.6/7.3	5.7/9.0	6.7/9.9	7.0/9.5	5.6/8.8	4.7/7.1	3.5/6.6	1.6/4.0	1.4/3.0	4.9/5.7

Direct Beam Solar Radiation for Concentrating Collectors (kWh/m²/day), Uncertainty ±8%

Tracker		Jan	Feb	Mar	Apr	May	June	July	Aug	Sept	Oct	Nov	Dec	Year
1-Axis, E-W Horiz Axis	Average	1.2	1.6	1.9	2.7	3.2	3.6	3.6	3.1	2.7	2.2	1.2	0.9	2.3
	Min/Max	0.4/2.1	0.9/2.6	1.3/3.3	1.5/3.5	2.2/4.5	2.8/5.1	2.7/4.8	2.0/4.3	1.8/3.4	1.5/3.7	0.4/2.2	0.3/1.5	2.1/2.7
1-Axis, N-S Horiz Axis	Average	0.8	1.4	2.1	3.3	4.1	4.7	4.7	4.1	3.2	2.1	0.9	0.5	2.7
	Min/Max	0.3/1.4	0.7/2.2	1.4/3.5	2.0/4.4	2.7/6.0	3.5/6.7	3.6/6.3	2.6/5.7	2.1/4.1	1.3/3.6	0.3/1.7	0.2/0.9	2.3/3.0
1-Axis, N-S Tilt=Latitude	Average	1.3	1.9	2.5	3.6	4.0	4.4	4.5	4.2	3.6	2.7	1.4	0.9	2.9
	Min/Max	0.4/2.2	1.0/3.0	1.7/4.3	2.1/4.7	2.7/5.9	3.4/6.3	3.5/6.0	2.6/5.8	2.4/4.6	1.8/4.7	0.4/2.5	0.3/1.6	2.6/3.3
2-Axis	Average	1.3	2.0	2.5	3.6	4.3	4.8	4.8	4.3	3.6	2.8	1.4	1.0	3.0
	Min/Max	0.4/2.4	1.1/3.1	1.7/4.3	2.1/4.8	2.8/6.2	3.7/6.8	3.7/6.5	2.7/6.0	2.4/4.6	1.8/4.7	0.4/2.6	0.3/1.7	2.7/3.5

Average Climatic Conditions

Element	Jan	Feb	Mar	Apr	May	June	July	Aug	Sept	Oct	Nov	Dec	Year
Temperature (°C)	-4.0	-2.7	2.9	8.7	14.4	19.8	22.2	21.3	17.7	11.6	5.9	-0.6	9.8
Daily Minimum Temp	-8.0	-7.1	-2.1	2.9	8.5	13.8	16.3	15.7	12.3	6.4	1.7	-4.2	4.7
Daily Maximum Temp	-0.1	1.7	7.9	14.4	20.3	25.7	28.0	26.9	23.1	16.7	10.0	3.0	14.8
Record Minimum Temp	-28.3	-26.1	-20.6	-12.2	-3.9	-0.6	5.0	3.3	0.0	-7.2	-16.1	-26.1	-28.3
Record Maximum Temp	22.8	20.6	28.3	31.1	33.3	40.0	39.4	38.9	38.3	32.2	27.8	25.0	40.0
HDD, Base 18.3°C	692	588	477	290	139	22	0	6	55	215	373	587	3445
CDD, Base 18.3°C	0	0	0	0	18	66	121	99	37	4	0	0	345
Relative Humidity (%)	73	73	70	66	67	69	70	73	74	71	72	74	71
Wind Speed (m/s)	5.5	5.1	5.3	5.0	4.3	4.1	3.8	3.6	3.9	4.3	5.0	5.2	4.6

Columbus, OH

WBAN NO. 14821

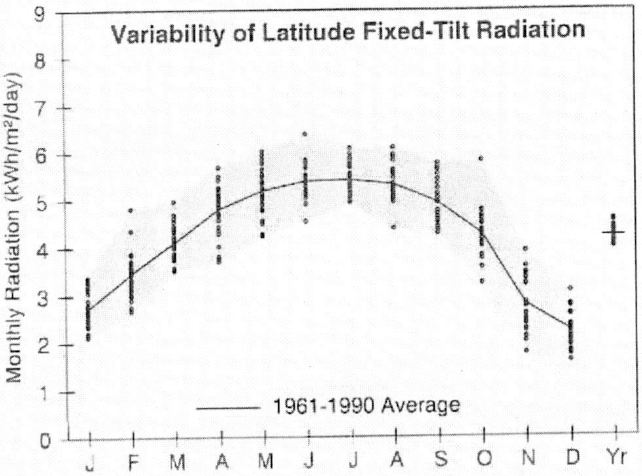

LATITUDE: 40.00° N
LONGITUDE: 82.88° W
ELEVATION: 254 meters
MEAN PRESSURE: 988 millibars

STATION TYPE: Secondary

Solar Radiation for Flat-Plate Collectors Facing South at a Fixed Tilt (kWh/m²/day), Uncertainty ±9%

Tilt (°)		Jan	Feb	Mar	Apr	May	June	July	Aug	Sept	Oct	Nov	Dec	Year
0	Average	1.8	2.5	3.5	4.6	5.5	6.0	5.9	5.3	4.3	3.1	1.9	1.5	3.8
	Min/Max	1.5/2.1	2.2/3.2	3.1/4.0	3.7/5.3	4.6/6.4	5.1/7.1	5.4/6.5	4.5/5.9	3.8/4.8	2.6/4.0	1.4/2.3	1.2/1.7	3.6/4.1
Latitude -15	Average	2.5	3.2	4.0	4.9	5.5	5.9	5.8	5.5	4.9	4.0	2.6	2.0	4.2
	Min/Max	2.0/3.0	2.6/4.4	3.5/4.8	3.8/5.8	4.5/6.4	4.9/6.9	5.3/6.5	4.6/6.3	4.3/5.7	3.1/5.4	1.7/3.5	1.5/2.7	4.0/4.6
Latitude	Average	2.7	3.4	4.1	4.8	5.2	5.4	5.4	5.3	4.9	4.3	2.8	2.2	4.2
	Min/Max	2.1/3.4	2.7/4.8	3.5/5.0	3.7/5.7	4.2/6.0	4.6/6.4	4.9/6.1	4.4/6.1	4.3/5.8	3.2/5.8	1.8/3.9	1.6/3.1	4.0/4.6
Latitude +15	Average	2.9	3.5	4.0	4.4	4.6	4.7	4.8	4.8	4.7	4.3	2.9	2.3	4.0
	Min/Max	2.2/3.6	2.7/5.1	3.4/4.9	3.4/5.3	3.8/5.3	4.0/5.5	4.3/5.3	4.0/5.6	4.1/5.5	3.2/5.9	1.8/4.1	1.6/3.3	3.7/4.3
90	Average	2.6	3.0	3.0	2.8	2.6	2.4	2.5	2.9	3.3	3.4	2.5	2.1	2.8
	Min/Max	1.9/3.5	2.2/4.7	2.5/3.7	2.2/3.3	2.2/2.9	2.2/2.7	2.4/2.8	2.4/3.3	2.9/3.9	2.5/4.8	1.4/3.7	1.4/3.1	2.6/3.0

Solar Radiation for 1-Axis Tracking Flat-Plate Collectors with a North-South Axis (kWh/m²/day), Uncertainty ±9%

Axis Tilt (°)		Jan	Feb	Mar	Apr	May	June	July	Aug	Sept	Oct	Nov	Dec	Year
0	Average	2.3	3.3	4.4	5.8	6.9	7.6	7.5	6.8	5.7	4.3	2.4	1.8	4.9
	Min/Max	1.9/2.9	2.5/4.5	3.7/5.4	4.3/7.2	5.4/8.6	6.3/9.3	6.6/8.6	5.5/8.1	4.8/6.7	3.1/5.8	1.5/3.3	1.4/2.5	4.6/5.4
Latitude -15	Average	2.8	3.8	4.8	6.1	7.0	7.5	7.5	7.1	6.2	4.9	2.9	2.3	5.3
	Min/Max	2.2/3.6	2.8/5.4	4.0/6.0	4.4/7.7	5.4/8.7	6.2/9.3	6.6/8.7	5.6/8.4	5.2/7.4	3.5/6.9	1.8/4.2	1.6/3.2	4.9/5.8
Latitude	Average	3.1	4.0	4.9	6.0	6.8	7.2	7.2	6.9	6.2	5.1	3.1	2.5	5.3
	Min/Max	2.3/3.9	2.9/5.8	4.1/6.2	4.3/7.6	5.2/8.4	5.9/8.9	6.4/8.4	5.4/8.3	5.2/7.5	3.7/7.3	1.8/4.5	1.7/3.5	4.9/5.8
Latitude +15	Average	3.2	4.0	4.8	5.8	6.4	6.7	6.8	6.6	6.0	5.1	3.2	2.5	5.1
	Min/Max	2.4/4.1	2.9/5.9	4.0/6.1	4.1/7.3	4.9/7.9	5.6/8.3	5.9/7.9	5.2/7.9	5.1/7.3	3.6/7.3	1.8/4.7	1.7/3.7	4.7/5.6

Solar Radiation for 2-Axis Tracking Flat-Plate Collectors (kWh/m²/day), Uncertainty ±9%

Tracker		Jan	Feb	Mar	Apr	May	June	July	Aug	Sept	Oct	Nov	Dec	Year
2-Axis	Average	3.2	4.1	4.9	6.1	7.1	7.7	7.6	7.1	6.2	5.2	3.2	2.6	5.4
	Min/Max	2.4/4.1	2.9/5.9	4.1/6.2	4.5/7.7	5.5/8.8	6.4/9.5	6.7/8.9	5.6/8.5	5.2/7.5	3.7/7.3	1.9/4.7	1.7/3.7	5.1/6.0

Direct Beam Solar Radiation for Concentrating Collectors (kWh/m²/day), Uncertainty ±8%

Tracker		Jan	Feb	Mar	Apr	May	June	July	Aug	Sept	Oct	Nov	Dec	Year
1-Axis, E-W Horiz Axis	Average	1.5	1.8	2.0	2.5	3.0	3.3	3.2	3.0	2.8	2.7	1.6	1.2	2.4
	Min/Max	0.7/2.4	1.0/3.0	1.3/3.1	1.3/3.6	1.8/4.3	2.3/4.8	2.5/4.2	1.9/4.0	2.0/3.7	1.5/4.4	0.7/2.9	0.6/2.2	2.1/2.9
1-Axis, N-S Horiz Axis	Average	1.1	1.6	2.3	3.2	3.9	4.4	4.3	4.0	3.4	2.6	1.2	0.8	2.7
	Min/Max	0.5/1.7	0.9/2.7	1.5/3.4	1.6/4.8	2.4/5.8	3.1/6.2	3.3/5.6	2.6/5.3	2.5/4.5	1.5/4.3	0.5/2.2	0.4/1.4	2.4/3.3
1-Axis, N-S Tilt=Latitude	Average	1.7	2.2	2.7	3.4	3.8	4.1	4.1	4.1	3.8	3.3	1.8	1.3	3.0
	Min/Max	0.7/2.6	1.2/3.6	1.7/4.1	1.7/5.1	2.4/5.6	2.9/5.9	3.1/5.3	2.6/5.4	2.8/5.1	1.9/5.5	0.7/3.2	0.6/2.3	2.6/3.6
2-Axis	Average	1.8	2.2	2.7	3.5	4.1	4.5	4.4	4.2	3.8	3.4	1.9	1.4	3.2
	Min/Max	0.8/2.8	1.2/3.7	1.7/4.1	1.7/5.1	2.5/6.0	3.1/6.4	3.3/5.7	2.7/5.6	2.8/5.1	1.9/5.5	0.8/3.4	0.7/2.5	2.8/3.8

Average Climatic Conditions

Element	Jan	Feb	Mar	Apr	May	June	July	Aug	Sept	Oct	Nov	Dec	Year
Temperature (°C)	-3.1	-1.3	4.9	10.6	16.2	20.7	22.9	21.9	18.6	12.1	6.1	-0.1	10.8
Daily Minimum Temp	-7.5	-6.0	-0.4	4.4	10.1	14.4	17.1	16.0	12.7	6.1	1.3	-4.1	5.3
Daily Maximum Temp	1.2	3.3	10.3	16.7	22.4	26.9	28.7	27.8	24.6	18.1	10.8	4.0	16.2
Record Minimum Temp	-28.3	-25.0	-21.1	-10.0	-3.9	1.7	6.1	3.9	-0.6	-6.7	-15.0	-27.2	-28.3
Record Maximum Temp	23.3	22.8	29.4	31.7	34.4	38.9	37.8	38.3	37.8	32.2	26.7	24.4	38.9
HDD, Base 18.3°C	665	551	415	233	104	13	0	7	45	201	368	570	3171
CDD, Base 18.3°C	0	0	0	0	38	83	143	119	53	6	0	0	443
Relative Humidity (%)	71	69	65	63	67	68	71	73	73	69	72	74	70
Wind Speed (m/s)	4.4	4.2	4.6	4.3	3.7	3.3	3.0	2.8	2.9	3.3	4.0	4.3	3.7

Dayton, OH

WBAN NO. 93815

LATITUDE: 39.90° N
LONGITUDE: 84.22° W
ELEVATION: 306 meters
MEAN PRESSURE: 981 millibars

STATION TYPE: Secondary

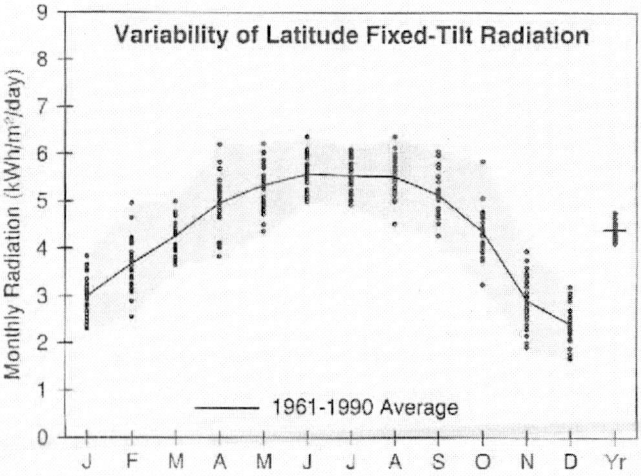

Solar Radiation for Flat-Plate Collectors Facing South at a Fixed Tilt (kWh/m²/day), Uncertainty ±9%

Tilt (°)		Jan	Feb	Mar	Apr	May	June	July	Aug	Sept	Oct	Nov	Dec	Year
0	Average	1.9	2.6	3.6	4.7	5.7	6.2	6.0	5.4	4.4	3.2	2.0	1.5	3.9
	Min/Max	1.6/2.3	2.1/3.3	3.2/4.0	3.8/5.7	4.7/6.6	5.6/7.1	5.4/6.6	4.6/6.2	3.8/5.0	2.6/4.0	1.5/2.3	1.2/1.8	3.8/4.2
Latitude -15	Average	2.7	3.4	4.2	5.1	5.7	6.1	6.0	5.7	5.1	4.1	2.7	2.1	4.4
	Min/Max	2.1/3.4	2.5/4.5	3.6/4.8	4.0/6.3	4.7/6.7	5.4/6.9	5.3/6.6	4.7/6.6	4.3/5.9	3.1/5.4	1.8/3.5	1.6/2.8	4.1/4.7
Latitude	Average	3.0	3.7	4.3	4.9	5.3	5.6	5.6	5.5	5.1	4.4	2.9	2.4	4.4
	Min/Max	2.3/3.8	2.6/4.9	3.7/5.0	3.8/6.2	4.4/6.2	5.0/6.4	4.9/6.1	4.5/6.4	4.3/6.1	3.2/5.8	1.9/3.9	1.7/3.2	4.1/4.8
Latitude +15	Average	3.1	3.8	4.1	4.6	4.7	4.8	4.9	5.0	4.9	4.4	3.0	2.5	4.2
	Min/Max	2.4/4.1	2.5/5.2	3.5/4.9	3.5/5.8	3.9/5.5	4.3/5.5	4.3/5.3	4.1/5.8	4.0/5.8	3.2/5.9	1.9/4.2	1.7/3.4	3.9/4.5
90	Average	2.9	3.2	3.1	2.9	2.6	2.5	2.6	3.0	3.4	3.5	2.6	2.3	2.9
	Min/Max	2.1/4.0	2.1/4.8	2.6/3.8	2.3/3.7	2.2/3.0	2.3/2.7	2.4/2.8	2.5/3.4	2.8/4.1	2.5/4.8	1.6/3.7	1.4/3.3	2.7/3.2

Solar Radiation for 1-Axis Tracking Flat-Plate Collectors with a North-South Axis (kWh/m²/day), Uncertainty ±9%

Axis Tilt (°)		Jan	Feb	Mar	Apr	May	June	July	Aug	Sept	Oct	Nov	Dec	Year
0	Average	2.5	3.5	4.6	6.1	7.2	7.9	7.8	7.1	5.9	4.4	2.6	2.0	5.1
	Min/Max	2.0/3.3	2.5/4.6	3.9/5.4	4.4/7.8	5.7/8.7	6.9/9.5	6.7/8.7	5.9/8.4	4.9/7.2	3.1/5.9	1.7/3.4	1.4/2.6	4.8/5.6
Latitude -15	Average	3.1	4.1	5.1	6.4	7.3	7.9	7.8	7.4	6.5	5.1	3.1	2.4	5.5
	Min/Max	2.4/4.1	2.8/5.6	4.3/6.0	4.6/8.4	5.7/8.9	6.9/9.5	6.7/8.8	6.0/8.8	5.3/7.9	3.6/6.9	1.9/4.3	1.7/3.3	5.1/6.1
Latitude	Average	3.4	4.3	5.1	6.3	7.1	7.6	7.5	7.3	6.5	5.3	3.3	2.6	5.5
	Min/Max	2.5/4.5	2.8/5.9	4.3/6.2	4.5/8.3	5.5/8.6	6.6/9.1	6.4/8.5	5.9/8.7	5.3/8.0	3.7/7.3	2.0/4.6	1.7/3.6	5.1/6.1
Latitude +15	Average	3.5	4.4	5.0	6.0	6.6	7.1	7.0	6.9	6.3	5.3	3.4	2.8	5.4
	Min/Max	2.6/4.7	2.8/6.1	4.2/6.1	4.3/8.0	5.1/8.1	6.1/8.5	6.0/7.9	5.6/8.3	5.1/7.8	3.6/7.3	2.0/4.8	1.8/3.8	4.9/5.9

Solar Radiation for 2-Axis Tracking Flat-Plate Collectors (kWh/m²/day), Uncertainty ±9%

Tracker		Jan	Feb	Mar	Apr	May	June	July	Aug	Sept	Oct	Nov	Dec	Year
2-Axis	Average	3.6	4.4	5.2	6.4	7.4	8.1	8.0	7.5	6.5	5.3	3.4	2.8	5.7
	Min/Max	2.6/4.7	2.9/6.1	4.3/6.2	4.6/8.4	5.8/9.0	7.0/9.7	6.8/8.9	6.0/8.9	5.3/8.0	3.7/7.3	2.0/4.8	1.8/3.9	5.2/6.3

Direct Beam Solar Radiation for Concentrating Collectors (kWh/m²/day), Uncertainty ±8%

Tracker		Jan	Feb	Mar	Apr	May	June	July	Aug	Sept	Oct	Nov	Dec	Year
1-Axis, E-W Horiz Axis	Average	1.8	2.1	2.2	2.7	3.2	3.6	3.5	3.3	3.1	2.8	1.7	1.4	2.6
	Min/Max	1.0/2.7	0.9/3.3	1.5/3.0	1.4/4.2	2.1/4.5	2.8/5.0	2.7/4.3	2.2/4.4	2.1/4.3	1.6/4.4	0.8/3.0	0.6/2.2	2.2/3.1
1-Axis, N-S Horiz Axis	Average	1.3	1.9	2.5	3.5	4.2	4.8	4.7	4.4	3.7	2.7	1.4	0.9	3.0
	Min/Max	0.7/2.0	0.9/3.0	1.7/3.3	1.8/5.4	2.8/5.9	3.7/6.6	3.6/5.7	3.1/5.8	2.7/5.2	1.5/4.3	0.6/2.3	0.4/1.5	2.6/3.6
1-Axis, N-S Tilt=Latitude	Average	2.0	2.5	2.9	3.7	4.1	4.5	4.5	4.5	4.2	3.5	2.0	1.5	3.3
	Min/Max	1.0/3.0	1.1/4.0	2.0/3.9	1.9/5.8	2.7/5.8	3.5/6.2	3.4/5.4	3.1/5.9	3.0/5.8	2.0/5.5	0.8/3.3	0.7/2.3	2.8/4.0
2-Axis	Average	2.1	2.6	2.9	3.8	4.4	4.9	4.8	4.6	4.2	3.5	2.1	1.6	3.5
	Min/Max	1.1/3.2	1.2/4.1	2.0/3.9	2.0/5.8	2.8/6.1	3.8/6.7	3.7/5.8	3.2/6.1	3.0/5.9	2.0/5.5	0.9/3.5	0.7/2.5	3.0/4.1

Average Climatic Conditions

Element	Jan	Feb	Mar	Apr	May	June	July	Aug	Sept	Oct	Nov	Dec	Year
Temperature (°C)	-3.3	-1.4	4.7	10.7	16.5	21.3	23.4	22.3	18.8	12.3	6.1	-0.2	10.9
Daily Minimum Temp	-7.8	-6.2	-0.6	4.7	10.6	15.1	17.4	16.3	12.8	6.4	1.3	-4.4	5.5
Daily Maximum Temp	1.2	3.3	10.0	16.6	22.5	27.6	29.4	28.3	24.7	18.1	10.7	3.9	16.4
Record Minimum Temp	-31.1	-26.7	-21.7	-9.4	-2.8	4.4	6.7	4.4	0.0	-6.1	-18.9	-28.9	-31.1
Record Maximum Temp	21.7	21.7	27.8	31.7	33.9	38.9	38.9	38.9	38.3	31.7	26.1	22.2	38.9
HDD, Base 18.3°C	672	554	422	230	103	6	0	3	41	197	368	575	3171
CDD, Base 18.3°C	0	0	0	0	46	96	158	128	54	10	0	0	492
Relative Humidity (%)	73	72	69	64	65	66	69	72	72	69	73	76	70
Wind Speed (m/s)	5.0	4.8	5.1	4.9	4.1	3.9	3.5	3.3	3.5	3.9	4.6	4.8	4.3

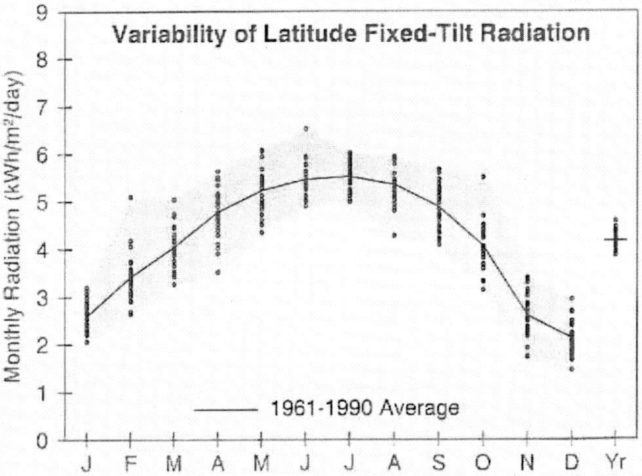

Mansfield, OH

WBAN NO. 14891

LATITUDE: 40.82° N
LONGITUDE: 82.52° W
ELEVATION: 395 meters
MEAN PRESSURE: 970 millibars

STATION TYPE: Secondary

Solar Radiation for Flat-Plate Collectors Facing South at a Fixed Tilt (kWh/m²/day), Uncertainty ±9%

Tilt (°)		Jan	Feb	Mar	Apr	May	June	July	Aug	Sept	Oct	Nov	Dec	Year
0	Average	1.7	2.5	3.4	4.6	5.5	6.1	6.0	5.3	4.2	3.0	1.8	1.4	3.8
	Min/Max	1.5/1.9	2.1/3.3	3.0/4.0	3.5/5.3	4.7/6.4	5.5/7.3	5.5/6.6	4.4/5.8	3.7/4.8	2.5/3.8	1.4/2.1	1.2/1.7	3.6/4.1
Latitude -15	Average	2.3	3.2	4.0	4.9	5.6	5.9	6.0	5.6	4.8	3.8	2.4	1.9	4.2
	Min/Max	1.9/2.8	2.5/4.6	3.3/4.9	3.6/5.8	4.6/6.5	5.3/7.1	5.4/6.5	4.5/6.2	4.1/5.6	3.0/5.1	1.7/3.0	1.4/2.6	3.9/4.6
Latitude	Average	2.6	3.4	4.0	4.8	5.2	5.5	5.5	5.3	4.9	4.1	2.6	2.1	4.2
	Min/Max	2.1/3.2	2.6/5.1	3.3/5.0	3.5/5.6	4.3/6.1	4.9/6.5	5.0/6.0	4.3/5.9	4.1/5.7	3.1/5.5	1.7/3.4	1.4/2.9	3.9/4.6
Latitude +15	Average	2.7	3.5	3.9	4.4	4.6	4.7	4.8	4.9	4.6	4.1	2.7	2.2	3.9
	Min/Max	2.1/3.4	2.6/5.3	3.1/4.9	3.2/5.3	3.9/5.4	4.3/5.6	4.4/5.3	3.9/5.4	3.9/5.4	3.1/5.6	1.7/3.5	1.4/3.2	3.6/4.3
90	Average	2.5	3.1	3.0	2.9	2.6	2.5	2.6	3.0	3.3	3.3	2.3	2.0	2.8
	Min/Max	1.9/3.1	2.2/4.9	2.4/3.8	2.2/3.5	2.3/3.0	2.3/2.8	2.4/2.8	2.4/3.3	2.7/3.9	2.5/4.6	1.4/3.2	1.2/3.0	2.5/3.0

Solar Radiation for 1-Axis Tracking Flat-Plate Collectors with a North-South Axis (kWh/m²/day), Uncertainty ±9%

Axis Tilt (°)		Jan	Feb	Mar	Apr	May	June	July	Aug	Sept	Oct	Nov	Dec	Year
0	Average	2.2	3.2	4.3	5.8	7.0	7.7	7.7	6.9	5.6	4.0	2.3	1.7	4.9
	Min/Max	1.8/2.6	2.5/4.7	3.3/5.4	4.0/6.9	5.4/8.7	6.6/9.9	6.8/8.7	5.2/7.9	4.6/6.5	3.1/5.5	1.5/2.9	1.3/2.3	4.5/5.5
Latitude -15	Average	2.7	3.8	4.7	6.1	7.1	7.7	7.7	7.1	6.1	4.7	2.7	2.1	5.2
	Min/Max	2.1/3.3	2.8/5.7	3.6/6.1	4.2/7.4	5.5/8.8	6.6/9.9	6.8/8.8	5.3/8.2	4.9/7.2	3.5/6.5	1.8/3.6	1.4/3.0	4.8/5.9
Latitude	Average	2.9	3.9	4.8	6.0	6.8	7.4	7.4	7.0	6.1	4.8	2.9	2.3	5.2
	Min/Max	2.2/3.6	2.9/6.1	3.6/6.2	4.1/7.3	5.3/8.6	6.3/9.5	6.6/8.4	5.2/8.1	4.9/7.3	3.5/6.8	1.8/3.9	1.5/3.3	4.8/5.9
Latitude +15	Average	3.0	4.0	4.7	5.7	6.4	6.9	7.0	6.6	5.9	4.8	3.0	2.4	5.0
	Min/Max	2.3/3.8	2.9/6.3	3.5/6.2	3.9/7.0	4.9/8.1	5.9/8.9	6.2/7.9	4.9/7.7	4.8/7.1	3.5/6.8	1.8/4.0	1.5/3.5	4.6/5.8

Solar Radiation for 2-Axis Tracking Flat-Plate Collectors (kWh/m²/day), Uncertainty ±9%

Tracker		Jan	Feb	Mar	Apr	May	June	July	Aug	Sept	Oct	Nov	Dec	Year
2-Axis	Average	3.0	4.0	4.8	6.1	7.2	7.9	7.9	7.2	6.1	4.9	3.0	2.4	5.4
	Min/Max	2.3/3.8	2.9/6.3	3.6/6.3	4.2/7.4	5.6/8.9	6.7/10.1	7.0/8.9	5.3/8.2	5.0/7.3	3.6/6.9	1.8/4.1	1.5/3.6	4.9/6.1

Direct Beam Solar Radiation for Concentrating Collectors (kWh/m²/day), Uncertainty ±8%

Tracker		Jan	Feb	Mar	Apr	May	June	July	Aug	Sept	Oct	Nov	Dec	Year
1-Axis, E-W Horiz Axis	Average	1.4	1.7	2.0	2.5	3.0	3.5	3.4	3.1	2.8	2.5	1.4	1.1	2.4
	Min/Max	0.7/2.2	1.0/3.1	0.9/3.1	1.2/3.5	1.8/4.5	2.6/5.2	2.7/4.2	1.7/4.0	1.9/3.6	1.3/4.0	0.5/2.3	0.3/2.0	2.0/3.0
1-Axis, N-S Horiz Axis	Average	1.0	1.5	2.2	3.2	4.0	4.5	4.5	4.0	3.3	2.4	1.1	0.7	2.7
	Min/Max	0.5/1.5	0.9/2.7	1.0/3.4	1.5/4.4	2.3/6.1	3.4/6.9	3.6/5.7	2.2/5.2	2.3/4.4	1.3/3.9	0.4/1.8	0.2/1.3	2.3/3.5
1-Axis, N-S Tilt=Latitude	Average	1.5	2.1	2.6	3.4	3.9	4.3	4.3	4.1	3.8	3.1	1.6	1.2	3.0
	Min/Max	0.8/2.4	1.2/3.7	1.2/4.1	1.6/4.7	2.3/5.9	3.2/6.5	3.4/5.4	2.3/5.3	2.6/5.0	1.6/5.0	0.6/2.6	0.3/2.1	2.5/3.8
2-Axis	Average	1.6	2.1	2.6	3.4	4.1	4.7	4.6	4.3	3.8	3.1	1.7	1.3	3.1
	Min/Max	0.9/2.6	1.2/3.8	1.2/4.1	1.7/4.8	2.4/6.3	3.5/7.1	3.7/5.8	2.3/5.5	2.6/5.0	1.7/5.0	0.6/2.7	0.4/2.3	2.6/4.0

Average Climatic Conditions

Element	Jan	Feb	Mar	Apr	May	June	July	Aug	Sept	Oct	Nov	Dec	Year
Temperature (°C)	-4.2	-2.8	3.1	9.1	14.9	19.9	22.3	21.3	17.7	11.5	5.3	-1.2	9.7
Daily Minimum Temp	-8.4	-7.3	-1.9	3.4	9.1	14.1	16.7	15.8	12.2	6.2	1.1	-5.2	4.7
Daily Maximum Temp	0.1	1.7	8.1	14.8	20.7	25.7	27.8	26.7	23.2	16.8	9.6	2.7	14.8
Record Minimum Temp	-30.0	-23.9	-21.1	-13.3	-3.9	2.8	6.1	4.4	0.6	-6.7	-16.7	-27.2	-30.0
Record Maximum Temp	17.2	19.4	27.8	30.0	33.3	38.3	37.8	36.1	33.9	29.4	25.6	22.8	38.3
HDD, Base 18.3°C	698	591	472	277	127	24	0	11	61	220	390	606	3477
CDD, Base 18.3°C	0	0	0	0	21	71	125	103	43	8	0	0	370
Relative Humidity (%)	76	74	71	66	68	70	71	74	74	70	74	78	72
Wind Speed (m/s)	5.9	5.2	5.6	5.5	4.9	4.2	3.8	3.7	4.0	4.5	5.2	5.8	4.9

Toledo, OH

WBAN NO. 94830

LATITUDE: 41.60° N
LONGITUDE: 83.80° W
ELEVATION: 211 meters
MEAN PRESSURE: 992 millibars

STATION TYPE: Secondary

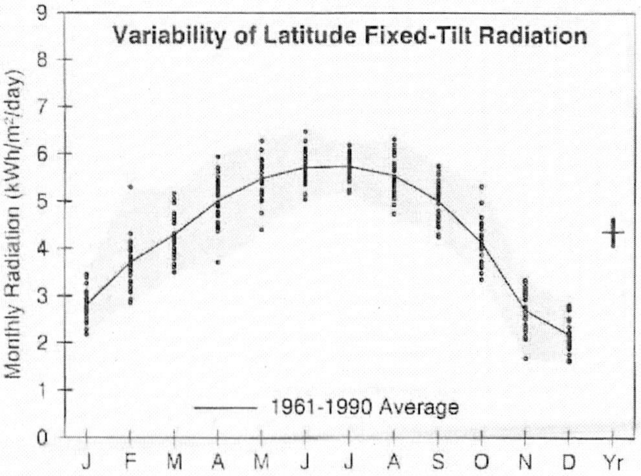

Solar Radiation for Flat-Plate Collectors Facing South at a Fixed Tilt (kWh/m²/day), Uncertainty ±9%

Tilt (°)		Jan	Feb	Mar	Apr	May	June	July	Aug	Sept	Oct	Nov	Dec	Year
0	Average	1.7	2.6	3.5	4.7	5.8	6.3	6.2	5.4	4.3	3.0	1.8	1.4	3.9
	Min/Max	1.5/2.0	2.2/3.4	3.1/4.0	3.7/5.4	4.7/6.6	5.6/7.2	5.6/6.7	4.8/6.1	3.8/4.8	2.6/3.6	1.4/2.1	1.1/1.6	3.7/4.1
Latitude -15	Average	2.5	3.4	4.2	5.1	5.8	6.2	6.2	5.8	5.0	3.9	2.5	2.0	4.4
	Min/Max	2.0/3.0	2.7/4.8	3.5/5.0	3.8/6.0	4.7/6.7	5.5/7.1	5.6/6.7	5.0/6.6	4.3/5.7	3.2/4.9	1.6/3.0	1.5/2.5	4.1/4.6
Latitude	Average	2.8	3.7	4.3	5.0	5.5	5.7	5.7	5.6	5.0	4.1	2.7	2.2	4.4
	Min/Max	2.2/3.5	2.9/5.3	3.5/5.2	3.7/5.9	4.4/6.3	5.0/6.5	5.2/6.2	4.7/6.3	4.2/5.8	3.3/5.3	1.7/3.3	1.6/2.8	4.1/4.6
Latitude +15	Average	3.0	3.8	4.2	4.6	4.9	5.0	5.0	5.0	4.8	4.1	2.8	2.3	4.1
	Min/Max	2.3/3.7	2.9/5.6	3.4/5.1	3.4/5.5	3.9/5.6	4.4/5.6	4.6/5.4	4.3/5.8	4.0/5.5	3.3/5.4	1.7/3.5	1.6/3.0	3.8/4.4
90	Average	2.8	3.4	3.2	3.1	2.8	2.6	2.8	3.1	3.4	3.4	2.4	2.1	2.9
	Min/Max	2.0/3.6	2.5/5.2	2.6/4.0	2.3/3.6	2.3/3.1	2.5/2.9	2.6/2.9	2.7/3.5	2.8/4.0	2.7/4.4	1.4/3.1	1.5/2.9	2.7/3.2

Solar Radiation for 1-Axis Tracking Flat-Plate Collectors with a North-South Axis (kWh/m²/day), Uncertainty ±9%

Axis Tilt (°)		Jan	Feb	Mar	Apr	May	June	July	Aug	Sept	Oct	Nov	Dec	Year
0	Average	2.3	3.4	4.6	6.1	7.4	8.1	8.0	7.1	5.7	4.0	2.3	1.8	5.1
	Min/Max	1.8/2.8	2.7/4.8	3.7/5.5	4.3/7.3	5.8/9.1	6.8/9.7	7.0/9.0	6.0/8.2	4.8/6.7	3.2/5.2	1.5/2.9	1.4/2.2	4.8/5.5
Latitude -15	Average	2.9	4.1	5.1	6.5	7.5	8.1	8.1	7.5	6.2	4.7	2.8	2.2	5.5
	Min/Max	2.2/3.6	3.1/5.9	4.0/6.3	4.5/7.8	5.9/9.3	6.7/9.7	7.0/9.1	6.2/8.6	5.2/7.4	3.7/6.3	1.7/3.6	1.6/2.9	5.1/5.9
Latitude	Average	3.2	4.3	5.2	6.4	7.3	7.8	7.8	7.3	6.3	4.9	3.0	2.4	5.5
	Min/Max	2.4/4.0	3.2/6.3	4.0/6.4	4.4/7.8	5.7/9.0	6.4/9.3	6.8/8.7	6.0/8.4	5.2/7.5	3.8/6.6	1.8/3.9	1.7/3.2	5.1/5.9
Latitude +15	Average	3.3	4.4	5.1	6.1	6.8	7.3	7.3	7.0	6.1	4.9	3.1	2.5	5.3
	Min/Max	2.5/4.2	3.2/6.5	3.9/6.4	4.2/7.4	5.3/8.5	6.0/8.7	6.3/8.2	5.7/8.0	5.0/7.3	3.8/6.6	1.8/4.0	1.7/3.3	4.9/5.8

Solar Radiation for 2-Axis Tracking Flat-Plate Collectors (kWh/m²/day), Uncertainty ±9%

Tracker		Jan	Feb	Mar	Apr	May	June	July	Aug	Sept	Oct	Nov	Dec	Year
2-Axis	Average	3.3	4.4	5.2	6.5	7.6	8.3	8.2	7.5	6.3	5.0	3.1	2.5	5.7
	Min/Max	2.5/4.2	3.2/6.5	4.0/6.4	4.5/7.8	5.9/9.4	6.9/9.9	7.2/9.2	6.2/8.6	5.2/7.5	3.9/6.6	1.8/4.0	1.7/3.4	5.3/6.1

Direct Beam Solar Radiation for Concentrating Collectors (kWh/m²/day), Uncertainty ±8%

Tracker		Jan	Feb	Mar	Apr	May	June	July	Aug	Sept	Oct	Nov	Dec	Year
1-Axis, E-W	Average	1.6	2.1	2.2	2.8	3.3	3.8	3.7	3.3	2.9	2.5	1.6	1.2	2.6
Horiz Axis	Min/Max	1.1/2.3	1.2/3.3	1.2/3.3	1.3/3.9	2.2/4.7	2.7/4.9	2.7/4.5	2.3/4.2	2.0/3.9	1.6/3.8	0.6/2.3	0.5/1.8	2.2/3.0
1-Axis, N-S	Average	1.1	1.8	2.5	3.5	4.4	5.0	4.9	4.3	3.4	2.4	1.2	0.8	2.9
Horiz Axis	Min/Max	0.8/1.6	1.0/2.8	1.3/3.5	1.6/4.8	2.9/6.3	3.5/6.4	3.6/6.0	3.1/5.5	2.4/4.6	1.5/3.7	0.4/1.7	0.3/1.2	2.6/3.4
1-Axis, N-S	Average	1.8	2.5	3.0	3.8	4.3	4.7	4.7	4.5	3.9	3.1	1.7	1.2	3.3
Tilt=Latitude	Min/Max	1.2/2.5	1.4/3.9	1.6/4.3	1.8/5.2	2.8/6.2	3.3/6.1	3.5/5.8	3.1/5.7	2.8/5.3	2.0/4.8	0.6/2.6	0.5/1.9	2.8/3.7
2-Axis	Average	1.9	2.5	3.0	3.8	4.6	5.1	5.0	4.6	3.9	3.2	1.8	1.4	3.4
	Min/Max	1.3/2.7	1.4/4.0	1.6/4.3	1.8/5.3	3.0/6.6	3.6/6.6	3.7/6.2	3.3/5.8	2.8/5.3	2.0/4.9	0.7/2.7	0.6/2.1	3.0/3.9

Average Climatic Conditions

Element	Jan	Feb	Mar	Apr	May	June	July	Aug	Sept	Oct	Nov	Dec	Year
Temperature (°C)	-5.3	-3.8	2.3	8.7	14.8	19.9	22.3	21.1	17.2	10.7	4.4	-2.3	9.2
Daily Minimum Temp	-9.5	-8.3	-2.9	2.4	8.2	13.3	15.9	14.7	10.8	4.4	-0.3	-6.4	3.6
Daily Maximum Temp	-1.0	0.8	7.5	14.9	21.4	26.6	28.6	27.4	23.6	16.9	9.2	1.8	14.8
Record Minimum Temp	-28.9	-25.6	-21.1	-13.3	-3.9	0.0	4.4	1.1	-3.3	-9.4	-16.7	-28.3	-28.9
Record Maximum Temp	16.7	20.0	26.7	31.1	35.0	40.0	39.4	36.7	36.7	32.8	25.6	20.0	40.0
HDD, Base 18.3°C	732	619	496	290	132	18	0	9	61	242	417	639	3655
CDD, Base 18.3°C	0	0	0	0	22	67	125	93	27	4	0	0	339
Relative Humidity (%)	74	73	70	66	66	69	72	76	76	72	76	79	72
Wind Speed (m/s)	5.0	4.7	5.1	4.9	4.2	3.8	3.3	3.1	3.4	3.8	4.5	4.6	4.2

Youngstown, OH

WBAN NO. 14852

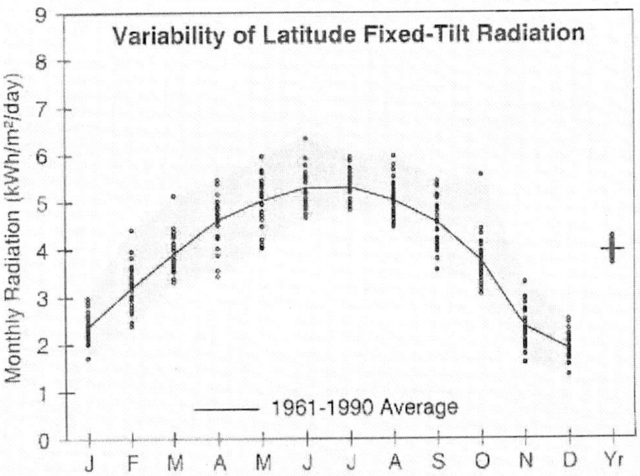

LATITUDE: 41.27° N
LONGITUDE: 80.67° W
ELEVATION: 361 meters
MEAN PRESSURE: 974 millibars

STATION TYPE: Secondary

Solar Radiation for Flat-Plate Collectors Facing South at a Fixed Tilt (kWh/m²/day), Uncertainty ±9%

Tilt (°)		Jan	Feb	Mar	Apr	May	June	July	Aug	Sept	Oct	Nov	Dec	Year
0	Average	1.6	2.4	3.3	4.4	5.3	5.9	5.8	5.0	4.0	2.8	1.7	1.3	3.6
	Min/Max	1.4/1.8	2.0/3.0	2.9/4.0	3.5/5.1	4.3/6.2	5.2/7.1	5.3/6.5	4.5/5.8	3.3/4.6	2.5/3.7	1.3/2.0	1.1/1.5	3.4/3.9
Latitude -15	Average	2.2	3.0	3.8	4.7	5.4	5.8	5.7	5.3	4.6	3.6	2.2	1.7	4.0
	Min/Max	1.6/2.7	2.3/4.0	3.3/4.9	3.6/5.6	4.3/6.3	5.1/6.9	5.2/6.4	4.7/6.2	3.6/5.4	3.0/5.1	1.5/3.0	1.3/2.2	3.7/4.3
Latitude	Average	2.4	3.2	3.9	4.6	5.0	5.3	5.3	5.0	4.6	3.8	2.4	1.9	3.9
	Min/Max	1.7/3.0	2.4/4.4	3.3/5.1	3.4/5.5	4.0/6.0	4.7/6.3	4.8/5.9	4.5/6.0	3.6/5.4	3.0/5.6	1.6/3.3	1.3/2.5	3.7/4.3
Latitude +15	Average	2.5	3.2	3.8	4.3	4.5	4.6	4.7	4.6	4.3	3.7	2.4	1.9	3.7
	Min/Max	1.7/3.2	2.3/4.6	3.2/5.1	3.2/5.1	3.6/5.3	4.1/5.5	4.3/5.2	4.1/5.4	3.3/5.2	3.0/5.7	1.6/3.4	1.3/2.7	3.5/4.0
90	Average	2.3	2.8	2.9	2.8	2.6	2.5	2.6	2.8	3.1	3.0	2.1	1.8	2.6
	Min/Max	1.5/3.0	2.0/4.3	2.4/3.9	2.1/3.4	2.2/3.0	2.3/2.8	2.4/2.8	2.5/3.3	2.4/3.7	2.4/4.7	1.3/3.1	1.1/2.5	2.4/2.9

Solar Radiation for 1-Axis Tracking Flat-Plate Collectors with a North-South Axis (kWh/m²/day), Uncertainty ±9%

Axis Tilt (°)		Jan	Feb	Mar	Apr	May	June	July	Aug	Sept	Oct	Nov	Dec	Year
0	Average	2.0	3.0	4.1	5.6	6.6	7.4	7.4	6.4	5.2	3.7	2.1	1.6	4.6
	Min/Max	1.5/2.5	2.3/4.0	3.5/5.6	4.0/6.8	5.0/8.1	6.2/9.4	6.5/8.6	5.4/7.9	4.0/6.3	3.0/5.5	1.4/2.7	1.2/2.0	4.2/5.0
Latitude -15	Average	2.4	3.5	4.6	5.9	6.7	7.4	7.4	6.7	5.7	4.3	2.5	1.9	4.9
	Min/Max	1.7/3.1	2.5/4.8	3.9/6.3	4.1/7.2	5.0/8.3	6.2/9.4	6.5/8.6	5.6/8.3	4.2/6.9	3.4/6.6	1.6/3.5	1.3/2.6	4.5/5.4
Latitude	Average	2.6	3.6	4.6	5.8	6.5	7.1	7.1	6.5	5.7	4.4	2.6	2.0	4.9
	Min/Max	1.8/3.4	2.6/5.2	3.9/6.5	4.0/7.2	4.8/8.0	5.9/9.0	6.3/8.3	5.5/8.2	4.2/7.0	3.5/6.9	1.6/3.7	1.3/2.8	4.5/5.4
Latitude +15	Average	2.7	3.7	4.5	5.5	6.1	6.6	6.7	6.2	5.5	4.4	2.6	2.1	4.7
	Min/Max	1.8/3.5	2.6/5.3	3.8/6.4	3.8/6.8	4.5/7.5	5.5/8.4	5.9/7.8	5.2/7.8	4.1/6.8	3.4/7.0	1.6/3.8	1.3/3.0	4.3/5.2

Solar Radiation for 2-Axis Tracking Flat-Plate Collectors (kWh/m²/day), Uncertainty ±9%

Tracker		Jan	Feb	Mar	Apr	May	June	July	Aug	Sept	Oct	Nov	Dec	Year
2-Axis	Average	2.7	3.7	4.6	5.9	6.8	7.6	7.6	6.7	5.7	4.5	2.7	2.1	5.1
	Min/Max	1.8/3.5	2.6/5.3	3.9/6.5	4.1/7.3	5.1/8.4	6.3/9.6	6.6/8.8	5.7/8.3	4.3/7.0	3.5/7.0	1.6/3.9	1.3/3.0	4.6/5.6

Direct Beam Solar Radiation for Concentrating Collectors (kWh/m²/day), Uncertainty ±8%

Tracker		Jan	Feb	Mar	Apr	May	June	July	Aug	Sept	Oct	Nov	Dec	Year
1-Axis, E-W Horiz Axis	Average	1.1	1.5	1.8	2.3	2.8	3.2	3.1	2.7	2.5	2.1	1.2	0.8	2.1
	Min/Max	0.3/1.9	0.7/2.4	1.1/3.3	1.1/3.5	1.5/4.0	2.1/4.9	2.3/4.1	1.9/3.9	1.4/3.4	1.4/4.1	0.5/2.3	0.2/1.6	1.8/2.6
1-Axis, N-S Horiz Axis	Average	0.8	1.3	2.0	3.0	3.6	4.2	4.2	3.6	3.0	2.1	0.9	0.5	2.4
	Min/Max	0.2/1.3	0.6/2.1	1.2/3.6	1.5/4.4	2.0/5.1	2.8/6.4	3.1/5.5	2.4/5.2	1.6/4.1	1.3/4.0	0.4/1.6	0.2/1.0	2.0/3.0
1-Axis, N-S Tilt=Latitude	Average	1.2	1.8	2.4	3.2	3.6	4.0	4.0	3.7	3.4	2.7	1.3	0.9	2.7
	Min/Max	0.3/2.1	0.9/2.9	1.5/4.4	1.6/4.7	2.0/5.0	2.6/6.1	3.0/5.3	2.5/5.3	1.9/4.7	1.7/5.1	0.6/2.5	0.3/1.7	2.2/3.2
2-Axis	Average	1.3	1.8	2.4	3.2	3.8	4.3	4.3	3.8	3.4	2.7	1.4	0.9	2.8
	Min/Max	0.4/2.2	0.9/3.0	1.5/4.4	1.6/4.8	2.1/5.3	2.8/6.6	3.2/5.7	2.6/5.5	1.9/4.7	1.7/5.2	0.6/2.6	0.3/1.8	2.3/3.4

Average Climatic Conditions

Element	Jan	Feb	Mar	Apr	May	June	July	Aug	Sept	Oct	Nov	Dec	Year
Temperature (°C)	-4.7	-3.4	2.4	8.5	14.2	19.0	21.3	20.4	16.8	10.7	5.0	-1.4	9.1
Daily Minimum Temp	-8.7	-7.8	-2.6	2.7	7.9	12.7	15.1	14.4	10.9	5.3	0.9	-5.1	3.8
Daily Maximum Temp	-0.7	1.1	7.4	14.3	20.4	25.2	27.4	26.4	22.6	16.1	9.1	2.2	14.3
Record Minimum Temp	-28.9	-25.6	-23.3	-11.7	-4.4	-1.1	5.6	0.0	-1.7	-6.7	-17.2	-24.4	-28.9
Record Maximum Temp	21.7	19.4	27.8	31.1	33.3	37.2	37.8	36.1	37.2	30.6	26.7	24.4	37.8
HDD, Base 18.3°C	713	608	494	295	149	34	4	12	72	242	400	612	3636
CDD, Base 18.3°C	0	0	0	0	19	54	96	78	25	4	0	0	276
Relative Humidity (%)	75	73	70	65	68	71	72	75	76	73	75	78	72
Wind Speed (m/s)	5.2	4.9	5.1	4.8	4.2	3.7	3.4	3.3	3.6	4.1	4.7	5.0	4.3

Oklahoma City, OK

WBAN NO. 13967

LATITUDE: 35.40° N
LONGITUDE: 97.60° W
ELEVATION: 397 meters
MEAN PRESSURE: 970 millibars

STATION TYPE: Secondary

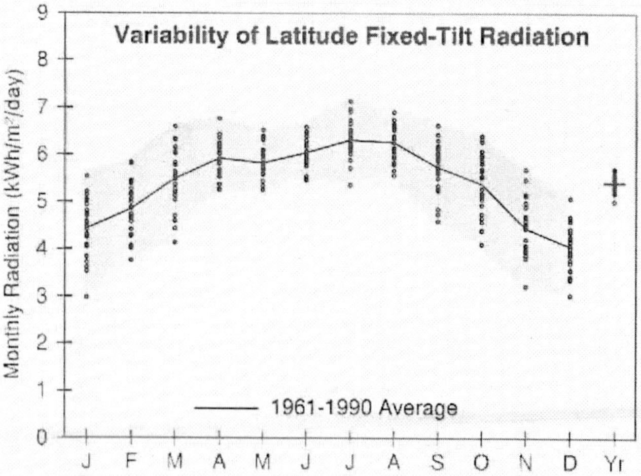

Solar Radiation for Flat-Plate Collectors Facing South at a Fixed Tilt (kWh/m²/day), Uncertainty ±9%

Tilt (°)		Jan	Feb	Mar	Apr	May	June	July	Aug	Sept	Oct	Nov	Dec	Year
0	Average	2.8	3.5	4.6	5.7	6.2	6.8	6.9	6.2	5.0	4.0	2.9	2.4	4.8
	Min/Max	2.2/3.2	3.0/4.0	3.6/5.3	5.0/6.4	5.6/6.9	6.1/7.4	5.8/7.8	5.6/6.9	4.2/5.7	3.3/4.6	2.4/3.5	2.0/2.8	4.6/4.9
Latitude -15	Average	3.9	4.4	5.3	6.0	6.2	6.6	6.8	6.5	5.6	5.0	3.9	3.5	5.3
	Min/Max	2.7/4.7	3.5/5.3	4.1/6.3	5.3/6.9	5.6/6.9	5.9/7.2	5.8/7.7	5.8/7.2	4.6/6.5	3.9/5.9	3.0/4.9	2.7/4.3	5.0/5.6
Latitude	Average	4.4	4.9	5.5	5.9	5.8	6.0	6.3	6.3	5.7	5.4	4.4	4.1	5.4
	Min/Max	3.0/5.5	3.8/5.8	4.1/6.6	5.2/6.8	5.2/6.5	5.4/6.6	5.3/7.1	5.6/6.9	4.6/6.6	4.1/6.4	3.2/5.7	3.0/5.1	5.0/5.7
Latitude +15	Average	4.7	5.0	5.4	5.5	5.2	5.2	5.5	5.7	5.5	5.5	4.7	4.4	5.2
	Min/Max	3.1/6.0	3.8/6.1	4.0/6.5	4.9/6.3	4.7/5.8	4.7/5.7	4.7/6.2	5.1/6.3	4.4/6.4	4.1/6.5	3.3/6.1	3.2/5.6	4.8/5.5
90	Average	4.3	4.1	3.8	3.2	2.5	2.3	2.4	3.0	3.6	4.2	4.1	4.0	3.4
	Min/Max	2.6/5.5	3.0/5.0	2.8/4.6	2.8/3.5	2.3/2.7	2.1/2.3	2.2/2.6	2.7/3.2	2.9/4.1	3.2/5.1	2.7/5.3	2.9/5.2	3.1/3.7

Solar Radiation for 1-Axis Tracking Flat-Plate Collectors with a North-South Axis (kWh/m²/day), Uncertainty ±9%

Axis Tilt (°)		Jan	Feb	Mar	Apr	May	June	July	Aug	Sept	Oct	Nov	Dec	Year
0	Average	4.0	4.9	6.2	7.6	8.1	8.9	9.3	8.5	6.9	5.8	4.2	3.6	6.5
	Min/Max	2.7/5.0	3.6/6.0	4.4/7.6	6.5/9.1	7.0/9.1	7.7/10.0	7.3/10.8	7.4/9.7	5.3/8.2	4.2/6.9	3.1/5.4	2.7/4.5	6.0/6.9
Latitude -15	Average	4.8	5.6	6.8	7.9	8.1	8.8	9.3	8.7	7.4	6.5	4.9	4.3	6.9
	Min/Max	3.1/6.1	4.1/6.9	4.8/8.4	6.8/9.5	7.0/9.2	7.7/9.9	7.3/10.8	7.6/9.9	5.7/8.8	4.7/7.8	3.5/6.4	3.2/5.5	6.4/7.4
Latitude	Average	5.3	5.9	7.0	7.8	7.9	8.4	8.9	8.6	7.5	6.8	5.3	4.8	7.0
	Min/Max	3.3/6.8	4.3/7.3	4.8/8.6	6.7/9.4	6.8/8.9	7.3/9.5	7.0/10.4	7.4/9.8	5.7/8.9	4.9/8.2	3.7/7.0	3.4/6.2	6.4/7.5
Latitude +15	Average	5.5	6.0	6.9	7.5	7.4	7.9	8.4	8.1	7.3	6.8	5.5	5.1	6.9
	Min/Max	3.4/7.1	4.3/7.5	4.7/8.5	6.4/9.1	6.4/8.4	6.8/8.9	6.6/9.8	7.1/9.3	5.6/8.7	4.9/8.3	3.8/7.3	3.6/6.6	6.2/7.3

Solar Radiation for 2-Axis Tracking Flat-Plate Collectors (kWh/m²/day), Uncertainty ±9%

Tracker		Jan	Feb	Mar	Apr	May	June	July	Aug	Sept	Oct	Nov	Dec	Year
2-Axis	Average	5.6	6.1	7.0	7.9	8.2	9.0	9.4	8.7	7.5	6.8	5.6	5.1	7.3
	Min/Max	3.4/7.2	4.3/7.5	4.9/8.6	6.8/9.5	7.1/9.3	7.8/10.2	7.4/11.0	7.6/10.0	5.7/8.9	4.9/8.3	3.8/7.4	3.6/6.7	6.6/7.7

Direct Beam Solar Radiation for Concentrating Collectors (kWh/m²/day), Uncertainty ±8%

Tracker		Jan	Feb	Mar	Apr	May	June	July	Aug	Sept	Oct	Nov	Dec	Year
1-Axis, E-W	Average	3.4	3.4	3.5	3.9	3.9	4.5	4.9	4.4	3.7	3.8	3.3	3.2	3.8
Horiz Axis	Min/Max	1.6/5.0	1.9/4.7	2.0/4.9	2.9/5.2	3.0/4.9	3.5/5.6	3.0/6.3	3.5/5.4	2.3/4.9	2.4/5.1	1.9/4.9	1.9/4.6	3.3/4.3
1-Axis, N-S	Average	2.7	3.3	4.2	5.1	5.2	5.9	6.5	5.9	4.7	4.1	2.8	2.4	4.4
Horiz Axis	Min/Max	1.3/4.0	1.8/4.6	2.2/5.7	3.7/6.9	3.9/6.5	4.6/7.4	4.1/8.4	4.7/7.3	2.9/6.2	2.4/5.4	1.7/4.2	1.5/3.5	3.8/4.9
1-Axis, N-S	Average	3.8	4.1	4.8	5.3	5.1	5.5	6.1	5.9	5.1	4.9	3.8	3.5	4.8
Tilt=Latitude	Min/Max	1.8/5.5	2.3/5.8	2.5/6.5	3.9/7.2	3.8/6.3	4.3/7.0	3.8/7.9	4.6/7.4	3.2/6.8	3.0/6.6	2.2/5.7	2.1/5.0	4.1/5.4
2-Axis	Average	4.1	4.2	4.8	5.3	5.4	6.0	6.6	6.1	5.1	5.0	4.0	3.8	5.0
	Min/Max	1.9/5.9	2.3/5.9	2.5/6.6	3.9/7.3	4.0/6.6	4.7/7.6	4.1/8.5	4.8/7.6	3.2/6.8	3.0/6.7	2.3/5.9	2.3/5.5	4.3/5.6

Average Climatic Conditions

Element	Jan	Feb	Mar	Apr	May	June	July	Aug	Sept	Oct	Nov	Dec	Year
Temperature (°C)	2.2	4.9	10.2	15.8	20.2	24.8	27.8	27.3	22.8	16.7	9.8	4.1	15.6
Daily Minimum Temp	-3.8	-1.3	3.6	9.3	14.3	18.9	21.4	20.9	16.8	10.2	3.7	-1.9	9.3
Daily Maximum Temp	8.2	11.2	16.7	22.2	26.2	30.7	34.1	33.6	28.8	23.1	15.8	9.9	21.7
Record Minimum Temp	-20.0	-19.4	-16.1	-6.7	2.8	8.3	11.7	10.6	2.2	-5.6	-11.7	-22.2	-22.2
Record Maximum Temp	26.7	28.9	33.9	37.8	40.0	40.6	42.8	43.3	38.9	35.6	30.6	30.0	43.3
HDD, Base 18.3°C	501	375	258	98	17	0	0	0	8	76	257	443	2033
CDD, Base 18.3°C	0	0	5	21	76	195	293	277	142	24	0	0	1033
Relative Humidity (%)	67	66	61	61	67	67	61	62	67	64	67	68	65
Wind Speed (m/s)	5.5	5.8	6.3	6.1	5.6	4.9	4.8	4.6	4.8	5.1	5.5	5.3	5.4

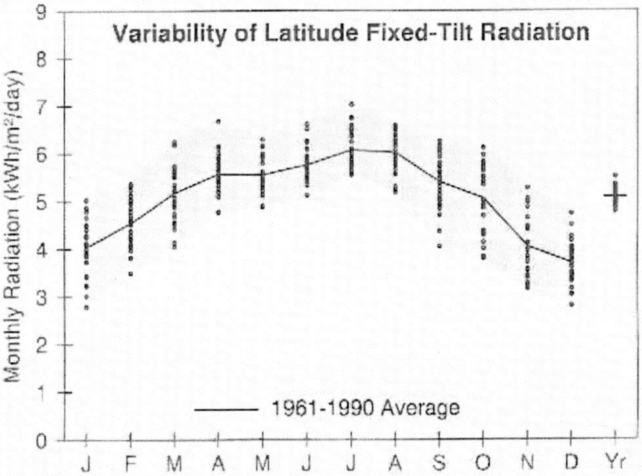

Tulsa, OK

WBAN NO. 13968

LATITUDE: 36.20° N
LONGITUDE: 95.90° W
ELEVATION: 206 meters
MEAN PRESSURE: 992 millibars

STATION TYPE: Secondary

Solar Radiation for Flat-Plate Collectors Facing South at a Fixed Tilt (kWh/m²/day), Uncertainty ±9%

Tilt (°)		Jan	Feb	Mar	Apr	May	June	July	Aug	Sept	Oct	Nov	Dec	Year
0	Average	2.5	3.3	4.3	5.3	5.9	6.4	6.7	6.0	4.7	3.8	2.7	2.2	4.5
	Min/Max	2.0/2.9	2.7/3.6	3.5/5.0	4.6/6.3	5.2/6.6	5.7/7.4	6.1/7.8	5.3/6.6	3.8/5.4	3.1/4.4	2.3/3.2	1.9/2.6	4.4/4.8
Latitude -15	Average	3.5	4.2	5.0	5.7	5.9	6.3	6.6	6.3	5.3	4.7	3.6	3.2	5.0
	Min/Max	2.5/4.3	3.3/4.8	4.0/6.0	4.9/6.8	5.2/6.7	5.6/7.2	6.0/7.6	5.4/6.8	4.1/6.1	3.6/5.6	2.9/4.6	2.5/4.0	4.8/5.4
Latitude	Average	4.0	4.5	5.2	5.6	5.5	5.8	6.1	6.0	5.4	5.1	4.0	3.7	5.1
	Min/Max	2.8/5.0	3.5/5.3	4.0/6.2	4.8/6.7	4.9/6.3	5.1/6.6	5.5/7.0	5.2/6.6	4.0/6.3	3.8/6.1	3.2/5.3	2.8/4.7	4.8/5.5
Latitude +15	Average	4.3	4.7	5.1	5.2	4.9	5.0	5.3	5.5	5.2	5.1	4.2	4.0	4.9
	Min/Max	2.9/5.4	3.5/5.6	3.9/6.2	4.4/6.2	4.3/5.6	4.5/5.7	4.8/6.1	4.7/6.0	3.8/6.0	3.7/6.2	3.2/5.6	2.9/5.2	4.5/5.3
90	Average	3.9	3.8	3.6	3.0	2.5	2.3	2.4	2.9	3.4	4.0	3.7	3.6	3.3
	Min/Max	2.4/4.9	2.8/4.6	2.8/4.4	2.6/3.5	2.3/2.7	2.1/2.4	2.3/2.6	2.6/3.1	2.5/3.9	2.8/4.9	2.7/5.0	2.6/4.8	2.9/3.5

Solar Radiation for 1-Axis Tracking Flat-Plate Collectors with a North-South Axis (kWh/m²/day), Uncertainty ±9%

Axis Tilt (°)		Jan	Feb	Mar	Apr	May	June	July	Aug	Sept	Oct	Nov	Dec	Year
0	Average	3.6	4.5	5.8	7.0	7.6	8.3	8.8	8.1	6.4	5.3	3.8	3.2	6.0
	Min/Max	2.5/4.5	3.4/5.4	4.4/7.2	5.7/8.8	6.3/8.8	7.1/10.0	8.0/10.8	6.7/9.3	4.7/7.6	3.9/6.5	3.0/4.9	2.5/4.1	5.7/6.7
Latitude -15	Average	4.4	5.2	6.3	7.3	7.6	8.3	8.8	8.3	6.9	6.0	4.4	3.9	6.5
	Min/Max	2.9/5.5	3.8/6.3	4.7/7.9	5.9/9.2	6.3/8.9	7.0/9.9	8.0/10.8	6.9/9.5	4.9/8.2	4.3/7.4	3.4/5.9	2.9/5.1	6.0/7.2
Latitude	Average	4.8	5.5	6.5	7.2	7.4	7.9	8.5	8.2	6.9	6.3	4.8	4.3	6.5
	Min/Max	3.1/6.1	4.0/6.7	4.8/8.1	5.8/9.1	6.1/8.6	6.7/9.5	7.7/10.4	6.7/9.4	4.9/8.3	4.4/7.8	3.6/6.4	3.1/5.7	6.1/7.2
Latitude +15	Average	5.0	5.6	6.4	6.9	7.0	7.4	8.0	7.8	6.8	6.3	5.0	4.5	6.4
	Min/Max	3.2/6.4	4.0/6.8	4.7/8.0	5.6/8.8	5.8/8.1	6.2/8.9	7.2/9.8	6.4/8.9	4.8/8.1	4.4/7.8	3.7/6.7	3.3/6.1	5.9/7.1

Solar Radiation for 2-Axis Tracking Flat-Plate Collectors (kWh/m²/day), Uncertainty ±9%

Tracker		Jan	Feb	Mar	Apr	May	June	July	Aug	Sept	Oct	Nov	Dec	Year
2-Axis	Average	5.0	5.6	6.5	7.3	7.7	8.5	9.0	8.4	6.9	6.4	5.0	4.6	6.7
	Min/Max	3.2/6.4	4.0/6.8	4.8/8.1	5.9/9.2	6.4/9.0	7.1/10.2	8.1/11.0	6.9/9.6	4.9/8.3	4.4/7.9	3.7/6.8	3.3/6.2	6.2/7.5

Direct Beam Solar Radiation for Concentrating Collectors (kWh/m²/day), Uncertainty ±8%

Tracker		Jan	Feb	Mar	Apr	May	June	July	Aug	Sept	Oct	Nov	Dec	Year
1-Axis, E-W Horiz Axis	Average	3.0	3.0	3.2	3.4	3.5	4.0	4.5	4.1	3.3	3.5	2.9	2.8	3.4
	Min/Max	1.5/4.3	1.8/4.2	1.8/4.6	2.3/5.0	2.4/4.5	2.9/5.7	3.7/6.3	2.9/5.1	1.7/4.4	1.9/4.7	1.9/4.3	1.6/4.2	3.0/4.1
1-Axis, N-S Horiz Axis	Average	2.4	2.9	3.7	4.4	4.7	5.3	6.0	5.5	4.2	3.6	2.4	2.1	3.9
	Min/Max	1.1/3.4	1.7/4.0	2.0/5.3	3.0/6.5	3.2/6.1	3.9/7.5	4.9/8.4	3.8/6.9	2.2/5.4	2.0/4.9	1.6/3.7	1.2/3.2	3.5/4.7
1-Axis, N-S Tilt=Latitude	Average	3.3	3.7	4.2	4.6	4.5	5.0	5.7	5.5	4.6	4.4	3.3	3.0	4.3
	Min/Max	1.6/4.7	2.2/5.1	2.4/6.1	3.1/6.8	3.1/5.9	3.6/7.0	4.7/8.0	3.8/6.9	2.4/6.0	2.4/6.0	2.2/5.0	1.8/4.6	3.8/5.2
2-Axis	Average	3.6	3.8	4.3	4.7	4.8	5.4	6.1	5.7	4.6	4.5	3.5	3.3	4.5
	Min/Max	1.7/5.1	2.2/5.2	2.4/6.1	3.1/6.9	3.3/6.2	3.9/7.6	5.0/8.6	3.9/7.2	2.4/6.0	2.4/6.1	2.3/5.3	1.9/5.0	4.0/5.4

Average Climatic Conditions

Element	Jan	Feb	Mar	Apr	May	June	July	Aug	Sept	Oct	Nov	Dec	Year
Temperature (°C)	1.8	4.6	10.3	16.4	20.7	25.4	28.5	27.5	22.9	16.8	9.9	3.8	15.7
Daily Minimum Temp	-3.9	-1.4	3.9	9.9	14.9	19.8	22.7	21.4	17.2	10.4	4.2	-1.7	9.8
Daily Maximum Temp	7.4	10.6	16.7	22.8	26.5	30.9	34.3	33.6	28.7	23.2	15.7	9.3	21.7
Record Minimum Temp	-22.2	-21.7	-19.4	-5.6	1.7	9.4	10.6	11.1	1.7	-3.3	-12.2	-22.2	-22.2
Record Maximum Temp	26.1	30.0	35.6	38.9	35.6	39.4	44.4	43.3	42.8	36.7	30.6	26.7	44.4
HDD, Base 18.3°C	513	384	254	84	23	0	0	0	11	80	252	449	2051
CDD, Base 18.3°C	0	0	6	26	97	212	315	284	149	32	0	0	1121
Relative Humidity (%)	67	65	62	61	69	69	64	65	70	66	67	69	66
Wind Speed (m/s)	4.6	4.8	5.3	5.4	4.9	4.4	4.2	4.1	4.1	4.3	4.7	4.5	4.6

Astoria, OR

WBAN NO. 94224

LATITUDE: 46.15° N
LONGITUDE: 123.88° W
ELEVATION: 7 meters
MEAN PRESSURE: 1017 millibars

STATION TYPE: Secondary

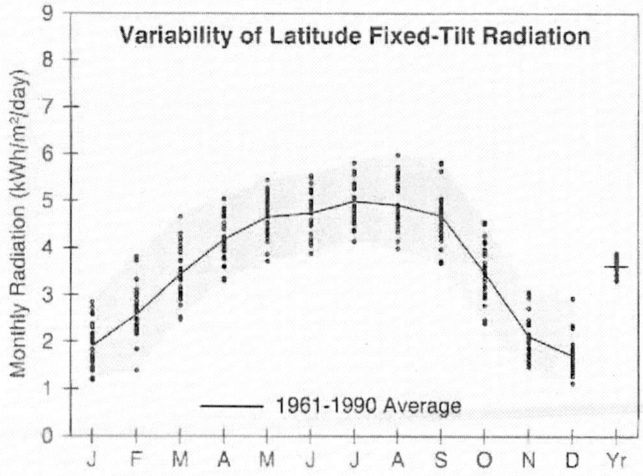

Solar Radiation for Flat-Plate Collectors Facing South at a Fixed Tilt (kWh/m²/day), Uncertainty ±9%

Tilt (°)		Jan	Feb	Mar	Apr	May	June	July	Aug	Sept	Oct	Nov	Dec	Year
0	Average	1.1	1.8	2.8	3.9	4.9	5.3	5.4	4.8	3.8	2.4	1.3	1.0	3.2
	Min/Max	0.9/1.4	1.3/2.3	2.3/3.4	3.3/4.5	4.0/5.6	4.4/6.1	4.6/6.2	4.0/5.6	3.3/4.5	1.9/2.9	1.1/1.6	0.9/1.3	3.0/3.4
Latitude -15	Average	1.7	2.4	3.4	4.3	5.0	5.2	5.4	5.1	4.7	3.3	2.0	1.5	3.7
	Min/Max	1.1/2.5	1.4/3.5	2.5/4.5	3.4/5.2	4.0/5.8	4.2/6.0	4.5/6.3	4.2/6.2	3.7/5.7	2.3/4.3	1.4/2.8	1.1/2.5	3.3/3.9
Latitude	Average	1.9	2.6	3.4	4.2	4.7	4.7	5.0	4.9	4.7	3.5	2.1	1.7	3.6
	Min/Max	1.2/2.8	1.4/3.8	2.5/4.7	3.3/5.0	3.7/5.4	3.9/5.5	4.1/5.8	4.0/6.0	3.7/5.8	2.4/4.6	1.5/3.1	1.1/2.9	3.3/3.9
Latitude +15	Average	2.0	2.6	3.3	3.8	4.1	4.1	4.4	4.4	4.5	3.5	2.2	1.8	3.4
	Min/Max	1.2/3.0	1.3/3.9	2.3/4.6	3.0/4.7	3.3/4.8	3.4/4.8	3.6/5.1	3.6/5.4	3.5/5.6	2.3/4.6	1.5/3.2	1.1/3.2	3.1/3.7
90	Average	1.8	2.2	2.6	2.7	2.7	2.5	2.7	3.0	3.4	2.9	1.9	1.6	2.5
	Min/Max	1.0/2.8	1.1/3.5	1.8/3.7	2.2/3.3	2.2/3.0	2.2/2.9	2.3/3.1	2.5/3.7	2.6/4.2	1.9/3.9	1.3/3.0	1.0/3.1	2.3/2.8

Solar Radiation for 1-Axis Tracking Flat-Plate Collectors with a North-South Axis (kWh/m²/day), Uncertainty ±9%

Axis Tilt (°)		Jan	Feb	Mar	Apr	May	June	July	Aug	Sept	Oct	Nov	Dec	Year
0	Average	1.5	2.3	3.5	4.9	6.1	6.4	6.7	6.1	5.2	3.3	1.7	1.3	4.1
	Min/Max	0.9/2.1	1.3/3.4	2.6/4.7	3.8/6.0	4.4/7.3	5.1/7.9	5.4/8.3	4.6/7.6	4.0/6.8	2.3/4.3	1.2/2.4	0.9/2.0	3.7/4.4
Latitude -15	Average	1.9	2.8	4.0	5.3	6.2	6.4	6.8	6.4	5.9	4.0	2.2	1.7	4.5
	Min/Max	1.2/3.0	1.4/4.3	2.8/5.5	3.9/6.6	4.5/7.6	5.1/8.0	5.4/8.4	4.8/8.1	4.4/7.8	2.6/5.3	1.5/3.2	1.1/3.0	4.0/4.9
Latitude	Average	2.1	2.9	4.1	5.2	6.0	6.1	6.6	6.3	5.9	4.1	2.3	1.8	4.5
	Min/Max	1.2/3.2	1.4/4.5	2.8/5.7	3.8/6.5	4.3/7.3	4.9/7.7	5.2/8.1	4.7/8.0	4.4/7.8	2.7/5.6	1.6/3.5	1.1/3.3	4.0/4.8
Latitude +15	Average	2.2	2.9	4.0	4.9	5.6	5.7	6.1	6.0	5.8	4.1	2.4	1.9	4.3
	Min/Max	1.2/3.4	1.4/4.6	2.7/5.6	3.6/6.2	4.0/6.9	4.5/7.2	4.8/7.6	4.4/7.6	4.2/7.7	2.6/5.6	1.6/3.6	1.2/3.5	3.9/4.7

Solar Radiation for 2-Axis Tracking Flat-Plate Collectors (kWh/m²/day), Uncertainty ±9%

Tracker		Jan	Feb	Mar	Apr	May	June	July	Aug	Sept	Oct	Nov	Dec	Year
2-Axis	Average	2.2	3.0	4.1	5.3	6.3	6.6	7.0	6.5	6.0	4.2	2.4	1.9	4.6
	Min/Max	1.2/3.4	1.4/4.6	2.8/5.7	4.0/6.6	4.5/7.7	5.3/8.2	5.6/8.6	4.9/8.2	4.4/7.9	2.7/5.6	1.6/3.6	1.2/3.6	4.2/5.0

Direct Beam Solar Radiation for Concentrating Collectors (kWh/m²/day), Uncertainty ±8%

Tracker		Jan	Feb	Mar	Apr	May	June	July	Aug	Sept	Oct	Nov	Dec	Year
1-Axis, E-W	Average	1.1	1.4	1.7	2.1	2.5	2.7	3.1	2.9	2.9	2.1	1.2	1.0	2.1
Horiz Axis	Min/Max	0.5/2.1	0.3/2.8	0.9/2.9	1.2/3.0	1.2/3.6	1.5/3.9	1.9/4.4	1.7/4.2	1.6/4.4	0.9/3.3	0.6/2.2	0.4/2.4	1.8/2.4
1-Axis, N-S	Average	0.7	1.2	1.8	2.5	3.3	3.4	3.9	3.6	3.3	1.9	0.8	0.6	2.3
Horiz Axis	Min/Max	0.3/1.4	0.3/2.3	0.9/2.9	1.5/3.7	1.5/4.7	2.0/5.0	2.5/5.7	2.0/5.3	1.8/5.2	0.8/3.0	0.4/1.5	0.2/1.4	1.9/2.6
1-Axis, N-S	Average	1.2	1.7	2.3	2.8	3.3	3.3	3.8	3.8	3.9	2.6	1.3	1.1	2.6
Tilt=Latitude	Min/Max	0.5/2.3	0.4/3.3	1.2/3.8	1.6/4.1	1.6/4.7	1.9/4.8	2.4/5.5	2.1/5.6	2.2/6.1	1.1/4.1	0.6/2.4	0.4/2.5	2.3/3.0
2-Axis	Average	1.3	1.7	2.3	2.8	3.5	3.6	4.1	4.0	3.9	2.7	1.4	1.1	2.7
	Min/Max	0.5/2.5	0.4/3.4	1.2/3.8	1.6/4.1	1.6/4.9	2.0/5.3	2.6/5.9	2.2/5.7	2.2/6.1	1.1/4.1	0.7/2.6	0.4/2.7	2.3/3.1

Average Climatic Conditions

Element	Jan	Feb	Mar	Apr	May	June	July	Aug	Sept	Oct	Nov	Dec	Year
Temperature (°C)	5.5	6.8	7.6	9.0	11.4	13.8	15.6	15.9	14.7	11.4	8.3	5.8	10.5
Daily Minimum Temp	2.2	2.9	3.4	4.6	7.1	9.7	11.3	11.4	9.4	6.7	4.6	2.6	6.3
Daily Maximum Temp	8.8	10.6	11.8	13.3	15.6	17.8	19.7	20.4	19.9	16.2	11.9	9.0	14.6
Record Minimum Temp	-11.7	-12.8	-5.6	-1.7	-1.1	2.8	3.9	3.9	0.6	-3.3	-9.4	-14.4	-14.4
Record Maximum Temp	19.4	22.2	22.8	28.3	30.6	33.9	37.8	35.6	35.0	29.4	21.7	17.8	37.8
HDD, Base 18.3°C	398	323	332	280	216	135	87	78	112	213	302	389	2866
CDD, Base 18.3°C	0	0	0	0	0	0	0	4	0	0	0	0	4
Relative Humidity (%)	83	82	81	79	79	80	80	82	81	83	83	84	81
Wind Speed (m/s)	4.3	4.2	4.1	4.1	3.9	4.1	4.0	3.8	3.5	3.5	4.1	4.1	4.0

Burns, OR

WBAN NO. 94185

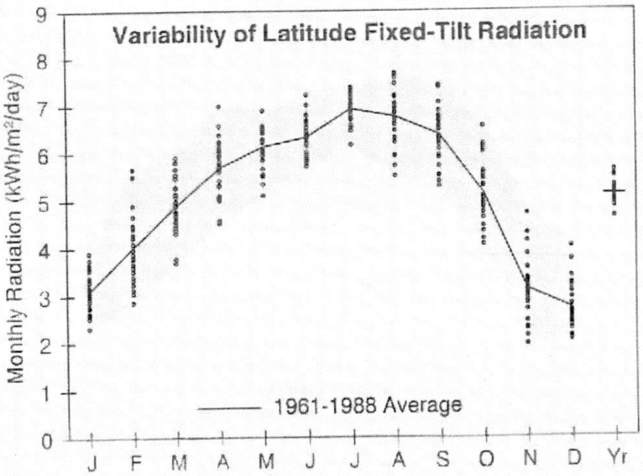

LATITUDE: 43.58° N
LONGITUDE: 119.05° W
ELEVATION: 1271 meters
MEAN PRESSURE: 874 millibars

STATION TYPE: Primary

Solar Radiation for Flat-Plate Collectors Facing South at a Fixed Tilt (kWh/m²/day), Uncertainty ±9%

Tilt (°)		Jan	Feb	Mar	Apr	May	June	July	Aug	Sept	Oct	Nov	Dec	Year
0	Average	1.8	2.6	3.8	5.2	6.4	7.1	7.5	6.5	5.1	3.4	1.9	1.5	4.4
	Min/Max	1.5/2.0	2.2/3.3	3.1/4.5	4.4/6.2	5.5/7.2	6.4/8.1	6.8/8.1	5.4/7.3	4.3/5.7	2.8/4.0	1.4/2.4	1.1/1.8	4.1/4.7
Latitude -15	Average	2.8	3.7	4.7	5.8	6.5	6.9	7.5	7.0	6.3	4.8	2.8	2.4	5.1
	Min/Max	2.1/3.4	2.7/5.1	3.7/5.7	4.7/7.1	5.5/7.3	6.2/7.9	6.7/8.0	5.8/8.0	5.3/7.2	3.8/6.0	1.9/4.1	1.9/3.4	4.7/5.6
Latitude	Average	3.1	4.0	4.9	5.7	6.1	6.3	6.9	6.8	6.4	5.2	3.1	2.7	5.1
	Min/Max	2.3/3.9	2.9/5.7	3.7/5.9	4.5/7.0	5.1/6.9	5.7/7.2	6.2/7.4	5.5/7.7	5.3/7.4	4.1/6.6	2.0/4.7	2.0/4.0	4.6/5.6
Latitude +15	Average	3.3	4.1	4.8	5.3	5.4	5.4	6.0	6.1	6.1	5.2	3.3	2.9	4.8
	Min/Max	2.4/4.2	2.8/5.9	3.5/5.8	4.2/6.5	4.5/6.1	4.9/6.2	5.3/6.3	5.0/6.9	5.0/7.2	4.1/6.7	2.0/5.0	2.1/4.3	4.3/5.3
90	Average	3.2	3.7	3.8	3.6	3.2	3.0	3.3	3.8	4.5	4.4	2.9	2.8	3.5
	Min/Max	2.3/3.9	2.4/5.4	2.8/4.6	2.9/4.4	2.7/3.5	2.8/3.3	2.9/3.4	3.2/4.3	3.6/5.3	3.3/5.7	1.7/4.6	2.0/4.2	3.1/3.9

Solar Radiation for 1-Axis Tracking Flat-Plate Collectors with a North-South Axis (kWh/m²/day), Uncertainty ±9%

Axis Tilt (°)		Jan	Feb	Mar	Apr	May	June	July	Aug	Sept	Oct	Nov	Dec	Year
0	Average	2.5	3.7	5.3	7.2	8.8	9.7	10.7	9.4	7.6	5.1	2.6	2.1	6.3
	Min/Max	2.0/3.1	2.7/5.2	3.8/6.4	5.6/8.9	7.0/10.1	8.4/11.4	9.6/11.8	7.2/11.1	6.3/9.1	4.0/6.6	1.8/3.9	1.6/3.0	5.6/6.8
Latitude -15	Average	3.3	4.5	6.0	7.7	9.0	9.8	10.9	10.0	8.6	6.2	3.4	2.8	6.8
	Min/Max	2.4/4.2	3.1/6.6	4.3/7.4	5.8/9.7	7.1/10.4	8.4/11.5	9.8/12.0	7.6/11.8	7.0/10.2	4.7/8.0	2.0/5.1	2.0/4.2	6.1/7.5
Latitude	Average	3.5	4.8	6.1	7.6	8.7	9.4	10.5	9.8	8.7	6.5	3.6	3.1	6.9
	Min/Max	2.6/4.6	3.2/7.0	4.3/7.6	5.8/9.6	6.9/10.1	8.0/11.1	9.4/11.6	7.4/11.6	7.0/10.4	4.9/8.4	2.1/5.6	2.2/4.7	6.1/7.6
Latitude +15	Average	3.7	4.8	6.1	7.3	8.2	8.8	9.9	9.3	8.5	6.5	3.7	3.2	6.7
	Min/Max	2.7/4.8	3.2/7.2	4.2/7.5	5.5/9.3	6.5/9.6	7.5/10.4	8.9/10.9	7.0/11.1	6.8/10.2	4.9/8.5	2.1/5.8	2.3/4.9	5.9/7.4

Solar Radiation for 2-Axis Tracking Flat-Plate Collectors (kWh/m²/day), Uncertainty ±9%

Tracker		Jan	Feb	Mar	Apr	May	June	July	Aug	Sept	Oct	Nov	Dec	Year
2-Axis	Average	3.7	4.9	6.2	7.7	9.1	10.0	11.1	10.0	8.7	6.5	3.7	3.3	7.1
	Min/Max	2.7/4.8	3.3/7.2	4.3/7.6	5.9/9.7	7.2/10.6	8.6/11.8	10.0/12.2	7.6/11.8	7.0/10.4	4.9/8.5	2.2/5.9	2.3/5.0	6.3/7.8

Direct Beam Solar Radiation for Concentrating Collectors (kWh/m²/day), Uncertainty ±8%

Tracker		Jan	Feb	Mar	Apr	May	June	July	Aug	Sept	Oct	Nov	Dec	Year
1-Axis, E-W Horiz Axis	Average	1.9	2.4	2.9	3.5	4.4	5.1	6.2	5.5	4.8	3.8	2.0	1.8	3.7
	Min/Max	0.9/3.1	0.8/4.4	1.6/4.0	1.9/5.2	2.7/5.5	3.8/6.6	5.3/7.1	3.3/6.9	3.5/6.1	2.3/5.5	0.6/3.9	0.9/3.4	3.2/4.3
1-Axis, N-S Horiz Axis	Average	1.3	2.1	3.2	4.6	5.9	6.9	8.3	7.3	5.9	3.7	1.5	1.1	4.3
	Min/Max	0.6/2.1	0.6/3.8	1.7/4.4	2.6/6.6	3.7/7.4	5.1/8.9	7.2/9.6	4.4/9.4	4.3/7.6	2.3/5.3	0.5/2.9	0.5/2.2	3.7/5.0
1-Axis, N-S Tilt=Latitude	Average	2.1	2.9	3.9	4.9	5.8	6.5	8.1	7.5	6.7	4.9	2.3	1.9	4.8
	Min/Max	1.0/3.4	0.9/5.3	2.1/5.4	2.8/7.2	3.6/7.3	4.8/8.4	7.0/9.3	4.6/9.7	4.9/8.7	2.9/7.0	0.7/4.4	0.9/3.6	4.0/5.6
2-Axis	Average	2.2	3.0	3.9	5.0	6.1	7.1	8.7	7.8	6.8	4.9	2.4	2.0	5.0
	Min/Max	1.0/3.7	0.9/5.5	2.1/5.4	2.8/7.3	3.8/7.7	5.2/9.2	7.5/10.0	4.7/10.0	4.9/8.7	3.0/7.1	0.7/4.6	1.0/3.9	4.2/5.8

Average Climatic Conditions

Element	Jan	Feb	Mar	Apr	May	June	July	Aug	Sept	Oct	Nov	Dec	Year
Temperature (°C)	-4.8	-1.4	2.4	6.0	10.4	14.4	19.0	17.9	12.8	7.2	0.9	-3.8	6.7
Daily Minimum Temp	-10.6	-7.1	-3.9	-1.7	2.2	5.3	8.4	7.2	2.4	-2.2	-5.6	-9.4	-1.2
Daily Maximum Temp	0.9	4.2	8.7	13.6	18.7	23.6	29.5	28.5	23.1	16.6	7.3	1.8	14.7
Record Minimum Temp	-32.8	-33.3	-15.6	-8.9	-5.6	-5.0	-3.9	-2.2	-8.3	-19.4	-25.0	-28.9	-33.3
Record Maximum Temp	12.8	17.2	23.3	28.9	33.9	35.0	37.2	38.3	36.1	30.0	22.0	14.4	38.3
HDD, Base 18.3°C	718	554	494	370	244	124	32	57	175	344	523	687	4325
CDD, Base 18.3°C	0	0	0	0	0	8	53	43	8	0	0	0	112
Relative Humidity (%)	78	75	65	55	51	49	38	40	46	57	74	79	59
Wind Speed (m/s)	2.1	2.2	3.0	3.5	3.4	3.0	3.0	2.3	2.5	2.3	2.2	2.4	2.7

Eugene, OR

WBAN NO. 24221

LATITUDE: 44.12° N
LONGITUDE: 123.22° W
ELEVATION: 109 meters
MEAN PRESSURE: 1004 millibars

STATION TYPE: Primary

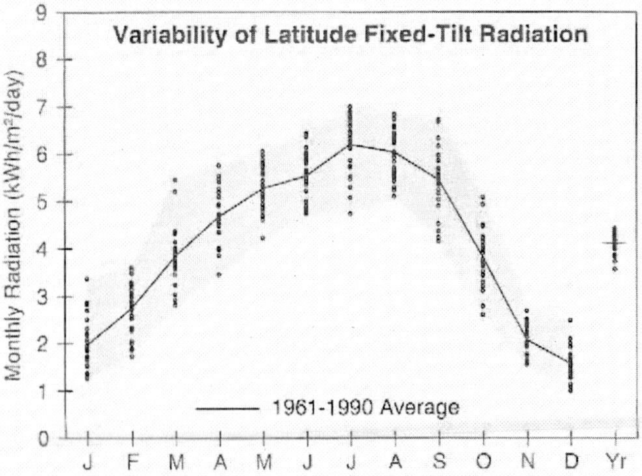

Solar Radiation for Flat-Plate Collectors Facing South at a Fixed Tilt (kWh/m²/day), Uncertainty ±9%

Tilt (°)		Jan	Feb	Mar	Apr	May	June	July	Aug	Sept	Oct	Nov	Dec	Year
0	Average	1.3	2.0	3.1	4.4	5.5	6.2	6.7	5.8	4.4	2.7	1.4	1.0	3.7
	Min/Max	0.9/1.7	1.4/2.4	2.5/4.0	3.5/5.2	4.6/6.3	5.3/7.2	5.2/7.6	5.0/6.5	3.5/5.2	2.1/3.3	1.1/1.7	0.7/1.3	3.3/4.0
Latitude -15	Average	1.8	2.6	3.8	4.8	5.6	6.0	6.7	6.3	5.4	3.6	1.9	1.4	4.2
	Min/Max	1.2/3.0	1.7/3.3	2.8/5.2	3.6/5.9	4.5/6.4	5.2/7.0	5.1/7.6	5.3/7.1	4.1/6.6	2.6/4.7	1.5/2.5	0.9/2.2	3.7/4.5
Latitude	Average	2.0	2.7	3.8	4.7	5.3	5.5	6.2	6.0	5.5	3.8	2.1	1.6	4.1
	Min/Max	1.3/3.4	1.7/3.6	2.8/5.4	3.5/5.8	4.2/6.1	4.7/6.4	4.7/7.0	5.1/6.8	4.2/6.7	2.6/5.1	1.5/2.7	1.0/2.5	3.6/4.4
Latitude +15	Average	2.0	2.8	3.7	4.3	4.6	4.8	5.4	5.5	5.2	3.8	2.1	1.6	3.8
	Min/Max	1.3/3.6	1.7/3.7	2.7/5.4	3.2/5.3	3.8/5.3	4.1/5.5	4.1/6.1	4.6/6.2	4.0/6.5	2.5/5.1	1.5/2.8	1.0/2.7	3.3/4.1
90	Average	1.8	2.3	2.9	3.0	2.8	2.7	3.1	3.5	3.9	3.2	1.8	1.5	2.7
	Min/Max	1.1/3.3	1.4/3.2	2.0/4.3	2.2/3.7	2.4/3.2	2.4/3.0	2.5/3.3	3.0/3.9	2.9/4.8	2.0/4.3	1.3/2.5	0.9/2.5	2.3/2.9

Solar Radiation for 1-Axis Tracking Flat-Plate Collectors with a North-South Axis (kWh/m²/day), Uncertainty ±9%

Axis Tilt (°)		Jan	Feb	Mar	Apr	May	June	July	Aug	Sept	Oct	Nov	Dec	Year
0	Average	1.6	2.5	4.0	5.6	7.2	8.1	9.2	8.1	6.3	3.7	1.8	1.2	5.0
	Min/Max	1.0/2.6	1.6/3.4	2.9/5.7	3.9/7.1	5.4/8.6	6.4/9.9	6.4/10.8	6.4/9.4	4.5/7.9	2.6/4.9	1.3/2.3	0.8/1.8	4.2/5.4
Latitude -15	Average	2.0	3.0	4.5	6.0	7.3	8.1	9.4	8.5	7.1	4.4	2.2	1.6	5.3
	Min/Max	1.3/3.5	1.8/4.1	3.1/6.6	4.1/7.6	5.5/8.9	6.4/9.9	6.4/11.0	6.7/10.0	5.0/8.9	2.8/6.0	1.6/2.9	1.0/2.5	4.5/5.8
Latitude	Average	2.1	3.1	4.6	5.9	7.1	7.8	9.1	8.4	7.1	4.6	2.3	1.7	5.3
	Min/Max	1.3/3.9	1.9/4.3	3.1/6.8	4.0/7.6	5.3/8.6	6.1/9.6	6.2/10.7	6.5/9.8	5.0/9.1	2.9/6.3	1.6/3.0	1.0/2.8	4.5/5.8
Latitude +15	Average	2.2	3.1	4.5	5.6	6.7	7.3	8.5	8.0	6.9	4.6	2.3	1.7	5.1
	Min/Max	1.3/4.0	1.8/4.3	3.0/6.7	3.8/7.3	4.9/8.1	5.7/8.9	5.8/10.0	6.2/9.4	4.9/8.9	2.8/6.3	1.6/3.1	1.0/2.9	4.3/5.6

Solar Radiation for 2-Axis Tracking Flat-Plate Collectors (kWh/m²/day), Uncertainty ±9%

Tracker		Jan	Feb	Mar	Apr	May	June	July	Aug	Sept	Oct	Nov	Dec	Year
2-Axis	Average	2.2	3.2	4.6	6.0	7.4	8.3	9.6	8.6	7.1	4.6	2.3	1.7	5.5
	Min/Max	1.4/4.1	1.9/4.4	3.2/6.8	4.1/7.7	5.5/9.0	6.6/10.2	6.6/11.3	6.7/10.0	5.1/9.1	2.9/6.3	1.7/3.1	1.0/3.0	4.6/6.0

Direct Beam Solar Radiation for Concentrating Collectors (kWh/m²/day), Uncertainty ±8%

Tracker		Jan	Feb	Mar	Apr	May	June	July	Aug	Sept	Oct	Nov	Dec	Year
1-Axis, E-W Horiz Axis	Average	1.1	1.4	2.0	2.5	3.2	3.9	5.0	4.4	3.8	2.4	1.1	0.8	2.6
	Min/Max	0.4/2.5	0.5/2.3	1.0/3.6	1.2/3.6	1.9/4.4	2.5/5.8	2.7/6.7	2.9/5.7	2.2/5.2	0.8/3.8	0.4/1.8	0.2/2.0	2.0/3.1
1-Axis, N-S Horiz Axis	Average	0.7	1.2	2.1	3.1	4.3	5.2	6.7	5.8	4.4	2.3	0.8	0.5	3.1
	Min/Max	0.3/1.7	0.4/2.0	1.0/3.8	1.5/4.6	2.4/5.8	3.3/7.6	3.5/8.8	3.7/7.6	2.6/6.2	0.8/3.5	0.3/1.3	0.1/1.2	2.3/3.6
1-Axis, N-S Tilt=Latitude	Average	1.1	1.7	2.6	3.4	4.2	4.9	6.5	6.0	5.1	3.0	1.2	0.8	3.4
	Min/Max	0.4/2.7	0.6/2.8	1.3/4.7	1.6/4.9	2.4/5.8	3.1/7.2	3.4/8.5	3.8/7.8	3.0/7.2	1.0/4.7	0.5/2.0	0.2/2.0	2.6/4.0
2-Axis	Average	1.2	1.7	2.6	3.4	4.5	5.4	7.0	6.2	5.1	3.0	1.2	0.9	3.5
	Min/Max	0.5/2.9	0.6/2.8	1.3/4.7	1.6/5.0	2.5/6.1	3.4/7.8	3.7/9.2	4.0/8.1	3.0/7.2	1.1/4.8	0.5/2.1	0.2/2.2	2.7/4.1

Average Climatic Conditions

Element	Jan	Feb	Mar	Apr	May	June	July	Aug	Sept	Oct	Nov	Dec	Year
Temperature (°C)	4.9	6.8	8.6	10.3	13.2	16.7	19.6	19.8	17.1	12.3	7.8	5.1	11.8
Daily Minimum Temp	1.8	2.8	3.8	4.8	6.9	9.8	11.6	11.8	9.6	6.4	4.3	2.2	6.3
Daily Maximum Temp	8.0	10.8	13.3	15.8	19.5	23.4	27.6	27.7	24.6	18.1	11.3	7.9	17.3
Record Minimum Temp	-20.0	-19.4	-6.7	-2.8	-2.2	0.0	3.9	3.3	0.0	-7.2	-11.1	-24.4	-24.4
Record Maximum Temp	19.4	21.7	25.0	30.0	33.9	37.8	40.6	42.2	39.4	34.4	24.4	20.0	42.2
HDD, Base 18.3°C	417	323	303	240	158	67	21	17	64	188	315	412	2526
CDD, Base 18.3°C	0	0	0	0	0	17	61	61	28	0	0	0	167
Relative Humidity (%)	87	85	80	76	74	70	64	65	70	81	88	89	77
Wind Speed (m/s)	3.4	3.5	3.7	3.5	3.3	3.4	3.5	3.3	3.3	2.9	3.5	3.4	3.4

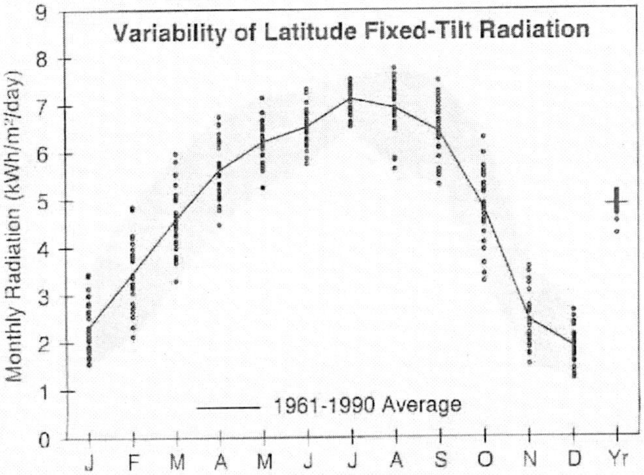

Medford, OR

WBAN NO. 24225

LATITUDE: 42.37° N
LONGITUDE: 122.87° W
ELEVATION: 396 meters
MEAN PRESSURE: 970 millibars

STATION TYPE: Primary

Solar Radiation for Flat-Plate Collectors Facing South at a Fixed Tilt (kWh/m²/day), Uncertainty ±9%

Tilt (°)		Jan	Feb	Mar	Apr	May	June	July	Aug	Sept	Oct	Nov	Dec	Year
0	Average	1.5	2.4	3.7	5.2	6.5	7.3	7.7	6.7	5.2	3.3	1.7	1.2	4.4
	Min/Max	1.1/1.9	1.8/3.0	2.9/4.4	4.3/6.1	5.5/7.4	6.4/8.2	7.1/8.2	5.4/7.4	4.5/5.9	2.6/4.0	1.2/2.1	0.9/1.5	4.0/4.5
Latitude -15	Average	2.1	3.2	4.5	5.7	6.6	7.1	7.7	7.2	6.3	4.5	2.3	1.7	4.9
	Min/Max	1.5/3.0	2.1/4.4	3.3/5.7	4.6/6.8	5.6/7.6	6.3/8.0	7.1/8.2	5.8/8.1	5.3/7.3	3.2/5.8	1.5/3.2	1.1/2.3	4.3/5.1
Latitude	Average	2.3	3.5	4.6	5.6	6.2	6.5	7.1	6.9	6.4	4.8	2.4	1.9	4.9
	Min/Max	1.5/3.5	2.1/4.8	3.3/6.0	4.5/6.7	5.2/7.1	5.8/7.3	6.5/7.5	5.6/7.8	5.3/7.5	3.3/6.3	1.5/3.6	1.2/2.6	4.3/5.1
Latitude +15	Average	2.4	3.5	4.5	5.2	5.5	5.6	6.2	6.3	6.2	4.8	2.5	2.0	4.5
	Min/Max	1.5/3.7	2.1/5.0	3.2/5.9	4.1/6.3	4.6/6.3	5.0/6.3	5.7/6.5	5.1/7.0	5.0/7.2	3.2/6.4	1.5/3.7	1.2/2.8	3.9/4.8
90	Average	2.2	3.0	3.4	3.5	3.1	2.9	3.2	3.8	4.4	4.0	2.2	1.8	3.1
	Min/Max	1.3/3.4	1.7/4.3	2.4/4.6	2.7/4.1	2.7/3.5	2.7/3.2	3.0/3.3	3.2/4.2	3.6/5.2	2.6/5.4	1.3/3.3	1.1/2.6	2.7/3.4

Solar Radiation for 1-Axis Tracking Flat-Plate Collectors with a North-South Axis (kWh/m²/day), Uncertainty ±9%

Axis Tilt (°)		Jan	Feb	Mar	Apr	May	June	July	Aug	Sept	Oct	Nov	Dec	Year
0	Average	1.9	3.1	4.9	6.9	8.7	9.8	11.0	9.5	7.6	4.7	2.1	1.5	6.0
	Min/Max	1.3/2.8	2.0/4.5	3.2/6.4	5.2/8.6	6.9/10.4	8.3/11.6	9.4/11.9	7.2/11.3	6.2/9.2	3.2/6.2	1.3/3.0	1.0/2.0	5.1/6.3
Latitude -15	Average	2.4	3.8	5.5	7.4	8.9	9.9	11.1	10.0	8.5	5.6	2.5	1.9	6.5
	Min/Max	1.5/3.7	2.2/5.5	3.6/7.4	5.5/9.3	7.1/10.7	8.3/11.6	9.5/12.1	7.6/11.9	6.8/10.3	3.7/7.5	1.5/3.8	1.2/2.7	5.5/6.8
Latitude	Average	2.6	4.0	5.6	7.3	8.7	9.5	10.7	9.9	8.6	5.9	2.7	2.0	6.5
	Min/Max	1.6/4.0	2.3/5.9	3.6/7.6	5.4/9.2	6.8/10.4	8.0/11.2	9.2/11.7	7.4/11.7	6.9/10.4	3.8/7.9	1.6/4.1	1.3/3.0	5.4/6.9
Latitude +15	Average	2.6	4.0	5.5	7.0	8.1	8.9	10.1	9.4	8.4	5.9	2.8	2.1	6.2
	Min/Max	1.6/4.2	2.2/6.0	3.5/7.5	5.1/8.9	6.4/9.8	7.4/10.5	8.6/11.0	7.1/11.1	6.7/10.2	3.8/8.0	1.6/4.2	1.3/3.1	5.2/6.7

Solar Radiation for 2-Axis Tracking Flat-Plate Collectors (kWh/m²/day), Uncertainty ±9%

Tracker		Jan	Feb	Mar	Apr	May	June	July	Aug	Sept	Oct	Nov	Dec	Year
2-Axis	Average	2.7	4.0	5.6	7.4	9.0	10.1	11.3	10.1	8.6	5.9	2.8	2.1	6.7
	Min/Max	1.6/4.2	2.3/6.0	3.6/7.6	5.5/9.3	7.2/10.8	8.5/11.9	9.7/12.4	7.6/11.9	6.9/10.4	3.8/8.0	1.6/4.2	1.3/3.2	5.6/7.1

Direct Beam Solar Radiation for Concentrating Collectors (kWh/m²/day), Uncertainty ±8%

Tracker		Jan	Feb	Mar	Apr	May	June	July	Aug	Sept	Oct	Nov	Dec	Year
1-Axis, E-W Horiz Axis	Average	1.3	2.1	2.7	3.5	4.5	5.4	6.5	5.6	4.8	3.4	1.3	1.0	3.5
	Min/Max	0.4/2.7	0.7/3.6	1.2/4.2	2.2/4.9	3.2/5.8	4.0/7.0	4.9/7.4	3.6/7.1	3.5/6.2	1.7/5.0	0.6/2.5	0.3/1.9	2.6/3.9
1-Axis, N-S Horiz Axis	Average	0.9	1.7	2.9	4.4	6.0	7.1	8.7	7.3	5.7	3.2	1.0	0.6	4.2
	Min/Max	0.3/1.9	0.5/3.1	1.2/4.6	2.6/6.3	4.1/7.8	5.1/9.3	6.4/10.0	4.6/9.6	4.1/7.7	1.6/4.8	0.4/1.8	0.2/1.2	3.1/4.5
1-Axis, N-S Tilt=Latitude	Average	1.4	2.4	3.5	4.7	5.9	6.8	8.4	7.6	6.6	4.2	1.5	1.1	4.5
	Min/Max	0.4/2.9	0.8/4.3	1.5/5.6	2.8/6.8	4.0/7.7	4.8/8.8	6.2/9.7	4.8/9.8	4.8/8.7	2.1/6.2	0.6/2.7	0.3/2.0	3.3/5.0
2-Axis	Average	1.5	2.5	3.5	4.8	6.2	7.4	9.0	7.8	6.6	4.2	1.6	1.1	4.7
	Min/Max	0.4/3.1	0.8/4.4	1.5/5.6	2.8/6.9	4.3/8.1	5.2/9.6	6.7/10.4	4.9/10.1	4.8/8.8	2.1/6.3	0.6/2.9	0.4/2.1	3.5/5.2

Average Climatic Conditions

Element	Jan	Feb	Mar	Apr	May	June	July	Aug	Sept	Oct	Nov	Dec	Year
Temperature (°C)	3.4	6.0	8.3	10.7	14.6	19.2	22.7	22.5	18.6	12.7	6.7	3.2	12.4
Daily Minimum Temp	-0.9	0.1	1.9	3.3	6.3	10.4	12.9	12.8	9.0	4.7	1.9	-0.4	5.2
Daily Maximum Temp	7.6	11.8	14.7	18.1	22.7	27.8	32.5	32.2	28.2	20.8	11.4	6.8	19.6
Record Minimum Temp	-19.4	-14.4	-8.9	-6.1	-2.2	-0.6	3.3	3.9	-1.7	-7.8	-12.2	-21.1	-21.1
Record Maximum Temp	21.7	25.0	30.0	33.9	39.4	42.8	46.1	45.6	43.3	37.2	23.9	22.2	46.1
HDD, Base 18.3°C	463	346	310	231	122	33	4	6	52	178	348	468	2562
CDD, Base 18.3°C	0	0	0	0	4	58	141	136	60	4	0	0	403
Relative Humidity (%)	84	78	71	65	61	55	48	51	57	70	84	87	67
Wind Speed (m/s)	1.7	2.1	2.4	2.5	2.5	2.7	2.6	2.4	2.1	1.7	1.7	1.6	2.1

North Bend, OR

WBAN NO. 24284

LATITUDE: 43.42° N
LONGITUDE: 124.25° W
ELEVATION: 5 meters
MEAN PRESSURE: 1018 millibars

STATION TYPE: Secondary

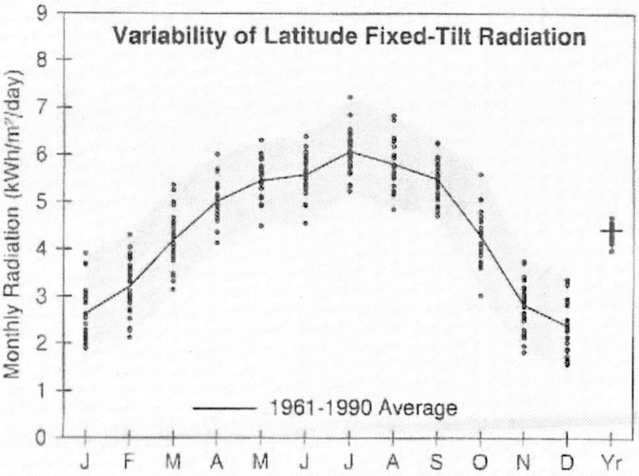

Solar Radiation for Flat-Plate Collectors Facing South at a Fixed Tilt (kWh/m²/day), Uncertainty ±9%

Tilt (°)		Jan	Feb	Mar	Apr	May	June	July	Aug	Sept	Oct	Nov	Dec	Year
0	Average	1.5	2.2	3.4	4.7	5.7	6.2	6.5	5.6	4.5	3.0	1.8	1.3	3.9
	Min/Max	1.3/1.9	1.7/2.6	2.8/4.1	4.0/5.4	4.8/6.5	5.1/7.1	5.6/7.7	4.8/6.4	4.0/5.0	2.4/3.6	1.4/2.1	1.0/1.6	3.6/4.0
Latitude -15	Average	2.4	3.0	4.1	5.1	5.8	6.1	6.5	6.0	5.4	4.0	2.6	2.1	4.4
	Min/Max	1.8/3.4	2.0/3.9	3.2/5.2	4.3/6.1	4.8/6.7	4.9/7.0	5.6/7.8	5.1/7.1	4.7/6.1	2.9/5.2	1.8/3.3	1.5/2.9	4.0/4.7
Latitude	Average	2.6	3.2	4.2	5.0	5.5	5.6	6.1	5.8	5.5	4.3	2.8	2.4	4.4
	Min/Max	1.9/3.9	2.1/4.3	3.2/5.4	4.1/6.0	4.5/6.3	4.6/6.4	5.2/7.2	4.8/6.8	4.7/6.3	3.0/5.6	1.8/3.8	1.6/3.4	4.0/4.7
Latitude +15	Average	2.7	3.3	4.1	4.6	4.8	4.8	5.3	5.3	5.2	4.3	2.9	2.5	4.2
	Min/Max	1.9/4.2	2.1/4.4	3.0/5.2	3.8/5.6	4.0/5.6	4.0/5.5	4.6/6.3	4.4/6.2	4.5/6.0	3.0/5.7	1.8/4.0	1.6/3.6	3.7/4.5
90	Average	2.5	2.8	3.1	3.1	2.9	2.7	3.0	3.3	3.8	3.6	2.6	2.3	3.0
	Min/Max	1.7/4.0	1.7/3.9	2.3/4.1	2.6/3.8	2.5/3.3	2.4/3.0	2.7/3.4	2.8/3.9	3.3/4.4	2.4/4.8	1.5/3.6	1.4/3.5	2.6/3.2

Solar Radiation for 1-Axis Tracking Flat-Plate Collectors with a North-South Axis (kWh/m²/day), Uncertainty ±9%

Axis Tilt (°)		Jan	Feb	Mar	Apr	May	June	July	Aug	Sept	Oct	Nov	Dec	Year
0	Average	2.1	2.9	4.4	6.0	7.2	7.8	8.4	7.3	6.2	4.1	2.4	1.8	5.1
	Min/Max	1.5/3.1	1.9/4.0	3.2/5.9	4.9/7.4	5.7/8.6	5.5/9.3	6.9/10.4	5.9/8.8	5.1/7.1	2.8/5.5	1.6/3.2	1.3/2.5	4.6/5.4
Latitude -15	Average	2.7	3.5	5.0	6.4	7.4	7.8	8.5	7.7	6.9	4.9	3.0	2.4	5.5
	Min/Max	1.9/4.2	2.2/4.9	3.5/6.7	5.2/8.0	5.8/8.9	5.5/9.4	7.0/10.6	6.2/9.4	5.7/8.0	3.3/6.7	1.9/4.1	1.6/3.5	4.9/5.9
Latitude	Average	3.0	3.7	5.0	6.3	7.2	7.5	8.2	7.6	7.0	5.1	3.2	2.7	5.5
	Min/Max	2.0/4.6	2.2/5.2	3.6/6.9	5.1/7.9	5.6/8.6	5.2/9.0	6.7/10.2	6.0/9.2	5.7/8.1	3.4/7.0	1.9/4.4	1.7/3.9	4.9/5.9
Latitude +15	Average	3.1	3.7	4.9	6.1	6.7	6.9	7.7	7.2	6.8	5.2	3.3	2.8	5.4
	Min/Max	2.0/4.8	2.2/5.3	3.5/6.8	4.8/7.6	5.2/8.1	4.8/8.4	6.2/9.6	5.7/8.8	5.6/7.9	3.3/7.1	1.9/4.6	1.7/4.1	4.7/5.8

Solar Radiation for 2-Axis Tracking Flat-Plate Collectors (kWh/m²/day), Uncertainty ±9%

Tracker		Jan	Feb	Mar	Apr	May	June	July	Aug	Sept	Oct	Nov	Dec	Year
2-Axis	Average	3.1	3.8	5.1	6.4	7.5	8.0	8.7	7.8	7.0	5.2	3.3	2.8	5.7
	Min/Max	2.1/4.9	2.3/5.3	3.6/6.9	5.2/8.0	5.8/9.0	5.6/9.6	7.1/10.8	6.2/9.4	5.8/8.1	3.4/7.1	1.9/4.6	1.7/4.2	5.1/6.1

Direct Beam Solar Radiation for Concentrating Collectors (kWh/m²/day), Uncertainty ±8%

Tracker		Jan	Feb	Mar	Apr	May	June	July	Aug	Sept	Oct	Nov	Dec	Year
1-Axis, E-W Horiz Axis	Average	1.7	1.9	2.3	2.8	3.4	3.8	4.5	4.0	3.7	2.8	1.8	1.6	2.9
	Min/Max	0.8/3.3	0.8/3.2	1.3/3.7	2.0/4.1	2.1/4.6	1.9/5.2	3.0/6.4	2.7/5.3	2.7/4.5	1.4/4.3	0.6/2.8	0.6/2.8	2.3/3.2
1-Axis, N-S Horiz Axis	Average	1.2	1.6	2.4	3.5	4.4	4.8	5.7	4.9	4.2	2.6	1.3	1.0	3.1
	Min/Max	0.5/2.3	0.6/2.7	1.3/4.1	2.5/5.0	2.7/5.9	2.3/6.7	3.8/8.1	3.3/6.6	3.1/5.2	1.2/4.1	0.4/2.2	0.4/1.8	2.6/3.5
1-Axis, N-S Tilt=Latitude	Average	1.8	2.2	3.0	3.8	4.4	4.6	5.6	5.1	4.9	3.5	2.0	1.7	3.5
	Min/Max	0.9/3.6	0.9/3.8	1.6/4.9	2.7/5.4	2.7/5.8	2.2/6.4	3.7/7.8	3.5/6.8	3.6/6.0	1.6/5.4	0.7/3.2	0.7/2.9	2.9/4.0
2-Axis	Average	2.0	2.3	3.0	3.8	4.6	5.0	6.0	5.3	4.9	3.5	2.1	1.8	3.7
	Min/Max	0.9/3.8	0.9/3.9	1.6/4.9	2.7/5.5	2.8/6.2	2.3/6.9	4.0/8.4	3.6/7.1	3.6/6.1	1.6/5.4	0.7/3.4	0.7/3.2	3.0/4.2

Average Climatic Conditions

Element	Jan	Feb	Mar	Apr	May	June	July	Aug	Sept	Oct	Nov	Dec	Year
Temperature (°C)	7.4	8.5	8.9	9.7	11.9	14.1	15.2	15.6	14.8	12.7	10.1	7.8	11.4
Daily Minimum Temp	3.8	4.7	5.1	5.8	8.1	10.4	11.4	11.7	10.3	8.2	6.2	4.3	7.5
Daily Maximum Temp	11.0	12.2	12.7	13.6	15.7	17.7	19.1	19.5	19.4	17.2	13.8	11.3	15.3
Record Minimum Temp	-7.2	-8.9	-2.8	-1.7	2.2	3.9	5.0	6.7	2.2	-1.7	-4.4	-10.6	-10.6
Record Maximum Temp	20.6	24.4	25.0	30.0	28.3	28.3	29.4	28.9	31.1	29.4	23.3	20.6	31.1
HDD, Base 18.3°C	338	276	293	258	200	128	97	87	108	174	248	326	2532
CDD, Base 18.3°C	0	0	0	0	0	0	0	0	0	0	0	0	0
Relative Humidity (%)	82	82	80	78	78	79	80	81	81	83	84	83	81
Wind Speed (m/s)	3.8	3.9	4.2	4.3	4.5	4.8	5.1	4.4	3.9	3.4	3.9	3.8	4.2

Pendleton, OR

WBAN NO. 24155

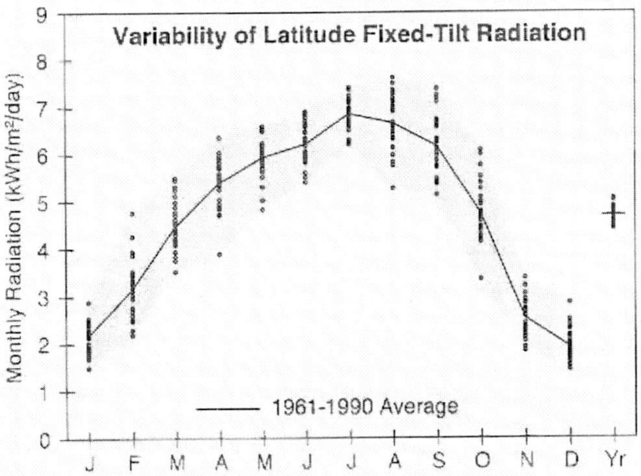

LATITUDE: 45.68° N
LONGITUDE: 118.85° W
ELEVATION: 456 meters
MEAN PRESSURE: 963 millibars

STATION TYPE: Secondary

Solar Radiation for Flat-Plate Collectors Facing South at a Fixed Tilt (kWh/m²/day), Uncertainty ±9%

Tilt (°)		Jan	Feb	Mar	Apr	May	June	July	Aug	Sept	Oct	Nov	Dec	Year
0	Average	1.4	2.1	3.4	4.9	6.2	6.9	7.4	6.3	4.8	3.0	1.6	1.1	4.1
	Min/Max	1.1/1.5	1.7/2.7	2.9/4.0	3.8/5.6	5.2/6.8	6.0/7.7	6.7/8.0	5.1/7.1	4.2/5.6	2.5/3.6	1.3/1.9	1.0/1.4	3.9/4.3
Latitude -15	Average	2.0	3.0	4.4	5.5	6.3	6.8	7.4	6.9	6.1	4.4	2.3	1.7	4.7
	Min/Max	1.5/2.5	2.1/4.3	3.5/5.3	4.0/6.4	5.2/7.0	5.9/7.5	6.7/8.0	5.5/7.9	5.1/7.2	3.2/5.6	1.7/3.0	1.4/2.5	4.5/5.0
Latitude	Average	2.2	3.2	4.5	5.4	5.9	6.2	6.8	6.7	6.2	4.7	2.5	1.9	4.7
	Min/Max	1.5/2.9	2.2/4.8	3.5/5.5	3.9/6.3	4.8/6.6	5.4/6.9	6.2/7.4	5.3/7.6	5.1/7.4	3.3/6.1	1.8/3.4	1.4/2.8	4.4/5.0
Latitude +15	Average	2.2	3.2	4.3	5.0	5.2	5.3	5.9	6.0	5.9	4.8	2.6	2.0	4.4
	Min/Max	1.5/3.0	2.1/4.9	3.4/5.4	3.6/5.9	4.3/5.8	4.7/5.9	5.4/6.4	4.8/6.9	4.9/7.1	3.3/6.2	1.8/3.5	1.4/3.1	4.1/4.7
90	Average	2.1	2.8	3.5	3.5	3.2	3.1	3.4	3.9	4.5	4.1	2.3	1.9	3.2
	Min/Max	1.3/2.9	1.8/4.4	2.6/4.4	2.5/4.1	2.7/3.5	2.8/3.3	3.2/3.6	3.2/4.4	3.7/5.4	2.8/5.3	1.6/3.3	1.3/3.0	3.0/3.5

Solar Radiation for 1-Axis Tracking Flat-Plate Collectors with a North-South Axis (kWh/m²/day), Uncertainty ±9%

Axis Tilt (°)		Jan	Feb	Mar	Apr	May	June	July	Aug	Sept	Oct	Nov	Dec	Year
0	Average	1.7	2.9	4.8	6.7	8.4	9.4	10.5	9.2	7.2	4.6	2.1	1.5	5.8
	Min/Max	1.3/2.2	2.0/4.2	3.8/6.0	4.7/8.0	6.7/9.6	7.8/10.9	8.9/11.6	7.0/10.9	5.7/8.9	3.2/5.9	1.5/2.8	1.2/2.1	5.3/6.1
Latitude -15	Average	2.2	3.5	5.5	7.2	8.6	9.5	10.7	9.8	8.2	5.6	2.7	1.9	6.3
	Min/Max	1.5/3.0	2.3/5.4	4.3/7.0	4.9/8.7	6.8/9.9	7.8/11.0	9.1/11.9	7.3/11.6	6.4/10.1	3.8/7.4	1.9/3.6	1.4/2.9	5.8/6.8
Latitude	Average	2.4	3.7	5.6	7.1	8.4	9.1	10.4	9.6	8.3	5.9	2.8	2.1	6.3
	Min/Max	1.5/3.3	2.4/5.8	4.3/7.2	4.8/8.7	6.6/9.6	7.5/10.6	8.8/11.5	7.2/11.5	6.5/10.3	3.9/7.8	2.0/3.9	1.5/3.2	5.8/6.8
Latitude +15	Average	2.4	3.7	5.5	6.9	7.9	8.6	9.8	9.2	8.1	5.9	2.9	2.2	6.1
	Min/Max	1.5/3.4	2.3/5.9	4.2/7.1	4.6/8.4	6.2/9.1	7.0/10.0	8.2/10.9	6.8/11.0	6.3/10.1	3.9/7.8	2.0/4.0	1.5/3.4	5.5/6.6

Solar Radiation for 2-Axis Tracking Flat-Plate Collectors (kWh/m²/day), Uncertainty ±9%

Tracker		Jan	Feb	Mar	Apr	May	June	July	Aug	Sept	Oct	Nov	Dec	Year
2-Axis	Average	2.5	3.7	5.6	7.3	8.8	9.7	10.9	9.8	8.3	5.9	2.9	2.2	6.5
	Min/Max	1.5/3.5	2.4/5.9	4.3/7.2	4.9/8.8	6.9/10.1	8.0/11.3	9.3/12.2	7.4/11.7	6.5/10.3	3.9/7.8	2.0/4.1	1.5/3.5	5.9/7.0

Direct Beam Solar Radiation for Concentrating Collectors (kWh/m²/day), Uncertainty ±8%

Tracker		Jan	Feb	Mar	Apr	May	June	July	Aug	Sept	Oct	Nov	Dec	Year
1-Axis, E-W Horiz Axis	Average	1.1	1.7	2.6	3.3	4.2	4.9	6.0	5.3	4.5	3.4	1.4	1.0	3.3
	Min/Max	0.4/2.0	0.6/3.5	1.6/3.8	1.7/4.5	2.8/5.2	3.6/6.4	4.6/7.1	3.3/6.8	3.0/6.1	1.7/5.0	0.6/2.3	0.5/2.0	2.9/3.8
1-Axis, N-S Horiz Axis	Average	0.7	1.5	2.9	4.2	5.6	6.5	8.1	7.0	5.4	3.2	1.0	0.6	3.9
	Min/Max	0.3/1.3	0.5/3.0	1.9/4.3	2.2/5.7	3.8/7.0	4.8/8.5	5.9/9.5	4.3/9.1	3.5/7.4	1.6/4.7	0.4/1.7	0.3/1.2	3.3/4.4
1-Axis, N-S Tilt=Latitude	Average	1.2	2.1	3.5	4.6	5.5	6.3	7.8	7.3	6.3	4.3	1.6	1.1	4.3
	Min/Max	0.4/2.2	0.8/4.2	2.3/5.2	2.4/6.2	3.8/7.0	4.6/8.1	5.8/9.3	4.5/9.5	4.1/8.6	2.2/6.3	0.6/2.6	0.5/2.1	3.7/4.9
2-Axis	Average	1.3	2.1	3.5	4.7	5.8	6.8	8.4	7.5	6.3	4.3	1.7	1.2	4.5
	Min/Max	0.4/2.3	0.8/4.3	2.3/5.2	2.4/6.3	4.0/7.3	5.0/8.9	6.2/9.9	4.7/9.8	4.1/8.6	2.2/6.4	0.7/2.8	0.5/2.3	3.8/5.1

Average Climatic Conditions

Element	Jan	Feb	Mar	Apr	May	June	July	Aug	Sept	Oct	Nov	Dec	Year
Temperature (°C)	0.8	4.0	7.1	10.2	14.4	19.0	22.7	22.2	17.3	11.3	5.3	1.3	11.3
Daily Minimum Temp	-2.7	-0.2	1.9	4.1	7.7	11.6	14.4	14.3	9.9	5.0	1.2	-2.3	5.4
Daily Maximum Temp	4.3	8.3	12.3	16.3	21.1	26.4	31.0	30.1	24.6	17.6	9.4	4.7	17.2
Record Minimum Temp	-30.0	-27.8	-12.2	-7.8	-3.9	1.7	5.6	4.4	-1.1	-11.7	-24.4	-28.3	-30.0
Record Maximum Temp	20.0	22.2	26.1	32.8	37.8	42.2	43.3	45.0	38.9	33.3	25.0	19.4	45.0
HDD, Base 18.3°C	543	401	348	245	126	39	8	13	81	217	392	529	2941
CDD, Base 18.3°C	0	0	0	0	3	59	144	133	49	0	0	0	389
Relative Humidity (%)	77	73	63	58	52	46	36	39	47	59	74	78	58
Wind Speed (m/s)	3.3	3.4	3.8	4.1	3.9	3.8	3.7	3.5	3.4	3.1	3.2	3.1	3.5

Portland, OR

WBAN NO. 24229

LATITUDE: 45.60° N
LONGITUDE: 122.60° W
ELEVATION: 12 meters
MEAN PRESSURE: 1017 millibars

STATION TYPE: Primary

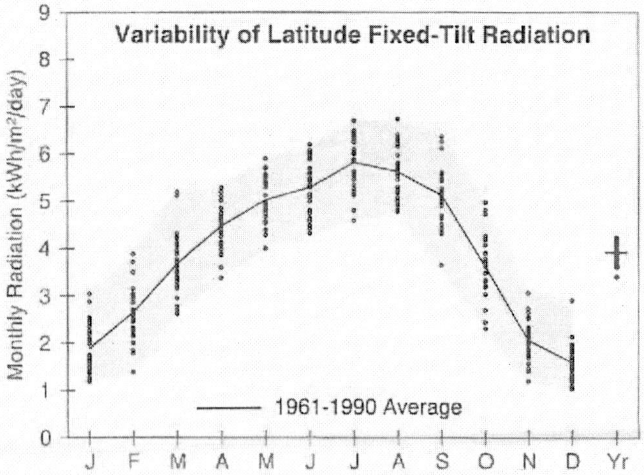

Solar Radiation for Flat-Plate Collectors Facing South at a Fixed Tilt (kWh/m²/day), Uncertainty ±9%

Tilt (°)		Jan	Feb	Mar	Apr	May	June	July	Aug	Sept	Oct	Nov	Dec	Year
0	Average	1.2	1.9	3.0	4.2	5.3	5.9	6.3	5.4	4.1	2.5	1.4	1.0	3.5
	Min/Max	0.9/1.5	1.4/2.4	2.4/3.8	3.4/4.8	4.3/6.2	4.8/6.9	5.0/7.2	4.6/6.3	3.2/4.9	1.8/3.2	0.9/1.7	0.7/1.3	3.1/3.8
Latitude -15	Average	1.7	2.5	3.6	4.6	5.4	5.8	6.3	5.9	5.1	3.4	1.9	1.4	4.0
	Min/Max	1.1/2.7	1.4/3.6	2.7/5.0	3.5/5.4	4.3/6.3	4.7/6.8	4.9/7.2	5.0/7.0	3.7/6.2	2.3/4.6	1.1/2.8	1.0/2.5	3.5/4.3
Latitude	Average	1.9	2.6	3.7	4.5	5.0	5.3	5.8	5.6	5.1	3.6	2.1	1.6	3.9
	Min/Max	1.2/3.0	1.4/3.9	2.6/5.2	3.4/5.3	4.0/5.9	4.3/6.2	4.6/6.7	4.8/6.7	3.6/6.4	2.3/5.0	1.2/3.1	1.0/2.9	3.4/4.2
Latitude +15	Average	1.9	2.7	3.5	4.1	4.4	4.6	5.1	5.1	4.9	3.6	2.1	1.6	3.6
	Min/Max	1.2/3.2	1.3/4.0	2.5/5.1	3.1/4.9	3.5/5.2	3.8/5.3	4.0/5.8	4.3/6.1	3.4/6.1	2.2/5.0	1.2/3.2	1.0/3.1	3.2/3.9
90	Average	1.8	2.3	2.8	2.9	2.8	2.7	3.1	3.4	3.7	3.0	1.9	1.5	2.6
	Min/Max	1.0/3.1	1.0/3.5	1.9/4.1	2.2/3.5	2.3/3.3	2.3/3.1	2.5/3.4	2.9/4.0	2.6/4.7	1.8/4.3	1.0/2.9	0.9/3.0	2.3/2.9

Solar Radiation for 1-Axis Tracking Flat-Plate Collectors with a North-South Axis (kWh/m²/day), Uncertainty ±9%

Axis Tilt (°)		Jan	Feb	Mar	Apr	May	June	July	Aug	Sept	Oct	Nov	Dec	Year
0	Average	1.5	2.4	3.8	5.2	6.6	7.4	8.3	7.3	5.7	3.4	1.7	1.2	4.5
	Min/Max	0.9/2.3	1.3/3.4	2.7/5.4	3.7/6.4	4.8/8.0	5.8/9.2	6.1/9.9	5.7/9.0	3.8/7.2	2.2/4.7	1.0/2.4	0.8/2.0	3.9/5.0
Latitude -15	Average	1.9	2.9	4.3	5.6	6.8	7.5	8.4	7.7	6.4	4.1	2.1	1.6	4.9
	Min/Max	1.1/3.2	1.4/4.3	2.9/6.3	3.9/6.9	4.9/8.3	5.8/9.3	6.1/10.1	6.1/9.6	4.2/8.2	2.5/5.8	1.2/3.2	1.0/2.9	4.2/5.4
Latitude	Average	2.0	3.0	4.3	5.5	6.5	7.2	8.1	7.5	6.5	4.3	2.2	1.7	4.9
	Min/Max	1.2/3.5	1.4/4.6	2.9/6.4	3.8/6.8	4.7/8.0	5.6/8.9	5.8/9.8	5.9/9.5	4.2/8.4	2.5/6.1	1.2/3.5	1.0/3.3	4.2/5.4
Latitude +15	Average	2.1	3.0	4.2	5.3	6.1	6.7	7.6	7.2	6.3	4.2	2.3	1.7	4.7
	Min/Max	1.2/3.6	1.3/4.6	2.8/6.3	3.6/6.5	4.4/7.5	5.2/8.3	5.5/9.2	5.6/9.0	4.0/8.2	2.5/6.1	1.2/3.6	1.0/3.4	4.0/5.2

Solar Radiation for 2-Axis Tracking Flat-Plate Collectors (kWh/m²/day), Uncertainty ±9%

Tracker		Jan	Feb	Mar	Apr	May	June	July	Aug	Sept	Oct	Nov	Dec	Year
2-Axis	Average	2.1	3.0	4.4	5.6	6.9	7.7	8.6	7.7	6.5	4.3	2.3	1.8	5.1
	Min/Max	1.2/3.7	1.4/4.7	2.9/6.4	3.9/6.9	5.0/8.4	6.0/9.5	6.2/10.3	6.1/9.7	4.2/8.4	2.5/6.1	1.3/3.6	1.0/3.5	4.3/5.6

Direct Beam Solar Radiation for Concentrating Collectors (kWh/m²/day), Uncertainty ±8%

Tracker		Jan	Feb	Mar	Apr	May	June	July	Aug	Sept	Oct	Nov	Dec	Year
1-Axis, E-W Horiz Axis	Average	1.0	1.4	1.9	2.2	2.8	3.4	4.2	3.8	3.3	2.2	1.1	0.8	2.3
	Min/Max	0.2/2.4	0.2/2.7	0.9/3.4	1.1/3.1	1.4/3.9	2.2/4.8	2.2/6.0	2.5/5.2	1.5/4.7	0.9/3.6	0.4/2.2	0.2/2.2	1.9/2.9
1-Axis, N-S Horiz Axis	Average	0.6	1.1	1.9	2.7	3.6	4.4	5.4	4.8	3.7	2.0	0.7	0.5	2.6
	Min/Max	0.2/1.5	0.2/2.2	0.9/3.7	1.3/3.8	1.8/5.1	2.9/6.3	2.9/8.1	3.1/6.7	1.6/5.3	0.8/3.3	0.3/1.4	0.1/1.3	2.1/3.2
1-Axis, N-S Tilt=Latitude	Average	1.1	1.6	2.4	3.0	3.6	4.2	5.3	5.0	4.4	2.7	1.2	0.9	3.0
	Min/Max	0.3/2.6	0.2/3.2	1.1/4.5	1.4/4.2	1.8/5.1	2.7/6.1	2.8/7.8	3.3/7.0	1.9/6.3	1.1/4.4	0.5/2.3	0.2/2.3	2.3/3.6
2-Axis	Average	1.2	1.7	2.4	3.0	3.8	4.5	5.7	5.2	4.4	2.7	1.3	0.9	3.1
	Min/Max	0.3/2.7	0.2/3.2	1.1/4.5	1.4/4.3	1.9/5.4	3.0/6.6	3.0/8.4	3.4/7.3	2.0/6.3	1.1/4.5	0.5/2.5	0.2/2.5	2.4/3.8

Average Climatic Conditions

Element	Jan	Feb	Mar	Apr	May	June	July	Aug	Sept	Oct	Nov	Dec	Year
Temperature (°C)	4.2	6.4	8.5	10.6	13.9	17.5	20.1	20.3	17.4	12.5	7.8	4.6	12.0
Daily Minimum Temp	0.9	2.3	3.7	5.2	8.3	11.6	13.6	13.8	11.1	7.2	4.2	1.6	6.9
Daily Maximum Temp	7.4	10.6	13.3	15.9	19.5	23.3	26.6	26.8	23.7	17.8	11.4	7.6	17.0
Record Minimum Temp	-18.9	-19.4	-7.2	-1.7	-1.7	3.9	6.1	6.7	1.1	-3.3	-10.6	-14.4	-19.4
Record Maximum Temp	17.2	21.7	26.7	30.6	37.8	37.8	41.7	41.7	40.6	33.3	22.8	17.8	41.7
HDD, Base 18.3°C	437	333	305	233	138	51	16	19	57	181	315	427	2512
CDD, Base 18.3°C	0	0	0	0	0	26	71	82	28	0	0	0	206
Relative Humidity (%)	81	78	75	72	69	66	63	65	69	78	82	83	73
Wind Speed (m/s)	4.4	4.2	3.8	3.4	3.4	3.4	3.5	3.3	3.1	3.0	4.0	4.2	3.6

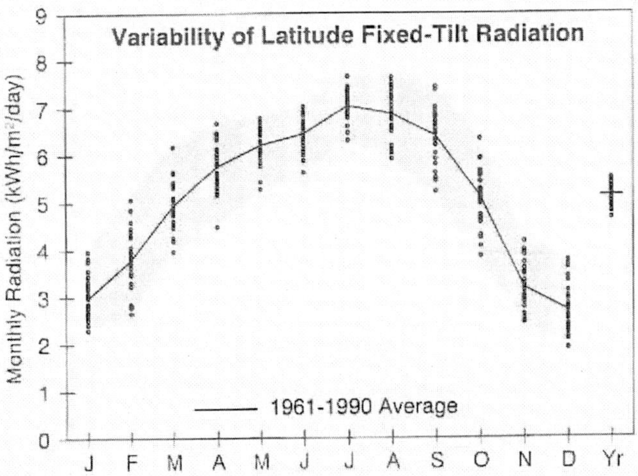

Redmond, OR

WBAN NO. 24230

LATITUDE: 44.27° N
LONGITUDE: 121.15° W
ELEVATION: 940 meters
MEAN PRESSURE: 909 millibars

STATION TYPE: Primary

Solar Radiation for Flat-Plate Collectors Facing South at a Fixed Tilt (kWh/m²/day), Uncertainty ±9%

Tilt (°)		Jan	Feb	Mar	Apr	May	June	July	Aug	Sept	Oct	Nov	Dec	Year
0	Average	1.7	2.5	3.8	5.3	6.5	7.2	7.6	6.6	5.1	3.3	1.9	1.4	4.4
	Min/Max	1.4/1.9	1.9/3.0	3.2/4.4	4.3/5.9	5.6/7.1	6.3/7.8	6.8/8.4	5.6/7.2	4.4/5.7	2.8/3.9	1.6/2.2	1.2/1.7	4.1/4.6
Latitude -15	Average	2.6	3.5	4.8	5.9	6.6	7.0	7.6	7.1	6.3	4.7	2.9	2.4	5.1
	Min/Max	2.1/3.4	2.5/4.6	3.9/5.9	4.6/6.8	5.6/7.2	6.1/7.7	6.8/8.3	6.1/7.9	5.2/7.3	3.7/5.8	2.3/3.7	1.7/3.2	4.7/5.5
Latitude	Average	3.0	3.8	4.9	5.7	6.2	6.4	7.0	6.9	6.4	5.1	3.2	2.7	5.1
	Min/Max	2.3/4.0	2.7/5.1	4.0/6.2	4.5/6.7	5.3/6.8	5.6/7.0	6.3/7.7	5.9/7.6	5.2/7.4	3.8/6.3	2.4/4.2	1.9/3.8	4.6/5.5
Latitude +15	Average	3.1	3.9	4.8	5.3	5.5	5.5	6.1	6.2	6.1	5.1	3.3	2.9	4.8
	Min/Max	2.4/4.2	2.7/5.2	3.8/6.1	4.1/6.2	4.6/5.9	4.8/6.0	5.5/6.6	5.4/6.9	4.9/7.2	3.8/6.5	2.5/4.4	2.0/4.1	4.3/5.2
90	Average	2.9	3.4	3.8	3.7	3.3	3.1	3.4	3.9	4.5	4.3	3.0	2.7	3.5
	Min/Max	2.2/4.0	2.3/4.6	3.0/4.9	2.8/4.2	2.9/3.5	2.8/3.3	3.1/3.6	3.5/4.3	3.6/5.3	3.2/5.5	2.2/4.0	1.9/3.9	3.1/3.8

Solar Radiation for 1-Axis Tracking Flat-Plate Collectors with a North-South Axis (kWh/m²/day), Uncertainty ±9%

Axis Tilt (°)		Jan	Feb	Mar	Apr	May	June	July	Aug	Sept	Oct	Nov	Dec	Year
0	Average	2.4	3.4	5.3	7.2	8.8	9.9	10.8	9.5	7.6	5.0	2.6	2.0	6.2
	Min/Max	1.9/3.1	2.4/4.6	4.1/6.5	5.5/8.5	7.2/10.0	8.2/11.1	9.0/12.5	7.7/10.9	6.1/9.0	3.7/6.2	2.0/3.4	1.5/2.7	5.5/6.7
Latitude -15	Average	3.1	4.2	6.0	7.7	9.1	9.9	11.0	10.1	8.5	6.1	3.4	2.7	6.8
	Min/Max	2.4/4.2	2.8/5.8	4.6/7.6	5.8/9.2	7.3/10.3	8.2/11.2	9.2/12.7	8.1/11.6	6.8/10.2	4.4/7.7	2.5/4.5	1.9/3.9	6.0/7.4
Latitude	Average	3.4	4.4	6.2	7.7	8.8	9.5	10.6	9.9	8.6	6.3	3.6	3.0	6.9
	Min/Max	2.5/4.7	3.0/6.2	4.7/7.9	5.7/9.2	7.1/10.0	7.9/10.8	8.9/12.3	8.0/11.4	6.8/10.4	4.5/8.1	2.7/4.9	2.0/4.3	6.0/7.5
Latitude +15	Average	3.5	4.5	6.1	7.4	8.3	8.9	10.0	9.5	8.4	6.4	3.7	3.2	6.7
	Min/Max	2.6/4.9	3.0/6.3	4.6/7.8	5.4/8.8	6.7/9.4	7.4/10.1	8.3/11.6	7.6/10.9	6.6/10.2	4.5/8.2	2.8/5.1	2.1/4.6	5.8/7.3

Solar Radiation for 2-Axis Tracking Flat-Plate Collectors (kWh/m²/day), Uncertainty ±9%

Tracker		Jan	Feb	Mar	Apr	May	June	July	Aug	Sept	Oct	Nov	Dec	Year
2-Axis	Average	3.6	4.5	6.2	7.8	9.2	10.2	11.2	10.1	8.6	6.4	3.8	3.2	7.1
	Min/Max	2.6/4.9	3.0/6.3	4.7/7.9	5.8/9.3	7.4/10.4	8.4/11.5	9.3/13.0	8.2/11.7	6.8/10.4	4.6/8.2	2.8/5.1	2.1/4.7	6.2/7.7

Direct Beam Solar Radiation for Concentrating Collectors (kWh/m²/day), Uncertainty ±8%

Tracker		Jan	Feb	Mar	Apr	May	June	July	Aug	Sept	Oct	Nov	Dec	Year
1-Axis, E-W Horiz Axis	Average	1.9	2.3	2.9	3.5	4.4	5.2	6.2	5.4	4.7	3.6	2.0	1.8	3.7
	Min/Max	1.3/3.3	1.1/3.8	1.9/4.5	2.1/4.8	2.7/5.4	3.7/6.4	4.6/7.7	3.8/6.8	3.1/6.1	2.1/5.2	1.2/3.3	0.8/3.2	3.0/4.2
1-Axis, N-S Horiz Axis	Average	1.3	1.9	3.2	4.5	5.9	6.9	8.3	7.2	5.6	3.5	1.5	1.1	4.3
	Min/Max	0.8/2.2	0.9/3.2	2.0/4.7	2.8/6.1	3.7/7.2	4.9/8.6	5.9/10.5	5.0/9.1	3.8/7.5	1.9/4.9	0.9/2.4	0.5/2.0	3.4/4.8
1-Axis, N-S Tilt=Latitude	Average	2.1	2.8	3.9	4.9	5.8	6.6	8.0	7.5	6.5	4.6	2.3	1.9	4.7
	Min/Max	1.4/3.6	1.3/4.6	2.4/5.9	3.0/6.6	3.7/7.2	4.7/8.2	5.7/10.2	5.2/9.4	4.4/8.6	2.6/6.5	1.4/3.7	0.9/3.3	3.8/5.4
2-Axis	Average	2.3	2.8	3.9	5.0	6.1	7.1	8.6	7.7	6.5	4.7	2.4	2.1	4.9
	Min/Max	1.5/3.8	1.3/4.7	2.4/5.9	3.0/6.7	3.9/7.6	5.1/8.9	6.2/11.0	5.4/9.7	4.4/8.6	2.6/6.6	1.4/3.9	0.9/3.6	4.0/5.6

Average Climatic Conditions

Element	Jan	Feb	Mar	Apr	May	June	July	Aug	Sept	Oct	Nov	Dec	Year
Temperature (°C)	-0.3	2.2	4.2	6.7	10.6	15.2	18.8	18.4	13.9	9.1	3.4	-0.1	8.5
Daily Minimum Temp	-5.9	-3.9	-3.2	-1.7	1.5	5.6	8.0	7.9	3.7	0.1	-2.6	-5.7	0.3
Daily Maximum Temp	5.1	8.4	11.5	15.1	19.6	24.8	29.6	28.8	24.1	17.9	9.4	5.4	16.7
Record Minimum Temp	-31.7	-27.8	-16.7	-11.1	-10.0	-3.3	-0.6	-0.6	-7.2	-15.0	-20.6	-31.7	-31.7
Record Maximum Temp	18.3	21.7	26.1	30.0	36.7	38.0	38.9	40.6	36.7	32.8	23.9	18.9	40.6
HDD, Base 18.3°C	579	451	439	348	241	107	44	53	146	288	447	572	3716
CDD, Base 18.3°C	0	0	0	0	0	14	60	55	13	0	0	0	142
Relative Humidity (%)	72	68	61	57	54	51	45	48	52	59	71	73	59
Wind Speed (m/s)	3.3	3.5	3.6	3.6	3.5	3.4	3.3	3.1	3.0	3.0	3.3	3.1	3.3

Salem, OR

WBAN NO. 24232

LATITUDE: 44.92° N
LONGITUDE: 123.02° W
ELEVATION: 61 meters
MEAN PRESSURE: 1011 millibars

STATION TYPE: Secondary

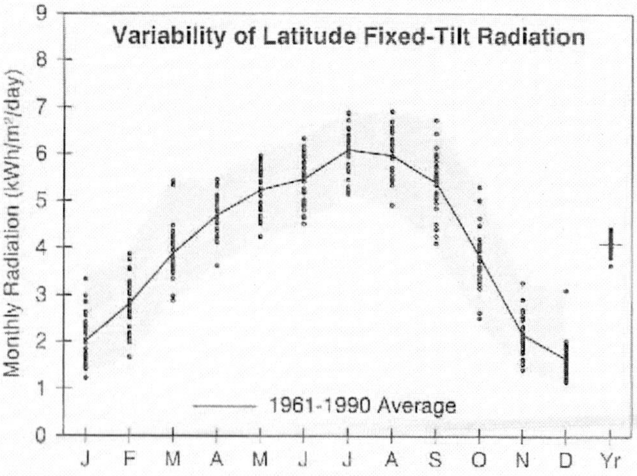

Solar Radiation for Flat-Plate Collectors Facing South at a Fixed Tilt (kWh/m²/day), Uncertainty ±9%

Tilt (°)		Jan	Feb	Mar	Apr	May	June	July	Aug	Sept	Oct	Nov	Dec	Year
0	Average	1.3	2.0	3.1	4.4	5.5	6.1	6.6	5.7	4.4	2.7	1.4	1.1	3.7
	Min/Max	1.0/1.7	1.5/2.5	2.5/3.9	3.5/5.0	4.5/6.2	5.1/7.0	5.5/7.4	4.9/6.5	3.5/5.2	2.0/3.3	1.1/1.8	0.9/1.4	3.4/3.9
Latitude -15	Average	1.8	2.7	3.8	4.8	5.6	6.0	6.6	6.2	5.3	3.6	2.0	1.5	4.2
	Min/Max	1.2/2.9	1.7/3.6	2.9/5.2	3.8/5.6	4.5/6.4	4.9/6.9	5.6/7.5	5.2/7.2	4.1/6.6	2.5/4.9	1.4/2.9	1.1/2.7	3.7/4.4
Latitude	Average	2.0	2.8	3.9	4.7	5.2	5.5	6.1	6.0	5.4	3.8	2.2	1.6	4.1
	Min/Max	1.2/3.3	1.7/3.9	2.9/5.4	3.6/5.5	4.2/6.0	4.5/6.3	5.2/6.9	4.9/6.9	4.1/6.7	2.5/5.3	1.4/3.3	1.2/3.1	3.6/4.4
Latitude +15	Average	2.1	2.8	3.7	4.3	4.6	4.7	5.3	5.4	5.1	3.8	2.2	1.7	3.8
	Min/Max	1.2/3.5	1.6/3.9	2.7/5.3	3.3/5.1	3.8/5.3	3.9/5.4	4.5/6.0	4.4/6.3	3.9/6.5	2.4/5.4	1.4/3.4	1.1/3.3	3.4/4.2
90	Average	1.9	2.4	2.9	3.0	2.9	2.8	3.1	3.5	3.9	3.2	1.9	1.5	2.8
	Min/Max	1.0/3.3	1.3/3.4	2.1/4.3	2.3/3.5	2.4/3.2	2.4/3.1	2.8/3.4	3.0/4.1	2.9/4.9	2.0/4.6	1.2/3.1	1.0/3.2	2.4/3.1

Solar Radiation for 1-Axis Tracking Flat-Plate Collectors with a North-South Axis (kWh/m²/day), Uncertainty ±9%

Axis Tilt (°)		Jan	Feb	Mar	Apr	May	June	July	Aug	Sept	Oct	Nov	Dec	Year
0	Average	1.6	2.5	4.0	5.6	7.0	7.9	8.9	7.9	6.1	3.7	1.8	1.3	4.9
	Min/Max	1.0/2.5	1.6/3.4	2.9/5.7	4.1/6.6	5.2/8.2	6.2/9.5	6.8/10.6	6.1/9.2	4.3/7.8	2.5/5.1	1.2/2.6	0.9/2.2	4.2/5.2
Latitude -15	Average	2.0	3.0	4.6	6.0	7.2	7.9	9.1	8.3	6.9	4.4	2.2	1.6	5.3
	Min/Max	1.2/3.5	1.7/4.3	3.2/6.6	4.3/7.1	5.3/8.4	6.1/9.6	6.9/10.8	6.4/9.8	4.8/8.9	2.8/6.3	1.4/3.4	1.1/3.2	4.6/5.7
Latitude	Average	2.2	3.2	4.6	5.9	7.0	7.6	8.7	8.2	7.0	4.5	2.4	1.8	5.3
	Min/Max	1.2/3.8	1.7/4.5	3.2/6.7	4.2/7.1	5.1/8.2	5.9/9.2	6.6/10.4	6.3/9.7	4.8/9.1	2.8/6.6	1.5/3.7	1.2/3.5	4.5/5.7
Latitude +15	Average	2.3	3.2	4.5	5.6	6.6	7.1	8.2	7.8	6.8	4.5	2.4	1.8	5.1
	Min/Max	1.2/4.0	1.7/4.6	3.1/6.7	4.0/6.8	4.8/7.7	5.5/8.6	6.2/9.8	5.9/9.2	4.7/8.9	2.8/6.6	1.5/3.9	1.2/3.7	4.3/5.5

Solar Radiation for 2-Axis Tracking Flat-Plate Collectors (kWh/m²/day), Uncertainty ±9%

Tracker		Jan	Feb	Mar	Apr	May	June	July	Aug	Sept	Oct	Nov	Dec	Year
2-Axis	Average	2.3	3.2	4.7	6.0	7.3	8.1	9.2	8.4	7.0	4.6	2.4	1.8	5.4
	Min/Max	1.2/4.0	1.7/4.6	3.2/6.8	4.3/7.2	5.4/8.6	6.3/9.8	7.0/11.0	6.4/9.9	4.9/9.1	2.9/6.6	1.5/3.9	1.2/3.8	4.6/5.9

Direct Beam Solar Radiation for Concentrating Collectors (kWh/m²/day), Uncertainty ±8%

Tracker		Jan	Feb	Mar	Apr	May	June	July	Aug	Sept	Oct	Nov	Dec	Year
1-Axis, E-W Horiz Axis	Average	1.1	1.5	2.0	2.4	3.1	3.6	4.6	4.1	3.5	2.3	1.1	0.8	2.5
	Min/Max	0.4/2.5	0.4/2.6	1.1/3.6	1.3/3.3	1.7/4.1	2.2/5.0	2.7/5.9	2.6/5.3	1.9/5.1	1.0/4.0	0.6/2.3	0.2/2.5	2.0/2.9
1-Axis, N-S Horiz Axis	Average	0.7	1.2	2.1	3.0	4.1	4.8	6.1	5.4	4.1	2.2	0.8	0.5	2.9
	Min/Max	0.2/1.7	0.3/2.1	1.1/3.8	1.6/4.1	2.2/5.3	2.9/6.6	3.5/7.9	3.3/7.0	2.0/6.0	0.9/3.7	0.4/1.6	0.1/1.4	2.3/3.3
1-Axis, N-S Tilt=Latitude	Average	1.2	1.7	2.6	3.3	4.0	4.6	5.9	5.6	4.8	2.9	1.3	0.9	3.2
	Min/Max	0.4/2.7	0.5/3.0	1.4/4.7	1.7/4.4	2.2/5.3	2.8/6.3	3.4/7.7	3.5/7.2	2.4/7.0	1.3/4.9	0.6/2.5	0.3/2.5	2.5/3.7
2-Axis	Average	1.3	1.8	2.6	3.4	4.3	5.0	6.4	5.8	4.8	2.9	1.3	1.0	3.4
	Min/Max	0.4/2.9	0.5/3.1	1.4/4.7	1.7/4.5	2.3/5.6	3.0/6.8	3.7/8.2	3.6/7.5	2.5/7.0	1.3/5.0	0.6/2.7	0.3/2.7	2.6/3.8

Average Climatic Conditions

Element	Jan	Feb	Mar	Apr	May	June	July	Aug	Sept	Oct	Nov	Dec	Year
Temperature (°C)	4.2	6.0	7.6	9.5	12.6	16.3	19.1	19.3	16.4	11.5	7.2	4.4	11.2
Daily Minimum Temp	0.4	1.2	2.0	3.2	5.7	9.1	10.5	10.8	8.4	5.1	3.1	0.9	5.0
Daily Maximum Temp	8.0	10.8	13.2	15.8	19.4	23.6	27.6	27.8	24.4	17.9	11.3	8.0	17.3
Record Minimum Temp	-23.3	-20.0	-11.1	-5.0	-3.9	0.0	2.8	2.2	-3.3	-5.0	-12.8	-24.4	-24.4
Record Maximum Temp	18.3	22.2	26.7	31.1	37.8	38.9	42.2	42.2	40.0	33.9	22.2	18.9	42.2
HDD, Base 18.3°C	437	346	332	265	179	78	23	29	73	212	333	431	2737
CDD, Base 18.3°C	0	0	0	0	0	18	45	58	16	0	0	0	137
Relative Humidity (%)	83	82	77	74	71	68	63	64	69	79	85	85	75
Wind Speed (m/s)	3.5	3.5	3.6	3.3	3.0	3.0	3.0	2.8	2.7	2.6	3.6	3.4	3.2

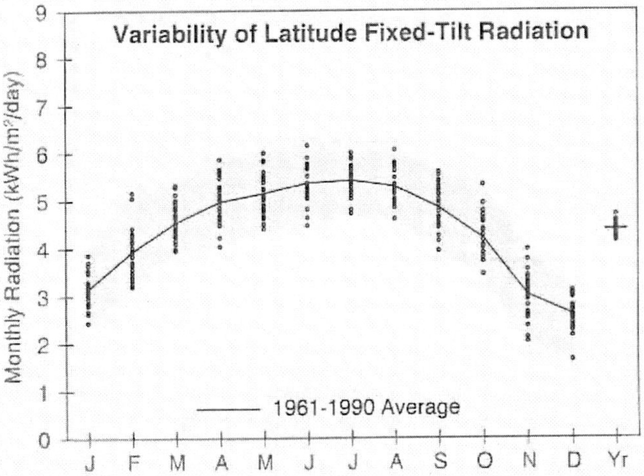

Allentown, PA

WBAN NO. 14737

LATITUDE: 40.65° N
LONGITUDE: 75.43° W
ELEVATION: 117 meters
MEAN PRESSURE: 1003 millibars

STATION TYPE: Secondary

Solar Radiation for Flat-Plate Collectors Facing South at a Fixed Tilt (kWh/m²/day), Uncertainty ±11%

Tilt (°)		Jan	Feb	Mar	Apr	May	June	July	Aug	Sept	Oct	Nov	Dec	Year
0	Average	1.9	2.7	3.7	4.7	5.4	6.0	5.9	5.2	4.2	3.1	2.0	1.6	3.9
	Min/Max	1.6/2.2	2.3/3.4	3.3/4.2	4.0/5.4	4.7/6.3	5.0/6.8	5.2/6.5	4.6/5.9	3.6/4.7	2.6/3.7	1.6/2.3	1.3/1.7	3.7/4.1
Latitude -15	Average	2.8	3.6	4.4	5.1	5.5	5.8	5.8	5.5	4.8	4.0	2.7	2.3	4.4
	Min/Max	2.2/3.4	3.0/4.7	3.9/5.1	4.2/6.0	4.7/6.4	4.8/6.7	5.1/6.5	4.8/6.3	4.0/5.5	3.3/5.0	1.9/3.5	1.6/2.7	4.2/4.7
Latitude	Average	3.1	3.9	4.5	5.0	5.2	5.4	5.4	5.3	4.9	4.2	3.0	2.6	4.4
	Min/Max	2.4/3.9	3.2/5.2	4.0/5.3	4.0/5.9	4.4/6.0	4.5/6.2	4.7/6.0	4.6/6.1	3.9/5.6	3.5/5.3	2.0/4.0	1.6/3.1	4.2/4.7
Latitude +15	Average	3.3	4.0	4.4	4.6	4.6	4.7	4.8	4.8	4.6	4.2	3.1	2.8	4.2
	Min/Max	2.5/4.2	3.3/5.4	3.8/5.2	3.7/5.5	3.9/5.3	3.9/5.3	4.2/5.3	4.2/5.5	3.7/5.4	3.4/5.4	2.0/4.2	1.7/3.3	3.9/4.5
90	Average	3.1	3.6	3.4	3.0	2.6	2.5	2.6	2.9	3.3	3.4	2.8	2.6	3.0
	Min/Max	2.4/4.0	2.8/5.0	2.8/4.0	2.5/3.6	2.3/2.9	2.2/2.7	2.4/2.8	2.6/3.3	2.6/3.8	2.7/4.4	1.7/3.8	1.4/3.2	2.8/3.2

Solar Radiation for 1-Axis Tracking Flat-Plate Collectors with a North-South Axis (kWh/m²/day), Uncertainty ±11%

Axis Tilt (°)		Jan	Feb	Mar	Apr	May	June	July	Aug	Sept	Oct	Nov	Dec	Year
0	Average	2.6	3.7	4.9	6.0	6.8	7.4	7.3	6.7	5.5	4.2	2.6	2.1	5.0
	Min/Max	2.1/3.2	3.0/4.9	4.1/5.8	4.7/7.5	5.6/8.3	5.8/8.7	6.2/8.5	5.8/7.8	4.3/6.4	3.4/5.4	1.8/3.4	1.4/2.5	4.7/5.4
Latitude -15	Average	3.3	4.4	5.4	6.4	6.9	7.4	7.4	6.9	6.0	4.8	3.2	2.7	5.4
	Min/Max	2.5/4.1	3.5/5.9	4.6/6.6	4.9/7.9	5.6/8.5	5.8/8.7	6.2/8.6	5.9/8.2	4.6/7.1	3.9/6.3	2.1/4.3	1.6/3.2	5.0/5.9
Latitude	Average	3.6	4.6	5.5	6.3	6.7	7.1	7.1	6.8	6.0	5.0	3.4	2.9	5.4
	Min/Max	2.7/4.5	3.7/6.3	4.6/6.7	4.8/7.9	5.4/8.3	5.5/8.4	5.9/8.3	5.8/8.0	4.6/7.1	4.0/6.6	2.2/4.7	1.7/3.6	5.0/5.9
Latitude +15	Average	3.8	4.7	5.4	6.0	6.2	6.6	6.6	6.5	5.9	5.1	3.5	3.1	5.3
	Min/Max	2.8/4.8	3.7/6.4	4.5/6.7	4.6/7.6	5.1/7.8	5.1/7.8	5.5/7.7	5.5/7.6	4.5/7.0	4.0/6.7	2.2/4.8	1.8/3.8	4.9/5.8

Solar Radiation for 2-Axis Tracking Flat-Plate Collectors (kWh/m²/day), Uncertainty ±11%

Tracker		Jan	Feb	Mar	Apr	May	June	July	Aug	Sept	Oct	Nov	Dec	Year
2-Axis	Average	3.8	4.7	5.6	6.4	7.0	7.5	7.5	7.0	6.0	5.1	3.6	3.1	5.6
	Min/Max	2.9/4.8	3.7/6.4	4.6/6.8	4.9/8.0	5.7/8.6	5.9/8.9	6.3/8.7	6.0/8.2	4.7/7.2	4.0/6.7	2.2/4.9	1.8/3.8	5.2/6.1

Direct Beam Solar Radiation for Concentrating Collectors (kWh/m²/day), Uncertainty ±11%

Tracker		Jan	Feb	Mar	Apr	May	June	July	Aug	Sept	Oct	Nov	Dec	Year
1-Axis, E-W Horiz Axis	Average	2.0	2.4	2.5	2.7	2.9	3.2	3.2	3.0	2.7	2.6	1.9	1.7	2.6
	Min/Max	1.4/2.8	1.5/3.6	1.9/3.6	1.6/3.8	2.0/4.2	1.9/4.4	2.2/4.1	2.2/3.9	1.7/3.5	1.7/3.9	0.8/3.0	0.6/2.4	2.2/3.0
1-Axis, N-S Horiz Axis	Average	1.5	2.1	2.8	3.5	3.8	4.2	4.2	3.9	3.2	2.5	1.4	1.1	2.9
	Min/Max	1.0/2.0	1.4/3.2	2.1/4.0	2.1/5.0	2.6/5.5	2.5/5.7	2.8/5.4	2.9/5.0	2.0/4.2	1.6/3.8	0.6/2.3	0.4/1.6	2.5/3.3
1-Axis, N-S Tilt=Latitude	Average	2.2	2.8	3.4	3.7	3.7	3.9	4.0	4.0	3.7	3.3	2.1	1.8	3.2
	Min/Max	1.5/3.0	1.8/4.3	2.5/4.7	2.2/5.3	2.5/5.4	2.4/5.3	2.7/5.2	3.0/5.2	2.3/4.8	2.1/4.9	0.9/3.3	0.7/2.5	2.8/3.7
2-Axis	Average	2.4	2.9	3.4	3.8	3.9	4.3	4.3	4.1	3.7	3.3	2.2	2.0	3.4
	Min/Max	1.6/3.2	1.9/4.4	2.5/4.7	2.3/5.4	2.7/5.7	2.6/5.8	2.9/5.6	3.1/5.3	2.3/4.8	2.1/5.0	1.0/3.5	0.7/2.7	2.9/3.8

Average Climatic Conditions

Element	Jan	Feb	Mar	Apr	May	June	July	Aug	Sept	Oct	Nov	Dec	Year
Temperature (°C)	-3.0	-1.5	4.1	9.8	15.7	20.8	23.4	22.3	18.2	11.8	6.2	-0.1	10.7
Daily Minimum Temp	-7.3	-6.2	-1.2	3.8	9.6	14.9	17.6	16.7	12.3	5.8	1.2	-4.2	5.3
Daily Maximum Temp	1.3	3.2	9.3	15.8	21.8	26.7	29.2	27.9	23.9	17.7	11.0	4.0	16.0
Record Minimum Temp	-24.4	-21.7	-18.3	-8.9	-2.2	3.9	8.9	5.0	-1.1	-6.1	-11.7	-22.2	-24.4
Record Maximum Temp	22.2	24.4	30.0	33.9	36.1	37.8	40.6	37.8	37.2	32.2	27.2	22.2	40.6
HDD, Base 18.3°C	661	556	441	255	102	6	0	4	46	207	365	572	3214
CDD, Base 18.3°C	0	0	0	0	21	79	157	128	41	4	0	0	429
Relative Humidity (%)	69	67	62	61	66	68	69	72	74	72	70	71	68
Wind Speed (m/s)	4.6	4.6	4.9	4.8	4.1	3.7	3.2	3.2	3.3	3.6	4.3	4.4	4.1

Bradford, PA

WBAN NO. 04751

LATITUDE: 41.80° N
LONGITUDE: 78.63° W
ELEVATION: 600 meters
MEAN PRESSURE: 940 millibars

STATION TYPE: Secondary

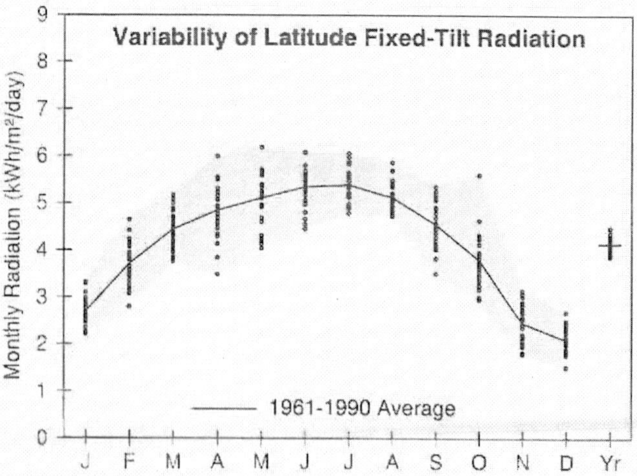

Solar Radiation for Flat-Plate Collectors Facing South at a Fixed Tilt (kWh/m²/day), Uncertainty ±9%

Tilt (°)		Jan	Feb	Mar	Apr	May	June	July	Aug	Sept	Oct	Nov	Dec	Year
0	Average	1.8	2.6	3.6	4.6	5.4	5.9	5.8	5.0	3.9	2.8	1.7	1.4	3.7
	Min/Max	1.5/2.0	2.2/3.0	3.2/4.0	3.5/5.5	4.4/6.4	5.0/6.7	5.2/6.6	4.7/5.7	3.2/4.5	2.4/3.7	1.4/2.0	1.2/1.6	3.5/3.9
Latitude -15	Average	2.4	3.4	4.3	5.0	5.4	5.8	5.8	5.3	4.5	3.6	2.3	1.9	4.2
	Min/Max	2.0/2.9	2.7/4.2	3.7/5.0	3.6/6.1	4.3/6.6	4.8/6.6	5.2/6.5	4.9/6.1	3.5/5.3	2.9/5.2	1.7/2.8	1.5/2.4	3.9/4.5
Latitude	Average	2.7	3.7	4.4	4.8	5.1	5.3	5.4	5.1	4.5	3.8	2.4	2.1	4.1
	Min/Max	2.2/3.3	2.8/4.6	3.8/5.2	3.5/6.0	4.0/6.2	4.4/6.1	4.8/6.1	4.7/5.9	3.5/5.3	3.0/5.6	1.8/3.1	1.5/2.7	3.9/4.5
Latitude +15	Average	2.8	3.8	4.4	4.5	4.5	4.6	4.7	4.6	4.3	3.8	2.5	2.2	3.9
	Min/Max	2.3/3.5	2.8/4.8	3.7/5.1	3.2/5.6	3.6/5.5	3.9/5.3	4.2/5.3	4.3/5.3	3.3/5.1	2.9/5.7	1.8/3.3	1.5/2.8	3.6/4.3
90	Average	2.7	3.5	3.6	3.0	2.7	2.5	2.6	2.9	3.1	3.1	2.2	2.1	2.8
	Min/Max	2.1/3.4	2.5/4.5	2.9/4.2	2.2/3.9	2.2/3.1	2.3/2.8	2.5/2.9	2.6/3.3	2.4/3.7	2.3/4.7	1.5/3.0	1.3/2.7	2.6/3.1

Solar Radiation for 1-Axis Tracking Flat-Plate Collectors with a North-South Axis (kWh/m²/day), Uncertainty ±9%

Axis Tilt (°)		Jan	Feb	Mar	Apr	May	June	July	Aug	Sept	Oct	Nov	Dec	Year
0	Average	2.3	3.5	4.7	5.9	6.8	7.5	7.4	6.5	5.1	3.7	2.1	1.7	4.8
	Min/Max	1.8/2.8	2.7/4.3	4.0/5.5	4.0/7.4	5.2/8.4	5.9/8.9	6.3/8.6	5.8/7.7	3.9/6.2	2.8/5.5	1.6/2.7	1.3/2.2	4.4/5.2
Latitude -15	Average	2.8	4.1	5.2	6.2	6.9	7.5	7.5	6.8	5.6	4.3	2.6	2.1	5.1
	Min/Max	2.2/3.5	3.0/5.2	4.4/6.3	4.1/7.9	5.3/8.6	5.8/8.9	6.3/8.6	6.1/8.0	4.2/6.8	3.2/6.6	1.8/3.3	1.5/2.8	4.8/5.6
Latitude	Average	3.0	4.3	5.4	6.1	6.7	7.2	7.2	6.6	5.6	4.5	2.7	2.3	5.1
	Min/Max	2.4/3.8	3.1/5.5	4.4/6.4	4.0/7.9	5.1/8.3	5.6/8.6	6.1/8.3	5.9/7.9	4.2/6.9	3.3/6.9	1.9/3.6	1.5/3.0	4.8/5.7
Latitude +15	Average	3.1	4.4	5.3	5.9	6.3	6.7	6.8	6.3	5.5	4.4	2.8	2.4	5.0
	Min/Max	2.5/4.0	3.2/5.7	4.3/6.4	3.8/7.6	4.7/7.8	5.2/8.0	5.7/7.8	5.6/7.5	4.0/6.7	3.2/7.0	1.9/3.7	1.6/3.1	4.6/5.5

Solar Radiation for 2-Axis Tracking Flat-Plate Collectors (kWh/m²/day), Uncertainty ±9%

Tracker		Jan	Feb	Mar	Apr	May	June	July	Aug	Sept	Oct	Nov	Dec	Year
2-Axis	Average	3.2	4.4	5.4	6.2	7.0	7.7	7.6	6.8	5.7	4.5	2.8	2.4	5.3
	Min/Max	2.5/4.0	3.2/5.7	4.4/6.5	4.1/8.0	5.3/8.7	6.0/9.1	6.4/8.8	6.1/8.1	4.2/6.9	3.3/7.0	1.9/3.7	1.6/3.2	4.9/5.8

Direct Beam Solar Radiation for Concentrating Collectors (kWh/m²/day), Uncertainty ±8%

Tracker		Jan	Feb	Mar	Apr	May	June	July	Aug	Sept	Oct	Nov	Dec	Year
1-Axis, E-W	Average	1.3	1.8	2.2	2.6	2.9	3.3	3.2	2.8	2.4	2.2	1.2	0.9	2.2
Horiz Axis	Min/Max	0.8/2.0	1.0/2.9	1.5/3.3	1.1/3.9	1.6/4.2	1.9/4.4	2.3/4.2	2.3/3.8	1.4/3.4	1.2/4.2	0.4/2.0	0.3/1.5	2.0/2.6
1-Axis, N-S	Average	0.9	1.5	2.4	3.3	3.8	4.3	4.3	3.7	2.9	2.1	0.9	0.6	2.6
Horiz Axis	Min/Max	0.5/1.3	0.9/2.5	1.7/3.6	1.4/4.9	2.1/5.5	2.5/5.8	3.0/5.5	2.9/4.9	1.7/4.0	1.1/4.0	0.4/1.5	0.2/1.0	2.3/3.0
1-Axis, N-S	Average	1.4	2.1	2.9	3.5	3.7	4.1	4.1	3.8	3.3	2.7	1.4	1.0	2.8
Tilt=Latitude	Min/Max	0.9/2.1	1.2/3.4	2.0/4.4	1.5/5.3	2.1/5.4	2.4/5.4	2.9/5.3	3.0/5.1	1.9/4.6	1.4/5.2	0.5/2.2	0.3/1.6	2.6/3.3
2-Axis	Average	1.5	2.2	2.9	3.6	3.9	4.4	4.4	3.9	3.3	2.7	1.5	1.1	3.0
	Min/Max	0.9/2.2	1.2/3.5	2.0/4.4	1.5/5.3	2.2/5.7	2.6/5.9	3.1/5.7	3.1/5.3	1.9/4.6	1.4/5.2	0.5/2.3	0.3/1.7	2.7/3.5

Average Climatic Conditions

Element	Jan	Feb	Mar	Apr	May	June	July	Aug	Sept	Oct	Nov	Dec	Year
Temperature (°C)	-7.1	-6.1	-0.4	5.7	11.5	16.1	18.4	17.6	13.8	8.1	2.3	-4.0	6.3
Daily Minimum Temp	-11.2	-10.9	-5.5	0.0	5.0	9.7	12.2	11.4	7.9	2.7	-1.8	-7.8	0.9
Daily Maximum Temp	-3.0	-1.3	4.6	11.4	17.9	22.4	24.7	23.7	19.7	13.5	6.3	-0.3	11.7
Record Minimum Temp	-32.8	-32.2	-27.8	-14.4	-7.8	-3.3	0.6	-3.3	-3.9	-12.2	-18.9	-28.9	-32.8
Record Maximum Temp	15.6	16.7	25.6	29.4	30.0	31.1	34.4	33.3	29.4	25.6	22.8	20.6	34.4
HDD, Base 18.3°C	787	683	582	378	217	81	38	54	142	317	482	692	4454
CDD, Base 18.3°C	0	0	0	0	5	13	42	31	6	0	0	0	96
Relative Humidity (%)	78	75	73	68	70	75	77	79	81	77	80	81	76
Wind Speed (m/s)	4.4	4.1	4.2	4.1	3.5	3.1	2.6	2.6	2.9	3.4	4.1	4.3	3.6

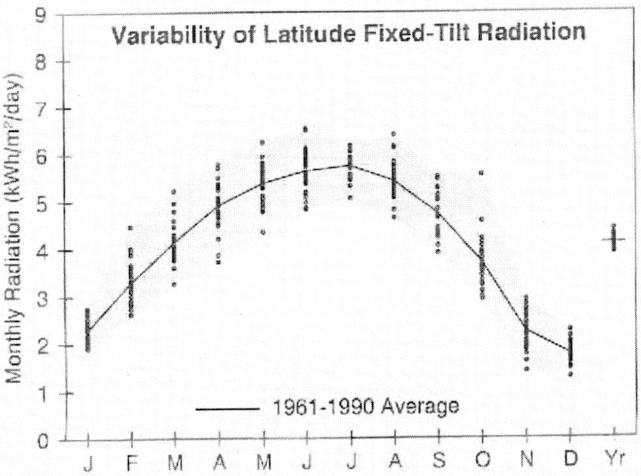

Erie, PA

WBAN NO. 14860

LATITUDE: 42.08° N
LONGITUDE: 80.18° W
ELEVATION: 225 meters
MEAN PRESSURE: 990 millibars

STATION TYPE: Secondary

Solar Radiation for Flat-Plate Collectors Facing South at a Fixed Tilt (kWh/m²/day), Uncertainty ±9%

Tilt (°)		Jan	Feb	Mar	Apr	May	June	July	Aug	Sept	Oct	Nov	Dec	Year
0	Average	1.6	2.4	3.4	4.6	5.7	6.3	6.2	5.3	4.1	2.7	1.6	1.3	3.8
	Min/Max	1.4/1.8	2.1/3.0	2.9/4.0	3.7/5.3	4.7/6.5	5.4/7.2	5.5/6.7	4.7/6.1	3.5/4.5	2.4/3.7	1.2/1.9	1.1/1.4	3.6/4.0
Latitude -15	Average	2.1	3.1	4.0	5.0	5.7	6.1	6.2	5.6	4.7	3.5	2.1	1.6	4.2
	Min/Max	1.8/2.5	2.5/4.1	3.3/5.0	3.9/5.9	4.7/6.7	5.3/7.1	5.4/6.7	4.9/6.6	3.9/5.4	2.9/5.1	1.4/2.7	1.3/2.0	4.0/4.4
Latitude	Average	2.3	3.3	4.1	4.9	5.4	5.6	5.8	5.4	4.7	3.7	2.2	1.8	4.1
	Min/Max	1.9/2.8	2.6/4.5	3.3/5.2	3.7/5.8	4.3/6.2	4.8/6.5	5.1/6.2	4.6/6.4	3.9/5.5	2.9/5.6	1.4/2.9	1.3/2.3	3.9/4.4
Latitude +15	Average	2.4	3.4	4.0	4.5	4.8	4.9	5.0	4.9	4.5	3.7	2.3	1.8	3.9
	Min/Max	2.0/2.9	2.6/4.7	3.1/5.1	3.4/5.4	3.9/5.5	4.2/5.6	4.4/5.4	4.2/5.8	3.7/5.3	2.8/5.7	1.4/3.0	1.3/2.4	3.6/4.1
90	Average	2.2	3.0	3.2	3.0	2.8	2.6	2.8	3.1	3.2	3.0	2.0	1.6	2.7
	Min/Max	1.8/2.8	2.3/4.3	2.3/4.0	2.4/3.6	2.3/3.1	2.4/2.9	2.6/3.0	2.6/3.6	2.7/3.8	2.2/4.7	1.1/2.7	1.1/2.2	2.5/2.9

Solar Radiation for 1-Axis Tracking Flat-Plate Collectors with a North-South Axis (kWh/m²/day), Uncertainty ±9%

Axis Tilt (°)		Jan	Feb	Mar	Apr	May	June	July	Aug	Sept	Oct	Nov	Dec	Year
0	Average	1.9	3.1	4.4	6.0	7.2	8.0	8.0	6.9	5.4	3.6	1.9	1.5	4.8
	Min/Max	1.6/2.3	2.4/4.1	3.5/5.6	4.2/7.1	5.5/8.7	6.6/9.6	6.8/8.8	5.9/8.4	4.2/6.4	2.8/5.5	1.3/2.5	1.1/1.8	4.6/5.2
Latitude -15	Average	2.3	3.6	4.9	6.3	7.4	8.0	8.1	7.2	5.9	4.2	2.3	1.8	5.2
	Min/Max	1.9/2.9	2.8/5.0	3.8/6.3	4.4/7.6	5.5/8.9	6.6/9.6	6.9/8.9	6.1/8.9	4.6/7.1	3.2/6.6	1.4/3.1	1.3/2.3	4.9/5.6
Latitude	Average	2.5	3.8	4.9	6.2	7.1	7.7	7.8	7.1	5.9	4.3	2.5	1.9	5.2
	Min/Max	2.0/3.1	2.9/5.3	3.8/6.5	4.3/7.6	5.3/8.6	6.3/9.2	6.6/8.6	6.0/8.7	4.6/7.2	3.3/6.9	1.4/3.3	1.3/2.5	4.9/5.6
Latitude +15	Average	2.6	3.8	4.9	6.0	6.7	7.2	7.3	6.7	5.8	4.3	2.5	1.9	5.0
	Min/Max	2.0/3.2	2.9/5.4	3.7/6.4	4.1/7.3	5.0/8.1	5.9/8.6	6.2/8.1	5.6/8.3	4.4/7.0	3.2/7.0	1.4/3.4	1.3/2.6	4.7/5.5

Solar Radiation for 2-Axis Tracking Flat-Plate Collectors (kWh/m²/day), Uncertainty ±9%

Tracker		Jan	Feb	Mar	Apr	May	June	July	Aug	Sept	Oct	Nov	Dec	Year
2-Axis	Average	2.6	3.8	5.0	6.3	7.5	8.2	8.3	7.3	6.0	4.4	2.5	1.9	5.3
	Min/Max	2.1/3.3	2.9/5.4	3.8/6.5	4.4/7.7	5.6/9.0	6.8/9.8	7.0/9.1	6.1/8.9	4.6/7.2	3.3/7.0	1.4/3.4	1.3/2.6	5.0/5.8

Direct Beam Solar Radiation for Concentrating Collectors (kWh/m²/day), Uncertainty ±8%

Tracker		Jan	Feb	Mar	Apr	May	June	July	Aug	Sept	Oct	Nov	Dec	Year
1-Axis, E-W Horiz Axis	Average	1.0	1.6	2.1	2.8	3.3	3.9	3.9	3.3	2.7	2.1	1.1	0.6	2.4
	Min/Max	0.5/1.5	0.9/2.5	1.4/3.5	1.4/3.8	2.1/4.5	2.7/5.1	2.7/4.7	2.5/4.6	1.8/3.6	1.2/4.2	0.3/1.9	0.1/1.3	2.1/2.8
1-Axis, N-S Horiz Axis	Average	0.7	1.3	2.3	3.5	4.4	5.0	5.1	4.2	3.2	2.0	0.8	0.4	2.7
	Min/Max	0.3/1.0	0.8/2.1	1.5/3.8	1.7/4.8	2.7/5.9	3.5/6.6	3.6/6.0	3.1/5.9	2.0/4.3	1.1/4.0	0.3/1.4	0.1/0.8	2.5/3.2
1-Axis, N-S Tilt=Latitude	Average	1.1	1.9	2.7	3.7	4.3	4.7	4.9	4.4	3.6	2.6	1.2	0.7	3.0
	Min/Max	0.5/1.6	1.0/3.0	1.8/4.6	1.8/5.2	2.6/5.8	3.3/6.3	3.5/5.8	3.2/6.1	2.3/4.9	1.5/5.2	0.4/2.1	0.1/1.3	2.7/3.5
2-Axis	Average	1.1	1.9	2.8	3.8	4.5	5.2	5.2	4.5	3.6	2.7	1.3	0.7	3.1
	Min/Max	0.5/1.7	1.1/3.0	1.8/4.6	1.8/5.2	2.8/6.2	3.6/6.8	3.7/6.3	3.3/6.3	2.3/4.9	1.5/5.3	0.4/2.2	0.1/1.4	2.8/3.7

Average Climatic Conditions

Element	Jan	Feb	Mar	Apr	May	June	July	Aug	Sept	Oct	Nov	Dec	Year
Temperature (°C)	-3.7	-3.4	2.2	7.8	13.8	19.2	21.8	21.2	17.8	11.8	6.2	-0.3	9.6
Daily Minimum Temp	-7.7	-7.8	-2.2	3.2	8.8	14.2	17.0	16.6	13.3	7.6	2.8	-3.8	5.2
Daily Maximum Temp	0.3	0.9	6.4	12.5	18.8	24.1	26.6	25.8	22.3	16.1	9.7	3.2	13.9
Record Minimum Temp	-26.7	-27.2	-22.8	-11.1	-3.3	0.0	6.7	2.8	0.6	-4.4	-13.9	-21.1	-27.2
Record Maximum Temp	18.3	19.4	27.2	31.7	32.2	37.8	37.2	33.9	34.4	31.1	26.7	23.9	37.8
HDD, Base 18.3°C	682	610	501	315	149	29	0	7	49	206	363	577	3488
CDD, Base 18.3°C	0	0	0	0	9	54	111	95	33	4	0	0	306
Relative Humidity (%)	74	75	72	68	69	71	72	74	74	71	72	75	72
Wind Speed (m/s)	6.1	5.5	5.6	5.3	4.8	4.5	4.2	4.2	4.6	5.2	6.1	6.2	5.2

Harrisburg, PA

WBAN NO. 14751

LATITUDE: 40.22° N
LONGITUDE: 76.85° W
ELEVATION: 106 meters
MEAN PRESSURE: 1005 millibars

STATION TYPE: Secondary

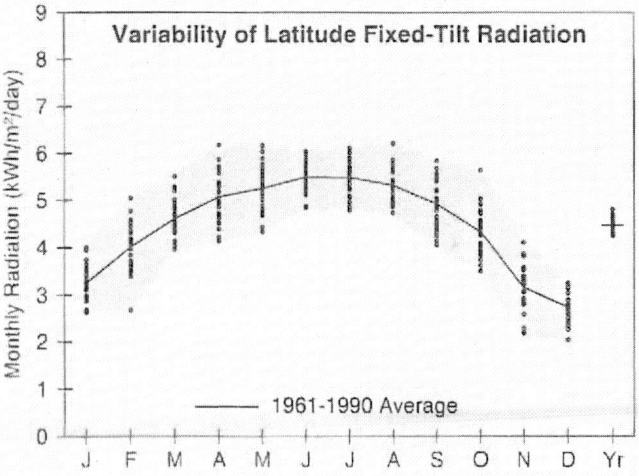

Solar Radiation for Flat-Plate Collectors Facing South at a Fixed Tilt (kWh/m²/day), Uncertainty ±9%

Tilt (°)		Jan	Feb	Mar	Apr	May	June	July	Aug	Sept	Oct	Nov	Dec	Year
0	Average	2.0	2.8	3.8	4.8	5.5	6.1	5.9	5.3	4.3	3.2	2.0	1.6	3.9
	Min/Max	1.7/2.3	2.2/3.3	3.3/4.3	4.1/5.7	4.7/6.4	5.4/6.7	5.2/6.6	4.7/6.0	3.7/4.8	2.8/3.8	1.6/2.4	1.4/1.8	3.7/4.2
Latitude -15	Average	2.9	3.7	4.5	5.2	5.6	6.0	5.9	5.5	4.9	4.1	2.9	2.4	4.5
	Min/Max	2.4/3.5	2.6/4.6	3.9/5.3	4.3/6.3	4.6/6.6	5.2/6.6	5.2/6.6	4.9/6.4	4.1/5.7	3.4/5.2	2.1/3.6	1.9/2.8	4.2/4.8
Latitude	Average	3.2	4.0	4.6	5.1	5.3	5.5	5.5	5.3	4.9	4.3	3.2	2.7	4.5
	Min/Max	2.6/4.0	2.7/5.1	4.0/5.5	4.1/6.2	4.3/6.2	4.8/6.0	4.8/6.1	4.7/6.2	4.1/5.8	3.5/5.6	2.2/4.1	2.0/3.3	4.2/4.8
Latitude +15	Average	3.5	4.1	4.5	4.7	4.7	4.8	4.8	4.8	4.7	4.4	3.3	2.9	4.3
	Min/Max	2.8/4.3	2.7/5.3	3.8/5.4	3.8/5.7	3.8/5.5	4.2/5.2	4.2/5.3	4.3/5.7	3.8/5.6	3.5/5.7	2.2/4.3	2.1/3.5	4.0/4.6
90	Average	3.2	3.6	3.4	3.0	2.6	2.5	2.6	2.9	3.3	3.5	2.9	2.7	3.0
	Min/Max	2.5/4.2	2.3/4.9	2.8/4.2	2.5/3.6	2.3/3.0	2.3/2.7	2.3/2.8	2.6/3.3	2.7/4.0	2.7/4.7	1.9/3.9	1.9/3.3	2.8/3.3

Solar Radiation for 1-Axis Tracking Flat-Plate Collectors with a North-South Axis (kWh/m²/day), Uncertainty ±9%

Axis Tilt (°)		Jan	Feb	Mar	Apr	May	June	July	Aug	Sept	Oct	Nov	Dec	Year
0	Average	2.8	3.8	5.0	6.2	7.0	7.6	7.5	6.8	5.6	4.3	2.8	2.2	5.1
	Min/Max	2.2/3.4	2.7/4.9	4.2/6.2	4.8/7.9	5.5/8.5	6.3/8.7	6.3/8.5	6.0/8.1	4.6/6.7	3.6/5.7	1.9/3.6	1.7/2.7	4.8/5.6
Latitude -15	Average	3.4	4.5	5.6	6.6	7.1	7.6	7.5	7.0	6.1	5.0	3.4	2.8	5.6
	Min/Max	2.7/4.3	3.0/5.8	4.7/6.9	5.0/8.4	5.6/8.6	6.3/8.6	6.3/8.6	6.2/8.5	4.9/7.4	4.0/6.7	2.3/4.5	2.0/3.4	5.2/6.1
Latitude	Average	3.7	4.8	5.7	6.5	6.8	7.3	7.2	6.9	6.2	5.2	3.6	3.1	5.6
	Min/Max	2.9/4.7	3.1/6.2	4.7/7.1	4.9/8.3	5.4/8.4	6.0/8.3	6.0/8.3	6.0/8.4	4.9/7.5	4.1/7.1	2.4/4.9	2.2/3.8	5.2/6.1
Latitude +15	Average	3.9	4.9	5.6	6.2	6.4	6.8	6.8	6.5	6.0	5.2	3.7	3.2	5.4
	Min/Max	3.0/5.0	3.1/6.3	4.6/7.0	4.7/8.0	5.0/7.9	5.6/7.8	5.6/7.8	5.7/7.9	4.7/7.3	4.1/7.1	2.4/5.1	2.3/4.0	5.1/6.0

Solar Radiation for 2-Axis Tracking Flat-Plate Collectors (kWh/m²/day), Uncertainty ±9%

Tracker		Jan	Feb	Mar	Apr	May	June	July	Aug	Sept	Oct	Nov	Dec	Year
2-Axis	Average	3.9	4.9	5.7	6.6	7.1	7.8	7.6	7.0	6.2	5.3	3.8	3.3	5.8
	Min/Max	3.1/5.0	3.1/6.3	4.7/7.1	5.0/8.4	5.7/8.7	6.4/8.9	6.4/8.7	6.2/8.5	4.9/7.5	4.1/7.1	2.4/5.1	2.3/4.0	5.4/6.3

Direct Beam Solar Radiation for Concentrating Collectors (kWh/m²/day), Uncertainty ±8%

Tracker		Jan	Feb	Mar	Apr	May	June	July	Aug	Sept	Oct	Nov	Dec	Year
1-Axis, E-W Horiz Axis	Average	2.1	2.4	2.6	2.9	3.0	3.4	3.3	3.0	2.8	2.7	2.0	1.8	2.7
	Min/Max	1.5/3.0	0.9/3.4	2.0/3.7	1.6/4.1	1.7/4.2	2.3/4.2	2.4/4.1	2.4/4.3	1.9/3.8	1.7/4.2	0.9/3.1	1.0/2.5	2.4/3.1
1-Axis, N-S Horiz Axis	Average	1.5	2.2	3.0	3.7	4.0	4.4	4.3	3.9	3.3	2.7	1.6	1.2	3.0
	Min/Max	1.1/2.1	0.9/3.1	2.2/4.2	2.2/5.4	2.3/5.6	3.0/5.5	3.1/5.4	3.0/5.5	2.2/4.5	1.8/4.2	0.7/2.5	0.6/1.7	2.7/3.5
1-Axis, N-S Tilt=Latitude	Average	2.3	2.9	3.5	3.9	3.9	4.2	4.1	4.0	3.8	3.4	2.3	1.9	3.4
	Min/Max	1.6/3.3	1.1/4.1	2.7/5.0	2.3/5.7	2.3/5.4	2.8/5.2	2.9/5.2	3.1/5.7	2.5/5.1	2.2/5.3	1.0/3.5	1.0/2.7	3.0/3.9
2-Axis	Average	2.5	3.0	3.5	4.0	4.1	4.5	4.4	4.2	3.8	3.5	2.4	2.1	3.5
	Min/Max	1.7/3.5	1.2/4.2	2.7/5.0	2.3/5.8	2.4/5.8	3.1/5.6	3.2/5.5	3.2/5.8	2.5/5.2	2.2/5.4	1.1/3.7	1.1/2.9	3.1/4.1

Average Climatic Conditions

Element	Jan	Feb	Mar	Apr	May	June	July	Aug	Sept	Oct	Nov	Dec	Year
Temperature (°C)	-1.9	-0.4	5.1	10.9	16.6	21.6	24.3	23.4	19.1	12.6	6.9	0.9	11.6
Daily Minimum Temp	-6.0	-4.8	0.0	5.1	10.6	15.9	18.7	17.9	13.6	7.0	2.3	-3.0	6.4
Daily Maximum Temp	2.2	4.0	10.2	16.7	22.5	27.3	29.9	28.8	24.6	18.2	11.4	4.8	16.7
Record Minimum Temp	-22.8	-20.6	-15.0	-7.2	-0.6	4.4	9.4	7.2	-1.1	-5.0	-10.6	-22.2	-22.8
Record Maximum Temp	22.8	23.9	30.0	33.9	36.1	37.8	41.7	38.3	38.9	36.1	28.9	23.9	41.7
HDD, Base 18.3°C	627	524	410	223	82	5	0	0	33	183	343	541	2971
CDD, Base 18.3°C	0	0	0	0	27	103	184	158	56	6	0	0	534
Relative Humidity (%)	64	63	61	59	65	68	69	72	74	70	69	66	67
Wind Speed (m/s)	3.8	3.8	4.1	4.1	3.3	3.0	2.8	2.6	2.7	3.0	3.5	3.6	3.4

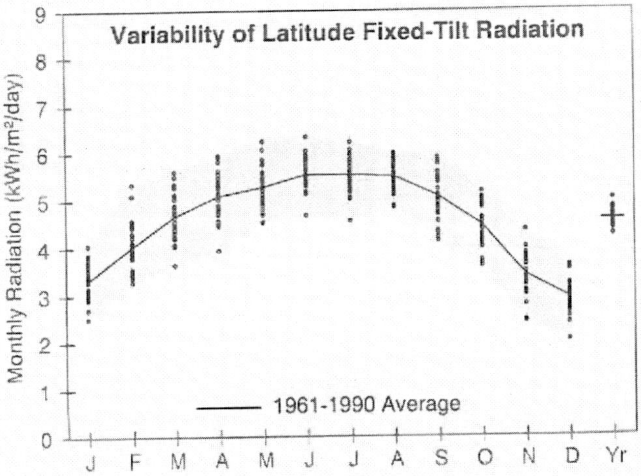

Philadelphia, PA

WBAN NO. 13739

LATITUDE: 39.88° N
LONGITUDE: 75.25° W
ELEVATION: 9 meters
MEAN PRESSURE: 1016 millibars

STATION TYPE: Secondary

Solar Radiation for Flat-Plate Collectors Facing South at a Fixed Tilt (kWh/m²/day), Uncertainty ±11%

Tilt (°)		Jan	Feb	Mar	Apr	May	June	July	Aug	Sept	Oct	Nov	Dec	Year
0	Average	2.0	2.8	3.8	4.8	5.5	6.1	6.0	5.4	4.4	3.2	2.1	1.7	4.0
	Min/Max	1.6/2.3	2.5/3.4	3.2/4.4	3.9/5.5	4.8/6.5	5.2/7.0	5.0/6.7	4.8/5.8	3.7/4.9	2.8/3.6	1.8/2.5	1.4/1.9	3.8/4.3
Latitude -15	Average	2.9	3.7	4.5	5.2	5.6	6.0	5.9	5.7	5.0	4.2	3.0	2.6	4.5
	Min/Max	2.3/3.5	3.1/4.8	3.6/5.4	4.1/6.0	4.8/6.6	5.1/6.9	4.9/6.7	5.0/6.2	4.1/5.8	3.4/4.8	2.3/3.8	1.8/3.1	4.2/4.9
Latitude	Average	3.3	4.0	4.7	5.1	5.3	5.5	5.5	5.5	5.1	4.4	3.4	2.9	4.6
	Min/Max	2.5/4.1	3.3/5.3	3.6/5.6	3.9/5.9	4.5/6.2	4.7/6.3	4.6/6.2	4.8/6.0	4.1/5.9	3.6/5.2	2.4/4.3	2.0/3.6	4.2/5.0
Latitude +15	Average	3.5	4.1	4.6	4.7	4.7	4.8	4.8	5.0	4.8	4.5	3.5	3.1	4.3
	Min/Max	2.6/4.4	3.3/5.6	3.5/5.5	3.6/5.5	4.0/5.5	4.1/5.4	4.0/5.4	4.4/5.4	3.9/5.7	3.6/5.2	2.5/4.6	2.1/3.9	4.0/4.8
90	Average	3.3	3.5	3.4	3.0	2.6	2.5	2.6	3.0	3.4	3.6	3.1	2.9	3.1
	Min/Max	2.3/4.2	2.8/5.1	2.6/4.1	2.4/3.5	2.3/3.0	2.2/2.7	2.3/2.8	2.7/3.2	2.7/4.0	2.8/4.2	2.1/4.1	1.8/3.6	2.8/3.4

Solar Radiation for 1-Axis Tracking Flat-Plate Collectors with a North-South Axis (kWh/m²/day), Uncertainty ±11%

Axis Tilt (°)		Jan	Feb	Mar	Apr	May	June	July	Aug	Sept	Oct	Nov	Dec	Year
0	Average	2.8	3.8	5.1	6.2	7.0	7.7	7.5	6.9	5.8	4.4	3.0	2.4	5.2
	Min/Max	2.1/3.5	3.1/5.2	3.9/6.1	4.7/7.6	5.7/8.5	6.1/9.1	5.8/8.6	5.8/7.8	4.7/6.7	3.5/5.2	2.1/3.8	1.7/2.9	4.9/5.7
Latitude -15	Average	3.5	4.5	5.7	6.6	7.1	7.7	7.5	7.2	6.3	5.2	3.6	3.0	5.7
	Min/Max	2.6/4.4	3.6/6.2	4.3/6.9	4.8/8.0	5.8/8.7	6.1/9.1	5.8/8.7	6.0/8.1	5.0/7.4	4.0/6.1	2.5/4.8	2.1/3.7	5.2/6.3
Latitude	Average	3.8	4.8	5.8	6.5	6.8	7.3	7.3	7.1	6.3	5.4	3.9	3.3	5.7
	Min/Max	2.8/4.8	3.8/6.6	4.3/7.0	4.7/8.0	5.6/8.4	5.8/8.7	5.6/8.4	5.8/7.9	5.0/7.5	4.2/6.3	2.7/5.2	2.2/4.1	5.2/6.3
Latitude +15	Average	4.0	4.9	5.7	6.2	6.4	6.8	6.8	6.7	6.2	5.4	4.1	3.5	5.6
	Min/Max	2.9/5.1	3.8/6.8	4.2/6.9	4.5/7.7	5.2/7.9	5.4/8.1	5.2/7.8	5.5/7.6	4.8/7.3	4.1/6.4	2.7/5.4	2.3/4.4	5.1/6.2

Solar Radiation for 2-Axis Tracking Flat-Plate Collectors (kWh/m²/day), Uncertainty ±11%

Tracker		Jan	Feb	Mar	Apr	May	June	July	Aug	Sept	Oct	Nov	Dec	Year
2-Axis	Average	4.0	4.9	5.8	6.6	7.2	7.8	7.7	7.3	6.4	5.4	4.1	3.5	5.9
	Min/Max	2.9/5.2	3.8/6.8	4.3/7.0	4.8/8.1	5.8/8.8	6.2/9.3	5.9/8.8	6.0/8.1	5.1/7.5	4.2/6.4	2.7/5.4	2.3/4.4	5.4/6.5

Direct Beam Solar Radiation for Concentrating Collectors (kWh/m²/day), Uncertainty ±11%

Tracker		Jan	Feb	Mar	Apr	May	June	July	Aug	Sept	Oct	Nov	Dec	Year
1-Axis, E-W Horiz Axis	Average	2.3	2.6	2.8	2.9	3.1	3.5	3.4	3.3	3.0	2.9	2.3	2.0	2.8
	Min/Max	1.5/3.0	1.6/3.8	1.7/3.8	1.5/4.0	2.1/4.3	2.5/4.6	2.0/4.2	2.3/4.0	2.0/3.8	1.9/3.6	1.2/3.4	1.1/2.8	2.5/3.4
1-Axis, N-S Horiz Axis	Average	1.7	2.3	3.1	3.7	4.0	4.6	4.4	4.2	3.6	2.8	1.8	1.4	3.1
	Min/Max	1.0/2.3	1.4/3.5	2.0/4.3	2.0/5.2	2.7/5.7	3.2/5.9	2.7/5.4	2.9/5.1	2.4/4.5	1.9/3.6	0.9/2.7	0.8/1.9	2.8/3.7
1-Axis, N-S Tilt=Latitude	Average	2.5	3.1	3.7	4.0	3.9	4.3	4.2	4.3	4.0	3.6	2.6	2.2	3.5
	Min/Max	1.6/3.3	1.9/4.6	2.3/5.0	2.1/5.5	2.7/5.5	3.0/5.6	2.5/5.2	3.0/5.2	2.7/5.1	2.4/4.5	1.3/3.8	1.2/3.0	3.1/4.2
2-Axis	Average	2.7	3.2	3.7	4.0	4.2	4.7	4.5	4.4	4.0	3.6	2.7	2.4	3.7
	Min/Max	1.7/3.6	2.0/4.7	2.3/5.0	2.2/5.6	2.8/5.8	3.3/6.1	2.7/5.6	3.1/5.4	2.7/5.1	2.4/4.6	1.4/4.0	1.3/3.3	3.3/4.3

Average Climatic Conditions

Element	Jan	Feb	Mar	Apr	May	June	July	Aug	Sept	Oct	Nov	Dec	Year
Temperature (°C)	-0.9	0.6	5.8	11.3	17.2	22.1	24.8	24.2	20.1	13.6	8.0	2.1	12.4
Daily Minimum Temp	-5.1	-4.0	0.7	5.6	11.5	16.6	19.6	19.1	14.8	8.0	3.1	-2.2	7.3
Daily Maximum Temp	3.3	5.0	10.9	17.0	22.8	27.6	30.1	29.2	25.3	19.1	12.8	6.3	17.4
Record Minimum Temp	-21.7	-20.0	-13.9	-7.2	-2.2	6.7	10.6	6.7	1.7	-3.9	-9.4	-17.2	-21.7
Record Maximum Temp	23.3	23.3	30.6	34.4	36.1	37.8	40.0	38.3	37.8	35.6	27.2	22.2	40.0
HDD, Base 18.3°C	596	498	389	210	68	3	0	0	18	157	310	503	2752
CDD, Base 18.3°C	0	0	0	0	32	116	202	181	71	9	0	0	612
Relative Humidity (%)	66	64	62	60	65	68	70	70	72	71	68	68	67
Wind Speed (m/s)	4.6	4.8	5.0	4.8	4.3	3.9	3.6	3.6	3.8	3.9	4.3	4.5	4.2

Pittsburgh, PA

WBAN NO. 94823

LATITUDE: 40.50° N
LONGITUDE: 80.22° W
ELEVATION: 373 meters
MEAN PRESSURE: 973 millibars

STATION TYPE: Primary

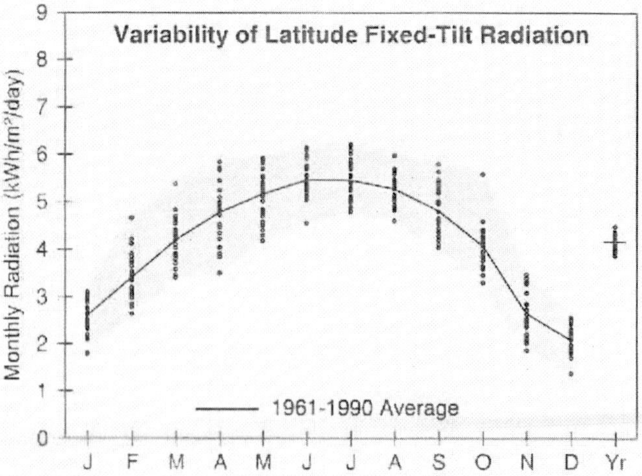

Solar Radiation for Flat-Plate Collectors Facing South at a Fixed Tilt (kWh/m²/day), Uncertainty ±9%

Tilt (°)		Jan	Feb	Mar	Apr	May	June	July	Aug	Sept	Oct	Nov	Dec	Year
0	Average	1.7	2.5	3.5	4.6	5.5	6.1	5.9	5.2	4.2	3.0	1.8	1.4	3.8
	Min/Max	1.4/2.0	2.2/3.1	3.0/4.2	3.5/5.4	4.5/6.2	5.1/6.8	5.2/6.7	4.7/5.8	3.7/4.9	2.5/3.8	1.5/2.1	1.2/1.6	3.6/4.0
Latitude -15	Average	2.4	3.2	4.1	4.9	5.5	5.9	5.9	5.5	4.8	3.8	2.4	1.9	4.2
	Min/Max	1.7/2.8	2.5/4.2	3.4/5.2	3.7/5.9	4.5/6.3	4.9/6.7	5.2/6.7	4.8/6.2	4.1/5.7	3.1/5.2	1.8/3.1	1.3/2.3	3.9/4.5
Latitude	Average	2.6	3.4	4.2	4.8	5.2	5.4	5.5	5.3	4.8	4.1	2.6	2.1	4.2
	Min/Max	1.8/3.1	2.6/4.7	3.4/5.4	3.5/5.8	4.2/5.9	4.6/6.1	4.8/6.2	4.6/6.0	4.0/5.8	3.3/5.6	1.9/3.5	1.4/2.5	3.9/4.5
Latitude +15	Average	2.7	3.5	4.1	4.4	4.6	4.7	4.8	4.8	4.6	4.1	2.7	2.2	3.9
	Min/Max	1.8/3.3	2.6/4.9	3.3/5.3	3.2/5.4	3.7/5.3	4.0/5.3	4.2/5.4	4.2/5.4	3.8/5.6	3.3/5.7	1.9/3.6	1.4/2.7	3.6/4.2
90	Average	2.5	3.0	3.1	2.9	2.6	2.5	2.6	2.9	3.2	3.3	2.3	2.0	2.7
	Min/Max	1.7/3.2	2.2/4.5	2.5/4.0	2.1/3.5	2.2/3.0	2.2/2.8	2.4/2.9	2.6/3.3	2.7/3.9	2.7/4.6	1.5/3.2	1.1/2.5	2.5/2.9

Solar Radiation for 1-Axis Tracking Flat-Plate Collectors with a North-South Axis (kWh/m²/day), Uncertainty ±9%

Axis Tilt (°)		Jan	Feb	Mar	Apr	May	June	July	Aug	Sept	Oct	Nov	Dec	Year
0	Average	2.2	3.2	4.4	5.7	6.8	7.5	7.4	6.7	5.5	4.0	2.3	1.7	4.8
	Min/Max	1.6/2.6	2.5/4.3	3.6/5.9	4.1/7.3	5.3/8.1	6.0/8.8	6.3/8.6	5.6/7.7	4.5/6.6	3.2/5.6	1.7/3.0	1.2/2.1	4.4/5.3
Latitude -15	Average	2.7	3.7	4.9	6.0	6.9	7.5	7.4	6.9	5.9	4.7	2.8	2.1	5.1
	Min/Max	1.8/3.3	2.8/5.2	3.9/6.6	4.2/7.7	5.3/8.2	5.9/8.8	6.3/8.7	5.7/8.1	4.8/7.2	3.7/6.6	1.9/3.7	1.3/2.6	4.7/5.7
Latitude	Average	2.9	3.9	5.0	6.0	6.6	7.2	7.2	6.8	6.0	4.8	2.9	2.3	5.1
	Min/Max	1.9/3.5	2.9/5.6	4.0/6.8	4.1/7.6	5.1/8.0	5.7/8.4	6.0/8.4	5.6/7.9	4.8/7.3	3.8/6.9	2.0/4.0	1.4/2.8	4.7/5.7
Latitude +15	Average	3.0	4.0	4.9	5.7	6.2	6.7	6.7	6.4	5.8	4.8	3.0	2.3	5.0
	Min/Max	2.0/3.7	2.9/5.7	3.9/6.7	3.8/7.3	4.8/7.5	5.3/7.8	5.6/7.8	5.3/7.5	4.7/7.1	3.8/6.9	2.0/4.2	1.4/3.0	4.5/5.5

Solar Radiation for 2-Axis Tracking Flat-Plate Collectors (kWh/m²/day), Uncertainty ±9%

Tracker		Jan	Feb	Mar	Apr	May	June	July	Aug	Sept	Oct	Nov	Dec	Year
2-Axis	Average	3.0	4.0	5.0	6.1	7.0	7.7	7.6	6.9	6.0	4.9	3.0	2.4	5.3
	Min/Max	2.0/3.7	2.9/5.7	4.0/6.8	4.2/7.7	5.3/8.3	6.1/8.9	6.4/8.8	5.7/8.1	4.9/7.3	3.9/7.0	2.0/4.2	1.4/3.0	4.8/5.8

Direct Beam Solar Radiation for Concentrating Collectors (kWh/m²/day), Uncertainty ±8%

Tracker		Jan	Feb	Mar	Apr	May	June	July	Aug	Sept	Oct	Nov	Dec	Year
1-Axis, E-W	Average	1.4	1.7	2.0	2.4	2.8	3.1	3.0	2.8	2.6	2.4	1.4	1.0	2.2
Horiz Axis	Min/Max	0.6/2.0	0.9/2.8	1.2/3.5	1.2/3.6	1.6/3.9	2.0/4.0	2.1/4.2	1.8/3.6	1.8/3.5	1.6/4.0	0.7/2.4	0.3/1.7	1.9/2.7
1-Axis, N-S	Average	1.0	1.5	2.2	3.1	3.6	4.1	4.0	3.7	3.1	2.3	1.1	0.7	2.5
Horiz Axis	Min/Max	0.4/1.5	0.8/2.4	1.4/3.9	1.5/4.6	2.1/5.1	2.7/5.2	2.8/5.5	2.4/4.7	2.1/4.1	1.6/3.9	0.5/1.8	0.2/1.1	2.2/3.1
1-Axis, N-S	Average	1.5	2.0	2.7	3.3	3.6	3.9	3.8	3.7	3.5	3.0	1.6	1.1	2.8
Tilt=Latitude	Min/Max	0.7/2.2	1.0/3.3	1.7/4.6	1.6/4.9	2.1/5.0	2.5/4.9	2.7/5.2	2.4/4.8	2.4/4.6	2.0/5.0	0.8/2.7	0.3/1.7	2.4/3.4
2-Axis	Average	1.6	2.0	2.7	3.3	3.8	4.2	4.1	3.9	3.5	3.0	1.7	1.2	2.9
	Min/Max	0.8/2.3	1.0/3.4	1.7/4.6	1.6/4.9	2.2/5.3	2.8/5.3	2.9/5.6	2.5/5.0	2.4/4.6	2.0/5.0	0.8/2.8	0.4/1.9	2.5/3.5

Average Climatic Conditions

Element	Jan	Feb	Mar	Apr	May	June	July	Aug	Sept	Oct	Nov	Dec	Year
Temperature (°C)	-3.3	-1.8	4.1	9.8	15.3	19.9	22.3	21.4	17.7	11.3	5.7	-0.3	10.2
Daily Minimum Temp	-7.5	-6.5	-1.2	3.8	9.1	13.8	16.4	15.7	11.9	5.7	1.2	-4.2	4.8
Daily Maximum Temp	0.9	2.7	9.4	15.7	21.4	26.1	28.1	27.1	23.5	16.9	10.2	3.7	15.5
Record Minimum Temp	-27.8	-24.4	-18.3	-10.0	-3.3	1.1	5.6	3.9	-0.6	-8.9	-18.3	-24.4	-27.8
Record Maximum Temp	20.6	20.6	27.8	31.7	32.8	36.7	39.4	37.8	36.1	30.6	27.8	23.3	39.4
HDD, Base 18.3°C	670	564	441	257	119	20	3	8	56	222	378	577	3316
CDD, Base 18.3°C	0	0	0	0	24	68	126	102	37	5	0	0	363
Relative Humidity (%)	70	67	64	60	63	66	69	71	72	68	70	72	68
Wind Speed (m/s)	4.7	4.5	4.7	4.5	3.9	3.5	3.2	3.0	3.2	3.6	4.2	4.6	4.0

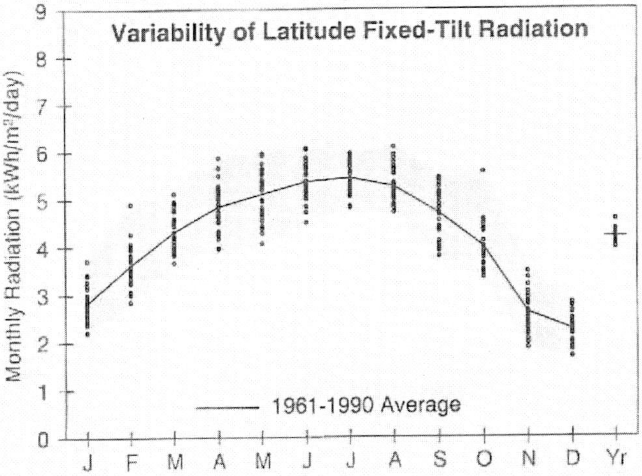

Wilkes-Barre, PA

WBAN NO. 14777

LATITUDE: 41.33° N
LONGITUDE: 75.73° W
ELEVATION: 289 meters
MEAN PRESSURE: 983 millibars

STATION TYPE: Secondary

Solar Radiation for Flat-Plate Collectors Facing South at a Fixed Tilt (kWh/m²/day), Uncertainty ±11%

Tilt (°)		Jan	Feb	Mar	Apr	May	June	July	Aug	Sept	Oct	Nov	Dec	Year
0	Average	1.8	2.5	3.6	4.6	5.4	6.0	5.9	5.2	4.1	2.9	1.8	1.4	3.8
	Min/Max	1.5/2.1	2.1/3.1	3.2/4.0	3.9/5.4	4.4/6.2	5.0/6.7	5.3/6.4	4.7/5.9	3.4/4.6	2.6/3.7	1.5/2.1	1.3/1.6	3.6/4.0
Latitude -15	Average	2.5	3.4	4.2	5.0	5.4	5.8	5.8	5.5	4.7	3.8	2.4	2.0	4.2
	Min/Max	2.0/3.3	2.7/4.4	3.7/4.9	4.1/6.0	4.3/6.3	4.9/6.6	5.2/6.4	4.9/6.3	3.8/5.4	3.2/5.1	1.8/3.1	1.6/2.5	4.0/4.5
Latitude	Average	2.8	3.6	4.3	4.8	5.1	5.4	5.4	5.3	4.7	4.0	2.6	2.2	4.2
	Min/Max	2.2/3.7	2.8/4.9	3.7/5.1	4.0/5.9	4.1/6.0	4.5/6.1	4.8/6.0	4.7/6.1	3.8/5.4	3.3/5.6	1.9/3.5	1.7/2.8	4.0/4.5
Latitude +15	Average	3.0	3.7	4.2	4.5	4.5	4.7	4.8	4.8	4.5	4.0	2.7	2.4	4.0
	Min/Max	2.3/4.0	2.9/5.2	3.5/5.0	3.6/5.4	3.6/5.3	3.9/5.2	4.2/5.2	4.3/5.5	3.6/5.2	3.3/5.7	1.8/3.6	1.7/3.0	3.7/4.3
90	Average	2.8	3.3	3.3	3.0	2.6	2.5	2.6	3.0	3.2	3.3	2.3	2.2	2.8
	Min/Max	2.1/3.9	2.5/4.8	2.6/3.8	2.5/3.6	2.2/3.0	2.3/2.7	2.4/2.8	2.7/3.3	2.6/3.7	2.7/4.7	1.5/3.3	1.5/2.9	2.6/3.1

Solar Radiation for 1-Axis Tracking Flat-Plate Collectors with a North-South Axis (kWh/m²/day), Uncertainty ±11%

Axis Tilt (°)		Jan	Feb	Mar	Apr	May	June	July	Aug	Sept	Oct	Nov	Dec	Year
0	Average	2.4	3.4	4.6	5.8	6.7	7.4	7.4	6.6	5.3	3.9	2.3	1.8	4.8
	Min/Max	1.9/3.1	2.6/4.6	4.0/5.6	4.5/7.3	5.1/8.2	5.7/8.8	6.2/8.4	5.9/8.0	4.2/6.2	3.3/5.5	1.7/3.0	1.4/2.2	4.6/5.3
Latitude -15	Average	2.9	4.0	5.1	6.2	6.8	7.4	7.4	6.9	5.8	4.6	2.7	2.3	5.2
	Min/Max	2.3/3.9	3.1/5.6	4.4/6.3	4.7/7.8	5.1/8.4	5.6/8.8	6.2/8.4	6.1/8.4	4.5/6.8	3.7/6.6	1.9/3.7	1.7/2.9	4.9/5.7
Latitude	Average	3.2	4.2	5.2	6.1	6.6	7.1	7.1	6.8	5.8	4.8	2.9	2.5	5.2
	Min/Max	2.4/4.3	3.2/6.0	4.4/6.5	4.7/7.7	4.9/8.1	5.4/8.5	6.0/8.1	6.0/8.2	4.5/6.9	3.9/6.9	2.0/4.0	1.8/3.2	4.9/5.8
Latitude +15	Average	3.3	4.3	5.1	5.8	6.2	6.6	6.7	6.4	5.6	4.8	3.0	2.6	5.0
	Min/Max	2.5/4.6	3.3/6.1	4.3/6.4	4.4/7.4	4.6/7.6	5.0/7.9	5.6/7.6	5.7/7.8	4.3/6.7	3.8/7.0	2.0/4.2	1.8/3.3	4.7/5.6

Solar Radiation for 2-Axis Tracking Flat-Plate Collectors (kWh/m²/day), Uncertainty ±11%

Tracker		Jan	Feb	Mar	Apr	May	June	July	Aug	Sept	Oct	Nov	Dec	Year
2-Axis	Average	3.3	4.3	5.2	6.2	6.9	7.6	7.6	7.0	5.8	4.8	3.0	2.6	5.4
	Min/Max	2.5/4.6	3.3/6.1	4.4/6.5	4.8/7.8	5.2/8.4	5.8/9.1	6.3/8.6	6.2/8.4	4.5/6.9	3.9/7.0	2.0/4.2	1.8/3.4	5.1/5.9

Direct Beam Solar Radiation for Concentrating Collectors (kWh/m²/day), Uncertainty ±11%

Tracker		Jan	Feb	Mar	Apr	May	June	July	Aug	Sept	Oct	Nov	Dec	Year
1-Axis, E-W Horiz Axis	Average	1.6	2.0	2.3	2.6	2.9	3.3	3.3	3.0	2.6	2.4	1.4	1.3	2.4
	Min/Max	1.0/2.5	1.2/3.3	1.5/3.3	1.5/3.8	1.5/4.1	1.9/4.5	2.3/4.1	2.4/4.2	1.6/3.4	1.6/4.1	0.6/2.4	0.6/1.9	2.1/2.9
1-Axis, N-S Horiz Axis	Average	1.1	1.7	2.5	3.3	3.7	4.2	4.3	3.9	3.0	2.3	1.1	0.8	2.7
	Min/Max	0.7/1.7	1.1/2.9	1.6/3.7	1.8/4.9	2.0/5.3	2.3/5.9	2.9/5.4	3.0/5.5	1.9/4.0	1.6/4.0	0.4/1.8	0.4/1.2	2.4/3.2
1-Axis, N-S Tilt=Latitude	Average	1.8	2.3	3.0	3.5	3.7	4.0	4.1	4.0	3.5	3.0	1.6	1.3	3.0
	Min/Max	1.1/2.7	1.5/3.9	1.9/4.4	2.0/5.2	1.9/5.2	2.2/5.6	2.8/5.1	3.1/5.6	2.2/4.6	2.0/5.2	0.6/2.7	0.6/2.0	2.7/3.6
2-Axis	Average	1.9	2.4	3.0	3.6	3.9	4.4	4.4	4.1	3.5	3.0	1.7	1.4	3.1
	Min/Max	1.1/2.9	1.5/4.0	1.9/4.4	2.0/5.3	2.0/5.5	2.4/6.1	3.0/5.5	3.2/5.8	2.2/4.6	2.1/5.2	0.7/2.9	0.7/2.2	2.8/3.8

Average Climatic Conditions

Element	Jan	Feb	Mar	Apr	May	June	July	Aug	Sept	Oct	Nov	Dec	Year
Temperature (°C)	-4.1	-2.9	2.7	8.9	14.9	19.6	22.1	21.1	17.0	10.8	5.2	-1.1	9.5
Daily Minimum Temp	-8.1	-7.2	-2.1	3.4	9.1	13.8	16.4	15.6	11.6	5.6	1.1	-4.8	4.6
Daily Maximum Temp	-0.1	1.4	7.5	14.3	20.7	25.3	27.7	26.5	22.4	16.1	9.3	2.6	14.5
Record Minimum Temp	-25.6	-26.7	-18.9	-10.0	-2.2	1.1	6.7	3.9	-1.1	-6.7	-12.8	-21.7	-26.7
Record Maximum Temp	19.4	21.7	28.3	33.3	33.9	35.0	37.8	34.4	35.0	28.3	26.7	18.9	37.8
HDD, Base 18.3°C	694	594	484	283	121	21	0	6	61	235	393	603	3495
CDD, Base 18.3°C	0	0	0	0	14	58	117	91	21	0	0	0	299
Relative Humidity (%)	70	68	64	60	64	70	71	73	75	71	72	73	69
Wind Speed (m/s)	3.9	3.9	4.2	4.2	3.8	3.5	3.2	3.1	3.3	3.5	3.8	3.9	3.7

Williamsport, PA

WBAN NO. 14778

LATITUDE: 41.27° N
LONGITUDE: 77.05° W
ELEVATION: 243 meters
MEAN PRESSURE: 998 millibars

STATION TYPE: Secondary

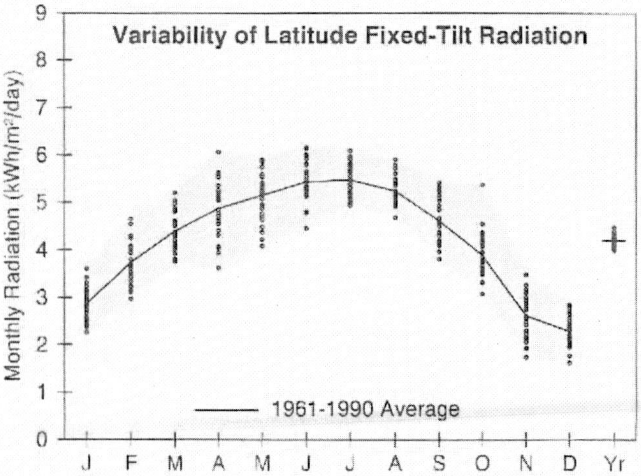

Solar Radiation for Flat-Plate Collectors Facing South at a Fixed Tilt (kWh/m²/day), Uncertainty ±11%

Tilt (°)		Jan	Feb	Mar	Apr	May	June	July	Aug	Sept	Oct	Nov	Dec	Year
0	Average	1.8	2.6	3.6	4.6	5.4	6.0	5.9	5.1	4.0	2.9	1.8	1.4	3.8
	Min/Max	1.5/2.1	2.2/3.0	3.2/4.1	3.7/5.5	4.4/6.2	5.0/6.8	5.3/6.5	4.7/5.7	3.4/4.5	2.4/3.6	1.4/2.1	1.3/1.6	3.6/3.9
Latitude -15	Average	2.6	3.4	4.3	5.0	5.5	5.9	5.9	5.4	4.6	3.7	2.4	2.1	4.2
	Min/Max	2.1/3.2	2.8/4.2	3.7/5.0	3.8/6.2	4.4/6.3	4.8/6.7	5.3/6.6	4.9/6.1	3.8/5.3	3.0/5.0	1.7/3.1	1.5/2.5	4.0/4.5
Latitude	Average	2.9	3.7	4.4	4.9	5.1	5.4	5.5	5.2	4.6	3.9	2.6	2.3	4.2
	Min/Max	2.3/3.6	3.0/4.7	3.8/5.2	3.6/6.1	4.1/5.9	4.4/6.2	4.9/6.1	4.7/5.9	3.8/5.4	3.1/5.4	1.7/3.5	1.6/2.8	4.0/4.5
Latitude +15	Average	3.0	3.8	4.3	4.5	4.6	4.7	4.8	4.8	4.4	3.9	2.7	2.4	4.0
	Min/Max	2.3/3.9	3.0/4.9	3.6/5.1	3.3/5.7	3.6/5.2	3.9/5.3	4.3/5.3	4.3/5.4	3.6/5.2	3.0/5.4	1.7/3.7	1.6/3.0	3.8/4.2
90	Average	2.8	3.4	3.3	3.0	2.6	2.5	2.6	3.0	3.1	3.1	2.3	2.2	2.8
	Min/Max	2.1/3.8	2.7/4.5	2.7/3.9	2.2/3.8	2.2/3.0	2.3/2.8	2.5/2.9	2.7/3.3	2.5/3.7	2.4/4.5	1.5/3.3	1.5/2.9	2.7/3.0

Solar Radiation for 1-Axis Tracking Flat-Plate Collectors with a North-South Axis (kWh/m²/day), Uncertainty ±11%

Axis Tilt (°)		Jan	Feb	Mar	Apr	May	June	July	Aug	Sept	Oct	Nov	Dec	Year
0	Average	2.4	3.5	4.7	5.9	6.7	7.5	7.4	6.4	5.1	3.7	2.2	1.8	4.8
	Min/Max	1.9/3.0	2.8/4.3	3.9/5.6	4.0/7.5	5.1/7.9	5.7/8.9	6.3/8.3	5.5/7.6	4.0/6.0	3.0/5.2	1.5/2.9	1.4/2.2	4.5/5.1
Latitude -15	Average	3.0	4.1	5.2	6.2	6.9	7.5	7.4	6.7	5.5	4.4	2.7	2.3	5.2
	Min/Max	2.3/3.8	3.2/5.2	4.3/6.3	4.2/8.0	5.1/8.1	5.6/8.9	6.3/8.4	5.7/8.0	4.3/6.6	3.4/6.3	1.8/3.7	1.6/2.9	4.9/5.6
Latitude	Average	3.2	4.3	5.3	6.1	6.6	7.2	7.1	6.6	5.6	4.5	2.9	2.5	5.2
	Min/Max	2.5/4.1	3.3/5.6	4.3/6.4	4.1/7.9	4.9/7.8	5.4/8.5	6.1/8.1	5.6/7.8	4.3/6.7	3.5/6.6	1.8/4.0	1.7/3.2	4.9/5.6
Latitude +15	Average	3.4	4.4	5.2	5.9	6.2	6.7	6.7	6.2	5.4	4.5	3.0	2.6	5.0
	Min/Max	2.5/4.4	3.3/5.7	4.2/6.3	3.8/7.7	4.6/7.4	5.0/7.9	5.6/7.6	5.2/7.4	4.1/6.5	3.5/6.6	1.8/4.1	1.7/3.4	4.7/5.4

Solar Radiation for 2-Axis Tracking Flat-Plate Collectors (kWh/m²/day), Uncertainty ±11%

Tracker		Jan	Feb	Mar	Apr	May	June	July	Aug	Sept	Oct	Nov	Dec	Year
2-Axis	Average	3.4	4.4	5.3	6.2	6.9	7.6	7.5	6.8	5.6	4.6	3.0	2.6	5.3
	Min/Max	2.6/4.4	3.3/5.7	4.3/6.4	4.2/8.0	5.2/8.2	5.8/9.1	6.4/8.5	5.7/8.0	4.3/6.7	3.5/6.6	1.8/4.2	1.7/3.4	5.0/5.8

Direct Beam Solar Radiation for Concentrating Collectors (kWh/m²/day), Uncertainty ±11%

Tracker		Jan	Feb	Mar	Apr	May	June	July	Aug	Sept	Oct	Nov	Dec	Year
1-Axis, E-W Horiz Axis	Average	1.7	2.1	2.4	2.6	2.9	3.4	3.3	2.9	2.5	2.3	1.4	1.3	2.4
	Min/Max	1.2/2.3	1.1/3.0	1.7/3.3	1.2/4.0	1.4/3.9	1.9/4.4	2.3/4.0	2.1/4.0	1.5/3.3	1.5/3.9	0.5/2.4	0.5/2.1	2.1/2.8
1-Axis, N-S Horiz Axis	Average	1.2	1.8	2.6	3.3	3.8	4.3	4.3	3.7	2.8	2.1	1.1	0.8	2.7
	Min/Max	0.8/1.6	1.0/2.6	1.8/3.6	1.4/5.0	1.9/5.0	2.4/5.7	3.0/5.2	2.6/5.1	1.6/3.8	1.4/3.7	0.4/1.8	0.3/1.3	2.4/3.1
1-Axis, N-S Tilt=Latitude	Average	1.8	2.5	3.1	3.6	3.7	4.1	4.1	3.8	3.3	2.8	1.6	1.4	3.0
	Min/Max	1.2/2.5	1.3/3.6	2.2/4.3	1.5/5.3	1.9/4.9	2.3/5.4	2.8/5.0	2.6/5.2	1.9/4.3	1.8/4.8	0.6/2.6	0.5/2.2	2.6/3.5
2-Axis	Average	2.0	2.5	3.1	3.6	3.9	4.4	4.4	3.9	3.3	2.8	1.7	1.5	3.1
	Min/Max	1.3/2.7	1.4/3.7	2.2/4.3	1.5/5.4	2.0/5.2	2.4/5.9	3.0/5.3	2.7/5.4	1.9/4.4	1.8/4.8	0.6/2.8	0.6/2.3	2.7/3.6

Average Climatic Conditions

Element	Jan	Feb	Mar	Apr	May	June	July	Aug	Sept	Oct	Nov	Dec	Year
Temperature (°C)	-3.8	-2.2	3.4	9.5	15.3	19.9	22.4	21.6	17.5	11.1	5.4	-0.6	9.9
Daily Minimum Temp	-8.3	-7.1	-1.9	3.4	8.8	13.7	16.4	15.8	11.7	5.3	0.9	-4.5	4.5
Daily Maximum Temp	0.7	2.6	8.7	15.6	21.7	26.1	28.4	27.3	23.2	16.9	9.9	3.3	15.4
Record Minimum Temp	-27.2	-25.0	-18.3	-9.4	-2.2	2.2	6.1	3.3	-2.2	-6.7	-13.3	-26.1	-27.2
Record Maximum Temp	20.6	21.7	30.6	33.3	35.0	38.9	39.4	37.8	38.9	32.8	28.3	19.4	39.4
HDD, Base 18.3°C	686	576	463	265	111	17	0	5	59	227	387	587	3382
CDD, Base 18.3°C	0	0	0	0	16	63	127	105	34	0	0	0	346
Relative Humidity (%)	70	68	65	61	67	73	74	77	79	75	73	72	71
Wind Speed (m/s)	4.0	3.9	4.1	4.2	3.5	3.2	2.8	2.7	2.8	3.0	3.7	3.9	3.5

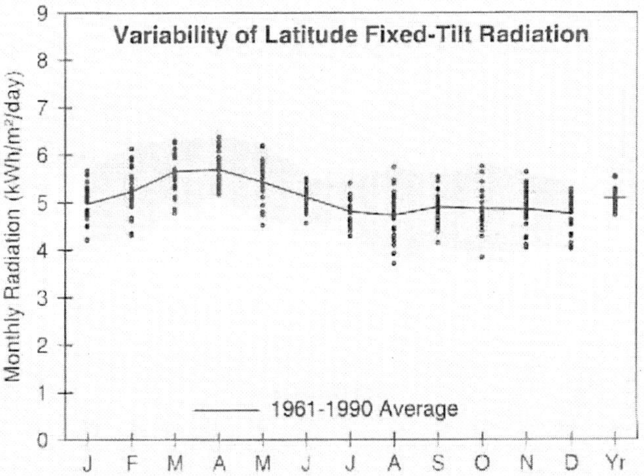

Guam, PI

WBAN NO. 41415

LATITUDE: 13.55° N
LONGITUDE: 144.83° E
ELEVATION: 110 meters
MEAN PRESSURE: 998 millibars

STATION TYPE: Primary

Solar Radiation for Flat-Plate Collectors Facing South at a Fixed Tilt (kWh/m²/day), Uncertainty ±9%

Tilt (°)		Jan	Feb	Mar	Apr	May	June	July	Aug	Sept	Oct	Nov	Dec	Year
0	Average	4.4	4.8	5.5	5.8	5.7	5.5	5.1	4.9	4.9	4.6	4.4	4.2	5.0
	Min/Max	3.8/5.0	4.1/5.6	4.7/6.1	5.2/6.5	4.7/6.6	4.8/5.9	4.5/5.8	3.9/5.9	4.2/5.5	3.6/5.4	3.7/5.0	3.6/4.6	4.7/5.4
Latitude -15	Average	4.3	4.8	5.4	5.8	5.7	5.5	5.1	4.9	4.8	4.6	4.3	4.1	4.9
	Min/Max	3.8/4.9	4.0/5.5	4.6/6.1	5.2/6.5	4.8/6.6	4.9/5.9	4.6/5.8	3.8/5.9	4.1/5.4	3.6/5.3	3.7/4.9	3.6/4.5	4.6/5.4
Latitude	Average	5.0	5.2	5.7	5.7	5.4	5.1	4.8	4.7	4.9	4.9	4.9	4.8	5.1
	Min/Max	4.2/5.7	4.3/6.1	4.8/6.3	5.2/6.4	4.5/6.2	4.6/5.5	4.3/5.4	3.7/5.7	4.1/5.5	3.8/5.8	4.0/5.6	4.0/5.3	4.7/5.5
Latitude +15	Average	5.3	5.4	5.6	5.3	4.9	4.5	4.3	4.4	4.7	4.9	5.1	5.1	5.0
	Min/Max	4.4/6.2	4.4/6.4	4.7/6.3	4.8/6.0	4.1/5.5	4.1/4.8	3.8/4.8	3.4/5.3	4.0/5.3	3.9/5.9	4.2/6.0	4.2/5.7	4.6/5.4
90	Average	3.8	3.3	2.6	1.7	1.5	1.5	1.5	1.5	2.0	2.8	3.5	3.8	2.5
	Min/Max	3.1/4.5	2.7/4.0	2.3/2.9	1.6/1.9	1.3/1.6	1.3/1.5	1.4/1.5	1.4/1.7	1.8/2.1	2.2/3.3	2.8/4.2	3.1/4.3	2.3/2.6

Solar Radiation for 1-Axis Tracking Flat-Plate Collectors with a North-South Axis (kWh/m²/day), Uncertainty ±9%

Axis Tilt (°)		Jan	Feb	Mar	Apr	May	June	July	Aug	Sept	Oct	Nov	Dec	Year
0	Average	5.7	6.2	7.0	7.4	7.2	6.8	6.2	5.8	5.9	5.7	5.6	5.4	6.2
	Min/Max	4.7/6.6	4.9/7.4	5.5/8.0	6.5/8.6	5.6/8.6	5.7/7.6	5.3/7.2	4.3/7.5	4.7/6.9	4.3/6.9	4.4/6.6	4.4/6.0	5.5/6.9
Latitude -15	Average	5.6	6.1	6.9	7.4	7.3	6.9	6.2	5.8	5.9	5.7	5.6	5.3	6.2
	Min/Max	4.7/6.6	4.9/7.4	5.5/8.0	6.5/8.6	5.6/8.6	5.7/7.6	5.3/7.2	4.3/7.5	4.7/6.9	4.3/6.9	4.4/6.5	4.4/5.9	5.5/6.9
Latitude	Average	6.1	6.4	7.1	7.3	7.0	6.6	6.0	5.7	6.0	5.9	5.9	5.8	6.3
	Min/Max	5.0/7.2	5.1/7.8	5.6/8.2	6.5/8.5	5.4/8.3	5.5/7.3	5.1/6.9	4.2/7.3	4.7/6.9	4.4/7.2	4.7/7.0	4.7/6.5	5.6/7.0
Latitude +15	Average	6.3	6.6	7.0	7.1	6.6	6.1	5.6	5.5	5.8	5.9	6.1	6.1	6.2
	Min/Max	5.1/7.5	5.2/8.0	5.5/8.1	6.2/8.2	5.1/7.8	5.1/6.7	4.8/6.5	4.0/7.0	4.5/6.8	4.4/7.3	4.8/7.3	4.9/6.8	5.5/6.9

Solar Radiation for 2-Axis Tracking Flat-Plate Collectors (kWh/m²/day), Uncertainty ±9%

Tracker		Jan	Feb	Mar	Apr	May	June	July	Aug	Sept	Oct	Nov	Dec	Year
2-Axis	Average	6.4	6.6	7.1	7.4	7.3	7.0	6.3	5.9	6.0	6.0	6.2	6.2	6.5
	Min/Max	5.2/7.6	5.2/8.0	5.6/8.2	6.6/8.6	5.7/8.7	5.8/7.8	5.4/7.4	4.3/7.5	4.7/7.0	4.5/7.3	4.8/7.4	4.9/6.9	5.8/7.3

Direct Beam Solar Radiation for Concentrating Collectors (kWh/m²/day), Uncertainty ±8%

Tracker		Jan	Feb	Mar	Apr	May	June	July	Aug	Sept	Oct	Nov	Dec	Year
1-Axis, E-W Horiz Axis	Average	3.1	2.9	3.0	3.1	3.2	3.0	2.4	2.0	2.1	2.3	2.8	2.9	2.7
	Min/Max	2.0/4.2	1.7/4.1	1.9/4.0	2.5/4.2	2.0/4.4	2.1/3.7	1.6/3.3	1.0/3.4	1.2/2.9	1.6/3.5	1.8/3.8	1.8/3.6	2.1/3.4
1-Axis, N-S Horiz Axis	Average	3.3	3.5	4.0	4.3	4.2	3.8	3.1	2.7	2.9	2.9	3.1	3.0	3.4
	Min/Max	2.1/4.4	2.1/4.9	2.4/5.3	3.5/5.8	2.6/5.8	2.6/4.7	2.0/4.2	1.4/4.4	1.6/3.8	1.9/4.3	2.0/4.2	1.9/3.7	2.5/4.2
1-Axis, N-S Tilt=Latitude	Average	3.6	3.7	4.1	4.2	4.0	3.6	2.9	2.6	2.9	3.1	3.3	3.3	3.4
	Min/Max	2.3/4.8	2.2/5.2	2.5/5.5	3.5/5.7	2.5/5.5	2.5/4.4	1.9/4.0	1.3/4.3	1.6/3.8	2.0/4.5	2.1/4.5	2.1/4.1	2.6/4.2
2-Axis	Average	3.8	3.8	4.1	4.3	4.2	3.9	3.1	2.7	2.9	3.1	3.5	3.6	3.6
	Min/Max	2.5/5.2	2.3/5.3	2.5/5.5	3.5/5.8	2.7/5.8	2.7/4.8	2.1/4.3	1.4/4.5	1.6/3.8	2.0/4.6	2.2/4.7	2.3/4.4	2.7/4.4

Average Climatic Conditions

Element	Jan	Feb	Mar	Apr	May	June	July	Aug	Sept	Oct	Nov	Dec	Year
Temperature (°C)	25.2	25.1	25.4	26.1	26.5	26.6	26.3	26.2	26.2	26.2	26.2	25.8	26.0
Daily Minimum Temp	21.7	21.6	21.7	22.3	22.6	22.7	22.4	22.3	22.3	22.4	22.8	22.6	22.3
Daily Maximum Temp	28.7	28.6	29.2	29.9	30.4	30.5	30.2	29.9	30.1	29.9	29.6	29.0	29.7
Record Minimum Temp	13.3	15.0	12.2	15.0	16.7	17.2	17.8	17.2	16.1	17.8	16.7	16.1	12.2
Record Maximum Temp	30.6	31.7	32.2	32.2	32.8	33.3	34.4	32.8	35.0	32.8	31.7	31.7	35.0
HDD, Base 18.3°C	0	0	0	0	0	0	0	0	0	0	0	0	0
CDD, Base 18.3°C	212	190	221	233	253	248	248	243	237	243	237	233	2797
Relative Humidity (%)	82	81	80	81	82	83	86	87	87	86	85	83	84
Wind Speed (m/s)	4.4	4.9	4.7	4.6	4.2	3.3	3.0	2.5	2.7	3.0	4.0	4.4	3.8

San Juan, PR

WBAN NO. 11641

LATITUDE: 18.43° N
LONGITUDE: 66.00° W
ELEVATION: 19 meters
MEAN PRESSURE: 1014 millibars

STATION TYPE: Primary

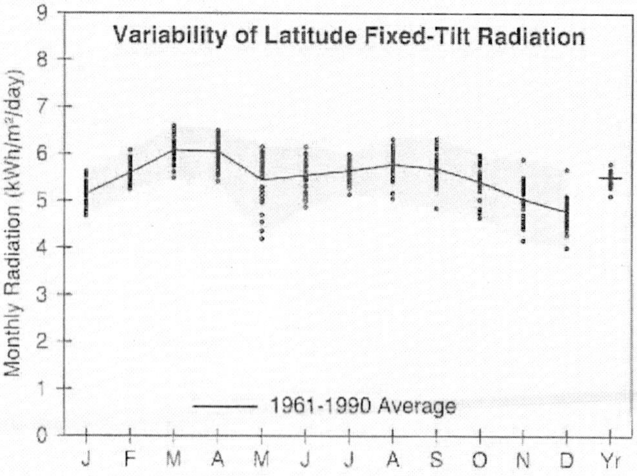

Solar Radiation for Flat-Plate Collectors Facing South at a Fixed Tilt (kWh/m²/day), Uncertainty ±9%

Tilt (°)		Jan	Feb	Mar	Apr	May	June	July	Aug	Sept	Oct	Nov	Dec	Year
0	Average	4.3	4.9	5.7	6.1	5.8	6.1	6.1	6.0	5.5	4.9	4.3	4.0	5.3
	Min/Max	4.0/4.6	4.7/5.3	5.2/6.2	5.4/6.5	4.4/6.6	5.3/6.8	5.5/6.5	5.2/6.5	4.8/6.1	4.3/5.4	3.7/4.9	3.4/4.5	4.9/5.5
Latitude -15	Average	4.5	5.1	5.8	6.1	5.7	6.0	6.0	6.0	5.6	5.0	4.5	4.1	5.4
	Min/Max	4.1/4.8	4.8/5.5	5.3/6.3	5.5/6.6	4.4/6.5	5.2/6.7	5.5/6.4	5.2/6.5	4.8/6.2	4.4/5.5	3.8/5.1	3.5/4.8	5.0/5.6
Latitude	Average	5.1	5.6	6.1	6.1	5.4	5.5	5.6	5.8	5.7	5.4	5.1	4.8	5.5
	Min/Max	4.7/5.6	5.3/6.1	5.5/6.6	5.4/6.5	4.2/6.2	4.9/6.2	5.1/6.0	5.1/6.3	4.8/6.3	4.7/6.0	4.2/5.9	4.0/5.7	5.1/5.8
Latitude +15	Average	5.5	5.8	6.0	5.7	4.9	4.8	5.0	5.3	5.5	5.5	5.4	5.2	5.4
	Min/Max	5.0/6.1	5.4/6.3	5.4/6.5	5.1/6.1	3.8/5.5	4.3/5.3	4.5/5.3	4.6/5.8	4.7/6.1	4.7/6.1	4.3/6.3	4.3/6.3	5.0/5.7
90	Average	4.2	3.9	3.1	2.0	1.5	1.5	1.5	1.7	2.5	3.3	3.9	4.1	2.8
	Min/Max	3.8/4.8	3.6/4.2	2.8/3.3	1.9/2.1	1.4/1.6	1.4/1.5	1.4/1.5	1.6/1.7	2.2/2.7	2.8/3.7	3.1/4.7	3.3/5.1	2.7/3.0

Solar Radiation for 1-Axis Tracking Flat-Plate Collectors with a North-South Axis (kWh/m²/day), Uncertainty ±9%

Axis Tilt (°)		Jan	Feb	Mar	Apr	May	June	July	Aug	Sept	Oct	Nov	Dec	Year
0	Average	5.6	6.4	7.4	7.8	7.2	7.6	7.6	7.5	7.0	6.3	5.5	5.1	6.8
	Min/Max	5.1/6.3	5.8/7.0	6.4/8.2	6.5/8.7	5.0/8.5	6.3/8.9	6.8/8.5	6.4/8.6	5.8/8.2	5.2/7.2	4.4/6.6	4.2/6.4	6.1/7.2
Latitude -15	Average	5.8	6.5	7.5	7.8	7.2	7.5	7.6	7.5	7.0	6.4	5.7	5.2	6.8
	Min/Max	5.2/6.5	5.9/7.2	6.5/8.3	6.5/8.7	5.0/8.4	6.3/8.9	6.7/8.5	6.4/8.6	5.8/8.2	5.3/7.3	4.5/6.8	4.3/6.5	6.1/7.3
Latitude	Average	6.3	6.9	7.7	7.8	6.9	7.2	7.3	7.4	7.1	6.7	6.1	5.7	6.9
	Min/Max	5.6/7.1	6.3/7.6	6.7/8.5	6.4/8.6	4.9/8.2	6.0/8.5	6.5/8.1	6.2/8.4	5.8/8.4	5.5/7.6	4.8/7.3	4.6/7.2	6.2/7.4
Latitude +15	Average	6.6	7.1	7.6	7.5	6.5	6.7	6.8	7.0	7.0	6.7	6.3	6.0	6.8
	Min/Max	5.8/7.4	6.4/7.8	6.6/8.4	6.2/8.3	4.6/7.7	5.6/7.9	6.0/7.6	5.9/8.0	5.7/8.2	5.5/7.7	4.9/7.7	4.8/7.7	6.2/7.4

Solar Radiation for 2-Axis Tracking Flat-Plate Collectors (kWh/m²/day), Uncertainty ±9%

Tracker		Jan	Feb	Mar	Apr	May	June	July	Aug	Sept	Oct	Nov	Dec	Year
2-Axis	Average	6.6	7.1	7.7	7.9	7.2	7.7	7.7	7.6	7.1	6.7	6.4	6.1	7.2
	Min/Max	5.9/7.5	6.4/7.8	6.7/8.5	6.5/8.8	5.1/8.5	6.4/9.1	6.8/8.6	6.4/8.6	5.9/8.4	5.5/7.7	5.0/7.7	4.9/7.8	6.5/7.7

Direct Beam Solar Radiation for Concentrating Collectors (kWh/m²/day), Uncertainty ±8%

Tracker		Jan	Feb	Mar	Apr	May	June	July	Aug	Sept	Oct	Nov	Dec	Year
1-Axis, E-W	Average	3.5	3.5	3.7	3.7	3.2	3.6	3.6	3.5	3.2	3.1	3.2	3.2	3.4
Horiz Axis	Min/Max	2.8/4.3	2.9/4.2	2.9/4.3	2.6/4.5	1.6/4.3	2.6/4.8	2.9/4.3	2.5/4.4	2.4/4.0	2.2/4.1	2.0/4.5	2.2/4.8	2.9/4.0
1-Axis, N-S	Average	3.5	4.0	4.8	4.9	4.1	4.5	4.6	4.6	4.2	3.7	3.3	3.0	4.1
Horiz Axis	Min/Max	2.8/4.3	3.2/4.8	3.6/5.6	3.4/6.1	2.1/5.5	3.3/6.2	3.7/5.6	3.3/5.9	3.1/5.4	2.6/4.9	2.0/4.6	2.1/4.7	3.5/4.8
1-Axis, N-S	Average	4.0	4.4	5.0	4.9	4.0	4.3	4.4	4.5	4.3	4.0	3.7	3.5	4.2
Tilt=Latitude	Min/Max	3.2/4.9	3.5/5.2	3.8/5.8	3.3/6.0	2.0/5.3	3.1/5.8	3.5/5.3	3.2/5.7	3.2/5.6	2.8/5.3	2.3/5.2	2.5/5.4	3.6/5.0
2-Axis	Average	4.3	4.5	5.0	5.0	4.2	4.6	4.7	4.6	4.3	4.0	3.9	3.8	4.4
	Min/Max	3.4/5.3	3.6/5.4	3.8/5.8	3.4/6.1	2.1/5.6	3.3/6.3	3.7/5.7	3.3/5.9	3.2/5.6	2.8/5.3	2.5/5.5	2.7/5.9	3.7/5.2

Average Climatic Conditions

Element	Jan	Feb	Mar	Apr	May	June	July	Aug	Sept	Oct	Nov	Dec	Year
Temperature (°C)	25.0	25.1	25.6	26.3	27.2	27.9	28.1	28.2	28.1	27.7	26.7	25.6	26.8
Daily Minimum Temp	21.6	21.4	22.0	22.7	23.6	24.5	24.9	24.8	24.6	24.2	23.3	22.4	23.3
Daily Maximum Temp	28.4	28.7	29.1	29.9	30.7	31.4	31.4	31.5	31.6	31.3	29.9	28.8	30.2
Record Minimum Temp	16.1	16.7	15.6	17.8	18.9	20.6	20.6	21.1	20.6	19.4	18.9	17.2	15.6
Record Maximum Temp	33.3	35.6	35.6	36.1	35.6	36.1	35.0	36.1	36.1	36.7	35.6	34.4	36.7
HDD, Base 18.3°C	0	0	0	0	0	0	0	0	0	0	0	0	0
CDD, Base 18.3°C	207	188	224	240	274	288	303	305	292	291	250	226	3088
Relative Humidity (%)	74	72	71	71	75	76	76	76	76	77	76	75	75
Wind Speed (m/s)	3.5	3.8	4.0	3.8	3.5	3.8	4.1	3.7	3.2	2.9	3.2	3.5	3.6

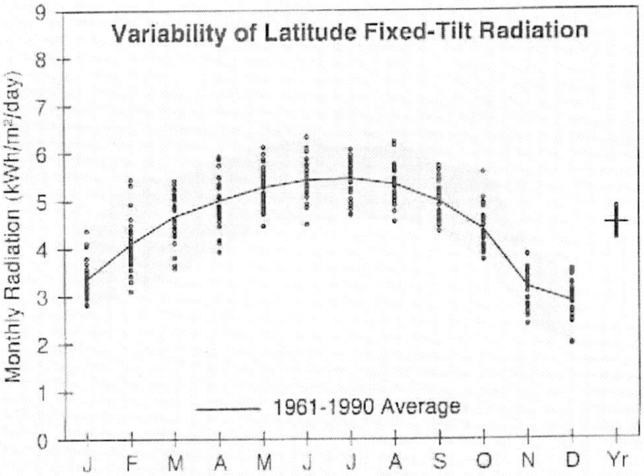

Providence, RI

WBAN NO. 14765

LATITUDE: 41.73° N
LONGITUDE: 71.43° W
ELEVATION: 19 meters
MEAN PRESSURE: 1014 millibars

STATION TYPE: Secondary

Solar Radiation for Flat-Plate Collectors Facing South at a Fixed Tilt (kWh/m²/day), Uncertainty ±11%

Tilt (°)		Jan	Feb	Mar	Apr	May	June	July	Aug	Sept	Oct	Nov	Dec	Year
0	Average	1.9	2.7	3.7	4.7	5.6	6.0	5.9	5.2	4.2	3.1	1.9	1.6	3.9
	Min/Max	1.7/2.3	2.3/3.3	3.1/4.1	3.9/5.5	4.7/6.4	5.0/7.0	5.2/6.5	4.6/5.9	3.8/4.8	2.8/3.7	1.7/2.2	1.3/1.8	3.7/4.1
Latitude -15	Average	3.0	3.7	4.5	5.1	5.6	5.9	5.9	5.5	4.9	4.1	2.9	2.5	4.5
	Min/Max	2.5/3.8	2.9/4.9	3.5/5.2	4.1/6.0	4.7/6.5	4.9/6.9	5.1/6.5	4.8/6.4	4.3/5.6	3.5/5.2	2.2/3.4	1.8/3.0	4.2/4.8
Latitude	Average	3.4	4.1	4.7	5.0	5.3	5.4	5.5	5.3	5.0	4.4	3.2	2.9	4.5
	Min/Max	2.8/4.4	3.1/5.4	3.6/5.4	3.9/5.9	4.5/6.1	4.5/6.3	4.7/6.1	4.6/6.2	4.3/5.7	3.7/5.6	2.4/3.8	2.0/3.5	4.2/4.8
Latitude +15	Average	3.6	4.2	4.6	4.6	4.7	4.7	4.8	4.8	4.7	4.4	3.3	3.1	4.3
	Min/Max	3.0/4.7	3.2/5.7	3.4/5.4	3.6/5.5	4.0/5.4	3.9/5.5	4.1/5.3	4.1/5.7	4.1/5.5	3.7/5.7	2.4/4.1	2.0/3.8	4.0/4.6
90	Average	3.4	3.7	3.5	3.0	2.7	2.6	2.7	3.0	3.4	3.6	3.0	2.9	3.1
	Min/Max	2.8/4.5	2.7/5.3	2.6/4.3	2.4/3.6	2.4/3.1	2.3/2.9	2.4/2.9	2.6/3.5	3.0/3.9	3.0/4.7	2.1/3.7	1.9/3.7	2.9/3.4

Solar Radiation for 1-Axis Tracking Flat-Plate Collectors with a North-South Axis (kWh/m²/day), Uncertainty ±11%

Axis Tilt (°)		Jan	Feb	Mar	Apr	May	June	July	Aug	Sept	Oct	Nov	Dec	Year
0	Average	2.7	3.8	5.0	6.1	7.0	7.5	7.4	6.7	5.6	4.3	2.7	2.2	5.1
	Min/Max	2.3/3.5	2.9/5.1	3.7/5.9	4.6/7.4	5.5/8.3	5.8/9.0	6.2/8.4	5.6/8.0	4.8/6.5	3.7/5.5	2.0/3.2	1.6/2.8	4.7/5.5
Latitude -15	Average	3.5	4.6	5.6	6.4	7.1	7.5	7.5	7.0	6.2	5.0	3.4	2.9	5.6
	Min/Max	2.9/4.7	3.4/6.3	4.1/6.7	4.8/7.9	5.6/8.5	5.7/9.1	6.2/8.5	5.8/8.4	5.2/7.3	4.3/6.6	2.5/4.2	2.0/3.7	5.1/6.0
Latitude	Average	3.9	4.9	5.8	6.4	6.9	7.2	7.2	6.9	6.2	5.3	3.6	3.2	5.6
	Min/Max	3.2/5.1	3.6/6.7	4.1/6.9	4.7/7.9	5.4/8.3	5.5/8.7	5.9/8.2	5.7/8.2	5.2/7.3	4.4/6.9	2.6/4.5	2.1/4.1	5.2/6.1
Latitude +15	Average	4.1	5.0	5.7	6.1	6.5	6.7	6.7	6.5	6.0	5.3	3.7	3.4	5.5
	Min/Max	3.3/5.4	3.6/6.9	4.0/6.8	4.4/7.6	5.0/7.8	5.1/8.1	5.5/7.7	5.4/7.8	5.0/7.1	4.4/7.0	2.6/4.7	2.2/4.4	5.0/6.0

Solar Radiation for 2-Axis Tracking Flat-Plate Collectors (kWh/m²/day), Uncertainty ±11%

Tracker		Jan	Feb	Mar	Apr	May	June	July	Aug	Sept	Oct	Nov	Dec	Year
2-Axis	Average	4.1	5.0	5.8	6.5	7.2	7.7	7.6	7.0	6.2	5.3	3.8	3.5	5.8
	Min/Max	3.4/5.5	3.6/6.9	4.2/6.9	4.8/8.0	5.7/8.6	5.8/9.3	6.3/8.7	5.9/8.4	5.3/7.3	4.5/7.0	2.7/4.7	2.2/4.4	5.3/6.3

Direct Beam Solar Radiation for Concentrating Collectors (kWh/m²/day), Uncertainty ±11%

Tracker		Jan	Feb	Mar	Apr	May	June	July	Aug	Sept	Oct	Nov	Dec	Year
1-Axis, E-W Horiz Axis	Average	2.4	2.6	2.7	2.8	3.1	3.4	3.3	3.1	2.9	2.8	2.1	2.0	2.8
	Min/Max	1.7/3.4	1.6/4.0	1.7/3.6	1.7/3.9	1.9/4.0	2.0/4.7	2.2/4.1	2.2/4.2	2.1/3.7	2.1/4.2	1.2/2.9	1.0/2.8	2.5/3.2
1-Axis, N-S Horiz Axis	Average	1.6	2.3	3.0	3.6	4.0	4.4	4.3	3.9	3.4	2.7	1.6	1.3	3.0
	Min/Max	1.2/2.3	1.4/3.5	1.8/4.0	2.1/5.0	2.4/5.3	2.6/6.0	2.9/5.3	2.9/5.4	2.4/4.4	2.1/4.0	0.9/2.2	0.7/1.8	2.6/3.5
1-Axis, N-S Tilt=Latitude	Average	2.6	3.1	3.6	3.8	4.0	4.2	4.1	4.0	3.9	3.5	2.4	2.1	3.4
	Min/Max	1.9/3.7	1.9/4.8	2.1/4.8	2.2/5.4	2.4/5.2	2.4/5.7	2.8/5.1	3.0/5.6	2.8/5.0	2.7/5.2	1.3/3.3	1.1/2.9	3.0/4.0
2-Axis	Average	2.7	3.2	3.6	3.9	4.2	4.5	4.4	4.2	3.9	3.6	2.5	2.3	3.6
	Min/Max	2.0/3.9	1.9/5.0	2.1/4.8	2.3/5.4	2.5/5.5	2.6/6.2	3.0/5.4	3.1/5.7	2.8/5.0	2.7/5.3	1.4/3.5	1.2/3.2	3.1/4.1

Average Climatic Conditions

Element	Jan	Feb	Mar	Apr	May	June	July	Aug	Sept	Oct	Nov	Dec	Year
Temperature (°C)	-2.3	-1.3	3.0	8.6	14.1	19.4	22.6	21.8	17.8	12.0	6.7	0.4	10.2
Daily Minimum Temp	-7.2	-6.2	-1.8	3.2	8.5	13.8	17.3	16.6	12.1	6.1	1.6	-4.2	5.0
Daily Maximum Temp	2.6	3.5	7.8	13.9	19.6	24.9	27.8	27.1	23.5	17.8	11.7	5.1	15.4
Record Minimum Temp	-25.0	-21.7	-17.2	-10.0	-1.7	5.0	8.9	4.4	0.6	-6.7	-14.4	-23.3	-25.0
Record Maximum Temp	18.9	22.2	26.7	36.7	34.4	36.1	38.9	40.0	37.8	30.0	25.6	21.1	40.0
HDD, Base 18.3°C	639	549	476	293	137	17	0	4	50	199	350	554	3269
CDD, Base 18.3°C	0	0	0	0	4	49	133	113	35	3	0	0	337
Relative Humidity (%)	64	63	63	61	67	70	71	72	73	70	69	67	68
Wind Speed (m/s)	4.9	5.1	5.4	5.4	4.8	4.4	4.2	4.1	4.2	4.2	4.7	4.8	4.6

Charleston, SC

WBAN NO. 13880

LATITUDE: 32.90° N
LONGITUDE: 80.03° W
ELEVATION: 12 meters
MEAN PRESSURE: 1017 millibars

STATION TYPE: Primary

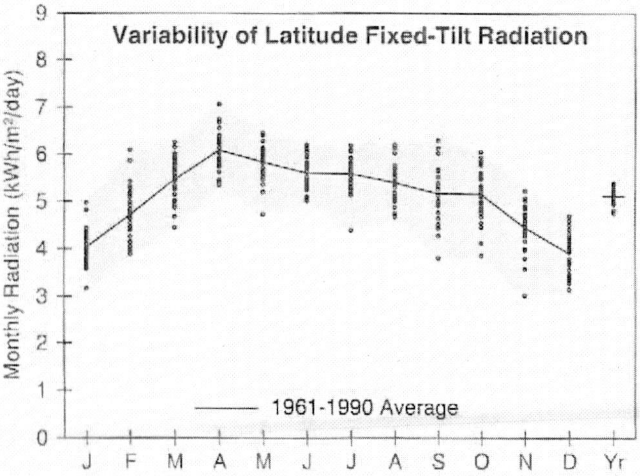

Solar Radiation for Flat-Plate Collectors Facing South at a Fixed Tilt (kWh/m²/day), Uncertainty ±9%

Tilt (°)		Jan	Feb	Mar	Apr	May	June	July	Aug	Sept	Oct	Nov	Dec	Year
0	Average	2.7	3.5	4.7	5.9	6.2	6.2	6.1	5.5	4.7	4.1	3.1	2.5	4.6
	Min/Max	2.3/3.2	3.1/4.3	3.9/5.3	5.2/6.8	5.0/6.8	5.5/6.9	4.8/6.8	4.8/6.3	3.7/5.6	3.2/4.6	2.3/3.5	2.2/2.9	4.3/4.8
Latitude -15	Average	3.5	4.3	5.3	6.2	6.2	6.1	6.0	5.6	5.1	4.8	3.9	3.4	5.0
	Min/Max	2.8/4.3	3.6/5.5	4.3/6.0	5.4/7.2	5.0/6.8	5.4/6.8	4.7/6.7	4.9/6.4	3.9/6.2	3.7/5.6	2.8/4.6	2.8/4.0	4.7/5.3
Latitude	Average	4.0	4.7	5.5	6.1	5.8	5.6	5.6	5.4	5.2	5.2	4.5	3.9	5.1
	Min/Max	3.2/5.0	3.9/6.1	4.4/6.3	5.3/7.1	4.7/6.5	5.0/6.2	4.4/6.2	4.7/6.2	3.8/6.3	3.9/6.1	3.0/5.2	3.2/4.7	4.8/5.4
Latitude +15	Average	4.3	4.9	5.4	5.7	5.2	4.9	4.9	4.9	5.0	5.2	4.7	4.2	4.9
	Min/Max	3.3/5.4	4.0/6.4	4.3/6.2	5.0/6.6	4.2/5.8	4.4/5.4	3.9/5.4	4.3/5.7	3.6/6.1	3.8/6.2	3.1/5.6	3.3/5.1	4.6/5.2
90	Average	3.7	3.9	3.6	3.1	2.4	2.1	2.2	2.5	3.1	3.9	4.0	3.8	3.2
	Min/Max	2.8/4.8	3.1/5.2	2.9/4.1	2.7/3.4	2.1/2.5	1.9/2.2	1.9/2.3	2.3/2.8	2.2/3.8	2.8/4.7	2.5/4.8	2.9/4.6	2.9/3.4

Solar Radiation for 1-Axis Tracking Flat-Plate Collectors with a North-South Axis (kWh/m²/day), Uncertainty ±9%

Axis Tilt (°)		Jan	Feb	Mar	Apr	May	June	July	Aug	Sept	Oct	Nov	Dec	Year
0	Average	3.8	4.9	6.3	7.8	7.9	7.7	7.5	6.8	6.0	5.5	4.3	3.5	6.0
	Min/Max	3.0/4.7	4.0/6.4	5.1/7.4	6.6/9.6	5.9/8.9	6.6/9.0	5.4/8.7	5.7/8.3	4.2/7.6	4.0/6.7	2.9/5.2	2.8/4.3	5.4/6.4
Latitude -15	Average	4.4	5.4	6.7	8.1	7.9	7.6	7.5	6.9	6.3	6.1	4.9	4.2	6.3
	Min/Max	3.4/5.5	4.4/7.3	5.4/8.0	6.8/9.9	5.9/8.9	6.5/8.9	5.4/8.7	5.7/8.4	4.3/8.1	4.3/7.4	3.3/6.0	3.3/5.1	5.7/6.7
Latitude	Average	4.8	5.8	6.9	8.0	7.7	7.3	7.2	6.8	6.4	6.3	5.3	4.6	6.4
	Min/Max	3.6/6.0	4.6/7.8	5.5/8.2	6.8/9.9	5.7/8.6	6.2/8.5	5.2/8.4	5.6/8.3	4.3/8.2	4.5/7.7	3.4/6.5	3.6/5.6	5.8/6.8
Latitude +15	Average	5.0	5.9	6.8	7.7	7.2	6.8	6.7	6.4	6.2	6.4	5.5	4.8	6.3
	Min/Max	3.7/6.3	4.7/8.0	5.4/8.1	6.5/9.5	5.3/8.1	5.8/7.9	4.8/7.8	5.3/7.8	4.1/8.0	4.5/7.8	3.5/6.7	3.7/6.0	5.7/6.7

Solar Radiation for 2-Axis Tracking Flat-Plate Collectors (kWh/m²/day), Uncertainty ±9%

Tracker		Jan	Feb	Mar	Apr	May	June	July	Aug	Sept	Oct	Nov	Dec	Year
2-Axis	Average	5.0	5.9	6.9	8.1	8.0	7.8	7.6	7.0	6.4	6.4	5.5	4.9	6.6
	Min/Max	3.8/6.4	4.7/8.0	5.5/8.2	6.8/10.0	5.9/9.0	6.6/9.1	5.5/8.8	5.8/8.4	4.3/8.2	4.5/7.8	3.5/6.8	3.8/6.1	6.0/7.1

Direct Beam Solar Radiation for Concentrating Collectors (kWh/m²/day), Uncertainty ±8%

Tracker		Jan	Feb	Mar	Apr	May	June	July	Aug	Sept	Oct	Nov	Dec	Year
1-Axis, E-W	Average	3.0	3.2	3.5	4.0	3.6	3.4	3.2	2.9	2.8	3.4	3.2	3.0	3.3
Horiz Axis	Min/Max	2.0/4.2	2.2/5.0	2.5/4.6	3.1/5.5	2.0/4.5	2.3/4.6	1.9/4.2	2.0/4.0	1.4/4.1	1.9/4.4	1.6/4.3	2.0/4.0	2.8/3.7
1-Axis, N-S	Average	2.5	3.2	4.2	5.3	4.8	4.4	4.2	3.7	3.4	3.6	2.8	2.3	3.7
Horiz Axis	Min/Max	1.6/3.4	2.2/5.0	3.0/5.6	4.0/7.4	2.6/6.0	3.0/6.0	2.4/5.6	2.6/5.3	1.6/5.2	2.0/4.8	1.4/3.9	1.5/3.1	3.1/4.1
1-Axis, N-S	Average	3.3	4.0	4.7	5.4	4.6	4.1	4.0	3.7	3.7	4.2	3.7	3.2	4.1
Tilt=Latitude	Min/Max	2.2/4.6	2.7/6.2	3.3/6.2	4.1/7.6	2.5/5.7	2.8/5.6	2.2/5.3	2.6/5.3	1.8/5.6	2.4/5.7	1.8/5.0	2.2/4.3	3.4/4.6
2-Axis	Average	3.6	4.1	4.7	5.5	4.9	4.5	4.2	3.8	3.7	4.3	3.9	3.5	4.2
	Min/Max	2.4/5.0	2.8/6.4	3.3/6.2	4.1/7.8	2.6/6.1	3.0/6.1	2.4/5.7	2.7/5.4	1.8/5.6	2.4/5.8	1.9/5.3	2.3/4.7	3.5/4.8

Average Climatic Conditions

Element	Jan	Feb	Mar	Apr	May	June	July	Aug	Sept	Oct	Nov	Dec	Year
Temperature (°C)	8.8	10.3	14.5	18.3	22.7	25.7	27.5	27.0	24.7	19.3	14.7	10.7	18.7
Daily Minimum Temp	3.2	4.4	8.6	12.2	17.2	20.6	22.6	22.3	19.9	13.5	8.4	4.8	13.2
Daily Maximum Temp	14.3	16.1	20.3	24.3	28.2	30.9	32.3	31.7	29.4	25.1	20.8	16.4	24.2
Record Minimum Temp	-14.4	-11.1	-9.4	-1.7	2.2	10.0	14.4	13.3	5.6	-2.8	-9.4	-13.3	-14.4
Record Maximum Temp	28.3	30.6	32.2	34.4	36.7	39.4	40.0	38.9	37.2	34.4	31.1	28.3	40.0
HDD, Base 18.3°C	304	230	133	37	0	0	0	0	0	41	129	244	1118
CDD, Base 18.3°C	8	4	14	35	134	222	284	269	190	72	19	6	1259
Relative Humidity (%)	70	67	68	68	72	75	77	79	78	74	73	72	73
Wind Speed (m/s)	3.9	4.3	4.4	4.3	3.8	3.7	3.5	3.3	3.4	3.6	3.6	3.7	3.8

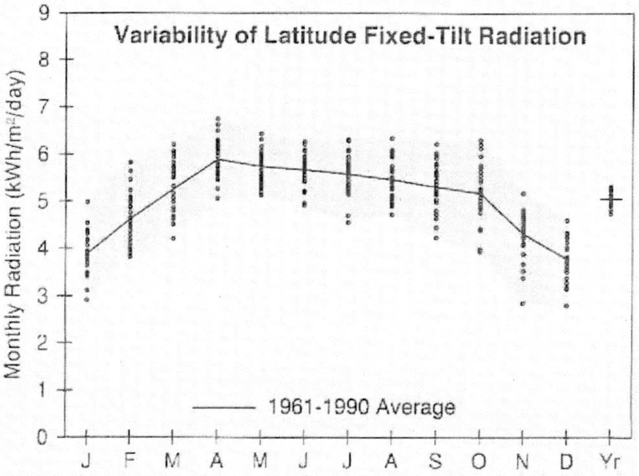

Columbia, SC

WBAN NO. 13883

LATITUDE: 33.95° N
LONGITUDE: 81.12° W
ELEVATION: 69 meters
MEAN PRESSURE: 1010 millibars

STATION TYPE: Secondary

Solar Radiation for Flat-Plate Collectors Facing South at a Fixed Tilt (kWh/m²/day), Uncertainty ±9%

Tilt (°)		Jan	Feb	Mar	Apr	May	June	July	Aug	Sept	Oct	Nov	Dec	Year
0	Average	2.6	3.4	4.5	5.7	6.1	6.3	6.1	5.5	4.8	4.0	2.9	2.4	4.5
	Min/Max	2.1/3.1	2.9/4.0	3.7/5.1	4.9/6.4	5.4/6.8	5.4/7.0	4.9/6.9	4.8/6.3	4.0/5.4	3.3/4.7	2.2/3.3	2.0/2.7	4.3/4.7
Latitude -15	Average	3.4	4.2	5.1	6.0	6.1	6.1	6.0	5.7	5.2	4.8	3.8	3.3	5.0
	Min/Max	2.7/4.3	3.5/5.2	4.1/5.9	5.2/6.8	5.4/6.8	5.3/6.8	4.9/6.8	4.9/6.6	4.3/6.1	3.8/5.8	2.7/4.5	2.5/3.9	4.7/5.2
Latitude	Average	3.9	4.6	5.3	5.9	5.7	5.7	5.6	5.5	5.3	5.2	4.3	3.8	5.1
	Min/Max	2.9/5.0	3.8/5.8	4.2/6.2	5.1/6.8	5.1/6.4	4.9/6.3	4.6/6.3	4.7/6.3	4.2/6.2	3.9/6.3	2.8/5.2	2.8/4.6	4.7/5.3
Latitude +15	Average	4.1	4.8	5.2	5.5	5.1	4.9	4.9	5.0	5.1	5.2	4.6	4.1	4.9
	Min/Max	3.0/5.4	3.9/6.1	4.1/6.1	4.7/6.3	4.6/5.7	4.3/5.4	4.0/5.5	4.3/5.8	4.0/6.0	3.9/6.4	2.9/5.5	2.9/5.0	4.5/5.2
90	Average	3.6	3.8	3.5	3.0	2.4	2.1	2.2	2.6	3.2	4.0	3.9	3.7	3.2
	Min/Max	2.6/4.8	3.0/5.0	2.8/4.3	2.7/3.4	2.2/2.6	2.0/2.2	2.0/2.4	2.4/2.9	2.5/3.8	2.9/4.9	2.4/4.8	2.5/4.6	2.9/3.4

Solar Radiation for 1-Axis Tracking Flat-Plate Collectors with a North-South Axis (kWh/m²/day), Uncertainty ±9%

Axis Tilt (°)		Jan	Feb	Mar	Apr	May	June	July	Aug	Sept	Oct	Nov	Dec	Year
0	Average	3.6	4.7	6.0	7.5	7.7	7.8	7.5	6.9	6.2	5.5	4.1	3.4	5.9
	Min/Max	2.7/4.6	3.8/6.1	4.7/7.2	6.3/9.0	6.5/8.8	6.4/9.1	5.6/8.8	5.7/8.3	4.8/7.4	4.2/6.9	2.7/4.9	2.4/4.1	5.5/6.2
Latitude -15	Average	4.2	5.3	6.5	7.8	7.8	7.8	7.5	7.1	6.5	6.1	4.8	4.0	6.3
	Min/Max	3.1/5.5	4.2/7.0	5.0/7.8	6.5/9.4	6.5/8.9	6.3/9.0	5.6/8.8	5.8/8.5	5.0/7.9	4.5/7.7	3.0/5.8	2.8/5.0	5.8/6.6
Latitude	Average	4.6	5.6	6.6	7.8	7.5	7.4	7.2	6.9	6.6	6.4	5.2	4.4	6.4
	Min/Max	3.3/6.1	4.4/7.5	5.1/8.0	6.4/9.3	6.3/8.6	6.1/8.6	5.4/8.5	5.7/8.3	5.0/8.0	4.7/8.1	3.2/6.3	3.1/5.5	5.8/6.8
Latitude +15	Average	4.8	5.7	6.5	7.5	7.1	6.9	6.7	6.6	6.4	6.4	5.3	4.7	6.2
	Min/Max	3.4/6.4	4.5/7.6	5.0/8.0	6.2/9.0	5.9/8.1	5.6/8.0	5.0/7.9	5.4/7.9	4.8/7.8	4.7/8.1	3.2/6.6	3.2/5.8	5.7/6.6

Solar Radiation for 2-Axis Tracking Flat-Plate Collectors (kWh/m²/day), Uncertainty ±9%

Tracker		Jan	Feb	Mar	Apr	May	June	July	Aug	Sept	Oct	Nov	Dec	Year
2-Axis	Average	4.8	5.8	6.6	7.8	7.8	7.9	7.6	7.1	6.6	6.4	5.4	4.7	6.6
	Min/Max	3.5/6.4	4.5/7.7	5.1/8.1	6.5/9.4	6.6/8.9	6.5/9.2	5.7/8.9	5.8/8.5	5.0/8.0	4.7/8.1	3.3/6.6	3.3/5.9	6.0/7.0

Direct Beam Solar Radiation for Concentrating Collectors (kWh/m²/day), Uncertainty ±8%

Tracker		Jan	Feb	Mar	Apr	May	June	July	Aug	Sept	Oct	Nov	Dec	Year
1-Axis, E-W Horiz Axis	Average	2.9	3.2	3.3	3.8	3.6	3.6	3.4	3.1	3.0	3.5	3.1	2.9	3.3
	Min/Max	1.8/4.2	2.2/4.8	2.2/4.4	2.7/5.1	2.6/4.6	2.5/4.8	2.0/4.6	2.2/4.2	2.0/4.1	2.0/4.9	1.4/4.2	1.6/3.9	2.8/3.7
1-Axis, N-S Horiz Axis	Average	2.3	3.1	3.9	5.0	4.8	4.7	4.3	4.0	3.7	3.6	2.7	2.2	3.7
	Min/Max	1.4/3.5	2.1/4.8	2.6/5.2	3.6/6.8	3.3/6.0	3.2/6.2	2.5/5.9	2.8/5.5	2.4/5.0	2.2/5.2	1.2/3.7	1.2/3.0	3.2/4.1
1-Axis, N-S Tilt=Latitude	Average	3.2	3.9	4.4	5.2	4.6	4.4	4.1	4.0	4.1	4.4	3.6	3.1	4.1
	Min/Max	2.0/4.7	2.6/5.9	2.9/5.9	3.7/7.0	3.2/5.8	3.0/5.8	2.4/5.6	2.8/5.5	2.6/5.4	2.6/6.2	1.6/4.9	1.7/4.3	3.5/4.5
2-Axis	Average	3.4	4.0	4.4	5.3	4.9	4.7	4.4	4.2	4.1	4.4	3.8	3.4	4.2
	Min/Max	2.1/5.1	2.7/6.1	3.0/5.9	3.8/7.1	3.3/6.1	3.3/6.3	2.5/6.0	2.9/5.7	2.6/5.4	2.6/6.3	1.7/5.1	1.9/4.6	3.6/4.7

Average Climatic Conditions

Element	Jan	Feb	Mar	Apr	May	June	July	Aug	Sept	Oct	Nov	Dec	Year
Temperature (°C)	6.6	8.2	12.9	17.2	21.6	25.2	27.1	26.5	23.4	17.4	12.6	8.3	17.3
Daily Minimum Temp	0.1	1.2	5.7	9.7	14.6	18.9	21.1	20.7	17.3	10.1	5.3	1.6	10.5
Daily Maximum Temp	12.9	15.2	20.1	24.7	28.6	31.6	33.1	32.3	29.5	24.6	19.9	14.9	23.9
Record Minimum Temp	-18.3	-15.0	-15.6	-3.3	1.1	6.7	12.2	11.7	4.4	-5.0	-11.1	-15.6	-18.3
Record Maximum Temp	28.9	28.9	32.8	34.4	38.3	41.7	41.7	41.7	38.3	38.3	32.2	28.3	41.7
HDD, Base 18.3°C	369	286	176	60	10	0	0	0	3	77	179	312	1472
CDD, Base 18.3°C	4	0	7	27	112	207	272	253	156	48	6	0	1092
Relative Humidity (%)	69	66	65	62	68	71	73	76	76	73	72	71	70
Wind Speed (m/s)	3.2	3.4	3.6	3.6	3.1	2.9	2.7	2.5	2.7	2.7	2.8	3.1	3.0

Greenville, SC

WBAN NO. 03870

LATITUDE: 34.90° N
LONGITUDE: 82.22° W
ELEVATION: 296 meters
MEAN PRESSURE: 983 millibars

STATION TYPE: Secondary

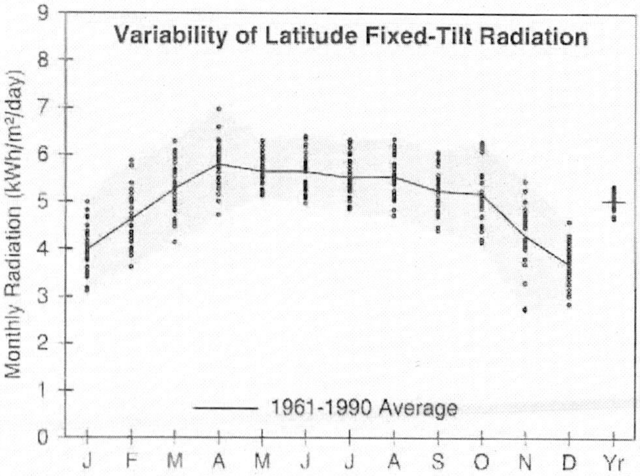

Solar Radiation for Flat-Plate Collectors Facing South at a Fixed Tilt (kWh/m²/day), Uncertainty ±9%

Tilt (°)		Jan	Feb	Mar	Apr	May	June	July	Aug	Sept	Oct	Nov	Dec	Year
0	Average	2.6	3.3	4.4	5.6	6.0	6.3	6.0	5.5	4.7	3.9	2.8	2.3	4.5
	Min/Max	2.2/3.0	2.8/4.0	3.7/5.1	4.6/6.6	5.5/6.7	5.6/7.2	5.3/6.8	4.8/6.3	4.1/5.3	3.3/4.6	2.1/3.4	1.9/2.6	4.2/4.7
Latitude -15	Average	3.5	4.2	5.1	5.9	6.0	6.1	5.9	5.7	5.2	4.8	3.8	3.2	5.0
	Min/Max	2.8/4.3	3.4/5.3	4.1/6.0	4.8/7.0	5.5/6.7	5.4/7.0	5.2/6.8	4.9/6.6	4.4/5.9	3.9/5.8	2.5/4.8	2.5/3.9	4.6/5.2
Latitude	Average	4.0	4.6	5.3	5.8	5.6	5.6	5.5	5.5	5.2	5.2	4.3	3.7	5.0
	Min/Max	3.1/5.0	3.6/5.9	4.1/6.3	4.7/7.0	5.1/6.3	5.0/6.4	4.8/6.3	4.7/6.3	4.4/6.1	4.2/6.3	2.7/5.4	2.8/4.6	4.7/5.3
Latitude +15	Average	4.2	4.8	5.2	5.4	5.0	4.9	4.8	5.1	5.0	5.2	4.5	4.0	4.8
	Min/Max	3.3/5.4	3.7/6.1	4.0/6.2	4.4/6.5	4.6/5.6	4.3/5.5	4.3/5.5	4.3/5.8	4.2/5.8	4.1/6.4	2.8/5.8	3.0/5.0	4.5/5.2
90	Average	3.8	3.9	3.6	3.1	2.4	2.2	2.2	2.7	3.2	4.0	3.9	3.6	3.2
	Min/Max	2.8/4.9	3.0/5.1	2.8/4.4	2.5/3.6	2.2/2.6	2.0/2.3	2.1/2.5	2.4/3.0	2.7/3.8	3.1/5.0	2.3/5.1	2.7/4.6	2.9/3.5

Solar Radiation for 1-Axis Tracking Flat-Plate Collectors with a North-South Axis (kWh/m²/day), Uncertainty ±9%

Axis Tilt (°)		Jan	Feb	Mar	Apr	May	June	July	Aug	Sept	Oct	Nov	Dec	Year
0	Average	3.7	4.7	6.0	7.4	7.6	7.9	7.5	7.1	6.1	5.5	4.0	3.3	5.9
	Min/Max	2.8/4.5	3.6/6.1	4.6/7.2	6.0/9.2	6.7/8.8	6.8/9.4	6.3/8.9	5.7/8.4	5.0/7.3	4.3/6.7	2.6/5.1	2.5/4.0	5.4/6.3
Latitude -15	Average	4.3	5.4	6.5	7.7	7.7	7.8	7.5	7.3	6.5	6.1	4.7	3.9	6.3
	Min/Max	3.3/5.5	4.0/7.0	4.9/7.9	6.2/9.6	6.7/8.9	6.7/9.4	6.3/8.9	5.8/8.6	5.3/7.8	4.8/7.6	2.9/6.1	2.9/4.9	5.8/6.7
Latitude	Average	4.7	5.7	6.7	7.7	7.4	7.5	7.2	7.1	6.6	6.4	5.1	4.3	6.4
	Min/Max	3.6/6.0	4.3/7.5	5.0/8.1	6.1/9.6	6.5/8.6	6.4/9.0	6.0/8.6	5.7/8.5	5.3/7.9	4.9/8.0	3.1/6.6	3.2/5.5	5.8/6.8
Latitude +15	Average	5.0	5.8	6.6	7.4	7.0	7.0	6.7	6.8	6.4	6.5	5.3	4.6	6.2
	Min/Max	3.7/6.4	4.3/7.7	4.9/8.1	5.9/9.2	6.1/8.1	6.0/8.3	5.6/8.0	5.3/8.1	5.1/7.7	4.9/8.1	3.1/6.9	3.3/5.8	5.7/6.7

Solar Radiation for 2-Axis Tracking Flat-Plate Collectors (kWh/m²/day), Uncertainty ±9%

Tracker		Jan	Feb	Mar	Apr	May	June	July	Aug	Sept	Oct	Nov	Dec	Year
2-Axis	Average	5.0	5.8	6.7	7.8	7.8	8.0	7.6	7.3	6.6	6.5	5.4	4.6	6.6
	Min/Max	3.7/6.4	4.3/7.7	5.0/8.1	6.2/9.7	6.8/9.0	6.8/9.6	6.4/9.1	5.8/8.7	5.3/7.9	5.0/8.1	3.2/6.9	3.4/5.9	6.0/7.0

Direct Beam Solar Radiation for Concentrating Collectors (kWh/m²/day), Uncertainty ±8%

Tracker		Jan	Feb	Mar	Apr	May	June	July	Aug	Sept	Oct	Nov	Dec	Year
1-Axis, E-W Horiz Axis	Average	3.1	3.3	3.4	3.9	3.7	3.7	3.5	3.3	3.1	3.6	3.2	2.9	3.4
	Min/Max	1.9/4.3	2.1/4.8	2.0/4.5	2.7/5.3	2.9/4.6	2.7/5.2	2.5/4.7	2.2/4.4	2.1/4.1	2.4/4.8	1.5/4.5	1.8/4.1	2.9/3.8
1-Axis, N-S Horiz Axis	Average	2.5	3.2	4.0	5.1	4.8	4.8	4.5	4.3	3.8	3.8	2.8	2.2	3.8
	Min/Max	1.6/3.4	2.0/4.8	2.4/5.3	3.6/7.0	3.8/6.1	3.6/6.8	3.3/6.1	2.8/5.8	2.6/5.1	2.5/5.1	1.3/3.8	1.3/3.1	3.3/4.3
1-Axis, N-S Tilt=Latitude	Average	3.4	4.1	4.6	5.2	4.6	4.5	4.2	4.3	4.1	4.6	3.7	3.1	4.2
	Min/Max	2.2/4.8	2.6/6.0	2.8/6.0	3.7/7.3	3.7/5.9	3.3/6.3	3.1/5.8	2.8/5.8	2.9/5.5	3.0/6.2	1.7/5.1	1.9/4.4	3.7/4.7
2-Axis	Average	3.6	4.2	4.6	5.3	4.9	4.9	4.6	4.5	4.2	4.6	3.9	3.4	4.4
	Min/Max	2.3/5.1	2.7/6.2	2.8/6.1	3.8/7.3	3.9/6.2	3.7/6.9	3.3/6.2	2.8/6.0	2.9/5.5	3.0/6.2	1.8/5.4	2.1/4.8	3.8/4.9

Average Climatic Conditions

Element	Jan	Feb	Mar	Apr	May	June	July	Aug	Sept	Oct	Nov	Dec	Year
Temperature (°C)	4.5	6.3	10.9	15.4	19.9	23.9	25.7	25.1	21.8	15.8	10.9	6.3	15.6
Daily Minimum Temp	-1.1	0.2	4.4	8.7	13.6	18.0	20.2	19.7	16.2	9.4	4.7	0.8	9.6
Daily Maximum Temp	10.1	12.4	17.5	22.2	26.3	29.7	31.2	30.4	27.3	22.2	16.9	11.8	21.5
Record Minimum Temp	-21.1	-13.3	-11.7	-3.9	-0.6	4.4	12.2	11.1	2.2	-3.9	-11.1	-15.0	-21.1
Record Maximum Temp	26.1	26.1	31.1	33.9	36.1	37.8	39.4	39.4	35.6	33.3	29.4	24.4	39.4
HDD, Base 18.3°C	429	336	229	98	27	0	0	0	4	97	223	374	1818
CDD, Base 18.3°C	0	0	0	12	77	167	227	208	107	20	0	0	818
Relative Humidity (%)	66	63	62	60	68	71	74	76	75	70	68	68	68
Wind Speed (m/s)	3.4	3.7	3.8	3.8	3.2	3.0	2.8	2.7	2.9	3.0	3.2	3.3	3.2

Huron, SD

WBAN NO. 14936

LATITUDE: 44.38° N
LONGITUDE: 98.22° W
ELEVATION: 393 meters
MEAN PRESSURE: 969 millibars

STATION TYPE: Secondary

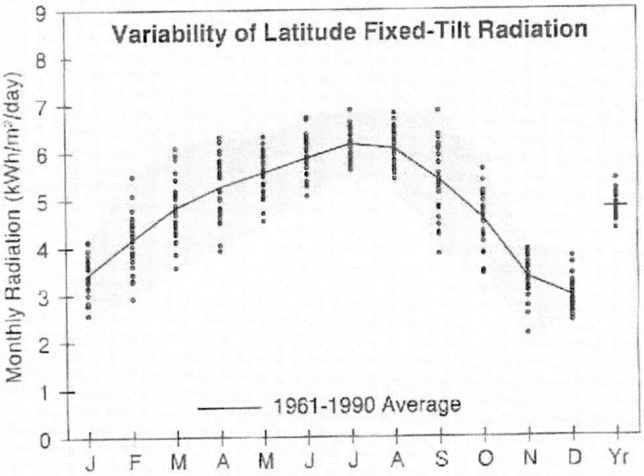

Solar Radiation for Flat-Plate Collectors Facing South at a Fixed Tilt (kWh/m²/day), Uncertainty ±9%

Tilt (°)		Jan	Feb	Mar	Apr	May	June	July	Aug	Sept	Oct	Nov	Dec	Year
0	Average	1.8	2.6	3.7	4.9	5.8	6.5	6.6	5.8	4.4	3.0	1.9	1.5	4.1
	Min/Max	1.5/2.0	2.1/3.2	3.0/4.4	3.8/5.7	4.8/6.6	5.7/7.4	6.1/7.4	5.3/6.5	3.5/5.3	2.5/3.5	1.5/2.0	1.3/1.6	3.8/4.4
Latitude -15	Average	3.0	3.8	4.7	5.4	6.0	6.4	6.7	6.3	5.3	4.3	3.0	2.6	4.8
	Min/Max	2.3/3.6	2.8/4.9	3.5/5.8	4.1/6.4	4.9/6.8	5.5/7.4	6.1/7.5	5.7/7.1	3.9/6.7	3.3/5.2	2.0/3.5	2.2/3.2	4.4/5.4
Latitude	Average	3.4	4.2	4.8	5.3	5.6	5.9	6.2	6.1	5.4	4.6	3.3	3.0	4.8
	Min/Max	2.6/4.1	2.9/5.5	3.6/6.1	3.9/6.3	4.5/6.3	5.1/6.7	5.6/6.9	5.4/6.8	3.8/6.9	3.4/5.6	2.2/3.9	2.4/3.8	4.3/5.4
Latitude +15	Average	3.7	4.3	4.8	4.9	4.9	5.1	5.4	5.5	5.2	4.6	3.5	3.2	4.6
	Min/Max	2.7/4.5	2.9/5.8	3.5/6.0	3.7/5.9	4.0/5.6	4.4/5.8	4.9/6.0	4.9/6.2	3.6/6.6	3.4/5.7	2.2/4.2	2.6/4.1	4.1/5.2
90	Average	3.6	3.9	3.9	3.4	3.0	2.9	3.1	3.5	3.8	3.9	3.2	3.1	3.4
	Min/Max	2.5/4.4	2.5/5.4	2.8/5.2	2.6/4.0	2.6/3.3	2.6/3.2	2.9/3.4	3.2/3.9	2.6/4.9	2.8/4.9	2.0/3.9	2.4/4.0	3.0/3.9

Solar Radiation for 1-Axis Tracking Flat-Plate Collectors with a North-South Axis (kWh/m²/day), Uncertainty ±9%

Axis Tilt (°)		Jan	Feb	Mar	Apr	May	June	July	Aug	Sept	Oct	Nov	Dec	Year
0	Average	2.7	3.8	5.1	6.5	7.8	8.7	9.0	8.1	6.3	4.4	2.8	2.2	5.6
	Min/Max	2.0/3.2	2.6/5.1	3.7/6.5	4.6/8.3	6.0/9.1	7.2/10.3	8.0/10.4	7.2/9.5	4.4/8.2	3.3/5.4	1.9/3.2	1.9/2.8	5.1/6.4
Latitude -15	Average	3.6	4.7	5.9	7.0	8.0	8.8	9.2	8.6	7.0	5.4	3.6	3.0	6.2
	Min/Max	2.6/4.4	3.1/6.4	4.1/7.6	4.8/8.9	6.1/9.4	7.2/10.4	8.1/10.6	7.5/10.1	4.7/9.3	3.9/6.7	2.2/4.3	2.5/4.0	5.5/7.1
Latitude	Average	4.0	5.0	6.0	6.9	7.7	8.5	8.9	8.4	7.1	5.6	3.9	3.4	6.3
	Min/Max	2.8/4.8	3.3/6.8	4.2/7.8	4.7/8.9	5.9/9.1	6.9/10.0	7.9/10.3	7.4/9.9	4.7/9.4	4.0/7.0	2.4/4.6	2.7/4.4	5.5/7.2
Latitude +15	Average	4.2	5.1	6.0	6.7	7.3	7.9	8.3	8.0	6.9	5.6	4.0	3.5	6.1
	Min/Max	2.9/5.1	3.3/7.0	4.1/7.8	4.5/8.5	5.5/8.6	6.5/9.3	7.4/9.7	7.0/9.4	4.5/9.2	4.0/7.1	2.4/4.8	2.8/4.7	5.4/7.1

Solar Radiation for 2-Axis Tracking Flat-Plate Collectors (kWh/m²/day), Uncertainty ±9%

Tracker		Jan	Feb	Mar	Apr	May	June	July	Aug	Sept	Oct	Nov	Dec	Year
2-Axis	Average	4.2	5.1	6.0	7.0	8.1	9.0	9.4	8.6	7.1	5.7	4.0	3.6	6.5
	Min/Max	3.0/5.2	3.3/7.0	4.2/7.9	4.8/9.0	6.2/9.5	7.4/10.6	8.3/10.9	7.6/10.1	4.7/9.4	4.1/7.1	2.4/4.8	2.9/4.7	5.7/7.5

Direct Beam Solar Radiation for Concentrating Collectors (kWh/m²/day), Uncertainty ±8%

Tracker		Jan	Feb	Mar	Apr	May	June	July	Aug	Sept	Oct	Nov	Dec	Year
1-Axis, E-W Horiz Axis	Average	2.4	2.6	2.7	3.2	3.7	4.4	4.8	4.4	3.6	3.1	2.3	2.1	3.3
	Min/Max	1.4/3.2	1.4/4.0	1.5/4.5	1.4/4.7	2.2/4.7	3.0/5.8	3.9/5.9	3.5/5.5	1.8/5.4	1.8/4.3	1.0/3.0	1.4/3.2	2.7/4.1
1-Axis, N-S Horiz Axis	Average	1.6	2.2	3.0	4.1	4.9	5.8	6.3	5.7	4.3	3.0	1.7	1.3	3.7
	Min/Max	0.9/2.1	1.1/3.5	1.5/4.8	1.8/6.1	3.0/6.3	4.0/7.6	5.2/7.9	4.6/7.3	2.2/6.5	1.8/4.0	0.7/2.3	0.9/2.0	3.0/4.5
1-Axis, N-S Tilt=Latitude	Average	2.6	3.1	3.7	4.4	4.9	5.6	6.1	5.9	5.0	3.9	2.6	2.2	4.2
	Min/Max	1.5/3.5	1.6/4.8	1.9/6.0	2.0/6.6	3.0/6.3	3.8/7.3	5.1/7.6	4.7/7.6	2.5/7.5	2.3/5.4	1.1/3.5	1.5/3.4	3.4/5.2
2-Axis	Average	2.8	3.2	3.7	4.5	5.2	6.0	6.6	6.1	5.0	4.0	2.7	2.4	4.4
	Min/Max	1.6/3.7	1.6/5.0	1.9/6.0	2.0/6.7	3.1/6.6	4.1/7.9	5.4/8.2	4.9/7.8	2.5/7.5	2.4/5.4	1.1/3.6	1.6/3.7	3.5/5.4

Average Climatic Conditions

Element	Jan	Feb	Mar	Apr	May	June	July	Aug	Sept	Oct	Nov	Dec	Year
Temperature (°C)	-10.4	-7.0	-0.1	7.9	14.2	19.9	23.6	22.1	16.0	9.2	0.2	-7.7	7.3
Daily Minimum Temp	-16.5	-12.7	-5.7	1.1	7.1	13.1	16.5	14.9	8.5	1.9	-5.7	-13.4	0.8
Daily Maximum Temp	-4.4	-1.3	5.6	14.8	21.3	26.8	30.6	29.3	23.4	16.4	6.1	-2.1	13.9
Record Minimum Temp	-38.3	-39.4	-31.1	-18.9	-8.3	0.0	2.8	2.2	-7.2	-12.8	-29.4	-34.4	-39.4
Record Maximum Temp	17.2	21.7	31.7	36.1	37.2	42.8	44.4	43.3	41.1	38.9	25.6	16.7	44.4
HDD, Base 18.3°C	892	709	570	312	142	27	6	12	96	284	543	808	4402
CDD, Base 18.3°C	0	0	0	0	15	75	168	129	26	0	0	0	413
Relative Humidity (%)	71	74	73	66	66	68	66	67	68	67	73	74	69
Wind Speed (m/s)	5.1	4.9	5.5	5.6	5.2	4.7	4.3	4.4	4.7	4.8	4.9	4.8	4.9

Pierre, SD

WBAN NO. 24025

LATITUDE: 44.38° N
LONGITUDE: 100.28° W
ELEVATION: 526 meters
MEAN PRESSURE: 953 millibars

STATION TYPE: Secondary

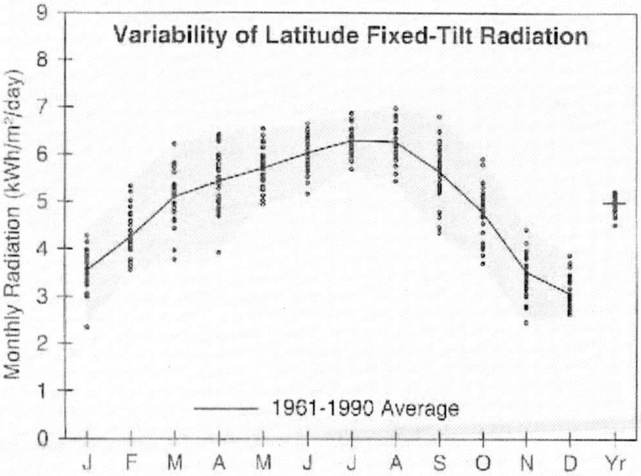

Solar Radiation for Flat-Plate Collectors Facing South at a Fixed Tilt (kWh/m²/day), Uncertainty ±9%

Tilt (°)		Jan	Feb	Mar	Apr	May	June	July	Aug	Sept	Oct	Nov	Dec	Year
0	Average	1.8	2.7	3.9	5.0	6.0	6.7	6.8	6.0	4.5	3.1	2.0	1.5	4.2
	Min/Max	1.4/2.0	2.4/3.0	3.2/4.4	3.9/5.8	5.2/6.8	5.8/7.4	6.2/7.4	5.3/6.6	3.8/5.3	2.6/3.6	1.6/2.2	1.4/1.7	4.0/4.3
Latitude -15	Average	3.1	3.9	4.9	5.6	6.1	6.6	6.8	6.5	5.6	4.5	3.2	2.7	4.9
	Min/Max	2.1/3.7	3.3/4.8	3.7/5.9	4.1/6.5	5.3/7.0	5.6/7.3	6.2/7.4	5.7/7.3	4.4/6.7	3.5/5.4	2.3/3.9	2.3/3.3	4.6/5.2
Latitude	Average	3.6	4.3	5.1	5.4	5.7	6.0	6.3	6.3	5.6	4.8	3.5	3.1	5.0
	Min/Max	2.3/4.3	3.5/5.3	3.8/6.2	3.9/6.4	4.9/6.6	5.2/6.7	5.7/6.9	5.4/7.0	4.3/6.8	3.7/5.9	2.5/4.4	2.6/3.9	4.5/5.2
Latitude +15	Average	3.8	4.4	5.0	5.0	5.1	5.2	5.5	5.7	5.4	4.8	3.7	3.3	4.7
	Min/Max	2.4/4.7	3.6/5.6	3.7/6.2	3.6/5.9	4.4/5.8	4.5/5.7	5.0/6.0	4.9/6.3	4.1/6.6	3.7/6.0	2.5/4.7	2.8/4.2	4.3/5.0
90	Average	3.7	4.0	4.1	3.5	3.1	2.9	3.1	3.6	4.0	4.1	3.4	3.2	3.6
	Min/Max	2.2/4.6	3.2/5.1	2.9/5.0	2.5/4.1	2.7/3.5	2.6/3.1	2.9/3.3	3.2/4.0	3.0/4.9	3.1/5.1	2.3/4.3	2.7/4.2	3.1/3.8

Solar Radiation for 1-Axis Tracking Flat-Plate Collectors with a North-South Axis (kWh/m²/day), Uncertainty ±9%

Axis Tilt (°)		Jan	Feb	Mar	Apr	May	June	July	Aug	Sept	Oct	Nov	Dec	Year
0	Average	2.8	3.9	5.4	6.8	8.0	9.0	9.3	8.4	6.6	4.7	2.9	2.3	5.9
	Min/Max	1.8/3.3	3.2/4.8	3.9/6.7	4.6/8.4	6.5/9.3	7.4/10.3	8.2/10.3	7.3/9.7	4.9/8.1	3.5/5.7	2.1/3.6	2.0/2.9	5.3/6.2
Latitude -15	Average	3.7	4.8	6.2	7.3	8.2	9.0	9.5	8.9	7.4	5.7	3.8	3.2	6.5
	Min/Max	2.4/4.6	4.0/6.1	4.4/7.8	4.8/9.1	6.7/9.6	7.4/10.4	8.3/10.6	7.7/10.3	5.5/9.2	4.2/7.0	2.6/4.8	2.7/4.0	5.8/6.8
Latitude	Average	4.1	5.1	6.4	7.2	8.0	8.7	9.1	8.8	7.5	5.9	4.1	3.5	6.5
	Min/Max	2.6/5.1	4.2/6.5	4.4/8.1	4.7/9.0	6.5/9.4	7.1/10.0	8.0/10.2	7.5/10.1	5.5/9.3	4.4/7.4	2.7/5.3	2.9/4.5	5.8/6.9
Latitude +15	Average	4.3	5.2	6.3	7.0	7.5	8.1	8.6	8.4	7.3	6.0	4.2	3.7	6.4
	Min/Max	2.7/5.3	4.2/6.7	4.3/8.0	4.4/8.7	6.1/8.9	6.7/9.4	7.5/9.6	7.1/9.7	5.3/9.1	4.4/7.5	2.8/5.5	3.1/4.8	5.6/6.8

Solar Radiation for 2-Axis Tracking Flat-Plate Collectors (kWh/m²/day), Uncertainty ±9%

Tracker		Jan	Feb	Mar	Apr	May	June	July	Aug	Sept	Oct	Nov	Dec	Year
2-Axis	Average	4.4	5.2	6.4	7.4	8.3	9.3	9.7	9.0	7.5	6.0	4.3	3.8	6.8
	Min/Max	2.7/5.4	4.3/6.7	4.5/8.1	4.8/9.1	6.8/9.8	7.6/10.7	8.5/10.8	7.7/10.4	5.5/9.3	4.4/7.5	2.8/5.5	3.1/4.9	6.0/7.2

Direct Beam Solar Radiation for Concentrating Collectors (kWh/m²/day), Uncertainty ±8%

Tracker		Jan	Feb	Mar	Apr	May	June	July	Aug	Sept	Oct	Nov	Dec	Year
1-Axis, E-W	Average	2.5	2.7	3.0	3.4	3.9	4.7	5.0	4.6	3.9	3.4	2.5	2.2	3.5
Horiz Axis	Min/Max	1.4/3.4	1.5/4.0	1.7/4.6	1.4/4.8	2.7/5.0	3.2/5.9	4.0/6.1	3.6/5.8	2.4/5.4	2.2/4.6	1.3/3.7	1.5/3.3	2.9/3.9
1-Axis, N-S	Average	1.7	2.3	3.3	4.3	5.2	6.1	6.6	6.1	4.7	3.2	1.8	1.4	3.9
Horiz Axis	Min/Max	0.8/2.2	1.2/3.4	1.8/5.0	1.8/6.2	3.6/6.7	4.3/7.8	5.3/8.0	4.8/7.6	2.8/6.4	2.0/4.4	0.9/2.7	0.9/2.0	3.3/4.3
1-Axis, N-S	Average	2.7	3.2	4.1	4.7	5.1	5.8	6.4	6.3	5.4	4.3	2.8	2.3	4.4
Tilt=Latitude	Min/Max	1.4/3.7	1.7/4.8	2.2/6.2	2.0/6.7	3.6/6.6	4.1/7.4	5.1/7.7	4.9/7.9	3.3/7.4	2.7/5.8	1.5/4.1	1.6/3.4	3.7/5.0
2-Axis	Average	2.9	3.3	4.1	4.8	5.4	6.4	6.9	6.5	5.4	4.3	2.9	2.5	4.6
	Min/Max	1.5/3.9	1.8/4.9	2.2/6.2	2.0/6.8	3.8/7.0	4.4/8.1	5.5/8.3	5.1/8.1	3.3/7.5	2.8/5.9	1.5/4.4	1.7/3.7	3.8/5.2

Average Climatic Conditions

Element	Jan	Feb	Mar	Apr	May	June	July	Aug	Sept	Oct	Nov	Dec	Year
Temperature (°C)	-8.4	-5.3	0.8	8.4	14.6	20.4	24.4	23.3	16.9	10.0	1.2	-6.2	8.3
Daily Minimum Temp	-14.1	-11.1	-5.2	1.6	7.6	13.2	16.7	15.5	9.2	2.8	-4.7	-11.6	1.7
Daily Maximum Temp	-2.7	0.4	6.7	15.2	21.6	27.5	32.2	31.1	24.6	17.2	7.0	-0.7	15.0
Record Minimum Temp	-35.6	-31.7	-26.1	-15.0	-5.6	3.3	5.6	4.4	-3.9	-10.0	-25.0	-33.9	-35.6
Record Maximum Temp	18.9	22.8	30.0	36.1	40.6	44.4	45.0	45.0	42.2	34.4	25.6	18.3	45.0
HDD, Base 18.3°C	828	661	544	298	132	24	4	9	82	258	515	759	4117
CDD, Base 18.3°C	0	0	0	0	17	86	193	163	39	0	0	0	498
Relative Humidity (%)	70	71	69	61	61	63	58	57	59	60	68	72	64
Wind Speed (m/s)	5.5	5.2	5.8	5.9	5.6	4.9	4.8	4.9	5.1	5.0	5.0	5.2	5.2

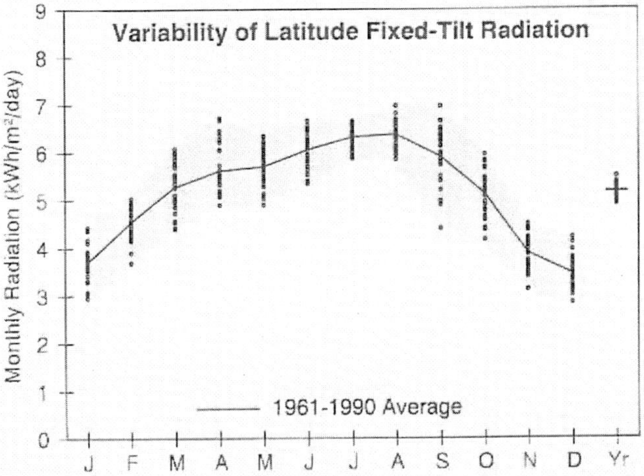

Rapid City, SD

WBAN NO. 24090

LATITUDE: 44.05° N
LONGITUDE: 103.07° W
ELEVATION: 966 meters
MEAN PRESSURE: 904 millibars

STATION TYPE: Secondary

Solar Radiation for Flat-Plate Collectors Facing South at a Fixed Tilt (kWh/m²/day), Uncertainty ±9%

Tilt (°)		Jan	Feb	Mar	Apr	May	June	July	Aug	Sept	Oct	Nov	Dec	Year
0	Average	1.9	2.8	4.0	5.1	6.0	6.7	6.8	6.1	4.7	3.3	2.1	1.6	4.3
	Min/Max	1.6/2.1	2.5/3.0	3.5/4.4	4.6/5.9	5.1/6.6	6.0/7.5	6.4/7.2	5.7/6.6	3.7/5.4	2.8/3.7	1.8/2.3	1.5/1.8	4.1/4.5
Latitude -15	Average	3.2	4.1	5.1	5.7	6.1	6.6	6.8	6.6	5.8	4.7	3.4	3.0	5.1
	Min/Max	2.6/3.8	3.4/4.5	4.3/5.8	5.0/6.8	5.2/6.8	5.8/7.3	6.3/7.2	6.1/7.2	4.4/6.8	3.9/5.5	2.8/3.9	2.5/3.6	4.9/5.4
Latitude	Average	3.7	4.5	5.3	5.6	5.7	6.0	6.3	6.4	5.9	5.1	3.9	3.4	5.2
	Min/Max	3.0/4.4	3.7/5.0	4.4/6.1	4.9/6.7	4.9/6.3	5.3/6.7	5.9/6.6	5.8/7.0	4.4/7.0	4.2/5.9	3.1/4.5	2.8/4.2	4.9/5.5
Latitude +15	Average	4.0	4.7	5.2	5.2	5.0	5.2	5.5	5.8	5.7	5.2	4.1	3.7	4.9
	Min/Max	3.1/4.8	3.8/5.3	4.3/6.0	4.5/6.2	4.3/5.6	4.6/5.7	5.1/5.7	5.3/6.3	4.2/6.7	4.2/6.1	3.3/4.8	3.0/4.6	4.7/5.2
90	Average	3.8	4.3	4.2	3.6	3.0	2.9	3.1	3.7	4.2	4.4	3.8	3.6	3.7
	Min/Max	2.9/4.7	3.4/4.9	3.5/5.0	3.1/4.5	2.7/3.3	2.6/3.1	2.9/3.2	3.4/4.0	3.1/5.0	3.5/5.2	3.0/4.4	3.0/4.4	3.5/3.9

Solar Radiation for 1-Axis Tracking Flat-Plate Collectors with a North-South Axis (kWh/m²/day), Uncertainty ±9%

Axis Tilt (°)		Jan	Feb	Mar	Apr	May	June	July	Aug	Sept	Oct	Nov	Dec	Year
0	Average	2.9	4.1	5.7	7.0	8.0	9.1	9.4	8.7	6.9	5.0	3.2	2.6	6.1
	Min/Max	2.3/3.5	3.3/4.7	4.6/6.6	5.9/8.8	6.5/9.3	7.8/10.6	8.6/10.1	7.8/9.7	5.0/8.4	4.0/5.8	2.6/3.7	2.2/3.1	5.7/6.6
Latitude -15	Average	3.9	5.2	6.5	7.6	8.3	9.2	9.5	9.2	7.8	6.1	4.2	3.6	6.8
	Min/Max	3.1/4.7	4.1/5.8	5.2/7.6	6.3/9.5	6.7/9.6	7.8/10.7	8.8/10.3	8.3/10.3	5.5/9.5	4.8/7.1	3.3/4.9	2.9/4.4	6.3/7.3
Latitude	Average	4.3	5.5	6.7	7.5	8.0	8.8	9.2	9.1	7.9	6.4	4.6	4.0	6.8
	Min/Max	3.4/5.2	4.3/6.2	5.3/7.8	6.3/9.5	6.5/9.4	7.5/10.3	8.4/9.9	8.1/10.1	5.6/9.6	5.0/7.5	3.6/5.3	3.2/4.9	6.4/7.4
Latitude +15	Average	4.5	5.6	6.6	7.2	7.6	8.3	8.7	8.6	7.7	6.5	4.7	4.2	6.7
	Min/Max	3.5/5.5	4.4/6.3	5.2/7.8	6.0/9.1	6.1/8.9	7.0/9.7	7.9/9.3	7.7/9.7	5.4/9.4	5.0/7.6	3.7/5.5	3.4/5.2	6.2/7.2

Solar Radiation for 2-Axis Tracking Flat-Plate Collectors (kWh/m²/day), Uncertainty ±9%

Tracker		Jan	Feb	Mar	Apr	May	June	July	Aug	Sept	Oct	Nov	Dec	Year
2-Axis	Average	4.6	5.6	6.7	7.6	8.4	9.4	9.7	9.3	7.9	6.5	4.8	4.3	7.1
	Min/Max	3.6/5.6	4.4/6.3	5.3/7.8	6.4/9.6	6.8/9.8	8.0/11.0	8.9/10.5	8.3/10.4	5.6/9.6	5.1/7.6	3.7/5.6	3.4/5.3	6.6/7.6

Direct Beam Solar Radiation for Concentrating Collectors (kWh/m²/day), Uncertainty ±8%

Tracker		Jan	Feb	Mar	Apr	May	June	July	Aug	Sept	Oct	Nov	Dec	Year
1-Axis, E-W Horiz Axis	Average	2.8	3.1	3.3	3.6	4.0	4.8	5.1	4.9	4.3	3.8	2.9	2.7	3.8
	Min/Max	1.8/3.8	2.0/3.9	2.4/4.4	2.7/5.2	2.9/5.1	3.6/6.1	4.6/5.7	4.1/5.8	2.5/5.6	2.7/4.7	1.9/3.7	1.8/3.8	3.4/4.3
1-Axis, N-S Horiz Axis	Average	1.9	2.6	3.6	4.6	5.3	6.4	6.8	6.4	5.1	3.6	2.1	1.7	4.2
	Min/Max	1.2/2.6	1.7/3.4	2.5/4.8	3.4/6.7	3.8/6.9	4.8/8.3	6.0/7.7	5.4/7.7	3.0/6.8	2.5/4.4	1.4/2.7	1.1/2.4	3.7/4.8
1-Axis, N-S Tilt=Latitude	Average	3.0	3.7	4.4	5.0	5.3	6.1	6.6	6.7	5.9	4.8	3.3	2.8	4.8
	Min/Max	2.0/4.2	2.4/4.7	3.1/5.9	3.7/7.3	3.8/6.9	4.5/7.9	5.8/7.4	5.6/7.9	3.5/7.8	3.3/5.9	2.2/4.2	1.9/4.0	4.3/5.5
2-Axis	Average	3.2	3.8	4.4	5.0	5.6	6.6	7.1	6.9	5.9	4.9	3.4	3.0	5.0
	Min/Max	2.1/4.4	2.5/4.8	3.1/5.9	3.7/7.4	4.0/7.3	4.9/8.5	6.3/8.0	5.8/8.2	3.5/7.9	3.3/6.0	2.3/4.4	2.0/4.4	4.5/5.7

Average Climatic Conditions

Element	Jan	Feb	Mar	Apr	May	June	July	Aug	Sept	Oct	Nov	Dec	Year
Temperature (°C)	-5.4	-2.9	1.2	7.3	12.9	18.2	22.3	21.4	15.6	9.3	1.6	-4.3	8.1
Daily Minimum Temp	-11.8	-9.3	-5.4	0.1	5.7	10.9	14.6	13.4	7.5	1.6	-5.1	-10.7	0.9
Daily Maximum Temp	1.0	3.4	7.7	14.4	20.1	25.4	30.1	29.5	23.6	16.9	8.2	2.0	15.2
Record Minimum Temp	-32.8	-31.1	-27.2	-17.2	-7.8	-0.6	3.9	3.3	-7.8	-18.9	-28.3	-34.4	-34.4
Record Maximum Temp	24.4	23.9	27.8	33.9	36.7	41.1	43.3	41.1	40.0	34.4	25.6	23.9	43.3
HDD, Base 18.3°C	736	596	532	332	173	57	10	17	117	281	503	703	4056
CDD, Base 18.3°C	0	0	0	0	4	54	134	114	34	0	0	0	339
Relative Humidity (%)	63	65	64	59	61	62	56	53	54	54	62	65	60
Wind Speed (m/s)	5.1	5.1	5.7	6.0	5.6	4.9	4.6	4.6	5.0	5.0	4.8	4.9	5.1

Sioux Falls, SD

WBAN NO. 14944

LATITUDE: 43.57° N
LONGITUDE: 96.73° W
ELEVATION: 435 meters
MEAN PRESSURE: 964 millibars

STATION TYPE: Secondary

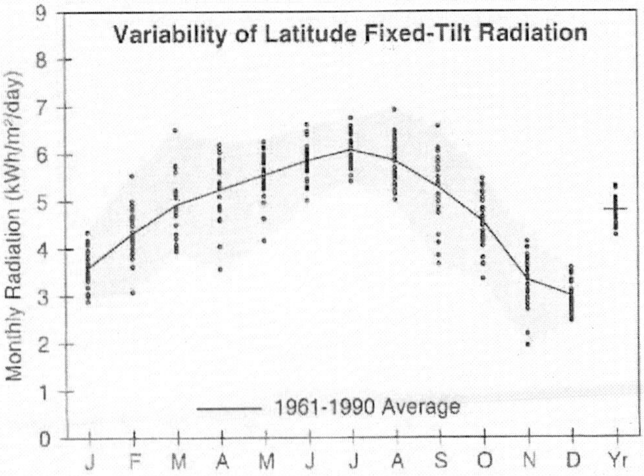

Solar Radiation for Flat-Plate Collectors Facing South at a Fixed Tilt (kWh/m²/day), Uncertainty ±9%

Tilt (°)		Jan	Feb	Mar	Apr	May	June	July	Aug	Sept	Oct	Nov	Dec	Year
0	Average	1.9	2.7	3.8	4.8	5.8	6.5	6.6	5.7	4.3	3.1	1.9	1.5	4.1
	Min/Max	1.7/2.1	2.2/3.2	3.2/4.8	3.5/5.7	4.5/6.5	5.6/7.3	5.9/7.3	5.0/6.6	3.3/5.2	2.4/3.5	1.5/2.2	1.4/1.8	3.8/4.4
Latitude -15	Average	3.1	3.9	4.7	5.4	5.9	6.4	6.6	6.1	5.2	4.3	3.0	2.6	4.8
	Min/Max	2.6/3.7	2.9/4.9	3.9/6.2	3.7/6.3	4.4/6.7	5.5/7.2	5.9/7.3	5.3/7.2	3.7/6.4	3.2/5.1	1.8/3.7	2.2/3.1	4.3/5.2
Latitude	Average	3.6	4.3	4.9	5.2	5.5	5.8	6.1	5.9	5.3	4.6	3.3	3.0	4.8
	Min/Max	2.9/4.3	3.1/5.5	3.9/6.5	3.6/6.2	4.2/6.3	5.0/6.6	5.4/6.8	5.0/6.9	3.7/6.6	3.4/5.5	1.9/4.2	2.5/3.6	4.3/5.3
Latitude +15	Average	3.8	4.5	4.8	4.8	4.9	5.1	5.3	5.3	5.1	4.6	3.5	3.2	4.6
	Min/Max	3.0/4.7	3.1/5.8	3.8/6.6	3.3/5.7	3.7/5.5	4.3/5.7	4.7/5.9	4.5/6.3	3.5/6.3	3.3/5.6	2.0/4.4	2.6/3.9	4.0/5.1
90	Average	3.7	4.1	3.9	3.3	2.9	2.8	3.0	3.4	3.7	3.9	3.2	3.1	3.4
	Min/Max	2.9/4.7	2.9/5.5	2.9/5.7	2.4/3.8	2.3/3.2	2.5/3.1	2.7/3.2	2.9/3.9	2.5/4.7	2.8/4.7	1.7/4.0	2.5/3.8	3.0/3.8

Solar Radiation for 1-Axis Tracking Flat-Plate Collectors with a North-South Axis (kWh/m²/day), Uncertainty ±9%

Axis Tilt (°)		Jan	Feb	Mar	Apr	May	June	July	Aug	Sept	Oct	Nov	Dec	Year
0	Average	2.9	3.9	5.2	6.5	7.7	8.6	8.9	7.8	6.2	4.5	2.8	2.3	5.6
	Min/Max	2.3/3.5	2.7/5.2	4.2/6.9	4.1/8.1	5.5/8.9	7.1/10.1	7.9/10.4	6.6/9.6	4.1/7.8	3.3/5.4	1.6/3.4	1.9/2.7	5.0/6.2
Latitude -15	Average	3.8	4.8	6.0	7.0	7.9	8.7	9.0	8.2	6.9	5.4	3.6	3.1	6.2
	Min/Max	3.0/4.7	3.3/6.4	4.6/7.9	4.2/8.7	5.6/9.2	7.1/10.2	8.0/10.6	6.9/10.2	4.5/8.8	3.9/6.6	2.0/4.5	2.5/3.8	5.4/6.9
Latitude	Average	4.2	5.1	6.1	6.9	7.6	8.3	8.7	8.1	6.9	5.6	3.9	3.4	6.2
	Min/Max	3.2/5.2	3.4/6.9	4.7/8.2	4.2/8.7	5.4/8.9	6.8/9.8	7.7/10.2	6.8/10.0	4.5/8.9	4.0/6.9	2.0/4.9	2.8/4.2	5.4/7.0
Latitude +15	Average	4.4	5.3	6.1	6.6	7.2	7.8	8.2	7.7	6.7	5.6	4.0	3.6	6.1
	Min/Max	3.4/5.5	3.5/7.1	4.7/8.2	4.0/8.3	5.1/8.4	6.4/9.2	7.2/9.6	6.4/9.6	4.3/8.8	4.0/6.9	2.1/5.1	2.9/4.4	5.3/6.9

Solar Radiation for 2-Axis Tracking Flat-Plate Collectors (kWh/m²/day), Uncertainty ±9%

Tracker		Jan	Feb	Mar	Apr	May	June	July	Aug	Sept	Oct	Nov	Dec	Year
2-Axis	Average	4.4	5.3	6.1	7.0	8.0	8.9	9.2	8.3	6.9	5.7	4.0	3.7	6.5
	Min/Max	3.4/5.5	3.5/7.1	4.7/8.2	4.3/8.8	5.7/9.3	7.3/10.5	8.2/10.8	6.9/10.3	4.5/8.9	4.1/6.9	2.1/5.1	2.9/4.5	5.6/7.3

Direct Beam Solar Radiation for Concentrating Collectors (kWh/m²/day), Uncertainty ±8%

Tracker		Jan	Feb	Mar	Apr	May	June	July	Aug	Sept	Oct	Nov	Dec	Year
1-Axis, E-W Horiz Axis	Average	2.6	2.8	2.9	3.2	3.7	4.3	4.6	4.1	3.5	3.2	2.3	2.1	3.3
	Min/Max	1.4/3.5	1.3/3.9	1.7/4.1	1.2/4.6	1.9/4.7	3.0/5.6	3.7/5.8	3.1/5.6	1.7/5.1	1.9/4.2	0.7/3.3	1.5/2.9	2.5/3.9
1-Axis, N-S Horiz Axis	Average	1.7	2.3	3.1	4.1	4.9	5.7	6.1	5.3	4.2	3.0	1.7	1.3	3.6
	Min/Max	0.9/2.4	1.0/3.5	1.8/4.4	1.5/6.0	2.7/6.2	4.0/7.4	4.9/7.8	4.1/7.4	2.1/6.2	1.9/4.0	0.5/2.4	0.9/1.8	2.9/4.3
1-Axis, N-S Tilt=Latitude	Average	2.8	3.3	3.8	4.5	4.8	5.4	5.9	5.5	4.8	4.0	2.6	2.2	4.1
	Min/Max	1.5/3.8	1.5/4.8	2.2/5.4	1.6/6.4	2.7/6.2	3.8/7.0	4.7/7.5	4.2/7.7	2.4/7.1	2.5/5.3	0.8/3.7	1.6/3.0	3.2/4.9
2-Axis	Average	3.0	3.4	3.8	4.5	5.1	5.9	6.3	5.7	4.8	4.0	2.7	2.4	4.3
	Min/Max	1.6/4.1	1.6/4.9	2.2/5.4	1.6/6.5	2.8/6.5	4.1/7.6	5.1/8.1	4.3/7.9	2.4/7.1	2.5/5.3	0.8/3.9	1.8/3.3	3.4/5.1

Average Climatic Conditions

Element	Jan	Feb	Mar	Apr	May	June	July	Aug	Sept	Oct	Nov	Dec	Year
Temperature (°C)	-10.1	-6.8	0.3	8.3	14.7	20.2	23.5	21.9	16.1	9.2	0.6	-7.6	7.5
Daily Minimum Temp	-15.9	-12.4	-5.2	1.6	7.7	13.4	16.8	15.2	9.3	2.2	-5.2	-13.0	1.2
Daily Maximum Temp	-4.3	-1.3	5.7	15.0	21.5	26.9	30.2	28.5	22.8	16.2	6.3	-2.2	13.8
Record Minimum Temp	-37.8	-35.0	-30.6	-15.0	-8.3	0.6	3.3	1.1	-5.6	-12.8	-27.2	-33.3	-37.8
Record Maximum Temp	18.9	21.1	30.6	34.4	37.8	43.3	42.2	42.2	40.0	34.4	24.4	16.1	43.3
HDD, Base 18.3°C	882	704	560	302	133	28	6	12	92	282	533	804	4338
CDD, Base 18.3°C	0	0	0	0	19	83	166	122	23	0	0	0	413
Relative Humidity (%)	72	73	72	65	63	65	65	68	70	67	73	76	69
Wind Speed (m/s)	4.9	4.9	5.5	5.8	5.2	4.7	4.4	4.4	4.6	4.7	5.0	4.8	4.9

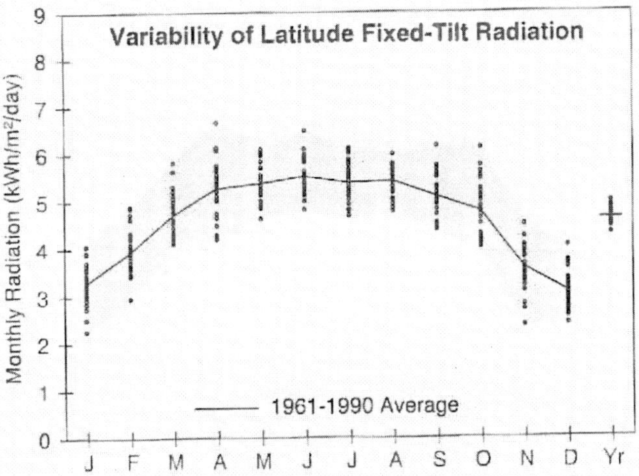

Bristol, TN

WBAN NO. 13877

LATITUDE: 36.48° N
LONGITUDE: 82.40° W
ELEVATION: 459 meters
MEAN PRESSURE: 964 millibars

STATION TYPE: Secondary

Solar Radiation for Flat-Plate Collectors Facing South at a Fixed Tilt (kWh/m²/day), Uncertainty ±9%

Tilt (°)		Jan	Feb	Mar	Apr	May	June	July	Aug	Sept	Oct	Nov	Dec	Year
0	Average	2.2	2.9	4.0	5.1	5.7	6.1	5.8	5.4	4.5	3.6	2.4	1.9	4.1
	Min/Max	1.8/2.5	2.4/3.4	3.6/4.7	4.2/6.3	5.0/6.5	5.3/7.2	5.1/6.6	4.8/5.9	4.0/5.3	3.2/4.4	1.9/2.8	1.7/2.3	3.9/4.4
Latitude -15	Average	2.9	3.6	4.6	5.4	5.7	6.0	5.8	5.6	5.0	4.5	3.2	2.7	4.6
	Min/Max	2.1/3.5	2.8/4.4	4.0/5.6	4.3/6.8	4.9/6.5	5.2/7.1	5.0/6.6	5.0/6.2	4.4/6.0	3.8/5.6	2.2/4.0	2.2/3.5	4.3/4.9
Latitude	Average	3.3	3.9	4.7	5.3	5.4	5.5	5.4	5.4	5.1	4.8	3.6	3.1	4.6
	Min/Max	2.3/4.1	3.0/4.9	4.1/5.8	4.2/6.7	4.6/6.1	4.8/6.5	4.7/6.1	4.8/6.0	4.4/6.2	4.0/6.1	2.4/4.5	2.4/4.0	4.3/5.0
Latitude +15	Average	3.5	4.0	4.6	4.9	4.8	4.8	4.7	5.0	4.9	4.8	3.7	3.3	4.4
	Min/Max	2.3/4.4	3.0/5.1	4.0/5.7	3.9/6.2	4.1/5.4	4.2/5.6	4.1/5.3	4.4/5.5	4.2/5.9	4.0/6.2	2.4/4.8	2.5/4.4	4.1/4.7
90	Average	3.1	3.3	3.3	2.9	2.4	2.2	2.3	2.7	3.2	3.8	3.2	3.0	3.0
	Min/Max	1.9/4.0	2.4/4.4	2.8/4.1	2.4/3.6	2.2/2.7	2.1/2.4	2.1/2.5	2.5/3.0	2.8/3.9	3.1/4.9	2.0/4.2	2.2/4.1	2.7/3.2

Solar Radiation for 1-Axis Tracking Flat-Plate Collectors with a North-South Axis (kWh/m²/day), Uncertainty ±9%

Axis Tilt (°)		Jan	Feb	Mar	Apr	May	June	July	Aug	Sept	Oct	Nov	Dec	Year
0	Average	3.0	3.9	5.2	6.6	7.2	7.7	7.3	6.8	5.9	4.9	3.3	2.7	5.4
	Min/Max	2.0/3.7	2.9/4.9	4.4/6.6	5.1/8.7	5.9/8.6	6.4/9.4	6.1/8.6	5.9/7.7	4.9/7.4	4.1/6.4	2.2/4.1	2.1/3.5	5.0/5.9
Latitude -15	Average	3.5	4.5	5.7	6.9	7.2	7.6	7.3	7.1	6.3	5.6	3.9	3.2	5.7
	Min/Max	2.3/4.4	3.2/5.7	4.8/7.2	5.2/9.2	5.9/8.6	6.4/9.4	6.1/8.6	6.1/8.0	5.3/8.0	4.6/7.3	2.5/5.0	2.5/4.3	5.3/6.3
Latitude	Average	3.8	4.7	5.8	6.8	7.0	7.3	7.0	6.9	6.3	5.8	4.2	3.5	5.8
	Min/Max	2.4/4.8	3.3/6.0	4.8/7.3	5.1/9.1	5.7/8.4	6.1/9.0	5.8/8.3	5.9/7.8	5.3/8.1	4.8/7.7	2.6/5.4	2.7/4.8	5.3/6.4
Latitude +15	Average	4.0	4.8	5.7	6.5	6.6	6.8	6.5	6.6	6.2	5.9	4.3	3.7	5.6
	Min/Max	2.5/5.1	3.4/6.2	4.8/7.3	4.9/8.8	5.4/7.9	5.7/8.4	5.4/7.8	5.6/7.5	5.2/7.9	4.8/7.8	2.6/5.6	2.8/5.1	5.2/6.2

Solar Radiation for 2-Axis Tracking Flat-Plate Collectors (kWh/m²/day), Uncertainty ±9%

Tracker		Jan	Feb	Mar	Apr	May	June	July	Aug	Sept	Oct	Nov	Dec	Year
2-Axis	Average	4.0	4.8	5.8	6.9	7.3	7.8	7.4	7.1	6.4	5.9	4.3	3.7	6.0
	Min/Max	2.5/5.2	3.4/6.2	4.9/7.4	5.2/9.2	6.0/8.8	6.5/9.6	6.2/8.8	6.1/8.0	5.4/8.1	4.8/7.8	2.6/5.7	2.8/5.2	5.5/6.6

Direct Beam Solar Radiation for Concentrating Collectors (kWh/m²/day), Uncertainty ±8%

Tracker		Jan	Feb	Mar	Apr	May	June	July	Aug	Sept	Oct	Nov	Dec	Year
1-Axis, E-W Horiz Axis	Average	2.2	2.4	2.8	3.2	3.3	3.6	3.3	3.2	3.0	3.2	2.4	2.2	2.9
	Min/Max	0.9/3.2	1.3/3.6	2.0/4.0	1.7/5.0	2.3/4.6	2.6/5.1	2.3/4.5	2.4/4.0	2.2/4.3	2.3/4.6	1.0/3.5	1.4/3.5	2.6/3.4
1-Axis, N-S Horiz Axis	Average	1.7	2.3	3.2	4.1	4.4	4.6	4.2	4.1	3.6	3.2	2.0	1.6	3.3
	Min/Max	0.7/2.5	1.3/3.4	2.3/4.8	2.3/6.5	3.0/6.1	3.4/6.6	2.9/5.8	3.0/5.1	2.7/5.3	2.3/4.8	0.9/2.9	1.0/2.6	2.9/3.9
1-Axis, N-S Tilt=Latitude	Average	2.4	3.0	3.7	4.3	4.2	4.3	4.0	4.1	4.0	4.0	2.8	2.3	3.6
	Min/Max	1.0/3.5	1.6/4.4	2.6/5.5	2.4/6.8	2.9/5.9	3.2/6.2	2.7/5.6	3.0/5.2	3.0/5.8	2.9/5.9	1.2/4.0	1.5/3.8	3.2/4.3
2-Axis	Average	2.6	3.1	3.7	4.4	4.5	4.7	4.3	4.3	4.0	4.0	2.9	2.5	3.7
	Min/Max	1.1/3.8	1.6/4.5	2.6/5.5	2.4/6.9	3.1/6.2	3.4/6.7	3.0/5.9	3.2/5.3	3.0/5.8	2.9/5.9	1.3/4.2	1.6/4.1	3.3/4.4

Average Climatic Conditions

Element	Jan	Feb	Mar	Apr	May	June	July	Aug	Sept	Oct	Nov	Dec	Year
Temperature (°C)	1.1	3.0	8.4	12.9	17.4	21.7	23.6	23.1	19.9	13.7	8.3	3.4	13.1
Daily Minimum Temp	-4.3	-2.9	1.9	6.1	10.9	15.5	17.8	17.3	13.7	6.8	2.2	-2.1	6.9
Daily Maximum Temp	6.5	8.9	14.9	19.7	24.0	27.9	29.2	28.9	26.2	20.6	14.6	8.9	19.2
Record Minimum Temp	-29.4	-20.6	-18.9	-6.1	-1.1	3.3	7.2	6.1	1.1	-6.7	-15.0	-22.8	-29.4
Record Maximum Temp	26.1	26.7	29.4	31.7	33.3	36.1	38.9	38.3	37.8	32.2	27.2	25.6	38.9
HDD, Base 18.3°C	534	429	307	166	67	3	0	0	24	156	300	462	2448
CDD, Base 18.3°C	0	0	0	0	39	105	162	148	73	13	0	0	540
Relative Humidity (%)	71	68	64	62	71	74	76	77	76	72	71	72	71
Wind Speed (m/s)	2.8	3.0	3.2	3.1	2.4	2.1	1.9	1.7	1.9	2.1	2.5	2.7	2.4

Chattanooga, TN

WBAN NO. 13882

LATITUDE: 35.03° N
LONGITUDE: 85.20° W
ELEVATION: 210 meters
MEAN PRESSURE: 994 millibars

STATION TYPE: Secondary

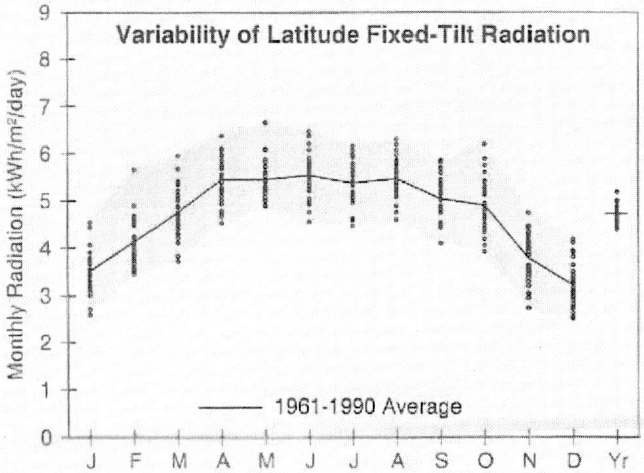

Solar Radiation for Flat-Plate Collectors Facing South at a Fixed Tilt (kWh/m²/day), Uncertainty ±9%

Tilt (°)		Jan	Feb	Mar	Apr	May	June	July	Aug	Sept	Oct	Nov	Dec	Year
0	Average	2.4	3.1	4.1	5.3	5.8	6.1	5.9	5.5	4.5	3.8	2.6	2.1	4.3
	Min/Max	2.0/2.8	2.8/3.9	3.4/4.9	4.4/6.0	5.2/7.1	5.1/7.2	4.9/6.7	4.7/6.2	3.8/5.2	3.2/4.6	2.1/3.0	1.9/2.5	4.0/4.6
Latitude -15	Average	3.1	3.8	4.6	5.6	5.8	6.0	5.8	5.7	5.0	4.6	3.4	2.8	4.7
	Min/Max	2.4/3.9	3.3/5.1	3.7/5.7	4.6/6.5	5.2/7.1	4.9/7.0	4.8/6.6	4.8/6.5	4.1/5.8	3.7/5.7	2.5/4.2	2.3/3.6	4.4/5.1
Latitude	Average	3.5	4.1	4.8	5.4	5.4	5.5	5.4	5.5	5.0	4.9	3.8	3.2	4.7
	Min/Max	2.6/4.5	3.4/5.7	3.7/6.0	4.5/6.4	4.9/6.7	4.6/6.5	4.5/6.2	4.6/6.3	4.1/5.9	3.9/6.2	2.7/4.7	2.5/4.2	4.4/5.2
Latitude +15	Average	3.7	4.2	4.6	5.1	4.9	4.8	4.7	5.0	4.8	4.9	4.0	3.4	4.5
	Min/Max	2.6/4.9	3.5/5.9	3.6/5.9	4.2/5.9	4.4/5.9	4.0/5.6	3.9/5.4	4.2/5.7	3.9/5.6	3.9/6.3	2.8/5.0	2.6/4.6	4.2/5.0
90	Average	3.2	3.4	3.2	2.9	2.4	2.2	2.2	2.7	3.1	3.8	3.4	3.0	3.0
	Min/Max	2.2/4.4	2.8/4.9	2.5/4.1	2.5/3.4	2.2/2.7	1.9/2.3	2.0/2.4	2.3/3.0	2.6/3.7	2.9/4.9	2.3/4.4	2.2/4.2	2.7/3.2

Solar Radiation for 1-Axis Tracking Flat-Plate Collectors with a North-South Axis (kWh/m²/day), Uncertainty ±9%

Axis Tilt (°)		Jan	Feb	Mar	Apr	May	June	July	Aug	Sept	Oct	Nov	Dec	Year
0	Average	3.2	4.2	5.3	6.8	7.3	7.6	7.2	6.9	5.8	5.1	3.5	2.8	5.5
	Min/Max	2.4/4.1	3.4/5.7	4.1/7.0	5.4/8.3	6.3/9.5	6.1/9.4	5.8/8.8	5.7/8.1	4.5/6.9	3.9/6.6	2.5/4.4	2.2/3.7	5.1/6.2
Latitude -15	Average	3.8	4.7	5.7	7.1	7.3	7.6	7.2	7.1	6.2	5.7	4.1	3.4	5.8
	Min/Max	2.7/4.9	3.8/6.6	4.3/7.6	5.5/8.6	6.4/9.6	6.1/9.3	5.8/8.8	5.8/8.4	4.7/7.3	4.3/7.4	2.8/5.3	2.6/4.5	5.4/6.6
Latitude	Average	4.1	4.9	5.9	7.0	7.1	7.2	6.9	6.9	6.2	5.9	4.4	3.7	5.9
	Min/Max	2.8/5.4	4.0/7.1	4.4/7.8	5.5/8.6	6.2/9.3	5.8/8.9	5.6/8.4	5.6/8.2	4.7/7.4	4.4/7.8	3.0/5.7	2.7/5.0	5.4/6.6
Latitude +15	Average	4.2	5.0	5.8	6.7	6.6	6.7	6.5	6.6	6.0	6.0	4.6	3.9	5.7
	Min/Max	2.9/5.7	4.0/7.3	4.3/7.7	5.2/8.3	5.8/8.7	5.4/8.3	5.2/7.9	5.3/7.8	4.6/7.3	4.4/7.9	3.0/5.9	2.8/5.3	5.3/6.5

Solar Radiation for 2-Axis Tracking Flat-Plate Collectors (kWh/m²/day), Uncertainty ±9%

Tracker		Jan	Feb	Mar	Apr	May	June	July	Aug	Sept	Oct	Nov	Dec	Year
2-Axis	Average	4.3	5.0	5.9	7.1	7.4	7.7	7.3	7.1	6.2	6.0	4.6	3.9	6.0
	Min/Max	2.9/5.7	4.0/7.3	4.4/7.8	5.6/8.7	6.4/9.7	6.2/9.5	5.9/8.9	5.8/8.4	4.7/7.4	4.5/7.9	3.0/6.0	2.8/5.4	5.6/6.9

Direct Beam Solar Radiation for Concentrating Collectors (kWh/m²/day), Uncertainty ±8%

Tracker		Jan	Feb	Mar	Apr	May	June	July	Aug	Sept	Oct	Nov	Dec	Year
1-Axis, E-W	Average	2.4	2.6	2.7	3.3	3.3	3.4	3.1	3.1	2.8	3.2	2.6	2.2	2.9
Horiz Axis	Min/Max	1.3/3.7	1.7/4.6	1.5/4.2	1.9/4.4	2.6/5.1	2.3/5.0	1.9/4.4	2.1/4.2	1.6/3.8	1.9/4.7	1.3/3.8	1.3/3.6	2.5/3.6
1-Axis, N-S	Average	1.9	2.5	3.2	4.2	4.3	4.4	4.1	4.0	3.4	3.3	2.2	1.7	3.3
Horiz Axis	Min/Max	1.0/2.9	1.6/4.4	1.7/5.1	2.5/5.9	3.4/6.8	3.0/6.5	2.5/5.8	2.8/5.4	1.9/4.5	1.9/4.9	1.1/3.2	0.9/2.7	2.8/4.1
1-Axis, N-S	Average	2.6	3.2	3.6	4.4	4.1	4.1	3.8	4.0	3.7	4.0	2.9	2.4	3.6
Tilt=Latitude	Min/Max	1.4/4.0	2.0/5.6	1.9/5.8	2.6/6.1	3.3/6.5	2.8/6.1	2.4/5.5	2.8/5.4	2.1/5.0	2.3/5.9	1.4/4.3	1.4/3.9	3.0/4.5
2-Axis	Average	2.8	3.3	3.7	4.5	4.4	4.5	4.1	4.1	3.7	4.0	3.1	2.6	3.7
	Min/Max	1.5/4.3	2.1/5.8	1.9/5.8	2.7/6.2	3.5/6.9	3.0/6.6	2.6/5.9	2.8/5.6	2.1/5.0	2.3/6.0	1.5/4.5	1.5/4.2	3.2/4.7

Average Climatic Conditions

Element	Jan	Feb	Mar	Apr	May	June	July	Aug	Sept	Oct	Nov	Dec	Year
Temperature (°C)	3.0	5.4	10.2	15.1	19.4	24.0	25.9	25.5	22.4	15.8	10.3	5.1	15.2
Daily Minimum Temp	-2.2	-0.5	3.9	8.2	13.0	17.7	20.2	19.8	17.0	9.4	4.4	-0.4	9.2
Daily Maximum Temp	8.2	11.3	16.6	21.9	25.8	30.2	31.7	31.2	27.8	22.2	16.2	10.4	21.1
Record Minimum Temp	-23.3	-17.2	-13.3	-3.3	1.1	5.0	10.6	10.0	2.2	-5.6	-15.6	-18.9	-23.3
Record Maximum Temp	25.6	26.1	30.6	33.9	37.2	40.0	41.1	40.6	38.9	34.4	28.9	25.6	41.1
HDD, Base 18.3°C	476	361	252	109	32	0	0	0	5	106	242	412	1993
CDD, Base 18.3°C	0	0	0	11	64	170	236	222	128	26	0	0	858
Relative Humidity (%)	71	68	66	64	71	73	75	76	77	75	73	73	72
Wind Speed (m/s)	3.0	3.2	3.4	3.2	2.6	2.3	2.2	2.1	2.2	2.2	2.6	2.9	2.7

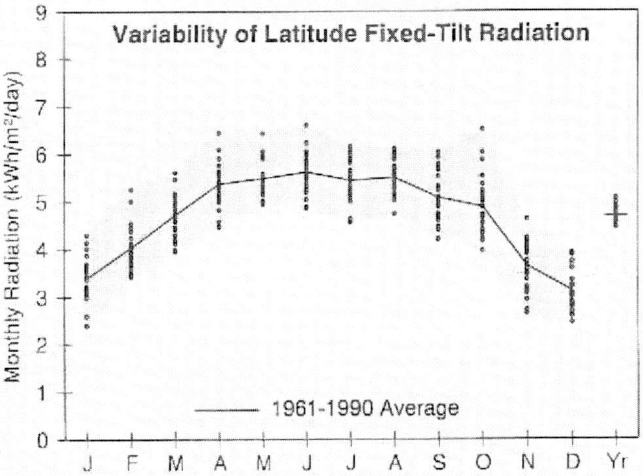

Knoxville, TN

WBAN NO. 13891

LATITUDE: 35.82° N
LONGITUDE: 83.98° W
ELEVATION: 299 meters
MEAN PRESSURE: 983 millibars

STATION TYPE: Secondary

Solar Radiation for Flat-Plate Collectors Facing South at a Fixed Tilt (kWh/m²/day), Uncertainty ±9%

Tilt (°)		Jan	Feb	Mar	Apr	May	June	July	Aug	Sept	Oct	Nov	Dec	Year
0	Average	2.3	3.0	4.0	5.2	5.8	6.2	5.9	5.5	4.5	3.7	2.5	2.0	4.2
	Min/Max	1.8/2.6	2.7/3.6	3.5/4.6	4.4/6.0	5.3/6.8	5.4/7.4	4.9/6.7	4.8/6.1	3.9/5.3	3.2/4.7	2.0/2.9	1.8/2.3	4.1/4.5
Latitude -15	Average	3.0	3.7	4.6	5.5	5.8	6.1	5.8	5.7	5.0	4.6	3.3	2.8	4.7
	Min/Max	2.2/3.7	3.2/4.7	3.9/5.4	4.6/6.5	5.3/6.8	5.3/7.2	4.9/6.6	4.9/6.3	4.2/5.9	3.8/6.0	2.5/4.1	2.3/3.4	4.5/5.0
Latitude	Average	3.4	4.0	4.7	5.4	5.5	5.6	5.4	5.5	5.1	4.9	3.7	3.1	4.7
	Min/Max	2.4/4.3	3.4/5.3	4.0/5.6	4.5/6.4	4.9/6.4	4.9/6.6	4.6/6.2	4.7/6.1	4.2/6.0	4.0/6.5	2.7/4.6	2.5/3.9	4.4/5.1
Latitude +15	Average	3.6	4.1	4.6	5.0	4.9	4.9	4.8	5.0	4.8	4.9	3.8	3.3	4.5
	Min/Max	2.5/4.6	3.5/5.4	3.8/5.5	4.1/6.0	4.4/5.7	4.3/5.7	4.0/5.4	4.3/5.6	4.0/5.8	3.9/6.7	2.7/4.9	2.5/4.2	4.2/4.8
90	Average	3.2	3.3	3.2	2.9	2.4	2.2	2.3	2.7	3.2	3.8	3.3	3.0	3.0
	Min/Max	2.0/4.2	2.7/4.5	2.7/3.8	2.5/3.5	2.3/2.7	2.0/2.4	2.1/2.5	2.4/3.0	2.6/3.8	3.0/5.2	2.2/4.3	2.2/3.9	2.7/3.2

Solar Radiation for 1-Axis Tracking Flat-Plate Collectors with a North-South Axis (kWh/m²/day), Uncertainty ±9%

Axis Tilt (°)		Jan	Feb	Mar	Apr	May	June	July	Aug	Sept	Oct	Nov	Dec	Year
0	Average	3.1	4.0	5.2	6.7	7.3	7.8	7.3	6.9	5.8	5.0	3.4	2.7	5.5
	Min/Max	2.1/3.8	3.3/5.2	4.3/6.5	5.3/8.2	6.3/9.1	6.4/9.8	5.7/8.7	5.8/7.9	4.6/7.2	4.1/6.9	2.4/4.3	2.2/3.4	5.2/6.1
Latitude -15	Average	3.6	4.6	5.7	6.9	7.4	7.7	7.3	7.1	6.2	5.7	4.0	3.3	5.8
	Min/Max	2.5/4.7	3.7/6.1	4.6/7.1	5.4/8.6	6.3/9.2	6.4/9.8	5.7/8.7	5.9/8.1	4.9/7.7	4.5/7.9	2.8/5.1	2.5/4.2	5.5/6.4
Latitude	Average	3.9	4.8	5.8	6.9	7.1	7.4	7.1	7.0	6.3	5.9	4.3	3.6	5.8
	Min/Max	2.6/5.1	3.8/6.5	4.6/7.2	5.3/8.6	6.1/8.9	6.1/9.4	5.4/8.4	5.8/8.0	4.9/7.8	4.7/8.3	2.9/5.6	2.7/4.6	5.5/6.5
Latitude +15	Average	4.1	4.9	5.7	6.6	6.7	6.9	6.6	6.7	6.1	6.0	4.4	3.7	5.7
	Min/Max	2.7/5.4	3.9/6.6	4.5/7.2	5.1/8.3	5.7/8.4	5.7/8.7	5.1/7.9	5.5/7.6	4.7/7.6	4.7/8.4	3.0/5.8	2.8/4.9	5.3/6.3

Solar Radiation for 2-Axis Tracking Flat-Plate Collectors (kWh/m²/day), Uncertainty ±9%

Tracker		Jan	Feb	Mar	Apr	May	June	July	Aug	Sept	Oct	Nov	Dec	Year
2-Axis	Average	4.1	4.9	5.8	7.0	7.4	7.9	7.4	7.2	6.3	6.0	4.4	3.8	6.0
	Min/Max	2.7/5.4	3.9/6.6	4.7/7.3	5.4/8.6	6.4/9.3	6.5/10.0	5.8/8.9	5.9/8.2	4.9/7.8	4.7/8.4	3.0/5.8	2.8/4.9	5.7/6.7

Direct Beam Solar Radiation for Concentrating Collectors (kWh/m²/day), Uncertainty ±8%

Tracker		Jan	Feb	Mar	Apr	May	June	July	Aug	Sept	Oct	Nov	Dec	Year
1-Axis, E-W Horiz Axis	Average	2.2	2.5	2.7	3.1	3.3	3.6	3.2	3.2	2.8	3.2	2.4	2.1	2.9
	Min/Max	1.1/3.4	1.4/3.9	1.8/3.8	1.8/4.4	2.6/4.8	2.5/5.4	1.9/4.5	2.1/4.0	1.8/4.0	2.2/5.1	1.3/3.6	1.3/3.1	2.6/3.5
1-Axis, N-S Horiz Axis	Average	1.8	2.4	3.1	4.1	4.4	4.6	4.2	4.1	3.4	3.3	2.0	1.6	3.2
	Min/Max	0.9/2.7	1.2/3.7	2.0/4.5	2.3/5.7	3.3/6.4	3.2/7.1	2.4/5.8	2.8/5.1	2.1/4.9	2.3/5.3	1.1/3.0	1.0/2.3	2.9/4.0
1-Axis, N-S Tilt=Latitude	Average	2.5	3.0	3.6	4.2	4.2	4.3	4.0	4.1	3.8	4.0	2.8	2.3	3.6
	Min/Max	1.3/3.8	1.6/4.8	2.3/5.2	2.4/6.0	3.2/6.2	3.0/6.7	2.3/5.5	2.8/5.2	2.3/5.4	2.7/6.5	1.5/4.1	1.4/3.4	3.2/4.3
2-Axis	Average	2.6	3.1	3.6	4.3	4.5	4.7	4.3	4.2	3.8	4.1	2.9	2.5	3.7
	Min/Max	1.3/4.0	1.7/4.9	2.3/5.2	2.5/6.0	3.4/6.5	3.3/7.2	2.5/5.9	2.9/5.3	2.3/5.4	2.8/6.5	1.6/4.3	1.5/3.7	3.3/4.5

Average Climatic Conditions

Element	Jan	Feb	Mar	Apr	May	June	July	Aug	Sept	Oct	Nov	Dec	Year
Temperature (°C)	2.2	4.5	9.4	14.2	18.6	22.9	24.8	24.4	21.2	14.7	9.3	4.5	14.2
Daily Minimum Temp	-3.3	-1.6	2.6	7.0	11.7	16.6	18.9	18.5	15.0	7.8	3.1	-1.1	7.9
Daily Maximum Temp	7.7	10.5	16.3	21.3	25.3	29.2	30.6	30.4	27.3	21.4	15.5	10.1	20.5
Record Minimum Temp	-31.1	-18.9	-17.2	-5.6	0.0	6.1	9.4	9.4	2.2	-3.9	-15.0	-21.1	-31.1
Record Maximum Temp	25.0	28.3	30.0	33.3	34.4	38.9	39.4	38.9	39.4	32.8	28.9	26.7	39.4
HDD, Base 18.3°C	499	387	276	130	52	0	0	0	10	132	272	429	2187
CDD, Base 18.3°C	0	0	0	5	59	137	200	189	95	18	0	0	703
Relative Humidity (%)	72	68	65	63	71	73	76	76	76	73	72	73	72
Wind Speed (m/s)	3.4	3.5	3.7	3.7	3.0	2.8	2.7	2.5	2.5	2.5	3.0	3.2	3.0

Memphis, TN

WBAN NO. 13893

LATITUDE: 35.05° N
LONGITUDE: 89.98° W
ELEVATION: 87 meters
MEAN PRESSURE: 1008 millibars

STATION TYPE: Secondary

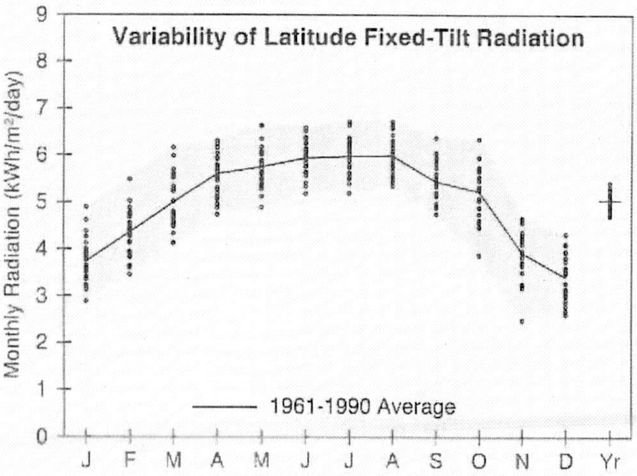

Solar Radiation for Flat-Plate Collectors Facing South at a Fixed Tilt (kWh/m²/day), Uncertainty ±9%

Tilt (°)		Jan	Feb	Mar	Apr	May	June	July	Aug	Sept	Oct	Nov	Dec	Year
0	Average	2.5	3.2	4.2	5.4	6.1	6.6	6.5	6.0	4.8	4.0	2.7	2.2	4.5
	Min/Max	2.1/2.9	2.8/3.8	3.6/5.0	4.6/6.0	5.2/7.0	5.8/7.4	5.7/7.3	5.3/6.6	4.3/5.5	3.1/4.6	2.0/3.0	1.8/2.5	4.3/4.8
Latitude -15	Average	3.3	4.0	4.8	5.7	6.1	6.4	6.4	6.2	5.4	4.9	3.5	3.0	5.0
	Min/Max	2.6/4.2	3.3/4.9	4.0/5.9	4.8/6.4	5.2/7.1	5.6/7.2	5.6/7.3	5.5/6.9	4.7/6.2	3.7/5.8	2.3/4.1	2.3/3.7	4.7/5.3
Latitude	Average	3.7	4.4	5.0	5.6	5.8	5.9	6.0	6.0	5.4	5.2	3.9	3.4	5.0
	Min/Max	2.9/4.9	3.5/5.5	4.1/6.2	4.7/6.3	4.9/6.6	5.2/6.6	5.2/6.7	5.3/6.7	4.7/6.4	3.9/6.3	2.5/4.7	2.6/4.3	4.7/5.4
Latitude +15	Average	4.0	4.5	4.9	5.2	5.1	5.1	5.2	5.5	5.2	5.3	4.1	3.6	4.8
	Min/Max	3.0/5.3	3.5/5.7	4.0/6.1	4.4/5.9	4.3/5.9	4.5/5.6	4.6/5.8	4.9/6.1	4.5/6.2	3.9/6.5	2.5/4.9	2.7/4.7	4.5/5.2
90	Average	3.5	3.6	3.4	3.0	2.4	2.2	2.3	2.9	3.4	4.1	3.5	3.3	3.1
	Min/Max	2.5/4.8	2.8/4.7	2.8/4.2	2.6/3.3	2.2/2.7	2.1/2.4	2.2/2.5	2.6/3.1	2.9/4.0	2.9/5.0	2.0/4.3	2.3/4.3	2.9/3.4

Solar Radiation for 1-Axis Tracking Flat-Plate Collectors with a North-South Axis (kWh/m²/day), Uncertainty ±9%

Axis Tilt (°)		Jan	Feb	Mar	Apr	May	June	July	Aug	Sept	Oct	Nov	Dec	Year
0	Average	3.4	4.4	5.6	7.1	7.9	8.5	8.4	7.8	6.4	5.5	3.7	3.0	6.0
	Min/Max	2.7/4.4	3.5/5.6	4.6/7.2	5.8/8.3	6.3/9.5	7.2/9.8	6.9/9.9	6.8/9.1	5.5/7.7	4.0/6.9	2.3/4.4	2.3/3.8	5.5/6.6
Latitude -15	Average	4.0	5.0	6.1	7.4	8.0	8.5	8.4	8.1	6.8	6.2	4.3	3.6	6.4
	Min/Max	3.1/5.4	3.9/6.4	4.9/7.9	5.9/8.7	6.4/9.6	7.1/9.8	6.9/9.8	6.9/9.3	5.9/8.2	4.4/7.8	2.6/5.2	2.7/4.7	5.9/7.0
Latitude	Average	4.4	5.3	6.2	7.3	7.7	8.1	8.1	7.9	6.9	6.5	4.6	3.9	6.4
	Min/Max	3.3/5.9	4.0/6.9	5.0/8.1	5.9/8.6	6.2/9.3	6.8/9.4	6.7/9.5	6.8/9.1	5.9/8.3	4.6/8.1	2.7/5.6	2.9/5.2	5.9/7.0
Latitude +15	Average	4.6	5.4	6.1	7.0	7.3	7.6	7.6	7.5	6.7	6.5	4.8	4.1	6.3
	Min/Max	3.4/6.2	4.1/7.0	4.9/8.0	5.6/8.3	5.8/8.8	6.3/8.7	6.2/8.9	6.5/8.7	5.8/8.2	4.5/8.2	2.7/5.9	3.0/5.4	5.7/6.9

Solar Radiation for 2-Axis Tracking Flat-Plate Collectors (kWh/m²/day), Uncertainty ±9%

Tracker		Jan	Feb	Mar	Apr	May	June	July	Aug	Sept	Oct	Nov	Dec	Year
2-Axis	Average	4.6	5.4	6.2	7.4	8.0	8.6	8.5	8.1	6.9	6.5	4.8	4.2	6.6
	Min/Max	3.4/6.3	4.1/7.0	5.0/8.1	6.0/8.7	6.4/9.7	7.2/10.0	7.0/10.0	7.0/9.3	5.9/8.3	4.6/8.2	2.8/5.9	3.0/5.5	6.1/7.3

Direct Beam Solar Radiation for Concentrating Collectors (kWh/m²/day), Uncertainty ±8%

Tracker		Jan	Feb	Mar	Apr	May	June	July	Aug	Sept	Oct	Nov	Dec	Year
1-Axis, E-W	Average	2.6	2.9	3.0	3.5	3.7	4.1	4.0	3.8	3.2	3.6	2.7	2.5	3.3
Horiz Axis	Min/Max	1.6/4.1	1.7/4.3	2.0/4.5	2.5/4.6	2.6/5.0	3.1/5.4	2.8/5.4	3.0/4.9	2.6/4.5	2.1/4.9	1.0/3.6	1.5/3.6	2.8/3.9
1-Axis, N-S	Average	2.1	2.8	3.6	4.6	5.0	5.4	5.3	5.0	4.0	3.7	2.3	1.8	3.8
Horiz Axis	Min/Max	1.3/3.2	1.7/4.2	2.4/5.3	3.2/6.1	3.4/6.7	4.0/7.1	3.7/7.1	3.9/6.5	3.2/5.5	2.2/5.2	0.9/3.1	1.1/2.8	3.3/4.5
1-Axis, N-S	Average	2.9	3.5	4.0	4.7	4.8	5.0	5.0	5.0	4.4	4.5	3.1	2.6	4.2
Tilt=Latitude	Min/Max	1.8/4.5	2.2/5.3	2.7/6.0	3.3/6.3	3.3/6.5	3.8/6.6	3.5/6.7	3.9/6.5	3.5/6.0	2.6/6.2	1.2/4.2	1.6/3.9	3.5/4.9
2-Axis	Average	3.1	3.6	4.1	4.8	5.1	5.5	5.4	5.2	4.4	4.6	3.3	2.9	4.3
	Min/Max	1.9/4.8	2.2/5.4	2.7/6.0	3.4/6.4	3.5/6.9	4.1/7.2	3.8/7.2	4.0/6.7	3.5/6.1	2.6/6.3	1.2/4.4	1.8/4.2	3.7/5.1

Average Climatic Conditions

Element	Jan	Feb	Mar	Apr	May	June	July	Aug	Sept	Oct	Nov	Dec	Year
Temperature (°C)	4.3	6.8	11.7	17.2	21.8	26.2	28.1	27.2	23.4	17.3	11.4	6.5	16.8
Daily Minimum Temp	-0.6	1.6	6.1	11.3	16.2	20.5	22.7	21.7	18.1	11.1	5.9	1.6	11.3
Daily Maximum Temp	9.2	11.9	17.3	22.9	27.2	31.8	33.5	32.7	28.8	23.5	16.8	11.4	22.3
Record Minimum Temp	-20.0	-23.9	-11.1	-1.7	3.3	8.9	11.1	8.9	2.2	-3.9	-12.8	-25.0	-25.0
Record Maximum Temp	25.6	27.2	29.4	34.4	37.2	40.0	42.2	40.6	39.4	35.0	29.4	27.2	42.2
HDD, Base 18.3°C	436	323	213	71	14	0	0	0	6	73	211	367	1712
CDD, Base 18.3°C	0	0	8	36	121	235	303	276	159	40	0	0	1177
Relative Humidity (%)	68	66	63	62	66	67	69	70	71	66	68	69	67
Wind Speed (m/s)	4.2	4.4	4.6	4.5	3.9	3.5	3.3	3.1	3.3	3.4	3.9	4.2	3.8

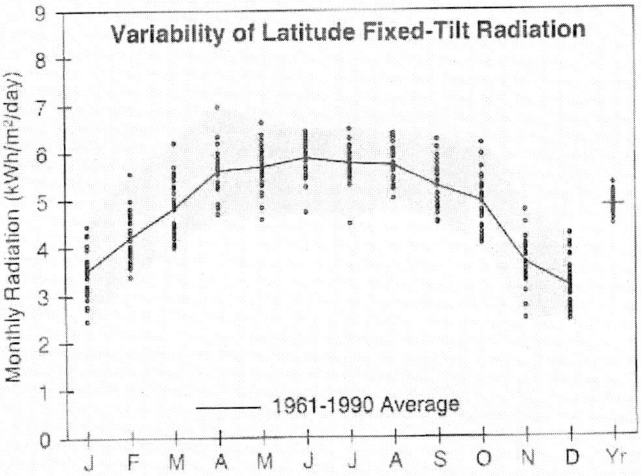

Nashville, TN

WBAN NO. 13897

LATITUDE: 36.12° N
LONGITUDE: 86.68° W
ELEVATION: 180 meters
MEAN PRESSURE: 996 millibars

STATION TYPE: Primary

Solar Radiation for Flat-Plate Collectors Facing South at a Fixed Tilt (kWh/m²/day), Uncertainty ±9%

Tilt (°)		Jan	Feb	Mar	Apr	May	June	July	Aug	Sept	Oct	Nov	Dec	Year
0	Average	2.3	3.1	4.1	5.4	6.0	6.5	6.3	5.7	4.7	3.8	2.5	2.0	4.4
	Min/Max	1.8/2.7	2.7/3.8	3.5/5.0	4.6/6.5	4.9/7.1	5.2/7.2	4.9/7.1	5.0/6.3	4.1/5.4	3.3/4.5	2.0/3.0	1.6/2.4	4.0/4.7
Latitude -15	Average	3.1	3.9	4.7	5.7	6.0	6.4	6.2	5.9	5.2	4.6	3.3	2.8	4.8
	Min/Max	2.3/3.9	3.2/5.0	3.9/5.9	4.8/7.0	4.9/7.1	5.1/7.0	4.8/7.0	5.2/6.6	4.5/6.2	3.9/5.7	2.3/4.2	2.2/3.6	4.4/5.2
Latitude	Average	3.5	4.2	4.8	5.6	5.7	5.9	5.8	5.7	5.3	4.9	3.6	3.1	4.9
	Min/Max	2.5/4.4	3.4/5.6	4.0/6.2	4.7/7.0	4.6/6.6	4.7/6.4	4.5/6.5	5.0/6.4	4.5/6.3	4.1/6.2	2.5/4.8	2.5/4.3	4.5/5.3
Latitude +15	Average	3.7	4.3	4.7	5.2	5.0	5.1	5.1	5.2	5.0	5.0	3.8	3.4	4.6
	Min/Max	2.5/4.8	3.4/5.8	3.9/6.1	4.3/6.5	4.1/5.8	4.2/5.6	4.0/5.7	4.6/5.8	4.3/6.0	4.1/6.3	2.5/5.0	2.6/4.6	4.3/5.1
90	Average	3.3	3.5	3.3	3.1	2.5	2.3	2.4	2.9	3.3	3.9	3.3	3.0	3.1
	Min/Max	2.1/4.3	2.8/4.8	2.7/4.3	2.6/3.8	2.1/2.8	2.1/2.5	2.1/2.6	2.5/3.1	2.8/4.0	3.1/4.9	2.0/4.4	2.2/4.3	2.9/3.4

Solar Radiation for 1-Axis Tracking Flat-Plate Collectors with a North-South Axis (kWh/m²/day), Uncertainty ±9%

Axis Tilt (°)		Jan	Feb	Mar	Apr	May	June	July	Aug	Sept	Oct	Nov	Dec	Year
0	Average	3.1	4.2	5.4	7.0	7.6	8.2	7.8	7.3	6.1	5.1	3.4	2.7	5.7
	Min/Max	2.2/3.9	3.3/5.5	4.4/7.1	5.7/8.9	6.0/9.5	5.9/9.3	5.7/9.4	6.3/8.5	5.0/7.5	4.1/6.5	2.3/4.5	2.2/3.6	5.1/6.3
Latitude -15	Average	3.8	4.8	5.8	7.3	7.7	8.1	7.8	7.5	6.5	5.8	4.0	3.3	6.0
	Min/Max	2.5/4.8	3.7/6.4	4.7/7.8	5.9/9.4	6.0/9.6	5.8/9.3	5.7/9.4	6.4/8.8	5.3/8.0	4.6/7.4	2.6/5.3	2.5/4.5	5.4/6.7
Latitude	Average	4.1	5.0	6.0	7.2	7.5	7.8	7.5	7.3	6.5	6.0	4.2	3.6	6.1
	Min/Max	2.7/5.2	3.9/6.9	4.8/8.0	5.8/9.3	5.8/9.3	5.6/8.9	5.5/9.1	6.3/8.6	5.3/8.1	4.8/7.8	2.7/5.8	2.7/5.1	5.5/6.8
Latitude +15	Average	4.2	5.1	5.9	6.9	7.0	7.3	7.0	6.9	6.3	6.1	4.4	3.8	5.9
	Min/Max	2.7/5.5	3.9/7.0	4.7/7.9	5.5/9.0	5.4/8.8	5.2/8.3	5.1/8.5	6.0/8.2	5.2/7.9	4.7/7.9	2.7/6.0	2.8/5.4	5.3/6.6

Solar Radiation for 2-Axis Tracking Flat-Plate Collectors (kWh/m²/day), Uncertainty ±9%

Tracker		Jan	Feb	Mar	Apr	May	June	July	Aug	Sept	Oct	Nov	Dec	Year
2-Axis	Average	4.3	5.1	6.0	7.3	7.8	8.3	7.9	7.5	6.5	6.1	4.4	3.8	6.3
	Min/Max	2.8/5.6	3.9/7.1	4.8/8.0	5.9/9.4	6.1/9.7	5.9/9.5	5.8/9.6	6.5/8.8	5.4/8.1	4.8/7.9	2.7/6.0	2.8/5.4	5.7/7.0

Direct Beam Solar Radiation for Concentrating Collectors (kWh/m²/day), Uncertainty ±8%

Tracker		Jan	Feb	Mar	Apr	May	June	July	Aug	Sept	Oct	Nov	Dec	Year
1-Axis, E-W Horiz Axis	Average	2.4	2.7	2.8	3.4	3.5	3.7	3.4	3.2	2.9	3.2	2.5	2.2	3.0
	Min/Max	1.3/3.5	1.5/4.4	1.9/4.4	2.3/5.0	2.3/5.2	1.9/4.8	1.8/4.9	2.5/4.3	2.2/4.2	2.2/4.6	1.1/3.8	1.4/3.7	2.6/3.6
1-Axis, N-S Horiz Axis	Average	1.8	2.5	3.3	4.4	4.6	4.8	4.4	4.2	3.6	3.3	2.0	1.6	3.4
	Min/Max	1.0/2.7	1.4/4.1	2.2/5.2	2.9/6.5	3.0/6.8	2.3/6.3	2.4/6.5	3.1/5.7	2.5/5.1	2.2/4.8	0.9/3.3	1.0/2.8	2.8/4.1
1-Axis, N-S Tilt=Latitude	Average	2.6	3.2	3.8	4.6	4.4	4.5	4.2	4.2	3.9	4.1	2.8	2.4	3.7
	Min/Max	1.4/3.8	1.8/5.3	2.5/5.9	3.0/6.8	2.9/6.6	2.2/5.8	2.3/6.1	3.2/5.8	2.8/5.6	2.7/5.8	1.2/4.4	1.5/4.0	3.2/4.6
2-Axis	Average	2.8	3.3	3.8	4.7	4.7	4.9	4.5	4.4	3.9	4.1	2.9	2.6	3.9
	Min/Max	1.5/4.1	1.9/5.5	2.5/5.9	3.1/6.9	3.1/7.0	2.4/6.3	2.4/6.6	3.3/5.9	2.8/5.6	2.7/5.9	1.3/4.6	1.6/4.3	3.3/4.8

Average Climatic Conditions

Element	Jan	Feb	Mar	Apr	May	June	July	Aug	Sept	Oct	Nov	Dec	Year
Temperature (°C)	2.3	4.7	10.1	15.1	19.8	24.2	26.3	25.6	22.1	15.8	10.0	4.7	15.1
Daily Minimum Temp	-3.1	-1.2	3.9	8.6	13.7	18.2	20.5	19.8	16.2	9.1	4.2	-0.6	9.1
Daily Maximum Temp	7.7	10.4	16.2	21.6	26.0	30.3	31.9	31.3	28.1	22.5	15.8	10.1	21.0
Record Minimum Temp	-27.2	-25.0	-16.7	-5.0	1.1	5.6	10.6	8.3	2.2	-3.3	-18.3	-23.3	-27.2
Record Maximum Temp	25.6	28.9	30.0	32.8	36.1	41.1	41.7	40.0	40.6	34.4	28.9	26.1	41.7
HDD, Base 18.3°C	496	383	261	107	33	0	0	0	12	108	250	422	2072
CDD, Base 18.3°C	0	0	6	11	79	177	246	226	125	29	0	0	898
Relative Humidity (%)	70	69	65	63	69	70	73	73	74	69	70	72	70
Wind Speed (m/s)	4.1	4.2	4.5	4.2	3.5	3.3	3.0	2.9	3.0	3.2	3.8	4.2	3.7

Abliene, TX

WBAN NO. 13962

LATITUDE: 32.43° N
LONGITUDE: 99.68° W
ELEVATION: 534 meters
MEAN PRESSURE: 953 millibars

STATION TYPE: Secondary

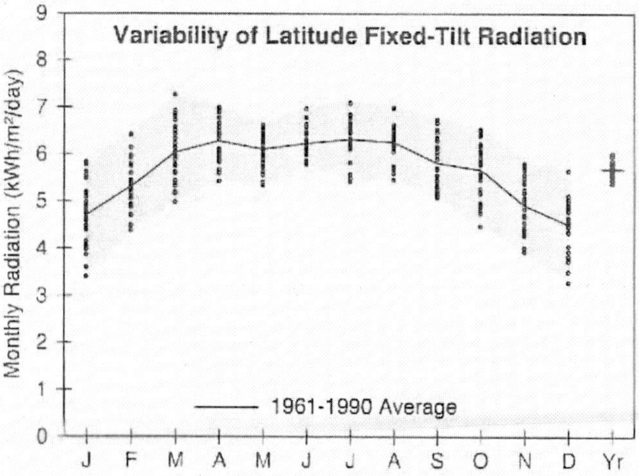

Solar Radiation for Flat-Plate Collectors Facing South at a Fixed Tilt (kWh/m²/day), Uncertainty ±9%

Tilt (°)		Jan	Feb	Mar	Apr	May	June	July	Aug	Sept	Oct	Nov	Dec	Year
0	Average	3.1	3.9	5.1	6.1	6.5	7.0	7.0	6.3	5.2	4.4	3.3	2.9	5.1
	Min/Max	2.5/3.6	3.4/4.6	4.4/6.0	5.3/6.8	5.7/7.1	6.5/7.9	6.0/7.9	5.5/7.1	4.6/5.9	3.7/4.9	2.9/3.8	2.3/3.3	4.9/5.3
Latitude -15	Average	4.1	4.8	5.8	6.4	6.5	6.8	6.8	6.5	5.7	5.3	4.3	3.9	5.6
	Min/Max	3.1/5.0	4.0/5.7	4.8/6.9	5.5/7.1	5.7/7.1	6.3/7.6	5.8/7.7	5.6/7.3	5.0/6.6	4.2/6.0	3.5/5.0	2.9/4.7	5.3/5.9
Latitude	Average	4.7	5.3	6.0	6.3	6.1	6.2	6.3	6.3	5.8	5.7	4.9	4.5	5.7
	Min/Max	3.4/5.8	4.4/6.4	5.0/7.3	5.4/7.0	5.3/6.6	5.8/7.0	5.4/7.1	5.4/7.0	5.1/6.7	4.5/6.5	3.9/5.8	3.3/5.7	5.4/6.0
Latitude +15	Average	5.1	5.5	6.0	5.8	5.4	5.4	5.5	5.7	5.6	5.8	5.2	4.9	5.5
	Min/Max	3.5/6.3	4.5/6.7	4.9/7.2	5.1/6.5	4.8/5.8	5.0/6.0	4.7/6.1	5.0/6.3	4.9/6.5	4.5/6.7	4.1/6.2	3.5/6.2	5.2/5.8
90	Average	4.4	4.4	4.0	3.1	2.3	2.0	2.2	2.7	3.4	4.3	4.4	4.4	3.5
	Min/Max	3.0/5.7	3.5/5.4	3.3/4.9	2.8/3.4	2.2/2.4	2.0/2.1	2.0/2.3	2.5/2.9	2.9/4.0	3.3/5.1	3.4/5.4	3.0/5.7	3.2/3.7

Solar Radiation for 1-Axis Tracking Flat-Plate Collectors with a North-South Axis (kWh/m²/day), Uncertainty ±9%

Axis Tilt (°)		Jan	Feb	Mar	Apr	May	June	July	Aug	Sept	Oct	Nov	Dec	Year
0	Average	4.5	5.6	7.1	8.2	8.6	9.2	9.3	8.5	7.1	6.2	4.8	4.2	7.0
	Min/Max	3.2/5.6	4.6/6.8	5.6/8.8	6.8/9.4	7.2/9.6	8.2/10.7	7.6/11.0	7.1/10.0	6.0/8.5	4.8/7.3	3.8/5.8	3.0/5.2	6.5/7.4
Latitude -15	Average	5.2	6.2	7.6	8.5	8.6	9.1	9.3	8.7	7.5	6.9	5.5	4.9	7.3
	Min/Max	3.7/6.6	5.1/7.7	6.0/9.5	7.0/9.8	7.2/9.6	8.1/10.6	7.6/11.0	7.2/10.2	6.3/9.0	5.3/8.1	4.3/6.7	3.5/6.2	6.9/7.9
Latitude	Average	5.7	6.6	7.8	8.4	8.4	8.7	8.9	8.6	7.6	7.2	6.0	5.4	7.4
	Min/Max	3.9/7.2	5.3/8.2	6.1/9.8	7.0/9.7	7.0/9.3	7.8/10.2	7.3/10.6	7.1/10.0	6.4/9.1	5.4/8.5	4.6/7.3	3.8/6.9	6.9/8.0
Latitude +15	Average	6.0	6.7	7.7	8.1	7.9	8.2	8.4	8.2	7.4	7.3	6.2	5.7	7.3
	Min/Max	4.0/7.6	5.4/8.4	6.0/9.8	6.7/9.3	6.5/8.8	7.3/9.5	6.8/9.9	6.7/9.6	6.2/8.9	5.4/8.6	4.8/7.6	3.9/7.4	6.8/7.9

Solar Radiation for 2-Axis Tracking Flat-Plate Collectors (kWh/m²/day), Uncertainty ±9%

Tracker		Jan	Feb	Mar	Apr	May	June	July	Aug	Sept	Oct	Nov	Dec	Year
2-Axis	Average	6.0	6.7	7.8	8.5	8.7	9.3	9.4	8.8	7.6	7.3	6.3	5.8	7.7
	Min/Max	4.1/7.7	5.4/8.4	6.1/9.8	7.1/9.8	7.3/9.7	8.3/10.9	7.7/11.2	7.2/10.3	6.4/9.1	5.5/8.6	4.8/7.7	4.0/7.5	7.2/8.2

Direct Beam Solar Radiation for Concentrating Collectors (kWh/m²/day), Uncertainty ±8%

Tracker		Jan	Feb	Mar	Apr	May	June	July	Aug	Sept	Oct	Nov	Dec	Year
1-Axis, E-W Horiz Axis	Average	3.7	3.9	4.1	4.3	4.4	4.9	4.9	4.4	3.8	4.1	3.8	3.7	4.2
	Min/Max	2.0/5.2	2.7/5.3	2.7/5.9	3.0/5.4	3.1/5.3	3.8/6.3	3.5/6.6	3.2/5.7	2.8/5.0	2.6/5.2	2.5/5.0	2.2/5.2	3.7/4.6
1-Axis, N-S Horiz Axis	Average	3.1	3.9	5.0	5.7	5.8	6.4	6.5	5.9	4.8	4.5	3.4	3.0	4.9
	Min/Max	1.7/4.4	2.8/5.4	3.2/7.2	4.1/7.3	4.2/7.1	5.1/8.3	4.6/8.8	4.2/7.8	3.5/6.5	2.9/5.7	2.3/4.5	1.8/4.2	4.3/5.4
1-Axis, N-S Tilt=Latitude	Average	4.2	4.8	5.6	5.9	5.6	6.0	6.2	5.9	5.2	5.3	4.4	4.0	5.3
	Min/Max	2.3/5.8	3.5/6.6	3.6/8.0	4.2/7.5	4.0/6.9	4.7/7.8	4.4/8.4	4.2/7.8	3.8/7.0	3.3/6.7	2.9/5.9	2.4/5.7	4.6/5.9
2-Axis	Average	4.5	4.9	5.6	6.0	5.9	6.5	6.6	6.1	5.2	5.4	4.6	4.4	5.5
	Min/Max	2.5/6.3	3.5/6.8	3.6/8.0	4.2/7.7	4.3/7.3	5.1/8.4	4.7/9.0	4.3/8.0	3.8/7.0	3.4/6.8	3.1/6.2	2.6/6.2	4.8/6.1

Average Climatic Conditions

Element	Jan	Feb	Mar	Apr	May	June	July	Aug	Sept	Oct	Nov	Dec	Year
Temperature (°C)	6.0	8.6	13.4	18.6	22.6	26.8	28.9	28.4	24.4	19.1	12.7	7.5	18.1
Daily Minimum Temp	-0.7	1.7	6.3	11.6	16.2	20.5	22.6	22.1	18.5	12.7	6.3	1.1	11.6
Daily Maximum Temp	12.7	15.4	20.5	25.4	29.1	33.0	35.1	34.7	30.4	25.5	19.1	13.9	24.6
Record Minimum Temp	-22.8	-21.7	-13.9	-3.9	2.2	8.3	12.8	12.8	1.7	-2.2	-10.0	-21.7	-22.8
Record Maximum Temp	31.7	32.2	36.1	37.2	41.7	42.8	43.3	42.8	41.1	39.4	33.3	31.7	43.3
HDD, Base 18.3°C	382	274	166	52	4	0	0	0	3	42	176	336	1436
CDD, Base 18.3°C	0	0	13	59	136	253	327	313	187	66	8	0	1362
Relative Humidity (%)	60	60	55	55	60	59	53	55	62	61	62	61	59
Wind Speed (m/s)	5.0	5.2	5.9	5.8	5.7	5.1	4.6	4.2	4.4	4.7	5.0	4.8	5.0

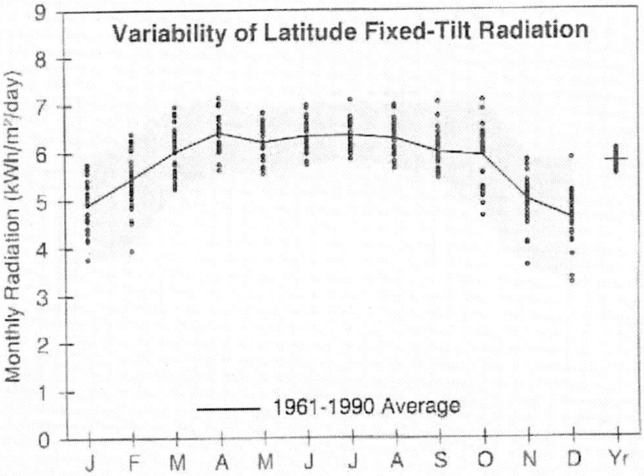

Amarillo, TX

WBAN NO. 23047

LATITUDE: 35.23° N
LONGITUDE: 101.70° W
ELEVATION: 1098 meters
MEAN PRESSURE: 891 millibars

STATION TYPE: Secondary

Solar Radiation for Flat-Plate Collectors Facing South at a Fixed Tilt (kWh/m²/day), Uncertainty ±9%

Tilt (°)		Jan	Feb	Mar	Apr	May	June	July	Aug	Sept	Oct	Nov	Dec	Year
0	Average	3.0	3.8	4.9	6.1	6.6	7.1	7.0	6.3	5.2	4.4	3.2	2.7	5.0
	Min/Max	2.5/3.3	3.0/4.3	4.4/5.6	5.4/6.7	5.9/7.3	6.4/7.9	6.4/7.8	5.7/7.0	4.9/6.0	3.7/5.0	2.6/3.6	2.1/3.1	4.8/5.2
Latitude -15	Average	4.2	4.9	5.8	6.5	6.6	6.9	6.9	6.5	5.9	5.5	4.4	3.9	5.7
	Min/Max	3.3/4.9	3.7/5.7	5.1/6.6	5.7/7.3	5.9/7.3	6.2/7.7	6.3/7.7	5.9/7.3	5.4/6.9	4.4/6.5	3.3/5.1	2.9/4.9	5.4/5.9
Latitude	Average	4.9	5.4	6.0	6.4	6.2	6.3	6.4	6.3	6.0	5.9	5.0	4.6	5.8
	Min/Max	3.8/5.8	3.9/6.4	5.2/7.0	5.6/7.2	5.6/6.8	5.8/7.0	5.9/7.1	5.7/7.0	5.5/7.1	4.7/7.1	3.6/5.8	3.2/5.9	5.5/6.1
Latitude +15	Average	5.3	5.7	5.9	6.0	5.5	5.5	5.5	5.7	5.8	6.0	5.3	5.0	5.6
	Min/Max	4.0/6.3	4.0/6.7	5.1/6.9	5.2/6.6	4.9/6.1	5.0/6.0	5.1/6.1	5.1/6.3	5.2/6.8	4.7/7.3	3.8/6.3	3.5/6.5	5.3/5.9
90	Average	4.8	4.7	4.2	3.4	2.6	2.2	2.4	2.9	3.7	4.7	4.7	4.6	3.7
	Min/Max	3.5/5.7	3.2/5.6	3.5/4.9	3.0/3.7	2.4/2.7	2.1/2.3	2.3/2.5	2.7/3.2	3.3/4.3	3.5/5.7	3.2/5.6	3.1/6.1	3.5/4.0

Solar Radiation for 1-Axis Tracking Flat-Plate Collectors with a North-South Axis (kWh/m²/day), Uncertainty ±9%

Axis Tilt (°)		Jan	Feb	Mar	Apr	May	June	July	Aug	Sept	Oct	Nov	Dec	Year
0	Average	4.5	5.6	7.0	8.4	8.8	9.5	9.4	8.6	7.3	6.4	4.8	4.1	7.0
	Min/Max	3.4/5.3	3.9/6.6	5.9/8.2	7.2/9.5	7.7/9.9	8.2/10.9	8.3/10.9	7.4/9.9	6.6/8.8	5.0/7.8	3.5/5.6	2.9/5.3	6.6/7.4
Latitude -15	Average	5.4	6.4	7.6	8.8	8.9	9.4	9.4	8.8	7.8	7.2	5.6	5.0	7.5
	Min/Max	4.1/6.5	4.4/7.7	6.4/9.0	7.5/10.0	7.7/9.9	8.1/10.8	8.3/10.8	7.6/10.2	7.0/9.5	5.6/8.9	4.0/6.7	3.4/6.5	7.0/7.9
Latitude	Average	6.0	6.8	7.8	8.7	8.6	9.0	9.0	8.7	7.9	7.6	6.1	5.5	7.6
	Min/Max	4.4/7.2	4.7/8.2	6.5/9.3	7.4/9.9	7.5/9.7	7.8/10.4	8.0/10.4	7.4/10.0	7.1/9.6	5.8/9.3	4.3/7.3	3.8/7.3	7.1/8.0
Latitude +15	Average	6.3	6.9	7.7	8.4	8.1	8.4	8.4	8.2	7.7	7.7	6.4	5.9	7.5
	Min/Max	4.6/7.6	4.7/8.4	6.4/9.2	7.1/9.6	7.1/9.1	7.2/9.7	7.5/9.8	7.0/9.5	6.9/9.4	5.8/9.4	4.4/7.6	3.9/7.7	7.0/7.9

Solar Radiation for 2-Axis Tracking Flat-Plate Collectors (kWh/m²/day), Uncertainty ±9%

Tracker		Jan	Feb	Mar	Apr	May	June	July	Aug	Sept	Oct	Nov	Dec	Year
2-Axis	Average	6.3	7.0	7.8	8.8	9.0	9.6	9.5	8.9	7.9	7.7	6.4	6.0	7.9
	Min/Max	4.6/7.6	4.7/8.4	6.5/9.3	7.5/10.0	7.8/10.1	8.3/11.1	8.4/11.0	7.6/10.2	7.1/9.6	5.8/9.4	4.4/7.7	4.0/7.9	7.4/8.3

Direct Beam Solar Radiation for Concentrating Collectors (kWh/m²/day), Uncertainty ±8%

Tracker		Jan	Feb	Mar	Apr	May	June	July	Aug	Sept	Oct	Nov	Dec	Year
1-Axis, E-W Horiz Axis	Average	4.0	4.1	4.2	4.6	4.6	5.2	5.0	4.6	4.2	4.6	4.0	3.9	4.4
	Min/Max	2.5/5.3	2.3/5.3	3.1/5.4	3.5/5.6	3.6/5.5	4.0/6.5	4.1/6.5	3.6/5.8	3.5/5.6	3.0/6.0	2.2/5.2	2.3/5.7	3.9/4.8
1-Axis, N-S Horiz Axis	Average	3.3	4.0	5.0	6.1	6.2	6.8	6.7	6.1	5.2	4.9	3.5	3.0	5.1
	Min/Max	2.0/4.3	2.2/5.3	3.5/6.4	4.7/7.5	4.9/7.4	5.2/8.6	5.3/8.6	4.7/7.9	4.5/7.0	3.2/6.5	2.0/4.5	1.7/4.4	4.5/5.4
1-Axis, N-S Tilt=Latitude	Average	4.5	5.0	5.7	6.3	6.0	6.4	6.3	6.1	5.7	5.9	4.7	4.3	5.6
	Min/Max	2.8/5.9	2.8/6.6	4.1/7.3	4.8/7.7	4.7/7.2	4.9/8.1	5.1/8.1	4.7/7.9	4.9/7.7	3.8/7.8	2.6/6.0	2.5/6.3	4.9/6.0
2-Axis	Average	4.8	5.2	5.7	6.4	6.3	6.9	6.8	6.3	5.7	6.0	4.9	4.6	5.8
	Min/Max	3.0/6.4	2.9/6.8	4.1/7.4	4.9/7.9	5.0/7.6	5.3/8.8	5.4/8.7	4.8/8.1	4.9/7.7	3.9/7.9	2.8/6.3	2.7/6.8	5.1/6.2

Average Climatic Conditions

Element	Jan	Feb	Mar	Apr	May	June	July	Aug	Sept	Oct	Nov	Dec	Year
Temperature (°C)	1.7	4.0	8.4	13.8	18.6	23.4	25.9	24.7	20.6	14.7	7.8	2.7	13.8
Daily Minimum Temp	-6.0	-3.6	0.4	5.6	10.9	15.9	18.6	17.7	13.6	6.9	0.2	-4.6	6.3
Daily Maximum Temp	9.4	11.6	16.4	21.9	26.2	30.9	33.2	31.7	27.7	22.5	15.4	10.1	21.4
Record Minimum Temp	-23.9	-25.6	-19.4	-10.0	-2.2	5.6	10.6	9.4	-1.1	-9.4	-17.8	-22.2	-25.6
Record Maximum Temp	27.2	31.1	34.4	36.7	38.9	42.2	40.6	41.1	38.9	35.0	30.6	27.2	42.2
HDD, Base 18.3°C	515	401	308	148	49	3	0	0	14	126	317	484	2366
CDD, Base 18.3°C	0	0	0	11	57	155	234	198	83	14	0	0	752
Relative Humidity (%)	58	59	52	48	54	56	53	58	61	56	59	59	56
Wind Speed (m/s)	5.9	6.2	6.8	6.8	6.6	6.2	5.8	5.3	5.7	5.8	5.9	5.7	6.0

Austin, TX

WBAN NO. 13958

LATITUDE: 30.30° N
LONGITUDE: 97.70° W
ELEVATION: 189 meters
MEAN PRESSURE: 994 millibars

STATION TYPE: Secondary

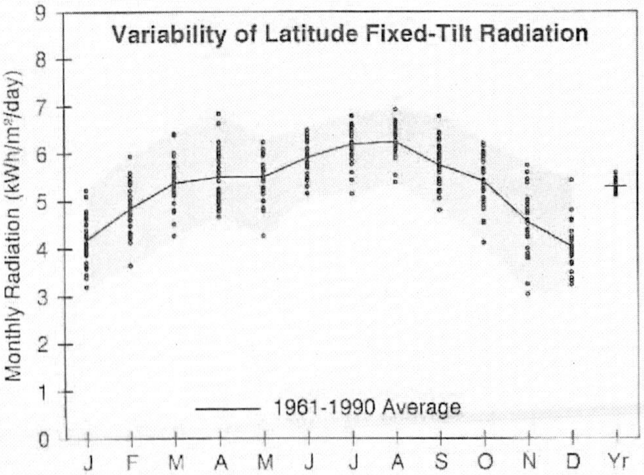

Solar Radiation for Flat-Plate Collectors Facing South at a Fixed Tilt (kWh/m²/day), Uncertainty ±9%

Tilt (°)		Jan	Feb	Mar	Apr	May	June	July	Aug	Sept	Oct	Nov	Dec	Year
0	Average	3.0	3.8	4.7	5.4	5.9	6.6	6.8	6.3	5.2	4.4	3.3	2.8	4.9
	Min/Max	2.5/3.5	3.1/4.4	3.9/5.5	4.7/6.6	4.6/6.7	5.7/7.3	5.6/7.5	5.5/7.0	4.5/6.1	3.5/4.9	2.5/3.9	2.4/3.4	4.7/5.1
Latitude -15	Average	3.7	4.4	5.2	5.6	5.8	6.4	6.7	6.5	5.7	5.0	4.1	3.5	5.2
	Min/Max	2.9/4.5	3.5/5.3	4.2/6.1	4.8/6.9	4.5/6.6	5.6/7.1	5.5/7.3	5.6/7.2	4.8/6.6	3.9/5.7	2.8/5.0	2.9/4.6	5.1/5.5
Latitude	Average	4.2	4.8	5.4	5.5	5.5	5.9	6.2	6.3	5.8	5.4	4.6	4.0	5.3
	Min/Max	3.2/5.2	3.7/5.9	4.3/6.4	4.7/6.8	4.3/6.2	5.2/6.5	5.2/6.8	5.4/6.9	4.8/6.8	4.1/6.2	3.0/5.8	3.2/5.4	5.1/5.6
Latitude +15	Average	4.4	5.0	5.3	5.1	4.9	5.1	5.4	5.7	5.5	5.5	4.8	4.3	5.1
	Min/Max	3.3/5.7	3.7/6.2	4.1/6.3	4.3/6.4	3.8/5.5	4.5/5.6	4.6/5.9	4.9/6.3	4.6/6.6	4.1/6.3	3.1/6.2	3.4/6.0	4.9/5.4
90	Average	3.8	3.8	3.4	2.7	2.1	1.9	2.0	2.6	3.3	4.0	4.0	3.8	3.1
	Min/Max	2.7/4.9	2.7/4.8	2.7/4.1	2.3/3.1	1.9/2.2	1.7/1.9	1.9/2.1	2.3/2.7	2.8/3.8	2.9/4.6	2.4/5.2	2.8/5.4	2.9/3.3

Solar Radiation for 1-Axis Tracking Flat-Plate Collectors with a North-South Axis (kWh/m²/day), Uncertainty ±9%

Axis Tilt (°)		Jan	Feb	Mar	Apr	May	June	July	Aug	Sept	Oct	Nov	Dec	Year
0	Average	4.1	5.1	6.2	6.9	7.4	8.4	8.8	8.3	6.9	6.0	4.6	3.8	6.4
	Min/Max	3.1/5.2	3.9/6.4	4.8/7.6	5.6/9.0	5.3/8.5	6.9/9.6	6.9/10.1	7.0/9.6	5.5/8.5	4.3/7.0	3.0/5.8	3.0/5.2	6.1/6.8
Latitude -15	Average	4.6	5.6	6.6	7.1	7.3	8.3	8.7	8.5	7.3	6.5	5.1	4.4	6.7
	Min/Max	3.4/5.9	4.1/7.1	5.0/8.1	5.8/9.3	5.2/8.5	6.9/9.5	6.8/10.1	7.1/9.7	5.7/8.9	4.6/7.6	3.3/6.6	3.3/6.1	6.3/7.2
Latitude	Average	5.0	5.9	6.7	7.0	7.1	8.0	8.4	8.3	7.3	6.8	5.5	4.8	6.7
	Min/Max	3.6/6.5	4.3/7.5	5.1/8.3	5.7/9.2	5.1/8.2	6.6/9.1	6.6/9.7	7.0/9.6	5.7/9.1	4.8/7.9	3.4/7.2	3.6/6.7	6.4/7.2
Latitude +15	Average	5.2	6.0	6.6	6.7	6.7	7.4	7.8	7.9	7.2	6.8	5.7	5.0	6.6
	Min/Max	3.7/6.8	4.3/7.7	5.0/8.3	5.4/8.8	4.7/7.7	6.1/8.4	6.2/9.0	6.7/9.1	5.6/8.9	4.8/8.0	3.5/7.5	3.8/7.2	6.3/7.1

Solar Radiation for 2-Axis Tracking Flat-Plate Collectors (kWh/m²/day), Uncertainty ±9%

Tracker		Jan	Feb	Mar	Apr	May	June	July	Aug	Sept	Oct	Nov	Dec	Year
2-Axis	Average	5.2	6.0	6.7	7.1	7.4	8.5	8.9	8.5	7.4	6.9	5.7	5.1	7.0
	Min/Max	3.7/6.9	4.3/7.7	5.1/8.4	5.8/9.3	5.3/8.6	7.0/9.7	6.9/10.2	7.2/9.8	5.8/9.1	4.8/8.0	3.5/7.5	3.8/7.3	6.6/7.5

Direct Beam Solar Radiation for Concentrating Collectors (kWh/m²/day), Uncertainty ±8%

Tracker		Jan	Feb	Mar	Apr	May	June	July	Aug	Sept	Oct	Nov	Dec	Year
1-Axis, E-W Horiz Axis	Average	3.0	3.3	3.3	3.3	3.4	4.2	4.5	4.3	3.6	3.7	3.3	3.0	3.6
	Min/Max	1.6/4.4	1.9/4.7	2.0/4.6	2.2/5.1	1.4/4.4	3.0/5.3	3.0/5.8	3.2/5.5	2.4/5.0	2.1/4.7	1.4/4.9	1.9/4.9	3.2/4.0
1-Axis, N-S Horiz Axis	Average	2.6	3.4	4.0	4.3	4.4	5.4	5.9	5.6	4.5	4.1	3.0	2.5	4.2
	Min/Max	1.4/3.9	2.0/4.9	2.4/5.6	2.9/6.7	1.9/5.7	3.9/6.9	3.9/7.7	4.1/7.2	2.9/6.4	2.2/5.3	1.4/4.5	1.5/4.1	3.7/4.7
1-Axis, N-S Tilt=Latitude	Average	3.4	4.0	4.4	4.3	4.2	5.0	5.6	5.6	4.8	4.7	3.8	3.3	4.4
	Min/Max	1.8/5.0	2.4/5.8	2.7/6.2	3.0/6.9	1.8/5.5	3.6/6.4	3.7/7.2	4.1/7.2	3.1/6.8	2.6/6.1	1.7/5.7	2.0/5.4	4.0/5.1
2-Axis	Average	3.6	4.1	4.4	4.4	4.5	5.5	6.0	5.7	4.9	4.8	4.0	3.6	4.6
	Min/Max	2.0/5.3	2.4/6.0	2.7/6.2	3.0/7.0	1.9/5.8	3.9/7.0	4.0/7.8	4.2/7.4	3.2/6.8	2.7/6.2	1.7/6.0	2.2/5.8	4.1/5.3

Average Climatic Conditions

Element	Jan	Feb	Mar	Apr	May	June	July	Aug	Sept	Oct	Nov	Dec	Year
Temperature (°C)	9.3	11.6	16.4	20.9	24.2	27.4	29.2	29.3	26.8	21.7	16.1	10.9	20.3
Daily Minimum Temp	3.7	5.6	10.6	15.4	19.2	21.9	23.3	23.3	21.0	15.6	9.9	5.1	14.6
Daily Maximum Temp	14.9	17.4	22.2	26.3	29.3	32.8	35.0	35.3	32.5	27.8	22.1	16.7	26.1
Record Minimum Temp	-18.9	-13.9	-7.8	1.7	6.1	11.7	17.8	16.1	5.0	0.0	-6.7	-15.6	-18.9
Record Maximum Temp	32.2	36.1	36.7	36.7	37.8	40.6	42.8	41.1	40.0	36.7	32.8	32.2	42.8
HDD, Base 18.3°C	286	196	97	16	0	0	0	0	0	9	97	237	938
CDD, Base 18.3°C	7	6	37	92	183	272	336	341	253	114	29	6	1676
Relative Humidity (%)	67	66	64	66	71	69	65	64	68	68	69	68	67
Wind Speed (m/s)	4.3	4.6	4.7	4.5	4.2	3.8	3.5	3.4	3.5	3.5	4.0	4.1	4.0

Brownsville, TX

WBAN NO. 12919

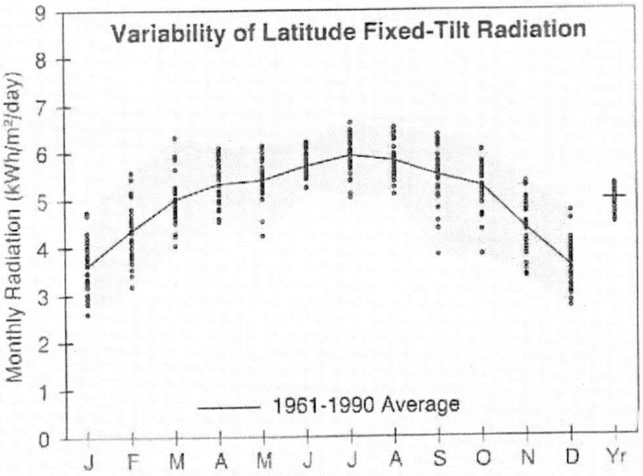

LATITUDE: 25.90° N
LONGITUDE: 97.43° W
ELEVATION: 6 meters
MEAN PRESSURE: 1015 millibars

STATION TYPE: Primary

Solar Radiation for Flat-Plate Collectors Facing South at a Fixed Tilt (kWh/m²/day), Uncertainty ±9%

Tilt (°)		Jan	Feb	Mar	Apr	May	June	July	Aug	Sept	Oct	Nov	Dec	Year
0	Average	2.9	3.7	4.6	5.3	5.8	6.4	6.5	6.0	5.2	4.5	3.4	2.7	4.8
	Min/Max	2.2/3.6	2.8/4.5	3.8/5.6	4.6/6.0	4.5/6.6	5.8/7.0	5.5/7.4	5.4/6.8	3.8/5.9	3.6/5.0	2.8/4.0	2.3/3.4	4.4/5.1
Latitude -15	Average	3.2	4.0	4.8	5.4	5.7	6.2	6.4	6.0	5.4	4.9	3.9	3.1	4.9
	Min/Max	2.4/4.1	3.0/5.0	3.9/6.0	4.6/6.2	4.4/6.5	5.6/6.7	5.4/7.2	5.3/6.8	3.9/6.2	3.7/5.6	3.1/4.7	2.5/4.1	4.5/5.2
Latitude	Average	3.6	4.3	5.0	5.3	5.4	5.7	5.9	5.8	5.5	5.3	4.4	3.6	5.0
	Min/Max	2.6/4.7	3.2/5.6	4.0/6.3	4.5/6.1	4.2/6.1	5.2/6.2	5.0/6.6	5.1/6.5	3.8/6.4	3.8/6.0	3.4/5.4	2.7/4.7	4.5/5.3
Latitude +15	Average	3.8	4.4	4.9	5.0	4.8	5.0	5.2	5.3	5.3	5.4	4.6	3.8	4.8
	Min/Max	2.7/5.1	3.2/5.8	3.9/6.3	4.2/5.7	3.8/5.4	4.6/5.4	4.5/5.8	4.7/5.9	3.6/6.2	3.8/6.2	3.4/5.7	2.8/5.1	4.2/5.1
90	Average	3.0	3.2	3.0	2.3	1.8	1.6	1.7	2.1	2.9	3.7	3.6	3.1	2.7
	Min/Max	2.1/4.2	2.3/4.3	2.4/3.8	2.1/2.6	1.7/1.9	1.5/1.8	1.5/1.9	1.9/2.3	2.1/3.3	2.5/4.2	2.5/4.6	2.2/4.4	2.3/2.9

Solar Radiation for 1-Axis Tracking Flat-Plate Collectors with a North-South Axis (kWh/m²/day), Uncertainty ±9%

Axis Tilt (°)		Jan	Feb	Mar	Apr	May	June	July	Aug	Sept	Oct	Nov	Dec	Year
0	Average	3.6	4.7	5.7	6.5	7.1	8.0	8.3	7.7	6.8	6.0	4.6	3.5	6.1
	Min/Max	2.6/4.9	3.3/6.3	4.4/7.6	5.4/7.5	5.0/8.4	7.0/9.1	6.6/9.8	6.8/9.2	4.3/8.2	4.1/6.9	3.4/5.7	2.6/4.7	5.3/6.6
Latitude -15	Average	3.9	5.0	6.0	6.6	7.1	7.9	8.2	7.8	6.9	6.3	4.9	3.8	6.2
	Min/Max	2.7/5.3	3.5/6.7	4.6/7.9	5.4/7.7	5.0/8.3	6.9/9.0	6.6/9.7	6.8/9.2	4.3/8.4	4.3/7.3	3.6/6.2	2.8/5.1	5.4/6.8
Latitude	Average	4.2	5.2	6.1	6.6	6.8	7.6	7.9	7.7	7.0	6.6	5.3	4.1	6.3
	Min/Max	2.9/5.8	3.6/7.1	4.7/8.1	5.4/7.6	4.8/8.1	6.6/8.6	6.3/9.4	6.7/9.1	4.3/8.5	4.3/7.7	3.8/6.7	3.0/5.7	5.4/6.9
Latitude +15	Average	4.3	5.3	6.0	6.3	6.4	7.1	7.4	7.3	6.9	6.7	5.4	4.3	6.1
	Min/Max	3.0/6.0	3.6/7.2	4.6/8.0	5.2/7.3	4.5/7.6	6.2/8.0	5.9/8.7	6.4/8.6	4.2/8.4	4.3/7.7	3.8/7.0	3.1/6.0	5.2/6.7

Solar Radiation for 2-Axis Tracking Flat-Plate Collectors (kWh/m²/day), Uncertainty ±9%

Tracker		Jan	Feb	Mar	Apr	May	June	July	Aug	Sept	Oct	Nov	Dec	Year
2-Axis	Average	4.4	5.3	6.1	6.6	7.1	8.1	8.4	7.8	7.0	6.7	5.5	4.4	6.5
	Min/Max	3.0/6.1	3.7/7.2	4.7/8.1	5.5/7.7	5.0/8.4	7.1/9.2	6.6/9.9	6.8/9.3	4.3/8.5	4.4/7.8	3.8/7.0	3.1/6.1	5.6/7.1

Direct Beam Solar Radiation for Concentrating Collectors (kWh/m²/day), Uncertainty ±8%

Tracker		Jan	Feb	Mar	Apr	May	June	July	Aug	Sept	Oct	Nov	Dec	Year
1-Axis, E-W Horiz Axis	Average	2.2	2.6	2.8	2.9	3.1	3.8	4.0	3.6	3.2	3.3	2.9	2.3	3.0
	Min/Max	1.3/3.5	1.4/4.2	1.9/4.2	1.9/3.8	1.6/4.2	3.0/4.7	2.6/5.5	2.8/4.9	1.0/4.6	1.2/4.2	1.0/4.1	1.2/3.7	2.4/3.6
1-Axis, N-S Horiz Axis	Average	2.0	2.8	3.4	3.7	4.0	4.9	5.2	4.9	4.1	3.8	2.8	2.0	3.6
	Min/Max	1.2/3.3	1.5/4.4	2.2/5.2	2.5/5.0	2.1/5.6	3.8/6.2	3.4/7.2	3.8/6.7	1.4/6.0	1.4/4.8	1.0/4.0	1.0/3.2	2.9/4.4
1-Axis, N-S Tilt=Latitude	Average	2.5	3.2	3.6	3.7	3.8	4.5	4.9	4.8	4.3	4.3	3.4	2.5	3.8
	Min/Max	1.5/4.0	1.8/5.2	2.4/5.6	2.5/5.0	2.0/5.3	3.6/5.8	3.2/6.8	3.7/6.6	1.4/6.3	1.6/5.4	1.3/4.8	1.3/4.0	2.9/4.5
2-Axis	Average	2.6	3.3	3.6	3.8	4.0	4.9	5.3	4.9	4.3	4.4	3.5	2.7	4.0
	Min/Max	1.6/4.3	1.8/5.3	2.4/5.6	2.5/5.1	2.1/5.7	3.9/6.3	3.4/7.3	3.8/6.8	1.5/6.3	1.6/5.5	1.3/5.1	1.4/4.4	3.0/4.7

Average Climatic Conditions

Element	Jan	Feb	Mar	Apr	May	June	July	Aug	Sept	Oct	Nov	Dec	Year
Temperature (°C)	15.2	16.9	20.4	24.1	26.6	28.3	29.2	29.2	27.7	24.3	20.4	16.7	23.2
Daily Minimum Temp	9.9	11.4	15.1	19.2	22.2	23.8	24.3	24.1	22.9	18.9	15.0	11.3	18.2
Daily Maximum Temp	20.5	22.3	25.8	28.9	31.0	32.8	34.1	34.2	32.4	29.6	25.7	22.1	28.3
Record Minimum Temp	-7.2	-5.6	0.0	3.3	11.1	15.6	20.0	17.2	12.8	4.4	0.6	-8.9	-8.9
Record Maximum Temp	33.9	34.4	41.1	38.9	38.9	38.9	38.9	38.9	40.0	35.6	36.1	34.4	41.1
HDD, Base 18.3°C	129	75	23	0	0	0	0	0	0	0	28	98	353
CDD, Base 18.3°C	33	34	88	173	257	300	336	336	280	184	89	48	2160
Relative Humidity (%)	79	77	75	75	77	75	73	74	76	75	76	78	76
Wind Speed (m/s)	4.7	5.3	5.9	6.0	5.7	5.1	5.0	4.6	4.2	4.1	4.7	4.8	5.0

Corpus Christi, TX

WBAN NO. 12924

LATITUDE: 27.77° N
LONGITUDE: 97.50° W
ELEVATION: 13 meters
MEAN PRESSURE: 1014 millibars

STATION TYPE: Secondary

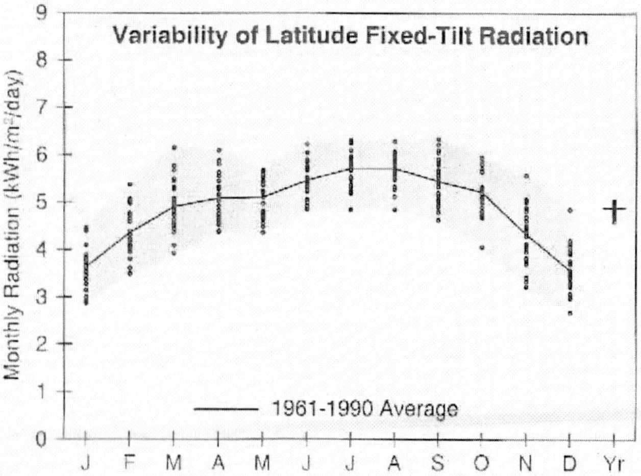

Solar Radiation for Flat-Plate Collectors Facing South at a Fixed Tilt (kWh/m²/day), Uncertainty ±9%

Tilt (°)		Jan	Feb	Mar	Apr	May	June	July	Aug	Sept	Oct	Nov	Dec	Year
0	Average	2.8	3.6	4.4	5.0	5.5	6.1	6.3	5.8	5.0	4.3	3.3	2.7	4.6
	Min/Max	2.4/3.3	3.0/4.2	3.6/5.4	4.4/6.0	4.7/6.1	5.4/7.0	5.2/6.9	5.0/6.4	4.4/5.8	3.5/4.8	2.6/4.0	2.2/3.3	4.3/4.7
Latitude -15	Average	3.2	4.0	4.7	5.2	5.4	5.9	6.1	5.9	5.3	4.8	3.9	3.1	4.8
	Min/Max	2.7/3.9	3.3/4.8	3.8/5.9	4.5/6.2	4.6/6.1	5.3/6.8	5.2/6.8	5.1/6.5	4.6/6.2	3.8/5.5	3.0/4.8	2.4/4.1	4.5/4.9
Latitude	Average	3.6	4.3	4.9	5.1	5.1	5.5	5.7	5.7	5.4	5.2	4.3	3.6	4.9
	Min/Max	2.9/4.5	3.5/5.4	3.9/6.2	4.4/6.1	4.3/5.7	4.8/6.2	4.8/6.3	4.8/6.3	4.6/6.3	4.1/5.9	3.2/5.6	2.7/4.8	4.6/5.0
Latitude +15	Average	3.8	4.5	4.8	4.7	4.6	4.8	5.0	5.2	5.2	5.3	4.6	3.8	4.7
	Min/Max	2.9/4.8	3.5/5.6	3.8/6.1	4.1/5.7	3.9/5.0	4.3/5.4	4.3/5.5	4.4/5.7	4.4/6.1	4.1/6.1	3.3/6.0	2.8/5.3	4.4/4.8
90	Average	3.1	3.3	3.0	2.3	1.8	1.7	1.8	2.2	3.0	3.7	3.7	3.2	2.7
	Min/Max	2.3/4.0	2.5/4.2	2.5/3.8	2.1/2.7	1.7/1.9	1.6/1.7	1.7/1.8	2.0/2.3	2.5/3.4	2.8/4.3	2.6/4.9	2.2/4.6	2.6/2.9

Solar Radiation for 1-Axis Tracking Flat-Plate Collectors with a North-South Axis (kWh/m²/day), Uncertainty ±9%

Axis Tilt (°)		Jan	Feb	Mar	Apr	May	June	July	Aug	Sept	Oct	Nov	Dec	Year
0	Average	3.6	4.7	5.6	6.3	6.8	7.7	8.0	7.6	6.7	5.9	4.5	3.5	5.9
	Min/Max	2.8/4.5	3.6/5.9	4.3/7.3	5.1/7.7	5.5/7.8	6.5/9.2	6.3/9.3	6.1/8.6	5.5/8.0	4.4/6.8	3.2/5.9	2.5/4.8	5.5/6.1
Latitude -15	Average	3.9	5.0	5.9	6.4	6.8	7.6	8.0	7.7	6.9	6.3	4.9	3.8	6.1
	Min/Max	3.0/5.0	3.8/6.4	4.4/7.7	5.1/7.9	5.5/7.8	6.4/9.1	6.3/9.3	6.2/8.6	5.7/8.3	4.7/7.3	3.4/6.5	2.7/5.3	5.6/6.3
Latitude	Average	4.2	5.2	6.0	6.3	6.5	7.3	7.7	7.5	7.0	6.5	5.2	4.2	6.1
	Min/Max	3.2/5.4	3.9/6.8	4.5/7.9	5.1/7.9	5.3/7.6	6.2/8.7	6.1/8.9	6.0/8.5	5.7/8.4	4.8/7.7	3.6/7.0	2.9/5.9	5.7/6.4
Latitude +15	Average	4.4	5.3	5.9	6.1	6.1	6.8	7.2	7.2	6.8	6.6	5.4	4.4	6.0
	Min/Max	3.2/5.7	4.0/6.9	4.4/7.8	4.8/7.6	5.0/7.1	5.8/8.1	5.7/8.3	5.7/8.1	5.6/8.2	4.8/7.8	3.7/7.3	3.0/6.3	5.6/6.3

Solar Radiation for 2-Axis Tracking Flat-Plate Collectors (kWh/m²/day), Uncertainty ±9%

Tracker		Jan	Feb	Mar	Apr	May	June	July	Aug	Sept	Oct	Nov	Dec	Year
2-Axis	Average	4.4	5.3	6.0	6.4	6.8	7.7	8.1	7.7	7.0	6.6	5.5	4.4	6.3
	Min/Max	3.3/5.7	4.0/6.9	4.5/7.9	5.1/7.9	5.5/7.9	6.6/9.3	6.4/9.4	6.2/8.7	5.7/8.4	4.9/7.8	3.8/7.4	3.0/6.4	5.9/6.6

Direct Beam Solar Radiation for Concentrating Collectors (kWh/m²/day), Uncertainty ±8%

Tracker		Jan	Feb	Mar	Apr	May	June	July	Aug	Sept	Oct	Nov	Dec	Year
1-Axis, E-W Horiz Axis	Average	2.3	2.7	2.8	2.8	2.9	3.6	3.9	3.6	3.2	3.4	3.0	2.4	3.1
	Min/Max	1.3/3.4	1.5/3.9	1.6/4.3	1.8/3.9	1.9/3.9	2.7/5.0	2.5/5.1	2.5/4.4	2.2/4.4	2.0/4.3	1.6/4.6	1.4/4.1	2.6/3.3
1-Axis, N-S Horiz Axis	Average	2.0	2.8	3.4	3.6	3.9	4.7	5.1	4.8	4.2	3.9	2.9	2.0	3.6
	Min/Max	1.2/3.1	1.6/4.1	1.9/5.2	2.3/5.1	2.5/5.2	3.5/6.5	3.3/6.7	3.3/5.8	2.9/5.7	2.3/4.9	1.5/4.4	1.1/3.5	3.1/3.9
1-Axis, N-S Tilt=Latitude	Average	2.6	3.3	3.7	3.7	3.7	4.3	4.8	4.7	4.4	4.4	3.5	2.6	3.8
	Min/Max	1.5/3.9	1.8/4.8	2.1/5.7	2.3/5.2	2.4/4.9	3.2/6.1	3.1/6.3	3.2/5.7	3.0/6.0	2.6/5.6	1.9/5.4	1.5/4.4	3.3/4.1
2-Axis	Average	2.8	3.4	3.7	3.7	3.9	4.7	5.1	4.9	4.4	4.5	3.7	2.8	4.0
	Min/Max	1.6/4.1	1.9/5.0	2.1/5.7	2.3/5.3	2.5/5.2	3.5/6.6	3.3/6.8	3.4/5.9	3.0/6.0	2.6/5.7	2.0/5.7	1.6/4.8	3.4/4.3

Average Climatic Conditions

Element	Jan	Feb	Mar	Apr	May	June	July	Aug	Sept	Oct	Nov	Dec	Year
Temperature (°C)	12.8	14.7	18.7	22.5	25.5	27.7	28.9	29.0	27.2	23.3	18.7	14.6	22.0
Daily Minimum Temp	7.4	8.9	12.9	17.3	20.8	23.0	23.8	23.9	22.4	17.7	13.1	9.1	16.7
Daily Maximum Temp	18.3	20.6	24.3	27.6	30.1	32.4	34.1	34.1	32.1	28.8	24.3	20.2	27.2
Record Minimum Temp	-10.0	-7.8	-4.4	0.6	8.3	14.4	17.8	17.8	10.0	3.9	-1.7	-10.6	-10.6
Record Maximum Temp	32.8	36.7	38.9	38.9	39.4	38.3	40.0	39.4	39.4	36.7	36.7	32.8	40.0
HDD, Base 18.3°C	188	118	56	5	0	0	0	0	0	0	53	145	564
CDD, Base 18.3°C	17	17	66	130	222	282	329	331	267	156	64	29	1911
Relative Humidity (%)	77	76	74	77	79	78	75	75	76	75	76	76	76
Wind Speed (m/s)	5.4	6.0	6.4	6.4	5.7	5.0	5.1	5.0	4.9	4.8	5.5	5.5	5.5

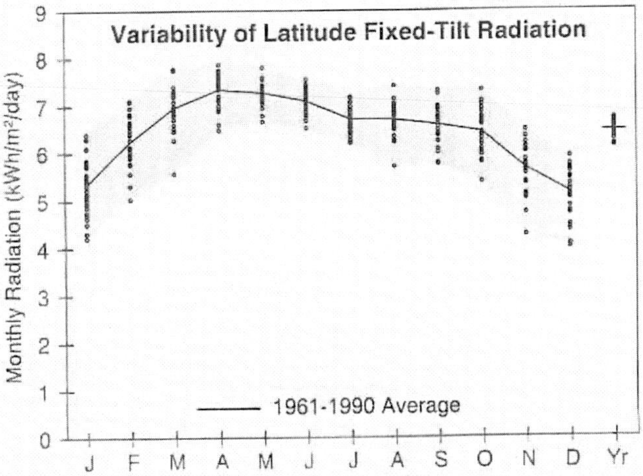

El Paso, TX

WBAN NO. 23044

LATITUDE: 31.80° N
LONGITUDE: 106.40° W
ELEVATION: 1194 meters
MEAN PRESSURE: 882 millibars

STATION TYPE: Primary

Solar Radiation for Flat-Plate Collectors Facing South at a Fixed Tilt (kWh/m²/day), Uncertainty ±9%

Tilt (°)		Jan	Feb	Mar	Apr	May	June	July	Aug	Sept	Oct	Nov	Dec	Year
0	Average	3.5	4.5	5.9	7.1	7.8	8.0	7.4	6.8	5.9	4.9	3.8	3.2	5.7
	Min/Max	2.9/4.0	3.8/5.1	4.9/6.5	6.3/7.6	7.2/8.5	7.3/8.7	6.8/7.9	5.8/7.5	5.2/6.4	4.2/5.5	3.1/4.2	2.7/3.6	5.5/5.9
Latitude -15	Average	4.6	5.6	6.6	7.4	7.8	7.7	7.2	6.9	6.4	5.9	4.9	4.4	6.3
	Min/Max	3.7/5.4	4.6/6.3	5.4/7.3	6.6/8.0	7.1/8.3	7.1/8.3	6.7/7.7	5.9/7.7	5.7/7.1	5.0/6.7	3.8/5.6	3.5/4.9	6.0/6.5
Latitude	Average	5.3	6.2	7.0	7.3	7.3	7.1	6.7	6.7	6.6	6.4	5.7	5.1	6.5
	Min/Max	4.2/6.4	5.0/7.1	5.6/7.8	6.5/7.9	6.7/7.8	6.5/7.5	6.2/7.2	5.7/7.4	5.8/7.3	5.4/7.3	4.3/6.5	4.0/5.9	6.1/6.7
Latitude +15	Average	5.8	6.5	6.9	6.8	6.4	6.0	5.8	6.1	6.3	6.6	6.1	5.6	6.2
	Min/Max	4.5/7.0	5.2/7.5	5.4/7.7	6.0/7.3	5.9/6.8	5.6/6.4	5.4/6.2	5.2/6.7	5.5/7.1	5.4/7.5	4.4/6.9	4.3/6.5	5.9/6.5
90	Average	5.1	5.2	4.6	3.5	2.5	2.1	2.2	2.8	3.8	4.9	5.2	5.1	3.9
	Min/Max	3.9/6.2	4.1/6.0	3.6/5.1	3.2/3.7	2.4/2.6	2.0/2.3	2.1/2.3	2.5/3.0	3.3/4.2	4.0/5.6	3.7/6.0	3.8/6.0	3.6/4.1

Solar Radiation for 1-Axis Tracking Flat-Plate Collectors with a North-South Axis (kWh/m²/day), Uncertainty ±9%

Axis Tilt (°)		Jan	Feb	Mar	Apr	May	June	July	Aug	Sept	Oct	Nov	Dec	Year
0	Average	5.1	6.6	8.3	9.9	10.7	10.8	9.6	9.0	8.1	7.1	5.6	4.8	8.0
	Min/Max	4.0/6.2	5.3/7.8	6.4/9.6	8.5/10.7	9.6/11.9	9.5/11.9	8.6/10.5	7.5/10.2	7.0/9.1	5.9/8.3	4.1/6.5	3.6/5.6	7.6/8.3
Latitude -15	Average	5.9	7.4	8.9	10.2	10.7	10.6	9.5	9.1	8.6	7.8	6.4	5.6	8.4
	Min/Max	4.7/7.2	5.9/8.7	6.9/10.4	8.8/11.1	9.6/12.0	9.4/11.8	8.6/10.4	7.7/10.4	7.4/9.7	6.5/9.2	4.6/7.5	4.2/6.6	8.0/8.8
Latitude	Average	6.5	7.9	9.2	10.1	10.4	10.2	9.2	9.0	8.7	8.2	7.0	6.3	8.6
	Min/Max	5.1/8.0	6.2/9.3	7.0/10.7	8.7/11.0	9.3/11.6	9.0/11.3	8.2/10.0	7.5/10.3	7.4/9.8	6.8/9.7	5.0/8.2	4.6/7.3	8.1/9.0
Latitude +15	Average	6.9	8.1	9.1	9.7	9.8	9.5	8.6	8.5	8.5	8.3	7.3	6.6	8.4
	Min/Max	5.3/8.4	6.3/9.6	6.9/10.6	8.4/10.6	8.7/10.9	8.4/10.5	7.7/9.4	7.2/9.8	7.2/9.6	6.8/9.8	5.1/8.5	4.9/7.8	7.9/8.9

Solar Radiation for 2-Axis Tracking Flat-Plate Collectors (kWh/m²/day), Uncertainty ±9%

Tracker		Jan	Feb	Mar	Apr	May	June	July	Aug	Sept	Oct	Nov	Dec	Year
2-Axis	Average	6.9	8.1	9.2	10.2	10.8	10.9	9.7	9.2	8.7	8.3	7.4	6.7	8.9
	Min/Max	5.3/8.6	6.3/9.6	7.0/10.7	8.8/11.1	9.7/12.1	9.6/12.1	8.7/10.6	7.7/10.5	7.4/9.8	6.8/9.8	5.2/8.6	4.9/8.0	8.3/9.3

Direct Beam Solar Radiation for Concentrating Collectors (kWh/m²/day), Uncertainty ±8%

Tracker		Jan	Feb	Mar	Apr	May	June	July	Aug	Sept	Oct	Nov	Dec	Year
1-Axis, E-W Horiz Axis	Average	4.5	4.9	5.2	5.6	6.1	6.3	5.2	4.8	4.7	4.9	4.7	4.4	5.1
	Min/Max	3.0/5.9	3.4/6.3	3.3/6.6	4.3/6.4	5.2/7.3	4.9/7.6	4.2/6.1	3.9/5.9	3.7/5.6	3.7/6.2	2.7/5.8	2.8/5.5	4.6/5.6
1-Axis, N-S Horiz Axis	Average	3.8	5.0	6.3	7.6	8.2	8.2	6.8	6.4	5.9	5.4	4.2	3.6	6.0
	Min/Max	2.7/5.1	3.5/6.5	4.0/8.2	5.8/8.8	7.0/9.9	6.4/10.0	5.4/8.1	5.1/7.8	4.7/7.1	4.0/6.9	2.4/5.3	2.2/4.5	5.4/6.5
1-Axis, N-S Tilt=Latitude	Average	5.0	6.1	7.0	7.7	7.9	7.7	6.4	6.3	6.3	6.3	5.4	4.8	6.4
	Min/Max	3.5/6.7	4.3/7.9	4.4/9.1	5.9/9.0	6.7/9.6	5.9/9.4	5.1/7.7	5.1/7.8	5.1/7.6	4.7/8.1	3.1/6.8	3.1/6.1	5.8/7.1
2-Axis	Average	5.4	6.3	7.1	7.9	8.4	8.3	6.9	6.5	6.4	6.4	5.7	5.2	6.7
	Min/Max	3.7/7.1	4.4/8.1	4.5/9.1	6.0/9.1	7.1/10.1	6.4/10.2	5.5/8.2	5.3/8.0	5.1/7.7	4.8/8.2	3.2/7.2	3.3/6.6	6.0/7.3

Average Climatic Conditions

Element	Jan	Feb	Mar	Apr	May	June	July	Aug	Sept	Oct	Nov	Dec	Year
Temperature (°C)	6.0	8.9	12.8	17.4	22.1	26.9	27.9	26.7	23.6	17.8	11.3	6.7	17.3
Daily Minimum Temp	-1.4	1.1	4.6	8.9	13.6	17.9	20.2	19.2	16.4	9.8	3.6	-0.7	9.4
Daily Maximum Temp	13.4	16.8	21.1	25.9	30.6	35.8	35.6	34.2	30.6	25.8	19.1	14.2	25.3
Record Minimum Temp	-22.2	-13.3	-10.0	-5.0	-0.6	7.8	13.9	13.3	5.0	-3.9	-17.2	-15.0	-22.2
Record Maximum Temp	26.7	28.3	31.7	36.7	40.0	43.9	44.4	42.2	40.0	35.6	30.6	26.7	44.4
HDD, Base 18.3°C	382	263	176	61	4	0	0	0	0	49	210	360	1504
CDD, Base 18.3°C	0	0	6	34	121	257	298	260	157	32	0	0	1163
Relative Humidity (%)	51	42	32	27	27	30	44	48	51	47	47	52	41
Wind Speed (m/s)	3.2	3.5	4.4	4.4	4.1	3.5	3.2	3.0	2.9	2.8	3.1	3.0	3.4

Fort Worth, TX

WBAN NO. 03927

LATITUDE: 32.83° N
LONGITUDE: 97.05° W
ELEVATION: 164 meters
MEAN PRESSURE: 996 millibars

STATION TYPE: Primary

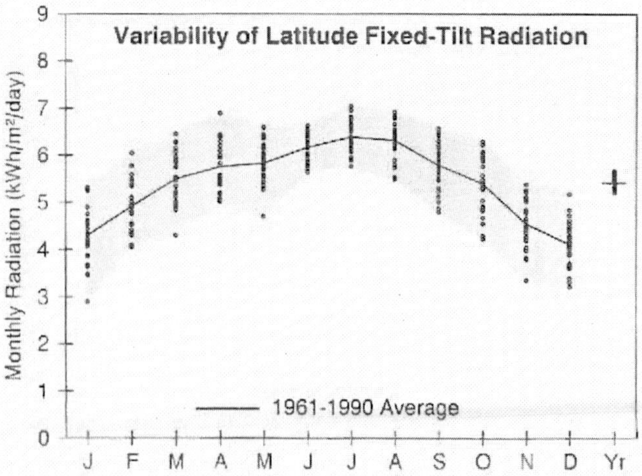

Solar Radiation for Flat-Plate Collectors Facing South at a Fixed Tilt (kWh/m²/day), Uncertainty ±9%

Tilt (°)		Jan	Feb	Mar	Apr	May	June	July	Aug	Sept	Oct	Nov	Dec	Year
0	Average	2.9	3.7	4.7	5.6	6.2	6.9	7.0	6.4	5.2	4.2	3.1	2.7	4.9
	Min/Max	2.2/3.3	3.2/4.3	3.9/5.4	4.9/6.6	5.1/7.0	6.3/7.5	6.3/7.8	5.6/7.0	4.5/5.8	3.5/4.8	2.6/3.6	2.3/3.1	4.7/5.1
Latitude -15	Average	3.8	4.5	5.3	5.9	6.2	6.7	6.9	6.5	5.7	5.0	4.0	3.5	5.3
	Min/Max	2.7/4.6	3.8/5.4	4.2/6.2	5.1/7.0	5.0/7.0	6.1/7.2	6.2/7.6	5.7/7.2	4.8/6.4	4.0/5.8	3.1/4.7	2.9/4.4	5.1/5.6
Latitude	Average	4.3	4.9	5.5	5.7	5.8	6.2	6.4	6.3	5.8	5.4	4.5	4.1	5.4
	Min/Max	2.9/5.3	4.1/6.0	4.3/6.4	5.0/6.9	4.7/6.6	5.6/6.6	5.8/7.0	5.5/6.9	4.8/6.6	4.2/6.3	3.3/5.4	3.2/5.2	5.2/5.7
Latitude +15	Average	4.6	5.1	5.4	5.3	5.2	5.3	5.6	5.7	5.6	5.5	4.8	4.4	5.2
	Min/Max	3.0/5.7	4.1/6.3	4.2/6.4	4.7/6.4	4.2/5.8	4.9/5.7	5.1/6.1	5.0/6.3	4.6/6.3	4.2/6.4	3.4/5.7	3.4/5.7	5.0/5.4
90	Average	4.0	4.0	3.6	2.9	2.3	2.1	2.3	2.8	3.4	4.1	4.1	4.0	3.3
	Min/Max	2.5/5.1	3.2/5.1	2.8/4.3	2.6/3.4	2.1/2.5	2.0/2.2	2.1/2.3	2.5/3.0	2.8/3.9	3.1/4.8	2.8/4.9	2.9/5.2	3.1/3.5

Solar Radiation for 1-Axis Tracking Flat-Plate Collectors with a North-South Axis (kWh/m²/day), Uncertainty ±9%

Axis Tilt (°)		Jan	Feb	Mar	Apr	May	June	July	Aug	Sept	Oct	Nov	Dec	Year
0	Average	4.0	5.1	6.3	7.3	7.9	8.9	9.2	8.4	6.9	5.8	4.4	3.8	6.5
	Min/Max	2.7/5.1	4.1/6.3	4.8/7.6	6.0/9.1	6.1/9.1	7.8/10.0	8.0/10.7	6.9/9.7	5.6/8.2	4.5/6.9	3.3/5.3	2.9/4.7	6.2/6.9
Latitude -15	Average	4.7	5.7	6.7	7.5	7.9	8.8	9.1	8.6	7.3	6.4	5.1	4.4	6.9
	Min/Max	3.0/5.9	4.6/7.1	5.1/8.2	6.2/9.5	6.0/9.1	7.8/9.9	7.9/10.6	7.1/9.9	5.8/8.7	4.9/7.7	3.6/6.1	3.4/5.6	6.5/7.3
Latitude	Average	5.1	6.0	6.9	7.4	7.7	8.4	8.8	8.5	7.4	6.7	5.5	4.9	6.9
	Min/Max	3.2/6.5	4.8/7.6	5.2/8.4	6.1/9.4	5.8/8.9	7.4/9.5	7.7/10.3	6.9/9.7	5.8/8.8	5.1/8.0	3.8/6.6	3.7/6.2	6.6/7.4
Latitude +15	Average	5.3	6.1	6.8	7.1	7.2	7.8	8.2	8.0	7.2	6.8	5.7	5.1	6.8
	Min/Max	3.3/6.9	4.8/7.8	5.1/8.4	5.8/9.0	5.5/8.3	6.9/8.9	7.2/9.6	6.6/9.3	5.7/8.6	5.1/8.1	3.9/6.9	3.8/6.6	6.4/7.2

Solar Radiation for 2-Axis Tracking Flat-Plate Collectors (kWh/m²/day), Uncertainty ±9%

Tracker		Jan	Feb	Mar	Apr	May	June	July	Aug	Sept	Oct	Nov	Dec	Year
2-Axis	Average	5.4	6.2	6.9	7.5	8.0	9.0	9.3	8.6	7.4	6.8	5.7	5.2	7.2
	Min/Max	3.3/6.9	4.9/7.8	5.2/8.4	6.2/9.5	6.1/9.3	7.9/10.1	8.1/10.9	7.1/9.9	5.9/8.8	5.1/8.1	3.9/6.9	3.8/6.7	6.8/7.6

Direct Beam Solar Radiation for Concentrating Collectors (kWh/m²/day), Uncertainty ±8%

Tracker		Jan	Feb	Mar	Apr	May	June	July	Aug	Sept	Oct	Nov	Dec	Year
1-Axis, E-W Horiz Axis	Average	3.2	3.4	3.4	3.5	3.7	4.4	4.7	4.2	3.6	3.7	3.4	3.2	3.7
	Min/Max	1.5/4.5	2.3/4.8	2.1/4.7	2.4/5.2	2.0/4.9	3.5/5.4	3.5/6.2	2.9/5.4	2.4/4.8	2.3/4.9	1.8/4.4	1.9/4.5	3.3/4.1
1-Axis, N-S Horiz Axis	Average	2.7	3.4	4.1	4.6	4.9	5.7	6.2	5.6	4.5	4.0	3.0	2.5	4.3
	Min/Max	1.2/3.8	2.3/4.7	2.5/5.7	3.2/6.8	2.7/6.5	4.5/7.2	4.7/8.3	3.8/7.4	3.0/6.1	2.5/5.3	1.7/4.0	1.5/3.5	3.8/4.7
1-Axis, N-S Tilt=Latitude	Average	3.6	4.2	4.6	4.7	4.7	5.3	5.9	5.6	4.9	4.7	3.9	3.5	4.6
	Min/Max	1.6/5.1	2.8/5.9	2.8/6.4	3.3/7.0	2.6/6.3	4.2/6.7	4.4/7.8	3.8/7.3	3.3/6.6	2.9/6.3	2.1/5.2	2.1/4.9	4.1/5.1
2-Axis	Average	3.9	4.3	4.6	4.8	5.0	5.8	6.3	5.8	4.9	4.8	4.1	3.8	4.8
	Min/Max	1.7/5.5	2.9/6.0	2.8/6.4	3.3/7.1	2.8/6.6	4.6/7.2	4.7/8.4	3.9/7.6	3.3/6.6	3.0/6.4	2.2/5.4	2.3/5.3	4.3/5.4

Average Climatic Conditions

Element	Jan	Feb	Mar	Apr	May	June	July	Aug	Sept	Oct	Nov	Dec	Year
Temperature (°C)	6.3	8.8	13.7	18.6	22.7	27.2	29.6	29.4	25.2	19.6	13.4	8.3	18.6
Daily Minimum Temp	0.4	2.7	7.6	12.6	17.0	21.1	23.4	23.1	19.4	13.2	7.4	2.4	12.6
Daily Maximum Temp	12.3	14.9	19.9	24.6	28.3	33.3	35.8	35.7	31.0	25.8	19.3	14.2	24.6
Record Minimum Temp	-15.6	-13.9	-9.4	-1.7	5.0	10.6	15.0	13.3	6.1	-1.7	-6.7	-18.3	-18.3
Record Maximum Temp	31.1	31.1	35.6	35.0	39.4	45.0	43.3	42.2	41.1	38.9	31.7	31.1	45.0
HDD, Base 18.3°C	372	269	159	42	0	0	0	0	0	28	153	314	1337
CDD, Base 18.3°C	0	3	16	50	137	267	349	343	207	66	6	3	1446
Relative Humidity (%)	68	66	64	65	70	66	60	59	67	66	67	68	65
Wind Speed (m/s)	5.1	5.4	5.8	5.7	5.2	4.6	4.3	4.0	4.2	4.4	4.9	4.9	4.9

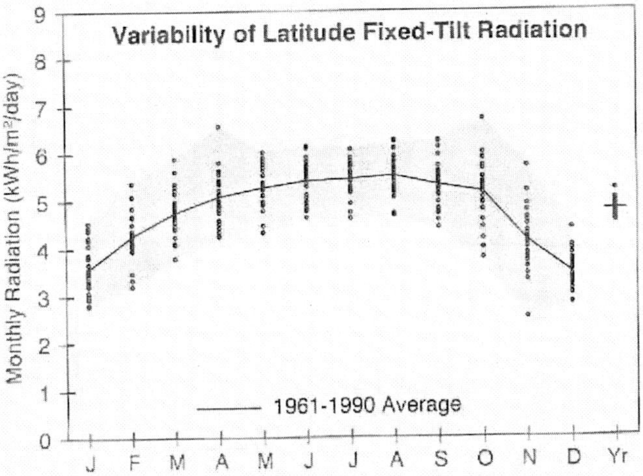

Houston, TX

WBAN NO. 12960

LATITUDE: 29.98° N
LONGITUDE: 95.37° W
ELEVATION: 33 meters
MEAN PRESSURE: 1014 millibars

STATION TYPE: Secondary

Solar Radiation for Flat-Plate Collectors Facing South at a Fixed Tilt (kWh/m²/day), Uncertainty ±9%

Tilt (°)		Jan	Feb	Mar	Apr	May	June	July	Aug	Sept	Oct	Nov	Dec	Year
0	Average	2.7	3.4	4.2	5.0	5.6	6.0	5.9	5.6	4.9	4.2	3.1	2.5	4.4
	Min/Max	2.2/3.2	2.8/4.1	3.6/5.1	4.3/6.4	4.6/6.4	5.1/6.9	5.0/6.7	4.8/6.3	4.2/5.7	3.3/5.2	2.2/3.9	2.2/2.9	4.2/4.8
Latitude -15	Average	3.2	3.9	4.6	5.2	5.6	5.9	5.8	5.7	5.2	4.8	3.7	3.0	4.7
	Min/Max	2.6/3.9	3.1/4.8	3.8/5.6	4.3/6.7	4.6/6.4	5.0/6.7	4.9/6.6	4.8/6.5	4.4/6.1	3.6/6.1	2.4/4.9	2.6/3.8	4.5/5.1
Latitude	Average	3.6	4.3	4.7	5.1	5.3	5.4	5.4	5.5	5.3	5.2	4.1	3.5	4.8
	Min/Max	2.8/4.5	3.2/5.4	3.8/5.9	4.2/6.6	4.3/6.0	4.6/6.1	4.6/6.1	4.7/6.3	4.4/6.3	3.8/6.7	2.5/5.7	2.8/4.4	4.5/5.2
Latitude +15	Average	3.8	4.4	4.6	4.7	4.7	4.7	4.8	5.0	5.1	5.2	4.3	3.7	4.6
	Min/Max	2.9/4.9	3.2/5.6	3.6/5.8	3.9/6.1	3.8/5.3	4.1/5.3	4.1/5.3	4.3/5.7	4.2/6.0	3.8/6.9	2.5/6.1	2.9/4.8	4.3/5.0
90	Average	3.1	3.3	3.0	2.5	2.0	1.8	1.9	2.4	3.0	3.8	3.5	3.2	2.8
	Min/Max	2.3/4.2	2.3/4.3	2.3/3.8	2.2/3.0	1.8/2.2	1.7/1.9	1.8/2.0	2.1/2.6	2.5/3.6	2.6/5.0	1.9/5.1	2.4/4.2	2.6/3.1

Solar Radiation for 1-Axis Tracking Flat-Plate Collectors with a North-South Axis (kWh/m²/day), Uncertainty ±9%

Axis Tilt (°)		Jan	Feb	Mar	Apr	May	June	July	Aug	Sept	Oct	Nov	Dec	Year
0	Average	3.5	4.5	5.4	6.3	7.0	7.6	7.5	7.2	6.4	5.7	4.1	3.3	5.7
	Min/Max	2.7/4.5	3.3/5.8	4.1/7.0	4.9/8.8	5.3/8.3	6.1/9.0	6.0/8.9	5.9/8.4	5.3/7.7	3.9/7.4	2.4/5.8	2.7/4.2	5.3/6.3
Latitude -15	Average	3.9	4.9	5.7	6.4	7.0	7.5	7.4	7.3	6.7	6.1	4.6	3.7	5.9
	Min/Max	3.0/5.1	3.5/6.3	4.3/7.4	5.0/9.0	5.3/8.3	6.0/8.9	6.0/8.8	6.0/8.5	5.5/8.0	4.2/8.1	2.6/6.6	3.0/4.8	5.6/6.5
Latitude	Average	4.2	5.1	5.8	6.4	6.8	7.2	7.2	7.2	6.7	6.4	4.9	4.0	6.0
	Min/Max	3.1/5.5	3.7/6.7	4.3/7.6	4.9/8.9	5.1/8.1	5.8/8.5	5.8/8.5	5.8/8.4	5.5/8.1	4.3/8.5	2.7/7.2	3.2/5.3	5.6/6.6
Latitude +15	Average	4.3	5.2	5.7	6.1	6.4	6.7	6.7	6.8	6.6	6.4	5.1	4.2	5.9
	Min/Max	3.2/5.8	3.7/6.9	4.2/7.6	4.7/8.6	4.8/7.6	5.4/7.9	5.4/7.9	5.6/8.0	5.3/8.0	4.3/8.6	2.7/7.5	3.3/5.6	5.5/6.5

Solar Radiation for 2-Axis Tracking Flat-Plate Collectors (kWh/m²/day), Uncertainty ±9%

Tracker		Jan	Feb	Mar	Apr	May	June	July	Aug	Sept	Oct	Nov	Dec	Year
2-Axis	Average	4.4	5.2	5.8	6.5	7.1	7.7	7.6	7.3	6.7	6.5	5.1	4.2	6.2
	Min/Max	3.2/5.8	3.7/6.9	4.3/7.6	5.0/9.0	5.3/8.4	6.1/9.1	6.1/8.9	6.0/8.6	5.5/8.1	4.3/8.6	2.7/7.5	3.3/5.7	5.8/6.8

Direct Beam Solar Radiation for Concentrating Collectors (kWh/m²/day), Uncertainty ±8%

Tracker		Jan	Feb	Mar	Apr	May	June	July	Aug	Sept	Oct	Nov	Dec	Year
1-Axis, E-W Horiz Axis	Average	2.3	2.7	2.7	2.8	3.1	3.4	3.3	3.2	3.1	3.4	2.8	2.3	2.9
	Min/Max	1.4/3.6	1.4/4.1	1.4/3.9	1.6/4.8	1.6/4.1	2.4/4.6	2.2/4.5	2.3/4.1	2.1/4.1	1.7/4.9	0.8/4.8	1.5/3.7	2.6/3.3
1-Axis, N-S Horiz Axis	Average	2.0	2.8	3.2	3.7	4.1	4.5	4.4	4.3	3.9	3.7	2.6	1.9	3.4
	Min/Max	1.2/3.2	1.5/4.3	1.7/4.8	2.1/6.5	2.1/5.4	3.0/6.0	3.0/6.0	3.1/5.4	2.8/5.3	1.9/5.4	0.8/4.5	1.2/3.1	3.1/3.9
1-Axis, N-S Tilt=Latitude	Average	2.6	3.3	3.6	3.7	3.9	4.2	4.1	4.3	4.1	4.3	3.3	2.5	3.7
	Min/Max	1.6/4.1	1.8/5.1	1.9/5.3	2.2/6.6	2.0/5.2	2.8/5.6	2.8/5.6	3.0/5.3	2.9/5.6	2.2/6.3	1.0/5.6	1.6/4.0	3.3/4.2
2-Axis	Average	2.8	3.4	3.6	3.8	4.1	4.6	4.4	4.4	4.2	4.3	3.4	2.8	3.8
	Min/Max	1.7/4.3	1.8/5.2	1.9/5.3	2.2/6.7	2.1/5.5	3.1/6.1	3.0/6.0	3.1/5.5	3.0/5.6	2.2/6.3	1.0/5.9	1.8/4.4	3.4/4.3

Average Climatic Conditions

Element	Jan	Feb	Mar	Apr	May	June	July	Aug	Sept	Oct	Nov	Dec	Year
Temperature (°C)	10.2	12.2	15.9	20.2	23.6	26.9	28.1	27.9	25.7	20.9	16.1	11.9	19.9
Daily Minimum Temp	4.3	5.9	10.0	14.5	18.0	21.4	22.4	22.2	19.9	14.2	9.8	5.7	14.1
Daily Maximum Temp	16.1	18.5	21.7	25.8	29.2	32.3	33.7	33.6	31.3	27.6	22.4	18.2	25.9
Record Minimum Temp	-11.1	-6.7	-5.6	-0.6	6.7	11.1	16.7	16.7	8.9	0.0	-7.2	-13.9	-13.9
Record Maximum Temp	28.9	32.8	32.8	35.0	36.1	39.4	40.0	41.7	38.9	35.6	31.7	28.3	41.7
HDD, Base 18.3°C	260	179	104	20	0	0	0	0	0	17	101	208	888
CDD, Base 18.3°C	9	6	28	75	164	257	303	298	220	97	34	10	1500
Relative Humidity (%)	75	73	73	74	75	75	75	75	76	74	75	75	75
Wind Speed (m/s)	4.0	4.2	4.4	4.4	3.9	3.6	3.2	3.0	3.3	3.4	3.8	3.9	3.8

Lubbock, TX

WBAN NO. 23042

LATITUDE: 33.65° N
LONGITUDE: 101.82° W
ELEVATION: 988 meters
MEAN PRESSURE: 904 millibars

STATION TYPE: Secondary

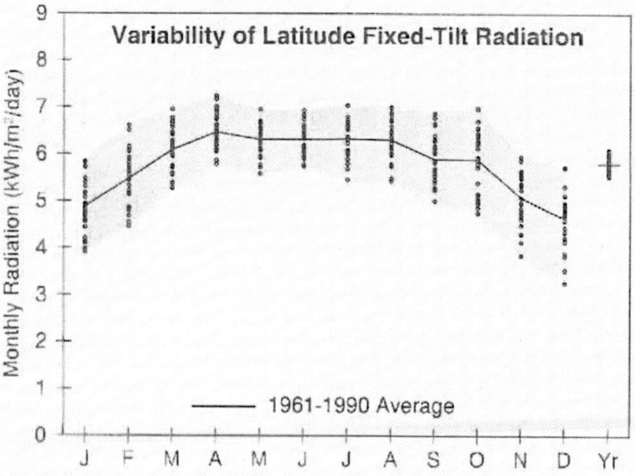

Solar Radiation for Flat-Plate Collectors Facing South at a Fixed Tilt (kWh/m²/day), Uncertainty ±9%

Tilt (°)		Jan	Feb	Mar	Apr	May	June	July	Aug	Sept	Oct	Nov	Dec	Year
0	Average	3.1	3.9	5.1	6.2	6.7	7.1	7.0	6.3	5.2	4.4	3.3	2.8	5.1
	Min/Max	2.7/3.5	3.4/4.6	4.5/5.7	5.6/6.9	5.9/7.4	6.4/7.8	6.0/7.8	5.5/7.0	4.5/6.0	3.7/5.1	2.8/3.7	2.2/3.3	4.9/5.3
Latitude -15	Average	4.2	5.0	5.8	6.6	6.7	6.9	6.8	6.5	5.8	5.4	4.4	3.9	5.7
	Min/Max	3.5/5.0	4.1/5.9	5.1/6.6	5.9/7.3	5.9/7.4	6.2/7.6	5.9/7.6	5.7/7.3	4.9/6.7	4.4/6.4	3.5/5.1	2.9/4.8	5.4/5.9
Latitude	Average	4.9	5.5	6.1	6.5	6.3	6.3	6.3	6.3	5.9	5.9	5.1	4.6	5.8
	Min/Max	3.9/5.8	4.5/6.6	5.3/6.9	5.8/7.2	5.6/6.9	5.7/6.9	5.4/7.0	5.4/7.0	5.0/6.9	4.7/7.0	3.8/5.9	3.3/5.7	5.5/6.1
Latitude +15	Average	5.3	5.7	6.0	6.0	5.6	5.4	5.5	5.7	5.7	6.0	5.4	5.0	5.6
	Min/Max	4.2/6.3	4.6/6.9	5.2/6.9	5.4/6.7	5.0/6.1	5.0/5.9	4.8/6.0	4.9/6.3	4.8/6.6	4.8/7.2	4.0/6.4	3.5/6.3	5.3/5.9
90	Average	4.7	4.6	4.1	3.3	2.5	2.1	2.2	2.8	3.5	4.6	4.7	4.6	3.6
	Min/Max	3.7/5.7	3.6/5.7	3.5/4.7	3.0/3.6	2.3/2.6	2.0/2.2	2.1/2.3	2.5/3.0	3.1/4.1	3.6/5.5	3.4/5.6	3.0/5.9	3.4/3.9

Solar Radiation for 1-Axis Tracking Flat-Plate Collectors with a North-South Axis (kWh/m²/day), Uncertainty ±9%

Axis Tilt (°)		Jan	Feb	Mar	Apr	May	June	July	Aug	Sept	Oct	Nov	Dec	Year
0	Average	4.6	5.7	7.1	8.5	8.9	9.4	9.3	8.6	7.2	6.4	4.9	4.2	7.1
	Min/Max	3.6/5.5	4.6/7.0	6.0/8.4	7.3/9.7	7.7/10.0	8.0/10.6	7.7/10.9	7.1/10.0	5.8/8.7	5.1/7.8	3.7/5.9	2.9/5.2	6.7/7.4
Latitude -15	Average	5.4	6.4	7.7	8.8	9.0	9.3	9.3	8.8	7.6	7.2	5.7	5.0	7.5
	Min/Max	4.2/6.6	5.1/8.0	6.4/9.1	7.6/10.1	7.8/10.1	7.9/10.5	7.7/10.9	7.2/10.2	6.2/9.3	5.6/8.8	4.3/6.9	3.4/6.3	7.1/7.9
Latitude	Average	5.9	6.8	7.9	8.7	8.7	8.9	8.9	8.6	7.7	7.5	6.2	5.6	7.6
	Min/Max	4.6/7.2	5.4/8.5	6.6/9.3	7.5/10.1	7.5/9.8	7.6/10.1	7.4/10.5	7.1/10.0	6.2/9.4	5.9/9.2	4.5/7.5	3.7/7.1	7.2/8.1
Latitude +15	Average	6.2	7.0	7.8	8.4	8.2	8.3	8.4	8.2	7.5	7.6	6.5	5.9	7.5
	Min/Max	4.8/7.6	5.5/8.7	6.5/9.3	7.2/9.7	7.1/9.2	7.1/9.4	6.9/9.9	6.7/9.6	6.0/9.2	5.9/9.3	4.7/7.8	3.9/7.5	7.0/7.9

Solar Radiation for 2-Axis Tracking Flat-Plate Collectors (kWh/m²/day), Uncertainty ±9%

Tracker		Jan	Feb	Mar	Apr	May	June	July	Aug	Sept	Oct	Nov	Dec	Year
2-Axis	Average	6.3	7.0	7.9	8.8	9.1	9.5	9.4	8.8	7.7	7.6	6.5	6.0	7.9
	Min/Max	4.8/7.7	5.5/8.7	6.6/9.4	7.6/10.2	7.8/10.2	8.1/10.8	7.8/11.1	7.3/10.3	6.2/9.4	5.9/9.4	4.7/7.9	4.0/7.7	7.4/8.3

Direct Beam Solar Radiation for Concentrating Collectors (kWh/m²/day), Uncertainty ±8%

Tracker		Jan	Feb	Mar	Apr	May	June	July	Aug	Sept	Oct	Nov	Dec	Year
1-Axis, E-W	Average	4.0	4.1	4.2	4.6	4.7	5.1	5.0	4.5	3.9	4.4	4.1	3.9	4.4
Horiz Axis	Min/Max	2.6/5.3	2.9/5.7	3.0/5.4	3.6/5.7	3.6/5.7	3.8/6.2	3.5/6.5	3.3/5.8	2.8/5.4	3.1/5.9	2.5/5.3	2.2/5.5	3.9/4.8
1-Axis, N-S	Average	3.3	4.1	5.1	6.1	6.2	6.7	6.6	6.1	5.0	4.8	3.6	3.1	5.1
Horiz Axis	Min/Max	2.1/4.4	2.8/5.7	3.6/6.5	4.7/7.6	4.8/7.6	5.0/8.1	4.7/8.7	4.3/7.8	3.5/7.0	3.3/6.5	2.2/4.7	1.7/4.3	4.5/5.5
1-Axis, N-S	Average	4.4	5.1	5.7	6.3	6.0	6.2	6.2	6.1	5.4	5.7	4.7	4.3	5.5
Tilt=Latitude	Min/Max	2.9/5.9	3.5/7.0	4.0/7.4	4.8/7.9	4.7/7.3	4.6/7.6	4.4/8.3	4.3/7.8	3.8/7.5	4.0/7.7	2.9/6.1	2.4/6.0	4.9/6.0
2-Axis	Average	4.8	5.2	5.7	6.4	6.4	6.8	6.7	6.2	5.4	5.8	5.0	4.6	5.7
	Min/Max	3.1/6.3	3.6/7.2	4.0/7.4	4.9/8.0	4.9/7.7	5.0/8.3	4.8/8.9	4.4/8.0	3.8/7.6	4.1/7.8	3.1/6.5	2.6/6.5	5.1/6.3

Average Climatic Conditions

Element	Jan	Feb	Mar	Apr	May	June	July	Aug	Sept	Oct	Nov	Dec	Year
Temperature (°C)	3.8	6.2	10.7	16.2	20.8	25.1	26.7	25.5	21.7	16.3	9.9	4.8	15.6
Daily Minimum Temp	-4.1	-1.9	2.4	8.2	13.2	17.9	20.0	19.0	15.2	8.9	2.5	-2.7	8.2
Daily Maximum Temp	11.6	14.2	18.9	24.1	28.4	32.2	33.3	32.0	28.3	23.7	17.3	12.3	23.1
Record Minimum Temp	-26.7	-22.2	-16.7	-5.6	-1.1	6.7	10.6	11.1	0.6	-6.7	-18.3	-18.9	-26.7
Record Maximum Temp	28.3	30.6	35.0	37.8	40.0	43.3	42.2	41.1	39.4	36.7	30.0	27.2	43.3
HDD, Base 18.3°C	451	341	243	89	14	0	0	0	9	86	253	420	1906
CDD, Base 18.3°C	0	0	6	24	90	203	258	222	111	23	0	0	938
Relative Humidity (%)	58	57	50	47	53	56	55	59	64	59	58	60	56
Wind Speed (m/s)	5.6	5.9	6.5	6.7	6.5	5.9	5.1	4.5	4.7	5.1	5.3	5.3	5.6

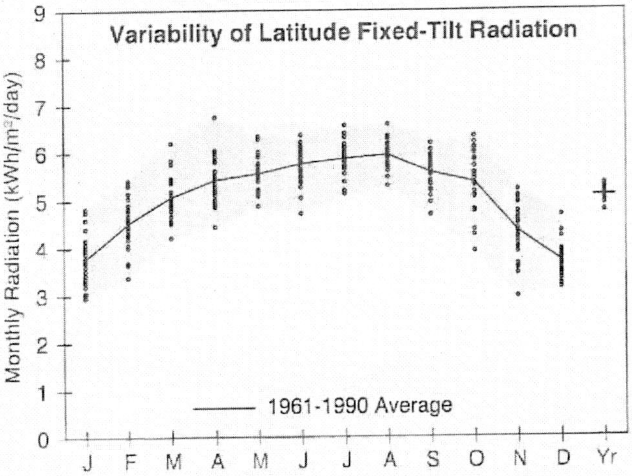

Lufkin, TX

WBAN NO. 93987

LATITUDE: 31.23° N
LONGITUDE: 94.75° W
ELEVATION: 96 meters
MEAN PRESSURE: 1006 millibars

STATION TYPE: Secondary

Solar Radiation for Flat-Plate Collectors Facing South at a Fixed Tilt (kWh/m²/day), Uncertainty ±9%

Tilt (°)		Jan	Feb	Mar	Apr	May	June	July	Aug	Sept	Oct	Nov	Dec	Year
0	Average	2.7	3.5	4.5	5.3	5.9	6.4	6.4	6.0	5.1	4.3	3.1	2.5	4.6
	Min/Max	2.3/3.2	2.9/4.0	3.8/5.3	4.4/6.5	5.2/6.8	5.2/7.1	5.6/7.2	5.4/6.6	4.4/5.6	3.4/4.9	2.4/3.6	2.3/3.0	4.4/4.8
Latitude -15	Average	3.4	4.1	4.9	5.5	5.9	6.2	6.3	6.2	5.5	5.0	3.9	3.2	5.0
	Min/Max	2.7/4.2	3.2/4.9	4.1/5.9	4.6/6.8	5.1/6.7	5.1/6.9	5.5/7.1	5.5/6.8	4.7/6.1	3.8/5.8	2.8/4.6	2.8/4.0	4.7/5.2
Latitude	Average	3.8	4.5	5.1	5.4	5.6	5.8	5.9	6.0	5.6	5.4	4.3	3.7	5.1
	Min/Max	3.0/4.8	3.4/5.4	4.2/6.2	4.4/6.8	4.9/6.3	4.7/6.4	5.1/6.6	5.3/6.6	4.7/6.2	3.9/6.3	3.0/5.2	3.1/4.7	4.8/5.3
Latitude +15	Average	4.0	4.6	5.0	5.0	4.9	5.0	5.2	5.4	5.4	5.4	4.6	4.0	4.9
	Min/Max	3.1/5.2	3.4/5.6	4.1/6.1	4.1/6.3	4.4/5.6	4.1/5.5	4.5/5.7	4.8/6.0	4.5/6.0	3.9/6.4	3.0/5.5	3.3/5.1	4.5/5.1
90	Average	3.4	3.6	3.3	2.7	2.2	1.9	2.1	2.6	3.2	4.0	3.8	3.5	3.0
	Min/Max	2.5/4.5	2.5/4.4	2.6/4.0	2.3/3.2	2.1/2.3	1.8/2.0	1.9/2.2	2.4/2.8	2.7/3.6	2.8/4.7	2.3/4.7	2.8/4.6	2.8/3.1

Solar Radiation for 1-Axis Tracking Flat-Plate Collectors with a North-South Axis (kWh/m²/day), Uncertainty ±9%

Axis Tilt (°)		Jan	Feb	Mar	Apr	May	June	July	Aug	Sept	Oct	Nov	Dec	Year
0	Average	3.6	4.7	5.8	6.8	7.4	8.1	8.1	7.8	6.6	5.8	4.2	3.4	6.0
	Min/Max	2.8/4.7	3.5/5.8	4.6/7.4	5.3/9.0	6.1/8.9	6.4/9.4	6.8/9.6	6.6/9.0	5.4/7.4	4.0/7.0	2.8/5.1	2.9/4.4	5.5/6.4
Latitude -15	Average	4.1	5.2	6.2	6.9	7.4	8.0	8.0	7.9	7.0	6.4	4.8	4.0	6.3
	Min/Max	3.1/5.4	3.7/6.5	4.8/7.9	5.4/9.3	6.1/8.9	6.3/9.3	6.8/9.5	6.7/9.1	5.6/7.8	4.3/7.7	3.1/5.8	3.3/5.1	5.8/6.7
Latitude	Average	4.4	5.5	6.3	6.9	7.2	7.7	7.8	7.8	7.0	6.6	5.2	4.3	6.4
	Min/Max	3.3/5.9	3.9/6.8	4.9/8.1	5.3/9.2	5.9/8.7	6.0/8.9	6.5/9.1	6.6/9.0	5.6/7.9	4.4/8.1	3.3/6.3	3.6/5.6	5.8/6.8
Latitude +15	Average	4.6	5.5	6.2	6.6	6.7	7.1	7.2	7.4	6.9	6.7	5.3	4.5	6.2
	Min/Max	3.4/6.2	3.9/7.0	4.8/8.0	5.1/8.9	5.5/8.1	5.6/8.3	6.1/8.6	6.2/8.6	5.5/7.7	4.4/8.2	3.3/6.6	3.7/6.0	5.7/6.7

Solar Radiation for 2-Axis Tracking Flat-Plate Collectors (kWh/m²/day), Uncertainty ±9%

Tracker		Jan	Feb	Mar	Apr	May	June	July	Aug	Sept	Oct	Nov	Dec	Year
2-Axis	Average	4.6	5.6	6.3	7.0	7.5	8.2	8.2	8.0	7.1	6.7	5.4	4.6	6.6
	Min/Max	3.5/6.2	3.9/7.0	4.9/8.1	5.4/9.3	6.1/9.0	6.4/9.5	6.9/9.6	6.7/9.2	5.7/7.9	4.5/8.2	3.3/6.7	3.8/6.1	6.0/7.0

Direct Beam Solar Radiation for Concentrating Collectors (kWh/m²/day), Uncertainty ±8%

Tracker		Jan	Feb	Mar	Apr	May	June	July	Aug	Sept	Oct	Nov	Dec	Year
1-Axis, E-W Horiz Axis	Average	2.6	2.9	3.0	3.1	3.3	3.8	3.8	3.7	3.3	3.6	3.0	2.6	3.2
	Min/Max	1.5/3.9	1.6/4.1	1.9/4.3	2.0/5.0	2.2/4.6	2.5/4.9	2.8/5.1	2.7/4.8	2.2/4.1	1.7/4.7	1.3/4.2	1.9/3.9	2.8/3.6
1-Axis, N-S Horiz Axis	Average	2.2	3.0	3.6	4.1	4.4	4.9	4.9	4.9	4.1	3.9	2.8	2.1	3.7
	Min/Max	1.3/3.3	1.6/4.2	2.2/5.3	2.6/6.7	2.8/6.1	3.3/6.4	3.6/6.7	3.5/6.3	2.7/5.1	1.8/5.2	1.2/3.8	1.5/3.2	3.2/4.3
1-Axis, N-S Tilt=Latitude	Average	2.9	3.6	4.0	4.2	4.2	4.6	4.7	4.8	4.4	4.6	3.5	2.9	4.0
	Min/Max	1.7/4.4	2.0/5.1	2.5/5.9	2.7/6.8	2.7/5.9	3.1/6.0	3.4/6.3	3.5/6.3	2.9/5.5	2.2/6.0	1.5/4.8	2.1/4.2	3.4/4.6
2-Axis	Average	3.1	3.7	4.0	4.2	4.4	5.0	5.0	5.0	4.4	4.6	3.7	3.1	4.2
	Min/Max	1.8/4.7	2.0/5.2	2.5/5.9	2.7/6.9	2.8/6.2	3.3/6.5	3.6/6.8	3.6/6.5	2.9/5.5	2.2/6.1	1.6/5.1	2.3/4.6	3.6/4.8

Average Climatic Conditions

Element	Jan	Feb	Mar	Apr	May	June	July	Aug	Sept	Oct	Nov	Dec	Year
Temperature (°C)	8.7	10.8	15.3	19.7	23.3	26.6	28.2	28.1	25.2	19.8	14.6	10.2	19.2
Daily Minimum Temp	2.7	4.3	8.6	13.2	17.3	20.8	22.3	21.8	19.2	12.7	8.0	3.8	12.9
Daily Maximum Temp	14.6	17.3	21.9	26.1	29.2	32.3	34.0	34.2	31.2	26.9	21.2	16.5	25.4
Record Minimum Temp	-13.3	-8.9	-7.2	-1.1	6.1	10.6	14.4	12.8	3.9	-1.0	-7.8	-16.1	-16.1
Record Maximum Temp	29.4	32.8	32.2	34.4	36.7	37.8	39.4	41.7	38.9	35.0	32.0	27.8	41.7
HDD, Base 18.3°C	306	215	119	30	0	0	0	0	0	27	129	259	1084
CDD, Base 18.3°C	6	5	24	70	153	247	307	302	207	73	17	6	1417
Relative Humidity (%)	74	71	70	72	75	75	74	73	75	73	75	76	74
Wind Speed (m/s)	3.2	3.4	3.6	3.5	3.1	2.6	2.4	2.4	2.6	2.6	3.0	3.3	3.0

Midland, TX

WBAN NO. 23023

LATITUDE: 31.93° N
LONGITUDE: 102.20° W
ELEVATION: 871 meters
MEAN PRESSURE: 916 millibars

STATION TYPE: Primary

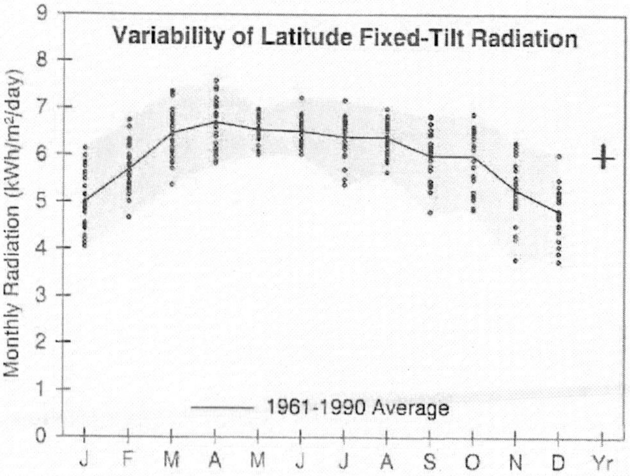

Solar Radiation for Flat-Plate Collectors Facing South at a Fixed Tilt (kWh/m²/day), Uncertainty ±9%

Tilt (°)		Jan	Feb	Mar	Apr	May	June	July	Aug	Sept	Oct	Nov	Dec	Year
0	Average	3.3	4.2	5.5	6.5	7.0	7.3	7.0	6.5	5.4	4.6	3.6	3.0	5.3
	Min/Max	2.8/3.8	3.6/4.8	4.8/6.1	5.7/7.3	6.4/7.4	6.7/8.2	5.9/8.0	5.7/7.1	4.4/6.1	3.9/5.2	2.7/4.1	2.6/3.6	5.2/5.5
Latitude -15	Average	4.3	5.1	6.2	6.8	7.0	7.1	6.9	6.6	5.9	5.5	4.6	4.1	5.8
	Min/Max	3.6/5.2	4.3/6.0	5.2/7.0	5.9/7.7	6.4/7.4	6.5/7.9	5.8/7.8	5.8/7.3	4.8/6.7	4.5/6.3	3.4/5.4	3.3/5.0	5.7/6.1
Latitude	Average	5.0	5.7	6.5	6.7	6.5	6.5	6.4	6.4	6.0	6.0	5.3	4.8	6.0
	Min/Max	4.0/6.1	4.7/6.7	5.4/7.4	5.8/7.6	6.0/7.0	6.0/7.2	5.4/7.2	5.6/7.0	4.8/6.8	4.8/6.9	3.8/6.3	3.8/6.0	5.8/6.2
Latitude +15	Average	5.4	5.9	6.4	6.2	5.8	5.6	5.5	5.8	5.8	6.1	5.6	5.2	5.8
	Min/Max	4.3/6.7	4.8/7.1	5.2/7.3	5.4/7.1	5.3/6.2	5.2/6.2	4.7/6.2	5.1/6.3	4.6/6.6	4.9/7.1	4.0/6.8	4.0/6.6	5.6/6.1
90	Average	4.7	4.7	4.2	3.2	2.4	2.0	2.2	2.7	3.5	4.5	4.8	4.7	3.6
	Min/Max	3.7/6.0	3.8/5.7	3.5/4.9	2.9/3.6	2.3/2.5	2.0/2.1	2.0/2.2	2.6/2.9	2.8/4.0	3.6/5.3	3.3/5.8	3.5/6.1	3.4/3.9

Solar Radiation for 1-Axis Tracking Flat-Plate Collectors with a North-South Axis (kWh/m²/day), Uncertainty ±9%

Axis Tilt (°)		Jan	Feb	Mar	Apr	May	June	July	Aug	Sept	Oct	Nov	Dec	Year
0	Average	4.8	6.0	7.7	8.8	9.3	9.6	9.4	8.7	7.4	6.6	5.2	4.5	7.3
	Min/Max	3.9/5.9	4.8/7.2	6.0/8.9	7.3/10.2	8.2/10.1	8.4/11.2	7.4/11.0	7.3/9.8	5.7/8.6	5.2/7.8	3.7/6.4	3.4/5.6	7.0/7.7
Latitude -15	Average	5.6	6.7	8.2	9.1	9.3	9.5	9.3	8.8	7.8	7.3	6.0	5.3	7.7
	Min/Max	4.4/6.9	5.3/8.1	6.4/9.6	7.6/10.5	8.3/10.2	8.3/11.1	7.4/10.9	7.5/10.0	6.0/9.1	5.7/8.6	4.2/7.3	4.0/6.7	7.4/8.1
Latitude	Average	6.1	7.1	8.4	9.0	9.0	9.1	9.0	8.7	7.8	7.6	6.5	5.8	7.9
	Min/Max	4.8/7.7	5.6/8.7	6.5/9.9	7.5/10.4	8.0/9.9	8.0/10.6	7.1/10.5	7.3/9.9	6.1/9.3	5.9/9.1	4.6/8.0	4.3/7.5	7.5/8.3
Latitude +15	Average	6.4	7.3	8.3	8.7	8.5	8.5	8.4	8.3	7.7	7.7	6.8	6.2	7.7
	Min/Max	4.9/8.1	5.7/8.9	6.4/9.8	7.2/10.1	7.5/9.3	7.4/9.9	6.6/9.8	7.0/9.4	5.9/9.1	6.0/9.2	4.7/8.3	4.6/7.9	7.4/8.2

Solar Radiation for 2-Axis Tracking Flat-Plate Collectors (kWh/m²/day), Uncertainty ±9%

Tracker		Jan	Feb	Mar	Apr	May	June	July	Aug	Sept	Oct	Nov	Dec	Year
2-Axis	Average	6.5	7.3	8.4	9.1	9.4	9.7	9.5	8.9	7.9	7.7	6.8	6.3	8.1
	Min/Max	5.0/8.2	5.8/8.9	6.5/9.9	7.6/10.6	8.3/10.3	8.5/11.3	7.5/11.1	7.5/10.1	6.1/9.3	6.0/9.2	4.8/8.4	4.6/8.1	7.8/8.6

Direct Beam Solar Radiation for Concentrating Collectors (kWh/m²/day), Uncertainty ±8%

Tracker		Jan	Feb	Mar	Apr	May	June	July	Aug	Sept	Oct	Nov	Dec	Year
1-Axis, E-W Horiz Axis	Average	4.0	4.3	4.6	4.7	4.9	5.2	4.9	4.5	4.0	4.5	4.2	4.1	4.5
	Min/Max	2.7/5.6	3.0/5.6	3.0/5.9	3.6/6.0	3.9/5.7	4.2/6.8	3.2/6.4	3.3/5.6	2.9/5.2	3.1/5.8	2.8/5.6	2.6/5.7	4.2/4.9
1-Axis, N-S Horiz Axis	Average	3.5	4.4	5.5	6.3	6.6	6.8	6.5	6.0	5.1	4.9	3.9	3.3	5.2
	Min/Max	2.3/4.8	3.0/5.7	3.5/7.2	4.7/8.1	5.2/7.7	5.3/8.9	4.2/8.6	4.5/7.6	3.6/6.8	3.3/6.5	2.5/5.1	2.1/4.6	4.9/5.7
1-Axis, N-S Tilt=Latitude	Average	4.6	5.3	6.2	6.5	6.3	6.4	6.2	6.0	5.5	5.8	4.9	4.5	5.7
	Min/Max	3.1/6.3	3.7/7.0	3.9/8.0	4.8/8.2	5.0/7.4	5.0/8.3	4.0/8.1	4.5/7.5	3.9/7.3	4.0/7.6	3.2/6.6	2.8/6.3	5.2/6.2
2-Axis	Average	4.9	5.5	6.2	6.6	6.7	6.9	6.6	6.2	5.5	5.9	5.2	4.8	5.9
	Min/Max	3.3/6.8	3.8/7.2	3.9/8.0	4.9/8.4	5.3/7.8	5.4/9.0	4.3/8.7	4.6/7.8	3.9/7.3	4.0/7.7	3.4/6.9	3.1/6.8	5.4/6.5

Average Climatic Conditions

Element	Jan	Feb	Mar	Apr	May	June	July	Aug	Sept	Oct	Nov	Dec	Year
Temperature (°C)	5.8	8.4	13.2	18.1	22.7	26.4	27.8	27.1	22.9	17.8	11.4	7.0	17.4
Daily Minimum Temp	-1.9	0.3	4.6	9.7	14.5	18.7	20.3	19.7	16.2	10.3	3.8	-0.7	9.6
Daily Maximum Temp	13.6	16.4	21.8	26.6	30.8	34.1	35.2	34.5	29.7	25.2	19.0	14.7	25.1
Record Minimum Temp	-22.2	-23.9	-12.8	-6.7	1.1	8.3	11.7	12.2	2.2	-3.3	-10.6	-18.3	-23.9
Record Maximum Temp	28.9	32.2	35.0	38.3	42.2	42.8	44.4	41.7	41.7	37.8	31.7	29.4	44.4
HDD, Base 18.3°C	388	278	172	60	6	0	0	0	6	59	209	351	1528
CDD, Base 18.3°C	0	0	12	53	140	243	293	272	144	42	3	0	1202
Relative Humidity (%)	57	55	46	45	51	53	51	54	61	60	59	58	54
Wind Speed (m/s)	4.8	5.1	5.8	5.9	5.8	5.6	5.0	4.7	4.6	4.6	4.8	4.7	5.1

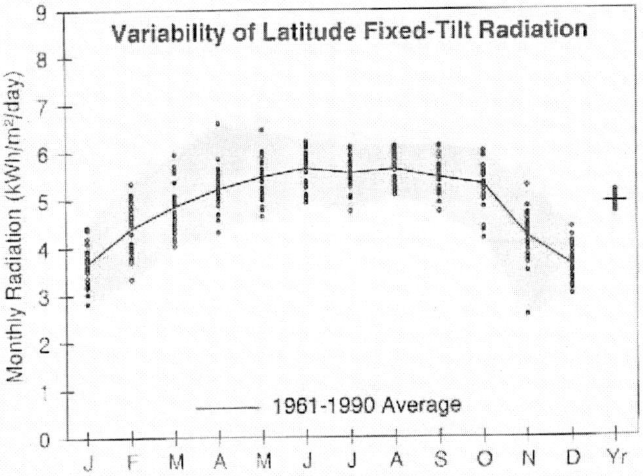

Port Arthur, TX

WBAN NO. 12917

LATITUDE: 29.95° N
LONGITUDE: 94.02° W
ELEVATION: 7 meters
MEAN PRESSURE: 1017 millibars

STATION TYPE: Secondary

Solar Radiation for Flat-Plate Collectors Facing South at a Fixed Tilt (kWh/m²/day), Uncertainty ±9%

Tilt (°)		Jan	Feb	Mar	Apr	May	June	July	Aug	Sept	Oct	Nov	Dec	Year
0	Average	2.7	3.5	4.3	5.2	5.8	6.3	6.1	5.7	5.0	4.3	3.1	2.6	4.6
	Min/Max	2.3/3.1	2.9/4.1	3.8/5.1	4.3/6.4	4.9/6.9	5.4/6.9	5.2/6.7	5.2/6.2	4.4/5.5	3.5/4.8	2.2/3.7	2.3/3.0	4.4/4.7
Latitude -15	Average	3.3	4.1	4.7	5.3	5.8	6.1	6.0	5.8	5.4	4.9	3.8	3.2	4.9
	Min/Max	2.6/3.9	3.2/4.8	4.0/5.7	4.4/6.7	4.9/6.9	5.3/6.7	5.1/6.6	5.3/6.3	4.7/6.0	4.0/5.6	2.5/4.6	2.7/3.8	4.7/5.1
Latitude	Average	3.7	4.4	4.9	5.2	5.5	5.7	5.5	5.6	5.5	5.3	4.2	3.6	4.9
	Min/Max	2.8/4.4	3.3/5.3	4.0/6.0	4.3/6.6	4.6/6.5	4.9/6.2	4.8/6.1	5.1/6.1	4.7/6.1	4.2/6.0	2.6/5.3	3.0/4.4	4.7/5.2
Latitude +15	Average	3.9	4.5	4.8	4.9	4.9	4.9	4.9	5.2	5.3	5.4	4.4	3.9	4.7
	Min/Max	2.9/4.7	3.3/5.6	3.9/5.9	4.0/6.2	4.2/5.7	4.3/5.4	4.2/5.3	4.7/5.6	4.6/5.9	4.2/6.1	2.6/5.6	3.1/4.8	4.5/5.0
90	Average	3.2	3.4	3.1	2.5	2.0	1.8	1.9	2.4	3.1	3.9	3.6	3.3	2.9
	Min/Max	2.3/4.0	2.5/4.3	2.5/3.8	2.2/3.0	1.9/2.2	1.7/1.9	1.8/2.0	2.2/2.5	2.8/3.5	2.9/4.5	1.9/4.7	2.6/4.2	2.7/3.0

Solar Radiation for 1-Axis Tracking Flat-Plate Collectors with a North-South Axis (kWh/m²/day), Uncertainty ±9%

Axis Tilt (°)		Jan	Feb	Mar	Apr	May	June	July	Aug	Sept	Oct	Nov	Dec	Year
0	Average	3.5	4.7	5.6	6.5	7.3	8.0	7.6	7.3	6.5	5.8	4.2	3.4	5.9
	Min/Max	2.8/4.3	3.5/5.8	4.3/7.2	4.9/8.7	5.8/9.1	6.6/8.9	6.4/8.8	6.4/8.3	5.5/7.6	4.5/6.7	2.5/5.3	2.8/4.2	5.5/6.3
Latitude -15	Average	4.0	5.1	5.9	6.7	7.3	7.9	7.6	7.4	6.8	6.3	4.7	3.9	6.1
	Min/Max	3.0/4.9	3.7/6.4	4.5/7.6	5.0/8.9	5.8/9.2	6.5/8.8	6.4/8.7	6.5/8.4	5.7/8.0	4.8/7.3	2.6/6.0	3.1/4.8	5.8/6.5
Latitude	Average	4.3	5.3	6.0	6.6	7.1	7.5	7.3	7.3	6.9	6.6	5.0	4.2	6.2
	Min/Max	3.2/5.3	3.8/6.8	4.6/7.8	5.0/8.8	5.6/8.9	6.3/8.4	6.2/8.4	6.4/8.2	5.7/8.1	5.0/7.7	2.7/6.5	3.4/5.3	5.8/6.6
Latitude +15	Average	4.4	5.4	5.9	6.3	6.7	7.0	6.8	6.9	6.7	6.7	5.2	4.4	6.0
	Min/Max	3.2/5.6	3.8/6.9	4.5/7.7	4.8/8.5	5.3/8.4	5.8/7.9	5.8/7.9	6.0/7.8	5.6/7.9	5.0/7.7	2.8/6.8	3.5/5.6	5.7/6.5

Solar Radiation for 2-Axis Tracking Flat-Plate Collectors (kWh/m²/day), Uncertainty ±9%

Tracker		Jan	Feb	Mar	Apr	May	June	July	Aug	Sept	Oct	Nov	Dec	Year
2-Axis	Average	4.5	5.4	6.0	6.7	7.4	8.0	7.7	7.5	6.9	6.7	5.2	4.5	6.4
	Min/Max	3.2/5.6	3.9/6.9	4.6/7.8	5.1/8.9	5.8/9.3	6.7/9.0	6.5/8.9	6.5/8.4	5.7/8.1	5.0/7.7	2.8/6.8	3.5/5.7	6.0/6.8

Direct Beam Solar Radiation for Concentrating Collectors (kWh/m²/day), Uncertainty ±8%

Tracker		Jan	Feb	Mar	Apr	May	June	July	Aug	Sept	Oct	Nov	Dec	Year
1-Axis, E-W Horiz Axis	Average	2.4	2.8	2.8	2.9	3.3	3.7	3.4	3.3	3.2	3.5	2.9	2.5	3.1
	Min/Max	1.3/3.4	1.5/4.1	1.6/4.2	1.7/4.8	1.8/4.9	2.8/4.6	2.5/4.3	2.5/4.1	2.2/4.2	2.3/4.4	0.9/4.3	1.8/3.6	2.8/3.5
1-Axis, N-S Horiz Axis	Average	2.1	2.9	3.4	3.9	4.4	4.9	4.5	4.4	4.0	3.9	2.7	2.1	3.6
	Min/Max	1.1/3.0	1.6/4.3	1.9/5.2	2.1/6.3	2.4/6.4	3.6/6.0	3.3/5.7	3.3/5.5	2.7/5.4	2.6/4.9	0.8/3.9	1.4/3.0	3.2/4.1
1-Axis, N-S Tilt=Latitude	Average	2.7	3.5	3.7	3.9	4.2	4.5	4.3	4.4	4.3	4.5	3.4	2.8	3.8
	Min/Max	1.5/3.9	1.9/5.1	2.1/5.8	2.2/6.4	2.3/6.2	3.4/5.6	3.1/5.4	3.2/5.5	2.9/5.7	2.9/5.6	1.0/4.9	1.9/3.9	3.4/4.4
2-Axis	Average	2.9	3.6	3.7	4.0	4.4	4.9	4.6	4.5	4.3	4.6	3.5	3.0	4.0
	Min/Max	1.6/4.1	2.0/5.2	2.1/5.8	2.2/6.5	2.5/6.5	3.7/6.1	3.4/5.8	3.3/5.6	2.9/5.7	3.0/5.7	1.1/5.2	2.1/4.3	3.6/4.6

Average Climatic Conditions

Element	Jan	Feb	Mar	Apr	May	June	July	Aug	Sept	Oct	Nov	Dec	Year
Temperature (°C)	10.5	12.4	16.3	20.5	24.0	27.1	28.2	28.1	25.9	20.9	16.2	12.4	20.2
Daily Minimum Temp	5.3	6.9	10.7	15.3	19.1	22.2	23.2	22.9	20.9	15.1	10.7	6.8	14.9
Daily Maximum Temp	15.7	17.9	21.9	25.7	28.9	31.9	33.3	33.2	30.7	26.8	21.8	17.9	25.5
Record Minimum Temp	-10.0	-6.7	-5.0	0.0	7.8	13.3	16.1	15.6	7.2	0.0	-5.6	-11.1	-11.1
Record Maximum Temp	27.2	29.4	30.6	34.4	36.1	37.8	39.4	41.7	37.8	35.0	31.1	28.9	41.7
HDD, Base 18.3°C	250	171	86	18	0	0	0	0	0	14	98	196	833
CDD, Base 18.3°C	7	6	24	83	176	262	307	303	227	96	35	12	1536
Relative Humidity (%)	79	77	76	77	79	79	81	80	79	77	78	79	78
Wind Speed (m/s)	4.6	4.9	5.0	5.1	4.4	3.8	3.2	3.1	3.6	3.8	4.4	4.5	4.2

San Angelo, TX

WBAN NO. 23034

LATITUDE: 31.37° N
LONGITUDE: 100.50° W
ELEVATION: 582 meters
MEAN PRESSURE: 949 millibars

STATION TYPE: Secondary

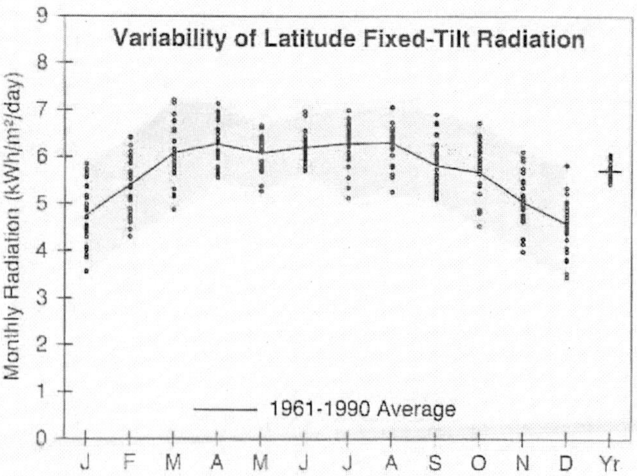

Solar Radiation for Flat-Plate Collectors Facing South at a Fixed Tilt (kWh/m²/day), Uncertainty ±9%

Tilt (°)		Jan	Feb	Mar	Apr	May	June	July	Aug	Sept	Oct	Nov	Dec	Year
0	Average	3.2	4.1	5.2	6.1	6.5	7.0	6.9	6.4	5.3	4.5	3.5	3.0	5.1
	Min/Max	2.6/3.7	3.4/4.6	4.3/6.0	5.4/6.9	5.6/7.1	6.4/7.9	5.6/7.8	5.3/7.2	4.7/6.1	3.8/5.1	3.0/4.0	2.4/3.5	4.9/5.4
Latitude -15	Average	4.1	4.9	5.8	6.4	6.5	6.7	6.8	6.5	5.7	5.3	4.4	3.9	5.6
	Min/Max	3.2/5.0	4.0/5.7	4.7/6.8	5.6/7.2	5.6/7.1	6.2/7.6	5.5/7.6	5.4/7.3	5.0/6.7	4.3/6.2	3.6/5.3	3.0/4.9	5.3/5.9
Latitude	Average	4.7	5.4	6.1	6.3	6.1	6.2	6.3	6.3	5.8	5.7	5.0	4.6	5.7
	Min/Max	3.5/5.8	4.3/6.4	4.9/7.2	5.6/7.1	5.3/6.7	5.7/7.0	5.1/7.0	5.3/7.1	5.1/6.9	4.5/6.7	4.0/6.1	3.4/5.8	5.4/6.1
Latitude +15	Average	5.1	5.6	6.0	5.8	5.4	5.3	5.5	5.7	5.6	5.8	5.4	5.0	5.5
	Min/Max	3.7/6.4	4.4/6.7	4.7/7.2	5.2/6.6	4.7/5.9	4.9/5.9	4.5/6.0	4.8/6.4	4.9/6.7	4.5/6.9	4.2/6.6	3.7/6.4	5.2/5.9
90	Average	4.4	4.4	4.0	3.0	2.3	1.9	2.1	2.7	3.4	4.3	4.5	4.4	3.4
	Min/Max	3.1/5.7	3.4/5.3	3.1/4.7	2.8/3.3	2.1/2.4	1.9/2.0	1.9/2.1	2.4/2.8	2.9/4.0	3.2/5.1	3.4/5.6	3.2/5.9	3.2/3.7

Solar Radiation for 1-Axis Tracking Flat-Plate Collectors with a North-South Axis (kWh/m²/day), Uncertainty ±9%

Axis Tilt (°)		Jan	Feb	Mar	Apr	May	June	July	Aug	Sept	Oct	Nov	Dec	Year
0	Average	4.6	5.7	7.1	8.1	8.4	9.1	9.2	8.6	7.1	6.3	5.0	4.3	7.0
	Min/Max	3.4/5.7	4.5/6.8	5.5/8.8	6.8/9.6	6.7/9.6	8.2/10.7	6.8/10.7	6.7/10.0	6.0/8.7	4.9/7.6	4.0/6.2	3.1/5.5	6.5/7.5
Latitude -15	Average	5.2	6.3	7.6	8.3	8.4	9.0	9.2	8.7	7.5	6.9	5.7	5.0	7.3
	Min/Max	3.8/6.6	4.9/7.6	5.8/9.4	7.1/9.9	6.7/9.6	8.1/10.6	6.8/10.7	6.8/10.2	6.3/9.2	5.3/8.4	4.4/7.1	3.6/6.5	6.9/7.9
Latitude	Average	5.7	6.7	7.8	8.3	8.2	8.6	8.8	8.6	7.6	7.2	6.2	5.5	7.4
	Min/Max	4.1/7.3	5.2/8.1	5.9/9.7	7.0/9.8	6.5/9.3	7.8/10.1	6.6/10.3	6.7/10.0	6.3/9.3	5.4/8.8	4.8/7.8	3.9/7.2	7.0/8.0
Latitude +15	Average	6.0	6.8	7.8	8.0	7.7	8.0	8.3	8.2	7.4	7.3	6.4	5.8	7.3
	Min/Max	4.2/7.7	5.2/8.4	5.8/9.6	6.7/9.5	6.1/8.8	7.2/9.4	6.1/9.6	6.4/9.5	6.2/9.1	5.4/8.9	4.9/8.1	4.1/7.7	6.8/7.9

Solar Radiation for 2-Axis Tracking Flat-Plate Collectors (kWh/m²/day), Uncertainty ±9%

Tracker		Jan	Feb	Mar	Apr	May	June	July	Aug	Sept	Oct	Nov	Dec	Year
2-Axis	Average	6.1	6.9	7.8	8.4	8.5	9.2	9.3	8.8	7.6	7.3	6.5	5.9	7.7
	Min/Max	4.3/7.8	5.3/8.4	5.9/9.7	7.1/9.9	6.8/9.7	8.3/10.8	6.9/10.9	6.8/10.2	6.3/9.3	5.5/8.9	4.9/8.2	4.2/7.8	7.2/8.3

Direct Beam Solar Radiation for Concentrating Collectors (kWh/m²/day), Uncertainty ±8%

Tracker		Jan	Feb	Mar	Apr	May	June	July	Aug	Sept	Oct	Nov	Dec	Year
1-Axis, E-W	Average	3.7	3.9	4.2	4.3	4.3	4.8	4.8	4.5	3.8	4.1	3.9	3.7	4.2
Horiz Axis	Min/Max	2.0/5.3	2.7/5.3	2.6/5.7	3.3/5.5	2.9/5.4	4.0/6.4	2.9/6.2	2.8/5.8	2.7/5.2	2.6/5.5	2.6/5.4	2.3/5.5	3.7/4.7
1-Axis, N-S	Average	3.2	4.0	5.1	5.6	5.7	6.3	6.4	6.0	4.8	4.5	3.6	3.0	4.9
Horiz Axis	Min/Max	1.7/4.6	2.7/5.3	3.2/7.0	4.2/7.4	3.7/7.2	5.2/8.4	3.8/8.4	3.8/7.8	3.4/6.8	2.9/6.1	2.4/5.0	1.8/4.4	4.3/5.5
1-Axis, N-S	Average	4.2	4.9	5.6	5.8	5.4	5.9	6.1	6.0	5.2	5.3	4.6	4.1	5.2
Tilt=Latitude	Min/Max	2.3/6.0	3.3/6.5	3.5/7.8	4.3/7.5	3.5/6.9	4.8/7.8	3.6/7.9	3.8/7.8	3.6/7.2	3.3/7.1	3.1/6.4	2.5/6.0	4.7/6.0
2-Axis	Average	4.5	5.0	5.6	5.8	5.7	6.4	6.5	6.1	5.2	5.4	4.8	4.5	5.5
	Min/Max	2.4/6.4	3.4/6.7	3.5/7.8	4.4/7.7	3.7/7.3	5.3/8.5	3.9/8.5	3.9/8.0	3.7/7.3	3.4/7.2	3.3/6.7	2.7/6.5	4.9/6.2

Average Climatic Conditions

Element	Jan	Feb	Mar	Apr	May	June	July	Aug	Sept	Oct	Nov	Dec	Year
Temperature (°C)	6.5	9.1	14.5	19.4	23.4	26.4	28.2	27.7	24.1	19.0	13.0	7.8	18.3
Daily Minimum Temp	-0.8	1.5	6.4	11.5	16.2	19.1	20.6	20.2	17.8	12.0	5.9	0.6	10.9
Daily Maximum Temp	13.8	16.7	22.6	27.3	30.8	33.7	35.7	35.2	30.4	26.0	20.1	15.0	25.6
Record Minimum Temp	-15.0	-18.3	-13.3	-3.9	1.7	8.9	13.3	12.2	2.8	-2.2	-10.6	-20.0	-20.0
Record Maximum Temp	32.2	32.8	36.1	39.4	41.7	43.3	43.9	42.8	41.7	37.8	33.9	32.8	43.9
HDD, Base 18.3°C	367	258	137	39	4	0	0	0	3	39	166	327	1341
CDD, Base 18.3°C	0	0	18	72	162	242	305	291	177	60	6	0	1333
Relative Humidity (%)	61	60	54	54	60	59	54	57	65	65	64	63	60
Wind Speed (m/s)	4.6	4.8	5.5	5.3	5.1	4.6	4.2	3.9	3.9	4.0	4.4	4.3	4.6

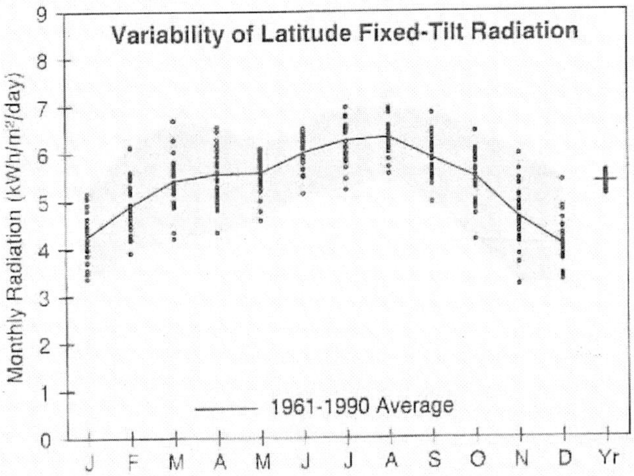

San Antonio, TX

WBAN NO. 12921

LATITUDE: 29.53° N
LONGITUDE: 98.47° W
ELEVATION: 242 meters
MEAN PRESSURE: 988 millibars

STATION TYPE: Primary

Solar Radiation for Flat-Plate Collectors Facing South at a Fixed Tilt (kWh/m²/day), Uncertainty ±9%

Tilt (°)		Jan	Feb	Mar	Apr	May	June	July	Aug	Sept	Oct	Nov	Dec	Year
0	Average	3.1	3.9	4.8	5.5	6.0	6.7	6.9	6.4	5.4	4.5	3.4	2.9	4.9
	Min/Max	2.6/3.5	3.3/4.6	3.8/5.8	4.3/6.4	4.9/6.5	5.7/7.3	5.6/7.7	5.6/7.1	4.7/6.2	3.6/5.1	2.7/4.1	2.5/3.5	4.7/5.2
Latitude -15	Average	3.7	4.5	5.2	5.7	5.9	6.5	6.7	6.6	5.8	5.1	4.1	3.5	5.3
	Min/Max	3.0/4.5	3.7/5.5	4.1/6.4	4.4/6.7	4.9/6.4	5.5/7.1	5.6/7.5	5.7/7.2	4.9/6.7	4.0/5.9	3.0/5.2	3.0/4.6	5.0/5.5
Latitude	Average	4.3	4.9	5.4	5.6	5.6	6.0	6.3	6.3	5.9	5.5	4.6	4.1	5.4
	Min/Max	3.4/5.2	3.9/6.1	4.2/6.7	4.3/6.6	4.6/6.1	5.1/6.5	5.2/7.0	5.6/6.9	5.0/6.8	4.2/6.5	3.2/6.0	3.3/5.4	5.1/5.6
Latitude +15	Average	4.5	5.0	5.3	5.2	5.0	5.2	5.5	5.8	5.7	5.6	4.9	4.4	5.2
	Min/Max	3.5/5.6	4.0/6.4	4.1/6.6	4.0/6.1	4.1/5.4	4.5/5.6	4.6/6.1	5.1/6.3	4.8/6.6	4.2/6.6	3.3/6.4	3.5/5.9	4.9/5.4
90	Average	3.8	3.8	3.4	2.7	2.1	1.9	2.0	2.6	3.3	4.0	4.0	3.8	3.1
	Min/Max	2.9/4.8	2.9/5.0	2.7/4.2	2.2/3.0	1.9/2.2	1.8/2.0	1.9/2.1	2.4/2.7	2.8/3.8	2.9/4.8	2.5/5.4	2.9/5.3	2.9/3.3

Solar Radiation for 1-Axis Tracking Flat-Plate Collectors with a North-South Axis (kWh/m²/day), Uncertainty ±9%

Axis Tilt (°)		Jan	Feb	Mar	Apr	May	June	July	Aug	Sept	Oct	Nov	Dec	Year
0	Average	4.2	5.2	6.3	6.9	7.3	8.4	8.7	8.3	7.1	6.1	4.7	3.9	6.4
	Min/Max	3.3/5.1	4.1/6.6	4.6/8.0	4.9/8.6	5.6/8.2	6.6/9.5	6.6/10.2	6.9/9.5	5.7/8.6	4.4/7.4	3.2/6.1	3.2/5.3	6.0/6.8
Latitude -15	Average	4.7	5.7	6.6	7.1	7.3	8.3	8.7	8.4	7.4	6.6	5.2	4.4	6.7
	Min/Max	3.6/5.8	4.4/7.3	4.9/8.5	5.0/8.8	5.6/8.2	6.5/9.4	6.6/10.1	7.0/9.6	5.9/9.0	4.7/8.0	3.5/6.9	3.5/6.0	6.2/7.1
Latitude	Average	5.1	6.0	6.8	7.0	7.1	7.9	8.4	8.3	7.5	6.9	5.6	4.8	6.8
	Min/Max	3.9/6.3	4.6/7.7	5.0/8.8	4.9/8.7	5.4/8.0	6.2/9.0	6.3/9.7	6.9/9.5	5.9/9.1	4.8/8.4	3.6/7.5	3.8/6.7	6.3/7.2
Latitude +15	Average	5.3	6.1	6.7	6.7	6.7	7.4	7.8	7.9	7.3	6.9	5.8	5.0	6.6
	Min/Max	4.0/6.7	4.6/7.9	4.9/8.7	4.7/8.4	5.0/7.5	5.8/8.4	5.9/9.1	6.6/9.0	5.8/8.9	4.8/8.5	3.7/7.8	4.0/7.1	6.2/7.0

Solar Radiation for 2-Axis Tracking Flat-Plate Collectors (kWh/m²/day), Uncertainty ±9%

Tracker		Jan	Feb	Mar	Apr	May	June	July	Aug	Sept	Oct	Nov	Dec	Year
2-Axis	Average	5.3	6.1	6.8	7.1	7.4	8.4	8.8	8.4	7.5	6.9	5.8	5.1	7.0
	Min/Max	4.0/6.7	4.7/7.9	5.0/8.8	5.0/8.8	5.6/8.3	6.6/9.6	6.7/10.3	7.1/9.7	5.9/9.1	4.9/8.5	3.7/7.9	4.0/7.2	6.5/7.4

Direct Beam Solar Radiation for Concentrating Collectors (kWh/m²/day), Uncertainty ±8%

Tracker		Jan	Feb	Mar	Apr	May	June	July	Aug	Sept	Oct	Nov	Dec	Year
1-Axis, E-W Horiz Axis	Average	3.0	3.3	3.3	3.2	3.3	4.0	4.4	4.1	3.6	3.7	3.3	3.0	3.5
	Min/Max	2.0/4.2	2.2/4.8	1.9/5.0	1.5/4.5	1.6/4.1	2.5/5.1	2.6/5.8	2.9/5.2	2.2/5.1	2.0/5.0	1.5/5.1	2.1/4.7	3.1/3.9
1-Axis, N-S Horiz Axis	Average	2.7	3.4	4.0	4.2	4.2	5.2	5.7	5.4	4.5	4.1	3.1	2.5	4.1
	Min/Max	1.7/3.8	2.3/5.0	2.3/6.1	1.9/6.0	2.1/5.3	3.1/6.7	3.3/7.6	3.8/6.8	2.8/6.4	2.2/5.7	1.5/4.8	1.7/4.0	3.6/4.6
1-Axis, N-S Tilt=Latitude	Average	3.4	4.1	4.4	4.2	4.1	4.8	5.3	5.3	4.8	4.7	3.9	3.2	4.4
	Min/Max	2.2/4.8	2.7/5.9	2.6/6.7	1.9/6.1	2.0/5.1	2.9/6.2	3.1/7.1	3.8/6.7	3.0/6.9	2.5/6.5	1.8/6.0	2.3/5.2	3.8/4.9
2-Axis	Average	3.7	4.2	4.4	4.3	4.3	5.3	5.7	5.5	4.8	4.8	4.1	3.5	4.5
	Min/Max	2.3/5.1	2.8/6.1	2.6/6.7	1.9/6.2	2.2/5.4	3.2/6.7	3.3/7.7	3.9/6.9	3.0/6.9	2.6/6.6	1.9/6.3	2.4/5.7	4.0/5.1

Average Climatic Conditions

Element	Jan	Feb	Mar	Apr	May	June	July	Aug	Sept	Oct	Nov	Dec	Year
Temperature (°C)	9.6	11.9	16.5	20.7	24.2	27.9	29.4	29.4	26.3	21.2	15.8	11.2	20.3
Daily Minimum Temp	3.3	5.2	9.8	14.7	18.7	22.6	23.9	23.6	20.7	14.9	9.3	4.9	14.3
Daily Maximum Temp	16.0	18.7	23.1	26.8	29.6	33.2	35.0	35.2	31.8	27.6	22.2	17.5	26.4
Record Minimum Temp	-17.8	-14.4	-7.2	-0.6	6.1	11.7	16.7	16.1	5.0	0.6	-6.1	-14.4	-17.8
Record Maximum Temp	31.7	36.1	37.8	37.8	39.4	40.6	41.1	42.2	39.4	37.2	34.4	32.2	42.2
HDD, Base 18.3°C	274	184	93	18	0	0	0	0	0	17	100	227	913
CDD, Base 18.3°C	4	6	36	89	181	287	344	343	238	106	23	7	1664
Relative Humidity (%)	67	65	63	66	71	69	65	65	68	67	68	68	67
Wind Speed (m/s)	3.9	4.2	4.5	4.4	4.4	4.3	4.2	3.8	3.8	3.8	3.8	3.8	4.1

Victoria, TX

WBAN NO. 12912

LATITUDE: 28.85° N
LONGITUDE: 96.92° W
ELEVATION: 32 meters
MEAN PRESSURE: 1012 millibars

STATION TYPE: Secondary

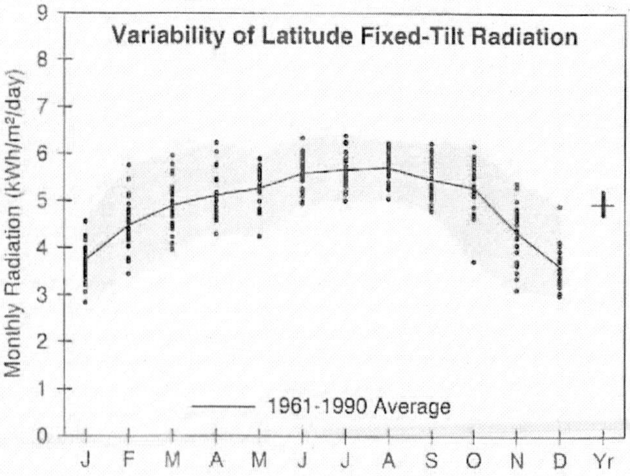

Solar Radiation for Flat-Plate Collectors Facing South at a Fixed Tilt (kWh/m²/day), Uncertainty ±9%

Tilt (°)		Jan	Feb	Mar	Apr	May	June	July	Aug	Sept	Oct	Nov	Dec	Year
0	Average	2.8	3.6	4.4	5.1	5.7	6.2	6.2	5.8	5.0	4.3	3.3	2.7	4.6
	Min/Max	2.4/3.3	3.0/4.4	3.7/5.2	4.4/6.1	4.6/6.3	5.5/7.1	5.4/7.1	5.2/6.4	4.5/5.7	3.3/4.9	2.6/3.9	2.3/3.3	4.4/4.8
Latitude -15	Average	3.3	4.1	4.8	5.2	5.6	6.1	6.1	5.9	5.4	4.9	3.9	3.2	4.9
	Min/Max	2.6/4.0	3.3/5.2	3.9/5.7	4.4/6.3	4.5/6.3	5.3/6.9	5.3/6.9	5.3/6.5	4.7/6.1	3.6/5.7	2.9/4.7	2.7/4.2	4.7/5.1
Latitude	Average	3.7	4.5	4.9	5.1	5.3	5.6	5.7	5.7	5.4	5.3	4.3	3.6	4.9
	Min/Max	2.8/4.6	3.4/5.8	4.0/6.0	4.3/6.2	4.3/5.9	4.9/6.3	5.0/6.4	5.1/6.2	4.8/6.2	3.7/6.2	3.1/5.4	3.0/4.9	4.7/5.2
Latitude +15	Average	3.9	4.6	4.8	4.8	4.7	4.9	5.0	5.2	5.2	5.4	4.6	3.8	4.7
	Min/Max	2.9/4.9	3.4/6.0	3.8/5.9	4.0/5.8	3.8/5.3	4.3/5.5	4.4/5.6	4.6/5.7	4.6/6.0	3.7/6.3	3.2/5.7	3.1/5.3	4.5/5.0
90	Average	3.2	3.4	3.0	2.4	2.0	1.8	1.9	2.4	3.0	3.8	3.7	3.3	2.8
	Min/Max	2.3/4.2	2.5/4.6	2.5/3.7	2.1/2.8	1.8/2.1	1.7/1.9	1.8/2.0	2.1/2.5	2.7/3.5	2.5/4.5	2.5/4.7	2.6/4.7	2.7/3.0

Solar Radiation for 1-Axis Tracking Flat-Plate Collectors with a North-South Axis (kWh/m²/day), Uncertainty ±9%

Axis Tilt (°)		Jan	Feb	Mar	Apr	May	June	July	Aug	Sept	Oct	Nov	Dec	Year
0	Average	3.7	4.7	5.6	6.3	7.0	7.8	7.9	7.5	6.6	5.9	4.4	3.5	5.9
	Min/Max	2.8/4.6	3.6/6.2	4.4/7.1	5.1/8.1	5.1/8.1	6.5/9.2	6.5/9.5	6.5/8.4	5.6/7.8	4.0/7.0	3.0/5.5	2.9/4.8	5.6/6.3
Latitude -15	Average	4.1	5.1	5.9	6.4	7.0	7.7	7.8	7.6	6.9	6.3	4.8	3.9	6.1
	Min/Max	3.0/5.1	3.8/6.8	4.6/7.5	5.2/8.3	5.1/8.1	6.4/9.1	6.5/9.4	6.6/8.5	5.8/8.1	4.2/7.6	3.3/6.2	3.2/5.4	5.8/6.5
Latitude	Average	4.4	5.4	6.0	6.3	6.7	7.4	7.5	7.5	6.9	6.6	5.2	4.2	6.2
	Min/Max	3.1/5.6	4.0/7.2	4.6/7.7	5.1/8.2	4.9/7.9	6.2/8.7	6.3/9.1	6.4/8.4	5.9/8.2	4.3/7.9	3.5/6.7	3.4/6.0	5.8/6.6
Latitude +15	Average	4.5	5.5	5.9	6.1	6.3	6.9	7.1	7.1	6.8	6.6	5.4	4.4	6.0
	Min/Max	3.2/5.8	4.0/7.4	4.5/7.6	4.8/7.9	4.6/7.4	5.7/8.1	5.9/8.5	6.1/8.0	5.7/8.0	4.2/8.0	3.5/6.9	3.5/6.3	5.7/6.4

Solar Radiation for 2-Axis Tracking Flat-Plate Collectors (kWh/m²/day), Uncertainty ±9%

Tracker		Jan	Feb	Mar	Apr	May	June	July	Aug	Sept	Oct	Nov	Dec	Year
2-Axis	Average	4.6	5.5	6.0	6.4	7.0	7.9	7.9	7.7	6.9	6.7	5.4	4.4	6.4
	Min/Max	3.2/5.9	4.0/7.4	4.6/7.7	5.2/8.3	5.2/8.2	6.6/9.3	6.6/9.6	6.6/8.6	5.9/8.2	4.3/8.0	3.5/7.0	3.5/6.4	6.0/6.8

Direct Beam Solar Radiation for Concentrating Collectors (kWh/m²/day), Uncertainty ±8%

Tracker		Jan	Feb	Mar	Apr	May	June	July	Aug	Sept	Oct	Nov	Dec	Year
1-Axis, E-W	Average	2.3	2.7	2.7	2.6	2.9	3.5	3.5	3.3	3.1	3.4	2.9	2.4	2.9
Horiz Axis	Min/Max	1.3/3.5	1.5/4.3	1.5/3.9	1.7/4.2	1.3/3.8	2.4/4.7	2.5/4.9	2.5/4.1	2.2/4.0	1.6/4.5	1.5/4.2	1.6/4.1	2.6/3.3
1-Axis, N-S	Average	2.1	2.9	3.3	3.4	3.9	4.6	4.6	4.5	4.0	3.8	2.7	2.0	3.5
Horiz Axis	Min/Max	1.1/3.2	1.6/4.5	1.8/4.9	2.2/5.5	1.8/5.1	3.1/6.1	3.2/6.6	3.4/5.5	2.9/5.3	1.7/5.1	1.3/4.0	1.4/3.4	3.0/3.8
1-Axis, N-S	Average	2.6	3.4	3.6	3.5	3.7	4.3	4.4	4.4	4.2	4.4	3.4	2.6	3.7
Tilt=Latitude	Min/Max	1.4/4.0	2.0/5.3	2.0/5.3	2.2/5.6	1.7/4.9	2.9/5.7	3.1/6.2	3.4/5.4	3.0/5.6	2.0/5.8	1.7/4.9	1.8/4.5	3.2/4.1
2-Axis	Average	2.8	3.5	3.6	3.6	3.9	4.6	4.7	4.5	4.2	4.4	3.6	2.8	3.9
	Min/Max	1.5/4.3	2.0/5.5	2.0/5.4	2.3/5.7	1.8/5.1	3.1/6.2	3.3/6.7	3.5/5.6	3.1/5.6	2.0/5.9	1.8/5.2	2.0/4.9	3.4/4.3

Average Climatic Conditions

Element	Jan	Feb	Mar	Apr	May	June	July	Aug	Sept	Oct	Nov	Dec	Year
Temperature (°C)	11.5	13.4	17.4	21.4	24.8	27.6	28.9	28.9	26.4	22.1	17.2	13.1	21.1
Daily Minimum Temp	5.8	7.4	11.6	16.1	19.8	22.6	23.7	23.4	21.3	16.1	11.3	7.3	15.6
Daily Maximum Temp	17.1	19.3	23.2	26.8	29.6	32.6	34.2	34.4	31.6	28.0	23.0	18.9	26.6
Record Minimum Temp	-10.0	-7.2	-6.1	0.6	9.4	15.0	16.7	17.2	9.4	2.2	-4.4	-12.8	-12.8
Record Maximum Temp	31.1	35.0	36.1	36.7	38.3	38.3	40.0	41.7	38.9	37.2	33.9	31.1	41.7
HDD, Base 18.3°C	224	148	71	8	0	0	0	0	0	8	82	179	720
CDD, Base 18.3°C	12	9	42	102	200	278	329	329	243	124	47	17	1732
Relative Humidity (%)	76	74	72	74	76	76	74	74	76	74	75	76	75
Wind Speed (m/s)	4.7	5.1	5.3	5.2	4.9	4.3	4.1	3.8	4.0	4.1	4.5	4.6	4.6

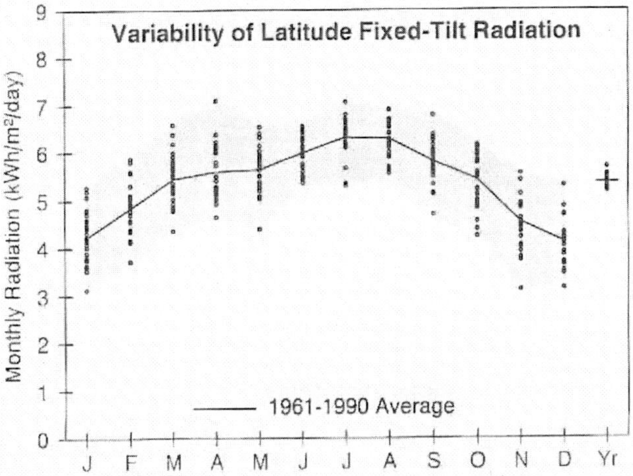

Waco, TX

WBAN NO. 13959

LATITUDE: 31.62° N
LONGITUDE: 97.22° W
ELEVATION: 155 meters
MEAN PRESSURE: 998 millibars

STATION TYPE: Secondary

Solar Radiation for Flat-Plate Collectors Facing South at a Fixed Tilt (kWh/m²/day), Uncertainty ±9%

Tilt (°)		Jan	Feb	Mar	Apr	May	June	July	Aug	Sept	Oct	Nov	Dec	Year
0	Average	2.9	3.7	4.7	5.5	6.0	6.7	6.9	6.4	5.2	4.3	3.2	2.7	4.9
	Min/Max	2.4/3.4	3.1/4.3	4.0/5.5	4.6/6.8	4.7/7.0	5.9/7.4	5.8/7.8	5.6/7.0	4.4/6.0	3.5/4.7	2.5/3.7	2.3/3.3	4.7/5.1
Latitude -15	Average	3.7	4.4	5.3	5.7	6.0	6.5	6.8	6.5	5.7	5.0	4.0	3.6	5.3
	Min/Max	2.9/4.5	3.5/5.3	4.3/6.3	4.8/7.2	4.7/6.9	5.8/7.2	5.7/7.7	5.8/7.2	4.7/6.6	4.0/5.7	2.9/4.8	2.8/4.5	5.1/5.6
Latitude	Average	4.2	4.8	5.4	5.6	5.6	6.0	6.3	6.3	5.8	5.4	4.5	4.1	5.4
	Min/Max	3.1/5.3	3.7/5.9	4.4/6.6	4.6/7.1	4.4/6.5	5.3/6.6	5.3/7.1	5.6/6.9	4.7/6.8	4.2/6.2	3.1/5.6	3.1/5.3	5.2/5.7
Latitude +15	Average	4.5	5.0	5.3	5.2	5.0	5.2	5.5	5.7	5.6	5.5	4.8	4.4	5.1
	Min/Max	3.2/5.7	3.8/6.1	4.2/6.5	4.3/6.6	3.9/5.8	4.7/5.6	4.7/6.1	5.1/6.3	4.5/6.6	4.2/6.3	3.2/5.9	3.3/5.8	5.0/5.5
90	Average	3.9	3.9	3.5	2.8	2.2	2.0	2.1	2.7	3.4	4.1	4.0	3.9	3.2
	Min/Max	2.6/5.0	2.8/4.9	2.8/4.3	2.4/3.3	2.0/2.4	1.8/2.0	2.0/2.2	2.5/2.8	2.7/3.9	3.0/4.7	2.5/5.1	2.8/5.3	3.0/3.4

Solar Radiation for 1-Axis Tracking Flat-Plate Collectors with a North-South Axis (kWh/m²/day), Uncertainty ±9%

Axis Tilt (°)		Jan	Feb	Mar	Apr	May	June	July	Aug	Sept	Oct	Nov	Dec	Year
0	Average	4.1	5.1	6.3	7.1	7.6	8.6	9.1	8.5	7.0	6.0	4.5	3.8	6.5
	Min/Max	3.0/5.1	3.9/6.3	4.9/7.8	5.7/9.5	5.4/9.2	7.4/9.8	7.3/10.8	7.3/9.6	5.6/8.6	4.5/6.8	3.1/5.6	2.9/5.0	6.1/7.0
Latitude -15	Average	4.6	5.6	6.7	7.3	7.6	8.6	9.1	8.7	7.4	6.5	5.1	4.4	6.8
	Min/Max	3.3/5.9	4.2/7.0	5.1/8.4	5.8/9.8	5.4/9.3	7.3/9.7	7.3/10.7	7.4/9.8	5.8/9.0	4.9/7.5	3.3/6.4	3.3/5.8	6.4/7.4
Latitude	Average	5.0	5.9	6.8	7.2	7.4	8.2	8.7	8.5	7.5	6.8	5.5	4.9	6.9
	Min/Max	3.5/6.5	4.4/7.4	5.2/8.6	5.8/9.8	5.2/9.0	7.0/9.3	7.0/10.3	7.3/9.7	5.8/9.1	5.1/7.9	3.5/7.0	3.6/6.5	6.5/7.4
Latitude +15	Average	5.3	6.0	6.8	6.9	7.0	7.6	8.2	8.1	7.3	6.9	5.7	5.1	6.7
	Min/Max	3.6/6.8	4.4/7.6	5.1/8.5	5.5/9.4	4.9/8.4	6.5/8.7	6.5/9.6	6.9/9.2	5.7/9.0	5.1/7.9	3.6/7.3	3.7/6.9	6.4/7.3

Solar Radiation for 2-Axis Tracking Flat-Plate Collectors (kWh/m²/day), Uncertainty ±9%

Tracker		Jan	Feb	Mar	Apr	May	June	July	Aug	Sept	Oct	Nov	Dec	Year
2-Axis	Average	5.3	6.0	6.9	7.3	7.7	8.7	9.2	8.7	7.5	6.9	5.7	5.2	7.1
	Min/Max	3.6/6.9	4.4/7.6	5.2/8.6	5.8/9.9	5.5/9.4	7.5/9.9	7.4/10.9	7.4/9.9	5.9/9.2	5.1/8.0	3.6/7.3	3.8/7.0	6.7/7.7

Direct Beam Solar Radiation for Concentrating Collectors (kWh/m²/day), Uncertainty ±8%

Tracker		Jan	Feb	Mar	Apr	May	June	July	Aug	Sept	Oct	Nov	Dec	Year
1-Axis, E-W Horiz Axis	Average	3.1	3.3	3.4	3.4	3.6	4.4	4.8	4.4	3.7	3.8	3.3	3.2	3.7
	Min/Max	1.6/4.5	1.9/4.7	2.1/4.9	2.3/5.5	1.6/5.0	3.2/5.4	3.2/6.3	3.4/5.3	2.5/5.1	2.4/4.7	1.5/4.7	1.9/4.7	3.3/4.2
1-Axis, N-S Horiz Axis	Average	2.6	3.3	4.1	4.4	4.8	5.7	6.3	5.9	4.7	4.1	3.0	2.6	4.3
	Min/Max	1.4/3.9	2.0/4.8	2.5/6.0	3.0/7.3	2.1/6.6	4.2/7.1	4.3/8.4	4.5/7.2	3.2/6.6	2.6/5.1	1.4/4.3	1.6/3.8	3.8/4.9
1-Axis, N-S Tilt=Latitude	Average	3.5	4.1	4.5	4.6	4.6	5.3	6.0	5.9	5.0	4.8	3.9	3.5	4.6
	Min/Max	1.8/5.1	2.4/5.8	2.8/6.6	3.1/7.5	2.0/6.3	3.9/6.6	4.0/7.9	4.5/7.1	3.4/7.0	3.1/6.0	1.8/5.5	2.1/5.2	4.1/5.3
2-Axis	Average	3.7	4.2	4.6	4.6	4.8	5.8	6.4	6.0	5.1	4.9	4.1	3.8	4.8
	Min/Max	1.9/5.5	2.5/6.0	2.8/6.7	3.1/7.6	2.2/6.7	4.3/7.2	4.3/8.5	4.6/7.4	3.4/7.0	3.1/6.1	1.8/5.8	2.3/5.6	4.3/5.5

Average Climatic Conditions

Element	Jan	Feb	Mar	Apr	May	June	July	Aug	Sept	Oct	Nov	Dec	Year
Temperature (°C)	7.3	9.7	14.6	19.5	23.5	27.5	29.8	29.8	25.9	20.3	14.3	9.1	19.3
Daily Minimum Temp	1.2	3.3	8.2	13.4	17.9	21.6	23.6	23.3	19.8	13.8	8.1	2.9	13.1
Daily Maximum Temp	13.4	16.0	20.9	25.6	29.1	33.3	36.0	36.2	32.0	26.8	20.4	15.2	25.4
Record Minimum Temp	-20.6	-15.6	-9.4	-2.8	2.8	11.1	15.6	13.9	4.4	-1.7	-8.3	-20.0	-20.6
Record Maximum Temp	31.1	33.3	37.8	38.3	38.9	42.8	42.2	44.4	41.1	38.3	33.3	32.8	44.4
HDD, Base 18.3°C	344	245	137	37	0	0	0	0	0	21	135	291	1211
CDD, Base 18.3°C	3	0	20	72	160	275	355	355	227	81	13	3	1564
Relative Humidity (%)	71	69	66	68	71	67	60	60	67	67	71	71	67
Wind Speed (m/s)	5.1	5.5	5.9	5.8	5.4	4.9	4.8	4.5	4.5	4.6	4.9	4.9	5.1

Wichita Falls, TX

WBAN NO. 13966

LATITUDE: 33.97° N
LONGITUDE: 98.48° W
ELEVATION: 314 meters
MEAN PRESSURE: 979 millibars

STATION TYPE: Secondary

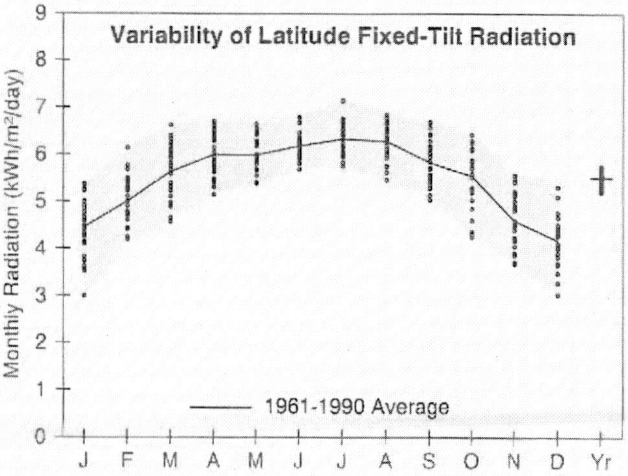

Solar Radiation for Flat-Plate Collectors Facing South at a Fixed Tilt (kWh/m²/day), Uncertainty ±9%

Tilt (°)		Jan	Feb	Mar	Apr	May	June	July	Aug	Sept	Oct	Nov	Dec	Year
0	Average	2.9	3.7	4.8	5.8	6.4	6.9	7.0	6.3	5.2	4.2	3.1	2.6	4.9
	Min/Max	2.3/3.3	3.2/4.3	4.0/5.4	5.0/6.4	5.8/7.0	6.4/7.7	6.3/7.9	5.5/6.9	4.5/5.8	3.4/4.7	2.7/3.5	2.1/3.1	4.7/5.1
Latitude -15	Average	3.9	4.6	5.4	6.1	6.4	6.7	6.8	6.5	5.7	5.1	4.1	3.6	5.4
	Min/Max	2.8/4.6	3.9/5.5	4.5/6.3	5.3/6.8	5.7/7.1	6.2/7.4	6.2/7.8	5.7/7.1	5.0/6.5	4.0/5.9	3.3/4.8	2.7/4.5	5.1/5.6
Latitude	Average	4.5	5.0	5.6	6.0	6.0	6.2	6.3	6.3	5.8	5.5	4.6	4.2	5.5
	Min/Max	3.0/5.4	4.2/6.1	4.6/6.6	5.2/6.7	5.4/6.6	5.7/6.8	5.8/7.1	5.5/6.8	5.0/6.7	4.3/6.4	3.7/5.6	3.0/5.3	5.2/5.8
Latitude +15	Average	4.8	5.2	5.5	5.6	5.3	5.3	5.5	5.7	5.6	5.6	4.9	4.5	5.3
	Min/Max	3.1/5.8	4.3/6.4	4.5/6.6	4.8/6.2	4.8/5.9	4.9/5.8	5.1/6.2	5.0/6.2	4.8/6.4	4.3/6.6	3.8/6.0	3.2/5.8	5.0/5.6
90	Average	4.2	4.2	3.8	3.1	2.4	2.2	2.3	2.9	3.5	4.3	4.2	4.1	3.4
	Min/Max	2.6/5.2	3.3/5.2	3.1/4.6	2.8/3.4	2.2/2.6	2.1/2.2	2.2/2.4	2.6/3.0	3.1/4.1	3.2/5.0	3.1/5.2	2.8/5.4	3.2/3.6

Solar Radiation for 1-Axis Tracking Flat-Plate Collectors with a North-South Axis (kWh/m²/day), Uncertainty ±9%

Axis Tilt (°)		Jan	Feb	Mar	Apr	May	June	July	Aug	Sept	Oct	Nov	Dec	Year
0	Average	4.1	5.2	6.5	7.7	8.3	9.1	9.3	8.5	7.1	6.0	4.4	3.8	6.7
	Min/Max	2.8/5.0	4.2/6.5	5.2/7.8	6.4/9.0	7.2/9.4	8.3/10.4	7.9/11.0	7.1/9.7	5.8/8.5	4.5/7.1	3.5/5.4	2.8/4.8	6.2/7.1
Latitude -15	Average	4.9	5.8	7.0	8.0	8.4	9.0	9.2	8.7	7.5	6.7	5.1	4.5	7.1
	Min/Max	3.2/6.0	4.7/7.4	5.5/8.5	6.6/9.3	7.2/9.4	8.2/10.3	7.9/11.0	7.2/10.0	6.2/9.1	5.0/8.0	4.0/6.4	3.2/5.9	6.6/7.5
Latitude	Average	5.3	6.2	7.2	7.9	8.1	8.6	8.9	8.6	7.6	7.0	5.6	5.0	7.2
	Min/Max	3.4/6.6	4.9/7.9	5.6/8.7	6.5/9.3	7.0/9.2	7.8/9.8	7.6/10.6	7.1/9.8	6.2/9.2	5.2/8.4	4.3/6.9	3.5/6.5	6.7/7.6
Latitude +15	Average	5.6	6.3	7.1	7.6	7.6	8.1	8.3	8.2	7.4	7.1	5.8	5.2	7.0
	Min/Max	3.5/6.9	5.0/8.1	5.5/8.6	6.3/8.9	6.6/8.6	7.3/9.2	7.1/9.9	6.8/9.3	6.1/9.0	5.2/8.4	4.4/7.2	3.6/6.9	6.5/7.4

Solar Radiation for 2-Axis Tracking Flat-Plate Collectors (kWh/m²/day), Uncertainty ±9%

Tracker		Jan	Feb	Mar	Apr	May	June	July	Aug	Sept	Oct	Nov	Dec	Year
2-Axis	Average	5.6	6.3	7.2	8.0	8.5	9.2	9.4	8.8	7.6	7.1	5.8	5.3	7.4
	Min/Max	3.5/7.0	5.0/8.1	5.7/8.7	6.6/9.4	7.3/9.5	8.4/10.5	8.0/11.2	7.3/10.0	6.2/9.2	5.2/8.5	4.4/7.3	3.6/7.1	6.9/7.8

Direct Beam Solar Radiation for Concentrating Collectors (kWh/m²/day), Uncertainty ±8%

Tracker		Jan	Feb	Mar	Apr	May	June	July	Aug	Sept	Oct	Nov	Dec	Year
1-Axis, E-W Horiz Axis	Average	3.4	3.5	3.7	3.9	4.1	4.7	4.9	4.4	3.9	4.0	3.5	3.3	3.9
	Min/Max	1.6/4.7	2.4/5.0	2.5/4.9	2.8/4.9	3.2/5.1	4.0/5.9	3.7/6.6	3.2/5.4	2.7/5.1	2.5/5.2	2.2/4.7	2.0/4.9	3.4/4.3
1-Axis, N-S Horiz Axis	Average	2.8	3.5	4.4	5.2	5.5	6.2	6.5	5.9	4.9	4.3	3.0	2.6	4.6
	Min/Max	1.3/3.9	2.4/5.1	3.0/5.8	3.6/6.6	4.2/6.7	5.3/7.8	4.8/8.8	4.2/7.4	3.4/6.6	2.7/5.6	1.8/4.2	1.5/3.8	4.0/5.1
1-Axis, N-S Tilt=Latitude	Average	3.8	4.3	4.9	5.4	5.3	5.8	6.1	5.9	5.3	5.1	4.0	3.6	5.0
	Min/Max	1.7/5.3	3.0/6.3	3.4/6.6	3.8/6.8	4.0/6.5	4.9/7.3	4.5/8.3	4.2/7.4	3.7/7.1	3.2/6.7	2.5/5.5	2.1/5.3	4.3/5.5
2-Axis	Average	4.1	4.4	4.9	5.4	5.6	6.3	6.6	6.1	5.3	5.2	4.2	4.0	5.2
	Min/Max	1.9/5.6	3.0/6.5	3.4/6.7	3.8/6.9	4.3/6.8	5.3/7.9	4.8/8.9	4.4/7.6	3.7/7.2	3.2/6.8	2.6/5.8	2.3/5.8	4.5/5.7

Average Climatic Conditions

Element	Jan	Feb	Mar	Apr	May	June	July	Aug	Sept	Oct	Nov	Dec	Year
Temperature (°C)	4.3	7.1	11.9	17.3	21.8	26.6	29.4	28.7	24.1	18.1	11.3	6.0	17.2
Daily Minimum Temp	-2.4	0.1	4.8	10.2	15.1	20.0	22.6	21.9	17.7	11.2	4.8	-0.7	10.4
Daily Maximum Temp	11.1	14.0	19.1	24.3	28.5	33.1	36.2	35.5	30.4	24.9	17.9	12.7	24.0
Record Minimum Temp	-20.6	-22.2	-13.3	-4.4	2.2	10.6	12.2	12.2	3.3	-3.9	-10.0	-21.7	-22.2
Record Maximum Temp	30.6	33.3	37.8	38.9	40.6	47.2	45.6	45.0	42.2	38.9	31.7	31.1	47.2
HDD, Base 18.3°C	434	316	209	69	6	0	0	0	3	58	213	382	1690
CDD, Base 18.3°C	0	0	11	38	112	247	344	321	177	51	0	0	1300
Relative Humidity (%)	66	66	61	62	66	64	56	58	67	66	68	68	64
Wind Speed (m/s)	5.1	5.4	5.9	5.9	5.7	5.2	5.0	4.6	4.7	4.9	5.1	4.9	5.2

Cedar City, UT

WBAN NO. 93129

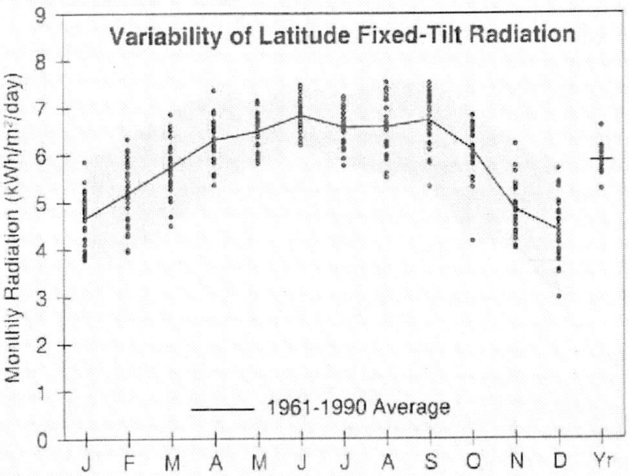

LATITUDE: 37.70° N
LONGITUDE: 113.10° W
ELEVATION: 1712 meters
MEAN PRESSURE: 829 millibars

STATION TYPE: Secondary

Solar Radiation for Flat-Plate Collectors Facing South at a Fixed Tilt (kWh/m²/day), Uncertainty ±9%

Tilt (°)		Jan	Feb	Mar	Apr	May	June	July	Aug	Sept	Oct	Nov	Dec	Year
0	Average	2.7	3.5	4.6	6.0	7.0	7.8	7.3	6.5	5.7	4.3	2.9	2.4	5.0
	Min/Max	2.3/3.1	2.9/4.0	3.9/5.4	5.2/6.8	6.3/7.6	7.0/8.6	6.3/8.0	5.5/7.4	4.7/6.3	3.2/4.7	2.6/3.5	1.9/2.8	4.7/5.5
Latitude -15	Average	4.0	4.7	5.5	6.4	7.0	7.5	7.1	6.8	6.6	5.6	4.2	3.7	5.8
	Min/Max	3.3/4.9	3.7/5.6	4.4/6.5	5.5/7.4	6.2/7.6	6.8/8.2	6.2/7.9	5.7/7.8	5.3/7.3	3.9/6.3	3.6/5.4	2.6/4.8	5.2/6.4
Latitude	Average	4.6	5.2	5.7	6.3	6.5	6.8	6.6	6.6	6.7	6.1	4.8	4.4	5.9
	Min/Max	3.8/5.9	4.0/6.3	4.5/6.9	5.4/7.4	5.8/7.2	6.2/7.5	5.8/7.2	5.5/7.5	5.3/7.5	4.2/6.8	4.0/6.2	3.0/5.7	5.3/6.6
Latitude +15	Average	5.0	5.4	5.7	5.9	5.7	5.9	5.7	6.0	6.5	6.2	5.1	4.8	5.7
	Min/Max	4.0/6.5	4.0/6.7	4.4/6.8	5.0/6.8	5.2/6.3	5.4/6.4	5.0/6.3	5.0/6.8	5.1/7.3	4.2/7.0	4.2/6.7	3.2/6.3	5.0/6.4
90	Average	4.7	4.7	4.2	3.6	2.8	2.6	2.6	3.3	4.3	5.0	4.6	4.6	3.9
	Min/Max	3.7/6.4	3.3/5.8	3.3/5.0	3.1/4.0	2.6/3.1	2.5/2.7	2.4/2.8	2.8/3.6	3.4/4.8	3.3/5.6	3.7/6.1	2.9/6.0	3.4/4.5

Solar Radiation for 1-Axis Tracking Flat-Plate Collectors with a North-South Axis (kWh/m²/day), Uncertainty ±9%

Axis Tilt (°)		Jan	Feb	Mar	Apr	May	June	July	Aug	Sept	Oct	Nov	Dec	Year
0	Average	4.2	5.2	6.6	8.4	9.6	10.8	10.1	9.2	8.4	6.6	4.5	3.8	7.3
	Min/Max	3.4/5.3	3.8/6.4	5.1/8.1	6.8/9.9	8.4/10.8	9.3/12.2	8.3/11.5	7.3/10.8	6.3/9.6	4.4/7.4	3.7/5.8	2.5/4.9	6.4/8.2
Latitude -15	Average	5.1	6.1	7.3	8.8	9.7	10.8	10.1	9.5	9.1	7.5	5.4	4.7	7.9
	Min/Max	4.1/6.6	4.4/7.6	5.5/9.0	7.1/10.4	8.5/10.9	9.3/12.2	8.4/11.5	7.6/11.2	6.8/10.5	4.9/8.5	4.4/7.2	3.1/6.3	6.8/8.9
Latitude	Average	5.6	6.5	7.5	8.7	9.4	10.3	9.7	9.4	9.3	7.9	5.9	5.3	8.0
	Min/Max	4.5/7.3	4.6/8.2	5.6/9.3	7.1/10.4	8.3/10.6	8.9/11.7	8.1/11.1	7.4/11.1	6.8/10.6	5.1/8.9	4.8/7.8	3.3/7.0	6.9/9.1
Latitude +15	Average	5.9	6.6	7.4	8.4	8.9	9.7	9.1	9.0	9.1	8.0	6.1	5.6	7.8
	Min/Max	4.7/7.8	4.6/8.4	5.5/9.2	6.8/10.0	7.8/10.1	8.3/10.9	7.6/10.4	7.1/10.6	6.6/10.4	5.1/9.1	5.0/8.2	3.5/7.4	6.7/9.0

Solar Radiation for 2-Axis Tracking Flat-Plate Collectors (kWh/m²/day), Uncertainty ±9%

Tracker		Jan	Feb	Mar	Apr	May	June	July	Aug	Sept	Oct	Nov	Dec	Year
2-Axis	Average	6.0	6.6	7.5	8.8	9.9	11.0	10.3	9.6	9.3	8.0	6.2	5.7	8.3
	Min/Max	4.7/7.8	4.7/8.4	5.6/9.3	7.2/10.5	8.6/11.1	9.5/12.5	8.5/11.7	7.6/11.3	6.8/10.6	5.1/9.1	5.0/8.2	3.6/7.6	7.1/9.4

Direct Beam Solar Radiation for Concentrating Collectors (kWh/m²/day), Uncertainty ±8%

Tracker		Jan	Feb	Mar	Apr	May	June	July	Aug	Sept	Oct	Nov	Dec	Year
1-Axis, E-W Horiz Axis	Average	3.7	3.8	3.8	4.5	5.1	6.2	5.5	5.1	5.1	4.8	3.9	3.7	4.6
	Min/Max	2.5/5.2	2.3/5.4	2.1/5.4	3.0/5.9	4.1/6.2	4.8/7.5	4.1/6.9	3.6/6.6	3.2/6.3	2.3/5.8	2.8/5.8	1.9/5.6	3.6/5.6
1-Axis, N-S Horiz Axis	Average	2.9	3.6	4.6	6.0	7.0	8.3	7.5	6.9	6.6	5.1	3.3	2.7	5.4
	Min/Max	2.0/4.1	2.0/5.2	2.6/6.5	4.1/7.9	5.6/8.5	6.3/10.1	5.5/9.4	4.8/9.0	4.0/8.1	2.5/6.1	2.3/4.9	1.4/4.1	4.2/6.5
1-Axis, N-S Tilt=Latitude	Average	4.2	4.7	5.3	6.3	6.8	7.8	7.1	7.0	7.2	6.3	4.5	4.0	5.9
	Min/Max	2.8/5.8	2.7/6.7	3.0/7.5	4.3/8.2	5.5/8.3	6.0/9.5	5.2/9.0	4.9/9.1	4.4/9.0	3.0/7.5	3.2/6.8	2.0/6.0	4.6/7.2
2-Axis	Average	4.5	4.8	5.3	6.4	7.2	8.5	7.7	7.2	7.2	6.4	4.7	4.4	6.2
	Min/Max	3.0/6.2	2.8/6.9	3.0/7.5	4.3/8.4	5.8/8.7	6.5/10.3	5.6/9.6	5.0/9.4	4.4/9.0	3.1/7.6	3.4/7.1	2.2/6.6	4.8/7.5

Average Climatic Conditions

Element	Jan	Feb	Mar	Apr	May	June	July	Aug	Sept	Oct	Nov	Dec	Year
Temperature (°C)	-1.4	1.4	4.4	8.6	13.6	19.3	23.4	22.2	17.2	10.9	4.3	-0.8	10.3
Daily Minimum Temp	-8.2	-5.5	-2.7	0.8	5.1	9.8	14.4	13.6	8.4	2.3	-2.9	-7.7	2.3
Daily Maximum Temp	5.4	8.3	11.6	16.4	22.2	28.7	32.3	30.8	26.1	19.4	11.5	6.1	18.2
Record Minimum Temp	-26.7	-30.6	-16.1	-12.2	-5.6	0.0	5.6	3.9	-4.4	-20.0	-20.0	-30.6	-30.6
Record Maximum Temp	19.4	22.2	23.9	27.8	33.3	38.3	39.4	37.2	33.9	29.4	23.3	18.9	39.4
HDD, Base 18.3°C	612	473	431	292	153	33	0	0	72	234	422	592	3312
CDD, Base 18.3°C	0	0	0	0	7	62	157	122	38	3	0	0	388
Relative Humidity (%)	66	62	56	46	40	31	35	40	41	47	58	65	49
Wind Speed (m/s)	2.7	3.1	3.7	4.0	3.9	3.8	3.6	3.4	3.4	3.0	3.3	2.7	3.4

Salt Lake City, UT

WBAN NO. 24127

LATITUDE: 40.77° N
LONGITUDE: 111.97° W
ELEVATION: 1288 meters
MEAN PRESSURE: 872 millibars

STATION TYPE: Primary

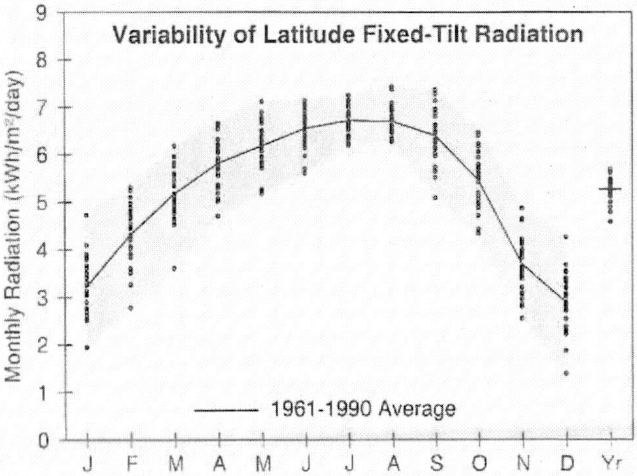

Solar Radiation for Flat-Plate Collectors Facing South at a Fixed Tilt (kWh/m²/day), Uncertainty ±9%

Tilt (°)		Jan	Feb	Mar	Apr	May	June	July	Aug	Sept	Oct	Nov	Dec	Year
0	Average	1.9	2.9	4.1	5.4	6.5	7.4	7.3	6.5	5.2	3.7	2.2	1.7	4.6
	Min/Max	1.5/2.4	2.2/3.3	3.0/4.7	4.5/6.1	5.5/7.5	6.2/8.1	6.7/7.9	6.0/7.2	4.3/5.9	3.1/4.2	1.8/2.7	1.1/2.1	4.2/4.9
Latitude -15	Average	2.9	4.0	5.0	5.9	6.6	7.2	7.3	7.0	6.3	5.0	3.3	2.5	5.2
	Min/Max	1.9/4.1	2.7/4.8	3.5/5.9	4.8/6.8	5.5/7.6	6.1/7.9	6.7/7.9	6.5/7.7	5.0/7.2	4.1/5.9	2.3/4.3	1.3/3.6	4.6/5.6
Latitude	Average	3.2	4.3	5.2	5.8	6.2	6.6	6.7	6.7	6.4	5.4	3.7	2.9	5.3
	Min/Max	2.0/4.7	2.8/5.3	3.6/6.2	4.7/6.7	5.2/7.1	5.6/7.1	6.2/7.2	6.3/7.4	5.1/7.3	4.3/6.5	2.5/4.9	1.4/4.3	4.6/5.7
Latitude +15	Average	3.4	4.4	5.1	5.4	5.5	5.6	5.8	6.1	6.1	5.5	3.9	3.1	5.0
	Min/Max	2.0/5.1	2.8/5.6	3.5/6.1	4.3/6.2	4.6/6.3	4.9/6.1	5.4/6.3	5.7/6.7	4.8/7.1	4.4/6.6	2.6/5.2	1.4/4.6	4.3/5.5
90	Average	3.2	3.9	3.9	3.5	3.0	2.8	2.9	3.6	4.3	4.5	3.5	2.9	3.5
	Min/Max	1.6/4.8	2.4/5.0	2.6/4.7	2.9/4.0	2.6/3.3	2.5/2.9	2.7/3.2	3.3/3.9	3.4/5.0	3.6/5.5	2.2/4.7	1.3/4.4	2.9/3.9

Solar Radiation for 1-Axis Tracking Flat-Plate Collectors with a North-South Axis (kWh/m²/day), Uncertainty ±9%

Axis Tilt (°)		Jan	Feb	Mar	Apr	May	June	July	Aug	Sept	Oct	Nov	Dec	Year
0	Average	2.7	4.0	5.6	7.3	8.8	10.0	10.1	9.2	7.6	5.5	3.2	2.3	6.4
	Min/Max	1.7/3.8	2.7/5.0	3.7/6.8	5.9/8.6	6.9/10.5	8.0/11.6	8.8/11.3	8.1/10.6	5.9/9.1	4.2/6.7	2.3/4.1	1.2/3.3	5.4/7.0
Latitude -15	Average	3.4	4.8	6.3	7.7	8.9	10.0	10.2	9.6	8.5	6.5	4.0	3.0	6.9
	Min/Max	2.0/5.1	3.0/6.1	4.1/7.7	6.2/9.2	7.1/10.8	8.0/11.6	8.9/11.4	8.5/11.1	6.4/10.0	4.9/7.9	2.7/5.3	1.4/4.5	5.9/7.6
Latitude	Average	3.7	5.1	6.5	7.7	8.7	9.6	9.8	9.4	8.6	6.8	4.3	3.3	7.0
	Min/Max	2.0/5.6	3.1/6.5	4.1/7.9	6.1/9.1	6.8/10.4	7.7/11.1	8.6/11.1	8.3/10.9	6.5/10.2	5.2/8.4	2.9/5.8	1.5/5.0	5.8/7.6
Latitude +15	Average	3.8	5.2	6.4	7.4	8.2	9.0	9.2	9.0	8.4	6.9	4.5	3.4	6.8
	Min/Max	2.1/5.9	3.1/6.7	4.1/7.9	5.8/8.8	6.4/9.9	7.1/10.4	8.0/10.4	7.9/10.4	6.3/10.0	5.2/8.4	2.9/6.0	1.5/5.3	5.6/7.5

Solar Radiation for 2-Axis Tracking Flat-Plate Collectors (kWh/m²/day), Uncertainty ±9%

Tracker		Jan	Feb	Mar	Apr	May	June	July	Aug	Sept	Oct	Nov	Dec	Year
2-Axis	Average	3.9	5.2	6.5	7.8	9.1	10.3	10.4	9.6	8.6	6.9	4.5	3.5	7.2
	Min/Max	2.1/5.9	3.1/6.7	4.2/8.0	6.2/9.2	7.1/10.9	8.2/11.9	9.0/11.7	8.5/11.1	6.5/10.2	5.2/8.5	2.9/6.1	1.5/5.4	6.0/7.9

Direct Beam Solar Radiation for Concentrating Collectors (kWh/m²/day), Uncertainty ±8%

Tracker		Jan	Feb	Mar	Apr	May	June	July	Aug	Sept	Oct	Nov	Dec	Year
1-Axis, E-W	Average	2.0	2.7	3.2	3.7	4.5	5.6	5.8	5.2	4.8	4.1	2.6	1.9	3.8
Horiz Axis	Min/Max	0.6/4.0	0.9/4.0	1.6/4.3	2.5/4.8	3.1/5.9	3.8/6.8	4.7/6.8	4.3/6.4	3.4/6.1	2.9/5.3	1.4/3.9	0.3/3.6	3.1/4.4
1-Axis, N-S	Average	1.4	2.4	3.6	4.8	6.1	7.4	7.6	6.9	5.8	4.1	2.0	1.2	4.5
Horiz Axis	Min/Max	0.5/2.8	0.9/3.5	1.7/4.9	3.3/6.2	4.1/8.0	5.0/9.2	6.3/9.1	5.5/8.6	4.1/7.5	2.7/5.4	1.1/3.0	0.2/2.4	3.6/5.1
1-Axis, N-S	Average	2.2	3.3	4.2	5.1	5.9	7.0	7.3	7.1	6.6	5.2	2.9	2.0	4.9
Tilt=Latitude	Min/Max	0.7/4.3	1.1/4.8	2.0/5.8	3.5/6.6	4.0/7.8	4.7/8.7	6.0/8.7	5.7/8.8	4.5/8.5	3.5/6.8	1.6/4.4	0.3/3.8	3.9/5.6
2-Axis	Average	2.3	3.3	4.3	5.2	6.3	7.6	7.9	7.3	6.6	5.3	3.1	2.2	5.1
	Min/Max	0.8/4.6	1.2/4.9	2.0/5.8	3.6/6.7	4.2/8.2	5.1/9.5	6.5/9.4	5.8/9.0	4.6/8.5	3.6/6.9	1.7/4.6	0.4/4.1	4.0/5.9

Average Climatic Conditions

Element	Jan	Feb	Mar	Apr	May	June	July	Aug	Sept	Oct	Nov	Dec	Year
Temperature (°C)	-2.3	1.2	5.4	9.8	14.9	20.6	25.5	24.2	18.4	11.8	4.9	-1.3	11.1
Daily Minimum Temp	-7.1	-4.1	-0.3	3.3	7.6	13.0	17.6	16.6	10.6	4.6	-0.6	-5.8	4.6
Daily Maximum Temp	2.4	6.4	11.2	16.3	22.2	28.2	33.4	31.9	26.2	18.9	10.4	3.2	17.6
Record Minimum Temp	-30.0	-34.4	-16.7	-10.0	-3.9	1.7	4.4	2.8	-2.8	-8.9	-25.6	-29.4	-34.4
Record Maximum Temp	16.7	20.6	25.6	29.4	33.9	40.0	41.7	40.0	37.8	31.7	23.9	19.4	41.7
HDD, Base 18.3°C	639	481	399	258	119	28	0	0	60	207	403	608	3203
CDD, Base 18.3°C	0	0	0	0	13	97	222	183	63	4	0	0	582
Relative Humidity (%)	74	70	60	53	49	41	36	39	46	56	66	74	55
Wind Speed (m/s)	3.5	3.8	4.3	4.4	4.2	4.2	4.3	4.3	4.2	3.8	3.8	3.4	4.0

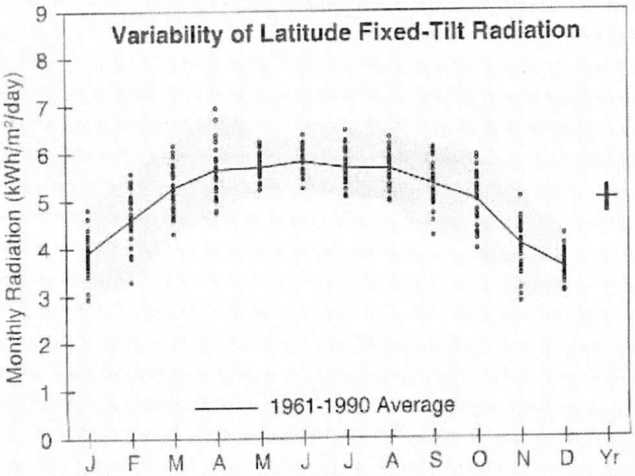

Lynchburg, VA

WBAN NO. 13733

LATITUDE: 37.33° N
LONGITUDE: 79.20° W
ELEVATION: 279 meters
MEAN PRESSURE: 984 millibars

STATION TYPE: Secondary

Solar Radiation for Flat-Plate Collectors Facing South at a Fixed Tilt (kWh/m²/day), Uncertainty ±9%

Tilt (°)		Jan	Feb	Mar	Apr	May	June	July	Aug	Sept	Oct	Nov	Dec	Year
0	Average	2.4	3.2	4.3	5.4	6.0	6.5	6.2	5.6	4.7	3.7	2.6	2.1	4.4
	Min/Max	2.0/2.7	2.6/3.7	3.7/4.9	4.6/6.5	5.5/6.6	5.8/7.1	5.5/7.0	5.0/6.2	3.9/5.2	3.2/4.3	2.0/3.0	1.9/2.3	4.2/4.5
Latitude -15	Average	3.4	4.2	5.1	5.8	6.1	6.3	6.1	5.9	5.3	4.6	3.6	3.1	5.0
	Min/Max	2.7/4.1	3.1/5.0	4.2/5.9	4.8/7.1	5.5/6.6	5.7/6.9	5.4/7.0	5.2/6.5	4.3/6.0	3.8/5.5	2.6/4.4	2.7/3.6	4.7/5.2
Latitude	Average	3.9	4.6	5.3	5.7	5.7	5.8	5.7	5.7	5.3	5.0	4.0	3.6	5.0
	Min/Max	2.9/4.8	3.3/5.6	4.3/6.2	4.7/6.9	5.2/6.2	5.3/6.4	5.1/6.5	5.0/6.3	4.2/6.1	4.0/5.9	2.8/5.1	3.0/4.3	4.7/5.2
Latitude +15	Average	4.2	4.8	5.2	5.3	5.1	5.0	5.0	5.2	5.1	5.0	4.3	3.8	4.8
	Min/Max	3.1/5.2	3.3/5.8	4.2/6.1	4.4/6.5	4.6/5.5	4.6/5.5	4.4/5.7	4.5/5.8	4.0/5.9	3.9/6.0	2.9/5.4	3.2/4.6	4.5/5.0
90	Average	3.8	4.0	3.7	3.2	2.6	2.4	2.5	2.9	3.4	4.0	3.7	3.5	3.3
	Min/Max	2.6/4.9	2.7/4.8	3.0/4.5	2.7/3.8	2.4/2.8	2.2/2.5	2.3/2.7	2.6/3.2	2.7/4.0	3.0/4.8	2.5/4.8	2.9/4.3	3.1/3.5

Solar Radiation for 1-Axis Tracking Flat-Plate Collectors with a North-South Axis (kWh/m²/day), Uncertainty ±9%

Axis Tilt (°)		Jan	Feb	Mar	Apr	May	June	July	Aug	Sept	Oct	Nov	Dec	Year
0	Average	3.4	4.6	5.9	7.2	7.7	8.2	7.8	7.2	6.2	5.1	3.7	3.0	5.8
	Min/Max	2.6/4.3	3.2/5.6	4.7/7.0	5.7/9.3	7.0/8.6	7.1/9.3	6.7/8.9	6.1/8.1	4.9/7.2	4.1/6.2	2.5/4.6	2.6/3.6	5.5/6.1
Latitude -15	Average	4.2	5.3	6.5	7.5	7.8	8.2	7.8	7.5	6.7	5.9	4.4	3.8	6.3
	Min/Max	3.1/5.3	3.6/6.5	5.1/7.8	5.9/9.8	7.0/8.8	7.1/9.2	6.7/8.9	6.2/8.4	5.2/7.8	4.5/7.1	3.0/5.6	3.2/4.6	5.9/6.6
Latitude	Average	4.6	5.6	6.6	7.4	7.6	7.8	7.5	7.3	6.7	6.1	4.8	4.1	6.4
	Min/Max	3.3/5.9	3.8/6.9	5.2/8.0	5.9/9.7	6.8/8.5	6.8/8.9	6.4/8.6	6.1/8.3	5.2/7.9	4.7/7.5	3.2/6.1	3.5/5.1	6.0/6.7
Latitude +15	Average	4.8	5.7	6.6	7.1	7.1	7.3	7.0	7.0	6.6	6.2	5.0	4.4	6.2
	Min/Max	3.4/6.2	3.8/7.1	5.1/7.9	5.6/9.4	6.4/8.0	6.3/8.3	6.0/8.0	5.8/7.9	5.0/7.7	4.7/7.5	3.3/6.4	3.6/5.4	5.9/6.6

Solar Radiation for 2-Axis Tracking Flat-Plate Collectors (kWh/m²/day), Uncertainty ±9%

Tracker		Jan	Feb	Mar	Apr	May	June	July	Aug	Sept	Oct	Nov	Dec	Year
2-Axis	Average	4.9	5.7	6.7	7.5	7.9	8.3	7.9	7.5	6.8	6.2	5.0	4.4	6.6
	Min/Max	3.5/6.3	3.8/7.1	5.2/8.0	6.0/9.8	7.1/8.9	7.2/9.4	6.8/9.1	6.3/8.5	5.2/7.9	4.7/7.6	3.3/6.4	3.7/5.5	6.2/6.9

Direct Beam Solar Radiation for Concentrating Collectors (kWh/m²/day), Uncertainty ±8%

Tracker		Jan	Feb	Mar	Apr	May	June	July	Aug	Sept	Oct	Nov	Dec	Year
1-Axis, E-W Horiz Axis	Average	2.9	3.2	3.4	3.6	3.7	3.9	3.6	3.4	3.2	3.4	2.9	2.7	3.3
	Min/Max	1.7/4.1	1.6/4.3	2.2/4.3	2.3/5.3	2.9/4.5	3.0/4.8	2.8/4.5	2.5/4.2	2.1/4.1	2.2/4.5	1.7/4.1	2.0/3.6	3.0/3.6
1-Axis, N-S Horiz Axis	Average	2.3	3.0	3.9	4.7	4.8	5.1	4.7	4.4	3.9	3.4	2.4	2.0	3.7
	Min/Max	1.4/3.2	1.5/4.1	2.5/5.0	2.9/7.0	3.8/5.9	4.0/6.3	3.6/5.7	3.2/5.4	2.6/5.1	2.2/4.6	1.3/3.3	1.4/2.6	3.3/4.0
1-Axis, N-S Tilt=Latitude	Average	3.2	3.9	4.5	4.9	4.7	4.8	4.5	4.5	4.3	4.2	3.3	2.9	4.1
	Min/Max	1.9/4.6	2.0/5.3	2.9/5.8	3.1/7.3	3.7/5.8	3.7/5.9	3.4/5.5	3.2/5.5	2.9/5.7	2.8/5.7	1.9/4.6	2.2/3.9	3.8/4.5
2-Axis	Average	3.5	4.0	4.5	5.0	4.9	5.2	4.8	4.6	4.3	4.3	3.5	3.2	4.3
	Min/Max	2.1/4.9	2.0/5.4	2.9/5.8	3.2/7.4	3.9/6.1	4.0/6.4	3.6/5.9	3.3/5.6	2.9/5.7	2.8/5.7	2.0/4.9	2.3/4.2	3.9/4.7

Average Climatic Conditions

Element	Jan	Feb	Mar	Apr	May	June	July	Aug	Sept	Oct	Nov	Dec	Year
Temperature (°C)	1.2	2.9	8.0	13.2	17.8	22.1	24.2	23.7	20.1	13.9	8.7	3.6	13.3
Daily Minimum Temp	-4.1	-2.6	1.9	6.6	11.4	16.0	18.4	18.0	14.3	7.6	2.9	-1.6	7.4
Daily Maximum Temp	6.4	8.4	14.1	19.8	24.2	28.1	30.0	29.4	25.8	20.2	14.5	8.7	19.1
Record Minimum Temp	-23.3	-17.8	-13.9	-6.7	-0.6	4.4	9.4	7.2	2.2	-6.1	-13.3	-20.0	-23.3
Record Maximum Temp	24.4	26.1	30.6	33.3	33.9	37.8	39.4	38.9	38.3	33.9	28.3	25.6	39.4
HDD, Base 18.3°C	531	432	321	157	53	3	0	0	18	150	288	458	2411
CDD, Base 18.3°C	0	0	0	0	37	115	183	166	69	12	0	0	582
Relative Humidity (%)	65	63	61	58	69	71	73	75	75	69	66	66	68
Wind Speed (m/s)	3.5	3.5	3.8	3.8	3.1	3.0	2.7	2.7	2.8	3.0	3.3	3.4	3.2

Norfolk, VA

WBAN NO. 13737

LATITUDE: 36.90° N
LONGITUDE: 76.20° W
ELEVATION: 9 meters
MEAN PRESSURE: 1017 millibars

STATION TYPE: Secondary

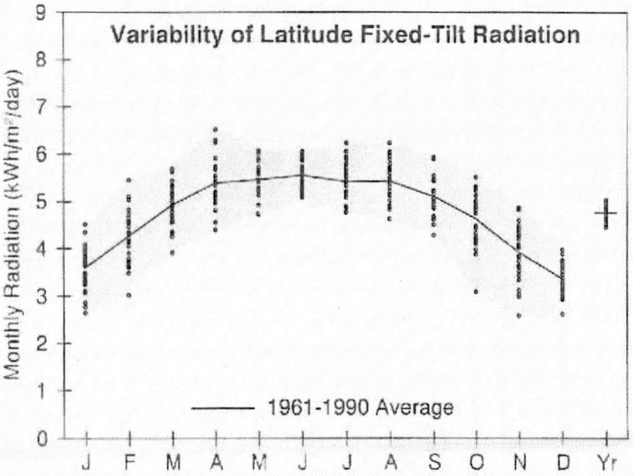

Solar Radiation for Flat-Plate Collectors Facing South at a Fixed Tilt (kWh/m²/day), Uncertainty ±9%

Tilt (°)		Jan	Feb	Mar	Apr	May	June	July	Aug	Sept	Oct	Nov	Dec	Year
0	Average	2.3	3.0	4.1	5.1	5.8	6.2	5.9	5.4	4.5	3.5	2.5	2.0	4.2
	Min/Max	1.9/2.6	2.4/3.6	3.4/4.6	4.3/6.1	5.0/6.4	5.6/6.7	5.2/6.7	4.6/6.1	3.9/5.1	2.6/4.0	1.9/2.9	1.7/2.3	4.0/4.4
Latitude -15	Average	3.2	3.9	4.8	5.5	5.8	6.0	5.8	5.6	5.0	4.3	3.5	2.9	4.7
	Min/Max	2.4/3.9	2.9/4.9	3.8/5.5	4.5/6.6	5.0/6.5	5.5/6.6	5.1/6.7	4.8/6.4	4.3/5.8	3.0/5.1	2.4/4.3	2.3/3.4	4.4/5.0
Latitude	Average	3.6	4.3	4.9	5.4	5.5	5.6	5.4	5.4	5.1	4.6	3.9	3.4	4.8
	Min/Max	2.6/4.5	3.0/5.4	3.9/5.7	4.4/6.5	4.7/6.1	5.1/6.1	4.8/6.2	4.6/6.2	4.3/5.9	3.1/5.5	2.6/4.9	2.6/4.0	4.4/5.0
Latitude +15	Average	3.8	4.4	4.8	5.0	4.8	4.8	4.8	4.9	4.9	4.7	4.1	3.6	4.6
	Min/Max	2.7/4.9	3.0/5.7	3.8/5.6	4.1/6.1	4.2/5.4	4.4/5.2	4.2/5.4	4.2/5.7	4.1/5.7	3.0/5.6	2.6/5.2	2.8/4.3	4.2/4.8
90	Average	3.4	3.6	3.4	3.0	2.5	2.3	2.3	2.8	3.3	3.6	3.6	3.3	3.1
	Min/Max	2.3/4.5	2.4/4.8	2.7/4.0	2.5/3.5	2.2/2.7	2.2/2.4	2.1/2.6	2.4/3.1	2.7/3.8	2.3/4.4	2.2/4.6	2.4/4.0	2.8/3.3

Solar Radiation for 1-Axis Tracking Flat-Plate Collectors with a North-South Axis (kWh/m²/day), Uncertainty ±9%

Axis Tilt (°)		Jan	Feb	Mar	Apr	May	June	July	Aug	Sept	Oct	Nov	Dec	Year
0	Average	3.2	4.2	5.5	6.7	7.3	7.7	7.4	6.9	5.9	4.8	3.6	2.9	5.5
	Min/Max	2.3/4.0	2.9/5.4	4.2/6.4	5.3/8.4	6.0/8.5	6.8/8.7	6.1/8.7	5.5/8.1	4.8/7.0	3.2/5.8	2.3/4.4	2.2/3.3	5.1/5.8
Latitude -15	Average	3.9	4.9	6.0	7.1	7.4	7.7	7.4	7.1	6.3	5.4	4.3	3.5	5.9
	Min/Max	2.8/4.9	3.3/6.3	4.6/7.1	5.4/8.9	6.0/8.6	6.7/8.6	6.1/8.7	5.7/8.4	5.1/7.6	3.5/6.6	2.7/5.3	2.7/4.2	5.5/6.3
Latitude	Average	4.2	5.1	6.2	7.0	7.1	7.4	7.1	7.0	6.4	5.6	4.6	3.9	6.0
	Min/Max	3.0/5.4	3.4/6.8	4.6/7.2	5.4/8.8	5.8/8.3	6.4/8.3	5.8/8.4	5.6/8.3	5.2/7.7	3.5/6.9	2.9/5.8	2.9/4.7	5.5/6.3
Latitude +15	Average	4.4	5.2	6.1	6.7	6.7	6.9	6.6	6.6	6.2	5.7	4.8	4.1	5.8
	Min/Max	3.0/5.7	3.4/6.9	4.5/7.2	5.1/8.5	5.5/7.9	6.0/7.7	5.4/7.8	5.3/7.9	5.0/7.5	3.5/7.0	2.9/6.1	3.0/4.9	5.4/6.2

Solar Radiation for 2-Axis Tracking Flat-Plate Collectors (kWh/m²/day), Uncertainty ±9%

Tracker		Jan	Feb	Mar	Apr	May	June	July	Aug	Sept	Oct	Nov	Dec	Year
2-Axis	Average	4.4	5.3	6.2	7.1	7.4	7.9	7.5	7.1	6.4	5.7	4.8	4.1	6.2
	Min/Max	3.0/5.8	3.4/6.9	4.7/7.3	5.5/8.9	6.1/8.7	6.9/8.8	6.2/8.9	5.7/8.4	5.2/7.7	3.6/7.0	2.9/6.1	3.1/5.0	5.7/6.5

Direct Beam Solar Radiation for Concentrating Collectors (kWh/m²/day), Uncertainty ±8%

Tracker		Jan	Feb	Mar	Apr	May	June	July	Aug	Sept	Oct	Nov	Dec	Year
1-Axis, E-W Horiz Axis	Average	2.6	2.9	3.1	3.3	3.4	3.6	3.4	3.2	3.0	3.0	2.9	2.5	3.1
	Min/Max	1.5/3.8	1.4/4.3	2.0/3.9	2.0/4.9	2.4/4.3	2.8/4.4	2.4/4.4	2.3/4.2	2.2/4.1	1.5/4.1	1.3/4.0	1.6/3.3	2.7/3.4
1-Axis, N-S Horiz Axis	Average	2.1	2.7	3.6	4.3	4.5	4.7	4.3	4.1	3.6	3.1	2.3	1.8	3.4
	Min/Max	1.1/3.0	1.3/4.1	2.3/4.5	2.6/6.3	3.1/5.7	3.6/5.8	3.0/5.9	2.8/5.5	2.6/5.0	1.5/4.2	1.0/3.3	1.1/2.3	3.0/3.8
1-Axis, N-S Tilt=Latitude	Average	2.9	3.5	4.1	4.5	4.3	4.4	4.1	4.2	4.0	3.8	3.2	2.7	3.8
	Min/Max	1.6/4.2	1.7/5.3	2.6/5.2	2.7/6.6	3.0/5.6	3.4/5.4	2.9/5.6	2.9/5.5	2.9/5.5	1.9/5.1	1.5/4.5	1.7/3.5	3.3/4.2
2-Axis	Average	3.1	3.6	4.1	4.6	4.6	4.8	4.4	4.3	4.0	3.8	3.4	2.9	4.0
	Min/Max	1.7/4.5	1.7/5.4	2.6/5.2	2.8/6.7	3.2/5.9	3.7/5.9	3.1/6.0	2.9/5.7	2.9/5.5	1.9/5.2	1.5/4.8	1.8/3.8	3.5/4.4

Average Climatic Conditions

Element	Jan	Feb	Mar	Apr	May	June	July	Aug	Sept	Oct	Nov	Dec	Year
Temperature (°C)	3.9	5.0	9.2	13.9	18.9	23.4	25.7	25.1	22.2	16.2	11.4	6.6	15.1
Daily Minimum Temp	-0.6	0.2	4.1	8.4	13.8	18.4	21.1	20.8	17.9	11.6	6.6	1.9	10.3
Daily Maximum Temp	8.5	9.8	14.4	19.4	24.1	28.3	30.2	29.5	26.4	20.8	16.2	11.2	19.9
Record Minimum Temp	-19.4	-13.3	-7.8	-2.2	2.2	7.2	12.2	9.4	7.2	-2.8	-6.7	-13.9	-19.4
Record Maximum Temp	25.6	27.2	31.1	36.1	37.8	38.3	39.4	40.0	37.2	35.0	30.0	26.7	40.0
HDD, Base 18.3°C	446	373	282	138	28	0	0	0	6	91	211	365	1942
CDD, Base 18.3°C	0	0	0	5	47	154	227	210	121	26	0	0	790
Relative Humidity (%)	66	66	65	63	69	71	73	75	74	72	69	67	69
Wind Speed (m/s)	5.1	5.4	5.6	5.3	4.8	4.5	4.1	4.0	4.4	4.6	4.8	5.1	4.8

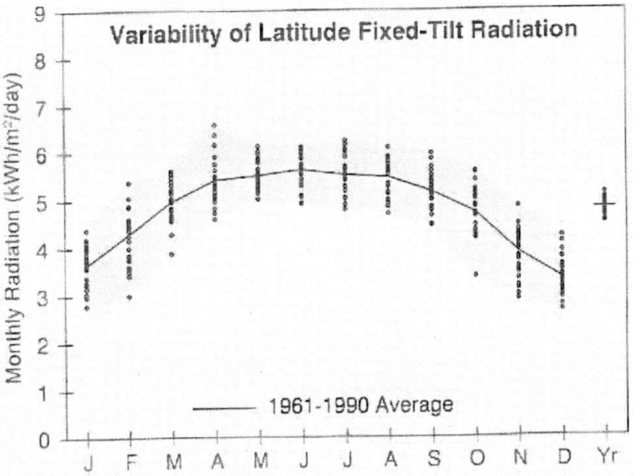

Richmond, VA

WBAN NO. 13740

LATITUDE: 37.50° N
LONGITUDE: 77.33° W
ELEVATION: 50 meters
MEAN PRESSURE: 1012 millibars

STATION TYPE: Secondary

Solar Radiation for Flat-Plate Collectors Facing South at a Fixed Tilt (kWh/m²/day), Uncertainty ±9%

Tilt (°)		Jan	Feb	Mar	Apr	May	June	July	Aug	Sept	Oct	Nov	Dec	Year
0	Average	2.3	3.0	4.1	5.2	5.8	6.3	6.0	5.4	4.5	3.5	2.5	2.0	4.2
	Min/Max	2.0/2.5	2.4/3.5	3.4/4.6	4.5/6.1	5.3/6.4	5.5/6.8	5.2/6.7	4.8/6.0	4.0/5.1	2.8/4.1	2.0/2.9	1.8/2.3	4.0/4.4
Latitude -15	Average	3.2	3.9	4.8	5.5	5.8	6.1	5.9	5.7	5.1	4.4	3.5	2.9	4.7
	Min/Max	2.5/3.8	2.9/4.8	3.8/5.4	4.7/6.7	5.3/6.5	5.3/6.6	5.2/6.7	4.9/6.3	4.4/5.9	3.3/5.2	2.7/4.3	2.4/3.6	4.5/5.0
Latitude	Average	3.6	4.3	5.0	5.4	5.5	5.6	5.5	5.5	5.2	4.7	3.9	3.3	4.8
	Min/Max	2.8/4.4	3.0/5.4	3.9/5.6	4.6/6.6	5.0/6.1	4.9/6.1	4.8/6.3	4.7/6.1	4.4/6.0	3.4/5.6	2.9/4.8	2.7/4.2	4.5/5.1
Latitude +15	Average	3.9	4.4	4.9	5.0	4.9	4.9	4.8	5.0	4.9	4.8	4.1	3.6	4.6
	Min/Max	2.9/4.7	3.0/5.6	3.7/5.5	4.3/6.1	4.5/5.4	4.3/5.3	4.2/5.5	4.3/5.6	4.2/5.8	3.3/5.7	3.0/5.2	2.8/4.6	4.3/4.9
90	Average	3.5	3.7	3.5	3.1	2.6	2.4	2.4	2.8	3.3	3.7	3.6	3.3	3.2
	Min/Max	2.5/4.4	2.4/4.7	2.7/4.1	2.7/3.7	2.4/2.7	2.2/2.5	2.2/2.7	2.5/3.1	2.8/3.9	2.5/4.5	2.6/4.6	2.5/4.3	2.9/3.4

Solar Radiation for 1-Axis Tracking Flat-Plate Collectors with a North-South Axis (kWh/m²/day), Uncertainty ±9%

Axis Tilt (°)		Jan	Feb	Mar	Apr	May	June	July	Aug	Sept	Oct	Nov	Dec	Year
0	Average	3.2	4.2	5.5	6.8	7.3	7.9	7.5	6.9	5.9	4.8	3.5	2.8	5.5
	Min/Max	2.5/3.8	2.9/5.3	4.3/6.3	5.5/8.5	6.5/8.5	6.6/8.9	6.3/8.6	5.7/8.0	4.9/7.0	3.5/5.8	2.6/4.4	2.2/3.5	5.1/6.0
Latitude -15	Average	3.9	4.9	6.1	7.1	7.4	7.8	7.5	7.1	6.4	5.5	4.2	3.5	6.0
	Min/Max	3.0/4.8	3.2/6.3	4.6/7.0	5.8/9.0	6.5/8.6	6.5/8.8	6.3/8.6	5.8/8.2	5.3/7.6	3.8/6.7	3.1/5.4	2.7/4.5	5.5/6.5
Latitude	Average	4.3	5.2	6.2	7.0	7.2	7.5	7.2	7.0	6.5	5.7	4.6	3.8	6.0
	Min/Max	3.2/5.3	3.3/6.7	4.6/7.1	5.7/8.9	6.3/8.3	6.3/8.5	6.0/8.3	5.7/8.1	5.3/7.7	3.9/7.0	3.3/5.8	3.0/5.0	5.5/6.5
Latitude +15	Average	4.5	5.3	6.1	6.7	6.8	7.0	6.7	6.7	6.3	5.8	4.8	4.1	5.9
	Min/Max	3.2/5.5	3.3/6.9	4.5/7.1	5.4/8.6	5.9/7.8	5.8/7.9	5.6/7.7	5.4/7.7	5.1/7.5	3.9/7.1	3.4/6.1	3.1/5.3	5.4/6.4

Solar Radiation for 2-Axis Tracking Flat-Plate Collectors (kWh/m²/day), Uncertainty ±9%

Tracker		Jan	Feb	Mar	Apr	May	June	July	Aug	Sept	Oct	Nov	Dec	Year
2-Axis	Average	4.5	5.3	6.2	7.1	7.5	8.0	7.6	7.2	6.5	5.8	4.8	4.1	6.2
	Min/Max	3.3/5.6	3.4/6.9	4.6/7.2	5.8/9.0	6.6/8.7	6.7/9.0	6.4/8.7	5.9/8.3	5.3/7.7	3.9/7.1	3.4/6.1	3.1/5.3	5.7/6.8

Direct Beam Solar Radiation for Concentrating Collectors (kWh/m²/day), Uncertainty ±8%

Tracker		Jan	Feb	Mar	Apr	May	June	July	Aug	Sept	Oct	Nov	Dec	Year
1-Axis, E-W Horiz Axis	Average	2.6	2.8	3.0	3.2	3.3	3.6	3.3	3.1	3.0	3.1	2.8	2.4	3.0
	Min/Max	1.5/3.5	1.2/4.3	1.7/3.7	2.1/4.6	2.5/4.3	2.6/4.5	2.4/4.2	2.2/4.0	2.2/3.8	1.5/4.1	1.7/3.9	1.6/3.5	2.6/3.5
1-Axis, N-S Horiz Axis	Average	2.0	2.7	3.5	4.2	4.3	4.7	4.3	4.0	3.6	3.1	2.3	1.7	3.4
	Min/Max	1.2/2.7	1.2/4.0	2.0/4.4	2.7/6.0	3.3/5.6	3.3/5.8	3.1/5.5	2.8/5.2	2.5/4.6	1.6/4.3	1.3/3.2	1.1/2.5	2.9/3.9
1-Axis, N-S Tilt=Latitude	Average	2.9	3.4	4.1	4.4	4.2	4.4	4.1	4.1	4.0	3.8	3.1	2.6	3.8
	Min/Max	1.7/3.9	1.5/5.2	2.4/5.1	2.9/6.3	3.2/5.4	3.1/5.5	2.9/5.2	2.8/5.2	2.8/5.1	2.0/5.2	1.9/4.4	1.6/3.7	3.2/4.3
2-Axis	Average	3.1	3.5	4.1	4.5	4.5	4.7	4.4	4.2	4.0	3.9	3.3	2.8	3.9
	Min/Max	1.9/4.2	1.5/5.3	2.4/5.1	2.9/6.4	3.4/5.7	3.4/5.9	3.1/5.6	2.9/5.4	2.8/5.1	2.0/5.3	2.0/4.7	1.8/4.1	3.3/4.5

Average Climatic Conditions

Element	Jan	Feb	Mar	Apr	May	June	July	Aug	Sept	Oct	Nov	Dec	Year
Temperature (°C)	2.1	3.7	8.9	14.1	18.9	23.3	25.6	24.9	21.1	14.8	9.8	4.5	14.3
Daily Minimum Temp	-3.5	-2.2	2.4	7.0	12.3	17.1	19.7	19.1	15.0	8.1	3.3	-1.2	8.1
Daily Maximum Temp	7.6	9.6	15.3	21.1	25.4	29.5	31.3	30.6	27.2	21.5	16.3	10.1	20.4
Record Minimum Temp	-24.4	-23.3	-11.7	-5.0	-0.6	4.4	10.6	7.8	1.7	-6.1	-12.2	-18.3	-24.4
Record Maximum Temp	26.7	28.3	33.9	35.6	37.8	40.0	40.6	38.9	39.4	37.2	30.0	26.7	40.6
HDD, Base 18.3°C	504	409	293	134	34	0	0	0	13	129	257	429	2202
CDD, Base 18.3°C	0	0	0	6	51	150	224	203	96	19	0	0	749
Relative Humidity (%)	68	66	63	61	70	72	75	77	77	74	69	69	70
Wind Speed (m/s)	3.7	3.9	4.2	4.1	3.6	3.4	3.1	2.9	3.0	3.2	3.4	3.6	3.5

Roanoke, VA

WBAN NO. 13741

LATITUDE: 37.32° N
LONGITUDE: 79.97° W
ELEVATION: 358 meters
MEAN PRESSURE: 976 millibars

STATION TYPE: Secondary

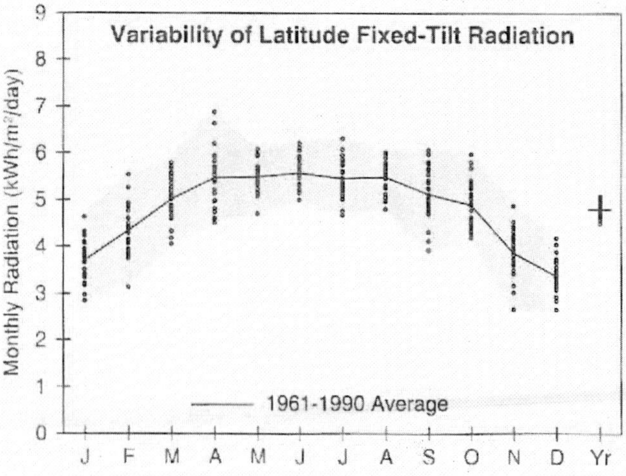

Solar Radiation for Flat-Plate Collectors Facing South at a Fixed Tilt (kWh/m²/day), Uncertainty ±9%

Tilt (°)		Jan	Feb	Mar	Apr	May	June	July	Aug	Sept	Oct	Nov	Dec	Year
0	Average	2.3	3.1	4.1	5.2	5.8	6.2	5.9	5.5	4.5	3.6	2.5	2.0	4.2
	Min/Max	2.0/2.7	2.5/3.7	3.5/4.6	4.4/6.4	5.1/6.4	5.6/6.9	5.1/6.8	4.9/5.9	3.7/5.2	3.3/4.3	1.9/2.9	1.8/2.3	4.0/4.4
Latitude -15	Average	3.3	4.0	4.8	5.6	5.8	6.0	5.9	5.7	5.1	4.6	3.5	2.9	4.8
	Min/Max	2.6/4.0	3.0/5.0	4.0/5.5	4.6/7.0	5.0/6.5	5.4/6.8	5.0/6.8	5.0/6.2	4.0/6.0	4.0/5.5	2.5/4.3	2.4/3.6	4.4/5.0
Latitude	Average	3.7	4.3	5.0	5.5	5.5	5.6	5.5	5.5	5.1	4.9	3.9	3.4	4.8
	Min/Max	2.8/4.6	3.1/5.5	4.1/5.8	4.5/6.9	4.7/6.1	5.0/6.2	4.7/6.3	4.8/6.0	3.9/6.1	4.2/6.0	2.7/4.9	2.7/4.2	4.5/5.1
Latitude +15	Average	3.9	4.5	4.9	5.1	4.9	4.8	4.8	5.0	4.9	4.9	4.1	3.6	4.6
	Min/Max	3.0/5.0	3.2/5.8	3.9/5.7	4.2/6.4	4.2/5.4	4.4/5.4	4.1/5.5	4.4/5.5	3.7/5.8	4.2/6.1	2.7/5.2	2.8/4.5	4.3/4.9
90	Average	3.6	3.7	3.6	3.1	2.5	2.3	2.4	2.8	3.3	3.9	3.6	3.3	3.2
	Min/Max	2.6/4.6	2.5/4.8	2.8/4.2	2.6/3.8	2.3/2.7	2.2/2.5	2.1/2.6	2.5/3.1	2.5/3.9	3.3/4.8	2.3/4.6	2.5/4.2	3.0/3.4

Solar Radiation for 1-Axis Tracking Flat-Plate Collectors with a North-South Axis (kWh/m²/day), Uncertainty ±9%

Axis Tilt (°)		Jan	Feb	Mar	Apr	May	June	July	Aug	Sept	Oct	Nov	Dec	Year
0	Average	3.3	4.3	5.5	6.8	7.3	7.8	7.4	7.0	5.9	5.0	3.5	2.9	5.6
	Min/Max	2.5/4.1	3.1/5.6	4.4/6.5	5.3/9.0	6.1/8.3	6.7/9.0	6.1/8.8	6.1/7.8	4.5/7.1	4.3/6.3	2.4/4.5	2.3/3.5	5.1/5.9
Latitude -15	Average	4.0	5.0	6.1	7.2	7.4	7.7	7.4	7.2	6.4	5.7	4.2	3.6	6.0
	Min/Max	3.0/5.1	3.5/6.6	4.8/7.2	5.5/9.5	6.1/8.4	6.7/9.0	6.1/8.9	6.2/8.1	4.7/7.7	4.9/7.2	2.8/5.4	2.8/4.5	5.5/6.4
Latitude	Average	4.3	5.2	6.2	7.1	7.2	7.4	7.2	7.1	6.4	6.0	4.6	3.9	6.1
	Min/Max	3.2/5.6	3.6/7.0	4.8/7.4	5.4/9.4	5.9/8.2	6.4/8.6	5.9/8.5	6.1/7.9	4.7/7.8	5.1/7.6	3.0/5.9	3.0/4.9	5.6/6.5
Latitude +15	Average	4.5	5.3	6.1	6.8	6.7	6.9	6.7	6.7	6.2	6.0	4.7	4.1	5.9
	Min/Max	3.3/5.9	3.6/7.1	4.7/7.4	5.1/9.1	5.5/7.7	5.9/8.1	5.5/8.0	5.8/7.5	4.5/7.6	5.1/7.6	3.0/6.2	3.1/5.3	5.4/6.3

Solar Radiation for 2-Axis Tracking Flat-Plate Collectors (kWh/m²/day), Uncertainty ±9%

Tracker		Jan	Feb	Mar	Apr	May	June	July	Aug	Sept	Oct	Nov	Dec	Year
2-Axis	Average	4.6	5.4	6.2	7.2	7.5	7.9	7.6	7.2	6.4	6.0	4.7	4.2	6.3
	Min/Max	3.3/6.0	3.6/7.1	4.8/7.5	5.5/9.5	6.2/8.5	6.8/9.2	6.2/9.0	6.2/8.1	4.7/7.8	5.1/7.6	3.0/6.2	3.1/5.3	5.8/6.7

Direct Beam Solar Radiation for Concentrating Collectors (kWh/m²/day), Uncertainty ±8%

Tracker		Jan	Feb	Mar	Apr	May	June	July	Aug	Sept	Oct	Nov	Dec	Year
1-Axis, E-W Horiz Axis	Average	2.7	2.9	3.1	3.3	3.3	3.5	3.3	3.2	3.0	3.3	2.7	2.5	3.1
	Min/Max	1.7/3.9	1.5/4.4	2.0/3.9	1.9/5.1	2.2/4.1	2.7/4.5	2.3/4.4	2.3/3.8	1.7/4.1	2.5/4.5	1.4/3.9	1.6/3.6	2.7/3.4
1-Axis, N-S Horiz Axis	Average	2.1	2.7	3.5	4.3	4.4	4.6	4.3	4.1	3.6	3.3	2.2	1.8	3.4
	Min/Max	1.3/3.0	1.5/4.2	2.3/4.6	2.4/6.7	2.9/5.4	3.5/5.9	3.1/5.7	3.1/5.0	2.2/4.9	2.6/4.7	1.1/3.2	1.2/2.6	3.0/3.9
1-Axis, N-S Tilt=Latitude	Average	2.9	3.5	4.1	4.5	4.3	4.3	4.1	4.2	4.0	4.1	3.1	2.7	3.8
	Min/Max	1.9/4.3	1.9/5.4	2.7/5.3	2.6/7.0	2.8/5.3	3.3/5.6	2.9/5.4	3.1/5.1	2.4/5.4	3.2/5.7	1.6/4.5	1.8/3.8	3.4/4.3
2-Axis	Average	3.2	3.6	4.1	4.6	4.5	4.7	4.4	4.3	4.0	4.2	3.3	2.9	4.0
	Min/Max	2.0/4.6	1.9/5.5	2.7/5.3	2.6/7.1	3.0/5.6	3.5/6.0	3.1/5.8	3.2/5.2	2.4/5.5	3.2/5.8	1.7/4.7	1.9/4.2	3.5/4.5

Average Climatic Conditions

Element	Jan	Feb	Mar	Apr	May	June	July	Aug	Sept	Oct	Nov	Dec	Year
Temperature (°C)	1.4	2.9	8.2	13.1	17.8	21.9	24.2	23.7	19.8	13.6	8.6	3.5	13.2
Daily Minimum Temp	-3.9	-2.7	2.1	6.6	11.4	15.7	18.2	17.7	13.8	7.1	2.8	-1.7	7.2
Daily Maximum Temp	6.6	8.5	14.3	19.6	24.3	28.3	30.2	29.6	25.8	20.1	14.4	8.7	19.2
Record Minimum Temp	-23.9	-17.2	-12.2	-6.7	-0.6	3.9	8.3	5.6	1.1	-5.6	-12.8	-20.0	-23.9
Record Maximum Temp	25.6	26.7	30.6	35.0	35.6	37.8	40.0	40.6	38.3	33.9	28.3	24.4	40.6
HDD, Base 18.3°C	526	431	313	161	58	3	0	0	22	157	292	460	2422
CDD, Base 18.3°C	0	0	0	4	43	111	183	166	67	11	0	0	584
Relative Humidity (%)	61	60	57	57	66	69	71	73	74	68	65	64	65
Wind Speed (m/s)	4.2	4.1	4.3	4.3	3.4	3.1	2.9	2.7	2.7	3.0	3.5	3.8	3.5

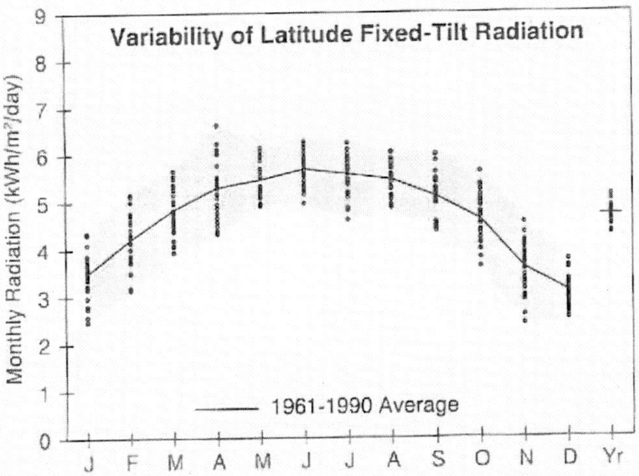

Sterling, VA

WBAN NO. 93738

LATITUDE: 38.95° N
LONGITUDE: 77.45° W
ELEVATION: 82 meters
MEAN PRESSURE: 1006 millibars

STATION TYPE: Primary

Solar Radiation for Flat-Plate Collectors Facing South at a Fixed Tilt (kWh/m²/day), Uncertainty ±9%

Tilt (°)		Jan	Feb	Mar	Apr	May	June	July	Aug	Sept	Oct	Nov	Dec	Year
0	Average	2.1	2.9	4.0	5.0	5.8	6.3	6.0	5.4	4.4	3.4	2.3	1.8	4.1
	Min/Max	1.6/2.5	2.3/3.4	3.3/4.5	4.2/6.2	5.2/6.4	5.5/6.9	5.0/6.7	4.9/5.9	4.0/5.1	2.8/4.0	1.8/2.7	1.5/2.1	3.7/4.4
Latitude -15	Average	3.1	3.9	4.7	5.4	5.8	6.1	6.0	5.7	5.0	4.3	3.2	2.7	4.7
	Min/Max	2.3/3.8	3.0/4.7	3.8/5.4	4.4/6.7	5.2/6.5	5.4/6.8	4.9/6.7	5.1/6.3	4.4/5.9	3.4/5.2	2.3/4.0	2.3/3.2	4.2/5.0
Latitude	Average	3.5	4.2	4.8	5.3	5.5	5.7	5.6	5.5	5.1	4.6	3.6	3.1	4.7
	Min/Max	2.5/4.3	3.1/5.2	3.9/5.7	4.3/6.6	4.9/6.1	4.9/6.3	4.6/6.2	4.8/6.0	4.4/6.0	3.6/5.6	2.4/4.5	2.5/3.7	4.3/5.1
Latitude +15	Average	3.7	4.3	4.7	4.9	4.8	4.9	4.9	5.0	4.9	4.6	3.7	3.3	4.5
	Min/Max	2.5/4.7	3.2/5.4	3.8/5.6	4.0/6.1	4.4/5.4	4.3/5.4	4.1/5.5	4.4/5.5	4.1/5.8	3.6/5.7	2.4/4.8	2.6/4.0	4.1/4.9
90	Average	3.4	3.7	3.5	3.1	2.6	2.5	2.6	2.9	3.3	3.7	3.3	3.0	3.1
	Min/Max	2.2/4.5	2.7/4.6	2.8/4.1	2.5/3.8	2.4/2.9	2.3/2.7	2.3/2.8	2.7/3.2	2.8/4.0	2.8/4.5	2.0/4.3	2.3/3.8	2.8/3.4

Solar Radiation for 1-Axis Tracking Flat-Plate Collectors with a North-South Axis (kWh/m²/day), Uncertainty ±9%

Axis Tilt (°)		Jan	Feb	Mar	Apr	May	June	July	Aug	Sept	Oct	Nov	Dec	Year
0	Average	3.0	4.1	5.3	6.5	7.2	7.8	7.4	6.8	5.8	4.6	3.2	2.5	5.4
	Min/Max	2.2/3.8	3.0/5.0	4.3/6.3	5.1/8.7	6.4/8.3	6.7/9.0	5.8/8.4	5.8/7.8	4.9/7.0	3.6/5.7	2.2/4.0	2.0/3.1	4.8/5.8
Latitude -15	Average	3.7	4.8	5.9	6.9	7.3	7.8	7.5	7.0	6.3	5.3	3.9	3.2	5.8
	Min/Max	2.6/4.7	3.5/6.0	4.7/7.1	5.3/9.2	6.5/8.5	6.6/8.9	5.8/8.4	5.9/8.1	5.2/7.6	4.1/6.7	2.5/5.0	2.6/4.0	5.2/6.4
Latitude	Average	4.0	5.0	6.0	6.8	7.1	7.4	7.2	6.9	6.3	5.6	4.2	3.5	5.8
	Min/Max	2.7/5.2	3.6/6.4	4.7/7.2	5.2/9.1	6.3/8.2	6.3/8.6	5.6/8.1	5.8/8.0	5.2/7.7	4.2/7.0	2.6/5.4	2.8/4.4	5.3/6.4
Latitude +15	Average	4.2	5.1	5.9	6.5	6.7	6.9	6.7	6.5	6.1	5.6	4.3	3.7	5.7
	Min/Max	2.8/5.5	3.7/6.5	4.6/7.1	4.9/8.8	5.9/7.7	5.9/8.0	5.2/7.6	5.4/7.6	5.1/7.5	4.2/7.0	2.6/5.6	2.9/4.6	5.1/6.3

Solar Radiation for 2-Axis Tracking Flat-Plate Collectors (kWh/m²/day), Uncertainty ±9%

Tracker		Jan	Feb	Mar	Apr	May	June	July	Aug	Sept	Oct	Nov	Dec	Year
2-Axis	Average	4.3	5.1	6.0	6.9	7.4	7.9	7.6	7.1	6.3	5.6	4.3	3.7	6.0
	Min/Max	2.8/5.5	3.7/6.5	4.7/7.2	5.3/9.2	6.5/8.6	6.8/9.1	5.9/8.6	5.9/8.1	5.3/7.7	4.3/7.0	2.7/5.7	2.9/4.7	5.5/6.6

Direct Beam Solar Radiation for Concentrating Collectors (kWh/m²/day), Uncertainty ±8%

Tracker		Jan	Feb	Mar	Apr	May	June	July	Aug	Sept	Oct	Nov	Dec	Year
1-Axis, E-W Horiz Axis	Average	2.4	2.7	2.9	3.1	3.2	3.4	3.2	3.0	2.9	3.0	2.4	2.2	2.9
	Min/Max	1.2/3.4	1.3/4.0	2.0/3.9	1.7/4.8	2.6/4.1	2.7/4.2	1.9/4.0	2.0/3.8	2.0/3.9	2.0/4.1	1.0/3.5	1.5/3.0	2.5/3.3
1-Axis, N-S Horiz Axis	Average	1.8	2.5	3.3	4.0	4.2	4.4	4.1	3.8	3.4	2.9	1.9	1.5	3.2
	Min/Max	0.9/2.5	1.3/3.6	2.2/4.5	2.2/6.3	3.4/5.4	3.5/5.5	2.5/5.3	2.5/5.0	2.5/4.7	2.0/4.1	0.8/2.8	1.1/2.1	2.8/3.7
1-Axis, N-S Tilt=Latitude	Average	2.7	3.3	3.8	4.2	4.1	4.2	3.9	3.9	3.9	3.7	2.7	2.3	3.6
	Min/Max	1.4/3.7	1.7/4.8	2.6/5.3	2.3/6.6	3.3/5.2	3.3/5.2	2.3/5.1	2.5/5.1	2.7/5.2	2.5/5.1	1.2/4.0	1.6/3.2	3.1/4.1
2-Axis	Average	2.8	3.4	3.8	4.3	4.3	4.5	4.2	4.0	3.9	3.8	2.9	2.5	3.7
	Min/Max	1.5/4.0	1.7/4.9	2.6/5.3	2.3/6.7	3.5/5.5	3.6/5.7	2.5/5.4	2.6/5.2	2.8/5.3	2.5/5.1	1.2/4.2	1.8/3.5	3.2/4.3

Average Climatic Conditions

Element	Jan	Feb	Mar	Apr	May	June	July	Aug	Sept	Oct	Nov	Dec	Year
Temperature (°C)	-0.8	0.9	6.2	11.5	16.8	21.7	24.2	23.4	19.5	12.8	7.4	1.9	12.1
Daily Minimum Temp	-6.1	-4.8	-0.1	4.6	10.0	15.1	17.8	17.1	13.0	5.8	1.2	-3.4	5.8
Daily Maximum Temp	4.5	6.6	12.6	18.4	23.5	28.2	30.6	29.7	26.1	19.8	13.7	7.2	18.4
Record Minimum Temp	-27.8	-25.6	-14.4	-8.3	-2.2	2.2	5.0	3.3	-1.1	-9.4	-12.8	-20.0	-27.8
Record Maximum Temp	23.9	26.1	31.7	33.3	36.1	37.8	40.0	40.0	37.2	32.2	28.9	25.6	40.0
HDD, Base 18.3°C	592	488	376	205	78	4	0	0	24	177	327	510	2781
CDD, Base 18.3°C	0	0	0	0	29	104	183	158	59	7	0	0	541
Relative Humidity (%)	68	66	63	62	70	72	73	74	75	73	70	69	70
Wind Speed (m/s)	3.7	3.8	4.1	3.9	3.4	3.0	2.8	2.6	2.7	3.0	3.4	3.5	3.4

Burlington, VT

WBAN NO. 14742

LATITUDE: 44.47° N
LONGITUDE: 73.15° W
ELEVATION: 104 meters
MEAN PRESSURE: 1003 millibars

STATION TYPE: Primary

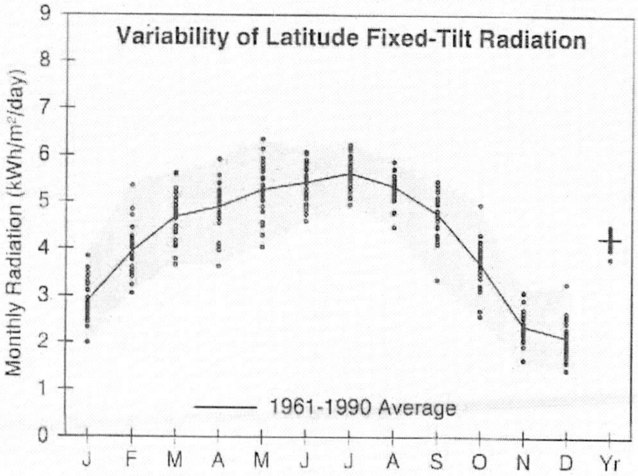

Solar Radiation for Flat-Plate Collectors Facing South at a Fixed Tilt (kWh/m²/day), Uncertainty ±9%

Tilt (°)		Jan	Feb	Mar	Apr	May	June	July	Aug	Sept	Oct	Nov	Dec	Year
0	Average	1.6	2.6	3.6	4.6	5.5	6.0	6.1	5.2	4.0	2.6	1.6	1.2	3.7
	Min/Max	1.2/2.0	2.1/3.1	3.0/4.3	3.5/5.3	4.3/6.6	5.2/6.8	5.4/6.7	4.4/5.7	3.1/4.4	2.0/3.3	1.2/1.8	0.9/1.6	3.4/3.9
Latitude -15	Average	2.6	3.6	4.5	5.0	5.6	5.9	6.1	5.6	4.7	3.5	2.2	1.9	4.3
	Min/Max	1.8/3.4	2.8/4.8	3.6/5.4	3.7/6.0	4.3/6.8	5.0/6.6	5.3/6.7	4.7/6.1	3.4/5.4	2.5/4.6	1.6/2.8	1.3/2.8	3.9/4.5
Latitude	Average	2.9	3.9	4.7	4.9	5.3	5.4	5.6	5.3	4.8	3.7	2.4	2.1	4.3
	Min/Max	2.0/3.8	3.0/5.3	3.7/5.6	3.6/5.9	4.0/6.3	4.6/6.1	4.9/6.2	4.5/5.9	3.4/5.4	2.6/4.9	1.6/3.1	1.4/3.3	3.8/4.5
Latitude +15	Average	3.1	4.1	4.6	4.5	4.6	4.7	4.9	4.8	4.5	3.7	2.4	2.2	4.0
	Min/Max	2.1/4.1	3.1/5.6	3.5/5.6	3.4/5.5	3.6/5.6	4.0/5.2	4.3/5.4	4.1/5.3	3.1/5.2	2.5/5.0	1.6/3.2	1.5/3.6	3.6/4.3
90	Average	3.0	3.8	3.9	3.2	2.9	2.7	2.9	3.2	3.4	3.1	2.2	2.2	3.0
	Min/Max	2.0/4.1	2.8/5.3	2.8/4.9	2.4/3.8	2.3/3.3	2.5/3.0	2.6/3.2	2.7/3.5	2.3/3.9	2.1/4.2	1.4/3.0	1.5/3.5	2.7/3.2

Solar Radiation for 1-Axis Tracking Flat-Plate Collectors with a North-South Axis (kWh/m²/day), Uncertainty ±9%

Axis Tilt (°)		Jan	Feb	Mar	Apr	May	June	July	Aug	Sept	Oct	Nov	Dec	Year
0	Average	2.3	3.6	5.0	6.0	7.1	7.7	7.9	6.9	5.4	3.6	2.0	1.7	4.9
	Min/Max	1.5/3.0	2.8/4.9	3.7/6.0	4.1/7.3	5.0/8.8	6.0/9.2	6.9/9.0	5.6/7.8	3.8/6.2	2.4/4.8	1.4/2.6	1.1/2.5	4.3/5.3
Latitude -15	Average	3.0	4.4	5.7	6.4	7.3	7.8	8.0	7.2	6.0	4.2	2.5	2.2	5.4
	Min/Max	2.0/4.1	3.3/6.1	4.2/6.9	4.4/7.9	5.1/9.1	5.9/9.3	7.0/9.1	5.8/8.3	4.0/7.0	2.8/5.8	1.7/3.4	1.4/3.4	4.7/5.8
Latitude	Average	3.3	4.6	5.8	6.3	7.0	7.4	7.7	7.1	6.0	4.4	2.6	2.4	5.4
	Min/Max	2.2/4.4	3.5/6.6	4.2/7.1	4.3/7.9	4.9/8.9	5.7/8.9	6.7/8.8	5.7/8.1	4.0/7.1	2.9/6.1	1.8/3.6	1.6/3.8	4.7/5.8
Latitude +15	Average	3.4	4.8	5.7	6.1	6.6	6.9	7.2	6.7	5.8	4.4	2.7	2.5	5.2
	Min/Max	2.3/4.7	3.6/6.7	4.2/7.1	4.1/7.5	4.6/8.3	5.3/8.3	6.3/8.3	5.4/7.7	3.8/6.9	2.9/6.1	1.8/3.7	1.6/4.0	4.5/5.6

Solar Radiation for 2-Axis Tracking Flat-Plate Collectors (kWh/m²/day), Uncertainty ±9%

Tracker		Jan	Feb	Mar	Apr	May	June	July	Aug	Sept	Oct	Nov	Dec	Year
2-Axis	Average	3.5	4.8	5.8	6.4	7.4	8.0	8.2	7.3	6.0	4.4	2.7	2.5	5.6
	Min/Max	2.3/4.7	3.6/6.7	4.3/7.1	4.4/7.9	5.2/9.2	6.1/9.5	7.1/9.3	5.8/8.3	4.0/7.1	3.0/6.2	1.8/3.8	1.6/4.1	4.9/6.0

Direct Beam Solar Radiation for Concentrating Collectors (kWh/m²/day), Uncertainty ±8%

Tracker		Jan	Feb	Mar	Apr	May	June	July	Aug	Sept	Oct	Nov	Dec	Year
1-Axis, E-W Horiz Axis	Average	1.7	2.2	2.5	2.8	3.2	3.5	3.7	3.2	2.8	2.2	1.3	1.2	2.5
	Min/Max	1.0/2.5	1.5/3.6	1.6/3.5	1.6/3.9	1.7/4.8	2.0/4.6	2.9/4.5	2.2/3.9	1.4/3.6	1.1/3.5	0.7/2.2	0.7/2.3	2.1/2.8
1-Axis, N-S Horiz Axis	Average	1.1	1.8	2.8	3.5	4.2	4.6	4.9	4.2	3.3	2.1	0.9	0.7	2.8
	Min/Max	0.6/1.7	1.3/3.1	1.7/3.7	2.0/4.9	2.1/6.2	2.6/6.2	3.9/5.9	2.8/5.2	1.7/4.2	1.0/3.3	0.5/1.7	0.4/1.4	2.3/3.1
1-Axis, N-S Tilt=Latitude	Average	1.8	2.6	3.4	3.8	4.2	4.4	4.7	4.3	3.8	2.8	1.4	1.2	3.2
	Min/Max	1.0/2.8	1.8/4.3	2.1/4.6	2.1/5.3	2.1/6.2	2.5/5.9	3.7/5.7	2.9/5.4	1.9/4.9	1.3/4.4	0.7/2.5	0.7/2.4	2.6/3.5
2-Axis	Average	2.0	2.7	3.4	3.9	4.4	4.8	5.1	4.5	3.8	2.8	1.5	1.3	3.3
	Min/Max	1.1/2.9	1.9/4.4	2.1/4.6	2.2/5.4	2.2/6.6	2.7/6.4	4.0/6.1	3.0/5.5	1.9/4.9	1.4/4.4	0.8/2.7	0.8/2.6	2.8/3.7

Average Climatic Conditions

Element	Jan	Feb	Mar	Apr	May	June	July	Aug	Sept	Oct	Nov	Dec	Year
Temperature (°C)	-8.7	-7.7	-0.7	6.6	13.5	18.4	21.4	19.9	14.9	8.8	2.7	-5.0	7.0
Daily Minimum Temp	-13.6	-12.8	-5.6	1.2	7.4	12.6	15.4	14.4	9.3	3.7	-1.3	-9.2	1.8
Daily Maximum Temp	-3.8	-2.5	4.1	12.0	19.6	24.3	27.3	25.5	20.6	13.9	6.7	-0.9	12.2
Record Minimum Temp	-34.4	-34.4	-28.9	-16.7	-4.4	0.6	3.9	1.7	-3.9	-9.4	-18.9	-32.2	-34.4
Record Maximum Temp	17.2	16.7	28.9	32.8	33.9	36.1	37.2	38.3	34.4	29.4	23.9	18.3	38.3
HDD, Base 18.3°C	839	728	591	352	157	32	3	16	111	296	470	723	4317
CDD, Base 18.3°C	0	0	0	0	7	36	98	66	9	0	0	0	216
Relative Humidity (%)	68	67	66	64	66	69	70	73	75	73	73	73	70
Wind Speed (m/s)	4.3	4.1	4.3	4.2	4.0	3.8	3.6	3.4	3.7	3.9	4.4	4.5	4.0

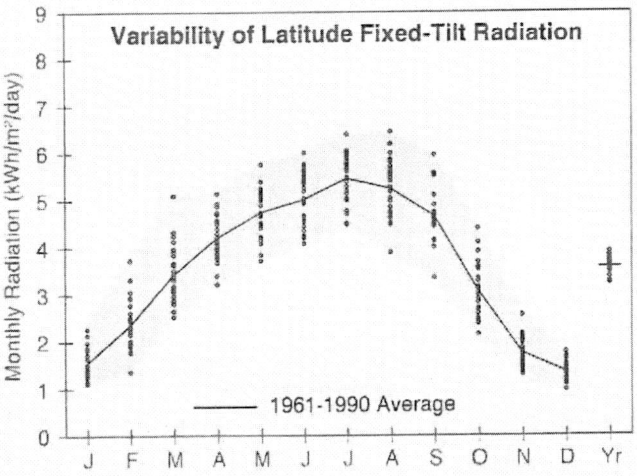

Olympia, WA

WBAN NO. 24227

LATITUDE: 46.97° N
LONGITUDE: 122.90° W
ELEVATION: 61 meters
MEAN PRESSURE: 1010 millibars

STATION TYPE: Secondary

Solar Radiation for Flat-Plate Collectors Facing South at a Fixed Tilt (kWh/m²/day), Uncertainty ±9%

Tilt (°)		Jan	Feb	Mar	Apr	May	June	July	Aug	Sept	Oct	Nov	Dec	Year
0	Average	1.0	1.7	2.8	4.0	5.0	5.6	5.9	5.1	3.8	2.2	1.2	0.9	3.3
	Min/Max	0.8/1.3	1.3/2.2	2.3/3.6	3.2/4.6	4.1/5.9	4.6/6.6	4.9/6.9	4.0/6.0	3.0/4.6	1.9/2.8	1.0/1.5	0.8/1.0	3.0/3.5
Latitude -15	Average	1.4	2.3	3.4	4.4	5.1	5.5	5.9	5.5	4.6	3.0	1.6	1.2	3.7
	Min/Max	1.1/2.1	1.4/3.4	2.5/4.9	3.4/5.3	4.0/6.1	4.4/6.5	4.8/6.9	4.1/6.7	3.4/5.9	2.1/4.1	1.2/2.3	0.9/1.6	3.3/4.0
Latitude	Average	1.5	2.4	3.4	4.2	4.7	5.0	5.5	5.2	4.6	3.1	1.7	1.3	3.6
	Min/Max	1.1/2.3	1.4/3.7	2.5/5.1	3.2/5.1	3.7/5.8	4.1/6.0	4.5/6.4	3.9/6.4	3.3/6.0	2.1/4.4	1.3/2.5	1.0/1.8	3.2/3.9
Latitude +15	Average	1.6	2.4	3.3	3.9	4.2	4.3	4.8	4.7	4.4	3.0	1.8	1.4	3.3
	Min/Max	1.1/2.4	1.3/3.8	2.4/5.0	2.9/4.8	3.3/5.1	3.5/5.2	3.9/5.6	3.5/5.8	3.1/5.7	2.0/4.4	1.3/2.6	0.9/1.8	3.0/3.6
90	Average	1.4	2.0	2.6	2.8	2.8	2.7	3.0	3.2	3.4	2.6	1.6	1.3	2.4
	Min/Max	1.0/2.2	1.1/3.4	1.9/4.1	2.1/3.4	2.2/3.3	2.3/3.1	2.6/3.4	2.5/4.0	2.4/4.4	1.7/3.8	1.1/2.4	0.8/1.7	2.2/2.7

Solar Radiation for 1-Axis Tracking Flat-Plate Collectors with a North-South Axis (kWh/m²/day), Uncertainty ±9%

Axis Tilt (°)		Jan	Feb	Mar	Apr	May	June	July	Aug	Sept	Oct	Nov	Dec	Year
0	Average	1.2	2.1	3.5	5.0	6.3	7.1	7.7	6.7	5.2	2.9	1.4	1.0	4.2
	Min/Max	0.9/1.8	1.3/3.2	2.6/5.1	3.6/6.1	4.7/7.9	5.3/8.8	5.8/9.5	4.5/8.4	3.6/7.0	2.1/4.1	1.1/2.0	0.8/1.3	3.7/4.6
Latitude -15	Average	1.6	2.5	4.0	5.3	6.5	7.1	7.9	7.1	5.8	3.5	1.8	1.3	4.5
	Min/Max	1.1/2.3	1.4/4.1	2.8/6.1	3.8/6.6	4.7/8.2	5.2/8.9	5.8/9.7	4.7/9.1	3.9/8.0	2.3/5.1	1.3/2.7	0.9/1.7	4.0/5.0
Latitude	Average	1.7	2.6	4.1	5.2	6.2	6.8	7.6	7.0	5.9	3.6	1.9	1.4	4.5
	Min/Max	1.1/2.5	1.4/4.3	2.8/6.3	3.7/6.6	4.5/7.9	5.0/8.6	5.6/9.4	4.5/8.9	3.8/8.1	2.3/5.3	1.4/2.8	1.0/1.9	4.0/5.0
Latitude +15	Average	1.7	2.6	4.0	5.0	5.9	6.3	7.1	6.6	5.7	3.5	1.9	1.4	4.3
	Min/Max	1.1/2.6	1.4/4.4	2.8/6.2	3.5/6.3	4.2/7.5	4.6/8.0	5.2/8.8	4.3/8.5	3.7/7.9	2.2/5.3	1.4/2.9	1.0/2.0	3.8/4.8

Solar Radiation for 2-Axis Tracking Flat-Plate Collectors (kWh/m²/day), Uncertainty ±9%

Tracker		Jan	Feb	Mar	Apr	May	June	July	Aug	Sept	Oct	Nov	Dec	Year
2-Axis	Average	1.7	2.7	4.1	5.3	6.6	7.3	8.0	7.2	5.9	3.6	1.9	1.5	4.7
	Min/Max	1.1/2.7	1.4/4.4	2.9/6.3	3.8/6.7	4.8/8.3	5.4/9.1	5.9/9.9	4.7/9.1	3.9/8.1	2.3/5.4	1.4/2.9	1.0/2.0	4.1/5.1

Direct Beam Solar Radiation for Concentrating Collectors (kWh/m²/day), Uncertainty ±8%

Tracker		Jan	Feb	Mar	Apr	May	June	July	Aug	Sept	Oct	Nov	Dec	Year
1-Axis, E-W Horiz Axis	Average	0.7	1.1	1.7	2.0	2.5	3.0	3.7	3.3	2.8	1.7	0.8	0.6	2.0
	Min/Max	0.2/1.4	0.3/2.6	0.8/3.3	1.0/2.9	1.3/3.8	1.5/4.5	2.1/5.2	1.4/4.7	1.2/4.5	0.6/3.0	0.5/1.5	0.3/1.0	1.6/2.4
1-Axis, N-S Horiz Axis	Average	0.4	0.9	1.8	2.5	3.3	3.9	4.8	4.2	3.2	1.5	0.6	0.4	2.3
	Min/Max	0.1/0.9	0.3/2.0	0.9/3.3	1.2/3.6	1.7/4.9	2.0/5.8	2.6/6.8	1.7/6.0	1.4/5.3	0.6/2.7	0.3/1.0	0.2/0.6	1.8/2.8
1-Axis, N-S Tilt=Latitude	Average	0.8	1.3	2.2	2.7	3.3	3.8	4.7	4.4	3.8	2.1	0.9	0.7	2.6
	Min/Max	0.2/1.5	0.4/3.0	1.1/4.3	1.3/3.9	1.7/4.9	1.9/5.6	2.6/6.7	1.9/6.3	1.6/6.3	0.8/3.7	0.5/1.6	0.3/1.0	2.1/3.1
2-Axis	Average	0.8	1.4	2.2	2.8	3.5	4.1	5.0	4.6	3.8	2.1	1.0	0.7	2.7
	Min/Max	0.2/1.6	0.4/3.1	1.1/4.3	1.4/4.0	1.8/5.2	2.1/6.1	2.8/7.2	1.9/6.5	1.6/6.3	0.8/3.8	0.5/1.7	0.3/1.1	2.1/3.2

Average Climatic Conditions

Element	Jan	Feb	Mar	Apr	May	June	July	Aug	Sept	Oct	Nov	Dec	Year
Temperature (°C)	3.3	5.1	6.6	8.6	11.8	14.8	17.2	17.4	14.4	9.8	5.8	3.4	9.8
Daily Minimum Temp	-0.2	0.4	0.9	2.3	5.0	8.0	9.6	9.7	7.1	3.8	1.7	0.1	4.0
Daily Maximum Temp	6.9	9.7	12.2	14.9	18.5	21.6	24.7	25.1	21.7	15.8	10.0	6.8	15.7
Record Minimum Temp	-22.2	-18.3	-12.8	-5.0	-3.9	-1.1	1.7	0.6	-3.9	-6.7	-18.3	-21.7	-22.2
Record Maximum Temp	17.2	22.8	24.4	30.6	35.6	38.3	39.4	40.0	36.7	32.2	23.3	17.8	40.0
HDD, Base 18.3°C	465	372	365	292	203	112	53	57	123	263	375	462	3142
CDD, Base 18.3°C	0	0	0	0	0	7	17	27	5	0	0	0	56
Relative Humidity (%)	88	85	80	75	73	72	71	72	78	85	88	89	80
Wind Speed (m/s)	3.3	3.2	3.3	3.4	3.3	3.1	2.9	2.8	2.7	2.7	3.3	3.1	3.1

Quillayute, WA

WBAN NO. 94240

LATITUDE: 47.95° N
LONGITUDE: 124.55° W
ELEVATION: 55 meters
MEAN PRESSURE: 1010 millibars

STATION TYPE: Secondary

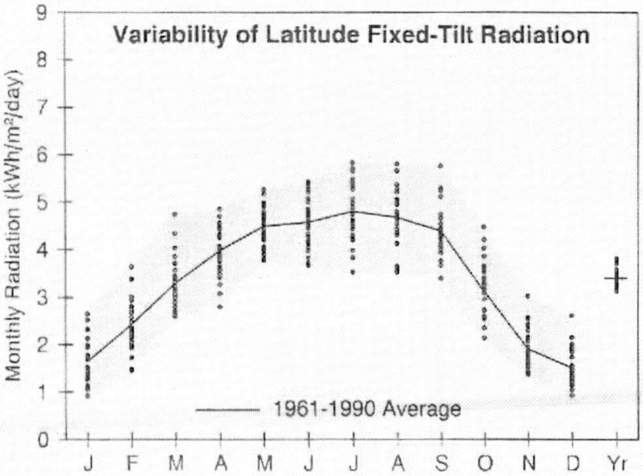

Solar Radiation for Flat-Plate Collectors Facing South at a Fixed Tilt (kWh/m²/day), Uncertainty ±9%

Tilt (°)		Jan	Feb	Mar	Apr	May	June	July	Aug	Sept	Oct	Nov	Dec	Year
0	Average	1.0	1.6	2.6	3.7	4.7	5.1	5.2	4.5	3.5	2.1	1.2	0.8	3.0
	Min/Max	0.7/1.3	1.2/2.0	2.2/3.3	2.8/4.3	4.1/5.4	4.2/5.9	4.0/6.2	3.6/5.4	3.0/4.4	1.7/2.7	1.0/1.5	0.7/1.1	2.8/3.2
Latitude -15	Average	1.5	2.3	3.2	4.1	4.8	5.0	5.2	4.9	4.4	3.0	1.8	1.4	3.5
	Min/Max	0.9/2.3	1.4/3.3	2.6/4.6	2.9/5.0	4.1/5.6	4.0/5.9	3.8/6.3	3.7/6.0	3.5/5.7	2.1/4.2	1.3/2.7	0.9/2.3	3.2/3.8
Latitude	Average	1.6	2.4	3.3	4.0	4.5	4.6	4.8	4.7	4.4	3.1	1.9	1.5	3.4
	Min/Max	0.9/2.7	1.5/3.6	2.6/4.7	2.8/4.8	3.8/5.3	3.7/5.4	3.5/5.8	3.5/5.8	3.4/5.8	2.1/4.5	1.4/3.0	0.9/2.6	3.1/3.8
Latitude +15	Average	1.7	2.4	3.2	3.7	4.0	4.0	4.2	4.2	4.2	3.1	2.0	1.6	3.2
	Min/Max	0.9/2.8	1.4/3.7	2.5/4.7	2.5/4.5	3.3/4.6	3.2/4.7	3.1/5.1	3.2/5.3	3.2/5.5	2.1/4.5	1.4/3.2	0.9/2.8	2.9/3.6
90	Average	1.6	2.1	2.6	2.7	2.7	2.5	2.7	3.0	3.2	2.6	1.8	1.5	2.4
	Min/Max	0.8/2.7	1.2/3.4	2.0/3.9	1.9/3.3	2.3/3.0	2.1/2.9	2.1/3.2	2.3/3.6	2.5/4.3	1.7/3.9	1.2/3.0	0.8/2.7	2.2/2.8

Solar Radiation for 1-Axis Tracking Flat-Plate Collectors with a North-South Axis (kWh/m²/day), Uncertainty ±9%

Axis Tilt (°)		Jan	Feb	Mar	Apr	May	June	July	Aug	Sept	Oct	Nov	Dec	Year
0	Average	1.3	2.1	3.4	4.7	5.9	6.3	6.5	5.8	4.9	2.9	1.5	1.1	3.9
	Min/Max	0.8/1.9	1.3/3.1	2.6/4.8	3.1/5.9	4.9/7.2	4.9/7.6	4.6/8.3	4.2/7.4	3.7/6.7	2.0/4.2	1.2/2.3	0.8/1.7	3.5/4.3
Latitude -15	Average	1.7	2.7	3.9	5.1	6.0	6.3	6.6	6.2	5.5	3.6	2.0	1.5	4.3
	Min/Max	0.9/2.7	1.5/4.1	3.0/5.7	3.2/6.4	4.9/7.5	4.8/7.8	4.6/8.5	4.3/8.0	4.1/7.7	2.3/5.3	1.4/3.2	0.9/2.6	3.9/4.8
Latitude	Average	1.8	2.8	4.0	5.0	5.8	6.0	6.4	6.1	5.6	3.7	2.1	1.6	4.2
	Min/Max	0.9/3.0	1.5/4.3	3.0/5.9	3.2/6.3	4.7/7.3	4.6/7.4	4.4/8.2	4.2/7.9	4.0/7.8	2.4/5.5	1.5/3.5	1.0/2.9	3.8/4.9
Latitude +15	Average	1.8	2.8	3.9	4.7	5.5	5.6	6.0	5.7	5.4	3.7	2.2	1.7	4.1
	Min/Max	0.9/3.1	1.5/4.4	2.9/5.8	3.0/6.1	4.4/6.8	4.2/6.9	4.1/7.7	4.0/7.5	3.9/7.6	2.3/5.5	1.5/3.6	1.0/3.1	3.7/4.7

Solar Radiation for 2-Axis Tracking Flat-Plate Collectors (kWh/m²/day), Uncertainty ±9%

Tracker		Jan	Feb	Mar	Apr	May	June	July	Aug	Sept	Oct	Nov	Dec	Year
2-Axis	Average	1.9	2.8	4.0	5.1	6.1	6.5	6.8	6.2	5.6	3.7	2.2	1.7	4.4
	Min/Max	0.9/3.2	1.5/4.4	3.0/5.9	3.3/6.4	5.0/7.6	4.9/7.9	4.7/8.7	4.4/8.1	4.1/7.8	2.4/5.6	1.5/3.6	1.0/3.2	4.0/5.0

Direct Beam Solar Radiation for Concentrating Collectors (kWh/m²/day), Uncertainty ±8%

Tracker		Jan	Feb	Mar	Apr	May	June	July	Aug	Sept	Oct	Nov	Dec	Year
1-Axis, E-W	Average	0.9	1.4	1.7	2.0	2.5	2.7	3.0	2.8	2.7	1.9	1.1	0.9	2.0
Horiz Axis	Min/Max	0.3/2.0	0.5/2.8	0.9/3.2	0.9/3.1	1.5/3.7	1.2/3.9	1.5/4.5	1.3/4.2	1.5/4.4	0.8/3.3	0.5/2.3	0.3/2.2	1.7/2.5
1-Axis, N-S	Average	0.6	1.1	1.8	2.5	3.2	3.4	3.8	3.5	3.1	1.7	0.7	0.5	2.2
Horiz Axis	Min/Max	0.2/1.2	0.4/2.2	1.0/3.2	1.1/3.9	2.0/4.8	1.7/5.0	2.0/5.9	1.6/5.3	1.7/5.1	0.7/3.0	0.3/1.5	0.2/1.2	1.9/2.7
1-Axis, N-S	Average	1.0	1.6	2.3	2.8	3.2	3.3	3.8	3.7	3.7	2.4	1.2	1.0	2.5
Tilt=Latitude	Min/Max	0.3/2.2	0.6/3.3	1.3/4.2	1.2/4.3	1.9/4.8	1.6/4.8	1.9/5.8	1.8/5.6	2.0/6.1	1.0/4.2	0.6/2.5	0.3/2.2	2.2/3.2
2-Axis	Average	1.1	1.7	2.3	2.8	3.4	3.6	4.0	3.8	3.7	2.4	1.3	1.0	2.6
	Min/Max	0.3/2.3	0.6/3.4	1.3/4.2	1.2/4.3	2.1/5.0	1.8/5.2	2.1/6.2	1.8/5.8	2.0/6.1	1.0/4.2	0.6/2.7	0.3/2.4	2.3/3.3

Average Climatic Conditions

Element	Jan	Feb	Mar	Apr	May	June	July	Aug	Sept	Oct	Nov	Dec	Year
Temperature (°C)	4.4	5.6	6.3	7.9	10.5	13.0	14.9	15.2	13.7	10.2	6.7	4.6	9.4
Daily Minimum Temp	0.9	1.6	1.8	2.9	5.4	8.1	9.7	9.8	7.9	5.2	2.9	1.3	4.8
Daily Maximum Temp	7.8	9.5	10.8	12.8	15.5	17.9	20.1	20.6	19.4	15.2	10.5	7.9	14.0
Record Minimum Temp	-13.9	-11.7	-7.2	-4.4	-1.7	0.6	3.3	2.2	-2.2	-4.4	-15.0	-13.9	-15.0
Record Maximum Temp	18.3	22.2	21.7	28.3	33.3	35.6	36.1	37.2	36.1	28.3	20.6	17.8	37.2
HDD, Base 18.3°C	432	358	374	313	243	160	107	100	142	252	348	426	3254
CDD, Base 18.3°C	0	0	0	0	0	0	0	3	0	0	0	0	3
Relative Humidity (%)	87	85	83	81	81	82	82	84	83	87	88	89	84
Wind Speed (m/s)	4.6	4.3	4.1	3.7	3.4	3.2	3.1	2.8	3.0	3.6	4.2	4.2	3.7

Seattle, WA

WBAN NO. 24233

LATITUDE: 47.45° N
LONGITUDE: 122.30° W
ELEVATION: 122 meters
MEAN PRESSURE: 1001 millibars

STATION TYPE: Primary

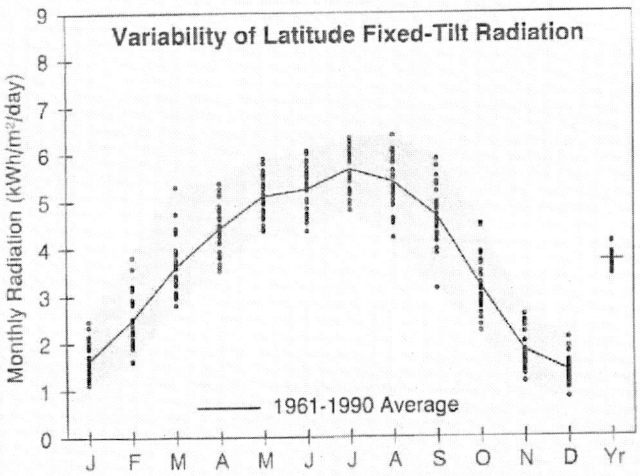

Solar Radiation for Flat-Plate Collectors Facing South at a Fixed Tilt (kWh/m²/day), Uncertainty ±9%

Tilt (°)		Jan	Feb	Mar	Apr	May	June	July	Aug	Sept	Oct	Nov	Dec	Year
0	Average	1.0	1.7	2.8	4.1	5.3	5.8	6.1	5.2	3.8	2.2	1.2	0.8	3.3
	Min/Max	0.8/1.2	1.2/2.1	2.4/3.6	3.5/4.8	4.6/6.1	4.8/6.7	5.2/6.7	4.2/5.9	2.9/4.5	1.8/2.8	0.9/1.4	0.6/1.0	3.2/3.6
Latitude -15	Average	1.5	2.3	3.5	4.6	5.4	5.7	6.1	5.6	4.7	3.0	1.7	1.3	3.8
	Min/Max	1.1/2.2	1.5/3.5	2.8/5.1	3.7/5.5	4.7/6.3	4.7/6.6	5.2/6.8	4.4/6.6	3.2/5.8	2.2/4.2	1.1/2.3	0.8/1.8	3.5/4.2
Latitude	Average	1.6	2.5	3.6	4.4	5.1	5.2	5.7	5.4	4.7	3.2	1.8	1.4	3.7
	Min/Max	1.1/2.5	1.6/3.8	2.8/5.3	3.5/5.4	4.4/5.9	4.4/6.1	4.8/6.3	4.2/6.4	3.1/5.9	2.2/4.5	1.1/2.6	0.8/2.1	3.4/4.1
Latitude +15	Average	1.7	2.5	3.5	4.1	4.5	4.5	4.9	4.9	4.5	3.2	1.8	1.4	3.5
	Min/Max	1.1/2.6	1.6/3.9	2.7/5.2	3.2/5.0	3.9/5.2	3.8/5.2	4.2/5.5	3.8/5.8	3.0/5.7	2.2/4.5	1.1/2.7	0.8/2.2	3.2/3.8
90	Average	1.5	2.2	2.8	3.0	3.0	2.8	3.1	3.4	3.4	2.7	1.7	1.3	2.6
	Min/Max	1.0/2.5	1.4/3.5	2.2/4.3	2.3/3.6	2.6/3.3	2.4/3.2	2.8/3.4	2.7/4.0	2.3/4.4	1.8/3.9	1.0/2.5	0.7/2.1	2.3/2.9

Solar Radiation for 1-Axis Tracking Flat-Plate Collectors with a North-South Axis (kWh/m²/day), Uncertainty ±9%

Axis Tilt (°)		Jan	Feb	Mar	Apr	May	June	July	Aug	Sept	Oct	Nov	Dec	Year
0	Average	1.2	2.2	3.7	5.2	6.7	7.3	8.0	6.9	5.1	3.0	1.5	1.0	4.3
	Min/Max	0.9/1.8	1.5/3.3	2.9/5.4	4.1/6.3	5.3/8.2	5.8/9.1	6.1/9.0	5.0/8.4	3.4/6.6	2.1/4.2	0.9/2.0	0.7/1.4	3.8/4.8
Latitude -15	Average	1.6	2.7	4.2	5.6	6.9	7.3	8.2	7.3	5.8	3.6	1.9	1.4	4.7
	Min/Max	1.1/2.5	1.7/4.3	3.2/6.5	4.3/7.0	5.3/8.6	5.8/9.3	6.2/9.2	5.3/9.0	3.7/7.6	2.5/5.2	1.1/2.7	0.8/2.1	4.1/5.3
Latitude	Average	1.8	2.8	4.3	5.5	6.7	7.0	7.9	7.2	5.9	3.7	2.0	1.5	4.7
	Min/Max	1.2/2.8	1.8/4.6	3.3/6.7	4.2/6.9	5.1/8.3	5.5/8.9	5.9/8.9	5.2/8.8	3.7/7.7	2.5/5.5	1.1/2.9	0.8/2.3	4.1/5.3
Latitude +15	Average	1.8	2.8	4.2	5.3	6.3	6.6	7.4	6.8	5.7	3.7	2.0	1.5	4.5
	Min/Max	1.2/2.9	1.7/4.6	3.2/6.6	4.0/6.6	4.7/7.8	5.1/8.3	5.5/8.4	4.9/8.4	3.6/7.5	2.4/5.5	1.1/3.0	0.8/2.4	3.9/5.1

Solar Radiation for 2-Axis Tracking Flat-Plate Collectors (kWh/m²/day), Uncertainty ±9%

Tracker		Jan	Feb	Mar	Apr	May	June	July	Aug	Sept	Oct	Nov	Dec	Year
2-Axis	Average	1.8	2.9	4.3	5.6	7.0	7.5	8.3	7.4	5.9	3.7	2.0	1.5	4.9
	Min/Max	1.2/2.9	1.8/4.7	3.3/6.7	4.3/7.0	5.4/8.7	5.9/9.5	6.3/9.4	5.3/9.0	3.8/7.7	2.5/5.5	1.1/3.0	0.8/2.4	4.2/5.5

Direct Beam Solar Radiation for Concentrating Collectors (kWh/m²/day), Uncertainty ±8%

Tracker		Jan	Feb	Mar	Apr	May	June	July	Aug	Sept	Oct	Nov	Dec	Year
1-Axis, E-W Horiz Axis	Average	0.9	1.4	1.9	2.3	3.0	3.3	4.0	3.5	2.9	1.8	1.0	0.8	2.2
	Min/Max	0.4/1.8	0.6/2.9	1.2/3.8	1.3/3.3	1.5/4.3	1.7/4.9	2.3/5.1	2.0/4.7	1.4/4.2	1.0/3.2	0.3/1.7	0.3/1.5	1.6/2.6
1-Axis, N-S Horiz Axis	Average	0.5	1.1	1.9	2.8	3.8	4.2	5.2	4.4	3.2	1.6	0.6	0.4	2.5
	Min/Max	0.2/1.1	0.5/2.3	1.1/3.8	1.6/4.0	1.7/5.7	2.1/6.4	2.8/6.5	2.4/5.9	1.6/4.8	0.7/2.8	0.2/1.2	0.2/0.8	1.8/3.0
1-Axis, N-S Tilt=Latitude	Average	0.9	1.6	2.5	3.1	3.9	4.1	5.1	4.7	3.9	2.2	1.1	0.8	2.8
	Min/Max	0.4/2.0	0.7/3.4	1.4/4.9	1.8/4.4	1.7/5.7	2.0/6.2	2.8/6.4	2.6/6.3	1.9/5.7	1.1/3.9	0.3/1.9	0.3/1.5	2.0/3.3
2-Axis	Average	1.0	1.7	2.5	3.1	4.1	4.4	5.5	4.8	3.9	2.3	1.1	0.9	2.9
	Min/Max	0.4/2.1	0.7/3.5	1.5/4.9	1.8/4.5	1.8/6.0	2.2/6.7	3.0/6.9	2.7/6.5	1.9/5.7	1.1/4.0	0.3/2.0	0.3/1.6	2.1/3.5

Average Climatic Conditions

Element	Jan	Feb	Mar	Apr	May	June	July	Aug	Sept	Oct	Nov	Dec	Year
Temperature (°C)	4.5	6.4	7.6	9.6	12.8	16.1	18.4	18.6	15.9	11.6	7.4	4.7	11.1
Daily Minimum Temp	1.8	3.0	3.6	5.1	7.9	11.1	12.9	13.2	11.1	7.7	4.5	2.1	7.0
Daily Maximum Temp	7.2	9.7	11.5	14.0	17.7	21.1	24.0	24.0	20.7	15.4	10.3	7.3	15.2
Record Minimum Temp	-17.8	-17.2	-11.7	-1.7	-2.2	3.3	6.1	6.7	1.7	-2.2	-14.4	-14.4	-17.8
Record Maximum Temp	17.8	21.1	23.9	29.4	33.9	35.6	37.2	37.2	36.7	31.7	23.3	17.2	37.2
HDD, Base 18.3°C	429	334	334	263	171	80	32	36	87	210	328	422	2727
CDD, Base 18.3°C	0	0	0	0	0	12	36	45	13	0	0	0	106
Relative Humidity (%)	78	75	74	71	69	67	66	68	73	79	80	80	73
Wind Speed (m/s)	4.1	4.0	4.0	4.0	3.7	3.7	3.5	3.4	3.4	3.5	4.0	4.0	3.8

Spokane, WA

WBAN NO. 24157

LATITUDE: 47.63° N
LONGITUDE: 117.53° W
ELEVATION: 721 meters
MEAN PRESSURE: 932 millibars

STATION TYPE: Secondary

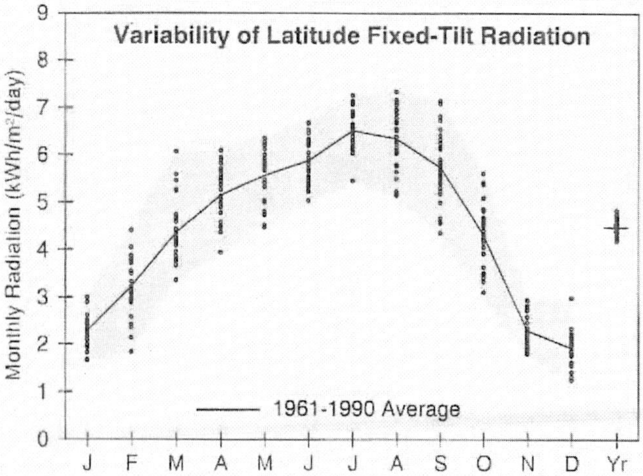

Solar Radiation for Flat-Plate Collectors Facing South at a Fixed Tilt (kWh/m²/day), Uncertainty ±9%

Tilt (°)		Jan	Feb	Mar	Apr	May	June	July	Aug	Sept	Oct	Nov	Dec	Year
0	Average	1.3	2.0	3.2	4.6	5.8	6.5	7.0	5.9	4.4	2.7	1.4	1.1	3.8
	Min/Max	1.0/1.6	1.5/2.5	2.7/4.2	3.8/5.3	4.8/6.5	5.7/7.4	5.9/7.7	5.0/6.8	3.7/5.2	2.2/3.2	1.2/1.6	0.8/1.4	3.7/4.1
Latitude -15	Average	2.1	3.0	4.2	5.3	6.0	6.4	7.0	6.6	5.7	4.0	2.1	1.7	4.5
	Min/Max	1.5/2.7	1.8/4.0	3.3/5.8	4.1/6.2	4.8/6.8	5.5/7.3	5.9/7.8	5.4/7.6	4.4/7.0	3.0/5.2	1.7/2.7	1.2/2.6	4.3/4.9
Latitude	Average	2.3	3.2	4.4	5.2	5.6	5.9	6.5	6.3	5.7	4.3	2.3	1.9	4.5
	Min/Max	1.7/3.0	1.8/4.4	3.4/6.1	3.9/6.1	4.5/6.3	5.0/6.7	5.4/7.3	5.2/7.3	4.4/7.1	3.1/5.6	1.8/3.0	1.3/3.0	4.2/4.8
Latitude +15	Average	2.4	3.3	4.3	4.8	4.9	5.1	5.6	5.7	5.5	4.4	2.4	2.0	4.2
	Min/Max	1.7/3.2	1.8/4.6	3.2/6.1	3.6/5.6	3.9/5.6	4.3/5.7	4.7/6.3	4.6/6.6	4.1/6.9	3.1/5.7	1.9/3.1	1.3/3.2	3.9/4.5
90	Average	2.3	3.0	3.5	3.4	3.2	3.1	3.5	3.9	4.3	3.8	2.2	2.0	3.2
	Min/Max	1.6/3.1	1.5/4.2	2.6/5.3	2.6/4.1	2.6/3.6	2.7/3.4	3.0/3.8	3.2/4.5	3.1/5.4	2.6/5.0	1.7/2.9	1.2/3.2	2.9/3.5

Solar Radiation for 1-Axis Tracking Flat-Plate Collectors with a North-South Axis (kWh/m²/day), Uncertainty ±9%

Axis Tilt (°)		Jan	Feb	Mar	Apr	May	June	July	Aug	Sept	Oct	Nov	Dec	Year
0	Average	1.8	2.8	4.6	6.4	7.9	8.9	10.0	8.7	6.6	4.1	1.9	1.4	5.4
	Min/Max	1.3/2.3	1.7/3.8	3.5/6.5	4.8/7.7	6.3/9.2	7.5/10.8	7.9/11.3	6.8/10.4	5.2/8.5	2.9/5.3	1.5/2.4	1.0/2.1	5.1/5.9
Latitude -15	Average	2.4	3.6	5.4	7.0	8.2	9.1	10.2	9.3	7.6	5.1	2.4	1.9	6.0
	Min/Max	1.6/3.1	1.9/5.0	4.0/7.7	5.1/8.4	6.4/9.5	7.6/10.9	8.0/11.6	7.2/11.2	5.7/9.9	3.5/6.8	1.9/3.2	1.2/3.0	5.6/6.5
Latitude	Average	2.5	3.7	5.5	6.9	8.0	8.7	9.9	9.2	7.7	5.3	2.6	2.1	6.0
	Min/Max	1.8/3.4	2.0/5.3	4.0/7.9	5.1/8.4	6.2/9.3	7.3/10.5	7.8/11.2	7.1/11.1	5.7/10.0	3.6/7.1	2.0/3.4	1.3/3.4	5.6/6.5
Latitude +15	Average	2.6	3.8	5.4	6.6	7.5	8.2	9.3	8.7	7.5	5.3	2.6	2.2	5.8
	Min/Max	1.8/3.5	2.0/5.4	3.9/7.9	4.8/8.1	5.8/8.8	6.8/9.9	7.3/10.6	6.7/10.6	5.5/9.8	3.6/7.2	2.0/3.5	1.4/3.6	5.4/6.3

Solar Radiation for 2-Axis Tracking Flat-Plate Collectors (kWh/m²/day), Uncertainty ±9%

Tracker		Jan	Feb	Mar	Apr	May	June	July	Aug	Sept	Oct	Nov	Dec	Year
2-Axis	Average	2.7	3.8	5.5	7.0	8.3	9.3	10.4	9.4	7.7	5.4	2.7	2.2	6.2
	Min/Max	1.8/3.6	2.0/5.4	4.1/7.9	5.2/8.5	6.5/9.7	7.8/11.2	8.2/11.9	7.3/11.3	5.7/10.0	3.7/7.2	2.0/3.5	1.4/3.6	5.7/6.7

Direct Beam Solar Radiation for Concentrating Collectors (kWh/m²/day), Uncertainty ±8%

Tracker		Jan	Feb	Mar	Apr	May	June	July	Aug	Sept	Oct	Nov	Dec	Year
1-Axis, E-W	Average	1.2	1.8	2.6	3.1	3.8	4.5	5.5	4.9	4.1	3.0	1.3	1.0	3.1
Horiz Axis	Min/Max	0.4/2.1	0.6/3.4	1.7/3.9	1.8/4.3	2.4/4.9	3.3/6.2	3.8/6.7	3.3/6.4	2.5/5.9	1.7/4.5	0.8/2.0	0.4/2.0	2.7/3.5
1-Axis, N-S	Average	0.7	1.5	2.8	4.0	5.1	6.0	7.4	6.5	4.9	2.8	0.9	0.6	3.6
Horiz Axis	Min/Max	0.2/1.3	0.5/2.7	1.7/4.2	2.4/5.4	3.3/6.7	4.5/8.3	5.0/8.9	4.4/8.6	3.1/7.1	1.5/4.2	0.5/1.4	0.3/1.1	3.1/4.0
1-Axis, N-S	Average	1.3	2.2	3.5	4.4	5.2	5.8	7.3	6.8	5.7	3.8	1.5	1.1	4.0
Tilt=Latitude	Min/Max	0.4/2.3	0.7/4.0	2.2/5.3	2.6/5.9	3.3/6.7	4.3/8.0	4.9/8.7	4.6/9.0	3.6/8.4	2.1/5.7	0.8/2.2	0.5/2.0	3.5/4.6
2-Axis	Average	1.4	2.2	3.5	4.5	5.5	6.3	7.8	7.0	5.8	3.9	1.6	1.1	4.2
	Min/Max	0.4/2.5	0.7/4.1	2.2/5.3	2.7/6.1	3.5/7.1	4.7/8.7	5.3/9.3	4.7/9.3	3.6/8.4	2.1/5.8	0.9/2.3	0.5/2.2	3.7/4.7

Average Climatic Conditions

Element	Jan	Feb	Mar	Apr	May	June	July	Aug	Sept	Oct	Nov	Dec	Year
Temperature (°C)	-2.7	0.7	3.7	7.7	12.2	16.7	20.4	20.2	14.9	8.5	1.7	-2.3	8.5
Daily Minimum Temp	-6.2	-3.4	-1.3	1.5	5.5	9.6	12.4	12.4	7.7	2.2	-1.8	-5.7	2.7
Daily Maximum Temp	0.7	4.8	8.7	13.9	18.8	23.7	28.4	28.1	22.2	14.8	5.2	1.0	14.2
Record Minimum Temp	-30.0	-27.2	-21.7	-8.3	-4.4	0.6	2.8	1.7	-4.4	-12.2	-29.4	-31.7	-31.7
Record Maximum Temp	15.0	16.1	21.7	32.2	35.6	37.8	39.4	42.2	36.7	30.0	19.4	13.3	42.2
HDD, Base 18.3°C	653	493	453	318	191	77	17	31	124	305	498	641	3801
CDD, Base 18.3°C	0	0	0	0	0	27	82	89	22	0	0	0	221
Relative Humidity (%)	83	79	70	61	58	54	44	45	54	67	83	85	65
Wind Speed (m/s)	4.2	4.3	4.5	4.7	4.4	4.3	4.1	3.8	3.9	3.8	4.2	3.7	4.2

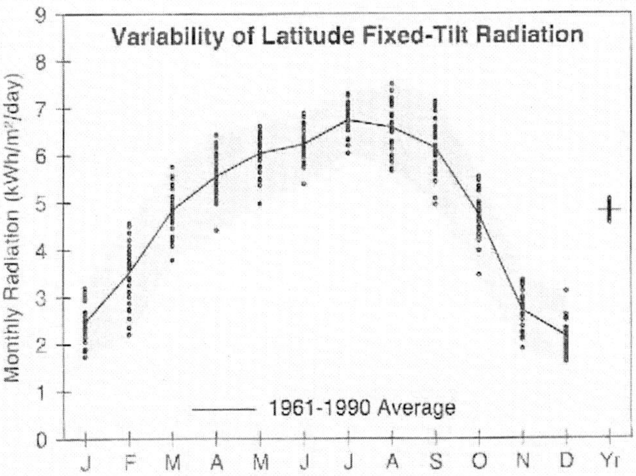

Yakima, WA

WBAN NO. 24243

LATITUDE: 46.57° N
LONGITUDE: 120.53° W
ELEVATION: 325 meters
MEAN PRESSURE: 978 millibars

STATION TYPE: Secondary

Solar Radiation for Flat-Plate Collectors Facing South at a Fixed Tilt (kWh/m²/day), Uncertainty ±9%

Tilt (°)		Jan	Feb	Mar	Apr	May	June	July	Aug	Sept	Oct	Nov	Dec	Year
0	Average	1.4	2.2	3.6	5.0	6.2	6.9	7.2	6.2	4.7	3.0	1.6	1.1	4.1
	Min/Max	1.2/1.6	1.7/2.5	3.0/4.1	4.1/5.6	5.3/6.8	6.0/7.7	6.5/7.8	5.4/6.9	4.0/5.3	2.5/3.3	1.3/1.8	1.0/1.3	4.0/4.3
Latitude -15	Average	2.2	3.3	4.7	5.7	6.4	6.8	7.3	6.8	6.0	4.4	2.5	1.9	4.8
	Min/Max	1.6/2.8	2.1/4.2	3.7/5.5	4.5/6.5	5.3/7.1	5.9/7.5	6.5/7.9	5.9/7.8	4.9/7.0	3.3/5.1	1.8/3.0	1.5/2.7	4.6/5.1
Latitude	Average	2.5	3.5	4.8	5.5	6.0	6.2	6.7	6.6	6.1	4.7	2.7	2.2	4.8
	Min/Max	1.7/3.2	2.2/4.6	3.8/5.8	4.4/6.4	5.0/6.6	5.4/6.9	6.0/7.3	5.7/7.5	4.9/7.1	3.5/5.5	1.9/3.3	1.6/3.1	4.6/5.0
Latitude +15	Average	2.6	3.6	4.7	5.1	5.3	5.3	5.8	5.9	5.9	4.8	2.8	2.3	4.5
	Min/Max	1.8/3.4	2.2/4.7	3.7/5.7	4.1/6.0	4.4/5.8	4.7/5.9	5.3/6.3	5.1/6.8	4.7/6.9	3.4/5.6	1.9/3.5	1.7/3.3	4.3/4.7
90	Average	2.5	3.2	3.8	3.7	3.4	3.2	3.5	4.0	4.5	4.1	2.6	2.2	3.4
	Min/Max	1.6/3.3	1.9/4.2	3.0/4.6	2.9/4.3	2.8/3.6	2.8/3.4	3.2/3.7	3.4/4.5	3.6/5.3	2.9/4.8	1.7/3.3	1.5/3.2	3.1/3.6

Solar Radiation for 1-Axis Tracking Flat-Plate Collectors with a North-South Axis (kWh/m²/day), Uncertainty ±9%

Axis Tilt (°)		Jan	Feb	Mar	Apr	May	June	July	Aug	Sept	Oct	Nov	Dec	Year
0	Average	1.9	3.1	5.1	6.9	8.6	9.5	10.3	9.0	7.1	4.5	2.2	1.6	5.8
	Min/Max	1.4/2.4	1.9/4.0	4.0/6.0	5.3/8.0	6.9/9.7	7.9/11.0	8.9/11.5	7.6/10.6	5.5/8.4	3.2/5.4	1.6/2.7	1.3/2.2	5.4/6.2
Latitude -15	Average	2.5	3.9	5.9	7.5	8.9	9.6	10.5	9.6	8.2	5.6	2.9	2.2	6.5
	Min/Max	1.8/3.4	2.3/5.2	4.6/7.2	5.7/8.8	7.1/10.0	7.9/11.2	9.1/11.8	8.1/11.4	6.2/9.7	3.9/6.7	2.0/3.6	1.6/3.2	6.0/6.8
Latitude	Average	2.8	4.1	6.1	7.4	8.6	9.2	10.2	9.5	8.2	5.8	3.1	2.4	6.5
	Min/Max	1.9/3.7	2.4/5.5	4.6/7.3	5.6/8.8	6.8/9.7	7.6/10.8	8.8/11.4	7.9/11.2	6.3/9.8	4.0/7.0	2.1/3.9	1.7/3.5	6.0/6.8
Latitude +15	Average	2.9	4.2	6.0	7.1	8.2	8.6	9.6	9.0	8.1	5.8	3.2	2.5	6.3
	Min/Max	1.9/3.9	2.4/5.6	4.5/7.3	5.3/8.5	6.4/9.2	7.1/10.1	8.3/10.8	7.5/10.8	6.1/9.6	4.0/7.1	2.1/4.0	1.8/3.8	5.8/6.6

Solar Radiation for 2-Axis Tracking Flat-Plate Collectors (kWh/m²/day), Uncertainty ±9%

Tracker		Jan	Feb	Mar	Apr	May	June	July	Aug	Sept	Oct	Nov	Dec	Year
2-Axis	Average	2.9	4.2	6.1	7.6	9.0	9.8	10.7	9.7	8.3	5.9	3.2	2.5	6.7
	Min/Max	1.9/3.9	2.4/5.6	4.6/7.3	5.7/8.9	7.2/10.1	8.1/11.5	9.3/12.1	8.1/11.5	6.3/9.8	4.1/7.1	2.1/4.0	1.8/3.8	6.2/7.1

Direct Beam Solar Radiation for Concentrating Collectors (kWh/m²/day), Uncertainty ±8%

Tracker		Jan	Feb	Mar	Apr	May	June	July	Aug	Sept	Oct	Nov	Dec	Year
1-Axis, E-W Horiz Axis	Average	1.4	2.2	3.0	3.5	4.4	5.0	5.8	5.2	4.5	3.4	1.7	1.3	3.5
	Min/Max	0.6/2.5	0.4/3.4	1.9/4.0	2.3/4.6	2.9/5.4	3.6/6.5	4.6/6.9	3.9/6.6	2.9/5.8	1.9/4.3	0.9/2.5	0.6/2.5	3.1/3.8
1-Axis, N-S Horiz Axis	Average	0.9	1.8	3.2	4.5	5.9	6.6	7.8	6.8	5.3	3.1	1.2	0.8	4.0
	Min/Max	0.3/1.6	0.4/2.8	2.1/4.2	2.8/5.7	4.0/7.3	4.7/8.7	6.1/9.3	5.2/8.8	3.3/6.9	1.7/4.1	0.6/1.8	0.4/1.5	3.5/4.4
1-Axis, N-S Tilt=Latitude	Average	1.5	2.6	4.0	4.9	5.8	6.4	7.6	7.1	6.3	4.2	1.9	1.4	4.5
	Min/Max	0.6/2.7	0.5/4.0	2.6/5.3	3.1/6.3	4.0/7.2	4.5/8.4	5.9/9.1	5.4/9.2	4.0/8.1	2.3/5.5	1.0/2.8	0.7/2.6	3.9/4.9
2-Axis	Average	1.7	2.7	4.0	5.0	6.2	6.9	8.1	7.4	6.3	4.3	2.0	1.5	4.7
	Min/Max	0.6/2.9	0.5/4.1	2.6/5.3	3.2/6.4	4.2/7.6	4.9/9.1	6.3/9.7	5.6/9.5	4.0/8.1	2.4/5.6	1.1/2.9	0.7/2.8	4.1/5.1

Average Climatic Conditions

Element	Jan	Feb	Mar	Apr	May	June	July	Aug	Sept	Oct	Nov	Dec	Year
Temperature (°C)	-1.3	2.4	6.1	9.7	13.9	18.1	21.1	20.6	15.9	9.9	3.7	-1.2	9.9
Daily Minimum Temp	-5.7	-3.1	-0.7	1.9	5.7	9.6	11.7	11.3	7.0	1.8	-1.7	-5.5	2.7
Daily Maximum Temp	3.1	8.0	12.9	17.3	22.0	26.6	30.4	29.8	24.9	18.0	9.1	3.1	17.1
Record Minimum Temp	-29.4	-31.7	-18.3	-6.7	-3.9	-1.1	1.1	1.7	-4.4	-11.7	-25.0	-27.2	-31.7
Record Maximum Temp	20.0	20.6	26.7	33.3	38.9	39.4	42.2	43.3	37.8	30.6	22.8	19.4	43.3
HDD, Base 18.3°C	608	445	379	260	142	50	11	21	94	260	440	606	3315
CDD, Base 18.3°C	0	0	0	0	4	43	95	90	22	0	0	0	254
Relative Humidity (%)	78	73	61	52	48	47	44	48	56	63	75	80	60
Wind Speed (m/s)	2.6	3.0	3.5	3.8	3.8	3.8	3.6	3.4	3.4	3.0	2.8	2.4	3.3

Eau Claire, WI

WBAN NO. 14991

LATITUDE: 44.87° N
LONGITUDE: 91.48° W
ELEVATION: 273 meters
MEAN PRESSURE: 983 millibars

STATION TYPE: Secondary

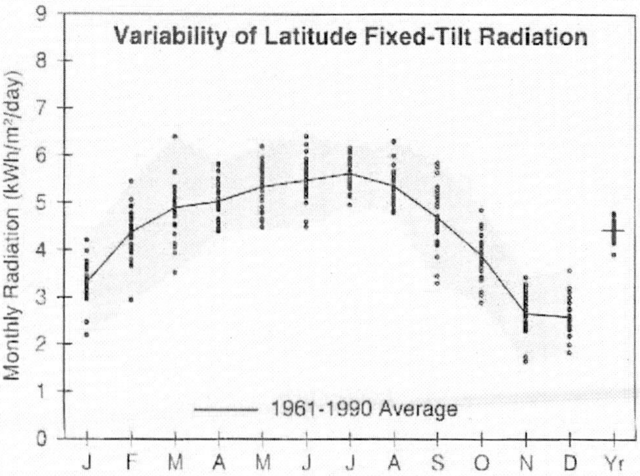

Solar Radiation for Flat-Plate Collectors Facing South at a Fixed Tilt (kWh/m²/day), Uncertainty ±9%

Tilt (°)		Jan	Feb	Mar	Apr	May	June	July	Aug	Sept	Oct	Nov	Dec	Year
0	Average	1.7	2.7	3.7	4.6	5.6	6.1	6.1	5.2	3.9	2.7	1.6	1.4	3.8
	Min/Max	1.4/2.0	2.2/3.0	3.0/4.5	4.2/5.2	4.8/6.4	5.1/7.1	5.4/6.6	4.7/6.0	3.1/4.6	2.2/3.1	1.3/1.9	1.1/1.7	3.5/4.1
Latitude -15	Average	2.9	4.0	4.7	5.1	5.7	6.0	6.1	5.6	4.7	3.7	2.4	2.3	4.4
	Min/Max	2.0/3.6	2.8/4.8	3.5/6.0	4.5/5.9	4.8/6.6	4.9/7.0	5.3/6.7	5.0/6.6	3.4/5.8	2.8/4.5	1.6/3.1	1.7/3.1	4.0/4.8
Latitude	Average	3.3	4.4	4.9	5.0	5.3	5.5	5.6	5.4	4.7	3.9	2.7	2.6	4.4
	Min/Max	2.2/4.2	2.9/5.4	3.5/6.4	4.4/5.8	4.5/6.2	4.5/6.4	4.9/6.2	4.8/6.3	3.3/5.8	2.9/4.8	1.6/3.4	1.8/3.6	3.9/4.8
Latitude +15	Average	3.6	4.6	4.8	4.6	4.7	4.7	4.9	4.8	4.5	3.9	2.8	2.8	4.2
	Min/Max	2.3/4.6	3.0/5.8	3.4/6.5	4.0/5.4	4.0/5.4	3.9/5.5	4.3/5.4	4.3/5.7	3.1/5.6	2.8/4.9	1.6/3.6	1.9/3.9	3.7/4.5
90	Average	3.5	4.3	4.1	3.3	2.9	2.8	2.9	3.2	3.3	3.3	2.5	2.7	3.2
	Min/Max	2.2/4.5	2.8/5.4	2.8/5.7	2.8/3.8	2.5/3.3	2.4/3.1	2.6/3.1	2.8/3.7	2.3/4.2	2.4/4.2	1.4/3.3	1.8/3.9	2.8/3.4

Solar Radiation for 1-Axis Tracking Flat-Plate Collectors with a North-South Axis (kWh/m²/day), Uncertainty ±9%

Axis Tilt (°)		Jan	Feb	Mar	Apr	May	June	July	Aug	Sept	Oct	Nov	Dec	Year
0	Average	2.6	3.9	5.1	6.2	7.4	7.9	8.1	7.1	5.3	3.8	2.2	2.0	5.1
	Min/Max	1.8/3.2	2.7/4.8	3.7/6.8	5.3/7.3	5.9/8.9	6.1/9.9	7.0/9.1	6.1/8.6	3.6/6.7	2.7/4.7	1.4/2.9	1.4/2.7	4.5/5.7
Latitude -15	Average	3.5	4.9	5.9	6.6	7.5	8.0	8.2	7.4	5.9	4.5	2.8	2.6	5.7
	Min/Max	2.3/4.4	3.2/6.2	4.1/7.9	5.6/7.9	6.0/9.1	6.1/9.9	7.1/9.3	6.4/9.1	3.8/7.6	3.2/5.7	1.7/3.8	1.9/3.7	4.9/6.3
Latitude	Average	3.8	5.2	6.1	6.5	7.3	7.7	7.9	7.3	6.0	4.7	3.0	2.9	5.7
	Min/Max	2.4/4.9	3.3/6.7	4.1/8.2	5.5/7.8	5.8/8.9	5.8/9.6	6.8/9.0	6.3/8.9	3.8/7.7	3.3/6.0	1.7/4.1	2.0/4.1	4.9/6.3
Latitude +15	Average	4.0	5.4	6.0	6.3	6.9	7.2	7.4	6.9	5.8	4.7	3.1	3.1	5.6
	Min/Max	2.5/5.2	3.3/6.9	4.1/8.2	5.2/7.5	5.5/8.4	5.4/9.0	6.4/8.5	5.9/8.5	3.6/7.5	3.3/6.0	1.7/4.2	2.1/4.4	4.7/6.2

Solar Radiation for 2-Axis Tracking Flat-Plate Collectors (kWh/m²/day), Uncertainty ±9%

Tracker		Jan	Feb	Mar	Apr	May	June	July	Aug	Sept	Oct	Nov	Dec	Year
2-Axis	Average	4.1	5.4	6.1	6.6	7.7	8.2	8.3	7.5	6.0	4.7	3.1	3.1	5.9
	Min/Max	2.5/5.2	3.3/6.9	4.1/8.2	5.6/7.9	6.1/9.3	6.3/10.2	7.2/9.5	6.4/9.1	3.8/7.7	3.3/6.1	1.7/4.2	2.1/4.4	5.1/6.5

Direct Beam Solar Radiation for Concentrating Collectors (kWh/m²/day), Uncertainty ±8%

Tracker		Jan	Feb	Mar	Apr	May	June	July	Aug	Sept	Oct	Nov	Dec	Year
1-Axis, E-W Horiz Axis	Average	2.2	2.7	2.7	2.9	3.4	3.7	3.8	3.4	2.8	2.4	1.6	1.7	2.8
	Min/Max	1.0/3.2	1.0/4.1	1.5/4.1	2.0/3.9	2.1/4.6	2.1/5.2	2.8/4.7	2.6/4.6	1.2/4.0	1.4/3.5	0.5/2.5	1.0/2.5	2.1/3.3
1-Axis, N-S Horiz Axis	Average	1.4	2.2	2.9	3.7	4.5	4.8	5.0	4.4	3.3	2.3	1.2	1.0	3.1
	Min/Max	0.7/2.0	0.8/3.4	1.6/4.3	2.5/4.9	2.9/6.1	2.8/7.0	3.7/6.4	3.4/6.1	1.4/4.7	1.2/3.3	0.3/1.9	0.6/1.5	2.4/3.8
1-Axis, N-S Tilt=Latitude	Average	2.3	3.2	3.6	4.0	4.5	4.6	4.9	4.6	3.8	3.1	1.8	1.7	3.5
	Min/Max	1.1/3.4	1.2/4.9	1.9/5.4	2.7/5.3	2.9/6.1	2.7/6.7	3.6/6.2	3.5/6.4	1.6/5.4	1.6/4.4	0.6/2.8	1.0/2.6	2.7/4.2
2-Axis	Average	2.5	3.3	3.6	4.1	4.7	5.0	5.2	4.8	3.8	3.1	1.9	1.9	3.7
	Min/Max	1.2/3.6	1.2/5.0	1.9/5.4	2.7/5.4	3.0/6.4	2.9/7.2	3.9/6.6	3.6/6.6	1.6/5.5	1.7/4.4	0.6/3.0	1.1/2.9	2.8/4.4

Average Climatic Conditions

Element	Jan	Feb	Mar	Apr	May	June	July	Aug	Sept	Oct	Nov	Dec	Year
Temperature (°C)	-11.8	-8.7	-1.3	7.2	14.1	19.1	21.9	20.3	15.0	8.6	0.2	-8.4	6.3
Daily Minimum Temp	-17.3	-14.6	-6.8	0.8	7.2	12.4	15.4	14.0	8.8	2.7	-4.6	-13.2	0.4
Daily Maximum Temp	-6.4	-2.9	4.1	13.6	20.8	25.7	28.4	26.7	21.1	14.6	4.9	-3.7	12.3
Record Minimum Temp	-38.3	-33.9	-36.1	-13.3	-5.0	1.1	6.7	3.9	-3.9	-11.1	-25.6	-35.0	-38.3
Record Maximum Temp	11.7	15.0	28.9	31.7	33.3	37.2	40.0	39.4	35.6	31.1	21.7	17.2	40.0
HDD, Base 18.3°C	935	758	609	333	154	33	8	16	107	302	543	830	4628
CDD, Base 18.3°C	0	0	0	0	21	56	120	78	7	0	0	0	282
Relative Humidity (%)	73	73	71	63	63	68	70	73	75	71	76	77	71
Wind Speed (m/s)	4.0	3.9	4.4	4.6	4.2	4.0	3.6	3.4	3.7	3.9	4.1	4.0	4.0

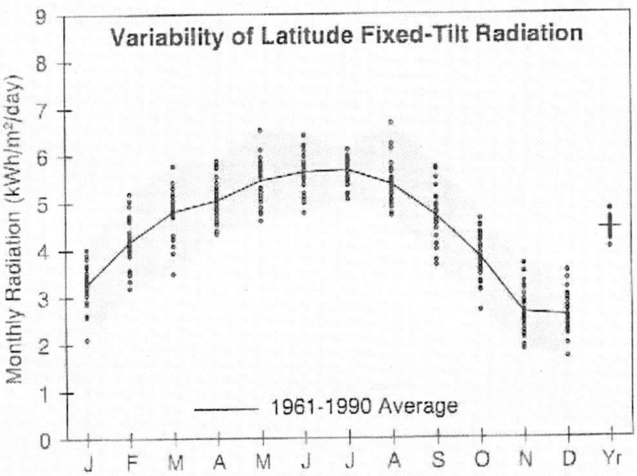

Green Bay, WI

WBAN NO. 14898

LATITUDE: 44.48° N
LONGITUDE: 88.13° W
ELEVATION: 214 meters
MEAN PRESSURE: 990 millibars

STATION TYPE: Secondary

Solar Radiation for Flat-Plate Collectors Facing South at a Fixed Tilt (kWh/m²/day), Uncertainty ±9%

Tilt (°)		Jan	Feb	Mar	Apr	May	June	July	Aug	Sept	Oct	Nov	Dec	Year
0	Average	1.7	2.6	3.7	4.7	5.7	6.3	6.1	5.2	3.9	2.7	1.6	1.4	3.8
	Min/Max	1.3/2.0	2.3/3.0	3.0/4.1	4.2/5.3	4.9/6.8	5.4/7.1	5.5/6.6	4.7/6.3	3.4/4.6	2.2/3.0	1.4/1.9	1.1/1.7	3.6/4.1
Latitude -15	Average	2.9	3.8	4.6	5.2	5.8	6.1	6.1	5.6	4.7	3.6	2.4	2.3	4.4
	Min/Max	1.9/3.5	3.0/4.7	3.4/5.5	4.5/6.0	4.9/7.0	5.2/7.0	5.4/6.6	4.9/6.9	3.7/5.6	2.6/4.3	1.8/3.3	1.5/3.0	4.1/4.8
Latitude	Average	3.3	4.2	4.8	5.0	5.4	5.6	5.7	5.4	4.7	3.8	2.6	2.6	4.4
	Min/Max	2.1/4.0	3.2/5.2	3.5/5.8	4.3/5.9	4.6/6.5	4.8/6.4	5.1/6.1	4.7/6.7	3.7/5.7	2.7/4.6	1.9/3.7	1.7/3.5	4.0/4.8
Latitude +15	Average	3.5	4.3	4.7	4.7	4.8	4.9	4.9	4.9	4.5	3.8	2.7	2.8	4.2
	Min/Max	2.2/4.3	3.3/5.4	3.4/5.8	4.0/5.5	4.1/5.8	4.1/5.5	4.4/5.3	4.3/6.1	3.4/5.5	2.6/4.7	1.9/3.9	1.8/3.8	3.8/4.6
90	Average	3.4	4.0	4.0	3.2	3.0	2.8	2.9	3.2	3.3	3.2	2.5	2.7	3.2
	Min/Max	2.1/4.3	3.0/5.1	2.8/4.9	2.7/3.8	2.6/3.5	2.5/3.1	2.6/3.1	2.8/3.9	2.5/4.2	2.2/3.9	1.6/3.6	1.6/3.8	2.8/3.5

Solar Radiation for 1-Axis Tracking Flat-Plate Collectors with a North-South Axis (kWh/m²/day), Uncertainty ±9%

Axis Tilt (°)		Jan	Feb	Mar	Apr	May	June	July	Aug	Sept	Oct	Nov	Dec	Year
0	Average	2.6	3.8	5.1	6.2	7.5	8.2	8.1	7.0	5.4	3.7	2.2	2.0	5.1
	Min/Max	1.7/3.1	2.8/4.7	3.6/6.2	5.2/7.3	6.1/9.3	6.6/9.8	7.1/8.9	6.0/9.0	4.1/6.6	2.7/4.6	1.6/2.9	1.3/2.6	4.6/5.7
Latitude -15	Average	3.4	4.7	5.8	6.6	7.7	8.2	8.2	7.4	6.0	4.4	2.8	2.6	5.7
	Min/Max	2.1/4.2	3.4/5.9	4.0/7.2	5.4/7.9	6.1/9.6	6.6/9.9	7.2/9.0	6.2/9.6	4.4/7.5	3.0/5.6	1.9/3.9	1.7/3.6	5.0/6.3
Latitude	Average	3.8	5.0	6.0	6.5	7.4	7.9	7.9	7.2	6.0	4.6	3.0	2.9	5.7
	Min/Max	2.3/4.7	3.6/6.3	4.1/7.4	5.3/7.8	5.9/9.3	6.3/9.5	6.9/8.7	6.1/9.4	4.4/7.6	3.1/5.8	2.0/4.3	1.8/4.0	5.0/6.3
Latitude +15	Average	3.9	5.1	5.9	6.3	7.0	7.4	7.4	6.9	5.8	4.6	3.1	3.1	5.5
	Min/Max	2.4/4.9	3.7/6.5	4.0/7.4	5.1/7.5	5.6/8.8	5.8/8.9	6.5/8.2	5.8/9.0	4.2/7.4	3.0/5.8	2.0/4.4	1.9/4.3	4.8/6.2

Solar Radiation for 2-Axis Tracking Flat-Plate Collectors (kWh/m²/day), Uncertainty ±9%

Tracker		Jan	Feb	Mar	Apr	May	June	July	Aug	Sept	Oct	Nov	Dec	Year
2-Axis	Average	4.0	5.1	6.0	6.6	7.8	8.4	8.4	7.4	6.0	4.6	3.1	3.1	5.9
	Min/Max	2.4/5.0	3.7/6.5	4.1/7.4	5.4/7.9	6.2/9.7	6.7/10.2	7.3/9.2	6.3/9.6	4.4/7.6	3.1/5.8	2.1/4.5	1.9/4.3	5.2/6.5

Direct Beam Solar Radiation for Concentrating Collectors (kWh/m²/day), Uncertainty ±8%

Tracker		Jan	Feb	Mar	Apr	May	June	July	Aug	Sept	Oct	Nov	Dec	Year
1-Axis, E-W Horiz Axis	Average	2.2	2.6	2.7	2.9	3.5	3.9	3.9	3.3	2.8	2.3	1.6	1.7	2.8
	Min/Max	1.0/3.0	1.4/3.6	1.4/3.8	2.1/4.0	2.2/5.0	2.5/5.4	2.9/4.5	2.6/5.0	1.5/4.0	1.2/3.3	0.7/2.8	0.9/2.6	2.3/3.3
1-Axis, N-S Horiz Axis	Average	1.4	2.1	2.9	3.7	4.6	5.2	5.1	4.4	3.3	2.2	1.2	1.0	3.1
	Min/Max	0.6/2.0	1.1/3.0	1.5/4.1	2.6/4.9	3.0/6.6	3.4/7.1	3.9/6.0	3.4/6.6	1.7/4.7	1.1/3.2	0.5/2.0	0.5/1.6	2.5/3.7
1-Axis, N-S Tilt=Latitude	Average	2.3	3.1	3.6	4.0	4.6	4.9	5.0	4.6	3.8	2.9	1.8	1.8	3.5
	Min/Max	1.1/3.3	1.6/4.3	1.8/5.0	2.8/5.4	3.0/6.5	3.2/6.8	3.7/5.8	3.5/6.8	2.0/5.5	1.5/4.2	0.8/3.1	1.0/2.8	2.8/4.2
2-Axis	Average	2.5	3.1	3.6	4.1	4.8	5.3	5.3	4.7	3.8	3.0	1.9	1.9	3.7
	Min/Max	1.1/3.5	1.7/4.4	1.8/5.0	2.8/5.5	3.1/6.9	3.5/7.4	4.0/6.3	3.6/7.0	2.0/5.6	1.5/4.3	0.9/3.2	1.1/3.0	2.9/4.4

Average Climatic Conditions

Element	Jan	Feb	Mar	Apr	May	June	July	Aug	Sept	Oct	Nov	Dec	Year
Temperature (°C)	-9.8	-7.6	-1.1	6.7	13.1	18.1	20.9	19.5	15.0	8.9	1.3	-6.6	6.6
Daily Minimum Temp	-14.6	-12.5	-5.9	1.1	6.5	11.9	14.9	13.8	9.3	3.6	-2.9	-10.8	1.2
Daily Maximum Temp	-5.1	-2.7	3.6	12.2	19.6	24.2	26.9	25.3	20.6	14.1	5.6	-2.4	11.8
Record Minimum Temp	-35.0	-32.2	-33.9	-13.9	-6.1	0.0	4.4	3.3	-4.4	-9.4	-22.8	-32.8	-35.0
Record Maximum Temp	10.0	12.8	25.0	31.7	32.8	36.7	37.2	37.2	35.0	31.1	22.2	16.7	37.2
HDD, Base 18.3°C	873	727	603	350	177	52	9	24	105	293	510	772	4494
CDD, Base 18.3°C	0	0	0	0	13	43	91	60	5	0	0	0	212
Relative Humidity (%)	74	74	73	67	66	69	71	75	76	74	77	77	73
Wind Speed (m/s)	4.8	4.6	4.8	4.9	4.4	4.0	3.5	3.4	3.8	4.2	4.6	4.6	4.3

La Crosse, WI

WBAN NO. 14920

LATITUDE: 43.87° N
LONGITUDE: 91.25° W
ELEVATION: 205 meters
MEAN PRESSURE: 992 millibars

STATION TYPE: Secondary

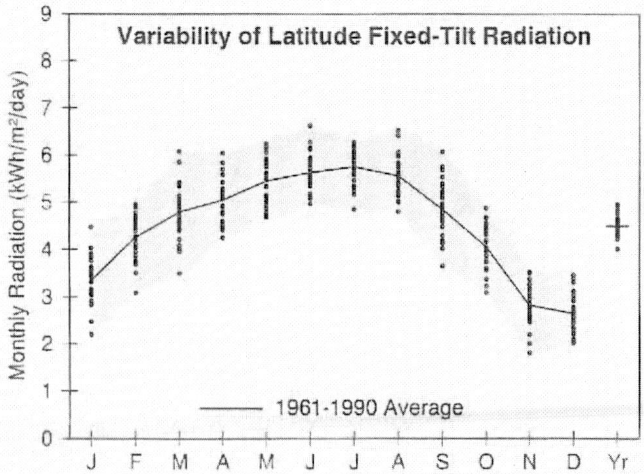

Solar Radiation for Flat-Plate Collectors Facing South at a Fixed Tilt (kWh/m²/day), Uncertainty ±9%

Tilt (°)		Jan	Feb	Mar	Apr	May	June	July	Aug	Sept	Oct	Nov	Dec	Year
0	Average	1.8	2.7	3.7	4.7	5.7	6.3	6.2	5.4	4.0	2.8	1.7	1.4	3.9
	Min/Max	1.4/2.1	2.3/3.0	3.0/4.4	4.1/5.4	4.9/6.5	5.6/7.3	5.3/6.7	4.8/6.2	3.4/4.8	2.4/3.2	1.4/2.0	1.2/1.7	3.6/4.2
Latitude -15	Average	2.9	3.9	4.6	5.2	5.8	6.1	6.2	5.8	4.8	3.8	2.6	2.3	4.5
	Min/Max	2.0/3.8	2.9/4.5	3.5/5.8	4.4/6.2	5.0/6.7	5.4/7.2	5.2/6.8	5.0/6.8	3.7/6.0	3.0/4.5	1.7/3.1	1.8/3.0	4.1/4.9
Latitude	Average	3.3	4.3	4.8	5.0	5.4	5.6	5.8	5.5	4.8	4.1	2.8	2.6	4.5
	Min/Max	2.2/4.5	3.1/4.9	3.5/6.1	4.3/6.0	4.7/6.2	5.0/6.6	4.8/6.3	4.8/6.5	3.7/6.1	3.1/4.9	1.8/3.5	2.0/3.5	4.0/5.0
Latitude +15	Average	3.6	4.4	4.7	4.7	4.8	4.9	5.0	5.0	4.6	4.1	2.9	2.8	4.3
	Min/Max	2.3/4.9	3.1/5.2	3.4/6.1	3.9/5.6	4.2/5.5	4.3/5.7	4.2/5.4	4.3/5.9	3.4/5.8	3.0/4.9	1.8/3.7	2.1/3.8	3.8/4.7
90	Average	3.5	4.1	3.9	3.2	2.9	2.8	2.9	3.2	3.4	3.4	2.6	2.7	3.2
	Min/Max	2.2/4.8	2.9/4.9	2.7/5.1	2.7/3.8	2.5/3.2	2.5/3.1	2.5/3.1	2.8/3.7	2.5/4.3	2.5/4.1	1.5/3.5	1.9/3.7	2.8/3.5

Solar Radiation for 1-Axis Tracking Flat-Plate Collectors with a North-South Axis (kWh/m²/day), Uncertainty ±9%

Axis Tilt (°)		Jan	Feb	Mar	Apr	May	June	July	Aug	Sept	Oct	Nov	Dec	Year
0	Average	2.7	3.9	5.1	6.2	7.5	8.2	8.2	7.2	5.5	3.9	2.3	2.0	5.2
	Min/Max	1.8/3.5	2.8/4.5	3.6/6.5	5.1/7.7	6.0/8.9	6.8/10.0	6.7/9.3	6.1/8.8	4.0/6.9	3.0/4.7	1.5/2.9	1.5/2.6	4.6/5.9
Latitude -15	Average	3.5	4.8	5.8	6.6	7.6	8.2	8.3	7.6	6.1	4.7	3.0	2.7	5.7
	Min/Max	2.3/4.7	3.3/5.7	4.0/7.5	5.4/8.3	6.2/9.1	6.8/10.1	6.7/9.5	6.3/9.3	4.3/7.8	3.5/5.7	1.8/3.8	2.0/3.6	5.0/6.5
Latitude	Average	3.9	5.1	5.9	6.5	7.4	7.9	8.0	7.4	6.1	4.9	3.2	3.0	5.8
	Min/Max	2.4/5.3	3.4/6.1	4.0/7.7	5.3/8.3	6.0/8.8	6.5/9.7	6.4/9.1	6.2/9.2	4.3/7.9	3.6/6.0	1.9/4.1	2.2/4.0	5.0/6.5
Latitude +15	Average	4.1	5.2	5.9	6.2	7.0	7.3	7.5	7.1	5.9	4.9	3.3	3.1	5.6
	Min/Max	2.5/5.6	3.5/6.2	3.9/7.7	5.1/7.9	5.6/8.3	6.1/9.1	6.0/8.6	5.9/8.8	4.1/7.8	3.5/6.0	1.9/4.3	2.3/4.2	4.8/6.4

Solar Radiation for 2-Axis Tracking Flat-Plate Collectors (kWh/m²/day), Uncertainty ±9%

Tracker		Jan	Feb	Mar	Apr	May	June	July	Aug	Sept	Oct	Nov	Dec	Year
2-Axis	Average	4.1	5.2	6.0	6.6	7.7	8.4	8.5	7.6	6.1	4.9	3.3	3.2	6.0
	Min/Max	2.5/5.6	3.5/6.2	4.1/7.7	5.4/8.4	6.3/9.2	7.0/10.4	6.8/9.6	6.4/9.4	4.3/7.9	3.6/6.1	1.9/4.3	2.3/4.3	5.2/6.7

Direct Beam Solar Radiation for Concentrating Collectors (kWh/m²/day), Uncertainty ±8%

Tracker		Jan	Feb	Mar	Apr	May	June	July	Aug	Sept	Oct	Nov	Dec	Year
1-Axis, E-W	Average	2.3	2.7	2.8	2.9	3.5	3.9	4.0	3.5	2.9	2.6	1.8	1.8	2.9
Horiz Axis	Min/Max	1.1/3.5	1.1/3.9	1.5/3.9	2.1/4.4	2.4/4.7	2.7/5.4	2.7/4.8	2.6/5.0	1.5/4.4	1.6/3.5	0.7/2.5	1.1/2.7	2.2/3.5
1-Axis, N-S	Average	1.5	2.3	3.0	3.7	4.6	5.1	5.2	4.6	3.4	2.4	1.3	1.1	3.2
Horiz Axis	Min/Max	0.7/2.3	0.9/3.3	1.5/4.3	2.6/5.6	3.1/6.2	3.5/7.1	3.6/6.5	3.4/6.5	1.8/5.2	1.5/3.3	0.5/1.8	0.7/1.7	2.5/3.9
1-Axis, N-S	Average	2.4	3.2	3.6	4.0	4.6	4.8	5.1	4.8	3.9	3.2	2.0	1.8	3.6
Tilt=Latitude	Min/Max	1.2/3.8	1.3/4.6	1.9/5.2	2.8/6.1	3.1/6.2	3.4/6.8	3.5/6.3	3.6/6.8	2.0/6.0	1.9/4.3	0.7/2.8	1.1/2.8	2.8/4.4
2-Axis	Average	2.6	3.3	3.7	4.1	4.9	5.3	5.4	4.9	3.9	3.3	2.1	2.0	3.8
	Min/Max	1.3/4.1	1.4/4.8	1.9/5.2	2.9/6.2	3.3/6.5	3.7/7.4	3.7/6.7	3.7/7.0	2.0/6.0	2.0/4.4	0.8/3.0	1.2/3.0	2.9/4.6

Average Climatic Conditions

Element	Jan	Feb	Mar	Apr	May	June	July	Aug	Sept	Oct	Nov	Dec	Year
Temperature (°C)	-9.8	-6.7	0.4	8.7	15.3	20.3	23.1	21.6	16.6	10.1	2.0	-6.5	7.9
Daily Minimum Temp	-14.8	-12.2	-4.7	2.8	8.9	13.9	16.8	15.5	10.7	4.6	-2.4	-10.8	2.4
Daily Maximum Temp	-4.7	-1.3	5.6	14.7	21.6	26.6	29.2	27.6	22.4	15.7	6.3	-2.2	13.4
Record Minimum Temp	-38.3	-37.8	-33.3	-13.9	-3.3	2.8	0.6	4.4	-2.2	-10.0	-22.8	-34.4	-38.3
Record Maximum Temp	13.9	17.8	28.9	33.9	34.4	38.9	40.0	40.6	37.8	33.9	23.9	17.8	40.6
HDD, Base 18.3°C	872	702	554	288	122	20	4	9	73	258	490	770	4162
CDD, Base 18.3°C	0	0	0	0	28	78	150	109	19	0	0	0	384
Relative Humidity (%)	73	72	71	64	65	70	72	75	77	71	75	77	72
Wind Speed (m/s)	3.9	3.7	4.1	4.4	4.0	3.7	3.4	3.3	3.6	3.9	4.1	3.8	3.8

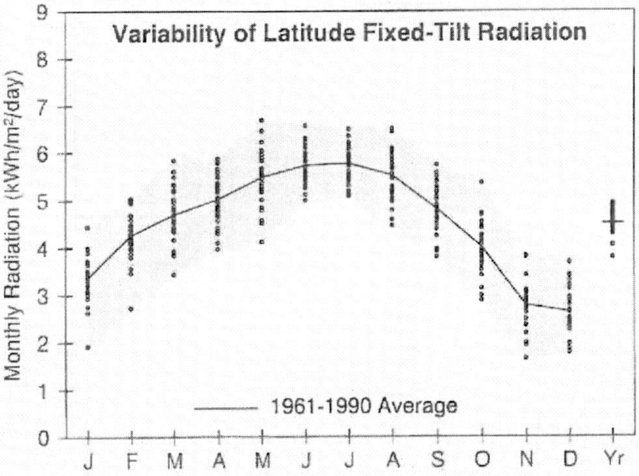

Madison, WI

WBAN NO. 14837

LATITUDE: 43.13° N
LONGITUDE: 89.33° W
ELEVATION: 262 meters
MEAN PRESSURE: 985 millibars

STATION TYPE: Primary

Solar Radiation for Flat-Plate Collectors Facing South at a Fixed Tilt (kWh/m²/day), Uncertainty ±9%

Tilt (°)		Jan	Feb	Mar	Apr	May	June	July	Aug	Sept	Oct	Nov	Dec	Year
0	Average	1.9	2.8	3.7	4.7	5.8	6.4	6.2	5.4	4.1	2.8	1.7	1.5	3.9
	Min/Max	1.3/2.2	2.2/3.2	3.0/4.3	3.9/5.4	4.3/7.0	5.6/7.3	5.6/7.0	4.4/6.2	3.4/4.7	2.2/3.5	1.2/2.2	1.0/1.9	3.5/4.3
Latitude -15	Average	3.0	3.9	4.5	5.1	5.8	6.2	6.2	5.7	4.8	3.8	2.5	2.3	4.5
	Min/Max	1.8/3.8	2.6/4.5	3.4/5.6	4.1/6.0	4.4/7.1	5.4/7.2	5.5/7.0	4.7/6.8	3.8/5.7	2.8/5.0	1.6/3.4	1.6/3.2	3.9/4.9
Latitude	Average	3.4	4.3	4.7	5.0	5.5	5.7	5.8	5.5	4.8	4.0	2.8	2.6	4.5
	Min/Max	1.9/4.4	2.7/5.0	3.4/5.8	4.0/5.9	4.1/6.7	5.0/6.6	5.1/6.5	4.5/6.5	3.8/5.7	2.9/5.4	1.7/3.8	1.8/3.7	3.8/4.9
Latitude +15	Average	3.6	4.4	4.6	4.6	4.8	4.9	5.0	5.0	4.6	4.0	2.9	2.8	4.3
	Min/Max	2.0/4.8	2.7/5.3	3.3/5.8	3.6/5.5	3.7/5.9	4.3/5.7	4.4/5.7	4.0/5.9	3.6/5.5	2.8/5.4	1.7/4.1	1.8/4.0	3.5/4.7
90	Average	3.5	4.0	3.7	3.2	2.9	2.8	2.9	3.2	3.4	3.3	2.6	2.7	3.2
	Min/Max	1.8/4.7	2.4/4.9	2.8/4.7	2.5/3.7	2.3/3.4	2.5/3.1	2.6/3.2	2.6/3.7	2.6/4.0	2.3/4.6	1.4/3.7	1.7/4.0	2.6/3.5

Solar Radiation for 1-Axis Tracking Flat-Plate Collectors with a North-South Axis (kWh/m²/day), Uncertainty ±9%

Axis Tilt (°)		Jan	Feb	Mar	Apr	May	June	July	Aug	Sept	Oct	Nov	Dec	Year
0	Average	2.7	3.9	5.0	6.1	7.4	8.1	7.9	7.0	5.4	3.9	2.3	2.0	5.1
	Min/Max	1.5/3.5	2.5/4.6	3.5/6.3	4.6/7.5	5.0/9.2	6.8/9.4	6.8/9.3	5.2/8.5	4.2/6.6	2.8/5.3	1.4/3.2	1.4/2.8	4.2/5.7
Latitude -15	Average	3.5	4.8	5.6	6.5	7.5	8.1	8.0	7.3	6.0	4.6	2.9	2.7	5.6
	Min/Max	1.9/4.7	2.8/5.8	3.8/7.3	4.8/8.0	5.1/9.4	6.8/9.4	6.8/9.4	5.4/9.0	4.6/7.4	3.2/6.4	1.7/4.2	1.8/3.8	4.5/6.3
Latitude	Average	3.9	5.0	5.8	6.4	7.3	7.8	7.7	7.1	6.0	4.8	3.2	3.0	5.7
	Min/Max	2.1/5.2	3.0/6.2	3.9/7.5	4.7/8.0	4.9/9.2	6.5/9.0	6.6/9.1	5.3/8.8	4.6/7.4	3.3/6.7	1.8/4.5	1.9/4.2	4.5/6.3
Latitude +15	Average	4.0	5.2	5.7	6.1	6.8	7.2	7.2	6.8	5.8	4.8	3.2	3.1	5.5
	Min/Max	2.1/5.5	3.0/6.3	3.8/7.4	4.4/7.7	4.6/8.6	6.1/8.4	6.1/8.5	4.9/8.4	4.4/7.2	3.3/6.7	1.8/4.7	1.9/4.5	4.3/6.2

Solar Radiation for 2-Axis Tracking Flat-Plate Collectors (kWh/m²/day), Uncertainty ±9%

Tracker		Jan	Feb	Mar	Apr	May	June	July	Aug	Sept	Oct	Nov	Dec	Year
2-Axis	Average	4.1	5.2	5.8	6.5	7.6	8.3	8.2	7.3	6.0	4.8	3.3	3.2	5.9
	Min/Max	2.1/5.6	3.0/6.3	3.9/7.5	4.8/8.1	5.1/9.6	6.9/9.6	7.0/9.6	5.4/9.0	4.6/7.4	3.3/6.8	1.8/4.7	2.0/4.6	4.6/6.5

Direct Beam Solar Radiation for Concentrating Collectors (kWh/m²/day), Uncertainty ±8%

Tracker		Jan	Feb	Mar	Apr	May	June	July	Aug	Sept	Oct	Nov	Dec	Year
1-Axis, E-W	Average	2.3	2.6	2.6	2.8	3.3	3.7	3.6	3.2	2.7	2.5	1.7	1.7	2.7
Horiz Axis	Min/Max	0.8/3.5	0.8/3.9	1.2/4.1	1.6/4.0	1.7/4.8	2.5/4.8	2.6/4.8	1.8/4.5	1.7/3.8	1.4/3.9	0.7/2.8	0.7/2.6	1.8/3.3
1-Axis, N-S	Average	1.5	2.2	2.8	3.5	4.4	4.8	4.7	4.0	3.2	2.3	1.3	1.1	3.0
Horiz Axis	Min/Max	0.5/2.3	0.6/3.4	1.2/4.4	1.9/5.2	2.1/6.4	3.2/6.3	3.3/6.3	2.0/5.8	2.0/4.5	1.4/3.8	0.5/2.1	0.4/1.6	1.9/3.6
1-Axis, N-S	Average	2.4	3.1	3.4	3.8	4.3	4.6	4.5	4.2	3.7	3.0	1.9	1.8	3.4
Tilt=Latitude	Min/Max	0.8/3.8	0.9/4.7	1.5/5.4	2.1/5.6	2.1/6.3	3.1/6.0	3.2/6.1	2.1/6.0	2.3/5.2	1.8/5.0	0.8/3.2	0.7/2.7	2.1/4.1
2-Axis	Average	2.6	3.1	3.4	3.9	4.6	5.0	4.9	4.3	3.7	3.1	2.0	2.0	3.5
	Min/Max	0.9/4.1	0.9/4.9	1.5/5.4	2.1/5.6	2.3/6.6	3.3/6.5	3.4/6.6	2.2/6.2	2.3/5.2	1.8/5.0	0.8/3.3	0.8/3.0	2.2/4.3

Average Climatic Conditions

Element	Jan	Feb	Mar	Apr	May	June	July	Aug	Sept	Oct	Nov	Dec	Year
Temperature (°C)	-8.9	-6.3	0.2	7.4	13.6	19.0	21.7	20.2	15.4	9.4	1.9	-5.7	7.3
Daily Minimum Temp	-13.8	-11.6	-5.0	1.2	6.8	12.3	15.3	13.8	9.0	3.2	-2.9	-10.3	1.5
Daily Maximum Temp	-4.0	-1.1	5.3	13.7	20.5	25.7	28.0	26.4	21.9	15.5	6.7	-1.2	13.1
Record Minimum Temp	-38.3	-33.3	-33.9	-17.8	-7.2	-0.6	2.2	1.7	-3.9	-10.6	-23.9	-31.7	-38.3
Record Maximum Temp	13.3	16.1	27.8	34.4	33.9	38.3	40.0	38.9	37.2	32.2	24.4	16.7	40.0
HDD, Base 18.3°C	844	691	563	327	163	38	7	21	93	277	493	746	4263
CDD, Base 18.3°C	0	0	0	0	17	58	110	78	7	0	0	0	269
Relative Humidity (%)	74	73	71	66	66	68	71	74	77	73	77	78	73
Wind Speed (m/s)	4.6	4.5	4.9	5.0	4.3	4.0	3.6	3.4	3.8	4.2	4.6	4.5	4.3

Milwaukee, WI

WBAN NO. 14839

LATITUDE: 42.95° N
LONGITUDE: 87.90° W
ELEVATION: 211 meters
MEAN PRESSURE: 991 millibars

STATION TYPE: Secondary

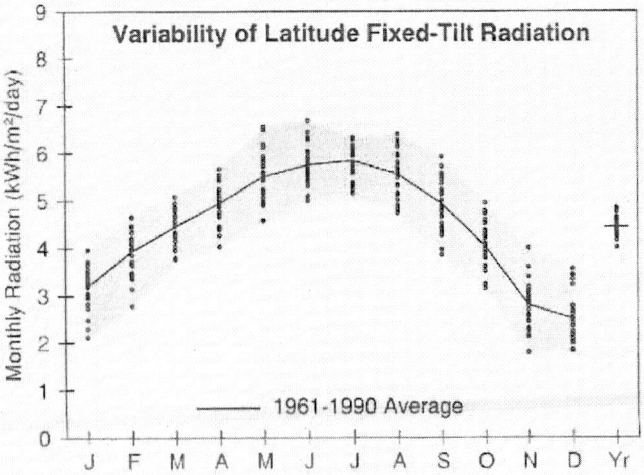

Solar Radiation for Flat-Plate Collectors Facing South at a Fixed Tilt (kWh/m²/day), Uncertainty ±9%

Tilt (°)		Jan	Feb	Mar	Apr	May	June	July	Aug	Sept	Oct	Nov	Dec	Year
0	Average	1.8	2.6	3.5	4.6	5.8	6.4	6.3	5.4	4.1	2.9	1.8	1.4	3.9
	Min/Max	1.4/2.1	2.2/3.0	3.2/3.9	3.9/5.2	4.9/6.8	5.6/7.4	5.6/6.8	4.7/6.1	3.5/4.8	2.4/3.3	1.4/2.1	1.2/1.7	3.7/4.2
Latitude -15	Average	2.8	3.6	4.3	5.1	5.9	6.2	6.3	5.8	4.9	3.8	2.5	2.2	4.5
	Min/Max	1.9/3.4	2.7/4.2	3.7/4.9	4.2/5.8	4.9/7.0	5.4/7.3	5.6/6.8	4.9/6.6	3.9/5.8	3.0/4.6	1.7/3.5	1.7/3.1	4.1/4.8
Latitude	Average	3.2	3.9	4.4	4.9	5.5	5.7	5.8	5.6	4.9	4.0	2.8	2.5	4.5
	Min/Max	2.1/4.0	2.8/4.7	3.8/5.1	4.0/5.7	4.6/6.6	5.0/6.7	5.2/6.3	4.7/6.4	3.8/5.9	3.2/4.9	1.8/4.0	1.8/3.6	4.0/4.8
Latitude +15	Average	3.4	4.1	4.4	4.6	4.9	5.0	5.1	5.0	4.7	4.0	2.9	2.7	4.2
	Min/Max	2.2/4.3	2.8/4.9	3.7/5.0	3.7/5.3	4.0/5.8	4.3/5.7	4.5/5.5	4.3/5.8	3.6/5.7	3.1/5.0	1.8/4.2	1.9/3.8	3.8/4.6
90	Average	3.3	3.7	3.5	3.1	2.9	2.7	2.9	3.2	3.4	3.3	2.6	2.6	3.1
	Min/Max	2.0/4.1	2.4/4.5	2.8/4.0	2.5/3.5	2.5/3.3	2.5/3.0	2.6/3.0	2.7/3.7	2.6/4.2	2.5/4.2	1.5/3.9	1.8/3.7	2.7/3.4

Solar Radiation for 1-Axis Tracking Flat-Plate Collectors with a North-South Axis (kWh/m²/day), Uncertainty ±9%

Axis Tilt (°)		Jan	Feb	Mar	Apr	May	June	July	Aug	Sept	Oct	Nov	Dec	Year
0	Average	2.6	3.6	4.7	6.0	7.5	8.2	8.3	7.2	5.6	3.9	2.3	2.0	5.2
	Min/Max	1.7/3.1	2.6/4.3	3.9/5.5	4.8/6.9	5.9/9.2	6.8/10.0	7.3/9.1	6.0/8.6	4.3/6.8	3.2/4.8	1.6/3.3	1.4/2.6	4.7/5.7
Latitude -15	Average	3.3	4.4	5.3	6.4	7.6	8.3	8.4	7.6	6.2	4.7	2.9	2.6	5.6
	Min/Max	2.1/4.1	3.0/5.3	4.3/6.3	5.1/7.4	6.0/9.5	6.8/10.1	7.3/9.3	6.3/9.0	4.7/7.6	3.6/5.8	1.8/4.3	1.8/3.6	5.0/6.3
Latitude	Average	3.7	4.6	5.4	6.3	7.4	7.9	8.1	7.4	6.2	4.9	3.2	2.8	5.7
	Min/Max	2.3/4.6	3.1/5.6	4.4/6.4	5.0/7.3	5.8/9.2	6.5/9.7	7.1/8.9	6.1/8.9	4.7/7.7	3.7/6.1	1.9/4.7	2.0/4.1	5.0/6.3
Latitude +15	Average	3.8	4.7	5.4	6.0	6.9	7.4	7.6	7.1	6.1	4.9	3.2	3.0	5.5
	Min/Max	2.4/4.8	3.1/5.7	4.3/6.3	4.8/7.0	5.4/8.7	6.1/9.1	6.6/8.4	5.8/8.5	4.5/7.5	3.7/6.1	1.9/4.9	2.1/4.3	4.8/6.1

Solar Radiation for 2-Axis Tracking Flat-Plate Collectors (kWh/m²/day), Uncertainty ±9%

Tracker		Jan	Feb	Mar	Apr	May	June	July	Aug	Sept	Oct	Nov	Dec	Year
2-Axis	Average	3.9	4.7	5.5	6.4	7.7	8.5	8.5	7.6	6.3	4.9	3.3	3.0	5.9
	Min/Max	2.4/4.9	3.1/5.8	4.4/6.4	5.1/7.4	6.1/9.6	7.0/10.4	7.5/9.4	6.3/9.1	4.7/7.7	3.7/6.1	1.9/4.9	2.1/4.4	5.2/6.5

Direct Beam Solar Radiation for Concentrating Collectors (kWh/m²/day), Uncertainty ±8%

Tracker		Jan	Feb	Mar	Apr	May	June	July	Aug	Sept	Oct	Nov	Dec	Year
1-Axis, E-W Horiz Axis	Average	2.1	2.3	2.4	2.8	3.5	4.0	4.0	3.5	3.0	2.5	1.7	1.6	2.8
	Min/Max	1.1/2.9	1.0/3.4	1.5/3.2	1.9/3.6	2.2/4.9	2.8/5.5	3.1/4.9	2.5/4.5	1.8/4.1	1.5/3.5	0.7/3.1	0.9/2.9	2.2/3.3
1-Axis, N-S Horiz Axis	Average	1.4	2.0	2.6	3.5	4.6	5.2	5.3	4.6	3.5	2.4	1.3	1.0	3.1
	Min/Max	0.7/1.9	0.9/3.0	1.6/3.5	2.5/4.4	2.9/6.4	3.7/7.2	4.1/6.3	3.3/5.9	2.2/4.8	1.5/3.3	0.5/2.2	0.6/1.8	2.5/3.7
1-Axis, N-S Tilt=Latitude	Average	2.3	2.8	3.2	3.8	4.5	4.9	5.1	4.7	4.0	3.1	1.9	1.7	3.5
	Min/Max	1.1/3.1	1.2/4.1	2.0/4.3	2.6/4.8	2.9/6.3	3.5/6.9	3.9/6.1	3.4/6.1	2.5/5.5	2.0/4.3	0.8/3.4	0.9/3.0	2.8/4.2
2-Axis	Average	2.4	2.8	3.2	3.8	4.8	5.3	5.5	4.9	4.0	3.2	2.0	1.8	3.7
	Min/Max	1.2/3.3	1.3/4.2	2.0/4.3	2.7/4.8	3.0/6.7	3.8/7.5	4.2/6.6	3.5/6.3	2.5/5.6	2.0/4.4	0.8/3.6	1.0/3.3	2.9/4.4

Average Climatic Conditions

Element	Jan	Feb	Mar	Apr	May	June	July	Aug	Sept	Oct	Nov	Dec	Year
Temperature (°C)	-7.3	-5.0	0.7	6.9	12.6	18.3	21.6	20.7	16.5	10.2	3.2	-4.2	7.8
Daily Minimum Temp	-11.3	-8.9	-3.2	2.1	7.1	12.8	16.7	16.0	11.6	5.4	-0.7	-8.1	3.3
Daily Maximum Temp	-3.3	-1.1	4.7	11.6	17.9	23.8	26.6	25.4	21.4	14.8	7.1	-0.4	12.4
Record Minimum Temp	-32.2	-28.3	-23.3	-11.1	-6.1	0.6	4.4	6.7	-2.2	-7.8	-20.6	-28.9	-32.2
Record Maximum Temp	16.7	18.3	27.8	32.8	33.9	38.3	38.3	39.4	36.7	31.7	25.0	17.2	39.4
HDD, Base 18.3°C	794	653	546	343	188	46	8	15	68	253	455	699	4069
CDD, Base 18.3°C	0	0	0	0	9	46	109	89	13	0	0	0	266
Relative Humidity (%)	72	72	71	69	68	70	72	75	75	72	75	76	72
Wind Speed (m/s)	5.5	5.3	5.5	5.5	4.9	4.6	4.3	4.2	4.5	5.0	5.3	5.3	5.0

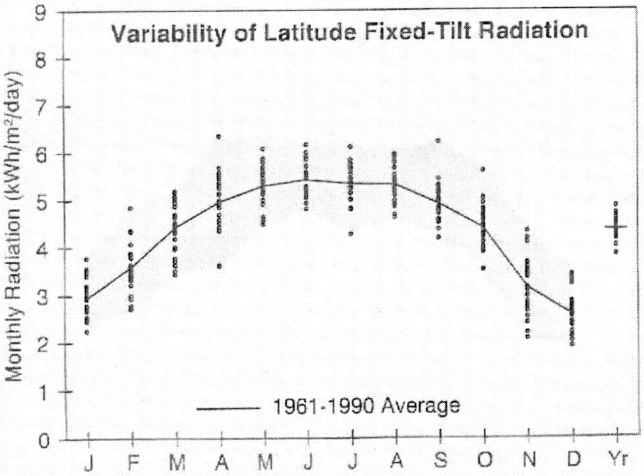

Charleston, WV

WBAN NO. 13866

LATITUDE: 38.37° N
LONGITUDE: 81.60° W
ELEVATION: 290 meters
MEAN PRESSURE: 983 millibars

STATION TYPE: Secondary

Solar Radiation for Flat-Plate Collectors Facing South at a Fixed Tilt (kWh/m²/day), Uncertainty ±9%

Tilt (°)		Jan	Feb	Mar	Apr	May	June	July	Aug	Sept	Oct	Nov	Dec	Year
0	Average	2.0	2.7	3.7	4.8	5.6	6.0	5.8	5.3	4.3	3.3	2.1	1.7	3.9
	Min/Max	1.7/2.3	2.2/3.3	3.1/4.2	3.7/5.9	4.8/6.4	5.3/6.9	4.7/6.6	4.7/5.8	3.8/5.2	2.8/4.0	1.7/2.6	1.5/2.0	3.5/4.3
Latitude -15	Average	2.7	3.3	4.3	5.1	5.6	5.9	5.7	5.5	4.9	4.1	2.9	2.3	4.4
	Min/Max	2.1/3.3	2.6/4.4	3.4/5.0	3.8/6.4	4.8/6.5	5.2/6.8	4.6/6.6	4.8/6.2	4.2/6.1	3.4/5.2	2.0/3.8	1.8/3.0	3.9/4.8
Latitude	Average	2.9	3.6	4.4	5.0	5.3	5.4	5.3	5.3	4.9	4.4	3.1	2.6	4.4
	Min/Max	2.3/3.8	2.7/4.8	3.4/5.2	3.6/6.3	4.5/6.1	4.8/6.2	4.3/6.1	4.6/6.0	4.2/6.2	3.5/5.6	2.1/4.3	1.9/3.4	3.8/4.9
Latitude +15	Average	3.1	3.7	4.3	4.6	4.7	4.7	4.7	4.9	4.7	4.4	3.3	2.7	4.1
	Min/Max	2.3/4.0	2.7/5.0	3.3/5.1	3.3/5.9	4.0/5.4	4.2/5.3	3.8/5.4	4.2/5.4	3.9/6.0	3.5/5.7	2.1/4.6	1.9/3.7	3.6/4.6
90	Average	2.8	3.1	3.2	2.9	2.5	2.4	2.4	2.8	3.2	3.5	2.8	2.5	2.8
	Min/Max	2.0/3.7	2.2/4.3	2.4/3.7	2.1/3.6	2.2/2.8	2.2/2.5	2.1/2.7	2.5/3.1	2.6/4.1	2.7/4.5	1.7/4.1	1.7/3.4	2.5/3.2

Solar Radiation for 1-Axis Tracking Flat-Plate Collectors with a North-South Axis (kWh/m²/day), Uncertainty ±9%

Axis Tilt (°)		Jan	Feb	Mar	Apr	May	June	July	Aug	Sept	Oct	Nov	Dec	Year
0	Average	2.6	3.5	4.8	6.1	7.1	7.5	7.2	6.7	5.6	4.4	2.8	2.2	5.0
	Min/Max	2.0/3.3	2.6/4.7	3.7/5.8	4.2/8.2	5.7/8.5	6.3/9.3	5.5/8.4	5.6/7.7	4.6/7.4	3.4/5.7	1.9/3.8	1.7/2.8	4.3/5.6
Latitude -15	Average	3.1	4.0	5.3	6.4	7.1	7.5	7.2	6.9	6.0	5.1	3.4	2.7	5.4
	Min/Max	2.3/4.1	2.9/5.6	3.9/6.4	4.2/8.7	5.8/8.6	6.2/9.2	5.4/8.5	5.7/8.0	4.9/8.1	3.9/6.6	2.1/4.7	1.9/3.6	4.6/6.1
Latitude	Average	3.3	4.2	5.3	6.3	6.9	7.2	6.9	6.7	6.1	5.3	3.6	2.9	5.4
	Min/Max	2.4/4.4	3.0/5.9	3.9/6.6	4.1/8.6	5.6/8.3	6.0/8.9	5.2/8.2	5.6/7.8	4.9/8.2	4.0/6.9	2.2/5.1	2.0/3.9	4.6/6.1
Latitude +15	Average	3.5	4.2	5.3	6.0	6.5	6.7	6.5	6.4	5.9	5.3	3.7	3.0	5.2
	Min/Max	2.5/4.6	3.0/6.0	3.8/6.5	3.9/8.3	5.2/7.8	5.5/8.3	4.9/7.6	5.3/7.4	4.7/8.0	4.0/7.0	2.2/5.3	2.1/4.2	4.5/6.0

Solar Radiation for 2-Axis Tracking Flat-Plate Collectors (kWh/m²/day), Uncertainty ±9%

Tracker		Jan	Feb	Mar	Apr	May	June	July	Aug	Sept	Oct	Nov	Dec	Year
2-Axis	Average	3.5	4.3	5.4	6.4	7.2	7.6	7.3	6.9	6.1	5.3	3.7	3.0	5.6
	Min/Max	2.5/4.7	3.0/6.0	4.0/6.6	4.3/8.7	5.8/8.7	6.4/9.4	5.6/8.6	5.7/8.0	4.9/8.2	4.1/7.0	2.2/5.4	2.1/4.2	4.7/6.3

Direct Beam Solar Radiation for Concentrating Collectors (kWh/m²/day), Uncertainty ±8%

Tracker		Jan	Feb	Mar	Apr	May	June	July	Aug	Sept	Oct	Nov	Dec	Year
1-Axis, E-W Horiz Axis	Average	1.7	1.9	2.4	2.7	3.1	3.3	3.1	3.0	2.7	2.7	1.9	1.6	2.5
	Min/Max	0.9/2.8	1.0/3.4	1.2/3.3	1.0/4.5	2.0/4.3	2.5/4.7	1.8/4.2	2.0/3.8	1.8/4.3	1.8/4.1	0.8/3.3	0.7/2.7	1.9/3.1
1-Axis, N-S Horiz Axis	Average	1.3	1.8	2.7	3.5	4.1	4.3	4.0	3.8	3.3	2.7	1.5	1.1	2.9
	Min/Max	0.6/2.1	0.9/3.1	1.4/4.0	1.3/5.9	2.6/5.7	3.2/6.3	2.4/5.4	2.5/5.0	2.2/5.3	1.8/4.0	0.6/2.6	0.5/1.9	2.2/3.5
1-Axis, N-S Tilt=Latitude	Average	1.9	2.3	3.1	3.7	4.0	4.1	3.8	3.9	3.7	3.4	2.2	1.7	3.2
	Min/Max	0.9/3.1	1.2/4.1	1.6/4.6	1.4/6.2	2.5/5.5	3.0/5.9	2.3/5.1	2.6/5.1	2.4/5.9	2.2/5.1	0.9/3.8	0.8/2.8	2.4/3.9
2-Axis	Average	2.0	2.4	3.1	3.8	4.2	4.4	4.1	4.0	3.7	3.5	2.3	1.8	3.3
	Min/Max	1.0/3.3	1.2/4.2	1.6/4.6	1.4/6.3	2.7/5.8	3.2/6.5	2.4/5.5	2.7/5.3	2.4/5.9	2.3/5.1	0.9/4.0	0.8/3.1	2.5/4.1

Average Climatic Conditions

Element	Jan	Feb	Mar	Apr	May	June	July	Aug	Sept	Oct	Nov	Dec	Year
Temperature (°C)	0.1	1.9	7.7	12.7	17.5	21.9	23.9	23.3	19.8	13.4	8.2	2.8	12.8
Daily Minimum Temp	-5.0	-3.5	1.7	6.0	10.8	15.4	18.0	17.4	13.6	6.8	2.4	-2.2	6.8
Daily Maximum Temp	5.1	7.4	13.7	19.3	24.2	28.4	29.8	29.1	26.0	20.1	14.1	7.8	18.8
Record Minimum Temp	-26.1	-21.1	-17.8	-7.2	-3.3	0.6	7.8	5.0	1.1	-8.3	-14.4	-24.4	-26.1
Record Maximum Temp	26.1	25.6	31.7	34.4	33.9	36.7	40.0	38.3	38.9	33.3	29.4	26.7	40.0
HDD, Base 18.3°C	567	459	329	173	72	6	0	0	24	166	303	482	2581
CDD, Base 18.3°C	0	0	0	3	46	112	174	153	69	14	0	0	573
Relative Humidity (%)	69	67	61	59	67	72	76	77	77	72	69	70	70
Wind Speed (m/s)	3.2	3.1	3.4	3.3	2.6	2.3	2.2	2.0	2.1	2.2	2.7	3.0	2.7

Elkins, WV

WBAN NO. 13729

LATITUDE: 38.88° N
LONGITUDE: 79.85° W
ELEVATION: 594 meters
MEAN PRESSURE: 947 millibars

STATION TYPE: Secondary

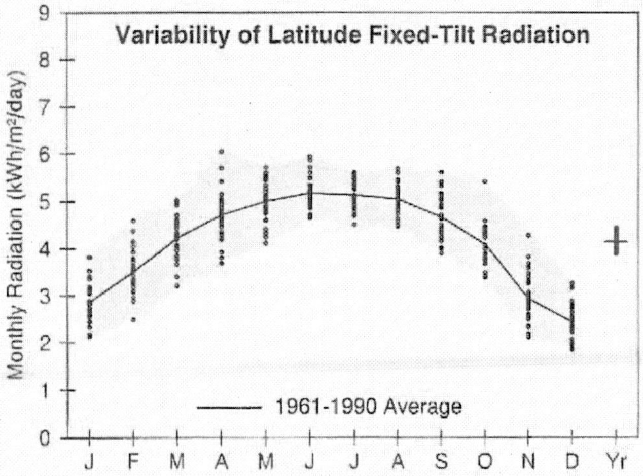

Solar Radiation for Flat-Plate Collectors Facing South at a Fixed Tilt (kWh/m²/day), Uncertainty ±9%

Tilt (°)		Jan	Feb	Mar	Apr	May	June	July	Aug	Sept	Oct	Nov	Dec	Year
0	Average	1.9	2.6	3.6	4.5	5.3	5.7	5.5	5.0	4.1	3.1	2.0	1.6	3.8
	Min/Max	1.7/2.2	2.1/3.1	3.0/4.1	3.8/5.7	4.4/6.0	5.2/6.6	4.9/6.0	4.5/5.6	3.6/4.8	2.7/3.8	1.6/2.6	1.4/1.9	3.6/4.0
Latitude -15	Average	2.6	3.3	4.1	4.8	5.3	5.6	5.5	5.2	4.6	3.9	2.7	2.2	4.2
	Min/Max	2.0/3.3	2.4/4.2	3.2/4.8	3.8/6.2	4.4/6.1	5.0/6.5	4.8/6.0	4.7/5.9	4.0/5.5	3.2/5.0	2.0/3.8	1.8/2.8	3.9/4.4
Latitude	Average	2.9	3.5	4.2	4.7	5.0	5.2	5.1	5.1	4.7	4.1	2.9	2.4	4.2
	Min/Max	2.1/3.8	2.5/4.6	3.2/5.0	3.7/6.1	4.1/5.7	4.7/5.9	4.5/5.6	4.5/5.7	3.9/5.6	3.4/5.4	2.1/4.3	1.9/3.3	3.9/4.4
Latitude +15	Average	3.0	3.6	4.1	4.4	4.4	4.5	4.5	4.6	4.5	4.1	3.0	2.6	3.9
	Min/Max	2.2/4.1	2.5/4.8	3.1/4.9	3.4/5.6	3.7/5.1	4.1/5.1	4.0/4.9	4.1/5.2	3.7/5.4	3.4/5.5	2.1/4.5	1.9/3.5	3.7/4.2
90	Average	2.7	3.1	3.1	2.8	2.4	2.3	2.4	2.7	3.1	3.2	2.6	2.4	2.7
	Min/Max	2.0/3.8	2.1/4.3	2.3/3.6	2.2/3.5	2.0/2.8	2.2/2.5	2.2/2.5	2.4/3.0	2.5/3.7	2.7/4.4	1.8/4.0	1.6/3.3	2.5/3.0

Solar Radiation for 1-Axis Tracking Flat-Plate Collectors with a North-South Axis (kWh/m²/day), Uncertainty ±9%

Axis Tilt (°)		Jan	Feb	Mar	Apr	May	June	July	Aug	Sept	Oct	Nov	Dec	Year
0	Average	2.5	3.4	4.5	5.7	6.6	6.9	6.6	6.1	5.1	4.0	2.6	2.1	4.7
	Min/Max	1.9/3.3	2.4/4.4	3.3/5.6	4.2/7.7	5.2/7.7	6.0/8.6	5.7/7.5	5.4/7.2	4.2/6.3	3.3/5.4	1.9/3.7	1.6/2.7	4.3/5.1
Latitude -15	Average	3.0	3.9	5.0	6.0	6.6	6.9	6.7	6.4	5.6	4.6	3.1	2.5	5.0
	Min/Max	2.2/4.1	2.7/5.2	3.5/6.2	4.3/8.2	5.2/7.8	5.9/8.5	5.7/7.5	5.5/7.5	4.5/6.9	3.8/6.3	2.2/4.6	1.9/3.4	4.7/5.4
Latitude	Average	3.2	4.0	5.0	5.9	6.4	6.6	6.4	6.2	5.6	4.8	3.3	2.7	5.0
	Min/Max	2.3/4.5	2.7/5.5	3.5/6.3	4.2/8.1	5.0/7.6	5.7/8.2	5.4/7.3	5.3/7.3	4.4/7.0	3.9/6.6	2.3/5.0	2.0/3.8	4.7/5.5
Latitude +15	Average	3.3	4.1	5.0	5.6	6.0	6.1	6.0	5.9	5.4	4.8	3.4	2.8	4.9
	Min/Max	2.3/4.7	2.7/5.7	3.4/6.2	4.0/7.8	4.6/7.1	5.3/7.6	5.1/6.8	5.0/7.0	4.3/6.8	3.9/6.7	2.3/5.2	2.0/4.0	4.5/5.3

Solar Radiation for 2-Axis Tracking Flat-Plate Collectors (kWh/m²/day), Uncertainty ±9%

Tracker		Jan	Feb	Mar	Apr	May	June	July	Aug	Sept	Oct	Nov	Dec	Year
2-Axis	Average	3.4	4.1	5.1	6.0	6.7	7.1	6.8	6.4	5.6	4.9	3.4	2.9	5.2
	Min/Max	2.4/4.8	2.7/5.7	3.5/6.3	4.3/8.2	5.2/7.9	6.1/8.7	5.8/7.7	5.5/7.5	4.5/7.0	4.0/6.7	2.3/5.2	2.0/4.0	4.8/5.6

Direct Beam Solar Radiation for Concentrating Collectors (kWh/m²/day), Uncertainty ±8%

Tracker		Jan	Feb	Mar	Apr	May	June	July	Aug	Sept	Oct	Nov	Dec	Year
1-Axis, E-W	Average	1.6	1.7	2.1	2.4	2.7	2.9	2.8	2.7	2.5	2.4	1.7	1.4	2.3
Horiz Axis	Min/Max	0.7/2.8	0.7/2.8	0.9/3.2	1.1/4.1	1.5/3.6	2.0/4.2	2.1/3.4	1.9/3.5	1.6/3.6	1.7/3.9	0.8/3.2	0.6/2.4	1.9/2.7
1-Axis, N-S	Average	1.2	1.6	2.4	3.1	3.6	3.8	3.5	3.3	2.9	2.4	1.3	0.9	2.5
Horiz Axis	Min/Max	0.5/2.1	0.7/2.6	1.0/3.5	1.4/5.3	2.0/4.7	2.6/5.5	2.6/4.4	2.5/4.4	1.9/4.2	1.7/3.8	0.6/2.5	0.4/1.7	2.1/3.0
1-Axis, N-S	Average	1.7	2.1	2.8	3.3	3.5	3.5	3.3	3.4	3.3	3.0	1.9	1.4	2.8
Tilt=Latitude	Min/Max	0.7/3.1	0.9/3.4	1.2/4.2	1.5/5.6	2.0/4.6	2.5/5.2	2.5/4.2	2.5/4.5	2.1/4.7	2.1/4.8	0.9/3.6	0.6/2.5	2.4/3.3
2-Axis	Average	1.8	2.2	2.8	3.3	3.7	3.8	3.6	3.5	3.3	3.1	2.0	1.6	2.9
	Min/Max	0.8/3.3	0.9/3.5	1.2/4.2	1.5/5.7	2.1/4.9	2.7/5.6	2.7/4.6	2.6/4.7	2.1/4.7	2.2/4.9	0.9/3.8	0.6/2.8	2.4/3.4

Average Climatic Conditions

Element	Jan	Feb	Mar	Apr	May	June	July	Aug	Sept	Oct	Nov	Dec	Year
Temperature (°C)	-2.7	-1.2	4.2	9.0	14.1	18.2	20.4	19.9	16.6	10.2	5.2	0.0	9.5
Daily Minimum Temp	-8.8	-7.6	-2.7	1.6	6.7	11.1	13.8	13.4	9.9	2.9	-1.3	-5.9	2.8
Daily Maximum Temp	3.3	5.2	11.2	16.4	21.6	25.3	26.9	26.3	23.1	17.6	11.6	5.9	16.2
Record Minimum Temp	-31.1	-30.0	-26.1	-16.1	-6.7	-3.9	0.0	1.1	-2.8	-11.7	-17.8	-31.1	-31.1
Record Maximum Temp	24.4	22.2	28.9	31.7	32.2	33.9	37.2	35.0	36.1	30.0	26.7	24.4	37.2
HDD, Base 18.3°C	653	548	437	280	138	34	6	13	73	254	395	568	3400
CDD, Base 18.3°C	0	0	0	0	7	31	70	61	19	3	0	0	192
Relative Humidity (%)	74	73	70	68	72	78	81	83	83	77	75	77	76
Wind Speed (m/s)	3.3	3.2	3.5	3.5	2.8	2.2	2.0	1.8	1.9	2.3	3.0	3.2	2.7

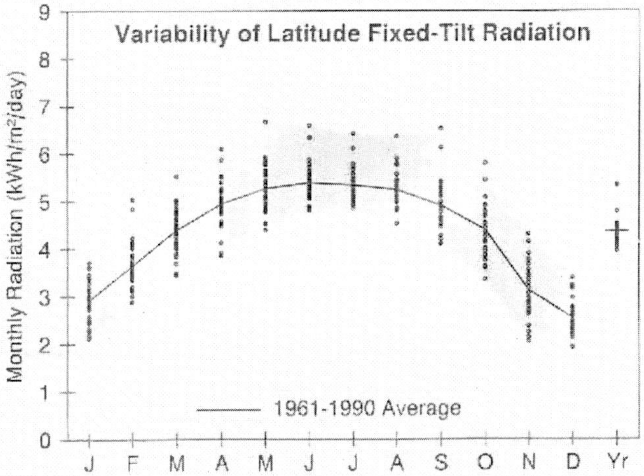

Huntington, WV

WBAN NO. 03860

LATITUDE: 38.37° N
LONGITUDE: 82.55° W
ELEVATION: 255 meters
MEAN PRESSURE: 988 millibars

STATION TYPE: Secondary

Solar Radiation for Flat-Plate Collectors Facing South at a Fixed Tilt (kWh/m²/day), Uncertainty ±9%

Tilt (°)		Jan	Feb	Mar	Apr	May	June	July	Aug	Sept	Oct	Nov	Dec	Year
0	Average	2.0	2.7	3.7	4.8	5.6	6.0	5.8	5.2	4.3	3.3	2.1	1.7	3.9
	Min/Max	1.7/2.3	2.3/3.4	3.1/4.4	3.9/5.7	4.7/7.0	5.3/7.4	5.3/7.0	4.6/6.2	3.7/5.5	2.7/4.1	1.6/2.6	1.4/2.0	3.7/4.7
Latitude -15	Average	2.6	3.4	4.3	5.1	5.6	5.9	5.7	5.4	4.9	4.1	2.8	2.3	4.4
	Min/Max	2.0/3.3	2.8/4.6	3.4/5.3	4.0/6.2	4.7/7.1	5.2/7.2	5.2/6.9	4.7/6.6	4.1/6.4	3.2/5.4	2.0/3.8	1.8/3.0	4.0/5.3
Latitude	Average	2.9	3.6	4.4	4.9	5.3	5.4	5.3	5.2	4.9	4.4	3.1	2.5	4.3
	Min/Max	2.1/3.7	2.9/5.0	3.4/5.5	3.9/6.1	4.4/6.7	4.8/6.6	4.8/6.4	4.5/6.4	4.1/6.5	3.4/5.8	2.0/4.3	1.9/3.4	3.9/5.3
Latitude +15	Average	3.1	3.7	4.3	4.6	4.7	4.7	4.7	4.8	4.7	4.4	3.2	2.7	4.1
	Min/Max	2.1/4.0	2.9/5.2	3.3/5.4	3.6/5.7	3.9/5.9	4.2/5.7	4.3/5.6	4.1/5.8	3.9/6.3	3.3/5.9	2.1/4.6	2.0/3.6	3.7/5.1
90	Average	2.8	3.1	3.1	2.8	2.5	2.3	2.4	2.8	3.2	3.5	2.8	2.4	2.8
	Min/Max	1.8/3.7	2.3/4.4	2.4/4.0	2.3/3.5	2.2/3.0	2.2/2.7	2.3/2.7	2.4/3.3	2.7/4.3	2.6/4.7	1.7/4.0	1.7/3.4	2.5/3.4

Solar Radiation for 1-Axis Tracking Flat-Plate Collectors with a North-South Axis (kWh/m²/day), Uncertainty ±9%

Axis Tilt (°)		Jan	Feb	Mar	Apr	May	June	July	Aug	Sept	Oct	Nov	Dec	Year
0	Average	2.6	3.5	4.8	6.1	7.1	7.5	7.3	6.6	5.6	4.5	2.8	2.1	5.1
	Min/Max	1.9/3.3	2.8/4.9	3.7/6.2	4.6/7.8	5.5/9.5	6.4/9.9	6.5/9.4	5.6/8.3	4.6/7.7	3.3/5.9	1.8/3.9	1.6/2.8	4.6/6.4
Latitude -15	Average	3.1	4.1	5.2	6.4	7.1	7.5	7.3	6.8	6.1	5.1	3.3	2.6	5.4
	Min/Max	2.2/4.0	3.1/5.8	4.0/6.9	4.7/8.2	5.6/9.6	6.4/9.9	6.4/9.5	5.8/8.6	4.9/8.4	3.7/6.9	2.1/4.8	1.9/3.5	4.8/6.9
Latitude	Average	3.3	4.3	5.3	6.3	6.9	7.2	7.0	6.7	6.1	5.3	3.6	2.8	5.4
	Min/Max	2.3/4.3	3.2/6.1	4.0/7.1	4.6/8.2	5.4/9.4	6.1/9.5	6.2/9.1	5.6/8.5	4.9/8.5	3.8/7.2	2.2/5.2	2.0/3.9	4.8/6.9
Latitude +15	Average	3.4	4.3	5.2	6.0	6.5	6.7	6.6	6.4	5.9	5.3	3.7	2.9	5.3
	Min/Max	2.3/4.5	3.2/6.3	3.9/7.0	4.4/7.9	5.0/8.8	5.7/8.9	5.8/8.6	5.3/8.1	4.7/8.3	3.8/7.3	2.2/5.4	2.1/4.1	4.7/6.8

Solar Radiation for 2-Axis Tracking Flat-Plate Collectors (kWh/m²/day), Uncertainty ±9%

Tracker		Jan	Feb	Mar	Apr	May	June	July	Aug	Sept	Oct	Nov	Dec	Year
2-Axis	Average	3.5	4.3	5.4	6.4	7.2	7.7	7.4	6.9	6.1	5.4	3.7	3.0	5.6
	Min/Max	2.3/4.6	3.2/6.3	4.0/7.1	4.7/8.3	5.7/9.8	6.5/10.1	6.6/9.6	5.8/8.6	4.9/8.5	3.8/7.3	2.2/5.4	2.1/4.2	5.0/7.2

Direct Beam Solar Radiation for Concentrating Collectors (kWh/m²/day), Uncertainty ±8%

Tracker		Jan	Feb	Mar	Apr	May	June	July	Aug	Sept	Oct	Nov	Dec	Year
1-Axis, E-W Horiz Axis	Average	1.7	2.0	2.4	2.7	3.1	3.3	3.1	2.9	2.7	2.8	1.9	1.5	2.5
	Min/Max	0.8/2.7	1.0/3.6	1.4/3.6	1.3/4.0	2.1/4.9	2.5/5.1	2.6/4.8	2.1/4.3	1.8/4.5	1.6/4.2	0.9/3.3	0.9/2.5	2.1/3.7
1-Axis, N-S Horiz Axis	Average	1.3	1.9	2.7	3.5	4.1	4.3	4.1	3.7	3.3	2.8	1.5	1.1	2.9
	Min/Max	0.6/2.0	1.0/3.3	1.6/4.2	1.8/5.3	2.7/6.6	3.2/6.7	3.4/6.4	2.8/5.5	2.2/5.4	1.6/4.1	0.6/2.7	0.6/1.8	2.4/4.3
1-Axis, N-S Tilt=Latitude	Average	1.9	2.5	3.2	3.7	4.0	4.1	3.9	3.8	3.7	3.5	2.2	1.6	3.2
	Min/Max	0.9/3.0	1.3/4.4	1.8/4.8	1.9/5.5	2.6/6.4	3.0/6.3	3.2/6.1	2.8/5.6	2.4/6.0	2.0/5.2	0.9/3.7	0.9/2.7	2.6/4.7
2-Axis	Average	2.0	2.5	3.2	3.8	4.2	4.4	4.2	3.9	3.7	3.5	2.3	1.8	3.3
	Min/Max	0.9/3.2	1.3/4.5	1.8/4.8	1.9/5.6	2.8/6.8	3.3/6.9	3.4/6.6	2.9/5.8	2.5/6.0	2.0/5.3	1.0/3.9	1.0/2.9	2.7/4.9

Average Climatic Conditions

Element	Jan	Feb	Mar	Apr	May	June	July	Aug	Sept	Oct	Nov	Dec	Year
Temperature (°C)	0.0	1.7	7.8	12.9	17.7	21.6	23.7	23.1	19.8	13.4	8.1	2.7	12.7
Daily Minimum Temp	-4.9	-3.3	1.9	6.6	11.3	15.8	18.3	17.7	13.9	7.2	2.8	-2.0	7.1
Daily Maximum Temp	4.8	6.8	13.6	19.3	24.0	27.4	29.1	28.4	25.6	19.6	13.3	7.4	18.3
Record Minimum Temp	-26.7	-21.1	-18.9	-6.7	-2.8	4.4	7.8	6.1	-0.6	-8.9	-13.3	-25.0	-26.7
Record Maximum Temp	23.3	26.1	30.0	33.3	33.9	37.8	38.9	37.8	36.1	30.0	27.8	26.7	38.9
HDD, Base 18.3°C	568	465	327	166	67	6	0	4	29	167	308	484	2592
CDD, Base 18.3°C	0	0	0	4	46	104	167	151	73	13	0	0	558
Relative Humidity (%)	70	67	62	59	68	73	76	77	77	71	69	71	70
Wind Speed (m/s)	3.4	3.4	3.6	3.5	2.8	2.6	2.4	2.3	2.4	2.6	3.2	3.3	3.0

Casper, WY

WBAN NO. 24089

LATITUDE: 42.92° N
LONGITUDE: 106.47° W
ELEVATION: 1612 meters
MEAN PRESSURE: 837 millibars

STATION TYPE: Secondary

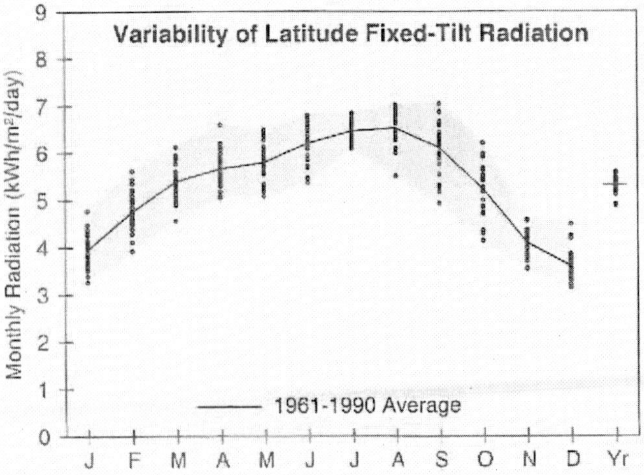

Solar Radiation for Flat-Plate Collectors Facing South at a Fixed Tilt (kWh/m²/day), Uncertainty ±9%

Tilt (°)		Jan	Feb	Mar	Apr	May	June	July	Aug	Sept	Oct	Nov	Dec	Year
0	Average	2.0	2.9	4.1	5.2	6.1	7.0	7.0	6.3	4.9	3.4	2.2	1.7	4.4
	Min/Max	1.8/2.2	2.6/3.2	3.6/4.5	4.7/5.9	5.4/6.7	6.0/7.6	6.6/7.5	5.4/6.7	4.2/5.6	2.9/3.9	2.0/2.4	1.6/2.0	4.1/4.6
Latitude -15	Average	3.4	4.3	5.2	5.8	6.2	6.8	7.0	6.8	6.0	4.8	3.6	3.1	5.2
	Min/Max	2.9/4.1	3.6/5.0	4.4/5.8	5.2/6.7	5.4/6.9	5.8/7.4	6.6/7.4	5.8/7.3	4.9/6.9	3.9/5.7	3.2/4.0	2.7/3.8	4.8/5.5
Latitude	Average	3.9	4.7	5.4	5.7	5.8	6.2	6.5	6.5	6.1	5.2	4.1	3.6	5.3
	Min/Max	3.3/4.8	3.9/5.6	4.6/6.1	5.1/6.6	5.1/6.5	5.4/6.8	6.1/6.8	5.5/7.0	4.9/7.0	4.1/6.2	3.5/4.6	3.1/4.5	4.9/5.6
Latitude +15	Average	4.3	4.9	5.3	5.3	5.1	5.3	5.6	5.9	5.9	5.3	4.3	3.9	5.1
	Min/Max	3.5/5.2	4.0/5.9	4.5/6.1	4.7/6.1	4.5/5.7	4.6/5.8	5.3/5.9	5.0/6.3	4.7/6.8	4.2/6.3	3.7/4.9	3.4/4.9	4.7/5.4
90	Average	4.1	4.5	4.3	3.6	3.0	2.8	3.0	3.7	4.3	4.5	4.0	3.8	3.8
	Min/Max	3.3/5.0	3.6/5.4	3.6/4.9	3.2/4.1	2.7/3.4	2.6/3.0	2.9/3.2	3.1/3.9	3.4/4.9	3.5/5.3	3.5/4.6	3.3/4.8	3.5/4.0

Solar Radiation for 1-Axis Tracking Flat-Plate Collectors with a North-South Axis (kWh/m²/day), Uncertainty ±9%

Axis Tilt (°)		Jan	Feb	Mar	Apr	May	June	July	Aug	Sept	Oct	Nov	Dec	Year
0	Average	3.2	4.4	5.9	7.2	8.2	9.5	9.7	9.0	7.3	5.2	3.5	2.7	6.3
	Min/Max	2.6/3.8	3.6/5.3	4.7/6.8	6.2/8.7	6.9/9.5	7.7/10.7	8.9/10.6	7.4/9.9	5.8/8.6	4.1/6.2	3.0/3.9	2.4/3.4	5.6/6.7
Latitude -15	Average	4.2	5.4	6.7	7.7	8.4	9.6	9.9	9.5	8.1	6.3	4.5	3.8	7.0
	Min/Max	3.5/5.2	4.4/6.6	5.4/7.8	6.6/9.4	7.1/9.7	7.7/10.8	9.1/10.8	7.7/10.5	6.3/9.7	4.9/7.5	3.8/5.1	3.2/4.7	6.2/7.5
Latitude	Average	4.6	5.8	6.9	7.6	8.2	9.2	9.5	9.3	8.2	6.6	4.9	4.2	7.1
	Min/Max	3.8/5.7	4.7/7.0	5.5/8.1	6.5/9.3	6.9/9.5	7.4/10.4	8.7/10.4	7.6/10.3	6.4/9.8	5.1/7.9	4.1/5.5	3.6/5.3	6.3/7.6
Latitude +15	Average	4.9	5.9	6.9	7.3	7.7	8.6	9.0	8.9	8.0	6.6	5.1	4.4	7.0
	Min/Max	4.0/6.0	4.7/7.2	5.4/8.0	6.3/8.9	6.5/8.9	6.9/9.7	8.2/9.8	7.2/9.9	6.2/9.6	5.1/8.0	4.3/5.8	3.8/5.7	6.2/7.4

Solar Radiation for 2-Axis Tracking Flat-Plate Collectors (kWh/m²/day), Uncertainty ±9%

Tracker		Jan	Feb	Mar	Apr	May	June	July	Aug	Sept	Oct	Nov	Dec	Year
2-Axis	Average	4.9	5.9	6.9	7.7	8.5	9.8	10.1	9.5	8.2	6.7	5.1	4.5	7.3
	Min/Max	4.0/6.1	4.8/7.2	5.5/8.1	6.6/9.4	7.2/9.9	7.9/11.1	9.2/11.0	7.8/10.6	6.4/9.8	5.2/8.0	4.3/5.8	3.9/5.8	6.5/7.8

Direct Beam Solar Radiation for Concentrating Collectors (kWh/m²/day), Uncertainty ±8%

Tracker		Jan	Feb	Mar	Apr	May	June	July	Aug	Sept	Oct	Nov	Dec	Year
1-Axis, E-W Horiz Axis	Average	3.1	3.4	3.5	3.6	4.1	5.2	5.5	5.1	4.5	3.9	3.1	2.9	4.0
	Min/Max	2.2/4.2	2.4/4.5	2.4/4.5	2.7/5.1	2.8/5.1	3.5/6.3	4.7/6.3	3.6/5.9	3.1/5.8	2.6/5.1	2.3/3.7	2.2/4.1	3.3/4.4
1-Axis, N-S Horiz Axis	Average	2.1	3.0	3.9	4.7	5.5	6.9	7.3	6.8	5.5	3.8	2.3	1.9	4.5
	Min/Max	1.5/2.9	2.1/4.0	2.6/5.1	3.4/6.6	3.8/6.9	4.6/8.4	6.2/8.4	4.9/8.0	3.7/7.1	2.5/4.9	1.7/2.8	1.4/2.7	3.6/5.0
1-Axis, N-S Tilt=Latitude	Average	3.3	4.1	4.7	5.1	5.4	6.6	7.0	7.0	6.3	5.0	3.5	3.1	5.1
	Min/Max	2.4/4.6	2.9/5.5	3.2/6.1	3.7/7.1	3.8/6.8	4.4/8.0	6.0/8.1	5.0/8.2	4.2/8.1	3.3/6.4	2.6/4.2	2.3/4.4	4.1/5.7
2-Axis	Average	3.6	4.2	4.8	5.2	5.8	7.1	7.5	7.2	6.3	5.0	3.7	3.3	5.3
	Min/Max	2.6/4.9	3.0/5.6	3.2/6.1	3.8/7.2	4.0/7.2	4.8/8.6	6.4/8.7	5.2/8.4	4.3/8.1	3.4/6.5	2.7/4.4	2.5/4.8	4.3/5.9

Average Climatic Conditions

Element	Jan	Feb	Mar	Apr	May	June	July	Aug	Sept	Oct	Nov	Dec	Year
Temperature (°C)	-5.3	-3.1	0.8	6.0	11.3	17.1	21.6	20.4	14.3	8.0	0.6	-4.6	7.3
Daily Minimum Temp	-11.1	-8.9	-5.7	-1.4	3.3	8.3	12.2	11.0	5.3	0.1	-5.7	-10.2	-0.2
Daily Maximum Temp	0.4	2.8	7.3	13.4	19.2	25.9	30.9	29.8	23.2	15.8	6.8	1.1	14.7
Record Minimum Temp	-40.0	-33.9	-29.4	-20.0	-8.9	-2.2	-1.1	0.6	-8.9	-19.4	-29.4	-40.6	-40.6
Record Maximum Temp	15.6	20.0	23.3	28.3	33.3	38.9	40.0	38.9	35.6	29.4	21.7	16.1	40.0
HDD, Base 18.3°C	734	599	543	370	219	77	6	20	139	321	532	709	4268
CDD, Base 18.3°C	0	0	0	0	0	39	106	84	19	0	0	0	247
Relative Humidity (%)	64	65	62	58	57	51	44	43	49	54	61	64	56
Wind Speed (m/s)	7.6	6.8	6.2	5.6	5.2	4.8	4.6	4.6	4.9	5.4	6.4	6.9	5.7

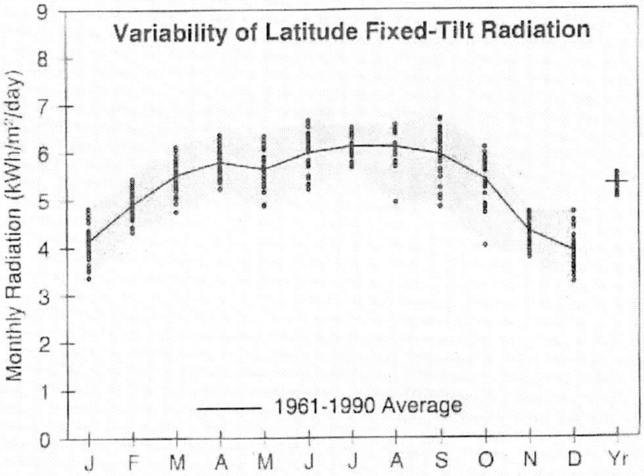

Cheyenne, WY

WBAN NO. 24018

LATITUDE: 41.15° N
LONGITUDE: 104.82° W
ELEVATION: 1872 meters
MEAN PRESSURE: 811 millibars

STATION TYPE: Secondary

Solar Radiation for Flat-Plate Collectors Facing South at a Fixed Tilt (kWh/m²/day), Uncertainty ±9%

Tilt (°)		Jan	Feb	Mar	Apr	May	June	July	Aug	Sept	Oct	Nov	Dec	Year
0	Average	2.2	3.1	4.2	5.3	6.0	6.7	6.7	5.9	4.9	3.6	2.4	1.9	4.4
	Min/Max	1.9/2.4	2.8/3.3	3.8/4.6	4.9/5.8	5.2/6.6	5.9/7.5	6.2/7.1	4.9/6.4	4.2/5.4	3.0/4.0	2.2/2.6	1.7/2.2	4.2/4.6
Latitude -15	Average	3.6	4.4	5.3	5.9	6.0	6.5	6.6	6.3	5.8	5.0	3.8	3.3	5.2
	Min/Max	3.0/4.1	3.9/4.9	4.6/5.8	5.3/6.5	5.2/6.7	5.7/7.3	6.1/7.1	5.1/6.8	4.8/6.6	3.8/5.6	3.3/4.2	2.8/4.0	4.9/5.4
Latitude	Average	4.1	4.9	5.5	5.8	5.6	6.0	6.1	6.1	6.0	5.4	4.3	3.9	5.3
	Min/Max	3.4/4.8	4.3/5.4	4.8/6.1	5.2/6.4	4.9/6.3	5.2/6.7	5.7/6.5	4.9/6.6	4.8/6.7	4.0/6.1	3.8/4.7	3.3/4.7	5.0/5.5
Latitude +15	Average	4.5	5.1	5.5	5.4	5.0	5.1	5.3	5.5	5.7	5.5	4.6	4.2	5.1
	Min/Max	3.6/5.2	4.5/5.7	4.7/6.1	4.9/5.9	4.3/5.6	4.5/5.7	4.9/5.6	4.5/5.9	4.6/6.5	4.0/6.2	4.0/5.0	3.5/5.2	4.8/5.3
90	Average	4.2	4.5	4.3	3.6	2.8	2.6	2.8	3.3	4.1	4.6	4.2	4.1	3.8
	Min/Max	3.3/4.9	4.0/5.2	3.7/4.9	3.3/4.1	2.5/3.1	2.4/2.8	2.6/2.9	2.8/3.5	3.2/4.6	3.3/5.2	3.7/4.7	3.4/5.1	3.5/3.9

Solar Radiation for 1-Axis Tracking Flat-Plate Collectors with a North-South Axis (kWh/m²/day), Uncertainty ±9%

Axis Tilt (°)		Jan	Feb	Mar	Apr	May	June	July	Aug	Sept	Oct	Nov	Dec	Year
0	Average	3.4	4.6	6.1	7.4	8.0	9.0	9.1	8.3	7.1	5.5	3.7	3.1	6.3
	Min/Max	2.8/4.0	4.1/5.2	5.1/7.0	6.6/8.3	6.6/9.1	7.5/10.5	8.4/9.9	6.3/9.2	5.6/8.2	4.1/6.3	3.3/4.1	2.5/3.7	5.8/6.7
Latitude -15	Average	4.5	5.7	6.9	7.9	8.1	9.0	9.2	8.7	7.8	6.6	4.8	4.1	6.9
	Min/Max	3.6/5.2	4.9/6.3	5.7/7.9	7.0/8.9	6.6/9.3	7.5/10.6	8.5/10.0	6.6/9.7	6.1/9.1	4.7/7.4	4.1/5.3	3.3/5.0	6.4/7.4
Latitude	Average	4.9	6.0	7.1	7.8	7.9	8.7	8.9	8.6	7.9	6.9	5.2	4.6	7.0
	Min/Max	3.9/5.8	5.2/6.8	5.8/8.1	6.9/8.8	6.4/9.0	7.2/10.1	8.2/9.7	6.4/9.5	6.2/9.2	4.9/7.8	4.5/5.7	3.7/5.6	6.5/7.5
Latitude +15	Average	5.2	6.2	7.0	7.5	7.4	8.1	8.3	8.2	7.7	6.9	5.4	4.9	6.9
	Min/Max	4.1/6.1	5.4/6.9	5.8/8.1	6.7/8.5	6.1/8.5	6.7/9.5	7.7/9.1	6.1/9.1	6.0/9.0	4.9/7.9	4.6/6.0	3.9/6.0	6.4/7.3

Solar Radiation for 2-Axis Tracking Flat-Plate Collectors (kWh/m²/day), Uncertainty ±9%

Tracker		Jan	Feb	Mar	Apr	May	June	July	Aug	Sept	Oct	Nov	Dec	Year
2-Axis	Average	5.2	6.2	7.1	7.9	8.2	9.3	9.3	8.8	7.9	6.9	5.4	4.9	7.3
	Min/Max	4.2/6.2	5.4/6.9	5.8/8.1	7.0/8.9	6.7/9.4	7.7/10.8	8.7/10.2	6.6/9.7	6.2/9.2	4.9/7.9	4.7/6.0	4.0/6.1	6.7/7.7

Direct Beam Solar Radiation for Concentrating Collectors (kWh/m²/day), Uncertainty ±8%

Tracker		Jan	Feb	Mar	Apr	May	June	July	Aug	Sept	Oct	Nov	Dec	Year
1-Axis, E-W Horiz Axis	Average	3.3	3.6	3.6	3.8	3.9	4.8	4.9	4.5	4.3	4.1	3.3	3.2	3.9
	Min/Max	2.3/4.3	2.8/4.3	2.5/4.5	3.0/4.6	2.6/4.8	3.4/6.1	4.3/5.7	2.9/5.2	2.9/5.3	2.4/4.9	2.6/3.9	2.3/4.3	3.5/4.3
1-Axis, N-S Horiz Axis	Average	2.3	3.2	4.1	5.0	5.2	6.4	6.5	6.0	5.2	4.1	2.6	2.1	4.4
	Min/Max	1.7/3.1	2.5/3.9	2.7/5.3	4.0/6.1	3.5/6.5	4.5/8.2	5.7/7.6	3.8/7.0	3.5/6.6	2.4/4.9	2.0/3.1	1.5/2.9	3.9/4.9
1-Axis, N-S Tilt=Latitude	Average	3.6	4.4	4.8	5.3	5.1	6.0	6.2	6.1	5.9	5.2	3.8	3.4	5.0
	Min/Max	2.6/4.7	3.4/5.3	3.2/6.2	4.2/6.5	3.5/6.3	4.3/7.8	5.5/7.3	3.9/7.1	4.0/7.4	3.1/6.3	3.0/4.5	2.4/4.6	4.4/5.5
2-Axis	Average	3.9	4.5	4.9	5.4	5.4	6.6	6.7	6.3	6.0	5.3	4.0	3.7	5.2
	Min/Max	2.7/5.1	3.5/5.4	3.2/6.2	4.3/6.6	3.7/6.7	4.6/8.4	5.9/7.8	4.1/7.4	4.0/7.4	3.1/6.3	3.2/4.7	2.6/5.0	4.6/5.7

Average Climatic Conditions

Element	Jan	Feb	Mar	Apr	May	June	July	Aug	Sept	Oct	Nov	Dec	Year
Temperature (°C)	-3.1	-1.5	0.9	5.8	11.1	16.3	20.2	19.1	14.1	8.3	1.8	-2.3	7.6
Daily Minimum Temp	-9.3	-7.7	-5.5	-1.1	4.1	9.1	12.6	11.6	6.5	1.1	-4.6	-8.5	0.7
Daily Maximum Temp	3.2	4.7	7.2	12.6	18.1	23.6	27.9	26.7	21.7	15.6	8.2	3.8	14.4
Record Minimum Temp	-33.9	-36.7	-29.4	-22.2	-8.9	-3.9	3.3	2.2	-13.3	-18.3	-25.6	-33.3	-36.7
Record Maximum Temp	18.9	21.7	23.3	27.8	32.2	37.8	37.8	35.6	33.9	28.3	22.8	20.6	37.8
HDD, Base 18.3°C	663	556	541	375	224	83	17	24	139	310	497	641	4070
CDD, Base 18.3°C	0	0	0	0	0	22	76	48	13	0	0	0	158
Relative Humidity (%)	53	55	56	54	56	53	51	51	52	50	54	54	53
Wind Speed (m/s)	7.1	6.6	6.5	6.5	5.8	5.1	4.6	4.6	5.0	5.4	5.9	6.5	5.8

Lander, WY

WBAN NO. 24021

LATITUDE: 42.82° N
LONGITUDE: 108.73° W
ELEVATION: 1696 meters
MEAN PRESSURE: 829 millibars

STATION TYPE: Primary

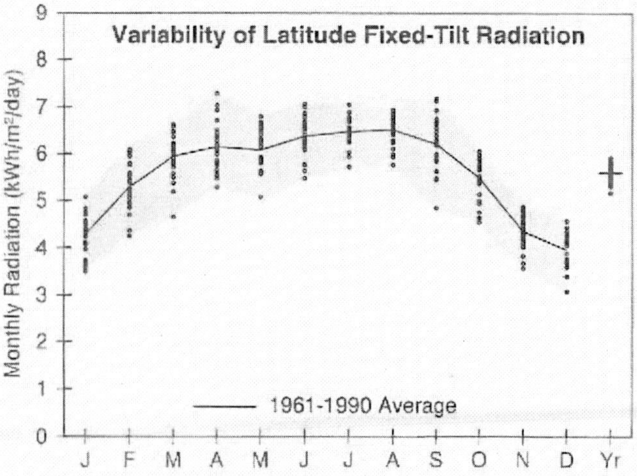

Solar Radiation for Flat-Plate Collectors Facing South at a Fixed Tilt (kWh/m²/day), Uncertainty ±9%

Tilt (°)		Jan	Feb	Mar	Apr	May	June	July	Aug	Sept	Oct	Nov	Dec	Year
0	Average	2.2	3.2	4.4	5.6	6.4	7.1	7.0	6.3	5.0	3.6	2.3	1.9	4.6
	Min/Max	1.9/2.4	2.7/3.6	3.7/4.8	4.9/6.5	5.3/7.1	6.1/7.9	6.2/7.6	5.6/6.7	4.2/5.7	3.1/3.9	2.1/2.6	1.7/2.0	4.3/4.8
Latitude -15	Average	3.7	4.8	5.7	6.2	6.5	7.0	7.0	6.8	6.1	5.1	3.8	3.4	5.5
	Min/Max	3.0/4.3	3.9/5.4	4.5/6.3	5.4/7.4	5.4/7.2	6.0/7.7	6.2/7.6	6.0/7.2	4.8/7.0	4.2/5.6	3.2/4.3	2.7/3.9	5.1/5.8
Latitude	Average	4.3	5.3	6.0	6.1	6.1	6.4	6.5	6.5	6.2	5.5	4.3	4.0	5.6
	Min/Max	3.5/5.1	4.3/6.1	4.7/6.6	5.3/7.3	5.1/6.8	5.5/7.1	5.7/7.0	5.8/6.9	4.8/7.2	4.5/6.1	3.6/4.9	3.1/4.6	5.2/5.9
Latitude +15	Average	4.6	5.6	5.9	5.7	5.4	5.5	5.6	5.9	6.0	5.6	4.6	4.3	5.4
	Min/Max	3.8/5.5	4.4/6.4	4.6/6.7	4.9/6.8	4.5/6.0	4.7/6.0	5.0/6.1	5.2/6.3	4.6/6.9	4.6/6.2	3.8/5.2	3.3/5.0	5.0/5.7
90	Average	4.5	5.1	4.8	4.0	3.2	2.9	3.1	3.6	4.3	4.7	4.3	4.3	4.1
	Min/Max	3.7/5.4	3.8/6.0	3.7/5.8	3.3/5.0	2.7/3.5	2.7/3.1	2.8/3.3	3.3/3.8	3.3/5.0	3.8/5.2	3.5/5.1	3.3/4.9	3.7/4.3

Solar Radiation for 1-Axis Tracking Flat-Plate Collectors with a North-South Axis (kWh/m²/day), Uncertainty ±9%

Axis Tilt (°)		Jan	Feb	Mar	Apr	May	June	July	Aug	Sept	Oct	Nov	Dec	Year
0	Average	3.4	4.9	6.5	7.7	8.5	9.8	9.8	9.0	7.4	5.5	3.7	3.0	6.6
	Min/Max	2.6/4.1	3.8/5.6	4.7/7.3	6.5/9.3	6.6/9.9	7.9/11.1	7.5/10.8	7.7/9.7	5.7/8.8	4.5/6.1	3.0/4.1	2.5/3.5	5.9/7.0
Latitude -15	Average	4.6	6.1	7.4	8.3	8.7	9.8	9.9	9.4	8.2	6.6	4.8	4.1	7.3
	Min/Max	3.5/5.5	4.7/7.1	5.4/8.5	6.9/10.0	6.7/10.2	7.9/11.2	7.6/10.9	8.1/10.3	6.2/9.9	5.4/7.4	3.8/5.4	3.2/4.8	6.5/7.8
Latitude	Average	5.0	6.5	7.6	8.3	8.5	9.4	9.5	9.3	8.3	6.9	5.2	4.6	7.4
	Min/Max	3.9/6.1	5.0/7.6	5.5/8.7	6.8/10.0	6.5/9.9	7.6/10.8	7.3/10.6	7.9/10.1	6.2/10.0	5.7/7.8	4.2/5.8	3.5/5.4	6.6/7.9
Latitude +15	Average	5.3	6.7	7.6	8.0	8.0	8.8	9.0	8.8	8.1	7.0	5.4	4.9	7.3
	Min/Max	4.2/6.5	5.1/7.8	5.4/8.7	6.5/9.6	6.1/9.4	7.1/10.1	6.8/9.9	7.6/9.6	6.0/9.8	5.7/7.9	4.3/6.1	3.7/5.7	6.5/7.8

Solar Radiation for 2-Axis Tracking Flat-Plate Collectors (kWh/m²/day), Uncertainty ±9%

Tracker		Jan	Feb	Mar	Apr	May	June	July	Aug	Sept	Oct	Nov	Dec	Year
2-Axis	Average	5.4	6.7	7.7	8.3	8.9	10.0	10.1	9.5	8.4	7.0	5.4	5.0	7.7
	Min/Max	4.2/6.6	5.2/7.8	5.5/8.7	6.9/10.0	6.8/10.3	8.1/11.5	7.8/11.1	8.1/10.3	6.2/10.0	5.7/7.9	4.3/6.2	3.8/5.8	6.8/8.2

Direct Beam Solar Radiation for Concentrating Collectors (kWh/m²/day), Uncertainty ±8%

Tracker		Jan	Feb	Mar	Apr	May	June	July	Aug	Sept	Oct	Nov	Dec	Year
1-Axis, E-W Horiz Axis	Average	3.2	3.7	3.9	3.9	4.2	5.2	5.3	5.0	4.5	4.1	3.3	3.2	4.1
	Min/Max	2.3/4.4	2.6/4.8	2.4/4.8	2.7/5.3	2.7/5.4	3.5/6.5	3.7/6.1	3.8/5.5	2.9/5.8	2.9/5.3	2.4/4.0	1.8/4.0	3.5/4.6
1-Axis, N-S Horiz Axis	Average	2.2	3.2	4.3	5.0	5.6	7.0	7.1	6.6	5.5	4.0	2.5	2.0	4.6
	Min/Max	1.5/3.1	2.1/4.2	2.5/5.4	3.5/6.9	3.5/7.3	4.7/8.7	4.6/8.2	4.9/7.5	3.5/7.2	2.8/5.2	1.8/3.0	1.1/2.6	3.8/5.1
1-Axis, N-S Tilt=Latitude	Average	3.5	4.5	5.3	5.4	5.6	6.6	6.9	6.9	6.3	5.2	3.7	3.3	5.3
	Min/Max	2.4/4.8	3.0/5.8	3.0/6.5	3.8/7.4	3.5/7.2	4.4/8.2	4.4/7.9	5.1/7.7	4.0/8.1	3.7/6.7	2.7/4.5	1.8/4.2	4.4/5.8
2-Axis	Average	3.8	4.6	5.3	5.5	5.9	7.2	7.4	7.1	6.3	5.3	3.9	3.6	5.5
	Min/Max	2.6/5.2	3.1/6.0	3.0/6.6	3.8/7.6	3.7/7.6	4.8/8.9	4.8/8.5	5.3/7.9	4.0/8.2	3.7/6.8	2.8/4.8	2.0/4.6	4.5/6.1

Average Climatic Conditions

Element	Jan	Feb	Mar	Apr	May	June	July	Aug	Sept	Oct	Nov	Dec	Year
Temperature (°C)	-6.9	-3.7	1.1	6.3	11.7	17.3	21.7	20.5	14.6	8.2	-0.4	-6.2	7.0
Daily Minimum Temp	-13.4	-10.2	-5.5	-0.7	4.4	9.3	13.3	12.1	6.8	0.9	-6.7	-12.6	-0.2
Daily Maximum Temp	-0.4	2.8	7.6	13.2	18.9	25.2	30.1	28.8	22.4	15.4	5.8	0.1	14.2
Record Minimum Temp	-38.3	-33.3	-26.7	-18.9	-7.8	-3.9	3.9	1.7	-12.2	-19.4	-27.8	-38.3	-38.3
Record Maximum Temp	17.2	20.0	24.4	27.8	32.8	37.8	38.3	38.3	34.4	29.4	21.1	17.8	38.3
HDD, Base 18.3°C	784	616	536	362	209	76	5	21	137	313	563	761	4383
CDD, Base 18.3°C	0	0	0	0	0	44	108	88	25	0	0	0	266
Relative Humidity (%)	64	62	56	52	50	45	40	40	46	52	62	65	53
Wind Speed (m/s)	2.4	2.5	3.1	3.5	3.6	3.4	3.4	3.3	3.1	2.7	2.5	2.5	3.0

Rock Springs, WY

WBAN NO. 24027

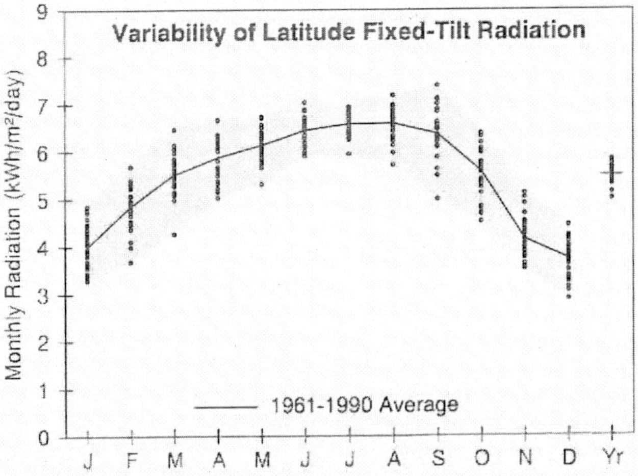

LATITUDE: 41.60° N
LONGITUDE: 109.07° W
ELEVATION: 2056 meters
MEAN PRESSURE: 794 millibars

STATION TYPE: Secondary

Solar Radiation for Flat-Plate Collectors Facing South at a Fixed Tilt (kWh/m²/day), Uncertainty ±9%

Tilt (°)		Jan	Feb	Mar	Apr	May	June	July	Aug	Sept	Oct	Nov	Dec	Year
0	Average	2.1	3.0	4.2	5.4	6.4	7.2	7.2	6.4	5.2	3.7	2.3	1.9	4.6
	Min/Max	1.9/2.4	2.5/3.4	3.5/4.7	4.8/6.0	5.6/7.1	6.6/7.9	6.5/7.6	5.6/6.9	4.3/5.9	3.2/4.1	2.1/2.7	1.7/2.1	4.3/4.8
Latitude -15	Average	3.5	4.4	5.3	6.0	6.5	7.0	7.1	6.9	6.3	5.2	3.7	3.2	5.4
	Min/Max	2.9/4.1	3.4/5.1	4.2/6.1	5.2/6.8	5.7/7.2	6.4/7.7	6.4/7.5	5.9/7.5	5.0/7.2	4.2/5.9	3.2/4.4	2.6/3.8	5.0/5.7
Latitude	Average	4.0	4.8	5.5	5.9	6.1	6.4	6.6	6.6	6.4	5.6	4.1	3.7	5.5
	Min/Max	3.3/4.8	3.7/5.7	4.3/6.5	5.0/6.7	5.3/6.7	5.9/7.0	5.9/6.9	5.7/7.2	5.0/7.4	4.5/6.4	3.5/5.1	2.9/4.4	5.0/5.8
Latitude +15	Average	4.3	5.1	5.5	5.5	5.4	5.5	5.7	6.0	6.1	5.7	4.4	4.1	5.3
	Min/Max	3.5/5.3	3.7/6.0	4.1/6.5	4.6/6.2	4.7/5.9	5.1/6.0	5.2/6.0	5.2/6.5	4.8/7.1	4.6/6.5	3.7/5.4	3.1/4.8	4.8/5.6
90	Average	4.2	4.6	4.3	3.6	3.1	2.8	3.0	3.6	4.4	4.7	4.0	3.9	3.8
	Min/Max	3.3/5.1	3.2/5.6	3.1/5.6	3.1/4.0	2.7/3.3	2.7/3.0	2.8/3.1	3.2/3.9	3.4/5.1	3.8/5.4	3.4/5.0	3.0/4.6	3.4/4.2

Solar Radiation for 1-Axis Tracking Flat-Plate Collectors with a North-South Axis (kWh/m²/day), Uncertainty ±9%

Axis Tilt (°)		Jan	Feb	Mar	Apr	May	June	July	Aug	Sept	Oct	Nov	Dec	Year
0	Average	3.3	4.6	6.1	7.5	8.8	10.0	10.0	9.1	7.7	5.7	3.6	2.9	6.6
	Min/Max	2.7/4.0	3.6/5.4	4.4/7.3	6.2/8.7	7.2/10.1	8.6/11.3	8.8/10.9	7.5/10.2	5.7/9.2	4.5/6.7	3.0/4.4	2.3/3.5	5.8/7.0
Latitude -15	Average	4.3	5.6	6.9	8.0	9.0	10.0	10.1	9.6	8.5	6.8	4.6	3.9	7.3
	Min/Max	3.5/5.3	4.2/6.6	5.0/8.3	6.6/9.4	7.3/10.3	8.6/11.4	8.9/11.1	7.8/10.7	6.3/10.2	5.3/8.0	3.8/5.7	2.9/4.7	6.4/7.7
Latitude	Average	4.7	6.0	7.1	8.0	8.7	9.6	9.8	9.4	8.6	7.1	4.9	4.4	7.4
	Min/Max	3.8/5.9	4.4/7.1	5.1/8.6	6.5/9.3	7.1/10.0	8.3/10.9	8.6/10.7	7.7/10.6	6.3/10.4	5.5/8.4	4.1/6.2	3.2/5.2	6.4/7.9
Latitude +15	Average	5.0	6.1	7.0	7.7	8.2	9.0	9.2	9.0	8.5	7.2	5.1	4.6	7.2
	Min/Max	4.0/6.2	4.5/7.3	4.9/8.6	6.2/9.0	6.7/9.5	7.8/10.3	8.1/10.1	7.3/10.1	6.1/10.2	5.5/8.4	4.2/6.5	3.4/5.6	6.3/7.8

Solar Radiation for 2-Axis Tracking Flat-Plate Collectors (kWh/m²/day), Uncertainty ±9%

Tracker		Jan	Feb	Mar	Apr	May	June	July	Aug	Sept	Oct	Nov	Dec	Year
2-Axis	Average	5.0	6.1	7.1	8.1	9.1	10.3	10.3	9.6	8.7	7.2	5.2	4.7	7.6
	Min/Max	4.0/6.3	4.5/7.3	5.1/8.6	6.6/9.4	7.4/10.4	8.8/11.7	9.1/11.3	7.8/10.8	6.3/10.4	5.5/8.5	4.3/6.5	3.4/5.7	6.6/8.1

Direct Beam Solar Radiation for Concentrating Collectors (kWh/m²/day), Uncertainty ±8%

Tracker		Jan	Feb	Mar	Apr	May	June	July	Aug	Sept	Oct	Nov	Dec	Year
1-Axis, E-W Horiz Axis	Average	3.0	3.4	3.6	3.8	4.4	5.3	5.4	5.0	4.7	4.2	3.1	2.9	4.1
	Min/Max	2.1/4.2	2.2/4.3	2.1/4.4	2.7/4.9	3.0/5.6	4.0/6.6	4.5/6.3	3.6/6.0	2.9/6.1	2.8/5.3	2.2/4.3	1.5/3.9	3.2/4.5
1-Axis, N-S Horiz Axis	Average	2.1	3.0	4.0	5.0	5.9	7.2	7.3	6.7	5.8	4.2	2.4	1.9	4.6
	Min/Max	1.5/3.0	2.1/3.9	2.2/5.1	3.4/6.4	4.0/7.6	5.3/8.9	6.0/8.6	4.7/8.2	3.5/7.7	2.8/5.5	1.7/3.3	0.9/2.6	3.6/5.2
1-Axis, N-S Tilt=Latitude	Average	3.3	4.1	4.8	5.3	5.8	6.8	7.0	6.9	6.6	5.4	3.5	3.1	5.2
	Min/Max	2.3/4.6	2.8/5.3	2.7/6.1	3.7/6.8	4.0/7.5	5.0/8.4	5.8/8.3	4.8/8.4	3.9/8.7	3.6/7.0	2.5/4.9	1.5/4.2	4.1/5.8
2-Axis	Average	3.5	4.2	4.8	5.4	6.1	7.4	7.5	7.1	6.6	5.5	3.7	3.3	5.4
	Min/Max	2.5/4.9	2.8/5.4	2.7/6.1	3.7/6.9	4.2/7.9	5.4/9.2	6.2/8.9	5.0/8.6	3.9/8.7	3.6/7.0	2.6/5.2	1.7/4.6	4.2/6.0

Average Climatic Conditions

Element	Jan	Feb	Mar	Apr	May	June	July	Aug	Sept	Oct	Nov	Dec	Year
Temperature (°C)	-6.4	-4.3	-0.4	4.8	10.2	15.7	20.0	18.8	13.1	6.8	-0.9	-5.9	5.9
Daily Minimum Temp	-11.6	-9.8	-6.3	-2.0	2.7	7.6	11.6	10.4	5.0	-0.4	-6.7	-11.2	-0.9
Daily Maximum Temp	-1.4	1.3	5.4	11.5	17.6	23.7	28.4	27.1	21.1	14.0	4.8	-0.7	12.7
Record Minimum Temp	-37.2	-32.8	-23.9	-13.3	-8.9	-1.7	2.2	0.6	-14.4	-14.4	-24.4	-33.9	-37.2
Record Maximum Temp	12.2	13.3	34.4	25.0	28.9	33.9	34.4	33.3	31.1	25.0	17.8	13.9	34.4
HDD, Base 18.3°C	768	633	581	407	253	103	12	37	168	357	577	751	4647
CDD, Base 18.3°C	0	0	0	0	0	23	64	51	10	0	0	0	148
Relative Humidity (%)	72	70	65	56	50	43	38	39	45	54	69	73	56
Wind Speed (m/s)	5.3	5.3	5.4	5.4	5.0	4.6	4.2	4.1	4.3	4.4	4.7	5.1	4.8

Sheridan, WY

WBAN NO. 24029

LATITUDE: 44.77° N
LONGITUDE: 106.97° W
ELEVATION: 1209 meters
MEAN PRESSURE: 878 millibars

STATION TYPE: Secondary

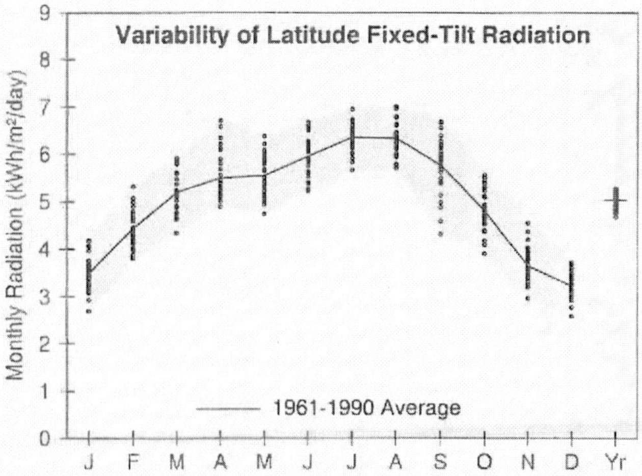

Solar Radiation for Flat-Plate Collectors Facing South at a Fixed Tilt (kWh/m²/day), Uncertainty ±9%

Tilt (°)		Jan	Feb	Mar	Apr	May	June	July	Aug	Sept	Oct	Nov	Dec	Year
0	Average	1.8	2.7	3.9	5.0	5.8	6.7	6.9	6.0	4.6	3.1	2.0	1.6	4.2
	Min/Max	1.6/2.1	2.4/3.1	3.4/4.3	4.6/5.9	5.0/6.6	5.8/7.4	6.1/7.5	5.5/6.6	3.7/5.2	2.7/3.4	1.8/2.3	1.4/1.7	3.9/4.4
Latitude -15	Average	3.1	4.0	5.0	5.6	5.9	6.5	6.9	6.6	5.7	4.5	3.2	2.8	5.0
	Min/Max	2.4/3.6	3.5/4.8	4.2/5.6	5.0/6.8	5.1/6.8	5.7/7.3	6.1/7.5	6.0/7.3	4.3/6.6	3.7/5.2	2.7/4.0	2.3/3.2	4.7/5.2
Latitude	Average	3.5	4.4	5.2	5.5	5.6	6.0	6.4	6.3	5.8	4.8	3.6	3.2	5.0
	Min/Max	2.7/4.2	3.8/5.3	4.3/5.9	4.9/6.7	4.7/6.4	5.2/6.7	5.7/6.9	5.7/7.0	4.3/6.7	3.9/5.6	3.0/4.6	2.6/3.7	4.7/5.3
Latitude +15	Average	3.7	4.6	5.1	5.1	4.9	5.1	5.5	5.7	5.5	4.9	3.8	3.5	4.8
	Min/Max	2.8/4.5	3.9/5.6	4.2/5.9	4.5/6.2	4.2/5.6	4.5/5.7	4.9/6.0	5.2/6.3	4.1/6.4	3.9/5.7	3.1/4.9	2.8/4.1	4.5/5.1
90	Average	3.6	4.2	4.2	3.6	3.0	2.9	3.2	3.7	4.1	4.1	3.6	3.4	3.6
	Min/Max	2.7/4.4	3.4/5.1	3.4/5.0	3.2/4.3	2.7/3.4	2.6/3.1	2.9/3.4	3.3/4.0	3.0/4.8	3.3/4.8	2.8/4.8	2.7/4.0	3.4/3.9

Solar Radiation for 1-Axis Tracking Flat-Plate Collectors with a North-South Axis (kWh/m²/day), Uncertainty ±9%

Axis Tilt (°)		Jan	Feb	Mar	Apr	May	June	July	Aug	Sept	Oct	Nov	Dec	Year
0	Average	2.7	4.0	5.5	6.9	7.8	9.1	9.6	8.7	6.8	4.7	3.0	2.4	5.9
	Min/Max	2.2/3.3	3.4/4.8	4.3/6.4	6.0/8.8	6.4/9.1	7.5/10.4	8.1/10.7	7.6/9.9	4.9/8.1	3.8/5.5	2.5/3.8	1.9/2.7	5.4/6.4
Latitude -15	Average	3.7	5.0	6.4	7.4	8.0	9.1	9.8	9.2	7.6	5.7	3.9	3.3	6.6
	Min/Max	2.8/4.4	4.2/6.1	5.0/7.5	6.4/9.6	6.5/9.4	7.5/10.5	8.3/10.9	8.0/10.5	5.4/9.2	4.5/6.8	3.1/5.0	2.6/3.9	6.0/7.0
Latitude	Average	4.0	5.3	6.5	7.4	7.8	8.8	9.5	9.0	7.7	6.0	4.3	3.7	6.7
	Min/Max	3.0/4.9	4.4/6.5	5.1/7.7	6.3/9.5	6.3/9.2	7.2/10.1	8.0/10.6	7.9/10.4	5.4/9.3	4.7/7.1	3.4/5.5	2.8/4.3	6.0/7.1
Latitude +15	Average	4.2	5.4	6.5	7.1	7.3	8.2	8.9	8.6	7.5	6.0	4.4	3.9	6.5
	Min/Max	3.1/5.2	4.5/6.7	5.0/7.7	6.0/9.2	5.9/8.7	6.7/9.5	7.5/10.0	7.5/9.9	5.3/9.1	4.7/7.2	3.5/5.7	3.0/4.6	5.9/6.9

Solar Radiation for 2-Axis Tracking Flat-Plate Collectors (kWh/m²/day), Uncertainty ±9%

Tracker		Jan	Feb	Mar	Apr	May	June	July	Aug	Sept	Oct	Nov	Dec	Year
2-Axis	Average	4.3	5.4	6.5	7.5	8.1	9.4	10.0	9.2	7.7	6.0	4.4	4.0	6.9
	Min/Max	3.2/5.2	4.5/6.7	5.1/7.7	6.4/9.6	6.6/9.6	7.7/10.8	8.4/11.2	8.1/10.6	5.5/9.3	4.7/7.2	3.5/5.8	3.0/4.7	6.2/7.3

Direct Beam Solar Radiation for Concentrating Collectors (kWh/m²/day), Uncertainty ±8%

Tracker		Jan	Feb	Mar	Apr	May	June	July	Aug	Sept	Oct	Nov	Dec	Year
1-Axis, E-W Horiz Axis	Average	2.3	2.7	3.1	3.4	3.7	4.7	5.3	4.8	4.1	3.4	2.5	2.3	3.5
	Min/Max	1.1/3.5	2.2/4.1	2.2/4.2	2.5/5.2	2.7/4.8	3.3/6.1	4.1/6.3	4.0/6.0	2.3/5.3	2.4/4.4	1.8/3.6	1.4/3.1	3.0/4.0
1-Axis, N-S Horiz Axis	Average	1.5	2.3	3.3	4.4	5.0	6.3	7.1	6.4	4.9	3.3	1.8	1.4	4.0
	Min/Max	0.8/2.3	1.8/3.4	2.2/4.5	3.3/6.8	3.5/6.5	4.4/8.1	5.3/8.4	5.1/8.0	2.8/6.6	2.2/4.2	1.2/2.6	0.8/1.9	3.4/4.5
1-Axis, N-S Tilt=Latitude	Average	2.5	3.3	4.1	4.7	5.0	6.0	6.9	6.7	5.7	4.3	2.8	2.4	4.5
	Min/Max	1.2/3.8	2.6/4.9	2.8/5.6	3.6/7.4	3.5/6.4	4.2/7.7	5.2/8.2	5.4/8.3	3.3/7.6	3.0/5.6	2.0/4.1	1.4/3.3	3.8/5.1
2-Axis	Average	2.7	3.4	4.1	4.8	5.3	6.6	7.4	6.9	5.7	4.4	3.0	2.6	4.7
	Min/Max	1.3/4.1	2.6/5.0	2.8/5.6	3.6/7.5	3.7/6.8	4.6/8.4	5.6/8.8	5.5/8.6	3.3/7.6	3.0/5.7	2.1/4.3	1.6/3.5	4.0/5.3

Average Climatic Conditions

Element	Jan	Feb	Mar	Apr	May	June	July	Aug	Sept	Oct	Nov	Dec	Year
Temperature (°C)	-6.2	-3.1	1.1	6.6	11.5	16.7	20.9	20.2	13.9	8.2	0.3	-5.2	7.1
Daily Minimum Temp	-13.1	-9.7	-5.8	-0.9	3.9	8.5	11.7	10.9	5.2	-0.2	-6.8	-12.1	-0.7
Daily Maximum Temp	0.6	3.5	7.9	13.9	19.1	24.9	30.1	29.6	22.7	16.5	7.4	1.7	14.8
Record Minimum Temp	-37.2	-35.6	-30.6	-18.9	-10.6	-2.8	1.7	1.1	-14.4	-22.8	-31.7	-38.3	-38.3
Record Maximum Temp	21.1	24.4	25.0	30.6	35.0	40.6	41.1	41.1	39.4	32.8	25.6	22.2	41.1
HDD, Base 18.3°C	761	601	536	353	212	87	16	35	149	315	542	730	4336
CDD, Base 18.3°C	0	0	0	0	0	38	94	93	18	0	0	0	244
Relative Humidity (%)	66	66	63	58	60	59	50	47	54	58	66	67	60
Wind Speed (m/s)	3.5	3.5	4.0	4.3	4.0	3.4	3.3	3.2	3.3	3.3	3.3	3.4	3.5